E D I T I O N 4

INTERVIEWING STRATEGIES FOR HELPERS
Fundamental Skills and Cognitive Behavioral Interventions

EDITION 4

INTERVIEWING STRATEGIES FOR HELPERS

Fundamental Skills and Cognitive Behavioral Interventions

Sherry Cormier
West Virginia University

Bill Cormier
Energy Flux, Inc., Annapolis, MD

Brooks/Cole Publishing Company

I(T)P® *An International Thomson Publishing Company*

Pacific Grove • Albany • Belmont • Bonn • Boston • Cincinnati • Detroit • Johannesburg • London • Madrid • Melbourne
Mexico City • New York • Paris • Singapore • Tokyo • Toronto • Washington

Sponsoring Editor: *Lisa Gebo*
Marketing Team: *Jean Thompson, Margaret Parks, Deanne Brown*
Editorial Assistants: *Lisa Blanton, Susan Carlson*
Production Editor: *Karen Ralling*
Manuscript Editor: *Patterson Lamb*
Permissions Editor: *Fiorella Ljunggren*
Interior Design: *Sharon Kinghan*

Interior Illustration: *John and Judy Waller*
Cover Art: *Sheryl Karas*
Cover Design: *Roy R. Neuhaus*
Art Editor: *Lisa Torri*
Typesetting: *Carlisle Communications*
Cover Printing: *R. R. Donnelley & Sons Company*
Printing and Binding: *R. R. Donnelley & Sons Company*

For more information, contact:

BROOKS/COLE PUBLISHING COMPANY
511 Forest Lodge Road
Pacific Grove, CA 93950
USA

International Thomson Publishing Europe
Berkshire House 168-173
High Holborn
London WC1V 7AA
England

Thomas Nelson Australia
102 Dodds Street
South Melbourne, 3205
Victoria, Australia

Nelson Canada
1120 Birchmount Road
Scarborough, Ontario
Canada M1K 5G4

International Thomson Editores
Campos Eliseos 385, Piso 7
Col. Polanco
11560 México D. F. México

International Thomson Publishing GmbH
Königswinterer Strasse 418
53227 Bonn
Germany

International Thomson Publishing Asia
221 Henderson Road
#05-10 Henderson Building
Singapore 0315

International Thomson Publishing Japan
Hirakawacho Kyowa Building, 3F
2-2-1 Hirakawacho
Chiyoda-ku, Tokyo 102
Japan

Printed in the United States of America

10   9   8   7   6   5   4

Library of Congress Cataloging-in-Publication Data
Cormier, L. Sherilyn (Louise Sherilyn)
    Interviewing strategies for helpers: fundamental skills and
  cognitive behavioral interventions / Sherry Cormier, Bill
  Cormier, —4th ed.
        p.   cm.
    Includes bibliographical references and index.
    ISBN 0-534-34916-1
    1. Counseling.   2. Helping behavior.   3. Interviewing.
  4. Cognitive therapy.   5. Behavior therapy.   I. Cormier, William H.
  (William Henry)   II. Title.
  BF637.C6C584 1998
  158'.3—dc21
                                            97-17465
                                        CIP

ABOUT THE AUTHORS

Sherry Cormier is a professor in the Department of Counseling, Rehabilitation Counseling and Counseling Psychology at West Virginia University, Morgantown, West Virginia. She also holds an adjunct appointment in the Department of Family Medicine at the West Virginia University School of Medicine. She is a licensed psychologist in the state of West Virginia. Her current research and practice interests are in counselor training issues and models, and counseling of girls and women. She also works with cognitive-behavioral, object-relations, and body-awareness approaches as well as Jungian and transpersonal psychology. She is the mother of two teenage daughters and a golden retriever dog.

Bill Cormier has been a teacher, clinician, consultant, and scholar for 26 years. He was Professor and Director of Clinical Training of the APA-accredited doctoral program in counseling psychology at West Virginia University. He held an adjunct appointment in Family Practice at the West Virginia University Medical Center. He is a licensed psychologist in West Virginia and Maryland. He was in general private practice from 1991 to 1996 with a specialty in behavioral medicine. His current interests are East/West approaches to psychotherapy and mind/body techniques. He has received extensive training in meditation, breath work, Tai Chi, and is a certified yoga instructor. Bill is currently affiliated with the Energy Flux Corporation that specializes in refocusing corporate energy and developing creative strategies for enhancing aptitudes. His avocational interests include boating, mountain biking, and fiction writing.

CONTENTS

PREFACE

When we wrote the previous edition in 1991, we relied on the axiom "If it's not broken, don't fix it!" That guidance seemed appropriate in that time and place, but today the world is very different; much has happened in the last seven years in the training for and delivery of counseling and psychological services as well as in the total environment. First, there is an increased awareness of and emphasis on multicultural issues and diversity as we move toward a global view of psychology and human services. Indeed, the review of literature for this edition showed that there has been a real explosion in this area. Integrating the multicultural focus into the book has been our primary emphasis for this edition. We have attempted to do this in a nonadversarial way that avoids stereotyping anyone by particular gender, race, ethnic group, and so on. We have tried to include divergent voices, including feminist and multicultural ones, to help increase awareness of inequities in power and instances of oppression, and how these impact us and our clients. At the same time we have tried not to blame particular groups of people for the oppression that has occurred and is still occurring. The emphasis on a multicultural model reflects our realization during this revision of the paradigm shift we were making in our own process and constructs. As Euro-American practitioners, we have been influenced and in ways constricted by our primary "mainstream" referent groups. Fortunately, in our colleagues, students, and reviewers, we have had models and mentors of increased diversity who have helped us to expand our cultural perspectives! We hope our increased awareness is reflected in this edition; we are, personally and professionally, the better for it and enriched by it.

A second area that has been rapidly changing during the course of this revision has been the transition to managed mental health care by various health maintenance organizations or HMOs. When we first began work on the new edition in 1994, HMOs, which were rapidly proliferating, were being welcomed as bringing a solution to the rising costs of health care services. Now, several years later, as we finish the chapters and ready the book for production, HMOs are increasingly under assault by both consumers and professionals. To some degree, these organizations are affecting how we practice and, therefore, what we write. In this fourth edition, you will find a revised chapter on outcome goals that also incorporates a pragmatic approach to practice evaluation (Chapter 10). You will also find a new chapter on treatment planning—a required activity for providers whose clients have managed care insurance and a critical element in provision of effective mental health services (Chapter 11). We believe that over the next decade there will continue to be great fluctuation in how health care services are provided and reimbursed.

Another trend we have observed over the last few years is the greater use of holistic treatment and health care services. Increasingly, people are recognizing that the mind and body are not two separate entities but are parts of a holistic and synergistic unity that affect the whole. The Public Broadcasting System has aired and published in book form an entire series devoted to Bill Moyers' exploration titled *Healing and the Mind*. Recognizing the mind-body connection, we have included in this edition another new chapter called *Breathing and Hatha Yoga* (Chapter 18).

We are also witnessing a renewed emphasis on the therapeutic relationship and the working alliance between therapist and client as well as increased awareness of the counselor as a person. Attachment issues are at the forefront of all of this. In a newly released book, *The Sibling Society,* Bly (1996) notes that society in the United States is abandoning its children and elders at an alarming pace; the result is that attachment issues abound as people from many diverse groups feel thrown away or disposed of. We have reworked the relationship chapter (Chapter 3), included a new chapter, *Knowing Yourself as a Counselor* (Chapter 2), and added an adult attachment model to our chapter on problem conceptualization (Chapter 8).

As a final change, we have updated many of the old references, including as much recent literature as possible. We have, however, retained old, valuable references for which there is no newer edition or older articles that are classics and provide an important historical perspective.

Over the years, some have commented on the length and detail of the book and have asked, "What is it like to put together a book like this?" Our first response is always, "It requires a lot of help." For this edition, we are indebted for their library research to Lynn Braun, Jeffrey Dulko, Joseph Kachik, and Paul Sundell, doctoral candidates in the Department of Counseling, Rehabilitation Counseling and Counseling Psychology at West Virginia University, and to Anne B. Drake and Victoria Railing for their word processing and manuscript preparation assistance. We also are grateful to Dr. Cynthia R. Kalodner, associate professor, Department of Counseling, Rehabilitation Counseling and Counseling Psychology at West Virginia University, for her revision of Chapter 19 on systematic desensitization. We also acknowledge the support of Dr. Jeffrey K. Messing, chairperson, Department of Counseling, Rehabilitation Counseling and Counseling Psychology at West Virginia University; and Dr. Jane H. Applegate, former dean of the College of Human Resources and Education at West Virginia University. We are very grateful to all the staff at Brooks/Cole, particularly to our former editor, Claire Verduin, who signed us to write the original edition of this book 20 years ago, and also to our current editor, Lisa Gebo, for her commitment and enthusiasm. We are indeed indebted to both these wonderful women for their "wise ways." We are also thankful for the support of our families and friends; you are too numerous to name, but you know who you are! Finally, the final form of this book as you, the reader, now see it would not have been possible without the superb efforts of the entire Brooks/Cole production team, headed by our production editor, Karen Ralling.

Finally, we wish to acknowledge with gratitude the contribution of our reviewers, who include the following: Kia J. Bentley, Virginia Commonwealth University; Terri Brown, Methodist College; Diane Coursol, Mankato State University; David R. Evans, University of Western Ontario; Ronn Johnson, University of San Diego; Norma S. C. Jones, Howard University; Theodore P. Remley, Jr., University of New Orleans; and Francine Shapiro. They probably do not fully realize the extent of their impact on the final form of this book. To all of you: Thank you. We could not have done this without each of your careful and detailed comments and suggestions.

Sherry Cormier
Bill Cormier

CHAPTER 1

ABOUT THIS BOOK

Imagine yourself as the helper in the following four situations. Try to see, hear, and sense what is happening to you.

A 14-year-old boy who is accused of setting fire to his family home walks in defiantly to see you. He has been "mandated" to see you by the judge. He sits down, crosses his arms and legs in front of him, and stares at the ceiling. He is silent after your initial greeting.

A young woman in her 20s walks in and can't hold back the tears and sobs. After a while, she talks about how upset she is feeling. In the last year, three of her close friends have died of acquired immune deficiency syndrome (AIDS); she has also lost her parents' support because she has told them she is a lesbian.

A Latino father and his teenage son come in together, but they are so at odds with each other that initially they refuse to be seen by you in the same room. According to the telephone intake report, they have repeatedly fought about the amount of freedom the son wants and the father is willing to give.

A middle-aged woman comes in. She has been escorted to your facility by her husband. She is so afraid to go out of her house that she does not drive anymore. In talking with her, you discover that she has confined herself to her home almost exclusively for the last year because of incapacitating anxiety attacks. Her husband has recently turned down a lucrative job offer to avoid having to move her into a new environment.

Now try to process exactly what it is like for you to imagine helping or counseling each of these four clients. How were you feeling? What thoughts were running through your head? How did you see or hear yourself responding? What things about yourself were you aware of that helped you in the interaction; what things hindered you? What skills did you utilize to deal with the client? What skills were you lacking? What did you observe about the client, and how did your observations affect your help giving? How did you know whether what you were doing was helpful?

Although responding to these kinds of questions may be difficult for you now, it will probably become easier as you go through the book and as you also acquire greater experience and more feedback. Specific purposes of the book are described in the following section.

PURPOSES OF THE BOOK

We hope that, in this book, you will find training experiences that facilitate personal growth, develop your counseling skills, and provide ways for you to evaluate your effectiveness. Personal growth is the most elusive and the most difficult to define of these three areas. Although it is beyond the scope of this book to focus primarily on self-development, you may engage in self-exploration as you go through certain parts of the book, particularly Chapters 2 and 3. We also encourage you to seek out additional experiences in which you can receive feedback from others about yourself, your strengths, and some behaviors that may interfere with counseling. These experiences might consist of individual or classroom activities and feedback, growth groups, and personal counseling. It is well documented that a counselor's warmth, empathy, and positive regard can contribute to client change. We feel that your demonstration of these relationship conditions will enhance the way you use the skills and strategies presented in this book.

We created the book with four specific purposes. First, we think it will help you acquire a repertory of counseling interview skills and strategies. The book focuses on *interview* skills and strategies as used in a helping relationship. It is directed (but not restricted) to applying skills within a counselor/client dyadic relationship. Although some of the skills and strategies may be used appropriately in group counseling, organizational interventions, or marriage and family counseling, the major focus of this book is on application of these skills with individuals. However, we would hope that you would develop proficiency in these other areas as well because ultimately your clients are affected by the context in which they live. The context includes family, neighborhood, work, regional environments, and countries.

In the first seven chapters of this book, we present what we call "fundamental skills." These include relationship conditions, nonverbal behavior, and verbal responses that are useful for practitioners of varying theoretical orientations. In the remaining chapters, our selection of models and strategies reflects a cognitive-behavioral framework. The intervention strategies we have chosen to include have some supporting database, although many of the existing research studies are analog ones (that is, conducted in simulated counseling settings) and the results may not always generalize to actual counseling situations.

In addition to the cognitive-behavioral "flavor" of the strategies, reference to skills and strategies based on other theoretical orientations is often mentioned throughout the book. This is because cognitive-behavioral therapies are increasingly broad-based in nature and focus (Goldfried, 1983) and also because of our own belief that skilled counselors are at least knowledgeable about, if not proficient in, more than one approach to working with client problems. For your benefit if you are not yet familiar with the concepts associated with various theoretical approaches to counseling and therapy, Table 1-1 presents a synopsis of four major forces in counseling theories.

Our second purpose is to assist you in identifying the potential applicability of many counseling strategies for different client problems. As Krumboltz and Thoresen (1976) point out, a variety of useful counseling methods are available for different problem areas. After you have finished the book, we hope you will be able to select and use appropriate counseling strategies when confronted with a depressed client, an anxious client, a nonassertive client, and so forth. We also hope you will be aware of cases in which approaches and strategies included in this book may not be very useful.

In addition to being able to identify appropriate counseling strategies for diverse client problems, you should also be able to apply counseling skills and strategies carefully and sensitively to a diverse group of clients. Helping you accomplish this is the third purpose of the book. Our world and therefore our range of clients is increasingly diverse and pluralistic. While you yourself may be a gay, Jewish man of middle socioeconomic level or an African American, straight, Protestant woman of upper socioeconomic level, your clients will invariably be somewhat different from you in terms of cultural background—*cultural* referring to factors such as age, culture, ethnicity, gender, language, disability, race,

gender expression, sexual orientation, and socioeconomic level. And even if you and your client share many similarities of background, he or she will still have unique characteristics that you must consider. Therefore, it is impossible to say that all persons who belong to a particular religious group or who hold a particular sexual orientation or are of a particular gender or ethnic origin will behave the same or will respond in the same way to counselors. Still, however, there are some general issues to consider about culture in working with clients from diverse backgrounds. For example, clients from Asian cultures often place less emphasis on the individual's needs and welfare and more emphasis on those of the group or family. As a result, counseling approaches such as assertion training that focus on the pursuit of individual rights may not be consistent with the values of *some* Asian clients. (We say *some,* because within the Asian American culture, as in any culture, there are areas of difference.) Similarly, counseling interventions that emphasize autonomy and objectivity rather than connectedness and subjectivity are reinforcing traditionally held *masculine* rather than *feminine* values. Although we cannot be prescriptive about the best way to work with all these cultural factors, it is important for us to recognize that they exist and that many of the counseling skills and strategies described in this book have been developed by Euro-American men in Western culture and may not be applicable to all clients. For these reasons, in the client descriptions used throughout the text as both model examples and practice ones, we often include cultural referent characteristics of the client such as race, ethnicity, sexual orientation, gender, age, physical/mental status, and so on.

Fourth, we hope to provide you with some ways to monitor and evaluate your behavior and the client's behavior during counseling. The recent emphasis on accountability requires each of us to explore the results of our helping activities more closely. Evaluation of counseling also assesses the extent to which the therapeutic goals are achieved.

Above all, we want to convey that the book is about *practical application* of selected skills and strategies. Our coverage of theoretical and research concepts is very limited because they are covered adequately in other texts.

AN OVERVIEW OF HELPING

A helping professional is someone who facilitates the exploration and resolution of issues and problems presented by a client, or the person seeking help. Helping interactions have four recognized components: (1) someone seeking help, (2) someone willing to give help who is also (3) capable of or

TABLE 1-1. Overview of four major forces in counseling and psychotherapy theory

	Theoretical system and relationship to foundational theories and family theory	Worldview	Major concepts and techniques
Multicultural counseling and therapy (the fourth force)	Foundational theories (empathic dimensions, microskills, decisional counseling, developmental counseling and therapy) explicitly and implicitly utilized as part of overall theoretical conception, but modified with cultural frames of reference. Family therapy concepts considered essential. Attneave's network therapy often an essential ingredient.	Counseling and therapy have been culturally encapsulated. The individual and family are based in the culture. The counselor or therapist needs to approach counseling with multicultural awareness. Many authors stress issues of development in the family and society. Seeks to integrate first-, second-, and third-force theories as part of worldview and counseling and therapy theory and case conceptualization.	As a newly evolving major theoretical group, the main point of agreement is that issues of culture, gender, and other multicultural issues need to take a central place in the helping process. Collaboration and network treatment planning are essential. Consciousness raising about ethnicity/race and gender issues often critical in the helping process.
Psychodynamic (the first force)	Foundational theories not explicitly considered, but post hoc examination shows that these concepts help explain the value of these orientations and makes their implementation more explicit. Family concepts not prominent, although attachment theory and the family unconscious are adding this emphasis. Bowen's intergenerational theory especially compatible. Historically, minimal attention to gender and multicultural issues.	The past is prelude to the present, and much of the past is held in the unconscious. Individuals are deeply influenced by the past, and we must understand this past if we are to facilitate individual growth. Sigmund Freud is major philosopher. The pragmatic and optimistic Bowlby stressed that we can facilitate growth through understanding and action. Taub-Bynum focused on family and cultural history playing themselves out in the individual.	These are the most complex set of theories available. The development of the person rests on early life experience. The interaction of person and environment is largely played out in the unconscious. Traditional Freudian theory emphasizes the Oedipal complex as central to development, whereas object relations and attachment theories focus on early infant and child experience as more important. Free association, dream analysis, and awareness of transference, countertransference, and projective identification are important.
Cognitive-behavioral: behavioral foundations (the second force)	Foundational theories often integrated into understanding and planning treatment. Decisional counseling and social skills portion of microskills a standard part of counseling and therapy.	Deeply rooted in the idea of progress and faith in science to solve human problems. B. F. Skinner often seen as major philosopher.	Through functional analysis, it is possible to understand the antecedents, resultant behavior, and consequences of the behavior. Many highly specific and proven techniques of

(continued)

TABLE 1-1. (continued)

	Theoretical system and relationship to foundational theories and family theory	Worldview	Major concepts and techniques
	Family concepts historically have not been important, but behavioral family approach illustrates how theory can be integrated.	Meichenbaum's more recent construction is more humanistic in orientation and provides a new integration of behaviorism with other theories. Cheek supplies a culturally-relevant view.	behavioral change available. Has had profound influence on popular cognitive-behavioral movement, particularly the work of Meichenbaum in social skills training.
Cognitive-behavioral: cognitive foundations (the second force)	Foundational theories tend to be implicit rather than explicit. Cognitive aspects of developmental counseling and therapy may help integrate this framework more closely with MCT, particularly action at the sensorimotor and systemic level, which is often missing in CBT. Family concepts historically have not been important, but are compatible.	Roots lie in stoic philosopher Epictetus—"We are disturbed not by events, but by the views we take of them." Attempt to integrate ideas about the world with action in the world. Ellis's rational-emotive therapy. Beck's cognitive therapy. Glasser's reality therapy.	Currently a popular theoretical orientation, as the system allows integration of many ideas from seemingly competing theories. Major focus is on thinking patterns and their modification, but maintains a constant emphasis on homework and taking new ideas out into the world and acting on them. Glasser's work is similar, but focuses very effectively on schools and youth in institutions.
Existential/humanistic (the third force)	The foundational concepts of empathy and the listening portion of microskills have been derived from this orientation. Family concepts historically have not been important, but are compatible. Whitaker's experiential family orientation is especially compatible. Has not consciously embraced MCT but is compatible.	The human task is to find meaning in a sometimes meaningless world. Rogers stresses the ability of the person to direct one's own life; Frankl, the importance of positive meanings; and Perls, that people are wholes, not parts, and can take direction of their own lives. Heidegger, Husserl, Binswanger, and Boss have been most influential at a basic philosophical level. Rogers's person-centered theory. Frankl's logotherapy. Perls's Gestalt therapy.	Each individual constructs the world uniquely. Rogers stresses the importance of self-actualization and careful listening to the client. Frankl emphasizes spirituality and a variety of specific techniques to facilitate the growth of meaning. Perls, with his many powerful techniques, may be described as the action therapist.

SOURCE: From *Counseling and Psychotherapy: A Multicultural Perspective*, 3rd. Ed., by A. E. Ivey, M. B. Ivey, and L. Simek-Morgan. Copyright © 1993 by Allyn and Bacon. Reprinted by permission.

trained to help (4) in a setting that permits help to be given and received (Hackney & Cormier, 1994, p. 2).

By some estimates, over 300 different forms of psychotherapy exist (Rossi, 1987). In all these therapies, the helpers have the following functions:

1. They initiate communication.
2. They engage in therapeutic work.
3. They have some general or specific criteria for solving the client's problem so they know when to end the therapy (Rossi, 1987, p. 100).

In this book, we describe skills and strategies associated with these three processes, and we conceptualize the processes as four primary stages of helping:

1. Relationship
2. Assessment and goal setting
3. Strategy selection and implementation
4. Evaluation and termination

The first stage of the helping process involves *establishing an effective therapeutic relationship* with the client. This part of the process is based primarily on client- or person-centered therapy (Rogers, 1951) and more recently on social influence theory (Strong & Claiborn, 1982) and psychoanalytic theory (Greenson, 1967; Gelso & Carter, 1994). The potential value of a sound relationship base cannot be overlooked, because the relationship is the specific part of the process that conveys the counselor's interest in and acceptance of the client as a unique and worthwhile person and builds sufficient trust for eventual self-disclosure and self-revelation to occur. For some clients, working with a counselor who stays primarily in this stage may be useful and sufficient. For other clients, the relationship part of therapy is necessary but not sufficient to help them with the kinds of choices and changes they seek to make. These clients need additional kinds of action or intervention strategies.

The second phase of helping, assessment and goal setting, often begins concurrently with or shortly after relationship building. In both stages, the counselor is interested mainly in helping clients *explore* themselves and their concerns. Assessment is designed to help both the counselor and the client obtain a better picture, idea, or grasp of what is happening with the client and what prompted the client to seek the services of a helper at this time. The information gleaned during the assessment phase is extremely valuable in planning strategies and also can be used to manage resistance. As the problems and issues are identified and defined, the counselor and client also work through the process of developing outcome goals. Outcome goals are the specific

results the client would like to occur as a result of counseling. Outcome goals also provide useful information for planning action strategies.

In the third phase of helping, strategy selection and implementation, the counselor's task is to facilitate client *understanding and related action.* Insight can be useful, but insight alone is far less useful than insight accompanied by a supporting plan that helps the client translate new or different understandings into observable and specific actions or behaviors. Toward this end, the counselor and client select and sequence a plan of action or intervention strategies that are based on the assessment data and are designed to help the client achieve the designated goals. In developing action plans, it is important to select plans that relate to the identified problems and goals and that also are not in conflict with the client's primary beliefs and values.

The last major phase of helping, evaluation, involves *assessing the effectiveness* of your interventions and the progress the client has made toward the desired goals. This kind of evaluation assists you in knowing when to terminate or when your action plans need revamping. Additionally, observable and concrete signs of progress are often quite reinforcing to clients, who can easily become discouraged during the change process.

Note that there is some flow and interrelationship among these four stages. In other words, all parts of these stages are present throughout the counseling process, although not with the same degree of emphasis. As Waehler and Lenox (1994) point out, "Counseling participants do not go through a discrete state of relationship building and then 'graduate' to undertaking assessment as stage models imply" (p. 19). Each stage is interconnected to the others so that even during the intervention phase the relationship process is still attended to. Similarly, evaluation and termination may be discussed early in the relationship.

Just as there are stages and processes associated with helping, there are also stages and themes of helpers as they enter the helping profession, seek training, encounter clients, and gain supervised experience. In an award-winning study now summarized in a book, Skovholt and Ronnestad (1995) explored the development of therapists and counselors over the life span of their careers as helpers. They found that helpers progress through a series of eight stages from the time they select counseling as a career to the point where they are experienced practitioners. These eight stages can be described as follows:

1. Conventional Stage
2. Transition to Professional Training Stage

3. Imitation of Experts Stage
4. Conditional Autonomy Stage
5. Exploration Stage
6. Integration Stage
7. Individuation Stage
8. Integrity Stage

As with the stages of helping, these stages are also interconnected and interrelated. At each stage there are certain themes that helpers are most concerned with. For example, an entry-level practitioner is necessarily more concerned with skill development than with individual identity as a helper. Entry-level helpers are also more likely to model the behaviors they see in expert teachers and supervisors. Further along the career path, different themes and concerns emerge. We point this out because we hear students who use this book lament their dependence on models and on practice, yet this dependence appears to be necessary in the overall developmental process of becoming an effective helper.

FORMAT OF THE BOOK

We have used a learning format designed to help you demonstrate and measure your use of the counseling competencies presented in this book. Each chapter includes a brief introduction, chapter objectives, content material interspersed with model examples, activities, and feedback, a postevaluation, and a role-play interview assessment. People who have participated in field-testing this book have found that using these activities has helped them to get involved and to interact with the content material. You can complete the chapters by yourself or in a class. If you feel you need to go over an exercise several times, do so! If part of the material is familiar, jump ahead. Throughout each chapter, your performance on the learning activities and self-evaluations will be a clue to the pace at which you can work through the chapter. To help you use the book's format to your advantage, we explain each of its components briefly.

Objectives

As we developed each chapter, we had certain goals in mind for the chapter and for you. For each major topic, there are certain concepts and skills to be learned. We feel the best way to communicate this is to make our intentions explicit. After a short chapter introduction, you will find a section called "Objectives." The list of objectives describes the kinds of things that can be learned from the chapter. Using objectives for learning is similar to using goals in counseling.

The objectives provide cues for your end results and serve as benchmarks by which you can assess your progress. As you will see in Chapter 10, an objective or goal contains three parts:

1. The behavior, or what is to be learned or performed
2. The level of performance, or how much or how often to demonstrate the behavior
3. The conditions of performance, or the circumstances or situations under which the behavior can be performed

Part 1 of an objective refers to what you should learn or demonstrate. Parts 2 and 3 are concerned with evaluation of performance. The evaluative parts of an objective, such as the suggested level of performance, may seem a bit hardnosed. However, setting objectives with a fairly high mastery level may result in more improved performance. In this book, the objectives are stated at the beginning of each chapter so you know what to look for and how to assess your performance in the activities and self-evaluations. If you feel it would be helpful to see some objectives now, take a look at the beginning of Chapter 2.

Learning Activities

Learning activities that reflect the chapter objectives are interspersed throughout each chapter. These learning activities, which are intended to provide both practice and feedback, consist of model examples, exercises, and feedback. There are several ways you can use the learning activities. Many of the exercises suggest that you write your responses. Your written responses may help you or your instructor check the accuracy and specificity of your work. Take a piece of paper and actually write the responses down. Or you may prefer to work through an activity covertly and just think about your responses.

Some exercises instruct you to respond covertly by imagining yourself in a certain situation, doing certain things. We feel that this form of mental rehearsal can help you prepare for the kinds of counseling responses you might use in a particular situation. Covert responding does not require any written responses. However, if it would help you to jot down some notes after the activity is over, go ahead. You are the best person to determine how to use these exercises to your advantage.

Many of the exercises, particularly in the first seven chapters, are based on cognitive self-instruction. The objective of this type of activity is to help you not only to acquire the skill in a rote manner but also to internalize it. Some research suggests that this may be an important addition to the more common elements of microtraining (modeling, rehearsal, feedback) found to be so helpful in skill acquisition (Morran, Kurpius, Brack, & Brack, 1995). The cognitive learning strategy is designed specifically to help you develop your own way to think about the skill or to "put it together" in a way that makes sense to you.

Another kind of learning activity involves a more direct rehearsal than the written or covert exercises. These "overt rehearsal" exercises are designed to help you apply your skills in simulated counseling sessions with a role-play interview. The role-play activities involve three persons or roles: a counselor, a client, and an observer. Each group should trade roles so that each person can experience the role play from these different perspectives. One person's task is to serve as the counselor and practice the skills specified in the instructions. The counselor role provides an opportunity to try out the skills in simulated counseling situations. A second person, the client, will be counseled during the role play.

We give one word of caution to whoever takes the client role. Assuming that "counselor" and "client" are classmates, or at least not close friends or relatives, each of you will benefit more when in the counselor's seat if the "client" shares a real concern. These concerns do not have to be issues of life or death. Often someone will say "I won't be a good client because I don't have a problem." It is hard to imagine a person who has no concerns. Maybe your role-play concern will be about a decision to be made, a relationship conflict, some uneasiness about a new situation, or feeling sorry for or angry with yourself or someone else. Taking the part of a client in these role-play exercises may require that you first get in touch with yourself.

The third person in the role-play exercise is the "observer." This is a very important role because it develops and sharpens observational skills that are an important part of effective counseling. The observer has three tasks to accomplish. First, this person should observe the process and identify what the client does and how the counselor responds. When the counselor is rehearsing a particular skill or strategy, the observer can also determine the strengths and limitations of the counselor's approach. Second, the observer can provide consultation at any point during the role play if it might facilitate the experience. Such consultation may occur if the counselor gets stuck or if the observer perceives that the counselor is practicing too many nonhelpful behaviors. In this capacity, we have often found it helpful for the observer to serve as a sort of alter ego for the counselor. The observer can then become involved in the role play to help give the counselor more options or better focus. It is important, however, not to take over for the counselor in such instances.

The third and most important task of the observer is to provide feedback to the counselor about his or her performance following the role play. The person who role-played the client may also wish to provide feedback.

Giving helpful feedback is itself a skill that is used in some counseling strategies. The feedback that occurs following the role play should be considered just as important as the role play itself. Although everyone involved in the role play will receive feedback after serving as the counselor, it is still sometimes difficult to "hear" negative feedback. Sometimes receptiveness to feedback will depend on the way the observer presents it. We encourage you to make use of these opportunities to practice giving feedback to another person in a constructive, useful manner. Try to make your feedback specific and concise. Remember, the feedback is to help the counselor learn more about the role play; it should not be construed as the time to analyze the counselor's personality or lifestyle.

Another learning activity involves having people learn the strategies as partners or in small groups by teaching one another. We suggest that you trade off teaching a strategy to your partner or the group. Person A might teach covert modeling to Person B, and then Person B will teach Person A muscle relaxation. The "student" can be checked out in role play. Person B would practice covert modeling (taught by A), and Person A would demonstrate the strategy learned from Person B. This method helps the teacher learn and teach at the same time. If the "student" does not master the skills, additional sessions with the "teacher" can be scheduled.

The Role of Feedback in Learning Activities. Most of the chapter learning activities are followed by some form of feedback. For example, if a learning activity involves identifying positive and negative examples of a counseling conversational style, the feedback will indicate which examples are positive and which are negative. We also have attempted in most of our feedback to give some rationale for the responses. In many of the feedback sections, several possible responses are included. Our purpose in including feedback is not for you to find out how many "right" and "wrong" answers you have given in a particular activity. The responses listed in the feedback sections should serve as a guideline for you to code and judge your own responses. With this in mind, we would like you to view the feedback sections as sources of information and alternatives. We hope you are not put off or discouraged if your responses are different from the ones in the feedback. We don't expect you to come up with identical responses; some of your responses may be just as good as or better than the ones given in the feedback. Space does not permit us to list all the possibly useful responses in the feedback for each learning activity.

Locating Learning Activities and Feedback Sections in the Text.   As we have indicated, each chapter contains a variety of learning activities and feedback for most but not all of the activities. Usually a learning activity directly follows the related content section. We have placed learning activities in this way (rather than at the end of a chapter) to give you an immediate opportunity to work with and apply that content area before moving ahead to new material. Feedback for each learning activity is usually given on the *following* spread. This is done to encourage you to work through the learning activity on your own without concurrently scanning the same page to see how we have responded. We believe this helps you work more independently and encourages you to develop and rely more on your own knowledge base and skills. A potential problem with this format is difficulty in finding a particular learning activity or its corresponding feedback section. To minimize this problem, each learning activity and its corresponding feedback section are numbered. For example, the first learning activity in the book with feedback is found on page 21; it is numbered 3. Its corresponding feedback section, found on page 23, is also labeled 3.

Postevaluation

A postevaluation can be found at the end of each chapter. It consists of questions and activities related to the knowledge and skills to be acquired in the chapter. Because you respond to the questions after completing a chapter, this evaluation is called *post;* that is, it assesses your level of performance *after* receiving instruction. The evaluation questions and activities reflect the conditions specified in the objectives. When the conditions ask you to identify a response in a written statement or case, take some paper and write down your responses to these activities. However, if the objective calls for demonstrating a response in a role play, the evaluation will suggest how you can assess your performance level by setting up a role-play assessment. Other evaluation activities may suggest that you do something or experience something to heighten your awareness of or information about the idea or skill to be learned.

The primary purpose of the postevaluation is to help you assess your competencies after completing the chapter. One way to do this is to check your responses against those provided in the feedback at the end of each postevaluation. If there is a great discrepancy, the postevaluation can shed light

on those areas still troublesome for you. You may wish to improve in these areas by reviewing parts of the chapter, redoing the learning activities, or asking for additional help from your instructor or a colleague.

Role-Play Evaluation

In actual counseling, you must demonstrate your skills orally—not write about them. To help you determine the extent to which you can apply and evaluate your skills, role-play evaluations are provided at the end of most chapters. Each role-play evaluation consists of a structured situation in which you are asked to demonstrate certain skills as a counselor with a role-play client. Your performance on the role-play interview can be assessed by using the role-play checklist at the end of the chapter. These checklists consist of steps and possible responses associated with a particular strategy. The checklist should be used only as a guideline. You should always adapt any helping strategy to the client and to the particular demands of the situation.

There are two ways to assess your role-play performance. You can ask your instructor, a colleague, or another person to observe your performance, using the checklist. Your instructor may even schedule you periodically to do a role-play "checkout" individually or in a small group. If you do not have anyone to observe you, assess yourself. Audiotape your interview and rate your performance on the checklist. Also ask your "client" for feedback. If you don't reach the criterion level of the objective on the first try, you may need some extra work. The following section explains the need for additional practice.

Additional Practice

You may find some skills more difficult than others to acquire the first time around. Often people are chagrined and disappointed when they do not demonstrate the strategy as well as they would like on their first attempt. We ask these individuals whether they hold similar expectation levels for their clients! You cannot quickly and simply let go of behaviors you don't find useful in counseling and acquire others that are more helpful. It may be unrealistic to assume you will always demonstrate an adequate level of performance on *all* the evaluations on the first go-round. Much covert and overt rehearsal may be necessary before you feel comfortable with skill demonstration in the evaluations. On some occasions, you may need to work through the learning activities and postevaluations more than once.

Some Cautions About Using This Format

Although we believe the format of this book will promote learning, we want you to consider several cautions about using it. As you will see, we have defined the skills and strategies in precise and systematic ways to make it easier for you to acquire and develop the skills. However, we do not intend that our definitions and guidelines be used like cookbook instructions. Perhaps our definitions and categories will give you some methodology for helping. But do not be restrained by this, particularly in applying your skills in the interview process. As you come to feel more comfortable with a strategy, we hope you will use the procedure creatively.

In following chapters, we present strategies or technological (body of knowledge) skills designed to produce therapeutic change or healing. We define healing as an internal process that relies on the client's inner resources. The healing is the process of seeing a concern or problem differently. Healing involves tapping the wisdom within each of us and achieving self-empowerment for the potential to act or respond differently. Counselors can facilitate healing when the technology (skills and strategies) of healing is used to help the client. The technical skills and strategies the counselor uses with the client may not be adequate. We believe that therapeutic change is enhanced when the technology of healing is used within the relationship of healing. The relationship of healing means that the counselor has the capacity for self-awareness and experiencing the moment-by-moment interchange with the client. In the relationship mode of healing, the counselor acts as a channel to facilitate the client's self-awareness, alteration of perception, and potential to respond differently. The counselor's self-awareness and the capacity of consciously being in the session with the client imply trust in the relating process that conveys to the client a therapeutic alliance.

One of the most difficult parts of learning counseling skills seems to be trusting the skills to work and not being preoccupied with your own performance. We are reminded of a story in *Time* (November 29, 1976) about the conductor of the Berlin Philharmonic, Herbert von Karajan. When asked why he didn't rely more on entry and cutoff cues in conducting a large orchestra, he replied, "My hands do their job because they have learned what to do. In the performance I forget about them" (p. 82).

Preoccupation with yourself, your skills, or a particular procedure reduces your ability to relate to and be involved with another person. At first, it is natural to focus on the skill or strategy because it is new and feels a little awkward or cumbersome. But once you have learned a particular skill or strategy, the skills will be there when you need them. Gradually, as you acquire your repertory of skills and strategies, you should be able to change your focus from the procedure to the person.

Remember, also, that counseling is a complex process composed of many interrelated parts. Although different counseling stages, skills, and strategies are presented in this book in separate chapters, in practice there is a meshing of all these components. As an example, the relationship does not stop or diminish in importance when a counselor and client begin to assess problems, establish goals, or implement strategies. Nor is evaluation something that occurs only when counseling is terminated. Evaluation involves continual monitoring throughout the counseling interaction. Even obtaining a client's commitment to use strategies consistently and to monitor their effects may be aided or hindered by the quality of the relationship and the degree to which client problems and goals have been defined clearly. In the same vein, keep in mind that most client problems are complex and multifaceted. Successful counseling may involve changes in the client's feelings, observable behavior, beliefs, and cognitions. To model some of the skills and procedures you will learn, we have included cases and model dialogues in most chapters. These are intended to illustrate one example of a way in which a particular procedure can be used with a client. However, the cases and dialogues have been simplified for demonstration purposes, and the printed words may not communicate the sense of flow and direction that is normally present in counselor/client interchanges. Again, with actual clients, you will encounter more dimensions in the relationship and in the client's concerns than are reflected in the chapter examples.

Our third concern involves the way you approach the examples and practice opportunities in this book. Obviously, reading an example or doing a role-play interview is not as real as seeing an actual client or engaging in a live counseling interaction. However, some practice is necessary in any new learning program. Even if the exercises seem artificial, they probably will help you learn counseling skills. The structured practice opportunities in the book may require a great deal of discipline on your part, but the degree to which you can generalize your skills from practice to an actual counseling interview may depend on how much you invest in the practice opportunities.

Options for Using the Book

We have written this book in its particular format because each component seems to play a unique role in the learning process. But we are also committed to the idea that each person must determine the most suitable individual method of learning. With this in mind, we suggest a number of ways to use this book. First, you can go through the book and use the entire format in the way it is described in this chapter. If you do this, we suggest you familiarize yourself carefully with the format as described here. If you want to use this format but do not understand it, it is not likely to be helpful. Another way to use the book is to use only certain parts of the format in any combination you choose. You may want to experiment initially to determine which components seem especially useful. For example, you might use the postevaluation but not complete the chapter learning activities. Finally, if you prefer a "straight" textbook format, you can read only the content of the book and ignore the special format. Our intent is for you to use the book in whatever way is most suitable for your learning strategies.

ONE FINAL THOUGHT

As you go through the book, you undoubtedly will get some feel for the particular ways to use each strategy. However, we caution you against using this book as a prescriptive device, like medicine handed over a counter automatically and without thought or imagination. Similarly, one counseling strategy may not work well for all clients. As your counseling experience accumulates, you will find that one client does not use a strategy in the same way, at the same pace, or with similar results as another client. In selecting counseling strategies, you will find it helpful to be guided by the documentation concerning the ways in which the strategy has been used. But it is just as important to remember that each client may respond in an idiosyncratic manner to any particular approach. Mahoney and Mahoney (1976) emphasize that counseling is a "personalized science, in which each client's problems are given due recognition for their uniqueness and potential complexity" (p. 100). Finally, remember that almost anybody can learn and perform a skill in a rote and mechanistic manner. But not everyone shows the qualities of sensitivity and ingenuity to give the skills his or her own unique touch.

CHAPTER 2

KNOWING YOURSELF AS A COUNSELOR

The therapeutic relationship is widely accepted today by persons of various theoretical orientations to counseling as an important part of the total helping process. According to Brammer, Shostrom, and Abrego (1989), the relationship is important not only because "it constitutes the principal medium for eliciting and handling significant feelings and ideas which are aimed at changing client behavior" but also because it often determines "whether counseling will continue at all" (pp. 74–75). Without an effective therapeutic relationship, client change is unlikely to occur. There is a body of research indicating that the *quality* of the relationship "can serve as a powerful positive influence on communication, openness, persuasibility, and ultimately, positive change in the client" (Goldstein & Higginbotham, 1991, p. 22). Part of the effectiveness of the relationship depends on the counselor's knowledge and use of himself or herself.

OBJECTIVES

After completing this chapter, you will be able to

1. Identify attitudes and behaviors about yourself that might facilitate or interfere with establishing a positive helping relationship, given a written self-assessment checklist.
2. Identify issues related to values, diversity, and ethics that might affect the development of a therapeutic relationship, given six written case descriptions.

CHARACTERISTICS OF EFFECTIVE HELPERS: SELF AWARENESS, INTERPERSONAL AWARENESS, AND ATTACHMENT THEORY

The most effective helper is one who not only has developed the expertise and technical skills associated with help giving, but also has recognized and worked with his or her own interpersonal issues. If you simply have excellent skills but are not aware of yourself, you are merely a good technician.

If you have good skills and you also know yourself, you are more than a technician; then you have the capacity to be healing. In this section of the chapter, we explore qualities and behaviors present in very effective helpers.

Self-Awareness

As you grow in the counseling profession and as you gain real-life counseling experience, you will become increasingly aware that you are affected by the clients with whom you work. Clients will trigger responses in you or push sensitive buttons that may cause you to react a certain way, to feel a certain way, or even to worry and lose sleep. This happens because we all carry wounds from our past and our family of origin. None of us came from a perfectly healthy past and we have all learned ways of dealing with ourselves and other persons; some of these ways are healthy and productive, and some are unhealthy and defensive. We are not perfect, without woundedness, and to be perfect is not necessary because it is often our woundedness that creates the most capacity for healing. The important point is that we are aware of our own wounds. We are cognizant of the wounds that are opened up in us by clients, and we acknowledge that we must take responsibility for healing these hurts. Otherwise, we may inadvertently have our clients heal our wounds for us; then, it is our needs rather than theirs that are being met and the relationship loses its healing capacity for *them*.

Day (1995) has pointed out the importance of examining our motives for even entering the helping profession. He cites three of the most common motivations:

1. To do for others what someone has done for me
2. To do for others what I wish had been done for me
3. To share with others certain insights I have acquired

Day (1995) claims that while these are common motives for becoming a helper, persons motivated in these ways "encounter several common problems that potentially convert their motivation to ill impact" (p. 109). Specifically, well-intentioned counselors with such motives can become frustrated and discouraged. These feelings can then affect their behavior with clients and lead to pride rather than humility, insistence rather than invitation, telling rather than listening, demanding rather than believing, and making or coercing rather than letting. Examining our motives for being a helper and the potential of these motives is another important facet of self-awareness.

There are three specific areas about yourself we invite you to explore in further depth: competence, power, and intimacy. (See also Learning Activity #1.)

Competence. Our attitudes about ourselves can significantly influence the way we behave. People who have negative views of themselves will "put themselves down" and will either seek out or avoid types of interactions with others that confirm their negative self-image. This has serious implications for counselors. If we don't feel competent or valuable as people, we may communicate this attitude to the client. Or if we don't feel confident about our ability to counsel, we may inadvertently structure the counseling process to meet our own self-image problems or to confirm our negative self-pictures. As Corey and Corey note, it is hard to be an effective helper, if you have a "fragile ego" (G. Corey & M. Corey, personal communication, September 29, 1992).

Power. Power can be misused in counseling in several ways. First, a counselor may try to be omnipotent. For this person, counseling is manageable only when it is controllable. Such a counselor may use a variety of maneuvers to stay in control, including persuading the client to do what the counselor wants, becoming upset or defensive if a client is resistant or hesitant, and dominating the content and direction of the interview. The counselor who needs to control the interview may be more likely to engage in a power struggle with a client.

In contrast, a counselor may be afraid of power and control. This counselor may attempt to escape from as much responsibility and participation in counseling as possible. Such a counselor avoids taking control by allowing the client too much direction and by not expressing opinions. In other words, risks are avoided or ignored.

Another way that unresolved power needs can influence counseling is seen in the "lifestyle converter." This person has very strong feelings about the value of one particular lifestyle. Such a counselor may take unwarranted advantage of the influence processes in a helping relationship by using counseling to convert the client to that lifestyle or ideology. Counseling, in this case, turns into a forum for the counselor's views and pet peeves. Finally, as Woodman (1992) points out, it is quite easy to become addicted to the power of helping.

Intimacy. A counselor's unresolved intimacy needs can also significantly alter the direction and course of counseling. Generally, a counselor who has trouble with intimacy may fear rejection or be threatened by closeness and affection. A counselor who is afraid of rejection may behave in ways that meet the need to be accepted and liked by the client. For example, the counselor may avoid challenging or confronting the client for fear the client may be "turned off." Or the counselor may subtly seek positive client feedback as a reassurance of being valued and liked. Negative client cues also may be ignored because the counselor does not want to hear expressions of client dissatisfaction.

A counselor who is afraid of intimacy and affection may create excessive distance in the relationship. The counselor may avoid emotional intimacy in the relationship by ignoring expressions of positive feelings from the client or by behaving in a gruff, distant, or aloof manner and relating to the client through the "professional role." These kinds of responses may create abandonment feelings in clients.

Interpersonal Awareness and Attachment Theory

At least 85% of the people who come for counseling or therapy have problems with relationships. Counselors help the clients explore these relationship concerns. In the therapeutic relationship, the counselor's role can be characterized as that of a caregiver, and the role of the client can be depicted as one of attachment. In the therapeutic relationship, then, the attachment system is activated by the counselor's position as a safe and secure place from which the client can explore vulnerable aspects of his or her attachment style, both inside and outside the counseling relationship (Pistole & Watkins, 1995, p. 463). Counselors need to be aware of their own style in relating to their clients. Without interpersonal awareness, counselors may threaten the alliance and healing that should occur in the therapeutic relationship. We feel that attachment theory offers a valuable model for the counselor as a way to conceptualize a broad range of issues in "adult development such as psychological functioning and identity formation, career dynamics and work behavior, relationship competence, parenting behavior, and adjustment to significant life transitions" (Lopez, 1995, p. 396). Before you explore and become aware of your own attachment style, we provide a very brief overview of attachment theory in the following section.

LEARNING ACTIVITY 1   *Exploring Personal Areas That Impact Your Counseling*

This activity, adapted in part from Marianne and Gerry Corey (1992), is designed to help you explore areas of yourself that will, in some fashion, impact your counseling. Take some time to consider these questions at *different points* in your development as a helper. There is no feedback for this activity, as the responses are yours and yours alone. You may wish to discuss your responses with a colleague or your own therapist.

1. Why are you attracted to the help-giving/counseling profession?
2. What was your role in your family when you were growing up, and what impact does it have on this attraction?
3. Who are you as a person?
4. What wounds or unfinished business do you carry with you into the counseling room?

5. In what way are you healing these wounds?
6. What do you notice or what do you think your work with clients will (does) trigger in you?
7. Do you honestly believe you *need* to be a counselor or you *want* to be a counselor?
8. Whom do you have unfinished business with?
9. How do you handle being in conflict? Being confronted? Being evaluated? What defenses do you use in these situations?
10. What are repetitive or chronic issues for you? How might these affect your work with clients?
11. What do you see in other people that you consistently do not like? See whether you can find these same qualities in yourself and "own" them as also belonging to you.

Attachment Theory: A Brief Overview.   Attachment means fidelity or affectionate regard or connection. Early patterns of connecting with our caregiver provide us with a style of perception, information processing, and interpersonal patterns of behaving that produce schema-consistent experiences. These experiences produce models for later adult social relations. Our cognitive or internal working models influence the way we regulate our emotions and our social competence.

According to Bowlby (1988), human beings are predisposed or programmed to seek and form attachment with others. Infants seek attachment and their caregivers respond to those innate needs. The attachment relationship between the infant and the caregiver creates a particular set-goal for proximity between them. The set-goal for proximity functions to keep the infant within safe range of the caregiver. When a dangerous situation seems unlikely, the attachment system recedes and allows the infant or child to learn through independent exploration of the environment. The particular set-goal for desired proximity can be threatened either by the child when exploring the environment or by the caregiver when separated from the child. As the child develops, the control system that was part of the caregiver–child relationship shifts to the child, who now has more control and autonomy for self-regulation. The theory assumes that in a *secure* attachment relationship, the child has trust and confidence that the sensitive and responsive caregiver will be available. When the caregiver displays inconsistent accessi-

bility, intrudes unpredictably or insensitively into the infant's activities, or rejects the infant's plea for physical contact, the infant (and later the child) can become anxious, clinging, hypervigilant, or prematurely independent. In this *insecure* style of attachment, elements of sensitivity, reliability, confidence, and trust are reduced in the relationship. This style of attachment relationship does not foster security for the infant or later the child and presents confusing messages about connection and fidelity.

The reciprocal quality of the early attachment relationship of the caregiver with the infant and the repeated attachment-related experiences create internal working models for the infant. These cognitive models provide the infant (and later the child) with a pattern of behaving and dealing with himself or herself and others. Working models consist of accumulated knowledge about the self, attachment figures or caregivers, and attachment relationships. Also, working models include cognitive, affective, and emotional defensive strategies, images, and ways to perceive one's self and one's relationships. The working models of self and others often function outside of awareness. The configuration of our cognitive working models for relating to self and others creates an attachment style. Bartholomew (1990; Bartholomew & Horowitz, 1991) has described two underlying dimensions of attachment style: models of the self (positive–negative) and models of others (positive–negative). These dimensions define four possible attachment styles, displayed in Figure 2-1. As you see in Figure 2-1, the quality of early

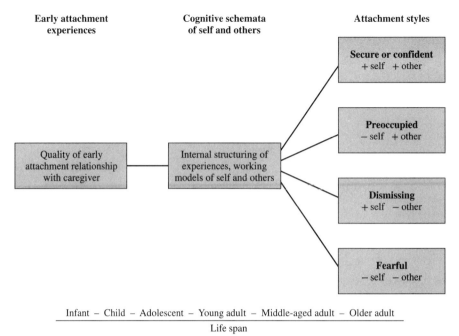

| Early attachment experiences | Cognitive schemata of self and others | Attachment styles |

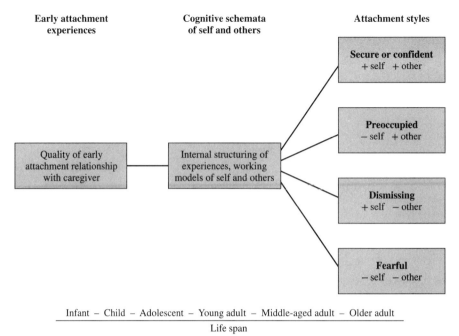

FIGURE 2-1. Overview of attachment theory

attachment experiences determines the style of internal working models of one's self and others. These experiences provide the cognitive schemata or working models of self and others during the earlier part of the life span. According to attachment theory (Bartholomew, 1990; Bartholomew & Horowitz, 1991), a person develops one of the following predominant attachment styles (see Figure 2-1): secure or confident—feels positive about self and positive about others; preoccupied—feels unsure or negative about self and positive about others; dismissing—feels confident and positive about self and untrusting or negative about others; and fearful or insecure—feels inadequate or negative about self and untrusting or negative about others.

Table 2-1 shows cognitive processes and internal working features of self and others for each attachment style. For each style, Table 2-1 includes the following cognitive processing categories: prototypical disclosure statements for each attachment style (Bartholomew & Horowitz, 1991), degree of confidence in self and others, infant experiences (Lopez, 1995), cognitive and behavioral features (Lopez, 1995), degree of comfort with closeness, concern for preoccupation with relationships, need for approval from others, relationships as secondary, trust of others, affect regulation within relationships, attachment system, self-disclosure, relationship commitment, self and other boundary disturbances, and independence/dependence balance.

We have some cautions about the use of these styles. First, it is not likely that one person uses precisely one attachment style. For example, a person may behave in a secure style with one person, but behave in a preoccupied fashion while interacting with another person. Second, a client who functions in a particular style might evoke unconsciously from a counselor a different style of interacting. In other words, the mutual social influence in building the therapeutic relationship and the reciprocal quality of the relationship need to be considered in applying attachment theory. Third, the models we use generally function outside our awareness. We are often unaware of the features and dynamics of our attachment style. A final caution is what we title the paradox of awareness. For example, a person may have secure cognitive working models of attachment, but the execution of attachment behaviors may reflect insecurity. Or this paradox can function so that a person's cognitive working models are insecure and produce anxiety, but the performance of attachment behaviors is secure. Counselors need to be aware of these cautions when working with client attachment or relationship concerns.

Despite the above cautions, attachment theory incorporates many other theories or descriptions of the way in which experience is encoded. These include (1) representation of interactions that have been generalized (Stern, 1985), (2) object relations or representations (Kohut, 1971, 1977), (3) psychosocial development of the ego across the life span

TABLE 2-1. Cognitive processes and internal working features of self and others for each attachment style

Cognitive processes and internal working features of self and others	Attachment styles			
	Secure	Preoccupied	Dismissing	Fearful
Prototype disclosure for each attachment style From Bartholomew & Horowitz, 1991.	"It is relatively easy for me to become emotionally intimate with others. I am comfortable depending on others and having others depend on me."	"I want to be completely emotionally intimate with others, but I often find that others are reluctant to get as close as I would like. I am uncomfortable being without close relationships, but I sometimes worry that others don't value me as much as I value them."	"I am comfortable without close emotional relationships. It is very important to me to feel independent and self-sufficient, and I prefer not to depend on others or have others depend on me."	"I am somewhat uncomfortable getting close to others. I want emotionally close relationships but I find it difficult to trust others completely, or to depend on them. I sometimes worry that I will be hurt if I allow myself to become too close to others."
Confidence of self and others	Positive model of self and others	Negative model of self and positive model of others	Positive model of self and negative model of others	Negative model of self and negative model of others
Infant experiences From Lopez, 1995, pp. 395–415.	Infant experiences caregiver as accessible, responsive. Attachment was secure Caregiver supported exploratory behavior while concurrently providing anxiety-reducing roles. High quality of caregiving enables infant to form an internal working model of self as worthy and competent. Model of others as responsive and dependable.	Caregiver is experienced as inconsistently responsive when needed. The condition of the relationship produces a model of self that is doubtful. Creates a model of others as potentially affirming but unreliable. Child's motivation, confidence, acts of exploration and mastery are conditional on the presence, support, and approval of attachment figures. Continued dependency of self on others retards development of affective self-regulation skills. Child is left vulnerable to stress and emotional change or instability.	Infant's efforts to get caregiver's protection, support, and caring were usually rejected. Results in working model of self as alone and unwanted. Model of others is rejecting and untrustworthy. Because of conditions of unresponsive relationships, attachment-seeking behaviors of infant become restrained. This model can result in compulsive self-sufficiency, chronic delinquency, and anti-social behavior.	Caregiver's patterns of interacting are experienced as disorganized and disoriented. These patterns result in a working model of the self as inadequate. Model of others exhibits erratic combinations of avoidant and anxious responses. Child is fearful and insecure because of inconsistent validation of self from caregiver. The same inconsistency provides anxiety when relating/attaching to others.

(continued)

Cognitive processes and internal working features of self and others	Attachment styles			
	Secure	Preoccupied	Dismissing	Fearful
Cognitive and behavioral features From Lopez, 1995, pp. 395–415.	Flexibility in shifting awareness and evaluation of self and others as needed. Flexibility permits thoughtful self-reflection. Ability to solicit, to receive, and to consider a wide range of social feedback. Ability to consider a wide range of viewpoints. Good at managing emotions. All of these abilities permit persons with this type of style to engage in constructive problem solving with others during periods of stress and instability. Capacity to draw on relationships for help and emotional support when needed.	Awareness and evaluation processes are less balanced between self and others. More focus and energy is directed toward others, which subtracts from objective self-awareness. Energy focused almost exclusively on others may encourage premature attributions of other's behavior. Persons with a preoccupied style depend on others to contain their distress and affirm their fragile selves. This process may overtax the supportive resources of their relationships and provoke their friends or partners to withdraw.	Awareness and evaluation processes are unbalanced in the direction of more energy directed toward self. This minimizes awareness of attachment-related needs. Persons with a dismissing style tend to withdraw from important others, to suppress disclosures of personal vulnerability, and to intensify involvement in impersonal activities such as work and/or hobbies. They are least likely to solicit social feedback or to accurately grasp the perspective of others. Persons with this style typically exhibit progressive alienation and self-absorption.	Awareness and evaluation processes are disorganized and disoriented. Persons with a fearful style have an impaired presentation of themselves that contributes to feelings of inadequacy and vulnerability. Their negative view of others encourages interpersonal disengagement and limits their awareness of others and access to corrective feedback. When stress occurs, they exhibit approach/avoidance behaviors. Others perceive them as unpredictable, and this undermines stable adult relationships. Without stable adult relationships, there is no opportunity for favorable revision of their negative working model, which is important for successful coping.
Comfort with closeness intimacy	Little or no discomfort with closeness	Some anxiety with closeness	Avoidant and anxious about closeness	Very fearful and insecure about closeness with others
Preoccupied with relationships	Very little preoccupation with relationships	Very preoccupied about relationships	Little preoccupation with relationships	Very preoccupied with relationships, more than other styles
Need for approval from others	Very little need for approval from others	Lots of need for approval from others	Little need for approval from others	A great deal of need for approval from others
Relationships as secondary	Relationships are extremely important, not secondary	Relationships are very important, not secondary	Relationships are secondary, self-activities are more important	Relationships are very important, but person is confused about how to maintain them
Trust	Very trustful of others	Less trusting	Sees others as untrustworthy	Has difficulty trusting others

LEARNING ACTIVITY   2   *Interpersonal Awareness*

Your effectiveness as a counselor or therapist depends on your awareness about yourself and the model you use in relating to others. This learning activity is placed at the beginning of the book because we feel that it is important for you to be aware of your style of interacting with others. This learning activity is an important start for your development as a counselor. As you gain more awareness, you may want to revise your own internal models about yourself and relating to others. Please complete this learning activity before continuing with the rest of the chapter.

I. This learning activity is designed to help you become aware of specific aspects of your life as an adult, by reflecting on your childhood. The questions are adapted from the Attachment Style Questionnaire by J. A. Feeney, P. Noller, and M. Hanrahan (1994). You may want to think about the questions before you answer them. Write the questions in a notebook or your journal with adequate space for each answer. Consider your reactions to the questions before recording them in writing. Take your time; don't rush. We feel it is important to bring into your awareness your responses to these questions. You may need a couple of days to finish this learning activity. Persevere; we feel this activity can provide you with new awareness about your early attachment experiences and their impact on your life. You may wish to discuss your responses with a colleague or your own therapist.

Think about your *relationship* with your parent(s) during your childhood; select some specific incidents that *reflect the quality and type of the relationship*. Choose a couple of incidents that depict your relationship with each parent. If you were raised by a single parent or caregiver, describe your relationship with that person. Use the incidents to help you respond to the questions in each of the following five areas.

1. *Preoccupied with relationship:* Think about the incidents you selected in relationship with each parent. Although these incidents may not apply directly, describe your *implied* reactions or feelings to the incidents with respect to the following statements: (a) I wanted to be close with my parent(s), but I felt they were reluctant to get close to me; (b) I worried that my parent(s) didn't care about me as much as I cared about them; (c) I worry now a lot about my relationships; (d) I got frustrated when my parent(s) were not available when I

needed them; (e) It was important to me to have close relationships with my parent(s) and others. As an adult, to what degree are you uncomfortable being without close relationships, but worry that others don't value you as much as you value them?

2. *Need for approval:* Now describe your *suggested* reactions or feelings to the selected incidents with respect to the following statements: (a) It was important for me to avoid doing things my parent(s) liked; (b) It was hard to make decisions unless I knew what my parent(s) thought; (c) I worried that I would not measure up to my parent(s)' expectations; (d) When I discussed my problems with my parent(s), I generally felt ashamed or foolish. As an adult, what is the degree of your need for approval?

3. *Discomfort with closeness:* Again, describe your *implied* reactions or feelings to the chosen incidents with respect to the following statements: (a) I preferred to depend on myself rather than on my parent(s); (b) I felt it was hard to trust my parents; (c) I preferred to keep to myself; (d) I wanted to get close to my parents, but I felt uneasy about it; (e) I felt comfortable depending on my parents. As an adult, to what degree do you trust others and feel comfortable depending on others?

4. *Relationships as secondary:* Again, describe your *suggested* reactions or feelings to the incidents with respect to the following statements: (a) To ask my parents for help was to admit failure; (b) I felt achieving things was more important than relating to my parents; (c) My relationship with my parent(s) was generally superficial; (d) My parent(s) were too busy with other activities to put much time into a relationship with me. To what degree do you, as an adult, view relationships as secondary?

5. *Confidence:* Finally, describe your *implied* reactions or feelings to the incidents with respect to the following statements: (a) My relationship with my parent(s) made me feel I was a worthwhile person; (b) I felt confident most of the time that my parent(s) would be there for me when I needed help; (c) My relationship with my parent(s) gave me confidence about relating to others; (d) I felt it was easy to get close to my parent(s). As an adult, to what degree do you feel confident in relating to and getting along with others?

(continued)

LEARNING ACTIVITY 2 *Continued*

There is no feedback for this learning activity. You will use your answers to increase your awareness of your own attachment style in Part II.

II. Adults with insecure attachment histories can revise their working models by examining their childhood experiences. Part I of this learning activity was designed to make you aware of your own attachment style, based on your reactions and feelings regarding parental interaction during childhood. Based on your responses in Part I, determine which attachment style generally is representative of your cognitive working model for relating to others. Use Table 2-1 to guide your answers to the following questions. When answering the following questions, it will be helpful to have images of particular people you relate to in your life. You may wish to discuss your responses with a colleague, your supervisor, or your own therapist.

1. Which attachment style disclosure statement in the first row of Table 2-1 generally fits you?
2. When you relate to others, where would you place yourself as far as confidence in yourself and others?
3. From your answers in Part I, decide which of the four descriptions of infant experiences best describes the experiences you remember.
4. Look at the four cognitive and behavioral features for each attachment style in Table 2-1. Which one of the four styles is generally representative of you? What type of people would cause you to use a more preoccupied, dismissing, or fearful style? With which type of people do you reflect a more secure model? Are you aware of people with whom you think you are secure, but with whom your behavior style reflects a preoccupied, dismissing, or fearful style? Regarding the other part of the awareness paradox, is there someone for whom you think your behavior is preoccupied or dismissing, but with whom your behavior actually reflects a more secure pattern of relating?
5. Generally, what style do your answers in Part I reflect your degree of comfort with closeness? With what type of person do you feel some anxiety about closeness? Avoidant? Fearful?
6. Are there times, situations, or people who make you feel overwhelmed or preoccupied with relationships?
7. With what types of people and/or situations do you feel the need for approval from others? How often is this need exhibited? Generally, what attachment style from others reflects your need for approval?
8. Where does your attachment style fit in regard to relationships that are primary? Secondary? What kinds of relationships are secondary for you? What types of relationships are primary?
9. Again reviewing your material and answers in Part I, what attachment style reflects your degree of trust of others? What persons are more trustful? Less trustful?
10. Where do you place yourself in managing emotions related to others? What types of people make you feel anxious or preoccupied? What type of person do you avoid? Are there people toward whom you adopt a fearful attachment style?
11. From your answers in Part I, where do you generally place your attachment style with respect to the balance between dependence and independence? What types of persons could shift you toward dependence? Toward independence?

We do not provide feedback for this learning activity. You may wish to ponder your answers throughout your development and practice of self-awareness.

(Erikson, 1968), (4) script theory of transactional analysis (Goulding & Goulding, 1978), (5) social learning theory (Bandura, 1977), (6) personal construct of cognitive psychology (Kelly, 1955; Neimeyer & Mahoney, 1995), (7) cognitive therapy (J. Beck, 1995; Baucom & Epstein, 1990), and (8) family and couple therapy (Boszormenyi-Nagy & Sparks, 1973; Bowen, 1978; Haley, 1987; Minuchin, 1974; Satir; 1972; Whitaker & Keith, 1981). Differences in adult attachment styles furnish the counselor with a conceptionalization of important variations of both adult affect regulation and social competence (Lopez, 1995, p. 404). We discuss the implications of attachment style for the therapeutic relationship in Chapter 3 and for the problem conceptualization process in Chapter 8. (See also Learning Activity #2.)

ISSUES AFFECTING HELPERS: VALUES, DIVERSITY, AND ETHICS

Although each therapeutic relationship is always defined somewhat idiosyncratically by each therapeutic dyad, certain issues will affect many of the therapeutic relationships you will encounter. These include (but are not limited to) values, diversity, and ethics.

Values

The word *value* denotes something we prize, regard highly, or prefer. Values are our feelings or attitudes about something and our preferred actions or behaviors. As an example, take a few minutes to think of (and perhaps list) five things you love to do. Now look over your list to determine how frequently and consistently you actually engage in each of these five actions. Your values are indicated by your frequent and consistent actions (Simon, 1995). If you say you value spending time with friends but you hardly ever do this, then other activities and actions probably have more value for you.

In interactions with clients, it is impossible to be "value-free." Values permeate every interaction. Counselors cannot be "scrupulously neutral" in their interactions with clients (Corey, Corey, & Callanan, 1993, p. 59). Okun (1997, p. 251) asserts that "in recent years, we have recognized that in any interpersonal relationship, whether a helping relationship or not, values are transmitted either directly or indirectly between the participants" (p. 233). Interviewers may unintentionally influence a client to embrace their values in subtle ways by what they pay attention to or by nonverbal cues of approval and disapproval (Corey et al., 1993). If clients feel they need the counselor's approval, they may act in ways they think will please the counselor instead of making choices independently according to their own value system.

Sometimes counselors push values on clients because of a loyalty to their own perspective or a belief that their values are better than the client's. There are several problems with this. First, when we push a value onto a client, the client may feel pushed, too—either pushed toward or pushed away. Second, when we impose our own perspective, we often miss seeing or hearing something of the client's real struggle.

Obviously, not all of our values have an impact on the helping process. For example, the counselor who values sailing can probably work with a client who values being a landlubber without any problem. However, values that reflect our ideas about "the good life," morality, ethics, lifestyle, roles, interpersonal living, and so forth have a greater chance of entering the helping process. The very fact that we have entered a helping profession suggests some of our values. There may be times when a referral is necessary because of an unresolved and interfering value conflict with a client. For example, a counselor who views rape as the most terrible and sexist act a person can perform might have difficulty counseling someone accused of rape. This counselor might tend to identify more with the rape victim than with the client. From an ethical viewpoint, if a counselor is unable to promote and

respect the welfare of a client, a referral may be necessary (American Counseling Association, 1995; American Psychological Association, 1992; National Association of Social Workers, 1996).

Lum (1996) asserts that although social work values are oriented toward client rights, the collective values of other cultures should also be addressed—values in the areas of family, spirituality and religion, and multicultural identity. We agree with Lum, and we have found that a significant issue for many counselors is the capacity to work with clients whose values and lifestyles are different from their own. Usually a counselor's value system and level of respect for different values are developed during formative years and are influenced by such facts as family, religious affiliation, cultural and ethnic background, and geographic location. As adults, we bring varying levels of acceptance for difference to the therapy situation. Generally, it is important to be able to expand our level of acceptance to include clients whose beliefs, values, lifestyle, and behaviors are quite different from our own. Otherwise we run the risk of regarding clients with different values or lifestyles as somehow less human than we are, or worse yet, as inferior to us. This denigration was the basis on which genocide was committed on a mass scale in the 1940s, and unfortunately, our world is still not free from these sorts of assumptions.

A first step is to acknowledge honestly what your struggle is with this issue rather than deny your blind spots, your ignorance, or your feelings of intolerance. Thus, if you are adamantly opposed to abortion, you acknowledge your opposition and also your struggle to work with clients who are pro-choice. We also believe it is particularly important for counselors to support the rights of those clients who because of their physical or mental abilities, age, gender, race, gender expression, or sexual orientation have traditionally been the recipients of prejudicial responses and often substandard treatment, ranging from negative stereotyping and emotional abuse to physical violence (Atkinson & Hackett, 1995).

For example, persons with disabilities suffer from economic and environmental discrimination and children with disabilities are more likely to be among the physically, emotionally, and sexually abused. The elderly also suffer from economic discrimination, negative stereotyping, and also abuse; elder abuse is growing at an alarming rate in the United States. And despite the advances of women in the 20th century, many have been stopped or reversed because of recent maneuvering of conservative political forces. Faludi (1991) points out as an example that the Equal Employment Opportunity Commission (EEOC), the agency responsible for enforcing anti-sex discrimination laws, has been almost dismantled. Although attitudes toward women's work roles have changed, American women still earn comparatively less in wages than their male counterparts, are much less likely than men to be selected for management positions, are more frequent victims of abuse and sexual harassment and assault than men, and are more likely to receive biased treatment from the mental health system.

Similar concerns exist with persons from racial/ethnic minorities who, despite civil rights legislation, continue to be the recipients of oppressed treatment and also harassment. It is a sad commentary on our times when a national sports magazine reports that an African American quarterback at a major state university is the only person on the entire squad to have the tires on his car slashed following the loss of a game. Unfortunately, harassment of persons of color appears to be on the upswing, if anything, and these groups continue to experience discrimination and exploitation in opportunities related to economics, education, and community inclusion, such as low rates of approval for bank loans, unfair treatment in employment, and so on. Even in the mental health field, there is an alarming equation of minority status and pathology. Stevenson and Renard (1993), for example, cite years of psychological research that assumes the behaviors of African Americans are deficient instead of different. Turner (1993) has noted the destructive effects of a body of research that has used a "deficit" theoretical model and also has assumed that the Euro-American middle class population in the United States represents the standard for measurement (p. 9).

Gay men and lesbian women also suffer from discriminatory treatment, including "social, legal, medical, psychological and religious" discrimination (Atkinson & Hackett, 1995, p. 100). Obviously, many persons in the world still view homosexuality as some pathogenic illness or see gay men and lesbian women as "perverted sinners." In addition to encountering homophobic attitudes from straight persons, lesbian women and gay men have to contend with the issue of "heterosexual privilege"—that is, the idea that heterosexuality is the more common and therefore the more acceptable sexual orientation. Generally, straight persons do not feel a need to hide their sexual orientation for fear of reprisal; for lesbian women and gay men, hiding or being "in the closet" for fear of the consequences is a pervading concern. In the mental health system, whereas treating the presenting problems of lesbian women and gay men is now customary, some counselors still make the condition of homosexuality the focus of therapy. As Coleman and Remafedi (1989) point out, gay adolescents may be at particular risk and may have the most difficulty receiving gay-affirmative counseling. While counselors often have more positive attitudes about gay men and lesbian women than do the public at large, much misinformation and stereotyping still appear to exist. These attitudes extend to variations in gender expression, as well. Instances of heterosexual bias have been reported in psychology training programs as well as in mental health systems (Pilkington & Cantor, 1996).

Diversity and Multicultural Competence

An area related to the counselor's values has to do specifically with the counselor's values about diversity. As we mentioned in Chapter 1, because our world and our society are pluralistic, we cannot expect always to encounter clients who are like us. From a cultural context there are many ways in which clients are diverse: *gender, gender expression, sexual orientation, race, ethnicity, socioeconomic level, religious or spiritual affiliation, disability status, age,* and so on. As Betancourt and López (1993) observe, "Because culture is so closely intertwined with concepts such as race, ethnicity, and social class . . . it is important to first define culture" (p. 630). These authors cite the definition of Triandis et al. (1980): Culture comprises a *physical* aspect such as distinctive buildings, government, roads, and so on as well as a *subjective* aspect that includes a "wide range of topics, such as familiar roles, communication patterns, affective styles, and values regarding personal control, individualism, collectivism, spirituality, and religiosity" (Betancourt & López, 1993, p. 630). Further, as Ho (1995) points out, subjective culture is a *culture*-level but not an *individual*-level construct. It is a useful way of describing a pattern of similar responses by those who belong to the culture, but it is not as useful in understanding worldviews of the individuals who are its members because two individuals, even from the same cultural group, may not share the same worldviews. Subjective culture focuses on "patterned regularities." It is thus of less use to counselors who, informed about these regularities, may make judgments about an individual client based on the knowledge of the cultural group, resulting in generalizations and stereotyping. Ho (1995) argues that counselors need to pay more attention to the cultural influences that exist for any given client and "that shape personality formation and various aspects of psychological functioning" (p. 5).

Ho's comment is applicable to variability within all cultural groups. As an example, consider that persons within various cultural groups may have preferences for how they are ethnically described. For example, a woman of one group may prefer to be called a Hispanic rather than a Latina

LEARNING ACTIVITY 3 *Personal Values*

This learning activity presents descriptions of six clients. If you work through this activity by yourself, we suggest that you imagine yourself counseling each of these clients. Try to generate a very vivid picture of yourself and the client in your mind. If you do this activity with a partner, you can role-play the counselor while your partner assumes the part of each client as described in the six examples. As you imagine or role-play the counselor, try to notice your feelings, attitudes, values, and behavior during the visualization or role-play process. After *each* example, stop to think about or discuss these questions:

1. What attitudes and beliefs did you have about the client?
2. Were your beliefs and attitudes based on actual or presumed information about the client?
3. How did you behave with the client?
4. What values are portrayed by your behavior?
5. Could you work with this person effectively?

There are no right or wrong answers. A reaction to this learning activity can be found in the feedback section.

Client 1
This client is a young woman who is having financial problems. She is the sole supporter of three young children. She earns her living by prostitution and pushing drugs. She states that she is concerned about her financial problems but can't make enough money from welfare or from an unskilled job to support her kids.

Client 2
You have been assigned a client who is charged with rape and sexual assault. The client, a man, tells you that he is not to blame for the incident because the victim, a woman, "asked for it."

Client 3
This client is concerned about general feelings of depression. Overweight and unkempt, the client is in poor physical condition and smokes constantly during the interview.

Client 4
The client is a young white woman who comes to you at the college counseling center. She is in tears because her parents have threatened to disown her on learning that her steady male partner is an African American.

Client 5
The client is a gay man who is angry with his gay lover because he found out from a mutual friend that his lover recently slept with another man.

Client 6
The client is an elderly man who confides in you that he is on two kinds of medicine for seizures and "thought control." He states that occasionally he believes people are out to get him. Although he hasn't been employed for some time, he is now thinking of returning to the workforce and wants your assistance and a letter of reference.

---

woman, a man may also wish to be known as a Hispanic rather than a Latino man. For another person, the Hispanic referent may feel offensive. It is best not to make any assumptions. Instead, ask the client or follow the client's lead in the ethnic description he or she uses. Waugh (1991) provides useful descriptions about various names in Table 2-2. The ethnic label a client chooses to use for self-identity is based on a number of cultural factors such as age, acculturation level, generation, political consciousness, country of birth, region of the country, and socioeconomic status (Baker & Krugh, 1996).

Other differences exist as well both between and within cultural subgroups. For example, women of some ethnic groups are still socialized as a group to be "nice" and not to express anger; some men, on the other hand, are socialized more to be strong and not to express tears or grief. However,

within any given group of women, some express anger easily, just as within any group of men, some can express sadness easily. Similarly, people of varying spiritual faiths have different spiritual beliefs, practices, and rituals. If you are of the Christian faith, you may choose to celebrate Christmas; if you are Jewish, you may choose to celebrate Hanukkah. Increasingly, some persons may decide not to honor a ritual from their particular religious affiliation. If you are Hindu or Buddhist, you most likely believe in reincarnation (as do some persons from other spiritual faiths) and also karmic theory—karmic referring to both good and bad deeds you do that follow you in future lives. You are also affected or shaped to some degree by your race and ethnic group affiliation. Euro-Americans have traditionally favored a "rugged individualism"; many of the counseling approaches reflect an emphasis on the importance of the individual,

TABLE 2-2. Facts about group names

People of color: Refers to everyone who isn't white. Has replaced Third World as term of preference.

White: Most commonly used term, a non-ethnic racial designation that has replaced Caucasian as preferred term.

Anglo: Refers only to those of English ancestry and is considered offensive by some, like the Irish, when used as an all-inclusive alternative for "white."

European American: An emerging term favored by some as a preferred alternative to white, Anglo and Caucasian, because it refers more broadly to the continent of origin and is culture-based rather than color- or race-based.

African American: Ramona H. Edelin, president of the National Urban Coalition, proposed it during an African American Summit in New Orleans in April, 1989. The Rev. Jesse Jackson's subsequent endorsement probably did more to popularize the term than anything else. Since adopted by major newspapers and prominent political leaders, such as Mayor David Dinkins of New York.

Negro: Still used by some blacks, especially in the South. The U.S. Census Bureau cited this reason for keeping it as a race category, with black, on the 1990 Census Form.

Black: Came into vogue during the black power movement of the late 1960s. Considered of great symbolic importance by many because it represented a casting-off of terminology—Negro—imposed by others. Used interchangeably with African American.

Hispanic: A term grouping all people of Spanish-speaking descent. UC-Berkeley scholar Margarita Melville says its first official sanction came with the 1968 presidential proclamation of National Hispanic Week. An ethnic advisory committee of the U.S. Census Bureau formed to recommend terminology in the 1970s recommended it as more inclusive than other terms, and it first appeared on the 1980 census form.

Latino/Latina: Generally preferred over Hispanic by Spanish-speaking people of the United States, especially in California because it emphasizes their Latin American origins, while Hispanic is seen by some as an officially imposed term connoting Spanish origins.

Chicano: A Southwestern term for working-class Mexican Americans that some consider pejorative. It gained wide political and popular favor among Mexican American activists during the 1960s Chicano civil rights movement. Still used as a term reflecting pride in the working-class roots of the Mexican American people.

Oriental: The East. Still used in many parts of the world, including Hawaii, to refer to people from Asia, or "the Orient." In disfavor among Asian Americans in many parts of the United States for reflecting an antiquated British perspective of the world.

Asian American: A now widely accepted term in the United States, especially in California, for people of Asian ancestry. Current usage is "Asian and Pacific Islanders" to include people from the Pacific Islands. Use of the term grew out of the Asian American Political Alliance that was formed by student activists on the UC-Berkeley campus in 1967 prior to the ethnic studies strike.

Native American/American Indian/Indian: All three are used to refer to indigenous peoples. Individuals often refer to themselves by tribal heritage. Native American may be in wider usage than the other two terms. The census uses Indian.

SOURCE: From sidebar accompanying "Ethnic Groups Change Names with the Times: Activists Discard Old Appellations in Ongoing Search for Self-Identity," by D. Waugh. *San Francisco Examiner (EX)*, September 2, 1991, 4th Ed., NEWS Section, p. A6. Copyright 1991 San Francisco Examiner Inc. Reprinted by permission.

individual rights, and individuation. Even this concept is shaped to some degree by learned gender roles. Recent theories of the psychology of women have emphasized the "self in relation" model (Miller, 1991)—that we are connected to and interdependent with each other, although the applicability of this model to women of color is less clear (Enns, 1993). Within-group variations of cultural differences have not received as much emphasis as between-group differences.

Cultural Syndromes and Cultural Biases in Counseling. Triandis (1996) notes that one way to differentiate cultural differences and regions is through the use of *cultural syndromes*. He defines a cultural syndrome as "a pattern of shared attitudes, beliefs, categorizations, self-definitions, norms, role definitions and *values* that are organized around a theme that can be identified among those who speak a particular language, during a specific historic period, and in a definable geographic region" (1996, p. 408). He notes that

Perhaps your visualizations or role plays revealed to you that you may have certain biases and values about sex roles, age, cultures, race, physical appearance, and rape. Some of your biases may reflect your past experiences with a person or an incident. Most people in the helping professions agree that some of our values are communicated to clients, even unintentionally. Try to identify any values or biases you hold now that could communicate disapproval to a client or could keep you from "promoting the welfare" of your client. With yourself, a peer, or an instructor, work out a plan to reevaluate your biases or to help you prevent yourself from imposing your values on clients.

examples of such cultural syndromes include *tightness* (number of norms and whether minor or major deviations from norms are punished or not), *activity-passivity* (competition, activity, and self-fulfillment versus reflective thought, cooperation, and leaving initiative to others), *honor* (sensitivity to affronts, the use of aggression for self-protection), *collectivism* (belonging to a family or being subordinated to the nation, personal goals of the collective), *individualism* (the self is independent of the group and personal goals have priority), and *vertical and horizontal relationships* (the degree to which social behavior is hierarchical or is more egalitarian). We point this out because these various cultural syndromes reflect different values and provide information about various cultural groups.

Pedersen (1987) has cited the emphasis on individualism and an overemphasis on independence as two of the more common culturally biased assumptions in many counseling approaches. Indeed, many counseling goals, processes, theories, and strategies are based on this cultural syndrome. Obviously, these goals and strategies are much less suitable to some cultural groups with different values and cultural syndromes. And, as Das (1995) points out, there is a dark side to the prevalence of "rugged individualism." When carried to an extreme, such individualism can corrode ties that bind people together and leave many successful Americans feeling alienated and without a sense of community. The emerging field of developmental social ecology (Bronfenbrenner, 1993), which explores the way in which the social milieu interacts with the individual, is, in some part, a reaction to the extreme emphasis on individualism. As counselors, we need to understand our own cultural values and their strengths and limitations, also, we must know something about the values of other cultural groups.

For other clients, such as some Asian clients, the individual is much less important than the group or family (Cimmarusti, 1996). Bringing honor to the group is greatly encouraged; behaving in a way that would reflect adversely on the group is considered shameful (Berg & Jaya, 1993; Sue & Sue, 1990). Sue (1994) notes that this concept of not "losing face" tends to keep problems such as mental health disorders out of public view. Losing face may contribute to an Asian American client's level of distress (Root, 1993). Duty and obligation are also important cultural values for Asian Americans, as is *Pakikisama*, the avoidance of direct confrontation in order to maintain harmony.

For some African Americans, the concept of family is also critical, particularly the concept of the extended family. Some research has shown that the "greatest source of life satisfaction among Black Americans is their family life" (Wilson & Stith, 1993, p. 102). Most African cultures value connectedness and close interpersonal space (Essandoh, 1995). Many African American families are also greatly influenced by their spiritual faith. Boyd-Franklin (1989) comments that the church and spiritual faith of Black Americans are so important that they must be integrated into the counseling process with many of these clients.

For many American Indian clients, connectedness is central. The extended family is paramount; Atkinson et al. (1993) refer to it as the single most important survival mechanism of this group. However, these bonds have been greatly disrupted by local, state, and federal laws targeted toward the dissolution of Native American nations. In addition, some American Indians have their own indigenous methods of healing that use a relationship with a healer as well as particular rituals or practices (LaFromboise, Trimble, & Mohatt, 1993).

Latinas/Latinos are a diverse group comprising persons from Mexico, Puerto Rico, Cuba, El Salvador, the Dominican Republic, and other Latin American groups. For many of these clients, family traditions and loyalty are very important, as is kinship with close friends because the extended family often includes persons not related by blood (Sue & Sue, 1990). Family members and issues are given priority; advice is sought from family leaders and also sometimes from a priest, folk healers, and *persona de confianza*, persons sanctioned by the family as trustworthy (Rosado & Elias, 1993). As acculturation increases, gender-role conflicts also become more apparent within the family structure (Espin, 1993; Smart & Smart, 1995b).

Having made these generalized assertions, we again want to emphasize that counselors consider the issue of diversity and culture in an individualized way for any given client. As Ho (1995) observes, "It is misleading to speak of, for instance, Native American culture as if it were a single monolithic entity—when in fact it is so rich in ethnic and linguistic diversity" (p. 10). When as counselors we are uninformed or misinformed about cultural syndromes and values and operate out of our own cultural biases, we run the risk of engaging in culturally insensitive and even injurious behaviors. For example, as Essandoh (1995) points out, boundary issues with clients who value connectedness are more diffuse and are often based on reciprocity. As a way of extending reciprocity, an African or Native American client, for example, may invite the counselor to speak at a ritual or attend a social gathering. Although this raises the ethical dilemma of multiple or dual relationships (see "Ethical Issues" later in this chapter), American counselors who interpret this as lack of good boundaries on the client's part are missing the intent. As another example, clients who favor connectedness will appear to be less individual or differentiated from their family and group. Counselors need to understand the implications of this behavior rather than label these clients immature and insist on greater self-expression, assertiveness, and so on. With clients who value vertical and hierarchical relationships, an egalitarian relationship with the counselor may not be valued by the client, may be confusing to the client, or be viewed as a sign of counselor incompetence (Essandoh, 1995). Some clients who value hierarchy may wait for the counselor to provide structure, direction, and suggestions. If the counselor is viewed as the authority figure by the client, the client may also be much more hesitant to speak up or speak out.

Sue (1994) points out that an alternative helping position that is less Eurocentric is to identify helping systems indigenous to other cultures and then to develop approaches that are consistent with the "intrinsic healing systems" of these cultures, such as the "*curranderismo,* the practitioner of Santeria, the acupuncturist, the Tai Chi Chuan teacher, or the Sufi of Islamic countries" (p. 295). When we do not seek to develop awareness about and knowledge of cultural factors, other than our own or other than the prevailing ones, we are being insensitive as helpers and also perpetuating culturally mistaken assumptions. Depression, for example, a major diagnostic category of the DSM-IV (*Diagnostic and Statistical Manual of Mental Disorders,* 4th ed.; American Psychiatric Association, 1994), does not appear to exist in some

non-Western cultures, and recommended "work" to allay grief is very different in Western culture than in some other cultural groups such as the Japanese, certain Native American tribes, and some Muslim groups (Stroebe, Gergen, Gergen, & Stroebe, 1992).

Acculturation and the Management of Diversity. *Acculturation* refers to the process by which individuals who belong to another cultural group accommodate to the attitudes, values, and behaviors of the host culture in which they live. The early models of acculturation assumed that some aspects of the indigenous culture are forgotten with time when aspects of the host culture are assimilated. Obviously, this rather unidimensional focus in acculturation is limited because it seems to assume that the host culture is preferred and has more intrinsic value than the indigenous culture. A more recent model of acculturation has been developed by Szapocznik and Kurtines (1980) and is referred to as *biculturalism.* In this model, the individual is affected in some mutually inclusive way by forces and values within both the indigenous culture and another culture. More recently, Szapocznik and Kurtines (1993) and Ho (1995) have expanded the bicultural model into a multicultural model that stresses the impact of multiple cultural environments upon an individual, recognizing the cultural pluralism that we discussed in Chapter 1. As an example of this, a Cuban American teenage boy may have been socialized to his Latino culture through his parents, to an African American culture through a friend, and to Euro-American culture through school. As Ho (1995) observes, this is an additive model in that people are able to gain competence in multiple cultures without either losing their cultural identity or having to choose one culture over another (p. 14). This model suggests better physical and psychological health for ethnic minority and ethnic majority persons (LaFromboise, Coleman, & Gerton, 1993).

Atkinson et al. (1993) point out that acculturation is related to the way that racial/ethnic minority clients view and respond to mental health services. They note that the research available suggests that "less acculturated racial/ethnic minorities are more likely to trust and express a preference for and a willingness to see an ethnically similar counselor than are their more acculturated counterparts" (p. 24). Acculturation and biculturalism seem to have important implications for the utilization of mental health services by racial/ethnic minority clients. Sue and Sue (1990) note that not only have racial/ethnic minority clients underutilized counseling services but also have tended to stop counseling often after only one contact with a therapist at a much higher rate than many Euro-American clients. Also, some ethnic minority clients receive lower quality and less frequent mental health services (Snowden & Cheung, 1990). Sue and Sue (1990) contend "that the reasons why minority-group individuals underutilize and prematurely terminate counseling/therapy lie in the biased nature of the services themselves" . . . which "are frequently antagonistic or inappropriate to the life experiences" of clients from diverse cultural groups (p. 7).

Smart and Smart (1995a) point out that acculturation is pervasive, intense, of lifelong duration, and often results in various kinds of stress. Coleman (1995) concurs that learning to cope with diversity is stressful in and of itself and is augmented when variables such as race and ethnicity are added in. He notes that clients will use various strategies varying from assimilation to separation to cope with acculturation stress and also to manage the counseling relationship. If the counselor and client use different strategies to cope with diversity, and the counselor is not aware of or sensitive to the client's strategy, the counselor runs the risk of utilizing culturally inappropriate processes, goals, and strategies.

Acculturation and the Psychology of Oppression.   Many clients have grown up in or have encountered in daily life an atmosphere of oppression—social, economic, and/or political. Clients who are not in positions of power are in effect part of an oppressed minority. Traditionally, power has been held by Euro-American, upper to middle socioeconomic-level, nondisabled, heterosexual males, and clients who are outside this circle have encountered varying degrees of oppression. Das (1995) points out that in a multicultural or pluralistic society, cultures do not exist in isolation but in effect influence one another. Therefore, contact with another culture leads to cultural change and a need for acculturation, although this acculturation "does not flow in a balanced way. Because of differences in power and status, people in the [United States] belonging to different racial and ethnic minorities are under greater pressure to acculturate to the more dominant (White) culture" (p. 49). (Note that this pressure also applies to other groups, such as gays and lesbians, women, and persons with physical and mental challenges.) As a result, not only are persons of the nondominant culture treated differentially and often denied equal access to opportunities, but they are also subject to both overt and covert forms of discrimination through oppression from the dominant group. (See Friere, 1972, *Pedagogy of the Oppressed.*) Oppression both widens the distance between various cultural groups and leads to conflict as well as intense anger. Our experience has shown that well-meaning counselors of the dominant cultural group sometimes do not understand much about oppression and its effects on clients—specifically, the anger that the legacy of

discrimination and oppression has created for some clients. As the writer, Michael Dorris, who is of French, Modoc Indian, and Irish ancestry, notes:

> I think those who are tolerant and well-meaning have to realize that there is an enormous well of anger that must be expressed before the tolerance can become reciprocal. And that makes people from the dominant group wary and impatient.
>
> There is a long hiatus before trust comes and before the other side, the side that has borne the brunt of intolerance, can tolerate back. To expect that experience to accelerate just because the oppressors finally see the light is, I think, tremendously naive.
>
> We have to recognize that there needs to be a period of healing. Utopia is not just around the bend. What happens next is that the side that has been making the rules, defining the history, explaining conquered people to themselves in language that is not their own, has got to just shut up and listen. Ask questions, but don't just ask questions, because even the questions are ethnocentric. We must listen to what people say and not secretly think that they don't know what they're talking about. (1995, p. 15)

It is important for counselors to understand the psychology of oppression because it is part of the context that the client brings to counseling and shapes part of the client's perceptions and views about the world, other people, and counseling itself. It is also a major factor in the way clients acquire a cultural identity.

Cultural Identity Development.   There appears to be a developmental process by which persons acquire an identity based on the particular racial/ethnic group to which they belong. Although a number of specific ethnic identity developmental models have been developed (Cross, 1971; Sue & Sue, 1971; Ruiz, 1990), Atkinson et al. (1993) have summarized the process by which various racial/ethnic populations develop an identity. This is referred to as the "minority identity development" or MID model (Atkinson et al., 1993, p. 27). This model is based on a common experience of oppression and discrimination and helps to explain how various ethnic groups see themselves and others in terms of their own culture and also the dominant culture. The model has five stages. Although it is considered more of a developmental process than a stage model, it is not irreversible, and not all persons move through it in the same way (Atkinson et al., 1993). These stages are summarized in Table 2-3.

Euro-Americans also experience racial identity development, although this development occurs differently for them (Helms, 1990a). The model showing their developmental process is presented in Table 2-4. At the lowest level, Euro-American individuals show disregard for differences in

TABLE 2-3. Summary of minority identity development model

Stage of minority development model	Attitude toward self	Attitude toward others of the same minority	Attitude toward others of different minority	Attitude toward dominant group
Stage 1: Conformity	Self-depreciating	Group-depreciating	Discriminatory	Group-appreciating
Stage 2: Dissonance	Conflict between self-depreciating and appreciating	Conflict between group-depreciating and group-appreciating	Conflict between dominant-held views of minority hierarchy and feelings of shared experience	Conflict between group-depreciating and group-appreciating
Stage 3: Resistance and Immersion	Self-appreciating	Group-appreciating	Conflict between feelings of empathy for other minority experiences and of culturo-centrism	Group depreciating
Stage 4: Introspection	Concern with basis of self-appreciation	Concern with nature of unequivocal appreciation	Concern with ethnocentric basis for judging others	Concern with the basis of group depreciation
Stage 5: Synergetic Articulation and Awareness	Self-appreciating	Group-appreciating	Group-appreciating	Selective appreciation

SOURCE: From *Counseling American Minorities: A Cross-Cultural Perspective,* 4th Ed., by Donald R. Atkinson, George Morten, and Derald Wing Sue. Copyright © 1993 Times Mirror Higher Education Group, Inc., Dubuque, Iowa. All Rights Reserved. Reprinted by permission.

TABLE 2-4. Summary of white racial identity attitudes as developed by Helms (1990a)

White racial identity attitudes	Characteristics	Solutions
Contact	Racial naivete; ignores racial differences; "people are people."	Avoid Blacks or befriend Blacks
Disintegration	Whiteness salient; internal standards of decency versus racial expectations/norms.	Retreat into white culture or overidentify with Blacks; paternalism.
Reintegration	"Idealized-white/negative-Black"; anger and hostility toward Blacks; stereotyping.	Maintain distance from Blacks or actively increase interactions with Blacks.
Pseudo-Independence	Sincere acceptance of racial differences; intellectualization; emotional distance.	Maintain few interactions with Blacks or actively increase interactions with Blacks.
Autonomy	Internalized positive White identity; emotional and intellectual integration of racial differences and similarities.	Value and seek cross-cultural relationships.

SOURCE: From "The Relationship between Racism and Racial Identity among White Americans: An Exploratory Investigation," by R. E. Carter, *Journal of Counseling and Development, 69,* 47. Copyright © 1990 ACA. Reprinted with permission. No further reproduction authorized without written permission of the American Counseling Association. Adapted by D. P. Pope-Davis, L. A. Menefee, and T. M. Ottavi from "The Comparison of White Racial Identity Attitudes among Faculty and Students: Implications for Professional Psychologists," *Professional Psychology, 24,* p. 445, Table 1. Copyright © 1993 by the American Psychological Association.

TABLE 2-5. Summary of identity formation process and therapeutic tasks

Stage	Client experiences and tasks	Therapist's tasks
First relationships (Coleman) Stable identity (Lewis) Identity acceptance (Cass)	Learning to function in a same-sex relationship; settling into new identity; *developing a "chosen family";* developing relationships that match own values; valuing lesbians; establishing a place in the subculture; choosing whom to be out to; learning how to pass as heterosexual and where to do so.	Continue intimacy work; educate couples in communication skills and in getting support for the relationship; help establish self-acceptance and outside support; help with establishing and maintaining relationships; help with decision making about coming out to others; help with incongruency about being accepted in some places and not in others; facilitate awareness of costs of hiding; help develop coping strategies for responding to others' negative reactions.
Identity pride (Cass)	*Valuing homosexuality as better than heterosexuality;* immersion in gay subculture.	Continue assisting in decisions about disclosure; facilitate expression of feelings about negative and positive responses received.
Integration (Coleman) Integration (Lewis) Identity synthesis (Cass)	*Incorporate public and private identities* into one self-image; accomplish other developmental tasks of adulthood, midlife, and old age; lesbianism recedes into other aspects of identity and demands less energy.	Assist with ongoing adaptation to identity; help with accepting costs and advantages of client's chosen degree of self-labeling and disclosure; help with ongoing development of social networks and personal relationships; facilitate view of lesbianism as one of many aspects of the self.

Based on Coleman (1982), Lewis (1984), Cass (1979), and Sophie (1982).

SOURCE: From *Psychotherapy with Lesbian Clients: Theory into Practice,* by K. L. Falco, pp. 100–101. Copyright 1991 by Brunner/Mazel.

race and culture; at the highest level, they recognize and accept commonalities and differences. Also, persons at this stage are actively working on behalf of oppressed groups.

Cass (1979) and others have developed a homosexual identity development model to describe the process by which gay men and lesbian women acquire their identity (see Table 2-5). Several models of women's identity development have been developed as well.

These models are useful for counselors in a number of ways. First, they help develop awareness of the role of oppression in the development of a client's cultural identity. As we discuss in Chapter 4, oppression is a major factor in the development of trust (and mistrust) for many racial/ethnic clients. Second, these models can help counselors understand identity development issues for clients from oppressed groups (Atkinson et al., 1993; Richardson & Helms, 1994).

Also, as Helms (1990b) has pointed out, certain combinations of cross-cultural counseling dyads have implications for both counseling process and outcome. For example, *parallel* dyads seem to be stable over time whereas in *progressive* dyads, racial tension and also the possibility of

change coexist. Overt racial conflict appears in *regressive* dyads and is even stronger in *crossed* dyads. These sorts of interactions can occur not only in the counseling relationship but in any setting in which there is a difference in perceived and real power in the dyad. Finally, all these identity development models are significant because prior to their existence, everyone had been assumed to develop a sense of identity in a similar fashion to everyone else—and in a fashion based on the experience of nonoppressed persons. This concept made the group in power the normative group, implying a set of standards and a model of "right" or "wrong" identity development using a majority group as the baseline. As Yoder and Kahn (1993) have observed, caution must be used so that no one group is held as the baseline from which other groups supposedly deviate or to which they conform (p. 848).

Cultural Competence and Guidelines for Working with Diverse Groups of Clients. Recent emphases in the training of counselors call for them to acquire a set of multicultural counseling competencies that include knowledge, attitudes and beliefs, and skills necessary to work effectively with diverse groups of clients (Das, 1995; Lum, 1996; Sue, Ivey, & Pedersen, 1996). A set of multicultural counseling competencies was developed in 1992 (Sue, Arredondo, & McDavis, 1992) and recently was operationalized (Arredondo et al., 1996). These proposed competencies are found in Appendix A.

There are many guidelines to consider in working effectively with issues of diversity. We have found an excellent summary of such guidelines in the work of Wilson and Stith on culturally sensitive therapy with Black clients. We have slightly modified their list, which follows, because we believe it is applicable to counseling with all clients from diverse groups.

1. Become aware of the historical and current experience of being a member of an oppressed group in America.
2. Consider value and cultural differences between various American ethnic groups and how your personal values influence the way you conduct therapy.
3. Consider how your personal values influence how you view both the presenting problem and the goals for therapy.
4. Include the value system of the client in the goal-setting process. Be sensitive to spiritual values and the value of the family and the church.
5. Be sensitive to and accepting of variations in family norms due to normal adaptations to stress.
6. Be aware of how ineffective verbal and nonverbal communication due to cultural variation in communication can lead to premature termination of therapy. Become familiar with nonstandard English and accept its use by clients.
7. Consider the client's problem in the large context. Include the extended family, other significant individuals, and larger systems in your thinking, if not in the therapy session.
8. Be aware of your client's racial identification, and do not feel threatened by your client's cultural identification with his or her own race.
9. Learn to acknowledge and to be comfortable with your client's cultural differences.
10. Consider the appropriateness of specific therapeutic models or interventions to specific clients. Do not apply interventions without considering the unique aspects of each client (Wilson & Stith, 1993, p. 109).

The importance of becoming familiar with the beliefs and practices of persons with backgrounds different from your own cannot be overestimated. If, for example, you are working with a client who believes in reincarnation, or who believes his or her grades have brought shame to the family, or who is being silent and nondisclosive as a sign of respect, or whose life is affected daily by a chronic and limiting disability, you run the risk of alienating the client if you are either ignorant or nonempathic about such issues.

Some members of the helping professions now believe in culturally sensitive therapy so strongly that it has become part of ethical codes of behavior and guidelines for practice. For example, an American Psychological Association (APA) task force has produced "Guidelines to Providers of Psychological Services to Ethnic, Linguistic, and Culturally Diverse Populations" (APA, 1993). The newest American Counselors Association (ACA) code of ethics contains a separate section on respecting diversity (ACA, 1995). In spite of these advances, a recent survey of APA members who had completed doctoral programs in clinical, counseling, and school psychology showed that very few respondents felt competent to provide adequate counseling services to ethnic minority and diverse clients (Allison, Crawford, Echemendia, Robinson, & Knepp, 1994). As Allison et al. (1994) note, "To expand the training of therapists beyond meeting the needs of YAVIS (i.e., young, attractive, verbal, intelligent, and successful) clients (Schoefield, 1964) and to prepare therapies to meet the mental health needs of an increasingly diverse population will require a continuing appreciation of the dimensions of human difference" (p. 793). At the end of this chapter, we provide a list of excellent multicultural resources as "Suggested Readings." We also have attempted to integrate a multicultural focus throughout this book.

Ethical Issues

The therapeutic relationship needs to be handled in such a way as to promote and protect the client's welfare. Indeed, as Brammer, Shostrom, and Abrego (1989, p. 81) observe, ethical handling of client relationships is a distinctive mark of the professional counselor/therapist. All professional groups of helpers have a code of ethics adopted by their profession, such as the ethical standards of the American Counseling Association (1995), the American Psychological Association (1992), and the National Association of Social Workers (1996) (see Appendixes B, C, and D). Marriage and family therapists, rehabilitation counselors, and health professionals also have their own sets of ethical standards. The counselor's value system is an important factor in determining ethical behavior. Behaving unethically can have consequences such as loss of membership in professional organizations and malpractice lawsuits. Of most consequence is the detrimental effect that unethical behavior can have on clients and on the therapeutic relationship.

LEARNING ACTIVITY **4** *Diversity*

This activity, adapted from Wilson and Stith (1993), is designed to help you expand your awareness of your own cultural group and of other groups as well. You may wish to do this with a partner or in a small group.

1. Describe your awareness of the historical and current experience of your own cultural group and any of the following groups that are not representative of your cultural group:
   African American/Black
   American Indian/Native American
   Asian American/Asian
   Euro-American/White
   Latino/Latina

2. Describe the value systems and variations in family norms for your own cultural group and also for any of the following groups that are not representative of your cultural group:
   African American/Black
   American Indian/Native American
   Asian American/Asian
   Euro-American/White
   Latino/Latina
3. Describe current examples of *stereotyping* and *oppression* for your own cultural group and also for any of the above groups that are not representative of your cultural group.

All student and practicing counselors and therapists should be familiar with the ethical codes of their profession. The following discussion highlights a few of the more critical issues and in no way is intended to be a substitute for careful scrutiny of existing ethical codes of behavior.

Client Welfare. Counselors are obligated to protect the welfare of their clients. In most instances, this means putting the client's needs first. It also means ensuring that you are intellectually and emotionally ready to give the best that you can to each client—or to see that the client has a referral option if seeing you is not in the client's best interests.

Confidentiality. Closely related to protecting client well-being is the issue of confidentiality. Counselors who breach client confidences can do serious and often irreparable harm to the therapeutic relationship. Counselors are generally not free to reveal or disclose information about clients unless they have first received written permission from the client. Exceptions vary from state to state but generally include the following as summarized by Vasquez (1993):*

1. If the client is a danger to self or others
2. If the client waives rights to privilege
3. If the client is a minor, and the therapist reasonably suspects child abuse

4. If the client is elderly or disabled and the therapist believes the client is a victim of physical abuse
5. If the client files suit against the therapist for breach of a duty
6. If the client is involved in court action and the client releases records

All these limits require a disclosure statement, which the client is given at the beginning of therapy.

On June 13, 1996, in the United States Supreme Court case *Jaffee v. Redmond,* the concept of confidentiality was explored by the Court. It ruled that communications between therapists and their clients are privileged and therefore protected from forced disclosure in cases arising under federal law. This ruling, which applies to federal courts, now brings them in line with courts in the 50 states and the District of Columbia, all of which have some type of therapist-client privilege.

A recent issue involving the limits of confidentiality has to do with whether the client who tests positive for the human immunodeficiency virus (HIV) is regarded as a danger to others. The question is whether the therapist has a duty to breach confidentiality and to warn another party of the client's HIV status. There are complex ethical, legal, and therapeutic issues surrounding this question, and although various critics have speculated on the issue, the clearest answers we have found are provided by Schlossberger and Hecker (1996), who draw the following two conclusions:

1. Although the therapist has a legal duty to warn others subject to an illegal danger [based on a case known as the

---

* SOURCE: From "The 1992 Ethics Code: Implications for the Practice of Psychotherapy," by M. Vasquez, *Texas Psychologist* (1993, June), *45*, 11. Copyright 1993 by the Texas Psychological Association. Reprinted by permission.

Tarasoff Case], the therapist has no duty to intervene when clients pose dangers that society, through law, grants them the right to pose. (p. 32)

2. Unless state law (criminal or tort) directly or indirectly generally requires seropositive clients to inform their partners, therapists have no legal duty to warn. (p. 33)

Confidentiality also has implications for effective multicultural practice. Lum (1996) points out that in a multicultural practice, the ethical value of confidentiality needs to be carefully explained to clients as variations about the understanding of this principle exist within different cultural groups.

Dual Relationships.   A dual relationship is one in which the counselor is in a therapeutic relationship with the client and simultaneously or consecutively also has another kind of relationship with that same person, such as an administrative, instructional, supervisory, social, sexual, or business relationship. Dual relationships are problematic because they reduce the counselor's objectivity, confuse the issue, and put the client in a position of diminished consent and potential abandonment. Counselors should avoid becoming involved in dual relationships. If such involvement is unavoidable, make use of the referral option so that two relationships are not carried on simultaneously or consecutively.

Sometimes dual relationships are unavoidable, such as in rural settings where therapists are more likely to know clients in other contexts and are less likely realistically to be able to refer clients elsewhere. Earlier we also pointed out that dual relationships may arise in instances in which a client's cultural and social values support multiple roles with a help-giving person. In instances when dual relationships cannot be avoided, it is important to seek face-to-face supervision and consultation about the issue. It is also important to document your discussion (Smith & Fitzpatrick, 1995). And under no circumstances should you agree to offer professional services when you can foresee clearly that a prior existing relationship would create harm or injury in any way. Obviously, sexual contact between therapist and client is never warranted under any circumstance and is explicitly proscribed by all the professional codes of ethics. Smith and Fitzpatrick (1995) also address the failure to maintain appropriate treatment boundaries by the use of excessive self-disclosure on the counselor's part. They note that this behavior is a common precursor to therapist-initiated sexual contact. They state: "Typically, there is a gradual erosion of treatment boundaries before sexual activity is initiated. Inappropriate therapist self-disclosure, more than any other kind of boundary violation, most frequently precedes therapist-patient sex" (p. 503).

Client Rights.   Establishing an effective therapeutic relationship entails being open with clients about their rights and options during the course of therapy. Nothing can be more damaging to trust and rapport than to have the client discover in midstream that the therapist is not qualified to help with a particular issue or that the financial costs of therapy are high or that therapy involves certain limitations or nonguarantees of outcomes. At the outset, the therapist should provide the client with enough information about therapy to help the client make informed choices (also called empowered consent). Usually this means discussing four general aspects of counseling with clients: (1) confidentiality and its limitations, (2) the procedures and goals of therapy and any possible side effects of change (such as anxiety, pain, or disruption of the status quo), (3) the qualifications and practices of the therapist, and (4) available resources and sources of help other than that particular therapist and other than traditional therapy (for example, self-help groups or indigenous healers).

Referral.   It is important for counselors to handle referral effectively and responsibly. Referring a client to another therapist may be necessary when, for one reason or another, you are not able to provide the service or care that the client requires or when the client wants another helper (Cheston, 1991). Careful referral, however, involves more than just giving the client the name of another counselor. The client should be given a choice among therapists who are competent and are qualified to deal with the client's problems. The counselor must obtain written client permission before discussing the case with the new therapist. And to protect against abandonment, the counselor should follow up on the referral to determine whether the appropriate contact was made. Also, referrals are mandated in situations in which the services of the therapist could be interrupted because of illness, death, relocation, financial limitations, or any other form of unavailability (Vasquez, 1994).

SUMMARY

Part of being an effective helper is knowing yourself so that you can use yourself as a therapeutic instrument—as a healer. This self-awareness enables you to respond to each client as a unique person, to develop acceptance for clients who have different values from you, and also to develop respect for clients who are from various cultural groups. Finally, your self-awareness should extend into your ethical code of behavior and allow you to put the client's welfare first.

POSTEVALUATION

Part One

According to Chapter Objective One, you will be able to identify attitudes and behaviors about yourself that could facilitate or interfere with establishing a positive helping relationship. In this activity, we present a Self-Rating Checklist. This checklist refers to characteristics of effective helpers. Your task is to use the checklist to assess yourself *now* with respect to these attitudes and behaviors. If you haven't yet had any or much contact with actual clients, try to use this checklist to assess how you believe you would behave in actual interactions. Identify any issues or areas you may need to work on in your development as a counselor. Discuss your assessment in small groups or with an instructor, colleague, or supervisor. There is no written feedback for this part of the postevaluation.

Self-Rating Checklist

Check the items that are most descriptive of you.

A. Competence Assessment

_____ 1. Constructive negative feedback about myself doesn't make me feel incompetent or uncertain of myself.

_____ 2. I tend to put myself down frequently.

_____ 3. I feel fairly confident about myself as a helper.

_____ 4. I am often preoccupied with thinking that I'm not going to be a competent counselor.

_____ 5. When I am involved in a conflict, I don't go out of my way to ignore or avoid it.

_____ 6. When I get positive feedback about myself, I often don't believe it's true.

_____ 7. I set realistic goals for myself as a helper that are within reach.

_____ 8. I believe that a confronting, hostile client could make me feel uneasy or incompetent.

_____ 9. I often find myself apologizing for myself or my behavior.

_____ 10. I'm fairly confident I can or will be a successful counselor.

_____ 11. I find myself worrying a lot about "not making it" as a counselor.

_____ 12. I'm likely to be a little scared by clients who would idealize me.

_____ 13. A lot of times I will set standards or goals for myself that are too tough to attain.

_____ 14. I tend to avoid negative feedback when I can.

_____ 15. Doing well or being successful does not make me feel uneasy.

B. Power Assessment

_____ 1. If I'm really honest, I think my counseling methods are a little superior to other people's.

_____ 2. A lot of times I try to get people to do what I want. I might get pretty defensive or upset if the client disagreed with what I wanted to do or did not follow my direction in the interview.

_____ 3. I believe there is (or will be) a balance in the interviews between my participation and the client's.

_____ 4. I could feel angry when working with a resistant or stubborn client.

_____ 5. I can see that I might be tempted to get some of my own ideology across to the client.

_____ 6. As a counselor, "preaching" is not likely to be a problem for me.

_____ 7. Sometimes I feel impatient with clients who have a different way of looking at the world than I do.

_____ 8. I know there are times when I would be reluctant to refer my client to someone else, especially if the other counselor's style differed from mine.

_____ 9. Sometimes I feel rejecting or intolerant of clients whose values and lifestyles are very different from mine.

_____ 10. It is hard for me to avoid getting into a power struggle with some clients.

C. Intimacy Assessment

_____ 1. There are times when I act more gruff than I really feel.

_____ 2. It's hard for me to express positive feelings to a client.

_____ 3. There are some clients I would really like to have as friends more than as clients.

_____ 4. It would upset me if a client didn't like me.

_____ 5. If I sense a client has some negative feelings toward me, I try to talk about it rather than avoid it.

_____ 6. Many times I go out of my way to avoid offending clients.

_____ 7. I feel more comfortable maintaining a professional distance between myself and the client.

_____ 8. Being close to people is something that does not make me feel uncomfortable.

_____ 9. I am more comfortable when I am a little aloof.

(continued)

POSTEVALUATION (continued)

_____ 10. I am very sensitive to how clients feel about me, especially if it's negative.

_____ 11. I can accept positive feedback from clients fairly easily.

_____ 12. It is difficult for me to confront a client.

Part Two

According to Chapter Objective Two, you will be able to identify issues related to values, ethics, and diversity that could affect the development of a therapeutic relationship, given six written case descriptions. In this activity, read each case description carefully; then identify in writing the major kind of issue reflected in the case by matching the type of issue with the case descriptions listed below. Feedback follows the postevaluation (p. 34).

Type of Issue

A. Values conflict
B. Values stereotyping
C. Ethics—breach of confidentiality
D. Ethics—client welfare and rights
E. Ethics—referral
F. Diversity issue

Case Description

_____ 1. You are counseling a client who is in danger of failing high school. The client states that he feels like a failure because all the other students are so smart. In an effort to make him feel better, you tell him about one of your former clients who also almost flunked out.

_____ 2. A 58-year-old man who is having difficulty adjusting to life without his wife, who died, comes to you for counseling. He has difficulty in discussing his concern or problem with you, and he is not clear about your role as a counselor and what counseling might do for him. He seems to feel that you can give him a tranquilizer. You tell him that you are not able to prescribe medication, and you suggest that he seek the services of a physician.

_____ 3. A fourth-grade girl is referred to you by her teacher. The teacher states that the girl is doing poorly in class yet seems motivated to learn. After working with the girl for several weeks, including giving a battery of tests, you conclude that she has a severe learning disability. After obtaining her permission to talk to her teacher, you inform her teacher of this and state that the teacher might as well not spend too much more time working on what you believe is a "useless case."

_____ 4. You are counseling a couple who are considering a trial separation because of constant marital problems. You tell them you don't believe separation or divorce is the answer to their problems.

_____ 5. A Euro-American counselor states in a staff meeting that "people are just people" and that he does not see the need for all this emphasis in your treatment facility on understanding how clients from diverse racial/ethnic/cultural backgrounds may be affected differentially by the therapy process.

_____ 6. A client comes into a mental health center and requests a counselor from his own culture. He also indicates that he would consider seeing a counselor who is not from his culture but who shares his worldview and has some notion of his cultural struggles. He is told that it shouldn't matter who he sees because all the therapists on the staff are value-free.

SUGGESTED READINGS

Ainsworth, M. D. S. (1989). Attachment beyond infancy. _American Psychologist, 44,_ 709–716.

Aquino, J., Russell, D. W., Cutrona, C. E., & Altmaier, E. (1996). Employment status, social support, and life satisfaction among the elderly. _Journal of Counseling Psychology, 43,_ 480–489.

Arredondo, P., Toporek, R., Brown, S. P., Jones, J., Locke, D. C., Sanchez, J., & Stadler, H. (1996). Operationalization of the multicultural counseling competencies. _Journal of Multicultural Counseling and Development, 24,_ 42–78.

Arthur, G. L., & Swanson, C. P. (1993). _Confidentiality and privileged communication._ Alexandria, VA: American Counseling Association.

Atkinson, D. R., Brown, M. T., Parham, T. A., Matthews, L. G., Landrum-Brown, J., & Kim, A. U. (1996). African American client skin tone and clinical judgments of African American and European American psychologists. _Professional Psychology, 27,_ 500–505.

Atkinson, D. R., Morten, G., & Sue, D. W. (1993). _Counseling American minorities_ (4th ed.). Madison, WI: Brown & Benchmark.

Atkinson, D. R., & Hackett, G. (1995). _Counseling diverse populations._ Madison, WI: Brown & Benchmark.

Axelson, J. (1993). *Counseling and development in a multicultural society* (2nd ed.). Pacific Grove, CA: Brooks/Cole.

Bartholomew, K. (1990). Avoidance of intimacy: An attachment perspective. *Journal of Social and Personal Relationships, 7,* 147–178.

Bartholomew, K., & Horowitz, L. M. (1991). Attachment styles among young adults: A test of a four category model. *Journal of Personality and Social Psychology, 61,* 226–244.

Betancourt, H., & Lopez, I. (1993). The study of culture, ethnicity, and race in American psychology. *American Psychologist, 48,* 629–637.

Bowlby, J. (1988). *A secure base: Parent-child attachments and healthy human development.* New York: Basic Books.

Buhrke, R. A., & Douce, L. A. (1991). Training issues for counseling psychologists in working with lesbian women and gay men. *The Counseling Psychologist, 19,* 216–234.

Cheston, S. E. (1991). *Making effective referrals.* New York: Gardner Press.

Coleman, H., Wampold, B., & Casali, S. (1995). Ethnic minorities' ratings of ethnically similar and European American counselors: A meta-analysis. *Journal of Counseling Psychology, 42,* 55–64.

Comas-Diaz, L., & Greene, B. (1994). *Women of color.* New York: Guilford.

Corey, G., Corey, M., & Callanan, P. (1993). *Issues and ethics in the helping professions* (4th ed.). Pacific Grove, CA: Brooks/Cole.

Corey, M., & Corey, G. (1993). *Becoming a helper* (2nd ed.). Pacific Grove, CA: Brooks/Cole.

Day, J. (1994). Obligation and motivation: Obstacles and resources for counselor well-being and effectiveness. *Journal of Counseling and Development, 73,* 108–110.

Dworkin, S. H., & Guiterrez, F. J. (1992). *Counseling gay men and lesbians.* Alexandria, VA: American Counseling Association.

Enns, C. (1993). Twenty years of feminist counseling and therapy. *The Counseling Psychologist, 21,* 3–87.

Elliott, D. M., & Guy, J. D. (1993). Mental health professionals versus non mental-health professionals: Childhood trauma and adult functioning. *Professional Psychology: Research and Practice, 24,* 83–90.

Feeney, J. A., Noller, P., & Hanrahan, M. (1994). Assessing adult attachment. In M. B. Sperling & W. H. Berman (Eds.), *Attachment in adults: Clinical and development perspectives* (pp. 128–152). New York: Guilford.

Fontaine, J., & Hammond, N. (1994). Twenty counseling maxims. *Journal of Counseling and Development, 73,* 223–226.

Fry, P. S. (1992). Major social theories of aging and their complications for counseling concepts and practice: A critical review. *The Counseling Psychologist, 20,* 246–329.

George, C., Kaplan, N., & Main, M. (1985). The Adult Interview. Unpublished protocol, Department of Psychology, University of California, Berkeley.

Good, G., Robertson, J., O'Neil, J., Fitzgerald, L., Stevens, M., DeBord, K., Bartels, K., & Braverman, D. (1995). Male gender role conflict: Psychometric issues and relations to psychological distress. *Journal of Counseling Psychology, 42,* 3–10.

Helms, J. (Ed.). (1990). *Black and white racial identity: Theory, research and practice.* New York: Greenwood Press.

Herlihy, B., & Corey, G. (1992). *Dual relationships in counseling.* Alexandria, VA: American Counseling Association.

Ho, David Y. F. (1995). Internalized culture, culturocentrism, and transcendence. *The Counseling Psychologist, 23,* 4–24.

Holtzen, D., Kenny, M., & Mahalik, J. (1995). Contributions of parental attachment to gay or lesbian disclosure to parents and dysfunctional cognitive processes. *Journal of Counseling Psychology, 42,* 350–355.

Ivey, A. E., Ivey, M. B., & Simek-Morgan, L. (1998). *Counseling and psychotherapy: A multicultural perspective.* Needham Heights, MA: Allyn & Bacon.

Kelly, A., Sedlacek, W., & Scales, W. (1994). How college students with and without disabilities perceive themselves and each other. *Journal of Counseling & Development, 73,* 178–182.

Liddle, B. (1996). Therapist sexual orientation, gender, and counseling practices as they relate to ratings of helpfulness by gay and lesbian clients. *Journal of Counseling Psychology, 43,* 394–401.

Lopez, F. G. (1995). Contemporary attachment theory: An introduction with implications for counseling psychology. *The Counseling Psychologist, 23,* 395–415.

Lum, D. (1996). *Social work practice and people of color.* Pacific Grove, CA: Brooks/Cole.

Main, M., Kaplan, N., & Cassidy, J. (1985). Security in infancy, childhood, and adulthood: A move to the level of representation. *Monographs for the Society for Research in Child Development, 50,* 66–104.

Marinelli, R., & Dell Orto, A. (Eds.). (1991). *The psychological and social impact of disability* (3rd ed.). New York: Springer.

McDowell, W. A., Bills, G. F., & Eaton, M. W. (1989). Extending psychotherapeutic strategies to people with disabilities. *Journal of Counseling & Development, 68,* 151–154.

Meara, N., Schmidt, L., & Day, J. (1996). A foundation for ethical decisions, policies, and character. *The Counseling Psychologist, 24,* 4–77.

Okun, B. F. (1997). *Effective helping* (5th ed.). Pacific Grove, CA: Brooks/Cole.

Payton, C. R. (1994). Implications of the 1992 (APA) Ethics code for diverse groups. *Professional Psychology, 25,* 317–320.

Pilkington, N. W. & Cantor, J. M. (1996). Perceptions of heterosexual bias in professional psychology programs: A survey of graduate students. *Professional Psychology, 27,* 604–612.

Pistole, M. C., & Watkins, Jr., C. E. (1995). Attachment theory, counseling process, and supervision. *The Counseling Psychologist, 23,* 457–478.

Ponteretto, J. G., Rieger, B. P., Barrett, A., & Sparks, R. (1994). Assessing multicultural counseling competence: A review of instrumentation. *Journal of Counseling and Development, 72,* 316–322.

Ridley, C., Mendoza, D., & Kanitz, B. (1994). Multicultural training: Re-examination, operationalization, and integration. *The Counseling Psychologist, 22,* 227–289.

Rodolfa, E., Hall, T., Holms, V., Pavena, A., Komatz, D., Antunez, M., & Hall, A. (1994). The management of sexual feelings in therapy. *Professional Psychology, 24,* 168–172.

Rothbard, J. C., & Shaver, P. R. (1994). Continuity of attachment across the life span. In M. B. Sperling & W. H. Berman (Eds.), *Attachment in adults: Clinical and development perspectives* (pp. 31–71). New York: Guilford.

Simon, S. (1995). *Values clarification* (2nd ed.). New York: Warner Books.

Skovholt, T., & Ronnestad, M. (1995). *The evolving professional self.* New York. Wiley.

Smart, J. F., & Smart, D. W. (1995). Acculturative stress: The experience of the Hispanic immigrant. *The Counseling Psychologist, 23,* 25–42.

Smith, D., & Fitzpatrick, M. (1995). Patient-therapist boundary issues. *Professional Psychology, 26,* 499–506.

Stanard, R., & Hazler, R. (1995). Legal and ethical implications of HIV and duty to warn for counselors. *Journal of Counseling & Development, 73,* 397–400.

Steenbarger, B. N. (1993). A multicontextual model of counseling. *Journal of Counseling & Development, 72,* 8–15.

Sue, D. W., & Sue, D. (1990). *Counseling the culturally different.* New York: Wiley.

Sue, D. W., Ivey, A., & Pedersen, P. (1996). *A theory of multicultural counseling and therapy.* Pacific Grove, CA: Brooks/Cole.

Swenson, L. C. (1997). *Psychology and law for the helping professions* (2nd ed.). Pacific Grove, CA: Brooks/Cole.

Szapocznik, J., & Kurtines, W. (1993). Family psychology and cultural diversity. *American Psychologist, 48,* 400–407.

Vasquez, M. (1993). The 1992 ethics code: Implications for the practice of individual psychotherapy. *Texas Psychologist, 45,* 11.

Wehrly, B. (1995). *Pathways to multicultural counseling competence.* Pacific Grove, CA: Brooks/Cole.

Whitman, J. (1995). Providing training about sexual orientation in counselor education. *Counselor Education & Supervision, 35,* 168–176.

FEEDBACK
POSTEVALUATION

II. Chapter Objective Two

1. C: Ethics—breach of confidentiality. The counselor broke the confidence of a former client by revealing his grade difficulties without his consent.

2. E: Ethics—referral. The counselor did not refer in an ethical or responsible way, because of failure to give the client names of at least several physicians or psychiatrists who might be competent to see the client.

3. B: Values stereotyping. The counselor is obviously stereotyping all kids with learning disabilities as useless and hopeless (the "label" is also not helpful or in the client's best interest)

4. A: Values conflict. Your values are showing: Although separation and divorce may not be your solution, be careful of persuading clients to pursue your views and answers to issues.

5. F: Diversity issue. The Euro-American counselor is obviously ignorant and also at Stage 1 (contact) of the white racial identity scale (racial naivete).

6. F and B: Diversity issue and values stereotyping. The response to this client ignores the client's racial identity status and also responds in a stereotypical way to his or her request.

INGREDIENTS OF AN EFFECTIVE
HELPING RELATIONSHIP

The quality of the therapeutic relationship remains the foundation on which all other therapeutic activities are built (Sexton & Whiston, 1994). The last several years have brought renewed acknowledgment of the significance of the most important components of that relationship (Gelso & Carter, 1994). Studies have been conducted to determine which factors in the relationship seem to be consistently related to positive therapeutic outcomes (Hill & Corbett, 1993; Sexton & Whiston, 1994; Steenbarger, 1994; Whiston & Sexton, 1993), and three have emerged:

1. *Facilitative Conditions:* Empathy, positive regard, and genuineness. Conditions, especially empathy, which, if present in the counselor and perceived by the client, contribute a great deal to the development of the relationship.
2. *Transference and Countertransference:* Issues of emotional intensity and objectivity felt by both client and counselor. Usually related to unfinished business with one's family of origin, yet triggered by and felt as a real aspect of the therapy relationship.
3. *A Working Alliance:* A sense in which both therapist and client work together in an active, joint effort toward particular goals and outcomes.

In this chapter, we describe these components of the therapeutic relationship and how they are utilized with clients. Also, we discuss the way such variables as client gender and cultural status may be differentially affected by these conditions.

OBJECTIVES

After completing this chapter you will be able to

1. Communicate the three facilitative conditions (empathy, genuineness, positive regard) to a client, given a role-play situation.

2. Identify issues related to transference and countertransference that might affect the development of the relationship and the working alliance, given four written case descriptions.

FACILITATIVE CONDITIONS

Facilitative conditions have roots in a counseling theory developed by Rogers (1951), called *client-centered* or *person-centered* therapy. Because this theory is the basis of these fundamental skills, we describe it briefly in this section.

The first stage of this theory (Rogers, 1942) was known as the *nondirective* period. The counselor essentially attended and listened to the client for the purpose of mirroring the client's communication. The second stage of this theory (Rogers, 1951) was known as the *client-centered* period. In this phase, the therapist not only mirrored the client's communication but also reflected underlying or implicit affect or feelings to help clients become more self-actualized or fully functioning people. (This is the basis of current concepts of the skill of empathy, discussed in the next section.)

In the most recent stage, known as *person-centered therapy* (Meador & Rogers, 1984; Raskin & Rogers, 1995), therapy is construed as an active partnership between two persons. In this current stage, emphasis is on client growth through *experiencing* of himself or herself and of the other person in the relationship.

Although client-centered therapy has evolved and changed, certain fundamental tenets have remained the same. One of these is that all people have an inherent tendency to strive toward growth, self-actualization, and self-direction. This tendency is realized when individuals have access to conditions (both within and outside therapy) that nurture growth. In the context of therapy, client growth is associated with high levels of three core, or facilitative, relationship conditions: *empathy* (accurate understanding), *respect* (positive regard), and *genuineness* (congruence) (Rogers, Gendlin,

Kiesler, & Truax, 1967). If these conditions are absent from the therapeutic relationship, clients may not only fail to grow, they may deteriorate (Berenson & Mitchell, 1974; Carkhuff, 1969a, 1969b; Truax & Mitchell, 1971). Presumably, for these conditions to enhance the therapeutic relationship, they must be communicated by the counselor *and* perceived by the client (Rogers, 1951, 1957).

Rogers's ideas have had an enormous impact on the evolution of the counseling relationship, partly because of his emphasis on the client's capacity for growth and partly because his ideas were consistent with the cultural context of the times: permissiveness and antiauthoritarianism (Gelso & Fretz, 1992; Mindess, 1988). However, as Lerman (1992) has observed, Rogers's humanistically oriented theory, while stressing growth, ignores the role of external influences and environmental constraints in a client's development. As she notes, studies of the lives of real women "have demonstrated that patriarchal institutions limit and severely constrict the possibilities for women" (p. 13). Ivey et al. (1993) also note this issue as a limitation in working with some clients from African American, Native American and Latin cultures, which focus less on the individual and more on the relationship.

Although Rogerian-based strategies for helping "are devoid of techniques that involve *doing* something to or for the client" (Gilliland, James, & Bowman, 1994), in later writings Rogers (1977) asserts that these three core conditions represent a set of skills as well as an attitude on the part of the therapist. In recent years, a variety of persons have developed concrete skills associated with these three core conditions; much of this development is based on accumulating research evidence (Carkhuff, 1969a, 1969b; Egan, 1994; Gazda et al., 1995; Ivey, 1994). This delineation of the core conditions into teachable skills has made it possible for people to learn how to communicate these core conditions to clients. As Wright and Davis (1994) note, aspects of the therapy relationship and technique are not separate domains but "are integrated aspects of a single process" (p. 29). In the following three sections, we describe these three important relationship conditions and associated skills in more detail.

Empathy, or Accurate Understanding

Empathy may be described as the ability to understand people from their frame of reference rather than your own. Responding to a client empathically may be "an attempt to think *with,* rather than *for* or *about* the client" (Brammer, Shostrom, & Abrego, 1989, p. 92). For example, if a client says "I've tried to get along with my father, but it doesn't work out. He's too hard on me," an empathic response would be something like "You feel discouraged about your unsuccessful attempts to get along with your father." In contrast, if you say something like "You ought to try harder," you are responding from your frame of reference, not the client's.

Empathy has received a great deal of attention from both researchers and practitioners over the years. Current concepts emphasize that empathy is far more than a single concept or skill. Empathy is believed to be a multistage process consisting of multiple elements (Barkham, 1988; Barrett-Lennard, 1981; Gladstein, 1983; Ottens, Shank, & Long, 1995). A useful review of empathy is provided by Duan and Hill (1996).

Ivey et al. (1993) distinguish between individual and multicultural empathy; the concept of multicultural empathy requires that we understand "different worldviews" from our own (p. 25). These authors have noted that the traditional concept of individual empathy focuses on a two-way interaction between the counselor and client, whereas cultural empathy involves four participants: the counselor and her or his cultural background and the client and his or her cultural background (Ivey, Ivey, & Simek-Downing, 1987). In cultural empathy, the counselor responds not only to the client's verbal and nonverbal messages but also to the historical-cultural-ethnic background of the client. Misunderstandings or breaches of empathy are often not just a function of miscommunications but of differences in understanding styles, nuances, and subtleties of various cultural beliefs, values, and use of language. As Sue and Sue (1990) note, "The cultural upbringing of many minorities dictates different patterns of communication" (p. 42) and "reared in a white middle-class society, counselors may (incorrectly) assume that certain behaviors or rules of speaking are universal and have the same meaning" (p. 52).

Current research has abandoned the "uniformity myth" (Kiesler, 1966) with respect to empathy and seeks to determine when empathic understanding is most useful for particular clients and problems and at particular stages in the counseling process. As Gladstein (1983, p. 178) observes, "In counseling/psychotherapy, affective and cognitive empathy can be helpful in certain stages, with certain clients, and for certain goals. However, at other times, they can interfere with positive outcomes." Generally, empathy is useful in influencing the quality and effectiveness of the therapeutic relationship. Empathy helps to build rapport and elicit information from clients by showing understanding, demonstrating civility (Egan, 1994), conveying that both counselor and client are working "from the same side," and fostering client goals related to self-exploration (Gladstein, 1983).

Rogers's theory of client-centered therapy and his view of the role of empathy in the therapeutic process assume that at the beginning of counseling, a client has a distinct and already fairly complete sense of himself or herself, referred to as a self-structure. This is often true of clients with more "neurotic" or everyday features of presenting problems; that is, they bring problems of living to the counselor and at the outset have an intact sense of themselves. Indeed, some recent research has suggested that empathy and the facilitative conditions in general may be more helpful for clients like this than for clients with a greater severity of presenting problems (Lambert & Bergin, 1992).

In contrast to the Rogerian view of the function of empathy—which is to help actualize the potential of an already established self-structure—is another view of empathy offered by the self-psychology theory of Kohut (1971b). This view assumes that many clients do not come into therapy with an established sense of self; that they lack a self-structure, and that the function of empathy in particular and of therapy in general is to build on the structure of the client's sense of self by completing a developmental process that was arrested at some time so that the person did not develop a whole sense of self. Empathy is conveyed to clients by reflective and additive verbal messages (Carkhuff, 1969a; Carkhuff & Pierce, 1975; Egan, 1994), by validating responses, by limit-setting responses, and by the provision of a safe-holding environment.

Verbal Means of Conveying Empathy.

Consider the following specific tools for conveying empathy:

- *Show desire to comprehend.* It is necessary not only to convey an accurate understanding from the client's perspective but also to convey your *desire* to comprehend from the client's frame of reference.

  Keeping in mind our discussion of cultural empathy, this desire includes an understanding not only of the individual, but also of the person's worldview, his or her environmental and sociopolitical context and cultural group. McGill (1992) offers the idea of the *Cultural Story* as a way to open communication and develop understanding about the client's cultural group. He states:

  > The cultural story refers to an ethnic or cultural group's origin, migration, and identity. Within the family, it is used to tell where one's ancestors came from, what kind of people they were and current members are, what issues are important to the family, what good and bad things have happened over time, and what lessons have been learned from their experiences. At the ethnic level, a cultural story tells the group's collective story of how to cope with life and how to respond to pain and trouble. It teaches people how to thrive in a multicultural society and what children should be taught so that they can sustain their ethnic and cultural story. (McGill, 1992, p. 340)

Your desire to comprehend is evidenced by statements indicating your attempts to make sense of the client's world and by clarification and questions about the client's experiences and feelings.

- *Discuss what is important to the client.* Show by your questions and statements that you are aware of what is most important to the client. Respond in ways that relate to the client's basic problem or complaint. This should be a brief statement that captures the thoughts and feelings of the client and one that is directly related to the client's concerns.
- Use verbal responses that *refer to client feelings.* One way to define empathy is through verbal statements that reflect the client's feelings. Use responses that convey your awareness of the client's feelings. Focus on the client's feelings by naming or labeling them. This is sometimes called *interchangeable* (Carkhuff, 1969a) or *basic* (Egan, 1994) empathy.
- Use verbal responses that bridge or *add on to implicit client messages.* Empathy also involves comprehension of the client's innermost thoughts and perspectives even when these are unspoken and implicit. According to Rogers (1977), "The therapist is so much inside the private world of the other that she can clarify not only the messages of which the client is aware but even those just below the level of awareness" (p. 11). The counselor bridges or adds on to client messages by conveying understanding of what the client implies or infers in order to add to the client's frame of reference or to draw out implications of the issue. This is sometimes called *additive* empathy (Carkhuff, 1969a) or *advanced* empathy (Egan, 1994). This skill involves abductive logic or reasoning to help counselors identify clues, develop hunches, and synthesize relevant information (Ottens, Shank, & Long, 1995).

Carkhuff and Pierce (1975) have developed a Discrimination Inventory that presents a scale for assessing both basic and additive empathy messages. On this scale, counselor responses are rated according to one of five levels; Level 3 is considered the *minimally* acceptable response. Level 3 responses on this scale correspond to Carkhuff and Pierce's concept of interchangeable empathy and Egan's (1994) concept of basic-level empathy; Level 4 corresponds to additive empathy (Carkhuff, 1969a) or advanced empathy (Egan); and

Level 5 represents facilitating action. The scale can be used either to discriminate among levels of responses or to rate levels of counselor communication. Here is an example of a verbal empathic response at each level of Carkhuff and Pierce's Discrimination Inventory.

Client: I've tried to get along with my father, but it doesn't work out. He's too hard on me.

Counselor at Level 1: I'm sure it will all work out in time [reassurance and denial].

or

You ought to try harder to see his point of view [advice].

or

Why can't you two get along? [question].

Level 1 is a question, reassurance, denial, or advice.

Counselor at Level 2: You're having a hard time getting along with your father.

Level 2 is a response to only the *content,* or cognitive portion, of the message; feelings are ignored.

Counselor at Level 3: You feel discouraged because your attempts to get along with your father have not been very successful.

Level 3 has understanding but no direction; it is a reflection of feeling and meaning based on the client's explicit message. In other words, a Level 3 response reflects both the feeling and the situation. In this response, "You feel discouraged" is the reflection of the feeling, and "because of not getting along" is the reflection of the situation.

Counselor at Level 4: You feel discouraged because you can't seem to reach your father. You want him to let up on you.

Level 4 has understanding and some direction. A Level 4 response identifies not only the client's feelings but also the client's deficit that is implied. In a Level 4 response, the client's deficit is personalized, meaning the client owns or accepts responsibility for the deficit, as in "You can't reach" in this response.

Counselor at Level 5: You feel discouraged because you can't seem to reach your father. You want him to let up on you. One step could be to express your feelings about this to your father.

A Level 5 response contains all of a Level 4 response plus at least one action step the person can take to master the deficit and attain the goal. In this example, the action step is "One step could be to express your feelings about this to your father." (See also Learning Activity #5.)

In using these verbal responses to convey empathy, it is important to remember that "it is by no means obvious that a particular client will consider reacting favorably," finding that your response conveys "understanding and interest." It may be just as likely that the client will react unfavorably; feeling the response presumes too much intimacy (Uhlemann, Lee, & Martin, 1994, p. 201).

Empathy and Validation of Clients' Experience. Both Rogers and Kohut developed their views on empathy from their work with various clients. For Kohut, the turning point was a client who came to each session with bitter accusations of him. As he stopped trying to explain and interpret her behavior and started to listen, he realized these accusations were her attempts to show him the reality of her very early childhood living with incapacitated caregivers who had been unavailable to her. Kohut surmised that clients show us their needs through their behavior in therapy, giving us clues to what they did not receive from their primary caretakers to develop an adequate sense of self and also to what they need to receive from the therapist.

It is important to remember that when a childhood need is not met or is blocked, it simply gets cut off but does not go away; it remains in the person in an often primitive form, which explains why some grown-up clients exhibit behaviors in therapy that can seem very childish. Kohut postulated five needs subsumed under headings he refers to as mirroring, idealizing, and twinship partnering (1984). These five are "the need to be admired, to have one's dreams, hopes, and fantasies recognized; the need to be favored or preferred; the need to be comforted, to have somebody listen and respond empathically; the need to be stimulated and to be attended to; and the need to be forgiven" (Okun, 1992b, p. 34). When these needs are chronically frustrated or repressed for the child, the child grows up with poor self esteem and an impaired self-structure. Also, the self is split into a *true self*—the capacity to relate to oneself and to others—and a *false self*—an accommodating self that exists mainly to deny one's true needs in order to comply with the needs of the primary caregivers (Winnicott, 1958). As Okun (1992b) notes, "The empathic therapeutic relationship becomes the single most curative variable in treatment. It provides the necessary context in which repressed parts of the ego or self and repressed unbearably painful feelings can surface, become understood, and integrated with the help of the therapist's [support]" (p. 21).

LEARNING ACTIVITY   **5**   *Verbal Empathy*

Using the description of Carkhuff and Pierce's (1975) Discrimination Inventory on pages 37–38, decide the level where each of the following counselor responses belongs:

Level 1—No understanding, no direction. Counselor response is a question, a denial or reassurance, or advice.

Level 2—No understanding, some direction. Counselor response highlights only *content* of client's message; feelings are ignored.

Level 3—Understanding present; direction absent. Counselor responds to both *content* or meaning and *feelings*.

Level 4—Both understanding and direction present. Helper responds to client *feelings* and identifies *deficit*.

Level 5—Understanding, direction, and action present. Counselor response includes all of Level 4 plus one *action* step.

After rating each response, explain your choice. An example is provided at the beginning of the activity. Feedback can be found after the learning activities.

Example

Client: I've become burned out with teaching. I've thought about changing jobs, but you know it's hard to find a good job now.

Counselor response: Teaching is no longer too satisfying to you.

This response is *Level 2. Response is only to the content or the situation of teaching. Client's feelings are ignored.*

Practice Statements

1. *Client:* I've always wanted to be a doctor, but I've been discouraged from this.

   Counselor: Oh, I'm sure this is something you could do if you really wanted to.
   This response is:
   Because:

2. *Client:* I've had such a rough semester. I don't know what I got myself into. I'm not sure where to go from here.

   Counselor: You feel perturbed about the way your semester turned out and confused because of this.
   This response is:
   Because:

3. *Client:* My teacher always picks on me.

   Counselor: Why do you suppose she picks on you?
   This response is:
   Because:

4. *Client:* I'm bored with my job. It's getting to be the same old thing. But what else is there to do?

   Counselor: You feel dissatisfied with your job because of the routine. You can't find anything in it that really turns you on. You want to find some more appealing work. One step is to list the most important needs a job meets for you and to identify how those needs could be met by certain jobs.
   This response is:
   Because:

5. *Client:* I don't understand why this accident happened to me; I've always led a good life; now this.

   Counselor: You feel resentful because you can't explain why this sudden accident happened to you. You want to at least figure out some reason that might make it seem more fair.
   This response is:
   Because:

6. *Client:* My parents are getting a divorce. I wish they wouldn't.

   Counselor: You feel upset because your parents are divorcing.
   This response is:
   Because:

7. *Client:* It just seems like each year goes by without our being able to have children.

   Counselor: You feel discouraged because you can't seem to get pregnant. You want to have a child very much.
   This response is:
   Because:

8. *Client:* I'm caught in the middle. I'm not able to move into public housing unless my partner leaves for good. But I'd also want my partner to continue to come and live with me at least some of the time.

   Counselor: Moving into public housing might prevent you and your partner from living together.
   This response is:
   Because:

9. *Client:* It's been hard for me to adjust since I've retired. The days seem so empty.

   Counselor: You feel useless because of all the time on your hands now. You can't find a way to fill up your days. You want to find some meaningful things to do. One step is to think of some ways you can continue using your work interests even though you are no longer employed.
   This response is:
   Because:

LEARNING ACTIVITY   6   *Validating Empathy*

Consider the following case descriptions of clients. What might be the effect on you on hearing this client's issue? What might you try to defend about yourself? How could you work with this to give a validating response to the client instead? Provide an example of such a response.

1. The client expresses a strong sexual interest in you and is mad and upset when she realizes you are not "in love" with her.

2. The client wants to be your favorite client and repeatedly wants to know how special he is to you.
3. The client is a man of Roman Catholic faith who wants to marry a Jewish woman. He is feeling a lot of pressure from his Latino parents, his priest, and his relatives to stop the relationship and go on with his life and find a woman of his own religious faith. He wants you to tell him what to do and is upset when you don't.

Kohut (1984) believes that empathy is at the core of providing a "corrective emotional experience" (in transactional analysis this is called *reparenting*) for clients. At the core of this is the therapist's acceptance of the client and his or her feelings. It means avoiding any sort of comment that may sound critical to clients. Because the lack of original empathic acceptance by caregivers has driven parts of the client's self underground, it is important not to repeat this process in counseling. Instead, the therapist needs to create an opposite set of conditions in which these previously buried aspects of the self can emerge, be accepted, and integrated (Kahn, 1991, pp. 96–97). The way to do this is to let clients know that the way they see themselves and their world "is not being judged but accepted as the most likely way for them to see it, given their individual history" (Kahn, 1991, p. 97).

This is known as a validating response. Validating responses are usually verbal messages from the counselor that mirror the *client's* experience. This sounds remarkably easy to do but often becomes problematic because of our own woundedness, which we discussed in Chapter 2. Too often we fail to validate the client because a button has been pushed in us and we end up validating or defending ourselves instead. Kahn (1991) provides a wonderful illustration of this:

Recently I had to close my office and see clients in a temporary place. One of my clients refused to meet me there because the parking was too difficult. She was angry and contemptuous that I would even ask her to do such a thing. I committed a whole list of Kohut sins. I told her that the parking was no harder there than anywhere else and that I assumed there was something else underlying her anger. She got angrier and angrier, and finally I did, too. It escalated into a near-disaster. Kohut would have felt his way into her situation and said with warmth and understanding, "I can really see how upsetting it is for you to have the stability of our seeing each other disturbed. I think that figuring out where you'd park really is difficult, and

I think there must be a lot of other upsetting things about our having to see each other somewhere else. And I can imagine some of those other things are even harder to talk about than the parking." Had she then kept the fight going, he might have said, "I think it must be really hard to have this move just laid on you without your having any say in the matter. It must seem like just one more instance where you get pushed around, where decisions get made for you, where you have to take it or leave it. It must be very hard."

Had I done that, I might have made it possible for her to explore other feelings—or I might have failed to do so. But whatever the outcome, she would have felt heard and understood. As it was, I became just one more in a long series of people telling her she was doing it wrong. (pp. 103–104)

The key to being able to provide validating responses is to be able to contain your own emotional reactions so that they do not get dumped out onto the client. This is especially hard to do when a client pushes your own buttons, and this is why working with yourself and your own "stuff" is so important. We discuss this further in the section on countertransference later in this chapter.

Empathy and Limit-Setting Responses.   As you can imagine, clients whose primary needs have not been met often present strong needs for immediate gratification to their counselors. As Kahn (1991) explains, "The most primitive side of clients wants to be gratified as *children;* that is, clients want to be hugged, to be told they're wonderful, to be reassured you will protect them and on and on" (p. 99). Part of providing an empathic environment is to reflect the client's wish or desire for this but not to actually provide the gratification. This constitutes a limit-setting response, and, combined with warmth and empathy, it contributes to the client's growth and also helps to create an atmosphere of safety. See whether you can tell the difference in the following two examples:

FEEDBACK 5
Verbal Empathy

1. Counselor response is at Level 1—no understanding and no direction. The response is a denial of client's concern and a form of advice.
2. Counselor response is at Level 3—understanding is present; direction is absent. Responds to client's feelings (you feel perturbed) and to content or situation (about the semester).
3. Counselor response is at Level 1—no understanding, no direction. Response is a question and ignores both the content and feelings of client's message.
4. Counselor response is at Level 5—understanding, direction, and action are all present. Response tunes in to client's feelings, identifies client's deficit, and identifies one action step (to list the important needs a job meets).
5. Counselor response is at Level 4—understanding and direction. Both client's feelings (you feel resentful) and client's deficit (you can't explain why) are included in counselor's response.
6. Counselor response is at Level 3—understanding is there; no direction. Counselor responds to client's feelings (you feel upset) and to the content or situation (your parents are divorcing).
7. Counselor response is at Level 4—understanding and direction. Response reflects client's feelings (you feel discouraged) and identifies her deficit (you can't seem to get pregnant).
8. Counselor response is at Level 2—some direction but no understanding. Response is only to content of client's message; feelings are ignored.
9. Counselor response is at Level 5—understanding, direction, and action are all there. Response picks up client's feelings and deficit and identifies one possible action step (to think of some ways).

Client: I don't think I do anything very well. No one else seems to think I'm real special either.

Example One

Counselor: Well you are so special to me . . .

Example Two

Counselor: I can see how that is a very hard thing for you—wanting to feel special and not feeling that way about yourself . . .

FEEDBACK 6
Validating Empathy

Here are some examples of validating responses. Discuss your emotional reactions and your response with a partner or in a small group.

1. I realize you are disappointed and upset about wanting me to have the same sort of feelings for you that you say you have for me.
2. It is important to you to feel very special to me. I understand this as your way of telling me something about what has been missing for you in your life.
3. I realize you feel caught between two things—your religious and cultural history and your love for this woman. You wish I could tell you what to do and you're upset with me that I won't.

In the first example, the counselor supplied reassurance and gratified the need, but in doing so, may have closed the door for more explanation of this issue by the client. In the second example, the counselor reflects the client's pain and also the client's desire or wish and leaves the door open for a client response.

Wells and Glickauf-Hughes (1986) note that "for clients with backgrounds of deprivation and/or neglect, limit setting is a needed art of caring, protective containment" (p. 462). Limits may need to be set on behaviors with clients prone to acting out—not only because the client is distressed but because the client's behavior violates your own limits as a counselor. If a client repeatedly shouts at you, you may set limits by saying something like "I'm aware of how much you shout at me during our sessions. I know this is your way of showing me something about what has happened to you and how awful you feel about it. Still, in order for me to work effectively with you, I want you to talk to me about your distress without shouting."

We want to emphasize that throughout our discussion of the concept of empathy from the viewpoint of Kohut (1971a, 1984) and self-psychology, we have intentionally used the word *caregivers.* This is consistent with our personal stance that among various ethnic and cultural groups, mothers, fathers, and often grandparents and extended family members represent the objects healthy children use to develop a cohesive sense of self; if self-esteem is impaired or the self needs are blocked, it is not the result of insufficient *mothering,* but rather insufficient and/or inconsistent caregiving in general. In Jungian terms, it is the absence of the feminine (or the "yang" of the "yin-yang") in both women and men and in predominantly patriarchal cultures that contributes to the lack of a "cherishing container" for the child (Woodman, 1993).

Empathy and the Holding Environment. The empathic mirroring and limit-setting responses we have described are often referred to as the provision of a therapeutic *holding environment* (Winnicott, 1958). A holding environment means that the therapist conveys in words and/or behavior that he or she knows and understands the deepest feelings and experience of the client and provides a safe and supportive atmosphere in which the client can experience deeply felt emotions. As Hackney and Cormier (1994) note, it means "that the counselor is able to allow and stay with or 'hold' the client's feelings instead of moving away or distancing from the feelings or the client. In doing this, the counselor acts as a *container;* that is, the counselor's comfort in exploring and allowing the emergence of client feelings provides the support to help the client contain or hold various feelings that are often viewed by the client as unsafe" (pp. 102–103). The therapist as a container helps the client to manage what might otherwise be experienced as overwhelming feelings by providing a structure and safe space in having to do so.

Teyber (1997) has noted that the effective holding environment provided by a therapist is usually dramatically different from what the young client is experiencing or what an adult client experienced while growing up. For example, if the child was sad, the parent may have responded by withdrawing, by denying the child's feelings, or by responding derisively (Teyber, 1997, p. 143). In all these parental reactions, the child's feeling was not heard, validated, or "contained"; as a result, the child learned over time to deny or avoid these feelings, (thus constituting the "false self" we described earlier). Children are developmentally unable to experience and manage feeling on their own without the presence of another person who can be emotionally present for them and receive and even welcome their feelings. If the parent was unable to help the child hold feelings in this manner, it will be up to the counselor to do so. In this way, the counselor allows clients to know that he or she can accept their painful feelings and still stay emotionally connected to them (Teyber, 1997, p. 147).

Josselson (1992) points out that "of all the ways in which people need each other, holding is the most primary and the least evident," starting with the earliest sensations infants experience—the sensation of being guarded by strong arms that keep them from falling and also help them to unfold as unique and separate individuals (p. 29). Not only is a child sufficiently nourished in such an environment, but just as important, the child also feels *real.* As she notes, in her seminal work on adult human relationships, this need for holding or groundedness does not disappear as we grow up, although the form of holding for adults may be with institutions, ideas, and words as much as with touch. Individuals who do not experience this sense of holding as children often grow up without a sense of groundedness in their own body as well as without a sense of self as a separate and unique person. Often their energy or "life force" is bound up and/or groundless and they may seek to escape their sense of nothingness by becoming attached to any number of addictions. Josselson (1992) provides an excellent description of how the process of therapy can support clients' growth through the provision of this sort of holding environment. She states:

> People often come to psychotherapy because they need to be held while they do the work of emotionally growing. They need a structure within which they can experience frightening or warded-off aspects of themselves. They need to know that this structure will not "let them down." They also need to trust that they will not be impinged upon by unwanted advice or by a therapist's conflicts or difficulties. Psychotherapy, because of clinicians' efforts to analyze what takes place, is one of the best understood of holding environments. Therapists "hold" patients as patients confront aspects of their memory and affective life that would be too frightening or overwhelming to face alone. (One of my patients once described her experience of therapy as my sitting with her while she confronts the monsters inside.) Therapists continue to hold patients even as patients rage at them in disappointment, compete with them, envy them, or yearn for them. Adequate holding continues despite the pain of relatedness. (p. 36)

Note that in various cultures this sort of holding environment is supplied by indigenous healers as well.

Genuineness

Genuineness means being oneself without being phony or playing a role. Although most counselors are trained to be *professionals,* a counselor can convey genuineness by being human and by collaborating with the client. Genuineness contributes to an effective therapeutic relationship by reducing the emotional distance between the counselor and client and by helping the client to identify with the counselor, to perceive the counselor as another person similar to the client. Genuineness has at least five components: supporting nonverbal behaviors, role behavior, congruence, spontaneity, and openness (see also Egan, 1994).

Supporting Nonverbal Behaviors.
Genuineness is communicated by the therapist's use of appropriate, or supporting, nonverbal behaviors. Nonverbal behaviors that convey genuineness include eye contact, smiling, and leaning toward the client while sitting. These two nonverbal behaviors, however, should be used discreetly and gracefully. For example, direct yet intermittent eye contact is perceived as more indicative of genuineness than is persistent gazing, which clients may interpret as staring. Similarly, continual smiling or leaning forward may be viewed as phony and artificial rather than genuine and sincere. As we mentioned during our discussion of empathy, when establishing rapport, the counselor should display nonverbal behaviors that parallel or match those of the client.

Role Behavior.
Counselors who do not overemphasize their role, authority, or status are likely to be perceived as more genuine by clients. Too much emphasis on one's role and position can create excessive and unnecessary emotional distance in the relationship. Clients can feel intimidated or even resentful.

The genuine counselor also is someone who is comfortable with himself or herself and with a variety of persons and situations and does not need to "put on" new or different roles to feel or behave comfortably and effectively. As Egan (1994, p. 55) observes, genuine counselors "do not take refuge in the role of counselor. Ideally, relating at many levels to others and helping are part of their lifestyle, not roles they put on or take off at will."*

Congruence.
Congruence means simply that the counselor's words, actions, and feelings match—they are consistent. For example, when a therapist becomes uncomfortable because of a client's constant verbal assault, she acknowledges this feeling of discomfort, at least to herself, and does not try to cover up or feign comfort when it does not exist. Counselors who are not aware of their feelings or of discrepancies between their feelings, words, and actions may send mixed, or incongruent, messages to clients—for example, saying, "Sure, go ahead and tell me how you feel about me" while fidgeting or tapping feet or fingers. Such messages are likely to be very confusing and even irritating to clients.

Spontaneity.
Spontaneity is the capacity to express oneself naturally without contrived or artificial behaviors. Spontaneity also means being tactful without deliberating about

everything you say or do. Spontaneity, however, does not mean that counselors need to verbalize every passing thought or feeling to clients, particularly negative feelings. Rogers (1957) suggests that counselors express negative feelings to clients only if the feelings are constant and persistent or if they interfere with the counselor's ability to convey empathy and positive regard.

Openness and Self-Disclosure.
Part of genuineness involves the ability to be open, to share yourself, to self-disclose. Because self-disclosure is a complex skill and should not be used indiscriminately, we discuss it in some detail in this section. (See also Learning Activity #7.)

Self-disclosure.
Self-disclosure may be defined as verbalized personal revelations made by the counselor to the client (Watkins, 1990, p. 478). Typically, counselors may choose to reveal something about themselves through verbal sharing of such information. Self-disclosure is not confined to verbal behavior, of course. As Egan (1994) points out, we always disclose information about ourselves through nonverbal channels and by our actions even when we don't intend to. This section, however, focuses on the purposeful use of verbal disclosure as a way to convey genuineness.

Using self-disclosure with a client can have several purposes. Counselor self-disclosure may generate an open and facilitative counseling atmosphere. In some instances, a disclosive counselor may be perceived as more sensitive and warm than a nondisclosive counselor (Nilsson, Strassberg, & Bannon, 1979). Counselor disclosure can reduce the role distance between a counselor and client (Egan, 1994). Counselor self-disclosure can also be used to increase the disclosure level of clients (Nilsson et al., 1979), to bring about changes in clients' perceptions of their behavior (Hoffman-Graff, 1977), and to increase client expression of feelings (McCarthy, 1982). Counselor self-disclosure may also help clients develop new perspectives needed for goal setting and action (Egan, 1994).

Ground rules.
Several ground rules may help a counselor decide what, when, and how much to disclose. As Nilsson et al. (1979, p. 399) observe, "The issue is far more complex than whether a counselor should or should not disclose. . . . Content, timing, and client expectation, are critical mediating variables determining the influence of counselor disclosure." One ground rule relates to the "breadth," the cumulative amount of information disclosed. Most of the evidence indicates that a moderate amount of disclosure has more positive effects than a high or low level (Edwards & Murdock, 1994). Some self-disclosure may indicate a desire for a close relationship and may increase the client's estimate

LEARNING ACTIVITY 7 *Self-Disclosure*

I. Respond to the following three client situations with a self-disclosing response. Make sure you reveal something about yourself. It might help you to start your statements with "I." Also try to make your statements similar in content and depth to the client messages and situations. An example is given first, and feedback is provided on page 46.

Example
The client is having a hard time stating specific reasons for seeking counseling. Your self-disclosing statement: "I'm reluctant at times to share something that is personal about myself with someone I don't know; I know it takes time to get started."

Now use your self-disclosure responses:

1. The client is feeling like a failure because "nothing seems to be going well."

Your self-disclosure:

2. The client is hinting that he or she has some concerns about sexual performance but does not seem to know how to introduce this concern in the session.

Your self-disclosure:

3. The client has started to become aware of feelings of anger for the first time and is questioning whether such feelings are legitimate or whether something is wrong with him or her.

Your self-disclosure:

II. In a conversation with a friend or in a group, use the skill of self-disclosure. You may wish to use the questions below as "starters." Consider the criteria listed in the feedback to assess your use of this response.

Preference Survey

1. What things or activities do you enjoy doing most?
2. What things or activities do you dislike?
3. What things or activities do you try to avoid?
4. When you're feeling down in the dumps, what do you do to get out of it?
5. What things or people do you think about most?
6. What things or people do you avoid thinking about?

---

of the helper's trustworthiness. Some self-disclosure can provide role modeling for clients from cultures with a low level of emotional expressiveness (Lum, 1996). Counselors who disclose very little could add to the role distance between themselves and their clients. At the other extreme, too much disclosure may be counterproductive. The counselor who discloses too much may be perceived as lacking in discretion, being untrustworthy, seeming self-preoccupied or needing assistance. A real danger in overdisclosing is the risk of being perceived as needing therapy as much as the client. This could undermine the client's confidence in the counselor's ability to be helpful. Also, as we noted in Chapter 2, excessive self-disclosure is the boundary violation that most likely precedes unethical sexual contact between therapist and client (Smith & Fitzpatrick, 1995). Excessive self-disclosure may represent a blurring of good treatment boundaries. Also, too much self-disclosure can lead clients who are from cultures unaccustomed to personal sharing to retreat (Lum, 1996).

Another ground rule concerns the duration of self-disclosure—the amount of time used to give information about yourself. Extended periods of counselor disclosure will consume time that could be spent in client disclosure. As one person reveals more, the other person will necessarily reveal

less. From this perspective, some conciseness in the length of self-disclosive statements seems warranted. Another consideration in duration of self-disclosure involves the capacity of the client to utilize and benefit from the information shared. As Egan (1994, p. 186) observes, counselors should avoid self-disclosing to the point of adding a burden to an already overwhelmed client.

A third ground rule to consider in using self-disclosure concerns the depth, or intimacy, of the information revealed. You should try to make your statements similar in content and mood to the client's messages. Ivey and Gluckstern (1984) suggest that the counselor's self-disclosure be closely linked to the client's statements. For example:

Client: I just feel so down on myself. My husband is so critical of me, and often I think he's right. I really can't do much of anything well.

Counselor (parallel): There have been times when I've also felt down on myself, so I can sense how discouraged you are. Sometimes, too, criticism from a male has made me feel even worse, although I'm learning how to value myself regardless of critical comments from my husband or a male friend.

Counselor (nonparallel): I've felt bummed out, too. Sometimes the day just doesn't go well.

Positive Regard

Positive regard, also called respect, means the ability to prize or value the client as a person with worth and dignity (Rogers, 1957). Communication of positive regard has a number of important functions in establishing an effective therapeutic relationship, including the communication of willingness to work with the client, interest in the client as a person, and acceptance of the client. Egan (1994) has identified four components of positive regard: having a sense of commitment to the client, making an effort to understand the client, suspending critical judgment, and expressing a reasonable amount of warmth.

Commitment.   Commitment means you are willing to work with the client and are interested in doing so. It is translated into such actions as being on time for appointments, reserving time for the client's exclusive use, ensuring privacy during sessions, maintaining confidentiality, and applying skills to help the client. Lack of time and lack of concern are two major barriers to communicating a sense of commitment.

Understanding.   Clients will feel respected to the degree that they *feel* the counselor is trying to understand them and to treat their problems with concern. Counselors can demonstrate their efforts to understand by being empathic, by asking questions designed to elicit information important to the client, and by indicating with comments or actions their interest in understanding the client and the client's cultural heritage and values.

Counselors also convey understanding with the use of specific listening responses such as paraphrasing and reflecting client messages (see also Chapter 6).

Nonjudgmental Attitude.   A nonjudgmental attitude is the counselor's capacity to suspend judgment of the client's actions or motives and to avoid condemning or condoning the client's thoughts, feelings, or actions. It may also be described as the counselor's acceptance of the client without conditions or reservations, although it does not mean that the counselor supports or agrees with all the client says or does. A counselor conveys a nonjudgmental attitude by warmly accepting the client's expressions and experiences without expressing disapproval or criticism. For example, suppose a client states "I can't help cheating on my wife. I love her, but I've got this need to be with other women." The counselor

who responds with regard and respect might say something like "You feel pulled between your feelings for your wife and your need for other women." This response neither condones nor criticizes the client's feelings and behaviors. In contrast, a counselor who states "What a mess! You got married because you love your wife. Now you're fooling around with other women" conveys criticism and lack of respect for the client as a unique human being. The experience of having positive regard for clients can also be identified by the presence of certain (covert) thoughts and feelings such as "I feel good when I'm with this person" or "I don't feel bothered or uncomfortable with what this person is telling me."

A question that counselors frequently face is how they can overcome personal and cultural biases to deal effectively with an individual who is perceived as unlikable, worthless, or offensive—such as a rapist, racist, or abuser of children.

A perspective on this is offered by Johanson and Kurtz (1991) who observe:

> It is difficult for therapists to help a rapist or racist if they have not made peace with the rapist or racist within themselves. If they do not do this first, their use of power will give rise to defenses against power in the offender, who will avoid change by playing whatever game is necessary to maintain the status quo. Therapists who recognize their own ability to manipulate, and who can employ power efficiently and dispassionately, are those with the best chance of transcending the manipulation and power plays of the offender. Therapists who can be honest and straightforward, communicating a sense of common human-beingness, are the ones who have the best chance of inviting offenders into the self-awareness of therapeutic processes. (p. 89)

Warmth.   According to Goldstein & Higginbotham (1991), without the expression of warmth, particular strategies and helping interventions may be "technically correct but therapeutically impotent" (p. 48). Warmth reduces the impersonal nature or sterility of a given intervention or treatment procedure. In addition, warmth begets warmth. In interactions with hostile or reluctant clients, warmth and caring can speak to the client's anger.

Nonverbal cues of warmth.   A primary way in which warmth is communicated is with supporting nonverbal behaviors such as voice tone, eye contact, facial animation and expressions, gestures, and touch. Johnson (1993) describes some nonverbal cues that express warmth or coldness (see Table 3-1). Remember that these behaviors may be interpreted differently by clients from various ethnic, racial, and cultural groups.

F E E D B A C K 7
Self-Disclosure

I. Here are some possible examples of counselor self-disclosure for these three client situations. See whether your responses are *similar;* your statements will probably reflect more of your own feelings and experiences. Are your statements fairly concise? Are they similar to the client messages in content and intensity?

1. "I, too, have felt down and out about myself at times."
   or
   "I can remember, especially when I was younger, feeling very depressed if things didn't turn out the way I wanted."
2. "For myself, I have sometimes questioned the adequacy of my sexual performance."
   or
   "I find it hard sometimes to start talking about really personal topics like sex."
3. "I can remember when I used to feel pretty afraid of admitting I felt angry. I always used to control it by telling myself I really wasn't angry."
   or
   "I know of times when some of my thoughts or feelings have seemed hard for me to accept."

II. Self-Disclosure Assessment

1. What was the amount of your self-disclosure in relation to the amount of the other person's—low, medium, or high?
2. What was the *total* amount of time you spent in self-disclosure?
3. Were your self-disclosure statements similar in content and depth to those expressed by the other person?
4. Did your self-disclosure detract from or overwhelm the other person?

An important aspect of the nonverbal dimension of warmth is touch (see also Chapter 5). In times of emotional stress, many clients welcome a well-intentioned touch. The difficulty with touch is that it may have a meaning to the client different from the meaning you intended to convey. In deciding whether to use touch, it is important to consider the level of trust between you and the client, whether the *client* may perceive the touch as sexual, the client's past history associated with touch (occasionally a client will associate touch with punishment or abuse and will say "I can't stand to be touched") and the client's cultural group (whether touch is

respectful and valued or not). To help you assess the probable impact of touch on the client, Gazda et al. (1995) recommend asking yourself the following questions:

1. How does the other person perceive this? Is it seen as genuine or as a superficial technique?
2. Is the other person uncomfortable? (If the other person draws back from being touched, adjust your behavior accordingly.)
3. Am I interested in the person or in touching the person? Whom is it for—me, the other person, or to impress those who observe?

Because of all the clients who present with trauma history, it is important to observe clear boundaries surrounding touch (Smith & Fitzpatrick, 1995). Check with the client and discuss these boundaries first.

Verbal responses associated with warmth and immediacy. Warmth can also be expressed to clients through selected verbal responses. One way to express warmth is to use enhancing statements (Ivey et al., 1987) that portray some positive aspect or attribute about the client, such as "It's great to see how well you're handling this situation," "You're really expressing yourself well," or "You've done a super job on this action plan." Enhancing statements offer positive reinforcement to clients and must be sincere, deserved, and accurate to be effective.

Another verbal response used to express warmth is immediacy. Immediacy is a characteristic of a counselor verbal response describing something *as it occurs* within a session. Immediacy involves self-disclosure but is limited to self-disclosure of *current* feelings or what is occurring at the present time in the relationship or the session. When persons avoid being immediate with each other over the course of a developing relationship, distance sets in and coldness can quickly evaporate any warmth formerly established. Egan (1990, p. 227) describes the potential impact on a relationship when immediacy is absent:

> People often fail to be immediate with one another in their interactions. For instance, a husband feels slighted by something his wife says. He says nothing and "swallows" his feelings. But he becomes a little bit distant from her the next couple of days, a bit more quiet. She notices this, wonders what is happening, but says nothing. Soon little things in their relationship that would ordinarily be ignored become irritating. Things become more and more tense, but still they do not engage in direct, mutual talk about what is happening. The whole thing ends as a game of "uproar" (see Berne, 1964)— that is, a huge argument over something quite small. Once

TABLE 3-1. Nonverbal cues of warmth and coldness

Nonverbal cue	Warmth	Coldness
Tone of voice	Soft	Hard
Facial expression	Smiling, interested	Pokerfaced, frowning, disinterested
Posture	Lean toward other; relaxed	Lean away from other; tense
Eye contact	Look into other's eyes	Avoid looking into other's eyes
Touching	Touch other softly	Avoid touching other
Gestures	Open, welcoming	Closed, guarding oneself, and keeping other away
Spatial distance	Close	Distant

SOURCE: From *Reaching Out: Interpersonal Effectivess and Self-Actualization*, 5th Ed., by D. W. Johnson, p. 163. Copyright © 1993 by Allyn and Bacon. Reprinted by permission.

they've vented their emotions, they feel both relieved because they've dealt with their emotions and guilty because they've done so in a somewhat childish way.

In using immediacy in counseling, the therapist reflects on a current aspect of (1) some thought, feeling, or behavior of the *counselor,* (2) some thought, feeling, or behavior of the *client,* or (3) some aspect of the *relationship.* Here are some examples of these three categories of immediacy.

1. *Counselor immediacy:* The counselor reveals his or her own thoughts or feelings in the counseling process as they occur "in the moment."
   "I'm glad to see you today."
   "I'm sorry, I am having difficulty focusing. Let's go over that again."
2. *Client immediacy:* The counselor provides feedback to the client about some client behavior or feeling as it occurs in the interview.
   "You're fidgeting and seem uncomfortable here right now."
   "You're really smiling now—you must be very pleased about it."
3. *Relationship immediacy:* The counselor reveals feelings or thoughts about how he or she experiences the relationship.
   "I'm glad that you're able to share that with me."
   "It makes me feel good that we're getting somewhere today."

Relationship immediacy may include references to specific "here and now" transactions or to the overall pattern or development of the relationship (Egan, 1994). For example, "I'm aware that right now as I'm talking again, you are looking away and tapping your feet and fingers. I'm wondering if you're feeling impatient with me or if I'm talking too much" (specific transaction). Consider another example in which immediacy is used to focus on the development and pattern of the relationship: "This session feels so good to me. I remember when we first started a few months ago and it seemed we were both being very careful and having trouble expressing what was on our minds. Today, I'm aware we're not measuring our words so carefully. It feels like there's more comfort between us."

Immediacy is not an end in and of itself but, rather, a means of helping the counselor and client work together better. If allowed to become a goal for its own sake, it can be distracting rather than helpful (Egan, 1990). Examples of instances in which immediacy might be useful include the following:

1. Hesitancy or "carefulness" in speech or behavior ("Mary, I'm aware that you [or I] seem to be choosing words very carefully right now—as if you [or I] might say something wrong").
2. Hostility, anger, resentment, irritation ("Joe, I'm feeling pretty irritated now because you're indicating you want me to keep this time slot open for you but you may not be able to make it next week. Because you have not kept your appointment for the last two weeks, I'm concerned about what might be happening in our relationship").
3. Attraction ("At first it seemed great that we liked each other so much. Now I'm wondering if we're so comfortable that we may be holding back a little and not sharing what's really on our minds").
4. Feeling of being "stuck"—lack of focus or direction ("Right now I feel like our session is sort of a broken

record. We're just like a needle tracking in the same groove without really making any music or going anywhere'').

5. Tension (''I'm aware there's some discomfort and tension we're both feeling now—about who we are as people and where this is going and what's going to happen'').

Immediacy can also be used to deal with the issues of transference and countertransference described in the next section of this chapter.

Immediacy has three purposes. One purpose is to bring out in the open something that you feel about yourself, the client, or the relationship that has not been expressed directly. Generally, it is assumed that covert, or unexpressed, feelings about the relationship may inhibit effective communication or may prevent further development of the relationship unless the counselor recognizes and responds to these feelings. This may be especially important for negative feelings. As Patterson and Welfel (1994) note, ''Immediacy responses provide opportunities for client and counselor to explore together any stresses in their relationship'' (p. 68). In this way, immediacy may reduce the distance that overshadows the relationship because of unacknowledged underlying issues.

A second purpose of immediacy is to generate discussion or to provide feedback about some aspects of the relationship or the interactions as they occur. This feedback may include verbal sharing of the counselor's feelings or of something the counselor sees going on in the interactive process. Immediacy is not used to describe every passing counselor feeling or observation to the client. But when something happens in the counseling process that influences the client's feelings toward counseling, then dealing openly with this issue has high priority. Usually it is up to the counselor to initiate discussion of unresolved feelings or issues (Patterson & Welfel, 1994). Immediacy can be a way to begin such discussion and, if used properly, can strengthen the counselor/client relationship and help the counselor and client work together more effectively.

Finally, immediacy is useful to facilitate client self-exploration and to maintain the focus of the interaction on the client or the relationship rather than on the counselor (McCarthy, 1982; McCarthy & Betz, 1978).

Steps in immediacy.   Immediacy is a complex set of skills. The first part of immediacy—and an important prerequisite of the actual verbal response—is awareness, or the ability to sense what is happening in the interaction (Egan, 1994). To do this, you must monitor the flow of the interaction to process what is happening to you, to the client, and to your developing relationship. Awareness also implies that you can

read the clues without a great number of decoding errors and without projecting your own biases and blind spots into the interaction. After awareness, the next step is to formulate a verbal response that somehow shares your sense or picture of the process with the client. The actual form of the response may vary and can include some of the listening or action responses we describe in Chapters 6 and 7. Regardless of the form, the critical feature of immediacy is its emphasis on the here and now—the present.

Turock (1980, p. 170) suggests some useful sentence stems for immediacy:

1. ''Right now I'm feeling _____ '' (counselor immediacy).
2. ''Even right now you're feeling _____ (feelings toward counselor) because _____ '' (client immediacy).
3. ''When I see (hear, grasp) you _____ (client's behavior or feelings), I _____ '' (counselor's behavior or feelings).

Ground rules.   Several rules can help counselors use immediacy effectively. First, the counselor should describe what she or he sees *as it happens.* If the counselor waits until later in the session or until the next interview to describe a feeling or experience, the impact is lost. In addition, feelings about the relationship that are discounted or ignored may build up and eventually be expressed in more intense or distorted ways. The counselor who puts off using immediacy to initiate a needed discussion runs the risk of having unresolved feelings or issues damage the relationship.

Second, to reflect the ''here-and-nowness'' of the experience, any immediacy statement should be in the present tense—''I'm feeling uncomfortable now,'' rather than ''I just felt uncomfortable.'' This models expression of current rather than past feelings for the client.

Further, when referring to your feelings and perceptions, own them—take responsibility for them—by using the personal pronoun *I, me,* or *mine,* as in ''I'm feeling concerned about you now'' instead of ''You're making me feel concerned.'' Expressing your current feelings with ''I'' language communicates that you are responsible for your feelings and observations, and this may increase the client's receptivity to your immediacy expressions.

Finally, as in using all other responses, the counselor should consider timing. Using a lot of immediacy in an early session may be overwhelming for some clients and can elicit anxiety in either counselor or client. As Gazda et al. (1995, p. 215) observe, ''It is highly desirable that a strong base relationship exist before using the dimension of immediacy.'' If a counselor uses immediacy and senses that this has threatened or scared the client, then the counselor should decide that the client is not yet ready to handle these feelings or issues. And not every feeling or observation a counselor has needs to be verbalized to a client. The session does not need to turn into a "heavy" discussion, nor should it resemble a confessional. Generally, immediacy is reserved for initiating exploration of the most significant or most influential feelings or issues. Of course, a counselor who never expresses immediacy may be avoiding issues that have a significant effect on the relationship.

There is some evidence that counselors tend to avoid immediacy issues even when raised directly by clients (Turock, 1980). Counselors who are not comfortable with their own self-image or who are struggling with intimacy issues in their own life are likely to have trouble with this skill or to try to avoid the use of it altogether. Unfortunately, this may result in the continuation of an unhealthy or somewhat stagnant therapeutic relationship (Turock, 1980).

EMOTIONAL OBJECTIVITY: TRANSFERENCE AND COUNTERTRANSFERENCE

The therapeutic relationship has the capacity to invoke great emotional intensity, often experienced by both the counselor and the client. To some extent, counselors need to become emotionally involved in the relationship. If they are too aloof or distant, clients will feel that the counselor is cold, mechanical, and noncaring. However, if counselors are too involved, they may scare the client away or may lose all objectivity and cloud their judgment. The degree of emotional objectivity and intensity felt by counselors can affect two relationship issues: transference and countertransference. (See Learning Activity #9.)

Transference

Transference is the

> process whereby clients project onto their therapists past feelings or attitudes toward significant people in their lives. . . . In transference, clients' unfinished business produces a distortion in the way they perceive and react to the therapist. They are rooted in past relationships but are now directed toward the therapist. (Corey et al., 1993, p. 41)

Transference can occur very easily with counselors of all theoretical orientations when the emotional intensity has become so great that the client loses his or her objectivity and starts to relate to the counselor as if he or she were some significant other person in the client's life.

Transference tends to occur regardless of the gender of the counselor (Kahn, 1991) and, according to Kohut (1984), may occur because in the presence of an empathic therapist, old unmet needs of the client resurface. Often the transference (positive or negative) is a form of reenactment of the client's familiar and old pattern or template of relating. The value of it is that through the transference clients may be trying to help us see how they felt at an earlier time when treated in a particular way. Transference often occurs when the therapist (usually inadvertently) does or says something that triggers unfinished business with the client, often with members of the client's family of origin—parents and siblings or significant others. Transference issues are thought to start at the beginning of therapy and become more intense as therapy continues. Counselors can make use of the transference, especially a negative one, by helping clients see that what they expect of us, they also expect of other people in their life; if, for example, a client wants to make the counselor "look bad," that is probably their intent with others as well.

Counselors can also work with transference issues in a helpful way by empathically reflecting on the client's desire or wish—for example, the wish to be loved, the wish to be important, the wish to control, and so on. Often the transference acted out by the client not only includes a reenactment of an earlier important relationship but also a replay of how the client wishes it were (Kahn, 1991). Transference also seems to be related to client patterns of attachment. As Mallinckrodt, Gantt, and Coble (1995) note, "Viewed from the perspective of attachment theory, transference may be understood as a misperception of the therapist and of the therapeutic relationship resulting from the client's use of long-established working models of self and others to resolve ambiguities in the new caregiving (therapeutic) attachment and to anticipate the motives and behavior of the new attachment figure (therapist)" (p. 316).

Countertransference

Countertransference includes feelings and attitudes the therapist has about the client. They may be realistic or characteristic responses, responses to transference, or responses to material and content that trouble the counselor (Kahn, 1991). As Kahn notes, a therapist's countertransference responses can be useful or damaging. He asserts that "at every moment deep characterological, habitual responses lie in wait, looking for an opportunity to express themselves as countertransference" (p. 121). Hurtful countertransference which comes from our own woundedness occurs when (1) we are blinded to an important area of exploration; (2) we focus on an issue

LEARNING ACTIVITY 8 *Immediacy*

I. For each of the following client stimuli, write an example of a counselor immediacy response. An example has been completed below, and feedback can be found on p. 52.

Example
The client has come in late for the third time, and you have some concern about this.

Immediacy response: "I'm aware that you're having difficulty getting here on time, and I'm feeling uncomfortable about this."

Now use immediacy in the following five situations:

1. Tears begin to well up in the client's eyes as he describes the loss of a close friend.
   Your immediacy response:
2. The client stops talking whenever you bring up the subject of her academic performance.
   Your immediacy response:
3. The client has asked you several questions about your competence and qualifications.
   Your immediacy response:
4. You experience a great deal of tension and caution between yourself and the client; the two of you seem to be treating each other with "kid gloves." You notice physical sensations of tension in your body, and signs of tension are also apparent in the client.
   Your immediacy response:
5. You and the client like each other a great deal and have a lot in common; lately you've been spending more time swapping life stories than focusing on or dealing with the client's presented concern of career indecision and dissatisfaction.
   Your immediacy response:

II. In a conversation with a close friend or in a group, use the sharing skill of immediacy. If possible, tape the conversation for feedback—or ask for feedback from the friend or the group. You should consider the criteria listed in the feedback in assessing your use of immediacy. You may wish to use the topics listed in the following Relationship Assessment Inventory as topics for discussion using immediacy.

Relationship Assessment Inventory

1. To what extent do we really know each other?
2. How do I feel in your presence?
3. How do you feel in my presence?
4. What areas do I have trouble sharing with you?
5. What is it about our relationship that makes it hard to share some things?
6. Do we both have a fairly equal role in maintaining our relationship, or is one of us dominant and the other passive?
7. How do we handle power and conflict in the relationship? Is one of us consistently "top dog" or "underdog"?
8. Do we express or avoid feelings of warmth and affection for each other?
9. How do our concepts of our sex roles affect the way we relate to each other?
10. How often do we give feedback to each other—and in what manner is it given?
11. How do we hurt each other?
12. How do we help each other?
13. Where do we want our relationship to go from here?

---

more of our own than pertaining to the client; (3) we use the client for vicarious or real gratification; (4) we emit subtle cues that "lead" the client; (5) we make interventions not in the client's best interest and, most important, (6) we adopt the roles the client wants us to play in his or her old script. Countertransference can be useful because it is a potential "generator of empathy" as long as the therapist can achieve an *optimal* distance from the feeling—that is, not be swamped or overwhelmed by it (Kahn, 1991, p. 127).

THE WORKING ALLIANCE

Sexton and Whiston (1994) point out that the therapeutic alliance has a "lengthy history" in psychotherapy, beginning with the work of Freud, although all therapeutic approaches considered the alliance to be very important (p. 35). The term *working alliance* was coined by Greenson (1967), who viewed the relationship as a sort of therapeutic collaboration and partnership—a sense in which both counselor and client are working together in a joint fashion, like rowing a boat. If only one person pulls the oars, the boat doesn't move as well through the water. Bordin (1979) expanded Greenson's work and noted specifically that this alliance comprises three parts:

1. Agreement on therapeutic *goals* (more is said on this in Chapter 10)
2. Agreement on therapeutic *tasks*
3. An *emotional bond* between client and therapist

LEARNING ACTIVITY 9   *Transference and Countertransference*

In a small group or with a partner, discuss the likely transference and countertransference reactions you discover in the following three cases. Feedback is provided.

1. The client is upset because you will not give her your home telephone number. She states that although you have a 24-hour on-call answering service, you are not really available to her unless you give her your home number.

2. You are an internship student and your internship is coming to an end. You have been seeing a client for weekly sessions during your year-long internship. As termination approaches, she becomes more and more anxious and angry with you and states that you are letting her down by forming this relationship with her and then leaving.

3. Your client has repeatedly invited you to his house for various social gatherings. Despite all you have said to him about "dual relationships," he says he still feels that if you really cared about him you would be at his parties.

In a recent innovative qualitative study, Bachelor (1995) examined the working alliance from the point of view of 34 clients to elucidate the characteristics of clients' perceptions of a good working alliance. Analyses provided by these clients yielded three distinct types of working alliances and specific therapy climates associated with each of these three types. The alliances were stable across three different phases of therapy: pretherapy, initial phase, and later phase. Based on these results, she conceptualizes three types of alliances:

1. Alliance as client-therapist *bond*
2. Alliance as improved client *self-understanding*
3. Alliance as client *collaboration*

However, Bachelor (1995) cautions that not all clients perceive the therapeutic alliance in the same way. She notes that clients perceive the positive alliance differentially, attaching significance to different components of it.

Gelso and Carter (1985, 1994) have expanded on Bordin's work as have others including Mallinckrodt (1991, 1993, 1996). Recent work on the alliance has been directed to ways in which ruptures in the alliance can be repaired (Safran, Crocker, McMain, & Murray, 1990). A recent meta-analysis found a stable and positive relationship between working alliance and therapy outcomes (such as client satisfaction and change) (Horvath & Symonds, 1991). Research also suggests that such an alliance needs to be founded early in therapy, may wax and wane over time but reemerge during times of crisis, and may be even more influential with more severe client issues (Sexton & Whiston, 1994). The working alliance is also affected by the kinds of bonds a client has established as a child with her or his parents and also by the client's social competence and also social support (Mallinckrodt, Coble, & Gantt, 1995; Mallinckrodt, 1996). These authors note that "therapists cannot rewrite a client's attachment history, but they can help a client acquire new social competencies. The therapist can become a stable quasi-attachment figure in the client's current life" (1995, p. 83). Recall from our discussion of attachment therapy in Chapter 2 that patterns of attachment are observed in infants and that these patterns are believed to govern adult attachment as well. Mallinckrodt, Gantt, and Coble (1995) have developed a client attachment to therapist scale (CATS) and have used this scale to identify patterns of attachment in therapy. Their results have implications for attachment theory and the working alliance. Clients who scored high on the CATS "securely attached" subscale perceived their therapist as responsive and accepting and reported a positive working alliance and good object-relations capacity. Clients who scored high on the CATS "preoccupied merger" subscale were more preoccupied with their therapist, wanted closer and more frequent therapist contact, had a number of serious object-relations deficits, and more readily formed a working alliance bond rather than coming to agreement in the goals and action phase of therapy. Clients who scored high on the "avoidant-fearful" CATS subscale tended to distrust their therapist, reported the poorest working alliance, and also had some object relation deficits. These clients appeared to long for an emotional connection but feared their ability to engage in such a connection and also feared rejection by the therapist.

Gelso and Carter (1994) point out that none of the relationship components we discuss in this chapter operate independently. For example, some part of the working alliance is influenced by both transference and countertransference, as a positive transference reaction may augment the therapy alliance and a negative transference reaction may erode it. Also, the stronger the alliance is, the safer the client

FEEDBACK 8
Immediacy

I. Here are some expressions of immediacy. See how these compare with yours.

1. "At this moment you seem to be experiencing this loss very intensely."
   or
   "I'm sensing now that it is very painful for you to talk about this."
2. "Every time I mention academic performance, like now, you seem to back off from this topic."
   or
   "I'm aware that, during this session, you stop talking when the topic of your grades comes up."
3. "You seem to be questioning now how qualified I am to help you."
   or
   "I'm wondering if it's difficult right now for you to trust me."
4. "I'm aware of how physically tight I feel now and how tense you look to me. I'm sensing that we're just not too comfortable with each other yet. We seem to be treating each other in a very fragile and cautious way right now."
5. "I'm aware of how well we get along and, because we have so much in common, how easy it is right now just to share stories and events instead of exploring your career concerns."

Are your immediacy responses in the present tense? Do you "own" your feelings and perceptions by using "I feel" rather than "You're making me feel"?

II. Immediacy Assessment

1. Did you express something personal about your feelings, the other person's feelings, or the relationship?
2. Were your immediacy statements in the present tense?
3. Did you use *I, me,* or *mine* when referring to *your* feelings and perceptions?
4. Did you express immediacy as your feelings occurred within the conversation?

FEEDBACK 9
Transference and Countertransference

1. The transference is the client's emotional reaction to not having you available to her at all times. Possible countertransference reactions on your part include frustration, anger, and feelings of failure.
2. The transference is the client's feelings of abandonment as termination approaches. Potential countertransference includes sadness, irritation, and pressure.
3. The transference is the client's expectation for you to be socially involved with him. Possible countertransference includes feelings of letting him down, being upset, impatience and so on.

Counselors obviously also need to pay attention to the ways in which this working alliance is formed with various clients, particularly clients from various cultural groups. As Berg and Jaya (1993) point out, proper attention to protocol may be very important in forming an initial alliance with some Asian American clients. They note that "paying proper respect to procedural rules is the first step in achieving a positive therapeutic alliance. *How* the client is shown respect is often more important than *what* the therapist does to help solve problems" (p. 33). Gender of clients may differentially affect the course of the working alliance. Ways to establish a productive working alliance with clients who have had few prior healthy attachments, have a history of sexual abuse, and strong fears of abandonment also need attention (Mallinckrodt, Coble, & Gantt, 1995). (See Learning Activity #10.)

SUMMARY

This chapter has described three major components of the therapy relationship: the facilitative conditions of empathy, genuineness, and positive regard; transference and countertransference; and the working alliance. These components contribute to both the effective process and outcome of therapy and are considered important by almost all theoretical approaches to counseling. None of these components operate independently from one another but, during the course of therapy, are connected to and influenced by each other. The components also are, to some degree, affected by client variables such as type and severity of problem, gender, and cultural affiliation.

usually feels in allowing and expressing the transferential reaction and pattern. The therapist's countertransference can also strengthen the alliance but therapists must be careful to ensure that their countertransference doesn't harm or injure the alliance.

LEARNING ACTIVITY 10 *Working Alliance*

With a partner or in a small group, discuss how the working alliance may be affected in working with various clients such as the following

1. Children
2. Adolescents
3. Elderly
4. Persons with disabilities
5. Men
6. Women
7. Clients of color
8. Gay, lesbian, and bisexual clients

POSTEVALUATION

I. According to the first objective of this chapter, you will be able to communicate the three facilitative conditions to a client, given a role-play situation. Complete this activity in triads, one person assuming the role of the counselor, another the role of client, and the third acting as the observer. The counselor's task is to communicate the behavioral aspects of empathy, genuineness, and positive regard to the client. The client can share a concern with the counselor. The observer will monitor the interaction, using the accompanying Checklist for Facilitative Conditions as a guide, and provide feedback after completion of the session. Each role play can last about 10–15 minutes. Switch roles so each person has an opportunity to be in each of the three roles. If you do not have access to another person to serve as an observer, find someone with whom you can engage in a role-played helping interaction. Tape-record your interaction and use the accompanying checklist as a guide to reviewing your tape.

Checklist for Facilitative Conditions
Counselor ___ Observer ___ Date ___ Instructions: Assess the counselor's communication of the three facilitative conditions by circling the number and word that best represent the counselor's overall behavior during this session.

Empathy
1. Did the counselor use verbal responses indicating a desire to comprehend the client?

1	2	3	4
A little	Somewhat	A great deal	Almost always

2. Did the counselor reflect *implicit,* or hidden, client messages?

1	2	3	4
A little	Somewhat	A great deal	Almost always

3. Did the counselor refer to the client's feelings?

1	2	3	4
A little	Somewhat	A great deal	Almost always

4. Did the counselor discuss what appeared to be important to the client?

1	2	3	4
A little	Somewhat	A great deal	Almost always

5. Did the counselor pace (match) the client's nonverbal behavior?

1	2	3	4
A little	Somewhat	A great deal	Almost always

6. Did the counselor show understanding of the client's historical-cultural-ethnic background?

1	2	3	4
A little	Somewhat	A great deal	Almost always

7. Did the counselor use verbal responses that validated the client's experience?

1	2	3	4
A little	Somewhat	A great deal	Almost always

Genuineness
8. Did the counselor avoid overemphasizing her or his role, position, and status?

1	2	3	4
A little	Somewhat	A great deal	Almost always

9. Did the counselor exhibit congruence, or consistency, among feelings, words, nonverbal behavior, and actions?

1	2	3	4
A little	Somewhat	A great deal	Almost always

10. Was the counselor appropriately spontaneous (for example, also tactful)?

1	2	3	4
A little	Somewhat	A great deal	Almost always

(continued)

POSTEVALUATION   (continued)

11. Did the counselor self-disclose, or share similar feelings and experiences at a moderate level?

1	2	3	4
A little	Somewhat	A great deal	Almost always

12. Did the counselor demonstrate supporting nonverbal behaviors appropriate for the client's culture?

1	2	3	4
A little	Somewhat	A great deal	Almost always

Positive Regard

13. Did the counselor demonstrate behaviors related to commitment and willingness to see the client (for example, starting on time, responding with intensity)?

1	2	3	4
A little	Somewhat	A great deal	Almost always

14. Did the counselor respond verbally and nonverbally to the client without judging or evaluating the client?

1	2	3	4
A little	Somewhat	A great deal	Almost always

15. Did the counselor convey warmth to the client with supporting nonverbal behaviors (soft voice tone, smiling, eye contact, touch) and verbal responses (enhancing statements and/or immediacy)?

1	2	3	4
A little	Somewhat	A great deal	Almost always

Observer comments: _____

_____

_____

_____

II. According to the second objective of the chapter you will be able to identify issues related to transference and countertransference that might affect the development of the relationship and the working alliance, given four written case descriptions. In this activity, read each case carefully, then identify in writing the transference/countertransference issue that is reflected in the written case description. Feedback follows.

1. You are leading a problem-solving group in a high school. The members are spending a lot of time talking about the flak they get from their parents. After a while, they start to "get the leader" and complain about all the flak they get from you.

2. You are counseling a person of the other sex who is the same age as yourself. After several weeks of seeing the client, you feel extremely disappointed and let down when the client postpones the next session.

3. You find yourself needing to terminate with a client but you are reluctant to do so. When the client presses you for a termination date, you find yourself overcome with sadness.

4. One of your clients is constantly writing you little notes and sending you cards basically saying what a wonderful person you are.

SUGGESTED READINGS

Bachelor, A. (1995). Clients' perceptions of the therapeutic alliance: A qualitative analysis. *Journal of Counseling Psychology, 42,* 323–337.

Brammer, L. M. (1993). *The helping relationship* (5th ed.). Needham Heights, MA: Allyn & Bacon.

Derlaga, V. J., Hendrick, S. S., Winstead, B. A., & Berg, J. H. (1991). *Psychotherapy as a personal relationship.* New York: Guilford Press.

Duan, C., & Hill, C. (1996). The current state of empathy research. *Journal of Counseling Psychology, 43,* 261–274.

Edwards, C. E., & Murdock, N. L. (1994). Characteristics of therapist self-disclosure in the counseling process. *Journal of Counseling and Development, 72,* 384–389.

Egan, G. (1994). *The skilled helper* (5th ed.). Pacific Grove, CA: Brooks/Cole.

Gelso, C. J., & Carter, J. A. (1994). Components of the psychotherapy relationship: Their interaction and unfolding during treatment. *Journal of Counseling Psychology, 41,* 296–306.

Gelso, C. J., Fassinger, R., Gomez, M., & Latts, M. (1995). Countertransference reactions to lesbian clients. *Journal of Counseling Psychology, 42,* 356–364.

Goldstein, A. P., & Higginbotham, H. N. (1991). Relationship-enhancement methods. In F. H. Kanfer & A. P. Goldstein (Eds.), *Helping people change* (4th ed., pp. 20–69). New York: Pergamon Press.

Hill, C. E., & Corbett, M. (1993). A perspective on the history of process and outcome research in counseling psychology. *Journal of Counseling Psychology, 40,* 3–24.

Johnson, D. W. (1997). *Reaching out: Interpersonal effectiveness and self-actualization* (6th ed.). Needham Heights, MA: Allyn & Bacon.

Josselson, R. (1992) *The space between us: Exploring the dimensions of human relationships.* San Francisco: Jossey-Bass.

Kahn, M. (1991). *Between therapist and client: The new relationship.* New York: W. H. Freeman.

Laidlaw, T., & Malmo, C. (Eds.). (1990). *Healing voices: Feminist approaches to therapy with women.* San Francisco: Jossey-Bass.

Mallinckrodt, B., Coble, H., & Gantt, D. (1995). Working alliance, attachment memories, and social competencies of women in brief therapy. *Journal of Counseling Psychology, 42,* 79–84.

Mallinckrodt, B., Gantt, D., & Coble, H. (1995). Attachment patterns in the psychotherapy relationship. *Journal of Counseling Psychology, 42,* 307–317.

Nickerson, K. J., Helms, J., & Terrell, F. (1994). Cultural mistrust, opinions about mental illness, and black students' attitudes toward seeking psychological help from white counselors. *Journal of Counseling Psychology, 41,* 378–385.

Ottens, A., Shank, G., & Long, R. (1995). The role of abductive logic in understanding and using advanced empathy. *Counselor Education and Supervision, 34,* 199–211.

Sexton, T. L., & Whiston, S. C. (1994). The status of the counseling relationship: An empirical review, theoretical implications, and research directions. *The Counseling Psychologist, 22,* 6–78.

Sexton, T. L., & Whiston, S. C. (Eds.). (1996). Counseling outcome research. Special feature of the *Journal of Counseling and Development, 74,* 588–623.

Steenbarger, B. N. (1994) Duration and outcome in psychotherapy: An integrative review. *Professional Psychology, 25,* 111–119.

Stevenson, H., & Renard, G. (1993). Trusting ole' wise owls: Therapeutic use of cultural strengths in African-American families. *Professional Psychology, 24,* 433–442.

Teyber, E. (1997). *Interpersonal process in psychotherapy.* Pacific Grove, CA: Brooks/Cole.

Toms, M. (1993). The eternal feminine: Mirror & container: An interview with Marion Woodman. *New Dimensions, 20,* 8–13.

Wells, M., & Glickauf-Hughes, C. (1986). Techniques to develop object constancy with borderline clients. *Psychotherapy, 23,* 460–468.

Whiston, S., & Sexton, T. (1993). An overview of psychotherapy outcome research: Implications for practice. *Professional Psychology, 24,* 43–51.

Wright, J., & Davis, D. (1994). The therapeutic relationship in cognitive-behavioral therapy: Patient perceptions and therapist responses. *Behavior Therapy, 1,* 25–45.

FEEDBACK
POSTEVALUATION

II. Chapter Objective Two

1. Transference. The group members seem to be transferring their angry feelings toward their parents onto you.
2. Countertransference. You are having a more than usually intense emotional reaction to this client (disappointment), which suggests that you are developing some affectionate feelings for the client and countertransference is occurring.
3. Countertransference: This is another example of your own countertransference—some emotional attachment on your part is making it hard for you to "let go" of this particular client (although termination does usually involve a little sadness for all parties).
4. Transference: The client is allowing herself to idealize you. At this point, this is a positive transference, although it could change.

RELATIONSHIP ENHANCEMENT VARIABLES
AND INTERPERSONAL INFLUENCE

In all human relationships, persons try to influence one another. The counseling relationship is no exception. That counselors do influence clients is inescapable and, according to Senour (1982, p. 346), the desire to avoid influence is "patently absurd [as] there would be no point to counseling if we had no influence on those with whom we work." Moreover, the influence process that operates in counseling is a two-way street. Clients also seek to influence their counselors. As Dorn (1984, p. 343) observes, "Although the client has sought counseling because of dissatisfaction with personal circumstances, this same client will attempt to influence the counselor's behavior."

Thus, the influence process in counseling and therapy is interpersonal—that is, between two persons—and reciprocal, or mutual. At the beginning of counseling, highly motivated clients perceive their counselors to be attractive and likable. The counselors of these same clients also perceive their clients to be quite interpersonally attractive. These counselors also are the ones who believe they have the greatest impact or influence on their clients.

The interpersonal and reciprocal nature of this influence process makes for "very intricate dynamics" during counseling (Dorn, 1984, p. 344). Dorn provides a useful example of how the influence process between the two parties occurs:

> Person A exhibits verbal and nonverbal behavior in an effort to have Person B respond in a specific manner. Person B responds, again with verbal and nonverbal behavior, and this behavior is immediate feedback to Person A about how successful his or her initial influence attempts were. Person A then assesses this feedback and compares it with his or her initial expectations. Person A then decides what behavior to exhibit next. Of course, Person B is simultaneously involved in the same process. (p. 344)

OBJECTIVES

1. Given written descriptions of six clients, match the client description with the corresponding client "test of trust."

2. Given a role play interaction, conduct a 30-minute *initial* interview in which you demonstrate both descriptive and behavioral aspects of attractiveness.

3. Given a role-play interaction, conduct a 30-minute *problem identification* interview in which you demonstrate verbal and nonverbal behaviors of expertness and trustworthiness.

STRONG'S MODEL OF COUNSELING AS INTERPERSONAL INFLUENCE

In 1968, Strong published what is now regarded as a landmark paper on counseling as a social influence process. He hypothesized that counselors' attempts to change clients precipitate dissonance in clients because of the inconsistency, or discrepancy, between the counselor's and the client's attitudes. This dissonance feels uncomfortable to clients, and they try to reduce this discomfort in a variety of ways, including discrediting the counselor, rationalizing the importance of their problem, seeking out information or opinions that contradict the counselor, attempting to change the counselor's opinion, or accepting the opinion of the counselor. Strong (1968) asserted that clients would be more likely to accept the counselor's opinions and less likely to discredit or refute the counselor if the clients perceive the counselor as expert, attractive, and trustworthy. These three helper characteristics (expertness, attractiveness, and trustworthiness) can also be called "relationship enhancers" (Goldstein & Higginbotham, 1991) because they have been identified as ways of making the therapeutic relationship more positive.

Strong (1968) suggested a two-stage model of counseling:

1. The counselor establishes a power base, or influence base, with the client through the three relationship enhancers of expertness, attractiveness, and trustworthiness. This influence base enhances the quality of the relationship and also encourages client involvement in counseling. This stage

of the model (drawing from social-psychology literature) assumes that counselors establish this influence base by drawing on power bases that can effect attitude change. Common power bases used by counselors include these:

- *Legitimate power:* power that occurs as a result of the counselor's role—a form that society at large views as acceptable and helpful.
- *Expert power:* power that results from descriptive and behavioral cues of expertness and competence.
- *Referent power:* power that results from descriptive and behavioral cues of interpersonal attractiveness, friendliness, and similarity between counselor and client (such as is found in "indigenous" helpers, for example).

2. The counselor actively uses this influence base to effect attitudinal and behavioral change in clients. In this second stage of the model, it is important that clients perceive the counselor as expert, attractive, and trustworthy, as it is the *client's* perception of these counselor characteristics that determines, at least in part, how much influence counselors will have with their clients.

During the last decade, an increasing number of research studies on Strong's social influence model have appeared, although much of the existing research consists of analog (not "in the field") studies limited to one or two contacts between persons (Heppner & Claiborn, 1989). A recent meta-analysis of Strong's model was conducted by Hoyt (1996).

THE INTERACTIONAL NATURE OF THE INFLUENCE PROCESS

As we noted at the beginning of this chapter, attempts at influence by counselor and client are interdependent and interrelated. In considering counselor attributes and behaviors (expertness, attractiveness, and trustworthiness) that contribute to influence, it is also important to consider client variables that may enhance or mediate counselor influence effects (Hoyt, 1996).

Although most of this chapter focuses on the three counselor characteristics that contribute most to the influence process in counseling, remember that certain client characteristics may also enhance or mediate the counselor's influence attempts. In other words, some clients may be more susceptible or less susceptible to counselor influence, depending on such things as

- Attractiveness and social competence
- Conceptual level and cognitive style
- "Myths," beliefs, and expectations about counseling

- Motivation
- Satisfaction with outcomes of counseling
- Level of commitment required to change target behaviors
- Gender, race, and cultural background

With respect to issues of gender, sexual orientation, and racial/ethnic status, clients who perceive their counselors as more dissimilar from themselves may feel less safe and be more guarded, especially in initial interactions. Counselors in these situations may need to work more diligently and consistently to establish an influence base with more dissimilar clients. As Sue and Sue (1990) observe, the history of oppression faced by many minority clients contributes to their sense of discomfort with majority counselors. They note that in these situations, "what a counselor says and does in the sessions can either enhance or diminish his/her credibility and attractiveness" (p. 81).

Additionally, as counseling is a *process* and involves distinctly different phases or stages, such as the ones mentioned in Chapter 1 (that is, relationship, assessment and goal setting, intervention and action, and evaluation and termination), it is also imperative to consider what kind of influence might be best suited for different phases of the process. For example, the descriptive aspects of expertness, such as role, reputation, education, and setting, are most useful and influential during the first part of counseling, in which you are trying to encourage the client to continue counseling by demonstrating your credibility. Yet, as counseling ensues, these external trappings are *not* sufficient unless accompanied by behavioral demonstrations of expertness or competence that indicate the counselor is skilled enough to handle the client's concerns successfully.

As Whiston and Sexton (1993) state,

Field studies indicated that factors that influence the counseling will vary on the stage of the counseling. Early in counseling, clients' ratings of counselor expertness, attractiveness, and trustworthiness may be high because of the *helping role* rather than because of any specific counselor behaviors (Heppner & Heesacker, 1982; LaCrosse, 1980). On the other hand, positive client ratings of counselor characteristics at the end of counseling were related to specific counselor behaviors (Heppner & Heesacker, 1982). [Further, in a review of 60 studies of the social influence model], Heppner and Claiborn (1989) "concluded that the interaction between the client and counselor is not static, and to understand the relationship it is important to consider events as distinct times or stages during the counseling that influence this interaction" (p. 45).

It is important, however, to note that almost all of these studies have used Euro-American participants as subjects. As Sue and Sue (1990) note, a caveat on the social influence model is that "findings that certain attributes contribute to a counselor's credibility and attractiveness may not be so perceived by culturally different clients. It is entirely possible that credibility, as defined by credentials indicating specialized training, may only mean to a Hispanic client that the White counselor has no knowledge or expertise in working with Hispanics" (p. 83).

The relationship of various aspects of counselor expertness, attractiveness, and trustworthiness to stages of counseling is depicted in summary form in Table 4-1. *Descriptive* cues associated with these three relationship enhancers refer to nonbehavioral aspects of the counselor such as demeanor, attire, and appearance, to situational aspects such as the office setting, and to the counselor's reputation inferred from introductions, prior knowledge, and the display of diplomas and certificates. *Behavioral* aspects of these three variables refer to the counselor's verbal and nonverbal behaviors or specific things the counselor says and does. In the remainder of the chapter, we describe the behavioral components of these three variables and provide examples.

COUNSELOR CHARACTERISTICS OR RELATIONSHIP ENHANCERS: EXPERTNESS, ATTRACTIVENESS, AND TRUSTWORTHINESS

Earlier we described the importance of three counselor characteristics for establishing and using an influence base with clients: expertness, attractiveness, and trustworthiness. These three characteristics are also related and, in fact, intercorrelated (Zamostny, Corrigan, & Eggert, 1981; Hoyt, 1996) to the extent that counselors who are perceived by clients as competent are also likely to be viewed as interpersonally attractive and trustworthy. These three social influence variables are related to client satisfaction (Zamostny et al., 1981; Heppner & Heesacker, 1983), changes in client self-concept ratings (Dorn & Day, 1985), and less likelihood of premature client termination in therapy (McNeil, May, & Lee, 1987). Hoyt (1996) has concluded that the relationship between these three social influence variables and measures of outcome is complex.

Expertness

Expertness, also known as "competence" (Egan, 1994), is the client's perception that the counselor will be helpful in resolving the client's concerns. Clients develop this perception from such things as the counselor's apparent level of skill, relevant education, specialized training or experience,

certificates or licenses, seniority, status, type of setting in which the counselor works, history of success in solving problems of others, and the counselor's ascribed role as a helper. Clients appear to formulate these perceptions from aspects of the counselor (language, attire, and so on) and of the setting (display of diplomas, certificates, professional literature, title) that are *immediately* evident to a client—that is, in initial contacts. Thus, in the initial stage of counseling, in which the main goal is to establish an effective relationship and build rapport, descriptive cues such as those mentioned above (see also Table 4-2) associated with expertness play a predominant part in helping the counselor to establish an influence base with the client.

Initially, the *role* of counselor also contributes to client perceptions of counselor competence. In our society, a "helper" role is viewed as socially acceptable and valuable. Counselors convey legitimate power or influence simply by the role they hold. Thus, the "counselor role carries considerable initial influence regardless of its occupant" (Corrigan, Dell, Lewis, & Schmidt, 1980, p. 425). Corrigan et al. believe that the legitimate power of our role is, in fact, so strong that demonstrated or behavioral cues of counselor expertness are masked in the initial stage of counseling because sufficient inherent power is ascribed to the role of a helper.

As shown in Table 4-1, in the initial stage of counseling, the counselor wants to create a favorable initial impression and also to encourage the client into counseling by communicating credibility (Whiston & Sexton, 1993). To some extent, such credibility will be conveyed for counselors by the inherent "power" of our roles as helpers. Additionally, counselors can seek to enhance evident and readily accessible descriptive cues associated with expertness by displaying diplomas, certificates, professional literature, titles, and so on. The helper's initial credibility is also enhanced when the counselor has acquired a positive reputation (based on past history of helping others to resolve their problems) and the client is aware of this reputation.

Role, reputation, and external or office "trappings," however, are insufficient to carry the counselor through anything but the initial phase of counseling. In subsequent phases, the counselor must show actual evidence of competence by his or her behavior. This is especially important in working with clients from diverse groups. Role and reputation may not establish credibility with clients of color; in fact, in some cases these external trappings may reduce it. As Sue and Sue (1990) note, "Behavior-expertness, or demonstrating your ability to help a client becomes critical in cross-cultural counseling" (p. 88). Behavioral expertness is

TABLE 4-1. Relationship of counselor expertness, attractiveness, and trustworthiness to stages of counseling

Stage of counseling	Purposes of influence efforts
Rapport and relationship (Stage 1)	
Descriptive aspects of *expertness:* education, role, reputation, setting	"Hook" client to continue counseling by communicating credibility
Physical demeanor and *attractiveness*	Create initial favorable impression
Interpersonal attractiveness conveyed by structuring	Reduce client anxiety, "check out" client expectations
Descriptive aspects of *trust*—role and reputation	Encourage client openness and self-expression
Assessment and goal setting (Stage 2)	
Behavioral aspects of *expertness:* Verbal and nonverbal attentiveness	Contribute to client understanding of self and of issues
Concreteness	Challenge client's language errors and omissions
Relevant and thought-provoking questions	Obtain specificity
Behavioral aspects of *attractiveness:*	
Responsive nonverbal behavior	Encourage relevant client self-disclosure and self-exploration
Self-disclosure	Convey likability and perceived similarity to client
Behavioral aspects of *trustworthiness:* nonverbal acceptance of client disclosures, maintaining of confidentiality, accurate paraphrasing, nondefensive reactions to client "tests of trust"	Convey yourself as trustworthy of client communications so client will feel comfortable "opening up" and self-disclosing
Intervention strategies and action steps (Stage 3)	
Behavioral aspects of *expertness:* directness, fluency, confidence in presentation, and delivery; interpretations	Use of selected skills and strategies and display of confidence to demonstrate ability to help client resolve problems and take necessary action
Behavioral aspects of *trustworthiness:* nonverbal dynamism, dependability and consistency of talk and actions, accurate and reliable information giving	Demonstrate dynamism, congruence, and reliability to encourage client to trust your suggestions and ideas for action; also to diffuse any resistance to action, especially if target behaviors require high level of commitment or change
Evaluation, termination, and follow-up (Stage 4)	
Behavioral areas of *expertness:*	
Relevant questions	Assess client progress and readiness for termination
Directness and confidence in presentation	Contribute to client confidence in maintenance of change through self directed efforts
Interpersonal attractiveness: structuring	Reduce client anxiety about termination and dissolution of therapeutic relationship
Trustworthiness: reputation or demonstrated lack of ulterior motives or personal gain, honesty, and openness	Increase client openness to dissolve relationship when appropriate and necessary

measured by the extent to which the counselor actually helps clients achieve their goals (Egan, 1994). Behavioral demonstrations of expertness are particularly crucial in the second and third stages of counseling (assessment, goal setting, and intervention). These stages require great skill or actual technical competence if a counselor is to make a thorough and accurate assessment of the client's problem, help the

client set realistic and worthwhile goals, and help the client take suitable action to reach those goals. Being able to accomplish these ends is extremely important because having charisma or being a "good guy" or a "good gal" will not get you through successive interactions with clients. As Corrigan et al. note, "In longer term counseling . . . continued evidence of a lack of expertise might negate the

TABLE 4-2. Descriptive and behavioral cues of expertness, attractiveness, and trustworthiness

Expertness	Attractiveness	Trustworthiness
Descriptive cues		
Relevant education (diplomas)	Physical attractiveness	Role as helper (regarded as trustworthy by society)
Specialized training or experience		
Certificates or licenses		Reputation for honesty and "straight-forwardness," lack of ulterior motives
Seniority		
Status		
Type of setting		
Display of professional literature		
Attire		
Reputation (past history of success in resolving problems of others)		
Socially validated role of helper		
Behavioral cues		
Nonverbal behaviors	*Nonverbal behaviors (responsive)*	*Nonverbal behaviors*
Eye contact	Eye contact	Nonverbal congruence
Forward body lean	Direct body orientation facing client	Nonverbal acceptance of client disclosures
Fluent speech delivery	Forward body lean	Nonverbal responsiveness/dynamism
	Smiling	
	Head nodding	
Verbal behaviors	*Verbal behaviors*	*Verbal behaviors*
Relevant and thought-provoking questions	Structuring	Accurate and reliable information giving
Verbal attentiveness	Moderate level of self-disclosure	Accurate paraphrasing
Directness and confidence in presentation	Content of self-disclosure similar to client experiences and opinions	Dependability and consistency of talk and actions
Interpretations		Confidentiality
Concreteness		Openness, honesty
		Reflection of "tests of trust" nondefensively

power conferred on a counselor by virtue of his/her role" (1980, p. 425).

Perceived expertness does not seem to be equivalent to counselor experience—that is, experienced counselors are not automatically perceived as more competent or expert than less experienced or even paraprofessional helpers (Heppner & Heesacker, 1982). Instead, expertness is enhanced by the presence or absence of selected nonverbal *and* verbal counselor behaviors that, together, interact to convey behavioral manifestations of competence (Barak, Patkin, & Dell, 1982). *Nonverbal* cues are especially important for enhancing the counselor's credibility (Hoyt, 1996). *Nonverbal* be-

haviors associated with the communication of expertness include these:

1. Eye contact
2. Forward body lean
3. Fluent speech delivery (see also Chapter 5)

These nonverbal behaviors appear to contribute to perceived expertness by conveying counselor attentiveness and spontaneity and lack of hesitancy in speech and presentation (see also Table 4-2).

Certain *verbal* behaviors seem to contribute to perceived expertness by establishing the counselor as a source of knowledge and skill and by promoting counselor credibility. These include the following:

1. Use of relevant and thought-provoking questions (see Chapter 7)
2. Verbal indications of attentiveness, such as verbal following, lack of interruptions, listening responses (see Chapter 6)
3. Directness and confidence in presentation
4. Interpretations (see Chapter 7)
5. Concreteness

Concreteness. Because the skill of concreteness is not presented in any other part of the book, we describe it in some detail in this section. What clients say to you is often an incomplete representation of their experience. Their words and language (sometimes called "surface structure") do not really represent their experience or the meaning of their communication (sometimes called "deep structure"). Not only is the language of clients an incomplete representation of their experience, it also is full of various sorts of gaps—three in particular:

Deletions—when things are left out, omitted.
Distortions—when things are not as they seem or are misconstrued.
Generalization—when a whole class of things or people is associated with one feeling or with the same meaning or when conclusions are reached without supporting data.

Because of these gaps, it is important for the counselor to use some linguistic tools to make meaning of the client's words and to fill in these gaps. The most efficient linguistic tool for achieving these two objectives is questions—not just any questions, but particular questions designed to extract exactness and concreteness from clients. These questions also help to ensure that you do not project your own sense of meaning onto the client, because your meaning may be irrelevant or inaccurate.

Consider the following example: A client says "I'm depressed." Therapist A responds by asking "About what, specifically?" Therapist B responds with "Depressed? Oh, yes, I know what that's like. How depressed?" The first therapist is likely to get a response from the client that leads to what depression is like for this client and eventually client responses that recover many missing pieces to the problem, as the client's initial statement is full of deletions, or omissions. The second therapist assumes that her sensory experience or meaning of the word *depressed* is the same as the client's and fails to determine how the client's model of reality (or depression) may differ from her own and even from those of other clients.

Concreteness is a way to ensure that general and common experiences and feelings such as depression, anxiety, anger, and so on are defined idiosyncratically for each client. Further, by requesting specific information from clients, you are relieved of having to search for your own equivalent meanings and interpretations. According to Lankton (1980, p. 52), "To translate a client's words into your own subjective experience, at best, results in valuable time and attention lost from the therapy session. At worst, the meaning you make of a client's experience may be wholly inaccurate."

Asking specific questions designed to elicit concreteness from clients is useful for assessing the client's current problems and also desired outcomes. Consequently, it is a facet of expertness that is particularly critical in the data-gathering and self-exploratory and understanding process characterized by the second stage of counseling—assessment and goal setting. Moreover, it helps you to identify client limitations and resources that could contribute to or mitigate against effective solutions to problems. Thus, it is also an important part of expertness in the third phase of counseling, in which action plans and intervention strategies are selected and applied.

As we mentioned earlier, client language contains linguistic errors known as deletions, distortions, and generalizations. Table 4-3 describes common categories of client incomplete linguistic communications and sample counselor responses designed to extract exactness and concreteness.

Expertness Is Not "One Up". It is extremely important to remember that expertness is not in any way the same as being dogmatic, authoritarian, or "one up." Expert helpers are those perceived as confident, attentive, and, because of background and behavior, capable of helping the client resolve problems and work toward goals. Helpers misuse this important variable when they come across as "the expert" or in a one-up position with clients. This posture may intimidate clients, who might then decide not to return for more counseling. In fact, particularly in initial sessions, helpers must do just the opposite: convey friendliness, equality, and likability. In later sessions, it is also important to deemphasize your influence efforts or make them inconspicuous in order to avoid engendering client resistance. In the next section, we describe how helpers exercise likability and friendliness through the variable of attractiveness.

Attractiveness

Attractiveness is inferred by clients from the counselor's apparent friendliness, likability, and similarity to the client. As we mentioned earlier, the counselor who is perceived as

TABLE 4-3. Categories and examples of client linguistic errors and sample counselor responses designed to elicit concreteness

Category	Description	Examples	Sample therapist responses
Deletions			
Simple deletion	Some object, person, or event is left out	"I am going" "I'm scared"	"Going where?" "Of what?"
Comparative deletion	Basis for using a comparative or superlative is deleted	"She is the best" "My brother is better than me"	"Best compared with whom?" "Best when (or where)?" "Better when (or how or where)?"
Referential index, lack of	Object or person being referred to is left out or is unspecified	"They're always in my way" "It makes me sick"	"Who, specifically, is always in your way?" "What, specifically, makes you sick?
Unspecified verb	Parts of the action are missing—for example, verb is introduced but not clarified	"He frustrates me" "I'm stymied"	"Specifically, how does he frustrate you?" "How, specifically, are you stymied?"
Modal operator of *necessity*	Assumption of no choice—"have to," "can't," "impossible," "necessary"	"I can't make sense of this list" "It's impossible to think straight"	"What stops you?" "What prevents you from thinking straight?"
Modal operator of *possibility*	Assumption of no choice—"should, "must not," "ought to"	"I should learn this list" "I must not neglect my studies"	"What would happen if you did not?" "What would happen if you did?"
Lost performative	Who is making a judgment or evaluation is omitted	"It is bad to neglect studying" "People should do better by each other"	"For whom it is bad?" "Bad in whose opinion?" "Who, specifically, should do better?" "Should do better in whose opinion?"

attractive by clients becomes an important source of referent power. The effects of attractiveness are apparently greatest when it is mutual—when the client likes the helper and the helper likes to work with the client (Heppner & Heesacker, 1982).

Attractiveness consists of both physical and interpersonal dimensions. Physical attractiveness is the primary descriptive cue associated with this relationship enhancer and, like the descriptive cues of expertness, appears to exert most influence in the *initial* stage of counseling—during relationship and rapport building, when impression formation by clients is based on relatively apparent and accessible cues. During later stages of counseling, the skills and competence, or behavioral manifestations, of expertness seem to outweigh

the effects of physical attractiveness. In one study, clients did not want to return for counseling with counselors having poor skills even if the counselors were perceived as physically attractive (Vargas & Borkowski, 1982). Zlotlow and Allen (1981, p. 201) conclude that "although the physically attractive counselor may have a head start in developing rapport with clients as a result of widely held stereotypes about good-looking people, this advantage clearly is not an adequate substitute for technical skill or social competence."

In the initial stage of counseling, counselors can utilize the potential benefit from the attractiveness stereotype by trying to maximize their physical attractiveness, appearance, and demeanor. Although there is obviously little we can do to alter certain aspects of our appearance short of plastic

TABLE 4-3. (*Continued*)

Category	Description	Examples	Sample therapist responses
Distortions			
Nominalization	Action (verb) made into a thing (noun)—tends to delete person's responsibility for the action	"I do not have freedom"	"How, specifically, do you not feel free? When, where, with whom?"
		"I want security"	"How do you want to be secure?"
Cause/effect	Assumption that one event *causes* another	"Your frowning makes me mad"	"Specifically, how does my frowning cause you to be mad?"
		"As long as my teacher is around, I feel happy"	"How, specifically, does the presence of your teacher cause you to be happy?"
Mind reading	Assumption of how the other person thinks or feels (inside) without specific evidence	"When you frown, I know you hate me"	"How, specifically, do you know that my frowning means I hate you?"
		"I know he doesn't love me"	"How, specifically, do you know this?"
Presuppositions	Some experience must be assumed for the statement to make sense	"You know I suffer" (Assumes that you know I suffer)	"How, specifically, do you suffer?"
		"My daughter is as stubborn as my husband" (Assumes husband is stubborn)	"How, specifically, does your husband seem stubborn to you?"
Generalizations			
Universal quantifiers	Generalization to whole class; "always," "never," "none," "all," "every"	"I always have trouble learning this kind of material"	"You *never* have learned any kind of material like this before, ever?"
		"I always lose arguments"	"Was there ever a time when you didn't lose an argument?"
Complex equivalence	Assuming that one experience means another (implied cause/effect)	"When he frowns, I know he hates me"	"Was there ever a time when he frowned and he loved you?"
			"Have you ever frowned at someone you love?"
		"My wife wants to work. She doesn't love me"	"Do you love your wife? And do you also work?"

SOURCE: Adapted from a list compiled by R. Rittenhouse (personal communication, June 1982) and Bandler and Grinder (1975). The authors appreciate their contribution to this table.

surgery, other aspects of our appearance, such as attire, weight, personal hygiene, and grooming, are under our control and can be used to enhance, rather than detract from, initial impressions that clients formulate of us.

Selected nonverbal and verbal behaviors convey interpersonal attractiveness and also are quite important during the first two stages of counseling—relationship/rapport and assessment and goal setting. Interpersonal attractiveness helps clients to open up and self-disclose by reducing client anxiety (through structuring and self-disclosure) and by creating the belief that this counselor is someone with whom the client wants to work.

Nonverbal behaviors that contribute to attractiveness include eye contact, body orientation (facing client), forward body lean, smiling, and head nodding (Hermansson, Webster, & McFarland, 1988; see also Table 4-2). These and other aspects of counselor nonverbal behavior are discussed more extensively in Chapter 5.

Verbal behaviors that contribute to attractiveness include self-disclosure and structuring, discussed below. These behaviors appear to enhance the relationship by creating positive expectations, reducing unnecessary anxiety, and increasing the perceived similarity between client and counselor.

LEARNING ACTIVITY 11 *Expertness*

I. Counselor Competence

A. With a partner or in small groups, describe the ideal counseling setting that would enhance most clients' initial impressions of counselor competence. Be very specific in your descriptions.

B. With your partner or in small groups, discuss any clients who might not view your ideal setting described above as indicative of counselor competence. Discuss the limitations of setting, role, and reputation as a means of enhancing the competence variable with these clients.

C. With your partner or in small groups, identify what you believe are the *three most important* things you can do behaviorally to enhance client perceptions of your competence. When you finish this part of the learning activity, you may want to share your descriptions of all three parts with another dyad or group, and vice versa.

II. Concreteness and Linguistic Errors

Twelve client statements are listed in this learning activity. Identify the category of the linguistic error contained in each statement (deletion, distortion, or generalization) and then write a sample counselor response that recovers the omission or challenges the distortion or generalization. An example is given. You may want to refer to Table 4-3 if you have difficulty. Feedback follows on page 66.

Example

1. a. Client statement: "I hate them."
   b. __✔__ Deletion   _____ Distortion
      _____ Generalization
   c. Counselor response: "Whom, specifically, do you hate?"
2. a. Client statement: "She upsets me."
   b. _____ Deletion   _____ Distortion
      _____ Generalization
   c. Counselor response:
3. a. Client statement: "I can't do this."
   b. _____ Deletion   _____ Distortion
      _____ Generalization
   c. Counselor response:

4. a. Client statement: "I know he thinks I'm dumb."
   b. _____ Deletion   _____ Distortion
      _____ Generalization
   c. Counselor response:
5. a. Client statement: "I always lose my cool in front of large groups."
   b. _____ Deletion   _____ Distortion
      _____ Generalization
   c. Counselor response:
6. a. Client statement: "The way you look makes me scared."
   b. _____ Deletion   _____ Distortion
      _____ Generalization
   c. Counselor response:
7. a. Client statement: "I'm sad."
   b. _____ Deletion   _____ Distortion
      _____ Generalization
   c. Counselor response:
8. a. Client statement: "I do not have independence."
   b. _____ Deletion   _____ Distortion
      _____ Generalization
   c. Counselor response:
9. a. Client statement: "My daughter wants to move out. I guess that means she doesn't like it here."
   b. _____ Deletion   _____ Distortion
      _____ Generalization
   c. Counselor response:
10. a. Client statement: "It is bad not to exercise."
    b. _____ Deletion   _____ Distortion
       _____ Generalization
    c. Counselor response:
11. a. Client statement: "It blows my top off."
    b. _____ Deletion   _____ Distortion
       _____ Generalization
    c. Counselor response:
12. a. Client statement: "I should do more work."
    b. _____ Deletion   _____ Distortion
       _____ Generalization
    c. Counselor response:

Self-Disclosure. With respect to attractiveness, three factors related to self-disclosure are worth reemphasizing:

1. Perceived attractiveness is related to a *moderate* level of helper self-disclosure (Edwards & Murdock, 1994). Too much or too little disclosure detracts from the client's perception of the helper as attractive.

2. The depth of personal material reflected in self-disclosure statements needs to be adapted to the stage of counseling and the degree of the therapeutic relationship. In early sessions, self-disclosure of a factual nature is more useful; in later sessions, more personal or self-involving disclosures are more helpful (McCarthy, 1982).

3. Attractiveness is enhanced when helpers self-disclose problems and concerns previously experienced that are *similar* to the client's present problem. Similarity of self-disclosure may also promote the credibility and com-

petence of the helper by suggesting that the counselor knows about and can understand the client's concern (Corrigan et al., 1980). Accordingly, "disclosure of any prior problem (now successfully resolved) may confer on the counselor some 'expertise in problem resolution' or credibility accorded to 'one who has also suffered' " (Corrigan et al., 1980, p. 425).

Structuring.   Another way to maximize perceived similarity between counselor and client is by the use of direct structuring. *Structuring* refers to an interactional process between counselors and clients in which they arrive at similar perceptions of the role of the counselor, an understanding of what occurs in the counseling process, and an agreement on which outcome goals will be achieved (Brammer, Shostrom, & Abrego, 1989; Day & Sparacio, 1980). Structuring enhances perceived counselor/client similarity and interpersonal attractiveness (Goldstein, 1971) and also fulfills an ethical obligation that requires counselors to inform clients of such things as the purposes, goals, techniques, and limitations of counseling (American Counseling Association, 1995).

Direct structuring means that the counselor actively and directly provides structure to the clients concerning the elements mentioned above. Direct structuring contributes to attractiveness by enhancing helper/client agreement on basic information and issues, thereby establishing some security in the relationship. Direct structuring is most important sometime in the first stage of counseling (relationship and rapport), in which ambiguity and anxiety and lack of information about counseling are likely to be greatest and the need to promote helper/client similarity is critical. As Goldstein and Higginbotham (1991) note:

> If, because of misinformation or lack of information about what to expect, the client later experiences events during these meetings with the helper that surprise or confuse him or her, negative feelings will result. Events that confirm his or her expectancies serve to increase his or her attraction to the helper. For example, many new psychotherapy patients come to therapy with expectations based primarily on their past experiences in what they judge to be similar relationships, such as what happens when they meet with their medical doctors. During those visits, the patient typically presented a physical problem briefly, was asked a series of questions by the physician, and was then authoritatively told what to do. Now, however, when the client with such "medical expectations" starts psychotherapy, he or she is in for some surprises. The client describes a psychological problem and sits back awaiting the helper's questions and eventual advice. The helper, unlike the general physician, wants the client to explore his or her feelings, to examine his or her history, to speculate about the causes of the problem. These are not the client's role expectations, and when such important expectations differ, the relationship clearly suffers. These are but a few of the several ways in which client and helper can differ in their anticipations of how each will behave. (pp. 26–27)

An example of the use of direct structuring with a new client in an initial interview follows:

Counselor: Mary, I understand this is the first time you have ever been to see a counselor. Is that accurate or not?

Client: Yes, it is. I've thought about seeing someone for a while, but I finally got the courage to actually do it just recently.

Counselor: I noticed you used the word *courage* as if perhaps you're feeling relieved you're here and also still somewhat uneasy about what happens in counseling.

Client: That's true. I'm glad I came, but I guess I'm also still a little unsure.

Counselor: One thing that might help with the uncertainty is to talk for a few minutes about what goes on in counseling, what my role and your role are, and the kinds of things you may want to talk about or work on. How does that sound?

Client: Great. I think that would help.

Counselor: OK. Many people come into counseling with something they need to get "off their chest"—at first sometimes they just need to talk and think about it. Later on it is usually important to also do something about the issue. My role is to help you identify, talk about, and understand issues of concern to you and then to help you take any action that seems important to resolve the issue or to take your life in a different direction. This process can take several months or longer. At first, it usually is a little hard to open up and share some personal things with someone you don't know, but one thing that might help you do this is to know that, short of feeling strongly like you are going to harm yourself or someone else, whatever you tell me is kept in this room between us. Now—what are your questions or reactions?

Direct structuring is also very useful during the last stage of counseling to ensure a smooth termination, to reduce client anxiety about dissolution of the therapeutic relationship, and to convey action expectations and information about what may happen after counseling terminates. Consider the following example as a counselor and client approach termination:

Counselor: Jim, we started seeing each other every week six months ago; the last two months you've been com-

FEEDBACK 11
Expertness

II. Concreteness and Linguistic Errors

2. b. Deletion (unspecified verb)
   c. "Specifically, how does she upset you?"
3. b. Deletion (modal operator of necessity)
   c. "What is stopping you?"
4. b. Distortion (mind reading)
   c. "How, specifically, do you know this?"
5. b. Generalization (universal quantifier "always")
   c. "You *never* have kept it together in front of a large group?"
6. b. Distortion (cause/effect)
   c. "How, specifically, does the way I look cause you to feel afraid?"
7. b. Deletion (simple omission)
   c. "About what?"
8. b. Distortion (nominalization)
   c. "How, specifically, do you not act (or behave) independently?"
9. b. Generalization (complex equivalence)
   c. "Have you ever left a place or a person you liked or loved?"
10. b. Deletion (lost performative)
    c. "For whom is it bad?"
11. b. Deletion (lack of referential index)
    c. "What, specifically, blows your top off?"
12. b. Deletion (modal operator of possibility)
    c. "What would happen if you didn't do more work?"

---

ing in every other week. Several times you've mentioned recently how good you feel and how your relationships with women are now starting to take off in a direction you want. It seems to me that after about one or two more contacts we will be ready to stop seeing each other because you are able to handle these issues on your own now. What is your reaction to this?

Client: That sounds about right. I do feel a lot more confident in the way my relationships are going. I guess it does seem a little strange to think of not coming in here.

Counselor: Yes. After you've been working together like we have, sometimes there's a little bit of strangeness or apprehension that accompanies the idea of finishing with counseling. However, I wouldn't suggest it at this time if I didn't feel very sure that you are ready to do this. It might help you to know that I'll be calling you several times after we finish to see how things are go-

ing, and of course, if anything comes up in the future you want to talk over, just give me a call.

According to Day and Sparacio (1980), structure is also helpful at major transition points in counseling, such as moving from one stage to another. This also reduces ambiguity, informs the client about any role and process changes in a different stage of therapy, and increases the likelihood that both counselor and client will approach the forthcoming stage with similar rather than highly discrepant perceptions.

To provide structure effectively with clients, consider the following ten guidelines for structuring suggested by Day and Sparacio (1980, pp. 248–249):

1. Structure should be negotiated or requested, not coerced. Clients should be given the opportunity to respond and react to structure as well as be able to modify it.
2. Structure, particularly restrictions and limitations, should not be applied for punitive reasons or in a punitive manner.
3. The counselor should be aware of his or her rationale for structuring and should explain the reasons at the time of structuring or be prepared to provide a rationale in response to the client's request for explanation.
4. The counselor should be guided by the client's readiness for structure and by the context of the relationship and process.
5. Too much or a too-rigid structure can be constraining for both the client and the counselor (Pietrofesa, Hoffman, Splete, & Pinto, 1978).
6. Ill-timed, lengthy, or insensitive structuring can result in client frustration or resistance and interrupt the continuity of the therapeutic process (Pietrofesa et al., 1978).
7. Unnecessary and purposeless recitation of rules and guidelines can imply that the counselor is more concerned with procedure than with helpfulness. In fact, a compulsive approach to structuring can be indicative of low levels of counselor self-assurance (Hansen, Rossberg, & Cramer, 1994).
8. The counselor must relate structure to the client's emotional, cognitive, and behavioral predisposition. For example, the highly independent individual or the isolate may be expected to resist what she or he interprets as personal threats or infringements. In such cases, structuring must be accomplished by sensitivity, tentativeness, and flexibility.
9. Structuring can "imply that the relationship will continue with this particular client. It may turn out that the counselor will decide not to work with this client, or that the client may not be suitable for this counselor. Hence, the client or counselor may feel too committed to the

LEARNING ACTIVITY **12** *Attractiveness*

I. Attributes of Attractive Persons

In a dyad or a small group, discuss the attributes of persons you know and consider to be "attractive" persons. Compile a written list of their attributes. Review your list to determine which attributes are descriptive ones, such as appearance and demeanor, and which attributes are behavioral—that is, things the person does. To what extent do the attributes of attractive people listed in your compilation generalize to effective helpers?

II. Structuring

In this activity, write an example of the use of direct structuring for each of the following four examples. Feedback follows on page 69.

1. Write an example of structuring in an initial interview with a client who has never seen a counselor before.
2. Write an example of structuring in an initial interview with a client who is new to you but has seen three other counselors.
3. Write an example of structuring prior to starting the termination phase with a client who has never been in counseling before.
4. Write an example of structuring prior to starting the termination phase with a client who has been in counseling several times.

---

relationship if it has been overstructured" (Brammer, Shostrom, & Abrego, 1989, p. 121).

10. Structure cannot replace or substitute for therapeutic competence. Structure is not a panacea. It is not the total solution to building a productive therapeutic relationship. Structure is complementary and supplementary to human relations, communications, diagnostic, and intervention skills.*

Trustworthiness

"Trust is the client's perception and belief that the counselor will not mislead or injure the client in any way"** (Fong & Cox, 1983, p. 163). According to the interpersonal influence model, trustworthiness is perceived by clients from such things as the counselor's role, reputation for honesty, demonstrated sincerity and openness, and lack of ulterior motives (Strong, 1968). Trust or mistrust is also greatly affected by the client's race, and ethnic and cultural affiliation.

Establishing Trust. In the initial stage of counseling (relationship and rapport), clients are also dependent on readily accessible descriptive cues to judge the trustworthiness of counselors. For example, many clients are likely to find counselors trustworthy, at least initially, because of the status of their role in society. According to Egan (1975, p. 111), "In our society, people who have certain roles are usually considered trustworthy until the opposite is demonstrated. . . . When exceptions do occur (as when a dentist is convicted of molesting a patient), the scandal is greater because it is unexpected." Clients also are more likely to perceive a counselor as trustworthy if she or he has acquired a reputation for honesty and for ethical and professional behavior. Likewise, a negative reputation can erode initial trust in a helper. Thus, many clients may put their faith in the helper initially on the basis of role and reputation and, over the course of counseling, continue to trust the counselor unless the trust is in some way abused. This is particularly true for clients from the dominant cultural group.

For clients who are not from the dominant cultural group, it may be the other way around. As LaFromboise and Dixon (1981, p. 135) observe, "A member of the minority group frequently enters the relationship suspending trust until the person proves that he/she is worthy of being trusted." For these and some other clients, counselors may have to earn initial trust. This is especially true as counseling progresses. Trust can be difficult to establish yet easily destroyed. Trust between counselor and client involves a series of "relationship interchanges" (Fong & Cox, 1983), takes time to develop fully, and is not a fixed phenomenon but changes constantly depending on the actions of both persons

---

*SOURCE: From "Structuring the Counseling Process," by R. W. Day and R. T. Sparacio, *The Personnel and Guidance Journal, 59,* pp. 246–249. Copyright 1980 by ACA. Reprinted with permission. No further reproduction authorized without written permission of the American Counseling Association.

**SOURCE: This and all other quotations from "Trust as an Underlying Dynamic in the Counseling Process: How Clients Test Trust, by M. L. Fong and B. G. Cox, *The Personnel and Guidance Journal, 62,* pp. 163–166, copyright 1983 by ACA, are reprinted with permission. No further reproduction authorized without written permission of the American Counseling Association.

(Johnson, 1993). Initial trust based on external factors such as the counselor's role and reputation must be solidified with appropriate actions and behaviors of helpers that occur during successive interactions (see also Table 4-2).

During the second stage of counseling, assessment and goal setting, trust is essential in order for the client to be open and revealing of very personal problems and concerns. Clients' self-exploration of problems during this phase can be limited by the amount of trust that has developed in the relationship prior to this time. Trust is also critical during the third and fourth stages of counseling. In the third stage (action/intervention), the client often has to set in motion the difficult and vulnerable process of change. Trust can provide the impetus necessary for the client to do so. Trust is also critical to the fourth stage of counseling (evaluation and termination). Effective termination ensues when the client trusts the counselor's decision to terminate, trusts that it is not too early (leaving the client hanging) or too prolonged (creating excessive dependency for the client), and trusts that the counselor is reliable and concerned enough to check in with the client on a periodic basis as a follow-up to therapy.

The behaviors that contribute most importantly to trustworthiness include counselor congruence, or consistency, of verbal and nonverbal behavior, nonverbal acceptance of client disclosures, and nonverbal responsiveness and dynamism (see also Chapter 5). Incongruence, judgmental or evaluative reactions, and passivity quickly erode any initial trust.

Important verbal behaviors (see also Table 4-2) contributing to trust include accurate paraphrasing (see also Chapter 6), dependability and consistency between talk and actions, confidentiality, openness and honesty, accurate and reliable information giving (see also Chapter 7), and nondefensive reflections/interpretations of clients' "tests of trust." This latter behavior is discussed in greater depth in the following section.

Client Tests of Counselor Trustworthiness.
According to Johnson (1993), trust between counselors and clients does not always develop automatically. Clients need to be assured that counseling will be structured to meet their needs and that the counselor will not take advantage of their vulnerability (Johnson, 1993). Often clients do not ask about these issues directly. Instead, they engage in subtle maneuvers to obtain data about the counselor's trustworthiness. Fong and Cox (1983) call these maneuvers "tests of trust" and liken them to trial balloons sent up to "see how they fly" before the client decides whether to send up the big one or the real one. Brothers (1995) also describes successful and failed tests of trust in the therapy relationship.

Counselors may be insensitive to such "tests of trust" and fail to identify that trust is the real concern of the client. Instead of responding to the trust issue, counselors may respond just to the content, the surface level of the message. Or the counselor may view the client as "defensive, resistant, or hostile" and respond negatively (Fong & Cox, 1983, p. 163). If the trust issue is unresolved, the relationship may deteriorate or even terminate with the counselor unaware that "the real issue was lack of trust" (Fong & Cox, 1983, p. 163).

Johnson (1993) observes that it is inappropriate to *never* trust or to *always* trust. We point this out because occasionally we have seen counselors who seem to expect clients to trust them automatically and feel offended if clients don't trust them or if clients "test" their trustworthiness. We believe that trust has to be earned and that it is useful for clients to be a little skeptical and guarded about revealing too much too soon to a relatively unknown therapist.

We take the position that it may be particularly unwise for clients of color to trust a counselor *initially* until the counselor behaviorally *demonstrates* trustworthiness and credibility—specifically until the counselor shows that (1) he or she will not recreate an oppressive atmosphere of any kind in the counseling interaction, (2) does not engage in discrimination, racist attitudes, and behaviors, and (3) does show some understanding and awareness of the client's racial and cultural affiliation.

Fong and Cox observe that some client statements and behaviors are used repeatedly by many clients as "tests of trust." They state that "the specific content of clients' questions and statements is unique to individual clients, but the general form that tests of trust take—for example, requesting information or telling a secret—are relatively predictable" (p. 164). These authors have identified six common types of client "tests of trust," which we describe as follows.

Requesting information (or "Can you understand and help me?").
Counselors need to be alert to whether client questions are searches for factual information or for counselor opinions and beliefs. Clients who ask questions like "Do you have children?" or "How long have you been married?" are probably looking for something in addition to the factual response. Most often they are seeking verification from you that you will be able to understand, to accept, and to help them with their particular set of concerns. In responding to such client questions, it is important to convey your understanding and acceptance of the clients' concerns and of their need to feel understood. For example, a counselor might say "Yes, I do have two children. I'm also wondering whether

FEEDBACK 12
Attractiveness

II. Structuring

1. "You're probably feeling uncertain about what to expect. It might help if we talked a little about what happens in counseling. You may have one or more things on your mind you want to talk about and work through. I'm here to help you do that and to do it in a safe and confidential place."

2. "You probably know what goes on in counseling generally. What I do might vary a little bit from the other counselors you've seen. I believe I'm here to listen to you, to help you understand some of your concerns, and to assist you in taking a course of action best for you to resolve these issues."

3. "I have the sense that you have really accomplished what you wanted to do. Let's take a few minutes to see . . . [reviews client's progress toward desired goals]. Since you also feel our work's about done, it seems like after two more sessions it will be time to close out our relationship for now. You may feel a little apprehensive about this, but I'll be calling you shortly after we finish to see how things are going. I imagine things will go quite smoothly for you. Once in a while, people find it's hard to start out on their own, but soon things start to fall into place. During these last three sessions, we'll also be working specifically on how you can take the things you've learned and done in our sessions out into your own world so things do fall into place for you out there."

4. "I believe after one or two more sessions, we're ready to finish. Since you've been through this before, do you have the same or a different opinion? [Time for client to respond.] One thing I want you to know about me is that I'll call you once or twice after we stop to see how things are going, and of course, feel free to call me, too, if something comes up you need to discuss.

---

you believe that the fact that I have children means I can better understand your concerns."

Telling a secret (or "Can I be vulnerable or take risks with you?"). Clients share secrets—very personal aspects of their lives—to test whether they can trust the counselor to accept them as they really are, to keep their communications confidential, and to avoid exploiting their vulnerability after they have disclosed very personal concerns. Usually, this secret is not even relevant to the client's presenting problem but, rather, is related to something the client does that has "embarrassment or shame attached to it" (Fong & Cox, 1983, p. 164). And "if the counselor becomes perceptively defensive in reaction to the client's revelation or makes some statement that seems to be judgmental, the client is almost certain to decide that it is unsafe to be vulnerable with this person. The level of trust drops. And further self-disclosure of any depth may not be forthcoming, at least for a very long time" (p. 164).

Counselors need to remember that clients who share secrets are really testing the waters to see how safe it is to self-disclose personal issues with you. Responding with non-verbal and verbal acceptance and listening assure clients that their private thoughts, feelings, and behaviors are safe with you. For example, suppose a client blurts out: "I had an affair several years ago. No one knows about this, not even my husband." The counselor must respond to the entire message, especially acknowledging the "risk" involved. "That is your way of saying to me that this is secret between you and me."

Asking a favor (or "Are you reliable?"). Clients may ask counselors to perform favors that may or may not be appropriate. According to Fong and Cox (1983, p. 165), "all requests of clients for a favor should be viewed, especially initially, as potential tests of trust." When clients ask you to lend them a book, see them at their home, or call their boss for them, whether you grant or deny the favor is not as important as how you handle the request and how reliably you follow through with your stated intentions. It is crucial to follow through on reasonable favors you have promised to do. For unreasonable favors, it is important to state tactfully but directly your reason for not granting the favor. Efforts to camouflage the real reason with an excuse or to grant an unreasonable favor grudgingly are just as damaging to trust as is failure to follow through on a favor (Fong & Cox, 1983, p. 165). For instance, if a client asks you to see her at her home in order to save her time and gas money, you might tactfully deny her favor by saying "Jane, I can certainly appreciate your need to save time and money. I would much prefer to continue to see you in the office, however, because it is easier for me to concentrate and listen to you without any distractions I'm not used to." Asking favors is generally an indication that the client is testing your reliability, dependability, honesty, and directness. A good rule of thumb to follow is "Don't promise more than you can deliver, and be sure to deliver what you have promised as promised." Consistency is especially important to establish with clients who are not from the dominant cultural group. As Sue and Sue (1990) state, "Generally, minority clients who enter counseling with a White therapist will tend to apply a consistency test to what the counselor says or does" (p. 85).

LEARNING ACTIVITY 13   *Trustworthiness*

I. Identification of Trust-Related Issues

With a partner or in a small group, develop responses to the following questions:

A. For clients belonging to the dominant cultural group or from racial/cultural backgrounds similar to your own:

1. How does trust develop during therapeutic interactions?
2. How is trust violated during therapeutic interactions?
3. How does it feel to have your trust in someone else violated?
4. What are ten things a counselor can do (or ten behaviors to engage in) to build trust? Of the ten, select five that are most important and rank-order these from 1 (most critical, top priority to establish trust) to 5 (least critical or least priority to establish trust).

B. Complete the same four questions above for clients of color or from a racial/cultural background distinctly different from your own.

II. Client Tests of Trust

Listed below are six client descriptions. For each description, (a) identify the content and process reflected in the test of trust, and (b) write an example of a counselor response that could be used appropriately with this type of trust test. You may wish to refer to Table 4-4. An example is completed. Feedback follows.

Example

1. The client asks whether you have seen other people before who have attempted suicide.

   a. Test of trust (content): request for information
      (process): can you understand and help me?
   b. Example of counselor response: "Yes, I have worked with other persons before you who have thought life wasn't worth living. Perhaps this will help you know that I will try to understand what this experience is like for you and will help you try to resolve it in your own best way."

2. The client's phone has been disconnected, and the client wants to know whether he can come ten minutes early to use your phone.

   a. Test of trust (content): _____
   _____

   (process): _____
   _____

   b. Example of counselor response: _____
   _____

3. The client wonders aloud whether you make enough money as a counselor that you would choose this occupation if you had to do it over again.

   a. Test of trust (content): _____
   _____

   (process): _____
   _____

   b. Example of counselor response: _____
   _____

4. The client states that she must be kind of stupid because she now has to repeat third grade when all the other kids in her class are going on to fourth grade.

   a. Test of trust (content): _____
   _____

   (process): _____
   _____

   b. Example of counselor response: _____
   _____

5. The client has changed the appointment time at the last minute four times in the last several weeks.

   a. Test of trust (content): _____
   _____

   (process): _____
   _____

   b. Example of counselor response: _____
   _____

6. The client is an Asian American male college student who indicates he is hesitant to speak openly and feels constrained by his concern that his family and friends do not discover he is coming to a counselor. He wonders whom you will tell about his visit to you.

   a. Test of trust (content): _____
   _____

   (process): _____
   _____

   b. Example of counselor response: _____
   _____
   _____
   _____

(continued)

LEARNING ACTIVITY **13** *Continued*

III.

Mr. Hernández is a 38-year-old Mexican American who works as a gardener. He and his family are making an inquiry at the Family Service Association agency regarding a problem that one of the children is having in school. Mr. Hernández speaks some English in his business, because much of his clientele is middle- and upper-class whites. From morning to evening, he drives his truck and maintains the yards and landscapes of many wealthy professionals who live in exclusive sections of the city. Mr. Hernández works hard and is friendly to his customers. During the holidays, many customers give him extra money and gifts for his family. When he returns home after a hard day of work, Mr. Hernández is tired. He has a few friends in the barrio who visit him in the evenings. He enjoys playing cards with them at home and drinking beer. Mr. Hernández is reluctant to talk about personal and family problems to outsiders. Rather, he confides in his wife on the rare occasions when he is deeply troubled over a situation.

Respond to the following:

It is important to enter the world of a person of color. For 30 minutes you are to become Mr. Hernández. Role-play him getting up in the morning before dawn, eating his breakfast, and leaving for a full day of gardening. Imagine his feelings about his work, his gardening skills, his conversations with some of his customers during the day, and his evenings at home with family and friends.

Over the course of several months, Mr. Hernández tries to cope with a family problem that involves his eldest son, but he and his wife are unable to solve the problem. What would you do if you were Mr. Hernández?

- Would you approach a formal social-service agency with your problem?
- How would you feel during the opening session?
- What kind of worker would you like to have help with your problem?
- What would you do to determine whether the worker is a person whom you can trust and in whom you are willing to confide?*

---

*From *Social Work Practice and People of Color* by D. Lum, p. 122. Copyright © 1996 by Brooks/Cole Publishing Company. Reprinted by permission.

Putting oneself down (or "Can you accept me?"). Clients put themselves down to test the counselor's level of acceptance. This test of trust is designed to help clients determine whether the counselor will continue to be accepting even of parts of themselves that clients view as bad, negative, or dirty. Often this test of trust is conveyed by statements or behaviors designed to shock the counselor, followed by a careful scrutiny of the counselor's verbal and nonverbal reactions. Counselors need to respond neutrally to client self-putdowns rather than condoning or evaluating the client's statements and actions. As Fong and Cox note,

> In responding to the client's self-putdowns, the counselor reflects to the client what the counselor has heard and then responds with statements of interest and acceptance. If the counselor makes the mistake of reacting either positively or negatively to the client's descriptions of their "bad" behavior early in the relationship, trust is unlikely to be built. Clients will see the counselor as potentially judgmental or opinionated. (1983, p. 165)

A client may say "Did you know I've had three abortions in the last three years? It's my own fault. I just get carried away and keep forgetting to use birth control." The counselor needs to respond with nonverbal acceptance and may say something like "You've found yourself with several unwanted pregnancies."

Inconveniencing the counselor (or "Do you have consistent limits?"). Clients often test trust by creating inconveniences for the counselor such as changing appointment times, canceling at the last minute, changing the location of sessions, or asking to make a phone call during the session. Counselors need to respond directly and openly to the inconvenience, especially if it occurs more than once or twice. When the counselor sets limits, clients may begin feeling secure and assured that the counselor is dependable and consistent. Setting limits often serves a reciprocal purpose: The clients realize they also can set limits in the relationship. As an example of this test of trust, consider the client who is repeatedly late to sessions. After three consecutive late starts, the counselor mentions "You know, Gary, I've realized that the last three weeks we've got off to quite a late start. This creates problems for me because if we have a full session, it throws the rest of my schedule off. Or if I stop at the designated time, you end up getting shortchanged of time. Can we start on time, or do we need to reschedule the appointment time?"

FEEDBACK
Trustworthiness

II. Client Tests of Trust

2. a. Test of Trust (content): Asking a favor (process): Are you reliable and open with me?
   b. Example response: "I know how difficult it can be to manage without a telephone. Unfortunately, I see someone almost up until the minute you arrive for your session, and so my office is occupied. There's a pay phone in the outer lobby of the building if you find you need to make a call on a particular day or time."

3. a. Test of trust (content): Questioning your motives (process): Do you really care, or are you just going through the motions?
   b. Example response: "Perhaps, Bill, you're feeling unsure about whether I see people like yourself for the money or because I'm sincerely interested in you. One way in which I really enjoy [value] working with you is . . ."

4. a. Test of trust (content): Putting oneself down (process): Can you accept me even though I'm not too accepting of myself right now?
   b. Example response: "You're feeling pretty upset right now that you're going to be back in the third grade again. I wonder if you're concerned, too, about losing friends or making new ones?"

5. a. Test of trust (content): Inconveniencing you (process): Do you have consistent limits?
   b. Example response: "Mary, I'm not really sure anymore when to expect you. I noticed you've changed your appointment several times in the last few weeks at the last minute. I want to be sure I'm here or available when you do come in, so it would help if you could decide on one time that suits you and then just one backup time in case the first time doesn't work out. If you can give some advance notice of a need to change times, then I won't have to postpone or cancel out on you because of my schedule conflicts."

6. a. Test of trust (content): Information request [process]: Do you know enough about my cultural background and affiliation for me to be disclosive with you?
   b. Example response: "I understand you are concerned right now about how much you can safely tell me. I respect your wish to keep the visit here just between us."

Questioning the counselor's motives (or "Is your caring real?"). As we mentioned earlier, one aspect of trustworthiness is sincerity. Clients test this aspect of trust by statements and questions designed to answer the question "Do you really care about me, or is it just your job?" Clients may ask about the number of other clients the counselor sees or how the counselor distinguishes and remembers all his clients or whether the counselor thinks about the client during the week (Fong & Cox, 1983). Fong and Cox observe that "unless counselors are alert to the fact that this is a form of testing trust, they may fail to respond adequately to the crucial issue; that is, the client's need to be seen as a worthwhile human being in the counselor's eyes and not just as a source of income for the counselor" (p. 166). For instance, suppose a client says to her counselor "I bet you get tired of listening to people like me all the time." The counselor may respond with something that affirms her interest in the client, such as "You're feeling unsure about your place here, wondering whether I really care about you when I see so many other persons. Suzanne, from you I've learned . . ." (follow through with a personal statement directly related to this client).

Tests of Trust in Cross-Cultural Counseling. Tests of trust may occur more frequently and with more emotional intensity in cross-cultural counseling. This is often because clients from non-dominant group membership have experienced past and current oppression, discrimination, and overt and covert racism. As a result, they may feel more vulnerable in interpersonal interactions, such as counseling, that involve self-disclosure and an unequal power base. During initial counseling interactions, clients with diverse backgrounds are likely to behave in ways that minimize their vulnerability and that maximize their self-protection (Sue & Sue, 1990). In U.S. culture, Euro-American counselors may be viewed automatically as members of the Establishment (Sue & Sue, 1990, p. 80). As Stevenson and Renard (1993) note, "The dynamics of hostile race relations still exist in our society. It is crucial that therapists question whether these relations are played out in the therapeutic context . . . sensitivity to oppression issues allows for the building of credibility for psychotherapists which becomes of supreme importance to cross-cultural relationships, especially in the early stages" (pp. 433–434). A therapist's disregard for issues of oppression only fuels a minority client's "legacy of mistrust" (Stevenson & Renard, 1993). Nickerson, Helms, and Terrell (1994) have recommended that clinicians must understand and address the impact of cultural mistrust during the entire therapeutic process, from intake to termination.

Table 4-4 presents a summary of these six tests of trust with sample client statements and helpful and nonhelpful counselor responses.

TABLE 4-4. Examples of client tests of trust and helpful and nonhelpful counselor responses

Test of trust	Client statement	Examples of nonhelpful responses	Example of helpful response
Requesting information (can you understand and help me?)	"Have you ever worked with anyone else who seems as mixed up as I am?"	"Yes, all the time" "No, not too often" "Once in a while" "Oh, you're not *that* mixed up"	"Many people I work with often come in feeling confused and overwhelmed. I'm also wondering whether you want to know that I have the experience to help you"
Telling a secret (can I be vulnerable with you?)	"I've never been able to tell anyone about this—not even my husband or my priest. But I did have an abortion several years ago. I just was not ready to be a good and loving mother"	"Oh, an abortion—really?" "You haven't even told your husband even though it might be his child too?"	"What you're sharing with me now is our secret, something between you and me"
Asking a favor (are you reliable?)	"Could you bring this information (or book) in for me next week?"	Promises to do it but forgets altogether or does not do it when specified	Promises to do it and does it when promised
Putting oneself down (can you accept me?)	"I just couldn't take all the pressure from the constant travel, the competition, the need to always win and be number one. When they offered me the uppers, it seemed like the easiest thing to cope with all this. Now I need more and more of the stuff"	"Don't you know you could hurt yourself if you keep going like this?" "You'll get hurt from this—is it really a smart thing to do?"	"The pressure has gotten so intense it's hard to find a way out from under it"
Inconveniencing the counselor (do you have consistent limits?)	"Can I use your phone again before we get started?"	"Of course, go ahead—feel free any time" "Absolutely not"	"Marc, the last two times I've seen you, you have started the session by asking to use my phone. When this happens, you and I don't have the full time to use for counseling. Would it be possible for you to make these calls before our session starts, or do we need to change our appointment time?"
Questioning the counselor's motives (is your caring real?)	"I don't see how you have the energy to see me at the end of the day like this. You must be exhausted after seeing all the other people with problems, too"	"Oh, no, I'm really not" "Yes, I'm pretty tired"	"You're probably feeling unsure about how much energy I have left for you after seeing other people first. One thing about you that helps me keep my energy up is . . ."

SUMMARY

In this chapter, we examined the social influence model of counseling. In this model, the counselor establishes an influence base with the client through the three relationship enhancers of expertness, attractiveness, and trustworthiness. The counselor then uses this influence base to effect client change.

Counselor characteristics contributing most to the influence process include expertness (or competence), attractiveness, and trustworthiness. Components of expertness include descriptive cues such as education and training, certificates and licenses, title and status, setting, reputation, and role. Behavioral cues associated with expertness include responsive nonverbal behaviors such as fluent speech delivery, nonverbal and verbal attentiveness, relevant and thought-provoking questions, interpretations, and concreteness. Nonverbal cues are especially significant.

Descriptive cues associated with attractiveness include physical attractiveness and demeanor. Behavioral cues of attractiveness are responsive nonverbal behavior, moderate level of counselor self-disclosure, similarity of the content of self-disclosure, and structuring.

Trustworthiness is based on one's role and reputation for honesty as well as nonverbal congruence, dynamism, and acceptance of client disclosures. Trustworthiness is also associated with accurate and reliable information giving, accurate paraphrasing, maintaining of confidentiality, openness and honesty, and nondefensive reactions to clients' "tests of trust."

Physical and interpersonal attractiveness and the role, reputation, and setting of the counselor contribute to early impressions of clients that the counselor is attractive, competent, and trustworthy. These aspects are most useful during the early sessions, in which the counselor strives to establish rapport and to motivate the client to continue with counseling. As therapy progresses, these aspects become less influential and must be substantiated by actual skills that demonstrate the counselor's competence and resourcefulness toward resolving client problems. Behavioral expressions of expertness and trustworthiness are particularly critical during all the remaining phases of counseling—assessment and goal setting, intervention and action, and evaluation and termination.

Although expertness, attractiveness, and trustworthiness are important ingredients of a therapeutic relationship, in cross-cultural counseling, counselors cannot safely assume the existence of these variables. As Sue and Sue (1990) state, "The counselor working with a minority client is likely to experience severe tests of his/her expertness and trustworthiness before serious counseling can proceed. The responsibility for proving to the client that you are a credible counselor is likely to be greater when working with a minority than a majority counselee. How you meet the challenge is important in determining your effectiveness as a cross-cultural counselor" (p. 91).

P O S T E V A L U A T I O N

Part One

Listed below are six written client descriptions. Your task is to match each description with the corresponding "test of trust" (Chapter Objective One). Feedback for this part follows on p. 77.

Test of Trust
    a. Information request
    b. Telling a secret
    c. Asking a favor
    d. Putting oneself down
    e. Inconveniencing the counselor
    f. Questioning the counselor's motives

Client Situation

1. The client asks you whether you get "burned out" or fatigued talking to people with problems all day.
2. The client says she has been sexually abused by her stepfather.
3. The client asks to borrow a book she sees on your desk.
4. The client wants to know whether you have been married before.
5. The client wants you to see him on the weekend.
6. The client says some people consider her a whore because she sleeps around a lot.

(continued)

Part Two

This part of the postevaluation is to be completed in triads; the first person assumes the role of counselor, another takes the role of client, and the third assumes the role of observer. Trade roles so that each person has an opportunity to try out each of the three roles once. If triads are not available, an instructor can also observe you, or you can audiotape or videotape your interview for additional assessment.

Instructions to Counselors

Your task is to conduct a 30-minute *initial interview* with a client in which you demonstrate descriptive and behavioral aspects of attractiveness listed in the Attractiveness Checklist that follows (Chapter Objective Two). Remember, too, the purposes of trying to enhance your perceived attractiveness in initial interviews: to reduce client anxiety, to be perceived as likable and friendly and similar to the client, and to increase the probability of client disclosure.

Instructions to Clients

Present a real or hypothetical concern to the counselor. Try to assume the role of a typical "new" client in an initial interview—somewhat apprehensive and a little reticent.

Instructions to Observers

Watch, listen, and assess the use of the counselor's physical and interpersonal cues associated with attractiveness. Use the Attractiveness Checklist that follows as a guide for your observation and feedback.

Attractiveness Checklist
I. Descriptive Cues

Instructions: Assess the counselor's degree of perceived attractiveness on these three items, using the following scale for rating: 1, not at all attractive; 2, minimally attractive; 3, somewhat attractive; 4, quite attractive; 5, very attractive.

1. Appearance

    1   2   3   4   5
2. Demeanor

    1   2   3   4   5
3. Grooming, hygiene

    1   2   3   4   5

II. Behavioral Cues

Instructions: Check "Yes" if the counselor demonstrated the following skills and behaviors; "No" if they were not demonstrated.

4. Use of structure

    Yes   No

5. Moderate level of self-disclosure

    Yes   No
6. Content of self-disclosure similar to client's concerns and experiences

    Yes   No
7. Disclosure of factual, nonintimate material (as this is an initial session)

    Yes   No
8. Responsive nonverbal behaviors
    a. Eye contact

        Yes   No
    b. Direct body orientation facing client

        Yes   No

Observer comments: _____

_____

_____

Part Three

This part of the postevaluation will also be conducted in triads so that each person can assume the roles of counselor, client, and observer. For continuity, you may wish to stay in the same triads you used in Part Two of the postevaluation and trade roles in the same sequence.

Instructions to Counselors

You will be conducting a 30-minute *problem identification interview*—one in which you assess or explore the client's primary problems or concerns. During this interview, your task is to demonstrate behaviors associated with expertness and trustworthiness listed on the Expertness and Trustworthiness Checklist that follows (Chapter Objective Three). Remember, too, that the purposes of trying to enhance your perceived expertness and trustworthiness during this stage of counseling are to contribute to the client's exploration and understanding of self and of issues, to work toward specificity and concreteness, and to encourage the client to share personal and relevant information with you.

Instructions to Clients

Be sure to have a particular "presenting problem" in mind to discuss with the counselor during this role play. It will be helpful if the problem is something real for you, although it doesn't have to be "heavy." In addition to discussing your "presenting problem," try also to ask several questions related to at least one of the following "tests of trust"— requesting information, telling a secret, asking a favor, putting yourself down, inconveniencing the counselor, or questioning the counselor's motives.

(continued)

POSTEVALUATION   (continued)

Instructions to Observers

Watch, listen, and assess the use of the counselor's behaviors associated with competence and trustworthiness. Use the Expertness and Trustworthiness Checklist that follows as a guide for your observation and feedback.

Expertness and Trustworthiness Checklist

I. Expertness

Instructions to observer: Check "Yes" if the counselor demonstrated the behavior; "No" if the counselor did not.

1. Did the counselor maintain eye contact with the client?
   Yes   No
2. Did the counselor lean toward the client during the interaction?
   Yes   No
3. Did the counselor talk fluently and without hesitation?
   Yes   No
4. Did the counselor use relevant and thought-provoking questions?
   Yes   No
5. Was the counselor attentive to the client?
   Yes   No
6. Was the counselor's presentation direct and confident?
   Yes   No
7. Did the counselor accurately interpret any implicit client messages?
   Yes   No
8. Did the counselor challenge any deletions, distortions, or generalizations apparent in the client's messages?
   Yes   No

II. Trustworthiness

9. Did the counselor convey nonverbal and verbal acceptance of the client's disclosures?
   Yes   No
10. Was the counselor's nonverbal behavior responsive and dynamic?
    Yes   No
11. Did the counselor engage in accurate paraphrasing of the client's messages?
    Yes   No
12. Did the counselor appear to safeguard and respect confidentiality of the client's communication?
    Yes   No
13. Did the counselor seem open, honest, and direct with the client?
    Yes   No
14. Was the information the counselor gave "checked out" (or promised to be checked out) for accuracy and reliability?
    Yes   No
15. Were the counselor's verbal messages consistent with overt actions or behaviors?
    Yes   No
16. Did the counselor respond to any client "tests of trust" appropriately and nondefensively?
    Yes   No
17. Did the counselor show some understanding and awareness of the client's racial/ethnic/cultural affiliation?
    Yes   No

Observer comments: _____
_____

SUGGESTED READINGS

Brothers, D. (1995). *Falling backwards: An exploration of trust and self experience.* N.Y.: Norton.

Dorn, F., & Day, B. J. (1985). Assessing change in self-concept: A social psychological approach. *American Mental Health Counselors Association Journal, 7,* 180–186.

Goldstein, A. P., & Higginbotham, H. N. (1991). Relationship-enhancement methods. In F. H. Kanfer & A. P. Goldstein (Eds.), *Helping people change* (4th ed., pp. 20–69). New York: Pergamon.

Heppner, P. P., & Claiborn, C. D. (1989). Social influence research in counseling. A review and critique. *Journal of Counseling Psychology, 36,* 365–387.

Hoyt, W. T. (1996). Antecedents and effects of perceived therapist credibility: A meta-analysis. *Journal of Counseling Psychology, 43,* 430–447.

McNeill, B., & Stoltenberg, C. D. (1989). Reconceptualizing social influence in counseling. The elaboration likelihood model. *Journal of Counseling Psychology 36,* 24–33.

Nickerson, K., Helms, J., & Terrell, F. (1994). Cultural mistrust, opinions about mental illness, and Black students' attitudes toward seeking psychological help from White counselors. *Journal of Counseling Psychology, 41,* 378–385.

Stevenson, H., & Renard, G. (1993). Trusting ole wise owls: Therapeutic use of cultural strengths in African-American families. *Professional Psychology, 24,* 433–442.

Sue, D. W., & Sue, D. (1990). *Counseling the culturally different* (2nd ed.). New York: Wiley.

Tracey, T. J. (1991). The structure of control and influence in counseling and psychotherapy: A comparison of several definitions and measures. *Journal of Counseling Psychology, 38,* 265–278.

Whiston, S. C., & Sexton, T. L. (1993). An overview of psychotherapy outcome research: Implications for practice. *Professional Psychology, 24,* 43–51.

FEEDBACK
POSTEVALUATION

Part One

1. f. Questioning your motives to see whether you really care
2. b. Telling you a secret, something she perhaps feels embarrassed about
3. c. Asking you a favor; in this case, it is probably a reasonable one
4. a. Requesting information overtly—but covertly wondering whether your personal life is together enough to help the client or whether you have enough significant life experiences similar to his own to help him
5. e. Trying to inconvenience you to see whether you have limits and how you set them and follow through on them
6. d. Putting herself down by revealing some part of herself she feels is "bad" and also something that will test your reaction to her.

CHAPTER 5

NONVERBAL BEHAVIOR

Nonverbal behavior plays an important role in our communication and relationships with others. In communicating, we tend to emphasize the spoken word. Yet much of the meaning of a message, 65% or more, is conveyed by our nonverbal behavior (Birdwhistell, 1970). Knapp and Hall (1992, p. 37) define nonverbal behavior as "all human communication events which transcend spoken or written words." Of course, many nonverbal behaviors are interpreted by verbal symbols. Nonverbal behavior is an important part of counseling because of the tremendous amount of information it communicates.

Counselors can learn much about a client by becoming sensitized to the client's nonverbal cues. Moreover, the counselor's nonverbal behavior has a great deal of impact on the client. One of the primary kinds of client verbal messages dealt with in counseling—the affective message—is highly dependent on nonverbal means of communication. Ekman and Friesen (1969a, p. 88) have noted that much of the information that can be gleaned from words of clients is derived from their nonverbal behavior.

Many emotions are expressed graphically by descriptions about the body. Smith (1985, pp. 60–61)* has provided an excellent list of some of these:

He holds his head high.
He has a tight jaw.
He does not hold his head straight.
His head is cocked.
He looks down his nose.
He keeps a stiff upper lip.
He does not look one in the eye.
He is down in the mouth.
He turns away.
He has shifty eyes.
He has a stiff neck.

He sticks his chest out.
His shoulders are stooped.
His arms are outstretched.
He is heavy-handed.
He puts his best foot forward.
He drags his heels.
He puts his finger on it.
He meets one with open arms.
He is tight-fisted.
He waved me away.
He is shady.
His shoulders are square.
He stoops low.
He sits tall.
He sits straight.
He is tied up in a knot.
He is backward.
He is forward.
He leans on people.
He wants to sit on it.
He is weak in the knees.
He puts his foot down.
He is a high stepper.

Five dimensions of nonverbal behavior with significant effects on communication are *kinesics, paralinguistics, proxemics, environmental factors,* and *time.* Body motion, or kinesic behavior, includes gestures, body movements, facial expressions, eye behavior, and posture (Knapp & Hall, 1992, p. 14). Associated with the work of Birdwhistell (1970), kinesics also involves physical characteristics that remain relatively unchanged during a conversation, such as body physique, height, weight, and general appearance. In addition to observing body motion, counselors must identify nonverbal vocal cues called paralanguage—the "how" of the message. Paralanguage includes voice qualities and vocalizations (Trager, 1958). Silent pauses and speech errors can be considered part of paralanguage as well (Knapp & Hall, 1992, p. 24). Also of interest to counselors is the area of

*SOURCE: This and all other quotations from *The Body in Psychotherapy*, by Edward W. L. Smith, copyright 1985 by McFarland & Company, Inc., are reprinted by permission of McFarland & Company, Inc., Publishers, Jefferson, NC 28640.

proxemics (Hall, 1966)—that is, one's use of social and personal space. As it affects the counseling relationship, proxemics involves the size of the room, seating arrangements, touch, and distance between counselor and client.

Perception of one's environment is another important part of nonverbal behavior because people react emotionally to their surroundings. Environments can produce effects on clients such as arousal or boredom and comfort or stress depending on the degree to which an individual tunes into or screens out relevant parts of the surroundings. A fifth aspect of nonverbal behavior involves perception and use of time. Time can be a significant factor in counseling. Time factors include promptness or delay in starting and ending sessions as well as the amount of time spent in communicating with a client about particular topics or events. All these aspects of nonverbal behavior are affected by cultural affiliation.

OBJECTIVES

1. From a list of client descriptions and nonverbal client behaviors, describe one possible meaning associated with each nonverbal behavior.
2. In an interview situation, identify as many nonverbal behaviors of the person with whom you are communicating as possible. Describe the possible meanings associated with these behaviors. The nonverbal behaviors you identify may come from any one or all of the categories of kinesics, or body motion; paralinguistics, or voice qualities; proxemics, or room space and distance; and the person's general appearance.
3. Demonstrate effective use of counselor nonverbal behaviors in a role-play interview.
4. Identify at least four out of five occasions for responding to client nonverbal behavior in an interview.

CLIENT NONVERBAL BEHAVIOR

An important part of a counselor's repertory is the capacity to discriminate various nonverbal behaviors of clients and their possible meanings. Recognizing and exploring client nonverbal cues is important in counseling for several reasons. First, clients' nonverbal behaviors are clues about their emotions. Even more generally, nonverbal behaviors are part of clients' expressions of themselves. As Perls states, "Everything the patient does, obvious or concealed, is an expression of the self" (1973, p. 75). Much of a client's nonverbal behavior may be obvious to you but hidden to the client. Passons (1975, p. 102) points out that most clients are more aware of their words than of their nonverbal behavior. Exploring

nonverbal communication may give clients a more complete understanding of their behavior.

Nonverbal client cues may represent more "leakage" than client verbal messages do (Ekman & Friesen, 1969a). Leakage is the communication of messages that are valid yet are not sent intentionally. Passons (1975) suggests that, because of this leakage, client nonverbal behavior may portray the client more accurately than verbal messages (p. 102). He notes that "nonverbal behaviors are generally more spontaneous than verbal behaviors. Words can be selected and monitored prior to being emitted. . . . Nonverbal behaviors, on the other hand, are not as easily subject to control" (p. 102). A client may come in and *tell* you one story and in nonverbal language convey a completely different story (Erickson, Rossi, & Rossi, 1976).

Knapp and Hall (1992, p. 7) point out that nonverbal and verbal behavior are interrelated. Recognizing the ways nonverbal cues support verbal messages can be helpful. Knapp and Hall identify six such ways:

1. *Repetition:* The verbal message is to "come in and sit down"; the hand gesture pointing to the room and chair is a nonverbal repeater.
2. *Contradiction:* The verbal message is "I like you," communicated with a frown and an angry tone of voice. Some evidence suggests that when we receive contradictory verbal and nonverbal messages, we tend to believe the nonverbal one.
3. *Substitution:* Often a nonverbal message is used in lieu of a verbal one. For example, if you ask someone "How are you?" and you get a smile, the smile substitutes for a "Very well today."
4. *Complementation:* A nonverbal message can complement a verbal message by modifying or elaborating the message. For example, if someone is talking about feeling uncomfortable and begins talking faster with more speech errors, these nonverbal messages add to the verbal one of discomfort.
5. *Accent:* Nonverbal messages can emphasize verbal ones and often heighten the impact of a verbal message. For example, if you are communicating verbal concern, your message may come through stronger with nonverbal cues such as furrow of the brows, frown, or tears. The kind of emotion one conveys is detected best by facial expressions. The body conveys a better description of the intensity of the emotion (Ekman, 1964; Ekman and Friesen, 1967).
6. *Regulation:* Nonverbal communication helps to regulate the flow of conversation. Have you ever noticed that when you nod your head at someone after he or she speaks, the person tends to keep talking? But if you look away and shift in body position, the person may stop

talking, at least momentarily. Whether we realize it, we rely on certain nonverbal cues as feedback for starting or stopping a conversation and for indicating whether the other person is listening [pp. 19–24].*

Identifying the relation between the client's verbal and nonverbal communication may yield a more accurate picture of the client, the client's feelings, and the concerns that have led the client to seek help. In addition, the counselor can detect the extent to which the client's nonverbal behavior and verbal behavior match or are congruent. Frequent discrepancies between the client's expressions may indicate lack of integration or some conflict (Passons, 1975).

Nonverbal behavior has received a great deal of attention in recent years in newspapers, magazine articles, and popular books. These publications may have value in increasing awareness of nonverbal behaviors. However, the meanings that have been attached to a particular behavior may

*SOURCE: Excerpt adapted from *Nonverbal Communication in Human Interaction*, 3rd Ed., by Mark L. Knapp and J. A. Hall. Copyright 1992 by Holt, Rinehart and Winston, Inc. Reprinted by permission of the publisher.

have become oversimplified. It is important to note that the meaning of nonverbal behavior will vary with people and situations (contexts). For example, water in the eyes may be a sign of happiness and glee for one person; for another, it may mean anger, frustration, or trouble with contact lenses. A person who has a lisp may be dependent; another may have a speech impediment. Twisting, rocking, or squirming in a seat might mean anxiety for one person and a stomach cramp for someone else. Further, nonverbal behaviors of one culture may have different or even opposite meanings in another culture. Watson (1970) reports significant differences among cultures in contact and noncontact nonverbal behaviors (distance, touch, eye contact, and so on). As an example, in some cultures, avoidance of eye contact is regarded as an indication of respect. We simply caution you to be careful not to assume that nonverbal behavior has the same meaning or effect for all. It is important to remember that much of what we know about client nonverbal behavior is extrapolated from research on "typical populations" or from analog studies, thus limiting the generalizability of the results. Examples of these are found in Table 5-1.

TABLE 5-1. Nonverbal attending patterns in European–North American culture compared with patterns of other cultures

Nonverbal dimension	European–North American pattern	Contrasting example from another culture
Eye contact	When listening to a person, direct eye contact is appropriate. When talking, eye contact is often less frequent.	Some African Americans may have patterns directly opposite and demonstrate more eye contact when talking and less when listening.
Body language	Slight forward trunk lean facing the person. Handshake is a general sign of welcome.	Certain Eskimo and Inuit groups in the Arctic sit side by side when working on personal issues. A male giving a female a firm handshake may be seen as giving a sexual invitation.
Vocal tone and speech rate	A varied vocal tone is favored, with some emotionality shown. Speech rate is moderate.	Many Latina/o groups have a more extensive and expressive vocal tone and may consider European–North American styles unemotional and "flat."
Physical space	Conversation distance is ordinarily "arm's length" or more for comfort.	Common in Arab and Middle-Eastern cultures is a six-to twelve-inch conversational distance, a point at which the European-American becomes uncomfortable.
Time	Highly structured, linear view of time. Generally "on time" for appointments.	Several South American countries operate on a more casual view of time and do not plan that specified, previously agreed-upon times for meetings will necessarily hold.

Note: It is critical to remember that individuals within a single cultural group vary extensively. *SOURCE:* From *Counseling and Psychotherapy: A Multicultural Perspective*, 3rd Ed., by A. E. Ivey, M. B. Ivey, and L. Simek-Morgan, p. 48. Copyright © 1993 by Allyn and Bacon. Reprinted by permission.

Kinesics

Kinesics involves eyes, face, head, gestures, body expressions, and movements.

Eyes.   Some cultural groups emphasize the importance of visual contact in interpersonal interactions. Therapists who are sensitive to the eye area of clients may detect various client emotions, such as the following:

Surprise: Eyebrows are raised so that they appear curved and high.
Fear: Brows are raised and drawn together.
Anger: Brows are lowered and drawn together. Vertical lines show up between the brows. The eyes may appear to have a "cold stare."
Sadness: Inner corners of the eyebrows are drawn up until the inner corners of the upper eyelids are raised.

Also significant to counselor/client interactions is eye contact (also called "direct mutual gaze"). Eye contact may indicate expressions of feeling, willingness for interpersonal exchange, or a desire to continue or stop talking. Lack of eye contact or looking away may signal withdrawal, embarrassment, or discomfort (Exline & Winters, 1965). Contrary to popular opinion, lack of eye contact does not seem to suggest deception or lack of truthfulness (Sitton & Griffin, 1981). People who generally avoid eye contact may nevertheless make eye contact when they seek feedback. Eye contact may also signal a desire to pause in the conversation or to say something (Knapp & Hall, 1992). The more shared glances there are between two persons, the higher is the level of emotional involvement and comfort. An averted gaze may serve to hide shame over expressing a particular feeling that is seen as culturally or socially taboo (Exline & Winters, 1965). Any kind of reduced eye movement, such as staring or fixated eyes, may signal rigidity or preoccupation in thought (Singer, 1975). Darting or rapid eye movement may mean excitation, anger, or poorly fitting contact lenses.

Excessive blinking (normal is 6 to 10 times per minute in adults) may be related to anxiety. During periods of attentiveness and concentration, blinking usually decreases in frequency. Moisture or tears in the eyes may have contrasting emotional meanings for different people. Eye shifts—away from the counselor to a wall, for example—may indicate that the client is processing or recalling material (Singer, 1975). Pupil dilation, which is an autonomic (involuntary) response, may indicate emotional arousal, attentiveness, and interest (Hess, 1975). Although pupil dilation seems to occur under conditions that represent positive interpersonal attitudes, little or no evidence supports the belief that the opposite (pupil constriction) is associated with negative attitudes toward people (Knapp & Hall, 1992).

In counseling, *more mutual gazing,* or eye contact, seems to occur in the following situations:

1. Greater physical distance exists between the counselor and client.
2. Comfortable, less personal topics are discussed.
3. Interpersonal involvement exists between the counselor and client.
4. You are listening rather than talking.
5. You are female.
6. You are from a culture that emphasizes visual contact in interaction.

Less gazing occurs when

1. The counselor and client are physically close.
2. Difficult, intimate topics are being discussed.
3. Either the counselor or the client is not interested in the other's reactions.
4. You are talking rather than listening.
5. You are embarrassed, ashamed, or trying to hide something.
6. You are from a culture that has sanctions on visual contact during some kinds of interpersonal interactions.

Unfortunately, counselors all too often equate avoidance of eye contact with disrespect, shyness, deception, and/or depression, but for some clients of color, less frequent eye contact is typical of their culture and is not a sign of any of the above.

Mouth.   Smiles are associated with the emotions of happiness and joy. Tight lips may mean stress, frustration, hostility, or anger. When a person has a quivering lower lip or is biting the lips, these signs may connote anxiety or sadness. An open mouth without speech may indicate surprise or difficulty in talking.

Facial Expressions.   The face of the other person may be the most important stimulus in an interaction because it is the primary communicator of emotional information (Ekman, 1982). Facial expressions are used to initiate or terminate conversation, provide feedback on the comments of others, underline or support verbal communication, and convey emotions. Most of the time, the face conveys multiple emotions (Ekman, 1982). For example, one emotion may be conveyed in one part of the face and another in a different area. It is rare for one's face to express only a single emotion at a time. More often than not, the face depicts a blend of varying emotions.

Different facial areas express different emotions. Happiness, surprise, and disgust may be conveyed through the lower face (mouth and jaw region) and the eye area, whereas sadness is conveyed with the eyes. The lower face and brows express anger; fear is usually indicated by the eyes. Although "reading" someone by facial cues alone is difficult, these cues may support other nonverbal indexes of emotion within the context of an interview.

Facial expressions conveying the basic emotions described above do *not* seem to vary much among cultures. In other words, primary or basic emotions such as surprise, anger, disgust, fear, sadness, and happiness do seem to be represented by the same facial expressions across cultures, although individual cultural norms may influence how much and how often such emotions are expressed (Mesquita & Frijda, 1992).

Head.   The movements of the head can be a rich source for interpreting a person's emotional or affective state. When a person holds his or her head erect, facing the other person in a relaxed way, this posture indicates receptivity to interpersonal communication. Nodding the head up and down implies confirmation or agreement. Shaking the head from left to right may signal disapproval or disagreement. Shaking the head with accompanying leg movements may connote anger. Holding the head rigidly may mean anxiety or anger, and hanging the head down toward the chest may reflect disapproval or sadness.

Shoulders.   The orientation of the shoulders may give clues to a person's attitude about interpersonal exchanges. Shoulders leaning forward may indicate eagerness, attentiveness, or receptivity to interpersonal communication. Slouched, stooped, rounded, or turned-away shoulders may mean that the person is not receptive to interpersonal exchanges. This posture also may reflect sadness or ambivalence. Shrugging shoulders may mean uncertainty, puzzlement, ambivalence, or frustration.

Arms and Hands.   The arms and hands can be very expressive of an individual's emotional state. Arms folded across the chest may signal avoidance of interpersonal exchange or reluctance to disclose. Anxiety or anger may be reflected in trembling and fidgety hands or clenching fists. Arms and hands that rarely gesture and are stiffly positioned may mean tension, anxiety, or anger. Relaxed, unfolded arms and hands gesturing during conversation can signal openness to interpersonal involvement or accentuation of points in conversation. The autonomic response of perspiration of the palms may reflect anxiety or arousal.

Legs and Feet.   If the legs and feet appear comfortable and relaxed, the person may be signaling openness to interpersonal exchange. Shuffling feet or a tapping foot may mean that the person is experiencing some anxiety or impatience or wants to make a point. Repeatedly crossing and uncrossing legs may indicate anxiety, depression, or impatience. A person who appears to be very "controlled" or to have "stiff" legs and feet may be uptight, anxious, or closed to an extensive interpersonal exchange.

Total Body and Body Movements.   Most body movements do not have precise social meanings. Body movements are learned and culture-specific. The body movements discussed in this section are derived from analyses of (and therefore most applicable to) White adults from middle and upper socioeconomic classes in the United States.

Body movements are not produced randomly. Instead, they appear to be linked to human speech. From birth, there seems to be an effort to synchronize body movements and speech sounds. In adults, lack of synchrony may be a sign of pathology. Lack of synchrony in body movements and speech between two persons may indicate an absence of listening behavior by both.

One of the most important functions of body movements is *regulation*. Various body movements regulate or maintain an interpersonal interaction. For example, important body movements that accompany the counselor's verbal greeting of a client include eye gaze, smiling, use of hand gestures, and a vertical or sideways motion of the head (Krivonos & Knapp, 1975). Body movements are also useful to terminate an interaction, as at the end of a counseling interview. Nonverbal exit or leave-taking behaviors accompanying a verbal summary statement include decreased eye gaze and positioning of your body near the exit. In terminating an interaction, particularly a therapeutic one, it is also important to display nonverbal behaviors that signify support, such as smiling, shaking the client's hand, touching the client on the arm or shoulder, and nodding your head. As Knapp and Hall (1992, p. 387) explain, "Supportiveness tends to offset any negativity which might arise from encounter termination signals while simultaneously setting a positive mood for the next encounter—that is, our conversation has terminated but our relationship hasn't."

Another way that body movements regulate an interaction involves *turn taking*—the exchange of speaker and listener roles within a conversation. Most of the time we take turns rather automatically. "Without much awareness of what we are doing, we use body movements, vocalizations, and some verbal behavior that often seems to accomplish this turn-taking with surprising efficiency" (Knapp & Hall, 1992, p. 382). Effective turn taking is important in a counseling interaction because it contributes to the perception that you and the client have a good relationship and that, as the counselor, you are a competent communicator. Conversely, ineffective turn taking may mean that a client perceives you as rude (too many interruptions) or as dominating (not enough talk time for the client).

In an innovative qualitative study, Reed, Patton, and Gold (1993) described turn-taking sequences of counselors conducting vocational test interpretation interviews. These authors note that through the use of turn-taking signals, counselors were able to do two things in these interviews: (1) connect test scores to client experiences, and (2) influence clients to see these connections (p. 153).

Duncan (1972, 1974) has found a variety of nonverbal behaviors, called *turn signals,* that regulate the exchange of speaking and listening roles. These turn signals are described in this section. *Turn yielding* occurs when the therapist (as speaker) wants to stop talking and expects a response from the client (as listener). To engage in effective turn yielding, the counselor can ask a question and talk more slowly, slow down the rate of speech, drawl on the last syllable of the last word, utter a "trailer" such as "you know," or use silence. Terminating body movements and gazing at the client will also indicate that it is now the client's turn to talk. If, after a lengthy silence, the client does not respond, more explicit turn-yielding cues may be used, such as touching the client or raising and holding up your eyebrows in expectation. Use of these more explicit nonverbal cues may be particularly important with a quiet, nontalkative client.

Turn maintaining occurs when the therapist (or client) wants to keep talking and not yield a turn to the other person, probably because an important idea is being expressed. Signals that indicate turn maintaining include talking louder and continuing or even increasing the gestures and other body movements accompanying your words. Counselors do not want to maintain their turns too often, or the client is likely to feel frustrated and unable to make a point. Conversely, overtalkative clients may try to take control of the interview by maintaining rather than yielding turns.

Turn requesting occurs when either the therapist or the client is listening and wants to talk. Turn requesting may be signified by an upraised index finger, often accompanied by an audible inspiration of breath and straightening or tightening of one's posture. The counselor can use turn-requesting signals more frequently with an overtalkative, rambling client who tends to hang on rather than give up his or her turn as speaker. In fact, to encourage a client to finish more

quickly, the counselor can use rapid head nods accompanied by verbalizations of pseudoagreement such as "Yes," "Mm-hmm," and "I see" (Knapp & Hall, 1992).

Turn denying occurs when we get a turn-yielding cue from the speaker but don't want to talk. For example, if the client is nontalkative or if the counselor wants the client to take more responsibility for the interaction, the counselor may choose to deny or give up a turn in order to prompt the client to continue talking. Turn-denying signals include a relaxed body posture, silence, and eye gaze. The counselor will also want to exhibit behaviors that show continuing involvement in the ideas expressed by the client by smiling, nodding, or using minimal verbal prompts such as "Mm-hmm."

Keep in mind that the sequence of turn taking may vary among cultures because of different learned patterns of communication. As Sue and Sue (1990) note,

> The cultural upbringing of many minorities dictates different patterns of communication that may place them at a disadvantage in counseling. Counseling initially demands that communication move from client to counselor. The client is expected to take the major responsibility for initiating conversation in the session, while the counselor plays a less active role.
>
> American Indians, Asian Americans, and Hispanics, however, function under different cultural imperatives that may make this difficult. These three groups may have been reared to respect elders and authority figures and "not to speak until spoken to." Clearly defined roles of dominance and deference are established in the traditional family. In the case of Asians, there is evidence to indicate that mental health is associated with exercising will power, avoiding unpleasant thoughts, and occupying one's mind with positive thoughts. Counseling is seen as an authoritative process in which a good counselor is more direct and active while portraying a father figure. A minority client who may be asked to initiate conversation may become uncomfortable and respond with only short phrases or statements. The counselor may be prone to interpret the behavior negatively, when in actuality it may be a sign of respect. (pp. 42–43)

In addition to regulation, body movements also serve the function of *adaptors.* Adaptors may include such behaviors as picking, scratching, rubbing, and tapping. In counseling, it is important to note the frequency with which a client uses nonverbal adaptors, because these behaviors seem to be associated with emotional arousal and psychological discomfort (Ekman, 1982). Body touching may reflect preoccupation with oneself and withdrawal from the interaction at hand. A client who uses adaptors frequently may be uncomfortable with the counselor or with the topic of discussion. The counselor can use the frequency of client adaptors as an index of the client's overall comfort level during counseling.

Another important aspect of a client's total body is his or her breathing. Changes in breathing rate (slower, faster) or depth (shallower, deeper) provide clues about comfort level, feelings, and significant issues. As clients relax, for example, their breathing usually becomes slower and deeper. Faster, more shallow breathing is more often associated with arousal, distress, discomfort, and anxiety.

Paralinguistics

Paralinguistics includes such extralinguistic variables as voice level (volume), pitch (intonation), rate of speech, and fluency of speech. Pauses and silence also belong in this category. Paralinguistic cues are those pertaining to *how* a message is delivered, although occasionally these vocal cues represent what is said as well.

Vocal cues are important in counseling interactions for several reasons. First, they help to manage the interaction by playing an important role in the exchange of speaker and listener roles—that is, turn taking. As you may recall from the discussion of body movements, certain vocalizations are used to yield, maintain, request, or deny turns. For example, decreased pitch is associated with turn yielding, whereas increased volume and rate of speech are associated with turn maintaining. Second, vocal characteristics convey data about a client's emotional states. You can identify the presence of basic emotions from a client's vocal cues if you are able to make auditory discriminations. In recognizing emotions from vocal cues, it is also important to be knowledgeable about various vocal characteristics of basic emotions. For example, a client who speaks slowly and softly may be feeling sad or may be reluctant to discuss a sensitive topic. Increased volume and rate of speech are usually signs of anger or happiness. Changes in voice level and pitch should be interpreted along with accompanying changes in topics of conversation and changes in other nonverbal behaviors.

Voice level may vary among cultures. Sue and Sue (1990) point out that some Americans have louder voice levels than people of other cultures. In counseling a client from a different cultural background, an American counselor should not automatically conclude that the client's lower voice volume indicates weakness or shyness (p. 57).

Vocal cues in the form of speech disturbances or aspects of *fluency* in speech also convey important information for therapists, as client anxiety or discomfort is often detected by identifying the type and frequency of client speech errors. Most speech errors become more frequent as anxiety and discomfort increase (Knapp & Hall, 1992).

Pauses and silence are another part of paralinguistics that can give the counselor clues about the level of arousal and anxiety experienced by the client. There are two types of pauses—filled and unfilled.

Filled pauses are those filled simply with some type of phonation such as "uh," or stutters, false starts, repetitions, and slips of the tongue (Knapp & Hall, 1992). Filled pauses are associated with emotional arousal and anxiety (Knapp & Hall, 1992). A client may be more likely to make a false start or a slip of the tongue when he or she is anxious or uncomfortable.

Unfilled pauses are those in which no sound occurs. Unfilled pauses occur to give the person time to interpret a message and to make a decision about past, present, or future responses. Unfilled pauses, or periods of silence, serve various functions in a counseling interview. The purpose of an unfilled pause often depends on whether the pause is initiated by the counselor or the client. Clients use silence to express emotions, to reflect on an issue, to recall an idea or feeling, to avoid a topic, or to catch up on the progress of the moment. Counselor-initiated silences are most effective when used with a particular purpose in mind, such as reducing the counselor's level of activity, slowing down the pace of the session, giving the *client* time to think, or transferring some responsibility to the client through turn yielding or turn denying. When therapists pause to meet their own needs, as, for example, because they are at a loss for words, the effects of silence may or may not be therapeutic. As Hackney and Cormier (1994) observe, in such instances, when the effect is therapeutic, the counselor is apt to feel lucky rather than competent.

Proxemics

Proxemics concerns the concept of environmental and personal space (Hall, 1966). As it applies to a counseling interaction, proxemics includes use of space relative to the counseling room, arrangement of the furniture, seating arrangements, and distance between counselor and client. Proxemics also includes a variable that seems to be very important to any human interaction—territoriality. Many people are possessive not only of their belongings but of the space around them. It is important for therapists to communicate nonverbal sensitivity to a client's need for space. A client who feels that his or her space or territory has been encroached on may behave in ways intended to restore the proper distance. Such behaviors may include looking away, changing the topic to a less personal one, or crossing one's arms to provide a "frontal barrier" (Knapp & Hall, 1992, p. 152).

In counseling, a distance of three to four feet between counselor and client seems to be the least anxiety-producing and most productive, at least for adult, middle-class Euro-Americans (Lecomte, Bernstein, & Dumont, 1981). For these Americans, closer distances may inhibit verbal productivity (Schulz & Barefoot, 1974), although females in this cultural group are generally more tolerant of less personal space than males, especially when interacting with other females. Disturbed clients also seem to require greater interaction distance. These spatial limits (three to four feet) may be inappropriate for clients of varying ages or cultures. The very young and very old seem to elicit interaction at closer distances. People from "contact" cultures (cultures where interactants face one another more directly, interact closer to one another, and use more touch and direct eye contact) may use different distances for interpersonal interactions than people from "noncontact" cultures (Watson, 1970). In short, unlike facial expressions, distance setting has no universals.

Sue and Sue (1990, p. 53) describe cross-cultural contrasts in proxemics as follows:

> For Latin Americans, Africans, Black Americans, Indonesians, Arabs, South Americans, and French, conversing with a person dictates a much closer stance than normally comfortable for Anglos. A Latin-American client may cause the counselor to back away because of the closeness taken. The client may interpret the counselor's behavior as indicative of aloofness, coldness, or a desire not to communicate. In some cross-cultural encounters, it may even be perceived as a sign of haughtiness and superiority. On the other hand, the counselor may misinterpret the client's behavior as an attempt to become inappropriately intimate, a sign of pushiness or aggressiveness. Both the counselor and the culturally different client may benefit from understanding that their reactions and behaviors are attempts to create the spatial dimension to which they are culturally conditioned.

An important use of proxemics in counseling is to note proxemic *shifts* such as increasing or decreasing space or moving forward or backward. Some evidence suggests that proxemic shifts signal important segments or hiatus points of an interaction, such as the beginning or ending of a topic or a shift to a different subject (Erickson, 1975). Proxemic shifts can give counselors clues about when the client is initiating a new topic, is finishing with a topic, or is avoiding a topic by changing the subject.

Another aspect of proxemics involves seating and furniture arrangement. In some cultures, most therapists prefer a seating arrangement with no intervening desk or objects, although many clients like to have the protective space or "body buffer" of a desk corner. Eskimos may prefer to sit side by side.

Seating and spatial arrangements are an important part of family therapy as well. Successful family therapists pay attention to family proxemics such as the following: How far apart do family members sit from each other? Who sits next to whom? Who stays closest to the therapist? Answers to these questions about family proxemics provide information about family rules, relationships, boundaries, alliances, roles, and so on.

A final aspect of proxemics has to do with touch. Although touch can be a powerful nonverbal stimulus, its effects in the counseling interaction have rarely been examined. A counselor-initiated touch may be perceived by the client as positive or negative, depending on the type of touch (expression of caring versus intimate gesture) and the context, or situation (supportive versus evaluative). In two studies, counseling touch consisting of handshakes and touches to the arm and back had a significant positive effect on the client's evaluation of counseling (Alagna, Whitcher, Fisher, & Wicas, 1979) and on the client's perceptions of the counselor's expertness (Hubble, Noble, & Robinson, 1981). According to Alagna et al. (1979, p. 471), these results

> point up the possibility of setting aside some of the reservations that practitioners experience when they think about physically contacting a client, since under no conditions did communication by touch lead to negative reactions. However, while the effect of touch in this experiment was consistently positive, it is obvious that under certain conditions (e.g., overtly sexual intent), touch in counseling could have negative effects.

Counselors should also be aware that ethical standards such as those adopted by the American Counseling Association (1995), the American Psychological Association (1992), and the National Association of Social Workers (1996) state that any form of sexual intimacy with clients, including sexuality in touch, is unethical.

Environment

Counseling and therapy take place in some surroundings, or environment—typically an office, although other indoor and outdoor environments can be used. The same surroundings can affect clients in different ways. Surroundings are perceived as arousing or nonarousing (Mehrabian, 1976). If a client reacts to an environment with low arousal and mild pleasure, the client will feel comfortable and relaxed. Environments need to be moderately arousing so that the client feels relaxed enough to explore her or his problems and to self-disclose. If the client feels so comfortable that the desire to work on a problem is inhibited, the therapist might consider increasing the arousal cues associated with the surroundings by moving the furniture around, using brighter colors, using more light, or even increasing vocal expressiveness. Therapists who talk louder and faster and use more expressive intonation patterns are greater sources of arousal for those around them (Mehrabian, 1976).

An important concept for considering the effects of environmental arousal on clients is *stimulus screening*—the extent to which a person characteristically screens out the less relevant parts of the environment and thereby effectively reduces the environmental load and the person's own arousal level (Mehrabian, 1976). This concept is useful for understanding different reactions of clients to the same room or office. Individuals who screen their environment well, or "screeners," select various parts of their surroundings to which they respond. As a result, they are more focused on key aspects of their surroundings because they screen out less relevant components. Nonscreeners, in contrast, are less selective in what they respond to in any environment. Consequently, nonscreeners experience locations as more complex and loaded; this reaction may result in too much arousal and even stress (Mehrabian, 1976). Overall, non-screeners are quite sensitive to the emotional reactions of others and to subtle changes in their environment and tend to react rather strongly to such changes.

Environment is also an important issue in working with clients from diverse cultural backgrounds. The idea of coming to an office with a scheduled appointment made in advance is a very Eurocentric notion. For some clients, a walk-in or drop-in visit may be more suitable. Also, for others, having sessions in an out-of-office environment may be desirable.

Time

Time has several dimensions that can affect a therapeutic interaction. One aspect has to do with the counselor's and client's perceptions of time and promptness or delays in initiating or terminating topics and sessions. Many clients will feel put off by delays or rescheduled appointments and, conversely, feel appreciated and valued when extra time is spent with them. Clients may communicate anxiety or resistance by being late or by waiting until the end of a session to bring up a significant topic. Perceptions of time also vary. Some persons have a highly structured view of time, so that being "on time" or ready to see the counselor (or client) is important. Others have a more casual view of time and do

---

LEARNING ACTIVITY 14   *Client Nonverbal Communication*

I. The purpose of this activity is to have you sample some nonverbal behaviors associated with varying emotions for different regions of the body. You can do this in dyads or in a small group. Act out each of the five emotions listed below, using your face, body, arms, legs, and voice.

1. Sadness, depression
2. Pleasure, satisfaction
3. Anxiety, agitation
4. Anger
5. Confusion, uncertainty

As an example, if the emotion to be portrayed were "surprise," you would show how your eyes, mouth, face, arms, hands, legs and feet, and total body might behave in terms of movement or posture, and you would indicate what your voice level and pitch would be like and how fluent your speech might be. After someone portrays one emotion, other members of the group can share how their nonverbal behaviors associated with the same emotion might differ.

II. This activity will help you develop greater sensitivity to nonverbal behaviors of clients. It can be done in dyads or triads. Select one person to assume the role of the communicator and another to assume the role of the listener. A third person can act as observer. As the communicator, recall recent times when you felt (1) very happy, (2) very sad, and (3) very angry. Your task is to retrieve that experience *nonverbally*. Do *not say* anything to the listener, and do *not* tell the listener in advance which of the three emotions you are going to recall. Simply decide which of the three you will recall and tell the listener when to begin. The listener's task is to *observe* the communicator, to *note* nonverbal behaviors and changes during the recall, and, from these, to *guess* which of the three emotional experiences the person was retrieving. After about three to four minutes, stop the interaction to process it. Observers can add behaviors and changes they noted at this time. After the communicator has retrieved one of the emotions, switch roles.

not feel offended or put off if the counselor is late for the appointment and do not expect the counselor to be upset when they arrive later than the designated time.

Time is also a concept that is greatly shaped by one's cultural affiliation. Traditional U.S. society is often characterized by a preoccupation with time as a linear product and as oriented to the future; in contrast, some American Indians and African Americans value a "present-day" time orientation, and Asian Americans and Hispanic Americans focus on both past and present dimensions of time. These differences in the way time is viewed and valued may contribute to discrepancies and misunderstandings in the pace and scheduling of counseling.

HOW TO WORK WITH CLIENT NONVERBAL BEHAVIOR

Many theoretical approaches emphasize the importance of working with client nonverbal behavior. For example, behavioral counselors may recognize and point out particular nonverbal behaviors of a client that constitute effective or ineffective social skills. A client who consistently mumbles and avoids eye contact may find such behaviors detrimental to establishing effective interpersonal relationships. Use of effective nonverbal behaviors also forms a portion of assertion training programs. In transactional analysis (TA), nonverbal behaviors are used to assess "ego states," or parts of one's personality used to communicate and relate to others. For example, the "critical" or controlling parent may be associated with a condescending and blaming voice tone, pointing fingers, frowning, hands on hips, and so forth. TA therapists also note how a client's nonverbal behavior may keep communication going (complementary transactions) or break communication down (crossed transactions). Client-centered therapists use client nonverbal behaviors as indicators of client feelings and emotions. Gestalt therapists help clients recognize their nonverbal behaviors in order to increase awareness of themselves and of conflicts or discrepancies. For example, a client may say "Yes, I want to get my degree" and at the same time shake his head no and lower his voice tone and eyes. Body-oriented therapists actively use body language as a tool for understanding hidden and unresolved "business," conflicts, and armoring. Adlerian counselors use nonverbal reactions of clients as an aid to discovering purposes (often hidden) of behavior and mistaken logic. Family therapists are concerned with a family's nonverbal (analogic) communication as well as verbal (digital) communication. A tool based on family nonverbal communication is known as "family sculpture" (Duhl, Kantor, &

Duhl, 1973). Family sculpture is a nonverbal arrangement of people placed in various physical positions in space to represent their relationship to one another. In an extension of this technique, family *choreography* (Papp, 1976), the sculptures or spatial arrangements are purposely moved to realign existing relationships and create new patterns.

Passons (1975) has described five ways of responding to client nonverbal behavior in an interview. His suggestions are useful because they represent ways of working with clients nonverbally that are consistent with various theoretical orientations. These five ways are the following:

1. Ascertain the congruence between the client's verbal and nonverbal behavior.
2. Note or respond to discrepancies, or mixed verbal and nonverbal messages.
3. Respond to or note nonverbal behaviors when the client is silent or not speaking.
4. Focus on nonverbal behaviors to change the content of the interview.
5. Note changes in client nonverbal behavior that have occurred in an interview or over a series of sessions.

Nonverbal communication is also a useful way for counselors to note something about the *appropriateness* of a client's communication style and to observe *how* something is said, not just *what* is said (Sue & Sue, 1990, p. 51). Such observations may be especially important in working with clients belonging to various ethnic/cultural groups. For example, in traditional Asian culture, *subtlety* in communication versus directness is considered a "prized art" (Sue & Sue, 1990, p. 51). These "social rhythms" of communication style will vary among race, culture, ethnicity, and gender of clients. (See also Learning Activity #15.)

Congruence Between Behaviors

The counselor can determine whether the client's verbal message is congruent with his or her nonverbal behavior. An example of congruence is the client's expressing confusion about a situation as she squints her eyes or furrows her brow. Another client may say, "I'm really happy with the way things have been working since I've been coming to see you," and accompany this statement with eye contact, relaxed body posture, and a smile. The counselor can respond in one of two ways to congruence between the client's verbal and nonverbal behaviors. A counselor might make a *mental* note of the congruence in behaviors. Or the counselor could ask the client to explain the meaning of the nonverbal behaviors. For example, the counselor could ask: "While you were saying this is a difficult topic for you, your eyes were moist, your head was lowered, and your hands were fidgety. I wonder what that means?"

Mixed Messages

The counselor can observe the client and see whether the client's words and nonverbal behavior are mixed messages. Contradictory verbal and nonverbal behavior would be apparent with a client who says, "I feel really [pause] excited about the relationship. I've never [pause] experienced anything like this before" while looking down and leaning away. The counselor has at least three options for dealing with a verbal/nonverbal discrepancy. The first is to note mentally the discrepancies between what the client says and the nonverbal body and paralinguistic cues with which the client delivers the message. The second option is to describe the discrepancy to the client, as in this example: "You say you are excited about the relationship, but your head was hanging down while you were talking, and you spoke with a lot of hesitation." (Other examples of confronting the client with discrepancies can be found in Chapter 7.) The third option is to ask the client, "I noticed you looked away and paused as you said that. What does that mean?"

Nonverbal Behavior During Silence

The third way a counselor can respond to the nonverbal behavior of the client is during periods of silence in the interview. Silence does not mean that nothing is happening! Also remember that silence has different meanings from one culture to another. In some cultures, silence is a sign of respect, not an indication that the client does not wish to talk more (Sue & Sue, 1990). The counselor can focus on client nonverbal behavior during silence by noting the silence mentally, by describing the silence to the client, or by asking the client about the meaning of the silence (Itai & McRae, 1994).

Changing the Content of the Interview

It may be necessary with some clients to change the flow of the interview, because to continue on the same topic may be unproductive. Changing the flow may also be useful when the client is delivering a lot of information or is rambling. In such instances, the counselor can distract the client from the verbal content by redirecting the focus to the client's nonverbal behavior.

For "unproductive" content in the client's messages, the counselor might say, "Our conversation so far has been dwelling on the death of your brother and your relationship with your parents. Right now, I would like you to focus on what we have been doing while we have been talking. Are you aware of what you have been doing with your hands?"

Such counselor distractions can be either productive or detrimental to the progress of therapy. Passons (1975) suggests that these distractions will be useful if they bring the client in touch with "present behavior." If they take the client away from the current flow of feelings, the distractions will be unproductive (p. 105). Passons also states that "experience, knowledge of the counselor, and intuition" all contribute to the counselor's decision to change the content of the interview by focusing on client nonverbal behavior.

Changes in Client Nonverbal Behavior

For some clients, nonverbal behaviors may be indexes of therapeutic change. For example, at the beginning of counseling, a client's arms may be folded across the chest. Later, the client may be more relaxed, with arms unfolded and hands gesturing during conversation. At the initial stages of counseling, the client may blush, perspire, and exhibit frequent body movement during the interview when certain topics are discussed. Later in counseling, these nonverbal behaviors may disappear and be replaced with a more comfortable and relaxed posture. Again, depending on the timing, the counselor can respond to nonverbal changes covertly or overtly.

This decision to respond to client nonverbal behavior covertly (with a mental note) or overtly depends not only on your purpose in focusing on nonverbal behavior but also on timing. Passons (1975) believes that counselors need to make overt responses such as immediacy (see Chapter 3) to client nonverbal behavior early in the therapeutic process. Otherwise, when you call attention to something the client is doing nonverbally after the tenth session or so, the client is likely to feel confused and bewildered by what is seen as a change in your approach. Another aspect of timing involves discriminating the likely effects of your responding to the nonverbal behavior with immediacy. If the immediacy is likely to contribute to increased understanding and continuity of the session, it may be helpful. If, however, the timing of your response interrupts the client's flow of exploration, your response may be distracting and interfering.

When responding to client nonverbal behavior with immediacy, it is helpful to be descriptive rather than evaluative and to phrase your responses in a tentative way. For example, saying something like "Are you aware that as you're talking with Gene, your neck and face are getting red splotches of color?" is likely to be more useful than evalua-

LEARNING ACTIVITY **15**   *Responding to Client Nonverbal Behavior*

I. The purpose of this activity is to practice verbal responses to client nonverbal behaviors. One person portrays a client (1) giving congruent messages between verbal and nonverbal behavior, (2) giving mixed messages, (3) being silent, (4) rambling and delivering a lot of information, and (5) portraying a rather obvious change from the beginning of the interview to the end of the interview in nonverbal behavior. The person playing the counselor responds verbally to each of these five portrayals. After going through all these portrayals with an opportunity for the role-play counselor to respond to each, switch roles. During these role plays, try to focus primarily on your responses to the other person's nonverbal behavior.

II. With yourself and several colleagues or members of your class to help you, use spatial arrangements to portray your role in your family and to depict your perceptions of your relationship to the other members of your family. Position yourself in a room and tell the other participants where to position themselves in relation to you and one another. (If you lack one or two participants, an object can

fill a gap.) After the arrangement is complete, look around you. What can you learn about your own family from this aspect of nonverbal behavior? Do you like what you see and feel? If you could change your position in the family, where would you move? What effect would this have on you and on other family members?

III. In a role-play interaction or counseling session in which you function as the therapist, watch for some significant nonverbal behavior of the client, such as change in breathing, shifts in eye contact, voice tone, and proxemics. (Do not focus on a small nonverbal behavior out of context with the spoken words.) Focus on this behavior by asking the client whether she or he is aware of what is happening to her or his voice, body posture, eyes, or whatever. Do not interpret or assign meaning to the behavior for the client. Notice where your focus takes the client.

IV. Contrast the areas of client nonverbal behavior we describe in this chapter—kinesics, paralinguistics, proxemics, environment, and time—for Euro-American and non-Euro-American clients.

---

tive and dogmatic comments such as "Why is your face getting red?" "You surely do have a red face," or "You're getting so red—you must feel very embarrassed about this."

COUNSELOR NONVERBAL BEHAVIOR

As a counselor, it is important for you to pay attention to your nonverbal behavior for several reasons. First, some kinds of counselor nonverbal behavior seem to contribute to a facilitative relationship; other nonverbal behaviors may detract from the relationship. For example, "high," or facilitative, levels of such nonverbal behaviors as direct eye contact and body orientation and relaxed body posture can contribute to positive client ratings of counselor empathy even in the presence of a low-level, or detracting, verbal message (Fretz, Corn, Tuemmler, & Bellet, 1979). In addition, the degree to which clients perceive you as interpersonally attractive and as having some expertise is associated with effective nonverbal skills (Claiborn, 1979).

Because much of the research on counselor nonverbal behavior has been done with ratings of videotapes and photographs, it is difficult to specify precisely what counselor nonverbal behaviors are related to counseling effectiveness. Table 5-2 lists presumed effective and ineffective uses of counselor nonverbal behaviors. In assessing this list, it is also

important to remember that the effects of various counselor nonverbal behaviors are related to contextual variables in counseling, such as type of client, verbal content, timing in session, and client's perceptual style (Hill, Siegelman, Gronsky, Sturniolo, & Fretz, 1981). Thus, clients who subjectively have a favorable impression of the therapist may not be adversely affected by an ineffective, or "low-level," nonverbal behavior such as tapping your finger or fiddling with your pen or hair (we still recommend you avoid such distracting mannerisms). Similarly, just engaging in effective use of nonverbal behaviors such as those listed in Table 5-2 may not be sufficient to alter the negative impressions of a particular client about yourself.

In addition to the use of effective nonverbal behaviors such as those listed in Table 5-2, there are three other important aspects of a therapist's nonverbal demeanor that affect a counseling relationship: sensitivity, congruence, and synchrony.

Sensitivity

Presumably, skilled interviewers are better able to send effective nonverbal messages (encoding) and are more aware of client nonverbal messages (decoding) than ineffective interviewers. There is some evidence that females of various cultures are better decoders—that is, more sensitive to other

TABLE 5-2 Effective and ineffective counselor nonverbal behavior

Ineffective use	Nonverbal mode of communication	Effective use
Doing any of these things will probably close off or slow down the conversation		These behaviors encourage talk because they show acceptance and respect for the other person
Distant or very close	Space	Approximately arm's length
Away	Movement	Toward
Slouching; rigid; seated leaning away	Posture	Relaxed but attentive; seated leaning slightly toward
Absent; defiant; jittery	Eye contact	Regular
You continue with what you are doing before responding; in a hurry	Time	Respond at first opportunity; share time with the client
Used to keep distance between the persons	Feet and legs (in sitting)	Unobtrusive
Used as a barrier	Furniture	Used to draw persons together
Does not match feelings; scowl; blank look	Facial expression	Match your own or other's feelings; smile
Compete for attention with your words	Gestures	Highlight your words; unobtrusive; smooth
Obvious; distracting	Mannerisms	None or unobtrusive
Very loud or very soft	Voice: volume	Clearly audible
Impatient or staccato; very slow or hesitant	Voice: rate	Average or a bit slower
Apathetic; sleepy; jumpy; pushy	Energy level	Alert; stay alert throughout a long conversation

SOURCE: From *Amity: Friendship in Action. Part 1: Basic Friendship Skills,* by Richard P. Walters. Copyright © 1980 by Richard P. Walters, Christian Helpers, Inc., Boulder, CO. Reprinted by permission.

persons' nonverbal cues—than are males (Sweeney, Cottle, & Kobayashi, 1980). Male therapists may need to ensure that they are not overlooking important client cues. All of us can increase our nonverbal sensitivity by opening up all our sensory channels. For example, people who tend to process information through auditory channels can learn to pay closer attention to visual cues, and those who process visually can sensitize themselves to voice cues.

Congruence

Counselor nonverbal behaviors in conjunction with verbal messages also have some consequences in the relationship, particularly if these messages are mixed, or incongruent. Mixed messages can be confusing to the client. For example, suppose a counselor says to a client, "I am really interested in how you feel about your parents," while the counselor's body is turned away from the client with arms folded across the chest. The effect of this inconsistent message on the

client could be quite potent. In fact, a *negative nonverbal* message mixed with a *positive verbal* one may have greater effects than the opposite (positive nonverbal and negative verbal). As Gazda et al. (1995, p. 87) point out, "When verbal and nonverbal messages are in contradiction, we usually believe the nonverbal message." Negative nonverbal messages are communicated by infrequent eye contact, body position rotated 45° away from the client, backward body lean (from waist up leaning back), legs crossed away from the client, and arms folded across the chest (Graves & Robinson, 1976). The client may respond to inconsistent counselor messages by increasing interpersonal distance and may view such messages as indicators of counselor deception (Graves & Robinson, 1976). Further, mixed messages may reduce the extent to which the client feels psychologically close to the counselor and perceives the counselor as genuine.

In contrast, congruence between counselor verbal and nonverbal messages is related to both client and counselor ratings of counselor facilitativeness (Hill et al., 1981; Reade &

LEARNING ACTIVITY 16 *Counselor Nonverbal Behavior*

The purpose of this activity is to have you experience the effects of different kinds of nonverbal behavior. You can do this in dyads or groups or outside a classroom setting.

1. Observe the response of a person you are talking with when
   a. You look at the person or have relaxed eye contact.
   b. You don't look at the person consistently; you avert your eyes with only occasional glances.
   c. You stare at the person.

   Obtain a reaction from the other person about your behavior.

2. With other people, observe the effects of varying conversational distance. Talk with someone at a distance of (a) 3 feet (about 1 meter), (b) 6 feet (2 meters), and (c) 9 feet (3 meters).
   Observe the effect these distances have on the person.
3. You can also do the same kind of experimenting with your body posture. For example, contrast the effects of two body positions in conversation: (a) slouching in seat, leaning back, and turning away from the person, compared with (b) facing the person, with a slight lean forward toward the person (from waist up) and with body relaxed.

LEARNING ACTIVITY 17 *Observation of Counselor and Client Nonverbal Behavior*

The purpose of this activity is to apply the material presented in this chapter in an interview setting. Using the Nonverbal Behavior Checklist at the end of the chapter, observe a counselor and determine how many behaviors listed on the checklist she or he demonstrates. In addition, in the role play, see how much you can identify about the client's nonverbal behaviors. Finally, look for evidence of synchrony (pacing) or dissynchrony between the two persons and congruence or incongruence for each person.

Smouse, 1980). The importance of counselor congruence, or consistency, among various verbal, kinesic, and paralinguistic behaviors cannot be overemphasized. Congruence between verbal and nonverbal channels seems especially critical when confronting clients (see Chapter 7) or when discussing personal, sensitive, or stressful issues (Reade & Smouse, 1980). A useful aspect of counselor congruence involves learning to match the *intensity* of your nonverbal behaviors with those of the client. For example, if you are asking the client to recall a time when she or he felt strong, resourceful, or powerful, it is helpful to convey these feelings by your own nonverbal behaviors. Become more animated, speak louder, and emphasize key words such as "strong" and "powerful." Many of us overlook one of our most significant tools in achieving congruence—our voice. Changes in pitch, volume, rate of speech, and voice emphasis are particularly useful ways of matching our experience with the experience of clients.

Synchrony

Synchrony is the degree of harmony between the counselor's and client's nonverbal behavior. In helping interactions, especially initial ones, it is important to match, or pace, the client's nonverbal behaviors. Pacing of body posture and other client nonverbal behaviors contributes to rapport and builds empathy (Maurer & Tindall, 1983). Synchrony does not mean that the counselor mimics every move or sound the client makes. It does mean that the counselor's overall nonverbal demeanor is closely aligned with or very similar to the client's. For example, if the client is sitting back in a relaxed position with crossed legs, the counselor matches and displays similar body posture and leg movements. Dissynchrony, or lack of pacing, is evident when, for example, a client is leaning back, very relaxed, and the counselor is leaning forward, very intently, or when the client has a very sad look on her face and the counselor smiles, or when the client speaks in a low, soft voice and the counselor responds in a strong, powerful voice. The more nonverbal patterns you can pace, the more powerful the effect will be. However, when learning this skill, it is too overwhelming to try to match many aspects of a client's nonverbal behavior simultaneously. Find an aspect of the client's demeanor, such as voice, body posture, or gestures, that feels natural and comfortable for you to match, and concentrate on synchronizing this one aspect at a time. (See Learning Activity #17.)

SUMMARY

The focus of this chapter has been on counselor and client nonverbal behavior. The importance of nonverbal communication in counseling is illustrated by the trust that both counselor and client place in each other's nonverbal messages. Nonverbal behavior may be a more accurate portrayal of our real selves. Most nonverbal behaviors are very spontaneous and cannot easily be faked. Nonverbal behavior adds significantly to our interpretation of verbal messages.

Five significant dimensions of nonverbal behavior were discussed in this chapter: kinesics (face and body expressions), paralinguistics (vocal cues), proxemics (space and distance), environment, and time. Although much popular literature has speculated on the meanings of "body lan-guage," in counseling interactions, counselors must remember that the meaning of nonverbal behavior varies with people, situations, and cultures and, further, cannot be easily interpreted without supporting verbal messages.

These categories of nonverbal behavior also apply to the counselor's use of effective nonverbal behavior in the interview. In addition to using nonverbal behaviors that communicate interest and attentiveness, counselors must ensure that their own verbal and nonverbal messages are congruent and that their nonverbal behavior is synchronized with, or matches, the client's nonverbal behavior. Congruence and synchrony are important ways of contributing to rapport and building empathy within the developing relationship.

POSTEVALUATION

Part One

Describe briefly one possible effect or meaning associated with each of the following 10 client nonverbal behaviors (Chapter Objective One). Speculate on the meaning of the client nonverbal behavior from the client description and context presented. If you wish, write your answers on a piece of paper. Feedback follows the evaluation.

Observed Client Nonverbal Behavior	Client Description (Context)
1. Lowered eyes—looking down or away	Client has just described incestuous relationship with father. She looks away after recounting the episode.
2. Pupil dilation	Client has just been informed that she will be committed to the state hospital. Her pupils dilate as she sits back and listens.
3. Quivering lower lip or lip biting	Client has just reported a recent abortion to the counselor. As she's finishing, her lip quivers and she bites it.
4. Nodding head up and down	Counselor has just described reasons for client to stop drinking. Client responds by nodding and saying "I know that."
5. Shrugging of shoulders	Counselor has just informed client that he is not eligible for services at that agency. Client shrugs shoulders while listening.
6. Fist clenching or holding hands tightly.	Client is describing recent argument with spouse. Her fists are clenched while she relates incident.
7. Crossing and uncrossing legs repeatedly	Counselor has just asked client whether he has been taking his medicine as prescribed. Client crosses and uncrosses legs while responding.
8. Stuttering, hesitations, speech errors	Client hesitates when counselor inquires about marital fidelity. Starts to stutter and makes speech errors when describing extramarital affairs.
9. Moving closer	As counselor self-discloses an episode similar to client's, client moves chair toward helper.
10. Flushing of face and appearance of sweat beads	Counselor has just confronted client about provocative clothing and posture. Client's face turns red and sweat appears on her forehead.

Part Two

Conduct a short interview as a helper and see how many client nonverbal behaviors of kinesics (body motion), paralinguistics (voice qualities), and proxemics (space) you can identify by debriefing with an observer after the session (Chapter Objective Two). Describe the possible effects or meanings associated with each behavior you identify. Con-

(continued)

fer with the observer about which nonverbal client behaviors you identified and which you missed.

Part Three

In a role-play interview in which you are the counselor, demonstrate effective use of your face and body, your voice, and distance/space/touch (Objective Three). Be aware of the degree to which your nonverbal behavior matches your words. Also attempt to pace at least one aspect of the client's nonverbal behavior, such as body posture or breathing rate and depth. Use the Nonverbal Behavior Checklist at the end of the chapter to assess your performance from a videotape or have an observer rate you during your session.

Part Four

Recall that there are five occasions for responding to client nonverbal behavior:

a. Evidence of congruence between the client's verbal and nonverbal behavior

b. A client's "mixed" (discrepant) verbal and nonverbal message

c. Client's use of silence

d. Changes in client's nonverbal cues

e. Focusing on client's nonverbal behavior to change or redirect the interview.

Identify four of the five occasions presented in the following client descriptions, according to Chapter Objective Four.

1. The client says that your feedback doesn't bother him; yet he frowns, looks away, and turns away.

2. The client has paused for a long time after your last question.

3. The client has flooded you with a great deal of information for the last five minutes.

4. The client says she feels angry about having to stay in the hospital. As she says this, her voice pitch gets louder, she clasps her hands together, and she frowns.

5. The client's face was very animated for the first part of the interview; now the client's face has a very serious look.

SUGGESTED READINGS

Ekman, P. (1993). Facial expression and emotion. *American Psychologist, 48,* 384–392.

Gazda, G., Asbury, F., Balzer, F., Childers, W., Phelps, R., & Walters, R. (1995). *Human relations development* (5th ed.). Needham Heights, MA: Allyn and Bacon.

Itai, G., & McRae, C. (1994). Counseling older Japanese American clients: An overview and observations. *Journal of Counseling and Development, 72,* 373–377.

Knapp, M. L., & Hall, J. (1992). *Nonverbal communication in human interaction* (3rd ed.). Orlando, FL: Holt, Rinehart and Winston.

Reed, J. R., Patton, M. J., & Gold, P. B. (1993). Effects of turn-taking sequences in vocational test interpretation interviews. *Journal of Counseling Psychology, 40,* 144–155.

Smith, E. L. (1985). *The body in psychotherapy.* Jefferson, North Carolina: McFarland and Co.

Sue, D. W., & Sue, D. (1990). *Counseling the culturally different* (2nd ed.). New York: Wiley.

FEEDBACK
POSTEVALUATION

Part One

Some of the possible meanings of these client nonverbal behaviors are as follows:

1. This client's lowering of her eyes and looking away probably indicates her *embarrassment* and *discomfort* in discussing this particular problem.

2. Dilation of this client's pupils probably signifies *arousal* and *fear* of being committed to the hospital.

3. In this example, the quivering of the client's lower lip and biting of the lip probably denote *ambivalence* and *sorrow* over her actions.

4. The client's head nodding indicates *agreement* with the counselor's rationale for remaining sober.

5. The client's shrugging of the shoulders may indicate *uncertainty* or *reconcilement.*

6. In this case, the client's fist clenching probably connotes *anger* with her spouse.

7. The client's crossing and uncrossing of his legs may signify *anxiety* or *discomfort.*

(continued)

FEEDBACK: POSTEVALUATION   (continued)

8. The client's hesitation in responding and subsequent stuttering and speech errors may indicate *sensitivity* to this topic as well as *discomfort* in discussing it.
9. In this case, the client's moving closer to the counselor probably indicates *intrigue* and *identification* with what the counselor is revealing.
10. The client's sweating and blushing may be signs of *negative arousal*—that is, *anxiety* and/or *embarrassment* with the counselor's confrontation about suggestive dress and pose.

Part Two
Have the observer debrief you for feedback or use the Nonverbal Behavior Checklist (below) to recall which nonverbal behaviors you identified.

Part Three
You or your observer can determine which desirable nonverbal behaviors you exhibited as a counselor, using the Nonverbal Behavior Checklist.

Part Four
The five possible occasions for responding to client nonverbal cues as reflected in the postevaluation examples are these:

1. b. Responding to a client's mixed message; in this case the client's frown, break in eye contact, and shift in body position contradict the client's verbal message.
2. c. Responding to client silence; in this example the client's pause indicates silence.
3. e. Responding to client nonverbal behaviors to redirect the interview focus—in this example, to "break up" the flood of client information.
4. a. Responding to congruence in client verbal and nonverbal messages; in this case, the client's nonverbal behaviors "match" her verbal report of feeling angry.
5. d. Responding to changes in client nonverbal cues—in this example, responding to the change in the client's facial expression.

Nonverbal Behavior Checklist
Name of Counselor _____
Name of Observer _____
Instructions: Using a videotaped or live interview, use the categories below as guides for observing nonverbal behavior. The checklist can be used to observe the counselor, the client, or both. The left-hand column lists a number of behaviors to be observed. The right-hand column has spaces to record a ✔ when the behavior is observed and to fill in any descriptive comments about it—for example, "Blinking—*excessive*" or "Colors in room—*high arousal*."

(continued)

	(✔)	Comments
I. Kinesics		
1. *Eyes*		
Eyebrows raised, lowered, or drawn together	____	_____
Staring or "glazed" quality	____	_____
Blinking—excessive, moderate, or slight	____	_____
Moisture, tears	____	_____
Pupil dilation	____	_____
2. *Face, mouth, head*		
Continuity or changes in facial expression	____	_____
Appropriate or inappropriate smiling	____	_____
Swelling, tightening, or quivering lips	____	_____
Changes in skin color	____	_____
Flushing, rashes on upper neck, face	____	_____
Appearance of sweat beads	____	_____
Head nodding	____	_____
3. *Body movements, posture, and gestures*		
Body posture—rigid or relaxed	____	_____
Continuity or shifts in body posture	____	_____
Frequency of body movements—excessive, moderate, or slight	____	_____
Gestures—open or closed	____	_____
Frequency of nonverbal adaptors (distracting mannerisms)—excessive, moderate, or slight	____	_____
Body orientation: direct (facing each other) or sideways	____	_____
Breathing—shallow or deep, fast or slow	____	_____
Continuity or changes in breathing depth and rate	____	_____
Crossed arms or legs	____	_____
II. Paralinguistics		
Continuity or changes in voice level, pitch, rate of speech	____	_____
Verbal underlining—voice emphasis of particular words/phrases	____	_____
Whispering, inaudibility	____	_____

(continued)

	(✔)	Comments
Directness or lack of directness in speech	____	_____
Speech errors—excessive, moderate, or slight	____	_____
Pauses initiated by counselor	____	_____
Pauses initiated by client	____	_____

III. Proxemics

Continuity or shifts in distance (closer, farther away)	____	_____
Use of touch (handshake, shoulder pat, back pat, and so on)	____	_____
Position in room—behind or next to object or person	____	_____

IV. Environment

Arousal (high or low) associated with

Furniture arrangement	____	_____
Colors	____	_____
Light	____	_____
Voice	____	_____
Overall room	____	_____

V. Time

Session started promptly or late	____	_____
Promptness or delay in responding to other's communication	____	_____
Amount of time spent on primary and secondary problems—excessive, moderate, or slight	____	_____
Continuity or changes in pace of session	____	_____
Session terminated promptly or late	____	_____

VI. Synchrony and Pacing

Synchrony or dissynchrony between nonverbal behaviors and words	____	_____
Pacing or lack of pacing between counselor and client nonverbal behavior	____	_____

VII. Congruence

Congruence or discrepancies:

Nonverbal—between various parts of the body	____	_____
Nonverbal/verbal—between nonverbal behavior and words	____	_____

VIII. Summary

Using your observations of nonverbal behavior and the cultural/contextual variables of the interaction, what conclusions can you make about the therapist? The client? The counseling relationship? Consider such things as emotions, comfort level, deception, desire for more exchange, and liking/attraction.

(continued)

CHAPTER 6

LISTENING RESPONSES

Communication is ever two-way. Listening is the other half of talking. Listening well is no less important than speaking well and it is probably more difficult. Could you listen to a 45-minute discourse without once allowing your thoughts to wander? Good listening is an art that demands the concentration of all your mental facilities. In general, people in the western world talk better than they listen. (Potter, 1965, p. 6)

Listening is a prerequisite for all other counseling responses and strategies. Listening should precede whatever else is done in counseling. When a counselor fails to listen, the client may be discouraged from self-exploring, the wrong problem may be discussed, or a strategy may be proposed prematurely.

We define listening as involving three processes: receiving a message, processing a message, and sending a message. These three processes are illustrated in Figure 6-1.

Each client message (verbal or nonverbal) is a stimulus to be received and processed by the counselor. When a client sends a message, the counselor receives it. Reception of a message is a covert process; that is, we cannot see how or what the counselor receives. Failure to receive all the message may occur when the counselor stops attending.

Once a message is received, it must be processed in some way. Processing, like reception, is covert, because it goes on within the counselor's mind and is not visible to the outside world—except, perhaps, from the counselor's nonverbal cues. Processing includes thinking about the message and pondering its meaning. Processing is important because a counselor's cognitions, self-talk, and mental (covert) preparation and visualization set the stage for overt responding (Morran, Kurpius, Brack, & Brack, 1995) . Errors in processing a message accurately often occur when counselors' biases or blind spots prevent them from acknowledging parts of a message or from interpreting a message without distortion. Counselors may hear what they want to hear instead of the actual message sent. Also, as Ivey, Gluckstern, and Ivey

(1993) note, clients tend to talk about what they perceive you are able and willing to listen to.

The third process of listening involves the verbal and nonverbal messages sent by a counselor. Sometimes a counselor may receive and process a message accurately but have difficulty sending a message because of lack of skills. Fortunately, you can learn to use listening responses to send messages. Problems in sending messages can be more easily corrected than errors in the covert processes of receiving and processing messages. We hope that you are already able to receive and process a message without difficulty. Of course, this is a big assumption! If you think your own covert processes in listening are in need of further development, this may be an area you will need to work on by yourself or with someone else.

This chapter is designed to help you acquire four verbal listening responses that you can use to send messages to a client: clarification, paraphrase, reflection, and summarization. As Sue and Sue (1990) note, "While breakdowns in communication often happen between members who share the same culture, the problem becomes exacerbated between people of different races or ethnic backgrounds" (p. 30).

OBJECTIVES

1. From a written list of 12 example counselor listening responses, accurately classify at least nine of them by type: clarification, paraphrase, reflection, or summarization.
2. From a list of three client statements, write an example of each of the four listening responses for each client statement.
3. In a 15-minute counseling interview in which you function as an observer, listen for and record five key aspects of client messages that form the basis of effective listening.
4. In a 15-minute role-play interview or a conversation in which you function as a listener, demonstrate at least two accurate examples of each of the four listening responses.

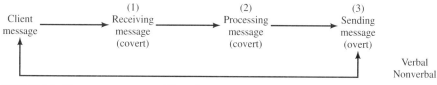

FIGURE 6-1. Three processes of listening.

LISTENING TO CLIENTS' STORIES

Ivey, Ivey, and Simek-Morgan (1993) note that listening helps to bring out the client's story with "minimal intrusions" on the counselor's part (p. 49). Listening is healing because it helps clients tell their stories. Clients' messages represent stories about themselves—narratives about their histories and current experiences from which clients construct their identities and infuse their lives with meaning and purpose (White & Epston, 1990). Good therapists listen to these stories to help clients recognize how these narratives create meaning and whether they help or hurt the development of the client's identity. Telling one's story can also provide emotional relief to clients, who have suffered trauma, even for those clients who are very young or old (Terr, 1990). For clients who are suffering a loss such as a separation or divorce, loss of a job, or loss of a significant other, stories provide a way to make sense of the bereavement (Sedney, Baker, & Gross, 1994). For dying clients, stories represent a life that has been lived and is now coming to an end, but may not yet have been told. Healing is particularly evident for clients who tell their story when "hidden difficulties" or "shame" have been involved (Ostaseski, 1994).

Stories are also almost universally present and relevant in various ethnic cultural groups. Helpers can listen to many things about clients' stories. Ivey, Ivey, and Simek-Morgan (1993) recommend listening to the *facts* of the story, the client's *feelings* about the story, and the way the client *organizes* the story. Sedney et al. (1994) recommend listening to how the story is started, the sequences in the story, and "hints of anger, regret, and what ifs," as well as the client's understanding of the story and the role she or he plays in it (p. 291). Significant omissions also may provide clues. Ostaseski (1994) comments that as helpers, we must simply trust that some insight will arise for clients just from the telling of the story. He concludes, "Often the story will deliver what is needed. So pay close attention to whatever you are presented with. Start with that. Take it. Believe it, and see where it leads you" (p. 11).

WHAT DOES LISTENING REQUIRE OF HELPERS?

When someone truly hears us, it is a special gift. We can all recall times when we felt wonderful—just because someone who means something to us stood or sat with us and really listened. Conversely, we can also remember instances in which we felt flustered because someone close to us was inattentive and distracted. Nichols (1995) has referred to listening as a "lost art," which he attributes partly to contemporary time pressures that distract our attention span and impoverish the quality of listening in our lives (p. 2). As a result, this lack of listening pervades our most prized relationships, contributes to interpersonal conflict, and leaves us with a sense of loss. Nichols (1995) observes that this loss is most severe when lack of listening occurs in relationships in which we counted on it to occur, such as in the therapy relationship.

In this chapter, we describe the use and purposes of four listening responses that, if acquired, will help you become a better listener. However, in addition to these responses, truly effective listening requires the capacity to be fully present to the client and free from distractions—both internal and external. In this way, the listening process provides the sort of "holding environment" we discussed in Chapter 3, and this requires a certain kind of *energy* on the helper's part—an energy that is highly involved and yet also quite contained.

Helpers who listen best usually have developed this sort of "mindfulness" about them that we discuss further in Chapter 16 in the meditation strategies. These helpers are also able to focus their energy very intently on the client with minimal intrusions—either from their own process or from the outer environment. This mindfulness quality is typically more highly developed in Eastern cultures, where persons, for example, rise at dawn to practice tai chi. Perhaps nowhere is this quality more evident than with those helpers who midwife dying persons. Ostaseski (1994), director of the San Francisco Zen Hospice project, discusses this process in the following way:

LEARNING ACTIVITY 18   *Cultivating the Listening Mind*

I. Prior to this, obtain three small objects to eat, such as raisins or M&Ms. Sit in a comfortable position, close your eyes, and focus on your teeth. If wandering thoughts come, let them flow by. Starting with one raisin, slowly lift it to your mouth. Chew it very slowly. Observe your arm lifting the raisin to your mouth. . . . Think about how your hand holds it . . . notice how it feels in your mouth. . . . Savor it as you chew it ever so slowly. While doing this . . . notice your tongue and throat as you very slowly swallow the raisin. . . .

Repeat this process with the next two. . . . Afterwards notice what you realize about eating and raisins. What do you usually tune out?

II. Lie still in a comfortable position. Scan your body starting with your toes and moving very slowly up to the top of your head. Direct your attention to a spot that feels most tense or painful. Put your hand on that spot. Leave it there for a few minutes. Breathe with it. Just notice what happens as you go into this part of your body with your breathing and your awareness. You are not trying to change anything; just be aware of this place and accept it. Stay with this for a little while and see what happens to this spot.

We sit at the bedside and we listen. We try to listen with our whole body, not just with our ears. We must perpetually ask ourselves, "Am I fully here? Or am I checking my watch or looking out the window?"

At the heart of it, all we can really offer each other is our full attention. When people are dying, their tolerance for bullshit is minimal. They will quickly sniff out insincerity. Material may arise that we don't particularly like or even strongly dislike. Just as we do in meditation, we need to sit still and listen, not knowing what will come next, to suspend judgment—at least for the moment—so that whatever needs to evolve will be able to do so. (p. 11)

If you feel this is a quality you need to develop further in yourself, we encourage you to practice Learning Activity #18, adapted from Kabat-Zinn (1993) on a daily basis.

FOUR LISTENING RESPONSES

This chapter presents four kinds of listening responses: clarification, paraphrase, reflection, and summarization. *Clarification* is a question, often used after an ambiguous client message. It starts with "Do you mean that . . ." or "Are you saying that . . ." along with a repetition or rephrasing of all or part of the client's previous message. Similar to a clarification is the *paraphrase,* defined as a rephrasing of the content part of the message, which describes a situation, event, person, or idea. In contrast, *reflection* is a rephrasing of the client's feelings, or the affect part of the message. Usually the affect part of the message reveals the client's feelings about the content; for example, a client may feel discouraged (affect) about not doing well in a class (content).

Summarization is an extension of the paraphrase and reflection responses that involves tying together and rephrasing two or more different parts of a message or messages.

To illustrate these four responses, we present a client message with an example of each:

Client, a 35-year-old Latina widow, mother of two young children: My whole life fell apart when my husband died. I keep feeling so unsure about my ability to make it on my own and to support my kids. My husband always made all the decisions for me and brought home money every week. Now I haven't slept well for so long, and I'm drinking more heavily—I can't even think straight. My relatives help me as they can but I still feel scared.

Counselor clarification: Are you saying that one of the hardest things facing you now is to have enough confidence in yourself?

Counselor paraphrase: Since your husband's death you have more responsibilities and decisions on your shoulders, even with support of relatives.

Counselor reflection: You feel concerned about your ability to shoulder all the family responsibilities now.

Counselor summarization: Now that your husband has died, you're facing a few things that are very difficult for you right now . . . handling the family responsibilities, making the decisions, trying to take better care of yourself, and dealing with fears that have come up as a result.

Table 6-1 presents the definitions and the *intended* or hypothesized purposes of the four counselor listening responses of clarification, paraphrase, reflection, and summari-

TABLE 6-1. Definitions and intended purposes of counselor listening responses

Response	Definition	Intended purpose
Clarification	A question beginning with, for example, "Do you mean that" or "Are you saying that" plus a rephrasing of the client's message	1. To encourage more client elaboration 2. To check out the accuracy of what you heard the client say 3. To clear up vague, confusing messages
Paraphrase (responding to content)	A rephrasing of the content of the client's message	1. To help the client focus on the content of his or her message 2. To highlight content when attention to feelings is premature or self-defeating
Reflection (responding to feelings)	A rephrasing of the affective part of the client's message	1. To encourage the client to express more of his or her feelings 2. To have the client experience feelings more intensely 3. To help the client become more aware of the feelings that dominate him or her 4. To help the client acknowledge and manage feelings 5. To help the client discriminate accurately among feelings
Summarization	Two or more paraphrases or reflections that condense the client's messages or the session	1. To tie together multiple elements of client messages 2. To identify a common theme or pattern 3. To interrupt excessive rambling 4. To review progress

zation. The counselor responses may not have the same results for all clients. For example, a counselor may find that reflecting feelings prompts some clients to discuss feelings, whereas other clients may not even acknowledge the counselor's statement (Uhlemann, Lee, & Martin, 1994). The point is that we are presenting some "modal" intentions for each counselor listening response; there are exceptions. The counselor responses will achieve their intended purposes most of the time. However, other dynamics within an interview may yield different client outcomes. Moreover, the effects of these verbal messages may vary depending on the nonverbal cues sent along with the message. It is helpful to have some rationale in mind for using a response. Keep in mind, however, that the influence a response has on the client may not be what you intended to achieve by selecting it. The guidelines in Table 6-1 should be used tentatively, *subject to modification by particular client reactions.*

The next three sections describe the listening responses and present model examples of each skill. Opportunities to practice each skill and receive feedback follow the examples.

Listening for Accuracy: The Clarification Response

Because most messages are expressed from the speaker's internal frame of reference, they may be vague or confusing. Messages that may be particularly confusing are those that include inclusive terms (*they* and *them*), ambiguous phrases

(*you know*), and words with a double meaning (*stoned, trip*) (Hein, 1980, p. 35). When you aren't sure of the meaning of a message, it is helpful to clarify it.

According to Hein (1980, p. 56), a clarification asks the client to elaborate on "a vague, ambiguous, or implied statement." The request for clarification is usually expressed in the form of a question and may begin with phrases such as "Are you saying this . . ." or "Could you try to describe that . . ." or "Can you clarify that . . ."

Purposes of Clarification. A clarification may be used to make the client's previous message explicit and to confirm the accuracy of your perceptions about the message. A clarification is appropriate for any occasion when you aren't sure whether you understand the client's message and you need more elaboration. A second purpose of clarification is to check out what you heard of the client's message. Particularly in the beginning stages of counseling, it is important to verify client messages before jumping to quick conclusions. The following example may help you see the value of the clarification response.

Client: Sometimes I just want to get away from it all.
Counselor: It sounds like you have to split and be on your own.
Client: No, it's not that. I don't want to be alone. It's just that I wish I could get out from under all this work I have to do.

In this example, the counselor drew a quick conclusion about the initial client message that turned out to be inaccurate. The session might have gone more smoothly if the counselor had requested clarification before assuming something about the client, as in the next example:

Client: Sometimes I just want to get away from it all.
Counselor: Could you describe for me what you mean by getting away from it all?
Client: Well, I just have so much work to do—I'm always feeling behind and overloaded. I'd like to get out from under that miserable feeling.

In this case, the clarification helped both persons to establish exactly what was being said and felt. Neither the client nor the counselor had to rely on assumptions and inferences that were not explored and confirmed. The skilled counselor uses clarification responses to determine the accuracy of messages as they are received and processed. Otherwise, inaccurate information may not be corrected and distorted assumptions may remain untested.

Steps in Clarifying.   There are four steps in clarifying for accuracy. First, identify the content of the client's verbal and nonverbal messages—what has the client told you? Second, identify any vague or confusing parts to the message that you need to check out for accuracy or elaboration. Third, decide on an appropriate beginning, or sentence stem, for your clarification, such as "Could you describe," "Could you clarify," or "Are you saying." In addition, use your voice to deliver the clarification as a question rather than a statement. Finally, remember to assess the effectiveness of your clarification by listening to and observing the client's response. If your clarification is useful, the client will elaborate on the ambiguous or confusing part of the message. If it is not useful, the client will clam up, ignore your request for clarification, and/or continue to make deletions or omissions. At this point, you can attempt a subsequent clarification or switch to an alternative response.

To help you formulate a clarification, decide when to use it, and assess its effectiveness, consider the following cognitive learning strategy:

1. What has this client told me?
2. Are there any vague parts or missing pictures to the message that I need to check out? If so, what? If not, decide on another, more suitable response.
3. How can I hear, see, or grasp a way to start this response?
4. How will I know whether my clarification is useful?

Notice how a counselor applies this cognitive learning strategy in clarifying the client's message given in the previous example:

Client: Sometimes I just want to get away from it all.
Counselor: 1. What has this client told me? That she wants to get away from something. [Asked and answered covertly]
2. Are there any vague parts or missing pictures in her message? If so, what? (If not, I'll decide on a more suitable response.)
Yes—I need to check out what she means by "getting away from it all."
3. Now, how can I begin a clarification response? I can see the start of it, hear the start of it, or grasp the start of it. Something like "Well, could you tell me, or could you describe . . ."
4. Now, how will I know that the response will be helpful? I'll have to see, hear, and grasp whether she elaborates or not. Let's try it . . .

Suppose that, at this juncture, the counselor's covert visualization or self-talk ends, and the following actual dialogue occurs:

Counselor clarification: Could you describe what you mean by "getting away from it all"?
Client response: Well, I just have so much work to do—I'm always feeling behind and overloaded. I'd like to get out from under that miserable feeling.

From the client's response, the counselor can determine that the clarification was effective because the client elaborated and added the missing parts or pictures from her previous message. The counselor can covertly congratulate himself or herself for not jumping ahead too quickly and for taking the time to check out the client's deletion and the resulting ambiguity.

The following learning activity gives you an opportunity to try out this cognitive learning strategy to develop the skill of clarification.

Listening for Content and Affect: Paraphrasing and Reflecting

In addition to clarifying the accuracy of client messages, the counselor needs to listen for information revealed in messages about significant situations and events in the client's life—and the client's feelings about these events. Ivey, Ivey, and Simek-Morgan (1993) talk about this as listening for (1) the main facts of the client's story and (2) the client's

LEARNING ACTIVITY **19** *Clarification*

In this learning activity, you are presented with three client practice messages. For each client message, develop an example of a clarification response, using the cognitive learning strategy described earlier and outlined in the following example. To internalize this learning strategy, you may wish to talk through these self-questions overtly (aloud) and then covertly (silently to yourself). The end product will be a clarification response that you can say aloud or write down or both. An example precedes the practice messages. Feedback follows on page 102.

Example

Client, a 15-year-old high school student: My grades have really slipped. I don't know why; I just feel so down about everything:

Self-question 1: What has this client told me? That she feels down and rather discouraged.

Self-question 2: Are there any vague parts or missing pictures to the message that I need to check out? If so, what? (If not, decide on a different response.) Yes, several—one is what she feels so down about. Another is what this feeling of being down is like for her.

Self-question 3: How can I hear, see, or grasp a way to start this response?
Well, "Are you saying there's something specific?" or "Can you describe this feeling. . . ?"

Self-question 4: Say aloud or write an actual clarification response:
"Are you saying there is something specific you feel down about?" or "Could you describe what this feeling of being down is like for you?"

Client Practice Messages

Client 1, a fourth-grader: I don't want to do this dumb homework anyway. I don't care about learning these math problems. Girls don't need to know this anyway.

Self-question 1: What has this client told me?
Self-question 2: Are there any vague parts or missing pictures I need to check out? If so, what?
Self-question 3: How can I hear, see, or grasp a way to start my response?
Actual clarification response: _____
_____

Client 2, a middle-aged man: I'm really discouraged with this physical disability now. I feel like I can't do anything the way I used to. Not only has it affected me in my job, but at home. I just don't feel like I have anything good to offer anyone.

Self-question 1: What has this client told me?
Self-question 2: Are there any vague parts or missing pictures I need to check out? If so, what?
Self-question 3: How can I hear, see, or grasp a way to start my response?
Actual clarification response: _____
_____

Client 3, an older person: The company is going to make me retire even though I don't want to. What will I do with myself then? I find myself just thinking over the good times of the past, not wanting to face the future at all. Sometimes retirement makes me so nervous I can't sleep or eat. My family suggested I see someone about this.

Self-question 1: What has this client told me?
Self-question 2: Are there any vague parts or missing pictures I need to check out? If so, what?
Self-question 3: How can I hear, see, or grasp a way to start my response?
Actual clarification response: _____
_____

feelings about his or her story. Each client message will express (directly or indirectly) some information about client situations or concerns and about client feelings. The portion of the message that expresses information or describes a situation or event is called the *content,* or the cognitive part, of the message. The cognitive part of a message includes references to a situation or event, people, objects, or ideas. Another portion of the message may reveal how the client feels about the content; expression of feelings or an emotional tone is called the *affective* part of the message

(Hackney & Cormier, 1994). Generally, the affect part of the verbal message is distinguished by the client's use of an affect or feeling word, such as *happy, angry,* or *sad.* However, clients may also express their feelings in less obvious ways, particularly through various nonverbal behaviors.

The following illustrations may help you distinguish between the content and affective parts of a client's verbal message.

Client, a 6-year-old first-grader: I don't like school. It isn't much fun.

FEEDBACK
Clarification

Client 1

1. What did the client say?
   That she doesn't want to do her math homework—that she thinks it's not important for girls.
2. Are there any vague parts or missing pictures?
   Yes—whether she really doesn't care about math or whether she's had a bad experience with it and is denying her concern.
3. Examples of clarification responses: "Are you saying that you really dislike math or that it's not going as well for you as you would like?"
   "Are you saying that math is not too important for you or that it is hard for you?"

Client 2

1. What did the client say?
   That he feels useless to himself and others.
2. Are there any vague parts or missing pictures?
   Yes—it's not clear exactly how things are different for him now and also whether it's the disability itself that's bothering him or its effects (inability to get around, reactions of others, and so on).
3. Examples of clarification responses: "Could you clarify exactly how things are different for you now than the way they used to be?"
   "Are you saying you feel discouraged about having the disability—or about the effects and constraints from it?"
   "Are you saying you feel differently about yourself now than the way you used to?"

Client 3

1. What did the client say?
   He is going to have to retire because of company policy. He doesn't want to retire now and feels upset about this. He's here at his family's suggestion.
2. Are there any vague parts or missing pictures?
   Yes—he says he feels nervous, although from his description of not eating and sleeping it may be sadness or depression. Also, is he here only because his family sent him or because he feels a need too? Finally, what specifically bothers him about retirement?
3. Examples of clarification responses: "Would you say you're feeling more nervous or more depressed about your upcoming retirement?"
   "Are you saying you're here just because of your family's feelings or because of your feelings too?" "Could you describe what it is about retiring that worries you?"

The first sentence ("I don't like school") is the affect part of the message. The client's feelings are suggested by the words "don't like." The second sentence ("It isn't much fun") is the content part of the message because it refers to a situation or an event in this child's life—not having fun at school.
Here is another example:

Client, a 20-year-old woman: How can I tell my boyfriend I want to break off our relationship? He will be very upset. I guess I'm afraid to tell him.

In this example, the first two sentences are the content because they describe the situation of wanting to break off a relationship. The third sentence, the affect part, indicates the client's feelings about this situation—being *afraid* to tell the boyfriend of her intentions.
See whether you can discriminate between the content and affective parts of the following two client messages:

Client 1, a young man: I just can't satisfy my wife sexually. It's very frustrating for me.

In this example, the content part is "I can't satisfy my wife sexually." The affect part, or the client's feelings about the content, is "It's very *frustrating* for me."

Client 2, an institutionalized man: This place is a trap. It seems like I've been here forever. I'd feel much better if I weren't here.

In the second example, the statements referring to the institution as a trap and being there forever are the content parts of the message. The statement of "feeling better" is the affect part.
The skilled counselor tries to listen for both content and affect parts of client messages because it is important to deal with significant situations or relationships *and* with the client's feelings about the situations. Responding to cognitive or affective messages will direct the focus of the session in different ways. At some points, the counselor will respond to content by focusing on events, objects, people, or ideas. At other times, the counselor will respond to affect by focusing on the client's feelings and emotions. Generally, the counselor can respond to content by using a paraphrase and can respond to affect with a reflection.

Paraphrase. A paraphrase is a rephrasing of the client's primary words and thoughts. Paraphrasing involves selective attention given to the cognitive part of the message—with the client's key ideas translated into *your own words.* Thus, an effective paraphrase is more than just "parroting" the words of the client. The rephrasal should be carefully worded to lead to further discussion or increased understanding on the part of the client. It is helpful to stress the most important words and ideas expressed by the client. Consider the following example:

Client: I know it doesn't help my depression to sit around or stay in bed all day.
Counselor: You know you need to avoid staying in bed or sitting around all day to help your depression.

In this example, the counselor merely "parroted" the client's message. The likely outcome is that the client may respond with a minimal answer such as "I agree" or "That's right" and not elaborate further or that the client may feel ridiculed by what seems to be an obvious or mimicking response. A more effective paraphrase would be "You are aware that you need to get up and move around in order to minimize being depressed."

Purposes of Paraphrasing.   There are several purposes in using the paraphrase at selected times in client interactions. First, use of the paraphrase tells clients that you have understood their communication. If your understanding is complete, clients can expand or clarify their ideas. Second, paraphrasing can encourage client elaboration of a key idea or thought. Clients may talk about an important topic in greater depth. A third reason for use of paraphrases is to help the client focus on a particular situation or event, idea, or behavior.

Sometimes, by increasing focus, paraphrasing can help get a client "on track." For example, accurate paraphrasing can help stop a client from merely repeating a "story" (Ivey, 1994).

A fourth use of paraphrase is to help clients who need to make decisions. As Ivey, Ivey, and Simek-Downing (1987, p. 73) observe, "Paraphrasing is often helpful to clients who have a decision to make, for the repetition of key ideas and phrases clarifies the essence of the problem." Paraphrasing is also useful when emphasizing content if attention to affect is premature or counterproductive.

Steps in Paraphrasing.   There are five steps in paraphrasing content. First, attend to and recall the message by restating it to yourself covertly—what has the client told you? Second, identify the content part of the message by asking yourself "What situation, person, object, or idea is discussed in this message?" Third, select an appropriate beginning, or sentence stem, for your paraphrase. Paraphrases can begin with many possible sentence stems. Try to select one that is likely to match the client's choice of sensory words. Table 6-2 provides examples of typical sensory words

TABLE 6-2. Examples of client sensory words and corresponding counselor phrases

Client sensory words		Corresponding counselor phrases
Visual		
see	bright	It seems like
clear	show	It appears as though
focus	colorful	From my perspective
picture	glimpse	As I see it
view	"now look"	I see what you mean
perspective		It looks like
Auditory		
listen	discuss	Sounds like
yell	should	As I hear it
tell	loud	What you're saying is
told	noisy	I hear you saying
talk	call	Something tells you
hear	"now listen"	You're telling me that
ears		
Kinesthetic		
feel	relaxed	You feel
touch	sense	From my standpoint
pressure	experience	I sense that
hurt	firm	I have the feeling that
pushy	"you know"	
grasp		

SOURCE: Adapted from Lankton, 1980.

that clients may use and corresponding counselor phrases. Next, using the sentence stem you selected, translate the key content or constructs into your own words and verbalize this into a paraphrase. Remember to use your voice as you deliver the paraphrase so it sounds like a statement instead of a question. Finally, assess the effectiveness of your paraphrase by listening to and observing the client's response. If your paraphrase is accurate, the client will in some way—verbally and/or nonverbally—confirm its accuracy and usefulness. Consider the following example of the way a counselor uses the cognitive learning strategy to formulate a paraphrase:

Client, a 40-year-old African American woman: How can I tell my husband I want a divorce? He'll think I'm crazy. I guess I'm just afraid to tell him. [Said in a level, monotone voice]

1. What has this client told me?
   That she wants a divorce and she's afraid to tell her husband, as he will think she's crazy.
2. What is the content of this message—what person, object, idea, or situation is the client discussing?

Wants divorce but hasn't told husband because husband will think she's crazy.

3. What is an appropriate sentence stem (one that matches the sensory words used by the client)?

Client uses the verb *tell* two times and *think* once, so I'll go with a stem such as "You think," "I hear you saying," or "It sounds like."

4. How can I translate the client's key content into my own words?

Want a divorce = break off, terminate the relationship, split.

5. How will I know whether my paraphrase is helpful?

Listen and notice whether the client confirms its accuracy.

Suppose that at this point the counselor's self-talk stopped and the following dialogue ensued:

Counselor paraphrase: It sounds like you haven't found a way to tell your husband you want to end the relationship because of his possible reaction. Is that right?

Client: Yeah—I've decided—I've even been to see a lawyer. But I just don't know how to approach him with this. He thinks things are wonderful.

At this point, the counselor can congratulate herself or himself for having formulated a paraphrase that has encouraged client elaboration and focus on a main issue. Learning Activity #20 gives you an opportunity to develop your own paraphrase responses.

Reflection of Feeling.   We have just seen that the paraphrase is used to restate the cognitive part of the message. Although the paraphrase and the reflection of feeling are not mutually exclusive responses, the reflection of feeling is used to rephrase the *affective* part of the message, the client's emotional tone. A reflection is similar to a paraphrase but different in that a reflection adds an emotional tone or component to the message that is lacking in a paraphrase. Here are two examples that may illustrate the difference between a paraphrase and a reflection of feeling.

Client: Everything is humdrum. There's nothing new going on, nothing exciting. All my friends are away. I wish I had some money to do something different.

Counselor paraphrase: With your friends gone and no money around, there is nothing for you to do right now.

Counselor reflection: You feel bored with the way things are for you right now.

Note the counselor's use of the affect word *bored* in the reflection response to tune into the feelings of the client created by the particular situation.

Purposes of Reflection.   Reflecting feelings has five intended purposes. First, this response, if used effectively and accurately, helps clients to feel understood. Clients tend to communicate more freely with persons whom they feel try to understand them. As Teyber (1997) observes, when understanding is present, "Clients begin to feel that they have been seen and are no longer invisible, alone, strange, or unimportant. At that moment, the client begins to perceive the therapist as someone who is different from most other people and possibly as someone who can help" (p. 49).

Reflection is also used to encourage clients to express more of their feelings (both positive and negative) about a particular situation, person, or whatever. Some clients do not readily reveal feelings because they have never learned to do so, and other clients hold back feelings until the therapist gives permission to focus on them. Expression of feelings is not usually an end in itself; rather, it is a means of helping clients and counselors understand the scope of the problem or situation. Most if not all the concerns presented by clients involve underlying emotional factors to be resolved (Ivey, Ivey, & Simek-Downing, 1987, p. 73). For example, in focusing on feelings, the client may become more aware of lingering feelings about an unfinished situation or of intense feelings that seem to dominate his or her reaction to a situation. Clients may also become aware of mixed, or conflicting, feelings. Ambivalence is a common way that clients express feelings about problematic issues. Teyber (1997) notes that two common affective constructions with mixed components include anger-sadness-shame and sadness-anger-guilt. In the first sequence, the primary feeling is often anger but it is a negative response to hurt or sadness. Often, the experiencing of the anger and sadness provokes shame. In the second sequence, the predominant feeling is sadness but it is often connected to anger that has been denied because the expression of it produces guilt. These affective sequences are typically acquired in childhood and are a result of both the rules and the interactions of the family of origin. These affective elements are also strongly influenced by cultural affiliation. As Sue and Sue (1990) note, in Western cultures, which emphasize individualism, the predominant affective reaction following wrongful behavior is *guilt*. However, in some non-Western cultures such as Asian, Hispanic, and Black, where the psychosocial unit is the family, group, or collective society, the primary affective reaction to wrongful behavior is not guilt but *shame*. Sue and Sue (1990) conclude that "guilt is an individual affect, while shame appears to be a group affect" (p. 36). A review of cultural variations in emotions can be found in Mesquita and Frijda (1992).

LEARNING ACTIVITY 20   *Paraphrase*

In this learning activity, you are presented with three client practice messages. For each client message, develop an example of a paraphrase response, using the cognitive learning strategy outlined in the example below. To internalize this learning strategy, you may wish to talk through these self-questions overtly (aloud) and then covertly (silently to yourself). The end product will be a paraphrase response that you can say aloud or write down or both. Feedback is given on page 106.

Example
Client, a middle-aged graduate student: It's just a rough time for me—trying to work, keeping up with graduate school, and spending time with my family. I keep telling myself it will slow down someday. [Said in a level, monotone voice]
Self-question 1: What has this client told me?
   That it's hard to keep up with everything he has to do.
Self-question 2: What is the content of this message— What person, object, idea, or situation is the client discussing?
   Trying to keep up with work, school, and family.
Self-question 3: What is an appropriate sentence stem?
   I'll try a stem like "It sounds like" or "There are."
Actual paraphrase response: It sounds like you're having a tough time balancing all your commitments *or* There are a lot of demands on your time right now.

Client Practice Statements
Client 1, a 30-year-old woman: My husband and I argue all the time about how to manage our kids. He says I always interfere with his discipline—I think he is too harsh with them. [Said in a level voice tone without much variation in pitch or tempo]
Self-question 1: What has this client told me?
Self-question 2: What is the content of this message? What person, object, idea, or situation is the client discussing?
Self-question 3: What is a useful sentence stem?
Actual paraphrase response: _____

_____

Client 2, a 6-year-old boy: I wish I didn't have a little sister. I know my parents love her more than me. [Said in slow, soft voice with downcast eyes]
Self-question 1: What has this client told me?
Self-question 2: What is the content of this message— what person, object, idea, or situation is this client discussing?
Self-question 3: What is a useful sentence stem?
Actual paraphrase response: _____

_____

Client 3, a college student: I've said to my family before, I just can't compete with the other students who aren't blind. There's no way I can keep up with this kind of handicap. I've told them it's natural to be behind and do more poorly. [Said in level, measured words with little pitch and inflection change]
Self-question 1: What has this client told me?
Self-question 2: What is the content of this message? What person, object, idea, or situation is the client discussing?
Self-question 3: What is a useful sentence stem?
Actual paraphrase response: _____

_____

A third purpose of reflection is to help clients manage feelings. Learning to deal with feelings is especially important when a client experiences intense emotions such as fear, dependency, or anger. Strong emotions can interfere with a client's ability to make a rational response (cognitive or behavioral) to pressure. Also, when clients are given permission to reveal and release feelings, often their energy and well-being is increased. As Hackney and Cormier (1994) note:

> This occurs because all of the deep primary feelings have a survival value (Kelley, 1979). For example, anger allows an individual to protect his or her rights and establish personal boundaries or limits (Kelley, 1979). The capacity to recognize and express anger is the basis for healthy assertiveness. As Kelley observed, "The person who cannot become angry,

> whose anger is deeply repressed, is severely handicapped. . . . These persons' assertions lack conviction and they are often at the mercy of or emotionally dependent on those who are capable of becoming angry" (1979, p. 25). Moreover, the expression of anger is useful in close relationships to "clear the air" and prevent chronic boredom and resentment from building up. Anger and disgust that are expressed and released prompt subsequent expression of love; pain and sadness that are expressed promote later expression of joy and pleasure; and fear that is discharged allows for greater trust. (pp. 92–93)*

_____

*SOURCE: From *Counseling Strategies and Interventions*, 4th Ed., by H. Hackney and L. S. Cormier, pp. 92-93. Copyright 1994 by Allyn & Bacon. Reprinted by permission.

FEEDBACK 20
Paraphrase

Client 1

Question 1. What has the client said?

That she and her husband argue over child rearing.

Question 2. What is the content of her message?

As a couple, they have different ideas on who should discipline their kids and how.

Question 3. What is a useful sentence stem?

Try "It sounds like" or "Your ideas about discipline are."

Actual paraphrase response: Here are examples; see whether yours are similar.

It sounds like you and your husband disagree a great deal on which one of you should discipline your kids and how it should be done *or* Your ideas about discipline for your kids are really different from your husband's, and this creates disagreements between the two of you.

Client 2

Question 1. What has this client said?

He believes his little sister is loved more by his folks than he is, and he wishes she weren't around.

Question 2. What is the content of his message?

Client feels "dethroned"—wishes the new "queen" would go away.

Question 3. What is a useful sentence stem?

I'll try "It seems that" or "I sense that."

Actual paraphrase response: Here are examples. What are yours like?

It seems that you'd like to be "number one" again in your family *or* I sense you are not sure of your place in your family since your little sister arrived.

Client 3

Question 1. What has this client said?

He is behind in school and is not doing as well as his peers because he is blind—a point he has emphasized to his family.

Question 2. What is the content of his message?

Client wants to impress on his family that to him his blindness is a handicap that interferes with his doing as much or as well as other students.

Question 3. What is a useful sentence stem?

"It sounds like," "I hear you saying," or "You'd like."

Actual paraphrase response: Here are some examples.

It sounds like it's very important to you that your family realize how tough it is for you to do well in your studies here *or* You'd like your family to realize how difficult it is for you to keep up academically with people who don't have the added problem of being blind.

A fourth use of reflection is with clients who express negative feelings about therapy or about the counselor. When a client becomes angry or upset with you or with the help you are offering, there is a tendency to take the client's remarks personally and become defensive. Using reflection in these instances "lessens the possibility of an emotional conflict, which often arises simply because two people are trying to make themselves heard and neither is trying to listen" (Long & Prophit, 1981, p. 89). The use of reflection in these situations lets clients know that the counselor understands their feelings in such a way that the intensity of the anger is usually diminished. As anger subsides, the client may become more receptive, and the counselor can again initiate action-oriented responses or intervention strategies.

Finally, reflection helps clients discriminate accurately among various feelings. Clients often use feeling words like *anxious* or *nervous* that, on occasion, mask deeper or more intense feelings (Ivey, Gluckstern, & Ivey, 1988). Clients may also use an affect word that does not really portray their emotional state accurately. It is common, for instance, for a client to say "It's my nerves" or "I'm nervous" to depict other feelings, such as resentment and depression. Other clients may reveal feelings through the use of metaphors (Ivey, Gluckstern, & Ivey, 1993). For example, a client may say, "I feel like the person who rolled down Niagara Falls in a barrel" or "I feel like I just got hit by a Mack truck." Metaphors are important indicators of client emotion; as Ivey, Gluckstern, and Ivey (1993) note, they suggest that much more is going on with the client than just the "surface expression" (p. 71). Accurate reflections of feeling help clients to refine their understanding of various emotional moods.

Ivey, Gluckstern, and Ivey (1993) note that processing emotions does not always occur easily with some clients. As they suggest,

> In White, North American and other cultures, men are expected to hold back their feelings. You aren't a "real man" if you allow yourself to feel emotion. While many men can and do express their feelings in the helping interview, some should not be pushed too hard in this area in the first phases of counseling. Later, with trust, exploration of feelings becomes more acceptable.
>
> In general, women in all cultures are more in touch with and more willing to share feelings than men. Nonetheless, this will vary with the cultural group. Some cultures (for example, Asian, Native American) at times pride themselves on their ability to control emotions. However, this may also be true with those of British and Irish extraction.
>
> African Americans and other minorities have learned over time that it may not be safe to share themselves openly with White Americans. In cross-cultural counseling situations, trust needs

to be built before you can expect in-depth discussion of emotions. (pp. 73–74)*

Steps in Reflecting Feelings.

Reflecting feelings can be a difficult skill to learn because feelings are often ignored or misunderstood (Long & Prophit, 1981). Reflection of feelings involves six steps that include identifying the emotional tone of the communication and verbally reflecting the client's feelings, using your own words.

The first step is to listen for the presence of feeling words, or affect words, in the client's messages. Positive, negative, and ambivalent feelings are expressed by one or more affect words falling into one of five major categories: anger, fear, conflict, sadness, and happiness. Table 6-3 presents a list of commonly used affect words at three levels of intensity. Becoming acquainted with such words may help you recognize them in client communications and expand your vocabulary for describing emotions.

A second way to identify the client's feelings is to watch the nonverbal behavior while the verbal message is being delivered. As you may remember from Chapter 5, nonverbal cues such as body posture, facial expression, and voice quality are important indicators of client emotion. In fact, nonverbal behavior is often a more reliable clue to client emotions because nonverbal behaviors are less easily controlled than words. Observing nonverbal behavior is particularly important when the client's feelings are implied or expressed very subtly.

After the feelings reflected by the client's words and nonverbal behavior have been identified, the next step involves verbally reflecting the feelings back to the client, using different words. The choice of words to reflect feelings is critical to the effectiveness of this skill. For example, if a client expresses feeling annoyed, interchangeable affect words would be *bothered, irritated,* and *hassled.* Words such as *angry, mad,* or *outraged,* however, probably go beyond the intensity expressed by the client. It is important to select affect words that accurately match not only the type of feeling but also its intensity; otherwise, the counselor makes an understatement, which can make a client feel ridiculed, or an overstatement, which can make a client feel put off or intimidated. Note the three major levels of intensity of affect words in Table 6-3—mild, moderate, and intense. You can also control the intensity of the expressed affect by the type of preceding adverb used—for example, *somewhat* (weak),

quite (moderate), or *very* (strong) *upset.* Study Table 6-3 carefully so that you can develop an extensive affect-word vocabulary. Overuse of a few common affect words misses the varied nuances of the client's emotional experience.

The next step in reflecting is to start the reflection statement with an appropriate sentence stem—if possible, one that matches the client's choice of sensory words. These are some sample reflections to match the visual modality:

"It *appears* that you are *angry* now."
"It *looks* like you are *angry* now."
"It is *clear* to me that you are *angry* now."

Sample reflections to match the auditory modality:

"It *sounds* like you are *angry* now."
"I *hear* you saying you are *angry* now."
"My *ears tell* me that you are *angry* now."

Sample reflections to match kinesthetic words:

"I can *grasp* your *anger.*"
"You are *feeling angry* now."
"Let's get in *touch* with your *anger.*"

If you don't know how the client processes information, or if the client uses visual, auditory, and kinesthetic words interchangeably, you can do the same by varying the sentence stems you select (refer also to Table 6-2).

The next step in reflecting is to add the context, or situation, around which the feelings occur. This takes the form of a brief paraphrase. Usually the context can be determined from the cognitive part of the client's message. For example, a client might say "I just can't take tests. I get so anxious I just never do well even though I study a lot." In this message, the affect is anxiety; the context is test taking. The counselor reflects the affect ("You feel uptight") *and* the context ("whenever you have to take a test").

The final step in reflecting feelings is to assess the effectiveness of your reflection after delivering it. Usually, if your reflection accurately identifies the client's feelings, the client will confirm your response by saying something like "Yes, that's right" or "Yes, that's exactly how I feel." If your response is off target, the client may reply with "Well, it's not quite like that" or "I don't feel exactly that way" or "No, I don't feel that way." When the client responds by denying feelings, it may mean your reflection was inaccurate or ill-timed. It is very important for counselors to decide when to respond to emotions. Reflection of feelings may be too powerful to be used *frequently* in the very early stage of counseling. At that time, overuse of this response may make the client feel uncomfortable, a situation that can result in

---

*SOURCE: From *Basic Attending Skills,* 3rd Ed., by A. E. Ivey, N. B. Gluckstern, and M. B. Ivey, pp. 73-74. Copyright 1993 by Microtraining Associates. Reprinted by permission.

TABLE 6-3. Words that express feelings

Relative intensity of words	Feeling category				
	Anger	**Conflict**	**Fear**	**Happiness**	**Sadness**
Mild feeling	Annoyed	Blocked	Apprehensive	Amused	Apathetic
	Bothered	Bound	Concerned	Anticipating	Bored
	Bugged	Caught	Tense	Comfortable	Confused
	Irked	Caught in a bind	Tight	Confident	Disappointed
	Irritated	Pulled	Uneasy	Contented	Discontented
	Peeved			Glad	Mixed up
	Ticked			Pleased	Resigned
				Relieved	Unsure
Moderate feeling	Disgusted	Locked	Afraid	Delighted	Abandoned
	Hacked	Pressured	Alarmed	Eager	Burdened
	Harassed	Torn	Anxious	Happy	Discouraged
	Mad		Fearful	Hopeful	Distressed
	Provoked		Frightened	Joyful	Down
	Put upon		Shook	Surprised	Drained
	Resentful		Threatened	Up	Empty
	Set up		Worried		Hurt
	Spiteful				Lonely
	Used				Lost
					Sad
					Unhappy
					Weighted
Intense feeling	Angry	Ripped	Desperate	Bursting	Anguished
	Boiled	Wrenched	Overwhelmed	Ecstatic	Crushed
	Burned		Panicky	Elated	Deadened
	Contemptful		Petrified	Enthusiastic	Depressed
	Enraged		Scared	Enthralled	Despairing
	Fuming		Terrified	Excited	Helpless
	Furious		Terror-striken	Free	Hopeless
	Hateful		Tortured	Fulfilled	Humiliated
	Hot			Moved	Miserable
	Infuriated			Proud	Overwhelmed
	Pissed			Terrific	Smothered
	Smoldering			Thrilled	Tortured
	Steamed			Turned on	

SOURCE: From *Helping Relationships and Strategies*, 2nd Ed., by D. Hutchins and C. Cole. Copyright 1992 by Brooks/Cole Publishing Company, a division of International Thomson Publishing Inc.

denial rather than acknowledgment of emotions. But do not ignore the potential impact or usefulness of reflection later on, when focusing on the client's feelings would promote the goals of the session. In the following example, notice the way a therapist uses a cognitive learning strategy (adapted from Richardson & Stone, 1981) to formulate a reflection of client feelings:

Client, a middle-aged man: You can't imagine what it was like when I found out my wife was cheating on me. I saw red! What should I do—get even—leave her—I'm not sure. [Said in loud, shrill, high-pitched voice, clenched fists]

1. What overt feeling words has this client used?
   None—except for the suggested affect phrase "saw red."

2. What feelings are implied in the client's voice and nonverbal behavior?

Anger, outrage, hostility?

3. What is a good choice of affect words that accurately describe this client's feelings at a similar level of intensity?

Furious, angry, vindictive, outraged.

4. What is an appropriate sentence stem that matches the sensory words used by the client?

From the client's use of words like "imagine" and "saw red," I'll try visual sentence stems like "It seems," "It appears," "It looks like."

5. What is the context, or situation, surrounding his feelings that I'll paraphrase?

Finding out his wife was cheating on him.

6. How will I know whether my reflection is accurate and helpful?

Watch and listen for the client's response—whether he confirms or denies the feeling of being angry and vindictive.

Actual examples of reflection:

It looks like you're very angry now about your wife's going out on you.

It appears that you're furious with your wife's actions.

It seems like you're both angry and vindictive now that you've discovered your wife has been going out with other men.

Suppose that, following the reflection, the client said "Yes, I'm very angry, for sure—I don't know about vindictive, though I guess I'd like to make her feel as crappy as I do." The client has confirmed the counselor's reflection of the feelings of anger and vindictiveness but has also given a clue that the word *vindictive* was too strong for the client to accept *at this time.* The counselor can congratulate himself or herself for having picked up on the feelings, noting that the word *vindictive* might be used again later, after the client has sorted through his mixed feelings about his wife's behavior.

Learning Activity #21 will give you an opportunity to try out the reflection-of-feeling response.

Listening for Themes: Summarization

Usually, after a client has expressed several messages or has talked for a while, her or his messages will suggest certain consistencies or patterns that we refer to as *themes*. Themes in client messages are expressed in topics that the client continually refers to or brings up in some way. The counselor can identify themes by listening to what the client repeats "over and over and with the most intensity" (Carkhuff, Pierce, & Cannon, 1977). The themes indicate what the client is trying to tell us and what the client needs to focus on in the counseling sessions. Ivey, Ivey, and Simek-Morgan (1993) refer to this as listening to the way the client organizes his or her story (p. 54). The counselor can respond to client themes by using a summarization response. For example, suppose you have been counseling a young man who, during the last three sessions, has made repeated references to homosexual relationships yet has not really identified this issue intentionally. You could use a summarization to identify the theme from these repeated references by saying something like "I'm aware that during our last few sessions you've spoken consistently about homosexual relationships. Perhaps this is an issue for you we might want to focus on."

As another example, suppose that in one session a client has given you several descriptions of different situations in which she feels concerned about how other people perceive her. You might discern that the one theme common to all these situations is the client's need for approval from others, or "other-directedness." You could use a summarization such as this to identify this theme: "One thing I see in all three situations you've described, Jane, is that you seem quite concerned about having the approval of other people. Is this accurate?"

Purposes of Summarization. One purpose of summarization is to tie together multiple elements of client messages. In this case, summarization can serve as a good feedback tool for the client by extracting meaning from vague and ambiguous messages. A second purpose of summarization is to identify a common theme or pattern that becomes apparent after several messages or sometimes after several sessions. Occasionally, a counselor may summarize to interrupt a client's incessant rambling or "storytelling." At such times, summarization is an important focusing tool that brings direction to the interview.

A fourth use of summarization is to moderate the pace of a session that is moving too quickly. In such instances, summaries provide psychological breathing space during the session. A final purpose of a summary is to review progress that has been made during one or more interviews.

A summarization can be defined as a collection of two or more paraphrases or reflections that condenses the client's messages or the session. Summarization "involves listening to a client over a period of time (from three minutes to a complete session or more), picking out relationships among key issues, and restating them back accurately to the client" (Ivey, Gluckstern, & Ivey, 1993, p. 92).

A summarization may represent collective rephrasings of either cognitive or affective data but most summarization

LEARNING ACTIVITY 21   *Reflection of Feelings*

In this learning activity, you are presented with three client practice messages. For each message, develop an example of a reflection-of-feeling response, using the cognitive learning strategy (Richardson & Stone, 1981) described earlier and outlined below. To internalize this learning strategy, you may wish to talk through these self-questions overtly (aloud) and then covertly (silently to yourself). The end product will be a reflection-of-feeling response that you can say aloud or write down or both. An example precedes the practice messages. Feedback is given after the learning activity (on page 112).

Example

Client, a 50-year-old steelworker now laid off: Now look, what can I do? I've been laid off over a year. I've got no money, no job, and a family to take care of. It's also clear to me that my mind and skills are just wasting away. [Said in a loud, critical voice, staring at the ceiling, brow furrowed, eyes squinting]

Self-question 1: What overt feeling words has the client used?
None.

Self-question 2: What feelings are implied in the client's nonverbal behavior?
Disgust, anger, upset, frustration, resentment, disillusionment, discouragement.

Self-question 3: What is a good choice of affect words that accurately describe the client's feelings at a similar level of intensity?
Seem to be two feelings—anger and discouragement. Anger seems to be the stronger emotion of the two.

Self-question 4: What is an appropriate sentence stem that matches the sensory words used by the client?
Use stems like "I see you" or "It's clear to me that you" or "From where I'm looking, you" to match the client phrases "now look" and "it's clear."

Self-question 5: What is the context, or situation, surrounding his feelings that I'll paraphrase?
Loss of job, no resources, no job prospects in sight.

Reflection-of-feeling response: I can see you're angry about being out of work and discouraged about the future *or* It looks like you're very upset about having your job and stability taken away from you.

Client Practice Statements

Client 1, an 8-year-old girl: I'm telling you I don't like living at home anymore. I wish I could live with my friend and her parents. I told my mommy that one day I'm going to run away, but she doesn't listen to me. [Said in level, measured words, glancing from side to side, lips drawn tightly together, flushed face]

Self-question 1: What overt feeling words has the client used?

Self-question 2: What feelings are implied in the client's nonverbal behavior?

Self-question 3: What are accurate and similar interchangeable affect words?

Self-question 4: What is a useful sentence stem that matches the sensory words used by the client?

Self-question 5: What is the context, or situation, concerning her feelings that I'll paraphrase?

Actual reflection response: _____
_____

Client 2, a middle-aged man in marital therapy: As far as I'm concerned, our marriage turned sour last year when my wife went back to work. She's more in touch with her work than with me. [Said in soft voice tone with downcast eyes]

Self-question 1: What overt feeling words did the client use?

Self-question 2: What feelings are implied in the client's nonverbal behavior?

Self-question 3: What are accurate and similar interchangeable affect words?

Self-question 4: What is a useful sentence stem that matches the sensory words used by the client?

Self-question 5: What is the context, or situation, surrounding his feelings that I'll paraphrase?

Actual reflection response: _____
_____

Client 3, an adolescent: Now look, we have too damn many rules around this school. I'm getting the hell out of here. As far as I can see, this place is a dump. [Said in loud, harsh voice]

Self-question 1: What overt feeling words has this client used?

Self-question 2: What feelings are implied in the client's nonverbal behavior?

Self-question 3: What are accurate and similar interchangeable affect words?

Self-question 4: What is a useful sentence stem that matches the sensory words used by the client?

Self-question 5: What is the context, or situation surrounding his feelings that I'll paraphrase?

Actual reflection response: _____
_____

responses will include references to both cognitive and affective messages, as in the following four examples:

1. Example of summarization to *tie together multiple elements* of a client message:

Client, a Native American medical student: All my life I thought I wanted to become a doctor and go back to work on my reservation. Now that I've left home, I'm not sure. I still feel strong ties there that are pulling me back. I hate to let my people down, yet I also feel like there's a lot out here I want to explore first.

Summarization: You're away from the reservation now and are finding so much in this different place to explore—at the same time, you're feeling your lifelong ties to your people and your dream you've had to go back as a doctor.

2. Example of summarization to *identify a theme:*

Client, a 35-year-old male: One of the reasons we divorced was because she always pushed me. I could never say no to her; I always gave in. I guess it's hard for me just to say no to requests people make.

Summarization: You're discovering that you tend to give in or not do what you want in many of your significant relationships, not just with your ex-wife.

3. Example of summarization *to regulate pace of session and to give focus:*

Client, a young woman: What a terrible week I had! The water heater broke, the dog got lost, someone stole my wallet, my car ran out of gas, and to top it all off, I gained five pounds. I can't stand myself. It seems like it shows all over me.

Summarization: Let's stop for just a minute before we go on. It seems like you've encountered an unending series of bad events this week.

4. Example of summarization *to review progress* (often used as termination strategy near end of session):

Counselor summary: Jane, we've got about five minutes left today. It seems like most of the time we've been working on the ways you find to sabotage yourself from doing things you want to do but yet feel are out of your control. This week I'd like you to work on the following homework before our next session. . . .

Steps in Summarizing.   Summarizing requires careful attention to and concentration on the client's verbal and nonverbal messages. Accurate use of this response involves good recall of client behavior, not only within a session but over time—across several sessions or even several months of therapy. Developing a summarization involves the following five steps:

1. Attend to and recall the message or series of messages by restating these to yourself covertly—what has the client been telling you, focusing on, working on? This is a key and difficult part of effective summaries because it requires you to be aware of many, varying verbal and nonverbal messages you have processed *over time.*
2. Identify any apparent patterns, themes, or multiple elements of these messages by asking yourself questions like "What has the client repeated over and over" or "What are the different parts of this puzzle?"
3. Select an appropriate beginning (sentence stem) for your summarization that uses the personal pronoun *you* or the client's name and matches one or more of the client's sensory words. (See Table 6-2.)
4. Next, using the sentence stem you've selected, select words to describe the theme or tie together multiple elements, and verbalize this as the summarization response. Remember to use your voice so that the summarization sounds like a statement instead of a question.
5. Assess the effectiveness of your summarization by listening for and observing whether the client confirms or denies the theme or whether the summary adds to or detracts from the focus of the session.

To help you formulate a summarization, consider the following cognitive learning strategy:

1. What was this client telling me and working on today and over time? That is, what are the *key content* and *key affect?*
2. What has the client repeated over and over today and over time? That is, what is the *pattern* or *theme?*
3. What is a useful sentence stem that matches the client's sensory words?
4. How will I know whether my summarization is useful?

Notice how a counselor applies this cognitive learning strategy in developing a summarization in the following example:

Client, a middle-aged male fighting alcoholism [he has told you for the last three sessions that his drinking is ruining his family life but he can't stop because it makes him feel better and helps him to handle job stress]: I know drinking doesn't really help me in the long run. And it sure doesn't help my family. My wife keeps threatening to leave. I know all this. It's hard to stay away from the booze. Having a drink makes me

FEEDBACK
Reflection of Feelings

Client 1

Question 1: What overt feeling words did the client use?
"Don't like."

Question 2: What feelings are implied in the client's non-verbal behavior?
Upset, irritation, resentment.

Question 3: What are interchangeable affect words?
Bothered, perturbed, irritated, upset.

Question 4: What sentence stem matches the client's sensory words?
"Seems like," "It sounds like," "I hear you saying that" match her words "tell" and "listen."

Question 5: What is the context, or situation, surrounding her feelings?
Living at home with her parents.

Actual examples of reflection: It sounds like you're upset about some things going on at your home now *or* I hear you saying you're bothered about your parents.

Client 2

Question 1: What overt feeling words did the client use?
No obvious ones except for phrases "turned sour" and "in touch with."

Question 2: What feelings are implied in the client's non-verbal behavior?
Sadness, loneliness, hurt.

Question 3: What are interchangeable affect words?
Hurt, lonely, left out, unhappy.

Question 4: What sentence stem matches the client's choice of sensory words?
"I sense," "You feel" match his phrases "turned sour" and "in touch with."

Question 5: What is the context, or situation, surrounding his feelings?
Wife's return to work.

Actual examples of reflection: You're feeling left out and lonely since your wife's gone back to work *or* I sense you're feeling hurt and unhappy because your wife seems so interested in her work.

Client 3

Question 1: What overt feeling words did the client use?
No obvious ones, but words like "damn," "hell," and "dump" suggest intensity of emotions.

Question 2: What feelings are implied in the client's non-verbal behavior?
Anger, frustration.

(continued)

Question 3: What are interchangeable affect words?
Angry, offended, disgusted.

Question 4: What sentence stem matches the client's sensory words?
Stems such as "It seems," "It appears," "It looks like," "I can see" match his words "now look" and "I can see."

Question 5: What is the context surrounding the feelings?
School rules.

Actual examples of reflection: It looks like you're pretty disgusted now because you see these rules restricting you *or* It seems like you're very angry about having all these rules here at school.

---

feel relieved. [Said in low, soft voice, downcast eyes, stooped shoulders]

Self-question 1: What has this client been telling me today and over time?
Key content: results of drinking aren't good for him or his family.
Key affect: drinking makes him feel better, less anxious.

Self-question 2: What has this client repeated over and over today and over time—pattern or theme?
That despite adverse effects and family deterioration, he continues to drink for stress reduction; that is, stress reduction through alcohol seems worth losing his family.

Self-question 3: What sentence stem matches the client's sensory words?
I'll try "You're feeling," "My sense of it," and so on to match his words "know" and "feel."

Suppose that at this time the counselor delivered one of the following summarizations to the client:

"Jerry, I sense that you feel it's worth having the hassle of family problems because of the good, calm feelings you get whenever you drink."

"Jerry, you feel that your persistent drinking is creating a lot of difficulties for you in your family, and I sense your reluctance to stop drinking in spite of these adverse effects."

"Jerry, I sense that, despite everything, alcohol feels more satisfying (rewarding) to you than your own family."

If Jerry confirms the theme that alcohol is more important now than his family, the counselor can conclude that the summarization was useful. If Jerry denies the theme or issue summarized by the counselor, the counselor can ask Jerry to clarify how the summarization was inaccurate, remembering that the summary may indeed be inaccurate or that Jerry may not be ready to acknowledge the issue at this time.

LEARNING ACTIVITY 22 *Summarization*

In this learning activity, you are presented with three client practice messages. For each message, develop a summarization response, using the cognitive learning strategy described earlier and outlined below. To internalize this learning strategy, you may wish to talk through these self-questions overtly (aloud) and then covertly—that is, silently to yourself. The end product will be a summarization response that you can say aloud or write down or both. An example precedes the practice messages. Feedback is given on page 114.

Example

Client, a 10-year-old girl:

At beginning of the session: I don't understand why my parents can't live together anymore. I'm not blaming anybody, but it just feels very confusing to me. [Said in low, soft voice with lowered, moist eyes]

Near the middle of the same session: I wish they could keep it together. I guess I feel like they can't because they fight about me so much. Maybe I'm the reason they don't want to live together anymore.

Self-question 1: What has this client been telling me and looking at today in terms of key *content* and key *affect?*
Key content: wants parents to stay together.
Key affect: feels sad, upset, responsible.

Self-question 2: What has the client repeated over and over today or over time—that is, pattern or theme? She's the one who's responsible for her parents' breakup.

Self-question 3: What is a useful sentence stem that matches the client's sensory words?
Try "I sense" or "You're feeling," to match her words such as "don't understand" and "feels."

Examples of summarization response: Joan, at the start of our talk today, you were feeling like no one person was responsible for your parents' separation. Now I sense you're saying that you feel responsible *or* Joan, earlier today you indicated you didn't feel like blaming anyone for what's happening to your folks. Now I'm sensing that you are feeling like you are responsible for their breakup.

Client Practice Messages

Client 1, a 30-year-old man who has been blaming himself for his wife's unhappiness: I really feel guilty about marrying her in the first place. It wasn't really for love. It was just a convenient thing to do. I feel like I've messed up her life really badly. I also feel obliged to her. [Said in low, soft voice tone with lowered eyes]

Self-question 1: What has this client been telling me and working on today?
Key content:
Key affect:

Self-question 2: What has this client repeated over and over today or over time in terms of patterns and themes?

Self-question 3: What is a sentence stem that matches the client's sensory words?

Summarization response: _____

_____

Client 2, a 35-year-old woman who focused on how her life has improved since having children: I never thought I would feel this great. I always thought being a parent would be boring and terribly difficult. It's not, for me. It's fascinating and easy. It makes everything worthwhile. [Said with alertness and animation]

Self-question 1: What has this client been telling me and working on today?
Key content:
Key affect:

Self-question 2: What has this client repeated over and over today or over time in terms of patterns and themes?

Self-question 3: What is a sentence stem that matches the client's sensory words?

Summarization response: _____

_____

Client 3, a 27-year-old woman who has continually focused on her relationships with men and her needs for excitement and stability:

First session: I've been dating lots and lots of men for the last few years. Most of them have been married. That's great because there are no demands on me. [Bright eyes, facial animation, high-pitched voice]

Fourth session: It doesn't feel so good anymore. It's not so much fun. Now I guess I miss having some commitment and stability in my life. [Soft voice, lowered eyes]

Self-question 1: What has this client been telling me and working on today?
Key content:
Key affect:

Self-question 2: What has this client repeated over and over today or over time in terms of patterns and themes?

Self-question 3: What is a sentence stem that matches the client's sensory words?

Summarization response: _____

_____

FEEDBACK 22
Summarization

Client 1

Question 1: What has the client told me?
Key content: He married for convenience, not love.
Key affect: Now he feels both guilty and indebted.

Question 2: What has the client repeated over and over now and before in terms of patterns and themes?
Conflicting feelings—feels a strong desire to get out of the marriage yet feels a need to keep relationship going because he feels responsible for his wife's unhappiness.

Question 3: What is an appropriate sentence stem that matches the client's sensory words?
Use stems such as "You're feeling," "My grasp of it is," "I sense that" to match his constant use of the verb "feel."

Examples of summarization response: I sense you're feeling pulled in two different directions. For yourself, you want out of the relationship. For her sake, you feel you should stay in the relationship *or* You're feeling like you've used her for your convenience and because of this you think you owe it to her to keep the relationship going *or* I can grasp how very much you want to pull yourself out of the marriage and also how responsible you feel for your wife's present unhappiness.

Client 2

Question 1: What has the client told me?
Key content: Children have made her life better, more worthwhile.
Key affect: Surprise and pleasure.

Question 2: What has the client said over and over in terms of patterns and themes?
Being a parent is uplifting and rewarding even though she didn't expect it to be. In addition, her children are very important to her. To some extent, they define her worth and value as a person.

Question 3: What is an appropriate sentence stem that matches the client's sensory words?
There are no real clear-cut sensory words exhibited in this message, so I may want to emphasize several in my response—for example, "It *seems* like you're *feeling*" or "I *hear feelings* of."

Examples of summarization response: It seems like you're feeling surprise, satisfaction, and relief about finding parenting so much easier and more rewarding than you had expected it would be *or* I hear feelings of surprise and pleasure in your voice as you reveal

(continued)

how great it is to be a parent and how important your children are to you *or* You seem so happy about the way your life is going since you've had children—as if they make you and your life more worthwhile.

Client 3

Question 1: What has the client told me?
Key content: She has been dating lots of men who have their own commitments.
Key affect: It used to feel great; now she feels a sense of loss and emptiness.

Question 2: What has she repeated over and over in terms of patterns and themes?
At first—feelings of pleasure, relief not to have demands in close relationships. Now, feelings are changing, feels less satisfied, wants more stability in close relationships.

Question 3: What is an appropriate sentence stem that matches the client's sensory words?
There are no clear-cut sensory words except for "feel" in the fourth session, so I'll vary my own words in my response and also include the word "feel."

Examples of summarization response: Lee Ann, originally you said it was great to be going out with a lot of different men who didn't ask much of you. Now you're also feeling it's not so great—it's keeping you from finding some purpose and stability in your life *or* In our first session, you were feeling "up" about all those relationships with noncommittal men. Now you're feeling like this is interfering with the stability you need and haven't yet found *or* At first it was great to have all this excitement and few demands. Now you're feeling some loss from lack of a more stable, involved relationship.

BARRIERS TO LISTENING

Egan (1994) discusses what he calls the "shadow side" of listening—that is, ways in which the listening process may fail. As he notes, active listening sounds good in theory but in practice is not without "obstacles and distractions" (p. 100).

We have observed three types of helpers who seem to have the most difficulty listening to clients. We identify these as follows:

1. *Frenetic* counselors: These counselors are so "hyper" and so much "in motion" (either internally, externally, or both) that they have great difficulty just sitting quietly and taking in clients' stories.

2. *Self-centered* counselors: These counselors are so in love with themselves and so "hell-bent" on getting their own ideas across that little opportunity exists for clients to tell their stories.

3. *Self-absorbed* counselors: These counselors often look physically present and attentive, yet there is so much "internal noise" going on inside that they aren't really emotionally available to hear clients.

Another obstacle to effective counseling is the ways the listening process itself may conflict with the values of some clients of color. As Atkinson et al. (1993) observe,

> Often, in order to encourage self-disclosure, the counseling situation is intentionally designed to be an ambiguous one, one in which the counselor listens empathically and responds only to encourage the client to continue talking. This lack of structure in the counseling process may conflict with need for structure that is a value in many cultures. Racial/ethnic minority clients frequently find the lack of structure confusing, frustrating, and even threatening. (p. 53)

Gender differences also play a role, as some men tend to have a more directive style, asking more questions, and doing more interrupting and problem solving than some women (Ivey, Gluckstern, & Ivey, 1993).

SUMMARY

We often hear these questions: "What good does all this listening do? How does just rephrasing client messages really help?" In response, we will reiterate the rationale for using listening responses in counseling.

1. Listening to clients is a very powerful reinforcer and may strengthen clients' desire to talk about themselves and their concerns. Not listening may prevent clients from sharing relevant information.

2. Listening to a client first may mean a greater chance of responding accurately to the client in later stages of counseling, such as problem solving. By jumping to quick solutions without laying a foundation of listening, you may inadvertently ignore the primary problem or propose inadequate and ill-timed action steps.

3. Listening encourages the client to assume responsibility for selecting the topic and focus of an interview. Not listening may meet your needs to find information or to solve problems. In doing so, you may portray yourself as an expert rather than a collaborator. Simply asking a series of questions or proposing a series of action steps in the initial phase of helping can cause the client to perceive you as the expert and can hinder proper development of client self-responsibility in the interview.

4. Good listening and attending skills model socially appropriate behavior for clients (Gazda et al., 1995). Many clients have not yet learned to use the art of listening in their own relationship and social contacts. They are more likely to incorporate these skills to improve their interpersonal relationships when they experience them firsthand through their contact with a significant other, such as a therapist.

All these guidelines have to consider both the gender and cultural affiliation of the client. Listening may have a differential impact depending on the client's gender and cultural affiliation.

Some counselors can articulate a clear rationale for listening but nevertheless cannot listen in an interview because of blocks that inhibit effective listening. Some of the most common blocks to listening are these:

1. The tendency to judge and evaluate the client's messages.
2. The tendency to stop attending because of distractions such as noise, the time of day, or the topic.
3. The temptation to respond to missing pieces of information by asking questions.
4. The temptation or the pressure put on yourself to solve problems or find answers.
5. The preoccupation with yourself as you try to practice the skills. This preoccupation shifts the focus from the client to you and actually reduces, rather than increases, your potential for listening.

Finally, effective listening requires an involved yet contained sort of energy that allows you to be fully present to the client. Listening is a process that does not stop after the initial session but continues throughout the entire therapeutic relationship with each client.

POSTEVALUATION

Part One

This part allows you to assess your performance on Chapter Objective One. On a sheet of paper, classify each of the counselor listening responses in the following list as a clarification, paraphrase, reflection of feelings, or summarization. If you identify 9 out of 12 responses correctly, you have met this objective. You can check your answers against those provided in the feedback that follows on page 119.

1. *Client,* an older, retired person: How do they expect me to live on this little bit of Social Security? I've worked hard all my life. Now I have so little to show for it—I've got to choose between heat and food.
   a. Can you tell me who exactly expects you to be able to manage on this amount of money?
   b. All your life you've worked hard, hoping to give yourself a secure future. Now it's very upsetting to have only a little bit of money that can't possibly cover your basic needs.
2. *Client:* I'm having all these horrendous images that keep coming at me. I always thought I had had a happy childhood but now I'm not so sure.
   a. Can you tell me what you mean by "horrendous images"?
   b. Recently you started to have some very scary memories about your past and it's made you question how great your childhood really was.
3. *Client:* I feel so nervous when I have to give a speech in front of lots of people.
   a. You feel anxious when you have to talk to a group of people.
   b. You would rather not have to talk in front of large groups.
4. *Client:* I always have a drink when I'm under pressure.
   a. Are you saying that you use alcohol to calm you down?
   b. You think alcohol has a calming effect on you.
5. *Client:* I don't know whether I've ever experienced orgasm. My husband thinks I have, though.
   a. Are you saying that you've been trying to have your husband believe you do experience orgasm?
   b. You feel uncertain about whether you've ever really had an orgasm, even though your husband senses that you have.
6. *Client:* I haven't left my house in years. I'm even afraid to hang out the clothes.
   a. You feel panicked and uneasy when you go outside the security of your house.

b. Because of this fear, you've stayed inside your house for a long time.

Part Two

Three client statements are presented. Objective Two asks you to verbalize or write an example of each of the four listening responses for each client statement.* In developing these responses, you may find it helpful to use the cognitive learning strategy you practiced earlier for each response. Feedback follows the evaluation.

Client 1, a 28-year-old woman: My life is a shambles. I lost my job, my friends never come around anymore. It's been months now, but I still can't seem to cut down. I can't see clearly. It seems hopeless. [Said in high-pitched voice, with crossed legs, lots of nervous "twitching" in hands and face]
Clarification:
Paraphrase:
Reflection:
Summarization:

Client 2, an African American high school sophomore: I can't seem to get along with my mom. She's always harassing me, telling me what to do. Sometimes I get so mad I feel like hitting her, but I don't, because it would only make the situation worse.
Clarification:
Paraphrase:
Reflection:
Summarization:

Client 3, a 54-year-old man: Ever since my wife died four months ago, I can't get interested in anything. I don't want to eat or sleep. I'm losing weight. Sometimes I just tell myself I'd be better off if I were dead, too.
Clarification:
Paraphrase:
Reflection:
Summarization:

Part Three

This part of the evaluation gives you an opportunity to develop your observation skills of key aspects of client behavior that must be attended to in order to listen effectively:

1. Vague or confusing phrases and messages
2. Expression of key content
3. Use of affect words
4. Nonverbal behavior illustrative of feeling or mood states
5. Presence of themes or patterns

---

*These three client messages can be put on audiotape with pauses between statements. Instead of reading the message, you can listen to the message and write or verbalize your responses during the pause.

(continued)

Objective Three asks you to observe these five aspects of a client's behavior during a 15-minute interview conducted by someone else. Record your observations on the Client Observation Checklist that follows. You can obtain feedback for this activity by having two or more persons observe and rate the same session—then compare your responses.

Client Observation Checklist

Name of counselor _____

Name of observer(s) _____

_____

Instructions: Given the five categories of client behavior to observe (left column), use the right column to record separate occurrences of behaviors within these categories as they occur during a short counseling interview.*

Observed category of behavior	Selected key client words and nonverbal behavior
1. Vague, confusing, ambiguous phrases, messages	1. _____ 2. _____ 3. _____ 4. _____ 5. _____
2. Key content (situation, event, idea, person)	1. _____ 2. _____ 3. _____ 4. _____ 5. _____
3. Affect words used	1. _____ 2. _____ 3. _____ 4. _____ 5. _____
4. Nonverbal behavior indicative of certain feelings	1. _____ 2. _____ 3. _____ 4. _____ 5. _____
5. Themes, patterns	1. _____ 2. _____ 3. _____ 4. _____ 5. _____

Observer impressions and comments _____

_____

_____

Part Four

This part of the evaluation gives you a chance to demonstrate the four listening responses. Objective Four asks you to conduct a 15-minute role-play interview in which you use at least two examples of each of the four listening responses. Someone can observe your performance, or you can assess yourself from an audiotape of the interview. You or the observer can classify your responses and judge their effectiveness using the Listening Checklist that follows. Try to select a listening response to use when you have a particular purpose in mind. Remember, in order to listen, it is helpful to

1. Refrain from making judgments
2. Resist distractions
3. Avoid asking questions
4. Avoid giving advice
5. Stay focused on the client

Obtain feedback for this activity by noting the categories of responses on the Listening Checklist and their judged effectiveness.

Listening Checklist

Name of counselor _____

Name of observer _____

Instructions: In the far left column, "Counselor response," summarize a few key words of each counselor statement, followed by a brief notation of the client's verbal and nonverbal response in the next column, "Client response." Then classify the message as a clarification, paraphrase, reflection of feeling, summarization, or other under the corresponding column. Rate the *effectiveness* of each counselor response in the far right column, "Effectiveness of response," on the following 1–3 scale.

    1 = not effective. Client ignored counselor message or gave indication that counselor message was inaccurate and "off target."

    2 = somewhat effective. Client gave some verbal or nonverbal indication that counselor message was partly right, accurate, "on target."

    3 = very effective. Client's verbal and nonverbal behavior confirmed that counselor response was very accurate, "on target," or "fit."

Remember to watch and listen for the *client's* reaction to the response for your effectiveness rating.

_____

*If observers are not available, audiotape or videotape your sessions and complete the checklist while reviewing the tape.

(continued)

POSTEVALUATION   (continued)

Counselor response (key words)	Client response (key words)	Type of counselor response					Effectiveness of response (determined by client response) Rate from 1 to 3 (3 = high)
		Clarification	Paraphrase	Reflection of feelings	Summarization	Other	
1.							
2.							
3.							
4.							
5.							
6.							
7.							
8.							
9.							
10.							
11.							
12.							
13.							
14.							
15.							
16.							
17.							
18.							
19.							
20.							

Observer comments and general observations _____

SUGGESTED READINGS

Atkinson, D. R., Morten, G., & Sue, D. W. (1993). *Counseling American minorities* (4th ed.). Madison, WI: Brown & Benchmark.

Brammer, L., Shostrom, E., & Abrego, P. (1989). *Therapeutic psychology* (5th ed.). Englewood Cliffs, NJ: Prentice Hall.

Carkhuff, R. R. (1993). *The art of helping VII*. Amherst, MA: Human Resource Development Press.

Egan, G. (1994). *The skilled helper* (5th ed.). Pacific Grove, CA: Brooks/Cole.

Hackney, H., & Cormier, L. S. (1994). *Counseling strategies and interventions* (4th ed.). Needham Heights, MA: Allyn & Bacon.

Ivey, A. E., Gluckstern, N., & Ivey, M. B. (1993). *Basic attending skills* (3rd ed.). North Amherst, MA: Microtraining Associates.

Nichols, M. P. (1995). *The lost art of listening*. New York: Guilford.

Teyber, E. (1997). *Interpersonal processes in psychotherapy*. Pacific Grove, CA: Brooks/Cole.

White, M., & Epston, D. (1990). *Narrative means to therapeutic ends*. New York: W. W. Norton.

F E E D B A C K
POSTEVALUATION

Part One
1. a. Clarification
    b. Summarization
2. a. Clarification
    b. Summarization
3. a. Reflection
    b. Paraphrase
4. a. Clarification
    b. Paraphrase
5. a. Clarification
    b. Reflection
6. a. Reflection
    b. Paraphrase

Part Three
Here are some examples of listening responses. See whether yours are similar:

Client statement 1

1. Clarification: "Can you describe what you mean by 'cutting down'?"
2. Paraphrase: "You seem to realize that your life is not going the way you want it to."
3. Reflection: "You appear frightened about the chaos in your life, and you seem uncertain of what you can do to straighten it out."
4. Summarization: "Your whole life seems to be falling apart. Your friends are avoiding you, and now you don't even have a job to go to. Even though you've tried to solve the problem, you can't seem to handle it alone. Coming here to talk is a useful first step in 'clearing up the water' for you."

Client statement 2

1. Clarification: "Can you describe what it's like when you don't get along with her?"
2. Paraphrase: "It appears that your relationship with your mom is deteriorating to the point that you feel you may lose control of yourself."
3. Reflection: "You feel frustrated and angry with your mom because she's always giving you orders."
4. Summarization: "It seems like the situation at home with your mom has become intolerable. You can't stand her badgering, and you feel afraid that you might do something you would later regret."

(continued)

Client statement 3

1. Clarification: "Are you saying that since the death of your wife, life has become so miserable that you occasionally contemplate taking your own life?"
2. Paraphrase: "Your life has lost much of its meaning since your wife's recent death."
3. Reflection: "It sounds like you're very lonely and depressed since your wife died."
4. Summarization: "Since your wife died, you've lost interest in living. There's no fun or excitement anymore, and further, you're telling yourself that it's not going to get any better."

ACTION RESPONSES

Listening responses involve responding to client messages primarily from the client's point of view, or frame of reference. There are times in the counseling process when it is legitimate to move beyond the client's frame of reference and to use responses that include more counselor-generated data and perceptions. These responses, which we have labeled *action responses,* are active rather than passive and reflect a counselor-directed more than a client-centered style. Whereas listening responses influence the client indirectly, action responses exert a more direct influence on the client (Ivey, 1994). Action responses are based as much on the counselor's perceptions and hypotheses as on the client's messages and behavior. We have selected four such action responses: probes, confrontation, interpretation, and information giving. The general purpose of action responses, according to Egan (1994), is to help clients see the need for change and action through a more objective frame of reference.

OBJECTIVES

1. From a written list of eight example counselor action responses, accurately classify at least six of them by type.
2. With a written list of three client statements, write an example of each of the four action responses for each client statement.
3. In a 30-minute counseling interview in which you are an observer, listen for and record five key aspects of client behavior that form the basis for action responding.
4. Conduct at least one 20-minute counseling interview in which you integrate relationship variables (Chapters 3 and 4), nonverbal behavior (Chapter 5), listening responses (Chapter 6), and action responses (Chapter 7).

ACTION RESPONSES AND TIMING

The most difficult part of using action responses is the timing, the point at which these responses are used in the interview. As you recall from Chapter 6, some helpers tend to

120

jump into action responses before listening and establishing rapport with the client. Listening responses generally reflect clients' understanding of themselves. In contrast, action responses reflect the *counselor's* understanding of the client. Action responses can be used a great deal in the interview as long as the counselor is careful to lay the foundation with attending and listening. The listening base can heighten the client's receptivity to a counselor action message. If the counselor voices his or her opinions and perceptions too quickly, the client may respond with denial, with defensiveness, or even with dropping out of counseling. When this happens, often the counselor needs to retreat to a less obtrusive level of influence and do more listening, at least until a strong base of client trust and confidence has been developed.

On the other hand, some clients from various cultural groups may actually be less defensive with a more active and directive counselor communication style because it is more consistent with their needs and values. Further, some clients of color may feel more comfortable with the use of directive and active skills because these responses provide them with data about "where the counselor is coming from, allowing the minority client to first test the waters" before self-disclosing too much (Sue & Sue, 1990, p. 71).

WHAT DOES ACTION REQUIRE OF HELPERS?

In the previous chapter, we discussed what listening requires of helpers. We noted that accurate and effective listening depends on the ability of helpers to be present to the client and to restrain some of their own energy and expressiveness. In contrast, action responses require the helper to be more expressive and more challenging. Egan (1994) describes the use of action responses as the responding to "sour notes" present in the client's communication and behavior. To use action responses effectively, helpers must first provide a supportive and safe environment by listening carefully, and then they must feel comfortable enough with themselves to provide feedback or amplification to the client that the client may not like. Counselors who have esteem issues of their own may find the use of action responses difficult as these responses carry the risk of upsetting a client by what is said or challenged. Ultimately, the use of effective action responses requires helpers to feel secure enough about themselves to have their own voice and to tolerate client disapproval and disagreement.

FOUR ACTION RESPONSES

We have selected four action responses to describe in this chapter: the probe, confrontation, interpretation, and information giving. A *probe* is an open or closed question or inquiry. A *confrontation* is a description of a client discrepancy. An *interpretation* is a possible explanation for the client's behavior. *Information giving* is the communication of data or facts about experiences, events, alternatives, or people.

Look at the way these four action responses differ in this illustration:

Client, a 35-year-old Latina widow, mother of two young children: My whole life fell apart when my husband died. I keep feeling so unsure about my ability to make it on my own and to support my kids. My husband always made all the decisions for me. Now, I haven't slept well for so long and I'm drinking more heavily—I can't even think straight. My relatives help me as much as they can but I still feel scared.

Counselor probe: What sorts of experiences have you had in being on your own—if any?

What feels most scary about this?

Counselor confrontation: It seems as if you're dealing with two things in this situation: first, the experience of being on your own for the first time, which feels so new and scary you're unsure you can do it, and second, the reality that, although your relatives help out, the responsibility for you and your children does now rest on your shoulders.

Counselor interpretation: When your husband was alive, you depended on him to take care of you and your children. Now it's up to you, but taking on this role is uncomfortable and also unfamiliar. Perhaps your increased drinking is a way to keep from having to face this. What do you think?

Counselor information giving: Perhaps you are still grieving over the loss of your husband. I'm wondering whether there are rituals in your culture as well as certain people who might be helpful to you in your loss.

Table 7-1 describes the definitions and intended purposes of these four action responses. Remember, these intended purposes are presented only as tentative guidelines, not as "the truth." The remainder of the chapter describes and presents model examples of these four skills. You will have an opportunity to practice each skill and receive feedback about your responses.

Probes

Probes, or questions, are an indispensable part of the interview process. Their effectiveness depends on the type of question and the frequency of their use. Questions have the potential for establishing a desirable or undesirable pattern of interpersonal exchange, depending on the skill of the therapist. Beginning interviewers err by assuming that a helping interview is a series of questions and answers or by asking the wrong kind of question at a particular time. These practices are likely to make the client feel interrogated rather than understood. Even more experienced counselors overuse this potentially valuable verbal response. Unfortunately, asking a question is all too easy to do during silence or when you are at a loss for words. Questions should not be asked unless you have a particular purpose for the question in mind. For example, if you are using a question as an open invitation to talk, realize that you are in fact asking the client to initiate a dialogue and allow the client to respond in this way.

Open and Closed Probes. Most effective questions are worded in an open-ended fashion, beginning with words such as *what, how, when, where,* or *who*. According to Ivey, Ivey, and Simek-Morgan (1993), the particular word used to begin an open-ended question is important. Research has shown that "what" questions tend to solicit facts and information, "how" questions are associated with sequence and process or emotions, and "why" questions produce reasons. Similarly, "when" and "where" questions solicit information about time and place, and "who" questions are associated with information about people. The importance of using *different* words in formulating open-ended questions is critical.

Open-ended probes have a number of purposes in different counseling situations (Hackney & Cormier, 1994; Ivey, 1994). These are

1. Beginning an interview
2. Encouraging the client to express more information
3. Eliciting examples of particular behaviors, thoughts, or feelings so that the counselor can better understand the conditions contributing to the client's problem

TABLE 7-1. Definitions and intended purposes of counselor action responses

Response	Definition	Intended purpose
Probe	Open-ended or closed question or inquiry	*Open-ended questions* 1. To begin an interview 2. To encourage client elaboration or to obtain information 3. To elicit specific examples of client's behaviors, feelings, or thoughts 4. To motivate client to communicate *Closed questions* 1. To narrow the topic of discussion 2. To obtain specific information 3. To identify parameters of a problem or issue 4. To interrupt an overtalkative client—for example, to give focus to the session
Confrontation	Description of client discrepancy	1. To identify client's mixed (incongruent) messages 2. To explore other ways of perceiving client's self or situation
Interpretation	Possible explanation of or association among various client behaviors	1. To identify the relation between client's implicit messages and behaviors 2. To examine client behavior from alternative view or with different explanation 3. To add to client's self-understanding as a basis for client action
Information giving	Verbal communication of data or facts	1. To identify alternatives 2. To evaluate alternatives 3. To dispel myths 4. To motivate clients to examine issues they may have been avoiding

4. Developing client commitment to communicate by inviting the client to talk and guiding the client along a focused interaction

In contrast to open-ended questions, closed, or focused, questions can be useful if the counselor needs a particular fact or seeks a particular bit of information. These questions begin with words such as *are, do, can, is, did* and can be answered with a yes, a no, or a very short response. As we see in Chapter 9, questions are a major tool for obtaining information during the assessment process.

These are examples of closed questions:

1. "Of all the problems we discussed, which bothers you the most?"
2. "Is there a history of depression in your family?"
3. "Are you planning to look for a job in the next few months?"

The purposes of closed questions include the following:

1. Narrowing the area of discussion by asking the client for a specific response
2. Gathering specific information
3. Identifying parameters of problems
4. Interrupting an overtalkative client who rambles or "storytells"

Closed questions must be used sparingly within an interview. Too many closed questions may discourage discussion and may subtly give the client permission to avoid sensitive or important topics.

Shainberg (1993) observes that "the point of a question is to open a person to [his or her] own process . . . to come in from a different angle than the client" (p. 87). She notes that it is all too easy for therapists to become sloppy or to lack creativity in formulating truly effective questions. For her, the difference between an effective and a poor question is whether the question enables the client to look at things in a new way or at a deeper level. She notes that "for many clients, being asked a good question is like having some new energy" (p. 88). Shainberg (1993) also describes other issues involved in using questions effectively:

1. The *frequency* of questions—more does not mean better.
2. The *timing* of a question—"stock" questions such as "How does this make you feel?" or "What was that like for you?" usually yield "pat" answers.

Examples of thought-provoking questions suggested by Shainberg can be found in Table 7-2.

Guidelines in the Use of Probes.   Probes will be used more effectively and efficiently if you remember some important guidelines for their use.

TABLE 7-2. Sample probes

"What sorts of experiences do you have with this issue?"
"Do you have some understanding of how painful this is—or is it painful?"
"This feeling you describe—does this happen in any particular part of your body?"
"Have you ever really let yourself experience this feeling?"
"What can you tell me about your readiness to heal?"
"What is scary (or painful, sad, etc.) about this?"
"What is the prize that will unlock the puzzle as things stand right now?"
"What is it that you don't want to see (feel or do)?"
"How can you come alive in your life?"
"Is there something from the past . . . some wound from the past that keeps you where you are right now?"
"What do you find yourself longing (wishing) for?"
"How is it that you create some of your own unhappiness?"
"What is your relationship with your own pain (fear, sadness)? Do you fight it, dread it, turn against yourself in it, or think someone else will fix it—or believe it will go on forever—or take good care of yourself when you have it?"
"What is your relationship with your family?"
"What is important to you? to your family? to your culture?"
"What is it time for right now in your life?"

SOURCE: Adapted from *Healing in Psychotherapy*, by D. Shainberg. Copyright 1993 by Gordon & Breach Science Publishers. Reprinted by permission.

First, develop questions that center on the concerns of the client. Effective questions arise from what the client has already said, not from the counselor's curiosity or need for closure.

Second, after a question, use a pause to give the client sufficient time to respond. Remember that the client may not have a ready response. The feeling of having to supply a quick answer may be threatening and may encourage the client to give a response that pleases the therapist.

Third, ask only one question at a time. Some interviewers tend to ask multiple questions (two or more) before allowing the client time to respond. We call this "stacking questions." It confuses the client, who may respond only to the least important of your series of questions.

Fourth, avoid accusatory or antagonistic questions. These are questions that reflect antagonism either because of the counselor's voice tone or because of use of the word *why*. You can obtain the same information by asking "what" instead of "why." Accusatory questions can make a client feel defensive.

Finally, avoid relying on questions as a primary response mode during an interview (an exception would be when doing an intake, a history, or an assessment session). Re-

member that in some cultural groups questions may seem offensive, intrusive, and lacking in respect. In any culture, consistent overuse of questions can create a number of problems in the therapeutic relationship, including creating dependency, promoting yourself as an expert, reducing responsibility and involvement by the client, and creating resentment (Gazda et al., 1995). The feeling of being interrogated may be especially harmful with "reluctant" clients. Questions are most effective when they provoke new insights and yield new information. To determine whether it is really necessary to use a question at any particular time during a session, ask the question covertly to yourself and see whether you can answer it for the client. If you can, the question is probably unnecessary, and a different response would be more productive.

Steps in the Use of Probes. There are four steps in formulating effective probes. First, determine the purpose of your probe—is it legitimate and therapeutically useful? Often, before probing for information, it is therapeutically useful to demonstrate first that you have heard the client's message. Listening before probing is particularly important when clients reveal strong emotions. It also helps clients to feel understood rather than interrogated. For this reason, before each of our example probes, we use a paraphrase or reflection response. In actual practice, this "bridging" of listening and action responses is very important. Second, depending on the purpose, decide what type of question would be most helpful. Remember that open-ended probes foster client exploration, while closed or focused questions should be reserved for times when you want specific information or you need to narrow the area of discussion. Make sure your question centers on concerns of the client, not issues of interest only to you. Finally, remember to assess the effectiveness of your questioning by determining whether its purpose was achieved. A question is not useful simply because the client answered or responded to it. Additionally, examine how the client responded and the overall explanation, inquiry, and dialogue that ensued as a result of particular questions.

These steps are summarized in the following cognitive learning strategy:

1. What is the purpose of my probe, and is it therapeutically useful?
2. Can I anticipate the client's answer?
3. Given the purpose, how can I start the wording of my probe to be most effective?
4. How will I know whether my probe is effective?

Notice how the counselor applies this cognitive learning strategy in the following example:

Client: I just don't know where to start. My marriage is falling apart. My mom recently died. And I've been having some difficulties at work.

Counselor: 1. What is the purpose of my probe—and is it therapeutically useful?
To get the client to focus more specifically on an issue of most concern to her.

2. Can I anticipate the client's answer?
No.

3. Given the purpose, how can I start the wording of my probe to be most effective?
"Which one of these?"
"Do you want to discuss _____?"

4. How will I know whether my probe is effective?
Examine the client's verbal and nonverbal response and resulting dialogue, as well as whether the purpose was achieved (whether client starts to focus on the specific concern).

Suppose that at this time the counselor's covert visualization or self-talk ends and the following dialogue ensues:

Counselor question: Things must feel overwhelming to you right now. [reflection] Of the three concerns you just mentioned, which one is of most concern to you now? [probe]

Client response: My marriage. I want to keep it together, but I don't think my husband does [accompanied by direct eye contact; body posture, which had been tense, now starts to relax].

From the client's verbal and nonverbal response, the therapist can conclude the question was effective because the client focused on a specific concern and did not appear to be threatened by the question. The therapist can now covertly congratulate herself or himself for formulating an effective question with this client.

Learning Activity #23 gives you an opportunity to try out this cognitive learning strategy in order to develop effective probes.

Confrontation

A confrontation is a verbal response in which the counselor describes discrepancies, conflicts, and mixed messages apparent in the client's feelings, thoughts, and actions. Confrontation has several purposes. One purpose is to help clients explore other ways of perceiving themselves or an issue, leading ultimately to different actions or behaviors. A second and major purpose of confrontation is to help the client become more aware of discrepancies or incongruities in thoughts, feelings, and actions. There are many instances within an interview in which a client says or does something that is inconsistent. For example, a client may say she doesn't want to talk to you because you are a male but then goes ahead and talks to you. In this case, the client's verbal message is inconsistent with her actual behavior. This is an example of an inconsistent, or mixed, message. The purpose of using a confrontation to deal with a mixed message is to describe the discrepancy or contradiction to the client. Often the client is unaware or only vaguely aware of the conflict before the counselor points it out. In describing the discrepancy, you will find it helpful to use a confrontation that presents or connects *both* parts of the discrepancy.

Six major types of mixed messages and accompanying descriptions of counselor confrontations are presented as examples (see also Egan, 1994; Ivey, 1994).

1. *Verbal and nonverbal behavior*

   a. The client says "I feel comfortable" (verbal message) and at the same time is fidgeting and twisting her hands (nonverbal message).
   Counselor confrontation: You say you feel comfortable, and you're also fidgeting and twisting your hands.

   b. Client says "I feel happy about the relationship being over—it's better this way" (verbal message) and is talking in a slow, low-pitched voice (nonverbal message).
   Counselor confrontation: You say you're happy it's over, and at the same time your voice suggests you have some other feelings, too.

2. *Verbal messages and action steps or behaviors*

   a. Client says "I'm going to call her" (verbal message) but reports the next week that he did not make the call (action step).
   Counselor confrontation: You said you would call her, and as of now you haven't done so.

   b. Client says "Counseling is very important to me" (verbal message) but calls off the next two sessions (behavior).
   Counselor confrontation: Several weeks ago you said how important counseling is to you; now I'm also aware that you called off our last two meetings.

3. *Two verbal messages* (stated inconsistencies)

   a. Client says "He's sleeping around with other people. I don't feel bothered [verbal message 1], but I think our

LEARNING ACTIVITY **23** *Probes*

In this learning activity, you are given three client practice statements. For each client message, develop an example of a probe, using the cognitive learning strategy described earlier and outlined below. To internalize this learning strategy, you may wish to talk through these self-questions overtly (aloud) and then covertly (silently to yourself). The end product will be a probe that you can say aloud or write down or both. An example precedes the practice messages. Feedback is at the end of the learning activity on page 126.

Example

Client 1, a middle-aged Latina woman: I just get so nervous. I'm just a bunch of nerves.
Self-question 1: What is the purpose of my probe—and is it therapeutically useful?
  To ask for examples of times when she is nervous. This is therapeutically useful because it contributes to increased understanding of the problem.
Self-question 2: Can I anticipate the client's answer? No.
Self-question 3: Given the purpose, how can I start the wording of my probe to be most effective? "When" or "what."
Actual probes: You say you're feeling pretty upset. [reflection] When do you feel this way? [probe] *or* What are some times when you get this feeling? [probe]

Client Practice Messages

The purpose of the probe is given to you for each message. Try to develop probes that relate to the stated purposes. Remember, too, to precede your probe with a listening response such as paraphrase or reflection.

Client 1, a retired Euro-American woman: To be frank about it, it's been pure hell around my house the last year.
Self-question 1: What is the purpose of my probe? To encourage client to elaborate on how and what has been hell for her.
Self-question 2: Can I anticipate the client's answer?
Self-question 3: Given the purpose, how can I start the wording of my probe to be most effective?
Actual probe(s): _____

_____

Client 2, a 40-year-old physically challenged man: Sometimes I just feel kind of blue. It goes on for a while. Not every day but sometimes.
Self-question 1: What is the purpose of my probe? To find out whether client has noticed anything that makes the "blueness" better.
Self-question 2: Can I anticipate the client's answer?
Self-question 3: Given the purpose, how can I start the wording of my probe to be most effective?
Actual probes(s): _____

_____

Client 3, a 35-year-old African American woman: I just feel overwhelmed right now. Too many kids underfoot. Not enough time for *me.*
Self-question 1: What is the purpose of my probe? To find out how many kids are underfoot and in what capacity client is responsible for them.
Self-question 2: Can I anticipate the client's answer?
Self-question 3: Given the purpose, how can I start the wording of my probe to be most effective?
Actual probe(s): _____

_____

relationship should mean more to him than it does" [verbal message 2].
  Counselor confrontation: First you say you feel OK about his behavior; now you're feeling upset that your relationship is not as important to him as it is to you.
  b. Client says "I really do love little Georgie [verbal message 1], although he often bugs the hell out of me" [verbal message 2].
  Counselor confrontation: You seem to be aware that much of the time you love him, and at other times you feel very irritated toward him, too.
4. *Two nonverbal messages* (apparent inconsistencies)
  a. Client is smiling (nonverbal message 1) and crying (nonverbal message 2) at the same time.

  Counselor confrontation: You're smiling and also crying at the same time.
  b. Client is looking directly at counselor (nonverbal message 1) and has just moved chair back from counselor (nonverbal message 2).
  Counselor confrontation: You're looking at me while you say this, and at the same time, you also just moved away.
5. *Two persons* (counselor/client, parent/child, teacher/student, spouse/spouse, and so on)
  a. Client's husband lost his job two years ago. Client wants to move; husband wants to stick around near his family.
  Counselor confrontation: Edie, you'd like to move. Marshall, you're feeling family ties and want to stick around.

FEEDBACK
Probes

Client 1

Sample probes based on defined purpose: It sounds like things have gotten out of hand. [paraphrase] What exactly has been going on that's been so bad for you? [probe] *or* How has it been like hell for you? [probe]

Client 2

Sample probes based on defined purpose: Now and then you feel kind of down. [reflection] What have you noticed that makes this feeling go away? [probe] *or* Have you noticed anything in particular that makes you feel better? [probe]

Client 3

Sample probes based on defined purpose: With everyone else to take care of, there's not much time left for you. [paraphrase] Exactly how many kids are underfoot? [probe] *or* How many kids are you responsible for? [probe]

---

   b. A woman presents anxiety, depression, and memory loss. You suggest a medical workup to rule out any organic dysfunction, and the client refuses.
   Counselor confrontation: Irene, I feel it's very important for us to have a medical workup so we know what to do that will be most helpful for you. You seem to feel very reluctant to have the workup done. How can we work this out?

6. *Verbal message and context or situation*

   a. A young child deplores her parents' divorce and states that she wants to help her parents get back together.
   Counselor confrontation: Juanita, you say you want to help your parents get back together. At the same time, you had no role in their breakup. How do you put these two things together?

   b. A young married couple have had severe conflicts for the past three years, and still they want to have a baby to improve their marriage.
   Counselor confrontation: The two of you have separated three times since I've been seeing you in therapy. Now you're saying you want to use a child to improve your relationship. Many couples indicate that having a child and being parents increases, rather than relieves, stress. How do you put this together?

Ground Rules for Confronting.   Confrontation needs to be offered in a way that helps clients examine the consequences of their behavior rather than defend their actions (Johnson, 1993). In other words, confrontation must be used carefully in order not to increase the very behavior or pattern that the therapist feels may need to be diminished or modified. The following ground rules may assist you in using this response to help rather than to harm. First, be aware of your own motives for confronting at any particular time. Although the word itself has a punitive or emotionally charged sound, confrontation in the helping process is not an attack on the client or an opportunity to badger the client (Patterson & Welfel, 1994). Confrontation is also not to be used as a way to ventilate or "dump" your frustration onto the client. It is a means of offering constructive, growth-directed feedback that is positive in context and intent, not disapproving or critical (Patterson & Welfel, 1994). To emphasize this, Egan (1994) uses the word *challenge* in lieu of *confront.* To avoid blame, focus on the incongruity as the problem, not on the person, and "allow your nonjudgmental stance to be reflected in your tone of voice and body language" (Ivey, 1994, p. 193). In describing the distortion or discrepancy, the confrontation should cite a *specific example* of the behavior rather than make a vague inference. A poor confrontation might be "You want people to like you, but your personality turns them off." In this case, the counselor is making a general inference about the client's personality and also is implying that the client must undergo a major "overhaul" in order to get along with others. A more helpful confrontation would be "You want people to like you, and at the same time you make frequent remarks about yourself that seem to get in the way and turn people off."

Moreover, before a counselor tries to confront a client, rapport and trust should be established. Confrontation probably should not be used unless you, the counselor, are willing to maintain or increase your involvement in or commitment to the counseling relationship (Johnson, 1993). The primary consideration is to judge what your level of involvement seems to be with each client and adapt accordingly. The stronger the relationship, the more receptive the client may be to a confrontation.

The *timing* of a confrontation is very important. As the purpose is to help the person engage in self-examination, try to offer the confrontation at a time when the client is likely to use it. The perceived ability of the client to act on the confrontation should be a major guideline in deciding when to confront (Johnson, 1993). In other words, before you jump in and confront, determine the person's attention level, anxiety level, desire to change, and ability to listen.

LEARNING ACTIVITY **24** *Mixed Messages*

An important part of developing effective confrontations is learning to identify accurately various client discrepancies and incongruities. In this learning activity, we present four client messages. For each message, use the list below to identify the type of mixed message you observe from the client's message, and identify concrete verbal/nonverbal cues that indicate the discrepancy. An example precedes the practice messages. Feedback is given after the learning activity (on page 128).

Types of mixed messages

1. Verbal and nonverbal behavior
2. Verbal messages and actions or behaviors
3. Two verbal messages
4. Two nonverbal messages
5. Two persons
6. Verbal messages and context

Example

Client: I'm very happy. [Said with lowered, moist eyes, stooped shoulders, impassive facial expression]

1. Identify the type of mixed message: #1—verbal and nonverbal behavior.
2. Identify any cues that are indicative of the mixed message: Client *says* she's happy, but eyes, shoulders, and face suggest sadness.

Client Practice Messages

Client 1, a teenage girl with an obvious limp: I'd like to start ballet or gymnastics. I'd like to be a star gymnast or ballerina.

1. Identify the type of mixed message: _____

2. Identify any cues that are indicative of the mixed message: _____

Client 2, a young man who has been talking very openly with you about some sexual concerns: A big part of the problem is that I can't talk to her about our sex life. I've just never been able to talk about sex.

1. Identify the type of mixed message: _____

2. Identify any cues that are indicative of the mixed message: _____

Client 3: My husband and I are thinking about having a baby. It would be a good way to fill this void we feel; to keep busy with someone who needs you all the time. It will be great as long as it isn't too confining. I don't want to be too tied down.

1. Identify the type of mixed message: _____

2. Identify any cues that are indicative of the mixed message: _____

Client 4: My partner is always interested in sex. She could just wear me out. She wants it two or three times a night. I've told her I think that's a little much, especially for two persons who have been together as long as we have.

1. Identify the type of mixed message: _____

2. Identify any cues that are indicative of the mixed message: _____

Appropriate use of timing also means that the helper does not confront on a "hit and run" basis (Johnson, 1993). Ample time should be given after the confrontation to allow the client to react to and discuss the effects of this response. For this reason, counselors should avoid confronting near the end of a therapy session.

It is also a good idea not to overload the client with confrontations that make heavy demands in a short time. The rule of "successive approximations" suggests that people learn small steps of behaviors gradually more easily than trying to make big changes overnight. Initially, you may want to confront the person with something that can be managed

fairly easily and with some success. Carkhuff (1987) suggests that two successive confrontations may be too intense and should be avoided.

Gender and cultural affiliations of clients also have an impact on the usefulness of the confrontation response. This response may be more suitable for Euro-American male clients, particularly manipulative and acting-out ones (Ivey, 1994). Some Asian, Latino, and Native American clients may view confrontation as "lacking in respect," "a crude and rude form of communication," and "a reflection of insensitivity" (Sue & Sue, 1990, p. 51). For *all* clients, irrespective of gender and racial/cultural affiliations, it is important to use

FEEDBACK
Mixed Messages

Client 1

1. #6—verbal messages and context.
2. Client says she wants to be a star in two sports that require considerable leg work and muscle dexterity; could be challenged by her obvious limp.

Client 2

1. #2—verbal message and behavior.
2. Client says he's never been able to talk about sex but has been talking about it openly with you.

Client 3

1. #3—two verbal messages.
2. Client says she wants to have a child to fill a void by keeping busy and also that she doesn't want to be too tied down or confined by a child.

Client 4

1. #5—two persons.
2. Client's partner wants frequent sex; client does not.

---

this response in such a way that the client views you as an *ally,* not an adversary (Patterson & Welfel, 1994).

Finally, acknowledge the limits of confrontation. Confrontation usually brings about client awareness of a discrepancy or conflict. Awareness of discrepancies is an initial step in resolving conflicts. Confrontation, as a single response, may not always bring about resolution of the discrepancy without additional discussion or intervention strategies such as role playing, role reversal, Gestalt dialoguing, and TA redecision work.

Client Reactions.   Sometimes counselors are afraid to confront because they are uncertain how to handle the client's reactions to the confrontation. Even clients who hear and acknowledge the confrontation may be anxious or upset about the implications. Generally, a counselor can expect four types of client reaction to a confrontation: denial, confusion, false acceptance, or genuine acceptance.

In a denial of the confrontation, the client does not want to acknowledge or agree to the counselor's message. A denial may indicate that the client is not ready or tolerant enough to face the discrepant or distorted behavior. Egan (1994,

pp. 174–175) lists some specific ways the client might deny the confrontation:

1. Discredit the counselor (for example, "How do you know when you don't even have kids?").
2. Persuade the counselor that his or her views are wrong or misinterpreted ("I didn't mean it that way").
3. Devalue the importance of the topic ("This isn't worth all this time anyway").
4. Seek support elsewhere ("I told my friends about your comment last week and none of them had ever noticed that").
5. Agree with the challenger but don't act on the challenge ("I think you're right. I should speak up and tell how I feel but I'm not sure I can do that").

At other times, the client may indicate confusion or uncertainty about the meaning of the confrontation. In some cases, the client may be genuinely confused about what the counselor is saying. This may indicate that your confrontation was not concise and specific. At other times, the client may use a lack of understanding as a smokescreen—that is, as a way to avoid dealing with the impact of the confrontation.

Sometimes the client may seem to accept the confrontation. Acceptance is usually genuine if the client responds with a sincere desire to examine her or his behavior. Eventually such clients may be able to catch their own discrepancies and confront themselves. But Egan (1994) cautions that false acceptance also can occur, which is another client game. In this case, the client verbally agrees with the counselor. However, instead of pursuing the confrontation, the client agrees only to get the counselor to leave well enough alone.

There is no set way of dealing with client reactions to confrontation. However, a general rule of thumb is to go back to the client-oriented listening responses of paraphrase and reflection. A counselor can use these responses to lay the foundation before the confrontation and return to this foundation after the confrontation. The sequence might go something like this:

Counselor: You seem to feel concerned about your parents' divorce. [reflection]
Client: Actually, I feel pretty happy—I'm glad for their sake they got a divorce. [said with low, sad voice—mixed message]
Counselor: You say you're happy, and at the same time, from your voice I sense that you feel unhappy. [confrontation]
Client: I don't know what you're talking about, really. [denial]

Counselor: I feel that what I just said has upset you.
[reflection]

Steps in Confronting.   There are three steps in developing effective confrontations. First, observe the client carefully to identify the type of discrepancy, or mixed message, that the client presents. Note the specific verbal cues, nonverbal cues, and behaviors that support the type of discrepancy. Second, summarize the different elements of the discrepancy. In doing so, use a statement that *connects* the parts of the conflict rather than disputes any one part, as the overall aim of confrontation is to resolve conflicts and to achieve integration. A useful summary is "On the one hand, you _____ , *and* on the other hand, _____ ." Note that the elements are connected with the word *and* rather than *but* or *yet*. Third, remember to assess the effectiveness of your confrontation. A confrontation is effective whenever the client acknowledges the existence of the incongruity or conflict.

To help you formulate a confrontation, consider the following cognitive learning strategy:

1. What discrepancy, or mixed message, do I see, hear, or grasp in this client's communication?
2. How can I summarize the various elements of the discrepancy?
3. How will I know whether my confrontation is effective?

Notice how a therapist uses this cognitive learning strategy for confrontation in the following example:

Client: It's hard for me to discipline my son. I know I'm too indulgent. I know he needs limits. But I just don't give him any. I let him do basically whatever he feels like doing. [Said in low, soft voice]

Counselor:

1. What discrepancy do I see, hear, or grasp in this client's communication?
   A discrepancy between two verbal messages and between verbal cues and behavior: client knows son needs limits but doesn't give him any.
2. How can I summarize the various elements of the discrepancy?
   Client believes limits would help son; at the same time, client doesn't follow through.
3. How will I know whether my confrontation is effective?
   Observe the client's response and see whether he acknowledges the discrepancy.

Suppose that at this point the therapist's self-talk or covert visualization ends and the following dialogue occurs:

Counselor confrontation: William, on the one hand, you feel like having limits would really help your son, and at the same time, he can do whatever he pleases with you. How do you put this together?
Client response: Well, I guess that's right. I do feel strongly he would benefit from having limits. He gets away with a lot. He's going to become very spoiled, I know. But I just can't seem to "put my foot down" or make him do something.

From the client's response, which confirmed the discrepancy, the counselor can conclude that the confrontation was initially useful (further discussion of the discrepancy seems necessary to help the client resolve the conflict between feelings and actions).

Learning Activity #25 gives you an opportunity to apply this cognitive learning strategy to develop the skill of confrontation.

Interpretation

Interpretation is a skill that involves understanding and communicating the meaning of a client's messages. In making interpretive statements, the counselor provides clients with a fresh look at themselves or with another explanation for their attitudes or behaviors (Ivey & Gluckstern, 1984). According to Brammer et al. (1989, p. 175), interpretation involves "presenting the client with a *hypothesis* about *relationships* or *meanings* among his or her behaviors." Johnson (1993) observes that interpretation is useful for clients because it leads to insight, and insight is a key to better psychological living and a precursor to effective behavior change.

Interpretive responses can be defined in a variety of ways (Brammer et al., 1989; Ivey & Gluckstern, 1984). An interpretation may vary to some degree according to your own perspective, your theoretical orientation, and what you decide is causing or contributing to the client's problems and behaviors (Johnson, 1993). We define an interpretation as a counselor statement that makes an association or a causal connection among various client behaviors, events, or ideas or presents a possible explanation of a client's behavior (including the client's feelings, thoughts, and observable actions). An interpretation differs from the listening responses (paraphrase, clarification, reflection, summarization) in that it deals with the *implicit* part of a message—the part the client does not talk about explicitly or directly. As Brammer et al. (1989) note, when interpreting, a counselor will often verbalize issues that the client may have felt only

LEARNING ACTIVITY 25   *Confrontation*

We give you three client practice statements in this learning activity. For each message, develop an example of a confrontation, using the cognitive learning strategy described earlier and outlined below. To internalize this learning strategy, you may wish to talk through these self-questions overtly (aloud) and then covertly (silently to yourself). The end product will be a confrontation that you can say aloud or write down or both. An example precedes the practice messages. Feedback follows the learning activity (on page 132).

Example

Client, a Latino male college student: I'd like to get through medical school with a flourish. I want to be at the top of my class and achieve a lot. All this partying is getting in my way and preventing me from doing my best work.

Counselor:

Self-question 1: What discrepancy do I see, hear, or grasp in this client's communication?
A discrepancy between verbal message and behavior; he says he wants to be at the top of his class and at the same time is doing a lot of partying.

Self-question 2: How can I summarize the various elements of the discrepancy?
He wants to be at the top of his class and at the same time is doing a lot of partying, which is interfering with his goal.

Actual confrontation response: You're saying that you feel like achieving a lot and being at the top of your class and also that you're doing a lot of partying, which appears to be interfering with this goal *or* Eduardo, you're saying that doing well in medical school is very important to you. You have also indicated you are partying instead of studying. How important is being at the top for you?

Client Practice Messages

Client 1, an Asian American graduate student: My wife and child are very important to me. They make me feel it's all worth it. It's just that I know I have to work all the time if I want to make it in my field, and right now I can't be with them as much as I'd like.

Self-question 1: What discrepancy do I see, hear, or grasp in this client's communication?

Self-question 2: How can I summarize the various elements of the discrepancy?

Actual confrontation response: _____

_____

Client 2, a 13-year-old African American girl: Sure, it would be nice to have mom at home when I get there after school. I don't feel lonely. It's just that it would feel so good to have someone close to me there and not to have to spend a couple of hours every day by myself.

Self-question 1: What discrepancy do I see, hear, or grasp in this client's communication?

Self-question 2: How can I summarize the various elements of the discrepancy?

Actual confrontation response: _____

_____

Client 3, a Euro-American high school student: My dad thinks it's terribly important for me to get all A's. He thinks I'm not working up to my potential if I get a B. I told him I'd much rather be well rounded and get a few B's and also have time to talk to my friends and play basketball.

Self-question 1: What discrepancy do I see, hear, or grasp in this client's communication?

Self-question 2: How can I summarize the various elements of the discrepancy?

Actual confrontation response: _____

_____

---

vaguely. Our concept of interpretation is similar to what Egan (1994) calls "advanced accurate empathy," which is a tool to help the client "move beyond the expressed message to partially expressed and implied messages."

There are many benefits and purposes for which interpretation can be used appropriately in a helping interview. First, effective interpretations can contribute to the development of a positive therapeutic relationship by reinforcing client self-disclosure, enhancing the credibility of the therapist, and communicating therapeutic attitudes to the client

(Claiborn, 1982, p. 415). Another purpose of interpretation is to identify causal relations or patterns between clients' explicit and implicit messages and behaviors. A third purpose is to help clients examine their behavior from a different frame of reference or with a different explanation in order to achieve a better understanding of the problem. Another important reason for using interpretation is to motivate the client to replace self-defeating or ineffective behaviors with more functional ones. Finally, almost all interpretations are offered to promote insight.

The frame of reference selected for an interpretation should be consistent with one's preferred theoretical orientation(s) to counseling: a psychodynamic therapist might interpret unresolved anxiety or conflicts; an Adlerian therapist might highlight the client's mistaken logic; a transactional analysis interviewer may interpret client games and ego states; a cognitive therapist might emphasize irrational and rational thinking; and a behavioral counselor may emphasize self-defeating or maladaptive behavior patterns. *Traditional* (or "old guard") client-centered counselors often refrained from interpreting, but recent client-centered therapists do interpret and often emphasize such themes as self-image and intimacy in their interpretations (Egan, 1994). Gestalt therapists consider interpretation a "therapeutic mistake" because it takes responsibility away from the client.

Here is an example that may help you understand the nature of the interpretation response more clearly. Note how the frame of reference or content varies with the counselor's theoretical orientation.

Client 1, a young woman: Everything is humdrum. There's nothing new going on, nothing exciting. All my friends are away. I wish I had some money to do something different.

1. *Interpretation from Adlerian orientation:* It seems as if you believe you need friends, money, and lots of excitement to make your life worthwhile and to feel good about yourself.
2. *Interpretation from TA perspective:* It seems as if you function best only when you can play and have a lot of fun. Your "Child" seems in control of so much of your life.
3. *Interpretation from cognitive or rational-emotive perspective:* It sounds as if you're catastrophizing—because you have no friends around now and no money, things are going to be terrible. Yet where is the proof or data for this? I suspect your feelings of boredom would change if you could draw a different and more logical conclusion about not having friends and money right now.
4. *Interpretation from behavioral orientation:* You seem to be saying that you don't know how to get along or have fun without having other people around. Perhaps recognizing this will help you learn to behave in more self-reliant ways.

In all the above examples, the counselors use interpretation to point out that the client is more dependent on things or other people than on herself for making her life meaningful. In other words, the counselor is describing a possible association, or relationship, between the client's explicit feelings of being bored and her explicit behavior of depending on others to alleviate the boredom. The counselor hopes this explanation will give the client an increased understanding of herself that she can use to create meaning and enjoyment in her life.

What Makes Interpretations Work?   During the last few years, a variety of research studies have explored the interpretive response, although conclusive evidence on this response is limited because of variations in definitions, designs, client differences, timing of the interpretation, and so on (Spiegel & Hill, 1989). Claiborn (1982) has proposed three ways that account for how interpretations work with clients. One is the *relationship* model, which assumes that interpretations work by enhancing the therapeutic relationship. Another is the *content* model, which assumes that the meaning and wording of the interpretive response effect subsequent change. The third is the *discrepancy* model; this assumes that the discrepancy between the counselor's ideas and the client's ideas motivates the client to change. Spiegel and Hill (1989) conclude that all three of these models are relevant in considering the use and impact of an interpretive response, as each model describes "an aspect that is operative in the process of intervention" (p. 123). Stated another way, all three models have some clinical relevance to offer. We describe this further in the next section.

Ground Rules for Interpreting.   From the relationship model, we learn that the overall quality of the therapeutic relationship affects the degree to which an interpretation is likely to be useful to the client. As Spiegel and Hill (1989) observe, "The relationship serves as both a *source* of interpretations and is also enhanced by them" (p. 126). Interpretations need to be offered in the context of a safe and empathic contact with the client. From the content model, the *quality* of the interpretation is as important as the *quantity*. As Spiegel and Hill (1989) note, "More is not always better" (p. 125). There are differences in what various authors have recommended as essential interpretive content; as we noted earlier, much of this depends on the theoretical orientation you use. Spiegel and Hill (1989) recommend organizing all possible types of interpretations into some coherent structure. We like the usefulness of their schema because it is general enough to be applied with varying theoretical orientations. It is summarized in Table 7-3.

Another aspect of the *content* involves making sure your interpretation is based on the client's actual message rather than your own biases and values projected onto the client.

TABLE 7-3. Categories for types of interpretations

1. *Obstacles to change,* in which the focus is on the client's failure to become involved in the treatment process or to work on the therapeutic tasks; in psychodynamic therapy, interpretations about defenses or resistance are often an essential prerequisite to therapeutic work on other conflicts.

2. *Self-awareness,* in which the focus is on how clients perceive self and others as well as how they come across to others; insight in this area may help clients understand their errors in perception that contribute to problems in self-esteem or interpersonal relationships.

3. *Awareness of feelings,* in which the focus is on feelings of which the client is not aware; Singer (1970) discussed the historical shift from Freud's original focus on interpretations as providing rational explanations (why the patient feels something) to interpretations as facilitating more experiential awareness (what the client is feeling).

4. *Clarification of unconscious elements,* in which the focus is on the client's unconscious impulses or conflicts, including transference, which helps the client recognize distortions in current relationships and leads to greater awareness of how these conflicts are expressed in behavior, feeling, and thought.

5. *Stressful life events,* in which the focus is on the client's awareness of the impact of stressful occurrences, which can help clients gain greater awareness of their situation as well as the factors that inhibit acknowledging their importance.

SOURCE: From "Guidelines for Research on Therapist Interpretation: Toward Greater Methodological Rigor and Relevance to Practice," by S. B. Spiegel and C. E. Hill, *Journal of Counseling Psychology, 36,* p. 125. Copyright 1989 by the American Psychological Association. Reprinted by permission.

FEEDBACK 25
Confrontation

Client 1

Question 1: There is a discrepancy between the client's verbal message that his wife and child are very important and his behavior, which suggests he doesn't spend as much time with them.

Question 2: He feels that his family is very valuable, but he also feels he must continually spend more time on his career than with his family.

Examples of confrontation responses: David, on the one hand, you feel your family is very important, and on the other, you feel your work takes priority over them. How do you put this together? *or* You're saying that your family makes things feel worthwhile for you. At the same time you're indicating you must

(continued)

make it in your field in order to feel worthwhile. How do these two things fit for you?

Client 2

Question 1: There is a discrepancy between two verbal messages—one denying she is lonely and the other indicating she doesn't want to be left alone.

Question 2: She doesn't feel lonely—and at the same time she wishes someone could be with her.

Examples of confrontation responses: Denise, you're saying that you don't feel lonely and also that you wish someone like your mom could be home with you. How do you put this together? *or* It seems as though you're trying to accept your mom's absence and at the same time still feeling like you'd rather have her home with you. I wonder if it does feel kind of lonely sometimes?

Client 3

Question 1: There is a discrepancy between two persons' views on this issue—the client and his father. A second discrepancy might be between the client's desire to please his father and to be well-rounded.

Question 2: The client feels getting all A's is not as important as being well rounded, whereas his father values very high grades. Also, for discrepancy #2, the client may want to please both his father and himself.

Examples of confrontation responses: Gary, you're saying that doing a variety of things is more important than getting all A's whereas your father believes that all A's should be your top priority.
or
Gary, you're saying you value variety and balance in your life; your father believes high grades come first.
or
Gary, you want to please your father and make good grades and, at the same time, you want to spend time according to your priorities and values.
(Note: Do not attempt to confront both discrepancies at once!)

This requires that you be aware of your own blind spots. As an example, if you have had a bad experience with marriage and are biased against people's getting or staying married, be aware of how this could affect the way you interpret client statements about marriage. If you aren't careful with your values, you could easily advise all marital-counseling clients away from marriage, a bias that might not be in the best interests of some of them. Try to be aware of whether you are interpreting to present helpful data to the client or only to show off your expertise. Make sure your interpretation is based on sufficient data, and offer the interpretation in a collaborative spirit, making the client an active participant.

A third aspect of the content concerns the way the counselor phrases the statement and offers it to the client. Although preliminary research suggests there is no difference between interpretations offered with absolute and with tentative phrasing (Milne & Dowd, 1983), we believe that in most cases the interpretation should be phrased tentatively, using phrases such as "perhaps," "I wonder whether," "it's possible that," or "it appears as though." Using tentative rather than absolute phrasing helps to avoid putting the counselor in a one-up position and engendering client resistance or defensiveness to the interpretation. The work of Jones and Gelso (1988) supports the use of tentative interpretations (versus absolute or questioning interpretations), at least in the early stages of counseling. After an interpretation, check out the accuracy of your interpretive response by asking the client whether your message fits. Returning to a clarification is always a useful way to determine whether you have interpreted the message accurately.

Finally, the content of an interpretation must also be congruent with the client's cultural affiliations. Because many of our counseling theories are based on Eurocentric assumptions, this can be a thought-provoking task. It is most important not to assume that simply because an interpretation makes sense to you, it will make the same sort of sense to a client whose racial, ethnic, and cultural background varies from your own.

The *discrepancy* model involves the *depth* of the interpretation you offer to the client. Depth is the degree of discrepancy between the viewpoint expressed by the counselor and the client's beliefs. Presenting clients with a viewpoint discrepant from their own is believed to facilitate change by providing clients with a reconceptualization of the problem (Claiborn, Ward, & Strong, 1981). An important question is to what extent the counselor's communicated conceptualization of the problem should differ from the client's beliefs. A study by Claiborn et al. (1981) addressed this concern. The results supported the general assumption that highly discrepant (that is, very deep) interpretations are more likely to be rejected by the client, possibly because they are unacceptable, seem too preposterous, or evoke resistance. In contrast, interpretations that are either congruent with or only slightly discrepant from the client's viewpoint are most likely to facilitate change, possibly because these are "more immediately understandable and useful to the clients" (Claiborn et al., 1981, p. 108; Claiborn & Dowd, 1985).

The depth of the interpretation also has some impact on the time at which an interpretation is offered—both within a

session and within the overall content of treatment. The client should show some degree of readiness to explore or examine himself or herself before you use an interpretation. Generally, an interpretation response is reserved for later, rather than initial, sessions because some data must be gathered as a basis for an interpretive response and because the typical client requires several sessions to become accustomed to the type of material discussed in counseling. The client may be more receptive to your interpretation if she or he is comfortable with the topics being explored and shows some readiness to accept the interpretive response. As Brammer et al. (1989, p. 180) note, a counselor usually does not interpret until the time when the client can almost formulate the interpretation for herself or himself.

Timing of an interpretation within a session is also important. If the counselor suspects that the interpretation may produce anxiety or resistance or break the client's "emotional dam," it may be a good idea to postpone it until the beginning of the next session (Brammer et al., 1989).

Client Reactions to Interpretation. Client reactions to interpretation may range from expression of greater self-understanding and release of emotions to less verbal expression and more silence.

Generally, the research on interpretation has not *systematically* explored differential client reactions. Research conducted in this area has yielded varying results (Spiegel & Hill, 1989). Based on these studies, Spiegel and Hill (1989) have speculated that an individual client's receptivity to an interpretation has to do with the "client's self-esteem, severity of disturbance, level of cognitive complexity, and psychological mindedness" (p. 1240). To their list, we would also add the client's cultural affiliation as an important moderating variable.

Although the concept of promoting insight, a goal of interpretation, is compatible for some Euro-American clients, for other clients, insight may not be so valued. As Sue and Sue (1990) note, "When survival on a day-to-day basis is important, it seems inappropriate for the counselor to use insightful processes" (p. 38). Also, some cultural groups simply do not feel the need to engage in contemplative reflection. Indeed, the very notion of thinking about oneself or one's problems too much is inconsistent for some clients who may have been taught not to dwell on themselves and their thoughts. As a character in Waller's (1995) book, *Border Music,* noted, too much reflection can do "violence to the soul" and "even worse things to the heart" (p. 26). Other clients may have learned to gain insight in a solitary manner, as in a "vision quest," rather than with another person such as a counselor. In actual practice, you can try to assess the client's receptivity by using a trial interpretation, bearing in mind that the client's initial reaction may change over time (Strupp & Binder, 1984).

If interpretation is met initially with defensiveness or hostility, it may be best to drop the issue temporarily and introduce it again later. Repetition is an important concept in the use of interpretations. As Brammer et al. observe, "Since a useful and valid interpretation may be resisted, it may be necessary for the counselor to repeat the interpretation at appropriate times, in different forms, and with additional supporting evidence" (1989, p. 182). However, don't push an interpretation on a resistant client without first reexamining the accuracy of your response (Brammer et al., 1989).

Steps in Interpreting.   There are five steps in formulating effective interpretations. First, listen for and identify the *implicit* meaning of the client's communication—what the client conveys subtly and indirectly. Second, formulate an interpretation that provides the client with a *slightly* different way to view the problem or issue. This alternative frame of reference should be consistent with your theoretical orientation and not too discrepant from the client's beliefs. Third, make sure your view of the issue, your frame of reference, is relevant to the client's cultural background, keeping in mind some of the precautions we addressed earlier. Next, select words in the interpretation that match the client's sensory words, or predicates. Finally, examine the effectiveness of your interpretation by assessing the client's reaction to it. Look for nonverbal "recognition" signs such as a smile or contemplative look as well as verbal and behavioral cues that indicate the client is considering the issue from a different frame of reference or that the client may not understand or agree with you.

To help you formulate an effective interpretation and assess its usefulness, consider the following cognitive learning strategy:

1. What is the implicit part of the client's message?
2. What is a slightly different way to view this problem or issue that is consistent with the theoretical orientation I am using with this client? (You can also refer to Table 7-3.)
3. Is my view of this issue culturally relevant for this client?
4. How will I know whether my interpretation is useful?

Notice how a therapist applies this cognitive learning strategy in the following example:

Client, Euro-American woman: I really don't understand it myself. I can always have good sex whenever we're not at home—even in the car. But at home it's never too good.

Counselor: 1. What is the implicit part of the client's message? That sex is not good or fulfilling unless it occurs in special, out-of-the-ordinary circumstances or places.
2. What is a slightly different way to view this problem or issue consistent with the theoretical orientation I am using with this client?
Behavioral: Adaptive response learned in special places hasn't generalized to home setting, perhaps because of different "setting events" there (for example, fatigue, lack of novelty).
Cognitive: Client is catastrophizing about "no good" sex when it occurs at home; continued reindoctrination prevents spontaneity, increases preoccupation with herself.
Psychodynamic: Problem with sex at home is indicative of unresolved conflict.
Transactional analysis: Sex is good in places that are novel and exciting, where the "Child" ego state can be predominant; not as good at home, where Child ego state is probably excluded.
Adlerian: Client has developed a lifestyle and accompanying mistaken logic that emphasizes novelty, excitement, out-of-the-ordinary events in order for things (like sex) to be good or great.
3. Is my view of this issue culturally relevant for this client? This client seems relatively comfortable in talking about and disclosing information about her sexual feelings and behaviors so almost any of the above interpretive responses would seem to fit for her. However, be careful not to make any assumptions about the client's sexual orientation. At this point we do not know whether this person is lesbian, bisexual, or straight.

Suppose that at this point the counselor's covert visualization or self-talk ends and the following dialogue ensues:

Counselor interpretation: Ann, I might be wrong about this—it seems that you get psyched up for sex only when it occurs in out-of-the-ordinary places where you feel there's a lot of novelty and excitement. Is that possible? [Note: This is an interpretation that could be consistent with a behavioral, Adlerian, or eclectic orientation.]
Client: [Lips part, slight smile, eyes widen] Well, I never thought about it quite that way. I guess I do need to feel like there are some thrills around when I do have sex—maybe it's that I find unusual places like the elevator a challenge.

At this point, the counselor can conclude that the interpretation was effective because of the client's nonverbal "recognition" behavior and because of the client's verbal response suggesting the interpretation was on target. The therapist might continue to help the client explore whether she needs thrills and challenge to function satisfactorily in other areas of her life as well.

Learning Activity #26 gives you an opportunity to try out the interpretation response.

Information Giving

There are many times in the counseling interview when a client may have a legitimate need for information. For instance, a client who reports being abused by her husband may need information about her legal rights and alternatives. A client who has recently become physically challenged may need some information about employment and about lifestyle adaptations such as carrying out domestic chores or engaging in sexual relationships. According to Selby and Calhoun (1980, p. 236), "Conveying information about the psychological and social changes accompanying a particular problem situation (. . . such as divorce) may be a highly effective addition to any therapeutic strategy." Note that information giving is an important tool of feminist therapy approaches. Feminist therapists, for example, may give information to clients about gender role stereotyping, the impact of cultural conditioning on gender roles, strategies for empowerment, and social/political structures that contribute to disempowerment (Valentich, 1986). Recent research has found that counselors tend to provide information about the counseling process and about facts concerning client behaviors, as well, when conducting consultation sessions (Lin, Kelly & Nelson, 1996).

We define information giving as the verbal communication of data or facts about experiences, events, alternatives, or people. As summarized in Table 7-1, there are four intended purposes of information giving in counseling. First, information is necessary when clients do not know their options. Giving information is a way to help them identify possible alternatives. For example, you may be counseling a pregnant client who says she is going to get an abortion because it is her only choice. Although she may eventually decide to pursue this choice, she should be aware of other options before making a final decision. Information giving is also helpful when clients are not aware of the possible outcomes of a particular choice or plan of action. Giving information can help them evaluate different choices and actions. For example, if the client is a minor and is not aware that she may need her parents' consent for an abortion, this information may influence her choice. In the preceding kinds of situations, information is given to counteract ignorance. Information giving can also be useful to correct invalid or unreliable data or to dispel a myth. In other words, information giving may be necessary when clients are misinformed about something. For example, a pregnant client may decide to have an abortion on the erroneous assumption that an abortion is also a means of subsequent birth control.

A final purpose of information giving is to help clients examine issues or problems they have been successfully avoiding (Egan, 1994). For example, a client who hasn't felt physically well for a year may be prompted to explore this problem when confronted with information about possible effects of neglected treatment for various illnesses.

Differences Between Information Giving and Advice. It is important to note that information giving differs from advice. In giving advice, a person usually recommends or prescribes a particular solution or course of action for the listener to follow. In contrast, information giving consists in presenting relevant information about the issue or problem, and the decision concerning the final course of action—if any—is made by the client. Consider the differences between the following two responses:

Client, a young mother: I just find it so difficult to refuse requests made by my child—to say no to her—even when I know they are unreasonable requests or could even be dangerous to her.

Counselor (advice giving): Why don't you start by saying no to her just on one request a day for now—anything that you feel comfortable with refusing—and then see what happens?

Counselor (information giving): I think there are two things we could discuss that may be affecting the way you are handling this situation. First, we could talk about what you feel might happen if you say no. We also need to examine how your requests were handled in your own family when you were a child. Very often as parents we repeat with our children the way we were parented—in such an automatic way we don't even realize it's happening.

In the first example, the counselor has recommended action that may or may not be successful. If it works, the client may feel elated and expect the counselor to have other magical solutions. If it fails, the client may feel even more discouraged and question whether counseling can really help her resolve this problem. Appropriate and effective information

LEARNING ACTIVITY 26   *Interpretation*

Three client practice statements are given in this learning activity. For each message, develop an example of an interpretation, using the cognitive learning strategy described earlier and outlined below. To internalize this learning strategy, you may want to talk through these self-questions overtly (aloud) and then covertly (silently to yourself). The end product will be an interpretation that you can say aloud or write down or both. An example precedes the practice messages. Feedback follows the learning activity (on page 138).

Example

Client, a young Asian American woman: I don't know what to do. I guess I just never thought I'd ever be asked to be a supervisor. I feel so content just being a part of the group I work with.

Self-question 1: What is the implicit part of the client's message?

That the client feels afraid to achieve more than she's presently doing.

Self-question 2: What is a slightly different way to view this problem or issue consistent with the theoretical orientation I am using with this client?

Behavioral—Client has acquired anxiety about job success, possibly because of lack of exposure to other successful female models and also because of lack of reinforcement from significant others for job achievement. It is also possible that she has been punished for being successful or achieving in the past, an experience that is maintaining her present avoidance behavior on this issue.

Cognitive—Client is indoctrinating herself with irrational or self-defeating thoughts of possible failure or loss of friends if she moves up the job ladder.

Psychodynamic—Client has anxiety about a job situation; suggests possible unresolved conflict and also identity issue.

Transactional analysis—Client is confronted with a decision that her Adapted Child ego state is afraid to make, possibly because in the role of supervisor she would have to function in more of her Parent ego state, which is probably excluded from her personality to some degree.

Adlerian—Client's lifestyle and family background haven't accommodated personal achievement; possibly client has relied on others, such as

husband, to take care of her rather than putting herself in roles where she is responsible for others.

Self-question 3: Is my view of this issue culturally relevant for this client?

The above theoretical interpretations do not take into account that with her Asian-American background, the client may feel more comfortable working in and for a collective group of people.

Actual interpretation response: Despite your obvious success on the job, you seem to be reluctant to move up to a position that requires you to work by yourself. There could be different reasons for this. I'm wondering whether you're responding in part to your cultural background, which stresses belonging to a group and working for the good of the group rather than promoting yourself.

Client Practice Statements

Client 1, a young, Native American woman: I can't stand to be touched any more by a man. And after I was raped, they wanted me to go see a doctor in this hospital. When I wouldn't, they thought I was crazy. I hope you don't think I'm crazy for that.

Self-question 1: What is the implicit part of the client's message?

Self-question 2: What is a slightly different way to view this issue consistent with my theoretical orientation?*

Self-question 3: Is my view of this issue culturally relevant for this client?

Actual interpretation response: _____

_____

_____

Client 2, a 50-year-old Jordanian man: Sure, I seemed upset when I got laid off several years ago. After all, I'd been an industrial engineer for almost 23 years. But I can support my family with my job supervising these custodial workers. So I should be very thankful. Then why do I seem down?

Self-question 1: What is the implicit part of this message?

Self-question 2: What is a slightly different way to view this problem or issue consistent with my theoretical orientation?*

Self-question 3: Is my view of this issue culturally relevant for this client?

Actual interpretation response: _____

_____

*If you have not yet had much exposure to counseling theories, you may find it useful just to reframe the client statement in a slightly different manner without trying to conceptualize it according to a counseling theory; refer also to Table 7-3.

(continued)

LEARNING ACTIVITY 26 *Continued*

Client 3, a young Euro-American man: I have a great time with Susie [his girlfriend], but I've told her I don't want to settle down. She's always so bossy and tries to tell me what to do. She always decides what we're going to do and when and where and so on. I get real upset at her.

Self-question 1: What is the implicit part of the client's message?

Self-question 2: What is a slightly different way to view this problem or issue consistent with my theoretical orientation?

Self-question 3: Is my view of this issue culturally relevant for this client?

Actual interpretation response: _____

_____

TABLE 7-4. The "when," "what," and "how" of information giving in helping

When—recognizing client's need for information	What—identifying type of information	How—delivery of information in interview
1. Identify information presently available to client.	1. Identify kind of information useful to client.	1. Avoid jargon.
2. Evaluate client's present information—is it valid? data-based? sufficient?	2. Identify reliable sources of information to validate accuracy of information.	2. Present all the relevant facts; don't protect client from negative information.
3. Wait for client cues of readiness to avoid giving information prematurely.	3. Identify any sequencing of information (option A before option B).	3. Limit amount of information given at one time; don't overload.
	4. Identify cultural relevance of information.	4. Ask for and discuss client's feelings and biases about information.
		5. Know when to stop giving information so action isn't avoided.
		6. Use paper and pencil to highlight key ideas or facts.

giving is presented as what the client *could* ponder or do, not what the client *should* do, and what the client *might* consider, not *must* consider.

Several dangers are associated with advice giving that make it a potential trap for counselors. First, the client may reject not only this piece of advice but any other ideas presented by the therapist in an effort to establish independence and thwart any conspicuous efforts by the counselor to influence or coerce. Second, if the client accepts the advice and the advice leads to an unsatisfactory action, the client is likely to blame the counselor and may terminate therapy prematurely. Third, if the client follows the advice and is pleased with the action, the client may become overly dependent on the counselor and expect, if not demand, more "advice" in subsequent sessions. Finally, there is always the possibility that an occasional client may misinterpret the advice and may cause injury to himself or herself or others in trying to comply with it.

Ground Rules For Giving Information.   Information giving is generally considered appropriate when the need for information is directly related to the client's concerns and goals and when the presentation and discussion of information are used to help the client achieve these goals.

To use information giving appropriately, a counselor should consider three major guidelines. These cover when to give information, what information is needed, and how the information should be delivered. Table 7-4 summarizes the "when," "what," and "how" guidelines for information giving in counseling. The first guideline, the "when," involves recognizing the client's need for information. If the client does not have all the data or has invalid data, a need exists.

To be effective, information must also be well timed. The client should indicate receptivity to the information before it is delivered. A client may ignore information if it is introduced too early in the interaction.

FEEDBACK
Interpretation

Client 1

Question 1: The implicit part of this message is that she doesn't see herself as crazy for reacting this way.

Question 2: Following are different theoretical ways to view this issue:

Behavioral: Following the client's exposure to being raped, she has developed generalized anxiety about being touched by all men.

Cognitive: She is making negative self-statements in the form of distortions about men in general and their intentions.

Psychodynamic: Her anxiety around men is symbolic of some unresolved earlier issue, triggered by the rape.

Transactional analysis: Her adapted Child ego state is in control, dominating her personality with feelings of anxiety.

Adlerian: Her feelings about men reflect not only the intolerable external trauma of having been raped but also her learned interpersonal "scripts" or beliefs in the form of overgeneralizations about men.

Question 3: Not only has the rape affected her feelings about men and seeking help from a doctor but, given her cultural background, she may feel more comfortable seeking help from a traditional healer rather than a physician.

Interpretation example: I'm guessing that not only has the rape affected your trust of other men—even doctors—but also your cultural background is having some effect, too. Could it be that you would respond differently if you could go to a traditional healer instead?

Client 2

Question 1: Implicit part of this client's message is that he's sort of an independent, self-made man whose independence, masculinity, and pride have been damaged by being laid off and by doing "menial" work.

Question 2: Alternative ways to view this issue, depending on your theoretical orientation, include the following:

Behavioral: Depression was precipitated and is maintained by loss of powerful positive, or reinforcing, contingencies (job loss, lack of satisfying current employment).

Cognitive: Depression is maintained by client's negative and irrational thoughts or self-statements about

his lack of human worth due to job loss and subsequent type of employment.

Psychodynamic: Depression is the result of his loss of masculinity.

Transactional analysis: Depressed feelings are the result of the adapted Child ego state, which probably became a greater part of his personality after the job loss, whereas other ego states formerly functional, such as nurturing parent, are being somewhat excluded now from his personality.

Adlerian: Client has developed lifestyle and image of self in which he is independent, intelligent, strong, and masculine. Self-image is now challenged by job loss and menial work, resulting in depression.

Question 3: All the above views of his issue are to some degree affected by what he has learned growing up in Jordan about the role of the male and specifically the husband and father.

Interpretation example: It sounds as though when you lost your job as an engineer, you also lost some parts of the role you have learned from your culture about being a man, a husband, and a father. Does that seem accurate?

Client 3

Question 1: Implicit part of this message is that client doesn't want to settle down with a controlling woman; he's also upset with himself for giving her so much control.

Question 2: Different ways to view this issue, depending on your theoretical orientation:

Behavioral: Client uses a relationship with a controlling female to maintain avoidance behavior against settling down.

Cognitive: Client allows himself to get upset with Susie because of all the negative things he tells himself about her and her need to control his life.

Psychodynamic: Client's need to be controlled by a domineering female suggests unresolved oedipal issue.

Transactional analysis: Client is both attracted to and repelled by Susie; he is hooked because he relates to her out of his Child ego state, which hooks or maintains her Parent ego state. At the same time, he dislikes feeling so "trapped" or controlled by her. He needs to find ways of relating to her from other ego states.

Adlerian: Client has developed lifestyle in which he feels adequate by giving up control to others. He may also have been parented by a controlling mother who made all his decisions for him.

Question 3: Any of the above theoretical orientations, which are more Eurocentric, would seem to fit for this particular client.

(continued)

(continued)

FEEDBACK #26   (continued)

Examples of interpretations:

1. You say you have great times with Susie although you told her you don't want to settle down. I may be wrong—I'm wondering whether you've got yourself hooked up with a controlling woman as a way to avoid settling down. [Consistent with behavioral framework.]

2. You seem to be both attracted to and turned off by Susie. I wonder what your relationship with your mother was like and whether you see any similarities between that relationship and the present one? [Consistent with psychodynamic and Adlerian frameworks.]

3. You seem to be blaming Susie for being so controlling and nurturing. Perhaps you haven't thought about how what you say to her might encourage her to be this way. Could it be that you relate to her as a child might relate to a parent? [Consistent with transactional analysis framework.]

The counselor also needs to determine what information is useful and relevant to clients. Generally, information is useful if it is something clients are not likely to find on their own and if they have the resources to act on the information. The counselor also needs to determine whether the information must be presented sequentially in order to make most sense to the client. Because clients may remember initial information best, presenting the most significant information *first* may be a good rule of thumb in sequencing information. Finally, in selecting information to give, be careful not to impose information on clients, who are ultimately responsible for deciding what information to use and act on. In other words, information giving should not be used as a forum for the counselor to subtly push his or her own values on clients (Egan, 1994).

One of the critical facets of giving information has to do with the cultural appropriateness of the information being given. As Lum (1996) observes, "Much cross-cultural contact involves communicating with people who do not share the same types of information" (p. 122). Also, some research suggests that people in different cultures vary in the types of information they attend to (Basic Behavioral Science Task Force of the National Advisory Mental Health Council, 1996). In the United States, counselors working with non-Euro-American clients too easily and too frequently provide them with information that is based on Eurocentric notions.

In an earlier chapter, we commented on the different ways in which grief and loss are handled across varying cultures. We also noted the role of certain indigenous healing rituals and healers in various cultures. For example, providing information to a sick client about the traditional physician and medical care setting in the United States may be useful if the client is Euro-American. However, for many non-Euro-American clients, such information may be so removed from their own cultural practices regarding health and illness that the information is simply not useful to them.

Other mismatches between Eurocentric and non-Eurocentric information abound in family therapy. Enmeshment—the concept of a family system lacking clear boundaries between and among individuals—is a prime example. In the United States, for some Euro-American families, enmeshment is considered a sign of pathology because in enmeshed families, the autonomy of individual members is considered hampered. However, for many Asian families and also for some rural Euro-American families, enmeshment is so completely the norm that any other structure of family living is foreign to them; in many of these families, the prevailing culture dictates that the good of the family comes before the individual members' needs and wishes (Berg & Jaya, 1993). Boundaries in these families are blurred in ways that do not usually occur in some Euro-American families. For example, young Asian children may be carried on their mothers' backs until they are three or four years of age; toilet training also occurs later, and often there is a practice of "co-sleeping" in which the children sleep in the same room as the parents (Itai & McRae, 1994). If the counselor assumes this behavior is pathological and gives the client information about becoming more "individuated" or "establishing clearer boundaries," the client may feel misunderstood and also greatly offended. Therefore, a very important question to be addressed in effective information giving is this: What cultural biases are reflected in the information I will give the client, based on the client's ethnic, racial, and cultural affiliations, and is this information culturally relevant and appropriate? If you are not careful to assess the assumptions reflected in the information you share with clients, not only may your information seem irrelevant but your credibility in the client's eyes may also be diminished.

In the interview itself, the actual delivery of information, the "how" of information giving, is crucial. The information should be discussed in a way that makes it usable to the client and encourages the client to hear and apply the information (Gazda et al., 1995). Moreover, information should be presented objectively. Don't leave out facts simply because they aren't pleasant. Watch out, too, for information overload. Most people cannot assimilate a great deal of information at one time. Usually, the more information you give the clients, the less they remember. Clients recall information best when you give no more than several pieces at one time.

Be aware that information differs in depth and may have an emotional impact on clients. Clients may not react emotionally to relatively simple or factual information such as information about a counseling procedure or an occupation or a résumé. However, clients may react with anger, anxiety, or relief to information that has more depth or far-reaching consequences, such as the results of a test. Ask about and discuss the client's reactions to the information you give. In addition, make an effort to promote client understanding of the information. Avoid jargon in offering explanations. Use paper and pencil as you're giving information to draw a picture or diagram highlighting the most important points, or give clients paper and pencil so they can write down key ideas. Remember to ask clients to verify their impression of your information either by summarizing it or by repeating it to you. Try to determine, too, when it's time to stop dealing with information. Continued information giving may reinforce a client's tendency to avoid taking action.

Steps in Information Giving.   There are six steps in formulating the what, when, and how of presenting information to clients. First, assess what information the client lacks about the issue or problem. Second, determine the cultural relevance of information you plan to share. Third, decide how the information can be sequenced in a way that facilitates client comprehension and retention. Fourth, consider how you can deliver the information in such a way that the client is likely to comprehend it. Keep in mind that in cross-cultural counseling situations, effective delivery requires you to communicate in language and style that the client can understand. Fifth, assess the emotional impact the information is likely to have on the client. Finally, determine whether your information giving was effective. Note client reactions to it and follow up on client use of the information in a subsequent session. Remember, too, that some clients may "store" information and act on it at a much later date—often even after therapy has terminated.

To facilitate your use of information giving, we have put these six steps in the form of questions that you can use as a cognitive learning strategy:

1. What information does this client lack about the problem or issue?
2. Based on the client's ethnic, racial, and cultural affiliations, is this information relevant and appropriate?
3. How can I best sequence this information?
4. How can I deliver this information so that the client is likely to comprehend it?
5. What emotional impact is this information likely to have on this client?
6. How will I know whether my information giving has been effective?

Consider the way a therapist uses this cognitive learning strategy in the first example of Learning Activity #27.

LEARNING ACTIVITY 27   *Information Giving*

In this learning activity, three client situations are presented. For each situation, determine what information the client lacks and develop a suitable information-giving response, using the cognitive learning strategy described earlier and outlined below. To internalize this learning strategy, you may want to talk through these self-questions overtly (aloud) and then covertly (silently to yourself). The end product will be an information-giving response that you can say aloud or write down or both. An example precedes the practice situations. Feedback follows the learning activity (on pages 148–149).

Example

The clients are a married couple in their 30s: Gus is a Euro-American man and his wife, Assani, is an Asian American woman. They disagree about the way to handle their four-year-old son. The father believes the boy is a "spoiled brat" and thinks the best way to keep him in line is to give him a spanking. The mother believes that her son is just a "typical boy" and the best way to handle him is to be understanding and loving. The couple admit that there is little consistency in the way the two of them deal with their son. The typical pattern is for the father to reprimand him and swat him while the mother stands, watches, comforts him, and often intercedes on the child's behalf.

Self-question 1: What information do these clients lack about this problem or issue?
Information about effective parenting and child-rearing skills.

(continued)

LEARNING ACTIVITY 27 *Continued*

Self-question 2: Based on the clients' ethnic, racial, and cultural affiliations, is this information relevant and appropriate?

I have to recognize that there are probably different cultural values brought to this parenting situation by the mother and dad. I'm going to have to find information that is appropriate to both value systems, such as the following:

1. All children need some limits at some times.
2. There is a hierarchy in parent/child relationships; children are taught to respect parents and vice versa.
3. Children function better when their parents work together on their behalf rather than disagreeing all the time, especially in front of the child.

Self-question 3: How can I best sequence this information?

Discuss #3 first—working together on the child's behalf—and note how each parent's approach reflects his or her own cultural background. Stress that neither approach is right or wrong, but that the approaches are different. Stress points of common agreement.

Self-question 4: How can I deliver this information so that the clients are likely to comprehend it?

Present the information in such a way that it appeals to the values of both parents. The mother values understanding, support, and nurturing whereas the father values authority, respect, and control.

Self-question 5: What emotional impact is this information likely to have on these clients?

If I frame the information positively, it will appeal to both parents. I have to be careful not to take sides or cause one parent to feel relieved while the other feels anxious, guilty, or put down.

Self-question 6: How will I know whether my information giving has been effective?

I'll watch and listen to their nonverbal and verbal reactions to it to see whether they support the idea and also follow up on their use of the information in a later session.

Example of information-giving response: You know, Assani and Gus, I sense that you are in agreement on the fact that you love your child and want what is best for him. So what I'm going to say next is based on this idea that you are both trying to find a way to do what is best for Timmy. In discussing how you feel about Timmy and his behavior—and this is most important—remember that Timmy will do better if you two can find a way to agree on parenting. I think part of your struggle is that you come from cultures where parenting is viewed in different ways. Perhaps we could talk first about these differences and then find areas where you can easily agree.

Client Practice Situations

Client 1 is a young Native American man who has had his driver's license taken away because of several arrests for driving under the influence of alcohol. He is irate because he doesn't believe drinking a six-pack of beer can interfere with his driving ability. After all, as he says, he has never had an accident. Moreover, he has seen many of his male relatives drive drunk for years without any problem. He believes that losing his license is just another instance of the White man's trying to take away something that justifiably belongs to him.

Self-question 1: What information does this client lack about the problem or issue?

Self-question 2: Based on the client's ethnic, racial, and cultural affiliations, is this information relevant and appropriate?

Self-question 3: How can I best sequence the information?

Self-question 4: How can I deliver the information so the client is likely to comprehend it?

Self-question 5: What emotional impact is the information likely to have on this client? Give an example of an information-giving response.

Client 2 is an African American male who has been ordered by the court to come in for treatment of heroin addiction. At one point in your treatment group, he talks about his drug use with several of his sexual partners. When you mention something about the risk of AIDS, his response is that it could never happen to him.

Self-question 1: What information does this client lack about the problem or issue?

Self-question 2: Based on the client's ethnic, racial, and cultural affiliations, is this information relevant and appropriate?

Self-question 3: How can I best sequence this information?

Self-question 4: How can I deliver the information so that the client is likely to comprehend it?

Self-question 5: What emotional impact is the information likely to have on this client?

(continued)

LEARNING ACTIVITY **27** *Continued*

Client 3 is a 35-year-old Euro-American woman with two teenage daughters. She is employed as an executive secretary in a large engineering firm. Her husband is a department store manager. She and her husband have had a stormy relationship for several years. She wants to get a divorce but is hesitant to do so for fear that she will be labeled a troublemaker and will lose her job. She is also afraid that she will not be able to support her daughters financially on her limited income. However, she indicates that she believes getting a divorce will make her happy and will essentially solve all her own internal conflicts.

Self-question 1: What information does this client lack about this problem or issue?
Self-question 2: Based on the client's ethnic, racial, and cultural affiliations, is this information relevant and appropriate?
Self-question 3: How can I best sequence the information?
Self-question 4: How can I deliver this information so that the client is likely to comprehend it?
Self-question 5: What emotional impact is the information likely to have on this client?

SUMMARY

Listening responses reflect clients' perceptions of their world. Action responses provide alternative ways for clients to view themselves and their world. A change in the client's way of viewing and explaining things may be one indication of positive movement in counseling. According to Egan (1994), counselor statements that move beyond the client's frame of reference are a "bridge" between listening responses and concrete change programs. To be used effectively, action responses require a great deal of counselor concern and judgment.

In the last two chapters, we have described these two different sorts of counselor communication styles. Part of the decision about the timing of these responses involves the counselor's awareness of the client's cultural affiliations. As Sue and Sue (1990) note, it is important for counselors to be able to shift their communication style to meet the unique cultural dimensions of every client.

SKILL INTEGRATION

In Chapters 3 and 4 you learned about important relationship conditions and variables such as empathy, genuineness, positive regard, competence, trustworthiness, and interpersonal attractiveness. In Chapter 5 you discovered valuable reasons for attending to and working with client nonverbal behavior as well as important aspects of your own nonverbal behavior, including kinesics, paralinguistics, proxemics, environment, and time variables. Synchrony, or matching, between counselor and client nonverbal behavior and congruence between your verbal and nonverbal messages were also emphasized. In Chapters 6 and 7 you acquired a base of various verbal responses to use in counseling interactions to facilitate client exploration, understanding, and action. These responses included clarification, paraphrase, reflection, summarization, probes, confrontation, interpretation, and information giving. And all of this has occurred within a context of cultural pluralism. You have learned, for example, that the impact of the facilitative conditions and the social influence factors will vary with clients from differing cultural/ethnic groups. Similarly, we have pointed out that various aspects of nonverbal behavior are culture bound and that the effectiveness of verbal communication styles also depends somewhat on the cultural affiliation of both counselor and client. You have also had various types of practice in which you have demonstrated each set of skills in role-play interactions. In actual counseling, these skills are blended together and used in a complementary fashion. In Part Four of the postevaluation, we structure a practice opportunity that simulates an actual initial helping interview with a client. The purpose of this activity is to help you put the skills together—that is, integrate them for yourself in some meaningful, coherent fashion. It is analogous to learning anything else that requires a set of skills for successful performance. To swim, for example, you have to learn first to put your face in the water, then to float, then to kick, then to move your arms in strokes, and finally to do it all at once. Initial attempts feel awkward, but out of such first steps evolve championship swimmers.

POSTEVALUATION

Part One

This part is designed to help you assess your performance on Objective One. Using the written list of client statements and counselor responses below, identify and write the type of action response—probe, confrontation, interpretation, or information giving—reflected in each counselor message. If you can accurately identify six out of eight responses, you have met this objective. You can check your answers against those provided in the feedback that follows the postevaluation.

1. *Client:* "The pressure from my job is starting to get to me. I'm always in a constant rush, trying to hurry and get several things done at the same time. There's never enough time."
   a. "What is it about your job that is causing you to feel so stressed?"
   b. "It's important you are aware of this. Continued anxiety and stress like this can lead to health problems if they go unchecked."

2. *Client:* "I'm tired of sitting home alone, but I feel so uptight when I ask a girl for a date."
   a. "You seem to be saying that you feel lonely and also that you're not willing to risk asking a girl to go out with you."
   b. "What makes you so anxious when you speak with girls?"

3. *Client:* "I don't know why I tolerate his abuse. I really don't love him."
   a. "On the one hand, you say that you don't love him, and on the other hand, you remain in the house and allow him to beat you. How do you put these two things together?"
   b. "You may be caught up in a vicious cycle about whether your feelings for him, even though they're not love, outweigh your regard for yourself. It might be helpful for you to know the process other women in your shoes go through before they finally get enough courage to leave for good."

4. *Client:* "I don't know why we ever got married in the first place."
   a. "What qualities attracted you to each other originally?"
   b. "You're having a difficult time right now, which has led you to question the entire marriage. I wonder whether you would react this way if this present problem weren't causing such distress."

Part Two

For each of the following three client statements, Objective Two asks you to verbalize or write an example of each of the four action responses. In developing these responses, it may be helpful to use the cognitive learning strategy you practiced earlier for each response. Example responses are given in the Postevaluation Feedback.

Client 1, a Euro-American parent: My house looks like a mess. I can't seem to get anything done with these kids always under my feet. I'm afraid that I may lose my temper and hit them one of these days.

Probe:

Confrontation:

Interpretation:

Information giving:

Client 2, an African American graduate student: I feel so overwhelmed. I've got books to read, papers to write. My money is running low and I don't even have a job. Plus my roommate is thinking of moving out.

Probe:

Confrontation:

Interpretation:

Information giving:

Client 3, a young, Native American man: "I haven't gotten hooked on this stuff. It doesn't make me feel high, though, just good. All my bad thoughts and all the pain go away when I take it. So why should I give it up?"

Probe:

Confrontation:

Interpretation:

Information giving:

Part Three

This part of the evaluation gives you an opportunity to develop your skills in observing key aspects of client behavior that must be attended to in order to develop effective and accurate action responses:

1. Issues and messages that need more elaboration, information, or examples
2. Discrepancies and incongruities
3. Implicit messages and themes
4. Distorted perceptions and ideas
5. Myths and inaccurate information

Objective Three asks you to observe these five aspects of client behavior during a 30-minute interview. Record your

(continued)

POSTEVALUATION (continued)

observations on the Client Observation Checklist that follows. You can obtain feedback for this activity by having two or more persons observe and rate the same session—then compare your responses.

Client Observation Checklist

Name of counselor _____

Name of observer(s) _____

_____

Instructions: Given the five categories of client behavior (left column), use the right column to record separate occurrences of behaviors within these categories as they occur during a 30-minute counseling interview.*

Observed category	Selected key client words and behavior
1. Issues and messages that need more elaboration, information, or examples	1. _____ 2. _____ 3. _____ 4. _____
2. Discrepancies and incongruities	1. _____ 2. _____ 3. _____ 4. _____
3. Implicit messages and themes	1. _____ 2. _____ 3. _____ 4. _____
4. Distorted perceptions and ideas	1. _____ 2. _____ 3. _____ 4. _____
5. Myths or inaccurate information	1. _____ 2. _____ 3. _____ 4. _____

Observer impressions and comments _____

_____

Part Four

To begin integrating your skills (Chapter Objective Four), conduct at least one role-play interview that is approximately 20 minutes long. You may want to consider this an initial helping interview. Your objective is to use as many of the verbal responses (listening, action) and the nonverbal

behaviors as seem appropriate within this time span. Also give some attention to the quality of your relationship with the client and the degree to which you can conduct the interview in a culturally competent way. Try to regard this interview as an opportunity to get involved with the person in front of you, not as just another practice. If you feel some discomfort you may wish to do several more interviews with different kinds of clients. To assess your interview, use the Interview Inventory that follows. You may wish to copy the inventory or superimpose a piece of paper over it for your ratings. After all the ratings are completed, look at your ratings in the light of these questions:

1. Which relationship variables were easiest for you to demonstrate? Hardest?
2. Examine the total number of the verbal responses you used in each category. Did you use responses from each category with the same frequency? Did most of your responses come from one category? Did you seem to avoid using responses from one category? If so, for what reason?
3. Was it easier to integrate the verbal responses or the nonverbal skills?
4. Which nonverbal skills were easiest for you to demonstrate? Which ones did you find most difficult to use in the interview?
5. What aspects of multicultural competence do you feel most comfortable with? What parts are still hard for you?
6. What do you see that you have learned about your counseling interview behavior so far? What do you think you need to improve?

Interview Inventory

Interview No. _____ Counselor _____

Client _____ Rater _____ Date _____

Instructions for rating: This rating form has four parts. Part One (Relationship Variables) measures aspects of establishing and enhancing a therapeutic relationship. Part Two (Verbal Behavior) assesses listening and action responses. Part Three (Nonverbal Behavior) evaluates your use of various nonverbal behaviors. Part Four (Multicultural Competence) assesses eight aspects of culturally competent interview behaviors. To use the Interview Inventory for rating, follow the instructions found on each part of the inventory.

Part One: Relationship Variables

Instructions: Using the 5-point scale, indicate which number on the scale best represents the counselor's behavior during the observed interaction. Circle the appropriate number on the chart on p. 145.

(continued)

Part Two: Verbal Behavior

Instructions: Check (✔) the type of verbal response represented by each counselor statement in the corresponding category on the rating form. At the end of the observation period, tally the total number of checks associated with each verbal response on the chart on p. 146.

Part Three: Nonverbal Behavior

Instructions: This part of the inventory lists a number of significant dimensions of nonverbal behavior. Check (✔) any that you observe and provide a brief description of the key aspects and appropriateness of the behavior. An example is given on the chart on p. 147.

Part Four: Cultural Impact of Your Helping Style

Using the 6-point scale, indicate the number on the scale that best represents the counselor's behavior during the observed interaction. Circle the appropriate number on the chart found on p. 147. Then, based on the observations made from your practice session and also your own observations, consider the following questions about your helping style, adapted from Sue and Sue (1990, p. 71):

1. What is my predominant counseling/communication style?
2. What does my style suggest about my values and biases regarding human behavior and people?
3. How might my nonverbal behaviors reflect stereotypes, fears, or preconceived ideas about various racial/ethnic groups?
4. In what way does my helping style hinder my ability to work effectively with a culturally different client?

Part One: Relationship Variables

1. Conveyed accurate understanding of the client.

1	2	3	4	5
Not at all	Minimally	Somewhat	A great deal	Almost always

2. Conveyed support and warmth without approving or disapproving of the client.

1	2	3	4	5
Not at all	Minimally	Somewhat	A great deal	Almost always

3. Focused on the person rather than on the procedure or on counselor's "professional role."

1	2	3	4	5
Not at all	Minimally	Somewhat	A great deal	Almost always

4. Conveyed spontaneity, was not "mechanical" when responding to client.

1	2	3	4	5
Not at all	Minimally	Somewhat	A great deal	Almost always

5. Responded to feelings and issues as they occurred within the session (that is, "here and now").

1	2	3	4	5
Not at all	Minimally	Somewhat	A great deal	Almost always

6. Displayed comfort and confidence in working with the client.

1	2	3	4	5
Not at all	Minimally	Somewhat	A great deal	Almost always

7. Responded with dynamism and frequency; was not "passive."

1	2	3	4	5
Not at all	Minimally	Somewhat	A great deal	Almost always

8. Displayed sincerity in intentions and responses.

1	2	3	4	5
Not at all	Minimally	Somewhat	A great deal	Almost always

9. Conveyed friendliness and goodwill in interacting with client.

1	2	3	4	5
Not at all	Minimally	Somewhat	A great deal	Almost always

(continued)

10. Informed client about expectations and what would or would not happen in session (that is, structuring).

1	2	3	4	5
Not at all	Minimally	Somewhat	A great deal	Almost always

11. Shared similar attitudes, opinions, and experiences about yourself with client when appropriate (that is, when such sharing added to, not detracted from, client focus).

1	2	3	4	5
Not at all	Minimally	Somewhat	A great deal	Almost always

12. Other significant relationship aspects _____
_____
_____
_____

Part Two: Verbal Behavior

	Listening responses					Action responses				
	Clarification	Paraphrase	Reflecting feeling	Summarization	Open question	Closed question	Focused question	Confrontation	Interpretation	Information giving
1										
2										
3										
•										
•										
•										
•										
•										
•										
•										
20										
Total										

(continued)

Part Three: Nonverbal Behavior

Behavior	Check (✔) if observed	Key aspects of behavior		Behavior	Check (✔) if observed	Key aspects of behavior
Example Body posture	✔	Tense, rigid until last part of session, then relaxed		*Example* Body posture	✔	Tense, rigid until last part of session, then relaxed
1. Eye contact				19. Time in responding to messages		
2. Facial expression				20. Time in ending session		
3. Head nodding				21. Autonomic response (for example, breathing, sweat, skin flush, rash)		
4. Body posture				22. Congruence/ incongruence between counselor verbal and nonverbal behavior		
5. Body movements				23. Synchrony/ dissynchrony between counselor/client nonverbal behavior		
6. Body orientation				24. Other		
7. Gestures						
8. Nonverbal adaptors						
9. Voice level and pitch						
10. Rate of speech						
11. Verbal underlining (voice emphasis)						
12. Speech errors						
13. Pauses, silence						
14. Distance						
15. Touch						
16. Position in room						
17. Environmental arousal						
18. Time in starting session						

Part Four: Multicultural Competence*

The counseling student:	Rarely		Sometimes		Consistently	
1. Displayed awareness of his or her own racial and cultural identity development and its impact on the counseling process.	1	2	3	4	5	6
2. Was aware of his or her own values, biases, and assumptions about other racial and cultural groups and did not let these biases and assumptions impede the counseling process.	1	2	3	4	5	6
3. Exhibited a respect for cultural differences among clients.	1	2	3	4	5	6
4. Was aware of the cultural values of each client as well as of the uniquenesses of each client within the client's racial and cultural group identification.	1	2	3	4	5	6
5. Was sensitive to nonverbal and paralanguage cross-cultural communication clues.	1	2	3	4	5	6

(continued)

POSTEVALUATION (continued)

The counseling student:	Rarely		Sometimes		Consistently	
6. Demonstrated the ability to assess the client's level of acculturation and to use this information in working with the client to implement culturally sensitive counseling.	1	2	3	4	5	6
7. Displayed an understanding of how race, ethnicity, and culture influence the treatment, status, and life chances of clients.	1	2	3	4	5	6
8. Was able to help the client sort out the degree to which the client's issues or problems are exacerbated by limits and regulations of the larger society.	1	2	3	4	5	6

*SOURCE: From *Pathways to Multicultural Counseling Competence: A Developmental Journey*, by B. Wehrly, p. 240. Copyright 1996 by Brooks/Cole Publishing Company, a division of International Thomson Publishing Inc.

SUGGESTED READINGS

Brammer, L. M., Shostrom, E. L., & Abrego, P. J. (1989). *Therapeutic psychology* (5th ed.). Englewood Cliffs, NJ: Prentice Hall.

Egan, G. (1994). *The skilled helper*. Pacific Grove, CA: Brooks/Cole.

Gazda, G. M., Asbury, F. S., Balzer, F. J., Childers, W. C., Phelps, R. E., & Walters, R. P. (1995). *Human relations development* (5th ed.). Needham Heights: Allyn & Bacon.

Hackney, H., & Cormier, L. S. (1994). *Counseling strategies and interventions*. Needham Heights, MA: Allyn & Bacon.

Hill, C. E. (1989). *Therapist techniques and client outcomes*. Newbury Park, CA: Sage.

Hill, C. E. (1992). An overview of four measures developed to test the Hill Process Model. *Journal of Counseling and Development, 70*, 728–739.

Ivey, A. E. (1994). *Intentional interviewing and counseling* (3rd ed.). Pacific Grove, CA: Brooks/Cole.

Jacobs, E. (1994). *Impact therapy*. Odessa, FL: Psychological Assessment Resources.

Johnson, D. W. (1997). *Reaching out: Interpersonal effectiveness and self actualization* (6th ed.). Boston: Allyn & Bacon.

Lin, Meei-Ju, Kelly, K. R. & Nelson, R. C. (1996). A comparative analysis of the interpersonal process in school-based counseling and consultation. *Journal of Counseling Psychology, 43*, 389–393.

Shainberg, D. (1993). *Healing in psychotherapy*. Langhorne, PA: Gordon and Breach Science Publishers.

Spiegel, S. B., & Hill, C. E. (1989). Guidelines for research on therapist interpretation. *Journal of Counseling Psychology, 36*, 121–129.

FEEDBACK 27
Information Giving

Client 1

Question 1: The client does not appear to know the general effects of drinking alcohol on judgment and reaction time. He also does not seem to know about the specific impact of a certain quantity of alcohol such as a six-pack of beer. He attributes the loss of his driver's license to a continuing cultural event (oppression by White people) rather than to his behavior of drinking too much alcohol to drive safely.

Question 2: This information is very relevant as the effects of drinking alcohol and driving are consistent across cultural groups in the United States. However, the counselor must also recognize that the client is engaging in a pattern he has observed while growing up in some of his male relatives, and so to him it seems culturally acceptable. Moreover, the loss of his license within the larger context of cultural oppression of his nation by the "White man" also is culturally consistent for him.

Question 3: First, it would be important to address this from his cultural view of it—that he sees it as an acceptable practice for many of the adult males he

(continued)

knows in his cultural group. After this, present information about the ways in which drinking alcohol can affect driving.

Question 4: Be factual and nonjudgmental in the presentation. Empathize with his feelings of loss and oppression. Providing a graphic display (such as a film) of what can happen to a person after drinking alcohol may be useful.

Question 5: His initial reaction may be denial or anger. He may need some time to trust you as a reliable source of information, especially if you are a White counselor.

Example of information-giving response: I realize this seems to you to be just another example of what White men do to people of your nation that is unjust and unfair. I also realize that you are following what you've seen many of your male relatives do. So I'm sure, based on all this, it does seem hard to believe that drinking a six-pack of beer can interfere with the way you drive. In fact, it can and does affect how you judge things and how quickly you react. So far, you're accident-free. I'm sure if you thought *you* could be in danger or could put someone else's life in danger from drinking and driving, you might think about it differently. Would you be willing to watch a short film clip with me?

Client 2

Question 1: Client may not be aware that he is in a high-risk category for AIDS because of his drug use with needles and multiple sexual partners. Or he may be aware of the danger but, in order to maintain his behavior, is in denial about it. He may also be unaware of the incidence of HIV infection in other African American men.

Question 2: This information is highly relevant.

Question 3: First, I might mention categories of high risk for HIV infection that he appears to belong to; then I might ascertain whether he is aware of these or is operating under certain myths that contribute to his denial.

Question 4: Be factual, concrete, and nonjudgmental.

Question 5: His feelings may range from fear to anger to more denial.

Example of information-giving response: Kevin, when you say this could never happen to you, it makes me wonder what you know about HIV. Do you know any Black men who have tested positive for HIV? And are you aware that the virus can be spread easily through shared needles and also through semen?

Client 3

Question 1: Client seems to lack certain legal information about possible management and consequences of

divorce. Also seems to lack information about possible psychological effects of divorce.

Question 2: Information that seems culturally relevant to her:

a. Getting a divorce rarely results in loss of job.

b. In most situations, the husband would be legally required to give financial support for the children.

c. Although divorce may make a person feel happy and relieved, it can also be unsettling, can result in temporary feelings of loss and depression, and is not an antidote for all other life issues.

Question 3: Present the need for legal information about her job status and child support, followed by other possible effects of divorce.

Question 4: Be factual and concrete. Possibly ask her to list pros and cons of divorce on paper.

Question 5: Her feelings could range from relief about the legal issues to disappointment that divorce is not a panacea.

Example of information-giving response: Leslie, in discussing your situation with you, there are a couple of things I want to mention. First, it might be useful for you to consider seeing a competent lawyer who specializes in divorce mediation. This person could give you detailed information about the legal effects and processes of a divorce. Usually, a person does not lose a job because of a divorce. Besides, in most instances, the husband is required to make support payments as long as the children are of minor age. I would encourage you to express these same concerns to the lawyer. The other thing I'd like to spend some time discussing is your belief that you will feel very happy after the divorce. That might be very true. It is also important to remember, though, that just the process of ending a relationship—even a bad relationship—can be very unsettling and can bring not only relief but often some temporary feelings of loss and maybe sadness.

F E E D B A C K
POSTEVALUATION

Part One

1. a. Probe
   b. Information giving
2. a. Confrontation
   b. Probe

(continued)

(continued)

FEEDBACK: POSTEVALUATION   (continued)

3. a. Confrontation
    b. Information giving
4. a. Probe
    b. Interpretation

Part Two

Here are some examples of action responses. Are yours similar?

Client statement 1

Probe: What exactly would you like to be able to accomplish during the day?

or

How could you keep the kids occupied while you do some of your housework?

or

When do you feel most like striking the children?

or

How could you control your anger?

Confrontation: On the one hand, you seem to be saying the kids are responsible for your difficulties, and on the other, it appears as if you feel you are the one who is out of control.

Interpretation: I wonder whether you would be able to accomplish what seems important to you even if the kids weren't always underfoot. Perhaps it's easy to use their presence to account for your lack of accomplishment.

Information giving: If you believe your problem would be solved by having more time alone, we could discuss some options that seemed to help other women in this situation—things to give you more time alone as well as ways to cope with your anger.

Client statement 2

Probe: How could you organize yourself better so that you wouldn't feel so overcome by your studies?

or

What kind of work might you do that would fit in with your class schedule?

or

How might you cope with these feelings of being so overwhelmed?

Confrontation: You've mentioned several reasons that you feel so overwhelmed now, and at the same time I don't think you mentioned anything you're doing to relieve these feelings.

Interpretation: You seem to feel so discouraged with everything that I imagine it would be easy now to feel justified in giving it all up, quitting grad school altogether.

(continued)

Information giving: Perhaps it would be helpful if we talked about some ways to help you with your time and money problems.

Client statement 3

Probe: What do you feel comfortable sharing with me about your pain?

Confrontation: You're telling me that you're pretty sure you're not hooked on this and at the same time you recognize it seems to medicate your pain. How do you put these two things together?

Interpretation: Even though you don't feel hooked on this substance, it seems as if using it helps you avoid certain things—do you think this is so?

Information giving: I'm wondering what you would think of the idea of our spending some time talking about other ways to deal with the pain—such as practices and rituals consistent with your own cultural and ethnic background.

CHAPTER 8

CONCEPTUALIZING AND UNDERSTANDING CLIENT PROBLEMS

Institutionalized patient: Why are people always out to get me?

Student: I can't even talk to my mom. What a hassle!

Physically challenged person: Ever since I had that automobile accident and had to change jobs, I don't seem to be able to get it together.

Older person: I never feel like I can do anything well anymore. And I feel so depressed all the time.

These client statements are representative of the types of concerns that clients bring to counselors every day. One thing these clients and others have in common is that their initial problem presentation is often vague. A counselor can translate vague client problems into specific problem statements by using certain skills associated with problem assessment. This chapter presents a conceptual framework that a counselor can use to assess client problems. Chapter 9 demonstrates a way for the counselor to implement this framework in the interview setting.

OBJECTIVES

After completing this chapter, you will be able to identify, in writing, using two client case descriptions,

1. The client's problem behaviors
2. Whether the problem behaviors are overt or covert
3. The antecedent contributing conditions
4. The consequences and secondary gains
5. The way each consequence influences the problem behaviors
6. The sociopolitical context of the problem

WHAT IS ASSESSMENT?

Problem assessment consists of procedures and tools used to collect and process information from which the entire counseling program is developed. Assessment has six purposes:

1. To obtain information on the client's presenting problem and on other, related problems.
2. To identify the controlling or contributing variables associated with the problem.
3. To determine the client's goals/expectations for counseling outcomes.
4. To gather baseline data that will be compared with subsequent data to assess and evaluate client progress and the effects of treatment strategies. This evaluation helps the practitioner decide whether to continue or modify the treatment plan or intervention strategy.
5. To educate and motivate the client by sharing your views of the problem with the client, increasing the client's receptivity to treatment, and contributing to therapeutic change through reactivity (that is, when behavior changes as a consequence of the assessment interview or procedure rather than as a result of a particular action or change strategy).
6. To use the information obtained from the client to plan effective treatment interventions and strategies. The information obtained during the assessment process should help to answer this well-thought-out question: "*What* treatment, by *whom,* is most effective for *this* individual with *that* specific problem and under *which* set of circumstances?" (Paul, 1967, p. 111)

This chapter focuses primarily on the first two purposes of assessment mentioned above: defining the problem and identifying the controlling variables associated with the problem. The next section presents several ways to conceptualize client problems.

METHODS OF CONCEPTUALIZING AND UNDERSTANDING CLIENT PROBLEMS

Interviewing the client and having the client engage in other assessment procedures are only part of the overall assessment process in counseling and therapy. Equally important is the therapist's own mental, or covert, activity that goes on during the process. The therapist typically gathers a great amount of information from clients during this stage of counseling. Unless the therapist can integrate and synthesize the data, they are of little value and use. The *counselor's* tasks during the assessment process include knowing what information to obtain and how to obtain it, putting it together in some meaningful way, and using it to generate clinical hunches, or hypotheses, about client problems, hunches that lead to tentative ideas for treatment planning. This mental activity of the counselor's is called "conceptualization"—which simply means the way the counselor thinks about the client's problem configuration. A recent study assessed the relationship of counselors' hypothesis formation skill levels and counseling effectiveness (Morran et al., 1994). The results of this study found that a higher level of hypothesis formation skill was associated with more positive client assessment of the counselor.

The assessment methods we describe later in this chapter and in Chapter 9, and our interview assessment model particularly, are based on a model of conceptualization we have used over the years in our teaching and in clinical practice. The origins of this model were first described by Kanfer and Saslow (1969). Before describing our model in detail, we would first like to describe two current models of client or case conceptualization proposed by Ainsworth and Bowlby (1991), and Lazarus (1976, 1989) that have influenced the development of our own clinical model of problem conceptualization.

The three models of problem conceptualization we present in this chapter have some distinct differences but are also similar in several respects. First, they represent a framework the therapist can use to develop hunches (educated guesses) about the client's presenting problem. Second, they recognize that problem behavior is usually multifaceted and affects how people think and feel as well as behave. Finally, they provide information about the problem that the therapist can use in selecting and planning relevant treatments. Although the major focus of the chapter is on the model we use for case conceptualization, we present two others because they are important historically in the development of case conceptualization models and because they enable the reader to look at client problems from more than one perspective.

The Attachment Theory Model of Problem Conceptualization

In Chapter 2 we presented an overview of attachment theory (Ainsworth, 1989; Bowlby, 1988; Lopez, 1995; Rothbard & Shaver, 1994). Based on variations of both adult affect regulation and social competence, a person can be classified in one of four predominant attachment styles (as illustrated in Table 2-1):

1. Secure: Positive model of self and others; can engage in constructive problem solving with others during periods of stress and instability.
2. Preoccupied: Negative cognitive or working model of self, but positive model of others; awareness and evaluation processes are less balanced, with more energy directed toward others, resulting in less affect regulation and objective self-awareness.
3. Dismissing: Positive cognitive schema or internal working model of self and negative model of others; awareness and evaluation processes are directed toward self, with minimized awareness of attachment-related needs.
4. Fearful: Cognitive or working models are negative for both self and others; fearful presentation of self contributes to feelings of vulnerability and inadequacy, and negative view of others encourages interpersonal disengagement and limits awareness of others.

In the interpersonal awareness learning activity in Chapter 2, you used the attachment theory model of problem conceptualization for your own attachment style. As another example, consider a 35-year-old, single, Euro-American male who reports having problems relating to women. He says that maintaining a long and trustful relationship with a woman is difficult for him because he feels worried and anxious that she might leave him. He reports feeling awkward because of his own vulnerability and inadequacy, and he fears making a commitment. The client says that he does not know how to behave or what to do when he is around a woman. He has no problems sexually, but he is unsure and confused about what behaviors he must engage in to create intimacy. He says that he has felt this way since he started dating when he was in high school. He is concerned now because he would like to find someone he feels comfortable with, make a commitment, and possibly start a family. The client states that he is looking for security, but he feels his behavior causes women not to stay in a relationship with him for very long periods of time. After breaking up a relationship, he goes for several months without dating or seeing anyone on a regular basis. During these periods of separation, he feels anxious and longing for not having a secure relationship. Table 8-1 shows the analysis of this case according to attachment theory. The conceptualization is based on some of the components of the client's internal working models as illustrated in Table 8-1, including confidence with self and others, trust, relationship

TABLE 8-1. Attachment theory used to conceptualize case

Components of internal working models	Fearful (Women)		Preoccupied (Supervisor)	
	Adult	**Child**	**Adult**	**Child**
Confidence with self & others	Feels inadequate with women, fears that they will leave him.	Mother was always interfering. Didn't want him to play sports. Wanted to know where he was going and what he was doing.	Doubts about self, anxious about pleasing his supervisor.	Wanted to please his father. His father was critical—said that he could do better. His performance conditional on father's presence.
Trust	Has difficulty trusting women because he fears they will leave him.	Mother had drinking problem and was erratic. At times she was available to him; other times she was not. Difficult to trust her.	Too much trust in his supervisor, overtaxes his need for support from his supervisor	Didn't have confidence or trust in himself; relied on parents for approval. Trusts supervisor more than self.
Relationship commitment	His feelings of vulnerability and inadequacy elicit fear of commitment.	His parents' inconsistency led to anxiety when relating to others. Feared loss makes commitment to women tentative.	Because he needs approval from his supervisor, he is overly committed to his supervisor. Cannot rely on his own competence.	Experienced his parents as inconsistently responsive when needed. Depended on them for approval. Their commitment seemed tentative.
Attachment system	His feared loss of women and his eager search for security in them activate a dysfunctional attachment style.	The model he developed as a child for others exhibits erratic combinations of avoidant and anxious responses.	Anxiety and worry about how his supervisor sees him creates an overly involved and demanding attachment system.	His parents created for him a model of self that was doubtful.
Self and other boundary disturbances	Very weak boundaries between himself and the women he dates.	His parents' patterns of disorganized and disoriented relating and allowing him to explore created weak boundaries between himself and others.	Being overly involved or enmeshed with his supervisor makes boundaries for self indistinct.	Parents created model of self which was doubtful. He needed approval, support, and presence of attachment figures.
Independence/ dependence balance	Avoidant and anxious responding with women creates a confusing model of independence and dependence.	Confusing parental pattern of interacting created unhealthy model for independence and dependence.	Overly dependent on supervisor for approval.	Need for attachment figure approval as an infant and child.

commitment, attachment system, self and other boundary disturbances, and independence/dependence balance. The analysis using these components of the model reveals that the client is classified with a *fearful* attachment style with respect to women (column 1 in Table 8-1). If we explore the antecedents of his childhood attachment quality, we find that his mother had a problem with alcohol abuse; she typically interfered in his activities as a child; she was always monitoring his activities without offering much support for self-exploration and individuation (see second column in

Table 8-1). In this case, the counselor conceptualizes the client's early childhood experiences as fearful, affected by the confusing interactions with his mother and his father's lack of approval.

It is often helpful to have clients explore two relationships in their present life to determine whether another attachment style is exhibited. Frequently, some people use a couple of attachment styles or models for relating to others. With the above example, the counselor could explore the client's relationship with his work supervisor. Table 8-1 illustrates in column 3 the components of the internal working models the client uses with his supervisor. The client reports that he does not feel independent enough to perform his job tasks without getting approval from his supervisor. He admits placing too much trust in his supervisor, and he does not feel he can rely exclusively on his own confidence to perform his job tasks. His system of attachment is overly enmeshed with his supervisor, and this dependence contributes to indistinct boundaries for himself. He is overly committed and dependent on supervisory approval. The components of his internal working models reflect a preoccupied style of attachment: The client sees himself as negative and the supervisor as positive. Table 8-1 (column 4) also illustrates his childhood quality of attachment that contributed to his adult internal working model with his supervisor. For example, the client's performance during childhood was conditional on the approval and presence of his father. He did not have confidence or trust in himself so he relied on his parents—maybe more on his father—for approval. Again, his boundaries were not distinguished. Finally, the quality of his parental attachment style created for him a lack of independence and contributed to a more preoccupied and dependent attachment style.

The model of problem conceptualization according to attachment theory stresses the contribution of early attachment experiences with the caregiver to adult internal working models of social competence and affect regulation. It is possible that early insecure attachment experiences can be changed to more secure attachment models in adulthood. Changes in the internal working models can occur alone and without help, or change can occur in a therapeutic relationship. One of the first goals of therapy for relationship concerns is to bring to the client's *awareness* the predominant models the client uses for relating to others. Then, the treatment plan would involve *replacing* current dysfunctional models with balanced and healthier internal work models for himself or herself and others. If counselors help to provide these changes, they must offer a secure place for clients to experience changes in their models within *and* outside the therapeutic relationship. To be helpful in the change process, counselors must be aware and secure in their own affect regulation and social competence because, as Bowlby (1988) contends, the therapy relationship is analogous to complementary attachment.

The Lazarus Model of Problem Conceptualization: The BASIC ID

According to Lazarus (1989), who is associated with broad-spectrum behavior therapy, there are seven modalities to explore in assessment and intervention with clients. To refer to these seven areas of assessment and treatment in abbreviated fashion, Lazarus uses the acronym BASIC ID. A brief discussion of each component of the BASIC ID follows. In using this model of conceptualization, remember that each modality described by Lazarus interacts with the other modalities and should not be treated in isolation.

B: Behavior.   Behavior includes simple and more complex psychomotor skills and activities such as smiling, talking, writing, eating, smoking, and having sex. In most clinical interviewing, the therapist has to infer what the client does or does not do on the basis of client self-report, although occasionally other measures of behavior can corroborate client verbal report. Lazarus (1989) notes the importance of being alert to behavioral excesses and deficits—things the client does too much or too little.

A: Affect.   Affect includes felt or reported feelings and emotions. According to Lazarus (1989) this is perhaps the most overworked area in psychotherapy and also one of the least understood. Included in this category would be presence or absence of particular feelings as well as hidden or distorted feelings.

S: Sensation.   Sensation includes five major senses with respect to sensory processing of information: visual (sight), kinesthetic (touch), auditory (hearing), olfactory (smell), and gustation (taste). Focus on sensory elements of experience is important to develop personal fulfillment. Sometimes, too, presenting complaints are described as felt body sensations such as stomach distress or dizziness (Lazarus, 1989). Therapists need to be alert to pleasant and unpleasant reported sensations as well as sensations of which clients seem unaware.

I: Imagery.   According to Lazarus, imagery comprises various mental pictures that exert influence on a client's life

(1989). For example, a husband who was nagged by what he called repetitive ideas that his wife was having an affair (apparently with no realistic basis) actually was troubled because he generated constant pictures or images of his wife in bed with another man. Lazarus (1989) believes that imagery is especially useful with clients who tend to overuse the cognitive modality and intellectualize their feelings.

C: Cognition. Cognitions are thoughts and beliefs, and Lazarus is most interested in exploring the client's mistaken beliefs—the illogical or irrational ones. He usually looks for three faulty assumptions that he believes are common and also potentially more damaging than others:

1. The tyranny of the SHOULD (Horney, 1950)—a belief that often can be inferred from the client's actions and behaviors as well as from self-report. This belief often places unreasonable demands on self and others.
2. Perfectionism—expecting infallibility, often not only of themselves but of others as well.
3. External attributions—the myths that clients verbalize when they feel they are the victims of outside persons or circumstances and have no control over or responsibility for what is happening to them.

I: Interpersonal Relationships. Many therapists (including Sullivan, Horney, and Fromm) have stressed the importance of interpersonal relationships, or "social interest" (Adler, 1964). Lazarus (1989) notes that problems in the way clients relate to others can be detected not only through self-report and role playing but also by observation of the therapist/client relationship. Assessment of this modality includes observing the way clients express and accept feelings communicated to them by others as well as the way they behave and react to others.

D: Drugs. Lazarus asserts that drugs represent an important nonpsychological modality to assess (and potentially treat) because neurological and biochemical factors can affect behavior, affective responses, cognitions, sensations, and so on. In addition to specific inquiries about psychotropic medications, assessment of this modality includes the following:

1. Overall appearance—attire, skin or speech disturbances, tics, psychomotor disorders.
2. Physiological complaints or diagnosed illnesses.
3. General health and well-being—physical fitness, exercise, diet and nutrition, avocational interests and hobbies, and leisure time pursuits.

This modality may often require consultation with or examination by a physician or other type of health professional.

Lazarus (1976) asserts that most therapists, including eclectic ones, fail to assess and treat these seven basic modalities. Instead, they deal with only one or two modalities, depending on their personal preferences and theoretical orientation, even though "durable results are in direct proportion to the number of specific modalities deliberately invoked by any therapeutic system" (p. 13).

The BASIC ID model of case conceptualization is applied to the following case and summarized in the modality profile (Lazarus, 1989) shown in Table 8-2.

TABLE 8-2. Modality profile of client case using BASIC ID (Lazarus, 1976, 1989)

Modality	Observations
B: Behavior	Passive responding; some withdrawal from conversation Slow rate of speech Frequent shrugging of shoulders Overeating
A: Affect	Alone—loneliness Unloved Denies concern or upset over weight
S: Sensation	Muscular tension—upper torso particularly
I: Imagery	Frequent fantasies of a move and different lifestyle Persistent dreams of being rescued
C: Cognition	Negative self-verbalizations and perceptions Self-perfectionistic standards Attributes problems to forces outside herself
I: Interpersonal relationships	Is exploited by ex-husband, daughters, boss Submissive in interactions with others
D: Drugs	Well groomed Well dressed 50–75 pounds overweight Articulate Stomach distress—weekly Good health—mostly sedentary activity Little leisure time

The client is a 35-year-old female who looks about 50 to 75 pounds overweight, though well groomed, well dressed, and articulate. The client states that she is in generally good health, does little exercise, works either on the job or at home, and has little free time. Free time is spent mainly in sedentary activities such as reading or watching TV. The client is divorced and has two school-age daughters. She does report occasional stomach distress—often as much as once or twice weekly. The client's presenting problem is overall "dissatisfaction with myself and my life." The client notes that she lives in a small town and has been unable to meet many available partners. She would like to have a good relationship with a male. She was divorced four years ago and states that her husband became interested in another woman and "took off." She says that she also has poor relationships with her two daughters, whom she describes as "irresponsible and lazy." On inquiry, it appears that the client is easily exploited and rather submissive in most of her relationships with significant others. In her job, she agrees to take work home with her even though she receives no overtime pay. She describes herself as feeling alone, lonely, and sometimes unloved or unlovable. She also reports that she often has thoughts that her life has been a failure and that she is not the kind of person she could be, although she portrays herself as a victim of circumstances (divorce, job, small town) beyond her control. However, she also reports rather frequent fantasies of moving and living in a different town and having a different job. She also describes repetitive dreams in which she can recall vividly the image of being rescued. She behaves very passively in the session—talks slowly, shrugs her shoulders, and occasionally withdraws from the conversation. Some muscular tension is apparent during the interview, particularly in her upper body. She states that overeating is a major problem, one that she attributes to not having her life go the way she wants it to and being unable to do much about it. At the same time, she appears to deny any concern about her weight, stating that if she's not worried about it, then it shouldn't matter to anyone else either.

In treatment planning, the first areas of focus would be the two modalities about which the client is most concerned—affective and interpersonal. If the interpersonal modality is selected as the initial area of focus, changes in this modality will likely lead to changes in the affective one also, as the client's feelings of loneliness are a direct result of lack of effective interpersonal relationships. From a feminist therapy perspective, part of the focus in the interpersonal and affective modalities would be on the way society's expectations of her as a woman have contributed to her feelings of distress and isolation. Skill training programs such as assertion training and social skills training are likely to be most effective in helping the client establish new relationships and avoid further exploitation in her present ones. Such skill training could also be directed toward some of her overt behaviors that may interfere with establishing new relationships, such as her speech rate and her style of responding in conversations. Although the client denies any concern about her weight, she may also allow her weight to prevent her from engaging in the very kind of social interactions and relationships she finds absent from her life. Strategies such as Gestalt dialoguing and TA redecision work may help her examine her conflicting feelings about being overweight. If and when she decides to make weight reduction a goal, cognitive strategies (such as cognitive restructuring, Chapter 15) aimed at modifying any problem-related cognitive misperceptions would be useful, as would behavioral strategies (such as self-management, Chapter 20) targeted toward helping her modify her overeating behavior and supporting environmental contingencies.

OUR ASSUMPTIONS ABOUT ASSESSMENT AND COGNITIVE BEHAVIOR THERAPY

Like the previously described case conceptualization models, our model of assessment in counseling and therapy is based on several assumptions about clients, problems, and behavior. These assumptions are drawn from the cognitive-behavioral approach to counseling. Cognitive behavior therapy includes a variety of techniques and strategies that are based on principles of learning and designed to produce constructive change in human behaviors. This approach was first developed in the 1950s under the term *behavior therapy* by, among others, Skinner, Wolpe, Lazarus, and Krumboltz. Early behavior therapists focused on the importance of changing clients' observable behavior. Since the 1950s, there have been significant developments in behavior therapy. Among the most important is the emergence of cognitive behavior therapy, which arose in the 1970s as a result of the work of such persons as Meichenbaum and Beck. Cognitive behavior therapy emphasizes the effects of private events such as cognitions, beliefs, and internal dialogue on resulting feelings and performance. This orientation to counseling now recognizes that both overt responding (observed behavior) and covert responding (feelings and thoughts) are important targets of change as long as they can be clearly specified (Rimm & Masters, 1979, p. 1). (See Learning Activity #28.)

LEARNING ACTIVITY 28 *Methods of Case Conceptualization*

Using the case of Mrs. Oliverio, described later in this Learning Activity, conceptualize the case according to the two models previously described: Attachment and BASIC ID. We provide specific questions below to consider for each model. You can do this exercise by yourself, although it may be a better learning experience if you work with it in small groups. You can then get feedback from your group or your instructor or supervisor. For additional work with these two models, you may also wish to apply the questions below to actual cases of your own or to the cases presented in the postevaluation at the end of this chapter.

1. Attachment theory

   a. Identify the client's degree of confidence about self and others.

   b. Identify the client's affect regulation within the relationships.

   c. Identify the extent of the client's preoccupation with relationships.

   d. Identify the level of trust the client has for others.

   e. Assess the degree of balance in the client's attachment system.

   f. Assess the degree of the client's independence/dependence balance.

   g. Based on your assessment of the above, identify the client's attachment style.

   h. Speculate on which of the above would be appropriate areas to address with the goal of changing the client's internal working model so as to resolve the issues the client presents.

2. BASIC ID (Lazarus) model:

   a. Identify the behavior exhibited by the client, particularly excesses and deficits.

   b. Identify the primary affect (feelings and emotions) reported by the client.

   c. Identify any major sensations or sensory experiences/processing reported by the client. Speculate on the client's primary sensory system.

   d. Identify the imagery or mental pictures that exert influence on the client.

   e. Identify the apparent cognitions (thoughts, beliefs) reported by the client.

   f. Assess the nature of the client's interpersonal relationships.

   g. Identify any physiological factors/complaints apparent in the problem.

   h. Speculate on which of these seven areas would be *primary* targets of change and which might be *secondary* targets in resolving the problems and issues this client presents.

The Case of Mrs. Oliverio

Mrs. Oliverio is a 28-year-old married woman who reports that an excessive fear that her husband will die has led her to seek therapy. She further states that because this is her second marriage, it is important for her to work out her problem so that it doesn't ultimately interfere with her relationship with her husband. However, her husband is a sales representative and occasionally has to attend out-of-town meetings. According to Mrs. Oliverio, whenever he has gone away on a trip during the two years of their marriage, she "goes to pieces" and feels "utterly devastated" because of recurring thoughts that he will die and not return. She states that this is a very intense fear and occurs even when he is gone on short trips, such as a half day or a day. She is not aware of any coping thoughts or behaviors she uses at these times. She indicates that she feels great as soon as her husband gets home. She states that this was also a problem for her in her first marriage, which ended in divorce five years ago. She believes the thoughts occur because her father died unexpectedly when she was 11 years old. Whenever her husband tells her he has to leave, or actually does leave, she reexperiences the pain of being told her father has died. She feels plagued with thoughts that her husband will not return and then feels intense anxiety. She is constantly thinking about never seeing her husband again during these anxiety episodes. According to Mrs. Oliverio, her husband has been very supportive and patient and has spent a considerable amount of time trying to reassure her and to convince her, through reasoning, that he will return from a trip. She states that this has not helped her to stop worrying excessively that he will die and not return. She also states that in the past few months her husband has canceled several business trips just to avoid putting her through all this pain.

Mrs. Oliverio also reports that this anxiety has resulted in some insomnia during the past two years. She states that as soon as her husband informs her that he must leave town, she has difficulty going to sleep that evening. When he has to be gone on an overnight trip, she reports, she doesn't sleep at all. She simply lies in bed and worries about her husband dying and also feels very frustrated that it is getting later and later and that she is still awake. She reports sleeping fairly well as long as her husband is home and a trip is not impending.

Mrs. Oliverio reports that she feels very satisfied with her present marriage except for some occasional times

(continued)

LEARNING ACTIVITY **28** *Continued*

when she finds herself thinking that her husband does not fulfill all her expectations. She is not sure exactly what her expectations are, but she is aware of feeling anger toward him after this happens. When she gets angry, she just "explodes" and feels as though she lashes out at her husband for no apparent reason. She reports that she doesn't like to explode at her husband like this but feels relieved after it happens. She indicates that her husband continues to be very supportive and protective in spite of her occasional outbursts. She suspects the anger may be her way of getting back at him for going away on a trip and leaving her alone. She also expresses feelings of hurt and anger since her father's death in being unable to find a "father substitute." She also reports feeling intense anger toward her ex-husband after the divorce—anger she still sometimes experiences.

Mrs. Oliverio has no children. She is employed in a responsible position as an executive secretary and makes $18,500 a year. She reports that she enjoys her work, although she constantly worries that her boss might not be pleased with her and that she could lose her job, even though her work evaluations have been satisfactory. She reports that another event she has been worried about is the health of her brother, who was injured in a car accident this past year. She further reports that she has an excellent relationship with her brother and strong ties to her church.

Clinical Hypothesis Formation
Using the case of Mrs. Oliverio,

1. Develop a clinical hypothesis describing Mrs. Oliverio and what you see as her major concern or issue.
2. Describe the aspects of this case that you believe support your hypothesis.
3. Formulate a list of questions you feel you would need answered to test the accuracy of your hypothesis.*

---

*SOURCE: These questions are adapted from the Clinical Hypothesis Rating Form developed by Morran et al. (1994) in "Relationship between Counselor's Clinical Hypotheses and Client Ratings of Counselor Effectiveness," by D. K. Morran, D. J. Kurpius, G. Brack, and T. G. Rozecki, *Journal of Counseling and Development, 72,* pp. 655-660. Copyright © 1994 by ACA. Reprinted with permission. No further reproduction authorized without written permission of the American Counseling Association.

Most Problem Behavior Is Learned

Problem (maladaptive) behavior is developed, maintained, and subject to alteration or modification in the same manner as normal (adaptive) behavior. Both prosocial and maladaptive, or self-defeating, behaviors are assumed to be developed and maintained either by external situational events or cues, by external reinforcers, or by internal processes such as cognition, mediation, and problem solving. For the most part, maladaptive behavior is not thought to be a function of physical disease or of underlying intrapsychic conflict. This fundamental assumption means that we do not spend a great deal of time sorting out or focusing on possible unresolved early conflicts or underlying pathological states. It does not mean, however, that we rule out or overlook possible organic and physiological causes of problem behavior. For example, clients who complain of "anxiety" and report primarily somatic (body-related) symptoms such as heart palpitations, stomach upset, chest pains, and breathlessness may be chronic hyperventilators (Lum, 1976), although hyperventilation can be considered only after the client has had a physical examination to rule out cardiopathy. Physical examinations also may reveal the presence of mitral valve heart dysfunction for some individuals who complain of "panic attacks." Other somatic symptoms suggesting anxiety, such as sweating, tachycardia, lightheadedness, and dizziness, could also result from organic disorders such as hypoglycemia, hyperthyroidism or other endocrine disorders, or a low-grade infection.

Physiological variables should always be explored, particularly when the results of the assessment do not suggest the presence of other specific stimuli eliciting the problem behavior. It is also important to recognize the need for occasional physiological management of psychological problems—for example, in the kinds of disorders mentioned above. Medications may be necessary in addition to psychological intervention. Antidepressants are typically recommended for some forms of depression, particularly the endogenous type as distinct from the more reactive (situational) type. They have been found helpful as a supplement to psychological treatment for some instances of agoraphobia, a disorder typified by a marked fear of being alone or in public places. Anxiety or panic attacks often are also managed with antidepressants but additionally with beta blockers and/or other antianxiety agents. Furthermore, a biological element, such as biochemical imbalance, seems to be present in many of the psychoses, such as schizophrenia, and these conditions usually require antipsychotic drugs to improve the client's overall level of functioning.

Causes of Problems and Therefore Treatments/Interventions Are Multidimensional

Rarely is a problem caused by only one factor, and rarely does a single, unidimensional treatment program work in actual practice. For example, with a client who reports depression, we may find evidence of organic contributing factors such as Addison's disease (dysfunction of the adrenal gland), of environmental contributing conditions such as being left by his wife after moving to a new town, and of internal contributing factors such as self-deprecatory thoughts and images. Causes and contributing conditions of most client problems are multiple and include overt behavior, environmental events and relationships with others, covert behavior such as beliefs, images, and cognitions, feelings and bodily sensations, and possibly physiological/organic conditions. Intervention is usually more effective when directed toward all these multiple factors. For the client described above, his endocrine balance must be restored and maintained, he must be helped to deal with his feelings of rejection and anger about his wife's departure, he needs to develop alternative resources and supports, including self-support, and he needs help in learning how to modify his self-deprecating thoughts and images. Additionally, he may benefit from problem-solving skills in order to decide the direction he wants his life to take. The more complete and comprehensive the treatment, the more successful the therapy tends to be, and also the less chance of relapse. According to Lazarus (1976, pp. 13–14),

> Comprehensive treatment at the very least calls for the correction of irrational beliefs, deviant behaviors, unpleasant feelings, intrusive images, stressful relationships, negative sensations, and possible biochemical imbalance. To the extent that problem identification (diagnosis) systematically explores each of these modalities, whereupon therapeutic intervention remedies whatever deficits and maladaptive patterns emerge, treatment outcomes will be positive and long-lasting. To ignore any of these modalities is to practice a brand of therapy that is incomplete.

Problems Are to Be Viewed Operationally

We suggest a way to view client problems that defines the client's present problem behaviors and some contributing problem conditions. This approach is called defining the problem "operationally," or "concretely." An operational problem definition functions like a measure, a barometer, or a "behavioral anchor." Operational definitions indicate some very specific problem behaviors; they do not infer vague traits or labels from the client's problem statement. Mischel (1973, p. 10) has contrasted these two approaches to problem conceptualization: "The emphasis is on what a person *does* in situations rather than on inferences about what attributes he *has* more globally."

Consider the following example of a way to view a client's problem operationally. In working with the "depressed" client, we would try to define precisely what the client means by "depressed" in order to avoid any misinterpretation of this self-report feeling statement. Instead of viewing the client's problem as "depression," we would try to specify some problem thoughts, feelings, actions, situations, and persons that are associated with the client's depression. We would find out whether the client experiences certain physiological changes during depression, what the client is thinking about while depressed, and what activities and behaviors occur during the depressed periods.

In other words, the therapist, in conjunction with the client, identifies a series of referents that are indicative of the state of being depressed, anxious, withdrawn, lonely, and so on. The advantage of viewing the problem this way is that vague phenomena are translated into specific and observable experiences. When this occurs, we not only have a better idea of what is happening with the client, but we also have made the problem potentially measurable, allowing us to assess therapy progress and outcome (see also Chapter 10).

Most Problems Occur in a Social Context and Are Functionally Related to Internal and External Antecedents and Consequences

Problems do not usually occur in a vacuum but are related to observable events (verbal, nonverbal, and motoric responses) and to less visible covert or indirect events (thoughts, images, moods and feelings, body sensations) that precipitate and maintain the problem. These internal and external events are called "antecedents" or "consequences." They are functionally related to the problem in that they exert control over it, so that a change in one of these variables often brings about a change in related variables. For example, a child's inability to behave assertively with his teacher may be a function of learned fears, lack of social skills, and the fact that he has moved, is in a new school, and also has his first male teacher. Changing one part of this overall problem—for example, helping him reduce and manage his fears—will exert an effect on all other variables in the situation.

In Chapter 10, in the discussion on goals, we learn that the therapist must be alert not only to the way different parts of the problem are related but also to the impact that change in one variable may have on the others. Occasionally a symptom may perform a very useful function for the client, and removing it could make things worse. For example, in the above illustration, add to the case the fact that the child had on one occasion been sexually abused by a male house intruder. The symptom of fear may be serving the function of protection in his relationships with unknown males. Removal of the fear without consideration of the other parts of the problem could make the presenting problem worse or could bring about the onset of other issues.

The functional relationship of behavior and the environment reflects a systemic/ecological view of human problems; it was articulated as early as 1979 and again in 1993 by Bronfenbrenner. In a social-ecological view of mental health treatment, the individual client and his or her environment are linked together, so that assessment includes not only an individual focus but a contextual focus, including key social settings, events, and resources (Rosado & Elias, 1993).

In the Bronfenbrenner style, counselors need to examine the social and cultural contexts of relationships among these key social settings, events, and resources. Whereas the ecological context of problems is important to consider for all clients, some writers have noted that the sociopolitical context surrounding the client is especially important for clients who feel marginalized, such as clients of color and women (Axelson, 1993; Brown, 1994; Rogoff & Chavajay, 1995). For example, Brown (1994) asserts that in feminist therapy, the first and foremost client is the cultural context; thus "feminist therapy concerns itself not simply with individual suffering but with the social and political meaning of both pain and healing. . . . Feminist therapy aims to deprivatize the lives of both therapists and the people with whom they work by asking, out loud and repeatedly, how each life and each pain are manifestations of processes operating in a larger social context" (p. 17). As an example, it is not enough just to help a female client with her stated feelings of "depression"; you must also explore how the culture's expectations of her gender contribute to her depression (Ivey, Ivey, and Simek-Morgan, 1993).

A similar focus on the cultural and political context with Native American clients has been described by LaFromboise and Low (1989) who state that

traditionally, Indian people live in relational networks that serve to support and nurture strong bonds of mutual assistance and affection. Many tribes still engage in a traditional system of collective interdependence, with family members responsible not only to one another but also the clan and tribe to which they belong. The Lakota Sioux use the term *tiospaye* to describe a traditional community way of life in which an individual's well-being remains the responsibility of the extended family. . . . When problems arise among Indian youth, they become problems of the community as well. The family, kin, and friends join together to observe the youth's behavior, draw the youth out of isolation, and integrate that person back into the activities of the group. (p. 121)

Ivey, Ivey, and Simek-Morgan (1993) point out that in working with some clients, such as the Lakota Sioux described in the above example, the therapist should extend the focus of the assessment from the individual to the *mitwelt*—that is, the immediate family, extended family, and community.

Stated another way, in a social-ecological view of mental health treatment, the individual client and his or her environment are linked together, so that assessment covers not only an individual focus but a contextual focus also, including key social settings, events, and resources (Rosado & Elias, 1993).

THE ABC MODEL OF BEHAVIOR

One way to identify the relationship between problem behavior and environmental events is with the ABC model (O'Leary & Wilson, 1987). The ABC model of behavior suggests that the behavior (B) is influenced by events that precede it, called antecedents (A), and by some types of events that follow behavior, called consequences (C). An antecedent (A) event is a cue or signal that can tell a person how to behave in a situation. A consequence (C) is defined as an event that strengthens or weakens a behavior. Note that these definitions of antecedents and consequences suggest that an individual's behavior is directly related to or influenced by certain events. For example, a behavior that appears to be caused by antecedent events such as anger may also be maintained or strengthened by consequences such as reactions from other people. Assessment interviews focus on identifying the particular antecedent and consequent events that influence or are functionally related to the client's defined problem behavior.

As a very simple example of the ABC model, consider a behavior (B) that most of us engage in frequently: talking. Our talking behavior is usually occasioned by certain cues, such as starting a conversation with another person, being asked a question, or being in the presence of a friend. Antecedents that might decrease the likelihood that we will talk may include worry about getting approval for what we say or how we answer the question or being in a hurry to get somewhere. Cultural norms can also serve as antecedents that occasion talking behavior—for example, in some cultures, there is a hierarchy from elders to youth and younger persons do not initiate conversation out of respect for the elders. Our talking behavior may be maintained by the verbal and nonverbal attention we receive from another person, which is a very powerful consequence, or reinforcer. Other positive consequences that might maintain our talking behavior may be that we are feeling good or happy and engaging in positive self-statements or evaluations about the usefulness or relevance of what we are saying. We may talk less when the other person's eye contact wanders, although the meaning of eye contact varies across cultures, or when he or she tells us more explicitly that we've talked enough. These are negative consequences (C) that decrease our talking behavior. Other negative consequences that may decrease our talking behavior could include bodily sensations of fatigue or vocal hoarseness that occur after we talk for a while, or thoughts and images that what we are saying is of little value to attract the interest of others. As you will see in the next three sections, not only do the components of problem behavior often vary among clients, but what functions as an antecedent or consequence for one person is often very different for someone else.

Behavior

Behavior includes things a client does as well as things a client thinks about. *Overt* behavior is behavior that is visible or could be detected by an observer, such as verbal behavior (talking), nonverbal behavior (for example, gesturing or smiling), or motoric behavior (engaging in some action such as betting, walking, or drinking). *Covert* behavior includes events that are usually internal—inside the client—and are not so readily visible to an observer, who must rely on client self-report and nonverbal behavior to detect such events. Examples of covert behavior include thoughts, beliefs, images, feelings, moods, and body sensations.

As we indicated earlier, problem behavior that clients report rarely occurs in isolated fashion. Most reported problems typically are part of a larger chain or set of behaviors. Moreover, each problem behavior mentioned usually has more than one component. For example, a client who complains of anxiety or depression is most likely using the label to refer to an experience consisting of an *affective* component (feelings, mood states), a *somatic* component (physiological and body-related sensation), a *behavioral* component (what the client does or doesn't do), and a *cognitive* component (thoughts, beliefs, images, or internal

dialogue). Additionally, the experience of anxiety or depression may vary for the client, depending on *contextual* factors (time, place, concurrent events, gender, culture, and sociopolitical climate), and on *relational* factors such as presence or absence of other people. All these components may or may not be related to a particular reported problem. For example, suppose our client who reports "anxiety" is afraid to venture out in public places except for home and work because of heightened anxiety and/or panic attacks. Her reported concern of anxiety seems to be part of a chain that starts with a cognitive component in which she thinks worried thoughts and produces images in which she sees herself alone and unable to cope or to get the assistance of others if necessary. The cognitive component leads to somatic discomfort and tension and to feelings of apprehension and dread. These three components work together to influence her overt behavior—for the last few years, she has successfully avoided almost all public places such as the grocery store, theater, or church, and she functions well only at home or at work. She consequently depends on the support of family and friends to help her function adequately in the home and at work and particularly on the few occasions when she attends public activities or uses public transportation.

It is important to determine the relative importance of each component of the reported problem behavior in order to select appropriate intervention strategies (see also Chapter 11). In Chapter 9 we describe ways to obtain descriptions of these various components of problem behavior with an interview assessment method. It is often valuable to list, in writing, the various components identified for any given problem behavior.

Antecedents

According to Mischel (1973), behavior is situationally determined. This means that given behaviors tend to occur only in certain situations. For example, most of us brush our teeth in a public or private bathroom rather than during a concert or a church service. Antecedents may elicit emotional and physiological reactions such as anger, fear, joy, headaches, or elevated blood pressure. Antecedents influence behavior by either increasing or decreasing its likelihood of occurrence. For example, a child in a first-grade class may behave differently at school than at home or differently with a substitute than with the regular teacher.

Antecedent events that occur immediately before a problem behavior exert influence on it. Events that are not in temporal proximity to the problem behavior can similarly increase or decrease the probability that the behavior will occur. Antecedents that occur in immediate temporal proximity to the problem behavior are technically called *stimulus events* (Bijou & Baer, 1976) and include any external or internal event or condition that either cues the behavior or makes it more or less likely to occur under that condition. Antecedents that are temporally distant from the problem are called *setting events* (Kantor, 1970) and include behavioral circumstances that the person has recently or previously passed through. Setting events may end well before the problem and yet, like stimulus events, still facilitate or inhibit its occurrence. Examples of setting events to consider in assessing client problems are the client's age, developmental stage, and physiological state; characteristics of the client's work, home, or school setting; multicultural factors; and behaviors that emerge to affect subsequent behaviors (Wahler & Fox, 1981). Both stimulus and setting antecedent conditions must be identified and defined individually for each client.

Antecedents also usually involve more than one source or type of event. Sources of antecedents may be *affective* (feelings, mood states), *somatic* (physiological and body-related sensations), *behavioral* (verbal, nonverbal, and motoric responses), *cognitive* (thoughts, beliefs, images, internal dialogue), *contextual* (time, place, multicultural factors, concurrent events), and *relational* (presence or absence of other people). For example, with our client who reported "anxiety," there may be a variety of antecedent sources that cue or occasion each aspect of the problem behavior, such as fear of losing control (cognitive/affective), negative self-statements and misperceptions of self and others (cognitive), awareness of apprehension-related body sensations, fatigue, and hypoglycemic tendencies (somatic), staying up late and skipping meals (behavioral), being in public places or needing to attend public functions (contextual), and absence of significant others such as friends and family (relational).

There are also a variety of antecedent sources that make components of the client's anxiety problem less likely to occur. These include feeling relaxed (affective), being rested (somatic), eating regularly (behavioral), decreased dependence on her husband (behavioral), decreased fear of separation from spouse (affective), positive appraisal of self and others (cognitive), expectation of being able to handle situations (cognitive), absence of need to go to public places or functions (contextual), and being accompanied to a public place by a significant other (relational).

The influence that antecedents have on our behavior may vary with each of us, depending on our learning history.

It is also important to keep in mind that antecedents are overt or covert events that in some way influence the problem behavior either by cuing it or by increasing or decreasing the likelihood that it will occur under certain conditions. In other words, not everything that precedes a behavior is automatically considered an antecedent—only those things that influence a behavioral response in some manner. Problem behavior may, however, also be affected by other situational factors (props) that are usually present in the problem situation but do not directly influence the behavior. This is especially observable if any of these situational factors changes dramatically (Goldiamond & Dyrud, 1967). For instance, a child's behavior in school may be at least temporarily affected if the child's only sibling is hospitalized for injuries received in an automobile accident or if the child's father, who has been a household spouse for ten years, starts to work full time outside the home.

During the assessment phase of counseling, it is important to identify those antecedent sources that facilitate desirable behaviors and those that are related to inappropriate responses. The reason is that, during the intervention (treatment) phase, it is important to select strategies that not only facilitate the occurrence of desirable behavior but also decrease the presence of cues for unwanted behavior. In Chapter 9 we describe and model ways to elicit information about antecedent sources and their effects on problem behavior with an interview assessment approach.

Consequences

The consequences of a behavior are events that follow a behavior and exert some influence on the behavior or are functionally related to the behavior. In other words, not everything that follows a behavior is automatically considered a consequence. For example, suppose you are counseling an overweight woman who tends occasionally to go on eating binges. She reports that, after a binge, she feels guilty, regards herself as even more unattractive, and tends to suffer from insomnia. Although these events are *results* of her eating-binge behavior, they are not consequences unless in some way they directly influence her binges, either by maintaining or by decreasing them. In this case, other events that follow the eating binges may be the real consequences. For instance, perhaps the client's binges are maintained by the enjoyment she gets from eating; perhaps they are temporarily decreased when someone else, such as her partner, notices her behavior and reprimands her for it or refuses to go out with her on their regular weekend splurge.

Consequences are categorized as positive or negative. Positive consequences can be referred to technically as *rewards* or *reinforcers;* negative ones can be labeled *punishers.* Like antecedents, the things that function as consequences will always vary with clients. By definition, positive consequences (rewarding events) will maintain or increase the behavior. Positive consequences often maintain or strengthen behavior by positive reinforcement, which involves the presentation of an overt or covert event following the behavior which increases the likelihood that the behavior will occur again in the future. People tend to repeat behaviors that result in pleasurable effects.

People also tend to engage in behaviors that have some payoffs, or value, even if the behavior is very dysfunctional (such payoffs are called *secondary gains*). For example, a client may abuse alcohol and continue to do so even after she loses her job or her family because she likes the feelings she gets after drinking and because the drinking helps her to avoid responsibility. Another client may continue to verbally abuse his wife despite the strain it causes in their relationship because the abusive behavior gives him a feeling of power and control. In these two examples, the problem behavior is often hard to change, because the immediate consequences make the person feel better in some way. As a result, the problem behavior is reinforced, even if its delayed or long-term effects are unpleasant. In other words, in these examples, the client "values" the behavior that he or she is trying to eliminate. Often the secondary gain, the payoff derived from a manifest problem, is a cover for more severe problems that are not always readily presented by the client. According to Fishman and Lubetkin (1983), it is important for therapists to be alert to this fact in order to focus on the core problem that, when ameliorated, will generalize to other problem areas as well. For example, consider a client who is overweight and wants to "lose weight" as her goal for therapy. Yet assessment of this presenting problem reveals that the client's obesity allows her to avoid looking for suitable employment and allows her to live at home with her parents. Successful therapy would need to be targeted not only to the manifest problem (weight and overeating) but also to the core, or underlying, problem that the weight masks—namely, avoidance of assuming responsibility for herself. Part of successful therapy would also involve acceptance of her own unique body image in a cultural context that values thinness. Similarly, the client described above who uses alcohol to avoid responsibility will need a treatment program targeted not only toward eliminating alcohol abuse but also toward changing her pattern of avoiding responsibil-

ity. As Fishman and Lubetkin note, many cognitive behavior therapists "are too wedded to the 'prima facie' problems that clients bring to therapy. We have observed from our own clinical experience that 'under material' may often be responsible for maintaining the manifest behavior" (1983, p. 27). Clients may not always know the reasons they engage in problem behavior. Part of therapy involves making reasons or secondary gains more explicit.

Positive consequences can also maintain behavior by negative reinforcement—removal of an unpleasant event following the behavior, increasing the likelihood that the behavior will occur again. People tend to repeat behaviors that get rid of annoying or painful events or effects. They also use negative reinforcement to establish *avoidance* and *escape* behavior. Avoidance behavior is maintained when an *expected* unpleasant event is removed. For example, staying at home stops agoraphobia fears. Avoidance of public places is maintained by removal of these expected fears. Escape behavior is maintained when a negative (unpleasant) event *already occurring* is removed or terminated. For example, abusive behavior toward a child temporarily stops the child's annoying or aversive behaviors. Termination of the unpleasant child behaviors maintains the parental escape behavior.

Negative consequences weaken or eliminate the behavior. A behavior is typically decreased or weakened (at least temporarily) if it is followed by an unpleasant stimulus or event (punishment), if a positive, or reinforcing, event is removed or terminated (response cost), or if the behavior is no longer followed by reinforcing events (operant extinction). As an example, an overweight man may maintain his eating binges because of the feelings of pleasure he receives from eating (a positive reinforcing consequence). Or his binges could be maintained because they allow him to escape from a boring work situation (negative reinforcing consequence). In contrast, his wife's reprimands or sarcasm or refusal to go out with him may, at least temporarily, reduce his binges (punishing consequence). Although using negative contingencies to modify behavior has many disadvantages, in real-life settings such as home, work, and school, punishment is widely used to influence the behavior of others. Therapists must be alert to the presence of negative consequences in a client's life and its effects on the client. Therapists must also be careful to avoid the use of any verbal or nonverbal behavior that may seem punitive to a client, because such behavior may contribute to unnecessary problems in the therapeutic relationship and subsequent client termination of (escape from) therapy.

Consequences also usually involve more than one source or type of event. Like antecedents, sources of consequences may be *affective, somatic, behavioral, cognitive, contextual,* and/or *relational.* For example, with our client who reports "anxiety," her avoidance of public places and functions is maintained because it results in a reduction of anxious feelings (affective), body tension (somatic), and worry (cognitive). Additional consequences that may help to maintain the problem may include avoidance of routine chores (behavioral) and increased attention from family and friends (relational).

It would be inaccurate to simply ask about whatever follows the problem behavior and automatically classify it as a consequence without determining its particular effect on the behavior. As Cullen (1983, p. 137) notes, "If variables are supposed to be functionally related to behavior when, in fact, they are not, then manipulation of those variables by the client or therapist will, at best, have no effect on the presenting difficulties or, at worst, create even more difficulties."

Occasionally students seem to confuse consequences as we present the concept in this chapter with the kinds of consequences that are often the results of problem behavior—for example, Julie frequently procrastinates on studying and, as a consequence, receives poor grades. Although poor grades are the result of frequent procrastination, they are not a consequence in the way we are defining it unless the poor grades in some way increase, decrease, or maintain the procrastination behavior. Otherwise, poor grades are simply the result of studying too little. One way to distinguish consequences from mere effects of problem behavior is to remember a rule of thumb termed "gradient of reinforcement." This term refers to the belief that consequences that occur soon after the behavior are likely to have a stronger impact than consequences that occur after a long time has elapsed (Hull, 1980). Poor grades are so far removed in time from daily studying (or lack of it) that they are unlikely to exert much influence on the student's daily study behavior.

During the assessment phase of counseling, it is important to identify those consequences that maintain, increase, or decrease both desirable and undesirable behaviors related to the client's problem. In the intervention (treatment) phase, this information will help you select strategies and approaches that will maintain and increase desirable behaviors and will weaken and decrease undesirable behaviors such as behavioral excesses and deficits. Information about consequences is also useful in planning treatment approaches that rely directly on the use of consequences to facilitate behavior change, such as self-reward (see also Chapter 20). In Chapter 9 we describe and model ways to elicit information about consequences and their effects on problem behavior with an interview assessment approach.

It is important to reiterate that antecedents, consequences, and components of the problem must be assessed and identified for each particular client. Two clients might complain of anxiety or "nerves," and the assessments might reveal very different components of the problem behavior and different antecedents and consequences. A multicultural focus here is also important; the problem behaviors, antecedents, and consequences may be affected by the client's cultural affiliations and social-political context. Also remember that there is often some overlap among antecedents, components of problem behavior, and consequences. For example, negative self-statements or irrational beliefs might function in some instances as both antecedents and consequences for a given component of the identified problem. Consider a college student who reports depression after situations with less than desired outcomes, such as asking a girl out and being turned down, getting a test back with a B or C on it, and interviewing for a job and not receiving a subsequent offer of employment. Irrational beliefs in the form of perfectionistic standards may function as an antecedent by cuing, or setting off, the resulting feelings of depression—for example, "Here is a solution that didn't turn out the way I wanted; it's awful; now I feel lousy." Irrational beliefs in the form of self-deprecatory thoughts may function as a consequence by maintaining the feelings of depression for some time even after the situation itself is over—for example, "When things don't turn out the way they should, I'm a failure."

DIAGNOSTIC CLASSIFICATION OF CLIENT PROBLEMS

Our emphasis throughout this chapter is on the need to conduct a thorough and precise assessment with each client to be able to define client problems in very concrete ways. In addition, counselors need to be aware that client problems can be organized in some form of diagnostic taxonomy (classification).

The official classification system used currently is found in the American Psychiatric Association's *Diagnostic and Statistical Manual of Mental Disorders,* fourth edition (*DSM-IV,* 1994). The reader is urged to consult the manual as well as the *DSM-IV Casebook* (Spitzer, Gibbon, Skodol, Williams, & First, 1994). Our interest is simply to summarize the basic diagnostic codes and categories found in *DSM-IV* so that the reader will not be caught off guard if a colleague or supervisor begins talking about "Axis I, II," and so on.

DSM-IV consists largely of descriptions of various mental and psychological disorders broken down into 17 major diagnostic classes, with additional subcategories within these major categories. Specific diagnostic criteria are provided for each category. These criteria are supposed to provide the practitioner with a way to evaluate and classify the client's problems. The particular evaluation system used by *DSM-IV* is called *multiaxial* because it consists of an assessment on five codes, or *axes:*

Axis I   Clinical disorders and other disorders that may be a focus of clinical attention
Axis II   Personality disorders and mental retardation
Axis III   General medical conditions
Axis IV   Psychosocial and environmental problems
Axis V   Global assessment of functioning

Axis I comprises the clinical disorders as well as any other disorders that the therapist decides are an important focus of clinical attention. Axis II is for reporting both personality disorders and mental retardation. The contributors to the *DSM-IV* note that these two listings were given a separate axis to help ensure that either condition would not be overlooked. Axis II may also be used to record information about the presence of client defensive mechanisms and maladaptive personality features that are present but not in sufficient strength to warrant a diagnosis of personality disorder. If no diagnosis on Axis II is present, the clinician uses the code of V71.09. Axis III is used to note current medical conditions of the client that are relevant to the understanding and/or management of the client's clinical disorders. For example, a client with hypothyroidism may suffer from some sort of depression (coded on Axis I) and the recording of hypothyroidism on Axis III notes the link between the two conditions.

Axis IV is used "for reporting psychosocial and environmental problems that may influence the diagnosis, treatment and prognosis of the mental disorder(s)" reported in Axis I and II (American Psychiatric Association, 1994, p. 29). These problems include nine general categories relating to negative life events, environmental and familial stresses, and lack of social support (identified by the therapist). Problems are grouped in the following categories:

Problems with primary support group
Problems related to the social environment
Educational problems
Occupational problems
Housing problems
Economic problems
Problems with access to health care services
Problems related to interaction with the legal system/crime
Other psychosocial and environmental problems

Axis V is use to report the therapist's assessment of the client's overall level of functioning. This rating is useful in planning treatment and assessing treatment goals. This evaluation is coded on a Global Assessment of Functioning (GAF). The assessment ranges from 0 inadequate information to 100 superior functioning. Descriptions for all other ratings are as follows:

91–100   Superior functioning in a wide range of activities; life's problems never seem to get out of hand.
81–90   Absent or minimal symptoms
71–80   Symptoms, if present, are transient and expectable reactions to stressors
61–70   Some mild symptoms
51–60   Moderate symptoms
41–50   Serious symptoms
31–40   Some impairment in reality testing or communication
21–30   Behavior considerably influenced by delusions or hallucinations or serious impairment in communication or judgment
11–20   Some danger of hurting self or others
 1–10   Persistent danger of severely hurting self or others

Examples of this multiaxial evaluation system can be found following the analyses of the client cases in this chapter. Table 8-3 describes the 17 major diagnostic categories of *DSM-IV* that are classified on Axis I and Axis II.

Taylor (1983) observes that, in spite of apparent conceptual and practical limitations of diagnosis, the process can aid therapists in assessing problem behaviors and in selecting appropriate interventions for treatment. For instance, knowledge about selected features of various types of clinical pathology, such as the usual age of the patient at the onset of some disorder or whether the disorder is more common in men or in women, can aid in assessment. A very useful addition to the *DSM-IV* is its routine inclusion of discussions of age, gender, and cultural implications of the various disorders. As an example, under panic attacks, it notes that in some cultures, a panic attack may involve an intense fear of witchcraft or magic, and under agoraphobia, it reports that in some cultural or ethnic groups, the participation of women in public life is restricted (see also Fodor, 1992).

Selected features of *DSM-IV* are useful for suggesting additional information about the problem behaviors and the controlling variables. For example, the operational criteria

TABLE 8-3. The 17 major *DSM-IV* classifications

Disorders usually first diagnosed in infancy, childhood, or adolescence—mental retardation Axis II, learning disorders, motor skills disorders, communication disorders, pervasive developmental disorders, attention-deficit and disruptive behavior disorders, feeding and eating disorders of infancy or early childhood, tic disorders, elimination disorders. Axis I

Delirium, dementia, and amnestic and other cognitive disorders—different types of delirium, dementia, and amnestic disorders. Axis I

Mental disorders due to a general medical condition not elsewhere classified—catatonic disorder and personality change. Axis I

Substance-related disorders—alcohol-related disorders; amphetamine (or amphetaminelike)-related disorders, caffeine-related disorders, cannabis-related disorders, cocaine-related disorders, hallucinogen-related disorders, inhalant-related disorders, nicotine-related disorders, opioid-related disorders, phencyclidine (or phencyclidinelike)-related disorders, sedative-, hypnotic-, or anxiolytic-related disorders, polysubstance related disorder. Axis I

Schizophrenia and other psychotic disorders—schizophrenia (paranoid, disorganized, catatonic, undifferentiated, or residual types), schizophreniform disorder, schizoaffective disorder, delusional disorder, brief psychotic disorder, shared psychotic disorder due to delusions or hallucinations, substance-induced psychotic disorder. Axis I

Mood disorders—depressive disorders, bipolar disorders. Axis I

Anxiety Disorders—panic disorder without agoraphobia, panic disorder with agoraphobia, agoraphobia without history of panic disorder, specific phobia, social phobia, obsessive-compulsive, posttraumatic stress disorder, acute stress disorder, generalized anxiety disorder. Axis I

Somatoform disorders—somatization disorder, undifferentiated somatoform disorder, conversion disorder, pain disorder, hypochondriasis, body dysmorphic disorder. Axis I

Factitious disorders—with predominantly psychological signs and symptoms, with predominantly physical signs and symptoms, with combined psychological and physical signs and symptoms. Axis I

Dissociate disorders—amnesia, fugue, identity and depersonalization disorders. Axis I

Sexual and gender identity disorders—sexual dysfunctions (due to desire, arousal, orgasmic or pain disorders), sexual dysfunction due to a general medical condition, paraphilias (due to exhibitionism, fetishism, frotteurism, pedophilia, sexual masochism, sexual sadism, transvestic fetishism, voyeurism, or paraphilia), gender identity disorders. Axis I

Eating disorders—anorexia nervosa, bulimia nervosa. Axis I

Sleep disorders—primary sleep disorders (dyssomnias, parasomnias), sleep disorders related to another mental disorder (insomnia, hypersomnia), other sleep disorders. Axis I

Impulse-control disorders not elsewhere classified—intermittent explosive disorder, kleptomania, pyromania, pathological gambling, trichotillomania. Axis I

Adjustment disorders—with depressed mood, with anxiety, with mixed anxiety and depressed mood, with disturbance of conduct, with mixed disturbance of emotions and conduct. Axis I

Personality disorders—paranoid, schizoid, schizotypal, antisocial, borderline, histrionic, narcissistic, avoidant, dependent, obsessive-compulsive. Axis II

Other conditions that may be a focus of clinical attention—psychological factors affecting medical condition, medication-induced movement disorders, other medication-induced disorder. **The following are classified with V codes:** relational problems, problems related to abuse or neglect, additional conditions that may be a focus of clinical attention (noncompliance with treatment, malingering, adult antisocial behavior, child antisocial behavior, borderline intellectual functioning, age-related cognitive decline, bereavement, academic problem, religious or spiritual problem, acculturation problem, phase of life problem, occupational problem, identity problem.) Axis I

found in *DSM-IV* often indicate further target behaviors associated with a particular disorder that should be assessed, and the associated features of a disorder often suggest controlling or contributing variables to be assessed. For instance, if a client describes behaviors related to depression, the therapist can use the operational criteria for major depressive episodes to ask about other target behaviors related to depression that the client may not mention. Therapists can also be guided by the associated features of this disorder to question the client about possible controlling variables typically associated with the disorder (for example, for depression, events such as life changes, loss of reinforcers, and family history of depression).

Nelson and Barlow (1981) also observe that diagnoses may be useful in suggesting treatments that have been found effective with similar problems. For example, clients with phobias typically benefit from modeling (see Chapters 12, 13, and 14) or from fear-reduction approaches such as systematic desensitization (see Chapter 19) and may also require antianxiety medication. A recently published report of a Division 12 (APA) task force describes empirically validated treatments for specific diagnostic categories such as desensitization for phobias and cognitive therapy for depression, anxiety, and pain. (We present a further discussion of this in Chapter 11.)

LIMITATIONS OF DIAGNOSIS: LABELS AND GENDER/MULTICULTURAL BIASES

Diagnostic classification presents certain limitations, and these are most apparent when a client is given a diagnostic classification without the benefit of a thorough and complete assessment. The most common criticisms of diagnosis are that it places labels on clients, often meaningless ones, and that the labels themselves are not well defined and do not describe what the clients do or don't do that makes them "histrionic" or "a conduct disorder" and so on.

Also, the process of making diagnoses using the current edition of the diagnostic and statistical manual has come under sharp criticism from members of feminist therapy groups, from persons of color, and from those who are advocates for clients of color (Sinacore-Guinn, 1995). For example, feminist therapists assert that the development of clinical disorders in women almost always involves a lack of both real and perceived power in their lives (Brown & Ballou, 1992). These therapists have noted that the concept of "distress," which permeates the traditional diagnostic classification system, reflects a "highly individualized phenomenon" and overlooks distress as "a manifestation of larger social and cultural forces" (Brown, 1992, p. 113). Root (1992) observes that "one of the contributions of the feminist perspective is to depathologize normal behavior" (p. 248). As an example, behavior that may be viewed in the traditional diagnostic classification as "regressive, signs of instability, or impaired traditional functioning, may be viewed from a feminist perspective as healthy strategies for staying alive and sane in dangerous and insane places" (Brown, 1992, p. 113), and as self-preservation behaviors (Root,

1992). Gender bias in the assignment of several different *DSM* disorders has been found in various studies (Becker & Lamb, 1994; Ford & Widiger, 1989). Brown (1992) argues that a feminist perspective of psychopathology must include the pathology of oppression. The theoretical foundations of psychiatry and psychology underpinning traditional diagnosis "have limited contexts and tend to be ahistorical . . . making invisible the experiences of large segments of the population who have been historically oppressed" (Root, 1992, p. 258). In this model, health is defined not just as an absence of distress "but also as the presence of nonoppressive attitudes and relationships toward other humans, animals, and the planet" (Brown, 1992, p. 112). A feminist model of psychopathology and health also examines the existence and the meaning of particular symptom patterns that may emerge with certain cultural groups. As Root (1992) observes, "For many minority groups, the repeated and/or chronic experience of traumatic events makes it difficult for the individual to believe in anything but unique vulnerability"; this sort of vulnerability "is reinforced in persons who are subject to repeated discrimination or threat, such as anti-gay/lesbian violence, racist-motivated violence, anti-Semitic violence, chronic torture experienced by many Southeast Asian refugees, and repeated interpersonal sexual assault and violence" (p. 244).

In addition to the social-political context described above, a feminist view of psychopathology and health also involves an examination of the social-political context surrounding an individual's expressed behavior. As Fodor (1992) notes, in an earlier time and in a different social context, not leaving the house was considered appropriate behavior for women; now this condition has become pathologized as "agoraphobia." Similarly, Ross (1990) observes that hearing voices or seeing ghosts is not viewed as a pathological symptom and a sign of psychosis in those cultures that consider such experiences to be indications of *divine favor*. A feminist conception of diagnosis includes cultural relativity and ascertains what is normal for *this* individual, in *this* particular time and place (Brown, 1992, p. 113). Feminist practitioners would ask a client, "What has happened to you?" rather than "What is wrong with you?"

Similar concerns about bias in diagnosis have been raised by cross-cultural researchers and practitioners. Snowden and Cheung (1990) found bias in diagnosis by practitioners when they were diagnosing clients of color. Atkinson, et al. (1996) found bias by the fact that Euro-American psychologists rated specific *DSM-IV* disorders as more appropriate for an African American female client than did the African American psychologists. They concluded that the differential diagnosis could be based on negative racial bias by white psychologists, positive racial bias by black psychologists, or biases on the parts of both groups of helpers (p. 504). They conclude that helpers must continually examine their rationale for their diagnoses of clients of color (p. 504). Mwaba and Pedersen (1990) found that in many cases clinicians misinterpret culturally appropriate behaviors as pathological.

Sinacore-Guinn (1995) cautions that within any cultural system, variables such as coping styles have specific meanings. In making gender- and culture-sensitive diagnoses, therapists must also ascertain the cultural meanings of symptoms, behaviors, and presenting issues for each client. As she notes, "Clients may be struggling with a structure within their culture or a conflict between two (or more) cultures of which they are a part. What may be misconstrued as an adjustment disorder may in fact be a cultural conflict in which the client is trying to negotiate and satisfactorily meet the demands of separate cultural systems" (p. 21). Immigration history is another example of a cultural variable that must be understood from the perspective of the client's cultural affiliations. As Sinacore-Guinn (1995) observes, "It is not uncommon in some cultures for children to be raised by a relative while one or both of the parents are establishing themselves, and thus ultimately the family, in a new country. If the counselor does not understand this situation from a cultural perspective, he or she may inappropriately attempt to diagnostically apply a Western notion of abandonment where it does not apply" (p. 22).

Sue (1991) points out that clinicians may make two kinds of errors in the assessment process with clients of color. First, they may assume that ethnic differences exist when they do not—or the reverse assumption that all clients will exhibit the same symptomatology. Sue (1991) recommends that in instances when the clinician is unfamiliar with a particular cultural group the aid of a consultant be sought. Sue and Sue (1990) point out that the history of oppression (described earlier by Brown for women) also affects resulting diagnoses made for clients of color who, because of this history, may be reluctant to self-disclose and, as a result, may be labeled *paranoid*. Sue and Sue (1990), like Brown, argue that diagnosis of clients of color must be understood from a larger social-political perspective. Otherwise, these clients may receive a diagnosis that overlooks the survival value of their behaviors in a racist society.

Itai and McRae (1994) observe that misdiagnoses can easily occur when English is not the primary language of the client. They also note that cultural practices that may seem psychotic to Euro-American counselors may be typical for some non-Euro-American clients. For example, an older Japanese American woman may say that she speaks with her dead husband. They also caution Euro-American counselors to be careful about assigning a diagnosis of certain personality disorders to older Japanese American clients. They note, for example, the Asian value of collectivity versus the Euro-American value of self-sufficiency. Because they value interdependence and collectivity, some Asian American clients tend to agree with most other persons, regardless of their own point of view. As Itai and McRae (1994) observe, "These behaviors do not suggest a diagnosis of Dependent Personality Disorder but show differences of cultural emphasis regarding interdependence, respect, and conformity" (p. 376). Bauermeister and others (1990) observe that the same phenomenon may occur in the diagnosis of ADHA or attention deficit-hyperactivity disorder for some Puerto Rican clients whose behaviors, although similar to some ADHD behaviors, simply reflect a different time and activity orientation, not pathology or overactivity.

In response to these concerns, the contributors to the *DSM-IV* recognize that accurate diagnosis can be challenging "when a clinician from one ethnic or cultural group uses the *DSM-IV* classification to evaluate an individual from a different ethnic or cultural group" (p. xiv). They observe that therapists who are not cognizant of the nuances of a client's cultural frame of reference "may incorrectly judge as psychopathology those normal variations in behavior, belief, or experience that are particular to the individual's culture" (p. vix). In the *DSM-IV,* in addition to very brief discussions of age, gender, and cultural features of the clinical disorders, there is an appendix that includes a glossary of 12 "culture-bound" syndromes—that is what the *DSM-IV* (APA, 1994) defines as "localized, folk, diagnostic categories that frame coherent meanings for certain repetitive, patterned, and troubling sets of experiences and observations" (p. 844). Although there is not usually an equivalent found in the *DSM-IV* clinical categories, associated relevant *DSM-IV* categories are also cross-listed with these syndromes. In addition to this glossary, the appendix also includes an outline for a supplemental "cultural formulation" to be used in addition to the multiaxial system. The categories to be included in the therapist's cultural formulation include (1) cultural identity/affiliation of the client, (2) any cultural explanations of the client's "illness," (3) cultural factors related to psychosocial environment and levels of functioning, (4) cultural elements of the relationship between the counselor and client, and (5) overall cultural assessment that may impact diagnosis and treatment of the client.

A broad-based perspective by which to conceptualize presenting problem behaviors from a cultural perspective has been developed by Sinacore-Guinn (1995). This model, which she calls "the diagnostic window" is depicted in Figure 8-1. It involves assessment of four categories:

1. *Cultural systems and structures* including "Community structure, family, schools, interaction styles, concepts of illness, life-stage development, coping patterns, and immigration history" (p. 21);
2. *Cultural values* including the five value orientations of *time, activity, relational orientation, person-nature orientation,* and *basic nature of people;*
3. *Gender socialization,* including cultural, ethnic, and racial variations in prescribed gender roles, meaning and attitudes assigned by clients to gender, and also preferred sexual orientation;
4. *Trauma,* including direct, indirect, and insidious trauma as well as the social environmental and sociopolitical context surrounding the trauma.

The counselor's task is to understand, in a culturally sensitive way, the client's presenting issues regarding these four areas that, as shown in Figure 8-1, fall outside the window. This understanding prevents an automatic labeling of issues in these areas as necessarily pathological, even though they may be troubling or conflict arousing for the client. It is *within* the window "that pathology exists and it is from the presence of pathology that the diagnosis is made" (Sinacore-Guinn, 1995, p. 25). Using this model, not all clients will have material inside the window. As Sinacore-Guinn (1995) asserts, "The key is for the counselor to carefully limit the frame of that window against overinclusion. Presenting problems that reside outside the window are not pathological and should be given a V code" (pp. 25–26).

Despite the apparent disadvantages of diagnosis, many counselors and therapists find themselves in field placement and work settings in which they are required to make a diagnostic classification of the client's problems. Often even clients request a diagnosis in order to receive reimbursement from their health insurance carrier for payment made for therapeutic services. This situation is becoming more common with the growth of managed care/HMOs (health maintenance organizations). We believe that when the *DSM-IV* system of classification needs to be used, it should be applied within the context of a complete multifactor and multicultural approach and should not be used as a substitute for an idiographic assessment of the specified problem events and behaviors.

AN ABC MODEL CASE

To assist you in conceptualizing client problems with an ABC model, we provide a case illustration followed by two practice cases for you to complete. The conceptual understanding you should acquire from this chapter will help you actually define client problems and contributing variables with an interview assessment, described in Chapter 9. The following hypothetical case will assist you in identifying the overt and covert ABCs of a client problem. Extensions of this case will be used as illustrations in remaining chapters of the book.

The Case of Joan

Joan is a 15-year-old Euro-American student completing her sophomore year of high school and presently taking a college preparatory curriculum. Her initial statement in the first counseling session is that she is "unhappy" and feels "dissatisfied" with this school experience but feels unable to do anything about it. On further clarification, Joan reveals that she is unhappy because she doesn't think she is measuring up to her classmates and that she dislikes being with these "top" kids in some of her classes, which are very competitive. She reports particular concern in one math class, which she says is composed largely of "guys" who are much smarter than

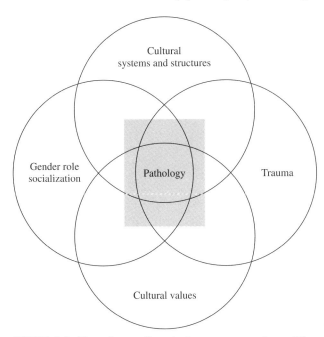

FIGURE 8-1. The diagnostic window. *SOURCE:* From "The Diagnostic Window: Culture- and Gender-Sensitive Diagnosis and Training," by A. L. Sinacore-Guinn, *Counselor Education and Supervision, 35,* pp. 18-31. © 1995 ACA. Reprinted with permission. No further reproduction authorized without written permission of the American Counseling Association.

she is. She states that she thinks about the fact that "girls are so dumb in math" rather frequently during the class. She reports that as soon as she is in this class, she gets anxious and "withdraws." She states that she sometimes gets anxious just thinking about the class, and when this happens, she gets "butterflies" in her stomach, her palms get sweaty and cold, and her heart beats faster. When asked what she means by "withdrawing," she says she sits by herself, doesn't talk to her classmates, and doesn't volunteer answers or go to the board. Often, when called on, she says nothing. As a result, she reports, her grades are dropping. She also states that her math teacher has spoken to her several times about her behavior and has tried to help her do better. However, Joan's nervousness in the class has resulted in her cutting the class whenever she can find any reason to do so, and she has almost used up her number of excused absences from school. She states that her fear of competitive academic situations has been a problem since junior high, when her parents started to compare her with other students and put "pressure" on her to do well in school so she could go to college. When asked how they pressure her, she says they constantly talk to her about getting good grades, and whenever she doesn't, they lash out at her and withdraw privileges, like her allowance. She reports that, during this year, since the classes are tougher and more competitive, school is more of a problem to her and she feels increasingly anxious in certain classes, especially math. Joan also states that sometimes she thinks she is almost failing on purpose to get back at her parents for their pressure. Joan reports that all this has made her dissatisfied with school and she has questioned whether she wants to stay in a college prep curriculum. She states that she has considered switching to a work-study curriculum so she can learn some skills and get a job after high school. However, she says she is a very indecisive person and does not know what she should do. In addition, she is afraid to decide this because if she changed curriculums, her parents' response would be very negative. Joan states that she cannot recall ever having made a decision without her parents' assistance. She feels they have often made decisions for her. She says her parents have never encouraged her to make decisions on her own, because they say she might not make the right decision without their help. Joan is an only child. She indicates that she is constantly afraid of making a bad or wrong choice.

Analysis of Case

Problem Situations.   There are three related but distinct problem areas for Joan. Her "presenting" problem is an academic one (note the corresponding diagnosis on the *DSM-IV* Axis I)—she feels anxious about her performance in competitive classes at school, primarily math, and is aware that her grades are dropping in this class. Second, she is generally unsure about her long-term goals and career choice and whether the college prep curriculum she is in is what she wants. (Note the diagnosis of identity problem on Axis I of the *DSM-IV* classification). She feels indecisive about this particular situation and regards herself as indecisive in many other situations as well. These two problems are exacerbated by her relationship with her parents. This is coded as a psychosocial and environmental problem on Axis IV of the *DSM-IV* classification. This coding notes the potential influence of this relational problem on the academic and decision-making/identity issues mentioned above. Next we analyze these two problems according to the ABC model we presented earlier in the chapter.

Analysis of School Problem

1. *Problem Behaviors*
   Joan's problem behaviors at school include
   A. Self-defeating labeling of her math class as "competitive" and of herself as "not as smart as the guys."
   B. Sitting alone, not volunteering answers in math class, not answering the teacher's questions or going to the board, and cutting class.

   Her self-defeating labels are a covert behavior; her sitting alone, not volunteering answers, and cutting class are overt behaviors.

2. *Context of Problem*
   A. *Antecedent Conditions*
      Joan's problem behaviors at school are elicited by anxiety about certain "competitive" classes, particularly math. Previous antecedent conditions would include verbal comparisons about Joan and her peers made by her parents and verbal pressure for good grades and withholding of privileges for bad grades by her parents. Note that these antecedent conditions do not occur at the same time. The antecedent of the anxiety in the "competitive" class occurs in close proximity to Joan's problem behaviors and is a "stimulus event." However, the verbal comparisons and parental pressure began several years ago and probably function as a "setting event."
   B. *Consequences*
      Joan's problem behaviors at school are maintained by

(1) An increased level of attention to her problems by her math teacher.

(2) Feeling relieved of anxiety through avoidance of the situation that elicits anxiety. By not participating in class and by cutting class, she can avoid putting herself in an anxiety-provoking situation.

(3) Her poorer grades, possibly because of two "payoffs," or secondary gains. (1) If her grades get too low, she may not qualify to continue in the college prep curriculum. This would be the "ultimate" way to avoid putting herself in competitive academic situations that elicit anxiety. (2) The lowered grades could also be maintaining her problem behaviors because she labels the poor grades as a way to "get back at" her parents for their pressure.

Analysis of Decision-Making Problem

1. *Problem Behaviors*
   Joan's problem behavior can be described as not making a decision for herself—in this case, about a curriculum change. Depending on the client, problems in making decisions can be either a covert or an overt problem. In people who have the skills to make a decision but are blocking themselves because of their "labels" or "internal dialogue" about the decision, the problem behavior would be designated as covert. In Joan's case, her indecisive behavior seems based on her past learning history of having many decisions either made for her or made with parental assistance. The lack of opportunities she has had to make choices suggests she has not acquired the skills involved in decision making. This would be classified as an overt problem.

2. *Context of Problem*
   A. *Antecedent Conditions*
      Joan's previous decision-making history is the primary antecedent condition. This consists of (1) having decisions made for her and (2) a lack of opportunities to acquire and use the skills of decision making.
   B. *Consequences*
      The consequences that seem to be maintaining her problem behavior of not deciding include
      (1) Getting help with her decisions, thereby avoiding the responsibility of making a choice.

(2) Anticipation of parental negative reactions (punishment) to her decisions through her self-talk.

(3) Absence of positive consequences or lack of encouragement for any efforts at decision making in the past.

(4) In the specific decision of a curriculum change, her low grades, which, if they get too bad, may help her avoid making a curriculum decision by automatically disqualifying her from the college prep curriculum.

C. *Social-political Context*
   This part of the assessment addresses the question of how Joan's presenting problems are a manifestation of the social-political context and structure in which she lives. Joan's concerns appear to be shaped by a context in which she has been reinforced (and punished) for what she does (or doesn't do). This pattern has led to a devaluing and uncertainty of who she is—and what she wants and needs. She appears to feel powerless in her current environment, partly, we suspect, because of the power her parents have exerted over her; partly because of the power exerted by a school system that emphasizes college-prep values; and partly because of lessons she has learned from her cultural groups about men, women, and achievement. Brown and Gilligan (1992), Pipher (1994), and others have noted the onset of adolescence as a turning point in girls' lives in which, in order to achieve and to relate, girls seem to lose a sense of their true self and become tentative in their expressiveness and unsure of their identity. Both the overt and covert problem behaviors we describe in #1 above, appear to be tools Joan is using to cope with this loss and sense of powerlessness as well as ways to attempt to increase the power she has and decrease the power held by other sources of authority.

DSM-IV DIAGNOSIS

Axis I:	V62.3 academic problem
	313.82 identity problem
Axis II	V71.09, no diagnosis
Axis III	None
Axis IV	Problems with primary support system
Axis V	GAS = 65 (current)

LEARNING ACTIVITY **29**   *ABCs of Problem Assessment*

To help you in conceptualizing a client's problem from the ABC model, we provide the following two cases. We suggest that you work through the first case completely before going on to the second one. After reading each case, by yourself or with a partner, respond to the questions following the case. Then check your responses with the feedback.

The Case of Ms. Weare and Freddie

Ms. Weare and her 10-year-old son, Freddie, have come to counseling at the referral of Family Services. Their initial complaint is that they don't get along with each other. Ms. Weare complains that Freddie doesn't dress by himself in the morning and this makes her mad. Freddie complains that his mother yells and screams at him frequently. Ms. Weare agrees she does, especially when it is time for Freddie to leave for school and he isn't dressed yet. Freddie agrees that he doesn't dress himself and points out that he does this just to "get mom mad." Ms. Weare says this has been going on as long as she can remember. She states that Freddie gets up and usually comes down to breakfast not dressed. After breakfast, Ms. Weare always reminds him to dress and threatens him that she'll yell or hit him if he doesn't. Freddie usually goes back to his room, where, he reports, he just sits around until his mother comes up. Ms. Weare waits until five minutes before the bus comes and then calls Freddie. After he doesn't come down, she goes upstairs and sees that he's not dressed. She reports that she gets very mad and yells "You're dumb. Why do you just sit there? Why can't you dress yourself? You're going to be late for school. Your teacher will blame me, since I'm your mother." She also helps Freddie dress. So far, he has not been late, but Ms. Weare says she "knows" he will be if she doesn't "nag" him and help him dress. On further questioning, Ms. Weare says this does not occur on weekends, only on school days. She states that, as a result of this situation, she feels very nervous and edgy after Freddie leaves for school, often not doing some necessary work because of this. Asked what she means by "nervous" and "edgy," she reports that her body feels tense and jittery all over. She indicates that this does not help her high blood pressure. She reports that since Freddie's father is not living at home, all the child rearing is on her shoulders. Ms. Weare also states that she doesn't spend much time with Freddie at night after school because she does extra work at home at night as she and Freddie "don't have much money."

DSM-IV Diagnosis for Ms. Weare
Axis I   V61.20 parent-child relational problem

Axis II   V71.09, no diagnosis
Axis III   401.9 (hypertension, essential)
Axis IV   None
Axis V   GAF = 75 (current)

Respond to these questions. Feedback follows the Learning Activity.

1. What problem behaviors does Freddie demonstrate in this situation?
2. Is each problem behavior you have listed overt or covert?
3. What problem behaviors does Ms. Weare exhibit in this situation?
4. Is each problem behavior you have listed overt or covert?
5. List one or more antecedent conditions that seem to bring about each of Freddie's problem behaviors.
6. List one or more antecedent conditions that seem to bring about each of Ms. Weare's problem behaviors.
7. List one or more consequences (including any secondary gains) that influence each of Freddie's problem behaviors. After each consequence listed, identify how the consequence seems to influence his behavior.
8. List one or more consequences that seem to influence each of Ms. Weare's behaviors. After each consequence listed, identify how the consequence seems to influence her behavior.
9. Identify aspects of the socio-political context that appear to impact Ms. Weare's behavior.

The Case of Mrs. Rodriguez

Mrs. Rodriguez is a 34-year-old Mexican American woman. She was brought to the emergency room by the police after her bizarre behavior in a local supermarket. According to the police report, Mrs. Rodriguez became very aggressive toward another shopper, accusing the man of "following me around and spying on me." When confronted by employees of the store about her charges, she stated "God speaks to me. I can hear His voice guiding me in my mission." On mental-status examination, the counselor initially notes Mrs. Rodriguez's unkempt appearance. She appears unclean. Her clothing is somewhat disheveled. She seems underweight and looks older than her stated age. Her tense posture seems indicative of her anxious state, and she smiles inappropriately throughout the interview. Her speech is loud and fast, and she constantly glances suspiciously around the room. Her affect is labile, fluctuating from anger to euphoria. On occasion, she looks at the ceiling and spontaneously

(continued)

LEARNING ACTIVITY 29 *Continued*

starts talking. When the counselor asks to whom she was speaking, she replies, "Can't you hear Him? He's come to save me!" Mrs. Rodriguez is alert and appears to be of average general intelligence. Her attention span is short. She reports no suicidal ideation and denies any past attempts. She does, however, express some homicidal feelings for those who "continue to secretly follow me around." When the family members arrive, the counselor is able to ascertain that Mrs. Rodriguez has been in psychiatric treatment on and off for the last ten years. She has been hospitalized several times in the past ten years during similar episodes of unusual behavior. In addition, she has been treated with several antipsychotic medicines for her problem. There is no evidence of organic pathology or any indication of alcohol or drug abuse. Her husband indicates that she recently stopped taking her medicine after the death of her sister and up until then had been functioning adequately during the past year with not much impairment.

DSM-IV Diagnosis

Axis I: 295.30, Schizophrenia, Paranoid Type
Axis II: V71.09, no diagnosis
Axis III: None
Axis IV: Problems with primary social support system (recent death of sister)
Axis V: GAF = 25 (current)

Respond to these questions. Feedback follows.

1. List several of the problem behaviors that Mrs. Rodriguez demonstrates.
2. Is each problem behavior you have listed overt or covert?
3. List one or more antecedents that seem to elicit Mrs. Rodriguez's problem behaviors.
4. List one or more consequences that appear to influence the problem behaviors, including any secondary gains. Describe how each consequence seems to influence the behavior.
5. Identify aspects of the socio-political context that impact her behavior.

SUMMARY

Assessment is the basis for development of the entire counseling program. Assessment has important informational, educational, and motivational functions in therapy. Although the major part of assessment occurs early in the counseling process, to some extent assessment, or identification of client concerns, goes on constantly during therapy.

An important part of assessment is the counselor's ability to conceptualize client problems. In this chapter we described the models of case or problem conceptualizations proposed by Ainsworth and Bowlby (1991) and Lazarus (1976, 1989). Conceptualization models help the counselor think clearly about the complexity of client problems.

The ABC model of assessment described in this chapter is based on several assumptions, including these:

1. Most problem behavior is learned, although some psychological problems may have organic (biological) causes.
2. Causes of problems are multidimensional.
3. Problems need to be viewed operationally, or concretely.
4. Problems occur in a social context and are affected by internal and external antecedents that are functionally related to or exert influence on the problem in various ways.

5. Components of the problem as well as sources of antecedents and consequences can be affective, somatic, behavioral, cognitive, contextual, and relational.

In addition to the need to identify components of the problem behavior and sources of antecedents and consequences, another part of assessment may involve a multiaxial diagnosis of the client. Current diagnosis is based on the *Diagnostic and Statistical Manual,* fourth edition, and involves classifying the disorders, medical conditions, and psychosocial and environmental problems, and making a global assessment of current functioning. Diagnosis can be a useful part of assessment. For example, knowledge about selected features of various types of clinical syndromes can add to understanding of the client's concern. Diagnosis, however, is not an adequate *substitute* for other assessment approaches and is not an effective basis for specifying goals and selecting intervention strategies unless it is part of a comprehensive treatment approach in which components of the problem are identified in a concrete, or operational, manner for each client. Research has shown that both assessment and diagnosis are subject to gender and cultural bias. The skilled practitioner conducts a multidimensional assessment process that includes an awareness of the current and historical sociopolitical context in which the client lives and also the client's gender and cultural referent groups.

FEEDBACK 28
ABCs of Problem Assessment

The Case of Ms. Weare and Freddie

1. Freddie's problem behavior is sitting in his room and not dressing for school.
2. This is an overt behavior, as it is visible to someone else.
3. Ms. Weare's problem behaviors are (a) feeling mad and (b) yelling at Freddie.
4. (a) Feeling mad is a covert behavior, as feelings can only be inferred. (b) Yelling is an overt behavior that is visible to someone else.
5. Receiving a verbal reminder and threat from his mother at breakfast elicits Freddie's behavior.
6. Ms. Weare's behavior seems to be cued by a five-minute period before the bus arrives on school days.
7. Two consequences seem to influence Freddie's problem behavior of not dressing for school. (a) He gets help in dressing himself; this influences his behavior by providing special benefits. (b) He gets some satisfaction from seeing that his mother is upset and is attending to him. This seems to maintain his behavior because of the attention he gets from her in these instances. A possible secondary gain is the control he exerts over his mother at these times. According to the case description, he doesn't seem to get much attention at other times from his mother.
8. The major consequence that influences Ms. Weare's behavior is that she gets Freddie ready on time and he is not late. This result appears to influence her behavior by helping her avoid being considered a poor mother by herself or someone else.
9. This parent-child relational problem is undoubtedly impacted by the fact that Ms. Weare is raising her son alone and appears to be living in a fairly isolated social climate with little social support. She also is the sole economic provider for Freddie, and her behavior and the child rearing are affected by her lack of financial re-

(continued)

sources. Overall she appears to feel disempowered in her ability to handle her parental and financial responsibilities. She has probably been influenced by a cultural norm in which she expects herself to do it all and feels uncomfortable about asking for help.

The Case of Mrs. Rodriguez

1. There are various problem behaviors for Mrs. Rodriguez: (a) disheveled appearance, (b) inappropriate affect, (c) delusional beliefs, (d) auditory hallucinations, (e) homicidal ideation, (f) noncompliance with treatment (medicine).
2. Disheveled appearance, inappropriate affect, and noncompliance with treatment are overt behaviors—they are observable by others. Delusions, hallucinations, and homicidal ideation are covert behaviors as long as they are not expressed by the client and therefore not visible to someone else. However, when expressed or demonstrated by the client, they become overt behaviors as well.
3. In this case, Mrs. Rodriguez's problem behaviors appear to be elicited by the cessation of her medication, which is the major antecedent. Apparently, when she stops taking her medicine an acute psychotic episode results.
4. This periodic discontinuation of her medicine and the subsequent psychotic reaction may be influenced by the attention she receives from the mental health profession, her family, and even strangers when she behaves in a psychotic, helpless fashion. Additional possible secondary gains include avoidance of responsibility and of being in control.
5. Identify the socio-political context of the problem. In this case it is important to note the potential influence of the cultural-ethnic affiliations of Mrs. Rodriguez. Ideas that may seem delusional in one culture may represent a common belief held by very many persons in another culture. Delusions with a religious thread may be considered a more typical part of religious experience in a particular culture, such as a sign of "divine favor." The skilled clinician would take this into consideration in the assessment before settling on a final diagnosis.

POSTEVALUATION

Read the case descriptions of Mr. Huang and of John that follow and then answer the following questions:

1. What are the client's problem behaviors?
2. Are the problem behaviors overt or covert?

3. What are the antecedent conditions of the client's concern?
4. What are the consequences of the problem behaviors? Secondary gains?

(continued)

5. In what way do the consequences influence the problem behaviors?
6. In what ways are the problem behaviors manifestations of the social-political context?

Answers to these questions are provided in the Feedback section that follows the Postevaluation.

The Case of Mr. Huang

A 69-year-old Asian American man, Mr. Huang, came to counseling because he felt his performance on his job was "slipping." Mr. Huang had a job in a large automobile company. He was responsible for producing new car designs. Mr. Huang revealed that he noticed he had started having trouble about six months before, when the personnel director came in to ask him to fill out retirement papers. Mr. Huang, at the time he sought counseling, was due to retire in nine months. (The company's policy made it mandatory to retire at age 70.) Until this incident with the personnel director and the completion of the papers, Mr. Huang reported, everything seemed to be "OK." He also reported that nothing seemed to be changed in his relationship with his family. However, on some days at work, he reported having a great deal of trouble completing any work on his car designs. When asked what he did instead of working on designs, he said "Worrying." The "worrying" turned out to mean that he was engaging in constant repetitive thoughts about his approaching retirement, such as "I won't be here when this car comes out" and "What will I be without having this job?" Mr. Huang stated that there were times when he spent almost an entire morning or afternoon "dwelling" on these things and that this seemed to occur mostly when he was alone in his office actually working on a car design. As a result, he was not turning in his designs by the specified deadlines. Not meeting his deadlines made him feel more worried. He was especially concerned that he would "bring shame both to his company and to his family who had always been proud of his work record." He was afraid that his present behavior would jeopardize the opinion others had of him, although he didn't report any other possible "costs" to him. In fact, Mr. Huang said that it was his immediate boss who had suggested, after several talks and after-work drinks, that he see a counselor. The boss also indicated the company would pay for Mr. Huang's counseling. Mr. Huang said that his boss had not had any noticeable reactions to his missing deadlines, other than reminding him and being solicitous, as evidenced in the talks and after-work drinks. Mr. Huang reported that he enjoyed this interaction with his boss and often wished he could ask his boss to go out to lunch with him. However, he stated that these meetings had all been at his boss's request. Mr. Huang felt somewhat hesitant about making the request himself. In the last six months, Mr. Huang had never received any sort of reprimand for missing deadlines on his drawings. Still, he was concerned with maintaining his own sense of pride about his work, which he felt might be jeopardized since he'd been having this trouble.

DSM-IV Diagnosis

Axis I: 309.24, Adjustment disorder with anxiety
Axis II: V71.09, no diagnosis
Axis III: None
Axis IV: Problems related to the social environment: adjustment to life cycle transition of retirement.
Axis V: GAS = 75 (current)

The Case of John

This is a complicated case with three presenting problems: (1) work, (2) sexual, and (3) alcohol. We suggest that you complete the analysis of ABCs (Questions 1–5 listed at the beginning of this postevaluation) *separately* for each of these three problems. Question 6 can be completed at the end of the third problem.

John, a 30-year-old African American business manager, has been employed by the same large corporation for two years, since his completion of graduate school. During the first counseling session, he reports a chronic feeling of "depression" with his present job. In addition, he mentions a recent loss of interest and pleasure in sexual activity, which he describes as "frustrating." He also relates a dramatic increase in his use of alcohol as a remedy for the current difficulties he is experiencing.

John has never before been in counseling and admits to feeling "slightly anxious" about this new endeavor. He appears to be having trouble concentrating when asked a question. He traces the beginning of his problems to the completion of his master's degree a little over two years ago. At that time, he states, "everything was fine." He was working part time during the day for a local firm and attending college during the evenings. He had been dating the same woman for a year and a half and reports a great deal of satisfaction in their relationship. Drinking occurred infrequently, usually only during social occasions or a quiet evening alone. On completion of his degree, John relates, "things changed. I guess maybe I expected too much too soon." He quit his job in the expectation of finding employment with a larger company. At first there were few offers, and he was beginning to wonder whether he had made a mistake. After several interviews, he was finally offered a job with a business firm that specialized in computer technology, an area in which John was intensely interested. He accepted and was immediately placed in a managerial position. Initially, John was comfortable and felt confident in his new occupation; however, as the weeks and months passed, the competitive nature of the job

(continued)

POSTEVALUATION   (continued)

began to wear him down. He relates that he began to doubt his abilities as a supervisor and began to tell himself that he wasn't as good as the other executives. He began to notice that he was given fewer responsibilities than the other bosses, as well as fewer employees to oversee. He slowly withdrew socially from his colleagues, refusing all social invitations. He states that he began staying awake at night obsessing about what he might be doing wrong. Of course, this lack of sleep decreased his energy level even further and produced a chronic tiredness and lessening of effectiveness and productivity at work. At the same time, his relationship with his girlfriend began to deteriorate slowly. He relates that "she didn't understand what I was going through." Her insistence that his sexual performance was not satisfying her made him even more apprehensive and lowered his self-esteem even further. After a time, his inhibition of sexual desire resulted in inconsistency in maintaining an erection throughout the sexual act. This resulted in an even greater strain on their relationship, so that she threatened to "call it quits" if he did not seek treatment for his problem. He reports that it was at this time that he began to drink more heavily. At first it was just a few beers at home alone after a day at the office. Gradually, he began

to drink during lunch, even though, he states, "I could have stopped if I had wanted to." However, his repeated efforts to reduce his excessive drinking by "going on the wagon" met with little success. He began to need a drink every day in order to function adequately. He was losing days at work, was becoming more argumentative with his friends, and had been involved in several minor traffic accidents. He states, "I think I'm becoming an alcoholic." John points out that he has never felt this low before in his life. He reports feeling very pessimistic about his future and doesn't see any way out of his current difficulties. He's fearful that he might make the wrong decisions, and that's why he's come to see a counselor at this time in his life.

DSM-IV Diagnosis

Axis I: 305.00 Alcohol abuse
            302.72 Male erectile disorder
            311.00 depressive disorder NOS (not otherwise specified)
Axis II: V71.09, no diagnosis
Axis III: None
Axis IV: Problems with primary social support system and occupational problems
Axis V: GAS = 55 (current)

SUGGESTED READINGS

Ainsworth, M.D.S. (1989). Attachment beyond infancy. *American Psychologist, 44,* 709–716.

American Psychiatric Association. (1994). *Diagnostic and statistical manual of mental disorders* (4th ed.). Washington, DC: Author.

Atkinson, D., Brown, M., Parham, T., Matthews, L., Landrum-Brown, J., & Kim, A. (1996). African American client skin tone and clinical judgments of African American and European American psychologists. *Professional Psychology, 27,* 500–505.

Axelson, J. (1993). *Counseling and development in a multicultural society* (2nd ed.). Pacific Grove, CA: Brooks/Cole.

Bartholomew, K. (1990). Avoidance of intimacy: An attachment perspective. *Journal of Social and Personal Relationships, 7,* 147–178.

Bartholomew, K., & Horowitz, L. M. (1991). Attachment styles among young adults: A test of a four category model. *Journal of Personality and Social Psychology, 61,* 226–244.

Becker, D., & Lamb, S. (1994). Sex bias in the diagnosis of borderline personality disorder and posttraumatic stress disorder. *Professional Psychology, 25,* 55–61.

Bellack, A. S., & Hersen, M. (Eds.). (1988). *Behavioral assessment.* New York: Pergamon Press.

Bowlby, J. (1988). *A secure base: Parent-child attachments and healthy human development.* New York: Basic Books.

Bradford, E., & Lyddon, W. (1994). Assessing adolescent and adult attachment. *Journal of Counseling and Development, 73,* 215–219.

Bronfenbrenner, U. (1993). The ecology of cognitive development. In R. H. Wozniak & K. W. Fischer (Eds.), *Development in context* (The Jean Piaget Symposium Series, pp. 3–44). Hillsdale, NJ: Erlbaum.

Brown, L. (1994). *Subversive dialogues: Theory in feminist therapy.* New York: Basic Books.

Brown, L., & Ballou, M. (Eds.). (1992) *Personality and psychopathology: Feminist reappraisals.* New York: Guilford.

Caplan, P. J. (1995). *They say you're crazy: How the world's most powerful psychiatrists decide who's normal.* New York: Addison-Wesley.

Fancher, R. T. (1995). *Cultures of healing: Correcting the image of American mental health care.* New York: Freeman.

Feeney, J. A., Noller, P., & Hanrahan, M. (1994). Assessing adult attachment. In M. B. Sperling & W. H. Berman (Eds.), *Attachment in adults: Clinical and development perspectives* (pp. 128–152). New York: Guilford.

Hohenshil, T. H. (1994). *DSM-IV:* What's new. *Journal of Counseling and Development, 73,* 105–107.

Lazarus, A. A. (1989). *The practice of multimodal therapy.* Baltimore, MD: Johns Hopkins University Press.

Lopez, F. G. (1995). Contemporary attachment theory: An introduction with implications for counseling psychology. *The Counseling Psychologist, 23,* 395–415.

Martin-Causey, T., & Hinkle, J. (1995). Multimodal therapy with an aggressive preadolescent. *Journal of Counseling and Development, 73,* 305–310.

Morran, D. K., Kurpius, D. J., Brack, G., & Rozecki, T. G. (1994). Relationship between counselor's clinical hypotheses and client ratings of counselor effectiveness. *Journal of Counseling and Development, 72,* 655–660.

Morrison, J. (1995). *DSM-IV made easy.* New York: Guilford.

Neighbors, H. W., Jackson, J. S., Campbell, L., & Williams, P. (1989). The influence of racial factors on psychiatric diagnosis: A review and suggestions for research. *Community Mental Health Journal, 25,* 301–311.

Parsons, R., Jorgensen, J., & Hernández, S. (1994). *The integration of social work practice.* Pacific Grove, CA: Brooks/Cole.

Pistole, M. C., & Watkins, C. E., Jr. (1995). Attachment theory, counseling process, and supervision. *The Counseling Psychologist, 23,* 457–478.

Rosado, J. W., Jr., & Elias, M. J. (1993). Ecological and psychocultural mediators in the delivery of services for urban, culturally diverse Hispanic clients. *Professional Psychology, 24,* 450–459.

Rothbard, J. C., & Shaver, P. R. (1994). Continuity of attachment across the life span. In M. B. Sperling & W. H. Berman (Eds.), *Attachment in adults: Clinical and development perspectives* (pp. 31–71). New York: Guilford.

Snowden, L. R., & Cheung, F. K. (1990). Use of inpatient mental health services by members of ethnic minority groups. *American Psychologist, 24,* 450–459.

Spitzer, R. L., Gibbon, M., Skodol, A. E., Williams, J. B., & First, M. B. (1994). *DSM-IV casebook.* Washington, DC: American Psychiatric Association.

Sue, S. (1991). Ethnicity and culture in psychological research and practice. In J. P. Goodchilds (Ed.), *Psychological perspectives in human diversity in America* (pp. 47–86). Washington, DC: American Psychological Association.

Teyber, E. (1997). *Interpersonal process in psychotherapy* (2nd ed.). Pacific Grove, CA: Brooks/Cole.

FEEDBACK POSTEVALUATION

The Case of Mr. Huang

1. Mr. Huang's self-reported problem behaviors include worry about retirement and not doing work on his automobile designs.

2. Worrying about retirement is a covert behavior. Not doing work on designs is an overt behavior.

3. One antecedent condition occurred six months ago, when the personnel director conferred with Mr. Huang about retirement and papers were filled out. This is an overt antecedent in the form of a setting event. The personnel director's visit seemed to elicit Mr. Huang's worry about retirement and his not doing his designs. A covert antecedent is Mr. Huang's repetitive thoughts about retirement, getting older, and so on. This is a stimulus event.

4. The consequences include Mr. Huang's being excused from meeting his deadlines and receiving extra attention from his boss.

5. Mr. Huang's problem behaviors appear to be maintained by the consequence of being excused from not meeting his deadlines, with only a "reminder." He is receiving some extra attention and concern from his boss, whom he values highly. He may also be missing deadlines and therefore not completing required car designs as a way to avoid or postpone retirement; that is, he may expect that if his designs aren't done, he'll be asked to stay longer until they are completed.

6. The anxiety that Mr. Huang is experiencing surrounding the transition from full-time employment to retirement is a fairly universal reaction to a major life change event. However, in addition to this, Mr. Huang is also affected by his cultural/ethnic affiliation in that he is concerned about maintaining pride and honor, not losing face or shaming the two groups he belongs to—his family and his company. This recognition is an important part of the assessment because it will also be a focus in the intervention phase.

The Case of John: Analysis of Work Problem

1. John's problem behaviors at work include (a) overemphasis on the rivalry that he assumes exists with his fellow administrators and resulting self-doubts about his competence compared with that of his peers and (b) missing days at work because of his feelings of depression as well as his alcohol abuse.

(continued)

2. His discrediting of his skills is a covert behavior, as is much of his current dejection. Avoiding his job is an overt behavior.

3. The antecedent conditions of John's difficulties at work are his apparent perceptions surrounding the competitiveness with his co-workers. These perceptions constitute a stimulus event. This apprehension has led him to feel inadequate and fosters his depressive symptomatology. It should be recognized that John's occupational difficulties arose only after he obtained his present job, one that requires more responsibility than any of his previous positions. Acquisition of this job and its accompanying managerial position is a setting event.

4. The consequences that maintain John's difficulties at work are (a) failing to show up for work each day and (b) alcohol abuse.

5. Failing to show up for work each day amounts to a variable-interval schedule of reinforcement, which is quite powerful in maintaining John's evasion of the workplace. A possible secondary gain of his absenteeism is the resulting decrease in his feelings of incompetence and depression. His abuse of alcohol provides him with a ready-made excuse to miss work whenever necessary or whenever he feels too depressed to go. It should be noted that alcohol as a drug is a central nervous system depressant as well. Alcohol abuse is also a common complication of depressive episodes.

Analysis of Sexual Problem

1. John's problem behavior is an apparent loss of interest in or desire for sexual activity, which is a significant change from his previous behavior. His feelings of excitement have been inhibited so that he is unable to attain or maintain an erection throughout the sexual act.

2. The inability to achieve and/or sustain an erection is an overt problem. We may also assume that whatever John is telling himself is somehow influencing his observable behavior. His self-talk is a covert problem.

3. There are apparently no organic factors contributing to the disturbance. Therefore, it appears likely that the antecedent conditions of John's current sexual problem are the anxiety and depression associated with the work situation.

4. The consequences maintaining John's sexual problem appear to be (a) the lack of reassurance from his girlfriend and (b) his current alcohol abuse. The girlfriend's ultimatum that he begin to regain his normal sexual functioning is creating psychological stress that will continue to prevent adequate sexual response. Although alcohol may serve as a relaxant, it also acts to physiologically depress the usual sexual response.

Analysis of Alcohol Problem

1. Problem behavior is frequent consumption of alcoholic beverages during the day as well as at night.

2. Although alcohol abuse is certainly an overt problem behavior, we might also assume that John is engaging in some self-defeating covert behaviors to sustain his alcohol abuse.

3. It is quite apparent that John's maladaptive use of alcohol occurred only after his difficulties with his job became overwhelming. It also appears to be linked to the onset of his sexual disorder. There is no history of previous abuse of alcohol or other drugs.

4. Consequences include the payoffs of avoidance of tension, responsibility, and depression related to his job as well as possible increased attention from others.

5. By abusing alcohol, John has been missing days at work and thus avoids the tension he feels with his job. Alcohol abuse is serving as a negative reinforcer. Moreover, his use of alcohol, which is a depressant, allows him to maintain his self-discrediting behavior, which, owing to the attention he derives from this, may also be maintaining the alcohol abuse. Finally, alcohol may also provide a ready-made excuse for his poor sexual functioning with his girlfriend.

6. John's issues appear to be affected by the sociopolitical climate in which he works, his particular life stage of development, and also probably by the history of oppression he has undoubtedly experienced as an African American male. John made a fairly rapid transition from being a student and working part time to being a manager in a competitive firm. This progress occurred at a time in which John was also concerned developmentally with issues of identity and intimacy. The difficulties he has encountered have challenged both his concept of himself and his intimate relationship with his girlfriend. Although his firm is competitive, his own sense of vulnerability and his mistrust of himself and of his colleagues no doubt has been impacted by his societal experiences of discrimination and oppression. (It is important to recognize that any cultural suspiciousness he feels has been for him an adaptive and healthy mechanism of coping with a host culture different from his own.)

   His increasing use of alcohol to self-medicate has further exacerbated both his work functioning and his sexual relationship with his girlfriend. His girlfriend appears to be responding to him with threats and intimidation, perhaps in an attempt to control or gain power in the relationship. Indeed, their relationship appears to lack a power base that is shared equally between both partners.

(continued)

CHAPTER 9

DEFINING CLIENT PROBLEMS WITH AN INTERVIEW ASSESSMENT

In Chapter 8 we described a number of important functions of the assessment process in therapy and noted that assessment is a way of identifying and defining a client's problems in order to make decisions about therapeutic treatment. A variety of tools or methods are available to the therapist that can help identify and define the range and parameters of client problems. These methods include standardized tests, such as interest and personality inventories; psychophysiological assessment, such as monitoring of muscle tension for chronic headaches with an electromyograph (EMG) machine; self-report checklists, such as assertiveness scales or anxiety inventories; observation by others, including observation by the therapist or by a significant person in the client's environment; self-observation, in which the client observes and records some aspect of the problem; imagery, in which the client uses fantasy and directed imagery to vicariously experience some aspect of the problem; role playing, in which the client may demonstrate some part of the problem in an in vivo yet simulated enactment; and direct interviewing, in which the client and counselor identify problems through verbal and nonverbal exchanges. All these methods are also used to evaluate client progress during therapy, in addition to their use in assessment for the purpose of collecting information about client problems. In this chapter we concentrate on direct interviewing, not only because it is the focus of the book but also because it is the one method readily available to all therapists without additional cost in time or money. We also, however, will mention ancillary use of some of the other methods of assessment named above. In actual practice, it is very important not to rely solely on the interview for assessment data but to use several methods of obtaining information about client problems.

OBJECTIVES

1. Given a written description of a selected client problem, outline in writing at least two questions for each of the 11 problem assessment categories that you would ask during an assessment interview with this person.

2. In a 30-minute role-play interview, demonstrate leads and responses associated with 9 out of 11 categories for assessing the problem. An observer can rate you, or you can rate your performance from a tape, using the Problem Assessment Interview Checklist at the end of the chapter. After the interview, identify orally or in writing some hypotheses about antecedent sources that cue the problem, consequences that maintain it, secondary gains, or "payoffs," and client resources, skills, and assets that might be used during intervention.

3. Given a written client case description, construct in writing a self-monitoring assessment plan for the client and an example of a log to use for self-recording the data.

4. Conduct a role-play interview in which you explain at least three parts of a self-monitoring assessment plan to a client (rationale, instructions, follow-up).

DIRECT ASSESSMENT INTERVIEWING

According to cognitive-behavioral literature, the interview is the most common behavioral assessment instrument (Nelson, 1983). "While elaborate behavioral and psychophysiological assessment procedures have been developed and evaluated, the assessment instrument most frequently employed in the clinical setting remains the behavioral interview" (Keane, Black, Collins, & Venson, 1982, p. 53). Nelson (1983) observes that the interview is the one assessment strategy used more consistently than any other procedure—perhaps because of its practicality in applied settings and its potential efficiency. Despite the overwhelming evidence confirming the popularity of the interview as an assessment tool, some persons believe it is the most difficult assessment approach for the therapist to enact. Successful assessment interviews require specific guidelines and training in order to obtain accurate and valid information from clients that will make a difference in treatment planning (Duley, Cancelli, Kratochwill, Bergan, & Meredith, 1983).

In this chapter we describe a structure and some guidelines to apply in assessment interviews in order to identify and define client problems. This chapter and other chapters in this book describe interview leads that, in applied settings, are likely to elicit certain kinds of client information. However, as Morganstern (1988) observes, little research on the effects of interview procedures has been conducted. The leads suggested in this chapter are supported more by practical considerations than by empirical data. As a result, you will need to be very attentive to the effects of using these questions with each client.

INTAKE INTERVIEWS AND HISTORY

Part of assessment involves eliciting information about the client's background, especially as it may relate to *current* problems or complaints. Past, or historical, information is not sought as an end in itself or because the therapist is necessarily interested in exploring or focusing on the client's "past" during treatment. Rather, it is used as a part of the overall assessment process that helps the therapist fit the pieces of the puzzle together concerning the client's presenting problems and current life difficulties. Often a client's current problems are precipitated and maintained by events found in the client's history. In no case is this more valid than with clients who have suffered trauma of one kind or another. For example, a 37-year-old woman came to a crisis center because of sudden onset of extreme anxiety. The interviewer noticed that she was talking in a "little girl" voice and was using gestures that appeared to be very childlike. The clinician commented on this and asked the client how old she felt right now. The client replied, "I'm seven years old," and went on to reveal spontaneously an incident in which she had walked into a room in an aunt's house and found her uncle fondling her cousin. No one had seen her and she had forgotten this until the present time. In cases such as this one, history may serve as a retrospective baseline measure for the client and may help to identify cognitive or historical antecedent conditions that still exert influence on the problem behavior and might otherwise be overlooked.

The process of gathering this type of information is called "history taking." In many agency settings, history taking occurs during an initial interview called an "intake interview." An intake interview is viewed as informational rather than therapeutic and, to underscore this point, is often conducted by someone other than the therapist assigned to see the client. In these situations, someone else, such as an intake worker, sees the client for an hour interview (shorter for children and adolescents), summarizes the information in writing, and passes the information along to the therapist. In other places, the therapists conduct their own intakes. For therapists who work either in private practice or in a school or agency in which intakes are not required, it is still a good idea to do some history taking with the client. A number of specific interview protocols for areas such as affective disorders, substance abuse, eating disorders, and attention deficit disorders are reported in Craig's (1989) book on diagnostic interviewing. Also, a sports psychology interview protocol has been developed by Taylor and Schneider (1992) and a suicide assessment protocol has been constructed by Sommers-Flanagan and Sommers-Flanagan (1995).

Various kinds of information can be solicited during history taking, but the most important areas are the following:

1. Identifying information about the client
2. General appearance and demeanor
3. History related to the presenting problems
4. Past psychiatric and/or counseling history
5. Educational and job history
6. Health (medical) history
7. Social/developmental history (including religious and cultural background and affiliations, predominant values, description of past problems, chronological/developmental events, military background, social/leisure activities, present social situation)
8. Family, marital, sexual history
9. Assessment of client communication patterns
10. Results of mental status; diagnostic summary

Table 9-1 presents specific questions or content areas to cover for each of these ten areas.

The sequence of obtaining this information in a history or intake interview is important. Generally, the interviewer begins with the least threatening topics and saves more sensitive topics (such as VI, VII, and VIII) until near the end of the session, when a greater degree of rapport has been established and the client feels more at ease about revealing personal information to a total stranger.

Handling Sensitive Subjects in the Interview Assessment Process

Morrison (1995) has pointed out that some important subjects that come up in intake and assessment interviews can be sensitive for both therapists and clients. Yet this potential sensitivity does not mean that such subjects should be overlooked or discarded. It does mean, however, that the helper should proceed with good judgment and seek consultation about when it is appropriate to assess these areas.

TABLE 9-1. History-taking interview content

I. *Identifying information*
Client's name, address, home, and work telephone numbers; name of another person to contact in case of emergency.

Age	Languages
Gender	Disabilities
Culture	Marital status
Ethnicity	Occupation
Race	Citizenship status

II. *General appearance*
Approximate height
Approximate weight
Brief description of client's dress, grooming, overall demeanor

III. *Presenting problems* (do for *each* problem or complaint that client presents)
Note the presenting complaint (quote client directly):
When did it start? What other events were occurring at that time?
How often does it occur?
What are thoughts, feelings, and observable behaviors associated with it?
Where and when does it occur most? least?
Are there any events or persons that precipitate it? make it better? make it worse?
How much does it interfere with the client's daily functioning?
What previous solutions/plans have been tried for the problem and with what result?
What made the client decide to seek help at this time (or, if referred, what influenced the referring party to refer the client at this time)?

IV. *Past psychiatric/counseling history*
Previous counseling and/or psychological/psychiatric treatment:
    Type of treatment
    Length of treatment
    Treatment place or person
    Presenting complaint
    Outcome of treatment and reason for termination
    Previous hospitalization
    Prescription drugs for emotional/psychological problems

V. *Educational/job history*
Trace academic progress (strengths and weaknesses) from grade school through last level of education completed
Relationships with teachers and peers
Types of jobs held by client
Length of jobs
Reason for termination or change
Relationships with co-workers
Training/education received for jobs
Aspects of work that are most stressful or anxiety-producing
Aspects of work that are least stressful or most enjoyable
Overall degree of current job satisfaction

VI. *Health/medical history*
Childhood diseases, prior significant illnesses, previous surgeries
Current health-related complaints or illnesses (for example, headache, hypertension)
Treatment received for current complaints: what type and by whom
Date of last physical examination and results
Significant health problems in client's family of origin (parents, grandparents, siblings)
Client's sleep patterns
Client's appetite level
Current medications (including such things as aspirin, vitamins, birth control pills, recreational substance use)
Drug and nondrug allergies

(continued)

TABLE 9-1. History-taking interview content (continued)

Client's typical daily diet, including caffeine-containing beverages/food; alcoholic beverages

Exercise patterns

VII. *Social/developmental history*

Current life situation (typical day/week, living arrangements, occupation and economic situation, contact with other people)

Social/leisure time activities, hobbies

Religious affiliation

Military background/history

Predominant values, priorities, and beliefs expressed by client

Significant chronological/developmental events noted by client:

    Earliest recollections

Significant events reported for the following developmental periods:

    Preschool (0–6 years)

    Childhood (6–13 years)

    Adolescence (13–21 years)

    Young adulthood (21–30 years)

    Middle adulthood (30–65 years)

    Late adulthood (65 years and over)

VIII. *Family, marital, sexual history*

Identifying data for client's mother and father

Ways in which mother rewarded and punished client

Ways in which father rewarded and punished client

Presence of physical and/or emotional abuse from parent, sibling, or other

Significant "parent tapes"

Activities client typically did with mother

How well client got along with mother

Activities client typically did with father

How well client got along with father

How well parents got along with each other

Identifying information for client's siblings (including those older and younger and client's birth order, or position in family)

Which sibling was most like client? least like client?

Which sibling was most favored by mother? father? least favored by mother? father?

Which sibling did client get along with best? worst?

History of previous psychiatric illness/hospitalization among members of client's family of origin

Use of substances in family of origin

Dating history

Engagement/marital history—reason for termination of relationship

Current relationship with intimate partner (how well they get along, problems, stresses, enjoyment, satisfaction, and so on)

Number and ages of client's children

Other people living with or visiting family frequently

Description of previous sexual experience, including first one

(note whether heterosexual, homosexual, or bisexual experiences are reported)

Present sexual activity—masturbation, intercourse, and so on; note frequency

Any present concerns or complaints about sexual attitudes or behaviors

Current sexual orientation

For female clients: obtain menstrual history (onset of first period, regularity of current periods, degree of stress and comfort before and during period)

Sexual contact and/or abuse by parent, sibling, or other

IX. Diagnostic summary (if applicable)

Axis I. Clinical disorders. *DSM-IV* Code

Axis II. Personality disorders and mental retardation

Axis III. General medical condition

Axis IV. Psychosocial and environmental problems (note: sociopolitical factors can be included here as well)

Axis V. Global assessment of functioning (0 to 100 scale)

For example, as one of our reviewers so aptly observed, it may be regarded as voyeuristic if a male counselor asks a young female presenting with an academic/career issue about her sexual practices and activity. On the other hand, if a client comes in and discusses problems in dating persons of the opposite sex and feelings of attraction to same sex people, not pursuing this is an important omission.

Specific subjects that may fall into the category of sensitive topics include questions about (1) *suicidal thoughts and behavior;* (2) *homicidal ideas and violent behavior;* (3) *substance use,* including alcohol, street drugs, and prescribed medications; (4) *sexual issues,* including sexual preference, sexual practices, and sexual problems; and (5) *physical, emotional, and sexual abuse,* both historic and current. As it is beyond the scope of this book to provide you with the specific information necessary to assess these five areas in an interview, we refer you to Morrison's (1995) excellent guide.

MENTAL-STATUS EXAMINATION

If, after conducting an initial interview, you are in doubt about the client's psychiatric status or suspicious about the possibility of an organic brain disorder, you may wish to conduct (or refer the client for) a mental-status examination. According to Kaplan and Sadock (1995), the mental-status exam is one that classifies and describes the areas and components of mental functioning involved in making diagnostic impressions and classifications. The major categories covered in a mental-status exam are general description and appearance of the client, mood and affect, perception, thought processes, level of consciousness, orientation to time, place, and people, memory, and impulse control. Additionally, the examiner may note the degree to which the client appeared to report the information accurately and reliably. Of these categories, disturbances in consciousness (which in-

volves ability to perform mental tasks, degree of effort, degree of fluency/hesitation in task performance) and orientation (whether or not clients know when, where, and who they are and who other people are) are usually indicative of organic brain impairment or disorders and require neurological assessment and follow-up as well. It is important for counselors and therapists to know enough about the functions and content of a mental-status exam to refer those clients who might benefit from this additional assessment procedure. A summary of the content of a mental-status exam is given in Table 9-2. For additional information about mental-status examinations and neurophysiological assessment, see Kaplan and Sadock (1995) and Morrison (1995).

History taking (and mental-status exams, if applicable) usually occur near the very beginning of counseling. After obtaining this sort of preliminary information about the client as well as an idea of the range of presenting complaints, you are ready to do some direct assessment interviewing with the client in order to define the parameters of problems and concerns more specifically. We present guidelines for assessment interviews in the next section, after Learning Activity #30.

ELEVEN CATEGORIES FOR ASSESSING CLIENT PROBLEMS

To help you acquire the skills associated with problem assessment interviews, we describe 11 categories of information you need to seek from each client. Most of this information is based on the case conceptualization models presented in Chapter 8. These 11 categories are illustrated and defined in the following list and subsections. They are also summarized in the Interview Checklist at the end of the chapter.

1. Explanation of *purpose* of assessment—presenting rationale for assessment interview to the client.

LEARNING ACTIVITY 30   *Intake Interviews and History*

To give you a sense of the process involved in doing an intake or history interview (if you don't already do lots of these on your job!), we suggest you pair up with someone else in your class and complete intake/history interviews with each other. Conduct a 30–45 minute session with one person serving as the counselor and the other taking the client's role; then switch roles. As the counselor, you can use the format in Table 9-1 as a guide. You may wish to jot

down some notes. After the session, it might be helpful to write a brief summary of the session, using the major categories listed in Table 9-1 as a way to organize your report. As the client, in this particular activity, rather than playing a role, be yourself. Doing so will allow you to respond easily and openly to the counselor's questions and both of you can more readily identify the way in which your particular history has influenced the current issues in your life.

TABLE 9-2. Summary of Mental-Status Exam

Information	Process
Mental-Status Exam	

Appearance
  Apparent age
  Ethnicity, race, and gender
  Body build, posture
  Nutrition
  Clothing: Neat? Clean? Style?
  Hygiene
  Hairstyle
Alertness: Full? Drowsy? Stupor? Coma?
General behavior
  Activity level
  Tremors?
  Mannerisms and stereotypes
  Facial expression
  Eye contact
  Voice
Attitude toward examiner
Mood
  Type
  Lability
  Appropriateness
  Intensity
Flow of thought
  Word associations
  Rate and rhythm of speech
Content of thought
  Delusions
  Hallucinations
  Anxiety
  Phobias
  Obsessions and compulsions
  Suicide and violence

Observed during history taking

(continued)

2. Identification of *range* of problems—using leads to help the client identify all the relevant primary and secondary issues to get the "big picture."

3. *Prioritization* and *selection* of issues and problems—using leads to help the client prioritize problems and select the initial area of focus.

4. Identification of *present problem behaviors*—using leads to help the client identify the six components of problem behavior(s): affective, somatic, behavioral, cognitive, contextual, and relational.

5. Identification of *antecedents*—using leads to help the client identify sources of antecedents and their effect on the problem behavior.

6. Identification of *consequences*—using leads to help the client identify sources of consequences and their influence on the problem behavior.

7. Identification of *secondary gains*—using leads to help the client identify underlying controlling variables that serve as payoffs to maintain the problem behavior.

8. Identification of *previous solutions*—using leads to help the client identify previous solutions or attempts to solve the problem and their subsequent effect on the problem.

9. Identification of *client coping skills*—using leads to help the client identify past and present coping or adaptive behavior and how such skills might be used in working with the present issue.

TABLE 9-2. Summary of Mental-Status Exam   (Continued)

Information	Process
Mental-Status Exam	
Orientation: Person? Place? Time?	"Now I'd like to ask some routine questions . . ."
Language: Comprehension, fluency, naming, repetition, reading, writing	
Memory: Immediate? Short term? Long term?	"How has your memory been? Do you mind if I test it?"
Attention and concentration	
Serial sevens	
Count backwards	
Cultural information	
Current events	
Five presidents	
Abstract thinking	
Proverbs	
Similarities and differences	
Insight	
Judgment	Closure
	Summarize findings
	Set next appointment
	"Do you have any questions for me?"

SOURCE: From *The First Interview: A Guide for Clinicians*, by J. Morrison, pp. 248–249. Copyright 1995 by The Guilford Press. Reprinted by permission.

10. Identification of the *client's perceptions* of the problem—using leads to help the client describe her or his understanding of the problem.
11. Identification of *problem intensity*—using leads and/or client self-monitoring to identify impact of problem on client's life, including (a) degree of problem severity and (b) frequency or duration of problem behaviors.

The first three categories—explanation of the purpose of assessment, identification of the range of problems, and prioritization and selection of problem concerns—are a logical starting place. First, it is helpful to give the client a rationale, a reason for conducting an assessment interview, before gathering information. Next, some time must be spent in helping the client explore all the relevant issues and prioritize problems to work on in order of importance, annoyance, and so on.

The other eight categories follow problem prioritization and selection. After the counselor and client have identified and selected the problems to work on, these eight categories of counselor leads are used to define and analyze parameters of the problem. The counselor will find that the order of the problem assessment leads varies among clients. A natural sequence will evolve in each interview, and the counselor will want to use the leads associated with these content categories in a pattern that fits the flow of the interview and follows the lead of the client. It is very important in assessment interviews not to impose your structure at the expense of the client. The amount of time and number of sessions required to obtain this information will vary with problems and with clients. It is possible to complete the assessment in one session, but with some clients, an additional interview may be necessary. Although the counselor may devote several interviews to assessment, the information gathering and hypothesis testing that go on do not automatically stop after these few sessions. Some degree of problem assessment continues throughout the entire therapy process.

Explaining the Purpose of Assessment

In explaining the purpose of problem assessment, the counselor gives the client a rationale for doing an assessment interview. The intent of this first category of problem assessment is to give the client a "set," or an expectation, of what will occur during the interview and why assessment is important to both client and counselor. We also usually tell clients that we may be asking more questions during this session than we have before or will in the future so that clients are not caught off guard if our interviewing style changes. One way the counselor can communicate the purpose of the assessment interview is "Today I'd like to focus on some concerns that are bothering you most. In order to find out exactly what you're concerned about, I'll be asking you for some specific kinds of information. This information will help both of us identify what you'd like to work on in counseling. How does this sound [or appear] to you?" After presenting the rationale, the counselor looks for some confirmation or indication that the client understands the importance of assessing problems. If client confirmation or understanding is not forthcoming, the counselor may need to provide more explanation before proceeding to other areas. It is also important in initial interviews with clients to create expectations that inspire hope (Lazarus, 1989). Most clients are so focused on their pain that they are unable to see, hear, or grasp much beyond it. So you need to be in touch not only with their pain but also with their potential, their possibilities, and their future.

Identifying the Range of Problems

In this category, the counselor uses open-ended leads to help clients identify all the major issues and concerns in their life now. Often clients will initially describe only one problem, and on further inquiry and discussion, the counselor discovers a host of other problems, some of which may be more severe or stressful or have greater significance than the one the client originally described. If the counselor does not try to get the "big picture," the client may reveal additional concerns either much later in the therapy process or not at all.

These are examples of range-of-problem leads:

"What are your concerns in your life now?"
"Could you describe some of the things that seem to be bothering you?"
"What are some present stresses in your life?"
"What situations are not going well for you?"
"Is there anything else that concerns you now?"

After using range-of-problem leads, the counselor should look for the client's indication of some general areas of concern or things that are troublesome for the client or difficult to manage. An occasional client may not respond affirmatively to these leads. Some clients may be uncertain about what information to share with the counselor. In such cases, the counselor may need to use a different approach from verbal questioning to elicit problem statements. For example, Lazarus (1989) has recommended the use of an

"Inner Circle" strategy to help a client disclose problem areas. The client is given a picture like this:

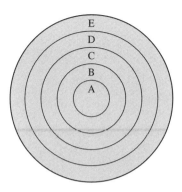

The counselor points out that topics in circle A are very personal, whereas topics in circle E are more or less public information. The counselor can provide examples of types of topics likely to be in the A circle, such as sexual concerns, feelings of hostility, marriage problems, and dishonesty. These examples may encourage the client to disclose personal concerns more readily. The counselor also emphasizes that good therapy takes place in the A and B circles and may say things like "I feel we are staying in Circle C" or "Do you think you've let me into Circle A or B yet?" (Lazarus, 1989). Sometimes the counselor may be able to obtain more specific problem descriptions from a client by having the client role-play a typical problem situation. Another client might provide more information by describing a fantasy or visualization about the problem. This last method has been used by Meichenbaum (1994), who asks the client "to run a movie through your head" in order to recall various aspects of the problem.

Exploring the range of problems is also a way to establish who is the appropriate client. A client may attribute the problem or the undesired behavior to an event or to another person. For instance, a student may say "That teacher always picks on me. I can never do anything right in her class." Because most clients seem to have trouble initially "owning" their role in their problem or tend to describe it in a way that minimizes their own contribution, the counselor will need to determine who is most invested in having the problem resolved and who is the real person requesting assistance. Often it is helpful to ask clients who feels it is most important for the problem to be resolved—the client or someone else. It is important for therapists not to assume that the person who arrives at their office is always the client. The client is the person who wants a change and who seeks assistance for it. In the example above, if the student had desired a change and had requested assistance, the student would be the client; if it were the teacher who wanted a change and requested assistance, the teacher would be the client. (Sometimes, however, the therapist gets "stuck" in a situation in which a family or a client wants a change and the person whose behavior is to be changed is "sent" to counseling as the client.)

The question of who is the appropriate client is also tricky when the problem involves two or more persons, such as a relationship, marital, or family problem. Many family therapists view family problems as devices for maintaining the status quo of the family and recommend that either the couple or the entire family be involved in counseling, rather than one individual. Although this is a great concept in theory, in practice it is sometimes difficult to implement.

Prioritizing and Selecting Problems

Rarely do clients or the results of assessment suggest only one area or problem that needs modification or resolution. Typically, a presenting problem turns out to be one of several unresolved issues in the client's life. For example, the assessment of a client who reports depression may also reveal that the client is troubled by her relationship with her teenage daughter and her obesity. History may reveal that this adult woman was also physically abused as a child. After the client describes all of her or his concerns, the counselor and client will need to select the problems that best represent the client's purpose for seeking counseling. The primary question to be answered by these leads is "What is the specific problem situation the client chooses to start working on?"

Prioritizing problems is an important part of assessment and goal setting. If clients try to tackle too many issues simultaneously, they are likely to soon feel overwhelmed and anxious and may not experience enough success to stay in therapy. Selection of the problem is the client's responsibility, although the counselor may help with the client's choice. If the client selects a problem that severely conflicts with the counselor's values, a referral to another counselor may be necessary. Otherwise, the counselor may inadvertently or purposely block the discussion of certain client problem areas by listening selectively to only those problems that the counselor can or wants to work with.

The following guidelines form a framework to help clients select and prioritize problems to work on:

1. Start with the presenting problem, the one that best represents the reason the client sought help. Fensterheim

(1983, p. 63) observes that relief of the presenting problem often improves the client's level of functioning and may then make other, related problems more accessible to treatment. Leads to use to help determine the initial or presenting problem include "Which issue best represents the reason you are here?" and "Out of all these problems you've mentioned, identify the one that best reflects your need for assistance."

2. Start with the problem that is primary or most important to the client to resolve. Often this is the one that causes the client the most pain or discomfort or annoyance or is most interfering to the client. Modifying the more important issues seems to lead to lasting change in that area, which may then generalize to other areas (Fensterheim, 1983). Responses to determine the client's most important priority include "How much happiness or relief would you experience if this issue were resolved?" "Of these concerns, which is the most stressful or painful for you?" "Rank-order these concerns, starting with the one that is most important for you to resolve to the one least important," and "How much sorrow or loss would you experience if you were unable to resolve this issue?"

3. Start with the problem or behavior that has the best chance of being resolved successfully and with the least effort. Some problems/behaviors are more resistant to change than others and require more time and energy to modify. Initially, it is important for the client to be reinforced for seeking help. One significant way to do this is to help the client resolve something that makes a difference without much cost to the client.

   Responses to determine what problems might be resolved most successfully include "Do you believe there would be any unhappiness or discomfort if you were successful at resolving this concern?" "How likely do you think we are to succeed in resolving this issue or that one?" and "Tell me which of these problems you believe you could learn to manage most easily with the greatest success."

4. Start with the problem that needs resolution before other problems can be resolved or mastered. Sometimes the presence of one problem sets off a chain of other ones; when this problem is resolved or eliminated, the other issues either improve or at least move into a position to be explored and modified. Often this problem is one that, in the range of elicited problems, is central or prominent.

Questions to ask to help determine the most central problem include "Out of all the problems we've discussed, which is the most predominant one?" and "Out of all the problems we've discussed, describe the one that, when resolved, would have the greatest impact on the rest of the issues."

If, after this process, the counselor and client still have difficulty prioritizing problems and selecting the initial area of focus, try the procedure recommended by Goldfried (1983). The client asks the following question about each identified problem: "What are the consequences of my *not* doing anything therapeutically to handle this particular problem?"

Understanding the Problem Behaviors

After selecting the initial area of focus, it is important to determine the components of the problem behavior. For example, if the identified problem is "not getting along very well with people at work," with an expected outcome of "improving my relationships with people at work," we would want to identify the client's *feelings* (affect), *body sensations* (somatic phenomena), *actions* (overt behavior), and *thoughts and beliefs* (cognitions) that occur during the problem situations at work. We would also explore whether these problematic feelings, sensations, actions, and thoughts occurred with *all* people at work or only *some* people (relationships) and whether they occurred only at work or in *other* settings, at what *times,* and under what *conditions* or with what *concurrent events* (context). Without this sort of exploration, it is impossible to define the problem operationally (concretely). Furthermore, it is difficult to know whether the client's work problems result from the client's actions or observable behaviors, from covert responses such as feelings of anger or jealousy, from cognitions and beliefs such as "When I make a mistake at work, it's terrible," from the client's transactions with significant others that suggest an "I'm not OK—they're OK" position, or from particular events that occur in certain times or situations during work, as during a team meeting or when working under a supervisor.

Without this kind of information about when and how the problem behavior is manifested, it would be very difficult and even presumptuous to select intervention strategies or approaches. The end result of this kind of specificity is that the problem is defined or stated in terms such that two or more persons can agree on when it exists. In the following sections, we describe specific things to explore for each of these six components and suggest some leads and responses to facilitate this exploration with clients.

Affect and Mood States.
Affective components of problem behavior include self-reported feelings or mood states, such as "depression," "anxiety," and "happiness." Feelings are generally the result of complex interactions among behavioral, physiological, and cognitive systems rather than unitary experiential processes (Basic Behavioral Science Research for Mental Health, 1995). Clients often seek therapy because of this component of the problem—that is, they feel bad, uptight, sad, angry, confused, and so on and want to get rid of such unpleasant feelings.

One category of things to ask the client about to get a handle on feelings or mood states is feelings about the problem behavior. After eliciting them, note the content (pleasant/unpleasant) and level of intensity. Example leads for this arc the following:

"How do you feel about this?"
"What kinds of feelings do you have when you do this or when this happens?"
"What feelings are you aware of?"

A second category is concealed or distorted feelings—that is, feelings that the client seems to be hiding from, such as anger, or a feeling like anger that has been distorted into hurt. Below are example responses for this:

"You seem to get headaches every time your husband criticizes you. What feelings are these headaches masking?"
"When you talk about your son, you raise your voice and get a very serious look on your face. What feelings do you have—deep down—about him?"
"You've said you feel hurt and you cry whenever you think about your family. What other feelings do you have besides hurt?"
"You've indicated you feel a little guilty whenever your friends ask you to do something and you don't agree to do it. Try on resentment instead of guilt. Try to get in touch with those feelings now."

The practitioner can always be on the lookout for concealed anger, which is the one emotion that tends to get "shoved under the rug" more easily than most. Distorted feelings that are common include reporting the feeling of hurt or anxiety for anger, guilt for resentment, and sometimes anxiety for depression, or vice versa. It is also important to be aware that exploration of the affective component may be very productive initially for clients who process information easily in a kinesthetic manner. For clients who do not, however, asking "How do you feel?" can draw a blank, uncomprehending look accompanied by an "I don't know what you mean" statement. Like any other response, "How do you feel?" is not equally productive with all clients and tends to be a tremendously overused lead in counseling sessions.

Somatic Sensations.   Closely tied to feelings are body sensations. Some clients are very aware of these "internal experiencings"; others are not. Some persons are so tuned into every body sensation that they become hypochondriacal, while others seem to be switched off "below the head" (Lazarus, 1989). Neither extreme is desirable. Somatic reactions are quite evident in problems such as sexual dysfunction, depression, and anxiety. Some persons may describe complaints in terms of body sensations rather than as feelings or thoughts—that is, as headaches, dizzy spells, back pain, and so on. Problem behavior can also be affected by other physiological processes, such as nutrition and diet, exercise and lifestyle, substance use, hormone levels, and physical illness. Usually, when this is the case, some form of physiological treatment is warranted as well as psychological intervention. Somatic sensations may be a symptom of abuse and trauma. The therapist will want to elicit information about physiological complaints, about lifestyle and nutrition, exercise, substance use, and so on and about other body sensations relating to the problem. Some of this information is gathered routinely during the health history portion of the intake interview.

Useful leads to elicit this component of the problem behavior include these:

"What goes on inside you when you do this or when this happens?"

"What are you aware of when this occurs?"

"What sensations do you experience in your body when this happens?"

"When this happens, are you aware of anything that feels bad or uncomfortable inside you—aches, pains, dizziness, and so on?"

Overt Behaviors or Motoric Responses.   Clients often describe a problem behavior in very nonbehavioral terms. In other words, they describe a situation or a process without describing their actions or specific behaviors within that event or process. For example, clients may say "I'm not getting along with my partner" or "I feel lousy" or "I have a hard time relating to authority figures" without specifying what they do to get along or not get along or to relate or not relate. In this part of the assessment interview, you are interested in finding out precisely what the client does and doesn't do related to the problem. Examples of overt behaviors might be compulsive handwashing, crying, excessive eating, stealing, and making deprecatory or critical comments about self or others.

When inquiring about the behavioral domain, the therapist will want to elicit descriptions of both the presence and absence of concrete overt behaviors connected to the problem—that is, what the client does and doesn't do. The therapist also needs to be alert to the presence of behavioral *excesses* and *deficits*. Excesses are things that the person does too much or too often or that are too extreme, such as binge eating, excessive crying, or assaultive behavior. Deficits are responses that occur too infrequently or are not in the client's repertory or do not occur in the expected context or conditions, such as failure to initiate requests on one's behalf, inability to talk to one's partner about sexual concerns and desires, or lack of physical exercise and body conditioning programs. The therapist may also wish to inquire about "behavioral opposites" (Lazarus, 1989) by asking about times when the person does *not* behave that way.

These are examples of leads to elicit information about overt behaviors and actions:

"Describe what happens in this situation."

"What do you mean when you say you're 'having trouble at work'?"

"What are you doing when this occurs?"

"What do you do when this happens?"

"What effect does this situation have on your behavior?"

"Describe what you did the last few times this occurred."

"If I were photographing this scene, what actions and dialogue would the camera pick up?"

Occasionally the counselor may want to supplement the information gleaned about behavior from the client's oral self-report with more objective assessment approaches, such as using role plays that approximate the problem or accompanying clients into their environments. These additional assessment devices will help therapists improve their knowledge of how the client does and doesn't act in the problematic situation. Additionally, when such observations are coupled with the interview data, the therapist can develop more reliable hunches about how the problem manifests itself and how the problem and the client may respond to treatment.

Cognitions, Beliefs, and Internal Dialogue.   In the last few years, therapists of almost all orientations have emphasized the relative importance of cognitions or symbolic processes (Bandura, 1969; Ellis, 1984) in contributing to, exacerbating, or improving problematic situations that clients bring to therapy. Unrealistic expectations of oneself and of others are often related to presenting problems, as are disturbing images, self-labeling and self-statements, and cognitive distortions. Gambrill (1977, pp. 112–113) elucidates various ways in which cognitions and symbolic processes influence problems:

Fear of elevators may be associated with vivid images of the elevator plunging to the ground. Anger may be fueled by the vivid reliving of perceived slights. Self-labeling may be involved in problematic reactions. A client may experience an unusual feeling state and consider it indicative of mental illness. He may become unduly attentive to the possible occurrence of such states, and may label ones that are only similar as also being indicators that he is "going crazy." A depressed client may engage in very few positive self-evaluative statements and a great number of punishing self-statements. Cognitive distortions may include arbitrary inference, in which conclusions are drawn in the absence of supporting evidence or in direct contradiction to such evidence; magnification, in which the meaning of an event is exaggerated; overgeneralization, in which a single incident is considered indicative of total incompetence; and dichotomous reasoning, in which polar differences are emphasized (Beck, 1970). . . . [In addition], it is assumed that there is a close relationship between the nature of self-statements and overt behavior, in that someone who copes well in a given situation has an internal dialog that is different from that of someone who does not cope well. Clients who complain of anxiety and depression often have an internal dialog consisting of anticipated bad consequences.*

When the cognitive component is a very strong element of the problem, part of the resulting treatment is usually directed toward this component and involves altering unrealistic ideas and beliefs, cognitive distortions and misconceptions, and dichotomous thinking.

Not all clients process cognitions or symbolic processes in the same way. Therefore, the therapist has to be sensitive to how this component may manifest itself with each client and respond accordingly. For example, some clients can easily relate to the term *irrational ideas;* others, particularly adolescents, seem to be offended by such terminology and prefer phrases like "clean up your thinking" (Baker, 1981). Clients who process information kinesthetically may have a great deal of difficulty exploring the cognitive component because they typically don't "think" this way. In contrast, people who process information visually may report cognitions as images or pictures. For example, if you ask "What do you *think* about when this happens?" the client may say "I *see* my wife getting into bed with someone else." Imagery may be a very useful supplemental assessment device with such clients. Clients who process in an auditory modality may report cognitions as "talking to myself" or "telling

myself" and can probably verbalize aloud for you a chain of internal dialogue connected to the problem.

Assessment of the cognitive component is accordingly directed toward exploring the presence of both irrational and rational beliefs and images related to the identified problem. Irrational beliefs will later need to be altered. Rational beliefs are also useful during intervention. Although irrational beliefs take many forms, the most damaging ones seem to be related to "shoulds" about oneself, others, relationships, work, and so on, "awfulizing" or "catastrophizing" about things that don't turn out as we expect, "perfectionistic standards" about ourselves and often projected onto others, and "externalization," the tendency to think that outside events are responsible for our feelings and problems. The therapist will also want to be alert for the presence of cognitive distortions and misperceptions, such as overgeneralization, exaggeration, and drawing conclusions without supporting data. Finally, it is important to note what clients do and do not "say" or "think" to themselves and how that relates to the identified problem.

Leads to use to assess these aspects of the cognitive component of the identified problem include the following:

"What beliefs [or images] do you hold that contribute to this problem? make it worse? make it better?

"Complete the following sentences for me—

    I should . . .

    People should . . .

    My husband [or mother, and so on] should . . .

    Work [or school] should . . .

    Sex should . . . "

"When something doesn't turn out the way you want or expect, how do you usually feel?"

"What data do you have to support these beliefs or assumptions?"

"What are you thinking about or dwelling on when this [problem] happens?"

"Can you describe what kinds of thoughts or images go through your mind when this occurs?"

"What do you say to yourself when this happens?"

"What do you say to yourself when it doesn't happen [or when you feel better, and so on]?"

"Let's set up a scene. You imagine that you're starting to feel a little upset with yourself. Now run through the scene and relate the images or pictures that come through your mind. Tell me how the scene changes [or relate the thoughts or dialogue—what you say to yourself as the scene ensues]."

"What are your mental commentaries on this situation?"

---

*This and all other quotations from this source are from *Behavior Modification: Handbook of Assessment, Intervention, and Evaluation* by E. Gambrill. Copyright © 1977 by Jossey-Bass. Reprinted by permission.

Context: Time, Place, and Concurrent Events.   Problem behaviors occur in a social context, not in a vacuum. Indeed, what often makes a behavior a "problem" is the context surrounding it or the way it is linked to various situations, places, and events. For example, it is not a problem to undress in your home, but the same behavior on a public street in many countries would be called "exhibitionism." In some other cultures, this same behavior might be more commonplace and would not be considered abnormal or maladaptive. Looking at the context surrounding the problem has implications not only for assessment but also for intervention, because a client's cultural background, lifestyle, and values can affect how the client views the problem and also the treatment approach to resolve it.

Assessing the context surrounding the problem is also important because most problems are "situation-specific"—that is, they are linked to certain events and situations, and they occur at certain times and places. For example, clients who say "I'm uptight" or "I'm not assertive" usually do not mean they are *always* uptight or nonassertive but, rather, in particular situations or at particular times. It is important that the therapist not reinforce the notion or belief in the client that the feeling, cognition, or behavior is pervasive. Otherwise, clients are even more likely to adopt the identity of the problem and begin to regard themselves as possessing a particular trait such as "nervousness," "social anxiety," or "nonassertiveness." They are also more likely to incorporate the problem into their lifestyle and daily functioning.

In assessing contextual factors associated with the problem, you are interested in discovering

1. Any *cultural, ethnic, and racial affiliations,* any particular *values* associated with these affiliations, and how these values affect the client's perception of the problem and of change.
2. *Situations* or *places* in which the problem usually occurs and situations in which it does not occur (*where* the problem occurs and where it does not).
3. *Times* during which the problem usually occurs and times during which it does not occur (*when* the problem occurs and when it does not).
4. *Concurrent events*—events that typically occur at or near the same time as the problem. This information is important because sometimes it suggests a pattern or a significant chain of events related to the problem that clients might not be aware of or may not report on their own.
5. Sociopolitical factors—that is, the overall Zeitgeist of the society in which the client lives, the predominant social

and political structures of this society, the major values of these structures, who holds power in these structures, and how all this affects the client.

These are example responses to elicit information about contextual components of the problem:

"What particular cultural and ethnic groups do you identify with?" [In addition to the more apparent ones, don't overlook ones such as Italian American, French Canadian, Polish American, and so on.] If so, how do the values of these groups affect the way you [think about, grasp] the problem?

"Describe some recent situations in which this problem occurred. What are the similarities in these situations? In what situations does this usually occur? Where does this usually occur?"

"Describe some situations when this problem does not occur."

"In what situations does this not occur?"

"When does this usually occur? not occur?"

"Can you identify certain times of the day [week, month, year] when this is more likely to happen? less likely?"

"Does the same thing happen at other times or in other places?"

"What else is going on when this problem occurs?"

"Describe a typical day for me when you feel 'uptight.' "

"Are you aware of any other events that normally occur at the same time as this problem?"

"How do you think the society in which you live has affected the onset and the development of this issue for you?"

In addition to the information obtained from the client's oral self-report during the assessment interviews, often both counselor and client can obtain a better idea of the context of the problem by having the client self-monitor such things as when and where the problem occurs and concurrent events.

Relationships and Significant Others.   Just as problems are often linked to particular times, places, and events, they are also often connected to the presence or absence of other people. People around the client can bring about or exacerbate a problem. Someone temporarily or permanently absent from the client's life can have the same effect. Assessing the client's relationships with others is a significant part of many theoretical orientations to counseling, including dynamic theories, Adlerian theory, family systems theory, and behavioral theory.

Interpersonal problems may occur because of a lack of significant others in the client's life or because of the way the client relates to others or because of the way significant others respond to the client. As we discussed in Chapters 2 and 8, the client's early attachment history is an important source of influence in current life problems. Childhood loss of attachment to a significant adult figure affects the child's subsequent relationship style, way of interacting with others, and even levels of brain chemicals (Bloom & Bills, 1995). Bradford and Lyddon (1994) point out that insecurely attached persons compared to securely attached persons exhibit lower career maturity, more symptoms of depression and disordered eating, greater psychological distress, and less satisfaction in work and relationship situations.

Other persons involved in the problem often tend to discount their role in it. It is helpful if the therapist can get a handle on what other persons are involved in the problem, how they perceive the problem, and what they might have to gain or lose from a change in the problem or the client. As Gambrill (1990) observes, such persons may anticipate negative effects of improvement in a problem and covertly try to sabotage the client's best efforts. For example, a husband may preach "equal pay and opportunity" yet secretly sabotage his wife's efforts to move up the career ladder for fear that she will make more money than he does or that she will find her new job opportunities more interesting and rewarding than her relationship with him. Other people can also influence a client's behavior by serving as role models (Bandura, 1969). People whom clients view as significant to them can often have a great motivational effect on clients in this respect.

Example leads to use to assess the relational component of the problem include the following:

"How many significant close relationships or friendships do you have in your life now?"

"What effects does this problem have on your relationships with significant others in your life?"

"What effects do these significant others have on this problem?"

"Who else is involved in this problem besides you? How are these persons involved? What would their reaction be if you resolved this issue?"

"From whom do you think you learned to act or think this way?"

"What persons *present* in your life now have the greatest positive impact on you? negative impact?"

"What persons *absent* from your life have the greatest positive impact on you? negative impact?"

"Whom do you know and respect who handles this issue in a way that you like?"

"What feelings do you have about your relationship?"

"How confident do you feel in relating to this person?"

"How much comfort do you have being close to this person?"

"How often do you need his or her approval?"

"What is the level of trust that you experience with this person?"

"How much anxiety and worry do you experience with this person?"

"How much do you feel this person is controlling your emotions?"

"How dependent are you on this person?"

"How often do you feel confused about how to act with this person?"

"How easy is it to regulate your own emotions when you are with or think about this person?"

"How much independence do you experience with this person?"

Identifying Antecedents

You may recall from Chapter 8 that there are usually certain things that happen before or after a problem that contribute to it. In other words, people are not born feeling depressed or thinking of themselves as inadequate. Other events may contribute to the problem by maintaining, strengthening, or weakening the problem behaviors, thoughts, or feelings. Much of the assessment process consists in exploring contributing variables that precede and cue the problem (antecedents) and things that happen after the problem (consequences) that, in some way, influence or maintain it.

To review our previous discussion of the ABC model, remember that, like problem behaviors, the sources of antecedents and consequences are varied and may be affective, somatic, behavioral, cognitive, contextual, or relational. Further, antecedents and consequences are likely to differ for each client. Antecedents are external or internal events that occasion or cue the problem behaviors and make them more or less likely to occur. Some antecedents occur immediately before the problem; other antecedents (setting events) may have taken place a long time ago.

In helping clients explore antecedents of the problem, you are particularly interested in discovering (1) what *current* conditions (covert and overt) exist *before* the problem that make it *more likely* to occur, (2) what *current* conditions (covert and overt) exist that occur *before* the problem that make it *less likely* to occur, and (3) what *previous* conditions, or setting events, exist that *still* influence the problem.

Example leads to identify antecedents follow and are categorized according to the six possible sources of antecedents, described in Chapter 8:

Affective

"What are you usually feeling before this happens?"

"When do you recall the first time you felt this way?"

"What are the feelings that occur before the problem and make it stronger or more constant?"

"What are the feelings that occur before the problem that make it weaker or less intense?"

"Are there any holdover feelings or unfinished feelings from past events in your life that still affect this problem?"

"What do you know of the roots of this feeling?"

Somatic

"What goes on inside you just before this happens?"

"Are you aware of any particular sensations in your body before this happens?"

"Are there any body sensations that occur right before this problem that make it weaker or less intense?"

"Is there anything going on with you physically—an illness or physical condition—or anything about the way you eat, smoke, exercise, and so on that affects or leads to this problem?"

Behavioral

"If I were photographing this, what actions and dialogue would I pick up before this happens?"

"Can you identify any particular behavior patterns that occur right before this happens?"

"What do you typically do before this happens?"

"Can you think of anything you do that makes this problem more likely to occur? Less likely to occur?"

Cognitive

"What kinds of pictures or images do you have before this happens?"

"What are your thoughts before this happens?"

"What are you telling yourself before this happens?"

"Can you identify any particular beliefs that seem to set the problem off?"

"What do you think about [see or tell yourself] before the problem occurs that makes it stronger or more likely to occur? Weaker or less likely to occur?"

Contextual

"Has this ever occurred at any other time in your life? If so, describe that."

"How long ago did this happen?"

"Where and when did this occur the first time?"

"How do you see those events related to your problem?"

"What things happened that seemed to lead up to this?"

"When did the problem start—what else was going on in your life at that time?"

"What were the circumstances under which the problem first occurred?"

"What was happening in your life when you first noticed this?"

"Are there any ways in which your cultural affiliation and values contribute to this problem? make it more likely to occur? less likely?"

"Are you aware of any events that occurred before this problem that in some way still influence it or set it off?"

"Do you see any particular aspects or structures in your society that have contributed to this issue?"

Relational

"Can you identify any particular people who seem to bring on this problem?"

"Are you usually with certain people right before or when this occurs?"

"Whom are you usually with right before this problem occurs?"

"Can you think of any person or of any particular reaction from a person that makes this problem more likely to occur? less likely?"

"Are there any people or relationships from the past that still influence or set off or lead to this problem in some way?"

"Do you think the way people in your society hold power has contributed to this issue? How?"

Identifying Consequences

Recall from Chapter 8 that consequences are external or internal events that influence the problem behavior by maintaining it, strengthening or increasing it, or weakening or decreasing it. Consequences occur after the problem behavior and are distinguished from results or effects of the problem behavior by the fact that they have direct influence on the problem by either maintaining or decreasing the problem in some way.

In helping clients explore consequences, you are interested in discovering both internal and external events that maintain and strengthen the problem behavior and also events that weaken or decrease it.

Example leads to identify consequences follow and are categorized according to the six sources of consequences described in Chapter 8.

Affective

"How do you feel after _____?"

"How does this feeling affect the problem (for example, keep it going, stop it)?"

"Are you aware of any particular feelings or emotions that you have after the problem that strengthen or weaken it?"

Somatic

"What are you aware of inside you just after this happens? How does this affect the problem?"

"Are there any body sensations that seem to occur after the problem that strengthen or weaken it?"

"Is there anything you can think of about yourself physically—illness, diet, exercise, and so on—that seems to occur after this problem? How does this affect the problem?"

Behavioral

"What do you do after this happens, and how does this make the problem worse? Better?"

"How do you usually react after this is over? In what ways does your reaction keep the problem going? Weaken it or stop it?"

"Can you identify any particular behavior patterns that occur after this? How do these patterns keep the problem going? Stop the problem?"

Cognitive

"What do you usually think about afterward? How does this affect the problem?"

"What do you picture after this happens?"

"What do you tell yourself after this occurs?"

"Can you identify any particular thoughts [or beliefs or self-talk] during or after the problem that make the problem better? Worse?"

"Are there certain thoughts or images you have afterward that either strengthen or weaken the problem?"

Contextual

"What happened after this?"

"When does the problem usually stop or go away? Get worse? Get better?"

"Where are you when the problem stops? Gets worse? Gets better?"

"Can you identify any particular times, places, or events that seem to keep the problem going? Make it worse or better?"

"Are there any ways in which your cultural affiliations and values seem to keep this problem going? Stop it or weaken it?"

"Do you think the particular social and political structures of your society maintain this problem? How?"

Relational

"Are you usually with certain people during and after the problem? When the problem gets worse? Better?"

"Can you identify any particular people who can make the problem worse? Better? Stop it? Keep it going?"

"Can you identify any particular reactions from other people that occur after the problem? In what ways do these reactions affect the issue?"

"How do you think this problem is perpetuated by the persons in your society who hold power?"

Identifying Secondary Gains: A Special Case of Consequences

As we mentioned in Chapter 8, occasionally clients have a vested interest in maintaining the status quo of the problem because of the payoffs that the problem produces. For example, a client who is overweight may find it difficult to lose weight, not because of unalterable eating and exercise habits, but because the extra weight has allowed him to avoid or escape such things as new social situations or sexual relationships and has produced a safe and secure lifestyle that he is reluctant to give up (Fishman & Lubetkin, 1983). A child who is constantly disrupting her school classroom may be similarly reluctant to give up such disruptive behavior even though it results in loss of privileges, because it has given her the status of "class clown," resulting in a great deal of peer attention and support.

It is always extremely important to explore with clients the payoffs, or secondary gains, they may be getting from having the problem, because often during the intervention phase such clients seem resistant. In these cases, the resistance is a sign the payoffs are being threatened. The most common payoffs include money, attention from significant others, immediate gratification of needs, avoidance of responsibility, security, and control.

Questions you can use to help clients identify possible secondary gains include these:

"The good thing about _____ is . . . "

"What happened afterward that was pleasant?"

"What was unpleasant about what happened?"

"Has your concern or problem ever produced any special advantages or considerations for you?"

"As a consequence of your concern, have you got out of or avoided things or events?"

"What are the reactions of others when you do this?"

"How does this problem help you?"

"What do you get out of this situation that you don't get out of other situations?"

"Do you notice anything that happens afterward that you try to prolong or to produce?"

"Do you notice anything that occurs afterward that you try to stop or avoid?"

"Are there certain feelings or thoughts occurring afterward that you try to prolong?"

"Are there certain feelings or thoughts occurring afterward that you try to stop or avoid?"

Exploring Previous Solutions

Another important part of the assessment interview is to explore what things the client has already attempted to resolve the problem and with what effect. This information is important for two reasons. First, it helps you to avoid recommendations for problem resolution that amount to "more of the same." Second, in many instances, solutions attempted by the client either create new problems or make the existing one worse.

Fisch, Weakland, and Segal (1982, pp. 13–14) explain how clients' attempted solutions are often responsible for the origin or persistence of problems:

> Problems begin from some ordinary life difficulty, of which there is never any shortage. This difficulty may stem from an unusual or fortuitous event. More often, though, the beginning is likely to be a common difficulty associated with one of the transitions regularly experienced in the course of life— marriage, the birth of a child, going to school, and so on. . . . Most people handle most such difficulties reasonably adequately—perfect handling is neither usual nor necessary— and thus we do not see them in our offices. But for a difficulty to turn into a problem, only two conditions need to be fulfilled: (1) the difficulty is mishandled, and (2) when the difficulty is not resolved, more of the same "solution" is applied. Then the original difficulty will be escalated, by a vicious-circle process, into a problem—whose eventual size and nature may have little apparent similarity to the original difficulty.

Leads to help the client identify previous solutions include the following:

"How have you dealt with this or other problems before? What was the effect? What made it work or not work?"

"How have you tried to resolve this problem?"

"What kinds of things have you done to improve this situation?"

"What have you done that has made the problem better? Worse? Kept it the same?"

"What have others done to help you with this?"

Identifying the Client's Coping Skills, Strengths, and Resources

When clients come to therapists, they usually are in touch with their pain and often only with their pain. Consequently, they are shortsighted and find it hard to believe that they have any internal or external resources that can help them deal with the pain more effectively. In the assessment interview, it is useful to focus not solely on the problems and pains but also on the person's positive assets and resources (which the pain may mask). This sort of focus is the primary one used by feminist therapists; that is, an emphasis is placed on the client's strengths rather than the client's weaknesses (Brown & Ballou, 1992). Recent cognitive behavioral therapists have also placed increasing emphasis on the client's self-efficacy (Wilson, 1995)—the sense of personal agency and the degree of confidence the client has that she or he can do something (Mahoney, 1989). This focus is also consistent with brief therapy models that recommend finding "at least one thing to like or respect about each client or his or her coping and call attention to it" (Cooper, 1995, p. 38). Therapists also should remember that, like many variables, coping skills are culture and gender specific; some men and women may not report using the same coping strategies, just as "effective coping" defined in one cultural system may be different in another one (Sinacore-Guinn, 1995).

Focusing on the client's positive assets achieves several purposes. First, it helps convey to clients that, in spite of the psychological pain, they do have internal resources available that they can muster to produce a different outcome. Second, it emphasizes wholeness—the client is *more* than just his or her "problem." Third, it gives you information on potential problems that may crop up during an intervention. Finally, information about the client's past "success stories" may be applicable to current problems. Such information is extremely useful in planning intervention strategies that are geared to using the kind of problem-solving and coping skills already available in the client's repertoire. As LeShan (1995) notes, focusing on what is right with a person or what has happened to the person rather than on what is wrong and how it can be fixed creates a paradigm shift for clients who can then begin to regard themselves with greater care.

Information to be obtained in this area includes the following:

1. Behavioral assets and problem-solving skills—at what times does the client display adaptive behavior instead of problematic behavior? Often this information can be obtained by inquiring about "opposites"—for example, "When don't you act that way?" (Lazarus, 1989).
2. Cognitive coping skills—such as rational appraisal of a situation, ability to discriminate between rational and irrational thinking, selective attention and feedback from distractions, and the presence of coping or calming "self-talk" (Meichenbaum & Cameron, 1983).

3. Self-control and self-management skills—including the client's overall ability to withstand frustration, to assume responsibility for self, to be self-directed, to control problematic behavior by either self-reinforcing or self-punishing consequences, and to perceive the self as being in control rather than being a victim of external circumstances (Lazarus, 1989).

The following leads are useful in identifying these kinds of client resources and assets:

"What skills or things do you have going for you that might help you with this concern?"
"Describe a situation when this concern or problem is not interfering."
"What strengths or assets can you use to help resolve this problem?"
"When don't you act this way?"
"What kinds of thoughts or self-talk help you handle this better?"
"When don't you think in self-defeating ways?"
"What do you say to yourself to cope with a difficult situation?"
"Identify the steps you take in a situation you handle well. What do you think about and what do you do? How could these steps be applied to the present issue?"
"In what situations is it fairly easy for you to manage or control this reaction or behavior?"
"Describe any times you have been able to avoid situations in which these problems have occurred."
"To what extent can you do something for yourself in a self-directed way without relying on someone else to prod you to do it?"
"How often do you get things done by rewarding yourself in some way?"
"How often do you get things done by punishing yourself in some way?"

Exploring the Client's Perception of the Problem

Most clients have their own perception of and explanation for their problem. It is important to elicit this information during an assessment session for several reasons. First, it adds to your understanding of the problem. The therapist can note which aspects of the problem are stressed and which are ignored during the client's assessment of the issue. Second, this process gives you valuable information about "patient position." *Patient position* refers to the client's strongly held beliefs and values—in this case, about the nature of the issue

or problem (Fisch et al., 1982). Usually clients allude to such "positions" in the course of presenting their perception of the problem. Ignoring the client's position may cause the therapist to "blunder into a strategy that will be met with resistance" (Fisch et al., 1982, p. 92). You can get clients to describe their view of the problem very concisely simply by asking them to give the problem a one-line title as if it were a movie, play, or book. Another way to elicit the client's perception of the problem that Lazarus (1989) recommends is to describe the problem in only one word and then to use the selected word in a sentence. For example, a client may say "guilt" and then "I have a lot of guilt about having an affair." The same client might title the problem "Caught Between Two Lovers." This technique also works extremely well with children, who typically are quick to think of titles and words without a lot of deliberation. It is also important to recognize the impact of culture, ethnicity, and race on clients' perceptions and reports of problems. As Rosado and Elias (1993) note, some Latino clients report the cause of problems in terms of external factors, supernatural forces, or both. Therapists must not minimize or ridicule such explanations; also, they should incorporate such explanations into the assessment and treatment process.

Leads to help clients identify and describe their view of the problem include these:

"What is your understanding of this issue?"
"How do you explain this problem to yourself?"
"What does the problem mean to you?"
"What is your interpretation [analysis] of this problem?"
"What else is important to you about the problem that we haven't mentioned?"
"Give the problem a title."
"Describe the issue with just one word."

Ascertaining the Frequency, Duration, and Severity of the Problem

It is also useful to determine the intensity of the problem. You want to check out how much the problem is affecting the client and the client's daily functioning. If, for example, a client says "I feel anxious," does the client mean a little anxious or very anxious? Is this person anxious all the time or only some of the time? And does this anxiety affect any of the person's daily activities, such as eating, sleeping, or working? There are two kinds of intensity to assess: the degree of problem intensity or severity and the frequency (how often) or duration (how long) of the problem.

Degree of Problem Intensity.   Often it is useful to obtain a client's subjective rating of the degree of discomfort, stress, or intensity of the problem. The counselor can use this information to determine how much the problem affects the client and whether the client seems to be incapacitated or immobilized by it. To assess the degree of problem intensity, the counselor can use leads similar to these:

"You say you feel anxious. On a scale from 1 to 10, with 1 being very calm and 10 being extremely anxious, where would you be now?"
"How strong is your feeling when this happens?"
"How has this interfered with your daily activities?"
"How would your life be affected if this issue were not resolved in a year?"

In assessing degree of intensity, you are looking for a client response that indicates how strong, interfering, or pervasive the problem seems to be.

Frequency or Duration of Problem Behaviors.   In asking about frequency and duration, your purpose is to have the client identify how long (duration) or how often (frequency) the problem behaviors occur. Data about how long or how often the problem occurs *before* a counseling strategy is applied are called baseline data. Baseline data provide information about the *present* extent of the problem. They can also be used later to compare the extent of the problem before and after a counseling strategy has been used (see also Chapter 10).

Leads to assess the frequency and duration of the problem behavior include the following:

"How often does this happen?"
"How many times does this occur?"
"How long does this feeling usually stay with you?"
"How much does this go on, say, in an average day?"

Some clients can discuss the severity, frequency, or duration of the problem behavior during the interview rather easily. However, many clients may be unaware of the number of times the problem occurs, how much time it occupies, or how intense it is. Most clients can give the counselor more accurate information about frequency and duration by engaging in self-monitoring of the problem behaviors with a written log. Use of logs to supplement the interview data is discussed later in this chapter.

Table 9-3 provides a review of the 11 categories of problem assessment. This table may help you conceptualize and summarize the types of information you will seek during assessment interviews.

TABLE 9-3. Review of 11 problem assessment categories

I. *Purpose* of assessment		
II. *Range* of problems		
III. *Prioritization* of problems		
IV. V, VI, VII. Identification of		

Antecedents	Problem behaviors	Consequences and secondary gains (payoffs)
Affective	Affective	Affective
Somatic	Somatic	Somatic
Behavioral	Behavioral	Behavioral
Cognitive	Cognitive	Cognitive
Contextual	Contextual	Contextual
Relational	Relational	Relational

VIII. *Previous solutions*
IX. *Coping skills*
X. *Client perceptions* of problem
XI. *Frequency, duration, severity* of problem

GENDER AND MULTICULTURAL FACTORS IN INTERVIEW ASSESSMENT

Assessment interviews based on the ABC model also may not be compatible with all clients, especially some women and some non-Euro-American clients. This model of an interview-based assessment reflects a sort of specificity that is both Eurocentric and androcentric. Some women and clients from some cultural groups may not think in these specific ways. They may have difficulty relating to these kinds of questions and have trouble in providing particular responses. The therapist who is sensitive to this difficulty will not regard such clients as resistant or uncooperative but rather will recognize that "many people have a style of cognitive processing that differs from white, middle-class masculine norms" (Kantrowitz & Ballou, 1992, p. 80). Belenky, Clinchy, Goldbeyer, and Tarule (1986) have found that women's cognitive processing styles or "ways of knowing" are often circular or nonlinear, intuitive, and based on connection. Jackson (1987) has noted a difference in the cognitive processing style of African Americans and Euro-Americans. He states that African Americans tend to use inferential reasoning based on contextual, interpersonal, and historical factors whereas Euro-Americans tend to rely on either inductive or deductive reasoning (p. 233).

Other clients may have difficulty responding to these sorts of interview leads because their cultural affiliation does not advocate the use of immediate self-disclosure, particularly to nonfamily members or strangers. Clients of color who have experienced discrimination and oppression from the host culture may also be reluctant to self-disclose too much in assessment interviews; and self-exploration may be discouraged in some families because it is considered an individual rather than a collective approach (Sue & Sue, 1990, p. 39).

Language is another factor that can affect the usefulness of the assessment interview. Often counselors in the United States will encounter clients for whom English is not the primary language. With them, the therapist should make efforts to allow them to speak in their primary language to enhance their feelings of ease and also to facilitate the assessment process.

LIMITATIONS OF INTERVIEW LEADS IN PROBLEM ASSESSMENT

The leads we present in this chapter are simply tools that the counselor can use to elicit certain kinds of client information. They are designed to be used as a road map to provide some direction for assessment interviews. However, the leads alone are an insufficient basis for problem assessment, because they represent only about half the process at most—the counselor responses. The other part of the process is reflected by the responses these leads generate from the client. A complete problem assessment includes not only asking the right questions but also synthesizing and integrating the client responses.

A useful way to synthesize client responses during assessment interviews is to continue to build on and use all the fundamental helping skills presented earlier in this book. Think of it this way: In an assessment interview, you are simply *supplementing* your basic skills with some specific leads designed to obtain certain kinds of information. Many of your leads will consist of open-ended questions or probes. However, even assessment interviews should not disintegrate into a question-and-answer or interrogation session. You can obtain information and give the information some meaning through other verbal responses, such as summarization, clarification, confrontation, and reflection. Demonstrating sensitivity is especially important because sometimes during assessment, a client may reveal or even reexperience very traumatic events and memories. Handling the assessment interview in an understanding and empathic way becomes critical. It is also extremely important to clarify and reflect the information the client gives you before jumping ahead to another question. The model dialogue that follows will illustrate this process. (See Learning Activity #31.)

MODEL DIALOGUE FOR PROBLEM ASSESSMENT: THE CASE OF JOAN

To help you identify how these problem assessment leads are used in an interview, a dialogue of the case of Joan (from Chapter 8) is given. An explanation of the counselor's response and the counselor's rationale for using it appear in italics before the responses.

Counselor response 1 is a **rationale** *to explain to the client the* **purpose** *of the assessment interview.*

1. *Counselor:* Joan, last week you dropped by to schedule today's appointment, and you mentioned you were feeling unhappy and dissatisfied with school. It might be helpful today to take some time just to explore exactly what is going on with you and school and anything else that concerns you. I'm sure there are ways we can work with this dissatisfaction, but first I think it would be helpful to both of us to get a better idea of what all the issues are for you now. Does this fit with where you want to start today?
   Client: Yeah. I guess school is the main problem. It's really bugging me.

Counselor response 2 is a lead to help Joan identify the **range** *of her concerns.*

2. *Counselor:* OK, you just said school is the *main* problem. From the way you said that and the way you look right now, I have the feeling school isn't the *only* thing you're concerned about in your life.
   Client: Well, you're right about that. I'm also kind of not getting along too well with my folks. But that's kind of related to this school thing, too.

In the next response, the counselor will simply **listen** *to Joan and synthesize what she's saying by using a* **paraphrase** *response.*

3. *Counselor:* So from your point of view, the school thing and the issue with your parents are connected.
   Client: Yeah, because I'm having trouble in some of my classes. There's too much competition. I feel the other kids are better than I am. I've thought about changing from this college prep program to the work-study program, but I don't know what to do. I don't like to make decisions anyway. At the same time, my folks put a lot of pressure on me to perform well, to make top grades. They have a lot of influence with me. I used to want to do well, but now I'm kind of tired of it all.

In the next response, the counselor continues to listen to Joan and **reflect her feelings.**

4. *Counselor:* It seems like you're feeling pretty overwhelmed and discouraged right now.
   Client: Yeah, I am. [Lowers head, eyes, and voice tone]

Counselor senses Joan has strong feelings about these issues and doesn't want to cut them off initially. **Instructs** *Joan to continue focusing on the feelings.*

5. *Counselor:* [Pause] Let's stay with these feelings for a few minutes and see where they take you.
   Client: [Pause; eyes fill with tears] I guess I just feel like all this stuff is coming down on me at once. I'd like to work something out, but I don't know how—or where, even—to start.

Counselor continues to **attend,** *to* **listen,** *and to* **reflect** *the client's current experience.*

6. *Counselor:* It seems like you feel you're carrying a big load on your shoulders—
   Client: Yeah.

In response 7, the counselor **summarizes** *Joan's concerns, followed by a lead to determine whether Joan has* **prioritized** *her problems.*

7. *Counselor:* I think before we're finished I'd like to come back to these feelings, which seem pretty strong for you now. Before we do, it might help you to think about not having to tackle everything all at once. You know you mentioned three different things that are bothering you—your competitive classes, having trouble making decisions, and not getting along with your parents. Which of these problems bothers you most?
   Client: I'm not really sure. I'm concerned right now about having trouble in my classes. But sometimes I think if I were in another type of curriculum, I wouldn't be so tense about these classes. But I'm sort of worried about deciding to do this.

Counselor response 8 is a **clarification.** *The counselor wants to see whether the client's interest in work-study is real or is a way to avoid the present problem.*

8. *Counselor:* Do you see getting in the work-study program as a way to get out of your present problem classes, or is it a program that really interests you?
   Client: It's a program that interests me. I think sometimes I'd like to get a job after high school instead of going to college. *But* I've been thinking about this for a year and I can't decide what to do. I'm not very good at making decisions on my own.

Counselor response 9 is a **summarization** *and* **instruction.** *The counselor goes back to the three problem areas mentioned in "Identifying the Range of Problems." Note that the counselor does not draw explicit attention to the client's last self deprecating statement.*

9. *Counselor:* Well, your concerns of your present class problems and of making this and other decisions are somewhat related. Your parents tie into this, too. Maybe you could explore all concerns and then decide later about what you want to work on first.
   Client: That's fine with me.

Counselor response 10 is a lead to **identify some present problem behaviors** *related to Joan's concern about competitive classes. Asking the client for examples can elicit specificity about what does or does not occur during the problem situation.*

10. *Counselor:* OK, what is an example of some trouble you've been having in your most competitive class?
    Client: Well, I withdraw in these classes. I've been cutting my math classes. It's the worst. My grades are dropping, especially in math class.

Counselor response 11 is a **problem behavior** *lead regarding the* **context** *of the problem to see whether the client's concern occurs at other* **times** *or other* **places.**

11. *Counselor:* Where else do you have trouble—in any other classes, or at other times or places outside school?
    Client: No, not outside school. And, to some degree, I always feel anxious in any class because of the pressures my parents put on me to get good grades. But my math class is really the worst.

Counselor response 12 is a lead to help the client identify **overt problem behaviors** *in math class (**behavioral** component of problem).*

12. *Counselor:* Describe what happens in your math class that makes it troublesome for you. [could also use imagery assessment at this point]
    Client: Well, to start with, it's a harder class for me. I have to work harder to do OK. In this class I get nervous whenever I go in it. So I withdraw.

Client's statement "I withdraw" is vague. So counselor response 12 is another **overt problem behavior** *lead to help the client specify what she means by "withdrawing." Note that since the counselor did not get a complete answer to this after response 8, the same type of lead is used again.*

13. *Counselor:* What do you do when you withdraw? [This is also an ideal place for a role-play assessment.]
    Client: Well, I sit by myself, I don't talk or volunteer answers. Sometimes I don't go to the board or answer when the teacher calls on me.

Now that the client has identified certain overt behaviors associated with the problem, the counselor will use a **covert problem behavior** *lead to find out whether there are any predominant* **thoughts** *the client has during the math class (**cognitive** component of problem).*

14. *Counselor:* What are you generally thinking about in this class?
    Client: What do you mean—am I thinking about math?

The client's response indicated some confusion. The counselor will have to use a more specific **covert problem behavior** *lead to assess cognition, along with some* **self-disclosure,** *to help the client respond more specifically.*

15. *Counselor:* Well, sometimes when I'm in a situation like a class, there are times when my mind is in the class and other times I'm thinking about myself or about something else I'm going to do. So I'm wondering whether you've noticed anything you're thinking about during the class?
    Client: Well, some of the time I'm thinking about the math problems. Other times I'm thinking about the fact that I'd rather not be in the class and that I'm not as good as the other kids.

The client has started to be more specific, and the counselor thinks perhaps there are still other thoughts going on. To explore this possibility, the counselor uses another **covert problem behavior** *lead in response 16 to assess* **cognition.**

16. *Counselor:* What else do you recall that you tell yourself when you're thinking you're not as good as other people?

    Client: Well, I think that I don't get grades that are as good as some other students'. My parents have been pointing this out to me since junior high. And in the math class I'm one of four girls. The guys in there are really smart. I just keep thinking how can a girl ever be as smart as a guy in math class? No way. It just doesn't happen.

The client identifies more specific problem-related thoughts and also suggests two possible antecedents—parental comparison of her grades and cultural stereotyping (girls shouldn't be as good in math as boys). The counselor's records show that the client's test scores and previous grades indicate that she is definitely not "dumb" in math. The counselor will **summarize** *this and then, in the next few responses, will focus on these and on other possible* **antecedents,** *such as the nervousness the client mentioned earlier.*

17. *Counselor:* So what you're telling me is that you believe most of what you've heard from others about yourself and about the fact that girls automatically are not supposed to do too well in math.

    Client: Yeah, I guess so, now that you put it like that. I've never given it much thought.

18. *Counselor:* Yes. It doesn't sound like you've ever thought about whether *you, Joan,* really feel this way or whether these feelings are just adopted from things you've heard others tell you.

    Client: No, I never have.

19. *Counselor:* That's something we'll also probably want to come back to later.

    Client: OK.

20. *Counselor:* You know, Joan, earlier you mentioned that you get nervous about this class. When do you notice that you feel this way—before the class, during the class, or at other times?

    Client: Well, right before the class is the worst. About ten minutes before my English class ends—it's right before math—I start thinking about the math class. Then I get nervous and feel like I wish I didn't have to go. Recently, I've tried to find ways to cut math class.

The counselor still needs more information about how and when the nervousness affects the client, so 21 is another **antecedent** *lead.*

21. *Counselor:* Could you tell me more about when you feel most nervous and when you don't feel nervous about this class?

    Client: Well, I feel worst when I'm actually walking to the class and the class is starting. Once the class starts, I feel better. I don't feel nervous about it when I cut it or at other times. However, once in a while, if someone talks about it or I think about it, I feel a little nervous.

The client has indicated that the nervousness seems to be more of an antecedent than a problem behavior. She has also suggested that cutting class is a consequence that maintains the problem, because she uses this to avoid the math class that brings on the nervousness. The counselor realizes at this point that the word **nervous** *has not been defined and goes back in the next response to a* **covert**

problem behavior *lead to find out what Joan means by* **nervous** *(affective component).*

22. *Counselor:* Tell me what you mean by the word *nervous*—what goes on with you when you're nervous?

    Client: Well, I get sort of a sick feeling in my stomach, and my hands get all sweaty. My heart starts to pound.

In the next response, the counselor continues to **listen** *and* **paraphrase** *to clarify whether the nervousness is experienced somatically.*

23. *Counselor:* So your nervousness really consists of things you feel going on inside you.

    Client: Yeah.

Next the counselor will use an **intensity** *lead to determine the* **severity** *of nervousness.*

24. *Counselor:* How strong is this feeling—a little or very?

    Client: Before class, very strong—at other times, just a little.

The client has established that the nervousness seems mainly to be exhibited in somatic forms and is more intense before class. The counselor will pursue the relationship between the client's nervousness and overt and covert problem behaviors described earlier to verify that the nervousness is an **antecedent.** *Another* **antecedent** *lead is used next.*

25. *Counselor:* Which seems to come first—feeling nervous, not speaking up in class, or thinking about other people being smarter than you?

    Client: Well, the nervousness. Because that starts before I get in the class.

The counselor will **summarize** *this pattern and confirm it with the client in the next response.*

26. *Counselor:* Let's see. So you feel nervous—like in your stomach and hands—before class and when math class starts. Then during class, on days you go, you start thinking about not being as smart in math as the guys and you don't volunteer answers or don't respond sometimes when called on. But after the class is over, you don't notice the nervousness so much. Is that right?

    Client: That's pretty much what happens.

The counselor has a clue from the client's previous comments that there are other antecedents in addition to nervousness that have to do with the client's problem behavior—such as the role of her parents. The counselor will pursue this in the next response, using an **antecedent** *lead.*

27. *Counselor:* Joan, you mentioned earlier that you have been thinking about not being as smart as some of your friends ever since junior high. When do you recall you really started to dwell on this?

    Client: Well, probably in seventh grade.

The counselor didn't get sufficient information about what happened to the client in the seventh grade, so another **antecedent** *lead will be used to identify this possible* **setting event.**

28. *Counselor:* Well, what things seemed to happen then when you began to compare yourself with others?

Client: Well, my parents said when you start junior high, your grades become really important in order to go to college. So for the last three or four years they have been telling me some of my grades aren't as good as other students'. Also, if I get a B, they will withhold a privilege, like my allowance.

The counselor has no evidence of actual parental reaction but will work with the client's report at this time, since this is how the client perceives parental input. If possible, a parent conference could be arranged later with the client's permission. The parents **seem** *to be using negative rather than positive consequences with Joan to influence her behavior. The counselor wants to pursue the relationship between the parents' input and the client's present behavior to determine whether parental reaction is eliciting part of Joan's present concerns and will use a lead to identify this as a possible* **antecedent.**

29. *Counselor:* How do you think this reaction of your parents' relates to your present problems in your math class?

Client: Well, since I started high school, they have talked more about needing to get better grades for college. And I have to work harder in math class to do this. I guess I feel a lot of pressure to perform—which makes me withdraw and just want to hang it up. Now, of course, my grades are getting worse, not better.

The counselor, in the next lead, will **paraphrase** *Joan's previous comment.*

30. *Counselor:* So the expectations you feel from your parents seem to draw out pressure in you.

Client: Yes, that happens.

In response 31, the counselor will explore another possible **antecedent** *that Joan mentioned before—thinking that girls aren't as good as boys in math.*

31. *Counselor:* Joan, I'd like to ask you about something else you mentioned earlier that I said we would come back to. You said one thing that you think about in your math class is that you're only one of four girls and that, as a girl, you're not as smart in math as a boy. Do you know what makes you think this way?

Client: I'm not sure. Everyone knows or says that girls have more trouble in math than boys. Even my teacher. He's gone out of his way to try to help me because he knows it's tough for me.

The client has identified a possible consequence of her problem behavior as teacher attention. The counselor will return to this later. First, the counselor is going to respond to the client's response that "everyone" has told her this thought. Counselors have a responsibility to point out things that clients have learned from stereotypes or irrational beliefs rather than actual data, as is evident in this case from Joan's academic record. Counselor will use **confrontation** *in the next response.*

32. *Counselor:* You know, studies have shown that when young women drop out of math, science, and engineering programs, they do so not because they're doing poorly, but because they don't believe they can do well.* It is evident to me from your records that you have a lot of potential for math.

Client: You mean I really could do as well in math as the guys?

Counselor response 33 is an **interpretation** *to help the client see the relation between overt and covert behaviors.*

33. *Counselor:* I don't see why not. But lots of times the way someone acts or performs in a situation is affected by how the person thinks about the situation. I think some of the reason you're having more trouble in your math class is that your performance is hindered a little by your nervousness and by the way you put yourself down.

In the next response, the counselor **checks out** *and* **clarifies** *the client's reaction to the previous interpretation.*

34. *Counselor:* I'm wondering now from the way you're looking at me whether this makes any sense or whether what I just said muddies the waters more for you?

Client: No, I guess I was just thinking about things. You mentioned the word *expectations*. But I guess it's not just that my parents expect too much of me. I guess in a way I expect too little of myself. I've never really thought of that before.

35. *Counselor:* That's a great observation. In a way the two sets of expectations are probably connected. These are some of the kinds of issues we may want to work on in counseling if this track we're on seems to fit for you.

Client: Yeah. OK, it's a problem.

The counselor is going to go back now to pursue possible consequences that are influencing the client's problem behavior. The next response is a lead to identify **consequences.**

36. *Counselor:* Joan, I'd like to go back to some things you mentioned earlier. For one thing, you said your teacher has gone out of his way to help you. Would you say that your behavior in his class has got you any extra attention or special consideration from him?

Client: Certainly extra attention. He talks to me more frequently. And he doesn't get upset when I don't go to the board.

Counselor response 37 will continue to explore the teacher's behavior as a possible **consequence.**

37. *Counselor:* Do you mean he may excuse you from board work?

Client: For sure, and I think he, too, almost expects me *not* to come up with the answer. Just like I don't expect myself to.

The teacher's behavior may be maintaining the client's overt problem behaviors in class by giving extra attention to her for her problems and by excusing her from some kinds of work. Declines in achievement in math and science among girls after seventh grade have been linked to teachers' differing expectations for girls and boys and their subsequent behavior in the classroom. Teachers often

---

*From studies conducted at Wellesley College's Center for Research on Women.

pay less attention to girls than boys in the classroom and when they do pay attention to girls, it is more in the form of protecting girls by solving problems for them (DeAngelis, 1994). A teacher conference may be necessary at some later point. The counselor, in the next two responses, will continue to use other leads to identify possible **consequences.**

38. *Counselor:* What do you see you're doing right now that helps you get out of putting yourself through the stress of going to math class?

    Client: Do you mean something like cutting class?

39. *Counselor:* I think that's perhaps one thing you do to get out of the class. What else?

    Client: Nothing I can think of.

The client has identified cutting class as one way to avoid the math class. The counselor, in the next response, will suggest another **consequence** *that the client mentioned earlier, though not as a way to get out of the stress associated with the class. The counselor will suggest that this consequence functions as a* **secondary gain,** *or* **payoff,** *in a tentative* **interpretation** *that is checked out with the client in the next three responses:*

40. *Counselor:* Well, Joan, you told me earlier that your grades were dropping in math class. Is it possible that if these grades—and others—drop too much, you'll automatically be dropped from these college prep classes?

    Client: That's right.

41. *Counselor:* I'm wondering whether one possible reason for letting your grades slide is that it is almost an automatic way for you to get out of these competitive classes.

    Client: How so?

42. *Counselor:* Well, if you became ineligible for these classes because of your grades, you'd automatically be out of this class and others that you consider competitive and feel nervous about. What do you think about that?

    Client: I guess that's true. And then my dilemma is whether I want to stay in this or switch to the work-study program.

In the next response, the counselor uses **summarization** *and ties together the effects of "dropping grades" to math class and to the earlier-expressed concern of a curriculum-change decision.*

43. *Counselor:* Right. And letting your grades get too bad will automatically mean that decision is made for you, so you can take yourself off the hook for making that choice. In other words, it's sort of a way that part of you has rather creatively come up with to get yourself out of the hassle of having to decide something you don't really want to be responsible for deciding about.

    Client: Wow! Gosh, I guess that might be happening.

44. *Counselor:* That's something you can think about. We didn't really spend that much time today exploring the issue of having to make decisions for yourself, so that will probably be something to discuss the next time we get together. I know you have a class coming up in about ten minutes, so there's just a couple more things we might look at.

    Client: OK—what next?

In the next few responses (45–52), the counselor continues to demonstrate **listening responses** *and to help Joan explore* **solutions** *she's tried already to resolve the problem. They look together at the* **effects** *of the use of the solutions Joan identifies.*

45. *Counselor:* OK, starting with the nervousness and pressure you feel in math class—is there anything you've attempted to do to get a handle on this problem?

    Client: Not really—other than talking to you about it and, of course, cutting class.

46. *Counselor:* So cutting class is the only solution you've tried.

    Client: Yeah.

47. *Counselor:* How do you think this solution has helped?

    Client: Well, like I said before—it helps mainly because on the days I don't go, I don't feel uptight.

48. *Counselor:* So you see it as a way to get rid of these feelings you don't like.

    Client: Yeah, I guess that's it.

49. *Counselor:* Can you think of any ways in which this solution has not helped?

    Client: Gee, I don't know. Maybe I'm not sure what you're asking.

50. *Counselor:* OK, good point! Sometimes when I try to do something to resolve a problem, it can make the issue better or worse. So I guess what I'm really asking is whether you've noticed that your "solution" of cutting class has in any way made the problem worse or in any way has even contributed to the whole issue?

    Client: [Pause] I suppose maybe in a way. [Pause] In that, by cutting class, I miss out on the work, and then I don't have all the input I need for tests and homework, and that doesn't help my poor grades.

51. *Counselor:* OK. That's an interesting idea. You're saying that when you look deeper, your solution also has had some negative effects on one of the problems you're trying to deal with and eliminate.

    Client: Yeah. But I guess I'm not sure what else I could do.

52. *Counselor:* At this point, you probably are feeling a little bit stuck, like you don't know which other direction or road to take.

    Client: Yeah, kind of like a broken record.

At this point, the counselor shifts the focus a little to exploration of Joan's **assets, strengths, and resources.**

53. *Counselor:* Well, one thing I sense is that your feelings of being so overwhelmed are sort of covering up the resources and assets you have within you to handle the issue and work it out. For example, can you identify any particular skills or things you have going for you that might help you deal with this issue?

    Client: [Pause]Well, are you asking me to brag about myself?

Clients often talk about their pain or limitations freely but feel reluctant to reveal their strengths, so in the next response, the counselor gives Joan a specific **directive** *and* **permission** *to talk about her* **assets.**

54. *Counselor:* Sure. Give yourself permission. That's certainly fine in here.

Client: Well, I am pretty responsible. I'm usually fairly loyal and dependable. It's hard to make decisions for myself, but when I say I'm going to do something, I usually do it.

55. *Counselor:* OK, great. So what you're telling me is you're good on follow-through once you decide something is important to you.

Client: Yeah. Mm-hmm. Also, although I'm usually uptight in my math class, I don't have the same feeling in my English class. I'm really doing well in there.

In response 56, the counselor will pick up on these "pluses" and use another **coping skills** *lead to have the client identify particular ways in which she handles positive situations, especially her English class. If she can demonstrate the steps to succeed in one class, this is useful information that can be applied in a different and problematic area. This topic is continued in response 57.*

56. *Counselor:* So there are some things about school that are going OK for you. You say you're doing well in your English class. What can you think of that you do or don't do to help you perform well in this class?

Client: Well, I go to class, of course, regularly. And I guess I feel like I do well in reading and writing. I don't have the hangup in there about being one of the few girls.

57. *Counselor:* So maybe you can see some of the differences between your English and math classes—and how you handle these. This information is useful because if you can start to identify the things you do and think about in English that make it go so well for you, then you potentially can apply the same process or steps to a more difficult situation, like math class.

Client: That sounds hopeful!

In the next few responses, the counselor tries to elicit **Joan's perception and assessment of the main issue.**

58. *Counselor:* Right. It is. I feel hopeful, too. Just a couple more things. Changing the focus a little now, could you think about the issues that you came in with today—and describe the main issue in one word?

Client: Ooh—that's a hard question!

59. *Counselor:* I guess it could be. Take your time. You don't have to rush.

Client: [Pause] Well, how about "can't?"

60. *Counselor:* OK, now, to help me get an idea of what that word means to you, use it in a sentence.

Client: Any sentence?

61. *Counselor:* Yeah. Make one up. Maybe the first thing that comes in your head.

Client: Well, "I can't do a lot of things I think I want to or should be able to do."

In response 62, the counselor uses a **confrontation** *to depict the incongruity revealed in the sentence Joan made up about her problem. This theme is continued in response 63.*

62. *Counselor:* OK, that's interesting too, because on the one hand, you're saying there are some things you *want* to do that aren't happening, and on the other hand, you're also saying there are

some things that aren't happening that you think you *should* be doing. Now, these are two pretty different things mixed together in the same sentence.

Client: Yeah. [Clarifies] I think the wanting stuff to happen is from me and the should things are from my folks and my teachers.

63. *Counselor:* OK, so you're identifying part of the whole issue as wanting to please yourself and others at the same time.

Client: M-m-hmm.

In response 64, the counselor identifies this issue as an extension of the **secondary gain** *mentioned earlier—avoiding deliberate decisions.*

64. *Counselor:* I can see how after a while that would start to feel like so much trouble that it would be easier to try to let situations or decisions get made for you rather than making a conscious or deliberate choice.

In the next two responses, the counselor explores the **context** *related to these issues and sets up some* **self-monitoring** *homework to obtain additional information. Note that this is a task likely to appeal to the client's dependability, which she revealed during exploration of* **coping skills.**

65. *Counselor:* That's something else we'll be coming back to, I'm sure. One last thing before you have to go. Earlier we talked about some specific times and places connected to some of these issues—like where and when you get in the rut of putting yourself down and thinking you're not as smart as other people. What I'd like to do is give you sort of a diary to write in this week to collect some more information about these kinds of problems. Sometimes writing these kinds of things down can help you start making changes and sorting out the issues. You've said that you're pretty dependable. Would doing this appeal to your dependability?

Client: Sure. That's something that wouldn't be too hard for me to do.

66. *Counselor:* OK, let me tell you specifically what to keep track of, and then I'll see you next week—bring this back with you. [Goes over instructions for Joan's log sheet—see section of chapter entitled "Client Self-Monitoring Assessment."]

At this time, the counselor also has the option of giving Joan a history questionnaire to complete and/or a brief self-report inventory to complete, such as an anxiety inventory or checklist.

NOTES AND RECORD KEEPING

Generally, some form of written record is started at the time the client requests an appointment. According to Mitchell (1991), clinicians must keep timely and accurately written records for three primary reasons: fiscal, clinical, and legal accountability. Increasingly with both government-funded agencies and private health maintenance organizations, written records are becoming a tool for billing as well as for

LEARNING ACTIVITY **31**   *Interview Assessment*

I. The following activity is designed to assist you in identifying problem assessment leads in an interview. You are given a counselor/client dialogue of the case of Ms. Weare and Freddie (Chapter 8). This dialogue consists of an interview with the mother, Ms. Weare. For each counselor response, your task is to identify and write down the type of problem assessment lead used by the counselor. You may find it helpful to use the Interview Checklist at the end of the chapter as a guide for this learning activity. There may be more than one example of any given lead in the dialogue. Also, responses from previous chapters (listening and action) may be used. Other basic verbal interview responses are also included. Feedback follows the Learning Activity (on page 206).

Dialogue with Ms. Weare and Counselor

1. *Counselor:* Hello, Ms. Weare. Could you tell me about some things going on now that are concerning you?
   Client: Not too much. Family Services sent me here.
2. *Counselor:* So you're here just because they sent you—or is there something bothering you?
   Client: Well, they don't think my kid and I get along too well. My kid is Freddie.
3. *Counselor:* What do you think about the way you and Freddie get along?
   Client: Well, I yell at him a lot. I don't like to do that but sometimes he gets me so mad. I don't like to, but he needs to learn.
4. *Counselor:* So there are times when you get real mad at Freddie and then you yell at him. You don't like to do this, but you see it as a way to help him learn right and wrong.
   Client: That's it. I don't like to, but there are times when he needs to know something. Like yesterday, I bought him new pants and he came home from school with a big hole in them. Now I just don't have money to keep buying him new pants.
5. *Counselor:* You just mentioned the incident with Freddie's pants. What are some other times that you get mad at Freddie?
   Client: Every morning. Freddie's in fifth grade now. But he still doesn't dress himself in the morning. I want to be a good mother and get him to school on time, and he tries to be late. He waits around not getting dressed.
6. *Counselor:* Any other times you can remember getting mad?

Client: Well, not too long ago he was playing outside and broke a window. I got mad then. But that doesn't happen every day like his not getting dressed does.
7. *Counselor:* So one thing that really bothers you is what goes on in the mornings. Could you tell me exactly what does happen each morning at your house?
   Client: Well, I call Freddie to get up and tell him to dress before he comes down for breakfast. He comes down all right—in his pajamas. I warn him after breakfast to get ready. Usually about five minutes before the bus comes, I'll go up. He'll just be sitting in his room! He's still not dressed. I'll yell at him and then dress him so he's not late.
8. *Counselor:* And your main feeling at this point is that you're mad. Anything else you feel?
   Client: No, just very mad.
9. *Counselor:* And what exactly do you do when you go upstairs and he's not dressed?
   Client: I yell at him. Then I help dress him.
10. *Counselor:* What kinds of things do you usually say to him?
    Client: I tell him he's dumb and he's going to be late for school, and that I have to make sure he won't be.
11. *Counselor:* You mentioned this happens in the morning. Does this situation go on every morning or only some mornings?
    Client: Just about every morning except weekends.
12. *Counselor:* When did these incidents seem to begin?
    Client: Ever since Freddie started going to school.
13. *Counselor:* So it appears that this has been going on for about five or six years, then?
    Client: Yes, I guess so.
14. *Counselor:* OK, now let's go back over this situation. You told me you remind Freddie every morning to get dressed. He never dresses by breakfast. You remind him again. Then, about five minutes before the bus comes, you go upstairs to check on him. When do you notice that you start to feel mad?
    Client: I think about it as soon as I realize it's almost time for the bus to come and Freddie isn't down yet. Then I feel mad.
15. *Counselor:* And what exactly do you think about right then?

(continued)

Client: Well, that he's probably not dressed and that if I don't go up and help him, he'll be late. Then I'll look like a bad mother if I can't get my son to school on time.

16. *Counselor:* So in a sense you actually go help him out so he won't be late. How many times has Freddie ever been late?

Client: Never.

17. *Counselor:* You believe that helping Freddie may prevent him from being late. However, your help also excuses Freddie from having to help himself. What do you think would happen if you stopped going upstairs to check on Freddie in the morning?

Client: Well, I don't know, but I'm his only parent. Freddie's father isn't around. It's up to me, all by myself, to keep Freddie in line. If I didn't go up and if Freddie was late all the time, his teachers might blame me. I wouldn't be a good mother.

18. *Counselor:* Of course, we don't *really* know what would happen if you didn't go up and yell at him or help him dress. It might be so different for Freddie after the first day or two he would dress himself. It could be that he thinks it's easier to wait and get your help than to dress himself. He might think that by sitting up there and waiting for you to help, he's getting a special advantage or attention from you.

Client: You mean like he's getting a favor from me?

19. *Counselor:* Sure. And when we find a way to get a favor from someone, we usually do as much as we can to keep getting the favor. Ms. Weare, I'd like to ask you about something else. Do you think maybe that you see helping Freddie out as a way to avoid having Freddie be late and then not having someone blame you for this?

Client: Sure. I'd rather help him than get myself in hot water.

20. *Counselor:* OK, so you're concerned about what you think might happen to you if he's late. You see getting him ready on time as a way to prevent you from getting the heat for him.

Client: Yes.

21. *Counselor:* How do you usually feel after these incidents in the morning are over?

Client: Well, it upsets me.

22. *Counselor:* OK, you feel upset. Do these feelings seem to make you want to continue or to stop helping Freddie?

Client: Probably to stop. I get worn out. Also, sometimes I don't get my work done then.

23. *Counselor:* So helping Freddie so he won't be late and you won't be blamed sort of makes you want to keep on helping him. Yet when you feel upset and worn out afterward, you're tempted to stop helping. Is this right?

Client: I guess that could be true.

24. *Counselor:* Gee, I imagine that all the responsibility for a 10-year-old boy would start to feel like a pretty heavy burden after a while. Would it be right to say that it seems like you feel very responsible for Freddie and his behavior?

Client: Yeah. I guess a lot of the time I do.

25. *Counselor:* Those may be feelings we'll want to talk about more. I'm also wondering whether there are any other things in your life causing you any difficulty now?

Client: No, this is about it.

26. *Counselor:* Ms. Weare, we've been talking a lot about some problem situations you've had with Freddie. Could you tell me about some times when the two of you get along OK?

Client: Well, on weekends we do. Freddie dresses himself whenever he gets up. I sleep later.

27. *Counselor:* What happens on weekends when the two of you get along better?

Client: Sometimes I'll take him to a movie or a game. And we eat all our meals together. Usually weekends are pleasant. He can be a good boy and I don't scream all the time at him.

28. *Counselor:* So you realize it is possible for the two of you to get along. How do you feel about my talking with Freddie and then with both of you together?

Client: That's OK.

II. To incorporate the interview leads into your verbal repertory. We suggest that you try a role-play interview of the case of Ms. Weare (Chapter 8) or the case of Mr. Huang (Chapter 8) with a triad. One person can take the role of the client (Ms. Weare or Mr. Huang); another can be the counselor. Your task is to assess the client's concerns using the interview leads described in this chapter. The third person can be the observer, providing feedback to the counselor during or following the role-play, using the Interview Checklist at the end of the chapter as a guide.

FEEDBACK
Interview Assessment

I. Identifications of the responses in the dialogue between Ms. Weare and the counselor are as follows:

1. Open-ended question (probe)
2. Clarification response
3. Open-ended question (probe)
4. Summarization response
5. Paraphrase response and problem behavior lead: exploration of context
6. Problem behavior lead: exploration of context
7. Paraphrase response and problem behavior lead: exploration of overt behavior
8. Reflection-of-feeling response and problem behavior lead: exploration of affect
9. Problem behavior lead: exploration of overt behavior
10. Problem behavior lead: exploration of overt behavior
11. Paraphrase and problem behavior lead: exploration of context
12. Antecedent lead: context
13. Clarification response
14. Summarization response and antecedent lead: affect
15. Problem behavior lead: exploration of cognitions
16. Paraphrase and probe responses
17. Consequences: overt behavior
18. Consequences: secondary gains for Freddie
19. Consequences: secondary gains for Ms. Weare
20. Summarization response and exploration of secondary gains for Ms. Weare
21. Consequences: affect
22. Consequences: affect
23. Summarization (of consequences)
24. Reflection-of-feeling and interpretation responses
25. Range-of-problems lead
26. Coping skills
27. Coping skills
28. Paraphrase and open-ended question

keeping track of client progress. Mitchell (1991) points out that if you don't have records to verify services rendered, both government-funding sources and insurance companies may not pay, or if they do, they may want their money returned until verification of services can be provided. Swenson (1997) notes that increasingly helpers are keeping records on computers, and while this makes record-keeping

easier, data stored in electronic form are copied "more easily and harder to protect than paper data" (p. 93). It is important to remember that duties to maintain confidentiality apply just as much to electronic data as to written records. Identifying data about the client are recorded initially, as well as appointment times, cancellations, and so on. The intake or initial history-taking session is recorded next. In writing up an intake or history, it is important to avoid labels, jargon, and inferences. If records were subpoenaed, such statements could appear inflammatory or slanderous. Be as specific as possible. Don't make evaluative statements or clinical judgments without supporting documentation. For example, instead of writing "This client is homicidal," you might write "This client reports engaging in frequent (at least twice daily) fantasies of killing an unidentified or anonymous victim," or instead of "The client is disoriented," consider "The client could not remember where he was, why he was here, what day it was, and how old he was."

It is also important to keep notes of subsequent treatment sessions and of client progress. These can be recorded on a standardized form such as the Sample Treatment Planning Form (Figure 9-1) or in narrative form. (See also Chapter 11 for further discussion.) Generally, treatment notes are brief and highlight only the major activities of each session and client progress and improvement (or lack of it). These notes are usually started during intakes, with additional information added from the assessment interviews. As therapy progresses, notations about goals, intervention strategies, and client progress are also included. Again, labels and inferences should always be avoided in written notes and records. Mitchell (1991) advocates the use of client participation in record keeping. He notes that reading your progress notes back to clients at the end of the sessions can encourage teamwork and reduce client stress and discontent.

It is also important to document in detail anything that has ethical or legal implications, particularly facts about case management. For example, with a client who reports depression and suicidal fantasies, it would be important to note that you conducted a suicide assessment and what its results were, that you consulted with your supervisor, and whether you did anything else to manage the case differently, such as seeing the client more frequently or setting up a contract with the client.

Many counselors, who are people-oriented persons, complain about paperwork. If you fall into this group, be aware of Mitchell's (1991) dictum: "Write with pride." Mitchell (1991) points out that often we are judged by the kind of written record and paper trail we leave behind us.

MENTAL HEALTH NETWORK: OUTPATIENT TREATMENT REPORT (OTR)   Provider:_____

Client Name:_____   Birthdate:_____   Age:_____   Sex: M   F

A. ASSESSMENT
   1. Presenting Problem (Client's Perspective): _____
   _____
   _____

   2. Precipitating Event(s) (Why Help-seeking Now?): _____
   _____
   _____

   3. Relevant Medical History (Medications, Drug/ETOH use, Illness, Injury, Surgery, etc.): _____
   _____
   _____

   4. Prior Psychiatric/Psychological Conditions & Treatments: _____
   _____

   5. Other Relevant History (Vocational/School, Relationship/Sexual, Social/Legal): _____

   6. Brief Mental Status Evaluation: (Check as necessary)

APPEARANCE/DRESS	INTELLIGENCE	JUDGMENT/INSIGHT	DELUS./HALLUCIN.	THOUGHT DISORDER	RECENT MEMORY	REMOTE MEMORY
__appropriate	__high	__intact	__absent	__absent	__intact	__intact
__inappropriate	__average	__impaired	__present	__present	__impaired	__impaired
__not assessed	__low	__not assessed	__not assessed	__not assessed	__not assessed	__not assessed

   7. Mood/Affect: (Describe) _____
   8. Suicide Assessment: (Risk, priors, plan) _____ Homicide Assessment: (Victim, violence, plan) _____
   9. Clinical Formulation (Explanation of symptoms; include strengths/resources, obstacles to treatment/hidden agendas): Please be specific yet brief and clear _____
   _____
   _____

   10. Code Nos. & Names DSM-4 Axis II: _____   Diagnostic Impressions: DSM-4 Axis I: _____

B. TREATMENT PLAN:
   1. Focused, Targeted, Behavior & Measurable GOALS (Prioritize) Specifically Addressing Presenting Problems(s): Use as many rows as needed.
   2. TYPE OF TREATMENT: Cognitive/Behavioral Interpersonal/Insight/Emotional Awareness Other:
   3. DURATION: Service dates this OTR: _____# Sessions expected for DISCHARGE: _____Discharge by (DATE): _____
   4. MODE: Individual   Couple   Family   Individual/Family combination   Medication management   Group (if available)   Other (__)
           (90844)   (90847)   (90862)   (90853)   CPT Code__

PROBLEM(S)	GOALS	MEASURABLE SUCCESS CRITERION	SELECTED INTERVENTIONS
1. _____	_____	_____	_____
2. _____	_____	_____	_____
3. _____	_____	_____	_____

Therapist Signature & Phone: _____   Lic. No.: _____   Date: _____

Client Comments:

FIGURE 9-1. Sample treatment planning form   *SOURCE:* Adapted from the Treatment Planning Form, by Mental Health Network, Inc., 771 Corporate Drive, Suite 410, Lexington, KY. Reprinted by permission.

CLIENT SELF-MONITORING ASSESSMENT

The data given by the client in the interview can be supplemented by client self-monitoring outside the interview. Self-monitoring can be defined as the process of observing specific things about oneself and one's interaction with others and the environment. In using self-monitoring as a problem assessment tool, the client is asked to record her or his observations in writing. These written recordings can be entered on a log or a daily record sheet.

One purpose of client self-monitoring is to help the counselor and client gain information about what actually occurs with respect to the problem in real-life settings. Another purpose is to validate the accuracy of the client's oral reports during the interviews. Client self-monitoring of problem situations and behaviors outside the interview should add more accuracy and specificity to the information discussed in the interview. As a result, client self-monitoring may accelerate treatment and enhance the client's expectations for change. Self-monitoring is also a useful way to test out hunches about the problem and to identify relations between classes of events such as thoughts, feelings, and behaviors.

As we mentioned earlier, a client can record observations on some type of written record, or log. Two types of logs can be used for different observations a client might make during problem definition. A *descriptive log* can be used to record data about identification and selection of problem concerns. A *behavior log* can be used to record information about the problem behaviors and their antecedents and consequences or the relation between these classes of events related to the problem.

Descriptive Logs

In an initial session with a client, a simple descriptive or exploratory log can be introduced to find out what is going on with the client, where, and when. Such a descriptive log could be set up as shown in Figure 9-2. The descriptive log is extremely useful when the client has difficulty identifying problem concerns or pinpointing problem situations. However, once the problem concerns have been identified and selected, a counselor and client may find that a behavior log is helpful as an interview adjunct for defining the ABCs of the problem.

Behavior Logs

The ABCs of a problem situation and the intensity of the problem can be clarified with client self-monitoring of the problem behaviors, the contributing conditions, and the frequency or duration of the problem behavior. All this information can be recorded in a behavior log, which is simply an extension of the descriptive log. Figure 9-3 is an example of a behavior log for our client Joan.

The client is also asked to observe and record how long (duration) or how often (frequency) the problem behaviors occur. Determining the level of the present problem serves as a baseline—that is, the rate or level of the problem *before* any counseling interventions have been started. The baseline is useful initially in helping establish the direction and level of change desired by the client. This information, as you will see in Chapter 10, is essential in establishing client goals. And as counseling progresses, these baseline data may help the client compare progress during and near the end of counseling with progress at the beginning of counseling.

In a behavior log, the defined problem behaviors are listed at the left. The client records the date, time, and place when these behaviors occur. To record contributing conditions, the client is asked to write down the behaviors and events that occur before and after the problem behaviors. This information helps to establish a pattern among the problem behaviors, things that cue or elicit the problem behaviors, and activities that maintain, strengthen, or weaken those behaviors.

DAILY RECORD SHEET					
Date	Time	Place	Activity	People	Observed behavior

FIGURE 9-2. A descriptive log

For Joan
Week of Nov. 6-13

(Problem behaviors) Behavior observing	Date	Time	Place	(Frequency/ duration) Number or amount	(Antecedents) What precedes behavior	(Consequences) What follows behavior
1. Thinking of self as not as smart as other students	Mon., Nov. 6	10:00 A.M.	Math class	IIII	Going into class, know have to take test in class	Leaving class, being with friends
	Tues., Nov. 7	10:15 A.M.	Math class	IIII IIII	Got test back with a B	Teacher consoled me
	Tues. Nov. 7	5:30 P.M.	Home	IIII II	Parents asked about test. Told me to stay home this weekend	Went to bed
	Thurs., Nov. 9	9:30 A.M.	English class	II	Thought about having to go to math class	Got to math class. Had substitute teacher
	Sun., Nov. 12	8:30 P.M.	Home	III	Thought about school tomorrow	Went to bed
2. a. Not volunteering answers	Tues., Nov. 7	10:05 A.M. 10:20	Math class	II	Felt dumb	Nothing
b. Not answering teacher questions	Thurs., Nov. 9	10:10 A.M. 10:20 10:40	Math class	III	Felt dumb	Nothing
c. Not going to board	Thurs., Nov. 9	10:30 A.M.	Math class	I	Teacher called on me	Nothing
	Fri., Nov. 10	10:10 A.M. 10:35 A.M.	Math class	II	Teacher called on me	Nothing
	Thurs., Nov. 9	10:45 A.M.	Math class	I	Didn't have a substitute teacher	Nothing
	Fri., Nov. 10	10:15 A.M.	Math class	I	Teacher asked girls to go up to board	Teacher talked to me after class
3. Cutting class	Wed., Nov. 8	9:55 A.M.	School	1 hour	Didn't want to hassle with class or think about test	Cut class. Played sick. Went to nurse's office for an hour

FIGURE 9-3. Example of behavior log

Uses of Logs

The success of written logs may depend on the client's motivation to keep a log as well as on the instructions and training given to the client about the log. Five guidelines may increase the client's motivation to engage in self-monitoring:

1. Establish a rationale for the log, such as "We need a written record in order to find out what is going on. This will help us make some decisions about the best way to handle your problem." A client is more likely to keep a log if he or she is aware of a purpose for doing so.

2. Provide specific, detailed instructions regarding how to keep the log. The client should be told *what, how, when,* and for *how long* to record. The client should be given an example of a model log to see how it may look. Providing adequate instructions may increase the likelihood that the client will record data consistently and accurately.

3. Adapt the type of log to the client's ability to do self-monitoring. At first, you may need to start with a very simple log that does not require a great deal of recording. Gradually, you can increase the amount of information the client observes and records. If a client has trouble keeping a written log, a substitute can be used, such as a tape recorder, golf wrist counter, or, for children, gold stars or pictures.

4. Adapt the log and the instructions to the client's problem and degree and type of problem. Clients exhibiting certain types of problems may experience predictable types of difficulties in implementing self-monitoring, particularly of covert, or cognitive, events. Hollon and Kendall (1981, pp. 350–351) summarize some of these reactions:

   Depressed clients either frequently report being overwhelmed by what seems to be an unmanageable task and/or fail to initiate or maintain monitoring because they do not anticipate that it can provide any help. Anxious clients (e.g., Beck & Emery, 1979) frequently avoid attending to cognitions because doing so seems to increase distress. Anorexic clients rarely report effects, per se. Rather, they list strings of inferential descriptions when asked to record how they feel. If asked to evaluate the validity of their beliefs, they are likely to respond with moralistic prescriptions, reminiscent of "New Year's resolutions" (Bemis, 1980). Obsessive clients, as might be expected, rarely get beyond listing their thoughts—frequently working long and hard to get it just "right."

5. Involve the client in discussing and analyzing the log within the interview. At first, the counselor can begin by putting together "hunches" about patterns of problem behavior and contributing conditions. As counseling progresses, the client can take a more active role in analyzing the log. Increasing the client's involvement in analyzing the log should serve as an incentive to the client to continue the time and effort required to collect the data.

The counselor should remember that the process of client self-monitoring can be reactive. In other words, the very act of observing oneself can influence that which is being observed. This reactivity may affect the data reflected in the log. Reactivity can be helpful in the overall counseling program when it changes the behavior in the desired direction. There are times when self-monitoring is used deliberately as a change strategy to increase or decrease a particular behavior (see Chapter 20).

Occasionally, when self-monitoring is used as an assessment tool, reactivity may cause an aspect of the problem behavior to get worse. This seems especially true when the client is monitoring negative affect, such as anxiety, anger, or depression (Hollon & Kendall, 1981). In such instances, sometimes the self-monitoring becomes a signal for the mood state, causing it to increase in frequency or intensity. If this happens, the use of self-monitoring as an *assessment* device should be discontinued, although using it as a change strategy and changing what and how the client self-records may create reactivity in the desired direction (see also Chapter 20).

The data obtained from client self-monitoring are used not only during the assessment process but also in establishing client goals. During assessment sessions, the self-monitoring data will help the client and counselor to determine the ABCs of the problem. These pretreatment data will be the starting place for the discussion of desired counseling outcomes (see Chapter 10).

WHEN IS "ENOUGH" ASSESSMENT ENOUGH?

Occasionally people will wonder whether assessment goes on forever—or when it stops! As we mentioned earlier, assessment is something of a continuous process during therapy, in that clients are people and can't—or shouldn't—be pigeonholed, and occasionally problems are shifted and redefined.

Shorter assessment periods are often necessary for clients in crisis or for practitioners in high-demand, high-caseload work settings. Regardless of the type of clients with whom you work or your work setting, we encourage you to view the assessment phase of counseling with healthy respect. Skipping or glossing over assessment can actually result in longer treatment and end up costing more of your time and of the client's money. A thorough assessment usually results in a more active and shorter course of treatment by eliminating a lot of guesswork, and may prevent client dropout from therapy.

Occasionally, even after a thorough assessment period, you may still feel that there are chunks of missing information or pieces to the puzzle. It is difficult to know whether additional assessment strategies are warranted in terms of time and cost and, most important, whether they will contribute to the effectiveness of treatment for this client. At this point, the counselor must make a deliberate therapeutic decision—to conduct additional assessments or to begin intervention strategies and work on what is known about the client. Fensterheim (1983) suggests three key questions to consider before making this decision:

1. Is enough information present to make a start on treatment?
2. How urgent is the need for immediate relief?
3. How pessimistic and despondent is the patient?

We would like to conclude this chapter with one caution: above all, remember that assessment is not an end in itself! The time, structure, and tools you use for assessment are of little value unless assessment has sufficient "treatment validity" (Nelson, 1983) or contributes to greater effectiveness of therapy treatment and outcome for clients.

SUMMARY

This chapter focused on the use of direct interviewing to assess client concerns. In many settings, initial assessment interviews often begin with an intake interview to gather information about the client's presenting problems and primary symptoms as well as information about such areas as previous counseling, social/developmental history, educational/vocational history, health history, and family, relationship, and sexual history. This interview often yields information that the counselor can use to develop hypotheses about the nature of the client's problems. History interviews also serve as a retrospective baseline of how the client was functioning before and what events contributed to the present difficulties and coping styles. For occasional clients, intakes or history interviews may be followed by a mental-status exam, which aids the therapist in assessing the client's psychiatric status.

The model presented in this chapter for direct assessment interviewing is based on the ABC model described in Chapter 8. Specifically, counselors are interested in defining six components of problem behavior—affective, somatic, behavioral, cognitive, contextual, and relational. They also seek to identify antecedent events that occur before the problem and cue it and consequent events that follow the problem and in some way influence it or maintain it. Consequences may include "payoffs," or secondary gains, which give value to the dysfunctional behavior and thus keep the problem going. Antecedent and consequent sources may also be affective, somatic, behavioral, cognitive, contextual, and relational. Gender and multicultural factors must also be considered in the assessment interview. Other important components of direct assessment interviewing include identifying previous solutions the client has tried for resolving the problem, exploring client coping skills and assets, exploring the client's perceptions of the issue, and identifying the frequency, duration, or severity of the problem.

In addition to direct assessment interviewing, other assessment tools include role playing, imagery, self-report measures, and self-monitoring. All these techniques can be useful for obtaining more specific information about the identified problems.

POSTEVALUATION

Part One
A client is referred to you with a presenting problem of "free-floating," or generalized (pervasive), anxiety. Outline the questions you would ask during an assessment interview with this client that pertain directly to her presenting component. Your objective (Objective One) is to identify at least 2 questions for each of the 11 problem assessment categories described in this chapter and summarized in Table 9-3. Feedback follows the postevaluation.

Part Two
Using the description of the above client, conduct a 30-minute role-play assessment interview in which your objective is to demonstrate leads and responses associated with at least 9 out of the 11 categories described for problem assessment (Chapter Objective Two). You can do this activity in triads in which one person assumes the role of counselor, another the "anxious" client, and the third
(continued)

POSTEVALUATION   (continued)

person the role of observer; trade roles two times. If groups are not available, audiotape or videotape your interview. Use the Interview Checklist at the end of the chapter as a guide to assess your performance and to obtain feedback.

After completing your interview, develop some hypotheses, or hunches, about the client. In particular, try to develop "guesses" about

1. Antecedent sources that cue or set off the anxiety, making its occurrence more likely
2. Consequences that maintain the anxiety, keep it going, or make it worse
3. Consequences that diminish or weaken the anxiety
4. Secondary gains, or payoffs, attached to the anxiety
5. Ways in which the client's "previous solutions" may contribute to the anxiety or make it worse
6. Particular strengths, resources, and coping skills of the client and how these might be best used during treatment/intervention
7. How the client's gender and culture impact the problem

You may wish to continue this part of the activity in triads or to do it alone, jotting down ideas as you proceed. At some point, it may be helpful to share your ideas with your group or your instructor.

Part Three
Devise a self-monitoring assessment procedure you could give to this client for homework, to obtain information about the time, place, frequency, duration, and severity of her anxious feelings. Write an example of a log you could give to her to obtain this information (Chapter Objective Three).

Part Four
Conduct a role-play interview with this client in which you assign the self-monitoring plan you devised as homework (Chapter Objective Four).

1. Provide a *rationale* to the client about the usefulness of the assignment.
2. Provide the client with detailed *instructions* about *what, how, when,* and for *how long* to monitor.
3. Follow-up—clarify and check out the client's understanding of your assignment.

If possible, continue with this activity in your triads and obtain feedback from the observer on the specificity and clarity of your instructions. If an observer is not available, tape your interview for self-assessment of its playback.

INTERVIEW CHECKLIST FOR ASSESSING CLIENT PROBLEMS

Scoring		Category of information	Examples of counselor leads or responses	Client response
Yes	No			
____	____	1. Explain purpose of assessment interview	"I am going to be asking you more questions than usual so that we can get an idea of what is going on. Getting an accurate picture about your concern (or problem) will help us to decide what we can do about it. Your input is important."	____ (check if client confirmed understanding of purpose)
____	____	2. Identify range of concerns and/or problems (if you don't have this information from history)	"What would you like to talk about today?" "How would you describe the things that are really bothering you now?" "What are some things that bug you?" "What specifically led you to come to see someone now?" "What things are not going well for you?" "Are there any other issues you haven't mentioned?"	____ (check if client described additional concerns)

(continued)

Scoring		Category of information	Examples of counselor leads or responses	Client response
Yes	No			
____	____	3. Prioritize and select primary or most immediate problem to work on	"What issue best represents the reason you are here?" "Of all these concerns, which one is most stressful (or painful) for you?" "Rank-order these concerns, starting with the one that is most important for you to resolve to the one least important." "Tell me which of these problems you believe you could learn to deal with most easily and with the most success." "Which one of the things we discussed do you see as having the best chance of being solved?" "Out of all the problems we've discussed, describe the one that, when resolved, would have the greatest impact on the rest of the issues."	____ (check if client selected problem to focus on)
____	____	4.0. Present problem behavior		____ (check if client identified the following components of problem)
		4.1. *Affective* aspects of problem: feelings, emotions, mood states	"What are you feeling when this happens?" "How does this make you feel when this occurs?" "What other feelings do you have when this occurs?" "What feelings is this problem hiding or covering up?" "How secure do you feel with this person?" "What feelings do you have about your relationship?"	____ (check if client identified feelings)
____	____	4.2. *Somatic* aspects of problem: body sensations, physiological responses, organic dysfunction and illness, medications	"What goes on inside you then?" "What do you notice in your body when this happens?" "What are you aware of when this happens?" "When this happens, are you aware of anything that goes on in your body that feels bad or uncomfortable—aches, pains, and so on?"	____ (check if client identified body sensations)
____	____	4.3. *Behavioral* aspects of problem: overt behaviors/actions (excesses and deficits)	"In photographing this scene, what actions and dialogue would the camera pick up?" "What are you doing when this occurs?" "What do you mean by 'not communicating'?" "Describe what you did the last few times this occurred."	____ (check if client identified overt behavior)

(continued)

POSTEVALUATION (continued)

Scoring		Category of information	Examples of counselor leads or responses	Client response
Yes	No			
____	____	4.4. *Cognitive* aspects of problem: automatic, helpful, unhelpful, rational, irrational thoughts and beliefs; internal dialogue; perceptions and misperceptions	"What do you say to yourself when this happens?" "What are you usually thinking about during this problem?" "What was going through your mind then?" "What kinds of thoughts can make you feel ____?" "What beliefs [or images] do you hold that affect this issue?" Sentence completions: I should ____, people should ____, it would be awful if ____, ____makes me feel bad.	____ (check if client identified thoughts, beliefs)
____	____	4.5. *Contextual* aspects of problem: time, place, or setting events	"Describe some recent situations in which the problem occurred. Where were you? When was it?" "When does this usually occur?" "Where does this usually occur?" "Does this go on all the time or only sometimes?" "Does the same thing happen at other times or places?" "At what time does this *not* occur? places? situations?" "What effect does your cultural/ethnic background have on this issue?" "What effects do the socio-political structures of the society in which you live have on this problem?"	____ (check if client identified time, places, other events)
____	____	4.6. *Relational* aspects of problem: other people	"What effects does this problem have on significant others in your life?" "What effects do significant others have on this problem?" "Who else is involved in the problem? How?" "How does this person influence your life?" "From whom do you think you learned to act or react this way?" "How many significant close relationships do you have in your life now?" "Whom do you know and respect who handles the issue the way you would like to?" "What persons *present* in your life now have the greatest positive impact on this problem? Negative impact?" "What about persons *absent* from your life?"	____ (check if client identified people)

(continued)

Scoring		Category of information	Examples of counselor leads or responses	Client response
Yes	No			
____	____	5.0. Antecedents— past or current conditions that cue, or set off, the problem		____ (check if client identified following antecedent sources)
____	____	5.1. *Affective* antecedents	"What are you usually feeling before this?" "When do you recall the first time you felt this way?" "What are the feelings that occur before the problem and make it more likely to happen? Less likely?" "Are there any holdover or unfinished feelings from past events in your life that still affect this problem? How?"	____ (feelings, mood states)
____	____	5.2. *Somatic* antecedents	"What goes on inside you just before this happens?" "Are you aware of any particular sensations or discomfort just before the problem occurs or gets worse?" "Are there any body sensations that seem to occur before the problem or when it starts that make it more likely to occur? Less likely?" "Is there anything going on with you physically—like illness or a physical condition or in the way you eat or drink— that leads up to this problem?"	____ (body sensations, physiological responses)
____	____	5.3. *Behavioral* antecedents	"If I were photographing this, what actions and dialogue would I pick up before this happens?" "Can you identify any particular behavior patterns that occur right before this happens?" "What do you typically do before this happens?"	____ (overt behavior)
____	____	5.4. *Cognitive* antecedents	"What kinds of pictures do you have before this happens?" "What are your thoughts before this happens?" "What are you telling yourself before this happens?" "Can you identify any particular beliefs that seem to set the problem off?" "What do you think about [or tell yourself] before the problem occurs that makes it more likely to happen? Less likely?"	____ (thoughts, beliefs, internal dialogue, cognitive schemes)

(continued)

POSTEVALUATION   **(continued)**

Scoring		Category of information	Examples of counselor leads or responses	Client response
Yes	No			
____	____	5.5. *Contextual* antecedents	"How long ago did this happen?" "Has this ever occurred at any other time in your life? If so, describe that." "Where and when did this occur the first time?" "How do you see those events as related to your problem?" "What things happened that seemed to lead up to this?" "What was happening in your life when you first noticed the problem?" "Are there any ways in which your cultural values and affiliations set off this problem? Make it more likely to occur? Less likely?" "How were things different before you had this concern?" "What do you mean, this started 'recently'?"	____ (time, places, other events)
____	____	5.6. *Relational* antecedents	"Are there any people or relationships from past events in your life that still affect this problem? How?" "Can you identify any particular people that seem to bring on this problem?" "Are you usually with certain people right before or when this problem starts?" "Are there any people or relationships from the past that trigger this issue in some way? Who? How?" "How do the people who hold power in your life trigger this issue?"	____ (other people)
____	____	6.0. Identify consequences— conditions that maintain and strengthen problem or weaken or diminish it		____ (check if client identified following sources of consequences)
____	____	6.1. *Affective* consequences	"How do you feel after this happens?" "How does this affect the problem?" "When did you stop feeling this way?" "Are you aware of any particular feelings or reactions you have after the problem that strengthen it? Weaken it?"	____ (feelings, mood states)

(continued)

Scoring		Category of information	Examples of counselor leads or responses	Client response
Yes	No			
____	____	6.2. *Somatic* consequences	"What are you aware of inside you—sensations in your body—just after this happens?" "How does this affect the problem?" "Are there any sensations inside you that seem to occur after the problem that strengthen or weaken it?" "Is there any physical condition, illness, and so on about yourself that seems to occur after this problem? If so, how does it affect the problem?"	____ (body or internal sensations)
____	____	6.3. *Behavioral* consequences	"What do you do after this happens, and how does this make the problem better? Worse?" "How do you usually react after this is over?" "In what ways does your reaction keep the problem going? Weaken it or stop it?" "Can you identify any particular behavior patterns that occur after this?" "How do these patterns keep the problem going? Stop it?"	____ (overt responses)
____	____	6.4. *Cognitive* consequences	"What do you usually think about afterward?" "How does this affect the problem?" "What do you picture after this happens?" "What do you tell yourself after this occurs?" "Can you identify any particular thoughts [beliefs, self-talk] that make the problem better? Worse?" "Are there certain thoughts or images you have afterward that either strengthen or weaken the problem?"	____ (thoughts, beliefs, internal dialogue)
____	____	6.5. *Contextual* consequences	"When does this problem usually stop or go away? Get worse? Get better?" "Where are you when the problem stops? Gets worse? Gets better?" "Can you identify any particular times, places, or events that seem to keep the problem going? Make it worse or better?" "Are there any ways in which your cultural affiliation and values seem to keep this problem going? Stop it or weaken it?"	____ (time, places, other events)

(continued)

POSTEVALUATION (continued)

Scoring		Category of information	Examples of counselor leads or responses	Client response
Yes	No			

Yes	No			
____	____	6.6. *Relational consequences*	"Can you identify any particular reactions from other people that occur following the problem?" "In what ways do their reactions affect the problem?" "Are you usually with certain people when the problem gets worse? Better?" "What happens to you after you have interacted with this person?" "Can you identify any particular people who can make the problem worse? Better? Stop it? Keep it going?" "How do the people who have power in your life situation perpetuate this problem?"	____ (other people)
____	____	7. Identify possible secondary gains from problem	"What happened afterward that was pleasant?" "What was unpleasant about what happened?" "Has your concern or problem ever produced any special advantages or considerations for you?" "As a consequence of your concern, have you got out of or avoided things or events?" "How does this problem help you?" "What do you get out of this situation that you don't get out of other situations?" "Do you notice anything that happens afterward that you try to prolong or to produce?" "Do you notice anything that occurs after the problem that you try to stop or avoid?" "Are there certain feelings or thoughts that go on after the problem that you try to prolong?" "Are there certain feelings or thoughts that go on after the problem that you try to stop or avoid?" "The good thing about ____ [problem] is . . . "	____ (check if client identified payoffs)
____	____	8. Identify solutions already tried to solve the problem	"How have you dealt with this or other problems before? What was the effect? What made it work or not work?" "How have you tried to resolve this problem?"	____ (check if client identified prior solutions)

(continued)

Scoring		Category of information	Examples of counselor leads or responses	Client response
Yes	No			
			"What kinds of things have you done to improve this situation?" "What have you done that has made the problem better? Worse? Kept it the same?" "What have others done to help you with this?"	
____	____	9. Identify client coping skills, strengths, resources	"What skills or things do you have going for you that might help you with this concern?" "Describe a situation when this concern or problem is not interfering." "What strengths or assets can you use to help resolve this problem?" "When don't you act this way?" "What kinds of thoughts or self-talk help you handle this better?" "When don't you think in self-defeating ways?" "What do you say to yourself to cope with a difficult situation?" "Identify the steps you take in a situation you handle well—what do you think about and what do you do? How could these steps be applied to the present problem?" "To what extent can you do something for yourself without relying on someone else to push you or prod you to do it?" "How often do you get things done by rewarding yourself in some way? By punishing yourself?"	____ (check if client identified assets, coping skills)
____	____	10. Identify client's description/ assessment of the problem (note which aspects of problem are stressed and which are ignored)	"What is your understanding of this issue?" "How do you explain this problem to yourself?" "What does the problem mean to you?" "What is your interpretation [analysis] of the problem?" "What else is important to you about the problem that we haven't mentioned?" "Sum up the problem in just one word." "Give the problem a title."	____ (check if client explained problem)

(continued)

POSTEVALUATION (continued)

Scoring		Category of information	Examples of counselor leads or responses	Client response
Yes	No			
____	____	11. Estimate frequency, duration, or severity of problem behavior/ symptoms (assign monitoring homework, if useful)	"How often [how much] does this occur during a day—a week?" "How long does this feeling stay with you?" "How many times do you ____ a day? a week?" "To what extent has this problem interfered with your life? How?" "You say sometimes you feel very anxious. On a scale from 1 to 10, with 1 being very calm and 10 being very anxious, where would you put your feelings?" "How has this interfered with other areas of your life?" "What would happen if the problem were not resolved in a year?"	____ (check if client estimated amount or severity of problem)

Yes	No	**Other skills**
____	____	12. The counselor listened attentively and recalled accurately the information given by the client.
____	____	13. The counselor used basic listening responses to clarify and synthesize the information shared by the client.
____	____	14. The counselor followed the client's lead in determining the sequence or order of the information obtained.

Observer comments _____

_____

_____

_____

SUGGESTED READINGS

Craig, R. J. (Ed.). (1989). *Clinical and diagnostic interviewing.* Northvale, NJ: Aronson.

Dana, R. H. (1993). *Multicultural assessment perspectives for professional psychology.* Needham Heights, MA: Allyn & Bacon.

Evans, D. R., Hearn, M. T., Uhlemann, M. R., & Ivey, A. E. (1993). *Essential interviewing: A programmed approach to effective communication* (4th ed.). Pacific Grove, CA: Brooks/Cole.

Gambrill, E. D. (1990). *Critical thinking in clinical practice.* San Francisco, CA: Jossey-Bass.

Lukas, S. (1993). *Where to start and what to ask: An assessment handbook.* New York: Norton.

Mitchell, R. W. (1991). *Documentation in counseling records.* Washington, DC: American Counseling Association.

Morganstern, K. P. (1988). Behavioral interviewing. In A. S. Bellack & M. Hersen (Eds.), *Behavioral assessment: A practical handbook* (3rd ed., pp. 86–118). New York: Pergamon Press.

Morrison, J. (1995). *The first interview: A guide for clinicians.* New York: Guilford.

Othmer, E., & Othmer, S. (1994). *The clinical interview using DSM-IV.* Vol. 1: *Fundamentals.* Washington, DC: American Psychiatric Press.

Pedersen, P. B., & Ivey, A. (1993). *Culture-centered counseling and interviewing skills.* New York: Praeger.

Steenbarger, B. (1993). A multicontextual model of counseling: Bridging brevity and diversity. *Journal of Counseling and Development, 72,* 8–15.

Wantz, D. W., & Morran, D. K. (1994). Teaching counselor trainees a divergent versus a convergent hypothesis-formation strategy. *Journal of Counseling and Development, 73,* 69–73.

F E E D B A C K
POSTEVALUATION

Part One

See whether the questions you generated are similar to the following ones:

Is this the only issue you're concerned about now in your life, or are there other issues you haven't mentioned yet? (Range of problems)

When you say you feel anxious, what exactly do you mean? (Problem behavior—affective component)

When you feel anxious, what do you experience inside your body? (problem behavior—somatic component)

When you feel anxious, what exactly are you usually doing? (problem behavior—behavioral component)

When you feel anxious, what are you typically thinking about [or saying to yourself]? (Problem behavior—cognitive component)

Try to pinpoint exactly what times the anxiety occurs or when it is worse. (Problem behavior—contextual component)

Describe where you are or in what situations you find yourself when you get anxious. (Problem behavior—contextual component)

Describe what other things are usually going on when you have these feelings. (Problem behavior—contextual component)

Can you tell me what persons are usually around when you feel this way? (Problem behavior—relational component)

Are there any feelings that lead up to this? (Antecedent source—affective)

What about body sensations that might occur right before these feelings? (Antecedent source—somatic)

Have you noticed any particular behavioral reactions or patterns that seem to occur right before these feelings? (Antecedent source—behavioral)

(continued)

Are there any kinds of thoughts—things you're dwelling on—that seem to lead up to these feelings? (Antecedent source—cognitive)

When was the first time you noticed these feelings? Where were you? (Antecedent source—contextual)

Can you recall any other events or times that seem to be related to these feelings? (Antecedent source—contextual)

Does the presence of any particular people in any way set these feelings off? (Antecedent source—relational)

Are you aware of any particular other feelings that make the anxiety better or worse? (Consequence source—affective)

Are you aware of any body sensations or physiological responses that make these feelings better or worse? (Consequence source—somatic)

Is there anything you can do specifically to make these feelings stronger or weaker? (Consequence source—behavioral)

Can you identify anything you can think about or focus on that seems to make these feelings better or worse? (Consequence source—cognitive)

At what times do these feelings diminish or go away? get worse? in what places? in what situations? (Consequence source—contextual)

Do certain people you know seem to react in ways that keep these feelings going or make them less intense? If so, how? (Consequence source—relational)

As a result of this anxiety, have you ever gotten out of or avoided things you dislike? (Consequence—secondary gain)

Has this problem with your nerves ever resulted in any special advantages or considerations for you? (Consequence—secondary gain)

What have you tried to do to resolve this issue? How have your attempted solutions worked out? (Previous solutions)

Describe some times and situations when you don't have these feelings or you feel calm and relaxed. What goes on that is different in these instances? (Coping skills)

How have you typically coped with other difficult situations or feelings in your life before? (Coping skills)

If you could give this problem a title—as if it were a movie or a book—what would that title be? (Client perceptions of problem)

How do you explain these feelings to yourself? (Client perceptions of problem)

How many times do these feelings crop up during a given day? (Frequency of problem)

How long do these feelings stay with you? (Duration of problem)

(continued)

FEEDBACK: POSTEVALUATION   (continued)

On a scale from 1 to 10, with 1 being not intense and 10
   being very intense, how strong would you say these
   feelings usually are? (Severity of problem)

Part Three

To start with, you may want to obtain information directly
related to time and place of anxiety occurrence. The log
might look something like this:

Date   Time   Place   Activity

The client would be asked to record anxious feelings
daily. Later, you may wish to add recording of information
about duration and severity. You could add these two
columns to the log:

How long   Intensity (1 to 10)

Although it might be valuable to ask the client also to
observe and record cognitions (thoughts, beliefs), there is
some evidence that many anxious clients fail to complete
this part of self-monitoring because attending to cognitions
is reactive and may cue, or increase, rather than decrease,
the anxiety (Hollon & Kendall, 1981).

CHAPTER 10

IDENTIFYING, DEFINING, AND EVALUATING OUTCOME GOALS

Pause for a few minutes to answer the following questions to yourself or with someone else.

1. What is one thing you would like to change about yourself?
2. Suppose you succeeded in accomplishing this change. How would things be different for you?
3. Does this outcome represent a change in yourself or for someone else?
4. What are some of the risks—to you or others—of this change?
5. What would be your payoffs for making this change?
6. What would you be doing, thinking, or feeling as a result of this change you would like to make for yourself?
7. In what situations do you want to be able to do this?
8. How much or how often would you like to be able to do this?
9. Looking at where you are now and where you'd like to be, are there some steps along the way to get from here to there? If so, rank them in an ordered list from "easiest to do now" to "hardest to do."
10. Identify any obstacles (people, feelings, ideas, situations) that might interfere with attainment of your goal.
11. Identify any resources (skills, people, knowledge) that you need to use or acquire to attain your goal.
12. How could you evaluate progress toward this outcome?

These steps reflect the process of identifying, defining, and evaluating goals for counseling. Goals represent desired results or outcomes and function as milestones of client progress. In this chapter we describe and model concrete guidelines you can use to help clients identify, define, and evaluate outcome goals for counseling.

OBJECTIVES

1. Identify a situation about you or your life that you would like to change. Identify and define one desired outcome for this issue, using the Goal-Setting Worksheet in the postevaluation as a guide.
2. Given a written client case description, describe the steps you would use with this client to explore and define desired outcome goals, with at least 11 of the 14 categories for selecting and defining goals represented in your description.
3. Demonstrate at least 11 of the 14 categories associated with identifying and defining outcome goals, given a role-play interview.
4. With yourself or another person or client, conduct an outcome evaluation of a real or a hypothetical counseling goal, specifying *when, what,* and *how* you will measure the outcome.

TREATMENT GOALS AND THEIR PURPOSES IN COUNSELING

Treatment goals represent results or outcomes described by clients and are a direct outgrowth of the problems identified during the assessment process. Goals have six important purposes in counseling. First, they provide some directions for counseling. Clearly defined goals reflect the areas of client concern that need most immediate attention. Establishing goals can also clarify the client's initial expectations of counseling. Goals may help both counselor and client anticipate more precisely what can and cannot be accomplished through counseling.

Although each theoretical orientation has its own direction for counseling, specifying goals individually for each client helps to ensure that counseling is structured specifically to meet the needs of *that* client. Clients are much more likely to support and commit themselves to changes that they create than changes imposed by someone else. Without goals, counseling may be directionless or may be based more on the theoretical biases and personal preferences of the counselor (Bandura, 1969, p. 70). Some clients may go through counseling without realizing that the sessions are devoid of direction or are more consistent with the counselor's preferences than the client's needs and aims. In other aspects of our lives, however, most of us would be quite aware of analogous situations. If we boarded an airplane destined for a place of our choice, and the airplane went around in circles or the pilots announced a change of destination that they desired, we would be upset and indignant.

Second, goals permit counselors to determine whether they have the skills, competencies, and interest for working with a particular client toward a particular outcome. Depending on the client's choice of goals and the counselor's values and level of expertise, the counselor decides whether to continue working with the client or to refer the client to someone else who may be in a better position to give services.

The third purpose of goals is their role in human cognition and problem solving. Goals facilitate successful performance and problem resolution because they are usually rehearsed in our working memory and because they direct our attention to the resources and components in our environment that are most likely to facilitate the solution of a problem (Dixon & Glover, 1984, pp. 128–129). This purpose of goals is quite evident in the performance of successful athletes who set goals for themselves and then use the goals not only as motivating devices but also as standards against which they rehearse their performance over and over, often cognitively or with imagery. Running backs, for example, constantly "see themselves" getting the ball and running downfield, over and past the goal line. Champion skiers are often seen closing their eyes and bobbing their heads in the direction of the course before the race. In the case of counseling goals, it is important for clients to be able to visualize and rehearse the target behaviors or end results reflected in their goals.

A fourth purpose of goals is to give the counselor some basis for selecting and using particular counseling strategies and interventions. The changes the client desires will, to some degree, determine the kinds of action plans and treatment strategies that can be used with some likelihood of success. Without an explicit identification of what the client wants from counseling, it is almost impossible to explain and defend one's choice to move in a certain direction or to use one or more counseling strategies. Without goals, the counselor may use a particular approach without any rational basis (Bandura, 1969, p. 70). Whether the approach will be helpful is left to chance rather than choice.

TABLE 10-1. Process and goals in counseling

Process	Goals
1. Appropriate	Appropriate
2. Appropriate	Inappropriate
3. Inappropriate	Appropriate
4. Inappropriate	Inappropriate

A fifth and most important purpose of goals is their role in an outcome evaluation of counseling. Goals can indicate the difference between what and how much the client is able to do now and what and how much the client would like to do in the future. With the ultimate goal in mind, the counselor and client can monitor progress toward the goal and compare progress before and after a counseling intervention. These data provide continuous feedback to both counselor and client. The feedback can be used to assess the feasibility of the goal and the effectiveness of the intervention.

Finally, goal-planning systems are useful because, like assessment procedures, they are often reactive; that is, clients make progress in change as a result of the goal-planning process itself. Goals induce an expectation for improvement, and purposeful goal setting contributes to a client's sense of hopefulness and well-being (Csikszentmihalyi, 1990). As Snyder (1995) notes, "Higher as compared to lower hope people have a greater number of goals, have more difficult goals, have success at achieving their goals, and perceive their goals as challenges" (p. 357).

CULTURAL ISSUES IN COUNSELING GOALS

Sue and Sue (1990) point out that "closely linked to the actual process of counseling are certain implicit or explicit goals . . . and that different cultural and subcultural groups may require different counseling processes and goals" (p. 162). They note that any given client may be exposed to one of four conditions in counseling, as shown in Table 10-1.

We have commented in earlier chapters about the cultural relevance (or lack of relevance) of counseling skills and processes for clients from varied cultural groups. In this chapter, we stress the importance of heeding Sue and Sue's (1990) concern about who determines the counseling goals and the relevance of these goals for the client. These authors provide the following example to distinguish between appropriate and inappropriate counseling goals.

A Black male student from the ghetto who is failing in school and getting into fights with other students can be treated by the counselor in a variety of ways. Sometimes such a student lacks the academic skills necessary to get good grades. The constant fighting is a result of peers' teasing him about his "stupidity." A counselor who is willing to teach the student study and test-taking skills as well as give advice and information may be using an appropriate process consistent with the expectations of the student. The appropriate goals defined between counselor and client, besides acquisition of specific skills, may be an elevation of grades.

Often times, a counseling strategy may be chosen by the counselor that is compatible with the client's life experiences, but the goals are questionable. Again, let us take the aforementioned example of the Black ghetto student. Here the counselor may define the goal as the elimination of "fighting behavior." The chosen technique may be behavior modification. Since the approach stresses observable behaviors and provides a systematic, precise, and structured approach to the "problem," much of the nebulousness and mystique of counseling is reduced for the Black student. Rather than introspection and self-analysis, which some people of color may find unappealing, the concrete tangible approach of behavioral counseling is extremely attractive.

While the approach may be a positive experience for many minorities, there is danger here regarding control and behavioral objectives. If the Black student is being tested and forced to fight because he is a minority-group member, then the goal of "stopping fighting behavior" may be inappropriate. The counselor in this situation may inadvertently be imposing his or her own standards and values on the client. The end goals place the problem in the hands of the individual rather than society, which produced the problems. (Sue & Sue, 1990, pp. 162–163)

Brown (1994) has offered a similar caveat with respect to gender aware therapy for clients. She suggests that the particular way the therapist conceptualizes the case will affect the client's choice of counseling goals. A feminist and a systemic interpretation—which places emphasis on the contribution of external sociopolitical factors—allows clients to have different choices about counseling outcomes. "Rather than compliantly taking on tasks assigned by the dominant culture—for example, 'Stop being depressed, become more productive'—a person furnished with this sort of knowledge may develop alternative goals for therapy such as 'Learn to get angry more often and see my connections to other people more clearly' " (p. 170).

Rosado and Elias (1993) point out that mainstream therapists often expect clients to develop long-range counseling goals (p. 454). They note that these sorts of goals are inconsistent with the mind-set of some clients from some cultural groups such as urban, low-income, and others. They recommend developing goals to meet "the pressing ecological need of these clients, even if they are of a short duration" (p. 454). As the authors note, financially indigent clients are concerned with survival issues and resolution of current problems. Similarly, Berg and Jaya (1993) note that for many Asian American clients, goals need to be developed that are short term, realistic, pragmatic, concrete, and oriented toward solutions of client problems.

When using a multicultural and gender-aware viewpoint to develop goals, the important point for counselors is to be aware of our own values and biases and to avoid deliberately or inadvertently steering the client toward goals that may reflect the norms and scripts of the mainstream culture rather than the client's expressed wishes (Burnett, Anderson, & Heppner, 1995; Chojnacki & Gelberg, 1995). A clear-cut example of this is in respecting a client's preferred sexual orientation when it is different from your own.

WHAT DO OUTCOME GOALS REPRESENT?

At the simplest level, an outcome goal represents what the client wants to happen as a result of the counseling process. Stated another way, outcome goals are an extension of the types of problems the client experiences. Outcome goals represent two major classifications of problems: *choice* and *change* (Dixon & Glover, 1984). In choice problems, the client has the requisite skills and resources for problem resolution but is caught between two or more choices, often conflicting ones. In these instances, the outcome goal represents a choice or a decision the client needs to make, such as "to decide between College A and College B" or "to choose either giving up my job and having lots of free time or keeping my job and having money and stability."

In change issues, the outcome goal is a change the client wants to make. The desired changes may be in overt behaviors or situations, covert behaviors, or combinations of the two. These outcome goals may be directed at eliminating something, increasing something, developing something, or restructuring something, but in all cases the change is expected to be an improvement over what currently exists. Furthermore, Lyddon (1995) distinguishes between *first-* and *second*-order changes that clients make. First-order changes consist primarily of relatively simple linear movements whereas second-order changes are complex, nonlinear, and radical. First-order changes represent more or less minor adaptations a client makes; second-order change is core or deep—that is, "a radical restructuring of a person's core self, mode of being, or worldview" (Hanna & Ritchie, 1995, p. 176). As an example of these two kinds of changes, consider a client who presents with alcohol and sexual addictions. The client is a 30-year-old male who drinks a six-pack of beer daily and has had sex with over 200 women in the last 20 years. His first sexual experience occurred at age 10 when he was "seduced" by a 21-year-old woman. He noted that this experience felt good so he has continued to seek out multiple sexual partners. History reveals that he is an only child who grew up with two alcoholic parents and witnessed multiple instances of domestic violence between the two. He is also a self-professed workaholic who has come to counseling because he has developed a severe case of hypertension and his physician has ordered him to slow down, work less, and cut back on his beer. He wants you to help him learn to take breaks from his work and to enjoy his leisure time when he does. He considers his childhood to be relatively happy and normal and shows no awareness of the relationship between his experiences in his family of origin and his current problems.

During the process of counseling, one of two things may happen: He may learn to cut back on his beer intake and on his work hours—a first-order change. It is also conceivable that he could begin to experience all the intense grief and rage that his addictions mask and experience change at a very deep, gut, or core level—a second-order change. However, it is important that his counselor honor the level of change *he* requests.

Prochaska, DiClemente, and Norcross (1992) have developed a transtheoretical model of client change. Their model suggests that a client experiences five stages of change in moving toward a particular outcome. These five steps are as follows:

1. *Precontemplation:* The client is unaware of a need for change or does not intend to change
2. *Contemplation:* The client is aware of a need for change and thinks seriously about it but has not decided to make it
3. *Preparation:* The client has decided to take some action in the near future and also has taken some action in the recent past that was not successful
4. *Action:* The client has begun to engage in successful action steps toward the desired outcome but has not yet attained the outcome
5. *Maintenance:* The client reaches his or her goals and now works both to prevent backslides and to consolidate changes made in the action phase

A concrete example of this stage model of change is weight loss. In the *precontemplation* phase, the client either doesn't recognize a need to lose weight or doesn't care; this person is content with his or her body image as it is. The client moves into the *contemplation* phase as he or she becomes aware of a heavier body image, starts to dislike it, and thinks about losing weight. Next, the client attempts some weight-loss program but is unsuccessful—but still plans to take additional action in the near future. This is the *preparation* stage. As the client finds successful ways of losing weight and commits to these, the *action* phase is initiated. Finally, as the client reaches the target weight loss, he or she engages in behaviors to maintain the current weight and to prevent regaining the weight lost; this is the *maintenance* phase.

Over the last 12 years, clinical research has provided support for this model. These five stages have been found to relate to client self-managed change, treatment intervention, treatment outcome effectiveness, and persistence in therapy. Prochaska et al. (1992) note that some clients move through these changes smoothly whereas others stumble at various stages and many recycle through the stages multiple times. Anyone who has followed the ups and downs of Oprah Winfrey's weight-loss program over the last few years can bear witness to this process. In areas such as weight loss and substance abuse, clients who are in or use these stages have more positive treatment outcomes than those who do not (Prochaska et al., 1992). Further, the research suggests that clients in the first two stages are not as ready for change as those in later stages and, if they start, they may terminate prematurely (Smith, Subich, & Kalodner, 1995).

This model has important implications for both goal setting and treatment selection. Clients in Stage 1, *precontemplation* (see Table 10-2), come to counseling at someone else's request or under some sort of pressure. Initially these clients may not participate in goal setting because they don't "own" the goal for change. The counselor's task is to help these clients begin to contemplate the problem and their role in it as well as the possibility of a new response. Being optimistic and providing a good rationale for interventions and change can help facilitate this process.

Clients in Stage 2, *contemplation,* may acknowledge the existence of a problem but may not see themselves as part of the solution, at least initially. The therapist must first help these clients explore the advantages and disadvantages of working toward a specific outcome so they can find outcomes that are realistic and feasible and are likely to result in more benefits than costs.

Clients in the stages of *preparation* and *action* acknowledge that there is a problem, see themselves as part

TABLE 10-2. Stages of change and corresponding interventions

Stages of change	Corresponding interventions
Precontemplation	Be optimistic.
	Provide strong rationale for interventions and change.
Contemplation	Educate client about change process.
	Weigh advantages/disadvantages of change.
Preparation	Define, work toward, and evaluate selected outcomes.
Action	Develop cognitions and skills to prevent relapse/setbacks prior to termination.
Maintenance	

of the solution, and are committed to working toward specific outcomes (see Table 10-2). In general, as Smith et al. (1995) observe, "Persons further along in the stages seem more likely to progress and benefit from therapy" (p. 35). With these clients, the counselor's task is to help them define their goals in specific ways so that progress toward outcomes can be determined. Outcome evaluation is important not only so that clients can see their actual progress but also to substantiate the effectiveness of counseling. Increasingly mental health providers working in agencies or receiving reimbursement from managed care companies are required to conduct outcome evaluations of counseling.

Clients in Stage 5, *maintenance,* often face difficulties in maintaining changes acquired in the prior stages. It is often easier to prepare and act than to maintain. As clients reach this stage and before they terminate counseling, it is important to work on beliefs and actions to equip them to maintain changes as well as to prevent setbacks and relapses. Maintenance goals and skills are especially important in areas such as substance abuse, mood disorders such as depression, and chronic mental health problems such as schizophrenia. An approach that seems particularly useful in working with clients at this last stage of the change model is Marlatt and Gordon's (1985) *Relapse Prevention* model. This model, which has been used frequently for relapse prevention in addictive behaviors (DeJong, 1994; MacKay & Marlatt, 1990), focuses on helping clients to

1. Identify high-risk situations for relapse
2. Acquire behavioral and cognitive coping skills
3. Attend to issues of balance in lifestyle

The model and processes we have described in this section are summarized in Table 10-2. As Cooper (1995)

notes, it is important for therapists to recognize what stage of change is reflected by each client as different kinds of interventions are required depending on the particular style. Counselors who are familiar with this change model can actively use it in therapy by first identifying the stage a given client is in and then by applying the appropriate methods of change to move the client from one stage to another.

In addition to understanding different kinds of change, counselors must heed Mahoney's (1991) conviction that the process of psychotherapeutic change is unbelievably complex. Mahoney (1991) observes that "bluntly, we are learning that psychological change is neither simple nor easy and yet that it is pervasive and relentless" (p. 259).

Mahoney (1991) also notes the role of human affect in producing change; "Some kinds of change are associated with relatively intense episodes of emotionality" (p. 259). Research in this area is in its infancy, although a recent study by Hanna and Ritchie (1995) described what appeared to be active ingredients of change as illustrated in 20 incidents of significant or second-order personal change in 18 persons. Twenty-four common factors derived from literature review were rated on a five-point Likert-type scale as to the role each factor had in the person's change, ranging from 1, not at all (2—somewhat a factor; 3—a definite factor; 4—a necessary condition) to 5, a sufficient condition. Of these 24 common factors, *insight* or *new understanding* was perceived by these participants as the most potent common ingredient of their change (average rating of 4.3). *Developing a new perspective* on the problem and *confronting or facing the problem* were also rated very highly (4.1 and 4.0, respectively). Also, *effort or will, a sense of necessity,* and the *willingness to experience anxiety or difficulty* were other common potent factors. Hanna and Ritchie (1995) conclude that although this is an exploratory study, these five variables seem to affect client involvement and motivation in pursuing treatment outcomes and also contribute to a stronger working alliance; "conversely, the degree of absence of such prerequisite variables may underlie client resistance" (p. 181). Stated another way, it is likely that persons in the first two stages of the Prochaska et al. (1992) model of change exhibit less of these variables than persons at the later three stages. Clients at Stages 1 and 2 may need education from the therapist about these common factors; for example, what does it mean to be willing to experience anxiety or difficulty along the journey to desired results? (Hanna & Ritchie, 1995, p. 181).

Occasionally, in addition to choice or change goals, clients want to maintain certain aspects of their life or certain behaviors at the same rate or in the same way. For example, a client may want to continue his sexual relationship with his wife (maintaining a behavior) the way it is even though he may wish to change other aspects of the relationship. Note that this is represented in the Prochaska et al. change model as Stage 5, Maintenance.

INTERVIEW LEADS FOR IDENTIFYING GOALS

In this section we discuss five categories of interview leads for identifying goals:

1. Providing a rationale
2. Eliciting outcome statements
3. Stating goals in positive terms
4. Determining what the goal is
5. Weighing advantages/disadvantages of the goal

The process of identifying goals and the associated interview leads are especially useful for clients at the first two stages of the Prochaska et al. change model: precontemplation and contemplation.

Providing a Rationale

The first step in identifying goals is to give the client a *rationale* for goals. This statement should describe goals, the purpose of having them, and the client's participation in the goal-setting process. The counselor's intent is to convey the importance of having goals as well as the importance of the client's participation in developing them. An example of what the counselor might say about the purpose of goals is "We've been talking about these two areas that bother you. It might be helpful now to discuss how you would like things to be different. We can develop some goals to work toward during our sessions. These goals will tell us what you want as a result of counseling. So today, let's talk about some things *you* would like to work on."

The counselor might also emphasize the role that goals play in resolving problems through attention and rehearsal. "Paulo, you've been saying how stuck you feel in your marriage and yet how hopeful you feel, too. If we can identify specifically the ways that you want to relate differently, this can help you attend to the things you do that cause difficulty as well as the things you know you want to handle differently." Occasionally, offering examples of how other persons, such as athletes or dancers, use goal setting to facilitate performance may be useful to clients.

After this explanation, the counselor will look for a client response that indicates understanding. If the client seems confused, the counselor will need to explain further the purposes of goals and their benefits to the client or to clarify and explore the client's confusion. As we have indicated, a strong rationale is especially important for clients at the early stages of the Prochaska et al. change model.

Eliciting Outcome Statements

Interview leads are used to help clients identify goals. Following are examples of leads that can help clients define goals and express them in outcome statements:

"Suppose some distant relative you haven't seen for a while sees you after counseling. What would be different then from the way things are now?"
"Assuming we are successful, what would you be doing or how would these situations change?"
"What do you expect to accomplish as a result of counseling? Is this a choice or a change?"
"How would you like counseling to benefit you?"
"What do you *want* to be doing, thinking, or feeling?"

The counselor's purpose in using these sorts of leads is to have the client identify some desired outcomes of counseling. The counselor is looking for some verbal indication of the results the client expects. If the client does not know of desired changes or cannot specify a purpose for engaging in counseling, some time should be spent in exploring this area before moving on. The counselor can assist the client in identifying goals in several ways: by assigning homework ("Make a list of what you can do now and what you want to do one year from now"), by using imagery ("Imagine being someone you admire. Who would you be? What would you be doing? How would you be different?"), by additional questioning ("If you could wave a magic wand and have three wishes, what would they be?"), or by self-report questionnaires or inventories. These sorts of leads and attitudes are useful to help clients contemplate changes in their life and behavior.

Stating the Goal in Positive Terms

An effective outcome goal is stated in *positive* rather than negative terms—as what the client *does* want to do, not what the client does not want to do. This direction is very important because of the role that goal setting plays in human cognition and performance, as mentioned earlier. When the goal is stated positively, clients are more likely to encode and rehearse the things they want to be able to *do* rather than the things they want to avoid or stop. For example, it is fairly easy to generate an image of yourself sitting down and watching TV. However, picturing yourself *not* watching TV is difficult. Instead of forming an image of not watching TV, you are likely to form images (or sounds) related to performing other activities instead, such as reading a book, talking to someone, or being in the TV room and doing something else.

The counselor will have to help clients "turn around" their initial goal statements, which are usually stated as something the person doesn't want to do, can't do, or wants to stop doing. Stating goals positively represents a self-affirming position and can be a helpful intervention for clients at the first two stages of the Prochaska et al. change model. If the client responds to the counselor's initial leads with a negative answer, the counselor can help turn this around by saying something like "That is what you *don't* want to do. Describe what you *do* want to do [think, feel]" or "What will you do instead, and can you see [hear, feel] yourself doing it every time?"

Determining What the Goal Is

As we mentioned earlier, clients at the first stage of the Prochaska change model, precontemplation, often want the goals to call for someone else to change rather than themselves—a teenager who says "I want my mom to stop yelling at me," a teacher who says "I want this kid to shut up so I can get some teaching done," or a husband who says "I want my wife to stop nagging." The tendency to project the desired change onto someone else is particularly evident in change problems that involve relationships with two or more persons.

Without discounting the client's feelings, the counselor needs to help get this tendency turned around. The client is the identified person seeking help and services and is the only person who can make a change. When two or more clients are involved simultaneously in counseling, such as a couple or a family, all identified clients need to contribute to the desired choice or change, not just one party or one "identified patient."

Who owns the change is usually directly related to the degree of *control* or *responsibility* that the client has in the situation and over the choice or change. For example, suppose you are counseling an 8-year-old girl whose parents are getting a divorce. The child says she wants you to help her persuade her parents to stay married. This goal would be very difficult for the child to attain, as she has no responsibility for her parents' relationship.

The counselor will need to use leads to help clients determine whether they or someone else owns the change

and whether anyone else needs to be involved in the goal-selection process. If the client steers toward a goal that requires a change by someone else, the counselor will need to point this out and help the client identify his or her role in the change process.

Interview Leads to Determine Who Owns the Change.
To help the client explore who owns the change, the counselor can use leads similar to the following ones:

"How much control do you have over making this happen?"
"What changes will this goal require of you?"
"What changes will this goal require someone else to make?"
"Can this goal be achieved without the help of anyone else?"
"To whom is this goal most important?"
"Who, specifically, is responsible for making this happen?"

The intent of these leads is to have the client identify a goal that represents choice or change for the client, not for others unless they are directly affected. If the client persists in selecting a goal that represents change for others rather than himself or herself, the counselor and client will have to decide whether to pursue this goal, to negotiate a reconsidered goal, or to refer the client to another helper, as we shall discuss shortly.

Weighing Advantages and Disadvantages of the Goal

It is important to explore the *cost/benefit* effect of all identified goals—that is, what is being given up (cost) versus what is being gained (benefit) from goal attainment (Dixon & Glover, 1984). We think of this step as exploration of *advantages,* or positive effects, and *disadvantages,* or negative effects, of goal attainment. Exploration of advantages and disadvantages helps clients assess the feasibility of the goal and anticipate the consequences; then they can decide whether the change is worth the cost to themselves or to significant others.

Oz (1995) points out that most clients already have considered what is attractive about given alternatives and in fact may be "stuck" either because they want to preserve the benefits without incurring any costs or because the choice involves a values conflict, often between self-denial and costs to a relationship or to their image (Oz, 1995, p. 81). In identifying goals, the client needs to be aware of possible risks and costs and whether he or she is prepared to take such risks if, in fact, they do occur.

Oz (1995) notes that although the cost factor most often affects a client's eventual choice, it is also the factor most

Desired changes	Immediate advantages	Long-term advantages	Immediate disadvantages	Long-term disadvantages
#1				
#2				
#3				

FIGURE 10-1. List for recording advantages and disadvantages of identified goals

often ignored by the client, and inadequate information about costs may lead to post-decisional regret (p. 79). The cost factor involved in change efforts also contributes frequently to client resistance to change. Advantages and disadvantages of particular goals may be emotional and cognitive as well as behavioral (Gambrill, 1977). Goal attainment may result in desirable or undesirable feelings and mood states; self-enhancing or self-defeating thoughts, images, and internal dialogue; and appropriate or inappropriate reactions and motoric responses.

Generally, the goals selected by clients should lead to benefits rather than losses. Advantages and disadvantages may be both short term and long term. The counselor helps the client identify various kinds of short- and long-term advantages and disadvantages associated with goal attainment and offers options to expand the client's range of possibilities (Gambrill, 1977). Sometimes it is helpful to write these in the form of a list that can be expanded or modified at any time, such as the one found in Figure 10-1. Weighing the advantages and disadvantages helps clients at the contemplation stage of the Prochaska et al. change model prepare for action.

Interview Leads for Advantages. Most clients can readily identify some positive consequences associated with their desired changes. Nevertheless, it is still a good idea with all clients to explore positive consequences of the change, for at least four reasons: to determine whether the advantages the client perceives are indicative of actual benefits; to point out other possible advantages for the client or for others that have been overlooked; to strengthen the client's incentive to change; and to determine to what degree the identified goal is relevant and worthwhile, given the client's overall functioning. These are examples of leads to explore advantages of client change:

"In what ways is it worthwhile for you to pursue counseling?"

"What do you see as the benefits of this change?"

"Who would benefit from this change and how?"

"What are some positive consequences that may result from this change?"

"How would attaining this goal help you?"

In using these leads, the counselor is looking for some indication that the client is pursuing a goal on the basis of the positive consequences the goal may produce. If the client overlooks some advantages, the counselor can describe them to add to the client's incentive to change.

If the client is unable to identify any benefits of change for herself or himself, this may be viewed as a signal for caution. Failure to identify advantages of change for oneself may indicate that the client is attempting to change at someone else's request or that the identified goal is not very feasible, given the "total picture." For instance, if a client wants to find a new job while she is also fighting off a life-threatening illness, the acquisition of a new job at this time may not be in the best interests of her desire to regain her health. Further exploration may indicate that another person is a more appropriate client or that other goals should be selected.

Interview Leads for Disadvantages. The counselor can also use leads to have the client consider some risks or side effects that might accompany the desired change. Some examples of leads the counselor might use to explore the risks or disadvantages of change are the following:

"How could this make life difficult for you?"

"Will pursuing this change affect your life in any adverse ways?"

"What might be some possible risks of doing this?"

"How would your life be changed if this happened?"

"What are some possible disadvantages of going in this direction? How willing are you to pay this price?"

"Who might disapprove of this action? How will that affect you?"

"What are some negative consequences this change might have for you—or for others?"

"How will this change limit or constrain you?"

"What new problems in living might pursuing this goal create for you?"

The counselor is looking for some indication that the client has considered the possible costs associated with the goal. If the client discounts the risks or cannot identify any, the counselor can use immediacy or confrontation to point out some disadvantages. However, the counselor should be careful not to persuade or coerce the client to pursue another alternative simply because the counselor believes it is better.

Decision Point: To Treat or Not

At this point in the process, the primary issue for the counselor is whether she or he can help this client. Most people agree that this is one of the biggest ethical and, to some extent, legal questions the counselor faces during the helping process. As Beutler and Clarkin (1990) observe, "The therapist must be aware that there are instances in which a decision *not* to treat at all is the optimal decision" (p. 24). They point out that therapy may be overused and unnecessary for well-adjusted persons experiencing a current life stressor. They also observe that therapy is unlikely to have much impact with clients who do not or cannot invest in the process. Similarly, some clients who are also high reactant persons are at greatest risk for negative effects from counseling. Occasionally, too, at the other extreme are clients who have become "addicted" to counseling; they move from one therapist to another and are really in need of a vacation from treatment.

The counselor and client will need to choose whether to continue with counseling and pursue the selected goals, to continue with counseling but reevaluate the client's initial goals, or to seek the services of another counselor. The particular decision is always made on an individual basis and is based on two factors: *willingness* and *competence* to help the client pursue the selected goals. Willingness involves your interest in working with the client toward identified goals and issues, your values, and your acceptance of the goals as worthwhile and important, given the overall functioning of the client. Competence involves your skills and knowledge and whether you are familiar with alternative intervention strategies and multiple ways to work with particular problems.

We offer the following ideas as food for thought in this area. First, as much as possible, be responsive to the *client's* requests for change (Gottman & Leiblum, 1974, p. 64) even if these goals do not reflect your theoretical biases or personal preferences. This sort of responsiveness has been described by Gottman and Leiblum (1974, p. 64): "If [the client] wants help in accepting his homosexuality, do not set up a treatment program designed to help him find rewards in heterosexuality. Respect for your client implies respect for his diagnosis of his needs and wishes."

Too often, either knowingly or unwittingly, counselors may lead a client toward a goal they are personally comfortable with or feel more competent to treat. As Gambrill observes, "Personal values and theoretical assumptions with no empirical basis have caused profound distress to many clients over the years" (1977, p. 1035).

Second, if you have a *major* reservation about pursuing selected goals, a referral might be more helpful to the client (Gottman & Leiblum, 1974, p. 43).

Referral may be appropriate in any of the following cases: if the client wants to pursue a goal that is incompatible with your value system; if you are unable to be objective about the client's concern; if you are unfamiliar with or unable to use a treatment requested by the client; if you would be exceeding your level of competence in working with the client; or if more than one person is involved and, because of your emotions or biases, you favor one person instead of another. Referral may be a better choice than continuing to work with the client in the midst of serious limitations or reservation. Referral is a way to provide an alternative counseling experience and, we hope, one that will leave the client with a positive impression of counseling.

In deciding to refer a client, the counselor does have certain responsibilities. From the initial counseling contacts, the counselor and client have entered into at least an unwritten contract. Once the counselor agrees to counsel a client, he or she assumes responsibility for the client (Van Hoose & Kottler, 1985). In deciding to terminate this "contract" by a referral, the referring counselor can be considered legally liable if the referral is not handled with due care—that is, ensuring that clients have choices of referral therapists (when available) and ensuring that referred therapists are considered competent and do not have a reputation for poor service or unethical practices (Van Hoose & Kottler, 1985). (See Learning Activity #32.)

MODEL DIALOGUE: THE CASE OF JOAN

To help you see how the leads for identifying goals are used with a client, the case of Joan, introduced in Chapter 8, is continued here as a dialogue in a counseling session directed

LEARNING ACTIVITY 32 *Decision Point*

For practice in thinking through the kinds of decisions you may face in the goal-setting process, you may want to use this learning activity. The exercise consists of three hypothetical situations. In each case, assume that you are the counselor. Read through the case. Then sit back, close your eyes, and try to imagine being in the room with the client and being faced with this dilemma. How would you feel? What would you say? What would you decide to do and why?

There are no right or wrong answers to these cases. You may wish to discuss your responses with another classmate, a co-worker, or your instructor.

Case 1

You are counseling a family with two teenage daughters. The parents and the younger daughter seem closely aligned; the elder daughter is on the periphery of the family. The parents and the younger daughter report that the older daughter's recent behavior is upsetting and embarrassing to them because she recently "came out of the closet" to disclose that she is a lesbian, and has begun to hang out with a few other lesbian young women in their local high school. The parents and younger daughter indicate that they think she is just going through a "phase." They state that they want you to help them get this girl "back in line" with the rest of the family and get her to adopt their values and socially acceptable behavior. What do you do?

Case 2

You are counseling a fourth-grader. You are the only counselor in this school. One day you notice that this boy seems to be all bruised. You inquire about this. After much hesitation, the child blurts out that he is often singled out on his way home by two big sixth-grade bullies who pick a fight, beat him up for a while, and then leave him alone until another time. Your client asks you to forget this information. He begs you not to say or do anything for fear of reprisal from these two bullies. He states he doesn't want to deal with this in counseling as he has come to see you about something else. What do you do?

Case 3

You are working with an elderly man whose relatives are dead. After his wife died six months ago, he moved from their family home to a retirement home. Although the client is relatively young (70) and is in good health and alert, the staff has requested your help because he seems to have become increasingly morbid and discouraged. In talking with you, he indicates that he has sort of given up on everything, including himself, because he doesn't feel he has anything to live for. Consequently, he has stopped going to activities, isolates himself in his room, and has even stopped engaging in self-care activities such as personal hygiene and grooming, leaving such things up to the staff. He indicates that he doesn't care to talk with you if these are the kinds of things you want to talk about. What do you do?

---

toward goal selection. Counselor responses are prefaced by an explanation (in italics).

*In response 1, the counselor starts out with a **review** of the last session.*

1. *Counselor:* Joan, last week we talked about some of the things that are going on with you right now that you're concerned about. What do you remember that we talked about?
   Client: Well, we talked a lot about my problems in school—like my trouble in my math class. Also about the fact that I can't decide whether or not to switch over to a vocational curriculum—and if I did my parents would be upset.

2. *Counselor:* Yes, that's a good summary. We did talk about a lot of things—such as the pressure and anxiety you feel in competitive situations like your math class and your difficulty in making decisions. I believe we mentioned also that you tend to go out of your way to please others, like your parents, or to avoid making a decision they might not like.
   Client: Mm-hmm. I tend to not want to create a hassle. I also just have never made many decisions by myself.

*In response 3, the counselor will move from problem definition to goal selection. Response 3 will consist of an **explanation** about goals and their **purpose.***

3. *Counselor:* Yes, I remember you said that last week. I've been thinking that since we've kind of got a handle on the main issues you're concerned about, today it might be helpful to talk about things you might want to happen—or how you'd like things to be different. This way we know exactly what we can be talking about and working on that will be most helpful to you. How does that sound?
   Client: That's OK with me. I mean, do you really think there are some things I can do about these problems?

*The client has indicated some uncertainty about possible change. The counselor will pursue this in response 4 and indicate more about the **purpose** of goals and possible effects of counseling for this person.*

4. *Counselor:* You seem a little uncertain about how much things can be different. To the extent that you have some control over a situation, it is possible to make some changes. Depending on

what kind of changes you want to make, there are some ways we can work together on this. It will take some work on your part, too. How do you feel about this?

Client: OK. I'd like to get out of the rut I'm in.

In response 5, the counselor will explore the ways in which the client would like to change. The counselor will use a lead to **identify client goals.**

5. *Counselor:* So you're saying that you don't want to continue to feel stuck. Exactly how would you like things to be different—say, in three months from now—from the way things are now?

Client: Well, I'd like to feel less pressured in school, especially in my math class.

The client has identified one possible goal, although it is stated in negative terms. In response 6, the counselor will help the client identify the goal in **positive terms.**

6. *Counselor:* OK, that's something you *don't* want to do. Can you think of another way to say it that would describe what you *do* want to do?

Client: Well, I guess I'd like to feel confident about my ability to handle tough situations like math class.

In the next response, the counselor **paraphrases** *Joan's goal and checks it out to see whether she has restated it accurately.*

7. *Counselor:* So you're saying you'd like to feel more positively about yourself in different situations—is that it?

Client: Yeah, I don't know if that is possible, but that's what I would like to have happen.

In responses 8–14, the counselor continues to help Joan **explore and identify desired outcomes.**

8. *Counselor:* Well, in a little while we'll take some time to explore just how feasible that might be. Before we do that, let's make sure we don't overlook anything else you'd like to work on—in what other areas is it important to you to make a change or to turn things around for yourself?

Client: Well, I'd like to start making some decisions for myself for a change, but I don't know exactly how to start.

9. *Counselor:* OK, that's part of what we'll do together—we'll look at how you can get started on some of these things. So far, then, you've mentioned two things you'd like to work toward—increasing your confidence in your ability to handle tough situations like math and starting to make some decisions by yourself without relying on help from someone else. Is that about it, or can you think of any other things you'd like to work on?

Client: Well, I guess it's related to making my own decisions, but I'd like to decide whether to stay in this curriculum or switch to the vocational one.

10. *Counselor:* So you're concerned also about making a special type of decision about school that affects you now.

Client: That's right. But I'm sort of afraid to, because I know if I decided to switch, my parents would have a terrible reaction when they found out about it.

11. *Counselor:* It seems that you're mentioning another situation that we might need to try to get a different handle on. As you mentioned last week, in certain situations, like math class or with your parents, you tend to back off and kind of let other people take over for you.

Client: That's true, and I guess this curriculum thing is an example of it. It's like a lot of things I do know what I want to do or say, but I just don't follow through. Like not telling my folks about my opinion about this college prep curriculum. Or not telling them how their harping at me about grades makes me feel. Or even in math class, just sitting there and sort of letting the teacher do a lot of the work for me when I really do probably know the answer or could go to the board.

12. *Counselor:* So what you're saying is that in certain situations with your folks or in math class, you may have an idea or an opinion or a feeling, yet you usually don't express it.

Client: Mm-hmm. Usually I don't because sometimes I'm afraid it might be wrong or I'm afraid my folks would get upset.

13. *Counselor:* So the anticipation that you might make a mistake or that your folks might not like it keeps you from expressing yourself in these situations?

Client: Yup, I believe so.

14. *Counselor:* Then is this another thing that you would like to work on?

Client: Yes, because I realize I can't go on withdrawing forever.

Because Joan has again stated the outcome in negative terms, in the next four responses (15, 16, 17, 18), the counselor helps Joan **restate the goal in positive terms.**

15. *Counselor:* OK, now again you're sort of suggesting a way that you don't want to handle the situation. You don't want to withdraw. Can you describe something you *do* want to do in these situations in a way that you could see, hear, or grasp yourself doing it each time the situation occurs?

Client: I'm not exactly sure what you mean.

16. *Counselor:* Well, for instance, suppose I want to lose weight. I could say "I don't want to eat so much, and I don't want to be fat." But that just describes not doing what I've been doing. So it would be more helpful to describe something I'm going to do instead, like "Instead of eating between meals, I'm going to go out for a walk, or talk on the phone, or create a picture of myself in my head as a thinner person."

Client: Oh, yeah, I do see what you mean. So I guess instead of withdrawing, I—well, what is the opposite of that? I guess I think it would be more helpful if I volunteered the answers or gave my ideas or opinions—things like that.

17. *Counselor:* OK, so you're saying that you want to express yourself instead of holding back. Things like expressing opinions, feelings, things like that.

Client: Yeah.

18. *Counselor:* OK, now we've mentioned three things you want to work on—anything else?

Client: No, I can't think of anything.

In the next response, the counselor asks Joan to **select one of the goals** *to work on initially. Tackling all three outcomes simultaneously could be overwhelming to a client.*

19. *Counselor:* OK, as time goes on and we start working on some of these things, you may think of something else—or something we've talked about today may change. What might be helpful now is to decide which of these three things you'd like to work on first.

Client: Gee, that's a hard decision.

In the previous response, Joan demonstrated in vivo one of her problems—difficulty in making decisions. In the next response, the counselor **provides guidelines** *to help Joan make a choice but is careful not to make the decision for her.*

20. *Counselor:* Well, it's one decision I don't want to make for you. I'd encourage you to start with the area you think is most important to you now—and also maybe one that you feel you could work with successfully.

Client: [Long pause] Can this change, too?

21. *Counselor:* Sure—we'll start with one thing, and if later on it doesn't feel right, we'll move on.

Client: OK, well, I guess it would be the last thing we talked about—starting to express myself in situations where I usually don't.

In the next response, the counselor will discuss the degree to which Joan believes the **change represents something she will do** *rather than someone else.*

22. *Counselor:* OK, sticking with this one area, it seems like these are things that you could make happen without the help of anyone else or without requiring anyone else to change too. Can you think about that for a minute and see whether that's the way it feels to you?

Client: [Pause] I guess so. You're saying that I don't need to depend on someone else; it's something I can start doing.

In the next response, the counselor shifts to exploring **possible advantages** *of goal achievement. Note that the counselor asks the client first to express her opinion about advantages; the counselor is giving her* in vivo *practice of one of the skills related to her goal.*

23. *Counselor:* One thing I'm wondering about—and this will probably sound silly because in a way it's obvious—but exactly how will making this change help you or benefit you?

Client: Mm—[Pause]—I'm thinking—well, what do you think?

In the previous response, the client shifted responsibility to the counselor and "withdrew," as she does in other anxiety-producing situations, such as math class and interactions with her parents. In the next response, the counselor **confronts** *this behavior pattern.*

24. *Counselor:* You know, it's interesting, I just asked you for your opinion about something, and instead of sharing it, you asked me to sort of handle it instead. Are you aware of this?

Client: Now that you mention it, I am. But I guess that's what I do so often it's sort of automatic.

In the next three responses (25, 26, 27), the counselor does some in vivo assessment of Joan's problems, which results in information that can be used later for **planning of subgoals and action steps.**

25. *Counselor:* Can you run through exactly what you were thinking and feeling just then?

Client: Well, just that I had a couple of ideas, but then I didn't think they were important enough to mention.

26. *Counselor:* I'm wondering if you also may have felt a little concerned about what I would think of your ideas.

Client: [Face flushes] Well, yeah. I guess it's silly, but yeah.

27. *Counselor:* So is this sort of the same thing that happens to you in math class or around your parents?

Client: Yeah—only in those two situations, I feel much more uptight than I do here.

In the next four responses, the counselor continues to explore **potential advantages** *for Joan of attaining this goal.*

28. *Counselor:* OK, that's real helpful because that information gives us some clues on what we'll need to do first in order to help you reach this result. Before we explore that, let's go back and see whether you can think of any ways in which making this change will help you.

Client: Well, I think sometimes I'm like a doormat. I just sit there and let people impose on me. Sometimes I get taken advantage of.

29. *Counselor:* So you're saying that at times you feel used as a result?

Client: Yeah. That's a good way to put it.

30. *Counselor:* OK, other advantages or benefits to you?

Client: Well, I'd become less dependent and more self-reliant. If I do decide to go to college, that's only two years away, and I will need to be a whole lot more independent then.

31. *Counselor:* OK, that's a good thought. Any other ways that this change would be worthwhile for you, Joan?

Client: Mm—I can't think of any. That's honest. But if I do, I'll mention them.

In the next responses (32–35), the counselor initiates exploration of **possible disadvantages** *of this goal.*

32. *Counselor:* OK, great! And the ones you've mentioned I think are really important ones. Now, I'd like you to flip the coin, so to speak, and see whether you can think of any disadvantages that could result from moving in this direction?

Client: Well, I can't think of any in math. Well, no, in a way I can. I guess it's sort of the thing to do there to act dumb. If I start expressing myself more, people might wonder what is going on.

33. *Counselor:* So you're concerned about the reaction from some of the other students?

Client: Yeah, in a way. Although there are a couple of girls in there who are pretty popular and also made the honor roll. So I don't think it's like I'd be a social outcast.

34. *Counselor:* It sounds, then, like you believe that is one disadvantage you could live with. Any other ways in which

doing this could affect your life in a negative way—or could create another problem for you?

Client: Well, I think a real issue there is how my parents would react if I started to do some of these things. I don't know. Maybe they would welcome it. But I sort of think they would consider it a revolt or something on my part and would want to squelch it right away.

35. *Counselor:* You seem to be saying you believe your parents have a stake in keeping you somewhat dependent on them.

Client: Yeah, I do.

*This is a difficult issue. Without observing her family, it would be impossible to say whether this is Joan's perception (and a distorted one) or whether the parents do play a role in this problem—and, indeed, from a diagnostic standpoint, family members are often significantly involved when one family member has a dependent personality. The counselor will thus **reflect both possibilities** in the next response.*

36. *Counselor:* That may or may not be true. It could be that you see the situation that way and an outsider like myself might not see it the same way. On the other hand, it is possible your parents might subtly wish to keep you from growing up too quickly. This might be a potentially serious enough disadvantage for us to consider whether it would be useful for all four of us to sit down and talk together.

Client: Do you think that would help?

*In the next two responses, the counselor and Joan continue to discuss potential **negative effects or disadvantages** related to this goal. Note that in the next response, instead of answering the client's previous question directly, the counselor shifts the responsibility to Joan and solicits her opinion, again giving her in vivo opportunities to demonstrate one skill related to the goal.*

37. *Counselor:* What do you think?

Client: I'm not sure. They are sometimes hard to talk to.

38. *Counselor:* How would you feel about having a joint session—assuming they were agreeable?

Client: Right now it seems OK. How could it help exactly?

*In the following response, the counselor changes from an **individual to a systemic focus,** since the parents may have an investment in keeping Joan dependent on them or may have given Joan an injunction "Don't grow up." The systemic focus avoids blaming any one person.*

39. *Counselor:* I think you mentioned it earlier. Sometimes when one person in a family changes the way she or he reacts to the rest of the family, it has a boomerang effect, causing ripples throughout the rest of the family. If that's going to happen in your case, it might be helpful to sit down and talk about it and anticipate the effects, rather than letting you get in the middle of a situation that starts to feel too hard to handle. It could be helpful to your parents, too, to explore their role in this whole issue.

Client: I see. Well, where do we go from here?

40. *Counselor:* Our time is about up today. Let's get together in a week and map out a plan of action.

(*Note:* The same process of goal identification would also be carried out in subsequent sessions for the other two outcome goals Joan identified earlier in this session.)

INTERVIEW LEADS FOR DEFINING GOALS

Most clients will select more than one goal. Ultimately, it may be more realistic for the client to work toward several outcomes. For example, in our model case, Joan has selected three terminal outcome goals: acquiring and demonstrating at least four initiating skills, increasing positive self-talk about her ability to function adequately in competitive situations, and acquiring and using five decision-making skills (see Joan's goal chart on p. 256). These three outcomes reflect the three core problems revealed by the assessment interview (Chapter 9). Selection of one goal may also imply the existence of other goals. For example, if a client states "I want to get involved in a relationship with a man that is emotionally and sexually satisfying," the client may also need to work on meeting men and her or his approach behaviors, developing communications skills designed to foster intimacy, and learning about what responses might be sexually satisfying for her or him.

At first, it is useful to have the client specify one or more desired goals for each separate problem. However, to tackle several outcome goals at one time would be unrealistic. The counselor should ask the client to choose one of the outcome goals to pursue first. After identifying an initial outcome goal to work toward, the counselor and client can define the three parts of the goal and identify subgoals. The next section of this chapter will introduce some counselor leads used to help the client define the outcome goals of counseling and will present some probable responses that indicate client responsiveness to the leads. These leads are particularly useful for clients moving from the preparation to the action phase of the change model developed by Prochaska et al. (1992).

Defining Behaviors Related to Goals

Defining goals involves specifying in operational or behavioral terms what the client (whether an individual, group member, or organization) is to *do* as a result of counseling. This part of an outcome goal defines the particular behavior the client is to perform and answers the question "*What* will the client do, think, or feel differently?" Examples of behavior outcome goals include exercising more frequently, asking for help from a teacher, verbal sharing of positive feelings about oneself, and thinking about oneself in positive ways. As you can see, both overt and covert behaviors, including thoughts and feelings, can be included in this part of the outcome goal as long as the behavior is defined by what it means for each client. Defining goals behaviorally makes the goal-setting process specific, and specifically defined goals are more likely to create incentives and guide performance than vaguely stated intentions (Bandura & Simon, 1977, p. 178). When goals are behaviorally or operationally defined, it is easier to evaluate the effects of counseling.

Interview Leads for Defining Goal Behavior. The following are some leads a counselor can use to identify the behavior part of a goal:

"When you say you want to _____, what do you see yourself doing?"

"What could I see you doing, thinking, or feeling as a result of this change?"

"You say you want to be more self-confident. What things would you be thinking and doing as a self-confident person?"

"Describe a good (and a poor) example of this goal."

It is important for the counselor to continue to pursue these leads until the client can define the overt and covert behaviors associated with the goal. This is not an easy task, for most clients talk about changes in vague or abstract terms. If the client has trouble specifying behaviors, the counselor can help with further instructions, information giving, or self-disclosing a personal goal. The counselor can also facilitate behavioral definitions of the goal by encouraging the client to use action verbs to describe what will be happening when the goal is attained (Dixon & Glover, 1984). As we mentioned earlier, it is important to get clients to specify what they *want* to do, not what they don't want or what they want to stop. The goal is usually defined sufficiently when the counselor can accurately repeat and paraphrase the client's definition.

Defining the Conditions of an Outcome Goal

The second part of an outcome goal specifies the conditions—that is, the *context* or *circumstances*—where the behavior will occur. This is an important element of an outcome goal for both the client and the counselor. The conditions suggest a particular *person* with whom the client may perform the desired behaviors or a particular *setting* and answers the question "*Where, when,* and *with whom* is the behavior to occur?" Specifying the conditions of a behavior sets boundaries and helps to ensure that the behavior will occur only in desired settings or with desired people and will not generalize to undesired settings. This idea can be illustrated vividly. For example, a woman may wish to increase the number of positive verbal and nonverbal responses she makes toward her partner. In this case, time spent with her partner would be the condition or circumstances in which the behavior occurs. However, if this behavior generalized to include all persons, it might have negative effects on the very relationship she is trying to improve.

Interview Leads for the Conditions of a Goal. Leads used to determine the conditions of the outcome goal include these:

"Where would you like to do this?"

"In what situations do you want to be able to do this?"

"When do you want to do this?"

"Whom would you be with when you do this?"

"In what situations is what you're doing now not meeting your expectations?"

The counselor is looking for a response that indicates where or with whom the client will make the change or perform the desired behavior. If the client gives a noncommittal response, the counselor may suggest client self-monitoring to obtain these data. The counselor can also use self-disclosure and personal examples to demonstrate that a desired behavior may not be appropriate in all situations or with all people.

Defining a Level of Change

The third element of an outcome goal specifies the level or *amount* of the behavioral change. In other words, this part answers "*How much* is the client to do or to complete in order to reach the desired goal?" The level of an outcome goal serves as a barometer that measures the extent to which the client will be able to perform the desired behavior. For example, a man may state that he wishes to decrease cigarette smoking. The following week, he may report that he did a better job of cutting down on cigarettes. However, unless he can specify how much he actually decreased smoking, both he and the counselor will have difficulty determining how much the client really completed toward the goal. In this case, the client's level of performance is ambiguous. In contrast, if he had reported that he reduced cigarette smoking by two cigarettes per day in one week, his level of performance could be determined easily. If his goal were to decrease cigarette smoking by eight cigarettes per day, this information would help to determine progress toward the goal. (See Learning Activity #33.)

Like the behavior and condition parts of an outcome goal, the level of change should always be established individually for each client. The amount of satisfaction derived from goal attainment often depends on the level of performance established (Bandura & Simon, 1977, p. 178). A suitable level of change will depend on such factors as the present level of the problem behavior, the present level of the desired behavior, the resources available for change, the client's readiness to change, and the degree to which other conditions or people are maintaining the present level of problem behavior.

As an example, suppose a client wants to increase the number of assertive opinions she expresses orally with her husband. If she now withholds all her opinions, her level of change might be stated at a lower level than that defined for another client who already expresses some opinions. And if the client's husband is accustomed to her refraining from giving opinions, this might affect the degree of change made, at least initially. The counselor's and client's primary concern is to establish a level that is manageable, that the client can attain with some success. Occasionally the counselor may encounter a client who always wants to achieve more change than is desirable or even possible. If the level is set too high, the desired behavior may not occur, thus ruling out chances for success and subsequent progress and rewards. As a general rule of thumb, it is better to err by moving too slowly and thus set the level too low rather than too high.

One way to avoid setting the level of a goal too high or making it too restrictive is to use a scale that identifies a series of *increasingly desired* outcomes for each given problem area. This concept, introduced by Kiresuk and Sherman (1968), is called "goal-attainment scaling" (GAS) and has been used increasingly in agencies that must demonstrate certain levels of client goal achievement in order to receive or maintain funding and third-party reimbursement. In goal-attainment scaling, the counselor and client devise five outcomes for a given problem and arrange these by level or extent of change on a scale in the following order (each outcome is assigned a numerical value): most unfavorable outcome (−2), less than likely expected outcome (−1), most likely or expected outcome (0), more than likely expected outcome (+1), most favorable outcome (+2). Table 10-3 shows an example of the use of GAS for a client with ulcerative

TABLE 10-3. Goal-attainment scale for client with ulcerative colitis

Date: 10/24/97	Frequency of colitis attacks
(−2) Most unfavorable outcome thought likely	One per day
(−1) Less than expected success with treatment	One every other day
(0) Expected level of treatment success	One per week
(+1) More than expected success with treatment	One every two weeks
(+2) Best expected success with treatment	None per month

SOURCE: Adapted from "Behavioral Treatment of Mucous Colitis," by K. J. Youell and J. P. McCullough, *Journal of Consulting and Clinical Psychology, 43,* pp. 740–745. Copyright © 1975 by the American Psychological Associaton.

colitis. A review of this GAS model and similar models is presented by Ogles, Lambert, and Masters (1996).

Leads to Identify the Level of Change. Here are some leads you can use to help identify the client's desired extent or level of change:

"How much would you like to be able to do this, compared with how much you're doing it now?"

"How often do you want to do this?"

"From the information you obtained during self-monitoring, you seem to be studying only about an hour a week now. What is a reasonable amount for you to increase this without getting bogged down?"

"You say you'd like to lose 40 pounds. Let's talk about a reasonable amount of time this might take and, to start with, what amount would be easy for you to lose just in the next 3 weeks."

"What amount of change is realistic, considering where you are right now?"

The counselor is looking for some indication of the present and future levels of the desired behavior. This level can be expressed by either the number of times or the amount the client wants to be able to do something. In some cases, an appropriate level may be only one, as when a client's outcome goal is to make one decision about a job change. The counselor can help the client establish an appropriate level of change by referring to the self-monitoring data collected during problem assessment. If the client has not engaged in monitoring, this is another point where it is almost imperative to have the client observe and record the present amounts of the problem behavior and the goal behavior. This information will give some idea of the present level of behavior, referred to as the base-rate or baseline level. This information is important because in setting the desired level, it should be contrasted with the present level of the overt or covert behaviors. As you may recall from Chapter 9, a client's data gathering is very useful for defining problems and goals and for monitoring progress toward the goals.

Level as an Indicator of Type of Problem and Goal.   The level reflected in an outcome goal reflects both the type of problem and the type of goal. From our earlier discussion about what outcome goals represent, recall that problems can be classified as either choice or change. In a choice problem, the level of the goal reflects a conflict to be resolved or a choice or decision to be made—for example, the client needs to decide on one of three options or decide between two different directions. In a change problem, the level reflected in the goal specifies both the direction and the type of change desired. In the example of the client who wants to be more assertive, if the client's present level of a specified assertive response is zero, then the goal would be to acquire the assertive skill. When the base rate of a behavior is zero, or when the client does not seem to have certain skills in her or his repertory, the goal is stated as acquiring a behavior. If, however, the client wants to improve or increase something that she or he can already do (but at a low level), the goal is stated as increasing a behavior. Increasing or acquiring overt and/or covert behaviors is a goal when the client's problem is a *response deficit,* meaning that the desired response occurs with insufficient intensity or frequency or in inappropriate form (Gambrill, 1977). Sometimes a client has an overt behavioral response in his or her repertoire, but it is masked or inhibited by the presence of certain feelings—in which case the goal would be directed toward the feelings rather than the overt behavior. In this instance, the problem stems from *response inhibition,* and the resulting goal is a disinhibition of the response, usually by the working through of the emotional reactions standing in the way.

In contrast, if the client is doing too much of something and wants to lower the present level, the goal is stated as decreasing a behavior and possibly, later on, eliminating it from the client's repertoire. Decreasing or eliminating overt and/or covert behaviors is a goal when the client's problem is a *response excess,* meaning that a response occurs so often,

so long, with such excessive intensity, or in socially inappropriate contexts that it is often annoying to the client and to others (Gambrill, 1977). In problems of response excesses, it is usually the frequency or amount of the response, rather than its form, that is problematic. It is almost always easier to work on developing or increasing a behavior (response increment or acquisition) than on stopping or decreasing a response (response decrement or elimination). This is another reason to encourage clients to state their goals in positive terms, working toward doing something or doing it more, rather than stopping something or doing it less.

Sometimes, when the client wants to eliminate something, she or he wishes to replace whatever is eliminated with a more appropriate or self-enhancing behavior. For instance, a client trying to lose weight may desire to replace junk-food snacks with low-calorie snacks. This client's goal is stated in terms of restructuring something about her or his environment—in this case, the type of snack eaten. Although this is an example of restructuring an overt behavior, restructuring can be cognitive as well. For example, a client may want to eliminate negative, self-defeating thoughts about difficulty in taking tests and replace these with positive, self-enhancing thoughts about the capacity to perform adequately in test-taking situations. Restructuring also often takes place during family counseling when boundaries and alliances between and among family members are shifted so that, for instance, a member on the periphery is pulled into the family, or triangles are broken up, or overinvolved alliances between two persons are rearranged. Restructuring overt or covert behaviors is a goal when the problem is *inadequate, inappropriate, or defective stimulus control,* meaning that the necessary supporting environmental conditions either are missing or are arranged in such a way as to make it impossible or difficult for the desired behavior to occur.

In some instances, the level of a goal reflects maintenance of a particular overt or covert response at the current rate or frequency or under existing conditions. As you recall from our earlier discussion of client change in this chapter, not all goals will reflect a discrepancy between the client's present and future behavior. Some goals may be directed toward maintaining a desired or satisfying situation or response (stage 5 of the Prochaska change model). Such goals may be stated as, for example, "to maintain my present amount (three hours daily) of studying," "to maintain the present balance in my life between work on weekdays and play on weekends," "to maintain the positive communication I have with my partner in our daily talks," or "to maintain my present level (two a day) of engaging in relaxation sessions." A maintenance goal suggests that the client's present level of behavior is satisfying and sufficient, at least at this particular time. A maintenance goal may help to put things in perspective by acknowledging the areas of the client's life that are going well. Maintenance goals are also useful and necessary when one of the change goals has been achieved. For example, if a client wanted to lose weight and has done so successfully, then the counselor and client need to be concerned about how the client can maintain the weight loss. As we mentioned earlier, often maintenance goals and programs are harder to achieve and take greater effort and planning than initial change attempts.

To summarize, the level stated by the outcome goal will usually reflect one of the categories of problems and goals summarized in Table 10-4. Because most clients have more than one outcome goal, a client's objectives may reflect more than one of these directions of change. Knowledge of the direction and level of change defined in the client's goals is important in selecting counseling strategies. For example, self-monitoring (see Chapter 20) is used differently depending on whether it is applied to increase or to decrease

TABLE 10-4. Categories of types of client problems and related goals

Type of problem	Type of goal
1. Choice	Decision between two or more alternatives
	Resolution of at least two conflicting issues
2. Change	
A. Response deficit	Response increment
	Response acquisition
B. Response inhibition	Disinhibition of response
	Working through of emotional reactions
C. Response excess	Response decrement
	Response elimination
D. Inadequate or inappropriate stimulus control	Response restructuring
3. Maintenance	Response maintenance at current frequency or amount or in current context

LEARNING ACTIVITY 33   *Defining Outcome Goals*

We have found that the most difficult part of developing goals with a client is specifying the three parts of the outcome goal. We believe the reason is that the concept is foreign to most of us and difficult to internalize. This is probably because, in our own lives, we think about small, very mundane goals. With more complex goals, we still don't assess the individual overt and covert behaviors to be changed, where and with whom change will occur, and the extent of the change. This learning activity is intended to help you create some *personal* meaning from these three parts of an outcome goal. If you feel comfortable with this, you are more likely to help a client define her or his goals.

1. During the next week, keep a log of any concerns, issues, or problems you're experiencing.
2. At the end of the week, go over your log and label each problem according to type: choice or change. If change, describe the problem as response deficit, response inhibition, response excess, or inadequate/inappropriate stimulus control (refer also to Table 10-4).
3. Select one problem you are interested in resolving and specify the corresponding type of goal (as listed in Table 10-4): decision or resolution of conflict, response increment, response acquisition, disinhibition or

working through a response, response decrement, response elimination, or response restructuring.
4. Define the goal by identifying

   a. What you'll be doing (overt behavior) and thinking and feeling (covert behavior) as a result of this goal—make sure you state this in positive terms.
   b. Where, when, and with whom you want to do this (conditions).
   c. How much or how often you want to do this (level). Your goal is probably defined sufficiently if an objective observer can paraphrase it accurately, stating exactly what, when, where, with whom, and how much you will be doing.

5. In addition to your definition above, create a goal-attainment scale for your outcome, using the example given in Table 10-3. Your scale should specify the following five levels of possible outcomes: most unfavorable outcome (−2), less than likely expected outcome (−1), most likely or expected outcome (0), more than likely expected outcome (+1), and most favorable outcome (+2).
6. Review your responses to this activity with a colleague, instructor, supervisor, or partner.

a response. One counseling strategy might be used appropriately to help a client acquire responses; yet another strategy may be needed to help a client restructure some responses. It is very important for the counselor and client to spend sufficient time on specifying the level of the goal, even if this process seems elusive and difficult.

Identifying and Sequencing Subgoals or Action Steps

All of us can probably recall times when we were expected to learn something so fast that the learning experience was overwhelming and produced feelings of frustration, irritation, and discouragement. The change represented by counseling goals can be achieved best if the process is gradual. Any change program should be organized into a sequence that guides the client through small steps toward the ultimate desired behaviors (Bandura, 1969, p. 74). In defining goals, this gradual learning sequence is achieved by breaking down the ultimate goal into a series of smaller goals called *subgoals* or *action steps*. Subgoals help clients move toward the solution of problems in a planned way (Dixon & Glover, 1984, p. 136). The subgoals are usually arranged in a hierarchy, so that the client completes subgoals at the bottom of the ranked list before the ones near the top. Although an overall outcome goal can provide a "general directive" for change, the specific subgoals may determine a person's immediate activities and degree of effort in making changes (Bandura & Simon, 1977, p. 178).

Sequencing goals into smaller subgoals is more likely to produce the desired results for two reasons. First, completion of subgoals may keep failure experiences to a minimum. Completing subgoals successfully will encourage the client and will help maintain the client's motivation to change (Bandura, 1969, p. 75). Second, arranging the ultimate goal into subgoals indicates that immediate, daily subgoals may be more potent than distant, weekly subgoals.

Subgoals identified may represent covert as well as overt behavior, since a comprehensive change program usually involves changes in the client's thoughts and feelings as well as in overt behaviors and environmental situations. Subgoals may arise out of treatment approaches or recommended ways to resolve a particular problem or, when formal procedures are not available, from more informal and common-sense ideas. In any event, they are always actions that move the client in the direction of the desired outcome goal (Carkhuff, 1993).

After subgoals are identified and selected, they are rank-ordered in a series of tasks—a hierarchy—according to

complexity and degree of difficulty and immediacy. Because some clients are put off by the word *hierarchy,* we use the term *goal pyramid* instead and pull out an 8½″ × 11″ sheet of paper that has a drawing of a blank pyramid on it, such as the one in Figure 10-2 (p. 241). A series of subgoal tasks may represent either increasing requirements of the same (overt or covert) behavior or demonstrations of different behaviors, with simpler and easier responses sequenced before more complex and difficult ones. The second criterion for ranking is immediacy. For this criterion, subgoals are ranked according to prerequisite tasks—that is, what tasks must be done before others can be achieved.

The sequencing of subgoals in order of complexity is based on learning principles called *shaping* and *successive approximations*. Shaping helps someone learn a small amount at a time, with reinforcement or encouragement for each task completed successfully. Gradually, the person learns the entire amount or achieves the overall result through these day-to-day learning experiences that successively approximate the overall outcome.

Steps in Identifying and Sequencing Subgoals. Identification and arrangement of subgoals are critical to the client's success with the outcome goal. The following steps are involved in this process.

First, the client identifies the *first* step he or she must take—that is, the first things that need to be done to move in the desired direction. The first step will be some action that is both comfortable and achievable (Gambrill, 1977).

Second, if the client progresses satisfactorily on the first step, additional intermediate steps that bridge the gap between the first step and the terminal goal are identified and ranked. (If the client does not progress on the first step, discuss this issue and consider revising the initial step.) Effective intermediate steps are ones that build on existing client assets and resources, do not conflict with the client's value system, are decided on and owned by the client, and represent immediate, daily, or short-term actions rather than weekly, distant, long-term actions (see also Bandura & Simon, 1977; Carkhuff, 1993; Gambrill, 1977).

There is no hard and fast rule concerning the number of intermediate steps identified, other than ensuring that the gap between adjacent steps is not too great. Each successive step gradually begins where the last step left off. The counselor also needs to make sure that each intermediate step requires only one basic action or activity by the client; if two or more activities are involved, it is usually better to make this two separate steps (Carkhuff, 1993).

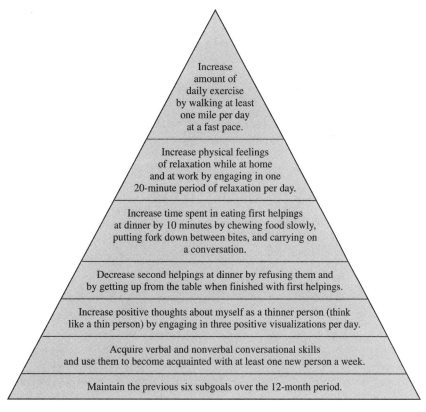

Terminal goal: To think, feel, and look like a thin person by losing 40 pounds over a 12-month period

Weekly goal: To think, feel, and look like a thin person by losing 1 pound per week

FIGURE 10-2. Goal pyramid subgoals for example client

As we mentioned earlier, intermediate steps are ranked on two aspects:

1. Degree of difficulty and complexity—"Which is easier; which is harder?" Less complex and demanding tasks are ranked ahead of others.
2. Immediacy—"What do I need to do before I can do this?" Prerequisite tasks are ranked before others.

Ranked steps are then filled in on the goal pyramid—usually in pencil, because in the process of moving through the hierarchy, the subgoals may need to be modified or rearranged.

Third, after all the steps have been identified and sequenced, the client starts to carry out the actions represented by the subgoals, beginning with the initial step and moving on. Usually, it is wise not to attempt a new subgoal until the client has successfully completed the previous one on the pyramid. Progress made on initial and subsequent steps provides useful information about whether the gaps between steps are too large or just right and whether the sequencing of steps is appropriate. As the subgoals are met, they become part of the client's current repertory that can be used in additional change efforts toward the terminal goals.

An example may clarify the process of identifying and sequencing subgoals for a client. Suppose you are working with a person who wishes to lose 40 pounds. Losing 40 pounds is not a goal that anyone can accomplish overnight or without small requisite changes along the way. First, the person will need to determine a reasonable weekly level of weight loss, such as 1 to 2 pounds. Next, you and the client will have to determine the tasks the client will need to complete to lose weight. These tasks can be stated as subgoals that the client can strive to carry out each day, starting with the initial subgoal, the one that feels most comfortable and easy to achieve, and working the way up the pyramid as each previous step is successfully completed and maintained in the client's repertory.

Although weight loss generally may include action steps such as alteration of eating levels, increase in physical activity, restructuring of cognitions and belief systems, and development of additional social skills, the exact tasks chosen by two or more clients who want to lose weight may be quite different. The therapist should be sensitive to such differences and not impose his or her method for solving the problem (such as weight loss) on the client. Similarly, each client will have a different idea of how subgoals will be best sequenced. In Figure 10-2 we illustrate how one particular client sequenced her identified subgoals on the goal pyramid. This client's rationale was that if she increased exercise and relaxation *first,* it would be easier to alter eating habits. For her, more difficult and also less immediate goals included restructuring her thoughts about herself and her body image and developing social skills necessary to initiate new relationships. This last subgoal she viewed as the most difficult one because her weight served partly to protect her from social distress situations. After all six subgoals are achieved, the final subgoal is simply to keep these actions going for at least a 12-month period. At the bottom of the pyramid, it would be important to discuss with her ways in which she can maintain the subgoals over an extended period of time. Note, too, in this example that her terminal goal is stated in positive terms—not "I don't want to be fat" but "I do want to feel, think, and look like a thin person." The subgoals represent actions she will take to support this desired outcome. Also note that all the subgoals are stated in the same way as the terminal outcome goal—with the definition of the behaviors to be changed, the level of change, and the conditions or circumstances of change so that the client knows what to do, where, when, with whom, and how much or how often.

Interview Leads for Identifying Subgoals.   In identifying subgoals, the counselor uses leads similar to the following to help the client determine appropriate subgoals or action steps:

"How will you go about doing [or thinking, feeling] this?"
"What exactly do you need to do to make this happen?"
"Let's brainstorm some actions you'll need to take to make your goal work for you."
"What have you done in the past to work toward this goal? How did it help?"
"Let's think of the steps you need to take to get from where you are now to where you want to be."

The counselor is always trying to encourage and support client participation and responsibility in goal setting, remembering that clients are more likely to carry out changes that they originate. Occasionally, however, after using leads like the ones above, some clients are unable to specify any desirable or necessary action steps or subgoals. The counselor may then have to use prompts, either asking the client to think of other people who have similar problems and to identify their strategies for action or providing a statement illustrating an example or model of an action step or subgoal (Dixon & Glover, 1984).

Interview Leads for Sequencing Subgoals.   General leads to use to sequence and rank subgoals include the following:

"What is your first step?"
"What would you be able to do most easily?"
"What would be most difficult?"
"What is most important for you to do now? least important?"
"How could we order these steps to maximize your success in reaching your goal?"
"Let's think of steps you need to take to get from where you are now to where you want to be—and arrange them in an order from what seems easiest to you to the ones that seem hardest for you."
"Can you think of some things you need to do before some other things as you make progress toward your goal?"

Identifying Obstacles

To ensure that the client can complete each subgoal step successfully, it is helpful to identify any *obstacles* that could interfere. Obstacles may include overt and/or covert behavior. *Potential* obstacles to check out with the client include the presence or absence of certain feelings or mood states, thoughts, beliefs and perceptions, other people, and situations or events. Another obstacle could be lack of knowledge or skill. Identification of lack of knowledge or skill is important if the client needs information or training before the subgoal action can be attempted (Gambrill, 1977).

Interview Leads to Identify Obstacles.   Clients are often not very aware of any factors that might interfere with completing a subgoal and may need prompts from the counselor, such as the following ones, to identify obstacles:

"What are some obstacles you may encounter in trying to take this action?"
"What people [feelings, ideas, situations] might get in the way of getting this done?"
"What or who might prevent you from working on this activity?"

LEARNING ACTIVITY 34 *Identifying and Sequencing Subgoals*

This learning activity is an extension of Learning Activity #33. Continue to work with the same goal you selected and defined in that activity. In this activity, we suggest the following steps:

1. First, identify the initial step you need to take to move toward your goal. Ask yourself:

   a. Does this step move directly toward the goal?
   b. Will I be able to take this step comfortably and successfully?

   Unless the answer to both questions is yes, redefine your initial step. Your initial step then becomes your first subgoal.

2. Generate a list of intermediate steps that bridge the gap between this initial step and your desired outcome. Consider where you will be after completing each step and what comes next. These intermediate steps become remaining subgoals. For each step in your list, consider:

a. Does the step represent only one major activity?
b. Is the step based on my existing assets and resources?
c. Does the step support most of my major values and beliefs?
d. Is the step something I want for myself?
e. Does the step represent immediate, short-term activities rather than distant, long-term ones?

If the answer to any of these questions for any particular step is no, rework that step.

3. Write each step on a 3 × 5 index card. Then assign a numerical rating to each card for degree of difficulty, using a 0 to 100 scale: 0 = least complex, least difficult; 100 = most complex, most difficult.

4. Rank-order your steps or subgoals by arranging your cards in order, starting with the one closest to zero and ending with the one closest to 100. This represents the sequence in which you will complete your subgoals.

"In what ways might you have difficulty completing this task successfully?"

"What information or skills do you need in order to complete this action effectively?"

Occasionally the counselor may need to point out apparent obstacles that the client overlooks. If significant obstacles are identified, a plan to deal with or counteract the effects of these factors needs to be developed. Often this is similar to an "antisabotage plan," in which the counselor and client try to predict ways in which the client might not do the desired activity and then work around the possible barriers. For example, suppose you explore obstacles with the client we described in the earlier section who wants to lose weight and become thin. Perhaps in exploring the first subgoal, walking at least one mile a day, she identifies two things that might keep her from doing this: rain and being alone. Ways to prevent these two factors from interfering with her walking might be to use an indoor facility and to walk with an exercise partner.

Identifying Resources

The next step is to identify *resources*—factors that will help the client complete the subgoal task effectively. Like obstacles, resources include overt and covert behaviors.

Potential resources to explore include feelings, thoughts and belief systems, people, situations, information, and skills. In this step, the counselor tries to help clients identify already present or developed resources that, if used, can make completion of the subgoal tasks more likely and more successful.

A specific resource involved in attaining desired outcomes is referred to by Bandura (1977, 1989) and others as *self-efficacy*. Self-efficacy has been the focus of a great deal of research in the last 15 years (Maddux, 1995). It involves two types of personal expectations that affect goal achievement; one is an *outcome* expectation and the other is an *efficacy* expectation. The outcome expectation has to do with whether and how much a client believes that engaging in particular behaviors will in fact produce the desired results. For example, in the weight-loss example given in Figure 10-2, the outcome expectation would be the extent to which this client believes that the actions represented by these subgoals will help her lose 40 pounds. The efficacy expectation has to do with the client's level of confidence regarding how well she or he can complete the behaviors necessary to reach the desired results. Again, in our example of the weight-loss client, the efficacy expectation has to do with the degree of confidence with which the client approaches the subgoals and his or her accompanying actions and behaviors. Of these two types of personal expectations, the *efficacy* ones seem to be the most important, thus strengthening the client's overall sense of perceived efficacy. (We discuss self-efficacy further in Chapter 20 on self-management interventions.)

Interview Leads for Identifying Resources. Possible leads include the following:

"Which resources do you have available to help you as you go through this activity [or action]?"
"What specific feelings [or thoughts] are you aware of that might make it easier for you to _____?"
"What kind of support system do you have from others that you can use to make it easier to _____?"
"What skills [or information] do you possess that will help you do _____ more successfully?"
"How much confidence do you have that you can do what it will take for you to _____?"
"To what extent do you believe that these actions will help you do _____?"

For example, the weight-loss client might identify a friend or other social support as a resource she could use for daily exercise, as well as her belief that exercise promotes wellness and good feelings.

EVALUATION OF OUTCOME GOALS

Therapists increasingly are under pressure to measure results with their clients. Mental health care providers who want to be chosen to be on panels of managed care companies are also under increasing pressure to provide data about their effectiveness. By some estimates, up to 88% of managed care companies are using outcome measurements to determine which providers serve on their panels (Hutchins, 1995). Practitioners who fail to measure outcomes will more than likely not be included. Also, and more important, there are ethical reasons to evaluate client outcomes. For example, the preamble of the 1992 APA Ethical Code states that "psychologists work to develop a valid and reliable body of scientific knowledge based on research" (p. 1599).

Doing outcome evaluations also guides treatment planning. As Resnick (1995) notes, "Outcomes measurements should be viewed as a mechanism to enhance and expand psychotherapeutic and psychological treatment. Using research-based measures to guide treatment, we can develop treatment paradigms, inform patients of likely outcome, and even suggest alternative courses of treatment" (p. 2). Such progress is already at work; Division 12 of APA has published

a task force report describing a number of empirically validated therapeutic treatments for specific problem areas. The Practice Directorate of APA is working toward building a national practice-research network that will consider both client characteristics and diagnostic categories and will enable therapists to compare their treatments and outcomes to those used with similar clients across the United States. We describe this further in the following chapter on treatment planning and selection. Many practitioners resist making outcome evaluations because of the time and complexity involved. As practitioners ourselves, we understand this concern, yet we hold fast to the premise that most of us cannot avoid putting off outcome evaluations too much longer. However, with the necessary monitoring and paperwork required by many insurance companies and also by government regulations, we have attempted to suggest ways to accomplish this that are brief, simple, and inexpensive. At the outset we wish to note that doing evaluations of client outcomes is not synonymous with empirical research and is subject to less rigor, less control, and more bias. Still, research and practice are and need to be interrelated (Kanfer, 1990; Marten & Heimberg, 1995; Strupp, 1989; Stricker, 1992). Research results need to guide practice and practice results need to inform research. Goldfried and Wolfe (1996) note that three sources of information currently exist to inform therapists about how to proceed clinically. These three sources include

1. Basic research on clinical disorders
2. Psychotherapy process research
3. Psychotherapy outcome research

Moreover, there are research designs such as single-case design (Galassi & Gersh, 1993; Jones, 1993) and case study designs (Moras, Telfer, & Barlow, 1993) that lend themselves particularly well to practice evaluations. Even so, objections and myths will always be raised about practice evaluations. Among these, Kazdin (1993) has noted the following:

1. Evaluating outcomes is unnecessary because change can be seen when it occurs.
2. Evaluation is an interference with effective treatment.
3. Clients are individuals and evaluation obscures individuality.
4. Evaluation minimizes or trivializes client's problems.
5. Client problems change so a systematic evaluation cannot be done.

Kazdin (1993) does an excellent job of refuting these myths and also points out that a data-based evaluation of outcomes is an important way to overcome inherent limitations and biases of human judgment (p. 40). In addition, our accountability as practitioners depends on "responsive evaluation" (p. 42).

RESPONSE DIMENSIONS OF OUTCOMES: WHAT TO MEASURE

The goal behaviors are evaluated by having the client assess the amount or level of the defined behaviors. Four dimensions commonly used to measure the direction and level of change in goal behaviors are frequency, duration, magnitude (intensity), and occurrence. A client may use one or a combination of these response dimensions, depending on the nature of the goal, the method of assessment, and the feasibility of obtaining particular data. The response dimensions should be individualized, particularly because they vary in the time and effort they cost the client.

Frequency

Frequency reflects the number (how many, how often) of overt or covert behaviors and is determined by obtaining measures of each occurrence of the goal behavior. Frequency counts are typically used when the goal behavior is discrete and of short duration. Panic episodes and headaches are examples of behaviors that can be monitored with frequency counts. Frequency data can also be obtained from comments written in a diary or daily journal. For example, the number of positive (or negative) self-statements before and after each snack or binging episode, reported in a daily diary, can be tabulated.

Sometimes frequency counts should be obtained as percentage data. For example, knowing the number of times a behavior occurred may not be meaningful unless data are also available on the number of *possible* occurrences of the behavior. Use percentage measures when it is important to determine the number of opportunities the client has to perform the target behavior as well as the number of times the behavior actually occurs. For example, data about the number of times an overweight client consumes snacks might be more informative if expressed as a percentage. In this example, the denominator would reflect the number of opportunities the person had to eat snacks; the numerator would indicate the number of times the person actually did snack. The advantage of percentage scores is that they indicate whether the change is a function of an actual increase or decrease in the number of times the response occurs or merely a function of an increase or decrease in the number of opportunities to perform the behavior. Thus, a percentage

score may give more accurate and more complete information than a simple frequency count. However, when it is hard to detect the available opportunities, or when it is difficult for the client to collect data, percentage scores may not be useful.

Duration

Duration reflects the length of time a particular response or collection of responses occurs. Duration measurement is appropriate whenever the goal behavior is not discrete and lasts for varying periods (Ciminero, Nelson & Lipinski, 1977, p. 198). Thinking about one's strengths for a certain period of time, the amount of time spent on a task or with another person, the period of time for depressive thoughts, and the amount of time that anxious feelings lasted are examples of behaviors that can be measured with duration counts. Duration may also involve time *between* an urge and an undesired response, such as the time one holds off before lighting up a cigarette or eating an unhealthy snack. It also can involve *elapsed* time between a covert behavior such as a thought or intention and an actual response, such as the amount of time before a shy person speaks up in a discussion (sometimes this is referred to as *latency*).

Frequency counts, percentage scores, and duration, can be obtained in one of two ways: continuous recording or time sampling. If the client can obtain data *each time* he or she engages in the goal behavior, then the client is collecting data continuously. Sometimes continuous recording is impossible, particularly when the goal behavior occurs very often or when its onset and termination are hard to detect. In such cases, a time-sampling procedure may be more practical. In time sampling, a day is divided into equal time intervals—90 minutes, 2 hours, or 3 hours, for example. The client keeps track of the frequency or duration of the goal behavior only during randomly selected intervals. In using time sampling, data should be collected during at least three time intervals each day and during *different* time intervals each day, so that representative and unbiased data are recorded. One variation of time sampling is to divide time into intervals and indicate the presence or absence of the target behavior for each interval in an "all or none" manner (Mahoney & Thoresen, 1974, p. 31). If the behavior occurred during the interval, a *yes* would be recorded; if it did not occur, a *no* would be noted. Time sampling is less precise than continuous recordings of frequency or duration of a behavior. Yet it does provide an estimate of the behavior and may be a useful substitute in monitoring high-frequency or nondiscrete target responses (Mahoney & Thoresen, 1974).

Intensity

The intensity or degree of the goal behavior can be assessed with a rating scale. For example, intensity of anxious feelings can be measured with ratings of 1 (not anxious) to 5 (panic). Cronbach (1990) suggests three ways of decreasing sources of error frequently associated with rating scales. First, the therapist should be certain that what is to be rated is well defined and specified in the client's language. For example, if a client is to rate depressed thoughts, the counselor and client specify, with examples, what constitutes depressed thoughts (such as "Nothing is going right for me," "I can't do anything right"). These definitions should be tailored to each client, on the basis of an analysis of the client's problem behavior and contributing conditions. Second, rating scales should be designed that include a description for each point on the scale. For example, episodes of anxious feelings in a particular setting can be rated on a 5-point scale, with 1 representing little or no anxiety, 2 equal to some anxiety, 3 being moderately anxious, 4 representing strong anxious feelings, and 5 indicating very intense anxiety. Third, rating scales should be unidirectional, starting with 0 or 1. Negative points (points below 0) should not be included. In addition, the therapist should consider the range of points in constructing the scale. There should be at least 4 points and no more than 7. A scale of less than 4 points may limit a person's ability to discriminate, whereas a scale that includes more than 7 points may not produce reliable ratings by the client because too many discriminations are required.

Occurrence

Occurrence refers to presence or absence of target behaviors. Checklists can be used to rate the occurrence of behaviors. They are similar to rating scales. The basic difference is the type of judgment one makes. On a rating scale, a person can indicate the degree to which a behavior is present; a checklist simply measures the presence or absence of a behavior. Checklists describe a cluster or collection of behaviors that a client may demonstrate.

Characteristics measuring occurrence of behaviors are very useful in providing the sort of outcome evaluations required by third-party payers who not only want to see data regarding symptoms/distress relief but also evidence of back-to-work functioning and productivity. For example, one checklist may measure the presence or absence of certain classroom behaviors of a teacher who has been treated for stress. Another may measure the presence or absence of behaviors at work for a person treated for substance abuse. A checklist can also be used in conjunction with frequency and duration counts and rating scales. As evaluative tools, checklists may be very useful, particularly when the reference points on the list are clearly defined and are representative of the particular performance domain being assessed.

CHOOSING OUTCOME MEASURES: HOW TO MEASURE OUTCOMES

A major factor facing the typical practitioner in evaluating outcomes is how to choose the most useful measures of outcome that are (1) psychometrically sound, (2) pragmatic and easy to use, (3) relevant to the client's stated goals, (4) relevant to the client's level of functioning, and (5) relevant to the client's gender and culture. From a recent survey of the literature, we have concluded that three pressing areas should be considered in evaluation of counseling outcomes: One is *client satisfaction*—that is, how satisfied the client is with you, the therapist, and with the overall results of therapy. A second is *clinical outcomes significance*—that is, whether there has been enough improvement toward the specified goals and in the overall functioning of the client to move the client from a dysfunctional to a functional level. (One study [Ankuta & Abeles, 1993] produced the interesting finding that a group of clients who showed such clinically significant symptom changes reported greater satisfaction and benefit from therapy than clients who changed only moderately or not at all.) The third variable is *cost effectiveness*—that is, whether the benefits of a particular therapy procedure outweigh the costs and if there are several available treatment procedures for a particular problem, which one is most time efficient. We discuss cost effectiveness in greater detail in the following chapter on treatment planning and selection.

In addition to examining these three areas of outcome, therapists should also evaluate outcomes at three different times:

1. At *intake* or the beginning of the therapy
2. At *termination* or the conclusion of therapy
3. At *follow-up*—some point in time one month to one year after termination of therapy

Client Satisfaction Measures

Several brief and easy to use measures of client satisfaction are available (see Table 10-5). An additional source of client satisfaction measures can be found in Fischer and Corcoran's (1994) excellent compendium, *Measures for Clinical*

TABLE 10-5. Measures of client satisfaction and clinical outcomes

Client satisfaction	Frequently used clinical outcome measures (Hutchins, 1995)
CSQ-8 Client Satisfaction Questionnaire.* Copyright © 1979, 1989, 1990. Used with written permission of Clifford Attkisson, Ph.D. UCSF, San Francisco. *Strupp Post-Therapy Client Questionnaire* (Strupp, Fox, & Lessler, 1969). *Session Evaluation Questionnaire* (Stiles & Snow, 1984).	**BASIS-32 (Behavior and Symptom Identification Scale)**—Also includes a depression scale. Available as a self-report or clinician report form. Information kit costs $20. Offers permission for individual providers and groups to copy free of charge. Licensing fee for software developers or consultants. Scoring available for a charge through the software company. Contact Evaluation Service Unit, McLean Hospital, 115 Mill St., Belmont, MA 02178-9106. (617) 855-2425. **GAF (Global Assessment of Functioning Scale)**—Part of the *DSM-IV Manual.* Contact American Psychiatric Press, Inc., American Psychiatric Association, 1400 K St., NW, Washington, DC 20005. (202) 682-6324. To order: 1 (800) 368-5777. **SF-36**—A health status questionnaire (also available in a 12-question format, SF-12). Cost: $25, which includes copyrighting instructions. Contact Medical Outcomes Trust, PO Box 1917, Boston, MA 02205-8516. Attn: Linda Birdsong. (617) 636-8098. **SCL-90-R**—Symptom checklist. Cost: $78 for a hand-scoring starter kit for 50 patients. Contact National Computer Systems, NCS Assessments, PO Box 1416, Minneapolis, MN 55440. (612) 939-5000. **Addiction Severity Index**—Free. Contact Treatment Research Institute, One Commerce Square, Suite 1020, 2005 Market Street, Philadelphia, PA 19103. (800) 335-9874. **SUDDS (Substance Use Disorder Diagnosis Schedule)**—Cost: $48.75, includes 25 schedules with hand-scoring kit. New version based on *DSM-IV* coming soon. Contact New Standards, Inc., 1080 Montreal Ave., Suite 300, St. Paul, MN 55116. (800) 755-6299. **Beck Depression Inventory-II**—The Psychological Corporation, 555 Academic Court, San Antonio, TX 78204-2498. (210) 299-1061; 1-800-228-0752. **Katz Activities of Daily Living**—Martin M. Katz, Clinical Research Branch, Resource Guild, National Institute of Mental Health, Chevy Chase, MD 20203. **FACES III**—David H. Olson. Family Social Science Department, University of Minnesota, 290 McNeal Hall, 1985 Beuford Avenue, St. Paul, MN 55108. (612) 625-7250. **Child Behavior Checklist and Youth Self-Report**—T. M. Achenbach, University Associates in Psychiatry, c/o Child Behavior Checklist, 1 South Prospect St., Burlington, VT 05401. (802) 656-8313. FAX: (802) 656-2602.

*The CSQ-8 is copyrighted and there is a cost for its use. Contact Clifford Attkisson, Ph. D., Professor of Medical Psychology, University of California at San Francisco, Millberry Union, 200 West, San Francisco, California 94143.

Practice. Client satisfaction with therapy reflects one aspect of the meaningfulness of change. In Strupp and Hadley's (1977) tripartite model of mental health, therapists' outcome is assessed by three parties: the clients, the therapist, and the public. Ankuta and Abeles (1993) note that "client satisfaction has a high level of face validity. It would be difficult to argue for an outcome criterion that could not stand the test of client satisfaction" (p. 73).

A measure of client satisfaction is considered a minimum part of an overall outcome evaluation; that is, it is better to have an indicator of client satisfaction than nothing at all. However, we believe this sort of measure needs to be supplemented with additional ones as some data suggest client satisfaction is tied to clinically significant change (Ankuta & Abeles, 1993) and also because some mental health consumers will not accurately judge the quality of service. As Eckert (1994) notes, "In the context of certain psychological conditions (e.g., paranoia, negative transference), perceptual distortion inevitably brews. In these cases, although the quality of client care may be excellent, the patient's evaluation may nevertheless be unfavorable" (p. 5).

Another limitation of client satisfaction is that this measure does not assess the particular outcomes specified by the counselor and client in the goal-setting process. For these reasons, Eckert (1994) argues, and we agree, that measures of clinical outcome should always be "multidimensional" (p. 5). In the next section we discuss ways to assess specific therapeutic goals and outcomes.

Goal-Related Outcome Measures

For the *researcher*, pragmatics is usually the least important criterion in choosing outcome measures. For the practitioner, who is faced with numbers of treatment plans to write and who may be paid according to the number of clients seen, pragmatics is usually the *most* important outcome. Given this consideration, we suggest two ways to proceed that are especially suitable to the demands of the typical practitioner's work environment. First, consider using the goal-attainment scaling system (Kiresuk & Sherman, 1968) (or some variation thereof) that we discussed in the section on defining goals. The GAS, which has been used extensively with a number of different client populations, simply requires you to take an outcome goal for any given client and construct a weighted scale of descriptions ranging from the most favorable result (+2) to the least favorable result (−2) with a sum of 0 representing the expected level of improvement. (You may wish to refer to Table 10-3 for a review of this process.) Using these numerical scores, levels of change in outcome goals can then be quantified by transforming these scores to standardized T scores (see also Kiresuk & Sherman, 1968.) As Marten and Heimberg (1995) observe, the advantage of the GAS system lies in "its ability to allow practitioners a means of evaluating treatment outcome in an idiographic manner by examining client changes within specific problem areas in a concrete and systematized way" (p. 49). The GAS is also relatively free of therapist bias about client "impairment." The GAS is constructed while the therapist and client are developing outcome goal statements and prior to the beginning of any treatment protocol or counseling intervention. A particular advantage of this method is that the GAS can be constructed within, rather than outside, the therapy situation and with the client's participation and assistance. It, therefore, requires almost no extra

TABLE 10-6. Forty-four problem areas addressed in *Measures for Clinical Practice* (Fischer & Corcoran, 1994)

abuse	narcissism
addiction and alcoholism	obsessive-compulsive
anger and hostility	parent-child relationship
anxiety and fear	perfectionism
assertiveness	phobias
beliefs	problem-solving
children's behaviors/ problems	procrastination
couple relationship	psychopathology and psychiatric symptoms
death concerns	rape
depression and grief	satisfaction with life
eating problems	schizotypal symptoms
family functioning	self-concept and esteem
geriatric	self-control
guilt	self-efficacy
health issues	sexuality
identity	smoking
interpersonal behavior	social functioning
locus of control	social support
loneliness	stress
love	suicide
marital/couple relationship	treatment satisfaction
mood	substance abuse

time from the therapist, reinforces the client's role in the change process, and also provides a quantifiable method of assessing outcome. Note that while the GAS example we present in Table 10-3 uses *frequency* as the level of change, you can also construct a GAS using duration and intensity as indicators of change. The GAS rating system is useful because it describes each point on the rating scale in a quantifiable way to eliminate ambiguity.

In addition to a numerical system of rating results, such as the GAS, the therapist should consider giving clients some sort of paper and pencil rapid-assessment instrument (RAI) they can complete to provide self-report data about symptom reduction and level of improvement. The most comprehensive compendium of such measures we have found is in two volumes of Fischer and Corcoran's work (1994). These authors provide descriptions of 318 RAIs for 44 problem areas (see Table 10-6), including a profile of the instrument, and information regarding norms, scoring, reliability, validity, and availability.

In selecting an RAI, it is important to choose one that has good psychometric properties; is easy to read, use, and score; and relates directly to the client's identified problems and symptoms at intake and stated outcome goals of counseling. For example, the Beck Depression Inventory-II (BDI-II) (Beck, Steer, & Brown, 1996) is frequently used with clients who are depressed at intake and want to become less depressed. It would not, however, be a suitable choice for someone who presents with a different problem such as anxiety, anger control, or marital dissatisfaction. Also, many of the psychometric properties of these RAIs have been normed on Caucasian clients, often middle-class college students. Caution must be applied when using some of these instruments with clients of color. If you cannot find a culturally relevant RAI, perhaps you should use the goal attainment scaling system instead. When possible, therapists should use RAIs that are as relevant as possible to the client's culture and gender. For example, if you are measuring the stress level of an African American woman, it is better to use the African-American Women's Stress Scale (Watts-Jones, 1990) than some other stress measure.

In addition to RAIs such as the Beck or the Lehrer and Woolfolk (1982) anxiety inventory, there are also RAIs that tap into general levels of client functioning and a range of symptoms. For example, recall from our discussion of the *DSM-IV* in Chapter 8 that there is a global assessment of functioning (GAF) scale used to report on Axis V. This scale can help to measure the impact of treatment and to track the progress of clients in global terms with the use of a single measure. It is rated on a 0 to 100 scale regarding the client's psychological, social, and occupational functioning but not physical or environmental limitations. It can be used for ratings at both intake and termination. (Refer to Chapter 8 for a visual display of this scale.) RAIs that cover a wide range of symptoms include such measures as the Derogatis (1983) Symptom Checklist SCL-90-R and the Behavior and Symptom Identification Scale (Basis-32). The more general RAIs are limited in that they do not directly measure the behaviors specified in the client's outcome goals. However, an advantage of a broader multifaceted RAI has to do with *clinical significance*. Within the last 10 years there has been a major movement in the evaluation of mental health services toward criteria that reflect *clinically* significant outcomes (Jacobson, Follette, & Revenstorf, 1984; Goldfried, Greenberg, & Marmer, 1990; Jacobson & Truax, 1991).

Clinical significance refers to the effect of a counseling treatment intervention on a single client and denotes improvement in client symptoms and functioning at a level comparable to that of the client's healthy peers (Jacobson et al., 1984). Clinically significant criteria are applied to each psychotherapy case and answer not only the question of the degree to which the client makes the specified changes, but

also address the *relevance* of those changes to the client's overall functioning and life style. In doing so, these criteria are considered to have a high degree of social validity (Kazdin, 1977). Increasingly, third-party payers are asking for outcome documentation from practitioners that not only includes symptom reduction but also the client's level of functioning in settings such as work, school, and home (Cavaliere, 1995).

Another outcome measure that can be used in a very idiographic way to evaluate goal-related client outcomes is self-monitoring. Self-monitoring is a process of observing and recording aspects of one's own covert or overt behavior. In evaluating goal behaviors, a client uses self-monitoring to collect data about the amount (frequency, duration, intensity) of the goal behaviors. The monitoring involves not only noticing occurrences of the goal behavior but also recording them with paper and pencil, mechanical counters, timers, or electronic devices. (In Chapter 20 we discuss the use of self-monitoring as a treatment/intervention strategy.) Self-monitoring is often necessary to collect information about frequency, duration, or intensity of target behaviors specified in a GAS. For example, in the GAS in Table 10-3, the client would need to self-monitor the frequency of colitis attacks during the day. Self-monitoring has a number of advantages as a way to collect data about client progress toward goal behaviors. Self-monitoring, or an ongoing account of what happens in a person's daily environment, can have more concurrent validity than some other data-collection procedures. In other words, self-monitoring may produce data that more closely approximate the goals of counseling than such measures as the rapid assessment inventories we discussed earlier. Moreover, the predictive validity of self-monitoring may be superior to that of other measurement methods, with the exception of direct observation. Self-monitoring can also provide a thorough and representative sample of the ongoing behaviors in the client's environment. And self-monitoring is relatively objective. Finally, self-monitoring is flexible. It can be used to collect data on covert and physiological indexes of change as well as more observable behaviors. However, self-monitoring should not be used by clients who cannot engage in observation because of the intensity or diagnostic nature of their problems or because of medication. Also, self-monitoring may have a high-cost for the client in the time and effort required to make such frequent records of the goal behavior.

Another consideration is that some clients may not monitor as accurately as others. To increase the accuracy and thus the reliability of a client's self-monitoring, consider these guidelines:

1. The behaviors to be observed should be defined clearly so there is no ambiguity about what is to be observed and reported. The counselor should spell out clearly the procedures for *what, where, when, how,* and *how long* to report the behaviors.

2. Any definition of the target behavior should be accompanied by examples so the client can discriminate instances of the observed behavior from instances of other behaviors. For example, a client should be instructed to observe particular responses associated with aggressiveness instead of just recording "aggressive behavior." In this case, the client might observe and record instances of raising his or her voice above a conversational tone, hitting another person, or using verbal expressions of hostility.

3. If possible, clients should be instructed to self-monitor in vivo when the behavior occurs rather than self-recording at the end of the day, when the client is dependent on retrospective recall. The accuracy of client reports may be increased by having the client record the target behaviors immediately rather than after a delay.

When to Measure Outcomes

There are several times during which a counselor and client can measure progress toward the goal behaviors. Generally, the client's performance should be assessed before counseling, during counseling or during application of a counseling strategy, immediately after counseling, and some time after counseling at a follow-up. Repeated measurements of client change may provide more precise data than only two measurement times, such as before and after counseling. Moreover, third-party payers are increasingly requiring that practitioners track outcomes for clients over a longer period of time, including and up to a one-year follow-up contact (Cavaliere, 1995). In some states, for example, Medicaid now requires five assessment outcome measures to be collected every 90 days.

A pretreatment assessment measures the goal behaviors before treatment. This period is a reference point against which therapeutic change in the client's goal behavior can be compared during and after treatment. The length of the pretreatment period can be three days, a week, two weeks, or longer. One criterion for the length of this period is that it should be long enough to contain sufficient data points to serve as a *representative sample* of the client's behavior. For example, with a depressed client, the counselor may ask the client to complete self-ratings on mood intensity at several different times during the next week or two. The counselor

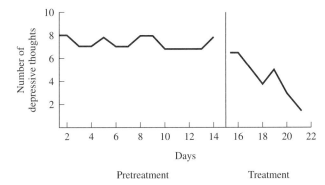

FIGURE 10-3. Graph of depressive thoughts

may also ask the client to self-monitor instances or periods of depression that occur during this time. This situation is graphed in Figure 10-3. Note that several data points are gathered to provide information about the stability of the client's behavior. If the counselor does not collect any pretreatment data at all, determining the magnitude or amount of change that has occurred after counseling is difficult because there are no precounseling data for comparison. The pretreatment period serves as a baseline of reference points showing where the client is before any treatment or intervention. In Figure 10-3, for example, you can see the number of depressed thoughts that occur for this client in the first two weeks of counseling but before any treatment plan or intervention is used. In the third week, when the counselor introduces and the client works with an intervention such as cognitive restructuring, the client continues to record the number of depressed thoughts. These records, as well as any recording after counseling, can be compared to the baseline data.

Baseline measurement may not be possible with all clients, however. The client's problem concern may be too urgent or intense to allow you to take the time for gathering baseline data. Baseline measurement is often omitted in crisis counseling. In a less urgent type of example, if a client reports "exam panic" and is faced with an immediate and very important test, the counselor and client will need to start working to reduce the test anxiety at once. In such cases, the treatment or counseling strategy must be applied immediately.

Assessment During Treatment Strategies

In therapy, the therapist and client monitor the effects of a designated treatment on the goal behaviors after collecting pretreatment data and selecting a counseling treatment strategy. The monitoring during the treatment phase of counseling is conducted by the continued collection of data about the client's performance of the goal behavior. The same types of data collected during the pretreatment period are collected during treatment. For example, if the client self-monitored the frequency and duration of self-critical thoughts during the pretreatment period, this self-monitoring would continue during the application of a helping strategy. Or if self-report inventories of the client's social skills were used during the pretreatment period, these same methods would be used to collect data during treatment. Data collection during treatment is a feedback loop that gives both the counselor and the client important information about the usefulness of the selected treatment strategy and the client's demonstration of the goal behavior. Figure 10-4 shows the data of a client who self-monitored the number of depressed thoughts experienced during the application of two counseling strategies: cognitive restructuring and stimulus control.

Posttreatment: Assessment after Counseling

At the conclusion of a counseling treatment strategy or at the conclusion of counseling, the counselor and client should conduct a posttreatment assessment to indicate in what ways and how much counseling has helped the client achieve the desired results. Specifically, the data collected during a posttreatment assessment are used to compare the client's demonstration and level of the goal behavior after treatment with the data collected during the pretreatment period and during treatment.

The posttreatment assessment may occur at the conclusion of a counseling strategy or at the point when counseling is terminated—or both. For instance, if a counselor is using cognitive restructuring (Chapter 15) to help a client reduce depressed thoughts, the counselor and client could collect data on the client's level of depressed thoughts after they have finished working with the cognitive restructuring strategy. This assessment may or may not coincide with counseling termination. If the counselor plans to use a second treatment strategy, then data would be collected at the conclusion of the cognitive restructuring strategy and prior to the use of another strategy. This example is depicted in Figure 10-4. Note that the client continued to self-monitor the number of depressed thoughts between the cognitive restructuring and stimulus control treatments and after stimulus control, when counseling was terminated.

Ideally, the same types of measures used to collect data before and during counseling should be employed in the posttreatment assessment. For instance, if the client self-monitored depressed thoughts before and during treatment,

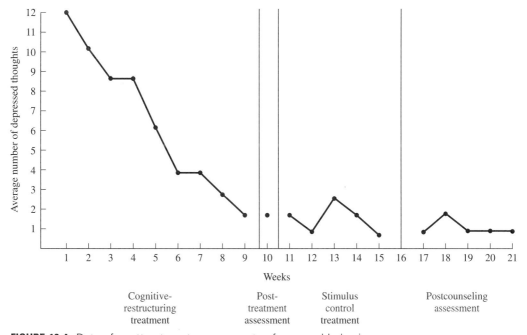

FIGURE 10-4. Data of posttreatment assessments of one goal behavior

then, as Figure 10-4 illustrates, self-monitoring data would also be collected during posttreatment assessment. If the counselor had also employed questionnaires such as the Beck Depression Inventory-II during the pretreatment period and treatment, these measures would be used during post-treatment data collection as well.

Follow-up Assessment

After the counseling relationship has terminated, some type of follow-up assessment should be conducted. A counselor can conduct both a short-term and a long-term follow-up. A short-term follow-up can occur three to six months after therapy. A long-term follow-up would occur one month to a year (or more) after counseling has been terminated. Generally the counselor should allow sufficient time to elapse before conducting a follow-up to determine the extent to which the client is maintaining desired changes without the counselor's assistance.

There are several reasons for conducting follow-up assessments. First, a follow-up can indicate the counselor's continued interest in the client's welfare. Second, a follow-up provides information that can be used to compare the client's performance of the goal behavior before and after counseling. Another important reason for conducting a follow-up is to determine how well the client is able to perform the goal behaviors in his or her environment without relying on the support and assistance of counseling. In other words, a follow-up can give some clues about the effectiveness of the counseling treatment or how well it has generalized to the client's actual environment. This reflects one of the most important evaluative questions to be asked: Has counseling helped the client to maintain desired behaviors and to prevent the occurrence of undesired ones in some self-directed fashion? As we have indicated, more and more third party payers are expecting therapists to provide some follow-up data on client outcomes.

Both short-term and long-term follow-ups can take several forms. The kind of follow-up a counselor conducts often depends on the client's availability to participate in a follow-up and the time demands of each situation. Here are some ways a follow-up can be conducted:

1. Invite the client in for a follow-up interview. The purpose of the interview is to evaluate how the client is coping with respect to his or her "former" concern or problem. The interview may also involve client demonstrations of the goal behavior in simulated role plays.
2. Mail an inventory or questionnaire to the client, seeking

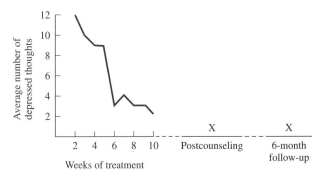

FIGURE 10-5. Graph of self-monitored follow-up data

information about her or his current status in relation to the original problem or concern. Be sure to include a stamped, self-addressed envelope.
3. Send a letter to the client asking about the current status of the problem.
4. Telephone the client for an oral report. The letter and telephone report could also incorporate the goal-attainment scale rating if that was used earlier.

These examples represent one-shot follow-up procedures that take the form of a single interview, letter, or telephone call. A more extensive (and sometimes more difficult to obtain) kind of follow-up involves the client's engaging in self-monitoring or self-rating of the goal behavior for a designated time period, such as two or three weeks. Figure 10-5 shows the level of depressed thoughts of a client at a six-month follow-up.

Some of the measures we described earlier are better suited for particular times. RAIs and client self-monitoring can be done most easily at pre- and posttreatment. Measures of client satisfaction and the GAF or similar ratings are more easily collected at posttreatment and follow-up measurement.

MODEL EXAMPLE: THE CASE OF JOAN

Throughout this book, we provide model illustrations of processes and strategies with our hypothetical client, Joan. We now provide sample illustrations of how a counselor could use the outcome evaluation procedures described in this chapter with Joan.

Joan's first outcome goal has been defined as "to acquire and demonstrate a minimum of four initiating skills, including four of the following: (1) asking questions and making reasonable requests, (2) expressing differences of opinion, (3) expressing positive feelings, (4) expressing negative feelings, and (5) volunteering answers or opinions in at least four situations a week with her parents and in math class." This overall goal can be assessed by a GAS (see model dialogue that follows). We would also give Joan a client satisfaction questionnaire at the end of counseling and at a six-month follow-up. Four subgoals are associated with the first goal:

1. To decrease anxiety associated with anticipation of failure in math class and rejection by parents from self-ratings of intensity of 70 to 50 on a 100-point scale during the next two weeks of treatment.
2. To increase positive self-talk and thoughts that "girls are capable" in math class and other competitive situations from zero or two times a week to four or five times a week over the next two weeks during treatment.
3. To increase attendance in math class from two or three times a week to four or five times a week during treatment.
4. To increase verbal participation and initiation in math class and with her parents from none or once a week to three or four times a week over the next two weeks during treatment. Verbal participation is defined as asking and answering questions with teacher or parents, volunteering answers or offering opinions, or going to the chalkboard.

The therapist and Joan need to establish the method of evaluating progress on each of the four subgoals and to determine the response dimension for each subgoal. We would recommend that a global self-report assessment inventory of anxiety (Lehrer & Woolfolk, 1982) be used to measure reductions in anxiety (subgoal 1). This RAI would be given at pre, post and follow-up times. For subgoal 1, Joan could also self-monitor intensity of anxiety associated with anticipated failure in math class and rejection from parents, on a scale ranging from 0 to 100. For subgoal 2, we recommend that Joan self-monitor her self-talk during math class and other competitive situations. She could be instructed to write (in vivo) her self-talk on note cards during baseline and treatment. Subgoal 3 is to increase her attendance in math class. Joan could keep a record of the days she attended class, and these data could be verified from the teacher's attendance records, with Joan's permission. For subgoal 4, verbal participation and initiation in math class and with her parents could be self-monitored (in vivo) by recording each time Joan performed these verbal responses. We illustrate this evaluation process in the following continuation of our model dialogue.

MODEL DIALOGUE: THE CASE OF JOAN

The counselor will start by **summarizing** *the previous session and by checking out whether Joan's goals have changed in any way. Goal setting is a flexible process, subject to revisions along the way.*

1. *Counselor:* OK, Joan, last week when we talked, just to recap, you mentioned three areas you'd like to work on. Is this still accurate, or have you added anything or modified your thinking in any way about these since we last met?
   Client: No, that's still where I want to go right now. And I still want to start with this whole issue of expressing myself and not worrying so much about what other people think. I've been doing a lot of thinking about that this week, and I think I'm really starting to see how much I let other people use me as a doormat and also control my reactions in a lot of ways.
2. *Counselor:* Yes, you mentioned some of those things last week. They seem to be giving you some incentive to work on this.
   Client: Yeah. I guess I'm finally waking up and starting to feel a little fed up about it.

In the next response, the counselor explains the **purpose** *of the session and solicits Joan's opinion, again giving her another opportunity to express her opinions.*

3. *Counselor:* Last week I mentioned it might be helpful to map out a plan of action. How does that sound to you? If it isn't where you want to start, let me know.
   Client: No, I do. I've been kind of gearing up for this all week.

In the next two responses, the counselor helps Joan define the **behaviors** *associated with the goal—what she will be doing, thinking, or feeling.*

4. *Counselor:* OK, last week when we talked about this area of change, you described it as wanting to express yourself more without worrying so much about the reactions of other people. Could you tell me what you mean by expressing yourself—to make sure we're on the same wavelength?
   Client: Well, like in math class, I need to volunteer the answers when I know them, and volunteer to go to the board. Also, I hesitate to ask questions. I need to be able to ask a question without worrying if it sounds foolish.
5. *Counselor:* OK, you've mentioned three specific ways in math class you want to express yourself. [makes a note] I'm going to jot these down on this paper in case we want to refer to these later. Anything else you can think of in math class?
   Client: No, not really. The other situation I have trouble with is with my folks.

"Trouble" is not very specific. Again, a **behavioral definition** *of the goal is sought in the next two responses.*

6. *Counselor:* OK, "trouble." Again, can you describe exactly how you'd like to express yourself when interacting with them?
   Client: Well, kind of the same stuff. Sometimes I would like to ask them a question. Or ask for help or something. But I don't. I almost never express my ideas or opinions to them, especially if I don't agree with their ideas. I just keep things to myself.
7. *Counselor:* So you'd like to be able to make a request, ask a question, talk about your ideas with them, and express disagreement.
   Client: Yeah. Wow—sounds hard.

*In the following response, the counselor prepares Joan for the idea of working in **small steps** and also explores **conditions (situations, people)** associated with the goal.*

8. *Counselor:* It will take some time, and we won't try to do everything at once. Just one step at a time. Now, you've mentioned two different situations where these things are important to you—math class and with your parents. I noticed last week there was one time when you were reluctant to express your opinion with me. Is this something you want to do in any other situations or with any other people?

   Client: Well, sure—in that it does crop up occasionally at different times or with different people, even friends. But it's worse in math and at home. I think if I could do it there, I could do it anywhere.

*In the next response, the counselor starts to explore the **level** or **desired extent of change**. The counselor is attempting to establish a **current base rate** in order to know how much the client is doing now.*

9. *Counselor:* OK, I'm making a note of this, too. Now could you estimate how often you express yourself in the ways you've described above *right now,* either in math class or with your folks, during the course of an average week?

   Client: You mean how many times do I do these things during the week?

10. *Counselor:* Yes.

    Client: Probably almost never—at least not in math class or at home. Maybe once or twice at the most.

*The counselor continues to help Joan identify a **practical** and **realistic** level of change.*

11. *Counselor:* OK, if you express yourself in one of these ways once or twice a week now, how often would you like to be doing this? Think of something that is also practical or realistic.

    Client: Mm. Well, I don't really know. Offhand, I'd guess about four or five times a week—that's about once a day, and that would take a lot of energy for me to be able to do that in these situations.

*At this point, the **behavior, conditions, and level of change** for this terminal goal are defined. The counselor asks Joan whether this definition is the way she wants it.*

12. *Counselor:* OK, I'll make a note of this. Check what I have written down—does it seem accurate? [Joan reads what is listed as the first terminal goal on her goal chart at the end of this dialogue.] (See p. 256.)

    Client: Yeah. Gosh, that sort of makes it official, doesn't it?

*This is the second time Joan has expressed a little hesitation. So the counselor will check out **her feelings** about the process in the next response.*

13. *Counselor:* Yes. What kinds of feelings are you having about what we're doing now?

    Client: Kind of good and a little scared too. Like do I really have what it takes to do this?

*In the next response, the counselor responds to Joan's concern. Joan has already selected this goal, yet if she has difficulty later on moving toward it, they will need to explore **what her present behavior is trying to protect**.*

14. *Counselor:* One thing I am sure of is that you do have the resources inside you to move in this direction as long as this is a direction that is important to you and one that is not necessary to protect any parts of you. If we move along and you feel stuck, we'll come back to this and see how you keep getting stuck at this point.

    Client: OK.

*Next the counselor introduces and **develops a goal attainment scale for this particular goal (responses 15–20)**.*

15. *Counselor:* OK, let's spend a little time talking about this particular goal we've just nailed down. I'd like to set up some sort of a system with you in which we rank what you expect or would like to happen, but also the best possible and the least possible success we could have with this goal. This gives us both a concrete target to work toward. How does that sound?

    Client: OK. What exactly do we do?

16. *Counselor:* Well, let's start with a range of numbers from −2 to +2. It will look like this: (draw the following numbers on a sheet of paper).

    −2
    −1
     0
    +1
    +2

    Now "0" represents an acceptable and expected level, one you could live with. What do you think it would be in this case?

    Client: Well, I guess maybe doing this at least twice a week—at least once in math and once at home. That would be better than nothing.

17. *Counselor:* OK, we'll put that down opposite "0". If that's acceptable, would you say that four per week you mentioned earlier is more than expected?

    Client: Yup.

18. *Counselor:* OK, let's put that down for +1 and how about eight per week for +2—that's sort of in your wildest dreams. How is that?

    Client: OK, let's go for it.

19. *Counselor:* OK, now if two per week is acceptable for you, what would be less than acceptable—one or zero?

    Client: Well, one is better than nothing—zero is just where I am now.

20. *Counselor:* So, let's put one per week with −1 and none per week with −2. Now we have a way to keep track of your overall progress on this goal. Do you have any questions about this?

    Client: No, it seems pretty clear.

(For a visual diagram of this GAS, see Table 10-7.)

*In the next response, the counselor introduces the idea of **subgoals** that represent small action steps toward the terminal goal and asks*

TABLE 10-7. Goal-attainment scale for Joan's first outcome goal

Date: 2/5/96	Frequency of verbal initiating skills
−2 Most unfavorable outcome	Zero per week
−1 Less than expected success	One per week—either with parents or in math class
0 Expected level	Two per week—at least one with parents and at least one in math class
+1 More than expected success	Four per week—at least two with parents and two in math class
+2 Best expected success	Eight or more per week—at least four with parents and four in math class

Joan to identify the **initial step.**

21. *Counselor:* OK. Another thing that I think might help with your apprehension is to map out a plan of action. What we've just done is to identify exactly where you want to get to—maybe over the course of the next few months. Instead of trying to get there all at once, let's look at different steps you could take to get there, with the idea of taking just one step at a time, just like climbing a staircase. For instance, what do you think would be your first step—the first thing you would need to do to get started in a direction that moves directly to this end result?
Client: Well, the first thing that comes to my mind is needing to be less uptight. I worry about what other people's reactions will be when I do say something.

In the next two responses, the counselor helps Joan define the **behavior and conditions associated with this initial subgoal,** *just as she did previously for the terminal goal.*

22. *Counselor:* OK, so you want to be less uptight and worry less about what other people might think. When you say other people, do you have any particular persons in mind?
Client: Well, my folks, of course, and to some degree almost anyone that I don't know too well or anyone like my math teacher, who is in a position to evaluate me.
23. *Counselor:* So you're talking mainly about lessening these feelings when you're around your folks, your math teacher, or other people who you think are evaluating you.
Client: Yes, I think that's it.

In response 24, the counselor is trying to establish the **current level of intensity** *associated with Joan's feelings of being uptight. She does this by using an* **imagery assessment** *in the interview.* **Self-reported ratings of intensity** *are used in conjunction with the imagery.*

24. *Counselor:* OK, now I'm going to ask you to close your eyes and imagine a couple of situations that I'll describe to you. Try to really get involved in the situation—put yourself there. If you feel any nervousness, signal by raising this finger. [Counselor shows Joan the index finger of her right hand and describes three situations—one related to parents, one related to math class, and one related to a job interview with a prospective employer. In all three situations, Joan raises her finger. After each situation, the counselor stops and asks Joan to rate the intensity of her anxiety on a 100-point scale, 0 being complete calm and relaxation and 100 being total panic.]

After the imagery assessment for base rate, the counselor asks Joan to **specify a desired level of change for this subgoal.**

25. *Counselor:* OK. Now, just taking a look at what happened here in terms of the intensity of your feelings, you rated the situation with your folks about 75, the one in math class 70, and the one with the employer 65. Where would you like to see this drop down to during the next couple of weeks?
Client: Oh, I guess about a 10.

It is understandable that someone with fairly intense anxiety wants to get rid of it, and it is possible to achieve that within the next few months. However, such goals are more effective when they are **immediate rather than distant.** *In the next two responses, the counselor asks Joan to* **specify a realistic level of change** *for the immediate future.*

26. *Counselor:* OK, that may be a number to shoot for in the next few months, but I'm thinking that in the next three or four weeks the jump from, say, 70 to 10 is pretty big. Does that gap feel realistic or feasible?
Client: Mm. I guess I was getting ahead of myself.
27. *Counselor:* Well, it's important to think about where you want to be in the long run. I'm suggesting three or four weeks mainly so you can start to see some progress and lessening of intensity of these feelings in a fairly short time. What number seems reasonable to you to shoot for in the short run?
Client: Well, maybe a 45 or 50.

At this point, the counselor and Joan continue to **identify other subgoals or intermediate steps** *between the initial goal and the terminal outcome.*

28. *Counselor:* OK, that seems real workable. Now, we've sort of mapped out the first step. Let's think of other steps between this first one and this end result we've written down here. [Counselor and client continue to generate possible action steps. Eventually they select and define the remaining three shown on Joan's goal chart.]

Assuming the remaining subgoals are selected and defined, the next step is to **rank-order** *or* **sequence** *the subgoals and* **list them in order on the goal pyramid.**

29. *Counselor:* OK, we've got the first step, and now we've mapped out three more. Consider where you will be after this first step is completed—which one of these remaining steps

comes next? Let's discuss it, and then we'll fill it in, along with this first step, on this goal pyramid, which you can keep so you know exactly what part of the pyramid you're on and when. [Counselor and Joan continue to rank-order subgoals, and Joan lists them in sequenced order on a goal pyramid.]

In response 30, the counselor points out that **subgoals may change in type or sequence.** *The counselor then shifts the focus to exploration of potential* **obstacles for the initial subgoal.**

30. *Counselor:* OK, now we've got our overall plan mapped out. This can change, too. You might find later on you may want to add a step or reorder the steps. Now, let's go back to your first step—decreasing these feelings of nervousness and worrying less about the reactions of other people. Since this is what you want to start working on this week, can you think of anything or anybody that might get in your way or would make it difficult to work on this?

Client: Not really, because it is mostly something inside me. In this instance, I guess I am my own worst enemy.

31. *Counselor:* So you're saying there don't seem to be any people or situations outside yourself that may be obstacles. If anyone sets up an obstacle course, it will be you.

Client: Yeah. Mostly because I feel I have so little control of those feelings.

The client has identified herself and her perceived lack of control over her feelings as **obstacles.** *Later on, the counselor will need to help Joan select and work with one or two* **intervention strategies.**

32. *Counselor:* So one thing we need to do is to look at ways you can develop skills and know-how to manage these feelings so they don't get the best of you.

Client: Yup. I think that would help.

Joan's goal chart

Terminal goal	*Related subgoals*
Goal 1	
(B) to acquire and demonstrate a minimum of four different initiating skills (asking a question or making a reasonable request, expressing differences of opinion, expressing positive feelings, expressing negative feelings, volunteering answers or opinions, going to the board in class) (C) in her math class and with her parents (L) in at least 4 situations a week	1. (B) to decrease anxiety associated with anticipation of failure (C) in math class or rejection by parents (L) from a self-rated intensity of 70 to 50 on a 100-point scale during the next 2 weeks 2. (B) to restructure thoughts or self-talk by replacing thoughts that "girls are dumb" with "girls are capable" (C) in math class and in other threatening or competitive situations (L) from 0–2 per day to 4–5 per day 3. (B) to increase attendance (C) at math class (L) from 2–3 times per week to 4–5 times per week 4. (B) to increase verbal participation skills (asking and answering questions, volunteering answers or offering opinions) (C) in math class and with her parents (L) from 0–1 times per day to 3–4 times per day
Goal 2	
(B) to increase positive perceptions about herself and her ability to function effectively (C) in competitive situations such as math class (L) by 50% over the next 3 months	1. (B) to eliminate conversations (C) with others in which she discusses her lack of ability (L) from 2–3 per week to 0 per week 2. (B) to increase self-visualizations in which she sees herself as competent and adequate to function independently (C) in competitive situations or with persons in authority (L) from 0 per day to 1–2 per day 3. (B) to identify negative thoughts and increase positive thoughts (C) about herself (L) by 25% in the next 2 weeks
Goal 3	
(B) to acquire and use five different decision-making skills (identifying an issue, generating alternatives, evaluating alternatives, selecting the best alternative, and implementing action) (C) at least one of which represents a situation in which significant others have given her their opinion or advice on how to handle it (L) in at least two different situations during a month	1. (B) to decrease thoughts and worry about making a bad choice or poor decision (C) in any decision-making situation (L) by 25% in the next 2 weeks 2. (B) to choose a course of action and implement it (C) in any decision-making situation (L) at least once during the next 2 weeks

Key: B = behavior; C = condition; L = level.

In the next response, the counselor explores **existing resources and support systems** *that Joan might use to help her work effectively with the subgoal.*

33. *Counselor:* OK, that's where I'd like to start in just a minute. Before we do, can you identify any people who could help you with these feelings—or anything else you could think of that might help instead of hinder you?

    Client: Well, coming to see you. It helps to know I can count on that. And I have a real good friend who is sort of the opposite of me, and she's real encouraging.

"Social allies" *are an important principle in effecting change, and the counselor uses this word in response 34 to underscore this point.*

34. *Counselor:* OK, so you've got at least two allies.

    Client: Yeah.

In response 35, the counselor helps Joan develop a way to continue the **self-ratings of the intensity** *of her nervous feelings. This gives both of them a* **benchmark to use in assessing progress and reviewing** *the adequacy of the first subgoal selected.*

35. *Counselor:* The other thing I'd like to mention is a way for you to keep track of any progress you're making. You know how you related these situations I described today? You could continue to do this by keeping track of situations in which you feel uptight and worry about the reactions of others—jot down a brief note about what happened and then a number on this 0 to 100 scale that best represents how intense your feelings are at that time. As you do this and bring it back, it will help both of us see exactly what's happening for you on this first step. Does that sound like something you would find agreeable?

    Client: Yeah—do I need to do it during the situation or after?

Clients are more likely to do **self-ratings or self-monitoring if it falls into their daily routine,** *so this is explored in the next response.*

36. *Counselor:* What would be most practical for you?

    Client: Probably after, because it's hard to write in the middle of it.

The counselor encourages Joan to make her notes soon after the situation is over. **The longer the gap, the less accurate or reliable the data might be.**

37. *Counselor:* That's fine; try to write it down as soon as it ends or shortly thereafter, because the longer you wait, the harder it will be to remember. Also, to get an idea of your current level of anxiety, I'd like you to take a few minutes at the end of our session today to fill out a paper and pencil form that asks you some questions about what you may be feeling. There are no right or wrong answers on this, so it's not like a test! I'll be asking you to do this again later on in the year and then we can compare the scores on both. Would you feel comfortable doing this?

Before the session ends, they have to work on the **obstacle** *Joan identified earlier—that she is her own worst enemy because her feelings are in control of her. At this point, some of the real nuts and bolts of counseling begin. The counselor will need to select an intervention strategy or theoretical approach to use with Joan in this*

instance. *(One such option, cognitive restructuring, is described and modeled in Chapter 15.)*

38. *Counselor:* Now, let's go back to that obstacle you mentioned earlier—that your feelings are in control of you . . .

Upon exploring this issue, the counselor will select treatment interventions to use with Joan. The process of treatment planning is described in the following chapter.

SUMMARY

The primary purpose of identifying goals is to convey to the client the responsibility and participation she or he has in contributing to the results of counseling. Without active client participation, counseling may be doomed to failure. The selection of goals should reflect *client* choices. Effective goals are consistent with the client's cultural identity and belief systems. The counselor's role is mainly to use leads that facilitate the client's goal selection. Together, the counselor and client explore whether the goal is owned by the client, whether it is realistic, and what advantages and disadvantages are associated with it. However, some value judgments by both counselor and client may be inevitable during this process. If the client selects goals that severely conflict with the counselor's values or exceed the counselor's level of competence, the counselor may decide to refer the client or to renegotiate this goal. If counselor and client agree to pursue the identified goals, these goals must be defined clearly and specifically. Throughout this process the client moves along a change continuum, ranging from contemplation to preparation and finally to action.

Well-defined goals make it easier to note and assess progress and also aid in guiding the client toward the desired goal(s). Goals are defined when you are able to specify the overt and covert behaviors associated with the goal, the conditions or context in which the goal is to be carried out or achieved, and the level of change. After the outcome goal is defined, the counselor and client work jointly to identify and sequence subgoals that represent intermediate action steps and lead directly to the goal. Obstacles that might hinder goal attainment and resources that may facilitate goal attainment are also explored.

As you go through the process of helping clients develop goals, remember that goal setting is a dynamic and flexible process (Bandura, 1969). Goals may change or may be redefined substantially as counseling progresses.

For these reasons, the outcome goals should always be viewed as temporary and subject to change. Client resistance at later stages in counseling may be the client's way of saying that the original goals need to be modified or redefined. The counselor who is committed to counseling to meet the client's needs will remember that, at any stage, the client always has the prerogative of changing or modifying directions.

As outcome goals are agreed on and defined, ways to evaluate progress toward these goals are also incorporated into the counseling process.

POSTEVALUATION

Part One

Objective One asks you to identify a problem for which you identify, define, and evaluate an outcome goal. Use the Goal-Setting Worksheet below for this process. You can obtain feedback by sharing your worksheet with a colleague, supervisor, or instructor.

Goal-Setting Worksheet

1. Identify a concern or problem.
2. State the desired outcome of the problem.
3. Assess the desired outcome (#2 above):
   a. Does it specify what you want to do? (If not, reword it so that you state what you want to do instead of what you don't want to do.)
   b. Is this something you can see (hear, grasp) yourself doing every time?
4. In what ways is achievement of this goal important to you? to others?
5. What will achieving this goal require of you? of others?
6. To what extent is this goal something you want to do? something you feel you should do or are expected to do?
7. Is this goal based on:
   ___rational, logical ideas?
   ___realistic expectations, ideas?
   ___irrational ideas and beliefs?
   ___logical thinking?
   ___perfectionistic standards (for self or others)?
8. How will achieving this goal help *you?* help significant others in your life?
9. What problems could achieving this goal create for you? for others?
10. If the goal requires someone else to change, is not realistic or feasible, is not worthwhile, or poses more disadvantages than advantages, rework the goal. Then move on to #11.
11. Specify exactly *what* you will be
    a. doing_____
    b. thinking_____
    c. feeling_____
    as a result of goal achievement. Be specific.
12. Specify your goal definition in #11 by indicating:
    a. *where* this will occur:_____
    b. *when* this will occur:_____
    c. *with whom* this will occur:_____
    d. *how much or how often* this will occur:_____

13. Develop a plan that specifies *how* you will attain your goal by identifying action steps included in the plan.
    a. _____
    b. _____
    c. _____
    d. _____
    e. _____
    f. _____
    g. _____
    h. _____
    i. _____
    j. _____
    k. _____
    l. _____
14. *Check* your list of action steps:
    a. Are the gaps between steps small? If not, add a step or two.
    b. Does each step represent only one major activity? If not, separate this one step into two or more steps.
    c. Does each step specify what, where, when, with whom, and how much or how often? If not, go back and define your action steps more concretely.
15. Use the goal pyramid on the next page to sequence your list of action steps, starting with the easiest, most immediate step on the top and proceeding to the bottom of the pyramid by degree of difficulty and immediacy or proximity to the goal.
16. For each action step (starting with the first), brainstorm what could make it difficult to carry out or could interfere with doing it successfully. Consider feelings, thoughts, places, people, and lack of knowledge or skills. Write down the obstacles in the space provided on page 259.
17. For each action step (starting with the first), identify existing resources such as feelings, thoughts, situations, people and support systems, information, skills, beliefs, and self-confidence that would make it more likely for you to carry out the action or complete it more successfully. Write down the resources in the space provided on page 259.
18. Identify a way to monitor your progress for completion of each action step.
19. Develop a plan to help yourself maintain the action steps once you have attained them.

(continued)

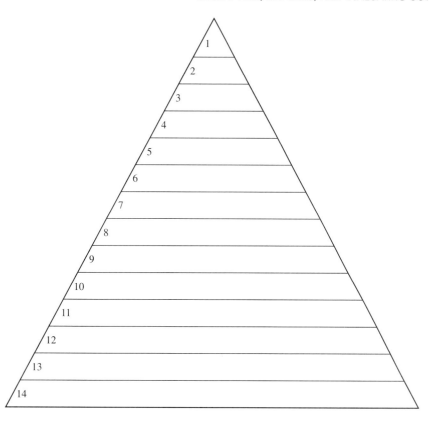

Goal pyramid

Obstacles   Action Steps   Resources

_____   1. _____
_____   2. _____
_____   3. _____
_____   4. _____
_____   5. _____
_____   6. _____
_____   7. _____
_____   8. _____
_____   9. _____
_____   10. _____
_____   11. _____
_____   12. _____

Part Two

In this part of the postevaluation, we describe a client case, the case of Manuel. Assuming that Manuel is your client, describe the steps you would go through to help him identify, define, and evaluate desired actions, given his stated problem (Chapter Objective Two). Try to include at least 11 of the 14 steps or categories we described in this chapter for identifying, defining, and evaluating outcome goals. You can do this orally with a partner, in small groups, or by yourself. If you do it by yourself, you may want to jot down your ideas in writing for someone else to look at. Feedback follows the postevaluation.

The Case of Manuel

Manuel Tréjos is a 52-year-old Latino man who is the manager of an advertising firm. He has been with the firm for 17 years and has another 12 years to go before drawing a rather lucrative retirement package. Over the last three

(continued)

POSTEVALUATION (continued)

years, however, Manuel has become increasingly dissatisfied with his job specifically and with the work world in general. He says he feels as if he would want nothing better than to quit but he and his wife want to build up a nest egg for their son's two young children, as their grandchildren are very important to them. He realizes that if he left the firm now, he would lose many of his retirement benefits. Manuel defines his problem as feeling burned out with the nine-to-five job routine. He wishes to have more free time but he also feels a sense of great responsibility as the head of his family to provide financial security.

Part Three

According to Objective Three, you will be able to demonstrate, in an interview setting, at least 11 of the 14 categories for identifying, defining, and evaluating client

outcome goals. We suggest you complete this part of the postevaluation in triads. One person assumes the role of the counselor and demonstrates helping the client with the goal-setting process in a 20- or 30-minute interview. The second person takes the role of the client. You may wish to portray the role and problem described for Manuel Tréjos in Part Two (if you choose to present something unfamiliar to the counselor, be sure to inform the counselor of your identified problem or concern before you begin). The third person assumes the role of the observer. The observer may act as the counselor's alter ego and cue the counselor during the role play if necessary. The observer also provides feedback to the counselor after the interview, using the Interview Checklist for Identifying, Defining, and Evaluating Goals that follows as a guide. If you do not have access to an observer, tape-record your interview so you can assess it yourself.

Review Checklist for Identifying, Defining, and Evaluating Goals

Directions: Determine which of the following leads or questions the counselor demonstrated. Check each counselor question or lead demonstrated. Also check whether the client answered the content of the counselor's question. Example counselor leads and questions are provided next to each item of the checklist. These are only suggestions; be alert to other responses used by the counselor.

Scoring	Category of information	Examples of counselor leads or questions	Client response
_____Yes_____No	1. Explain the purpose and importance of having goals or positive outcomes to the client.	"Let's talk about some areas you would like to work on during counseling. This will help us to do things that are related to what you want to accomplish."	___indicates understanding
_____Yes_____No	2. Determine *positive* changes desired by client ("I would like" versus "I can't").	"What would you like to be doing [thinking, feeling] differently?" "Suppose some distant relative you haven't seen for a while comes here in several months. What would be different then from the way things are now?" "Assuming we are successful, what do you want to be doing, or how would this change for you?" "In what ways do you want to benefit from counseling?"	___identifies goal in positive terms
_____Yes_____No	3. Determine whether the goal selected represents changes owned by the client rather than someone else ("I want to talk to my mom without yelling at her," rather than "I	"How much control do you have to make this happen?" "What changes will this require of you?" "What changes will this require someone else to make?" "Can this be achieved without the help of anyone else?"	___identifies who owns the goal

(continued)

Scoring	Category of information	Examples of counselor leads or questions	Client response
	want my mom to stop yelling at me").	"To whom is this change most important?"	
_____Yes_____No	4. Identify advantages (positive consequences) to client and others of goal achievement.	"In what ways is it worthwhile to you and others to achieve this?" "How will achieving this goal help you?" "What problems will continue for you if you don't pursue this goal?" "What are the advantages of achieving this change—for you? others?" "Who will benefit from this change—and how?"	__identifies advantages
_____Yes_____No	5. Identify disadvantages (negative consequences) of goal achievement to client and others.	"What new problems in living might achieving this goal pose for you?" "Are there any disadvantages to going in this direction?" "How will achieving this change affect your life in adverse ways?" "How might this change limit or constrain you?"	__identifies disadvantages
_____Yes_____No	6. Identify whether, as the helper, you can pursue counseling with this particular client.	"These are things I am able to help you work with." "I feel uncomfortable working with you on this issue because of my own personal values [or lack of knowledge]. I'd like to give you the names of several other counselors. . . ." "This would be hard for me to help you with because it seems as if you're choosing something that will restrict you and not give you any options. Let's talk more about this."	__responds to counselor's decision
_____Yes_____No	7. Identify what the client will be doing, thinking, or feeling in a concrete, observable way as a result of goal achievement ("I want to be able to talk to my mom without yelling at her," rather than "I want to get along with my mom").	"What do you want to be able to do [think, feel] differently?" "What would I see you doing [thinking, feeling] after this change?" "Describe a good and a poor example of this goal."	__specifies overt and covert behaviors
_____Yes_____No	8. Specify under what conditions and what situations goals will be achieved: when, where, and with whom ("I want to be able to talk to my mom at home during the next month without yelling at her").	"When do you want to accomplish this goal?" "Where do you want to do this?" "With whom?" "In what situations?"	__specifies people and places

(continued)

POSTEVALUATION   **(continued)**

Scoring	Category of information	Examples of counselor leads or questions	Client response
_____Yes_____No	9. Specify how often or how much client will do something to achieve goal ("I want to be able to talk to my mom at home during the next month without yelling at her at least once a day").	"How much [or how often] are you doing this [or feeling this way] now?" "What is a realistic increase or decrease?" "How much [or how often] do you want to be doing this to be successful at your goal?" "What amount of change is realistic, considering where you are right now?"	___specifies amount
_____Yes_____No	10. Identify and list small action steps the client will need to take to reach the goal (that is, break the big goal down into little subgoals). *List of Action Steps* 1. 2. 3. 4. 5. 6. 7. 8. 9. 10.	"How will you go about doing [thinking, feeling] this?" "What exactly do you need to do to make this happen?" "Let's brainstorm some actions you'll need to take to make your goal work for you." "What have you done in the past to work toward this goal?" "How did it help?" "Let's think of the steps you need to take to get from where you are now to where you want to be."	___lists possible action steps
_____Yes_____No	11. Sequence the action steps on the goal pyramid (a hierarchy) in terms of a. degree of difficulty (least to most) b. immediacy (most to least immediate)	"What is your first step?" "What would you be able to do most easily?" "What would be most difficult?" "What is your foremost priority?" "What is most important for you to do soon? least important?" "How could we order these steps to maximize your success in reaching your goal?" "Let's think of the steps you need to take to get from where you are now to where you want to be and arrange them in an order from what seems easiest to you to the ones that seem hardest." "Can you think of some things you need to do before some other things as you make progress toward this outcome?"	___assists in rank-ordering

Least difficult, most immediate

1
2
3
4
5
6
7
8
9
10

Most difficult, least immediate

(continued)

Scoring	Category of information	Examples of counselor leads or questions	Client response
_____Yes_____No	12. Identify any people, feelings, or situations that could prevent the client from taking action to reach the goal.	"What are some obstacles you may encounter in trying to take this action?" "What people [feelings, ideas, situations] might get in the way of getting this done?" "In what ways could you have difficulty completing this task successfully?" "What do you need to know to take this action?" or "What skills do you need to have?"	___identifies possible obstacles
_____Yes_____No	13. Identify any resources (skill, knowledge, support) that client needs to take action to meet the goal.	"What resources do you have available to help you as you complete this activity?" "What particular thoughts or feelings are you aware of that might make it easier for you to _____?" "What kind of support system do you have from others that you can use to make it easier to _____?" "What skills [or information] do you possess that will help you do this more successfully?"	___identifies existing resources and supports
_____Yes_____No	14. Develop a plan to evaluate progress toward the goal.	"Would it be practical for you to rate these feelings [count the times you do this] during the next two weeks? This information will help us determine the progress you are making." "Let's discuss a way you can keep track of how easy or hard it is for you to take these steps this week."	___agrees to monitor in some fashion

Part Four

Objective Four asks you to conduct an outcome evaluation with yourself, another person, or a client, specifying *what* will be measured, *when* it will be measured, and *how*. You may wish to do so using the following guidelines.

1. Define and give examples of a desired goal behavior.
2. Specify what type of data you or the other person will collect (for example, verbal reports, frequency, duration, ratings, or occurrence of the behavior).
3. a. Identify the methods to be used to collect these data (such as self-monitoring, inventories, self-rating).
   b. For *each* method to be used, describe very specifically the instructions you or the client would need to use this method.
4. Collect data on the goal behaviors at least several times before implementing any treatment (change) strategy.
5. Following pretreatment data collection, implement some treatment strategy for a designated time period.
6. Collect data after treatment, and compare changes with the pretreatment data. Graph your data or visually inspect the data. What do your data suggest about the effectiveness of your treatment? Share your results with a partner, colleague, or instructor.

SUGGESTED READINGS

Ankuta, G., & Abeles, N. (1993). Client satisfaction, clinical significance, and meaningful change in psychotherapy. *Professional Psychology, 24,* 70–74.

Bandura, A. (1989). Human agency in social cognitive theory. *American Psychologist, 44,* 1175–1184.

Bisman, C. (1994). *Social work practice.* Pacific Grove, CA: Brooks/Cole.

Clement, P. W. (1994). Quantitative evaluation of 26 years of private practice. *Professional Psychology, 25,* 173–176.

Csikszentmihalyi, M. (1990). *Flow: The psychology of optimal experience.* New York: Harper Perennial.

Fischer, J., & Corcoran, K. (1994). *Measures for clinical practice* (2nd ed.). New York: Free Press.

Galassi, J. P., & Gersh, T. L. (1993). Myths, misconceptions, and missed opportunity: Single-case designs and counseling psychology. *Journal of Counseling Psychology, 40,* 525–531.

Goldfried, M. R., Greenberg, L. S., & Marmer, C. (1990). Individual psychotherapy: Process and outcome. *Annual Review of Psychology, 41,* 659–688.

Gordon, J., & Shontz, F. (1990). Representative case research: A way of knowing. *Journal of Counseling and Development, 69,* 62–66.

Hanna, F. J., & Puhakka, K. (1991). When psychotherapy works: Pinpointing an element of change. *Psychotherapy, 28,* 598–607.

Hanna, F. J., & Ritchie, M. H. (1995). Seeking the active ingredients of psychotherapeutic change: Within and outside the context of therapy. *Professional Psychology, 26,* 176–183.

Jacobson, N. S., & Truax, P. (1991). Clinical significance: A statistical approach to defining meaningful change in psychotherapy research. *Journal of Consulting and Clinical Psychology, 59,* 12–19.

Kanfer, F. H. (1990). The scientist-practitioner connection: A bridge in need of constant attention. *Professional Psychology, 21,* 264–270.

Kazdin, A. E. (1993). Evaluation in clinical practice: Clinically sensitive and systematic methods of treatment delivery. *Behavior Therapy, 24,* 11–45.

Lambert, M. J., Ogles, B., & Masters, K. (1992). Choosing outcome assessment devices: An organizational and conceptual scheme. *Journal of Counseling & Development, 70,* 527–532.

Lyddon, W. J. (1990). First- and second-order change . . . Implications for rationalist and constructivist cognitive therapies. *Journal of Counseling & Development, 69,* 122–127.

Marten, P. A., & Heimberg, R. G. (1995). Toward an integration of independent practice and clinical research. *Professional Psychology, 26,* 48–53.

Moras, K., Telfer, L. A., & Barlow, D. H. (1993). Efficacy and specific effects data on new treatments: A case study strategy with mixed anxiety-depression. *Journal of Counseling and Clinical Psychology, 61,* 412–420.

Oz, S. (1995). A modified balance-sheet procedure for decision making in therapy: Cost-cost comparison. *Professional Psychology, 25,* 78–81.

Prochaska, J. O., DiClemente, C., & Norcross, J. (1992). In search of how people change: Applications to addictive behaviors. *American Psychologist, 47,* 1102–1114.

Satterfield, W. A., Buelow, S. A., Lyddon, W. J., & Johnson, T. J. (1995). Client stages of change and expectations about counseling. *Journal of Counseling Psychology, 42,* 476–478.

Snyder, C. R. (1995). Conceptualizing, measuring and nurturing hope. *Journal of Counseling and Development, 73,* 355–360.

Stricker, G. (1992). The relationship of research to clinical practice. *American Psychologist, 47,* 543–549.

Strupp, H. H. (1988). What is therapeutic change. *Journal of Cognitive Psychotherapy, 2,* 75–82.

Strupp, H. H. (1989). Psychotherapy: Can the practitioner learn from the research? *American Psychologist, 44,* 717–724.

Todd, D. M., Jacobus, S. I., & Boland, J. (1992). Uses of a computer database to support research-practice integration in a training clinic. *Professional Psychology, 23,* 52–58.

VandenBos, G. R. (Ed.). (1996). Outcome assessment of psychotherapy (Special Issue). *American Psychologist, 51* (10).

FEEDBACK
POSTEVALUATION

Part Two

1. First, explain to Manuel the *purpose and importance* of developing goals.
2. Help Manuel state the goal or desired change in *positive terms.*
3. Help Manuel determine whether the goal he is moving toward represents *changes owned by him* and whether such factors are under his control. Probably,

(continued)

deciding to give up his job and/or take a leave of absence would be changes under his control.

4. Help Manuel identify *advantages* or *benefits* to be realized by achieving his goal. He seems to be thinking about increased leisure time as a major benefit. Are there others?

5. Help Manuel identify *disadvantages* or *possible costs* of making the desired change. He has mentioned loss of retirement benefits as one cost and subsequent loss of a nest egg for his grandchildren as another. Do the perceived benefits outweigh the costs? What effect would leaving his job have on his wife and family? Is this consistent with his cultural identity and his beliefs?

6. If Manuel's goal looks as if it will have too many costs, explore other options with him, leaving the final decision about goals up to him. At this point, you will need to *decide whether you are able to help* him pursue his goal.

7. Assuming you will continue to work with Manuel, help him *define his goal behaviorally* by specifying exactly what he will be doing, thinking, and feeling as a result of goal achievement.

8. Further specification of the goal includes *where, when,* and *with whom* this will occur.

9. It will also include *how much* or *how often* the goal will occur. An option that might be useful for Manuel is to develop and scale five possible outcomes, ranging from the most unfavorable one to the most expected one to the best possible one (goal attainment scaling).

10. Help Manuel explore and *identify action steps or subgoals* that represent small approximations toward the overall goal. Help him choose action steps that are practical, are based on his resources, and support his values and culture.

11. Help Manuel sequence the action steps according to *immediacy and difficulty* so he knows what step he will take first and what step will be his last one.

12. Explore any *obstacles* that could impede progress toward the goal, such as the presence or absence of certain feelings, ideas, thoughts, situations, responses, people, and knowledge and skills.

13. Explore existing *resources* that could help Manuel complete the action steps more successfully. Like examination of obstacles, exploration of resources also includes assessing the presence or absence of certain feelings, ideas, thoughts, situations, responses, people, knowledge, skills, beliefs, and confidence in pursuing desired outcomes.

14. Help Manuel develop a *plan to review completion of the actions steps* and *evaluate progress toward the goal,* including a way to monitor and reward himself for progress and a plan to help him maintain changes.

CHAPTER 11

TREATMENT PLANNING AND SELECTION

An increasingly important part of therapeutic activity involves what is known as treatment planning. Treatment planning involves specification of mode, duration, and types of treatment to use with a client. It is initiated after the problems are assessed and after the outcome goals and means for evaluating outcomes are defined. A treatment plan specifies which kinds of counseling interventions will help the client reach his or her stated goals (type of treatment), how long this process will take (duration of treatment), and the specific format or way in which the intervention will be delivered to the client (mode of treatment).

Treatment planning is not just a discrete activity that occurs at one point in time. It is a continuous process and often the initial plan is readjusted whenever there is new information or the client or therapist becomes "stuck" (Caspar, 1995). Treatment planning has always been an important part of the counseling process and in the last several years its importance has at least doubled because it is required for third-party reimbursement from many managed care systems. In addition to this recent thrust, we advocate the use of treatment plans for several reasons. First, the plan is to be developed conjointly with the client as a collaborator. This process often increases the client's emotional investment in the counseling process. Second, the use of treatment planning helps to ensure the likelihood that the best combination of interventions for a given client with particular outcome goals will be used. Third, treatment planning helps to keep counseling on track; that is, it promotes a structure that ensures the client's needs are being considered and met as well as possible.

OBJECTIVES

1. For a given client case description and corresponding treatment plan, identify the following:

   a. Ways in which the selected treatment interventions used by the counselor conflicted with the client's cultural values and worldview.

   b. Recommended type, duration, and mode of treatment you would use as the counselor.

2. For a given client case, using a sample treatment planning form, develop in writing a treatment plan that specifies the type, duration, and mode of treatment.

FACTORS AFFECTING TREATMENT SELECTION

Treatment planning is based on a number of important factors. According to Beutler and Clarkin (1990), "Effective treatment is a consequence of a sequence of fine-grained decisions" about a number of treatment variables that are linked together and contribute "synergistically" to effective client change (pp. 20–21). These factors include client characteristics and preferences, counselor preferences and flexibility, and documentation and practice guidelines.

Client Characteristics and Preferences

"The characteristics that the patient brings to the treatment experience are the single most powerful sources of influence in the benefit to be achieved by treatment" (Beutler & Clarkin, 1990, p. 31). Banken and Wilson (1992) found, for example, that the effectiveness of several different treatment methods for depression suggests that the specific content of the therapies may be less important than nonspecific client factors, such as the client's readiness and motivation for change, belief systems, resources, and self-efficacy. Clients' incentive to change affects their use of a treatment plan. A client might indicate readiness to work by giving verbal permission, by demonstrating awareness of positive consequences of change, and by doing at least some covert work or hard thinking between sessions. Sometimes a client's readiness to pursue the outcomes is indicated by a shift in one part of his or her behavior. For example, the client may become more disclosive or may do more initiating in the interview.

Another client may demonstrate readiness for action by starting to assert his or her right to begin the session on time.

A recently developed and researched theoretical framework for the assessment of client readiness for change is the transtheoretical model of change described in Chapter 10 on outcome goals (Prochaska & DiClemente, 1982). Results of various research studies conducted on this model suggest that particular stages and processes of change are related to client readiness for change (McConnaughy, Prochaska, & Velicer, 1983; Prochaska & DiClemente, 1982, 1985; Prochaska, DiClemente, & Norcross, 1992). A recent study by Smith, Subich, and Kalodner (1995) found that clients who terminate therapy prematurely were at the precontemplation (first) stage of change rather than at the contemplation or action stage. The Prochaska and DiClemente (1982) model of change also found ten active processes of client change that cut across various theoretical approaches. These ten processes and sample corresponding interventions are listed in Table 11-1. Again, in the study referred to earlier by Smith et al. (1995), premature and nonpremature terminators from counseling were distinguishable in these processes of change. Clients who stayed in

therapy and did not terminate prematurely used more of these change processes more frequently than clients who dropped out of therapy early. You can examine Table 11-1 to identify example interventions for these change processes.

In addition to assessing client readiness, Caspar (1995) asserts that a therapist needs a good understanding of the client's motivational structure to plan for effective treatment strategies (p. 220). We need to remember that clients' motivational structures are not just internal, dynamic mechanisms but reflect cultural values as well. For example, one client, who has had neglectful parents, may wish to avoid personal rejection by the therapist and one way to do so is to be accepting of the therapist's treatment plan. Another client may be equally accepting of the therapist's treatment plan but for a different reason—in this case, to avoid rejection and to save face with her family and cultural group.

As therapists, we need to be aware that often (perhaps more often than not) clients will have competing motivations, and as a therapist you do a "dance" around these. Consider the case of a college sophomore whose father died when she was ten. She is the only child. She describes her mother as overprotective and portrays her mother as someone

TABLE 11-1. The Transtheoretical Model: Ten change processes and corresponding interventions

Process	Definitions: Sample interventions
Self-reevaluation	Assessing how one's problem affects physical environment: empathy training, documentaries
Consciousness-raising observations	Increasing information about self and problem: confrontations, interpretations, bibliotherapy
Dramatic relief	Experiencing and expressing feelings about one's problems solutions: psychodrama, grieving losses, role playing
Environmental reevaluation	Assessing how one's problem affects physical environment: empathy training, documentaries
Self disclosure and trust (helping relationship)	Being open and trusting about problems with someone who cares: therapeutic alliance, social support, self-help groups
Reinforcement (contingency) management	Rewarding oneself or being rewarded by others for making changes: contingency contracts, overt and covert reinforcement, self-reward
Counterconditioning	Substituting alternatives for problem behaviors: relaxation, desensitization, assertion, positive self-statements
Stimulus control	Avoiding or countering stimuli that elicit problem behaviors: restructuring one's environment (e.g., removing alcohol or fattening foods), avoiding high-risk cues, fading techniques
Self and social liberation (sociopolitical)	Increasing alternatives for nonproblem behaviors available in society; acting as an advocate for rights of repressed; empowering; policy interventions

SOURCE: Adapted from *The Clinician's Handbook,* 4th Ed., by R. G. Meyer and S. Deitsch, p. 16. Copyright © 1996 by Allyn and Bacon. Adapted by permission.

LEARNING ACTIVITY 35 — Client Characteristics and Treatment Planning

We are indebted to Caspar (1995) for the use of this particular learning activity, which may help to put you in touch with some of the effects particular client characteristics may have on treatment planning. You can do this with a partner or in a small group. Feedback follows on p. 270.

Task 1. Client A has an important plan to get rid of his stuttering problems in therapy. As a therapist, you have the complementary plan of helping him with this endeavor. As part of this plan, you must find out in precisely which situations the client stutters and what the subjective meaning of these situations is to the client. Please construct concrete behavior for this information gathering. Formulate concrete therapist statements to elicit the desired situational information. You may make additional assumptions about the client and about the situations, but you should

not make up too unusual a case. Within about 5 to 10 minutes you should have collected a workable repertory for this. If possible, discuss the results in your group. Otherwise, try to evaluate the proposals yourself.

Task 2. Client M has an extremely important plan not to admit weaknesses. To do so would be incompatible with his ideal of a strong, well-functioning man. The very fact that he is seeing a therapist is an ongoing violation of his plan. During the sessions he seems to be making a special effort to be a "strong man," equal in strength to the therapist. Please write down as concrete a list as possible of behaviors the therapist should refrain from to avoid getting into conflict with the client's plan of not admitting weaknesses. (Caspar, 1995, pp. 235–236)

---

who is constantly trying to arrange her life for her. She indicates that the only area of her life she feels she has any real control over is her personal exercise, and she finds herself exercising compulsively. She also feels "blue" for no reason apparent to her. Over the course of your sessions with her, you notice that she keeps asking you for your opinions and suggestions and for various sorts of information; she seems to want to rely heavily on what you say. In this case, her competing motivations are between dependence and control. As her therapist, you must honor both. It is also not uncommon for bicultural clients to have competing motivations between their two (or more) cultures. An example is a case of a young woman who is an Asian American student in a predominantly Caucasian high school. Although she was born and has grown up in the United States, her parents came here from Vietnam. She reports feeling pulled in two directions all the time. Exploration of this concern reveals that she wants to be like her high school friends—meaning she wants to assert her freedom and independence and make her own decisions; yet she is afraid that doing so will mean facing dishonor and shame from her parents and her relatives. She feels caught between the mainstream culture of her school and friends, which emphasizes individual autonomy and personal freedom, and the Asian culture of her family and relatives, which emphasizes belonging to a collective culture and honoring these values and wishes. Again, as a therapist, you must recognize both these different motivations and wishes.

The notion of "patient position"—the strongly held values and beliefs—is important because these are likely to

influence the client's supporting or rejecting the use of a particular approach. If the therapist is insensitive to the client's beliefs and values and forges ahead with a strategy that runs counter to them, the client is likely either to resist the procedure or to fail to really invest herself or himself in working with it. Occasionally, a client will reject a proposed strategy quite openly: "That sounds pretty silly to me!" "How in the world would that help?" or "I didn't even try it." Other clients convey their disapproval far more subtly, requiring the counselor from the very beginning to be alert to the client's cues, often nonverbal, for "yes" and "no." For instance, some clients' lack of acceptance of a proposed procedure may be communicated by a change in face color, eye or mouth movements, or breathing patterns.

As Caspar (1995) notes, treatment plans need to be complementary with the way clients generally shape their lives and specifically with the clients' level of autonomy. If the plan enhances the clients' autonomy, it will be more workable than if it restricts the clients' autonomy (p. 214). Effective treatment plans also build on client resource areas of strength and situations that are unproblematic as well as on clients' social and environmental support systems (Caspar, 1995, p. 233).

Counselor Preferences and Flexibility

Beutler and Clarkin (1990) observe that the therapist's capacity to be flexible and adaptive to the idiosyncratic needs of each client is more powerful in treatment selection than the right match between therapist and client. As Caspar (1995) states, the more possible ways of proceeding a therapist can choose from, the higher the probability that he or she can achieve a favorable effect with a maximum of benefits and a minimum of costs (p. 215).

This sort of flexibility requires therapists to have a high level of awareness of themselves, of each client, and of the interactional pattern between the two. Some of this awareness is developed by intuition or instinct, but some of it requires analysis, planning, and consultation with another therapist or supervisor.

Counselor preferences for particular theories and interventions also play a role in treatment selection. Often counselors select interventions on the basis of how they have been trained, what they feel most comfortable with, and which strategies they feel competent to carry out. Sometimes, too, counselors are encouraged to "stretch" a bit in their plans and interventions by their supervisors! From an ethical standpoint, it is important to practice within your limits of training and experience. If you have been trained only in cognitive change approaches, it is unlikely that you would be able to implement a Gestalt intervention without additional training and experience. All too often, however, treatment decisions are made on the basis of the counselor's allegiance to a favorite counseling treatment orientation (refer to Table 1-1 for a synopsis of these) rather than on the kinds of problems presented and outcomes desired by the client. You know this is happening when a counselor treats all clients with the same theoretical orientation. The importance of having multitheoretical perspectives in effective treatment planning cannot be emphasized enough. To use a golf analogy, if you have only one club in your golf bag to use in various situations, you are much more limited in your game. If you feel hampered in this way by your own training and/or experience, you have the option of referring a client who needs a particular kind of treatment outside the realm of your own competence or of expanding your own training and supervised experience to include more diverse perspectives.

At the same time, regardless of how much training and experience you have, counselor preferences should not be the only or even the primary consideration in treatment planning and selection. In addition to your preferences and training and the client's characteristics and preferences, it is also important to consider the available literature supporting the use of particular kinds of interventions for particular client problems.

Documentation and Practice Guidelines

Varying amounts of data exist for different counseling procedures. These data can help you determine, for your treatment plan, the ways in which the strategy has been used successfully and with what types of client problems. All the strategies presented in the remainder of this book have some empirical support.

Increasingly, however, documentation about therapeutic interventions is used as a basis for newly emerging *practice guidelines* (see Clinton, McCormick, & Besteman, 1994). Practice guidelines discuss ways to implement given therapeutic procedures. Practice guidelines are defined as "systematically developed statements to assist practitioner and patient decisions about appropriate health care for specific clinical circumstances" (AHCPR, 1990, p. 3). Practice guidelines are currently available for panic disorder and suicide management (Fishelman, 1991; Gottleib, 1991) and also depression (AHCPR, 1993; Antonuccio, Danton, & DeNelsky, 1995; Jacobson & Hollon, 1996; Karasu, Docherty, Gelenberg, Kupfer, Merriam, & Shadoan, 1993); guidelines are in progress for various other problems including anxiety and phobias, marital therapy, eating disorders, child behavior problems, pain control, and smoking cessation. AHCPR guidelines are developed by a multidisciplinary process that includes representatives of affected provider groups. These guidelines are based on empirical data and provider review and are to be revised every 18 to 36 months. A recent summary of guidelines for depression, panic, anxiety, and phobias based on client self-report data was made available to the public (*Consumer Reports,* 1995).

According to the American Psychiatric Association (1993), the question of whether the research evidence is adequate to warrant a treatment's implementation is especially relevant to the current environment in which professional groups and others are making *clinical* recommendations based on the research literature. Practice guidelines are developing as a result of the proliferation of health maintenance organizations and managed health care.

In spite of several inherent weaknesses of these guidelines (described later), some practitioners believe that disregarding documentation of various psychotherapeutic strategies may result in the exclusion of *all* psychotherapeutic interventions by third party payers. The worst possible result (short of an uprising of all concerned and indignant mental health professionals)* would be a public whose only reimbursed treatment option for emotional problems would

---

*At the time of this writing we are aware of at least two groups of mental health providers who have formed their own HMOs: the American Mental Health Alliance based in Boston, and the Island Behavioral Health Association based in Long Island, New York.

Now imagine that Clients A and M are the same person! First, check to see whether you have already made proposals for Task 1 that do not conflict with the constraints described in Task 2. Then try to construct additional behaviors that are in line with the constraints from both tasks (Caspar, 1995, p. 236).

---

be medication. Hollon (1996) points out that while medication "tends to out-perform psychotherapy" in symptom reduction of more severe disorders, therapy also accomplishes results that medication alone does not (p. 1028).

The best case scenario is that the quality of mental health care is improved by ensuring that the most appropriate treatments are used for particular emotional/psychological problems and with particular groups of clients. As an example, Division 12 (Clinical Psychology) of the American Psychological Association recently established a task force to define empirically validated psychotherapeutic treatments and to educate practitioners, third party payers, and the public about these interventions. In a report adopted in 1993, the members of this task force reported on data showing some interventions to be superior to others, depending on the nature of the disorder. For example, relaxation (described in Chapter 18) was less effective than cognitive therapy (described in Chapter 15) for panic disorder (Beck, Sokol, Clark, Berchick, & Wright, 1992). Based on data collected, this task force published two lists of empirically validated treatment interventions: (1) well-established treatments (Table 11-2), and (2) probably efficacious treatments (Table 11-3). The intent of the report is not to declare any one particular theoretical orientation effective in general but to indicate particular approaches that are effective for particular problems. As Hollon (1996) notes, "It is a continuing source of consternation to outcome researchers that those approaches that have shown the most promise in controlled trials are so rarely used in applied settings" (p. 1028).

Another excellent source of treatment protocols for a number of psychological disorders pertinent to adults, such as depression, anxiety, eating disorders, alcoholism, couple distress, and borderline personality disorder, can be found in Barlow's (1994) *Clinical Handbook of Psychological Disorders*. Note that the lists provided in Tables 11-2 and 11-3 are derived from the 1993 task force report and that a new task force is already working on supplements to these treatments.

The studies used in the APA Task Force report are referred to as *efficacy* studies, meaning there is a comparison of one type of treatment with another type of treatment under carefully controlled experimental conditions. Those data also have been summarized in a book by Seligman (1994), who notes that cognitive therapy and interpersonal therapy are useful for unipolar depression, cognitive therapy is useful for pain disorders, systematic desensitization relieves specific phobia, meditation and breathing relieve anxiety, and cognitive therapy is useful for bulimia. As Seligman (1995) notes, "Because treatment in efficacy studies is derived under tightly controlled conditions to carefully selected patients, sensitivity is maximized and efficacy studies are very useful for deciding whether one treatment is better than another treatment for a given disorder" (p. 966).

Note also that practice guidelines have resulted in a proliferation of treatment manuals (see Luborsky & DeRubeis, 1984). Treatment manuals are used to standardize the application of a particular therapeutic procedure. Treatment manuals often describe interventions that have been used in empirical studies. Wilson and Agras (1992) speculated that this "small revolution" (Luborsky & DeRubeis, 1984) will continue to grow. A list of currently available treatment manuals from the Division 12 Task Force can be found in Appendix E. Wilson (1996) has reported on the uses and misuses of treatment manuals in actual practice.

Efficacy studies in the area of marriage and family therapy (MFT) have also been reviewed and reported by Pinsof and Wynne (1995), who found that MFT is more effective than no therapy at all for many disorders; more clinically and cost effective than individual therapy for a few disorders; and is a necessary, but not sufficient, treatment for very severe disorders. A summary of their results is found in Table 11-4. One of the more important implications of their review for mental health treatment planning is that "the more severe, pervasive and disruptive the disorder, the greater the need to include *multiple* components of effective treatments" (p. 605).

In addition to these *efficacy* or research studies, *effectiveness* studies are also useful in generating data about effective treatments. Effectiveness studies are large-scale, in-the-field surveys of actual clients and the results they report. Seligman (1995) argues that the effectiveness study is a better way to assess how therapy is actually conducted because it taps into critical variables that cannot be measured

TABLE 11-2. APA Division 12 Task Force report on empirically validated treatments

Well-established treatments	Citation for efficacy evidence
Beck's cognitive therapy for depression	Dobson (1989)
Behavior medication for developmentally disabled individuals	Scotti, Evans, Meyer, & Walker (1991)
Behavior modification for enuresis and encopresis	Kupfersmid (1989); Wright & Walker (1978)
Behavior therapy for headache and for irritable bowel syndrome	Blanchard, Schwartz, & Radnitz (1987); Blanchard, Andrasik, Ahles, Teders, & O'Keefe (1980)
Behavior therapy for female orgasmic dysfunction and male erectile dysfunction	LoPiccolo & Stock (1986); Auerback & Kilmann (1977)
Behavioral marital therapy	Azrin et al. (1980); Jacobson & Follette (1985)
Cognitive behavior therapy for chronic pain	Keefe, Dunsmore, & Burnett (1992)
Cognitive behavior therapy for panic disorder with and without agoraphobia	Barlow, Craske, Cerny, & Klosko (1989); Clark et al. (1994)
Cognitive behavior therapy for generalized anxiety disorder	Butler, Fennell, Robson, & Gelder (1991); Borkovec et al. (1987); Chambless & Gillis (1993)
Exposure treatment for phobias (agoraphobia, social phobia, simple phobia) and PTSD	Mattick, Andrews, Hadzi-Pavlovic, & Christensen (1990); Trull et al. (1988); Foa, Rothbaum, Riggs, & Murdock (1991)
Exposure and response prevention for obsessive-compulsive disorder	Marks & O'Sullivan (1988); Steketee, Foa, & Grayson (1982)
Family education programs for schizophrenia	Hogarty et al. (1986); Falloon et al. (1985)
Group cognitive behavioral therapy for social phobia	Heimberg, Dodge, Hope, Kennedy, & Zollo (1990); Mattick & Peters (1988)
Interpersonal therapy for bulimia	Fairburn, Jones, Peveler, Hope, & O'Conner (1993); Wilfley et al. (1993)
Klerman and Weissman's interpersonal therapy for depression	DiMascio et al. (1979); Elkin et al. (1989)
Parent training programs for children with oppositional behavior	Wells & Egan (1988); Walter & Gilmore (1973)
Systematic desensitization for simple phobia	Kazdin & Wilcoxon (1976)
Token economy programs	Liberman (1972)

Criteria for empirically validated treatments: Well-established treatments

I. At least two good group design studies, conducted by different investigators, demonstrating efficacy in one or more of the following ways:
   A. Superior to pill or psychological placebo or to another treatment.
   B. Equivalent to an already established treatment in studies with adequate statistical power (about 30 per group: cf. Kazdin & Bass, 1989).

OR

II. A large series of single-case design studies demonstrating efficacy and the following characteristics:
   A. Used good experimental designs
   B. Compared the interventions to another treatment as in IA

Further criteria for both I and II

III. Studies must be conducted using standardized treatment manuals as protocols.
IV. Characteristics of the client samples must be clearly specified.

TABLE 11-3. APA Division 12 Task Force report on probably efficacious treatments

Probably efficacious treatments	Citation for efficacy evidence
Applied relaxation for panic disorders	Ost (1988); Ost & Westling (1991)
Brief psychodynamic therapies	Piper, Azim, McCallum, & Joyce (1990); Shefler & Dasberg (1989); Thompson, Gallagher, & Breckenridge (1987); Winston et al. (1991); Woody, Luborsky, McLellan, & O'Brien (1990)
Behavior modification for sex offenders	Marshall, Jones, Ward, Johnston, & Barbaree (1991)
Dialectical behavior therapy for borderline personality disorder	Linehan et al. (1991)
Emotionally focused couples therapy	Johnson & Greenberg (1985)
Habit reversal and control techniques	Azrin, Nunn, & Frantz (1980); Azrin, Nunn, & Frantz-Renshaw (1980)
Lewinsohn's psychoeducational treatment for depression	Lewinsohn, Hoberman, & Clarke (1989)

Criteria for empirically validated treatments: Probably efficacious treatments

I. Two studies showing the treatment is more effective than a waiting-list control group.

OR

II. Two studies otherwise meeting the well-established treatment criteria I, III, and IV, but both conducted by the same investigator. Or one good study demonstrating effectiveness by these same criteria.

OR

III. At least two good studies demonstrating effectiveness but flawed by heterogeneity of the client samples.

OR

IV. A small series of single-case design studies otherwise meeting the well-established treatment criteria II, III, and IV.

NOTE: Treatments that have not been established as probably effective are considered experimental, although treatments not appearing here should not be discarded. The task force may have overlooked a treatment that cannot at present be evaluated because of insufficient evidence.

---

in an efficacy study. A large-scale effectiveness study reported by *Consumer Reports* (1995, November) found that clients benefited significantly from therapy, that no one specific therapy approach was better than another, that longer-term therapy was better than short-term, and that clients whose length of therapy or choice of therapist was limited by insurance or managed care did worse. There have been a host of reactions to the *Consumer Reports* study; a useful critique and summary can be found in a special issue of *American Psychologist* (VandenBos, 1996). Seligman (1995) concludes that ultimately treatments should be validated by a combination of efficacy and effectiveness techniques.

In considering practice guidelines, empirically validated treatments, and standardized treatment manuals, it is critical to note that all this documentation is in the infancy stage with respect to significant client demographic variables, particularly for clients of color. We have obtained and reviewed most of the treatment manuals listed in Appendix E and have

found very little information available with respect to ethnocultural aspects of the problem and the treatment procedure. Fortunately, other sources are now beginning to compile more ethnocultural data. For example Marsella, Friedman, Gerrity and Scarsfield (1996) have published a book on ethnocultural aspects of posttraumatic stress syndrome. Still, however, attention to ethnocultural factors in practice guidelines is relatively sparse.

Earlier, we mentioned a comprehensive review of psychotherapy and medication for the treatment of bipolar depression (Antonuccio et al., 1995). These authors pointed out that issues of ethnicity and gender in almost all the studies were "sorely neglected" (1995, p. 380). This omission is problematic. Because most of the samples used are typically 90% Caucasian, it is difficult to specify how the results apply to clients of color. Also, in the studies of depression, because 65% to 80% of all subjects are women, it is unclear how the results will generalize to men. From a treatment perspective, approximately 75% of all antide-

TABLE 11-4 Outcomes for marriage therapy (MT) and/or family therapy (FT) versus no treatment and individual treatment

---

Disorders for which marriage therapy (MT) or family therapy (FT) was superior to no treatment

Adult schizophrenia (FT)

Outpatient depressed women in distressed marriages (MT)

Marital distress and conflict (MT)

Adult alcoholism and drug abuse (FT/MT)

Adult hypertension (MT)

Elderly dementia (FT)

Adult obesity (MT)

Cardiovascular risk factors in adults (FT)

Adolescent conduct disorder (FT)

Anorexia in young adolescent girls (FT)

Adolescent drug abuse (FT)

Child conduct disorders (FT)

Aggression and noncompliance in ADHD children (FT)

Childhood autism (FT)

Chronic physical illness in children (asthma, diabetes, etc.) (FT)

Child obesity (FT)

Cardiovascular risk factors in children (FT)

Disorders for which marriage and family therapy (MFT) was superior to individual treatment

Adult schizophrenia

Depressed outpatient women in distressed marriages

Marital distress

Adult alcoholism and drug abuse

Adolescent conduct disorders

Adolescent drug abuse

Anorexia in young adolescent females

Childhood autism

Various chronic physical illnesses in adults and children

Additionally, involving the family in engaging alcoholic adults in treatment is more efficacious than just working with the individual adult.

SOURCE: Based on data from Pinsof and Wynne (1995).

pressants prescribed are given to women. This observation raises several concerns. First, it appears that prescribing pills to women is easier than looking for sociopolitical factors that contribute to increased evidence of depression in this group. Second, it seems that women are disproportionately exposed to the risks of taking these prescribed medications, especially because they are more likely to experience unpleasant side effects than men (Munoz, Hollon, McGrath, Rehm, & VandenBos, 1994).

When evaluating the results and recommendations of research about practice, consider asking the following questions recommended by Brown (1994):

What are the writer's assumptions about norms, values, and appropriate behaviors?

Is the researcher measuring a way of being that is usual and normative within a culture or a response to the researcher's own dominant status?

TABLE 11-5. Sample treatment options for insomnia based on the *DSM-IV* categories

Many insomniacs have significant anxiety, depression, and/or obsessive-compulsive features that contribute to the insomnia, as well as to other aspects of functioning. If these exist, the reader is referred to those chapters for relevant treatment recommendations (Harsh and Ogilvie, 1995).

Chemotherapy is commonly used for insomnia, sometimes with a very specific focus—for example, oxycodone for restless-leg syndrome. Benzodiazepines, especially those with a short (diazepam) or intermediate (triazolam or temazepam) absorption rate, are the preferred hypnotic-sleeping drug for most clients. Antihistamines can be useful in the short run, but tolerance develops quickly. Numerous other agents are available for specific client needs (e.g., sedative antidepressants, barbiturates, chloral hydrate, etc.), but most have some problematic side effects if used for any length of time, and many quickly produce physical dependence. Indeed, any chemotherapy directed toward the sleep disorder as a whole must be administered with caution (Montplaisir and Godbout, 1991): It can be especially useful in breaking the insomnia cycle (e.g., if prescribed for no more than a week), but if prescribed for any length of time loses its effectiveness, brings on the risk of drug dependence and/or addiction, and can generate a rebound effect of even greater insomnia. Blood levels of melatonin, a natural hormone produced in the pineal gland, have been related to sleep-pattern disruptions. Melatonin at a dosage of 2–3 mg administered one to two hours before bedtime appears to restore disrupted sleep-wake cycle insomnia in as little as a week and to be especially useful for the elderly.

Various behavioral techniques have been found to be effective (Harsh and Ogilvie, 1995; Hauri, 1991; Montplaisir and Godbout, 1991), but the most efficient techniques are probably cognitive therapy, relaxation training, and biofeedback where the emphasis is on muscle-tension release. Ancillary techniques include (1) initiating a regular program of vigorous exercise, although it is best to avoid vigorous exercise late in the evening—ideally, do aerobic exercise three to five hours before bed; (2) avoiding large or late meals or, conversely, going to bed very hungry—a high carbohydrate snack is helpful; (3) avoiding napping, especially long or late naps during the day—keep any naps to forty-five minutes or less; (4) go to bed only when sleepy, and limit allotted time in bed to estimated need, usually no more than seven hours—this creates an actual, controlled sleep debt and helps to break the cycle of rumination in bed; (5) cutting down on caffeine or heavy smoking or alcohol use, especially later in the evening; (6) ceasing to fret over inconsistencies in the sleep pattern—give yourself some "quiet time" before going to bed, ideally one to two hours, and tell yourself as you prepare for bed that you will sleep well; (7) keeping the bedroom as quiet, dark, and comfortable as possible—most find a temperature of 65 degrees to be optimal—and possibly using a "white noise" generator or audio tapes to introduce masking or restful sound; and (8) relearning a more appropriate bedtime routine by (a) going to bed only when feeling tired, (b) awakening at the same time each morning, (c) avoiding all nonsleep-related activities (within reason) in the bedroom, and (d) leaving the bed in twenty or thirty minutes if sleep has not occurred, to do something like read a book, and not returning until again feeling drowsy. As the philosopher Friedrich Nietzsche said in *Thus Spake Zarathustra,* "It is no small art to sleep: To achieve it one must stay awake."

SOURCE: From *The Clinician's Handbook,* 4th Ed., by R. G. Meyer and S. E. Deitsch, pp. 150–151. Copyright © 1996 by Allyn and Bacon. Reprinted by permission.

Is the meaning ascribed to the behaviors centered in the culture, or does it evolve from a biased perspective?

These questions are even more critical when the researchers are members of the mainstream culture.

DECISION RULES IN PLANNING FOR TYPE, DURATION, AND MODE OF TREATMENT

As early as 1976, Shaffer proposed the use of diagnostic cues and patterns or *decision rules* to use in client treatment planning. A decision rule is a series of mental questions or heuristics that the counselor constantly asks himself or herself during interviews in order to match techniques to clients and their identified concerns. More recent formulations of this sort of therapist-client matching have been discussed by Beutler and Clarkin (1990), Santiago-Rivera (1995), Hackney and Cormier (1996), and Meyer and Deitsch (1996).

The Meyer and Deitsch (1996) model is anchored in the *DSM-IV.* This model describes various diagnoses, links these diagnoses with specific assessment and standardized test data, and then describes recommended treatment procedures for each diagnostic category. For an example, we provide their treatment options for insomnia in Table 11-5. (We chose this as an example because students often report developing sleeping problems as a function of graduate school!)

The Santiago-Rivera (1995) model is one that integrates various dimensions of culture including level of acculturation, culturally sensitive types of treatment, and availability of culturally relevant resources (see Figure 11-1). Her model is especially useful in that it has the greatest focus on cultural variables of any of the four models.

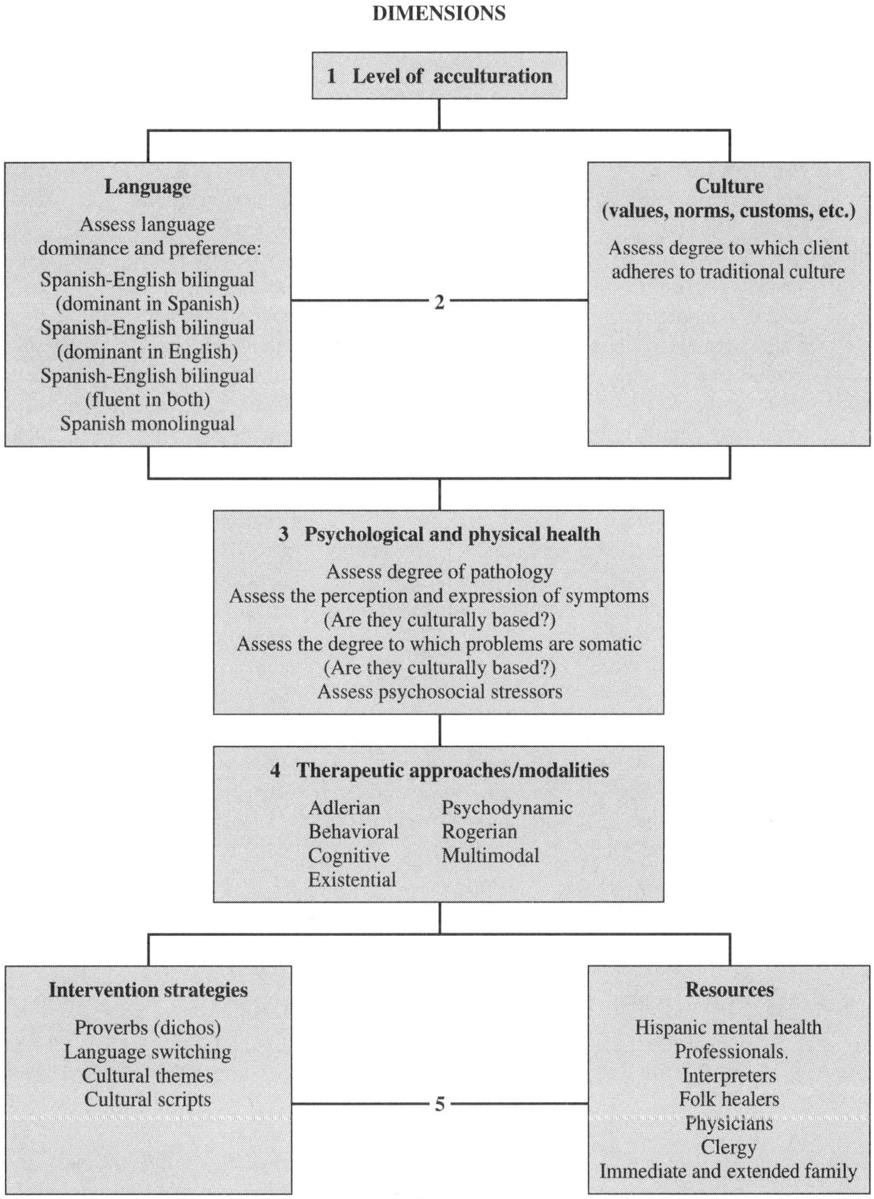

DIMENSIONS

1 **Level of acculturation**

Language

Assess language
dominance and preference:

Spanish-English bilingual
(dominant in Spanish)
Spanish-English bilingual
(dominant in English)
Spanish-English bilingual
(fluent in both)
Spanish monolingual

Culture
(values, norms, customs, etc.)

Assess degree to which client
adheres to traditional culture

2

3 **Psychological and physical health**

Assess degree of pathology
Assess the perception and expression of symptoms
(Are they culturally based?)
Assess the degree to which problems are somatic
(Are they culturally based?)
Assess psychosocial stressors

4 **Therapeutic approaches/modalities**

Adlerian   Psychodynamic
Behavioral   Rogerian
Cognitive   Multimodal
Existential

Intervention strategies

Proverbs (dichos)
Language switching
Cultural themes
Cultural scripts

5

Resources

Hispanic mental health
Professionals,
Interpreters
Folk healers
Physicians
Clergy
Immediate and extended family

FIGURE 11-1. Santiago-Rivera model of culturally sensitive treatment selection
SOURCE: From "Developing a Culturally Sensitive Treatment Modality for Bilingual Spanish-Speaking Clients," by A. L. Santiago-Rivera, *Journal of Counseling and Development, 74,* p. 14.
Copyright © 1995 ACA. Reprinted with permission. No further reproduction authorized without written permission of the American Counseling Association.

The Hackney and Cormier (1996) model matches types of problems in five categories—affective, cognitive, behavioral, systemic, and cultural, and particular manifestations of these problems—to corresponding types of interventions and theoretical orientations. Their model is depicted in Table 11-6.

The Beutler and Clarkin (1990) model is an integrated one that considers client characteristics, the nature of the client's problem, and the treatment objectives although it does not consider cultural variables as thoroughly as it could. Their model is summarized in Table 11-7. (For the full complexity of their model, we refer you to their book.)

TABLE 11-6. Treatment strategies and corresponding manifestations of client problems

Affective	Cognitive	Behavioral	Systemic	Cultural
Person-centered therapy; Gestalt therapy; body awareness therapies; psychodynamic therapies; experiential therapies: Active listening; empathy; positive regard; genuineness; awareness techniques; empty chair; fantasy; dreamwork; bioenergetics; biofeedback; core energetics; radix therapy; free association; transference analysis; dream analysis; focusing techniques.	**Rational-emotive therapy; Beck's cognitive therapy; transactional analysis; reality therapy:** A-B-C-D-E analysis; homework assignments; counter-conditioning; bibliotherapy; media-tapes; brainstorming; identifying alternatives; reframing; egograms; script analysis; problem definition; clarifying interactional sequences; coaching; defining boundaries; shifting triangulation patterns; prescribing the problem (paradox).	**Skinner's operant conditioning; Wolpe's counter-conditioning; Bandura's social learning; Lazarus' multimodal therapy:** Guided imagery; role-playing;, self-monitoring; physiological recording; behavioral contracting; assertiveness training; social skills training; systematic desensitization; contingency contracting; action planning; counter-conditioning.	**Structural therapy; strategic family therapy; intergenerational systems:** Instructing about subsystems; enmeshment and differentiation; addressing triangulation, alliances and coalitions; role restructuring; clarifying interactional systems; reframing; prescribing the problem (paradox); altering interactional sequences; genogram analysis; coaching; defining boundaries; shifting triangulation patterns.	**Multicultural counseling; cross-cultural counseling:** meta-theoretical, multimodal, culturally based interventions; focus on worldviews, cultural orientation, cultural identity; liberation and empowerment perspectives; culturally sensitive language, metaphors, rituals, practices, and resources, collaboration, networking, consciousness raising, advocacy.
Manifestations Emotional expressiveness and impulsivity; instability of emotions; use of emotions in problem-solving and decision-making; sensitivity to self and others; receptive to feelings of others.	*Manifestations* Intellectualizing; logical rational, systematic behavior; reasoned; computer-like approach to problem-solving and decision-making; receptive to logic, ideas, theories, concepts, analysis, and synthesis.	*Manifestations* Involvement in activities; strong goal orientation; need to be constantly doing something; receptive to activity, action, getting something done; perhaps at expense of others.	*Manifestations* Enmeshed or disengaged relationships; rigid relationship boundaries and rules; dysfunctional interaction patterns.	*Manifestations* Level of acculturation; type of worldview; level of cultural identity; bi- or trilingual; presenting problems are, to some degree, culturally based.

SOURCE: From *The Professional Counselor,* 3rd Ed., by H. Hackney and L. S. Cormier, p. 135. Copyright © 1996 by Allyn and Bacon. Reprinted by permission.

Type of Treatment

To summarize the Beutler and Clarkin (1990) model, client problems can be characterized in one of two ways: those problems involving symptom distress and those problems involving symbolic conflicts. The easiest way to distinguish between the two is to consider this question: "Are these problems simple habits maintained by the environment or are they symbolized expressions of unresolved conflictual experiences?" (Beutler & Clarkin, 1990, p. 226). Problems that are environment specific and clearly related to antecedents and consequences are symptom problems. For example, a teenage girl is referred to you by her parents who are convinced that during the last few months she has become more "distant and withdrawn" from them; specifically, they have noticed this behavior since she developed a relationship with an older boy.

On the other hand, recurrent symptom patterns that do not seem to be functionally related to specific environmental antecedents and consequences and continue to be evoked in situations that have little resemblance to the originally evolving situation are indicative of underlying conflict (Beutler & Clarkin, 1990, p. 226). For example, another teenage girl is referred to you and comes to see you with her parents who are concerned with their daughter's "distance and withdrawal" not only from them but also from her younger

TABLE 11-7. Beutler and Clarkin (1990) model of systematic treatment selection

Nature of problem	Targets of change	Corresponding treatments
Symptom Distress	Behaviors (1st order change)	1. Social skills training. 2. In vivo or in vitro exposure to avoided events 3. Graded practice 4. Reinforcement (Beutler & Clarkin, 1990, p. 244)
	Cognitions (1st order change)	1. Identification of cognitive errors 2. Evaluation of risk or degree of distortion 3. Questioning of dysfunctional assumptions and beliefs 4. Self-monitoring 5. Self-instruction 6. Practice alternative thinking 7. Testing of new assumptions (Beutler & Clarkin, 1990, pp. 244–245)
Symbolic Conflicts	Feelings (2nd order change)	1. Focus on sensory states 2. Reflection of feelings 3. Two-chair work on emotional "splits" 4. One- and two-chair work related to unfinished business 5. Structured imagery 6. Gestalt dream work 7. Reflective mirroring of the hidden self 8. Enacting emotional opposites 9. Free association to sensory cues 10. Physical expression and release exercises (Beutler & Clarkin, 1990, p. 248)
	Unconscious Conflicts (2nd order change)	1. Free association 2. Dream interpretation 3. Encouragement of transferential projections 4. Interpretation of resistance and defense 5. Analysis of hidden motives through assessment of common mistakes or slips 6. Free fantasy explorations 7. Discussion of early memories 8. The construction and analysis of genograms 9. Two-chair work on intrapersonal "splits" (Beutler & Clarkin, 1990, p. 249)

brother. However, they note that this behavior is nothing new; it has been a pattern since they adopted her at age 2 along with her infant brother. Prior to age 2, she was left alone for several days in an abandoned house by her biological mother. She describes herself as a loner and doesn't have any real friends.

In symptom-based client problems, the targets of change are observed behavioral excesses or deficits and dysfunctional cognitions. In conflict-based client problems, the targets of change are feelings that are masked by the symptoms and underlying unconscious conflicts. Table 11-8 provides examples of these four targets of change for two selected problems: depression and agoraphobia.

For problems that are primarily symptom based and involve changes in altering behaviors and cognitions, recommended treatment strategies involve behavioral and cognitive interventions such as modeling, graded practice, cognitive restructuring, self-monitoring, and so on (see Table 11-7), depending on the degree to which symptoms are overt or covert. Overt, external symptoms are more responsive to behavioral strategies; covert symptoms are more responsive to cognitive therapies (Beutler & Clarkin, 1990). Behavioral and cognitive therapies also appear to achieve better results with clients who are likely to externalize their distress.

For problems that are primarily conflict based and involve changes in altering feelings, recommended treatment strategies include interventions for enhancing emotional and sensory outcomes (Greenberg & Goldman, 1988), such as reflection of feelings, Gestalt two-chair work and dream work, imagery, and body expression and related activities (see Table 11-7). Recommended treatment strategies for

TABLE 11-8. Targets of change for two client problems

Depression

Observed symptoms (e.g., depression) social skills deficits associated with the presenting problem (e.g., lack of communication skills)—cognitions directly related to symptoms (e.g., negative self-attributions)—feelings and affects masked by the symptoms (e.g., primary emotions such as anger at a significant other)—unconscious conflicts (e.g., mourning for a lost mother-child relationship).

Agoraphobia

Observed symptoms (e.g., inability to go in a crowded building)—coping skill deficits (e.g., inability to control arousal level)—cognitions associated with symptoms (e.g., anticipations of death)—emotions associated with or covered by symptom (e.g., fear of losing a significant other)—unconscious conflicts (e.g., conflicts over seeking dependency).

SOURCE. From *Systematic Treatment Selection,* by L. Beutler and J. Clarkin, p. 226. Copyright © 1990 by Brunner/Mazel, Inc. Reprinted with permission.

resolving unconscious conflict include interventions for addressing unconscious experience and recurrent interpersonal patterns as well as hidden motives (Luborsky, 1984; Strupp & Binder, 1984), such as interpretation, confrontation, early recollections, genograms, and two-chair work (see Table 11-7). Therapies of this sort, which address the client's level of unconscious motives and feelings, appear to work better with clients who cope with stress by internalizing.

Duration of Treatment

Note that there is a hierarchy in this table and model of treatment selection in that the initial focus of treatment should be briefer and less complex—that is, symptom oriented rather than conflict oriented. Beutler and Clarkin (1990) conclude, "The initial assumption should be that any problem is a simple habit or transient adjustment problem. Only when it can be demonstrated with reasonable reliability and confidence that the symptoms represent a link in a recurrent and debilitating pattern of interpersonal relationships should the focus of treatment be upon the symbolically represented conflict rather than upon the symptom alone" (p. 233).

Beutler and Clarkin (1990) argue that beginning a treatment plan with a symptom-focused treatment is useful because there is no evidence that using a narrow band of treatment to deal with a complex problem yields negative results, but using a broad base of treatment (such as conflict focused) to treat a more simple problem is overkill. They acknowledge that although "symptom patterns may recur when more narrowly focused treatments are applied alone to complex problems,...mixing treatments of varying breadths actually may allay this recurrence" (Beutler & Clarkin, 1990, p. 234). They point out that as the targets of change increase and the client's strengths and resources

decline, a longer duration of treatment does become important. There is some evidence that the greater the total amount of treatment and the longer the duration of treatment, the greater the benefit to the client (*Consumer Reports,* 1995; Seligman, 1995). However, some settings and some managed care companies have policies that limit the duration of therapy to a few sessions, depending on the client's diagnosis. Clients who present with transient situational problems are good candidates for crisis intervention, but we cannot assume that all clients who come for therapy are automatically good candidates for brief or time-limited therapy. Brief therapy seems to work better with clients who may drop out of therapy early, such as clients who are at the precontemplation stage of the transtheoretical model of change and who have unidimensional symptoms or conflictual problems. Also, briefer therapy may better fit the preferences and belief systems of some clients of color. The primary indicators for brief symptom-focused and brief conflict-focused therapies are summarized in Table 11-9.

Mode of Treatment

Mode of treatment refers to the specific way the interventions are delivered to the client:

1. Individual treatment
2. Couple and/or family treatment
3. Group treatment
4. Medication

All these modes have certain advantages. Individual therapy promotes greater privacy, self-disclosure, sharing, individualized attention, and identification with the therapist than the other modes (Beutler & Clarkin, 1990, p. 123). Couples therapy allows for the direct observation of interaction between the partners; includes both parties, including

TABLE 11-9. Indicators for brief treatment

Indications for the brief conflict-focused therapies
Indications
1. Focal intrapsychic conflict involving separations, oedipal issues, narcissistic injury, or stress-response syndromes.
2. Goal is character change in one focal area.

Enabling factors
1. Client has had at least one significant relationship in early childhood.
2. Client relates quickly, flexibly, and openly to counselor.
3. Client can focus on a central conflict in the evaluation.
4. Client is willing to examine feelings and behavior.
5. Client has relatively high ego strength as evidenced by educational, work, and sexual performance.
6. Client is motivated to change behavior and understand self better.
7. Client is intelligent and able to communicate verbally thoughts, feelings, and fantasies.

Indications for the brief symptom-focused therapies
Indications
1. Anxiety disorders; agoraphobia, panic disorder, generalized anxiety, hyperventilation, examination anxiety, interpersonal anxiety.
2. Depression of mild to moderate severity.
3. Deficits in social skills.
4. Sexual dysfunctions.
5. Compulsive rituals.

Enabling factors
1. Time-limited orientation may be more suited for some clients of color.

SOURCE: From *Systematic Treatment Selection,* by L. Beutler and J. Clarkin, p. 147. Copyright © 1990 by Brunner/Mazel, Inc. Reprinted with permission.

both "sides" of conflict and collusion; and allows for the development of mutual support, communication, and conflict resolution sources (Beutler & Clarkin, 1990, p. 123). These same advantages are extended to family therapy in which both parents and children are involved. A group mode allows for extensive modeling, support, and feedback from others. Couples, group, and family interventions are good settings for providing education and skills training. Individual, couple/family, and group modes of treatment do not typically involve adverse side effects. Medication management involves the use of psychotropic drugs to help alleviate and/or manage psychological disorders for which there is a clear biochemical imbalance. Generally, these disorders include psychotic conditions and endogenous or major depression and bipolar disorders. If medication management seems warranted, the client will also need to be evaluated and followed by a physician. Clients rarely learn new ways of problem solving and coping with *just* medication, so some mode of psychotherapy in addition is often warranted. Also, medications pose greater risks and adverse side effects to clients than the other modes of treatment.

The mode of treatment is often indicated by the nature of the client's problem. As Beutler and Clarkin (1990)

observe, "In general, if the symptom of conflict reflects a transient or uncomplicated pattern that is under the control or direction of the individual patient, with little confounding from the current family environment, then it can be dealt with in either individual or group therapy. If, however, the symptom or conflict is significantly confounded by the current marital/family interpersonal environment, then family/marital therapy format may be the format of choice. Likewise, if the symptom or conflict is interpersonal in nature, extends beyond the home environment, is easily observed to be destructive in group interactions, group therapy may be the format of choice" (pp. 119–120).

Specific descriptions of clients and client problems suitable for various modes of treatment are presented in Table 11-10. Although the guidelines in this table are not necessarily inclusive of all the conditions leading to specific modes of treatment, one advantage of having guidelines such as these (and indeed an advantage of even doing treatment planning) is that such guides may counteract the tendencies of some helpers to develop plans and select and use interventions on the basis of how they have been trained rather than on the basis of the client's identified problems and goals. This unidimensional focus leads to unidimensional treatment. So, for example, someone who received training primarily

TABLE 11-10. Relationship of client problems and mode of therapy treatment

Mode	Client characteristics and problems
Individual	1. The client's symptoms or problematic ways of relating interpersonally are based on internal conflict and a coping style that manifests itself in repetitive life patterns that transcend the particulars of the current interpersonal environment. 2. The client is an adolescent or young adult who is striving for autonomy from family of origin. 3. Problems or difficulties are of such an embarrassing nature that the privacy of individual treatment is required for the clients to feel safe.
Couple	1. The partners are committed to each other and present with symptoms/conflicts that occur almost exclusively within the coupleship. 2. One partner has an individual symptom—for example, agoraphobia or depression—that is maintained or exacerbated by the couple's interaction patterns. 3. There is a need to involve one partner in an effective treatment program for the other (e.g., one person suffers from anorexia or obesity, or from phobias, and the partner is needed to assist in behavioral treatment, increase treatment compliance, and provide general support). 4. The couple's relationship suggests the presence of some role inflexibility.
Family	1. Family problems are presented as such without any one family member designated as the identified client; problems are predominantly within the relationship patterns. 2. Family presents with current structured difficulties in intrafamilial relationships, with each person contributing collusively or openly to the reciprocal interaction problems. 3. Acting-out behavior is exhibited by an adolescent member. 4. Improvement of one family member has led to symptoms or signs of deterioration in another. 5. Chronic mental disorder (e.g., schizophrenia) is present in one family member; there is a need for the family to cope with the condition.
Group	A. Heterogeneous Groups (clients in group have different problems) 1. Client's most pressing problems occur in current interpersonal relationships, both outside and inside family situations. Examples could include these: (a) client is lonely and wishes to get closer to others; has social and work inhibitions, excessive shyness (b) client has an inability to share; manifests selfishness and exhibitionism; needs excessive admiration; has difficulty perceiving and responding to the needs of others (c) client is excessively argumentative; oppositional toward authority; shows passive-aggressive traits (d) client is excessively dependent; relatively unable to individuate from family of origin; has difficulties with self-assertion (e) client has an externalizing coping style in interpersonal situations; tends to act immediately on feelings 2. The client may not have predominant interpersonal problems, but there may be other reasons to refer to heterogeneous group, such as these: (a) becomes intensely involved with individual therapist and cannot maintain self-observation (b) is extremely intellectualized and may benefit by being confronted about this defensive style B. Homogeneous Groups (clients in group share a specific and common problem) 1. specific impulse problems such as obesity, alcoholism, addictions, gambling, violence 2. problems adapting to and coping with acute, environmental stressors such as cardiac ailments, divorce, iliostomy, terminal illness 3. problems associated with a specific but transient developmental phase of life such as child rearing, geriatrics 4. specific symptom constellations such as phobias, schizophrenia, bipolar disorder

SOURCE: From *Systematic Treatment Selection*, by L. Beutler and J. Clarkin, pp. 126–131. Copyright © 1990 by Brunner/Mazel, Inc. Reprinted with permission.

in couples and family or system-based approaches may view all presenting problems and corresponding interventions as family oriented, whereas someone else who has been trained solely in individual counseling may miss or overlook important group and systemic parts of the problem as well as other nonindividual interventions. The importance of having a multidimensional treatment perspective cannot be overemphasized! Recall from an earlier discussion of efficacy studies in the area of marriage and family therapy that the more severe and pervasive the problem, the more critical is the use of multiple modes of treatment.

Another consideration in choosing mode of treatment has to do with cost effectiveness. Although we discuss this in the following section in greater detail, the individual mode requires the most time from the therapist and client; the group mode requires the least, with couples and family therapy following somewhere between as two or more clients are seen by a therapist in a single session. Couples treatment and family treatment also have been found to be more cost effective for standard inpatient and/or residential treatment for schizophrenia, delinquency, and severe adolescent conduct disorders, and adolescent and adult substance abuse (Pinsof & Wynne, 1995).

Cost Effectiveness of Treatment

As mentioned in the previous section, an increasingly important factor in selecting types, duration, and mode of treatment involves cost effectiveness (Miller & Magruder, in press). According to Herron, Javier, Primavera, & Schultz (1994), the cost of therapy "reflects specific conceptions of mental health involving a range of available goals . . . the current controversy is the latest development in society's historical uncertainty about the value and, therefore, the cost of different levels of mental health" (p. 106). As these authors point out, current funding policies tend to take the view that the amount of psychotherapy to be covered by insurance is whatever is deemed enough to restore or maintain a client's mental health "at the level of necessity," that is, "the absence of symptoms that prevent individuals from carrying out the tasks necessary for maintaining their life" (p. 106). One of the implications of this "standard of care" from a *cost* viewpoint is that there is little concern with either *improvement* or *prevention*. As Herron et al. (1994) note, "Levels of functioning below the acceptable range are considered grounds for psychotherapy but the need for a higher level is considered unnecessary," at least from the point of view of managed mental health care (p. 106).

Herron et al. (1994) identified two areas that all of us must be concerned with in considering the costs and benefits of therapy. One is the incorrect but often used assumption that equates usage with effectiveness. Many health maintenance organizations limit mental health coverage to 20 sessions or fewer per year and assume that this number of visits represents a "safe harbor" for even those people needing more than the average amount of therapy. Yet as Herron and colleagues (1994) point out, data exist that dispute this assumption. One very comprehensive study suggested that 52 sessions per year was the best number for maximum effectiveness for the largest number of people (Howard, Kopta, Krause, & Orlinsky, 1986). The American Psychological Association supports the 52-session concept as one that is both cost *and* clinically effective (Welch, 1992). Obviously, there will be clients who need more therapy and, lacking in either their own resources or in access to insurance coverage, will be deprived of needed treatment. Indigent clients and some clients of color are at greater risk of not having access to mental health services than are those in more favorable circumstances. It is difficult to make sweeping generalizations about an effective duration for *all* clients; indeed, this is why treatment plans are developed for each individual client.

A second issue is that cost containment should be better balanced with concern for quality care (Newman & Tejeda, 1996). Managed care companies have relied on utilization review to limit both the number of treatment sessions and the kind of provider. As a result, some providers who charge higher fees or use more treatment sessions have been excluded from provider panels; worse yet, some clients have been pushed out of therapy too soon.

Eckert (1994) notes that "denial of appropriate access to clinical care has resulted not only in psychological suffering for patients but also in higher overall costs associated with administration, litigation, and recommended treatment" (p. 3). Some managed care companies now recognize that utilization review may cost more than it saves and are now using a process instead called *provider profiling*. Provider profiling involves collecting data on the most cost-effective therapists and giving them the most referrals. Therapists are "scored" on such things as level of client satisfaction, level of clinical outcomes, and time and cost reported to achieve these outcomes.

An assessment of cost effectiveness of therapy treatment addresses such questions as these:

1. With this particular client who has this set of identified problems and goals, is treatment A more cost effective than treatment B?

2. Do the benefits of this particular treatment outweigh the costs?

3. How I can achieve the best possible outcomes with this client in the least amount of time?

For the typical practitioner, cost effectiveness means the following:

1. Describe alternative treatment or intervention or modalities available for each client's stated goal.

2. Generate a list of both costs and benefits for each treatment modality.

   Costs include such things as the therapist's or clinic's fee for provision of the service, costs borne by the client including any lost time from work, and also any overhead costs associated with the therapist's practice or work setting. Costs also involve an estimation of the number of sessions that would be required to use treatment A and/or treatment B.

   Benefits include such things as changes in the client's overall level of functioning, and level of improvement in symptom reduction and any changes in the client's overall quality of life. Benefits also include an assessment of the impact of treatment on significant others, significant settings in the client's life, family, school, work and so on. Obviously, time spent on any given treatment modality with one modality taking less time to achieve the same results is also a benefit.

3. Select and use the treatment modality that is the most cost and clinically effective and where the likelihood of helping clients reach their goals is greatest; that is where the benefits outweigh the costs.

GENDER AND MULTICULTURAL ISSUES IN TREATMENT PLANNING AND SELECTION

At the outset, practitioners need to realize the origins of psychotherapeutic treatment strategies and interventions and the implications of these origins for treatment planning with clients who feel "marginalized" or out of the mainstream. Most of the strategies we describe in the second half of this book are drawn from theoretical positions developed by a founding father or fathers. As a result, often the most widely used therapeutic strategies reflect the dominant values of the mainstream culture defined by "whiteness, middle-class position, youth, able-bodiedness, heterosexuality and maleness" (Brown, 1994, p. 63). Clients who fall outside these characteristics are more likely to feel marginalized. Another important point is to recognize that some traditional psychotherapeutic techniques found to be successful with Euro-American clients may be culturally contra-indicated for clients who fall outside these descriptions. Typically, traditional treatment planning and selection have not addressed issues of race, gender, and social class—particularly poverty and international perspectives (Brown, 1994). Two important questions for practitioners to address in overall treatment planning are these:

1. Do the usual or recommended treatment approaches oppress a client even more?

2. What should therapy look like and offer when treating "multiply oppressed" clients? (Brown, 1994).

If such questions as these are not addressed, practitioners run a greater risk of having Eurocentric and androcentric ideas reflected in their treatment plans.

Another way to look at this is to assess the client's worldview and then to consider how that worldview affects the client's receptivity to counseling in general and to specific treatment interventions. "Worldview" is defined as our basic perceptions and understanding of the world (Treviño, 1996). Sue and Sue propose that there are two major dimensions of worldview that impact treatment planning. One is locus of control and the other is locus of responsibility. Locus of control (Rotter, 1966) can be either internal or external. In internal locus of control, people believe they can shape their own destiny by their beliefs and actions. In external locus of control, they believe their destiny is shaped by fate or luck and occurs independently of their own actions. Locus of responsibility, derived from attribution theory (Jones et al., 1972), also can be internal or external. If internal, people feel they are responsible for their success and failure. If external, they feel that the system is responsible for their problems and their successes.

Essandoh (1995) provides a good cross-cultural example of how attribution style may vary among clients of color. He notes that clients in some cultures which value interdependence, humility, and collectivism over independence, selfishness, and individualism, attribute success to external factors (such as God, ancestral spirits, significant others, counselors) and failure to external factors (such as evil spirits, bad luck, God's will). He points out that this attribution usually reflects a desire to be modest and does not suggest depression, low self-esteem, or problems in assertiveness. He concludes that when problems are externalized, the counselor "should communicate respect for this different worldview (if [the problems] are attributed to evil spirits especially) and offer an alternative worldview and possibly some suggestions on how these problems could be solved" (p. 357).

IC-IR	IC-ER
I.	*III.*
I'm OK and have control over myself.	I'm OK and have control, but need a chance
Society is OK, and I can make it in the system.	Society is not OK, and I know what's wrong and seek to change it.

EC-IR	EC-ER
II.	*IV.*
I'm OK, but my control comes best when I define myself according to the definition of the dominant culture.	I'm not OK and don't have much control; I might as well give up or please everyone.
Society is OK the way it is; it's up to me.	Society is not OK and is the reason for my plight; the bad system is all to blame.

FIGURE 11-2. Sue's cultural identity quadrants   *SOURCE:* From *Counseling and Development in a Multicultural Society,* 2nd Ed., by J. A. Axelson, p. 399. Copyright 1993 by Brooks/Cole Publishing Company, a division of International Thomsom Publishing Inc.

Figure 11-2 is derived from transactional analysis and shows an analysis of Sue and Sue's (1990) quadrants of four possible dimensions of locus of control and locus of responsibility. The type of client that we described earlier in this chapter—white, middle class, young, able-bodied, heterosexual and male—is most likely to fall in the upper left quadrant (I) IC-IR, reflecting the "rugged individualism" of the mainstream culture in the United States. Treatment approaches such as cognitive therapy and problem solving that are person centered, reflect an individual focus, and are rational in orientation fall into this quadrant as do self-help groups. These treatment approaches are likely to work best with clients who subscribe to dimensions of both internal control and internal responsibility.

Clients who fall in the other three categories are likely in some way to feel marginalized and to encounter a number of specific problem situations in their relationships with the dominant society. Lum (1996, p. 115) has described five specific treatment strategies relevant to clients with the kinds of worldview depicted in these three categories. These five kinds of interventions* are described as follows:

- *Liberation* (vs. oppression) is the client's experience of release or freedom from oppressive barriers and control when change occurs. For some, it accompanies personal growth and decision-making: The client has decided no longer to submit to oppression. In other cases, liberation occurs under the influence of environmental change—for example, the introduction of a job-training program or the election of an ethnic mayor who

makes policy, legislative, and program changes on behalf of people of color.

- *Empowerment* (vs. powerlessness) is a process in which persons who belong to a stigmatized social category can be assisted to develop and increase skills in the exercise of interpersonal influence. Claiming the human right to resources and well-being in society, individuals experience power by rising up and changing their situational predicament. The first step to empowerment is to obtain information about resources and rights. Then, choosing an appropriate path of action, the client participates in a situation in which his or her exercise of power confers palpable benefits. Practical avenues of empowerment include voting, influencing policy, or initiating legislation on the local level.

- *Parity* (vs. exploitation) relates to a sense of equality. For people of color, parity entails being of equal power, value, and rank with others in society and being treated accordingly. Its focal theme is fairness and the entitlement to certain rights. Parity is expressed in terms of resources that guarantee an adequate standard of living, such as entitlement programs (Social Security, Medicare), income maintenance, and medical care.

- *Maintenance of culture* (vs. acculturation) asserts the importance of the ideas, customs, skills, arts, and language of a people. By tracing the history of an ethnic group, counselor and client can identify moments of crisis and challenge through which it survived and triumphed. Applying such lessons of history to the present inspires the client to overcome obstacles and provides a source of strength on which the client may draw. Maintenance of culture secures the client's identity as an ethnic individual.

- *Unique personhood* (vs. stereotyping) is an intervention strategy by which stereotypes are transcended. Functional casework asserts the view that each person is unique in the helping relationship and that there is something extraordinary in each individual. When people of color act to gain freedom

---

*SOURCE: From *Social Work Practice and People of Color: A Process-Stage Approach,* 3rd Ed., by Doman Lum. Copyright © 1996 by Brooks/Cole Publishing Company, a division of International Thomson Publishing Inc.

from social stereotypes, they assert their unique personhood and discover their humanity.

In the bottom left quadrant (II) EC-IR, we find clients who reflect the U.S. mainstream culture's definition of self-responsibility (IR) but feel very little personal control in their lives (EC). These are likely to be clients who feel caught between the system of the mainstream culture and some other cultural affiliation of their own that impacts their destiny, such as being a woman; having a mental or physical challenge; being elderly, poor, or gay/lesbian and bisexual; and/or being of a different racial/ethnic origin than Euro-American. Although they share aspects of two or more cultures, these individuals have become more acculturated to aspects of the mainstream culture than to their original culture and tend to deny the impact of their other cultural affiliations. They are likely to be in subordinate positions of a dominant-subordinate power structure, yet either do not realize this or believe it is their fault. Women who are battered by their male partners are an example as are clients with severe trauma in their histories. Non-Euro-American clients who hold this worldview even often prefer a Euro-American helper because of rejection of their own race and culture (Sue & Sue, 1990).

Treatment approaches that are likely to be helpful with clients with this worldview are approaches that are more intuitive and subjective rather than rational and objective, help clients recognize the effects of their acculturation to the mainstream culture, help them recognize the effects of a dominant-submissive hierarchy, and also help them develop a voice, agency, and resources about themselves—specifically about reclaiming their strengths and achievements, about their own needs and desires, and their own experiences and expressions of personal authority and self-worth (Young-Eisendrath, 1984). The therapist must also operate within a working alliance based on collaboration and mutual respect, and embrace a position of antidenomination when working with clients who hold an EC-IR worldview. Using Lum's (1996) model, *maintenance of culture* and *liberation* are especially useful with clients of this worldview.

Reframing the issues presented by these clients as social-contextual problems and helping clients separate their own personal identities from these issues can also be helpful (Young-Eisendrath, 1984). As Sue and Sue (1990) note, "A culturally encapsulated White counselor who does not understand the sociopolitical dynamics of the client's concerns may unwittingly perpetuate the conflict" (p. 150).

The relationship and communication skills we describe in Chapters 2 through 6 are especially useful for EC-IR clients, particularly the use of empathy and listening responses as ways of consciously validating the client's voice and experiences. As Young-Eisendrath (1984) notes, "Consensual validation is the foundation of human sanity in that it is the means by which we know the truthfulness of an experience and by which we replenish our self-esteem" (p. 60). Thinking of this another way, it does little good to give the client a *voice* if the therapist does not have an *ear* or a way to understand the newly emerging perspectives a client in the EC-IR mode may be trying to articulate (Young-Eisendrath, 1984).

At the upper right of the quadrant (III) are clients who hold the opposite worldview—IC, ER—in the sense that they believe in their own personal ability to shape their lives if given the opportunities (IC). They also believe that the "system" blocks opportunities because of oppression, stereotyping, and exploitation (Sue & Sue, 1990). In contrast to the EC-IR clients, IC-ER clients typically identify with and take pride in aspects of their own cultural identity, even if it is different from what is espoused by the mainstream. These are clients who have worked through levels of their racial, sexual, or gender identity (as we described earlier in Chapter 2) sufficiently to now reclaim and honor the characteristics of their cultural affiliations and gender. Sue and Sue (1990) predict that as clients who vary from the nondominant culture become increasingly conscious of aspects of their own cultural identity, there will be more and more clients with this IC-ER worldview. The authors contend that clients with this worldview will be the least trusting and most difficult for typical Euro-American counselors who hold the IC-IR views of the dominant culture. These clients are likely to terminate counseling even before the treatment planning stage occurs because of their dissatisfaction with the counselor and the ways in which counseling is conducted that does not "fit" for them. These clients are also more likely to challenge counselor statements or actions that appear to be oppressive or stereotypical. In working with clients who hold an IC-ER view, counselors can expect to deal with more of the "tests of trust" described in Chapter 4. These clients may also be more reluctant to be open and disclosive, particularly with a counselor who holds an IC-IR worldview.

Clients with an IC-ER belief are also more likely to view the problems as being outside themselves and as the product of "the Establishment"; they will be less likely to establish an alliance with a counselor who holds an IR worldview and sees problems as residing within the client.

These clients are much more likely to expect and utilize treatment approaches that are directive in style, action oriented in technique, and systems-oriented in focus. As Sue and Sue (1990) note, with IC-ER clients, "demands for the counselor to take external action on the part of the client will be strong . . . while most of us have been taught not to intervene externally on behalf of the client, all of us must look seriously at the value base of this dictate" (p. 157). With IC-ER clients the action-oriented skills we described in Chapter 7 and also many behavioral interventions are quite useful. Using Lum's (1996) model, *parity, unique personhood,* and *liberation* are also especially useful interventions for these clients.

In the fourth quadrant, in the lower right (IV), are clients who hold an EC-ER worldview. These clients are the ones most likely to feel hopeless, discouraged, and disempowered as they are high in both external control and system blame, believing there is little they can do to overcome severe external obstacles such as discrimination and exploitation (Sue and Sue, 1990). Their high EC response may indicate that they have truly given up and have developed what Seligman (1982) refers to as "learned helplessness," the idea that people "exposed to prolonged noncontrol in their lives develop expectation of helplessness in later situations" (p. 151). The high EC response also can indicate that these clients have simply learned that the best way to deal with their disempowerment is to placate the powerful. "Passivity in the face of oppression is the primary reaction of the placater" (Sue & Sue, 1990, p. 151). Both learned helplessness and placating behaviors represent survival strategies by the people who adopt them; direct expression of anger or healthy assertiveness is viewed as too risky because of the potential for punishing consequences from some aspect of the environment (Taylor, Gilligan, & Sullivan, 1995). When slavery existed, if African Americans did not show deferential behavior to their white "masters" they were severely punished. Unfortunately "holdover" aspects of this demeaning and objectionable social phenomenon still exist at a conscious level in many current interactions between Euro-American and non-Euro-American races. At some level we are all constantly influenced as well by *unconscious* stereotyping and bias about race, ethnicity, sex roles, sexual orientation, and so on. Women, too, have learned the risks of direct expression of anger in a society in which power is held largely by men. As Young-Eisendrath (1984) notes, "When a woman, in any social context, is insistent, angry, or convinced of her authority, she is often interpreted as domineering, overwhelming, or overcontrolling. Rarely is she simply understood to be angry or authoritative" (p. 72). EC-ER clients may be overly

polite and deferential with counselors and, unlike IC-ER clients, may not challenge or confront aspects of a counselor's bias because they perceive such behavior as either too risky or as not worth the effort. They are also unlikely to become directly and openly angry with you, even when you have said or done something that in some way violates them. They may act as though they believe your ideas are of value but, once outside the session, they may discard your ideas as unworkable or untenable for them in their particular life space. In planning treatment approaches with EC-ER clients, first you must recognize the survival values of their behaviors so you do not perceive these clients as "lacking in courage and ego strength" (Sue & Sue, 1990, p. 152). As with EC-IR clients, validation and respect form the foundation for all treatment approaches. Validation of their anger and resentment and also of their strengths and successes is very useful. In addition, the use of interventions that teach these clients new coping strategies is also very helpful (Sue & Sue, 1990). In Lum's (1996) model, *empowerment* interventions are particularly useful with these clients.

In applying this model to treatment planning, remember several caveats, as noted by Sue and Sue (1990, pp. 157–158). First, each worldview has something to offer that is useful and positive; for example, "the individual responsibility and achievement orientation of quadrant I, biculturalism and cultural flexibility of quadrant II, ability to compromise and adapt to life conditions of quadrant III, and collective action and social concern of quadrant IV need not be at odds with one another" (Sue & Sue, 1990, p. 157). The counselor's task is to plan treatment approaches that help clients to integrate aspects of the worldview that will increase their effectiveness and well-being. Second, research on this model is in a developing stage, so some of the observations we have made are tentative (Treviño, 1996). Third, the four styles described are conceptual and in reality "most people represent mixes of each rather than a pure standard" (Sue & Sue, 1990, p. 158). Thus, in practice, counselors will provide treatment planning to clients who hold both primary and secondary views about control and responsibility.

In general, we offer the following guidelines for developing multicultural treatment plans:

1. Make sure your treatment plan is *culturally* as well as clinically *literate* and relevant. Consider the cultural illiteracy of the suggestion that an Asian parent discipline his or her child by the American parenting technique of "grounding" or "time out." "For Asian families, being excluded from the family is the worst possible punishment one can

endure; exclusion is extremely rare. Therefore, when children misbehave, they are threatened with banishment from the family and told to get out. Of course, the children have to fight to stay in the family. Once in, they never leave; sons bring their wives in and expect the parents to help raise the grandchildren" (Berg & Jaya, 1993, p. 32). In other words, make sure the plan reflects the values and worldviews of the *client's* cultural identity, not your own.

2. Make sure your treatment plan addresses the needs and impact of the client's *social system* as well as the individual client, including (but not limited to) oppressive conditions within the system. For example, consider the case of a Native American youth you are working with for substance use. Whereas some cognitive-behavioral interventions may be useful, they are not sufficient unless accompanied by an exploration of the client's cultural context and what and how oppressive acts may contribute to the use of substances. Freire (1972) in the classic book, *The Pedagogy of the Oppressed,* states that an important purpose of counseling is *conscientizacào,* the development of "critical consciousness"—specifically, the development of awareness of oneself in a social context.

3. Make sure your treatment plan addresses the relevant *indigenous* practices and *supports* of your client. In other words, make sure that your treatment plan considers the role of important *subsystems* and *resources* in the client's life such as the family structure and external support systems. Networking with these resources of extended family, local community, and spiritual practices is useful, as numerous studies have shown collaboration between the therapist and local support systems helps the therapist develop a treatment plan that is more "culturally syntonic" (Rosado & Elias, 1993, p. 454).

4. Make sure your treatment plan addresses the client's view of *health* and *recovery* and *ways of solving problems.* The client's spirituality may play an important role in this regard; folk beliefs, mythology and supernatural forces may all be significant factors.

5. Assess for and consider the client's level of acculturation and language dominance and preference in planning treatment. Incorporate the use of culturally relevant themes, scripts, folk tales, proverbs, and metaphors and also the possibilities of "language switching" as culture-specific treatment interventions into your overall treatment plan (Santiago-Rivera, 1995).

6. Make sure that the length of your proposed treatment matches the needs and *time perspective* held by the client. As Rosado and Elias (1993) note, clients must survive before they can thrive (p. 452). Pragmatic strategies,

tangible coping skills, and short-term treatments may be favored by some clients. Rosado and Elias (1993) recommend tailoring treatment plans to the ecological needs of the client even if this means that treatment is of shorter duration. "It is more desirable to have the client return several times a year for specific issues than to drop out and never to return for much-needed mental health issues" (Rosado & Elias, 1993, p. 454).

We have found a very useful case illustration describing the design of helping interventions from a multicultural framework in the work of Cheatham, Ivey, Ivey, and Simek-Morgan (1993). The case involves work with a low-income Puerto Rican woman who suffers from *ataques de nervios* (physiological reactions that may be typically related to trauma and grief in the Puerto Rican culture). Notice when you read this case how skillfully the counselor weaves aspects of treatment planning into and with the client's cultural identity. We provide you with an opportunity to do this on your own in the learning activity on p. 288.

CASE STUDY: A PUERTO RICAN WOMAN SUFFERING FROM *ATAQUES DE NERVIOS**

The client colleague is a single parent, twenty-five years of age with two children. (As once was common in Puerto Rico) she has been sterilized with only minimal information given to her before she gave consent. She has suffered physical abuse both as a child and in more recent relationships. The following is an example of how multicultural counseling and therapy might use cultural identity development theory to facilitate *conscientizacào,* the generation of critical consciousness and effective treatment planning.

Acceptance—Diagnostic signs. The client enters counseling hesitatingly as her *ataques de nervios* are increasing in frequency. A physician has referred the client to you believing that the fainting spells are psychological in origin as no physical reasons can be found. As you talk with the client, you discover that she blames herself for the failures in her life. She comments that she is "always choosing the wrong man," and she states she should have been sterilized sooner and thus fewer children would be born.

Acceptance—Helping interventions and producing dissonance. Your intervention at this stage is to listen, but following Freire (1972, pp. 114–116), you can seek to help her codify or

*SOURCE: From "Multicultural Counseling and Therapy" by H. Cheatham, A. Ivey, M. B. Ivey, and L. Simek-Morgan. In *Counseling and Psychotherapy: A Multicultural Perspective,* 3rd Ed., by A. Ivey, M. B. Ivey, and L. Simek-Morgan, pp. 114–115. Copyright © 1993 by Allyn and Bacon. Reprinted by permission.

make sense of her present experience. You use guided imagery [see Chapter 13] as you help her review critical life events—the scenes around sterilization, the difficulties of economic survival when surrounded by others who have wealth, and actual discrimination against Puerto Ricans in nearby factories. . . . Through listening, the move to a more critical consciousness is begun. But, at the same time, your client colleague needs help. You may see that she has sufficient food and shelter; you may help her find a job. You may teach her basic stress management and relaxation [see Chapter 17], but especially you listen and learn [see Chapter 6].

Naming and resistance—Diagnostic signs. At this point, your client is likely to become very angry, for the responsibility or "fault" which she believed was hers is now seen as almost totally in the oppressive environment. Her eyes may flash as she talks about "them." An emotional release may occur as she becomes aware that the decision for sterilization was not truly hers, but imposed by an authoritarian physician. The woman is likely to seek to strike back wherever possible against those who she feels have oppressed her. In the early stages of naming, she may fail to separate people who have truly victimized her from those who have merely stood by and said nothing.

Naming and resistance—Interventions to help and to produce dissonance. Early in this stage, you are very likely to do a lot of listening. You may find it helpful to teach the client culturally-appropriate assertiveness training and anger management. There may be a delayed anger reaction to traditional sex roles. Later, this client may profit from reality therapy. However, the therapy must be adapted to her relational Puerto Rican heritage. You may support constructive action on her part to change oppressive situations. In the later stages of work with her, you may want to help her see that much of her consciousness and being depends on her *opposition* to the status quo and that she has given little attention to her own real needs and wishes. (At this point, identifying and naming contradictions between self and society may be especially important.)

Reflection and redefinition—Diagnostic signs. It gets very tiring to spend one's life in total anger toward society and others. The consciousness-raising theories find that at this stage that clients often retreat to their own gender and/or cultural community to reflect on what has happened to them and to others. Responsibility is now seen as more internal in nature, but keen awareness of external issues remains. You may note that the client colleague at this stage is less interested in action and more interested in understanding self and culture. There may be a great interest in understanding and appreciating her Puerto Rican heritage and how it plays itself out in North America.

Reflection and redefinition—Interventions to help and produce dissonance. Teaching clients the cultural identity development theories can be useful for them at this stage in that they help explain issues of development in culture. In addition, culturally appropriate theories such as . . . feminist theory may

be especially helpful, although they are useful at all consciousness levels. Cognitive-behavioral, psychodynamic, and person-centered theories may be used if adapted to the culture and needs of the person. (The reflective consciousness is still considered a form of naive consciousness by Freire, as much of the emphasis is on the individual with insufficient attention given to systemic roots of difficulties.)

Multiperspective integration—Diagnostic signs. The client draws from all previous stages as appropriate to the situation. At times, she may accept situations; at other times, be appropriately aggressive and angry, and later withdraw and reflect on herself and her relationships to others and society. She is likely to be aware of how her physical symptoms of *ataques de nervios* were a logical result of the position of women in her culture. She is able to balance responsibility between herself and society. At the same time, she does not see her level of *conscientizacào* as "higher" than others. She respects alternative frames of reference.

Multiperspective integration—Interventions to help and produce dissonance. You as helper may ask the client to join with you and your group to attack some of the issues that "cause" emotional personal and financial difficulty. The Puerto Rican woman may establish a family planning clinic with accurate information on the long-term effects of sterilization or she may establish a day-care center. The woman is clearly aware of how her difficulties developed in a system of relationships, and she balances internal and external responsibility for action. In terms of introducing dissonance, your task may require helping her with time management, stress management, and balancing the many possible actions she encounters. You may also arrange to see that she has accurate feedback from others about her own life and work. (You do not merely encourage her to work to transform the system. You also work with her to facilitate the process. You and your client colleague are now working together to produce cultural change in oppressive conditions.)

THE PROCESS OF TREATMENT PLANNING AND EMPOWERED CONSENT

In our opinion, the choice of appropriate counseling strategies is a joint decision in which both counselor and client are actively involved. We believe it is a misuse of the inherent influence of the helping process for the counselor to select a strategy or to implement a treatment plan independent of the client's input.

During the last decade, we have also witnessed an increasing consumerism movement in counseling and therapy, which has led to the following changes:

1. The client needs to be an active, rather than passive, participant in treatment planning.
2. The client's rights need to be made explicit.

LEARNING ACTIVITY 36

Gender and Multicultural Factors in Treatment Planning and Selection

In this activity, you are given a case description of a low-income woman living in a rural area of the United States. Your task is to read the case carefully and identify the potential multicultural factors that are present in working with this woman using the six guidelines we discuss on pp. 285–286. After this, with a partner or in a small group, identify a treatment plan implementing various interventions that address the multicultural factors you have identified above. Consider the type, duration, and mode of treatment based on the client's identified problems and stated goals. Feedback follows on p. 290.

The Case of Jane

Jane Wiggins is a 34-year-old Euro-American woman living in an isolated rural area in the Appalachian Mountains. She has been referred for counseling to the nearest mental health center because she sought treatment at the local health care clinic following a rape.

She is very suspicious of the counselor, who is a Caucasian man, and she talks reluctantly and without much eye contact. Gradually she reveals that she is married, unemployed, has no children, and lives with her husband who receives Social Security disability payments because of his very poor health. She has lived in this area all her life. She indicates that she has been followed for the last year by a White man who also grew up in the area. She knows him by name and sight only. In addition to following her, he has also sent her numerous letters and has made phone calls containing lewd and suggestive remarks.

She indicates that she and her husband went to the sheriff's office several times to complain, but their complaints were never followed up. It appears that this man may be related to a deputy in the sheriff's office. She indicates that several weeks ago her husband had gone over the hill to a neighbor's house to visit and, unknown to her, had left the door of their home unlocked. The man who has been following her apparently was around, noticed the husband's departure, and came into her house and raped her. She said she had told no one other than the neighbor and her husband because she feels so ashamed. She indicates she is a very religious person and has been reluctant to go to church or to confide in her minister for fear of what the church people may say. She has also been reluctant to tell her parents and sister, who live nearby, for the same reason. She sees no point in reporting the rape to the authorities because they dismissed her earlier reports. She feels a lot of guilt about the rape because she believes she should have been able to prevent it. She has been doing a lot of praying about this. As a few sessions go by, she gradually becomes more open with the counselor and seems to indicate she is willing to come back for "as long as it takes" for her to deal with her guilt and sadness; as a result of counseling, she would like to be able to feel happy again and not be so consumed by guilt. She finally discloses she would feel more comfortable talking these things out with another woman, as this is really a "female problem."

---

3. The treatment planning process needs to be demystified. This demystification can occur by having the counselor develop the treatment plan jointly with the client.

4. The client must consent to treatment. This step is important for all clients, regardless of the setting in which they are helped. Counselors also must usually take special precautions with clients of "low power," such as minors and institutionalized persons, to ensure that their rights are not violated and that treatment programs are not implemented without their participation and consent. Occasionally, some therapists argue that they are withholding information about treatment in order to base a therapeutic strategy on "confusion." Limited data, however, as well as ethical and legal principles, suggest that each client has the right to choose services and strategies appropriate to his or her needs.

We believe the counselor is acting in good faith to protect clients' rights and welfare by providing the following kinds of information to clients about strategies:

1. A description of *all* relevant and potentially useful treatment approaches for this particular client with this particular problem

2. A rationale for each procedure

3. A description of the therapist's role in each procedure

4. A description of the client's role in each procedure

5. Discomforts or risks that may occur as a result of the procedure

6. Benefits expected to result from the procedure

7. The estimated time and cost of each procedure

The therapist also needs to state that he or she will try to answer any questions the client has now or later about the procedure and that the client is always free to discontinue participation in the procedure at any time. If the client is a minor, consent must be obtained from a parent or legal guardian, just as consent must be obtained from a guardian or legal representative if the client has been declared mentally incompetent to give consent.

This sort of therapy contract was initially proposed in the 1970s by feminist therapists as a way to help empower clients by giving them active roles as consumers of mental health services. By the 1990s, informed consent had become part of mainstream practice.

But as Brown (1994) points out, the impetus for providing informed consent has shifted from empowering the client to protecting the therapist from litigation. As a result, the primary reasons for even providing consent information to clients—"respect, relationality and empowerment"—have come to be disregarded and amended in contemporary practice (Brown, 1994, p. 182). We like Brown's use of the term *empowered* consent rather than informed consent because, as she points out, the word *informed* raises such questions as "Who is being informed. How? By and about what or whom? Under what conditions of freedom of choice or 'friendly' coercion is consent being given?" (p. 180). Empowered consent provides complete and meaningful disclosure in an interactive way that supports the client's freedom of choice (Brown, 1994). Moreover, as Brown (1994) notes, providing information and consent is not a one-step activity but an ongoing process throughout the therapeutic relationship.

In Chapter 9 and in Figure 11-3 we provide an illustration of a sample outpatient treatment form. This form is representative of many that are currently used by practitioners in various human services settings. Section A, Assessment, incorporates the various assessment information we described in Chapters 8 and 9. Section B, Treatment Plan, incorporates the material on defining and evaluating outcome goals and planning treatments that we have described in this chapter and in Chapter 10. Section C contains three treatment update reports in which progress on outcome goals is noted, based on evaluation of these goals, as well as changes, obstacles, and request for some type of continued services. Section D, Discharge Report, occurs at termination and indicates the degree to which the outcome goals for counseling have been met. We believe the process of treatment planning is most effective when the therapist actively uses such a form during the assessment, goal-setting, evaluation, and treatment selection sessions such as the one depicted in Figure 11-3. Having the client participate in the completion of this form and including a place on the form for

client comments helps to ensure that the client plays an active role in the therapy process.

MODEL DIALOGUE: THE CASE OF JOAN

In this dialogue, the counselor will explore and help Joan plan some of the treatment strategies they could use to work with the first subgoal on Joan's goal chart (Chapter 10) for Terminal Outcome Goal #1. This dialogue is a continuation of the ones described in Chapters 9 and 10. In this session, Joan and the counselor will explore strategies that could help Joan decrease her nervousness about math class and anticipation of rejection from her parents. Note that all three strategies suggested are based on Joan's diagnostic pattern of specific, or focal, anxiety, as opposed to generalized anxiety.

In the initial part of the interview, the counselor will summarize the previous session and will introduce Joan to the idea of **exploring** *treatment strategies.*

1. *Counselor:* Last week, Joan, we talked about some of the things you would like to see happen as a result of counseling. One of the things you indicated was pretty important to you was being able to be more initiating. You had mentioned things like wanting to be able to ask questions or make responses, express your opinions, and express your feelings. We had identified the fact that one thing that keeps you from doing these things more often is the apprehension you feel in certain situations with your parents or in math class. There are several ways we might deal with your apprehension. I thought today we might explore some of the procedures that may help. These procedures are things we can do together to help you get where you want to be. How does that sound?
   Client: It's OK. So we'll find a way, maybe, that I could be less nervous and more comfortable at these times.

In the second response, the counselor tries to explain to Joan what treatment planning involves and the importance of **Joan's input.**

2. *Counselor:* Yes. One thing to keep in mind is that there are no easy answers and there is not necessarily one right way. What we can do today is explore some ways that are typically used to help people be less nervous in specific situations and try to come up with a way that *you* think is most workable for you. I'll be giving you some information about these procedures for your input in this decision.
   Client: OK.

In responses 3 and 4, the counselor suggests possible strategies for Joan to consider. The counselor also explains how one intervention strategy, relaxation, **is related to Joan's concerns and can help her achieve her goal.**

3. *Counselor:* From my experience, I believe that there are a couple of things that might help you manage your nervousness to the point where you don't feel as if you have to avoid the

F E E D B A C K   36
Gender and Multicultural Factors in Treatment Planning and Selection

Consider these guidelines about your plan for Jane:

1. To what extent is your treatment plan *"culturally* literate" —that is, it matches the *client's* values and worldview? In this case, have you considered that Jane is a low-income Euro-American woman who has lived in the same rural community all her life surrounded by family and friends from church. She holds herself responsible for the rape and also feels powerless to get help from local authorities who in fact have been unresponsive to her requests for assistance.

2. How has your plan addressed the impact of the client's *social system* including oppressive conditions within that system? In this case, have you noted that Jane is a poor, White woman living with a husband on disability due to poor health in a rural area near a small town and that their complaints to local authorities have not been taken seriously, compounded by the fact that the man who raped her is related to a deputy in the sheriff's office. In essence, she feels disempowered and silenced by the system. Also, Jane views the rape at this time as a "female" problem and does not yet really see it as an act of social violence and power.

3. In what ways has your plan addressed any relevant *indigenous practices, supports,* and *resources?* In this case, Jane reports herself to be a very spiritual person who relies on the power of prayer to help her through tough times. (Are you familiar with the book *Healing Words* by Dossey, 1993?) How has your plan considered the role of important *subsystems* in the client's life? For this client, the most important subsystems are her family and her church. However, with the exception of her husband who supports her, she feels cut off from both these subsystems because of the nature of the problem, her own views about it, and her fears of her friends' and family's reactions.

4. Does your plan reflect the client's view of *health, recovery,* and *ways of solving problems?* For example, with Jane, spirituality seems to play an important role in the way she solves problems. Gender also appears to be an issue: Jane seems to feel that a female counselor would be better equipped to help her with this issue (consider a referral to a woman counselor and/or to a women's support group).

5. Have you considered the *level of acculturation* and any *language preferences* she has? (Also the use of culturally relevant themes, scripts, proverbs, metaphors.)

(continued)

Implied in this framework is considering the role of Jane's general history; the geographic location in which she lives and how long she has lived there; the type of setting; her socioeconomic status, age, gender, role; and the specific effects of all of this on her language use and comprehension.

In Jane's case, she is a relatively young White woman on a limited and fixed income and has lived all her life in the same area—an isolated rural area near a small town in a section of the United States referred to as Appalachia. These demographics make for some interesting contradictions: She feels safe living in this region, enough to keep doors unlocked, yet she was raped. The area is small enough to know who lives in it and who is a stranger, yet she has no support from the local sheriff's office because the assailant in this case (rather than the victim) is a relative of a deputy.

Influenced by the societal and cultural norms of the area, she feels ashamed about what has happened to her. All these things are likely to impact your plan—for example, she may be mistrustful of you because you are a male and an outsider. She also may view you as part of a social system that, in her eyes, is similar to the sheriff's office.

The themes of cultural mistrust and gender-linked shame can be addressed in the types of treatment you use with Jane.

6. How does the proposed *length* of your plan meet the needs and the perspective of Jane? To what extent does her willingness to return for additional sessions depend on the gender of the counselor? How will her income and her husband's disability status affect her ability to come in for more sessions? Does your agency offer free services or a sliding scale for low-income clients? If not, how can you be her advocate so she can receive the number of sessions she needs?

---

situation. First of all, when you're nervous, you're tense. Sometimes when you're tense, you feel bad or sick or just out of control. One thing we could do is to teach you some relaxation methods [Chapter 17]. The relaxation can help you learn to identify when you're starting to feel nervous, and it can help you manage this before it gets so strong you just skip class or refuse to speak up. Does this make sense?

Client: Yes, because when I really let myself get nervous, I don't want to say anything. Sometimes I force myself to, but I'm still nervous and I don't feel like it.

4. *Counselor:* That's a good point. You don't have the energy or desire to do something you're apprehensive about. Sometimes, for some people, just learning to relax and control your nervousness might be enough. If you want to try this first and it

MENTAL HEALTH NETWORK: OUTPATIENT TREATMENT REPORT (OTR)   Provider:_____

Client Name:_____   Birthdate:_____   Age:_____   Sex: M   F

A. ASSESSMENT
   1. Presenting Problem (Client's Perspective): _____
   _____
   _____

   2. Precipitating Event(s) (Why Help-seeking Now?): _____
   _____
   _____

   3. Relevant Medical History (Medications, Drug/ETOH use, Illness, Injury, Surgery, etc.): _____
   _____
   _____

   4. Prior Psychiatric/Psychological Conditions & Treatments: _____
   _____

   5. Other Relevant History (Vocational/School, Relationship/Sexual, Social/Legal): _____

   6. Brief Mental Status Evaluation: (Check as necessary)

APPEARANCE/DRESS	INTELLIGENCE	JUDGMENT/INSIGHT	DELUS./HALLUCIN.	THOUGHT DISORDER	RECENT MEMORY	REMOTE MEMORY
__appropriate	__high	__intact	__absent	__absent	__intact	__intact
__inappropriate	__average	__impaired	__present	__present	__impaired	__impaired
__not assessed	__low	__not assessed	__not assessed	__not assessed	__not assessed	__not assessed

   7. Mood/Affect: (Describe) _____
   8. Suicide Assessment: (Risk, priors, plan) _____ Homicide Assessment: (Victim, violence, plan) _____
   9. Clinical Formulation (Explanation of symptoms; include strengths/resources, obstacles to treatment/hidden agendas): Please be specific yet brief and clear _____
   _____
   _____

   10. Code Nos. & Names DSM-4 Axis II: _____   Diagnostic Impressions: DSM-4 Axis I: _____

B. TREATMENT PLAN:
   1. Focused, Targeted, Behavior & Measurable GOALS (Prioritize) Specifically Addressing Presenting Problems(s): Use as many rows as needed.
   2. TYPE OF TREATMENT: Cognitive/Behavioral Interpersonal/Insight/Emotional Awareness Other:_____
   3. DURATION: Service dates this OTR: _____# Sessions expected for DISCHARGE: _____Discharge by (DATE): _____
   4. MODE: Individual   Couple   Family   Individual/Family combination   Medication management   Group (if available)   Other (___)
        (90844)   (90847)   (90862)   (90853)   CPT Code __

PROBLEM(S)	GOALS	MEASURABLE SUCCESS CRITERION	SELECTED INTERVENTIONS
1. _____	_____	_____	_____
2. _____	_____	_____	_____
3. _____	_____	_____	_____

Therapist Signature & Phone: _____   Lic. No.: _____   Date: _____

Client Comments: (continued)

FIGURE 11-3. Sample treatment planning form   *SOURCE:* Adapted from the Treatment Planning Form, by Mental Health Network, Inc., 771 Corporate Drive, Suite 410, Lexington, KY. Reprinted by permission.

TREATMENT UPDATE REPORT #1. Service Dates since Intake OTR: _____ Pt. Name: _____

C. GOALS (As indicated on reverse)   PROGRESS ON CRITERION   COMMENTS

                                       YES   SOME   NO

1. _____   ____   _____   ____   _____
2. _____   ____   _____   ____   _____
3. _____   ____   _____   ____   _____

Signifiant changes or interferences: _____

_____

Requested treatment services: _____

Therapist Signature: _____ Lic. No. _____ Date: _____

Update 1 Therapist Phone: _____ C.M. Signature: _____ Cert. Vst. _____ Den. _____ Date: _____

TREATMENT UPDATE REPORT #2. Service Dates since Update #1: _____ No. Sessions to date: _____

GOALS (As indicated on reverse)   PROGRESS ON CRITERION   COMMENTS

                                       YES   SOME   NO

1. _____   ____   _____   ____   _____
2. _____   ____   _____   ____   _____
3. _____   ____   _____   ____   _____

Signifiant changes or interferences: _____

_____

Requested treatment services: _____

Therapist Signature: _____ Lic. No. _____ Date: _____

Update 2 Therapist Phone: _____ C.M. Signature: _____ Cert. Vst. _____ Den. _____ Date: _____

TREATMENT UPDATE REPORT #3. Service Dates since Update #2: _____ No. Sessions to date: _____

GOALS (As indicated on reverse)   PROGRESS ON CRITERION   COMMENTS

                                       YES   SOME   NO

1. _____   ____   _____   ____   _____
2. _____   ____   _____   ____   _____
3. _____   ____   _____   ____   _____

Signifiant changes or interferences: _____

_____

Requested treatment services: _____

Therapist Signature: _____ Lic. No. _____ Date: _____

Update 3 Therapist Phone: _____ C.M. Signature: _____ Cert. Vst. _____ Den. _____ Date: _____

D. DISCHARGE REPORT: Service Dates since Update #3: _____ Final Session Date: _____ Total Sessions: _____

GOALS (As indicated on reverse/above)   SUCCESS CRITERION MET   *Discharge Reason*

                                       YES   SOME   NO   COMMENTS   __Goals met

1. _____   ____   ____   ____   _____   __Ineffective (pt.)
2. _____   ____   ____   ____   _____   __Ineffective (tx.)
3. _____   ____   ____   ____   _____   __Patient moved
                                                                            __Ineligibility
                                                                 __Patient dropped

Therapist Signature & Phone: _____ Lic. No.: _____ Date: _____

FIGURE 11-3. Continued

292

helps you be less nervous to the point where you can be more initiating, then that's fine. However, there are some other things we might do also, so I'd like you to know about these action plans, too.

Client: Like what?

The counselor proposes an additional intervention strategy in response 5 and indicates how this procedure can help Joan decrease her nervousness by **describing how it is also related to Joan's problem and goal.**

5. *Counselor:* Well, one procedure has a very interesting name—it's called "stress inoculation" [Chapter 16]. You know when you get a shot like a polio inoculation, the shot helps to prevent you from getting polio. Well, this procedure helps you to prevent yourself from getting so overwhelmed in a stressful situation, such as your math class or with your folks, that you want to avoid the situation or don't want to say anything.

Client: Is it painful like a shot?

The counselor provides more information about what stress inoculation would involve from Joan in terms of the **time, advantages, and risks of the procedure;** *this information should help Joan assess her preferences.*

6. *Counselor:* No, not like that, although it would involve some work on your part. In addition to learning the relaxation training I mentioned earlier, you would learn how to cope with stressful situations—through relaxing your body and thinking some thoughts that would help you handle these difficult or competitive situations. When you are able to do this successfully with me, you would start to do it in your math class and with your folks. Once you learned the relaxation, it would take several sessions to learn the other parts. The advantage of stress inoculation is that it helps you learn how to cope with rather than avoid a stressful situation. Of course, it does require you to practice the relaxation and the coping thoughts on your own, and this takes some time each day. Without this sort of daily practice, this procedure may not be that helpful.

Client: It does sound interesting. Have you used it a lot?

The counselor indicates some **information and advantages** *about the strategy based on the* **counselor's experience** *and use of it with others.*

7. *Counselor:* I believe I tend to use it, or portions of it, whenever I think people could benefit from learning to manage nervousness and not let stressful situations control them. I know other counselors have used it and found that people with different stresses can benefit from it. It has a lot of potential if you're in a situation where your nervousness is getting the best of you and where you can learn to cope with the stress. Another advantage of this procedure is that it is pretty comprehensive. By that I mean it deals with different parts of a nervous reaction—like the part of you that gets sweaty palms and butterflies in your stomach, the part of you that thinks girls are dumb in math or girls don't have much to say, and then the part

of you that goes out of your way to avoid these sticky situations. It's kind of like going shopping and getting a whole outfit—jeans, shirt, and shoes—rather than just the shoes or just the shirt.

Client: Well, it sounds OK to me. I also like the idea of the relaxation that you mentioned earlier.

The counselor moves on in response 8 to describe another possible treatment strategy, explains what this involves and how it might help Joan manage her nervousness, and **relates the use of the procedure to her problem and goal.**

8. *Counselor:* There's also another procedure called "desensitization" that is a pretty standard one to help a person decrease anxiety about situations [Chapter 19]. It is a way to help you desensitize yourself to the stress of your math class.

Client: Well, how exactly does that work—to desensitize yourself to something?

The counselor explains how this strategy helps Joan decrease her nervousness and **explains elements, advantages, and risks of this strategy.**

9. *Counselor:* It works on the principle that you can't be relaxed and nervous at the same time. So, after teaching you how to relax, then you imagine situations involving your math class—or with your folks. However, you imagine a situation only when you're relaxed. You practice this way to the point where you can speak up in class or with your folks without feeling all the nervousness you do now. In other words, you become desensitized. Most of this process is something we would do together in these sessions and is an advantage over something requiring a lot of outside work on your part.

Client: Does that take a long time?

The counselor gives Joan some information about the **time** *or* **duration** *and the* **mode** *involved.*

10. *Counselor:* This may take a little longer than the other two procedures. This procedure has helped a great many people decrease their nervousness about specific situations—like taking a test or flying. Of course, keep in mind that any change plan takes some time.

Client: It sounds helpful.

The counselor points out more of the **time factors** *involved in these procedures.*

11. *Counselor:* We would be working together in these individual sessions for several months.

Client: That's OK. I have study hall during this period and I usually just talk to my friends then anyway.

In response 12, the counselor indicates **his or her preferences** *and provides information about* **documentation.**

12. *Counselor:* I'd like us to make the decision together. I feel comfortable with all of these things I've mentioned. Also, all three of these procedures have been found to be pretty effective in dealing with the different fears of many people who are

concerned about working on their nervousness in situations so it isn't a handicap. In fact, for some of these procedures there are even guidelines I can give you so you can practice these on your own.

Client: I'm wondering exactly how to decide where to go from here.

In responses 13 and 14, the counselor elicits information about **client preferences.**

13. *Counselor:* Well, perhaps if we reviewed the action plans I've mentioned and go over them, you can see which one you feel might work best for you, at least now. We can always change something at a later point. How does that sound?

Client: Good. There's a lot of information and I don't know if I remember everything you mentioned.

14. *Counselor:* OK. Well, we talked first about relaxation as something you could learn here and then do on your own to help you control the feelings and physical sensations of nervousness. Then we discussed stress inoculation, which involves giving you a lot of different skills to use to cope with the stressful situations in your math class. The third plan, desensitization, involves using relaxation first but also involves having you imagine the scenes related to your math class and to interactions with your parents. This procedure is something we would work on together, although the relaxation requires daily practice from you. What do you think would be most helpful to you at this point?

Client: I think maybe the relaxation might help, since I can practice with it on my own. It also sounds like the simplest to do, not so much in time but just in what is involved.

In the last response, the counselor pursues the option that Joan has been leaning toward during the session, thus building on **client preferences.**

15. *Counselor:* That's a good point. Of the three procedures I mentioned, relaxation training is probably the easiest and simplest to learn to use. You have also mentioned once or twice before in our session that you were intrigued with this idea, so it looks as if you've been mulling it over for a little while and it still sounds appealing and workable to you. If so, we can start working with it today.

SUMMARY AND INTRODUCTION TO THE TREATMENT STRATEGY CHAPTERS

Most clients will present complex problems with a diverse set of counseling goals. Addressing these will require a set of interventions and combinations of strategies designed to work with all the major target areas of a person's functioning. Both counselor and client should be active participants in developing a treatment plan and selecting treatment strategies that are appropriate for the client's problem and desired outcomes. The strategies reflected by the overall treatment plan should be relevant to the client's gender and culture, and sufficient to deal with all the important target areas of change and matched, as well as possible, to the response components of the defined problem. After the strategies have been selected, the counselor and client will continue to work together to implement the procedures.

In the following chapters (12–20) we describe a number of treatment strategies that are primarily (though not exclusively) cognitive-behavioral. We do so because we agree with Beutler and Clarkin (1990) that initial treatment choice is usually symptom rather than conflict focused and also because many of these strategies are empirically validated or probably efficacious. However, we would be remiss if we did not point out that cognitive-behavior therapy in general has only recently begun to address issues of complex cultural and gender influences and identities of clients served by these strategies. In the following chapters we include a section on the use of the strategy with diverse populations. However, as recently as 1995, Hays concluded that this list is quite small because of a "dearth of cognitive-behavior therapy research with minority groups" (p. 313). Lum (1996) has also concluded that until recently, culturally diverse social work practice has also been too sparse.

The questions addressed by Brown (1994) about treatment planning for clients who fall outside the mainstream have also been raised in critiques about the multicultural application of cognitive-behavioral strategies. Kantrowitz and Ballou (1992) note that there is nothing *inherent* in cognitive-behavioral therapy approaches that enhance sensitivity to gender, role, and class issues. In a similar vein, Hays (1995) has noted that "cognitive-behavior therapy does not exclude the consideration of socio-cultural influences, but because it has not been explicit about the impact of racism and other forms of oppression in clients, these forces are easily overlooked, particularly by therapists of dominant cultural groups" (p. 311). As an example of this, there is often a lack of attention to cultural differences, including aspects of oppression and aspects of strengths in a cognitive-behavioral assessment. We have attempted to incorporate this in the model we present in Chapters 8 and 9 and exhort each of you to make explicit attempts to consider such factors in all phases of the therapy process—from the working alliance, to assessment, to selecting and evaluating goals, to planning, implementing, and evaluating treatment.

Cognitive-behavioral approaches have also been criticized because they at least implicitly support and often perpetuate the values and standards of mainstream society, such as assertiveness, independence, verbal ability, and individual change (the IC-IR worldview). These values and standards may be reflected in the goals of treatment and methods of treatment as well as cognitive styles, interactional patterns, and worldviews espoused by therapists to clients. Several issues arise from this. First, an individual client may feel blamed when in fact her problems have been created by unjust social situations. In an example provided by Kantrowitz and Ballou (1992), consider the situation of a female client who has been sexually harassed in her place of work. Her female cognitive-behavioral therapist suggests assertiveness training as the recommended treatment approach. As these authors note, "Although there is nothing wrong with improving assertiveness skills, by focusing on the woman's skill development and schema change, neither the external factors—aggressive sexuality and boundary violation—nor the dominant social norm of women's duty to protect themselves are articulated and questioned" (p. 79).

Moreover, the quality of assertiveness and the value attached to it reflect a Eurocentric perspective. As Hays (1995) so aptly points out, "Although there are ethnic groups including many Native American, Latino, Asian and Arab cultures, in which the quality of respect is more highly valued than that of assertiveness, psychotherapists have yet to develop 'respect training' for individuals who lack this quality in their interactions with others" (p. 313). Moreover, teaching a client from a nondominant group to be more assertive may also pose significant risks for that person (Hays, 1995). Assertiveness is also inconsistent with the specified beliefs of some clients.

Still, cognitive-behavioral approaches have potential strengths for use with culturally diverse clients (Hays, 1995). These include the emphasis on the uniqueness of the individual and the adaptation of therapy to meet this uniqueness, a focus on client empowerment, commitment to therapist-client collaboration, use of direct and pragmatic treatment focus, and emphasis on conscious (versus unconscious) processes and specific (versus abstract) behaviors and treatment protocols (Hays, 1995; Kantrowitz & Ballou, 1992). We agree with Hays (1995) who asserts that overall, "the key to multicultural applications of cognitive-behavior therapy lies in the need for explicit attention to cultural influence and minority populations that have traditionally been ignored" (p. 313).

POSTEVALUATION

Part One

We describe the case and treatment plan of David Chan (Sue & Sue, 1990, p. 259) in this section. After you read the case, identify (1) ways in which the selected treatment interventions used by the counselor conflicted with the client's cultural values and worldview, and (2) recommended type, duration, and mode of treatment you would follow as David's counselor (Chapter Objective One). Feedback follows the postevaluation.

David Chan is a 21-year-old student majoring in electrical engineering. He first sought counseling because he was having increasing study problems and was receiving failing grades. These academic difficulties became apparent during the first quarter of his senior year and were accompanied by headaches, indigestion, and insomnia. Since he had been an excellent student in the past, David felt that his lowered academic performance was caused by illness. However, a medical examination failed to reveal any organic disorder.

During the initial interview, David seemed depressed and anxious. He responded to inquiries with short, polite statements and would seldom volunteer information about himself. He avoided any statements that involved feelings and presented his problem as strictly an education one.

Although he never expressed it directly, David seemed to doubt the value of counseling and needed much reassurance and feedback about his performance in the interview.

After several sessions, the counselor was able to discern one of David's major concerns. David did not like engineering and felt pressured by his parents to go into this field. Yet, he was afraid to take responsibility for changing this decision without their approval; felt dependent on his parents, especially for bringing honor to them; and was afraid to express the anger he felt toward them. Using the Gestalt "empty chair technique," the counselor had David pretend that his parents were seated in empty chairs opposite him. The counselor encouraged him to express his true feelings toward them. Initially, he found this very difficult to do, but David was able to ventilate some of his true feelings under constant encouragement by the counselor. Unfortunately, the following sessions with David proved unproductive in that he seemed more withdrawn and guilt-ridden than ever.

Questions

1. How did the counseling intervention used by this counselor (Gestalt empty chair) conflict with this client's traditional Asian culture values and worldview?

(continued)

POSTEVALUATION   (continued)

2. If you were David's counselor, what type, duration, and mode of treatment would you use?

Part Two

Chapter Objective Two asks you to develop in writing a treatment plan for a given client case using the Sample Treatment Planning Form (Figure 11-3), specifying the type, duration, and mode of treatment. If you currently have a client caseload of your own, we suggest you do this for one of your actual clients and consult with your supervisor or a colleague or instructor after you and your client have completed the Treatment Planning Form. If you are a student and do not yet have clients, we suggest you use the case of Joan described in Chapter 8. Complete Part A, Assessment, based on the model dialogue information in Chapter 9. Complete Part B, Treatment Planning, based on the model dialogue information in Chapter 10 and in this chapter. Do Part B for Joan's first problem and goal *only* listed on her goal chart in Chapter 10 on p. 256 so that you have a description for #1 only of the problem, goal, measurable outcome, and selected intervention on 2-B of the Treatment Planning Form. Feedback for the case of Joan follows the postevaluation.

SUGGESTED READINGS

American Psychiatric Association. (1993). Practice guidelines for major depressive disorder in adults. *American Journal of Psychiatry, 150,* 1–26.

Antonuccio, D. O., Danton, W. G., & Denelsky, G. Y. (1995). Psychotherapy versus medication for depression: Challenging the conventional wisdom with data. *Professional Psychology, 26,* 574–585.

Axelson, J. A. (1993). *Counseling and development in a multicultural society,* (2nd ed.). Pacific Grove, CA: Brooks/Cole.

Barlow, D. H. (Ed.). (1994). *Clinical handbook of psychological disorders* (2nd ed.). New York: Guilford.

Beutler, L., & Clarkin, J. (1990). *Systematic treatment selection.* New York: Brunner/Mazel.

Brown, L. S. (1994). *Subversive dialogues: Theory in feminist therapy.* New York: Basic Books.

Caspar, F. (1995). *Plan analysis: Toward optimizing psychotherapy.* Seattle, WA: Hogrefe & Huber.

Chambless, D. (1993). *Division 12 of APA Task Force on Promotion and Dissemination of Psychological Procedures.* Unpublished report, American University, Washington, DC.

Clinton, J., McCormick, K., & Besteman, J. (1994). Enhancing clinical practice: The role of practice guidelines. *American Psychologist, 49,* 30–33.

Elkin, I., Shea, M., Watkins, J., Imber, S., Stosky, S., Collins, J., Glass, D., Pilkonis, P., Leber, W., Docherty, J., Fiester, S., & Parloff, M. (1989). National Institute of Mental Health Treatment of Depression Collaborative Research Program. *Archives of General Psychiatry, 46,* 971–982.

Field, M. J., & Lohr, K. N. (Eds.). (1992). *Guidelines for clinical practice: From development to use.* Washington, DC: National Academy Press.

Hackney, H., & Cormier, L. S. (1996). *The professional counselor* (3rd ed.). Boston: Allyn & Bacon.

Hanna, F. J., & Ritchie, M. H. (1995). Seeking the active ingredients of psychotherapeutic change: Within and outside the context of therapy. *Professional Psychology, 26,* 176–183.

Hays, P. A. (1995). Multicultural applications of cognitive-behavior therapy. *Professional Psychology: Research and Practice, 26,* 309–315.

Hollon, S. (1996). The efficacy and effectiveness of psychotherapy relative to medications. *American Psychologist, 51,* 1025–1030.

Inglehart, A. P., & Becerra, R. M. (1995). *Social services and the ethnic community.* Needham Hts, MA: Allyn & Bacon.

Kantrowitz, R. E., & Ballou, M. (1992). A feminist critique of cognitive-behavioral therapy. In L. S. Brown & M. Ballou (Eds.), *Personality and psychopathology: Feminist reappraisals.* (pp. 70–87). New York: Guilford.

Lazarus, A. A. (1989). *The practice of multimodal therapy.* Baltimore, MD: Johns Hopkins University Press.

Logan, M. L., Freeman, E. M., & McRoy, R. G. (1990). *Social work practice with Black families.* New York: Longman.

Meyer, R. G. & Deitsch, S. (1996). *The clinicians handbook: Integrated diagnostics, assessment, and intervention in adult and adolescent psychopathology.* (4th ed.). Boston, MA: Allyn/Bacon.

Miller, N., & Magruder, K. (Eds.). (in press). *The cost-effectiveness of psychotherapy: A guide for practitioners, researchers, and policy-makers.* New York: Wiley.

Miranda, J. (Ed.). (1993). Understanding diversity: Special edition, *The Behavior Therapist, 16* (9), 226–241.

Pinsof, W., & Wynne, L. (Eds.). (1995). Family therapy effectiveness: Current research and theory. Special issue, *Journal of Marital and Family Therapy, 21* (4).

Reid, W. H. (1995). *Treatment of the DSM-IV psychiatric disorders.* New York: Brunner/Mazel.

Rosado, J. W., & Elias, M. J. (1993). Ecological and psychocultural mediators in the delivery of service for urban, culturally diverse Hispanic clients. *Professional Psychology, 23,* 450–459.

Sanderson, W. C., & Woody, S. (Eds.). (1995). *Manuals for empirically validated treatments: A project of the Division 12 APA's psychological interventions.* Unpublished manuscript, Albert Einstein College of Medicine, New York.

Santiago-Rivera, A. (1995). Developing a culturally sensitive treatment modality for bilingual Spanish-speaking clients. *Journal of Counseling and Development, 74,* 12–17.

Seligman, M. (1994). *What you can change and what you can't.* New York: Knopf.

Seligman, M. (1995). The effectiveness of psychotherapy: The Consumer Reports study. *American Psychologist, 50,* 965–974.

Smith, K. J., Subich, L. M., & Kalodner, C. (1995). The transtheoretical model's stages and processes of change and their relation to premature termination. *Journal of Counseling Psychology, 45,* 34–39.

Sobell, L., Toneattio, T., & Sobell, M. (1994). Behavioral assessment and treatment planning for alcohol, tobacco, and other drug problems. *Behavior Therapy, 25,* 533–580.

Steenbarger, B. (1993). A multicultural model of counseling: Bridging brevity and diversity. *Journal of Counseling and Development, 72,* 8–15.

Steketee, G. (1994). Behavioral assessment and treatment planning with obsessive-compulsive disorder. *Behavior Therapy, 25,* 613–634.

Thompson, C. L., & Rudolph, L. B. (1996). *Counseling children* (4th ed.). Pacific Grove, CA: Brooks/Cole.

Treviño, J. G. (1996). Worldview and change in cross-cultural counseling. *The Counseling Psychologist, 24,* 198–215.

VandenBos, G. R. (Ed.). (1996). Outcome assessment of psychotherapy (Special issue.). *American Psychologist, 51* (10).

FEEDBACK
POSTEVALUATION

Part One

1. The counselor used a Gestalt empty chair counseling intervention. This intervention, which encouraged David to express feelings toward his parents, clashes with the traditional Asian value of respecting your elders (parents) and refraining from arguing with them even when you don't agree with their views.

2. As David's counselor it is important first to clarify *David's* expectations for counseling because they may be different from your own. David may value a type of treatment intervention that is direct and contains helpful suggestions. It is also important to select a type of treatment that honors and respects David's traditional Asian cultural values—restraint of feelings, restraint of open discussion of problems with a stranger, and collective/family responsibility. Interventions that focus on behaviors and cognitions rather than feelings are going to be more consistent with David's cultural beliefs. Because of this focus, the *duration* of treatment is likely to be short term. However, interventions that require individual self-assertion and expression of feelings are not useful because these clash with his worldview. It is also important to honor and validate the bicultural conflict David is experiencing between obeying his parents' wishes for an engineering career and his own values and preferences. Because of the nature of the conflict and also the Asian value of belonging to a family, family consultation as a *mode* of treatment rather than individual counseling may be helpful. However, in working with David and his parents it would be important not to force self-disclosure, to respect the family's hierarchy, and to try to arrive at a resolution that is mutually acceptable to all parties.

Part Two
If you used the case of Joan, check your responses with the ones on the form on the next page.

(continued)

MENTAL HEALTH NETWORK: OUTPATIENT TREATMENT REPORT (OTR)   Provider:_____(Your Name)_____

Client Name:____Joan____   Birthdate:_____   Age:____16____   Sex: M   (F)

A. ASSESSMENT

1. Presenting Problem (Client's Perspective): _Unhappy with school, expecially math. Feels anxious about "competition."_
She withdraws socially. Dropping grades. Cuts class. Feels pressure from parents about grades and college. Feels afraid
to speak up. Describes herself as indecisive.

2. Precipitating Event(s) (Why Help-seeking Now?): _She is doing poorly in school, especially math and is getting "flak"_
from parents.

3. Relevant Medical History (Medications, Drug/ETOH use, Illness, Injury, Surgery, etc.): _None—no medication,_
illness, surgery, etc. No report of drug use, no prescription meds.

4. Prior Psychiatric/Psychological Conditions & Treatments: _None_

5. Other Relevant History (Vocational/School, Relationship/Sexual, Social/Legal): _School homework ok until this year._
An only child. Parents have made decisions for her. She has typical number of friends.

6. Brief Mental Status Evaluation: (Check as necessary)

APPEARANCE/DRESS	INTELLIGENCE	JUDGMENT/INSIGHT	DELUS./HALLUCIN.	THOUGHT DISORDER	RECENT MEMORY	REMOTE MEMORY
x appropriate	**x** high	**x** intact	__absent	__absent	__intact	__intact
__inappropriate	__average	__impaired	__present	__present	__impaired	__impaired
__not assessed	__low	__not assessed	**x** not assessed	**x** not assessed	**x** not assessed	**x** not assessed

7. Mood/Affect: (Describe) _Anxious, somewhat reticent_

8. Suicide Assessment: (Risk, priors, plan) _____ Homicide Assessment: (Victim, violence, plan) _____

9. Clinical Formulation (Explanation of symptoms; include strengths/resources, obstacles to treatment/hidden agendas): Please be specific yet brief and clear _____
Recent problems in school appear to provide her with a way to be more assertive with parents and to resolve "crisis" with
them in terms of her own identity and autonomy.

10. Code (Nos. & Names) DSM-IV Axis II: ___V71.09___   Diagnostic Impressions: DSM-IV Axis I: _V62.3 313.82_

B. TREATMENT PLAN:

1. Focused, Targeted, Behavior & Measurable GOALS (Prioritize) Specifically Addressing Presenting Problems(s): Use as many rows as needed.

2. TYPE OF TREATMENT: (Cognitive/Behavioral) Interpersonal/Insight/Emotional Awareness Other:_____

3. DURATION: Service dates this OTR: _Feb. 21, 1997_ # Sessions expected for DISCHARGE: _13 sessions_ Discharge by (DATE): _June 8, 1997_

4. MODE: (Individual (90844)) Couple Family (90847) Individual/Family combination Medication management (90862) Group (if available) (90853) Other (___) CPT Code __

PROBLEM(S)	GOALS	MEASURABLE SUCCESS CRITERION	SELECTED INTERVENTIONS
1. _anxiety over self assertion_	_Use of four initiating skills at least four times a week_	1. _A plus one or above rating on a 5-point goal attainment scale._	_Relaxation_ _Stress inoculation_ _Systematic desensitizaton_
2. _____	_____	2. _Reduction of anxiety from 70 to 50 on a 100 pt. SUDS scale._	_____
3. _____	_____		

Therapist Signature & Phone: _____   Lic. No.: _____   Date: _____

Client Comments:

SOURCE: Adapted from the Treatment Planning Form, by Mental Health Network, Inc., 771 Corporate Drive, Suite 410, Lexington, KY. Reprinted by permission.

CHAPTER 12

SYMBOLIC MODELING AND PARTICIPANT MODELING

Picture the following series of events. A young girl is asked what she wants to be when she grows up. Her reply: "A doctor, just like my mom." Think of a child who points a toy gun and says "Bang, bang, you're dead" after watching a police program on television. Think of people flocking to stores to buy clothes that reflect the "outdoor" or "leisure look," or "warm-up suit look," which has been described and featured in some magazines. Imagine a television news program showing pictures of teenagers breaking glass doors and windows of stores, and entering and stealing whatever they can carry in 30 seconds or less. Several days later, after the story is shown on television, there is an increase in the number of teenager robberies similar to the ones reported on the news. Consider a person reporting dysfunctional patterns of behaving similar to his or her own that were displayed by a parent when he or she was a child. All these events are examples of a process called imitation, copying, mimicry, vicarious learning, identification, observational learning, or modeling. Rosenthal and Steffek (1991, p. 70) define modeling as "the processes by which information guides an observer (often without messages conveyed through language), so that conduct is narrowed from 'random' trial-and-error toward an intended response. By intended response, we mean that much of the practice takes place covertly, through information-processing, decision-making, and evaluative events *in advance* of visible or audible overt performance."

There are several ways people can learn through modeling. A person can acquire new behaviors from live or symbolic modeling. Modeling can help a person perform an already acquired behavior in more appropriate ways or at more desirable times. Modeling can also extinguish client fears. Modeling procedures have been used to help clients acquire parent and caregiver behaviors, change misbehavior, reduce fears, reduce anxiety disorders, decrease stress, reduce weight and smoking, and change perspective about themselves. Modeling strategies have also been used in preventive and behavioral medicine with patients using the health care system (Rosenthal & and Steffek, 1991).

Matson (1985) differentiates modeling, imitation, and observational learning. Modeling encompasses the other two. Imitation is observing and then displaying a series of behaviors or responses. Observational learning occurs by observing others but not necessarily imitating the behaviors exactly (Matson, 1985, p. 150). Modeling is a process of observing an individual or group (observation learning) and imitating similar behaviors. The model acts as a stimulus for thoughts, beliefs, feelings, and actions of the observer.

Rosenthal and Steffek (1991) advise that "it is not enough to acquire the skills to earn desired outcomes. People must also gain enough self-efficacy (confidence) that they can perform the needed acts despite stresses, dangers, moments of doubts, and can persevere in the face of setbacks" (p. 75). (Refer to Chapter 20 for a discussion of the importance of self-efficacy for implementing any treatment strategy.) They also emphasize the importance of restructuring vulnerable thoughts with positive coping statements (see Chapter 15) and other cognitive aids to support the perseverance of the desired behavior (Rosenthal & Steffek, 1991). Higher self-efficacy is achieved by enhancing confidence for each component of the series or graduated hierarchy of behaviors to be acquired. Also, self-efficacy enhancement serves as an induction and supportive aid in facilitating behavioral changes.

We present symbolic and participant modeling in this chapter; covert modeling is presented in Chapter 13 and cognitive modeling is described in Chapter 14. Table 12-1 illustrates some of the different conditions in which modeling has been used. Table 12-2 is a sample of the recent research about symbolic and participant modeling. As you can see, symbolic modeling has been used with AIDS prevention, altruistic behavior, anger and aggression, gender stereotyping modification, and surgery. Research about participant modeling has dealt with such concerns as anxiety, fear, phobia, self-injections, sexual abuse of children, stuttering, surgery, verbally abusive behavior, and women's self-defense skills.

TABLE 12-1. Types of modeling

Observational

Symbolic Model—behavior to be imitated is presented as film, videotape, written matter, or cartoons

Covert Model—person imagines others or himself or herself engaging in the behavior to be imitated (Chapter 13)

Cognitive Modeling—desired thoughts or self-talk is modeled out loud or symbolically (Chapter 14)

Participant Modeling

The desired behavior is modeled in vivo and the client is guided through performance of the behavioral sequence

SOURCE: Adapted from "Modeling Methods," by T. L. Rosenthal and B. D. Steffek, In *Helping People Change*, 4th Ed., by F. H. Kanfer and A. P. Goldstein (Eds.). Copyright 1991 by Pergamon Press.

OBJECTIVES

1. Develop and try out a script for one symbolic model with a role play or an actual client. After completing the script, evaluate it on the Checklist for Developing Symbolic Models at the end of the chapter.
2. Describe how you would apply the four components of participant modeling in a simulated client case.
3. Demonstrate at least 14 out of 17 steps associated with participant modeling with a role-play client.

SYMBOLIC MODELING

In symbolic modeling, the model is presented through written materials, audio- or videotapes, films, or slide tapes. Symbolic models can be developed for an individual client or can be standardized for a group of clients. Cartoon characters can be used as models to teach decision-making skills to children. These characters can be presented in a self-contained set of written materials and a cassette audiotape. Counselors may find that developing a standardized model is more cost effective because it can reach a greater number of clients. For instance, a school counselor who sees many students with deficits in information-seeking skills could develop one tape that could be used by many of these students.

In this section, we present some suggestions for developing self-instructional symbolic modeling procedures. A self-instructional model contains demonstrations of the target behavior, opportunities for client practice, and feedback. In developing a self-instructional symbolic modeling procedure, the counselor will have to consider the following elements: the characteristics of the consumers who will use the model;

the goal behaviors to be modeled or demonstrated; the media to be used; the content of the script; and the field testing of the model. These five steps are summarized in the Checklist for Developing Symbolic Models at the end of the chapter.

Characteristics of Consumers

The first consideration in developing a symbolic model is to determine the characteristics of the people for whom the model is designed. For example, the age, gender, cultural practices, racial characteristics, and problems of the people who will use the procedure should be assessed. For example, to determine problem situations, Sarason and Sarason (1981) conducted extensive interviews to assess what social skills are needed for low-achieving high school students. The authors interviewed teachers, counselors, students attending the school, former students who had dropped out, and employers who typically hired the students (Sarason & Sarason, 1981, p. 910).

The characteristics of the symbolic model should be similar to those of the people for whom the procedure is designed. The counselor should also consider the degree of variation that may exist in these characteristics among the users of the symbolic model. Including several persons as models (using multiple models) can make a symbolic model more useful for a variety of clients. For example, Sarason and Sarason's models were high school students who "represented several racial/ethnic groups—Western European, Mexican-American, Filipino, and black" (1981, p. 911). Gilbert and colleagues (1982) used as peer models two children age 6 and 8—a Black male and White female—who were trained for self-injection of insulin. One model may be satisfactory in some situations. Peterson and Shigetomi (1981) presented a film, *Ethan Has an Operation* (produced by Melamed & Siegel, 1975), that showed a 7-year-old White male as a model. Also, an 8-year-old Caucasian boy was used as a coping and a mastery model with children who were pedodontic patients (Klorman, Hilpert, Michael, LaGana, & Sveen, 1980). Finally, a six-year-old leukemia patient who came to an oncology clinic for bone marrow aspiration and a spinal tap was used to model positive coping behavior and to describe her thoughts and feelings (Jay, Elliot, Ozolins, Olson, & Pruitt, 1985). Finally, Malgady, Rogler, and Costantino (1990) used Puerto Rican *cuentos* or folk tales and their characters to promote self-esteem, ethnic pride, and anxiety reduction with Puerto Rican children.

In some instances, former clients may serve as appropriate symbolic models on audio- and videotapes. Reeder and Kunce (1976) used ex-addict paraprofessional staff members

TABLE 12-2. Symbolic and participant modeling research

Symbolic modeling	

AIDS prevention

Maibach, E., & Flora, J. A. (1993). Symbolic modeling and cognitive rehearsal: Using video to promote AIDS prevention self-efficacy. Special Issue: The role of communication in health promotion. *Communication Research, 20,* 517–545.

Altruistic behavior

Yoshimi, Y., Yonezawa, S., Sugiyama, K., & Matsui, H. (1989). The effects of the viewpoints and observers' traits on the observational learning of altruistic behavior. *Japanese Journal of Psychology, 60,* 98–104.

Anger and aggression

Feindler, E. L., Ecton, R. B., Kingsley, D., & Dubey, D. R. (1986). Group anger-control training for institutionalized psychiatric male adolescents. *Behavior Therapy, 17,* 109–123.

Larson, J. D. (1992). Anger and aggression management techniques through the Think First curriculum. *Journal of Offender Rehabilitation, 18,* 101–117.

Gender stereotyping modification

Katz, P. A., & Walsh, P. V. (1991). Modification of children's gender-stereotyped behavior. *Child Development, 62,* 338–351.

Peer counseling training

Romi, S., & Teichman, M. (1995). Participant and symbolic modelling training programmes. *British Journal of Guidance and Counselling, 23,* 83–94.

Surgery

Anderson, E. A. (1987). Preoperative preparation for cardiac surgery facilitates recovery, reduces psychological distress, and reduces the incidence of acute postoperative hypertension. *Journal of Consulting and Clinical Psychology, 55,* 513–550.

Participant modeling	

Anxiety

Hughes, D. (1990). Participant modeling as a classroom activity. *Teaching of Psychology, 17,* 238–240.

Fear

Samson, D., & Rachman, S. (1989). The effect of induced mood on fear reduction. *British Journal of Clinical Psychology, 28,* 227–238.

Phobia

Arnow, B. A., Taylor, C. B., Agras, W. S., & Telch, M. J. (1985). Enhancing agoraphobia treatment outcome by changing couple communication pattern. *Behavior Therapy, 16,* 452–467.

Mattick, R. P., Peters, L., & Clarke, J. C. (1989). Exposure and cognitive restructuring for social phobia. *Behavior Therapy, 29,* 3–23.

Ritchie, E. C. (1992). Treatment of gas mask phobia. *Military Medicine, 157,* 104–106.

Sanders, M. R., & Jones, L. (1990). Behavioural treatment of injection, dental and medical phobias in adolescents: A case study. *Behavioral Psychotherapy, 18,* 311–316.

Zaragoza, R. E., & Navarro-Humanes, J. F. (1988). Behavioral treatment of blood phobia: Participant modeling vs. gradual "in vivo" exposure. *Analisis y Modificacion de Conducta, 14,* 119–134.

Self-injections

Erasmus, U. (1992). Behavioral treatment of needle phobia. *Tijdschrift voor Psychotherapie, 18,* 335–347.

Sexual abuse of children

Wurtele, S. K., Marrs, S. R., & Miller-Perrin, C. L. (1987). Practice makes perfect? The role of participant modeling in sexual abuse prevention programs. *Journal of Consulting and Clinical Psychology, 55,* 599–602.

Stuttering

Bhargava, S. C. (1988). Participant modeling in stuttering. *Indian Journal of Psychiatry, 30,* 91–93.

Surgery

Faust, J., Olson, R., & Rodriguez, H. (1991). Same-day surgery preparation: Reduction of pediatric patient arousal and distress through participating modeling. *Journal of Consulting and Clinical Psychology, 59,* 475–478.

Verbally aggressive behavior

Vaccaro, F. J. (1990). Application of social skills training in a group of institutionalized aggressive elderly subjects. *Psychology and Aging, 5,* 369–378.

Women's self-defense skills

Ozer, E. M., & Bandura, A. (1990). Mechanisms governing empowerment effects: A self-efficacy analysis. *Journal of Personality and Social Psychology, 58,* 472–486.

and "advanced" residents of a drug-abuse treatment program as the models for their six video-model scenarios. The models in each scenario displayed a coping attitude while performing the various skills required for achieving the goal behaviors associated with one of six problem areas. For example:

> The model was initially shown as being pessimistic and ineffective in the given problem area. The model would then reflect upon his problem and discuss it with a peer or staff member. Following reflection and discussion, the model would try out new problem-solving behaviors. As the scenarios progressed, the model would progressively display more independence in solving problems, becoming less dependent upon the advice of the others. (p. 561)

Goal Behaviors to Be Modeled

The goal behavior, or what is to be modeled, should be specified. A counselor can develop a series of symbolic models to focus on different behaviors, or a complicated pattern of behavior can be divided into less complex skills. For instance, Reeder and Kunce (1976) developed scenarios for their video models for six problem areas: accepting help from others, capitalizing on street skills, job interviewing, employer relations, free-time management, and new lifestyle adjustment. Craigie and Ross (1980) employed actors who modeled appropriate and inappropriate target behaviors for pretherapy training programs to encourage alcohol detoxification patients to seek treatment. Webster-Stratton (1981a, 1981b) used videotaped vignettes of parent models who displayed appropriate parent behaviors (nurturant, playful) and inappropriate behaviors (rigid and controlling) to train mothers. Webster-Stratton, Kolpacoff, and Hollinsworth (1988) used self-administered videotape therapy for families with conduct-problem children. Gilbert et al. (1982) used videotaped models who gave information about self-injection of insulin, described feelings about the procedure, modeled appropriate coping statements, and used self-instructions and self-praise statements (p. 189). Gresham and Nagle (1980) had female and male 9- and 10-year-olds model on videotape such social skills as participation, cooperation, communication, friendship making, and initiating and receiving positive and negative peer interaction. Sarason and Sarason (1981) used models who displayed social and cognitive skills for the following situations: "job interview, resisting peer pressure, asking for help in school, asking questions in class, getting along with the boss, dealing with frustration on the job, cutting class, asking for help at work, and getting along with parents" (p. 911).

Whether one model or a series of models is developed, the counselor should structure the model around three questions: What behaviors are to be acquired? Should these behaviors or activities be divided into a sequence of less complex skills? How should the sequence of skills be arranged?

Media

In an attempt to help you acquire counseling skills, we have presented written symbolic models throughout the book in the form of modeled examples, practice exercises, and feedback. Any of these modeled examples could be filmed, audio- or videotaped, or presented on slide tape. The choice of the medium will depend on where, with whom, and how symbolic modeling will be used. Written, filmed, audiotaped, and videotaped symbolic models can be checked out for the client and used independently in a school, in an agency, or at home. We have found that audiotaped models (cassettes) are economical and extremely versatile. However, in some instances, audiotapes may not be as effective because they are not visual. Written models can serve as a bibliotherapeutic procedure (reading) by portraying a person or situation similar to the client and the desired goal. However, a self-instructional written symbolic model differs from traditional bibliotherapy procedures by including additional components of self-directed practice and feedback. In other words, self-instructional symbolic models can be administered by the client without therapist contact.

Content of the Presentation

Regardless of the medium used to portray the modeled presentation, the counselor should develop a script to reflect the content of the modeling presentations. The script should include five parts: instructions, modeling, practice, feedback, and a summarization. (See Learning Activity #37.)

Instructions. Instructions should be included for each behavior or sequence of behaviors to be demonstrated. Brief but explicit and detailed instructions presented before the model will help the client identify the necessary components of the modeled display. Instructions provide a rationale for the modeling and cues to facilitate attention to the model. The instructions can also describe the type of model portrayed, such as "The person you are going to see or hear is similar to you."

Modeling. The next part of the script should include a description of the behavior or activity to be modeled and possible dialogues of the model engaging in the goal behavior or activity. This part of the script should present complex patterns of behavior in planned sequences of skills.

Practice.   The effects of modeling are likely to be greater when presentation of the modeled behavior is followed by opportunities to practice. In symbolic modeling, there should be opportunities for clients to practice what they have just read, heard, or seen the model do.

Feedback.   After the client has been instructed to practice and sufficient time is allowed, feedback in the form of a description of the behavior or activity should be included. The client should be instructed to repeat the modeling and practice portions again if the feedback indicates some trouble spots.

Summarization.   At the conclusion of a particular scenario or series, the script should include a summary of what has been modeled and the importance for the client of acquiring these behaviors.

Field-Testing the Model

It is a good idea to check out the script before you actually construct the symbolic model from it. You can field-test the script with some colleagues or people from the target or client group. The language, the sequencing, the model, practice time, and feedback should be examined by the potential consumer before the final symbolic model is designated ready for use. If at all possible, a pilot program should be designed for the initial use of the symbolic model. Data from the field testing can be used to make any necessary revisions before the finished product is used with a client.

PARTICIPANT MODELING

Participant modeling consists of modeled demonstration, guided practice, and successful experiences (Bandura, 1977, 1986). Participant modeling assumes that a person's successful performance is an effective means of producing change. By successfully performing a formerly difficult or fearful response, a person can achieve potentially enduring changes in behavior. Participant modeling has been used to reduce avoidance behavior and the person's associated feelings about fearful activities or situations. For example, imagine an outside house painter who develops acrophobia. Participant modeling could be used to help the painter gradually climb "scary" heights by dealing directly with the anxiety associated with being in high places. In participant modeling with phobic clients, successful performance in fearful activities or situations helps the person learn to cope with the feared situation.

Another application of participant modeling is with people who have behavioral deficits or who lack such skills as social communication, assertiveness, child management, or physical fitness. Some of these skills might be taught as preventive measures in schools or community agencies. For example, parents can be taught child-management skills by modeling and practicing effective ways of dealing with and communicating with their children.

There are four major components of participant modeling: rationale, modeling, guided participation, and successful experiences (homework). These components are essentially the same whether participant modeling is used to reduce fearful avoidance behavior or to increase some behavior or skill. As you can see from the Interview Checklist for Participant Modeling at the end of the chapter, each component includes several parts. For each component, we present a description, a hypothetical counselor/client dialogue illustrating the implementation and use of the participant modeling strategy, and a learning activity.

LEARNING ACTIVITY 37 *Symbolic Modeling*

Your client is a young, biracial woman whose mother is Filipino and whose father is Mexican. Both her parents are émigrés to the United States. Your client is the only child of these parents. She is applying for a job at a local child care center and is interested in learning some child care skills as she has never had siblings or babysitting experience.

1. Describe the type of model you would select, including age, gender, race, a coping or mastery model, and skills.

2. Develop an outline for a script you would use for the audiotaped model. Include in the script instructions to the client; a description of the model; a brief example of one modeled scenario; an example of a practice opportunity; feedback about the practice; and a summarization of the script. Feedback follows.

FEEDBACK 37
Symbolic Modeling

1. Keep in mind that your client is both biracial and bicultural. You would want to select other young female models—ideally one who is Filipino or at least Asian American and one who is a Mexican or Latina woman. A mastery model may be too discouraging; initial modeling by coping models is preferred. The skills presented by the models would be similar to what your client wants to acquire—child care skills. Using models who have already had some child care experience would be useful.

Treatment Rationale

Here is an example of a rationale the counselor might give for participant modeling:

> This procedure has been used to help other people overcome fears or acquire new behaviors. [rationale] There are three primary things we will do. First, you will see some people demonstrating _____. Next, you will practice this with my assistance in the interview. Then we'll arrange for you to do this outside the interview in situations likely to be successful for you. This type of practice will help you perform what is now difficult for you to do. [overview] Are you willing to try this now? [client's willingness]

Modeling

The modeling component of participant modeling consists of five parts:

1. The goal behaviors, if complex, are divided into a series of subtasks or subskills.
2. The series of subskills is arranged in a hierarchy.
3. Models are selected.
4. Instructions are given to the client before the modeled demonstration.
5. The model demonstrates each successive subtask with as many repetitions as necessary.

Dividing the Goal Behaviors. Before you (the counselor) or someone else models the behavior to be acquired by the client, determine whether the behavior should be divided. Complex patterns of behavior should be divided into subskills or tasks and arranged by small steps or by a graduated series of tasks in a hierarchy. Dividing patterns of behavior and arranging them in order of difficulty may ensure that the client can perform initial behaviors or tasks. This is a very important step in the participant modeling strategy because you want the client to experience success in performing what is modeled. Start with a response or a behavior the client can perform.

For our house painter who fears heights, the target behavior might be engaging in house painting at a height of 30 feet off the ground. This response could be divided into subtasks of painting at varying heights. Each task might be elevated by several feet at a time.

Arranging the Subskills or Tasks in a Hierarchy. The counselor and client then arrange the subskills or subtasks in a hierarchy. The first situation in the hierarchy is the least difficult or threatening; other skills or situations of greater complexity or threat follow. Usually, the first behavior or response in the hierarchy is worked with first. After each of the subtasks has been successfully practiced one at a time, the client can practice all the subskills or tasks. With a nonassertive client, the counselor and client may decide it would be most helpful to work on eye contact first, then speech errors, then response latency, and finally all these behaviors at once.

In phobic cases, the content and arrangement of each session can be a hierarchical list of feared activities or objects. First, work with the situation that poses the least threat or provokes the least fear for the client. For our acrophobic house painter, we would begin with a situation involving little height and gradually progress to painting at greater heights.

Selecting a Model. Before implementing the modeling component, an appropriate model should be selected. At times, it may be most efficient to use the therapist as the model. However, as you may recall from earlier in this chapter, therapeutic gains may be greater when multiple models are used who are somewhat similar to the client. For example, phobia clinics have successfully employed participant modeling to extinguish phobias by using several formerly phobic clients as the multiple models.

Prior Instructions to the Client. Immediately before the modeled demonstration, to draw the client's attention to the model, the counselor should instruct the client about what will be modeled. The client should be told to note that the model will be engaging in certain responses without experiencing any adverse consequences. With a nonassertive client, the counselor might say something like "Notice the way this person looks at you directly when refusing to type your paper." With the house painter, the counselor might say "Look to see how the model moves about the scaffolding easily at a height of five feet."

Modeled Demonstrations.   In participant modeling, a live model demonstrates one subskill at a time. Often, repeated demonstrations of the same response are necessary. Multiple demonstrations can be arranged by having a single model repeat the demonstration or by having several models demonstrate the same activity or response. For example, one model could show moving about on the scaffolding without falling several times, or several models could demonstrate this same activity. When it is feasible to use several models, you should do so. Multiple models lend variety to the way the activity is performed and believability to the idea that adverse consequences will not occur. With clients who are biracial and/or bicultural, multiple models are particularly important.

Guided Participation

After the demonstration of the behavior or activity, the client is given opportunities and necessary guidance to perform the modeled behaviors. Guided participation or performance is one of the most important components of learning to cope, to reduce avoidance of fearful situations, and to acquire new behaviors. People must experience success in using what has been modeled. The client's participation in the counseling session should be structured in a nonthreatening manner.

Guided participation consists of the following five steps (Bandura, 1976a, p. 262):

1. Client practice of the response or activity with counselor assistance
2. Counselor feedback
3. Use of various induction aids for initial practice attempts
4. Fading of induction aids
5. Client self-directed practice

Each of these steps is described and illustrated.

Client Practice.   After the model has demonstrated the activity or behavior, the client is asked to do what has been modeled. The counselor has the client perform each activity or behavior in the hierarchy. The client performs each activity or behavior, starting with the first one in the hierarchy, until he or she can do this skillfully and confidently. It is quite possible that, for an occasional client, there does not need to be a breakdown of the behaviors or activities. For these clients, guided practice of the entire ultimate goal behavior may be sufficient without a series of graduated tasks.

Our house painter would practice moving about on a ladder or scaffolding at a low height. Practices would continue at this height until the painter could move about easily and comfortably; then practices at the next height would ensue.

Counselor Feedback.   After each client practice attempt, the counselor provides verbal feedback to the client about his or her performance. There are two parts to the feedback: (1) praise or encouragement for successful practice and (2) suggestions for correcting or modifying errors. With the house painter, the counselor might say "You seem comfortable at this height. You were able to go up and down the ladder very easily. Even looking down didn't seem to bother you. That's really terrific."

Use of Induction Aids.   Induction aids are supportive aids arranged by the counselor to assist a client in performing a feared or difficult response. Many people consider successful performance a good way to reduce anxiety. However, most people are just not going to participate in something they dread simply because they are told to do so. For instance, can you think of something you really fear, such as holding a snake, riding in an airplane or a boat, climbing a high mountain, or getting in a car after a severe accident? If so, you probably realize you would be very reluctant to engage in this activity just because at this moment you read the words *do it*. However, suppose we were to be there and hold the snake first, and hold the snake while you touch it, and then hold its head and tail while you hold the middle, then hold its head while you hold the tail, and so on. You might be more willing to do this or something else you fear under conditions that incorporate some supportive aids. To help our acrophobic painter reduce fear of heights, an initial induction aid might be joint practice. If actual practice with a ladder or scaffold is possible, nothing may be more supportive than having the model up on the scaffold with the painter or standing directly behind or in front of the painter on a ladder. This also functions as a type of protective aid. Of course, this scenario requires a model who is not afraid of heights. In the experience of one couple, the one who is nonacrophobic induces the other to climb lighthouses, landmarks, hills, and other such "scenic views" by going first and extending a hand. This type of induction aid enables both of them to enjoy the experience together. As a result, the fears of one person have never interfered with the pleasures of the other, because continued practice efforts with some support have reduced the fear level substantially.

Induction aids can be used in the counseling session, but they should also be applied in settings that closely approximate the natural setting. If at all possible, the counselor or a model should accompany the client into the "field," where the client can witness further demonstrations and can participate in the desired settings. For example, teaching assertive behavior to a client in the interview must be supplemented with modeling and guided participation in the natural setting in which the final goal behavior is to occur. It is doubtful that a counselor would be equipped with scaffolds so that our acrophobic house painter could practice the modeled activities at different heights. The counselor could use covert rehearsal instead of overt practice. Our point is that the counselor who uses live participant modeling procedures must be prepared to provide aids and supports that help the client practice as closely as possible the desired goal behavior. If this cannot be done, the next best alternative is to simulate those activities as closely as possible in the client's real situation.

Fading of Induction Aids. Induction aids can be withdrawn gradually. With a nonassertive client, the use of four induction aids initially might be gradually reduced to three, two, and one. Or, with the painter, a very supportive aid, such as joint practice, could be replaced by a less supportive aid, such as verbal coaching. The gradual withdrawal of induction aids bridges the gap between counselor-assisted and client-directed practice.

Client Self-Directed Practice. At some point, the client should be able to perform the desired activities or responses without any induction aids or assistance. A period of client self-directed practice may reinforce changes in the client's beliefs and self-evaluation and may lead to improved behavioral functioning. Therefore, the counselor should arrange for the client to engage in successful performance of the desired responses independently unassisted. Ideally, client self-directed practice would occur both within the interview and in the client's natural setting. For example, the house painter would practice moving on the ladder or scaffold alone. Client self-directed practice is likely to be superior to therapist-directed practice.

In addition to application of the participant modeling procedures in the counseling sessions, facilitating the transfer of behavior from the training session to the natural environment should be an integral part of counseling. Generalization of desired changes can be achieved by success or by reinforcing experiences that occur as part of a transfer-of-training program.

Success, or Reinforcing, Experiences

The last component of the participant modeling procedure is success (reinforcing) experiences. Clients must experience success in using what they are learning. Further, as Bandura points out, psychological changes "are unlikely to endure unless they prove effective when put into practice in everyday life" (1976a, p. 248). Success experiences are arranged by tailoring a transfer-of-training program for each client. In an adequate transfer-of-training program, the client's new skills are used first in low-risk situations in the client's natural environment or in any situation in which the client will probably experience success or favorable outcomes. Gradually, the client extends the application of the skills to natural situations that are more unpredictable and involve a greater threat.

To summarize, success experiences are arranged through a program that transfers skill acquisition from the interview to the natural setting. This transfer-of-training program involves the following steps:

1. The counselor and client identify situations in the client's environment in which the client desires to perform the target responses.
2. These situations are arranged in a hierarchy, starting with easy, safe situations in which the client is likely to be successful and ending with more unpredictable and risky situations.
3. The counselor accompanies the client into the environment and works with each situation on the list by modeling and guided participation. Gradually the counselor's level of participation is decreased.
4. The client is given a series of tasks to perform in a self-directed manner.

Bandura (1976a) concludes that participant modeling achieves results, given adequate demonstration, guided practice, and positive experiences. One advantage of participant modeling is that "a broad range of resource persons," such as peers or former clients, can serve as therapeutic models (p. 249). Bandura also points out that participant modeling helps clients to learn new responses under "lifelike conditions." As a result, the problems of transfer of learning from the interview to the client's real-life environment are minimized.

MULTICULTURAL APPLICATIONS OF SOCIAL MODELING

The primary areas in which we have seen applications of social modeling with diverse populations have been substance use and prevention and parenting/family/child behavior issues (see Table 12-3). Testing a longitudinal model of smoking initiation and maintenance of 233 Black seventh-grade students, Botvin, Baker, Botvin, and Dusenbury (1993) found that social modeling—in the form of friends' smoking—was the most important early factor in the smoking *initiation*

TABLE 12-3. Multicultural research on social modeling

Altering smokeless tobacco use	Riley, W., Barenie, J., Mabe, P., & Myers, D. (1990). Smokeless tobacco use in adolescent females: Prevalence and psychosocial factors among racial/ethnic groups. *Journal of Behavioral Medicine, 13,* 207–220. Riley, W., Barenie, J., Mabe, P., & Myers, D. (1991). The role of race and ethnic status on the psychosocial correlates of smokeless tobacco use in adolescent males. *Journal of Adolescent Health, 12,* 15–21.
Substance use/substance prevention	Botvin, G., Baker, E., Botvin, E., & Dusenbury, L. (1993). Factors promoting cigarette smoking among Black youth: A causal modeling approach. *Addictive Behaviors, 18,* 397–405. Mail, P. (1995). Early modeling of drinking behavior by Native American elementary school children playing drunk. *International Journal of the Addictions, 30,* 1187–1197.
HIV/AIDS risk/chemical dependence	Picucci, M. (1992). Planning an experiential weekend workshop for lesbians and gay males in recovery. *Journal of Chemical Dependency Treatment, 5,* 119–139. Rhodes, F., & Humfleet, G. (1993). Using goal-oriented counseling and peer support to reduce HIV/AIDS risk among drug users not in treatment. *Drugs and Society, 7,* 185–204.
Parenting	Hurd, E., Moore, C., & Rogers, R. (1995). Quiet success: Parenting strengths among African Americans. *Families in Society, 76,* 434–443. Kliewer, W., & Lewis, H. (1995). Family influences on coping processes in children and adolescents with sickle cell disease. *Journal of Pediatric Psychology, 20,* 511–525. Middleton, M., & Cartledge, G. (1995). The effects of social skills instruction and parental involvement on the aggressive behavior of African American males. *Behavior Modification, 19,* 192–210. Reyes, M., Routh, D., Jean-Gilles, M., & Sanfilippo, M. (1991). Ethnic differences in parenting children in fearful situations. *Journal of Pediatric Psychology, 16,* 717–726.
Prosocial behavior in children	Reichelova, E., & Baranova, E. (1994). Training program for the development of prosocial behavior in children. *Psycholigia a Patopsychologia Dietata, 29,* 41–50.
Skill training in children	Dowrick, P., & Raeburn, J. (1995). Self-modeling: Rapid skill training for children with physical disabilities. *Journal of Developmental and Physical Disabilities, 7,* 25–37.
Family violence	Williams, O. (1994). Group work with African American men who batter: Toward more ethnically sensitive practice. *Journal of Comparative Family Studies, 25,* 91–103.
Child safety	Alvarez, J., & Jason, L. (1993). The effectiveness of legislation, education, and loaners for child safety in automobiles. *Journal of Community Psychology, 21,* 280–284.
Anxiety reduction	Malgady, R., Rogler, T., & Costantino, G. (1990). Culturally sensitive psychotherapy for Puerto Rican children and adolescents: A program of treatment outcome research. *Journal of Consulting and Clinical Psychology, 58,* 704–712.
Pain	Neill, K. (1993). Ethnic pain styles in acute myocardial infarction. *Western Journal of Nursing Research, 15,* 531–543.
Breast self-examination	Anderson, R., & McMillion, P. (1995). Effects of similar and diversified modeling on African American women's efficacy expectations and intentions to perform breast self-examination. *Health Communication, 7,* 327–343.

process, although perceived smoking norms and intrapersonal factors were more important in smoking *maintenance*.

Modeling is also a strong predictor in beginning the use of smokeless tobacco among three groups of adolescent females and males: African American, Caucasian, and Native American (Riley et al., 1990, 1991). Mail (1995) has suggested that substance use prevention programs include modeling as a major component, but that all prevention programs be culturally adapted.

The use of culturally sensitive helpers as models in both HIV/AIDS prevention and in group treatment for lesbian and gay males in chemical dependency recovery also has been noted by Rhodes and Humfleet (1993) and Picucci (1992), as well as in group work with African American men and family violence (Williams, 1994). Helpers who are culturally similar to clients are likely to be of greatest support and value. In a recent ethnographic study of poor African American, Caucasian, and Latina adolescent girls, Taylor, Gilligan, and Sullivan (1995) found that the most important connections with adults reported by these girls were with adults from cultural backgrounds that were similar to or the same as their own. Anderson and McMillion (1995) used similar models with African American women to increase their confidence and intentions to perform breast self-examinations.

Parental modeling is also an important factor in a variety of ways. In an excellent article, Hurd, Moore, and Rogers (1995) examined strengths among 53 African American parents. In describing the values and behaviors they imparted to their children, a high frequency of positive role modeling emerged as a significant factor as well as parental involvement and support from other adults. Parental modeling has also emerged as a critical factor in the way African American children and adolescents with sickle cell disease cope with their illness (Kliewer & Lewis, 1995). Reyes, Routh, Jean-Gilles, and Sanfilippo (1991) have noted several differences among various ethnic groups in the use of parental modeling and have urged therapists to be alert to both historical and cultural trends when using modeling approaches with a diverse group of parents.

As a treatment/intervention strategy, modeling has been an important component of child safety educational programs for African American, Asian American, Caucasian, and Hispanic parents (Alvarez & Jason, 1993) and in prosocial behavior training for Slovak school-age children (Reichelova & Baranova, 1994). An excellent example of multicultural symbolic modeling with young Puerto Rican children involves the use of Puerto Rican *cuentos* or folk tales. The characters in these tales are used as therapeutic peer models depicting beliefs, values, and target behaviors that children can first attend to and then identify with and imitate (Malgady, Rogler, & Costantino, 1990). For bicultural children, stories can be developed that bridge two or more different cultures, such as Puerto Rican and American. With Puerto Rican adolescents, biographical stories of heroic Puerto Ricans have been used to expose the teenagers to successful adult models in their own culture (Malgady et al., 1990).

Guidelines for Using Modeling with Diverse Groups of Clients

When you are developing a social modeling intervention for culturally diverse clients, consider the following guidelines:

1. Make sure your live or symbolic model is culturally compatible with the client's background. For example, if you are developing a modeling program for Black youth with substance use issues, a model who is also Black, relatively young, and familiar with substance use issues will be more effective than an older, White model who knows little about or has had scant experience with substance use issues.

2. Make sure that the *content* of your modeled presentation is culturally relevant to the client and not simply a reflection of what may be your or society's Eurocentric values. For example, the concept of social skills training is more relevant to many middle-class clients than to poorer clients and some clients of color. Similarly, a model of assertion training is more relevant to many Caucasian clients and less applicable to many Asian American clients. Modeling of boundary work is more relevant to some Caucasian clients and less relevant to some Native American clients.

3. Keep in mind the cultural differences in the way people attend to, learn from, and utilize modeling approaches. Determine whether your particular client or group of clients will find value in even having or seeing a model.

Malgady et al. (1990) found complex treatment effects when using culturally sensitive models with Puerto Rican children and adolescents. They noted that client process variables such as the client's cognitive responses and the client's familial context can affect treatment outcome.

MODEL DIALOGUE: PARTICIPANT MODELING

Here is an example of the use of participant modeling with our client Joan. The participant modeling will be used to help

Joan perform the four behaviors in math class that she typically avoids. The rationale for the counselor responses is set off by the italicized comments that precede the responses.

Session 1

In the first response, the counselor will provide a **rationale** *about the strategy and a brief* **overview** *of the procedure.*

1. *Counselor:* This procedure has been of help to other people who have trouble in classroom participation. We'll take each of the ways you would like to participate, and either I myself or maybe one of your classmates will show you a way to do this, then help you practice it. At first we'll just practice here. Then gradually you'll try this out in your other classes and, of course, finally in your math class. What do you think about this?

   Client: It's OK with me. It's just something I know I can do but I don't because I'm a little nervous.

The counselor will pick up on Joan's previous response and use it to provide an **additional rationale** *for the participant modeling strategy.*

2. *Counselor:* And nervousness can keep you from doing something you want. This procedure helps you to learn to do something in small steps. As you accomplish each step, your confidence in yourself will increase and your nervousness will decrease.

   Client: I think that will definitely help. Sometimes I just don't believe in myself.

In response 3, the counselor ignores Joan's previous self-effacing comment. The counselor instead begins with the **modeling component** *by reviewing the ways Joan wanted to increase selected initiating skills in math class.*

3. *Counselor:* You know, last week I believe we found some ways that you would like to increase your participation in math class. And I think we arranged these in an order, starting with the one that you thought was easiest for you now, to the one that was hardest for you. Can you recall these things and this order?

   Client: Yes, I believe it was like this: answering questions, going to the board, volunteering answers, and then volunteering opinions or ideas.

The counselor asks the client **whether additional activities** *need to be added or* **whether the hierarchy order** *needs to be rearranged.*

4. *Counselor:* OK, after thinking about it for a week, have you thought of any other ways you'd like to participate—or do you think this order needs to be rearranged at all?

   Client: Well, one more thing—I would like to be able to work the problems on my own after I ask Mr. Lamborne for help. That's where I want to begin. He usually takes over and works the problems for me.

In response 5, the counselor will explore **a potential model** *for Joan and obtain Joan's input about this decision.*

5. *Counselor:* OK, one thing we need to do now is to consider who might model and help you with these activities. I can do it, although if you can think of a classmate in math who participates

the way you want to, this person could assist you when you try this out in your class. What do you think?

   Client: Is it necessary to have someone in the class with me? If so, I think it would be less obvious if it were someone already in the class.

The counselor picks up on Joan's discomfort about the counselor's presence in her class and **suggests another classmate as the model.**

6. *Counselor:* Well, there are ways to get around it, but it would be more helpful if someone could be there in your class, at least initially. I think you would feel more comfortable if this person were another classmate rather than me. If there is someone you could pick who already does a good job of participating, I'd like to talk to this person and have him or her help during our next sessions. So try to think of someone you like and respect and feel comfortable around.

   Client: Well, there's Debbie. She's a friend of mine, and she hardly ever gets bothered by answering Mr. Lamborne's questions or going to the board. I could ask her. She'd probably like to do something like this. She works independently, too, on her math problems.

The counselor provides **a rationale** *for how Joan's friend will be used as the model so that Joan understands how her friend will be involved. Note that Joan's reaction to this is solicited. If Joan were uncomfortable with this, another option would be explored.*

7. *Counselor:* OK, if you could ask her and she agrees, ask her to drop by my office. If that doesn't work out, stop back and we'll think of something else. If Debbie wants to do this, I'll train her to help demonstrate the ways you'd like to participate. At our session next week, she can start modeling these things for you. How does that sound?

   Client: OK. It might be kind of fun. I've helped her with her English themes before, so now maybe she can help with this.

The counselor encourages the idea of these two friends' providing **mutual help** *in the next response.*

8. *Counselor:* That's a good idea. Good friends help each other. Let me know what happens after you talk to Debbie.

After session 1, Joan stopped in to verify that Debbie would be glad to work with them. The counselor then arranged a meeting with Debbie to explain her role in the participant modeling strategy. Specifically, Debbie practiced modeling the other four participation goals Joan had identified. The counselor gave Debbie instructions and feedback so that each behavior was modeled clearly and in a coping manner. The counselor also trained Debbie in ways to assist Joan during the guided-participation phase. Debbie practiced this, with the counselor taking the role of Joan. In these practice attempts, Debbie also practiced using various induction aids that she might use with Joan, such as joint practice, verbal coaching, and graduated time intervals and difficulty of task. Debbie also practiced having the counselor (as Joan) engage in self-directed practice. Classroom simulations of success experiences were also rehearsed so Debbie could learn her role in arranging for actual success experiences with Joan. When Debbie felt comfortable with her part in the strategy, the next session with Joan was scheduled.

Session 2

In response 1, the counselor gives **instructions to Joan about what to look for** *during the modeled demonstration. Note that the counselor also points out the* **lack of adverse consequences** *in the modeling situation.*

1. *Counselor:* It's good to see you today, Joan. I have been working with Debbie, and she is here today to do some modeling for you. What we'll do first is to work with one thing you mentioned last week, telling Mr. Lamborne you want to work the problems yourself after you ask him for an explanation. Debbie will demonstrate this first. So I'll play the part of Mr. Lamborne and Debbie will come up to me and ask me for help. Note that she tells me what she needs explained, then firmly tells Mr. Lamborne she wants to try to finish it herself. Notice that this works out well for Debbie—Mr. Lamborne doesn't jump at her or anything like that. Do you have any questions?
   Client: No, I'm ready to begin. [Modeling ensues.]
   Debbie (as model): Mr. Lamborne, I would like some explanation about this problem. I need you to explain it again so I can work it out all right.
   Counselor (as Mr. Lamborne): OK, well, here is the answer . . .
   Debbie (as model, interrupts): Well, I'd like to find the answer myself, but I'd like you to explain this formula again.
   Counselor (as Mr. Lamborne): OK, well, here's how you do this formula . . .
   Debbie (as model): That really helps. Thanks a lot. I can take it from here. [Goes back to seat.]

After the modeling, the counselor **asks Joan to react** *to what she saw.*

2. *Counselor:* What reactions did you have to that, Joan?
   Client: Well, it looked fairly easy. I guess when I do ask him for help, I have a tendency just to let him take over. I am not really firm about telling him to let me finish the problem myself.

The counselor picks up on Joan's concern and **initiates a second modeled demonstration.**

3. *Counselor:* That's an important part of it—first being able to ask for an additional explanation and then being able to let him know you want to apply the information and go back to your seat and try that out. It might be a good idea to have Debbie do this again—see how she initiates finishing the problem so Mr. Lamborne doesn't take over.
   Client: OK.
   [Second modeled demonstration ensues.]

In response 4, the counselor asks Joan for her opinion about **engaging in a practice.**

4. *Counselor:* How ready do you feel now to try this out yourself in a practice here?
   Client: I believe I could.

Before the first practice attempt, the counselor will introduce **one induction aid, verbal coaching,** *from Debbie.*

5. *Counselor:* OK. Now I believe one thing that might help you is to have Debbie sort of coach you. For instance, if you get stuck or start to back down, Debbie can step in and give you a cue or a reminder about something you can do. How does that sound?
   Client: Fine. That makes it a little easier.

The first practice attempt begins.

6. *Counselor:* OK, let's begin. Now I'll be Mr. Lamborne and you get up out of your seat with the problem.
   Client: Mr. Lamborne, I don't quite understand this problem.
   Counselor (as Mr. Lamborne): Well, let me just give you the answer; you'll have one less problem to do then.
   Client: Well, uh, I'm not sure the answer is what I need.
   Counselor (as Mr. Lamborne): Well, what do you mean?
   Debbie (intervenes to prompt): Joan, you might want to indicate you would prefer to work out the answer yourself, but you do need another explanation of the formula.
   Client: Well, I'd like to find the answer myself. I do need another explanation of the formula.
   Counselor (as Mr. Lamborne): OK. Well, it goes like this . . .
   Client: OK, thanks.
   Debbie: Now be sure you end the conversation there and go back to your seat.

The counselor will **assess Joan's reactions** *to the practice.*

7. *Counselor:* OK, what did you think about that, Joan?
   Client: It went pretty well. It *was* a help to have Debbie here. That is a good idea.

In the next response, the counselor **provides positive feedback** *to Debbie and to Joan.* **Another practice is initiated;** *this also serves as an* **induction aid.**

8. *Counselor:* I think she helps, too. You seemed to be able to start the conversation very well. You did need a little help in explaining to him what you wanted and didn't want. Once Debbie cued you, you were able to use her cue very effectively. Perhaps it would be a good idea to try this again. Debbie will only prompt this time if she really needs to.
   [Second practice ensues; Debbie's amount of prompting is decreased.]

The counselor explores the idea of **a self-directed practice.**

9. *Counselor:* That seemed to go very smoothly. I think you are ready to do this again without any assistance. How does that sound?
   Client: I think so, too.

After obtaining an affirmative response from Joan, the counselor asks Debbie to leave the room. Just Debbie's physical presence could be a protective condition for Joan, which is another induction aid, so Debbie leaves to make sure the **self-directed practice occurs.**

LEARNING ACTIVITY **38** *Participant Modeling*

This activity is designed to be completed for a behavior of yours that you wish to change. You will need a partner to complete this activity.

1. Select a skill that you wish to acquire.
2. Define the skill by describing what you would be doing, thinking, and/or feeling differently. Decide whether the skill is so broad that it needs to be divided into a series of subskills. If so, identify these and arrange them on a hierarchy in order of difficulty.
3. Ask your partner to model or demonstrate the skill for you. (You can also arrange to observe other people you know and respect who might be likely to use similar skills in naturally occurring circumstances.)
4. With the help of your partner, prepare for your own initial practice of the skill or of the first subskill on the hierarchy. Your partner should facilitate your initial practice attempts with at least one or two induction aids, such as joint practice or verbal coaching. With successive practice attempts, these induction aids will gradually be removed. Your partner also needs to provide feedback after each practice.
5. With your partner, identify actual situations in which you want to apply your newly acquired skill. Rehearse such attempts and identify any induction aids that may be necessary in your initial efforts at skill application in these situations.
6. Call or see your partner to report on how you handled rehearsal efforts in step 5. Identify whether you need additional modeling, practice, or induction aids.

---

10. *Counselor:* I'm going to ask Debbie to leave the room so you'll be completely on your own this time.
    [Self-directed practice ensues.]

Next the counselor cues Joan to provide herself with **feedback** *about her self-directed practice.*

11. *Counselor:* How did you feel about that practice, Joan?
    Client: Well, I realized I was relying a little on Debbie. So I think it was good to do it by myself.

The counselor notes the link between self-directed performance and confidence and starts to work on **success experiences** *outside counseling.*

12. *Counselor:* Right. At first it does help to have someone there. Then it builds your confidence to do it by yourself. At this point, I think we're ready to discuss ways you might actually use this in your class. How does that sound?
    Client: Fine. A little scary, but fine.

The counselor introduces the idea of **Debbie's assistance as an initial aid in Joan's practice outside the session.**

13. *Counselor:* It's natural to feel a little apprehensive at first, but one thing we will do to help you get started on the right foot is to use Debbie again at first.
    Client: Oh, good. How will that work?

In response 14, the counselor **identifies a hierarchy of situations** *in Joan's math class. Joan's first attempts will be assisted by Debbie to ensure success at each step.*

14. *Counselor:* Well, apparently math is the only class where you have difficulty doing this, so we want to work on your using this in math class successfully. Since Debbie is in the class with you, instead of going up to Mr. Lamborne initially by yourself, first you can go with her. In fact, she could initiate the request for help the first time. The next time you could both go up and you could initiate it. She could prompt you or fill in. Gradually, you would get to the point where you would go up by yourself. But we will take one step at a time. *Client:* That really does help. I know the first time it would turn out better if she was there, too.

15. *Counselor:* Right. Usually in doing anything new, it helps to start slowly and feel good each step of the way. So maybe Debbie can come in now and we can plan the first step.

Debbie will model and guide Joan in using these responses in their math class. Next, the entire procedure will be repeated to work with the other initiating skills Joan wants to work on.

SUMMARY

The two modeling strategies presented in this chapter can be used to help clients acquire new responses or extinguish fears. These modeling strategies promote learning by providing a model to demonstrate the goal behaviors for the client. The way the model is presented differs slightly among the modeling procedures. Symbolic modeling uses media for modeled presentations; participant modeling usually employs a live modeling demonstration. As an intervention, modeling may not be as salient across all diverse cultural groups and subgroups. When modeling is an appropriate intervention for a diverse group of clients, the type of model (live or symbolic) should be culturally similar to the client and the content of the modeled presentation should be culturally relevant.

P O S T E V A L U A T I O N

Part One

Objective One asks you to develop a script for a symbolic model. Your script should contain

1. Examples of the modeled dialogue
2. Opportunities for practice
3. Feedback
4. Summarization

Use the Checklist for Symbolic Models as a guide.

Part Two

Objective Two asks you to describe how you would apply the four components of participant modeling with a hypothetical client case. Describe how you would use the four components of participant modeling (rationale, modeling, guided practice, and success experiences) to help a client acquire verbal and nonverbal skills necessary to initiate social contacts with someone he or she wants to ask out.

Part Three

Objective Three asks you to demonstrate 14 out of 17 steps of participant modeling with a role-play client. The client might take the role of someone who is afraid to perform certain responses or activities in certain situations. You can assess yourself, using the Interview Checklist for Participant Modeling below. Feedback for the postevaluation follows.

Checklist for Developing Symbolic Models

Instructions: Determine whether the following guidelines have been incorporated into the construction of your symbolic model.

Check if completed:

_____ 1. Determine what clients will use the symbolic modeling procedure and identify their characteristics.

        _____ Age

        _____ Gender

        _____ Ethnic origin, cultural practices, and race

        _____ Coping or mastery model portrayed

        _____ Possesses similar concern or problem to that of client group or population

_____ 2. Goal behaviors to be modeled by client have been enumerated.

_____ 3. Medium is selected (for example, written script, audiotape, videotape, film, slide tape).

_____ 4. Script includes the following ingredients:

        _____ Instructions

        _____ Modeled dialogue

        _____ Practice

        _____ Written feedback

        _____ Written summarization of what has been modeled, with its importance for client

_____ 5. Written script has been field-tested.

Interview Checklist for Participant Modeling

Instructions: Determine which of the following leads the counselor used in the interview. Check the leads used.

Item	Examples of counselor leads
I. Rationale about Strategy	
_____ 1. Counselor provides rationale about participant modeling strategy.	"This procedure has been used with other people who have concerns similar to yours. It is a way to help you overcome your fear of _____ or to help you acquire these skills."
_____ 2. Counselor provides brief description of components of participant modeling.	"It involves three things. I'll model what you want to do, you'll practice this with my assistance, and then you'll try this out in situations that at first will be pretty easy for you so you can be successful."
_____ 3. Counselor asks for client's willingness to use strategy.	"Would you be willing to try this now?"

(continued)

Item	Examples of counselor leads

II. Modeling

_____ 4. Counselor and client decide whether to divide goal behaviors into a series of subtasks or skills.

"Well, let's see . . . Right now you hardly ever go out of the house. You say it bothers you even to go out in the yard. Let's see whether we can identify different activities in which you would gradually increase the distance away from the house, like going to the front door, going out on the porch, out in the yard, out to the sidewalk, to the neighbor's house, and so on."

_____ 5. If goal behaviors are divided (step 4), these subskills are arranged by counselor and client in a hierarchy according to order of difficulty.

"Perhaps we can take our list and arrange it in an order. Start with the activity that is easiest for you now, such as going to the door. Arrange each activity in order of increasing difficulty."

_____ 6. Counselor and client determine and select appropriate model.

"I could help you learn to do this, or we could pick someone whom you know or someone similar to yourself to guide you through this. What are your preferences?"

_____ 7. Counselor provides instructions to client before demonstration of what to look for.

"Notice that when the doorbell rings, I will go to the door calmly and quickly and open it without hesitation. Also notice that after I go to the door, I'm still calm; nothing has happened to me."

_____ 8. Model demonstrates target response at least once; more demonstrations are repeated if necessary.

"OK, let me show this to you again."

_____ If hierarchy is used, first skill is modeled, followed successively by all others, concluding with demonstration combining all subskills.

"Now that I've done the first scene, next I'll show you stepping out on the porch. Then we'll combine these two scenes."

III. Guided Participation

_____ 9. Client is asked to perform target response. If a hierarchy is used, first skill in hierarchy is practiced first, successfully followed by second, third, and so on.

"This time you try going to the door when the doorbell rings. I will assist you as you need help."

_____ 10. After each practice, model or counselor provides feedback consisting of positive feedback and error corrections.

"That was quite smooth. You were able to go to the door quickly. You still hesitated a little once you got there. Try to open the door as soon as you see who is there."

_____ 11. Initial practice attempts of each skill by client include a number of induction aids, such as

"I'm going to assist you in your first few practices."

_____ a. Joint practice with model or counselor

"Let's do it together. I will walk with you to the door."

_____ b. Verbal and/or physical coaching or guiding by model or counselor

"Let me give you a suggestion here. When you open the door, greet the person there. Find out what the person wants."

_____ c. Repeated practice of one subtask until client is ready to move on to another.

"Let's work on this a few more times until you feel really comfortable."

_____ d. Graduated time intervals for practice (short to long duration)

"This time we'll increase the distance you have to walk to the door. Let's start back in the kitchen."

(continued)

POSTEVALUATION (continued)

Item	Examples of counselor leads
_____ e. Arrangement of protective conditions for practice to reduce likelihood of feared or undesired consequences	"We'll make sure someone else is in the house with you."
_____ f. Graduated levels of severity of threat or complexity of situation	"OK, now we've worked with opening the door when a good friend is there. This time let's pretend it's someone you are used to seeing but don't know as a friend, like the person who delivers your mail."
_____ 12. In later practice attempts, number of induction aids is gradually reduced.	"I believe now that you can do this without my giving you so many prompts."
_____ 13. Before moving on, client is able to engage in self-directed practice of all desired responses.	"This time I'm going to leave. I want you to do this by yourself."

IV. Success Experiences (Homework)

Item	Examples of counselor leads
_____ 14. Counselor and client identify situations in client's environment in which client desires to perform target responses.	"Let's list all the actual times and places where you want to do this."
_____ 15. Situations are arranged in hierarchy from easy with least risk to difficult with greater risk.	"We can arrange these in an order. Put the easiest one first, and follow it by ones that are successively harder or more threatening for you."
_____ 16. Starting with easiest and least risky situation, counselor (or model) and client use live or symbolic modeling and guided practice in client's real-life environment. Steps 4–11 are repeated outside session until gradually counselor (or model) reduces level of assistance.	"Starting with the first situation, we're going to work with this when it actually occurs. At first I'll assist you until you can do it on your own."
_____ 17. Client is assigned a series of related tasks to carry out in a self-directed manner.	"Now you're ready to tackle this situation without my help. You have shown both of us you are ready to do this on your own."

SUGGESTED READINGS

Anderson, E. A. (1987). Preoperative preparation for cardiac surgery facilitates recovery, reduces psychological distress, and reduces the incidence of acute postoperative hypertension. *Journal of Consulting and Clinical Psychology, 55,* 513–550.

Arnow, B. A., Taylor, C. B., Agras, W. S., & Telch, M. J. (1985). Enhancing agoraphobia treatment outcome by changing couple communication pattern. *Behavior Therapy, 16,* 452–467.

Bhargava, S. C. (1988). Participant modeling in stuttering. *Indian Journal of Psychiatry, 30,* 91–93.

Botvin, G., Baker, E., Botvin, E., & Dusenbury, W. (1993). Factors promoting cigarette smoking among Black youth: A causal modeling approach. *Addictive Behaviors, 30,* 397–405.

Erasmus, U. (1992). Behavioral treatment of needle phobia. *Tijdschrift voor Psychotherapie, 18,* 335–347.

Faust, J., Olson, R., & Rodriguez, H. (1991). Same-day surgery preparation: Reduction of pediatric patient arousal and distress through participating modeling. *Journal of Consulting and Clinical Psychology, 59,* 475–478.

Feindler, E. L., Ecton, R. B., Kingsley, D., & Dubey, D. R. (1986). Group anger-control training for institutionalized psychiatric male adolescents. *Behavior Therapy, 17,* 109–123.

Hughes, D. (1990). Participant modeling as a classroom activity. *Teaching of Psychology, 17,* 238–240.

Katz, P. A., & Walsh, P. V. (1991). Modification of children's gender-stereotyped behavior. *Child Development, 62,* 338–351.

Larson, J. D. (1992). Anger and aggression management techniques through the Think First curriculum. *Journal of Offender Rehabilitation, 18,* 101–117.

Maibach, E., & Flora, J. A. (1993). Symbolic modeling and cognitive rehearsal: Using video to promote AIDS prevention self-efficacy. Special Issue: The role of communication in health promotion. *Communication Research, 20,* 517–545.

Mail, P. (1995). Early modeling of drinking behavior by Native American elementary school children playing drunk. *International Journal of Addictions, 30,* 1187–1197.

Matson, J. (1985). Modeling. In A. S. Bellack & M. Hersen (Eds.), *Dictionary of behavior therapy techniques,* (pp. 150–151). New York: Pergamon Press.

Mattick, R. P., Peters, L., & Clarke, J. C. (1989). Exposure and cognitive restructuring for social phobia. *Behavior Therapy, 29,* 3–23.

Ozer, E. M., & Bandura, A. (1990). Mechanisms governing empowerment effects: A self-efficacy analysis. *Journal of Personality and Social Psychology, 58,* 472–486.

Reyes, M., Routh, D., Jean-Gilles, M., & Sanfilippo, M. (1991). Ethnic differences in parenting children in fearful situations. *Journal of Pediatric Psychology, 16,* 717–726.

Romi, S., & Teichman, M. (1995). Participant and symbolic modelling training programmes. *British Journal of Guidance and Counseling, 23,* 83–94.

Rosenthal, T. L., & Steffek, B. D. (1991). Modeling methods. In F. H. Kanfer & A. P. Goldstein (Eds.), *Helping people change* (4th ed., pp. 70–121). New York: Pergamon Press.

Samson, D., & Rachman, S. (1989). The effect of induced mood on fear reduction. *British Journal of Clinical Psychology, 28,* 227–238.

Sanders, M. R., & Jones, L. (1990). Behavioural treatment of injection, dental and medical phobias in adolescents: A case study. *Behavioural Psychotherapy, 18,* 311–316.

Vaccaro, F. J. (1990). Application of social skills training in a group of institutionalized aggressive elderly subjects. *Psychology and Aging, 5,* 369–378.

Wurtele, S. K., Marrs, S. R., & Miller-Perrin, C. L. (1987). Practice makes perfect? The role of participant modeling in sexual abuse prevention programs. *Journal of Consulting and Clinical Psychology, 55,* 599–602.

Yoshimi, Y., Yonezawa, S., Sugiyama, K., & Matsui, H. (1989). The effects of the viewpoints and observers' traits on the observational learning of altruistic behavior. *Japanese Journal of Psychology, 60,* 98–104.

Zaragoza, R. E., & Navarro-Humanes, J. F. (1988). Behavioral treatment of blood phobia: Participant modeling vs. gradual "in vivo" exposure. *Analisis y Modificacion de Conducta, 14,* 119–134.

FEEDBACK
POSTEVALUATION

Part One

Check the contents of your script outline with item 4 on the Checklist for Developing Symbolic Models on page 312.

Part Two

Here is a brief description of how you might use participant modeling to help your client.

Rationale

First, you would explain to your client that the procedure can help him or her acquire the kinds of skills he or she will need to initiate social contacts with someone to ask out. You would also tell him or her that the procedure involves modeling, guided practice, and success experiences. You might emphasize that the procedure is based on the idea that change can occur when the desired activities are learned in small steps with successful performance emphasized at each step.

Modeling

You and your client would explore the verbal and nonverbal responses that might be involved in approaching people and asking them to lunch, for a drink, and so on. For example, these skills might be divided into making a verbal request, speaking without hesitation or errors, and speaking in a firm, strong voice. After specifying all the components of the desired response, you and your client would arrange them in a hierarchy according to order of difficulty.

You and your client would select a culturally appropriate model—yourself or an aide. The model selected would demonstrate the first response in the hierarchy (followed by all the others) to the client. Repeated demonstrations of any response might be necessary.

(continued)

FEEDBACK: POSTEVALUATION   (continued)

Guided participation

After the modeled demonstration of a response, you would ask your client to perform it. The first attempts would be assisted with induction aids administered by you or the model. For example, you might verbally coach the client to start with a short message and gradually increase it. After each practice, you would give your client feedback, being sure to encourage his or her positive performance and make suggestions for improvement. Generally, your client would practice each response several times, and the number of induction aids would be reduced gradually. Before moving on to practice a *different* response, your client should be able to perform the response in a self-directed manner without your presence or support.

Success experiences

You and your client would identify situations in his or her environment in which the client would like to use the learned skills. In this case, most of the applications would be in social situations. Some of these situations involve more risk than others. The situations should be arranged in order, from the least to the most risky. The client would work on the least risky situation until he or she was able to do that easily and successfully before going on. Ideally, it would help to have the counselor or model go along with the client to model and guide. If the model was one of the client's colleagues, this would be possible. If this was not possible, the client could telephone the counselor just before the "contact" to rehearse and to receive coaching and encouragement.

Part Three

Use the Interview Checklist for Participant Modeling to assess your performance or to have someone else rate you.

CHAPTER 13

GUIDED IMAGERY AND COVERT MODELING

With some client problems, a counselor may find that it is impossible or unrealistic to provide live or symbolic models or to have the client engage in overt practice of the goal behaviors. In these cases, it may be more practical to employ strategies that use the client's imagination for the modeling or rehearsal. This chapter describes two therapeutic procedures that rely heavily on client imagery: guided imagery and covert modeling. In both these strategies, scenes are developed that the client visualizes or rehearses by imagination. It has been assumed that the client must be able to generate strong, vivid images to use these procedures, but we do not know at present to what extent the intensity of the client's imagery correlates with therapeutic outcomes.

Often people use the terms *emotive imagery, visualization,* and *guided imagery* interchangeably. The tenth edition (1993) of Merriam-Webster's Collegiate Dictionary defines visualization as the "formation of mental images; the act or process of interpreting in visual terms or of putting into visible form" (p. 1321). Visualization of mental images or pictures can be spontaneous or guided. All visualizations or images have an emotive, emotional, or mental connection. We experience events, people, or things in our environment through our senses, which influence our conditioned limbic (emotional) system and our perceptions. This chapter describes two therapeutic procedures that rely heavily on client imagery: guided imagery and covert modeling. In both strategies, mental scenes are developed that the client visualizes or rehearses by imagination. Some people have trouble initially generating strong, vivid images. We have found that most clients have the capacity to evoke vivid images, and with help from the therapist they can elicit their visualization potential.

OBJECTIVES

1. Given seven examples of counselor leads, identify which of the five steps of the guided imagery procedure are presented in each counselor lead. You should be able to identify accurately at least six out of seven examples.
2. Demonstrate 10 out of 13 steps of guided imagery with a role-play client, using the Interview Checklist for Guided Imagery at the end of the chapter to assess your performance.
3. Describe how you would apply the five components of covert modeling, given a simulated client case.
4. Demonstrate at least 22 out of 28 steps associated with covert modeling with a role-play client, using the Interview Checklist for Covert Modeling at the end of the chapter to assess your use of this strategy.

CLIENT IMAGERY: ASSESSMENT AND TRAINING

In both guided imagery and covert modeling, assessing the client's potential for engaging in imagery or fantasy is essential. To some extent, the success of these two strategies may depend on the client's capacity to generate vivid images. Some clients may be turned off by imagery; others may have difficulty picturing vivid scenes in their minds.

Here is a way to assess the intensity and clarity of client images. First, before using guided imagery or covert modeling, you can ask clients to recall some recent event they enjoyed or an event that made them feel relaxed and pleasant. Tell clients to close their eyes, take a couple of deep breaths to relax, and describe the enjoyable event to you. After the clients describe the event, ask them to rate its vividness, using 4 for a very clear image, 3 for a moderately clear image, 2 for a fairly clear image, 1 for an unclear image, and 0 for an indiscernible image. You can assess the vividness of clients' imagery using the rating scale displayed in Table 13-1. A score of 60 or more means that clients probably can evoke vivid images very easily. A total score lower than 30 might suggest that they need more training to evoke more vivid images. The vividness scale can be self-administered

TABLE 13-1. Example of a vividness imagery scale

The Imagery Vividness Scale

Let's try a simple test. You will be asked to picture certain images. If your image is "very clear" give it a rating of 4; if "moderately clear" give it a 3; a "fairly clear" image rates a 2; and an "unclear" image rates a 1. If you cannot form an image, or if it is "very unclear" or "indiscernible," give it 0. After reading each item, close your eyes, picture it as clearly as you can, and then record your own rating. *Authors' addition:* Then add the ratings together. If your score is 60 or greater, you can form images quite easily and vividly. If your score is 30 or lower, more training in imagery may be needed before using this guided imagery intervention.

Think about a very close relative or friend:

		Rating
1.	See him/her standing in front of you.	( )
2.	Imagine him/her laughing.	( )
3.	Picture his/her eyes.	( )
4.	Picture of bowl of fruit.	( )
5.	Imagine driving down a dry, dusty road.	( )
6.	See yourself throwing a ball.	( )
7.	Picture your childhood home.	( )
8.	See a white, sandy beach.	( )
9.	Imagine looking into a shop window.	( )
10.	See a blank television screen.	( )
11.	Imagine the sound of a barking dog.	( )
12.	Imagine the sound of an exploding firecracker.	( )
13.	Feel the warmth of a hot shower.	( )
14.	Imagine feeling the texture of rough sandpaper.	( )
15.	Picture yourself lifting a heavy object.	( )
16.	Imagine yourself walking up a steep stairway.	( )
17.	Imagine the taste of lemon juice.	( )
18.	Think of eating ice cream.	( )
19.	Imagine the smell of cooking cabbage.	( )
20.	Imagine yourself smelling a rose.	( )

SOURCE: From *In the Mind's Eye: The Power of Imagery for Personal Enrichment*, by A. Lazarus, pp. 9–11. Copyright © 1984 by The Guilford Press. Reprinted by permission.

or the counselor can read the instructions and items to the client and obtain a rating for each item. We have found that reading the instructions and items to clients is less disruptive because they do not have to open their eyes to read each item of the scale.

To train clients to add more vividness to imagery or mental scenes, the counselor and the clients can also develop practice scenes that clients can use to generate images. For example, the counselor might instruct a client to "visualize a scene or an enjoyable event that makes you feel relaxed and

pleasant. Select an event that you enjoy and feel good about—relax your body and mind, and try to be aware of all your sensations while visualizing the image of the event." We have found that instructing clients about sensations associated with the scene is a powerful induction aid for facilitating their vividness of the scene. The counselor can provide sensory cues by asking: "What objects are pictured in the scene? "What colors do you see?" "How is the light in the scene—is it dark or light?" "What is the degree of temperature that you feel in the scene?" "What odors do you experience?" "What sounds do you hear?" "What sensation of taste do you experience in your mouth?" "What sensations do you experience in your body?" When clients have input in selecting and developing the details of a practice scene, they develop more self-efficacy or confidence in using imagery. As counselors assess clients' ability to visualize or when they train clients to have vivid images, they should suggest selecting a scene or an event that was very enjoyable and pleasurable. We are much more aware of the sensations we experience in pleasurable events or situations. Table 13-2 illustrates dimensions of the senses and some examples of sensory experiences.

If clients have difficulty imagining specific details, the counselor can train them to engage in more detailed imagery. If this seems too time-consuming or if clients are reluctant to use or are uninterested in using imagery, a strategy that does not involve imagery may be more appropriate. If a client has good feelings about imagery and can adapt to practicing it, then the counselor and client may decide to continue with either guided imagery or covert modeling, depending on the client's problem and goal behavior.

MULTICULTURAL APPLICATIONS OF IMAGERY

To date, imagery has been used in fairly limited ways with diverse groups of clients. Herring and Meggert (1994) used imagery processes as a counseling strategy with Native American children. Andrada and Korte (1993) used imagery scenes with auditory, tactile, verbal, visual, and taste stimuli to enhance reminiscing among elderly Hispanic men and women living in a nursing home. Brigham (1994) describes specific imagery scenes for use with clients who are HIV positive.

There are several guidelines to consider in using imagery with diverse groups of clients. First, the models of imagery used across cultures are likely to be different. Gaines and Price-Williams (1990) and Brigham (1994) have noted that the imaginative model of Euro-Americans is indi-

TABLE 13-2. Sensations to enhance vividness of imagery

Senses	Dimensions	Experiences
Seeing	Colors, lightness, darkness, depth of field	Perception of proximity, distance, movement, stillness, physical objects, events, animals, people, nature
Hearing	Noise, sound, direction, pitch, loudness	Perception of sound: euphonious, unpleasant, from people, music, objects, nature
Smelling	Molecules of substances—odors, scent, or aroma	Airborne molecules drawn into the nose from substances, such as plants, animals, people, objects, nature, traffic, cooking, perfume, cologne, smoke, smog, moisture, pollution, fragrance of flowers, air
Tasting	Sour, bitter, sweet, salty	"Gruesome scene left bad taste in my mouth," or "the ocean breeze left a pleasant taste" Taste and smell provide experience of flavor
Touching	Pressure, pain, temperature	Perception of touch: moisture and feelings of warmth or coolness, usually experienced as skin sensations
Moving and positioning of head and body	Motion, body's position in space	Which end is up? Different movements provide feedback on body's position

vidualistic while the imaginative model for some other cultural groups is collective.

Specific mode and content of imagery scenes appear to be correlated with a person's philosophical and cultural settings (Gaines & Price-Williams, 1990). Therefore, imagery must be viewed within the context of the client's notion of both self and culture. Also, the imagery used must be culturally relevant to the client. For example, a Native American client may choose to visualize the use of a "power shield" to ward off bad spirits and to enter into harmony with nature. Other clients may choose to visualize themselves as "peace warriors."

If healing symbols are used in imagery, they need to be culturally specific or at least universally recognized. One such symbol is the mandala, a circle of wholeness and sacred space (Brigham, 1994). Myths, folk tales, and legends indigenous to a client's culture also can be used in imagery. For example, Brigham (1994) uses the compact disc *Skeleton Woman* in which the group Flesh and Bone have set music to the legend made so well known in *Women Who Run with the Wolves* (Estes, 1992). The Skeleton Woman tale can be read and then the client can imagine some aspect of this legend.

GUIDED IMAGERY

In using the guided imagery procedure, a person focuses on positive thoughts or images while imagining a discomforting or anxiety-arousing activity or situation. By focusing on positive and pleasant images, the person is able to "block" the painful, fearful, or anxiety-provoking situation. One can think of blocking in emotive imagery as a process that takes advantage of the difficulty of focusing on pleasant thoughts and on anxiety, pain, or tension at the same time. This is difficult because these emotions are incompatible.

Self-initiated imagery has been used to help a variety of different types of concerns. Table 13-3 presents a sample of research illustrating the variety of uses for guided imagery. As you see, the uses of guided imagery are very diverse, ranging from helping people control allergic responses to ragweed, to helping individuals cope with loss and grief, to strengthening the immune function, to reducing migraine headaches, to improving a person's problem solving ability. In recent years, guided imagery has become a very important part of sports psychology (Suinn, 1986; Martin & Hall, 1995). Also, like many other strategies presented in this edition, guided imagery has been used to complement other

TABLE 13-3. Guided imagery research

Allergy

Cohen, R. E., Creticso, P. S., & Norman, P. S. (1993–1994). The effects of guided imagery (GI) on allergic subjects' responses to ragweed-pollen nasal challenge: An exploratory investigation. *Imagination, Cognition and Personality, 13,* 259–269.

Breast cancer

Gruber, B. L., Hersh, S. P., Hall, N. R., & Waletsky, L. R. (1993). Immunological responses of breast cancer patients to behavioral interventions. *Biofeedback and Self-Regulation, 18,* 1–22.

Cancer

Baider, L., Uziely, B., & Kaplan-DeNour, A. (1994). Progressive muscle relaxation and guided imagery in cancer patients. *General Hospital Psychiatry, 16,* 340–347.

Chemical dependency

Avants, S. K., Margolin, A., & Singer, J. L. (1994). Self-reevaluation therapy: A cognitive intervention for the chemically dependent patient. *Psychology of Addictive Behaviors, 8,* 214–222.

Cassel, R. N., Hoey, D., & Riley, A. D. (1991). Guided imagery with subliminal stimulus in a mind-body-health program for chemical dependency rehabilitation (New Beginnings basic program). Special Issue: Special recognition to Dr. Russell N. Cassel. *Psychology: A Journal of Human Behavior, 27,* 3–9.

Children

Myrick, R. D., & Myrick, L. S. (1993). Guided imagery: From mystical to practical. Special Issue: Counseling and children's play. *Elementary School Guidance and Counseling, 28,* 62–70.

Complementary role of imagery

Overholser, J. C. (1991). The use of guided imagery in psychotherapy: Modules for use with passive relaxation training. *Journal of Contemporary Psychotherapy, 21,* 159–172.

Cultural exploration

Strongylou, N., & Woodard, V. (1993). Exploring images of the Greek-Cypriot woman through drama therapy. *Arts in Psychotherapy, 20,* 161–165.

Emotionally charged memories and states

Edwards, D. J. (1990). Cognitive therapy and the restructuring of early memories through guided imagery. *Journal of Cognitive Psychotherapy, 4,* 33–50.

Rosenthal, T. L. (1993). To soothe the savage beast. *Behaviour Research and Therapy, 31,* 439–462.

Grief

Brown, J. C. (1990). Loss and grief: An overview and guided imagery intervention model. *Journal of Mental Health Counseling, 12,* 434–445.

Hemoptysis

Palan, B. M., & Lakhani, J. D. (1991). Converting the "threat" into a "challenge": A case of stress-related hemoptysis managed with hypnosis. *American Journal of Clinical Hypnosis, 33,* 241–247.

HIV

Auerbach, J. E., Olsen, T. D., & Solomon, G. F. (1992). A behavioral medicine intervention as an adjunctive treatment for HIV-related illness. Special Issue: Biopsychosocial aspects of HIV infection. *Psychology and Health, 6,* 325–334.

Immune function

Zachariae, R., Kristensen, J. S., Hokland, P., & Ellegaard, J. (1990). Effect of psychological intervention in the form of relaxation and guided imagery on cellular immune function in normal healthy subjects: An overview. *Psychotherapy and Psychosomatics, 54,* 32–39.

Zachariae, R., Hansen, J. B., Andersen, M., & Jinquan, T. (1994). Changes in cellular immune function after immune specific guided imagery and relaxation in high and low hypnotizable healthy subjects. *Psychotherapy and Psychosomatics, 61,* 74–92.

Learned helplessness

Weisenberg, M., Gerby, Y., & Mikulincer, M. (1993). Aerobic exercise and chocolate as means for reducing learned helplessness. *Cognitive Therapy and Research, 17,* 579–592.

Migraine headache

Ilacqua, G. E. (1994). Migraine headaches: Coping efficacy of guided imagery training. *Headache, 34,* 99–102.

---

treatment strategies such as muscle relaxation, desensitization, eye movement desensitization and reprocessing, goal setting, and problem solving. The use of visualization in muscle relaxation training can create as great a physiological response as the actual experience of tightening and relaxing the muscles (Overholser, 1990, 1991).

Many people experience thoughts and emotions in the form of images or mental pictures. Judith Beck (1995) teaches clients to access their distressing images as a technique in problem identification. She might say to the client, "When you had that thought or emotion, what image or picture did you have in your head?" (p. 230). If the counselor and client can identify the distressing image, they can create a new image and later restructure the thought or emotions associated with the distressing image. Guided imagery has become very popular as a complementary strategy in medicine. Jeanne Achterberg (1985) in her *Imagery in Healing* describes the many uses of imagery in medicine. Finally,

TABLE 13-3. Guided imagery research (continued)

Music with imagery

Bonny, H. L. (1989). Sound as symbol: Guided imagery and music in clinical practice. *Music Therapy Perspectives, 6,* 7–10.

Mayer, J., Allen, J., & Beauregard, K. (1995). Mood inductions for four specific moods: A procedure employing *guided imagery* vignettes with music. *Journal of Mental Imagery, 19,* 151–159.

Pain

Dunne, P. W., Sanders, M. R., Rowell, J. A., & McWhirter, W. R. (1991). An evaluation of cognitive-behavioural techniques in the management of chronic arthritic pain in men with hemophilia. *Behaviour Change, 8,* 70–78.

Panic attacks

Der, D. F., & Lewington, P. (1990). Rational self-directed hypnotherapy: A treatment for panic attacks. *American Journal of Clinical Hypnosis, 32,* 160–167.

Personal meaning

Farr, C. (1990). A study of the guided imagery process: Awareness and the discovery of personal meaning. *Canadian Journal of Counselling, 24,* 45–52.

Rancour, P. (1991). Guided imagery: Healing when curing is out of the question. *Perspectives in Psychiatric Care, 27,* 30–33.

Physically challenged

Short, A. E. (1992). Music and imagery with physically disabled elderly residents: A GIM adaptation. *Music Therapy, 11,* 65–98.

Problem solving

Koziey, P. W. (1990). Patterning language usage and themes of problem formation/resolution. *Canadian Journal of Counselling, 24,* 230–239.

Psychodynamic therapy

Feinberg, M., Beverly, B., & Oatley, K. (1990). Guided imagery in brief psychodynamic therapy: Outcome and process. *British Journal of Medical Psychology, 63,* 117–129.

Respiratory problems

Connolly, M. J. (1993). Respiratory rehabilitation in the elderly patient. *Reviews in Clinical Gerontology, 3,* 281–294.

Smoking cessation

Wynd, C. A. (1992). Personal power imagery and relaxation techniques used in smoking cessation programs. *American Journal of Health Promotion, 6,* 184–189, 196.

Sports psychology

Martin, K., & Hall, C. (1995). Using mental imagery to enhance intrinsic motivation. *Journal of Sports and Exercise Psychology, 17,* 54–65.

Suinn, R. (1986). *Seven steps to peak performance: The mental training manual for athletes.* Lewiston, NY: Huber.

Strain injury

Nicol, M. (1993). Hypnosis in the treatment of repetitive strain injury. *Australian Journal of Clinical and Experimental Hypnosis, 21,* 121–126.

Stress reduction

Prerost, F. J. (1993). A strategy to enhance humor production among elderly persons: Assisting in the management of stress. *Activities, Adaptation and Aging, 17,* 17–24.

Weinburger, R. (1991). Teaching elderly stress reduction. *Journal of Gerontological Nursing, 17,* 23–27.

Test anxiety

Fernandez, D., & Allen, G. J. (1989). Test anxiety, emotional responding under guided imagery, and self-talk during an academic examination. *Anxiety Research, 2,* 15–26.

Vomiting

Torem, M. S. (1994). Hypnotherapeutic techniques in the treatment of hyperemesis gravidarum. *American Journal of Clinical Hypnosis, 37,* 1–11.

Watson, M., & Marvell, C. (1992). Anticipatory nausea and vomiting among cancer patients: A review. *Psychology and Health, 6,* 97–106.

---

there are commercial audiotapes that offer ways for effective visualization, techniques for improving vividness of mental imagery, visualization and achieving goals, using one's inner adviser, and imagery exercises that contribute to alertness and mindfulness (Pulos, L. 1996).

Guided imagery involves five steps: a rationale, assessment of the client's imagery potential, development of imagery scenes, practice of scenes, and homework. See the Interview Checklist for Guided Imagery at the end of the chapter for some examples of counselor leads associated with these steps.

Treatment Rationale

The following illustration of the purpose and overview of guided imagery can be used with people who have anxiety about a medical procedure and for relieving discomfort of such a procedure:

Here is how guided imagery works while you are having the procedure. Often people have a lot of anxiety about the procedure just before they go to the hospital. This is normal and natural. It is often very difficult to shake this belief.

Usually the belief intensifies anxiety about the procedure. When some discomfort occurs, the result is even more anxiety.

Now, since the anxiety magnifies the effect of the medical procedure, the more anxious you become, the more actual pain you will probably experience. You can say the anxiety you experience escalates or intensifies the pain. This vicious circle happens often with patients undergoing this procedure. You can reverse the anxiety-pain cycle by using a technique called guided imagery. It works like this: You visualize a scene or an event that is pleasant for you and makes you feel relaxed while the procedure is occurring. You cannot feel calm, relaxed, secure—or whatever other emotion the scene evokes—and anxious at the same time. You cannot experience feelings of anxiety and relaxation or calmness at the same time. These two different emotions are incompatible with each other.

So, you select a scene that you can get into and that makes you feel relaxed, calm, and pleasant. You visualize the image or scene you have selected while the medical procedure is being performed. The imagery blocks the anxiety that can lead to increased discomfort while the procedure is being performed. People who have used guided imagery have reported that visualizing and holding a pleasant scene or image raises their threshold of pain. Using imagery while the medical procedure is performed on you eliminates anxiety-related discomfort and also has a dulling effect on what might be experienced as pain.

The rationale ends with a request for the client's willingness to try the strategy.

Assessment of Client's Imagery Potential

Because the success of the guided imagery procedure may depend on the amount of positive feelings and calmness a client can derive from visualizing a particular scene, it is important for the counselor to get a feeling for the client's potential for engaging in imagery. The counselor can assess the client's imagery potential by the methods discussed at the beginning of this chapter: a self-report questionnaire, a practice scene with client narration, or counselor "probes" for details.

Development of Imagery Scenes

If the decision is made to continue with the guided imagery procedure, the client and counselor will then develop imagery scenes. They should develop at least two, although one scene might be satisfactory for some clients. The exact number and type of scenes developed will depend on the nature of the concern and the individual client.

Two basic ingredients should be included in the selection and development of the scene. First, the scenario should promote calmness, tranquility, or enjoyment for the client. Examples of such client scenes might include skiing on a beautiful slope, hiking on a trail in a large forest, sailing on a sunny lake, walking on a secluded beach, listening to and watching a symphony orchestra perform a favorite composition, or watching an athletic event. The scenes can involve the client as an active participant or as a participant observer or spectator. For some clients, the more active they are in the scene, the greater the degree of their involvement.

The second ingredient of the scene should be as much sensory detail as possible, such as sounds, colors, temperature, smell, touch, and motion. There may be a high positive correlation between the degree and number of sensations a scene elicits for the client and the amount or intensity of pleasant and enjoyable sensations the client experiences in a particular imagery scene. The counselor and client should decide on the particular senses that will be incorporated into the imagery scenes. Remember that the scene needs to be culturally relevant for the client.

As an example, the following imagery scene can be used with a client who experiences discomfort about a medical procedure. Note the sensations described in the scene instructions:

> Close your eyes, sit back, take a couple of deep breaths (pause), and relax. With your eyes closed, sitting back in the chair and feeling relaxed, visualize yourself on a beautiful ocean beach. Visualize a few puffy clouds scattered throughout a dazzling blue sky. Notice the blue-green ocean water with the white caps of the surf rolling in toward the beach. See yourself wearing a bathing suit, and feel bright, warm rays of sun on your body. Take a deep breath and experience the fresh and clean air. Hear the waves gently rolling in onto the beach and the water receding to catch the next wave. Smell the salt and moisture in the air. Experience the touch of a gentle breeze against your body. See yourself unfolding a beach towel, and feel the texture of the terry cloth material as you spread it on the sandy beach. Notice the sand on your feet; experience the warmth of the sand and feel the relaxing and soothing sensations the sand provides. Now visualize yourself walking toward the surf and standing in ankle-deep water, experiencing the wetness of the ocean. You experience the water as warm and comfortable. You are all alone, and you walk out in the surf up to your waist, then up to your chest. You hear the waves and see the sun glistening off the blue-green water. Smell the salt in the air as you surface dive into the oncoming surf. Experience the warmth of the water on your body as you swim out just beyond where the surf is breaking. See yourself treading water, and experience the gentle movement of going up and down on the surface of the water. Notice how warm the sun and water feel, and how relaxing it is just to linger there in the water.

Experience the gentle motions of your arms and legs moving beneath the surface. Picture yourself slowly swimming back to the beach, and catching a wave and riding it in. Visualizing yourself standing up and walking slowly toward your beach towel. Feel the warm sun and the air temperature of about 90 degrees on your wet body. You reach the beach towel, and you stretch out and lie down. You are alone; there is no else in the water or on the beach. You feel the warmth of the sun on the front of your body. Every muscle in your body feels totally relaxed. You look up at the sky watching the large billowy clouds drifting by.

Practice of Imagery Scenes

After the imagery scenes have been developed, the client is instructed to practice using them. There are two types of practice. In the first type, the client is instructed to focus on one of the scenes for about 30 seconds and to picture as much detail as was developed for that scene and feel all the sensations associated with it. The counselor cues the client after the time has elapsed. After the client uses the scene, the counselor obtains an impression of the imagery experience—the client's feelings and sensory details of the scene. If other scenes are developed, the client can be instructed to practice imagining them. Variations on this type of practice might be to have the client use or hold a scene for varying lengths of time.

The second type of practice is to have the client use the scenes in simulated anxious, tense, fearful, or painful situations. The counselor and client should simulate the problem situations while using the imagery scenes. Practice under simulated conditions permits the counselor and client to assess the effectiveness of the scenes for blocking out the discomfort or for reducing the anxiety or phobic reaction. Simulated situations can be created by describing vividly the details of an anxiety-provoking situation while the client uses a scene. For example, the counselor can describe the pleasant scene while interweaving a description of the discomforting situation. The counselor can simulate painful situations by squeezing the client's arm while the client focuses on one of the scenes. Or, to simulate labor contractions, the labor coach squeezes the woman's thigh while she focuses on a pleasant image. Another simulation technique is to have clients hold their hands in ice water for a certain length of time. Simulated practice may facilitate generalization of the scene application to the actual problem situation. After the simulated practices, the counselor should assess the client's reactions to using the scene in conjunction with the simulated discomfort or anxiety.

Homework and Follow-Up

The client is instructed to apply the guided imagery in vivo—that is, to use the scenes in the fearful, painful, tense, or anxious situation. The client can use a homework log to record the day, time, and situation in which guided imagery was used. The client can also note reactions before and after using guided imagery with a five-point scale, 1 representing minimum discomfort and 5 indicating maximum discomfort. The client should be told to bring the homework log to the next session or to a follow-up session.

MODEL EXAMPLE: GUIDED IMAGERY

In this model example, we are going to deviate from our usual illustrations of hypothetical cases and present a narrative account of how the two of us used guided imagery before and during labor for the birth of our two children.

1. *Rationale*: First, we discussed a rationale for using guided imagery during labor in conjunction with the breathing and relaxation techniques (see Chapter 17) we had learned in our prepared-childbirth class. We decided before labor started that we would try guided imagery at a point during labor when the breathing needed to be supplemented with something else. We also worked out a finger-signaling system to use during contractions so Sherry could inform Bill whether to continue or stop with the imagery scenes, depending on their effectiveness.

2. *Assessment of imagery potential*: We also discussed whether Sherry was able to use fantasy effectively enough to concentrate during a labor contraction. We tested this out by having Bill describe imagery stimuli and having Sherry imagine these and try to increase use of all sensations to make the imagery scenes as vivid as possible.

3. *Development of imagery scenes*: Together we selected two scenes to practice with before labor and to use during labor. One scene involved being on a sailboat on a sunny, warm day and sailing quite fast with a good breeze. We felt this scene would be effective because it produced enjoyment and also because it seemed to evoke a lot of sensory experience. The second scene involved being anchored at night on the boat on a warm night with a soft breeze. Because both these scenes represented actual experiences, we felt they might work better than sheer fantasy.

4. *Practice of imagery scenes*: We knew that much of the success of using guided imagery during labor would depend on the degree to which we worked with it before labor. We practiced with our imagery scenes in two ways. First, Sherry imagined these scenes on her own, sometimes in conjunction with her self-directed practice in breathing and relaxation, and sometimes just as something to do—for instance, in a boring situation. Second, Sherry evoked the scenes deliberately while Bill simulated a contraction by tightly squeezing her upper arm.

LEARNING ACTIVITY **39** *Guided Imagery*

This learning activity is designed to help you learn the process of guided imagery. It will be easier to do this with a partner, although you can do it alone if a partner is not available. You may wish to try it with a partner first and then by yourself.

Instructions for Dyadic Activity

1. In dyads, one of you can take the helper role; the other takes the part of the one helped. Switch roles after the first practice with the strategy.
2. The helper should give an explanation about the guided imagery procedure.
3. The helper should determine the potential for imagination of the one being helped: ask the person to imagine several pleasant scenes and then probe for details.
4. The two together should develop two imagery scenes the one being helped can vividly imagine. Imagination of these scenes should produce pleasant, positive feelings and should be culturally relevant to the helpee.
5. The person should practice imagining these scenes—as vividly and as intensely as possible.

6. He or she should practice imagining a scene while the helper simulates a problem situation. For example, the helper can simulate an anxiety-provoking situation by describing it while the other engages in imagery, or the helper can simulate pain by squeezing the other person's arm during the imagination.

Instructions for Self-Activity

1. Think of two scenes you can imagine very vividly. These scenes should produce positive or pleasant feelings for you and be culturally relevant for you. Supply as many details to these scenes as you can.
2. Practice imagining these scenes as intensely as you can.
3. Next, practice imagining one of these scenes while simulating a problem (discomforting) situation such as grasping a piece of ice in your hands or holding your hands under cold water. Concentrate very hard on the imagery as you do so.
4. Practice this twice daily for the next three days. See how much you can increase your tolerance for the cold temperature with daily practice of the imagery scene.

---

5. *Homework: In vivo* We had a chance to apply guided imagery during labor itself. We started to use it during the active phase of labor—about midway through the time of labor, when the contractions were coming about every 2 minutes. In looking back, we felt it was a useful supplement to the breathing and relaxation typically taught in the Lamaze childbirth method. Sherry felt that a lot of the effectiveness had to do with the soothing effect of hearing Bill's vocal descriptions of the scenes—in addition to the images she produced during the scene descriptions.

COVERT MODELING

Covert modeling is a procedure in which the client imagines a model performing behaviors by means of instructions. A live or symbolic performance by a model is not necessary. Instead, the client is directed to imagine someone demonstrating the desired behavior. Covert modeling has been used to train athletes acquire desired behaviors. Bill Cormier has used covert modeling to train patients and students to acquire Yoga postures or positions. Uhlemann and Koehn (1989) used covert modeling and overt modeling to acquire empa-

thetic responding. Covert modeling was used with disturbed special education students, third through ninth grades, in Finland (Lindh, 1987).

Covert modeling has several advantages: The procedure does not require elaborate therapeutic or induction aids; scenes can be developed to deal with a variety of problems; the scenes can be individualized to fit the unique concerns of the client; the client can practice the imagery scenes alone; the client can use the imagery scenes as a self-control procedure in problem situations; and covert modeling may be a good alternative when live or filmed models cannot be used or when it is difficult to use overt rehearsal in the interview.

Some questions about certain aspects of covert modeling remain unanswered, such as the importance of the identity of the model, the role of reinforcing consequences, and the type and duration of imagery scenes best used in the procedure. We have tried to point out the possible alternatives in our description of the components of the covert modeling strategy. The five major components of covert modeling are rationale about the strategy, practice scenes, developing treatment scenes, applying treatment scenes, and homework.

Within each of these five components are several substeps. If you would like an overview of the procedure, see the Interview Checklist for Covert Modeling at the end of the chapter.

Treatment Rationale

After the counselor and client have reviewed the problem behaviors and the goal behaviors for counseling, the counselor presents the rationale for covert modeling. Here is an explanation for using covert modeling that Bill uses when he teaches Yoga to an individual or with people in a class:

> In developing your own practice at home, it is helpful to rehearse or practice the different Yoga postures with visualization of the form for each posture. Imagining each posture can help you perform the posture in your daily practice. For example, you can imagine yourself doing each position or posture for the Sun Salutation or Warm Up. Seeing yourself perform each position in your imagination can strongly influence your confidence and behavior when you actually perform the Sun Salutation.

Another rationale provides an illustration of the way Bill describes the covert modeling process to Yoga students or clients:

> After we go through each position, I'll ask you to close your eyes and try to imagine, as clearly as possible, that you are observing yourself performing the first position of the Sun Salutation. Then, I will instruct you to visualize each of the other eight positions, one at a time. Try to use your senses in your visualization of each position. I will instruct you to focus and to experience the senses as you visualize going into each position. After we complete visualizing the last position, I will ask you some questions concerning your feelings about the sequence and how clearly you imagined each position.

Practice Scenes

After providing a rationale to the client, the counselor may decide to try out the imagery process with several practice scenes. For most clients, imagining a scene may be a new experience and may seem foreign. Kazdin (1976b, p. 478) suggests that practice scenes may help to familiarize clients with the procedure and sensitize them to focus on specific details of the imagery. Use of practice scenes also helps the counselor assess the client's potential for carrying out instructions in the imagination.

The practice scenes usually consist of simple, straightforward situations that are unrelated to the goal behaviors. For example, if you are using covert modeling to help a client acquire job-interview skills, the practice scenes would be unrelated to job-seeking responses. You might use some of the following as practice scenes:

1. Imagine watching a person at a golf match on the eighteenth hole of a beautiful golf course on a gorgeous day.
2. Imagine someone hiking to the top of a mountain with a panoramic view.
3. Imagine watching a comedian at a comedy club.
4. Imagine someone taking a walk on a beautiful day.

In using practice scenes with a client, the counselor usually follows six steps.

1. The counselor instructs the client to close his or her eyes and to sit back and relax. The client is instructed to tell the counselor when he or she feels relaxed. If the client does not feel relaxed, the counselor may need to decide whether relaxation procedures (Chapter 17) should be introduced. The effect of relaxation on the quality of imagery in covert modeling has not been evaluated. However, live and symbolic modeling may be facilitated when the client is relaxed.

2. The counselor describes a practice scene and instructs the client to imagine the scene and to raise an index finger when the scene has been imagined vividly. The practice scenes are similar to the four previous examples. The counselor reads the scene or instructs the client about what to imagine.

3. The counselor asks the client to open his or her eyes after the scene is vividly imagined (signal of index finger) and to describe the scene or to narrate the imagined events.

4. The counselor probes for greater details about the scene—the clothes or physical features of a person in the imagery, the physical setting of the situation, the amount of light, colors of the furniture, decorative features, noises, or smells. This probing may encourage the client to focus on details of the imagery scene.

5. The counselor may suggest additional details for the client to imagine during a subsequent practice. Working with practice scenes first can facilitate the development of the details in the actual treatment scenes.

6. Usually each practice scene is presented several times. The number of practice scenes used before moving on to developing and applying treatment scenes will depend on several factors. If the client feels comfortable with the novelty of the imagined scenes after several presentations,

the counselor might consider this a cue to stop using the practice scenes. Additionally, if the client can provide a fairly detailed description of the imagery after several practice scenes, this may be a good time to start developing treatment scenes. If a client reports difficulty in relaxing, the counselor may need to introduce relaxation training before continuing. For a client who cannot generate images during the practice scenes, another modeling strategy may be needed in lieu of covert modeling.

Developing Treatment Scenes

The treatment scenes used in covert modeling are developed in conjunction with the client and grow out of the desired client outcomes or goals. The scenes consist of a variety of situations in which the client wants to perform the target response in the real-life environment. If a client wants to acquire effective job-interview skills, the treatment scenes are developed around job-interview situations.

Five things should be considered in the development of treatment scenes: the model characteristics, whether to use individualized or standardized scenes, whether to use vague or specific scenes, the ingredients of the scenes, and the number of scenes. It is important for the client to help in the development of treatment scenes because client participation can provide many specifics about situations in which the goal behavior is to be performed.

Model Characteristics. Similarity between the model and the client contributes to client change. Models of the same sex and age as the client may be more effective. Also, clients who imagine several models may show more change than clients who imagine only one model. Recall that clients are more likely to learn from a model who is similar to them in gender, race, and ethnicity. Coping models also may be generally more effective in covert modeling than mastery models. A coping model who self-verbalizes his or her anxiety and uses covert self-talk for dealing with fear may enhance and facilitate the behaviors to be acquired.

One of the most interesting questions about the covert model is the identity of the model: clients who imagine *themselves* as the model and clients who imagine *another person* as the model. We believe that for most people, imagining themselves may be more powerful. However, there are not sufficient data to indicate who the model should be in the covert modeling procedure. We suspect that the answer varies with clients and suggest that you give clients the option of deciding whether to imagine themselves or

another person as the model. For biracial clients, this may mean asking them which part of their culture they identify with the most. One key factor may involve the particular identity the client can imagine most easily and comfortably. For clients who feel some stress at first in imagining themselves as models, imagining someone else might be introduced first and later replaced with self-modeling. Using yourself as the model is a good option when culturally similar models are not readily available.

Individualized versus Standardized Scenes. The treatment scenes used in covert modeling can be either standardized or individualized. Standardized scenes cover different situations in everyday life and are presented to a group of clients or to all clients with the same target responses. For example, a counselor can use a series of standardized scenes describing situations of job interviewing behavior. Individualized scenes represent situations specifically tailored to suit an individual client. For example, one nonassertive client may need scenes developed around situations with strangers; another may need scenes that involve close friends. Generally, treatment scenes should be individualized for those who have unique concerns and who are counseled individually, as some standardized scenes may not be relevant for a particular client.

Specificity of Scenes. Another consideration in developing treatment scenes is the degree of specificity of instruction that should be included. Some clients may benefit from very explicit instructions about the details of what to imagine. Other clients may derive more gains from covert modeling if the treatment scenes are more general, allowing the client to supply the specific features. A risk of not having detailed instructions is that some clients will introduce scene material that is irrelevant or detracts from the desired outcomes. We suggest this decision should consider the client's preferences.

Here is an example of a fairly general treatment scene for a prison inmate about to be released on parole who is seeking employment:

> Picture yourself (or someone else like you) in a job interview. The employer asks why you didn't complete high school. You tell the employer you got in some trouble, and the employer asks what kind of trouble. You feel a little uptight but tell her you have a prison record. The employer asks why you were in prison and for how long. The employer then asks what makes you think you can stay out of trouble and what you have been doing while on parole. Imagine yourself saying that you have been looking for work while on parole and have been thinking about your life and what you want to do with it.

The generality of the treatment scene in this example assumes that the client knows what type of response to make and what details to supply in the scene.

A more detailed treatment scene would specify more of the actual responses. For example:

> Picture yourself (or someone else) in a job interview and imagine the following dialogue. The employer says, "I see that you have only finished three years of high school. You don't intend to graduate?" Picture yourself saying (showing some anxiety): "Well, no, I want to go to vocational school while I'm working." The employer asks: "What happened? How did you get so far behind in school?" Imagine yourself (or someone else) replying: "I've been out of school for awhile because I've been in some trouble." Now imagine the employer is showing some alarm and asks: "What kind of trouble?" You decide to be up front as you imagine yourself saying: "I want you to know that I have a prison record." As the employer asks: "Why were you in prison?" imagine yourself feeling a little nervous but staying calm and saying something like: "I guess I was pretty immature. Some friends and I got involved with drugs. I'm on parole now. I'm staying away from drugs and I'm looking hard for a job. I really want to work."

Remember, the degree of specificity of each scene will depend largely on the client, the problem or concern, and the goals for counseling.

Ingredients of the Scene.   Three ingredients are required for a treatment scene in the covert modeling procedure: a description of the situation or context in which the behavior is to occur, a description of the model demonstrating the behavior to be acquired, and a depiction of some favorable outcome of the goal behavior.

> *Situation:* Imagine yourself playing tennis on a bright sunny day. You are playing in a tournament. Your opponent hits the ball to your forehand and you return it. On your opponent's return, the ball is hit low to your backhand.

> *Demonstrating the desired behavior:* You see yourself put both hands on the racket, pull your arms back ready to hit the low ball. See yourself hit the ball across the net, and inside the line. Notice the bodily sensation you experience while hitting the ball across the net.

Below is a covert modeling scene that includes a favorable outcome for an adult who wants to stop smoking. This is an example of a scene:

> Imagine yourself in a restaurant having a drink with some friends before your reservation to be seated for dinner. All your friends in the group are smoking, and the smell of the smoke makes you want a cigarette. You have been drinking with them for about ten minutes, and one of them offers you a cigarette. In the past, you would have taken a cigarette if you did not have any of your own. Now cope with the situation in your imagination. Imagine yourself feeling the urge to smoke, but see yourself refuse and focus on what the group is discussing.

Inclusion of a favorable consequence as a scene ingredient is based on research indicating that if a client sees a model rewarded for the behavior or feels good about the outcome of the behavior, the client is more likely to perform the response. Moreover, specifying a possible favorable outcome to imagine may prevent a client from inadvertently imagining an unfavorable one. Clients who receive covert modeling treatment scenes that are resolved favorably are more likely to engage in the desired goal behaviors than clients who imagined scenes without any positive consequences.

We believe that the favorable outcome in the scene should take the form of some action initiated by the client or of covert self-reinforcement or praise. For example, the favorable outcome illustrated in the scene for the "stop smoking" client was the client's self-initiated action of refusing to take a cigarette. We prefer that the action be initiated by the client or model instead of someone else in the scene because, in a real situation, it may be too risky to have the client rely on someone else to deliver a favorable outcome in the form of a certain response. We cannot guarantee that clients will always receive favorable responses from someone else in the actual situation.

Another way to incorporate a favorable outcome into a treatment scene is to include an example of client (or model) self-reinforcement or praise. For instance, models might congratulate themselves by saying "That is terrific. I am proud of myself for what I said to the hotel clerk." A favorable consequence in the form of model or client self-praise is self-administered. Again, in a real-life situation, it may be better for clients to learn to reward themselves than to expect external encouragement that might not be forthcoming.

The person who experiences the favorable outcomes will be the same person the client imagines as the model. If the client imagines someone else as the model, then the client would also imagine that person initiating a favorable outcome or reinforcing himself or herself. Clients who imagine themselves as the models would imagine themselves receiving the positive consequences. There is very little actual evidence on the role of reinforcement in covert modeling. Some of the effectiveness of adding favorable consequences to the treatment scene may depend on the identity of the covert model and the particular value of the consequences for the client, all of which may vary depending on the client's gender, age, and culture.

Number of Scenes.   The counselor and client can develop different scenes that portray the situation in which the client experiences difficulty or wants to use the newly acquired behavior. Multiple scenes can depict different situations in which the desired behavior is generally appropriate. The number of scenes the therapist and client develop will depend on the client and his or her problem. Although there is no set number of scenes that should be developed, certainly several scenes provide more variety than only one or two.

Applying Treatment Scenes

After all the scenes have been developed, the counselor can apply the treatment scenes by having the client imagine each scene. The basic steps in applying the treatment scenes are these:

1. Arranging the scenes in a hierarchy
2. Instructing the client before scene presentation
3. Presenting one scene at a time from the hierarchy
4. Presenting a scene for a specified duration
5. Obtaining the client's reactions to the imagined scene
6. Instructing the client to develop verbal summary codes and/or to personalize each treatment scene
7. Presenting each scene at least twice with the aid of the counselor or tape recorder
8. Having the client imagine each scene at least twice while directing himself or herself
9. Selecting and presenting scenes from the hierarchy in a random order

Hierarchy.   The scenes developed by the counselor and client should be arranged in a hierarchy for scene presentation. The hierarchy is an order of scenes beginning with one that is relatively easy for the client to imagine with little stress. More difficult or stressful scenes would be ranked by the client.

Instructions.   It may be necessary to repeat instructions about imagery to the client if a great amount of time has elapsed since using the practice scenes. The counselor might say:

> Close your eyes, take a couple of deep breaths, and relax. I want you to imagine as vividly and clearly as possible that you are observing yourself doing all the Sun Salutation postures.

Tune in and use all your senses. For example, try to actually hear my voice giving instructions for each posture; see the colors of the clothing you are wearing; see yourself performing each posture. What odor or smell do you experience? Notice the sensations you experience in your body. What sensations do you feel in a particular muscle group of your body for each posture, and what taste do you have in your mouth? In a moment, I will ask you some questions concerning your feelings about the entire sequence of the scene and how clearly you imagined it.

If a person other than the client is the model, the client is instructed to picture someone of his or her own age, gender, and culture whom he or she knows. The client is told that the person who is pictured as the model will be used in all the treatment scenes. The counselor also instructs the client to signal by raising an index finger as soon as the scene is pictured clearly and to hold the scene in imagery until the counselor signals to stop.

Sequence of Scene Presentation.   Initially, the first scene in the hierarchy is presented to the client. Each scene is presented alone. When one scene has been covered sufficiently, the next scene in the hierarchy is presented. This process continues until all scenes in the hierarchy have been covered.

Duration of Scenes.   There are no general ground rules for the amount of time to hold the scene in imagery once the client signals. For some clients, a longer duration may be more beneficial; for others, a shorter duration may be. We feel that the choice will depend on the counselor's personal preference and experience with the covert modeling procedure, the nature of the client's problem, the goal behavior for counseling, and—perhaps most important—the client's input about the scene duration. After one or two scenes have been presented, the counselor can query the client about the suitability of the scene duration. Generally, a scene should be held long enough for the client to imagine the three scene ingredients vividly without rushing. We have found that visualizing a scene is perceived as much longer in time than it is in "real" time.

Client Reactions to the Scene.   After the client has imagined a particular scene, the counselor queries the client about how clearly it was imagined. The client is asked to describe feelings during particular parts of the scene. The counselor should also ask whether the scene was described too rapidly or the duration of the scene was adequate for the client to imagine the scene ingredients clearly. These questions enable the counselor and client to modify aspects of a scene before it is presented the second time. Client input about possible scene revision can be very helpful. If particular episodes of the scene elicit intense feelings of anxiety, the content of the scene or the manner of presentation can be revised. Perhaps the order of the scenes in the hierarchy needs rearrangement.

Another way to deal with the client's unpleasant feelings or discomfort with a scene is to talk it over. If the client feels stressful when the model (or the self) is engaging in the behavior or activity in the scene, examine with the client what episode in the scene is producing the discomfort. In addition, if the client is the model and has difficulty performing the behavior or activity, discuss and examine the block. Focus on the adaptive behavior or the coping with the situational ingredient of the scene rather than on the anxiety or discomfort.

After each scene presentation, the counselor should assess the rate of delivery for the scene description, the clarity of the imagery, and the degree of unpleasantness of the scene for the client. Perhaps if the client has a great deal of input in developing the scenes, the level of discomfort produced will be minimized. In addition, the intensity of the imagined scene can be enhanced by using verbal summary codes or by personalizing the scene.

Verbal Summary Codes and Personalization.

Verbal summary codes are brief descriptions about the behavior to be acquired and the context in which the behavior is to occur *in the client's own words*. Verbal coding of the modeling cues can facilitate acquisition and retention of the behaviors to be modeled and may maintain client performance during and after treatment by helping clients encode desired responses in their working memory. The verbalizations (or verbal summary codes) of the scene provide an alternative representational process to imagery or covert modeling. The therapist instructs the client to use his or her descriptions of the behavior and the situation. We recommend that clients rehearse using verbal summary codes with *practice* scenes and receive feedback from the therapist about the descriptions of the practice scenes. The practice should occur before presentation of the *treatment* scenes. Then the treatment scenes are presented and the client is instructed to develop his or her own verbal summary codes (descriptions of behavior and situation) for the scene. Have the client "try out" the treatment scene on the first presentation *without* the use of the summary code. On the second presentation of the scene, instruct the client to use the summary code and to say aloud exactly what it is.

Personalization of treatment scenes is another technique that can enhance covert modeling. After the scene has been presented once as developed, then the client is instructed *to change the treatment scene in any way as long as the model responses to be acquired are represented in the scene*. As with verbal summary codes, the client is asked to rehearse personalizing (individualizing) or elaborating a scene using a practice scene, and he or she receives feedback about the elaboration. The counselor should encourage the client to use variations within the context of the situation presented by the scene. Variations include more details about the model responses and the situation in which the responses are to occur. The client is asked to elaborate the scene the second time the treatment scene is presented. Elaboration may lead to more client involvement because the scenes are individualized.

Remember to have the client experience imagining a scene first without instructions to use verbal summary codes or to personalize the scene. Then, on the second presentation of the treatment scene, the client is instructed to use one of these techniques. To verify that the client is complying with the instructions, the therapist can instruct the client to say aloud the verbal summary code or elaboration being used.

Counselor-directed Scene Repetition.

After presenting the first scene and making any necessary revisions, the counselor may want to repeat the scene a couple of times. The number of scene repetitions may be dictated by the degree of comfort the client experiences while imagining the scene and the complexity of the activities or behaviors the client images. A complex series of motor skills, for example, may require more repetitions; and engaging in some situations may require repetition until the client feels reasonably comfortable. Again, make the decision about the number of scene repetitions on the basis of client input: ask the client. If you use the verbal summary codes or personalization of the scene, remember to instruct the client to use the technique during later repetitions of each scene.

Client-directed Scene Repetition.

In addition to counselor-directed scene practice, the client should engage in self-directed scene practice. Again, the number of client practices is somewhat arbitrary, although perhaps two is a minimum. Generally, the client can repeat imagining the scenes alone until he or she feels comfortable in doing so. The client can either use the verbal summary codes without saying the codes aloud or can personalize the scenes. Overt rehearsal of the scene can facilitate acquisition and retention of the imagined behaviors. The client should be instructed to overtly enact (rehearse) the scene with the therapist after the second or third time each scene is presented.

Random Presentation of Scenes.   After all the scenes in the hierarchy have been presented adequately, the counselor can check out the client's readiness for self-directed homework practice by presenting some of the scenes in random order. This random presentation guards against any "ordering" effect that the hierarchy arrangement may have had in the scene presentation.

Homework and Follow-Up

Self-directed practice in the form of homework is perhaps the most important therapeutic ingredient for generalization. If a person can apply or practice the procedure outside the counseling session, the probability of using the "new" behavior or of coping in the actual situation is greatly enhanced. Homework can consist in having clients identify several situations in their everyday lives in which they could use the desired responses.

In arranging the homework tasks, the counselor and client should specify how often, how long, what times during the day, and where practice should occur. The counselor should also instruct the client to record the daily use of the modeling scenes on log sheets. The counselor should verify whether the client understands the homework and should arrange for a follow-up after some portion of the homework is completed.

MODEL DIALOGUE: COVERT MODELING

Here is an example of a covert modeling dialogue with our client Joan to help her increase initiating skills in her math class. (See Learning Activity #40.)

In response 1, the counselor briefly describes the **rationale** *and gives an* **overview** *of the strategy.*

1. *Counselor:* Joan, one way we can help you work on your initiating skills in math class is to help you learn the skills you want through practice. In this procedure, you will practice using your imagination. I will describe situations to you and ask you to imagine yourself or someone else participating in the way described in a situation. How does that sound?
   Client: OK. You mean I imagine things like daydreaming?

Further **instructions** *about the strategy are provided in counselor response 2.*

2. *Counselor:* It has some similarities. Only instead of just letting your mind wander, you will imagine some of the skills you want to use to improve your participation in your math class.
   Client: Well, I'm a pretty good daydreamer, so if this is similar, I will probably learn from it.

In response 3, the counselor initiates the idea of using **practice scenes** *to determine Joan's level and style of imagery.*

3. *Counselor:* Well, let's see. I think it might help to see how easy or hard it is for you to actually imagine a specific situation as I describe it to you. So maybe we could do this on a try-out basis to see what it feels like for you.
   Client: OK, what happens?

In response 4, the counselor instructs Joan to **sit back and relax before imagining the practice scene.**

4. *Counselor:* First of all, just sit back, close your eyes, and relax. [Gives Joan a few minutes to do this.] You look pretty comfortable. How do you feel?
   Client: Fine. It's never too hard for me to relax.

In response 5, the counselor instructs Joan **to imagine the scene vividly and to indicate this by raising her finger.**

5. *Counselor:* OK, now, Joan, I'm going to describe a scene to you. As I do so, I want you to imagine the scene as vividly as possible. When you feel you have a very strong picture, then raise your index finger. Does that seem clear?
   Client: Yes.

The counselor will offer a practice scene next. Note that the **practice scene** *is simple and relatively mundane. It asks Joan only to imagine another person.*

6. *Counselor:* OK, imagine that someone is about to offer you a summer job. Just picture a person who might offer you a job like this. [Gives Joan time until Joan raises her index finger.]

In response 7, the counselor asks Joan **to describe what she imagined.**

7. *Counselor:* OK, Joan, now open your eyes. Can you tell me what you just imagined?
   Client: Well, I pictured myself with a middle-aged man who asked me if I wanted to lifeguard this summer. Naturally I told him yes.

Joan's imagery report was specific in terms of the actions and dialogue, but she didn't describe much about the man, so the counselor **will probe for more details.**

8. *Counselor:* OK, fine. What else did you imagine about the man? You mentioned his age. What was he wearing? Any physical characteristics you can recall?
   Client: Well, he was about 35 [a 16-year-old's impression of "middle age" is different from a 30-, 40-, or 50-year-old person's definition], he was wearing shorts and a golf shirt—you see, we were by the pool. That's about it.

Joan was able to describe the setting and the man's dress but no other physical characteristics, so the counselor **will suggest that Joan add this to the next practice attempt.**

9. *Counselor:* OK, so you were able to see what he was wearing and also the setting where he was talking to you. I'd like to try another practice with this same scene. Just imagine everything

you did before, only this time try to imagine even more details about how this man actually looks. [Counselor presents the same scene, which goes on until Joan raises her finger.]

In response 10, the counselor will again **query Joan about the details of her imagery.**

10. *Counselor:* OK, let's stop. What else did you imagine this time about this person or the situation?

Client: Well, he was wearing white shorts and a blue shirt. He was a tall man and very tanned. He had dark hair, blue eyes, and had sunglasses on. He was also barefoot. We were standing on the pool edge. The water was blue and the sun was out and it felt hot.

In response 11, the counselor will **try to determine how comfortable Joan is with imagery** *and whether more practice scenes are necessary.*

11. *Counselor:* OK, that's great. Looks like you were able to imagine colors and temperature—like the sun feeling hot. How comfortable do you feel now with this process?

Client: Oh, I like it. It was easier the second time you described the scene. I put more into it. I like to imagine this anyway.

In response 12, the counselor decides Joan can move ahead and **initiates development of treatment scenes.**

12. *Counselor:* Well, I believe we can go on now. Our next step is to come up with some scenes that describe the situations you find yourself in now with respect to participation in math class.

Client: And then I'll imagine them in the same way?

The counselor sets the stage to **obtain all the necessary information to develop treatment scenes.** *Note the emphasis in response 13 on Joan's* **participation** *in this process.*

13. *Counselor:* That's right. Once we work out the details of these scenes, you'll imagine each scene as you just did. Now we have sort of a series of things we need to discuss in setting up the scenes in a way that makes it easiest for you to imagine, so I'll be asking you some questions. Your input is very valuable here to both of us.

Client: OK, shoot.

In response 14, the counselor **asks Joan whether she would rather imagine herself or someone else** *as the model.*

14. *Counselor:* Well, first of all, in that practice scene I asked you to imagine someone else. Now, you did that, but you were also able to picture yourself from the report you gave me. In using your class scenes, which do you feel would be easiest and least stressful for you to imagine—yourself going through the scene or someone else, maybe someone similar to you, but another person? [Gives Joan time to think.]

Client (pauses): Well, that's hard to say. I think it would be easier for me to imagine myself, but it might be a little less stressful to imagine someone else . . . [Pauses again.] I think I'd like to try using myself.

In the next response, the counselor **reinforces Joan's choice and also points out the flexibility of implementing the procedure.**

15. *Counselor:* That's fine. And besides, as you know, nothing is that fixed. If we get into this and that doesn't feel right and you want to imagine someone else, we'll change.

Client: Okey-dokey.

In response 16, the counselor **introduces the idea of a coping model.**

16. *Counselor:* Also, sometimes it's hard to imagine yourself doing something perfectly to start with, so when we get into this, I might describe a situation where you might have a little trouble but not much. That may seem more realistic to you. What do you think?

Client: That seems reasonable. I know what you mean. It's like learning to drive a car. In Driver's Ed, we take one step at a time.

In response 17, the counselor **will pose the option of individualizing the scenes** *or using* **standardized scenes.**

17. *Counselor:* You've got the idea. Now we have another choice also in the scenes we use. We can work out scenes just for you that are tailored to your situation, or we can use scenes on a cassette tape I have that have been standardized for many students who want to improve their class-participation skills. Which sounds like the best option to you?

Client: I really don't know. Does it really make a difference?

It is not that uncommon for a client not to know which route to pursue. In the next response, the counselor will **indicate a preference** *and check it out with Joan.*

18. *Counselor:* Probably not, Joan. If you don't have a choice at this point, you might later. My preference would be to tailor-make the scenes we use here in the session. Then, if you like, you could use the taped scenes to practice with at home later on. How does that sound to you?

Client: It sounds good, like maybe we could use both.

In responses 19 and 20, the counselor asks Joan to **identify situations in which Joan desires to increase these skills.** *This is somewhat a review of goal behavior described in Chapter 10.*

19. *Counselor:* Yes, I think we can. Now let's concentrate on getting some of the details we need to make up the scenes we'll use in our sessions. First of all, let's go over the situations in math class in which you want to work on these skills.

Client: Well, it's some of those things we talked about earlier, like being called on, going to the board, and so on.

Next the counselor **explores whether Joan prefers a very general description or a very specific one.** *Sometimes this makes a difference in how the person imagines.*

20. *Counselor:* OK, Joan, how much detail would you like me to give you when I describe a scene—a little detail, to let you fill in the rest, or do you want me to describe pretty completely what you should imagine?

Client: Maybe somewhere in between. I can fill in a lot, but I need to know what to fill in.

In response 21, the counselor **is asking about the specific situations** *in which Joan has trouble participating in her math class.*

21. *Counselor:* OK, let's fill out our description a little more. We're talking about situations you confront in your math class. I remember four situations in which you wanted to increase these skills—you want to answer more when Mr. Lamborne calls on you, volunteer more answers, go to the board, and tell Mr. Lamborne you want to work the problems yourself after you ask for an explanation. Any others, Joan?
Client: I can't think of any offhand.

In responses 22 through 27; the counselor asks Joan **to identify the desired behaviors for these situations.** *Again, much of this is a review of identifying outcome goals (Chapter 10).*

22. *Counselor:* OK, so we've got about four different situations. Let's take each of these separately. For each situation, can we think of what you would like to do in the situation—like when Mr. Lamborne calls on you, for instance?
Client: Well, I'd like to give him the answer instead of saying nothing or saying "I don't know."

23. *Counselor:* OK, good. And if you did give him the answer—especially when you do know it—how would you feel?
Client: Good, probably relieved.

24. *Counselor:* OK. Now what about volunteering answers?
Client: Well, Mr. Lamborne usually asks who has the answer to this; then he calls on people who raise their hand. I usually never raise my hand even when I do know the answer, so I need to just raise my hand and, when he calls on me, give the answer. I need to speak clearly, too. I think sometimes my voice is too soft to hear.

25. *Counselor:* OK, now, how could you tell Mr. Lamborne to let you work out the problems yourself?
Client: Well, just go up to him when we have a work period and tell him the part I'm having trouble with and ask him to explain it.

26. *Counselor:* So you need to ask him for just an explanation and let him know you want to do the work.
Client: Yup.

27. *Counselor:* OK, now, what about going to the board?
Client: Well, I do go up. But I always feel like a fool. I get distracted by the rest of the class so I hardly ever finish the problem. Then he lets me go back to my seat even though I didn't finish it. I need to concentrate more so I can get through the entire problem on the board.

Now that the content of the scenes has been developed, the counselor asks Joan to **arrange the four scenes in a hierarchy.**

28. *Counselor:* OK, so we've got four different situations in your math class where you want to improve your participation in some way. Let's take these four situations and arrange them in an order. Could you select the situation that right now is easiest

for you and least stressful to you and rank the rest in terms of difficulty and degree of stress?
Client: Sure, let me think. . . . Well, the easiest thing to do out of all of these would be to tell Mr. Lamborne I want to work out the problems myself. Then I guess it would be answering when he calls on me and then going to the board. I have a lot of trouble with volunteering answers, so that would be hardest for me.

The counselor emphasizes the **flexibility of the hierarchy** *and provides* **general instructions to Joan about how they will work with these scenes.**

29. *Counselor:* OK. Now, this order can change. At any point if you feel it isn't right, we can reorder these situations. What we will do is to take one situation at a time, starting with the easiest one, and I'll describe it to you in terms of the way you want to handle it and ask you to imagine it. So the first scene will involve your telling Mr. Lamborne what you need explained in order to work the problems yourself.
Client: So we do this just like we did at the beginning?

The counselor will **precede the scene presentation with very specific instructions** *to Joan.*

30. *Counselor:* Right. Just sit back, close your eyes, and relax. . . . [Gives Joan a few minutes to do so.] Now remember, as I describe the scene, you are going to imagine yourself in the situation. Try to use all your senses in your imagination—in other words, get into it. When you have a very vivid picture, raise your index finger. Keep imagining the scene until I give a signal to stop. OK?
Client: Yeah.

The counselor **presents the first scene in Joan's hierarchy slowly** *and with ample pauses to give Joan time to generate the images.*

31. *Counselor:* OK, Joan, picture yourself in math class. . . . [Pause] Mr. Lamborne has just finished explaining how to solve for x and y. . . . Now he has assigned problems to you and has given you a work period. . . . You are starting to do the problems and you realize there is some part of the equation you can't figure out. You take your worksheet and get up out of your seat and go to Mr. Lamborne's desk. You are telling Mr. Lamborne what part of the equation you're having trouble with. You explain to him you don't want him to solve the problem, just to explain the missing part. Now you're feeling good that you were able to go up and ask him for an explanation. [The counselor waits for about 10 seconds after Joan signals with her finger.]

The counselor **signals Joan to stop** *and in responses 32 through 35* **solicits Joan's reactions** *about the imagery.*

32. *Counselor:* OK, Joan, open your eyes now. What did you imagine?
Client: Well, it was pretty easy. I just imagined myself going up to Mr. Lamborne and telling him I needed more explanation but that I wanted to find the answers myself.

LEARNING ACTIVITY 40 *Covert Modeling*

As you may recall from reading the goals and subgoals of Ms. Weare (Chapter 10), one of her subgoals was to arrange a school conference with Freddie's teacher. Ms. Weare was going to use the conference to explain her new strategy in dealing with Freddie and request help and cooperation from the school. Specifically, Ms. Weare would point out that one thing that may happen initially might be an increase in Freddie's tardiness at school. Assume that Ms. Weare is hesitant to initiate the conference because she is unsure about what to say during the meeting. Describe how you could use covert modeling to help Ms. Weare achieve this subgoal. Describe specifically how you would use (1) a rationale, (2) practice scenes, (3) development of treatment scenes, (4) application of treatment scenes, and (5) homework to help Ms. Weare in this objective. Feedback is provided; see whether some of your ideas are similar.

Feedback follows on page 334.

---

33. *Counselor:* OK, so you were able to get a pretty vivid picture?
    Client: Yes, very much so.
34. *Counselor:* What were your feelings during this—particularly as you imagined yourself?
    Client: I felt pretty calm. It didn't really bother me.
35. *Counselor:* OK, so imagining yourself wasn't too stressful. Did I give you enough time before I signaled to stop?
    Client: Well, probably. Although I think I could have gone on a little longer.

On the basis of Joan's response about the length of the first scene, the counselor **will modify the length during the next presentation.**

36. *Counselor:* OK, I'll give you a little more time the next time.

Before the counselor presents the treatment scenes the second time, the counselor explores whether the client would like to use **verbal summary codes** *or to* **personalize the treatment scenes.**

37. *Counselor:* Joan, there are two techniques that you can use to enhance your imagery scene of Mr. Lamborne's math class. One technique is to describe briefly the behavior you want to do and the situation in Mr. Lamborne's class when you will perform the behavior. All that you are doing is just describing the scene in your own words. This process can help you remember what to do. With the other technique, you can change the scene or elaborate on the scene in any way as long as you still imagine engaging in the behaviors you want to do. Do you have any questions about these two techniques?
    Client: You think these techniques might help me imagine the scene better?
38. *Counselor:* That's right. Is there one technique you think might be more helpful to you?
    Client: Yes, I think that for me to describe the scene in my own words might work better for me. It might help me to remember better what to do.
39. *Counselor:* OK, for the first scene, what verbal summary or description would you use?

Client: Well—after Mr. Lamborne explains how to solve for x and y and assigns problems, I might find something I can't figure out. I get out of my seat and go up to Mr. Lamborne and tell him I need more explanation but I want to find the answer myself.
40. *Counselor:* That's great, Joan!

The counselor **presents the same scene again.** *Usually each scene is presented* **a minimum of two times** *by the counselor or on a tape recorder. If the client has chosen one or both treatment-scene enhancement techniques, instruct the client on the technique with each scene.*

41. *Counselor:* Let's try it again. I'll present the same scene, and I'll give you more time after you signal to me you have a strong picture. [Presents the same scene again and checks out Joan's reactions after the second presentation.]

After the counselor-presented scenes, the counselor **asks Joan to self-direct her own practice.** *This also occurs a minimum of two times on each scene.*

42. *Counselor:* You seem pretty comfortable now in carrying out this situation the way you want to. This time instead of my describing the scene orally to you, I'd like you just to go through the scene on your own—sort of a mental practice without my assistance.
    Client: OK. [Pauses to do this for several minutes.]
43. *Counselor:* OK, how did that feel?
    Client: It was pretty easy even without your instructions, and I think I can see where I can actually do this now with Mr. Lamborne.

The other scenes in the hierarchy are worked through in the same manner.

44. *Counselor:* Good. Now we will work through the other three scenes in the same manner, sticking with each one until you can perform your desired behaviors in your imagination pretty easily. [The other three situations in the hierarchy are worked through.]

FEEDBACK
Covert Modeling

1. *Rationale*

   First, you would explain that covert modeling could help Ms. Weare find ways to express herself and could help her practice expressing herself before having the actual conference. Second, you would briefly describe the strategy, emphasizing that she will be practicing her role and responses in the school conference, using her imagination.

2. *Practice Scenes*

   You would explain that it is helpful to see how she feels about practicing through her imagination. You would select several unrelated scenes, such as imagining someone coming to her home, imagining an old friend calling her, or imagining a new television show about a policewoman. You would present one practice scene and instruct Ms. Weare first to close her eyes, imagine the scene intensely, and signal to you with her finger when she has a strong picture in her mind. After this point, you could tell her to open her eyes and to describe the details of what she imagined. You might suggest additional details and present the same scene again or present a different scene. If Ms. Weare is able to use imagery easily and comfortably, you could move on to developing the actual treatment scenes.

3. *Developing Treatment Scenes*

   At this point, you would seek Ms. Weare's input about certain aspects of the scenes to be used as treatment scenes. Specifically, you would decide who would be used as the model, whether individualized or standardized scenes would be used, and whether Ms. Weare felt she could benefit from general or specific scenes. Our preference would be to use pretty specific, individualized scenes in which Ms. Weare imagines herself as the model, as she will ultimately be carrying out the action. Next, you should specify the three ingredients of the scenes: (1) the situation in which the behaviors should occur, (2) the behaviors to be demonstrated, and (3) a favorable outcome. For example, the scenes could include Ms. Weare calling the teacher to set up the conference, beginning the conference, explaining her strategy in the conference, and ending the conference. Specific examples of things she could say would be included in each scene. Favorable outcomes might take the form of covert self-praise or of relief from stressful, anxious feelings.

4. *Applying Treatment Scenes*

   After all the treatment scenes have been developed,

(continued)

Ms. Weare would arrange them in a hierarchy from least to most difficult. Starting with the first scene in the hierarchy, you would again instruct Ms. Weare about how to imagine. After the first scene presentation, you would obtain Ms. Weare's reactions to the clearness of her imagery, the duration of the scene, and so on. Any needed revisions could be incorporated before a second presentation of the same scene. You would also encourage Ms. Weare to develop a verbal summary code for each scene after the initial presentation of that scene. You would present each scene to Ms. Weare several times; then have her self-direct her own scene-imagining several times. After all the scenes in the hierarchy had been covered adequately, Ms. Weare would be ready for homework.

5. *Homework*

   You would instruct Ms. Weare to continue to practice the scenes in her imagination outside the session. A follow-up should be arranged. You should be sure that Ms. Weare understands how many times to practice and how such practice can benefit her. Ms. Weare might record her practice sessions on log sheets. She could also call in and verbalize the scenes, using a phone-mate.

---

45. *Counselor:* Well, how do you feel now that we've gone over every scene?

    Client: Pretty good. I never thought that my imagination would help me in math class!

After the hierarchy has been completed, the counselor **picks scenes to practice at random.** *This is a way to see how easily the client can perform the scene when it is not presented in the order of the hierarchy.*

46. *Counselor:* Well, sometimes imagining yourself doing something helps you learn how to do it in the actual situation. Now I'd like to just pick a scene here at random and present it to you and have you imagine it again. [Selects a scene from the hierarchy at random and describes it.]

    Client: That was pretty easy, too.

The counselor **initiates homework practice** *for Joan.*

47. *Counselor:* OK, I just wanted to give you a chance to imagine a scene when I took something out of the order we worked with today. I believe you are ready to carry out this imagination practice on your own during the week.

    Client: Is this where I use the tapes?

The **purpose of homework** *is explained to Joan.*

48. *Counselor:* Sure. This tape has a series of scenes dealing with verbal class participation. So instead of needing me to describe a scene, the tape can do this. I'd like you to practice with this daily, because daily practice will help you learn to participate more quickly and easily.

    Client: So I just go over a scene the way we did today?

The counselor instructs Joan on **how to complete the homework practice.**

49. *Counselor:* Go over the scenes one at a time—maybe about four times for each scene. Make your imagination as vivid as possible. Also, each time you go over a scene, make a check on your log sheets. Indicate the time of day and place where you use this—also, the length of each practice. And after each practice session, rate the vividness of your imagery on this scale: 1 is not vivid and 5 is very vivid. How about summarizing what you will do for your homework?

    Client: Yes. I just do what we did today and check the number of times I practice each scene and assign a number to the practice according to how strongly I imagined the scene.

At the termination of the session, the counselor **indicates that a follow-up on the homework** *will occur at their next meeting.*

50. *Counselor:* Right. And bring your log sheets in at our next meeting and we'll go over this homework then. OK? We had a really good session today. You worked hard. I'll see you next Tuesday.

SUMMARY

Guided imagery and covert modeling are procedures that may be useful when media modeling and live modeling are not feasible. These two strategies can be used without elaborate therapeutic aids or expensive equipment. Both strategies involve imagery, which makes the procedures quite easy for a client to practice in a self-directed manner. The capacity of clients to generate vivid images may be important for the overall effectiveness of guided imagery and covert modeling. Assessing client potential to engage in imagery is a necessary prerequisite before using either of these procedures. Assuming that clients can produce clear images, counselors may use guided imagery to help them deal with fears or discomfort or teach them covert modeling to promote desired responses. Although imagery has been used in limited ways with diverse clients, it is an intervention that can be helpful if the imagery is adapted to and relevant for the client's culture.

POSTEVALUATION

Part One

According to Objective One, you should be able to identify accurately six out of seven examples of guided imagery steps represented in written examples of counselor leads. For each of the following seven counselor leads, write down which part of guided imagery the counselor is implementing. More than one counselor lead may be associated with any part of the procedure, and the leads given here are not in any particular order. The five major parts of emotive imagery are as follows:

1. Rationale
2. Determining the client's potential to use imagery
3. Developing imagery scenes
4. Imagery-scene practice training
5. Homework and follow-up

Feedback follows the Postevaluation on pages 342–343.

1. "Can you think of several scenes you could imagine that give you calm and positive feelings? Supply as many details as you can. You can use these scenes later to focus on instead of the anxiety."
2. "It's important that you practice with this. Try to imagine these scenes at least several times each day."
3. "This procedure can help you control your anxiety. By imagining very pleasurable scenes, you can block out some of the fear."
4. "Let's see whether you feel that it's easy to imagine something. Close your eyes, sit back, and visualize anything that makes you feel relaxed."
5. "Now, select one of these scenes you've practiced. Imagine this very intensely. I'm going to apply pressure to your arm, but just focus on your imaginary scene."
6. "What we will do, if you feel that imagination is easy for you, is to develop some scenes that are easy for you to visualize and that make you feel relaxed. Then we'll practice having you focus on these scenes while also trying to block out fear."
7. "Now I'd like you just to practice these scenes we've developed. Take one scene at a time, sit back, and relax. Practice imagining this scene for about 30 seconds. I will tell you when the time is up."

Part Two

Objective Two asks you to demonstrate 10 out of 13 steps of guided imagery with a role-play client. You or an observer can rate your performance, assisted by the Interview Checklist for Guided Imagery following this Postevaluation.

Part Three

Objective Three asks you to describe how you would use the five components of covert modeling with a simulated client case. Use the case of Mr. Huang (Chapter 8) and his stated goal of wanting to decrease his worrying about retirement and increase his positive thoughts about retiring, particularly in his work setting. Describe how you would use a rationale, practice scenes, developing treatment scenes, applying treatment scenes, and homework to help Mr. Huang do this. Feedback follows the Postevaluation.

(continued)

POSTEVALUATION   (continued)

Part Four

Objective Four asks you to demonstrate at least 22 out of 28 steps associated with covert modeling with a role-play client. The client might take the part of someone who wants to acquire certain skills or to perform certain activities. Use the Interview Checklist for Covert Modeling on pages 338–341 to help you assess your interview.

Interview Checklist for Guided Imagery

Instructions: In a role-play counselor/client interview, determine which of the following counselor leads or questions were demonstrated. Indicate by a check the leads used by the counselor. A few examples of counselor leads are presented in the right column.

Item	Examples of counselor leads
I. Rationale	
___ 1. Counselor describes purpose of guided imagery.	"The procedure is called guided imagery because you can emote pleasant thoughts or images in situations that evoke fear, pain, tension, anxiety, or routine boredom. The procedure helps you block your discomfort or reduce the anxiety that you experience in the problem situation. The technique involves focusing on imaginary scenes that please you and make you feel relaxed while in the uncomfortable situation. This procedure works because it is extremely difficult for you to feel pleasant, calm, happy, secure, or whatever other emotion is involved in the scene and anxious (tense, fearful, stressed) at the same time. These emotions are incompatible."
___ 2. Counselor gives an overview of procedure.	"What we will do first is to see how you feel about engaging in imagery and look at the details of the scene you used. Then, we will decide whether guided imagery is a procedure we want to use. If we decide to use it, we will develop scenes that make you feel calm and good and generate positive feelings for you. We will practice using the scenes we have developed and try to rehearse using those scenes in a simulated fashion. Later, you will apply and practice using the scene in the real situation. Do you have any questions about my explanation?"
___ 3. Counselor assesses client's willingness to try strategy.	"Would you like to go ahead and give this a try now?"
II. Assessment of Client's Imagery Potential	
___ 4. Counselor instructs client to engage in imagery that elicits good feelings and calmness.	"Close your eyes, sit back, and relax. Visualize a scene or event that makes you feel relaxed and pleasant. Select something you really enjoy and feel good about. Try to be aware of all your sensations in the scene."
___ 5. After 30 seconds to a minute, the counselor probes to ascertain the sensory vividness of the client's imagined scene (colors, sounds, movement, temperature, smell). Counselor asks client's feelings about imagery and about "getting into" the scene (feeling good with imaginal process).	"Describe the scene to me." "What sensations did you experience while picturing the scene?" "What temperature, colors, sounds, smells, and motions did you experience in the scene?" "How do you feel about the imagery?" "How involved could you get with the scene?"

(continued)

Item	Examples of counselor leads
___ 6. Counselor discusses with client the decision to continue or discontinue guided imagery. Decision is based on client's attitude (feelings about imagery) and imaginal vividness.	"You seem to feel good with the imagery and are able to picture a scene vividly. We can go ahead now and develop some scenes just for you." "Perhaps another strategy that would reduce tension without imagery would be better as it is hard for you to 'get into' a scene."

III. Development of Imagery Scenes

Item	Examples of counselor leads
___ 7. Counselor and client develop at least two scenes that promote positive feelings for client, involve many sensations (sound, color, temperature, motion, and smell), and are culturally relevant.	"Now I would like to develop an inventory of scenes or situations that promote calmness, tranquility, and enjoyment for you. We want to have scenes that will have as much sensory detail as possible for you, so that you can experience color, smell, temperature, touch, sound, and motion. Later, we will use the scenes to focus on instead of anxiety, so we want to find scenes for you that are also consistent with and meaningful to you culturally. What sort of scenes can you really get into?"

IV. Practice of Imagery Scene

Item	Examples of counselor leads
___ 8. Counselor instructs client to practice focusing on the scene for about 30 seconds.	"Take one of the scenes, close your eyes, sit back, and relax. Practice or hold this scene for about 30 seconds, picturing as much sensory detail as possible. I will cue you when the time is up."
___ 9. Counselor instructs client to practice focusing on scene with simulated discomfort or anxiety.	"Let us attempt to simulate or create the problem situation and to use the scenes. While I squeeze your arm to have you feel pain, focus on one of the imagery scenes we have developed." "While I describe the feared situation or scene to you, focus on the scene."
___10. Counselor assesses client's reaction after simulated practice.	"How did that feel?" "What effects did my describing the discomforting situation [my application of pain] have on your relaxation?" "Rate your ability to focus on the scene with the discomfort." "How comfortable did you feel when you imagined this fearful situation then?"

V. Homework and Follow-up

Item	Examples of counselor leads
___11. Counselor instructs client to apply guided imagery in vivo.	"For homework, apply the guided imagery scenes to the discomforting situation. Focus on the scene as vividly as possible while you are experiencing the activity or situation."
___12. Counselor instructs client to record use of guided imagery and to record level of discomfort or anxiety on log sheets.	"After each time you use guided imagery, record the situation, the day, the time, and your general reaction on this log. For each occasion that you use imagery, record also your level of discomfort or anxiety, using a 5-point scale, with 5 equal to maximum discomfort and 1 equal to minimum discomfort."
___13. Counselor arranges a follow-up session.	"Let's get together again in two weeks to see how your practice is going and to go over your homework log."

Observer comments: _____

(continued)

POSTEVALUATION   **(continued)**

Interview Checklist for Covert Modeling

Instruction: Determine which of the following leads the counselor used in the interview. Check the leads used.

Item	Examples of counselor leads
I. Rationale	
___ 1. Counselor describes purpose of strategy.	"This strategy can help you learn how to discuss your prison record in a job interview. I will coach you on some things you could say. As we go over this, gradually you will feel as if you can handle this situation when it comes up in an actual interview."
___ 2. Counselor provides overview of strategy.	"We will be relying on your imagination a lot in this process. I'll be describing certain scenes and asking you to close your eyes and imagine that you are observing the situation I describe to you as vividly as you can."
___ 3. Counselor confirms client's willingness to use strategy.	"Would you like to give this a try now?"
II. Practice Scenes	
___ 4. Counselor instructs client to sit back, close eyes, and relax in preparation for imagining practice scenes.	"Just sit back, relax, and close your eyes."
___ 5. Counselor describes a practice scene unrelated to goal and instructs client to imagine scene as counselor describes it and to raise index finger when scene is vividly imagined.	"As I describe this scene, try to imagine it very intensely. Imagine the situation as vividly as possible. When you feel you have a vivid picture, raise your index finger."
___ 6. After client indicates vivid imagery, counselor instructs client to open eyes and describe what was imagined during scene.	"OK, now let's stop—you can open your eyes. Tell me as much as you can about what you just imagined."
___ 7. Counselor probes for additional details about scene to obtain a very specific description from client.	"Did you imagine the color of the room? What did the people look like? Were there any noticeable odors around you? How were you feeling?"
___ 8. Counselor suggests ways for client to attend to additional details during subsequent practice.	"Let's do another scene. This time try to imagine not only what you see but what you hear, smell, feel, and touch."
___ 9. Counselor initiates additional practices of one scene or introduces practice of new scene until client is comfortable with the novelty and is able to provide a detailed description of imagery.	"Let's go over another scene. We'll do this for a while until you feel pretty comfortable with this."
___10. After practice scenes, counselor does one of the following:	
___a. Decides to move on to developing treatment scenes.	"OK, this seems to be going pretty easily for you, so we will go on now."
___b. Decides that relaxation or additional imagery training is necessary.	"I believe before we go on it might be useful to try to help you relax a little more. We can use muscle relaxation for this purpose."
___c. Decides to terminate covert modeling because of inadequate client imagery.	"Judging from this practice, I believe another approach would be more helpful where you can actually see someone do this."

(continued)

Item	Examples of counselor leads

III. Developing Treatment Scenes

___11. Counselor and client decide on appropriate characteristics of model to be used in treatment scenes, including

    ___a. Identity of model (client or someone else) — "As you imagine this scene, you can imagine either yourself or someone else in this situation. Which would be easier for you to imagine?"

    ___b. Coping or mastery model — "Sometimes it's easier to imagine someone who doesn't do this perfectly. What do you think?"

    ___c. Single or multiple models — "We can have you imagine just one other person—someone like you—or several other people."

    ___d. Specific characteristics of model to maximize client/model similarity — "Let's talk over the specific type of person you will imagine, someone similar to you."

___12. Counselor and client specify — "We have two options in developing the scenes you will imagine. We can discuss different situations and develop the scenes just to fit you, or we can use some standardized scenes that might apply to anyone with a prison record going through a job interview. What is your preference?"

    ___a. Individualized scenes

    ___b. Standardized scenes

___13. Counselor and client decide to use either — "On the basis of these situations you've just described, I can present them to you in one of two ways. One way is to give you a general description and leave it up to you to fill in the details. Or I can be very detailed and tell you specifically what to imagine. Which approach do you think would be best for you?"

    ___a. General descriptions of scenes

    ___b. Specific, detailed descriptions of scenes

___14. Counselor and client develop specific ingredients to be used in scenes. Ingredients include — "Let's decide the kinds of things that will go in each scene."

    ___a. Situations or context in which behaviors should occur — "In the scene in which you are interviewing for a job, go over the type of job you might seek and the kind of employer who would be hard to talk to."

    ___b. Behaviors and coping methods to be demonstrated by model — "Now, what you want to do in this situation is to discuss your record calmly, explaining what happened and emphasizing that your record won't interfere with your job performance."

    ___c. Favorable outcome of scene, such as

        ___1. Favorable client self-initiated action — "At the end of the scene you might want to imagine you have discussed your record calmly without getting defensive."

        ___2. Client self-reinforcement — "At the end of the scene, congratulate yourself or encourage yourself for being able to discuss your record."

___15. Counselor and client generate descriptions of multiple scenes. — "OK, now, the job interview is one scene. Let's develop other scenes where you feel it's important to be able to discuss your record—for example, in establishing a closer relationship with a friend."

IV. Applying Treatment Scenes

___16. Counselor and client arrange multiple scenes in a hierarchy for scene presentation according to — "Now I'd like you to take these six scenes we've developed and arrange them in an order. Start with the situation that you feel most comfortable with and that is easiest for you to discuss your record in now. End with the situation that is most difficult and gives you the most discomfort or tension."

    ___a. Client degree of discomfort in situation

    ___b. Degree of difficulty or complexity of situation

(continued)

POSTEVALUATION   **(continued)**

Item	Examples of counselor leads
__17. Counselor precedes scene presentation with instructions to client, including	"I'm going to tell you now what to do when the scene is presented."
__a. Instructions to sit back, relax, close eyes	"First, just sit back, close your eyes, and relax."
__b. Instructions on whom to imagine	"Now come up with an image of the person you're going to imagine, someone similar to you."
__c. Instructions to imagine intensely, using as many senses as possible	"As I describe the scene, imagine it as vividly as possible. Use all your senses—sight, smell, touch, and so on."
__d. Instructions to raise index finger when vivid imagery occurs	"When you start to imagine very vividly, raise your finger."
__e. Instructions to hold imagery until counselor signals to stop	"And hold that vivid image until I tell you when to stop."
__18. Counselor presents one scene at a time, by describing the scene orally to client or with a tape recorder.	"Here is the first scene. . . . Imagine the employer is now asking you why you got so far behind in school. Imagine that you are explaining what happened in a calm voice."
__19. Duration of each scene is determined individually for client and is held until client imagines model performing desired behavior as completely as possible (perhaps 20–30 seconds).	"You should be able to imagine yourself saying all you want to about your record before I stop you."
__20. After first scene presentation, counselor solicits client reactions about	
__a. Rate of delivery and duration of scene	"How did the length of the scene seem to you?"
__b. Clearness and vividness of client imagery	"How intense were your images? What did you imagine?"
__c. Degree of discomfort or pleasantness of scene	"How did you feel while doing this?"
__21. On basis of client reactions to first scene presentation, counselor does one of the following:	
__a. Presents scene again as is	"I'm going to present this same scene again."
__b. Revises scene or manner of presentation before second presentation	"Based on what you've said, let's change the type of employer. Also, I'll give you more time to imagine this the next time."
__c. Changes scene order in hierarchy and presents another scene next	"Perhaps we need to switch the order of this scene and use this other one first."
__d. Precedes another presentation of scene with relaxation or discussion of client discomfort	"Let's talk about your discomfort."
__22. Imagery enhancement techniques explained to client:	
__a. Verbal summary codes	"You can briefly describe the scene in your own words, which can help you remember the behaviors to perform in the situation."
__b. Personalization or elaboration of treatment scene	"You can change or elaborate on the scene in any way as long as you still imagine the behavior you want to do."
__23. Each scene is presented a minimum of two times by counselor or on tape recorder.	"OK, now I'm going to present the same scene one or two more times."

(continued)

Item	Examples of counselor leads
___24. Following counselor presentations of scene, client repeats scene at least twice in a self-directed manner.	"This time I'd like you to present the scene to yourself while imagining it, without relying on me to describe it."
___25. After each scene in hierarchy is presented and imagined satisfactorily, counselor presents some scenes to client in a random order, following steps 18–20.	"Now I'm just going to pick a scene at random and describe it while you imagine it."

V. Homework

___26. Counselor instructs client to practice scenes daily outside session and explains purpose of daily practice.	"During the week, I'd like you to take these cards where we've made a written description of these scenes and practice the scenes on your own. This will help you acquire this behavior more easily and quickly."
___27. Instructions for homework include	
___a. What to do	"Just go over one scene at a time—make your imagination as vivid as possible."
___b. How often to do it	"Go over this five times daily."
___c. When and where to do it	"Go over this two times at home and three times at school."
___d. A method for self-observation of homework completion	"Each time you go over the scene, make a tally on your log sheet. Also, after each practice session, rate the intensity of your imagery on this scale."
___28. Counselor arranges for a follow-up after completion of some amount of homework.	"Bring these sheets next week so we can discuss your practices and see what we need to do as the next step."

Observer comments: _____

SUGGESTED READINGS

Achterberg, J. (1985). *Imagery in healing: Shamanism and modern medicine.* Boston: Shambhala.

Andrada, P., & Korte, A. (1993). *En aquellas tiempos:* A reminiscing group with Hispanic elderly. *Journal of Gerontological Social Work, 20,* 25–42.

Baider, L., Uziely, B., & Kaplan-DeNour, A. (1994). Progressive muscle relaxation and guided imagery in cancer patients. *General Hospital Psychiatry, 16,* 340–347.

Beck, J. S. (1995). *Cognitive therapy: Basics and beyond.* New York: Guilford.

Brigham, D. (1994). *Imagery for getting well.* New York: Norton.

Brown, J. C. (1990). Loss and grief: An overview and guided imagery intervention model. *Journal of Mental Health Counseling, 12,* 434–445.

Cohen, R. E., Creticso, P. S., & Norman, P. S. (1993–1994). The effects of guided imagery (GI) on allergic subjects' responses to ragweed-pollen nasal challenge: An exploratory investigation. *Imagination, Cognition and Personality, 13,* 259–269.

Fernandez, D., & Allen, G. J. (1989). Test anxiety, emotional responding under guided imagery, and self-talk during an academic examination. *Anxiety Research, 2,* 15–26.

Fezler, W. (1989). *Creative imagery: How to visualize in all five senses.* New York: Simon & Schuster.

Ilacqua, G. E. (1994). Migraine headaches: Coping efficacy of guided imagery training. *Headache, 34,* 99–102.

Lazarus, A. (1984). *In the mind's eye: The power of imagery for personal enrichment.* New York: Guilford.

Lindh, R. (1987). Suggestive covert modeling as a method with disturbed pupils. *Jyvaskyla Studies in Education, Psychology and Social Research, 60,* 1–194.

Myrick, R. D., & Myrick, L. S. (1993). Guided imagery: From mystical to practical. Special Issue: Counseling and children's play. *Elementary School Guidance and Counseling, 28,* 62–70.

Overholser, J. C. (1991). The use of guided imagery in psychotherapy: Modules for use with passive relaxation training. *Journal of Contemporary Psychotherapy, 21,* 159–172.

Rosenthal, T. (1993). To soothe the savage beast. *Behaviour Research and Therapy, 31,* 439–462.

Sheikh, A. A. (Ed.). (1986). *Anthology of imagery techniques.* Milwaukee, WI: American Imagery Institute.

Simonton, O. C., & Hensen, R. (1992). *The healing journey.* New York: Bantam.

Suinn, R. (1986). *Seven steps to peak performance: The mental training manual for athletes.* Lewiston, NY: Huber.

Uhlemann, M. R., & Koehn, C. V. (1989). Effects of covert and overt modeling on the communication of empathy. *Canadian Journal of Counselling, 23,* 372–381.

Wynd, C. A. (1992). Personal power imagery and relaxation techniques used in smoking cessation programs. *American Journal of Health Promotion, 6,* 184–189, 196.

FEEDBACK
POSTEVALUATION

Part One

1. Instructing the client to *develop imagery scenes.* These are used as the scenes to focus on to block the unpleasant sensation.
2. Part of *homework*—in vivo application of imagery.
3. *Rationale*—giving the client a reason for guided imagery.
4. The counselor is determining *the client's potential to use imagery.*
5. *Imagery-scene practice*—with a pain-provoking situation.
6. *Rationale*—the counselor is giving an overview of the procedure.
7. *Imagery-scene practice*—the client is trained to imagine the scenes very vividly before using them in simulation of anxiety-provoking situations.

Part Two

Rate your performance with the Interview Checklist for Guided Imagery found after the Postevaluation.

Part Three

Rationale

First you would give Mr. Huang an explanation of covert modeling. You would briefly describe the process to him

(continued)

and explain how using his imagination to "see" someone doing something could help him perform his desired responses.

Practice Scenes

Next you would present a couple of unrelated practice scenes. You would instruct Mr. Huang to close his eyes, relax, and imagine the scene as you describe it. When Mr. Huang signals he is imagining the scene, you would stop and query him about what he imagined. You might suggest additional details for him to imagine during another practice scene. Assuming Mr. Huang feels relaxed and can generate vivid images, you would go on to develop treatment scenes.

Developing Treatment Scenes

You and Mr. Huang would specify certain components to be included in the treatment scenes, including the identity of the model (Mr. Huang or someone else), type of model (coping or mastery), single or multiple models, and specific characteristics of the model to maximize client/model similarity. Next you would decide whether to use individualized or standardized scenes; perhaps in Mr. Huang's case, his own scenes might work best. You would also need to decide how detailed the scene should be. In Mr. Huang's case, a scene might include some examples of positive thoughts and allow room for him to add his own. You and Mr. Huang would generate a list of scenes to be used, specifying the following:

1. The situation (which, for him, would be at work when the negative thoughts crop up)
2. The behavior and coping methods he would acquire (stopping interfering thoughts, generating positive thoughts about retirement, and getting back to his project at work)
3. Favorable outcomes (for Mr. Huang, this might be being able to get his work done on time, which would help him avoid shame and maintain pride in his work)

Applying Treatment Scenes

You and Mr. Huang would arrange the scenes in order—starting with a work situation in which his thoughts are not as interfering and proceeding to situations in which they are most interfering. Starting with the first scene, you would give Mr. Huang specific instructions on imagining. Then you would present the scene to him and have him hold the scene in imagination for a few seconds after he signaled a strong image. After the scene presentation, you would get Mr. Huang's reactions to the scene and make any necessary revisions in duration, scene content, order in the hierarchy, and so on. At this time Mr. Huang could either develop a verbal summary code or personalize the scene by changing or elaborating on it in some way. The

(continued)

same scene would be presented to Mr. Huang at least one more time, followed by several practices in which he goes through the scene without your assistance. After you had worked through all scenes in the hierarchy, you would present scenes to him in a random order.

Homework

After each scene presentation in the session, you would instruct Mr. Huang to practice the scenes daily outside the session. A follow-up on this homework should be arranged.

Part Four

Assess your interview or have someone else assess it, using the Interview Checklist for Covert Modeling on pages 338–341.

COGNITIVE MODELING AND PROBLEM SOLVING

Most systems of therapy recognize the importance of overt behavior change *and* cognitive and affective, or covert, behavior change. In recent years, increasing attention and effort have been directed toward developing and evaluating procedures aimed at modifying thoughts, moods, emotions, attitudes, and beliefs. These procedures come under the broad umbrella of cognitive therapy or cognitive behavior modification. Several assumptions are made about cognitive-change procedures. One of the basic assumptions is that a person's thoughts and beliefs can contribute to maladaptive behavior. Another is that maladaptive behaviors can be altered by dealing directly with the person's beliefs, attitudes, or thoughts. In many instances, a client's unreasonable self-standards and negative self-thoughts can diminish the power of a treatment program. Attention to the client's beliefs and expectations may be necessary for other therapeutic strategies to be successful.

Two cognitive change procedures are presented in this chapter: cognitive modeling and self-instructional training and problem solving. Cognitive restructuring is described in Chapter 15, and in Chapter 16 stress inoculation and reframing are presented. Meditation, presented in Chapter 17, is another procedure for enhancing cognitive change. All these cognitive procedures are efforts to eliminate "cognitive pollution."

OBJECTIVES

1. Using a simulated client case, describe how you would apply the seven components of cognitive modeling and self-instructional training.
2. Demonstrate 16 out of 21 steps of cognitive self-instructional modeling with a role-play client, using the Interview Checklist for Cognitive Modeling at the end of the chapter to rate your performance.
3. Identify which step of the problem-solving strategy is reflected in each of ten counselor responses, accurately identifying at least eight of the ten examples.

4. Demonstrate 16 out of 19 steps of problem solving in a role-play interview, using the Interview Checklist for Problem Solving at the end of the chapter to assess your performance.

COGNITIVE MODELING WITH COGNITIVE SELF-INSTRUCTIONAL TRAINING

Cognitive modeling is a procedure in which counselors show people what to say to themselves while performing a task. Cognitive modeling and self-instructional training have been applied to a variety of client concerns. Table 14-1 presents a list of research studies about cognitive modeling and self-instructional training. These procedures have been used in numerous populations—with children who have attention deficit and hyperactivity disorder, for controlling anger and hostility, as preparation for cardiac catheterization, for school-based consultation, in helping students with learning disabilities, for facilitating problem solving, in enhancing self-efficacy, with preservice and active teachers to reduce anxiety and increase perception of control, in decreasing test anxiety and worry for ninth grade girls, and for training and supervision. We have found only one application of cognitive modeling with a multicultural population. Hains (1989) used cognitive modeling with self-guiding verbalizations to teach anger-control skills to Black, White, and Hispanic male juvenile offenders. After training, 75% of the participants had an increased use of self-instruction and thinking-ahead statements during both provoking incidents and interpersonal conflicts.

Cognitive modeling with a self-instructional training strategy consists of five steps:

1. The counselor serves as the model (or a symbolic model can be used) and first performs the task while talking aloud to himself or herself.
2. The client performs the same task (as modeled by the counselor) while the counselor instructs the client aloud.

TABLE 14-1. Cognitive modeling and self-instructional training research

Attention deficit hyperactivity disorder

Westby, C. E., & Cutler, S. K. (1994). Language and ADHD: Understanding the bases and treatment of self-regulatory deficits. Special Issue: ADD and its relationship to spoken and written language. *Topics in Language Disorders, 14,* 58–76.

Anger and hostility

Normand, D., & Robert, M. (1990). Modeling of anger/hostility control with preadolescent Type A girls. *Child Study Journal, 20,* 237–262.

Cardiac catheterization

Anderson, K. O., & Masur, F. T. (1989). Psychological preparation for cardiac catheterization. *Advances, 6,* 8–10.

Consultation

Gutkin, T. B. (1993). Cognitive modeling: A means for achieving prevention in school-based consultation. *Journal of Educational and Psychological Consultation, 4,* 179–183.

Learning disability

Van Reusen, A. K., & Head, D. N. (1994). Cognitive and metacognitive interventions: Important trends for teachers of students who are visually impaired. *Review, 25,* 153–162. Simmonds, E. P. (1990). The effectiveness of two methods for teaching a constraint-seeking questioning strategy to students with learning disabilities. *Journal of Learning Disabilities, 23,* 229–232.

Problem solving

Gorrell, J. (1993). Cognitive modeling and implicit rules: Effects on problem-solving performance. *American Journal of Psychology, 106,* 51–65.

Self-efficacy

Gist, M. E. (1989). The influence of training method on self-efficacy and idea generation among managers. *Personnel Psychology, 42,* 787–805.

Gorrell, J., & Capron, E. W. (1989). Cognitive modeling effects on preservice teachers with low and moderate success expectations. *Journal of Experimental Education, 57,* 231–244.

Teaching

Hazaressingh, N. A., & Bielawski, L. L. (1991). The effects of cognitive self-instruction on student teachers' perceptions of control. *Teaching and Teacher Education, 7,* 383–393. Lauth, G. W., & Wiedl, K. H. (1989). Cognitive teaching methods for special education: Development of approaches for intervention and assessment in Germany. Special Theme: Cognitive teaching methods for special education. *International Journal of Disability, Development and Education, 36,* 187–202. Payne, B. D., & Manning, B. H. (1990). The effect of cognitive self-instructions on preservice teachers' anxiety about teaching. *Contemporary Educational Psychology, 15,* 261–267.

Test anxiety

Sud, A. (1993). Efficacy of two short term cognitive therapies for test anxiety. *Journal of Personality and Clinical Studies, 9,* 39–46.

Training and supervision

Morran, D. K., Kurpius, D. J., Brack, C. J., & Brack, G. (1995). A cognitive-skills model for counselor training and supervision. *Journal of Counseling & Development, 73,* 384–389. Nutt-Williams, E., & Hill, C. E. (1996). The relationship between self-talk and therapy process variables for novice therapists. *Journal of Counseling Psychology, 43,* 170–177.

---

3. The client is instructed to perform the same task again while instructing himself or herself aloud.
4. The client whispers the instructions while performing the task.
5. Finally, the client performs the task while instructing himself or herself covertly.

Note that cognitive modeling is reflected in step 1, whereas in steps 2 through 5, the client practices self-verbalizations while performing a task or behavior. The client's verbalizations are faded from an overt to a covert level.

We propose that cognitive modeling and self-instructional training should be implemented with seven steps as guidelines:

1. A rationale about the procedure
2. Cognitive modeling of the task and of the self-verbalizations

Client practice in the following form:

3. Overt external guidance
4. Overt self-guidance
5. Faded overt self-guidance
6. Covert self-guidance
7. Homework and follow-up

Each of these steps is explained in the following section. Illustrations are also provided in the Interview Checklist for Cognitive Modeling at the end of the chapter.

Treatment Rationale

Here is an example of the counselor's rationale for cognitive modeling:

> It has been found that some people have difficulty in performing certain kinds of tasks. Often the difficulty is not because they don't have the ability to do it but because of what they say or think to themselves while doing it. In other words, a person's 'self-talk' can get in the way or interfere with performance. For instance, if you get up to give a speech and you're thinking 'What a flop I'll be,' this sort of thought may affect how you deliver your talk. This procedure can help you perform something the way you want to by examining and coming up with some helpful planning or self-talk to use while performing [rationale]. I'll show what I am saying to myself while performing the task. Then I'll ask you to do the task while I guide or direct you through it. Next, you will do the task again and guide yourself aloud while doing it. The end result should be your performing the task while thinking and planning about the task to yourself [overview]. How does this sound to you? [client willingness]

After the rationale has been presented and any questions have been clarified, the counselor begins by presenting the cognitive model.

Model of Task and Self-Guidance

First, the counselor instructs the client to listen to what the counselor says to herself or himself while performing the task. Next, the counselor models performing a task while talking aloud.

Questions
1. What has to be done?
2. Answers question in form of planning what to do.
3. Self-guidance and focused attention.
4. Self-reinforcement.
5. Coping self-evaluative statements with error correction options.

Dialogue
1. "Okay, what is it I have to do?"
2. "You want me to copy the picture with different lines."
3. "I have to go slow and be careful. Okay, draw the line down, down, good; then to the right, that's it; now down some more and to the left."
4. "Good. Even if I make an error I can go on slowly and carefully. Okay, I have to go down now."
   "Finished. I did it."
5. "Now back up again. No, I was supposed to go down. That's okay. Just erase the line carefully."

As this example indicates, the counselor's modeled self-guidance should include five parts. The first part of the verbalization asks a question about the nature and demands of the task to be performed. The purposes of the question are to compensate for a possible deficiency in comprehending what to do, to provide a general orientation, and to create a cognitive set. The second part of the modeled verbalization answers the question about what to do. The answer is designed to model cognitive rehearsal and planning to focus the client's attention on relevant task requirements. Self-instruction in the form of self-guidance while performing the task is the third part of the modeled verbalization. The purpose of self-guidance is to facilitate attention to the task and to inhibit any possible overt or covert distractions or task irrelevancies. In the example, modeled self-reinforcement is the fourth part and is designed to maintain task perseverance and to reinforce success. The last part in the modeled verbalization contains coping self-statements to handle errors and frustration, with an option for correcting errors. The example of the modeled verbalization used by Meichenbaum and Goodman (1971) depicts a coping model. In other words, the model does make an error in performance but corrects it and does not give up at that point. See whether you can identify these five parts of modeled self-guidance in Learning Activity #41 on p. 347.

Overt External Guidance

After the counselor models the verbalizations, the client is instructed to perform the task (as modeled by the counselor) while the counselor instructs or coaches. The counselor coaches the client through the task or activity, substituting the personal pronoun *you* for *I* (for example, "What is it that *you* . . . , *you* have to wheel your chair . . . , *you* have to be careful"). The counselor should make sure that the coaching contains the same five parts of self-guidance that were previously modeled: question, planning, focused attention, coping self-evaluation, and self-reinforcement. Sometimes in the client's real-life situation, other people may be watching when the client performs the task— as could be the case whenever the client in a wheelchair appears in public. If the presence of other people appears to interfere with the client's performance, the counselor might say "Those people may be distracting you. Just pay attention to what you are doing." This type of coping statement can be included in the counselor's verbalizations when using overt external guidance in order to make this part of the procedure resemble what the client will actually encounter.

LEARNING ACTIVITY **41** *Modeled Self-Guidance*

The following counselor verbalization is a cognitive model for a rehabilitation client with a physical challenge who is learning how to use a wheelchair. Identify the five parts of the message: (1) questions of what to do, (2) answers to the question in the form of planning, (3) self-guidance and focused attention, (4) coping self-evaluative statements, and (5) self-reinforcement. Feedback for this activity follows.

"What do I have to do to get from the parking lot over the curb onto the sidewalk and then to the

building? I have to wheel my chair from the car to the curb, get over the curb and onto the sidewalk, and then wheel over to the building entrance. Okay, wheeling the chair over to the curb is no problem. I have to be careful now that I am at the curb. Okay, now I've just got to get my front wheels up first. They're up now. So now I'll pull up hard to get my back wheels up. Whoops, didn't quite make it. No big deal—I'll just pull up very hard again. Good. That's better, I've got my chair on the sidewalk now. I did it! I've got it made now."

Overt Self-Guidance

The counselor next instructs the client to perform the task while instructing or guiding himself or herself aloud. The purpose of this step is to have the client practice the kind of self-talk that will strengthen attention to the demands of the task and will minimize outside distractions. The counselor should attend carefully to the content of the client's self-verbalizations. Again, as in the two preceding steps, these verbalizations should include the five component parts, and the client should be encouraged to use his or her own words. If the client's self-guidance is incomplete or if the client gets stuck, the counselor can intervene and coach. If necessary, the counselor can return to the previous steps—either modeling again or coaching the client while the client performs the task (overt external guidance). After the client completes this step, the counselor should provide feedback about parts of the practice the client completed adequately and about any errors or omissions. Another practice might be necessary before moving on to the next step: faded overt self-guidance.

Faded Overt Self-Guidance

The client next performs the task while whispering (lip movements). This part of cognitive modeling serves as an intermediate step between having the client verbalize aloud, as in overt self-guidance, and having the client verbalize silently, as in the next step, covert self-guidance. In other words, whispering the self-guidance is a way for the client to approximate successively the end result of the procedure: thinking the self-guidance steps while performing them. In our own experience with this step, we have found that it is necessary to explain this to an occasional client who seems hesitant or concerned about whispering. If a client finds the

whispering too foreign or aversive, he or she might prefer to repeat overt self-guidance several times and finally move directly to covert self-guidance. If the client has difficulty performing this step or leaves out any of the five parts, an additional practice may be required before moving on.

Covert Self-Guidance

Finally, clients perform the task while guiding or instructing covertly, or "in their heads." It is very important for clients to instruct themselves covertly after practicing the self-instructions overtly. After the client does this, the counselor might ask for a description of the covert self-instructions. If distracting or inhibiting self-talk has occurred, the counselor can offer suggestions for more appropriate verbalizations or self-talk and can initiate additional practice. Otherwise, the client is ready to use the procedure outside the session.

Homework and Follow-up

Assigning the client homework is essential for generalization to occur from the interview to the client's environment. The therapist should instruct the client to use the covert verbalizations while performing the desired behaviors alone, outside the counseling session. The homework assignment should specify what the client will do, how much or how often, and when and where the homework is to be done. The counselor should also provide a way for the client to monitor and reward himself or herself for completion of homework. A follow-up on the homework task should also be scheduled.

These seven components of cognitive modeling are modeled for you in the following dialogue with our client Joan. Again, this strategy is used as one way to help Joan achieve her goal of increasing her verbal participation in math class.

FEEDBACK
Modeled Self-Guidance

Question: "What do I have to do to get from the parking lot over the curb onto the sidewalk and then to the building?"

Answers with planning: "I have to wheel my chair from the car to the curb, get onto the curb and onto the sidewalk, and then wheel over to the building entrance."

Self-guidance and focused attention: "Okay, wheeling the chair over to the curb is no problem. I have to be careful now that I am at the curb. Okay, now I've just got to get my front wheels up first. They're up now. So now I'll pull up hard to get my back wheels up."

Coping self-evaluation and error-correction option: "Whoops, didn't quite make it. No big deal—I'll just pull up very hard again."

Self-reinforcement: "Good. That's better, I've got my chair on the sidewalk now. I did it! I've got it made now."

MODEL DIALOGUE: COGNITIVE MODELING WITH COGNITIVE SELF-INSTRUCTIONAL TRAINING

In response 1, the counselor introduces the possible use of cognitive modeling to help Joan achieve the goal of increasing initiating skills in her math class. The counselor is giving a **rationale** *about the strategy.*

1. *Counselor:* One of the goals we developed was to help you increase your participation level in your math class. One of the ways we might help you do that is to use a procedure in which I demonstrate the kinds of things you want to do—and also I will demonstrate a way to think or talk to yourself about these tasks. So this procedure will help you develop a plan for carrying out these tasks, as well as showing you a way to participate. How does that sound?
   Client: OK. Is it hard to do?

In response 2, the counselor provides an **overview** *of the procedure, which is also a part of the rationale.*

2. *Counselor:* No, not really, because I'll go through it before you do. And I'll sort of guide you along. The procedure involves my showing you a participation method and, while I'm doing that, I'm going to talk out loud to myself to sort of guide myself. Then you'll do that. Gradually, we'll go over the same participation method until you do it on your own and can think to

yourself how to do it. We'll take one step at a time. Does that seem clear to you?
Client: Well, pretty much. I've never done anything like this, though.

In response 3, the counselor determines **Joan's willingness** *to try out the procedure.*

3. *Counselor:* Would you like to give it a try?
   Client: Sure—I'm willing.

In responses 4 and 5, the counselor sets the stage for **modeling of the task** *and accompanying* **self-guidance** *and* **instructs the client in what will be done and what to look for in this step.**

4. *Counselor:* We mentioned there were at least four things you could do to increase your initiating skills—asking Mr. Lamborne for an explanation only, answering more of Mr. Lamborne's questions, going to the board to do problems, and volunteering answers. Let's just pick one of these to start with. Which one would you like to work with first?
   Client: Going to the board to work algebra problems. If I make a mistake there, it's visible to all the class.

5. *Counselor:* Probably you're a little nervous when you do go to the board. This procedure will help you concentrate more on the task than on yourself. Now, in the first step, I'm going to pretend I'm going to the board. As I move out of my chair and up to the board, I'm going to tell you what I'm thinking that might help me do the problems. Just listen carefully to what I say, because I'm going to ask you to do the same type of thing afterward. Any questions?
   Client: No, I'm just waiting to see how you handle this. I'll look like Mr. Lamborne. His glasses are always down on his nose and he stares right at you. It's unnerving.

In responses 6 and 7, the counselor **initiates and demonstrates** *the task with accompanying* **self-guidance.** *Note, in the modeled part of response 7, the* **five components of the self-guidance process.** *Also note that a simple problem has been chosen for illustration.*

6. *Counselor:* OK, you do that. That will help set the scene. Why don't you start by calling on me to go to the board?
   Client (as teacher): Joan, go to the board now and work this problem.

7. *Counselor* (gets out of seat, moves to imaginary board on the wall, picks up the chalk, verbalizing aloud): What is it I need to do? He wants me to find y. OK, I need to just go slowly, be careful, and take my time. OK, the problem here reads $4x + y = 10$, and x is 2.8. OK, I can use x to find y. [Counselor asks *question* about task.] OK, I'm doing fine so far. Just remember to go slowly. OK, y has to be $10 - 4x$. If x is 2.8, then y will be $10 - 4$ multiplied by 2.8. [Counselor focuses *attention* and uses *self-guidance.*] Let's see, 4×2.8 is 10.2. Oops, is this right? I hear someone laughing. Just keep on going. Let me refigure it. No, it's 11.2. Just erase 10.2 and put in $y = 10 - 11.2$. OK, good. If I keep on going slowly, I can catch any error and redo it. [Counselor uses *coping self-evaluation* and makes *error correction.*] Now it's simple. $10 - 11.2$ is -1.2 and y is -1.2. Good, I

LEARNING ACTIVITY 42

Cognitive Modeling with Cognitive Self-Instructional Training

You may recall from the case of Ms. Weare and Freddie (Chapter 8) that Ms. Weare wanted to eliminate the assistance she gave Freddie in getting ready for school in the morning. One of Ms. Weare's concerns is to find a successful way to instruct Freddie about the new ground rules—mainly that she will not help him get dressed and will not remind him when the bus is five minutes away. Ms. Weare is afraid that after she tells Freddie, Freddie will either pout or talk back to her. She is concerned that she will not be able to follow through with her plan or will not

be firm in the way she delivers the rules to him. (a) Describe how you would use the seven major components of cognitive modeling and self-instructional training to help Ms. Weare to do this, and (b) write out an example of a cognitive modeling dialogue that Ms. Weare could use to accomplish this task. Make sure this dialogue contains the five necessary parts of the self-guidance process: question, answer, focused attention, self-evaluation, and self-reinforcement. Feedback follows.

---

did it, I'm done now and I can go back to my seat. [Counselor *reinforces self.*]

In responses 8 and 9, the counselor initiates **overt external guidance:** *the client performs the task while the counselor continues to verbalize aloud the self-guidance, substituting* **you** *for* **I** *as used in the previous sequence.*

8. *Counselor:* OK, that's it. Now let's reverse roles. This time I'd like you to get up out of your seat, go to the board, and work through the problem. I will coach you about what to plan during the process. OK?

   Client: Do I say anything?

9. *Counselor:* Not this time. You just concentrate on carrying out the task and thinking about the planning I give you. In other words, I'm just going to talk you through this the first time.

   Client: OK, I see.

In response 10, the counselor **verbalizes self-guidance while the client performs** *the problem.*

10. *Counselor:* OK, I'll be Mr. Lamborne. I'll ask you to go to the board, and then you go and I'll start coaching you (as teacher): Joan, I want you to go to the board now and work out this problem: If $2x + y = 8$ and $x = 2$, what does y equal? [Joan gets up from chair, walks to imaginary board, and picks up chalk.] (as counselor): OK, first you write the problem on the board. $2x + y = 8$ and $x = 2$. Now ask yourself "What do I have to do with this problem?" OK, now answer yourself [question].

      You need to find the value of y [answer to question]. OK, just go slowly, be careful, and concentrate on what you're doing. You know $x = 2$, so you can use x to find y. Your first step is to subtract $8 - 2x$. You've got that up there. OK, you're doing fine—just keep going slowly [focuses attention and uses self-guidance].

      $8 - 2$ multiplied by 2, you know is $8 - 4$. Someone is laughing at you. But you're doing fine, just keep thinking about what you're doing. $8 - 4$ is 4, so $y = 4$ [coping self-evaluation].

      Now you've got y. That's great. You did it. Now you can go back to your seat [self-reinforcement].

In response 11, the counselor **assesses the client's reaction** *before moving on to the next step.*

11. *Counselor:* OK, let's stop. How did you feel about that?

    Client: Well, it's such a new thing for me. I can see how it can help. See, usually when I go up to the board I don't think about the problem. I'm usually thinking about feeling nervous or about Mr. Lamborne or the other kids watching me.

In response 12, the counselor reiterates the **rationale** *for the cognitive modeling procedure.*

12. *Counselor:* Yes, well, those kinds of thoughts distract you from concentrating on your math problems. That's why this kind of practice may help. It gives you a chance to work on concentrating on what you want to do.

    Client: I can see that.

In responses 13 and 14, the counselor instructs the client to perform the task while verbalizing to herself **(overt self-guidance).**

13. *Counselor:* This time I'd like you to go through what we just did—only on your own. In other words, you should get up, go to the board, work out the math problem, and as you're doing that, plan what you're going to do and how you're going to do it. Tell yourself to take your time, concentrate on seeing what you're doing, and give yourself a pat on the back when you're done. How does that sound?

    Client: OK, I'm just going to say something similar to what you said the last time—is that it?

14. *Counselor:* That's it. You don't have to use the same words. Just try to plan what you're doing. If you get stuck, I'll step in and give you a cue. Remember, you start by asking yourself what you're going to do in this situation and then answering yourself. This time let's take the problem $5x + y = 10$; with $x = 2.5$, solve for y.

    Client: (gets out of seat, goes to board, writes problem): What do I need to do? I need to solve for y. I know $x = 2.5$. Just think about this problem. My first step is to subtract $10 - 5x$. 5 multiplied by 2.5 is 12.5. So I'll subtract $10 - 12.5$. [Counselor laughs; Joan turns around.] Is that wrong?

FEEDBACK 42
Cognitive Modeling with Cognitive Self-Instructional Training

a. Description of the seven components:

1. *Rationale.* First, you would explain to Ms. Weare how cognitive modeling could help her in instructing Freddie and what the procedure would involve. You might emphasize that the procedure would be helpful to her in both prior planning and practice.

2. *Model of task and self-guidance.* In this step, you would model a way Ms. Weare could talk to Freddie. You need to make sure that you use language that is relevant and acceptable to Ms. Weare. Your modeling would include both the task (what Ms. Weare could say to Freddie) and the five parts of the self-guidance process.

3. *Overt external guidance.* Ms. Weare would practice giving her instructions to Freddie while you coach her on the self-guidance process.

4. *Overt self-guidance.* Ms. Weare would perform the instructions while verbalizing aloud the five parts of the self-guidance process. If she gets stuck or if she leaves out any of the five parts, you can cue her. This step also may need to be repeated before moving on.

5. *Faded overt self-guidance.* Assuming Ms. Weare is willing to complete this step, she would perform the instructions to give Freddie while whispering the self-guidance to herself.

6. *Covert self-guidance.* Ms. Weare would practice giving the instructions to Freddie while covertly guiding herself. When she is able to do this comfortably, you would assign homework.

7. *Homework.* You would assign homework by asking Ms. Weare to practice the covert self-guidance daily and arranging for a follow-up after some portion had been completed.

b. Example of a model dialogue:

"OK, what is it I want to do in this situation [question]? I want to tell Freddie that he is to get up and dress himself without my help, that I will no longer come up and help him even when it's time for the bus to come [answer]. OK, just remember to take a deep breath and talk firmly and slowly. Look at Freddie. Say "Freddie, I am not going to help you in the morning. I've got my own work to do. If you want to get to school on time, you'll need to decide to get yourself ready" [focused attention and self-guidance]. Now, if he gives me flak, just stay calm and firm. I won't back down [coping self-evaluation]. That should be fine. I can handle it [self-reinforcement].

Counselor: Check yourself but stay focused on the problem, not on my laughter.

Client: Well, $10 - 12.5$ is -2.5 $y = -2.5$. Let's see if that's right. $5 \times 2.5 = 12.5 - 2.5 = 10$. I've got it. Yeah.

In response 15, the counselor **gives feedback** *to Joan about her practice.*

15. *Counselor:* That was really great. You only stumbled one time—when I laughed. I did that to see whether you would still concentrate. But after that, you went right back to your work and finished the problem. It seemed pretty easy for you to do this. How did you feel?

Client: It really was easier than I thought. I was surprised when you laughed. But then, like you said, I just tried to keep going.

In responses 16, 17, and 18, the counselor instructs Joan on how to **perform the problem while whispering instructions** *to herself* **(faded overt self-guidance).**

16. *Counselor:* This time we'll do another practice. It will be just like you did the last time, with one change. Instead of talking out your plan aloud, I just want you to whisper it. Now you probably aren't used to whispering to yourself, so it might be a little awkward at first.

Client (laughs): Whispering to myself? That seems sort of funny.

17. *Counselor:* I can see how it does. But it is just another step in helping you practice this to the point where it becomes a part of you—something you can do naturally and easily.

Client: Well, OK. I guess I can see that.

18. *Counselor:* Well, let's try it. This time let's take a problem with more decimals, since you get those, too. If it seems harder, just take more time to think and go slowly. Let's take $10.5x + y = 25$, with $x = 5.5$.

Client: (gets out of seat, goes to board, writes on board, whispers): OK, what do I need to do with this problem? I need to find y. This has more decimals, so I'm just going to go slowly. Let's see, $25 - 10.5x$ is what I do first. I need to multiply 10.5 by 5.5. I think it's 52.75. [Counselor laughs.] Let's see, just think about what I'm doing. I'll redo it. No, it's 57.75. Is that right? I'd better check it again. Yes, it's OK. Keep going. $25 - 57.75$ is equal to -32.75, so $y = -32.75$. I can check it—yes, 10.5×5.5 is $57.75 - 32.75 = 25$. I got it!

Counselor **gives feedback** *in response 19.*

19. *Counselor:* That was great, Joan—very smooth. When I laughed, you just redid your arithmetic rather than turning around or letting your thoughts wander off the problem.

Client: It seems like it gets a little easier each time. Actually, this is a good way to practice math, too.

In responses 20 and 21, the counselor gives Joan instructions on how to **perform the problem while instructing herself covertly (covert self-guidance).**

20. *Counselor:* That's right. Not only for what we do in here, but even when you do your math homework. Now, let's just go

through one more practice today. You're really doing this very well. This time I'd like you to do the same thing as before—only this time I'd like you to just think about the problem. In other words, instead of talking out these instructions, just go over them mentally. Is that clear?

Client: You just want me to think to myself what I've been saying?

21. *Counselor:* Yes—just instruct yourself in your head. Let's take the problem $12x - y = 36$, with $x = 4$. Solve for y. [Joan gets up, goes to the board, and takes several minutes to work through this.]

In response 22, the counselor **asks the client to describe what happened during covert self-guidance** *practice.*

22. *Counselor:* Can you tell me what you thought about while you did that?

Client: Well, I thought about what I had to do, then what my first step in solving the problem would be. Then I just went through each step of the problem, and then after I checked it, I thought I was right.

In response 23, the counselor **checks to see whether another practice is needed** *or whether they can move on to homework.*

23. *Counselor:* So it seemed pretty easy. That is what we want you to be able to do in class—to instruct yourself mentally like this while you're working at the board. Would you like to go through this type of practice one more time, or would you rather do this on your own during the week?

Client: I think on my own would help right now.

In response 24, the counselor sets up Joan's **homework assignment** *for the following week.*

24. *Counselor:* OK. I think it would be helpful if you could do this type of practice on your own this week—where you instruct yourself as you work through math problems.

Client: You mean my math homework?

In response 25, the counselor **instructs Joan on how to do homework,** *including what to do, where, and how much to do.*

25. *Counselor:* Well, that would be a good way to start. Perhaps you could take seven problems a day. As you work through each problem, go through these self-instructions mentally. (Do this at home.) Does that seem clear?

Client: Yes, I'll just work out seven problems a day the way we did here for the last practice.

In response 26, the counselor instructs Joan **to observe her homework completion on log sheets** *and* **arranges for a follow-up** *of homework at their next session.*

26. *Counselor:* Right. One more thing. On these blank log sheets, keep a tally of the number of times you actually do this type of practice on math problems during the day. This will help you keep track of your practice. And then next week bring your log sheets with you and we can go over your homework.

PROBLEM-SOLVING THERAPY

Problem-solving therapy or problem-solving training emerged in the late 1960s and early 1970s as a trend in the development of intervention and prevention strategies for enhancing competence in specific situations. D'Zurilla (1988) defines problem solving as a "cognitive-affective-behavioral process through which an individual (or group) attempts to identify, discover, or invent effective means of coping with problems encountered in every day living" (p. 86). Rose and LeCroy (1991) describe problem solving as a strategy whereby "the client learns to systematically work through a set of steps for analyzing a problem, discovering new approaches, evaluating those approaches, and developing strategies for implementing those approaches in the real world" (p. 439). Problem-solving therapy or training has been used with children, adolescents, and adults as a treatment strategy, a treatment-maintenance strategy, or a prevention strategy. As a treatment strategy, problem solving has been used alone or in conjunction with other treatment strategies presented in this book.

The list in Table 14-2 shows that problem-solving therapy has been used effectively for a wide range of purposes. It has helped inpatient and aggressive adolescent boys, cancer patients, families, children with learning and behavior problems, and single parents. It has been used successfully to improve clients' communication skills, dietary management, conflict resolution, parenting skills, stress management, and effectiveness in applying solutions. Using problem-solving therapy, clients have reduced their depression, health complaints and expectancies, interpersonal dependency, substance abuse, and worry.

Heppner (1988) has developed a problem-solving inventory: the scales of the inventory include problem-solving confidence (or self-assurance while engaging in problem activities), tendency to approach or avoid problem-solving activities, and the extent to which individuals believe they are in control of their behaviors and emotions while solving problems. D'Zurilla and Nezu (1990) have also developed an inventory that focuses on social problem solving.

Perceptions and attitudes about problem solving can play either a facilitative or a disruptive role. If perceptions and attitudes have a facilitative role, the client is motivated to learn and engage in problem-solving behaviors. Clients who are unmotivated or avoid dealing with problems because of their perception of problems will not want to learn the problem-solving strategy. In these cases, the therapist will have to first help the client deal with these perceptions and attitudes. The therapist also can help the client engage in some effective coping activities or behaviors that can facilitate the

TABLE 14-2. Problem solving research

Adolescents

Foxx, R. M., Kyle, M. S., Faw, G. D., & Bittle, R. G. (1989). Teaching a problem solving strategy to inpatient adolescents: Social validation, maintenance, and generalization. *Child and Family Behavior Therapy, 11,* 71–88.

Aggressive boys

Guervremont, D. C., & Foster, S. L. (1993). Impact of social problem-solving training on aggressive boys: Skill acquisition, behavior change, and generalization. *Journal of Abnormal Child Psychology, 21,* 13–27.

Behavior problems

Hains, A. A., & Fouad, N. A. (1994). The best laid plans . . .: Assessment in an inner-city high school. Special Issue: Multicultural assessment. *Measurement and Evaluation in Counseling and Development, 27,* 116–124.

Cancer

Fawzy-Fawzy, I., Cousins, N., Fawzy, N. W., & Kemeny, M. E. (1990). A structured psychiatric intervention for cancer patients: I. Changes over time in methods of coping and affective disturbance. *Archives of General Psychiatry, 47,* 720–725.

Children

Crammond, B., Martin, C. E., & Shaw, E. L. (1990). Generalizability of creative problem solving procedures to real-life problems. *Journal for the Education of the Gifted, 13,* 141–155.
Erwin, P. G., & Ruane, G. E. (1993). The effects of a short-term social problem solving programme with children. *Counseling Psychology Quarterly, 6,* 317–323.
Shure, M. B. (1993). I can problem solve (ICPS): Interpersonal cognitive problem solving for young children. Special Issue: Enhancing young children's lives. *Early Child Development and Care, 96,* 49–64.
Spivack, G., & Shure, M. B. (1989). Interpersonal Cognitive Problem Solving (ICPS): A competence-building primary prevention program. *Prevention in Human Services, 6,* 151–178.

Communication behaviors

Firestien, R. L. (1988). Creative problem solving and communication behavior in small groups. *Creativity Research Journal, 1,* 106–114.
Firestien, R. L. (1990). Effects of creative problem solving training on communication behaviors in small groups. *Small Group Research, 21,* 507–521.

Decisional conflict resolving

Clarke, K. M., & Greenberg, L. S. (1986). Differential effects of the gestalt two-chair intervention and problem solving in resolving decisional conflict. *Journal of Counseling Psychology, 33,* 11–15.

Depression

Nezu, A. M., & Perri, M. G. (1989). Problem-solving therapy for unipolar depression: An initial dismantling investigation. *Journal of Consulting and Clinical Psychology, 57,* 408–413.
Nezu, A. M., Nezu, C. M., & Perri, M. G. (1990). Psychotherapy for adults within a problem-solving framework: Focus on depression. *Journal of Cognitive Psychotherapy, 4,* 247–256.

Dietary self-management

Goodall, T. A., Halford, W. K., & Mortimer, R. (1993). Problem solving training to enhance dietary self-management in a diabetic patient. *Behavioural Psychotherapy, 21,* 147–155.

Families

Bentley, K. J., Rosenson, M. K., & Zito, J. M. Promoting medication compliance: Strategies for working with families of mentally ill people. *Social Work, 35,* 274–277.
Nangle, D. W., Carr-Nangle, R. E., & Hansen, D. J. (1994). Enhancing generalization of a contingency-management intervention through the use of family problem-solving training: Evaluation with a severely conduct-disordered adolescent. *Child and Family Behavior Therapy, 16,* 65–76.

Health

Elliott, T. R., & Marmarosh, C. L. (1994). Problem-solving appraisal, health complaints, and health-related expectancies. *Journal of Counseling & Development, 72,* 531–537.

Interpersonal dependency

Overholser, J. C., & Fine, M. A. (1994). Cognitive-behavioral treatment of excessive interpersonal dependency: A four-stage psychotherapy model. *Journal of Cognitive Psychotherapy, 8,* 55–70.

Learning and behavior problems

Coleman, M., Wheeler, L., & Webber, J. (1993). Research on interpersonal problem-solving training: A review. *RASE Remedial and Special Education, 14,* 25–37.

(continued)

---

problem-solving process. If the perceptual, attitudinal, and emotional components of behavior are dealt with, problem-solving training or therapy will be more than an intellectual exercise (D'Zurilla, 1988, pp. 116–117).

Many clients prefer to ignore or to avoid a problem because they believe the problem will probably go away by itself. Although some problems may simply disappear, others will *not* vanish if ignored or avoided. In fact, some may get worse and can become antecedents to more problems if the client does not solve the initial difficulty. The role of the therapist is to help the client take responsibility for solving problems and commit to spending the time and energy needed to solve them by changing the client's attitudes and perceptions about these issues.

TABLE 14-2. Problem solving research (continued)

Marital therapy

Baker, M. P., Blampied, N. M., & Haye, L. (1989). Behavioural marital therapy for alcoholics: Effects on communication skills and marital satisfaction. Special Issue: Behavioural marital therapy. *Behaviour Change, 6,* 178–186.

Jacobson, N. S., Schmaling, K. B., & Holtzworth-Munroe, A. (1987). Component analysis of behavioral marital therapy: 2-year follow-up and prediction of relapse. *Journal of Marital and Family Therapy, 13,* 187–195.

Noller, P., & Guthrie, D. (1989). Assessment and modification of marital communication. Special Issue: *Behavioural Marital Therapy, 6,* 124–136.

Upton, L. R., & Jensen, B. J. (1991). The acceptability of behavioral treatment for marital problems: A comparison of behavioral exchange and communication skills training procedures. *Behavior Modification, 15,* 51–63.

Mental retardation

Nezu, C. M., Nezu, A. M., & Arean, P. (1991). Assertiveness and problem-solving training for mildly mentally retarded persons with dual diagnoses. *Research in Developmental Disabilities, 12,* 371–386.

O'Reilly, M. F., & Chadsey-Rusch, J. (1992). Teaching a social skills problem-solving approach to workers with mental retardation: An analysis of generalization. *Education and Training in Mental Retardation, 27,* 324–334.

Park, H. S., & Gaylord-Ross, R. (1989). A problem solving approach to social skills training in employment settings with mentally retarded youth. Special Issue: Supported employment. *Journal of Applied Behavior Analysis, 22,* 373–380.

Parenting skills

Cunningham, C. E., Davis, J. R., Bremner, R., & Dunn, K. W. (1993). Coping modeling problem solving versus mastery modeling: Effects on adherence, in-session process, and skill acquisition in a residential parent-training problem. *Journal of Consulting and Clinical Psychology, 61,* 871–877.

Doll, B., & Kratochwill, T. R. (1992). Treatment of parent-adolescent conflict through behavioral technology training: A case study. *Journal of Educational and Psychological Consultation, 3,* 281–300.

Gammon, E. A., & Rose, S. D. (1991). The Coping Skills Training Program for parents of children with developmental disabilities: An experimental evaluation. *Research on Social Work Practice, 1,* 244–256.

Kazdin, A. E., Siegel, T. C., & Bass, D. (1992). Cognitive problem-solving skills training and parent management training in the treatment of antisocial behavior in children. *Journal of Consulting and Clinical Psychology, 60,* 733–747.

Spaccarelli, S., Cotler, S., & Penman, D. (1992). Problem-solving skills training as a supplement to behavioral parent training. *Cognitive Therapy and Research, 16,* 1–17.

Single parenting

Pfiffner, L. J., Jouriles, E. N., Brown, M. M., & Etscheidt, M. A. (1990). Effects of problem-solving therapy on outcomes of parent training for single-parent families. *Child and Family Behavior Therapy, 12,* 1–11.

Solution effectiveness

Yoman, J., & Edelstein, B. A. (1993). Relationship between solution effectiveness ratings and actual solution impact in social problem solving. *Behavior Therapy, 24,* 409–430.

Stress management

D'Zurilla, T. J. (1990). Problem-solving training for effective stress management and prevention. *Journal of Cognitive Psychotherapy, 4,* 327–354.

Heppner, P. P., Baumgarder, A. H., Larson, L. M., & Petty, R. E. (1988). The utility of problem-solving training that emphasizes self-management principles. Special Issue: Stress counseling. *Counselling Psychology Quarterly, 1,* 129–143.

Substance abuse

Platt, J. J., & Hermalin, J. A. (1989). Social skills deficit intervention for substance abusers. Special Issue: Society of Psychologists in Addictive Behaviors comes of age. *Psychology of Addictive Behaviors, 3,* 114–133.

Worry

Dugas, M. J., Letarte, H., Rheaume, J., & Freeston, M. H. (1995). Worry and problem solving: Evidence of a specific relationship. *Cognitive Therapy and Research, 19,* 109–120.

MULTICULTURAL APPLICATIONS OF PROBLEM SOLVING

Of all of the intervention strategies we describe in this book, problem solving has been the one most widely used with diverse groups of clients. We suspect this is in part because it is a strategy that focuses on direct action and observable, attainable results. Problem solving has been used with diverse groups of clients in several areas, including the development of educational and academic achievement skills and the development of coping skills and conflict management skills; it has also been incorporated as a component of various prevention programs including HIV prevention, tobacco use prevention, and violence prevention (see Table 14-3).

Armour-Thomas, Bruno, and Allen (1992) examined the nature of higher-order thinking in academic problem-solving situations involving 77 African American and 30 Latino high

TABLE 14-3. Multicultural problem solving research

Developing educational and academic achievement skills	Armour-Thomas, E., Bruno, K., & Allen, B. (1992). Toward an understanding of higher-order thinking among minority students. *Psychology in the Schools, 29,* 273–280. Bell, Y., Brown, R., & Bryant, A. (1993). Traditional and culturally relevant presentations of a logical reasoning task and performance among African-American students. *Western Journal of Black Studies, 17,* 173–178. Mehan, H., Hubbard, L., & Villanueva, I. (1994). Forming academic identities: Accommodations without assimilation among involuntary minorities. *Anthropology and Education Quarterly, 25,* 91–117. Pollard, D. (1993). Gender, achievement, and African-American students' perceptions of their school experience. *Educational Psychologist, 28,* 341–356.
Developing coping skills	Rao, R., & Kramer, L. (1993). Stress and coping among mothers of infants with sickle cell condition. *Children's Health Care, 22,* 169–188. Yang, B., & Clum, G. (1994). Life stress, social support, and problem-solving skills predictive of depression symptoms, hopelessness, and suicide ideation in an Asian student population: A test of a model. *Suicide and Life Threatening Behavior, 24,* 127–139.
Promoting conflict management skills	Watson, D., Bell, P., & Chavez, E. (1994). Conflict handling skills used by Mexican American and White non-Hispanic students in the educational system. *High School Journal, 78,* 35–39.
As a component of prevention programs	
1. **HIV prevention**	Bracho-de-Carpio, A., Carpio-Cedraro, F., & Anderson, L. (1990). Hispanic families and teaching about AIDS: A participatory approach at the community level. *Hispanic Journal of Behavioral Sciences, 12,* 165–176. Kelly, J., & St. Lawrence, J. (1990). The impact of community-based groups to help persons reduce HIV infection risk behaviors. *AIDS Care, 2,* 25–36. St.-Lawrence, J., Brasfield, T., Jefferson, K., Alleyne, E., O'Bannon, R., & Shirley, A. (1995). Cognitive-behavioral intervention to reduce African American adolescents' risk for HIV infection. *Journal of Consulting and Clinical Psychology, 63,* 221–237.
2. **Tobacco use prevention**	Moncher, M., & Schinke, S. (1994). Group intervention to prevent tobacco use among Native American youth. *Research on Social Work Practice, 4,* 160–171.
3. **Violence prevention**	Hammond, R., & Yung, B. (1991). Preventing violence in at-risk African American youth. *Journal of Health Care for the Poor and Underserved, 2,* 359–373.

school students. They found that both planning and monitoring were important components of solving academic problems for these students, especially in the area of mathematics. In a study conducted by Pollard (1993), African American male and female students who were successful achievers in school used more active problem-solving strategies than did their less successful peers.

Two studies have been conducted that specifically focused on culturally relevant educational processes and problem solving. In one of these, Bell, Brown, and Bryant (1993) compared the performance of 195 African American college students on a problem-solving task. Students were given both a traditional, analytic presentation of the task and a culturally relevant, holistic presentation. The students who received the culturally relevant presentation did better on the task than those given the traditional presentation. These authors note the impact of cultural factors on problem solving and also stress the importance for African American students of exposure to models of education that are Afrocentric rather than Eurocentric in nature.

In the second culturally relevant study of problem solving, Mehan, Hubbard, and Villaneuva (1994) describe a program that has been implemented in the San Diego high schools. In this program, African American and Latino students who took college preparatory classes were also offered a special elective class emphasizing collaborative instruction, writing, and problem solving. The program is called AVID—Advancement via Individual Determination. The students who went through the AVID program enrolled in four-year colleges at a level well above the national average. An interesting result of the AVID program is that students in the program reported they were able to achieve at these schools without compromising their ethnic identity. However, they did this in a way that is reported frequently by students of color involved in mainstream schools: they supported each other and bypassed racist teachers and counselors. Essentially, they did not look to the school for their sense of self-worth. Although the AVID program supports the use of problem solving for African American and Latino youth, it does not appear to solve the issue of cultural relevance of mainstream culture schools for students of color.

Other studies have explored nonacademic uses of problem solving. In one of these, Rao and Kramer (1993) interviewed African American mothers of infants with sickle cell anemia and infants with sickle cell trait to explore coping skills the mothers used to care for their children. Mothers of both groups reported using social support, positive reappraisal, and planned problem solving most frequently to cope with stressors related to parenting their children.

Yang and Clum (1994) explored the relationship between a problem-solving model and a social-support model and stress, depression, hopelessness, and suicidal ideation among 88 Asian international students (aged 18–40 years) living in the United States. The problem-solving model was related to stress, depression, hopelessness, and suicidal ideation. Specifically, students who were depressed and hopeless had deficits in problem-solving skills, suggesting that training in these skills for the international students could help them cope with stress and depression.

Watson, Bell, and Chavez (1994) explored differences in problem solving in the area of conflict-handling skills among Mexican American and White non-Hispanic high school students and drop-outs. The White students generally used a competitive problem-solving approach to conflict management whereas the Mexican American students and drop-outs used a more cooperative problem-solving approach. These authors recommend that school systems provide opportunities for collaborative problem solving for all students, including students of various racial and ethnic groups.

In another area, a major thrust in the use of problem-solving training with diverse clients has been in HIV prevention. Bracho-de-Carpio, Carpio-Cedraro, and Anderson (1990) reported the use of an HIV educational prevention program based on problem solving with Hispanic families. Kelly and St. Lawrence (1990) described the usefulness of a problem-solving training model for HIV prevention with gay men. Schinke, Botvin, Orlandi, and Schilling (1990) have developed a culturally sensitive HIV prevention program incorporating problem solving with HIV facts and elements of ethnic pride. Their program is designed for African American and Hispanic American adolescents and can be used in both school and nonschool settings. An interactive microcomputer-based training approach using their model is also available (Schinke, Orlandi, Gordon, & Weston, 1989).

In an excellent empirical study of HIV prevention programs, 246 African American adolescents (28% male and 72% female) were given either education (EC) training only or behavioral skills training (BST) that involved correct condom use, sexual assertion, refusal, information, self-management, risk recognition, and problem solving (St. Lawrence et al., 1995). For example, participants identified problem situations they had encountered in the past as well as ones they anticipated in the future and practiced practical problem-solving skills to deal with these situations. The outcomes in lowering the participants' risk for HIV infection were evaluated on multiple measures at both post-intervention and at a one-year follow-up. These indicated significantly greater benefit for the youths in the skills training program than in the information only intervention. Although these results held for both female and male participants, some gender differences emerged. These authors conclude that because HIV risk behavior is influenced by a variety of "cognitive, interpersonal and situational determinants, multifaceted intervention approaches" are needed (p. 236). Also, for such training to be most effective with diverse groups of clients they must be "developmentally appropriate and culturally relevant" (p. 235).

Moncher and Schinke (1994) developed a group intervention for tobacco use prevention for Native American youth. The intervention included information on bicultural competence, coping skills, and problem-solving skills. Hammond and Yung (1991) developed a violence prevention program called PACT (Positive Adolescents Choices Training) targeted specifically at Black adolescents. In a pilot study of the PACT program, the participants' communication, problem-solving, and negotiation skills were improved.

Guidelines for Using Problem Solving with Diverse Groups of Clients

Problem solving has been shown to be an effective intervention strategy with diverse groups of clients in different ways, but there are some guidelines to consider in enhancing its effectiveness. First, there is evidence that the specific nature of problem-solving skills varies according to gender, race, and ethnic group. It is unwise to assume that diverse groups of clients will solve problems in the same way. The traditional Eurocentric and Androcentric model of problem solving that is individualistic is used widely by some male and Caucasian clients whereas many females and clients of color prefer a more collaborative and cooperative approach. As a result, both intervention and prevention programs that incorporate problem-solving training must be adapted for client characteristics such as age, gender, race, and ethnic affiliation to make problem solving both developmentally appropriate and culturally relevant.

Second, problem-solving training for clients from diverse cultures will be more effective for these clients when the training is conducted in a culturally sensitive manner—that is, in a way that respects the rituals and traditions of the client's culture, that promotes relevant cultural identity such as sexual and/or racial identity and also ethnic pride, that develops biracial and/or bicultural competence, and that helps these clients to acquire problem-solving skills without having to assimilate the norms of the mainstream culture.

SIX STAGES OF PROBLEM SOLVING

We propose the following stages enumerated by D'Zurilla (1986):

1. *Treatment rationale:* to provide the purpose and overview of problem-solving strategy.
2. *Problem orientation:* to assess the client's problem-solving coping style; to educate clients about maladaptive and facilitative problem-solving coping skills; to determine cognitive and emotional obstacles to problem solving, and then to train the client to overcome these, attacking the problem from many different vantage points and assessing the time, energy, and commitment needed to resolve the difficulty.
3. *Problem definition and formulation:* to help the client gather relevant and factual information for understanding the problems, to identify problem-focused and/or emotion-focused components of the issues, and to identify problem-solving goals for each one.
4. *Generation of alternative solutions:* to instruct the client to think about different ways to handle each problem goal and to use the deferment of judgment, quantity, and variety principles.
5. *Decision making:* to instruct the client to screen the list of alternative solutions, to evaluate (judge and compare) solution outcomes for each problem goal, and to select solution plans.
6. *Solution implementation and verification:* to encourage the client to carry out several alternative solutions simultaneously; to have the client self-monitor, evaluate, and reinforce the implementation of the solutions; and to help the client troubleshoot and recycle the problem-solving strategy if the solutions do not work.

Each of these parts is described in this section. A detailed description of these six components can be found in the Interview Checklist for Problem Solving at the end of the chapter.

Treatment Rationale

The rationale used in problem-solving therapy is that it attempts to strengthen the client's belief that problem solving can be an important coping skill for dealing with a variety of concerns requiring effective functioning. The following is an example of the rationale for problem-solving treatment.

Each of us is faced with both minor and important problems. Some problems are routine, such as trying to decide what to wear or what movie to see. Other problems are more stressful, such as dealing with a difficult relationship. One way to enhance responsibility and self-control is to learn techniques for solving our problems. To take the time, energy, and commitment necessary to solve or deal with problems immediately may relieve future frustration and pain created by a problem.

Here is an example of an overview of the procedure.

You will learn how to become aware of how you see a problem. You will look at how much control, time and effort, and commitment you feel you have for solving the problem.

We will need to gather information to understand and define the problems. Also, we'll need to look at what might prevent you from solving the problem. It is important that we explore a variety of solutions for solving the problem. After obtaining several solutions, you will decide which solutions feel most reasonable to implement simultaneously. Finally, you will implement the solution plans. (Pause) I'm wondering what your thoughts and feelings are regarding what I have described. Do you have any questions?

The rationale and treatment overview must be relevant for the client's culture.

Problem Orientation

After giving a rationale to the client, the counselor asks the client to describe how she or he typically solves problems. The counselor determines whether the client has a maladaptive or facilitative problem-solving style and then helps the client distinguish between these two coping styles. People with maladaptive coping styles blame themselves or others for the problem. These people often believe that something is wrong with them and may feel abnormal, hopeless, depressed, stupid, or unlucky (D'Zurilla, 1988). Maladaptive coping styles are often exhibited in persons who either minimize the benefits of problem solving or who maximize or exaggerate the losses that may occur from failure to solve the problem successfully. Individuals with poor problem-solving skills often perceive the problem as insolvable and so avoid dealing with it. Also, poor problem solvers may feel inadequate or incompetent and they prefer having someone else produce a solution (D'Zurilla, 1988). Some people have difficulty solving problems because they either never learned how or they feel the difficulty is too overwhelming.

The role of the counselor is to help the client engaging in a maladaptive problem-solving style to change his or her perception about problem solving. However, the counselor must do this in a culturally sensitive way, as there are differences in problem-solving styles across gender, age, race, ethnicity, and religious affiliation. It is critical not to equate maladaptive problem solving with non-Eurocentric world views. Instead of viewing problems as a threat or a personal inadequacy, the counselor helps the client see them as an opportunity for personal growth and self-improvement (D'Zurilla, 1988). Clients may feel that it is easier not to solve the problem and wait for things to get better. Counselors need to help these clients to realize that if a problem is not solved, it may very well come back to haunt them later (Peck, 1978). Clients need to believe that there is a solution to the problem and that they have the capacity and

self-control to find the solution independently and successfully (D'Zurilla, 1988). An expectation that one can cope with and solve a problem successfully will produce an ability to do so.

Problem solving takes time and energy. It is sometimes easier to avoid dealing with problems when they are influenced more by a person's feelings than by his or her reason. People often respond to problems impulsively and do not take time and effort to think about viable solutions. Problem solving requires time, energy, and commitment—a delay of gratification. The counselor needs to assess the client's willingness to spend time and energy, to be committed, and to delay gratification, and the counselor may have to motivate the client to make the necessary commitment to solve the problem.

Another component of problem orientation is discussing how the client's cognitions and emotions affect problem solving. Clients may be unmotivated to work on the problem because of how they think about it. Also, poor cognitions or self-talk such as, "it is their fault," "it will go away," or "I can't work on it" inhibits the motivation to work on a problem. The purpose of the therapy is to instruct and to train the client in positive coping methods, with the intent to overcome cognitive and/or emotional obstacles to problem solving (D'Zurilla, 1986, 1988). Strategies such as cognitive restructuring, reframing, stress inoculation, meditation, muscle relaxation or systematic desensitization, breathing exercises, and Yoga (see Chapters 15, 16, 17, 18 and 19) may help a client deal with cognitive and/or emotional barriers to problem solving. Once the sources of the client's resistance to problem solving have been minimized, the client is ready for the next phase of the problem-solving strategy.

Problem Definition and Formulation

The purpose of the defining and formulating step in problem solving is for the counselor to help the client gather as much objective information about the problem as possible. In cases where a client has a distorted cognitive view or perception of the problem, the counselor may have to use rational-emotive therapy or cognitive restructuring (see Chapter 15). The counselor explains that problem-solving strategy is a skill and a practical approach whereby a person attempts to identify, explore, or create effective and adaptive ways of coping with everyday problems (D'Zurilla, 1986). A problem can be viewed as a discrepancy between how a present situation is being experienced and how that situation should or could be experienced (D'Zurilla, 1988). The client needs to obtain relevant information about the problem by identifying the obstacles that are creating the discrepancy or are preventing effective responses for reducing the discrepancy. It is also important to examine antecedent conditions or unresolved problems that may be contributing to or causing the present problem or concern. Counselors may want to use the techniques presented in Chapter 9 for this step of problem definition.

Some therapists make an important distinction between problem-focused and emotion-focused problem definition (D'Zurilla, 1988). Problems that have problem-focused definitions center on problem-solving goals, with the purpose of changing the problem situation for the better. Problems with an emotion-focused definition concentrate on changing the client's personal reactions to the problem (D'Zurilla, 1988). Alternatively, D'Zurilla (1988) suggests that if the problem situation is assessed as *changeable,* the problem-focused definition should be emphasized in therapy. If problem-focused problems are *unchangeable,* the counselor helps the client deal with the client's reaction to the problem (D'Zurilla, 1988). In some cases, the client's problem may be first assessed as problem-focused but later the counselor and client discover the problem is unchangeable and the therapeutic focus then becomes changing the personal reaction. It has been our experience that it is best to include both problem-focused and emotion-focused goals in defining the client's problem. Again, problem-focused and emotion-focused definitions of a problem and corresponding goals seem to have some gender and cultural variation.

After the problem has been identified and defined (see Chapter 9), the counselor and the client set realistic emotion-focused and/or problem-focused goals (see Chapter 10). A goal is defined as what the client would like to have happen as a consequence of solving the problem. The goals should be realistic and attainable, and they should specify the type of behavior, level of behavior, and conditions under which the goal will facilitate solving the problem. The counselor should help the client identify obstacles that might interfere with problem-solving goals. Finally, the counselor should help the client to understand that the complexities involved in most problem situations usually require attacking the problem from many different vantage points (Nezu & Nezu, 1989). Establishing problem-solving goals will help the client with the next step in the therapy: creating alternative solutions.

Generation of Alternative Solutions

The purpose of the next stage of problem solving is to have the client generate as many alternative solutions to the problem as possible. The counselor instructs the client to think of *what* she or he could do, or *ways* to handle the problem. The counselor also instructs the client not to worry about *how* to go about making a plan work or how to put the solution into effect; that will come later. The client is instructed to allow her or his imagination to think of a great variety of new and original solutions, no matter how ridiculous the solution may seem. According to D'Zurilla (1986, 1988), the greater the quantity of alternative solutions that the client produces, the greater will be the quality of the solutions available for the client to choose. Similarly, when the client defers judgment or critical evaluation of the solutions, solutions of greater quality will be produced by the client.

After generating this list of alternative solutions to the problem, the client is asked to identify the number of different strategies represented. If too few strategies are represented, the client is instructed to generate more strategy solutions or more solutions for a specific strategy. This "free-wheeling" or brainstorming process is intended to filter out functional fixity, practicality, and feasibility in generating solutions. If there are several goals for the problem, the counselor encourages the client to generate several alternative solutions for each problem goal, the rationale being that most problems are complicated and a simple alternative is often inadequate.

Decision Making

The purpose of the decision-making step is to help the client decide on the best solution to solve the problem by judging and comparing all the alternatives. The client is first instructed to screen the list of available alternatives and to eliminate any solution that may be a risk or is not feasible (D'Zurilla, 1988). The best solutions are the ones that maximize benefits and minimize costs for the client's personal, immediate, and long-term welfare. The client is instructed to anticipate outcomes for each alternative (D'Zurilla, 1988) and then is asked to evaluate each solution using the following four criteria: (1) will the problem be solved, (2) what will be the status of the client's emotional well-being, (3) how much time and effort will be needed to solve the problem, and (4) what effect will it have on the client's overall personal and social well-being? (D'Zurilla, 1988). When working with diverse groups of clients, it is important not to impose your own values and culture on this process.

After the client selects and evaluates all the alternative solutions, D'Zurilla (1988) recommends that the client be instructed to answer the following three questions: (1) can the problem be solved, (2) is more information needed before a solution can be selected and implemented, and (3) what solution or combination of solutions should be used to solve the problem? If the problem cannot be solved with one of the existing solutions, the counselor may have to help the client redefine the problem and/or gather more information about the problem. If the previous three questions have been satisfactorily answered, the client is ready to implement the solution, as long as the chosen solution is consistent with the goal for solving the problem.

During the first five stages of the problem-solving strategy, the counselor assumes a directive role with the client to ensure a thorough application of the following steps: the problem orientation, problem definition, generation of alternatives, and decision making. The counselor assumes a less directive role with the client during the solution implementation stage. The therapeutic goal during the last stage of the strategy is to have the client become more responsible.

Solution Implementation and Verification

The purpose of solution implementation and verification in problem solving is to test the chosen solutions to the problem-solving goals and to verify whether these solutions solve the problem. The client simultaneously implements as many solutions as possible. According to D'Zurilla (1988), there are four parts to verifying whether the solution plan is working. The first part is to implement the chosen solution. If there are obstacles (behavioral deficits, emotional concerns, or dysfunctional cognitions) to implementing the solution, the client can acquire performance skills, defuse affective concerns, and restructure cognition to remove the obstacles.

Second, the client can use self-monitoring techniques (see Chapter 20) to assess the effects of the chosen solutions for solving the problem. The counselor instructs the client to keep a daily log or journal of the self-talk or emotional reactions to the chosen solutions. The self-talk or statements recorded in the journal can be rated on a scale where 5 = extremely negative, 0 = neutral, and −5 = extremely positive and the accompanying affect state also can be recorded, such as loved, depressed, frustrated, guilty, happy, or neutral—which can increase the level of emotional awareness.

Third, the client assesses whether the chosen solutions achieve the desired goals for solving the problem. This self-evaluation process is assessed in relationship to the solution in the following areas: (1) problem resolution; (2) emotional well-being; (3) time and effort exerted; (4) the ratio of total benefit to cost (D'Zurilla, 1988).

Finally, if the chosen solution meets all the criteria, the client engages in some form of self-reward (see Chapter 20) for having successfully solved the problem. However, if the chosen solutions do not solve the problem, the counselor and client try to pinpoint trouble areas and retrace the problem-solving steps. Some common trouble areas that clients have are emotional reaction to the problem, inadequate definition of the problem, unresolved antecedent issues to the problem, problem-focused instead of emotion-focused definition, and unsatisfactory solution choices.

D'Zurilla (1988) offers three cautions about the problem-solving strategy. One concern is the possible failure of the counselor to recognize when other strategies are more appropriate. A client with a serious concern or severe maladaptive behavior will require other strategies. For example, a depressed person may require intensive cognitive restructuring (see Chapter 15) before problem-solving therapy could be considered an appropriate strategy (D'Zurilla, 1988). The second caution is the danger of viewing problem-solving therapy as a "rational," "intellectual," or "head-trip therapy or exercise" rather than as a coping strategy that involves all three components of behavior, cognition, and affect (D'Zurilla, 1988). The problem-solving strategy should be viewed as an overall or general system for therapy that must include the emotional, behavioral, and cognitive modes of a person and be culturally relevant as well. The third caution is the potential failure of counselors to recognize that rapport with the client or a positive therapeutic relationship is a necessary condition for successful therapy (D'Zurilla, 1988). The ingredients of effective therapy (see Chapter 3) and the variables that enhance the therapeutic relationship (see Chapter 4) are important for successful application of any strategy; the problem-solving strategy is no exception. Sometimes problem solving can be difficult for clients who are not accustomed to thinking of long-range effects, such as many adolescents and some clients with severe trauma histories who don't allow themselves to think much into the future. Bly (1996) contends that many Euro-Americans in the United States are shortsighted in problem solving and lack the *vertical gaze*—that is, the Native American custom of looking ahead to the possible effects of a given solution for the next seven generations.

Finally, Nezu, Nezu, and Perri (1989, pp. 133–136) offer some guidelines for implementing the problem-solving strategy.

1. Training in problem solving should *not* be presented in a mechanistic manner.
2. The therapist should attempt to individualize the strategy and make it relevant to the specific client concern.

3. Homework and in vivo practice of the problem-solving components are crucial.
4. The therapist should be caring and sensitive to the client's concerns and feelings.
5. The judicious use of humor by the therapist can be an effective therapeutic tool.
6. The therapist needs continuously to ensure that an accurate assessment of the problem has been obtained.
7. The therapist should encourage the client to implement as many solutions as possible during treatment.
8. The therapist must make a thorough evaluation of the patient's abilities and limitations in order to implement a solution alternative. The evaluation would also include how much control the client has over the problem situation.
9. The therapist must determine whether the client's problem deals with problem-focused coping or emotion-focused coping or both.

The role of the counselor throughout the problem-solving process is to educate the client about the problem-solving strategy and to guide the client through the problem-solving steps. As we mentioned before, the counselor is less directive with the client during the last stage of the problem-solving process in order to help the client become more independent and take responsibility for applying the chosen problem solutions and verifying the effectiveness of the chosen solutions. The counselor can help the client to maintain these problem-solving skills and to generalize them to other concerns. The counselor also can assist the client in anticipating obstacles to solving strategies and can prepare the client for coping with them. The client should be able to cope fairly well if he or she takes the time to *examine objectively* her or his orientation to the problem, to carefully define the problem, to generate a variety of alternative solutions, and to make a decision about solution alternatives that are compatible with goals or desired outcomes. Solution implementation may be easier if the first four stages of the problem-solving process have been thoroughly and objectively processed. (See also Learning Activity #43).

MODEL EXAMPLE: PROBLEM-SOLVING THERAPY

In this model example, we present a narrative account of how problem-solving therapy might be used with a 49-year-old male client. Yoshi, an air traffic controller, has reported that he would like to decrease the stress he experiences in his job.

He believes that decreasing this stress will help his ulcer and help him cope better with his job. In addition to the physical signs of stress (insomnia), Yoshi also reports that he worries constantly about making mistakes in his job. He has also thought about taking early retirement, which would relieve him of the stress and worry.

Rationale

First, we explain to Yoshi that all of us face problems and sometimes we feel stuck because we don't know how to handle a problem. We tell Yoshi that solving problems as they occur can prevent future discomfort. We provide Yoshi with an overview of problem-solving therapy, telling him that we'll need to look at how he sees the problem and what obstacles there are to solving the problem. We tell him that we will need to define the problem, think of many different solutions for solving the problem, select several solutions to solve the problem, and try out the solutions and see how well the solutions solve the problem. We emphasize that problem solving is a collaborative, cooperative process. Finally, we confirm Yoshi's willingness to use problem-solving therapy and answer any questions he may have about the procedure.

Problem Orientation

We determine how Yoshi typically solves problems. We ask him to give an example of a problem he has had in the past and how and what he did to solve it. Then we describe for Yoshi the difference between maladaptive and facilitative problem solving. We explain to him that most people inherently have problem-solving ability but something blocks the use of it. We tell him that problem-solving therapy removes the blocks or obstacles that are maladaptive for good problem solving. We explain that healthy problem solving is the capacity to view problems as an opportunity. If Yoshi is encountering cognitive or emotional obstacles in his problem-solving attempts, we would introduce appropriate strategies to help remove them. Finally, we assess how much time, energy, and commitment Yoshi has for solving the problem.

Problem Definition and Formulation

We briefly describe the problem-solving strategy for Yoshi. We explain to him that we need to gather information about the problem, such as his thoughts and feelings about the problem, what unresolved issues contribute to the problem, how intense the problem is, what has been done to solve the problem, and when and where the problem occurs. We ask Yoshi what other information is needed to define the problem. If he has distorted views or perceptions of the problem, we would have to help him reframe his perception of the problem. We have to determine whether Yoshi's problem is problem-focused or emotion-focused or both. For example, we can probably change his emotional and cognitive reaction to the work situation and help him reduce the stress but he cannot change the job requirements unless he leaves or retires. We help Yoshi identify problem-solving goals or what he would like to have happen so that the problem would be solved. For Yoshi, one of the most attainable and realistic goals might be to reduce job stress.

Generation of Alternative Solutions

Yoshi is instructed to generate as many alternative solutions as possible for solving the problem. We inform him not to worry about how to go about making the alternatives work or how to put the solution into effect. Also, he is instructed to defer judgment about how good or feasible his ideas or solutions are until later, to generate as many alternatives as he can think of (because quantity produces quality), and to be creative and to think of nontraditional and unconventional solutions as well.

Decision Making

We instruct Yoshi to screen the list of alternatives and to use the following criteria for evaluating each solution: will the problem be solved with this solution; what will be the status of Yoshi's emotional well-being; how much time and effort will be needed to use the alternative solutions; and what will be Yoshi's overall personal and social well-being using each of these alternative solutions. Yoshi is reminded that it is important to evaluate each solution by answering these four criteria questions. Finally, he is instructed to select the best solutions that are compatible with the problem-solving goals and that "fit" best with his own culture.

Solution Implementation and Verification

We instruct Yoshi to try his chosen solutions for solving the problem. We also instruct him to self-monitor the alternative solutions he chose to solve the problem to determine their effectiveness. Suppose he chose to reduce his stress in the workplace by using meditation as one solution. We instruct Yoshi to self-monitor by keeping a written log or journal of the effectiveness of meditation using the following criteria questions: How effective is meditation in reducing job stress? How well does he feel emotionally about the meditative experience; is the time and effort spent with daily meditations worth it? Are there more benefits for using meditation than costs and what are your thoughts, feelings, and behavior in relationship to the solution implementation? He is instructed to complete the self-monitoring or journal each day just before bedtime. Yoshi is instructed to rate each of the criteria questions on a five-point scale with descriptive words for each point on the scale. We tell him that he needs to reward himself after successfully solving the problem (reducing the stress) by selecting rewarding things or activities. Also, we tell him to determine the best time to receive something or to engage in rewarding activity. If, for example, the meditation did not contribute to solving the problem, we instruct Yoshi to look at trouble areas that might be obstacles to solving the problem, such as his emotional reactions to the problem, the fact that the problem may not be well defined, or unresolved issues that may be contributing to the problem.

SUMMARY

Cognitive modeling is a procedure designed to help people learn how to use self-talk to enhance performance. In this strategy, implicit or covert responses as well as overt responses are modeled. Problem-solving therapy or training provides clients with a formalized system for viewing problems more constructively. As a treatment strategy, problem solving can be used alone or in conjunction with other treatment strategies presented in this book. Problem-solving has been used in a number of ways with diverse groups of clients. Some differences in problem-solving styles are apparent in client characteristics such as age, gender, race, and ethnicity. Culturally sensitive problem-solving training also includes elements of ethnic pride and bicultural competence.

In the next chapter we see how clients can be taught to alter self-defeating thoughts by learning to replace these with incompatible coping thoughts and skills. The strategy of cognitive restructuring, presented in Chapter 15, is directed toward replacing or reformatting self-defeating thoughts and schemata.

LEARNING ACTIVITY 43   *Problem Solving*

This learning activity provides an opportunity to try out problem solving. Think of a problem that you have, and apply the steps to problem solving to your problem. Do this in a quiet place when you will not be interrupted.

1. Determine how you solve problems. How is your approach impacted by your gender and culture?
2. Assess your problem-solving style. How does it reflect your world view(s)?
3. What cognitive and/or emotional obstacles do you have that might be barriers to solving problems?

4. How much time, energy, and commitment do you have for solving the problem?
5. Define your problem and determine whether your problem is problem focused or emotion focused or both.
6. Generate solutions for solving the problem. Be sure to think of a variety of solutions that fit your culture.
7. Select the best solutions using the criteria in #13 of the Interview Checklist for Problem Solving.
8. Implement your solutions to the problem or at least think about how to implement the solutions and think of a method for verifying the effectiveness of each solution.

POSTEVALUATION

Part One

Describe how you would use the seven components of cognitive modeling and self-instructional training to help Mr. Huang (from Chapter 8) initiate social contacts with his boss (Objective One). These are the seven components:

1. Rationale
2. Model of task and self-guidance
3. Overt external guidance
4. Overt self-guidance
5. Faded overt self-guidance
6. Covert self-guidance
7. Homework and follow-up

Feedback follows the evaluation.

Part Two

Objective Two asks you to demonstrate at least 16 out of 21 steps of the cognitive self-instructional modeling procedure with a role-play client. You can audiotape your interview or have an observer assess your performance, using the Interview Checklist for Cognitive Modeling that follows the Postevaluation.

Part Three

Objective Three asks you to identify accurately the steps of the problem-solving therapy represented by at least eight out of ten examples of counselor interview responses. For each of the following counselor responses, identify on paper which step of the problem-solving procedure is being used. There may be more than one counselor response associated with a step. The six major steps of problem solving are as follows:

1. Rationale for problem solving
2. Problem orientation
3. Problem definition and formulation
4. Generation of alternative solutions
5. Decision making
6. Solution implementation and verification

Feedback for the Postevaluation follows on page 371.

1. "Self-monitoring involves keeping a diary or log about your thoughts, feelings, and behaviors."
2. "To help you assess each solution, you can answer several questions about how effective the solution will be in solving the problem."
3. "Be creative and free-wheeling. Let your imagination go. Whatever comes into your mind write it down."
4. "What goals do you want to set for your emotional or personal reaction to the problem?"
5. "When you have concerns or problems, give me an example of the problem and describe how you typically solve it."
6. "Solving problems as they occur can prevent future discomfort."
7. "Most people have an ability to solve problems but often they block the use of it and become poor problem solvers."
8. "What unresolved issues may be contributing to the problem? When does the problem occur? Where does it occur?"
9. "Look over your list of solutions and see how much variety is on your list; think of new and original ones."
10. "You need to think about what you can reward yourself with after you complete this step."

(continued)

Part Four

Objective Four asks you to demonstrate 16 out of 19 steps associated with the problem-solving strategy in a role-play interview. You can audiotape your interview or have an observer rate it, using the Interview Checklist for Problem Solving on pp. 366–370.

Interview Checklist for Cognitive Modeling

Instructions: Determine which of the following leads the counselor used in the interview. Check each of the leads used. Some examples of counselor leads are provided in the right column; however, these are only suggestions.

Item	Examples of counselor leads
I. Rationale about Strategy	
_____ 1. Counselor provides a rationale for the strategy.	"This strategy is a way to help you do this task and also plan how to do it. The planning will help you perform better and more easily."
_____ 2. Counselor provides overview of strategy.	"We will take it step by step. First, I'll show you how to do it and I'll talk to myself aloud while I'm doing it so you can hear my planning. Then you'll do that. Gradually, you'll be able to perform the task while thinking through the planning to yourself at the same time."
_____ 3. Counselor checks client's willingness to use strategy.	"Would you like to go ahead with this now?"
II. Model of Task and Self-Guidance	
_____ 4. Counselor instructs client in what to listen and look for during modeling.	"While I do this, I'm going to tell you orally my plans for doing it. Just listen closely to what I say as I go through this."
_____ 5. Counselor engages in modeling of task, verbalizing self-guidance aloud, using language relevant to the client.	"OK, I'm walking in for the interview. [Counselor walks in.] I'm getting ready to greet the interviewer and then wait for his cue to sit down" [sits down].
_____ 6. Self-guidance demonstrated by counselor includes five components:	
_____ a. *Question* about demands of task	"Now what is it I should be doing in this situation?"
_____ b. *Answers* question by planning what to do	"I just need to greet the person, sit down on cue, and answer the questions. I need to be sure to point out why they should take me."
_____ c. *Focused attention* to task and *self-guidance* during task	"OK, just remember to take a deep breath, relax, and concentrate on the interview. Just remember to discuss my particular qualifications and experiences and try to answer questions completely and directly."
_____ d. *Coping self-evaluation* and, if necessary, *error correction*	"OK, now, if I get a little nervous, just take a deep breath. Stay focused on the interview. If I don't respond too well to one question, I can always come back to it."
_____ e. *Self-reinforcement* for completion of task	"OK, doing fine. Things are going pretty smoothly."
III. Overt External Guidance	
_____ 7. Counselor instructs client to perform task while counselor coaches.	"This time you go through the interview yourself. I'll be coaching you on what to do and on your planning."
_____ 8. Client performs task while counselor coaches by verbalizing self-guidance, changing *I* to *you*. Counselor's verbalization includes the five components of self-guidance:	"Now just remember you're going to walk in for the interview. When the interview begins, I'll coach you through it."

(continued)

POSTEVALUATION   **(continued)**

Item	Examples of counselor leads
III. Overt External Guidance (continued)	
_____a. Question about task	"OK, you're walking into the interview room. Now ask yourself what it is you're going to do."
_____b. Answer to question	"OK, you're going to greet the interviewer. [Client does so.] Now he's cuing you to sit down." [Client sits.]
_____c. Focused attention to task and self-guidance during task	"Just concentrate on how you want to handle this situation. He's asking you about your background. You're going to respond directly and completely."
_____d. Coping self-evaluation and error correction	"If you feel a little nervous while you're being questioned, just take a deep breath. If you don't respond to a question completely, you can initiate a second response. Try that now."
_____e. Self-reinforcement	"That's good. Now remember you want to convey why you should be chosen. Take your time to do that. [Client does so.] Great. Very thorough job."
IV. Overt Self-Guidance	
_____ 9. Counselor instructs client to perform task and instruct self aloud.	"This time I'd like you to do both things. Talk to yourself as you go through the interview in the same way we have done before. Remember, there are five parts to your planning. If you get stuck, I'll help you."
_____10. Client performs task while simultaneously verbalizing aloud self-guidance process. Client's verbalization includes five components of self-guidance:	
_____a. Question about task	"Now what is it I need to do?"
_____b. Answer to question	"I'm going to greet the interviewer, wait for the cue to sit down, then answer the questions directly and as completely as possible."
_____c. Focused attention and self-guidance	"Just concentrate on how I'm going to handle this situation. I'm going to describe why I should be chosen."
_____d. Coping self-evaluation and error correction	"If I get a little nervous, just take a deep breath. If I have trouble with one question, I can always come back to it."
_____e. Self-reinforcement	"OK, things are going smoothly. I'm doing fine."
_____11. If client's self-guidance is incomplete or if client gets stuck, counselor either	
_____a. Intervenes and cues client or	"Let's stop here for a minute. You seem to be having trouble. Let's start again and try to . . ."
_____b. Recycles client back through step 10	"That seemed pretty hard, so let's try it again. This time you go through the interview and I'll coach you through it."
_____12. Counselor gives feedback to client about overt practice.	"That seemed pretty easy for you. You were able to go through the interview and coach yourself. The one place you seemed a little stuck was in the middle, when you had trouble describing yourself. But overall, it was something you handled well. What do you think?"

(continued)

Item	Examples of counselor leads
V. Faded Overt Self-Guidance	
_____13. Counselor instructs client on how to perform task while whispering.	"This time I'd like you to go through the interview and whisper the instructions to yourself as you go along. The whispering may be a new thing for you, but I believe it will help you learn to do this."
_____14. Client performs task and whispers simultaneously.	"I'm going into the room now, waiting for the interviewer to greet me and to sit down. I'm going to answer the questions as completely as possible. Now I'm going to talk about my background."
_____15. Counselor checks to determine how well client performed.	
_____a. If client stumbled or left out some of the five parts, client engages in faded overt practice again.	"You had some difficulty with _____. Let's try this type of practice again."
_____b. If client performed practice smoothly, counselor moves on to next step.	"You seemed to do this easily and comfortably. The next thing is . . ."
VI. Covert Self-Guidance	
_____16. Counselor instructs client to perform task while covertly (thinking only) instructing self.	"This time while you practice, simply *think* about these instructions. In other words, instruct yourself mentally or in your head as you go along."
_____17. Client performs task while covertly instructing.	Only client's actions are visible at this point.
_____18. After practice (step 17), counselor asks client to describe covert instructions.	"Can you tell me what you thought about as you were doing this?"
_____19. On basis of client report (step 18)	
_____a. Counselor asks client to repeat covert self-guidance.	"It's hard sometimes to begin rehearsing instructions mentally. Let's try it again so you feel more comfortable with it."
_____b. Counselor moves on to homework.	"OK, you seemed to do this very easily. I believe it would help if you could apply this to some things that you do on your own this week. For instance. . . ."
VII. Homework	
_____20. Counselor instructs client on how to carry out homework. Instructions include	"What I'd like you to do this week is to go through this type of mental practice on your own."
_____a. What to do	"Specifically, go through a simulated interview where you mentally plan your responses as we've done today."
_____b. How much or how often to do the task	"I believe it would help if you could do this two times each day."
_____c. When and where to do it	"I believe it would be helpful to practice at home first, then practice at school [or work]."
_____d. A method for self-monitoring during completion of homework	"Each time you do this, make a check on this log sheet. Also, write down the five parts of the self-instructions you used."
_____21. Counselor arranges for a face-to-face or telephone follow-up after completion of homework assignment.	"Bring in your log sheets next week or give me a call at the end of the week and we'll go over your homework then."

*Observer Comments:*_____

(continued)

POSTEVALUATION   (continued)

Interview Checklist for Problem Solving

Instructions: Determine whether the counselor demonstrated each of the leads listed in the checklist. Check which leads were used.

Item	Examples of counselor leads
I. Rationale for Problem Solving	
_____ 1. Counselor explains purpose of problem-solving therapy in a way that is consistent with the client's culture.	"All of us are faced with little and big concerns or problems. Sometimes we feel stuck because we don't know how to handle a problem. This procedure can help you identify and define a problem and examine ways of solving the problem. You can be in charge of the problem instead of the problem being in charge of you. Solving problems as they occur can prevent future discomfort."
_____ 2. Counselor provides brief overview of procedure in a way that is consistent with the client's culture.	"There are five steps we'll do in using this procedure. Most problems are complex and to solve the problem we'll need to handle the problem from many different perspectives. First, we'll need to look at how you see the problem. We'll examine what are unhelpful and helpful problem-solving skills. Another part of this step is to explore how to overcome thoughts and feelings that could be obstacles to solving the problem. We'll also need to see how much time and energy you are willing to use to solve the problem. Second, we will define the problem by gathering information about it. Third, we'll want to see how many different solutions we can come up with for solving the problem. Next, we'll examine the solutions and decide which one to use. Finally, you will try out the chosen solutions and see how well they solve the problem. (Pause) What are your thoughts about what I have described? Do you have any questions?"
II. Problem Orientation	
_____ 3. Counselor determines how the client solves problems.	"When you have concerns or problems, give me an example of the problem and describe how you typically solve it."
_____ 4. Counselor describes the difference between maladaptive and facilitative problem solving, recognizing variations across gender, age, and culture and world views.	"Most people have the ability to solve problems but often they block the use of it. Problem-solving therapy helps to remove blocks or obstacles and helps bring important issues into focus. Problem-solving therapy provides a formalized system for viewing problems differently. People who don't solve problems very well may feel inadequate or incompetent to solve their problem. Often these people want to avoid the problem or want someone else to solve it. People sometimes feel it is easier not to solve the problem and things will get better. At times poor problem solvers feel hopeless, depressed, or unlucky. If you feel like a poor problem solver, we'll have to consider ways that make you feel like you are in charge. You can solve problems; they are a part of daily

(continued)

Item	Examples of counselor leads

II. Problem Orientation (continued)

living. There are usually a variety of solutions to every problem and you have the capacity to find the solution. It can be helpful to think of problems as an opportunity."

_____ 5. Counselor determines what cognitive and emotional obstacles the client might have as barriers to solving the problem.

"When you think about your problem, what thoughts do you have concerning the problem? What are you usually thinking about during this problem? Do you have any 'shoulds' or beliefs concerning the problem? What feelings do you experience when thinking about the problem? Are there any holdover or unfinished feelings from past events in your life that still affect the problem? How do your thoughts and feelings affect the problem and your ability to solve it? You may not be aware of it, but think about some past issues or unfinished business you may have as we do problem solving."

(If there are any obstacles, the counselor introduces a strategy or strategies [for example, Rational Emotive Therapy, cognitive restructuring, reframing, meditation, muscle relaxation, etc.] to help the client remove cognitive or emotional obstacles to problem solving.)

_____ 6. Counselor assesses the client's time, energy, and commitment to solving the problem.

"Any problem usually takes time, effort, and commitment to solve. But it is often important to solve the problem now rather than wait and solve it later—or not at all. It is important to know how committed you are to solving the problem. (Wait for the answer.) Also, solving a problem takes time; do you feel you have enough time to work on the problem? (Pause, wait for answer.) Thinking about and working on a problem can take a lot of energy. How energized do you feel about working on this problem?" (Pause for answer.)

III. Problem Definition and Formulation

_____ 7. Counselor describes the problem-solving strategy for the client in a culturally relevant way.

"People have problems or concerns. Some problems are minor and some are major. Problem-solving strategy is a skill and practical approach. People use problem solving to identify, explore, or create effective ways of dealing with everyday problems."

_____ 8. Counselor helps the client gather information about the problem. (The steps of problem identification presented in Chapter 9 can be used for this step of problem solving.)

"We want to gather as much information about the problem as we can. What type of problem is it? What thoughts do you have when the problem occurs? What feelings do you experience with the problem? How often does the problem occur or is it ongoing? What unresolved issues may be contributing to the problem? Who or what other people are involved with the problem? When does the problem occur? Where does it occur? How long has the problem been going on? How intense is the problem? What have you done to solve the problem? What obstacles can you identify that prevent you from solving the problem? Tell me, how do you see the problem?"

(continued)

POSTEVALUATION   (continued)

Item	Examples of counselor leads
III. Problem Definition and Formulation (continued)	
_____ 9. Counselor determines whether the client's problem is problem-focused or emotion-focused or both.	"What other information do we need to define the problem? What is your definition of the problem?" "From the way you have defined the problem, how can the problem be changed? What aspects of the problem can be changed? What emotional reactions do you have about the problem? How would you like to change your personal/emotional reaction to the problem? There may be some things about the problem you cannot change. Some problem situations are unchangeable. If there are aspects of the problem that are changeable, we will work on those things that can be changed. One thing we can change is your emotional or personal reaction to the problem."
_____10. Counselor helps the client identify culturally relevant problem-solving goals. (The steps in goal setting presented in Chapter 10 can be used for this step.)	"Now that we have identified and defined the problem, we need to set some goals. A goal is what you would like to have happen so that the problem would be solved. The goals should be things you can do or things that are attainable and realistic." "How many obstacles are there that prevent you from setting problem-solving goals? How can you remove these obstacles? What goals do you want to set for your emotional or personal reaction to the problem? What behaviors do you want to change? How much or what level of behavior is going to change? Under what condition or circumstance will the behavior change occur? What goals do you want to set for things that are changeable in the problem situation? What behaviors or goals do you want to set for yourself, the frequency of these behaviors, and in what problem conditions? These goals will help us in the next step of problem solving."
IV. Generation of Alternative Solutions	
_____11. Counselor presents the guidelines for generating alternative solutions.	"We want to generate as many alternative solutions for solving the problem as possible. We do this because problems are often complicated and a single alternative is often inadequate. We need to generate several alternative solutions for each problem-solving goal."
_____a. What options	"Think of what you could do or ways to handle the problem. Don't worry about how to go about making your plan work or how to put the solution into effect—you'll do that later."
_____b. Defer judgment	"Defer judgment about your ideas or solutions until later. Be loose and open to any idea or solution. You can evaluate and criticize your solutions later."
_____c. Quantity	"Quantity breeds quality. The more alternative solutions or ideas you can think of, the better. The more alternatives you produce, the more quality solutions you'll discover."

(continued)

Item	Examples of counselor leads
IV. Generation of Alternative Solutions (continued) _____d. Variety	"Be creative and free-wheeling. Let your imagination go. Whatever comes into your mind write it down. Allow yourself to think of a variety of unusual or unconventional solutions as well as more traditional or typical ones. Look over your list of solutions and see how much variety there is. If there is little variety on your list, generate more and think of new and original solutions."
V. Decision Making _____12. Counselor instructs the client to screen the list of alternative solutions.	"Now you need to screen and look over your list of alternative solutions for solving the problem. You want to look for the _best_ solutions. The best solutions are the ones that maximize benefits and minimize costs for your personal, social, immediate, and long-term welfare."
_____13. Counselor provides criteria evaluating each solution.	"To help you assess _each_ solution, you answer the following four questions: 1. Will the problem be solved with this particular solution? 2. By using this solution, what will be the status of my emotional well-being? 3. If I use this solution, how much time and effort will be needed to solve the problem? 4. What will be my overall personal and social well-being if I use this solution? Remember that it is important to evaluate _each_ solution by answering these four questions."
_____14. Counselor instructs the client to make a decision and select the best solutions compatible with problem-solving goals, and the client's culture.	"Select as many solutions that you think will work or solve the problem. Answer these questions: 1. Can the problem be solved reasonably well with these solutions? 2. Is more information about the problem needed before these solutions can be selected and implemented? Decide whether the solutions fit with the problem-solving goals.

(If the answer to questions one and two are Yes and No, respectively, move on to the next step. Answers of No to question one and Yes to question two may require recycling by redefining the problem, gathering more information, and determining problem obstacles.)

VI. Solution Implementation and Verification _____15. Counselor instructs the client to carry out chosen solutions.	"For the last stage of problem solving, try out the solutions you have chosen. If there are obstacles to trying out the solutions, we'll have to remove them. You can use several alternative solutions at the same time. Use as many solutions as you can."
_____16. Counselor informs client about self-monitoring strategy (described in Chapter 20).	"We'll need to develop a technique for you to see whether the solution solves the problem. Self-monitoring involves keeping a diary or log about your thoughts, feelings, and behavior. You can record these behaviors as

(continued)

POSTEVALUATION   **(continued)**

Item	Examples of counselor leads
VI. Solution Implementation and Verification (continued)	you implement your chosen solutions. We'll need to discuss what responses you'll record, when you'll record, and method of recording."
_____17. Counselor instructs the client to use these criteria to assess whether the solution achieves the desired goal for solving the problem.	"You'll need to determine whether your solution solves the problem. One way to do this is to ask yourself the following:
_____a. Problem resolved	"Did the solutions solve the problem?"
_____b. Emotional well-being	"How is your emotional well-being after you used the solutions?"
_____c. Time and effort exerted	"Was the time and effort you exerted worth it?"
_____d. Ratio of total benefits to total costs	"Were there more benefits for using the solutions than costs?"
_____18. Counselor instructs the client about self-reward (see Chapter 20 for a description of the self-reward strategy).	"You need to think about how you can reward yourself after successfully solving the problem. What types of things or activities are rewarding to you? When would be the best time to receive something or to engage in a rewarding activity?"
_____19. Counselor instructs the client on what to do if solutions do not solve the problem.	"When the solutions do not solve the problem, we need to look at some trouble areas that might be obstacles to solving the problem. What is your emotional reaction to the problem? The problem may not be defined well. There may be old unresolved problems contributing to the present problem."

SUGGESTED READINGS

D'Zurilla, T. J. (1986). *Problem-solving therapy: A social competence approach to clinical intervention.* New York: Springer.

D'Zurilla, T. J. (1988). Problem-solving therapies. In K. S. Dobson (Ed.), *Handbook of cognitive-behavioral therapies* (pp. 85–135). New York: Guilford.

D'Zurilla, T. J., & Nezu, A. M. (1990). Development and preliminary evaluation of the Social Problem-Solving Inventory (SPSI). *Psychological Assessment: A Journal of Consulting and Clinical Psychology, 2,* 156–163.

Gorrell, J. (1993). Cognitive modeling and implicit rules: Effects on problem-solving performance. *American Journal of Psychology, 106,* 51–65.

Gutkin, T. B. (1993). Cognitive modeling: A means for achieving prevention in school-based consultation. *Journal of Educational and Psychological Consultation, 4,* 179–183.

Nezu, A. M., & Nezu, C. M. (Eds.). (1989). Clinical decision making in behavior therapy: A problem-solving perspective. Champaign, IL: Research Press.

Nezu, A. M., Nezu, C. M., & Perri, M. B. (1989). *Problem-solving therapy for depression: Theory, research and clinical guidelines.* New York: Wiley.

Paye, B. D., & Manning, B. H. (1990). The effect of cognitive self-instructions on preservice teachers' anxiety about teaching. *Contemporary Educational Psychology, 15,* 261–267.

Rose, S. D., & LeCroy, C. W. (1991). Group methods. In F. H. Kanfer & A. P. Goldstein (Eds.), *Helping people change* (pp. 422–453). New York: Pergamon Press.

Schinke, S., Botvin, G., Orlandi, M., & Schilling, R. (1990). African American and Hispanic American adolescents, HIV infection, and prevention intervention. *AIDS Education and Prevention, 2,* 305–312.

St. Lawrence, J., Brasfield, T., Jefferson, K., Alleyne, E., O'Bannon, R., & Shirley, A. (1995). Cognitive-behavioral interventions to reduce African American adolescents' risk for HIV infection. *Journal of Consulting and Clinical Psychology, 63,* 221–237.

Watson, D., Bell, P., & Chavez, E. (1994). Conflict handling skills used by Mexican American and White non-Hispanic students in the educational system. *High School Journal, 78,* 35–39.

Yang, B., & Clum, G. (1994). Life stress, social support, and problem-solving skills predictive of depression symptoms, hopelessness, and suicide ideation in an Asian student population: A test of a model. *Suicide and Life Threatening Behavior, 24,* 127–139.

FEEDBACK
POSTEVALUATION

Part One

1. *Rationale*
   First, you would explain the steps of cognitive modeling and self-instructional training to Mr. Huang. Then you would explain how this procedure could help him practice and plan the way he might approach his boss.

2. *Model of task and self-guidance*
   You would model for Mr. Huang a way he could approach his boss to request a social contact. You would model the five parts of the self-guidance process: (1) the question about what he wants to do, (2) the answer to the question in the form of planning, (3) focused attention on the task and guiding himself through it, (4) evaluating himself and correcting errors or making adjustments in his behavior in a coping manner, and (5) reinforcing himself for adequate performance. In your modeling it is important to use language that is relevant to Mr. Huang.

3. *Overt external guidance*
   Mr. Huang would practice making an approach or contact while you coach him through the five parts of self-guidance as just described.

(continued)

4. *Overt self-guidance*
   Mr. Huang would practice making a social contact while verbalizing aloud the five parts of the self-guidance process. If he got stuck, you could prompt him, or you could have him repeat this step or recycle step 3.

5. *Faded overt self-guidance*
   Mr. Huang would engage in another practice attempt, only this time he would whisper the five parts of the self-guidance process.

6. *Covert self-guidance*
   Mr. Huang would make another practice attempt while using the five parts of the self-guidance process covertly. You would ask him afterward to describe what happened. Additional practice with covert self-guidance or recycling to step 4 or 5 might be necessary.

7. *Homework*
   You would instruct Mr. Huang to practice the self-guidance process daily before actually making a social contact with his boss.

Part Two

Rate an audiotape of your interview or have an observer rate you, using the Interview Checklist for Cognitive Modeling on pages 363–365.

Part Three

1. Solution Implementation and Verification
2. Decision Making
3. Generation of Alternative Solutions
4. Problem Definition and Formulation
5. Problem Orientation
6. Rationale for Problem Solving
7. Problem Orientation
8. Problem Definition and Formulation
9. Generation of Alternative Solutions
10. Solution Implementation and Verification

Part Four

Rate an audiotape of your interview or have someone else rate it, using the Interview Checklist for Problem Solving on pages 366–370.

CHAPTER

COGNITIVE THERAPY AND COGNITIVE RESTRUCTURING

Since the first edition of this book almost 20 years ago, considerable changes have emerged in the concept and practice of cognitive therapy. Cognitive therapy includes a variety of techniques and approaches that assume problematic emotions and behaviors to be the result of how clients perceive and interpret events. Three levels of cognition appear to play a significant role in producing emotional and behavioral difficulties: (1) automatic thoughts, (2) schemas or underlying assumptions, and (3) cognitive distortions (Beck, 1967). Clinical improvement depends on changes in these three levels of cognition, or *cognitive restructuring.*

OBJECTIVES

1. Identify and describe the six components of cognitive restructuring from a written case description.
2. Teach the six major components of cognitive restructuring to another person, or demonstrate these components in a role-play interview.

DEVELOPMENTS IN COGNITIVE THERAPY

Mahoney (1995) suggests that "the major conceptual developments in the cognitive psychotherapies over the past three decades have been (a) the differentiation of rationalist and constructivist approaches to cognition; (b) the recognition of social, biological, and embodiment issues; (c) the reappraisal of unconscious processes; (d) an increasing focus on self and social systems; (e) the reappraisal of emotional and experiential processes; and (f) the contribution of the cognitive psychotherapies to the psychotherapy integration movement" (p. 6). Some of the contributors to these developments are listed in Table 15-1. This table gives a historical context to the development of cognitive therapy.

Albert Ellis (1962) founded a form of cognitive therapy based on the cognitive control of irrational thinking. Some of the additional foundation for the therapeutic application of

cognitive behavior modification started with the covert conditioning described by Homme (1965) and Cautela's (1966) covert sensitization. The work of Homme and Cautela laid the base from which others developed the cognitive strategies described in Chapter 13 about guided imagery and the desensitization procedure presented in Chapter 19 of this edition. Meichenbaum and Goodman's (1971) self-instructional training demonstrated the influence between verbal self-instruction and behavior described in the cognitive modeling section of the previous chapter. In 1967, Beck presented his thesis that emotional disorders are the consequence of distorted thinking or unrealistic cognitive appraisal of life events. A description of cognitive restructuring similar to the ideas of A. T. Beck (1967, 1976) and J. Beck (1995) is presented in this chapter. The anxiety management training developed by Suinn and Richardson (1971) is based on the process of reciprocal inhibition. Clients are taught to use relaxation (Chapter 17, meditation and muscle relaxation) to control their anxious feelings. The problem-solving therapy we described in Chapter 14 (D'Zurilla & Goldfried, 1971) is a cognitive form of self-control in which the client explores response alternatives for coping with problem situations. Spivack, Platt, and Shure (1976) developed an interpersonal cognitive problem-solving treatment approach based on the same skills first identified by D'Zurilla and Goldfried. Mahoney and Thoresen (1974) contributed self-control to the literature of cognitive therapies in general and specifically to self-management, presented in Chapter 20. Finally, the symbolic and participant modeling approaches discussed in Chapter 12 and self-efficacy described in the last section of Chapter 20 are based on the social learning theory of Bandura (1986). This chapter describes an intervention considered the cornerstone of cognitive-behavioral approaches: *cognitive restructuring.* More recently, it has been referred to as *cognitive replacement,* or the shift from old mental pictures and rules to new ones. More recent discussions of the practice of cognitive therapy are provided by Beck (1995), Goldfried (1995), Leahy (1996), and Mahoney (1995).

TABLE 15-1. Contributors to the development of cognitive therapies

Therapy	Author and work
Cognitive therapies	
Rational-Emotive Therapy	Ellis, A. (1962). *Reason and emotion in psychotherapy.* New York: Stuart.
Covert Conditioning	Homme, L. (1965). Perspectives in psychology: XXIV. Control of coverants, the operants of the mind. *Psychological Record, 15,* 501–511.
Covert Sensitization	Cautela, J. R. (1966). The treatment of compulsive behavior by covert sensitization. *Psychological Record, 16,* 33–41.
Self-Instructional Training	Meichenbaum, D. H., & Goodman, J. (1971). Training impulsive children to talk to themselves. *Journal of Abnormal Psychology, 77,* 127–132.
Cognitive Therapy	Beck, A. T. (1967). *Depression: Causes and treatment.* Philadelphia: University of Pennsylvania Press.
Anxiety Management Training	Suinn, R. M., & Richardson, F. (1971). Anxiety management training: A nonspecific behavior therapy program for anxiety control. *Behavior Therapy, 2,* 498–510.
Problem-Solving Therapy	D'Zurilla, T. J., & Goldfried, M. R. (1971). Problem-solving and behavior modification. *Journal of Abnormal Psychology, 78,* 107–126.
	Spivack, G., Platt, J. J., & Shure, M. B. (1976). *The problem-solving approach to adjustment.* San Francisco: Jossey-Bass.
Self-Control	Mahoney, M. J., & Thoresen, C. E. (1974). *Self-control: Power to the person.* Monterey, CA: Brooks/Cole.
Stress Inoculation Training	Meichenbaum, D., Turk, D., & Burstein, S. (1975). The nature of coping with stress. In I. G. Sarason & C. D. Spielberger (Eds.), *Stress and anxiety: Vol. II.* New York: Wiley.
Social Learning	Bandura, A. (1985). *Social foundations of thought and actions: A social cognitive theory.* Englewood Cliffs, NJ: Prentice Hall.

USES OF COGNITIVE RESTRUCTURING

Cognitive restructuring (CR) has its roots in the elimination of distorted or invalid inferences (Beck, Rush, Shaw, & Emery, 1979), disputing irrational thoughts or beliefs (Ellis, 1979), and promoting rule-governed behavior (described by Meichenbaum, 1977). In many ways, cognitive restructuring is considered an essential component of almost every cognitive-behavioral procedure. A sample of the cognitive restructuring research is listed in Table 15-2. Inspection of this list of research reveals a variety of concerns to which cognitive restructuring was applied. We have listed several studies using CR to help control generalized anxiety, anxiety of elementary school children, and test and math anxiety. CR has also been used in a wide range of situations involving, among others, arthritis patients, athletes in competition, cancer patients, crime victims, and adult children of alcoholic parents (see Table 15-2).

MULTICULTURAL APPLICATIONS OF COGNITIVE THERAPY AND COGNITIVE RESTRUCTURING

In recent years, the use of cognitive therapy with diverse groups of clients has received increased attention, although as Hays (1995) notes, such multicultural applications are still too few in number. Critiquing cognitive therapy from a multicultural viewpoint, Hays (1995) observed that the values embraced in this approach are those supported by the

TABLE 15-2. Cognitive restructuring research

Anger

Deffenbacher, J. L., Thwaites, G. A., Wallance, T. L., & Oetting, E. R. (1994). Social skills and cognitive-relaxation approaches to general anger reduction. *Journal of Counseling Psychology, 41,* 386–396.

Deffenbacher, J. L., Oetting, E. R., Huff, M. E., & Thwaites, G. A. (1995). Fifteen-month follow-up of social skills and cognitive-relaxation approaches to anger reduction. *Journal of Counseling Psychology, 42,* 400–405.

Anxiety

Barlow, D. H., Rapee, R. M., & Brown, T. A. (1992). Behavioral treatment of generalized anxiety disorder. *Behavior Therapy, 23,* 551–570.

Hiley-Young, B. (1990). Facilitating cognitive-emotional congruence in anxiety disorders during self-determined cognitive change: An integrative model. *Journal of Cognitive Psychotherapy, 4,* 225–236.

Strumpf, J. A., & Fodor, I. (1993). The treatment of test anxiety in elementary school-age children: Review and recommendations. *Child and Family Behavior Therapy, 15,* 19–42.

Sud, A. (1993). Efficacy of two short term cognitive therapies for test anxiety. *Journal of Personality and Clinical Studies, 9,* 39–46.

Vance, W. R., & Watson, T. S. (1994). Comparing anxiety management training and systematic rational restructuring for reducing mathematics anxiety in college students. *Journal of College Student Development, 35,* 261–266.

Arthritis

Basler, H. D., & Rehfisch, H. P. (1991). Cognitive-behavioral therapy in patients with ankylosing spondylitis in a German self-help organization. *Journal of Psychosomatic Research, 35,* 345–354.

Athletic competition

Greenspan, M. J., & Feltz, D. L. (1989). Psychological interventions with athletes in competitive situations: A review. *Sport Psychologist, 3,* 219–236.

Bulimia nervosa

Hsu, L. G., Santhouse, R., & Chesler, B. E. (1991). Individual cognitive behavioral therapy for bulimia nervosa: The description of a program. *International Journal of Eating Disorders, 10,* 273–283.

Cancer

Cella, D. F. (1990). Health promotion in oncology: A cancer wellness doctrine. *Journal of Psychosocial Oncology, 8,* 17–31.

Crime victims

Davis, R. C. (1991). A crisis intervention program for crime victims. *Response to the Victimization of Women and Children, 14*(79, No. 2), 7–11.

Children of alcoholics

Ferstein, M. E., & Whiston, S. C. (1991). Utilizing RET for effective treatment of adult children of alcoholics. *Journal of Rational Emotive and Cognitive Behavior Therapy, 9,* 39–49.

Cultural differences

Giannini, A. J., Quinones-Delvalle, R. M., & Blackshear, G. The use of cognitive restructuring in cross-cultural therapy. *Psychiatric Forum, 15,* 30–32.

Depression

Pace, T. M., & Dixon, D. N. (1993). Changes in depressive self-schemata and depressive symptoms following cognitive therapy. *Journal of Counseling Psychology, 40,* 288–294.

Emerson, P., West, J. D., & Gintner, G. G. (1991). An Adlerian perspective on cognitive restructuring and treating depression. *Journal of Cognitive Psychotherapy, 5,* 41–53.

Epilepsy

Upton, D., & Thompson, P. J. (1992). Effectiveness of coping strategies employed by people with chronic epilepsy. *Journal of Epilepsy, 5,* 119–127.

Gambling

Sharpe, L., & Tarrier, M. (1992). A cognitive-behavioral treatment approach for problem gambling. *Journal of Cognitive Psychotherapy, 6,* 193–203.

Sylvain, C., & Ladouceur, R. (1992). Corrective cognition and gambling habits in players of video poker. *Canadian Journal of Behavioural Science, 24,* 479–489.

Headache

Mehta, M. (1992). Biobehavioral intervention in recurrent headaches in children. *Headache Quarterly, 3,* 426–430.

Lascelles, M. A., Cunningham, S. J., McGrath, P. J., & Sullivan, M. J. (1989). Teaching coping strategies to adolescents with migraine. *Journal of Pain and Symptom Management, 4,* 135–145.

Insomnia

Chambers, M. J., & Alexander, S. D. (1992). Assessment and prediction of outcome for a brief behavioral insomnia treatment program. *Journal of Behavior Therapy and Experimental Psychiatry, 23,* 289–297.

Lifestyle transitions

Brammer, L. M. (1992). Coping with life transitions. Special Section: Counselling and health concerns. *International Journal for the Advancement of Counselling, 15,* 239–253.

Marital therapy

Halford, W. K., Sanders, M. R., & Behrens, B. C. (1993). A comparison of the generalization of behavioral marital therapy and enhanced behavioral marital therapy. Special Section: Couples and couple therapy. *Journal of Consulting and Clinical Psychology, 61,* 51–60.

(continued)

TABLE 15-2. Cognitive restructuring research (continued)

Marital therapy

Baucom, D. H., Sayers, S. L., & Sher, T. G. (1990). Supplementing behavioral marital therapy with cognitive restructuring and emotional expressiveness training: An outcome investigation. *Journal of Consulting and Clinical Psychology, 58,* 636–645.

Memory and belief

Lachman, M. E., Weaver, S. L., Bandura, M., & Elliott, E. (1992). Improving memory and control beliefs through cognitive restructuring and self-generating strategies. *Journal of Gerontology, 47,* P293–P299.

Claridge, K. E. (1992). Restructuring memories of abuse: A theory-based approach. *Psychotherapy, 29,* 243–252.

Obsessive-compulsive disorder

Sookman, D., Pinard, G., & Beauchemin, N. (1994). Multidimensional schematic restructuring treatment for obsessions: Theory and practice. *Journal of Cognitive Psychotherapy, 8,* 175–194.

Pain

Basler, H. D. (1993). Group treatment for pain and discomfort. Special Issue: Psychosocial aspects of rheumatic diseases. *Patient Education and Counseling, 20,* 167–175.

Grant, L. D., & Haverkamp, B. E. (1995). A cognitive-behavioral approach to chronic pain management. *Journal of Counseling & Development, 74,* 25–31.

Spence, S. H., & Kennedy, E. (1989). The effectiveness of a cognitive behavioral treatment approach to work-related upper limb pain. *Behaviour Change, 6,* 12–23.

Subramanian, K. (1994). Long-term follow-up of a structured group treatment for the management of chronic pain. Special Issue: Empirical research on the outcomes of social work with groups. *Research on Social Work Practice, 4,* 208–223.

Panic

Barlow, D. H. (1992). Cognitive-behavioral approaches to panic disorder and social phobia. Annual Meeting of the American Psychiatric Association: Integrated treatment of panic disorder and social phobia, Washington, DC. *Bulletin of the Menninger Clinic, 56,* 14–28.

Craske, M. G., Brown, T. A., & Barlow, D. H. (1991). Behavioral treatment of panic disorder: A two-year follow-up. *Behavior Therapy, 22,* 289–304.

Hajjar, M. (1989). Combination of clomipramine (anafranil) and cognitive restructuring: Interventions in a chronic case of panic anxiety. *Medical Psychotherapy: An International Journal, 2,* 189–192.

Parent training

Gammon, E. A., & Rose, S. D. (1991). The Coping Skills Training Program for parents of children with developmental disabilities: An experimental evaluation. *Research on Social Work Practice, 1,* 244–256.

Perfectionistic behavior

Halgin, R. P., & Leahy, P. M. (1989). Understanding and treating perfectionistic college students. *Journal of Counseling and Development, 68,* 222–225.

Phobia

Ball, S. G., & Otto, M. W. (1994). Cognitive-behavioral treatment of choking phobia: 3 case studies. *Psychotherapy and Psychosomatics, 62,* 207–211.

Heard, P. M., Dadds, M. R., & Conrad, P. (1992). Assessment and treatment of simple phobias in children: Effects on family and marital relationships. *Behaviour Change, 9,* 73–82.

Psychosomatic complaints

Hellman, C. J., Budd, M., Borysenko, J., & McClelland, D. C. (1990). A study of the effectiveness of two group behavioral medicine interventions for patients with psychosomatic complaints. *Behavioral Medicine, 16,* 165–173.

Relapse with alcoholics

Watson, L. (1991). Paradigms of recovery: Theoretical implications for relapse prevention in alcoholics. *Journal of Drug Issues, 21,* 839–858.

Self-esteem

Horan, J. (1996). Effects of computer-based cognitive restructuring on rationally mediated self-esteem. *Journal of Counseling Psychology, 43,* 371–375.

Smoking

Haaga, D. A. F., & Allison, M. L. (1994). Thought suppression and smoking relapse: A secondary analysis of Haaga (1989). *British Journal of Clinical Psychology, 33,* 327–331.

Stress

Hains, A. A. (1992). Comparison of cognitive-behavioral stress management techniques with adolescent boys. *Journal of Counseling and Development, 70,* 600–605.

Tallant, S., Rose, S. D., & Tolman, R. M. (1989). New evidence for the effectiveness of stress management training in groups. Special Issue: Empirical research in behavioral social work. *Behavior Modification, 13,* 431–446.

Tolman, R. M., & Rose, S. D. (1989). Teaching clients to cope with stress: The effectiveness of structured group stress management training. Special Issue: Advances in group work research. *Journal of Social Service Research, 13,* 45–66.

Worry

Robinson, E. L. (1989). The relative effectiveness of cognitive restructuring and coping desensitization in the treatment of self-reported worry. *Journal of Anxiety Disorders, 3,* 197–207.

Writing

Nedate, Y., & Tagami, F. (1994). Effects of instruction through writing on subjective well-being modification: When adopting a cognitive restructuring approach. *Japanese Journal of Counseling Science, 27,* 21–26.

status quo of the mainstream culture. As an example, the emphasis on self-control that fits with the Euro-American value of personal autonomy may be empowering for some clients but may also "imply placing blame on the individual for problems that are previously a result of unjust social conditions" (Hays, 1995, p. 311; Ivey, Ivey, & Simek-Morgan, 1993).

Cognitive therapy and cognitive restructuring have also been critiqued by feminist therapists (Brown, 1994; Kantrowitz & Ballou, 1992). In the early days of feminism, cognitive-behavioral therapy was seen as admirable "for teaching women new ways of behaving" (Brown, 1994, p. 55). This was in the era when individual change was viewed as the primary solution for societal problems. Current models of feminist therapy, multicultural therapy, and ecological therapy do not agree with the assertion that if you change an oppressed individual's way of thinking, the presenting problems will automatically be resolved. Further, as Kantrowitz and Ballou (1992) suggest, the "rational thinking" orientation of cognitive therapy and cognitive restructuring reinforces worldviews and cognitive processes that are stereotypically both Euro-American and masculine; other worldviews and cognitive processes are at the least overlooked and at the worst, devalued. Thus, cognitive processing styles of some women and some persons of color may be rejected. In addition, as Brown (1994) observes, a useful response to the cognitive therapy notion of "irrational thinking" is the suggestion "that for some people, in some places and at some times, these supposedly universally pathological conditions might be reasonable, or even life saving" (p. 61). Moreover, cognitive therapy and cognitive restructuring, as a cornerstone of the procedure, challenge one's beliefs and thoughts. As Kantrowitz and Ballou (1992) suggest, this challenging may not fit with some culture and gender socialization patterns (p. 81). For example, Asians "have been taught to create emotional harmony and avoid conflict in accordance with their cultural norms" (Kantrowitz & Ballou, 1992, p. 81).

In light of all this, Hays (1995) does assert that cognitive therapy can be useful with diverse groups of clients if cultural adaptations in the intervention are made. We have found useful adaptations of cognitive restructuring with diverse groups of clients. Several of these are listed in Table 15-3.

Ahijevych and Wewers (1993) have suggested the use of cognitive restructuring as an intervention for nicotine-dependent African American women who want to stop smoking. Addressing a different problem, Hatch and Paradis (1993) have developed a 12-week group treatment incorpo-rating cognitive restructuring, breathing, and relaxation for African American women with panic disorder. They found that the women in the group particularly valued audiovisual aids and self-help material but noted the small number of African Americans represented in television programs as well as among the group therapists who could be role models. An African American woman who had completed successful treatment for panic attacks was invited to speak to the group. Addressing racial issues and providing access to other African Americans for both education and support seem to be critical parts of this cognitive intervention.

Earlier, we addressed the relevance of cognitive restructuring for Asians. Iwamasa (1993) contends that cognitive restructuring can be culturally compatible for some Asian American clients, especially those who are well educated and achievement oriented. He notes that part of the appeal of this intervention is that it is structured (versus unstructured), emphasizes thoughts and behaviors, and does not require the Asian American client to oppose the traditional Asian value of revealing very personal and/or familial difficulties to strangers.

Johnson and Ridley (1992) have described an adaptation of cognitive restructuring for clients who sought "Christian counseling." The clients were encouraged to challenge problematic beliefs by using Biblical scriptures as the basis for the disputation. These authors contended that cognitive restructuring can be adapted to culture-specific values of some Christian clients.

Increasingly, cognitive restructuring is being used with the elderly, especially those with major depressive disorders (Thompson, 1996). This use of the technique is especially important because many older people have serious side effects with some antidepressant medications. The best attempts at using cognitive restructuring with older persons have presented the intervention as an educational rather than a therapeutic experience (Freiberg, 1995), with sensitivity to the older clients' fears and biases about disclosing problems and dealing with their beliefs about being old (Arean, 1993). Group modes of cognitive therapies are especially useful because they provide greater social involvement and support (Arean, 1993; Freiberg, 1995). Also, the delivery of the intervention may need to be modified depending on the client's hearing and seeing abilities and other special needs (Thompson, 1996). Cognitive therapy has also been used to improve rehabilitation outcomes in elderly persons (Lopez & Mermelstein, 1995) and to facilitate life-review work of the elderly (Weiss, 1995).

Cognitive restructuring also has been used with gay men and lesbian women. Ussher (1990) and Kuehlwein (1992)

TABLE 15-3. Multicultural cognitive therapy research

African American clients	Ahijevych, K., & Wewers, M. (1993). Factors associated with nicotine dependence among African American women cigarette smokers. *Research in Nursing and Health, 16,* 283–292. Hatch, M., & Paradis, C. (1993). Panic disorder with agoraphobia. A focus on group treatment with African Americans. *The Behavior Therapist, 16,* 240–242. Haley, W., Roth, D., Coleton, M., & Ford, G. (1996). Appraisal, coping and social support as mediators of well-being in Black and White family caregivers of patients with Alzheimer's disease. *Journal of Consulting and Clinical Psychology, 64,* 121–129.
Asian American clients	Iwamasa, G. Y. (1993). Asian Americans and cognitive behavioral therapy. *The Behavior Therapist, 16,* 233–235.
Christian clients	Johnson, W. B., & Ridley, C. R. (1992). Brief Christian and non-Christian rational-emotive therapy with depressed Christian clients: An exploratory study. *Counseling and Values, 36,* 220–229.
Clients with physical challenges	Ellis, A. (1997). Using rational emotive behavior therapy techniques to cope with disability. *Professional Psychology, 28,* 17–22.
Elderly clients	Arean, P. A. (1993). Cognitive behavioral therapy with older adults. *The Behavior Therapist, 16,* 236–239. Dick, L. P., & Gallagher-Thompson, D. (1995). Cognitive therapy with the core beliefs of a distressed, lonely caregiver. *Journal of Cognitive Psychotherapy, 9,* 215–227. Lopez, M., & Mermelstein, R. (1995). A cognitive-behavioral program to improve geriatric rehabilitation outcome. *Gerontologist, 35,* 696–700. Thompson, L. W. (1996). Cognitive-behavioral therapy and treatment for late-life depression. *Journal of Clinical Psychiatry, 57,* 29–37. Weiss, J. (1995). Cognitive therapy and life-review therapy. *Journal of Mental Health Counselors, 17,* 157–172.
Gay men	Kuehlwein, K. T. (1992). Working with gay men. In A. Freemen & F. M. Dattillio (Eds.), *Comprehensive casebook of cognitive therapy* (pp. 249–255). New York: Plenum Ussher, J. (1990). Cognitive behavioral couples therapy with gay men referred for counseling in an AIDS setting: A pilot study. *AIDS-Care, 2,* 43–51.
Latino clients	Organista, K., Dwyer, E. V., & Azocar, F. (1993). Cognitive behavioral therapy with Latino outpatients. *The Behavior Therapist, 16,* 229–232.
Lesbian clients	Wolfe, J. L. (1992). Working with gay women. In A. Freeman & F. M. Dattillio (Eds.), *Comprehensive casebook of cognitive therapy* (pp. 249–255). New York: Plenum.
Low-income clients	Miranda, J., & Dwyer, E. V. (1993). Cognitive behavioral therapy for disadvantaged medical patients. *The Behavior Therapist, 16,* 226–228.
Native American clients	Renfrey, G. S. (1992). Cognitive-behavior therapy and the Native American client. *Behavior Therapy, 23,* 321–340.

have used cognitive restructuring to help gay male clients examine and correct internalized heterosexist beliefs and thoughts. Wolfe (1992) has described the use of cognitive restructuring with a lesbian client dealing with both parental and social discrimination as a result of her sexual orientation.

Organista, Dwyer, and Azocar (1993) describe some very specific modifications of cognitive restructuring for Latino clients. First, it is important to have a linguistic match between predominantly Spanish-speaking clients and their therapists. Second, Organista and colleagues have found consistent issues and theories for Latino clients related to marriage and family and acculturation stress. Culture and gender-related issues are common in Latina women and often produce depression, partly because of the culture's emphasis for Latina women on *marianismo*—a cultural trait that values a Latina's role in the family as one who is self-sacrificing and willing to endure suffering—and also because of the concept of *guadar*—to hold in rather than to express anger. These authors do use cognitive restructuring, but with Latino clients, they recommend a one-step rather than a multistep disputational process to challenge errors in thinking such as "yes, but."

Miranda and Dwyer (1993) discuss the use of cognitive therapy with low-income medical patients. They also note the importance of having bilingual and multicultural treatment staff available. They use treatment manuals to teach cognitive restructuring and recommend that the content and reading level be adapted to the client. For example, rather than using the term *generalization,* they use the phrase "thinking all bad things means everything will be bad" (Miranda & Dwyer, 1993, p. 227). They also note the importance of dealing with psychosocial stressors as well as cognitions. They found that group rather than individual cognitive treatment is more effective for most of their clients.

Renfrey (1992) discusses the use of cognitive therapy with Native American clients. Renfrey (1992) notes that current mental health needs of many Native Americans are great, largely because of acculturation and deculturation stressors brought on by the European American culture. Native Americans' mental health needs also are underserved. Renfrey (1992) recommends that a therapist collaborate with traditional healers in the Native American community. He believes that culturally sensitive cognitive therapy can be useful because it is specific and direct, involves homework, and focuses on altering present actions rather than emotional states. He recommends making an initial assessment of the client's acculturation status prior to any intervention because

this variable will affect treatment process and outcome. At the very least, cognitive restructuring must be offered in a way that promotes bicultural competence, so that treatment begins with enhancement of the Native American's traditional identity and focuses on skills to help the client meet the demands of the indigenous culture, the mainstream culture, and the transculture. (Renfrey, 1992, p. 330).

Guidelines for Using Cognitive Restructuring with Diverse Groups of Clients

As Hays (1995) has noted, cognitive therapy and cognitive restructuring are potentially quite applicable to diverse groups of clients, depending on the way the procedure is implemented and on the "therapist's sensitivity to diverse perspectives" (p. 312). We offer the following guidelines for using cognitive restructuring in a culturally sensitive way.

First, be very careful about the language you use when describing client cognitions. Although we don't recommend the use of the terms *rational* and *irrational* with any client, we consider these terms—and even others like *maladaptive* or *dysfunctional*—to be particularly inappropriate for women, gays, lesbians, clients of color, and all others who feel marginalized from the mainstream culture. These terms can only further diminish a sense of self-efficacy and marginalization.

Second, present a rationale for cognitive restructuring that is educational rather than therapeutic to help remove the stigma some clients of some cultural groups may have learned about mental health treatment. A didactic approach that includes specific and direct homework assignments is useful.

Third, adapt the language presented in cognitive restructuring to the client's primary language; age; educational level; and hearing, seeing, and reading abilities. Avoid jargon. Consider streamlining the procedure, focusing on one or two steps rather than multistep processes such as challenging self-defeating thoughts. Provide examples of skills and coping thoughts that are bicultural and transcultural.

Fourth, utilize and/or collaborate with bilingual, ethnically similar helpers and/or traditional healers who can help you address issues of psychosocial stressors, race, and discrimination. Remember that for clients who feel marginalized, addressing these issues is as important as addressing issues of internal cognition. Also, consider the usefulness for some of these clients of cognitive restructuring offered in a group rather than an individual setting.

SIX COMPONENTS OF COGNITIVE RESTRUCTURING

Our presentation of cognitive restructuring reflects the research presented in Table 15-2 and the suggested readings at the end of this chapter as well as our own adaptations of it based on clinical usage. We present cognitive restructuring in six major parts:

1. Rationale: purpose and overview of the procedure
2. Identification of client thoughts during problem situations
3. Introduction and practice of coping thoughts
4. Shifting from self-defeating to coping thoughts
5. Introduction and practice of positive or reinforcing self-statements
6. Homework and follow-up

Each of these parts is described in this section. A detailed description of these six components can be found in the Interview Checklist for Cognitive Restructuring at the end of the chapter and in Learning Activity #44.

Treatment Rationale

The rationale used in cognitive restructuring attempts to strengthen the client's belief that "self-talk" can influence performance and particularly that self-defeating thoughts or negative self-statements can cause emotional distress and can interfere with performance. Examples used should be relevant to the client's gender and culture.

Rationale.   The following general performance anxiety rationale can be used with clients with different concerns. You can fashion the rationale based on the specific client problem.

> One of our goals is for you to become aware of your thoughts or what you say to yourself that seems to maintain your anxiety when you are doing this activity (or while you are performing). Once we have identified these automatic thoughts, we can replace or change them. The thoughts about your performance are probably contributing to your anxiety. The performance situation may create these automatic thoughts, or perhaps the feelings you have about the situation create these thoughts. In either case, the thoughts or the feelings create physiological responses in your body and these responses as well as your feelings and thoughts influence your performance. When we become aware of these automatic thoughts, we can deal with them by changing what you think about.

Overview.   Here is an example of an overview of the procedure:

> We will learn how to deal with your automatic thoughts by becoming aware of when the thoughts occur or discovering what you say to yourself and what these internal self-statements are. Awareness of self-defeating automatic thoughts is one of the first steps in changing and decreasing the anxiety while performing. Once we know what the self-defeating statements are, they can act as a red flag or ring a bell for you to shift your self-talk to more self-enhancing performance statements. In other words, we will generate self-enhancing thoughts to shift to when you become aware of the automatic self-defeating thoughts. By shifting to the self-enhancing statements, your physiological and emotional responses will also become self-enhancing, and this will help you perform with less self-defeating anxiety. We will learn how to shift to self-enhancing statements while you are performing, and before and after your performances.

Contrast of Self-Defeating and Self-Enhancing Thoughts. In addition to providing a standard rationale such as the one just illustrated, the cognitive restructuring procedure should be prefaced by some contrast between self-enhancing, or rational, thoughts and self-defeating, or irrational, thoughts. This explanation may help clients discriminate between their own self-enhancing and self-defeating thoughts during treatment. Many clients who could benefit from cognitive restructuring are all too aware of their self-defeating thoughts but are unaware of or unable to generate self-enhancing thoughts. Providing a contrast may help them see that they can develop more realistic thinking styles.

One way to contrast these two types of thinking is to model some examples of both positive, enhancing self-talk and negative, defeating self-talk. These examples can come out of your personal experiences or can relate to the client's problem situations. Again, providing culturally relevant examples is important. The examples might occur *before, during,* or *after* a problem situation. For example, you might say to the client that in a situation that makes you a little uptight, such as meeting a person for the first time, you could get caught up in very negative thoughts:

Before meeting:
"What if I don't come across very well?"
"What if this person doesn't like me?"
"I'll just blow this chance to establish a good relationship."
During meeting:
"I'm not making a good impression on this person."
"This person is probably wishing our meeting were over."
"I'd just like to leave and get this over with."
"I'm sure this person won't want to see me after this."

After meeting:

"Well, that's a lost cause."

"I can never talk intelligently with a stranger."

"I might as well never bother to set up this kind of meeting again."

"How stupid I must have sounded!"

In contrast, you might demonstrate some examples of positive, self-enhancing thoughts about the same situation:

Before meeting:

"I'm just going to try to get to know this person."

"I'm just going to be myself when I meet this person."

"I'll find something to talk about that I enjoy."

"This is only an initial meeting. We'll have to get together more to see how the relationship develops."

During meeting:

"I'm going to try to get something out of this conversation."

"This is a subject I know something about."

"This meeting is giving me a chance to talk about _____ ."

"It will take some time for me to get to know this person, and vice versa."

After meeting:

"That went OK; it certainly wasn't a flop."

"I can remember how easy it was for me to discuss topics of interest to me."

"Each meeting with a new person gives me a chance to see someone else and explore new interests."

"I was able just to be myself then."

Influence of Self-Defeating Thoughts on Performance. The last part of the rationale for cognitive restructuring should be an *explicit* attempt to point out how self-defeating thoughts or negative self-statements are unproductive and can influence emotions and behavior. You are trying to convey to the client that we are likely to believe and to act on whatever we tell ourselves. However, it is also useful to point out that, in some situations, people don't *literally* tell themselves something. In many situations, our thoughts are so well learned that they reflect our core beliefs or schemas. For this reason, you might indicate that you will often ask the client to monitor or log what happens during actual situations between counseling sessions.

The importance of providing an adequate rationale for cognitive restructuring cannot be overemphasized. If one begins implementing the procedure too quickly, or without the client's agreement, the process can backfire. One way to prevent difficulty in implementing the procedure is to enhance the client's self-efficacy. The counselor can do this by practicing with the client in the therapy session sufficiently that the client is comfortable with the shifting facets of the procedure. With repeated practice, the client can gain enough experience that the self-enhancing thoughts become almost as automatic as the self-defeating ones. Practice helps loosen the grip on perseverative, self-defeating thoughts and enables the client to formulate experiences more realistically (Beck, 1993). Also, repeated practice can enhance the client's self-efficacy with the procedure. The counselor should not move ahead until the client's commitment to work with the strategy is obtained.

Identification of Client Thoughts in Problem Situations

Assuming that the client accepts the rationale provided about cognitive restructuring, the next step involves an analysis of the client's thoughts in anxiety-provoking or distressing situations. Both the range of situations and the content of the client's thoughts in these situations should be explored.

Description of Thoughts in Problem Situations. Within the interview, the counselor should query the client about the particular distressing situations encountered and the things the client thinks about before, during, and after these situations. The counselor might say something like "Sit back and think about the situations that are really upsetting to you. What are they?" and then "Can you identify exactly what you are thinking about or telling yourself before you go to _____? What are you thinking during the situation? And afterward?"

In identifying negative or self-defeating thoughts, the client might be aided by a description of possible cues that compose a self-defeating thought. The counselor can point out that a negative thought may have a "worry quality" such as "I'm afraid . . . ," or a "self-oriented quality" such as "I won't do well." Negative thoughts may also include elements of catastrophizing ("If I fail, it will be awful") or exaggerating ("I *never* do well" or "I *always* blow it"). Clients can identify the extent to which self-defeating thoughts contribute to situational anxiety by asking themselves, "Do I (1) make unreasonable demands of myself, (2) feel that others are evaluating my performance or actions, and (3) forget that this is only one small part of my life?"

Modeling of Links between Events and Emotions. If clients have trouble identifying negative thoughts, Guidano (1995) suggests that they engage in what he calls the movieola technique. Clients are instructed to run scenes of the situation in their heads. "Then, as if [he or she] were in an editing room, the client is instructed to 'pan' the scenes, going back and forth in slow motion, thereby allowing the client to 'zoom in' on a single scene, to focus on particular aspects" of the scene (p. 157). The counselor may need to point out that the thoughts are the link between the situation and the resulting emotion and ask the client to notice explicitly what this link seems to be. If the client is still unable to identify thoughts, the counselor can model this link, using either the client's situations or situations from the counselor's life. For example, the counselor might say this:

> Here is one example that happened to me. I was a music major in college, and several times a year I had to present piano recitals that were graded by several faculty members and attended by faculty, friends, and strangers. Each approaching recital got worse—I got more nervous and more preoccupied with failure. Although I didn't realize it at the time, the link between the event of the recital and my resulting feelings of nervousness was things I was thinking that I can remember now—like "What if I get out there and blank out?" or "What if my arms get so stiff I can't perform the piece?" or "What if my shaking knees are visible?" Now can you try to recall the specific thoughts you had when you felt so upset about _____ ?

Identification of Client Schemas Underlying Distorted Thoughts.

Recent developments in cognitive therapy stress the role of examining a client's *schema,* defined by Goldfried (1995) as a "cognitive representation of one's past experiences with situations or people" (p. 55). Schemas are thought to provide a way of understanding the distortions in automatic thoughts and beliefs that clients make in processing information about themselves, other persons, and their environments. For example, if a client has a schema of being inferior and abandoned, she or he will use selective attention to focus on information related to failure, isolation, and rejection, and will dismiss information related to success, relational connection, and acceptance (Leahy, 1996). It is believed that schemas about self and others grow out of object representations in early childhood, specifically the early attachment styles (Bowlby, 1988) described in Chapter 2 (particularly the preoccupied, dismissing, and fearful styles).

The social-political context of a child's life can also influence the developing schema. (Consider this in reference to the various worldviews discussed in Chapter 11.) For example, children who suffer or witness traumatic events and instances of discrimination and oppression probably develop different schemas than children who are not impacted in those ways. In a longitudinal study, Nolen-Hoeksema, Girgus, and Seligman (1992) find that stressful life events caused children to develop a negative cognitive style that was predictive of later depression if additional stressors occurred.

Leahy (1996) notes that challenging a client's schema can be difficult in that it was established in childhood, when the client was operating at a preoperational level of processing "marked by egocentrism, centration, magical thinking and moral realism" (p. 192). Still he believes it is useful to explore and understand the particular schemas that underlie the client's distorted thoughts and beliefs. He provides some sample questions to illustrate this identification process:

How did your parents (siblings, peers, teachers) teach you that you are _____? (Fill in the word that best describes the client's schema.)

When you learned the schema, you were 5 years old. Do you think it is wise to guide your life by what a 5-year-old thinks?

What evidence is there that you are not _____ ? Or that you are _____ ? (Fill in the word that best describes the client's schema.)

What is the consequence of demanding this of yourself? Is it ever OK to be helpless? To fail? To depend on others? To be disapproved of?

How would you challenge your mother and father, now that you are an adult, if they were to describe you as _____? (Fill in the word that best describes the client's schema.) (Leahy, 1996, p. 194).

Client Modeling of Thoughts.

The counselor can also have the client identify situations and thoughts by monitoring and recording events and thoughts outside the interview in the form of homework. An initial homework assignment might be to have the client observe and record for one week at least three self-defeating statements and emotions a day in the stressful situation. For each day of the week, the client could record on a daily log the self-defeating self-statements and emotions for each situation in which these statements were noted (see Figure 15-1).

Using the client's data, the counselor and client can determine which of the thoughts were self-enhancing and productive and which were self-defeating and unproductive. The counselor should try to have the *client* discriminate

Date: _____	Week: _____	
Situation	*Emotions*	*Self-Defeating Statements*
1.	1.	1.
2.	2.	2.
3.	3.	3.

FIGURE 15-1. Example of daily log

between the two types of statements and identify why the negative ones are unproductive. The identification serves several purposes. First, it is a way to determine whether the client's present repertory consists of both positive and negative self-statements or whether the client is generating or recalling only negative thoughts. These data may also provide information about the client's degree of distress in a particular situation. If some self-enhancing thoughts are identified, the client becomes aware that alternatives are already present in his or her thinking style. If no self-enhancing thoughts are reported, this is a cue that some specific attention may be needed in this area. The counselor can demonstrate how the client's unproductive thoughts can be changed by showing how self-defeating thoughts can be restated more constructively.

Introduction and Practice of Coping Thoughts

At this point in the procedure, there is a shift in focus from the client's self-defeating thoughts to other kinds of thoughts that are incompatible with the self-defeating ones. These incompatible thoughts may be called coping thoughts, coping statements, or coping self-instructions. They are developed for each client. There is no attempt to have all clients accept a common core of rational beliefs, as is often done in rational-emotive therapy.

Introduction and practice of coping statements is, as far as we know, crucial to the overall success of the cognitive restructuring procedure. Rehearsal of coping statements, by itself, is almost as effective as the combination of identifying negative statements and replacing these with incompatible coping thoughts.

Explanation and Examples of Coping Thoughts. The purpose of coping thoughts should be explained clearly. The client should understand that it is difficult to think of failing at an experience (a self-defeating thought) while concentrating on just doing one's best, regardless of the outcome (a coping thought). The counselor could explain the purpose and use of coping thoughts like this:

> So far we've worked at identifying some of the self-defeating things you think during _____. As long as you're thinking about those kinds of things, they can make you feel anxious. But as soon as you replace these with coping thoughts, then the coping thoughts take over, because it is almost impossible to concentrate on both failing at something and coping with the situation at the same time. The coping thoughts help you to manage the situation and to cope if you start to feel overwhelmed.

The counselor should also model some examples of coping thoughts so that the client can clearly differentiate between a self-defeating and a coping thought. Some examples of coping thoughts to use *before* a situation might be these:

"I've done this before, and it is never as bad as I think."
"Stay calm in anticipating this."
"Do the best I can. I'm not going to worry how people will react."
"This is a situation that can be a challenge."
"It won't be bad—only a few people will be there."

Examples of coping thoughts to use *during* a situation include these:

"Focus on the task."
"Just think about what I want to do or say."
"What is it I want to accomplish now?"
"Relax so I can focus on the situation."
"Step back a minute, take a deep breath."
"Take one step at a time."
"Slow down, take my time, don't rush."
"OK, don't get out of control. It's a signal to cope."

If you go back and read over these lists of coping examples, you may note some subtle differences among them. There can be four types of coping statements. (1) *Situational coping statements* help the client reduce the potential level of threat or severity of the anticipated situation. Examples of this first type are "It won't be too bad," or "Only a few people will be watching me." (2) Other coping statements refer more to the plans, steps, or behaviors the person will need to demonstrate during the situation, such as "concentrate on what I want to say or do," "think about the task," or "what do I want to accomplish?" These are called *task-oriented coping statements.* (3) *Coping with being overwhelmed* is another type such as: "keep cool," "stay calm," or "relax, take a deep breath." (4) A fourth type of coping statement, which we call *positive self-statements,* is used to have clients reinforce or encourage themselves for having coped. These include such self-instructions as "Great, I did it," or "I managed to get through that all right." Positive self-statements can be used during a stressful situation and especially after the situation. The use of positive self-statements in cognitive restructuring is described in more detail later in this chapter.

In explaining about and modeling potential coping thoughts, the counselor may want to note the difference between *coping* and *mastery* thoughts. Coping thoughts are ones that help a client deal with or manage a situation, event, or person adequately. Mastery thoughts are ones that are directed toward helping a person "conquer" or master a situation in almost a flawless manner. For some clients, mastery self-instructions may function as perfectionistic standards that are, in reality, too difficult to attain. For these clients, use of mastery thoughts can make them feel more pressured rather than more relieved. For these reasons, we recommend that counselors avoid modeling the use of mastery self-statements and also remain alert to clients who may spontaneously use mastery self-instructions in subsequent practice sessions during the cognitive restructuring procedure.

Client Examples of Coping Thoughts.   After providing some examples, the counselor should ask the client to think of additional coping statements. The client may come up with self-enhancing or positive statements she or he has used in other situations. The client should be encouraged to select coping statements that feel most natural. Clients can identify coping thoughts by discovering convincing counterarguments for their unrealistic thoughts.

Client Practice.   Using these client-selected coping statements, the counselor should ask the client to practice verbalizing coping statements aloud. This is very important because most clients are not accustomed to using coping statements. Such practice may reduce some of the client's initial discomfort and can strengthen confidence in being able to produce different "self-talk." In addition, clients who are "formally" trained to practice coping statements systematically may use a greater variety of coping thoughts, may use more specific coping thoughts, and may report more consistent use of coping thoughts in vivo.

At first, the client can practice verbalizing the individual coping statements she or he selected to use before and during the situation. Gradually, as the client gets accustomed to coping statements, the coping thoughts should be practiced in the natural sequence in which they will be used. First, the client would anticipate the situation and practice coping statements before the situation to prepare for it, followed by practice of coping thoughts during the situation—focusing on the task and coping with feeling overwhelmed.

It is important for the client to become actively involved in these practice sessions. The counselor should try to ensure that the client does not simply rehearse the coping statements by rote. Instead, the client should use these practices to try to internalize the meaning of the coping statements. One way to encourage more client involvement and self-assertion in these practice attempts is to suggest that the client pretend he

or she is talking to an audience or a group of persons and needs to talk in a persuasive, convincing manner to get his or her point across.

Shifting from Self-Defeating to Coping Thoughts

After the client has identified negative thoughts and has practiced alternative coping thoughts, the counselor introduces rehearsal of shift from self-defeating to coping thoughts during stressful situations. Practice of this shift helps the client use a self-defeating thought as a cue for an immediate switch to coping thoughts.

Counselor Demonstration of Shift.   The counselor should model this process before asking the client to try it. This gives the client an accurate idea of how to practice this shift. Here is an example of a counselor modeling for a high school student who constantly "freezes up" in competitive situations.

> OK, I'm sitting here waiting for my turn to try out for cheerleader. Ooh, I can feel myself getting very nervous. [anxious feeling] Now, wait, what am I so nervous about? I'm afraid I'm going to make a fool of myself. [self-defeating thought] Hey, that doesn't help. [cue to cope] It will take only a few minutes, and it will be over before I know it. Besides, only the faculty sponsors are watching. It's not like the whole school. [situation-oriented coping thoughts]
>
> Well, the person before me is just about finished. Oh, they're calling my name. Boy, do I feel tense. [anxious feelings] What if I don't execute my jumps? [self-defeating thought] OK, don't think about what I'm not going to do. OK, start out, it's my turn. Just think about my routine—the way I want it to go. [task-oriented coping thoughts]

Client Practice of the Shift.   After the counselor demonstration, the client should practice identifying and stopping self-defeating thoughts and replacing them with coping thoughts. The counselor can monitor the client's progress and coach if necessary. Rehearsal of this shift involves four steps:

1. The client imagines the stressful situation or carries out his or her part in the situation by means of a role play.
2. The client is instructed to recognize the onset of any self-defeating thoughts and to signal this by raising a hand or finger.
3. Next, the client is told to stop these thoughts, or reframe these thoughts.
4. After the self-defeating thought is stopped, the client immediately replaces it with the coping thoughts. The client should be given some time to concentrate on the

coping thoughts. Initially, it may be helpful for the client to verbalize coping thoughts; later, this can occur covertly.

As the client seems able to identify, stop, and replace the self-defeating thoughts, the counselor can gradually decrease the amount of assistance. Before homework is assigned, the client should be able to practice and carry out this shift in the interview setting in a completely self-directed manner.

Introduction and Practice of Reinforcing Self-Statements

The last part of cognitive restructuring involves teaching clients how to reinforce themselves for having coped. This is accomplished by counselor modeling and client practice of positive, or reinforcing, self-statements. Many clients who could benefit from cognitive restructuring report not only frequent self-defeating thoughts but also few or no positive or rewarding self-evaluations. Some clients may learn to replace self-defeating thoughts with task-oriented coping ones and feel better but not satisfied with their progress. The purpose of including positive self-statements in cognitive restructuring is to help clients learn to praise or congratulate themselves for signs of progress. Although the counselor can provide social reinforcement in the interview, the client cannot always be dependent on encouragement from someone else when confronted with a stressful situation.

Purpose and Examples of Positive Self-Statements.
The counselor should explain the purpose of reinforcing self-statements to the client and provide some examples. An explanation might sound like this:

> You know, Joan, you've really done very well in handling these situations and learning to stop those self-defeating ideas and to use some coping thoughts. Now it's time to give yourself credit for your progress. I will help you learn to encourage yourself by using rewarding thoughts, so that each time you're in this situation and you cope, you also give yourself a pat on the back for handling the situation and not getting overwhelmed by it. This kind of self-encouragement helps you to note your progress and prevents you from getting discouraged.

Then the counselor can give some examples of reinforcing self-statements:

"Gee, I did it."
"Hey, I handled that OK."
"I didn't let my emotions get the best of me."
"I made some progress, and that feels good."
"See, it went pretty well after all."

Client Selection of Positive Self-Statements.
After providing examples, the counselor should ask the client for additional positive statements. The client should select those statements that feel suitable. This is particularly important in using reinforcing statements, because the reinforcing value of a statement may be very idiosyncratic.

Counselor Demonstration of Positive Self-Statements.
The counselor should demonstrate how the client can use a positive self-statement after coping with a situation. Here is an example of a counselor modeling the use of positive self-statements during and after a stressful situation. In this case, the client was an institutionalized adolescent who was confronting her parents in a face-to-face meeting.

> OK, I can feel them putting pressure on me. They want me to talk. I don't want to talk. I just want to get the hell out of here. [self-defeating thought] Slow down, wait a minute. Don't pressure yourself. Stay cool. [coping with being overwhelmed] Good. That's better. [positive self-statement]
> Well, it's over. It wasn't too bad. I stuck it out. That's progress. [positive self-statement]

Client Practice of Positive Self-Statements.
The client should be instructed to practice using positive self-statements during and after the stressful situation. The practice occurs first within the interview and gradually outside the interview with in vivo assignments.

Homework and Follow-Up

Although homework is an integral part of every step of the cognitive restructuring procedure, the client ultimately should be able to use cognitive restructuring whenever it is needed in actual distressing situations. The client should be instructed to use cognitive restructuring in vivo but cautioned not to expect instant success. Clients can be reminded of the time they have spent playing the old tape over and over in their heads and of the need to make frequent and lengthy substitutions with the new tape. The client can monitor and record the instances in which cognitive restructuring was used over several weeks.

The counselor can facilitate the written recording by providing a homework log sheet that might look something like Figure 15-2. The client's log data can be reviewed at a follow-up session to determine the number of times the client is using cognitive restructuring and the amount of progress that has occurred. The counselor can also use the follow-up session to encourage the client to apply the procedure to stressful situations that could arise in the future. This may encourage the client to generalize the use of cognitive

Directions: When you notice your mood getting worse, ask yourself, "What's going through my mind right now?" and as soon as possible jot down the thought or mental image in the Automatic Thought column.

Date/time	Situation	Automatic thought(s)	Emotion(s)	Adaptive response	Outcome
	1. What actual event or stream of thoughts, or daydreams or recollection led to the unpleasant emotion? 2. What (if any) distressing physical sensations did you have?	1. What thought(s) and/or image(s) went through your mind? 2. How much did you believe each one at the time (0 to 100%)?	1. What emotion(s) (sad/anxious/angry/ etc.) did you feel at the time? 2. How intense (0–100%) was the emotion?	1. (optional) What cognitive distortion did you make? 2. Use questions at bottom to compose a response to the automatic thought(s). 3. How much do you believe each response?	1. How much do you now believe each automatic thought? 2. What emotion(s) do you feel now? How intense (0–100%) is the emotion? 3. What will you do (or did you do)?
Friday 2/23 10 AM Tuesday 2/27 12 PM Thursday 2/29 5 PM	Talking on the phone with Donna Studying for my exam Thinking about my economics class tomorrow Noticing my heart beating fast and my trouble concentrating	She must not like me any more 90% I'll never learn this 100% I might get called on and I won't give a good answer 80% What's wrong with me?	Sad 80% Sad 95% Anxious 80% Anxious 80%		

Questions to help compose an alternative response: (1) What is the evidence that the automatic thought is true? Not true? (2) Is there an alternative explanation? (3) What's the worse that could happen? Could I live through it? What's the best that could happen? What's the most realistic outcome? (4) What's the effect of my believing the automatic thought? What could be the effect of my changing my thinking? (5) What should I do about it? (6) If _____[friend's name] was in the situation and had this thought, what would I tell him/her?

FIGURE 15-2. Example of homework log sheet *SOURCE: From Cognitive Therapy: Basics and Beyond,* by J. S. Beck, p. 126. Copyright 1995 by The Guilford Press. Reprinted by permission.

restructuring to situations other than those that are presently considered problematic.

Occasionally, a client's level of distress may not diminish even after repeated practice of restructuring self-defeating thoughts. In some cases, negative self-statements do not precede or contribute to the person's strong feelings. Some emotions may be classically conditioned and therefore treated more appropriately by a counterconditioning procedure, such as systematic desensitization (see Chapter 19). However, even in classically conditioned fears, cognitive processes may also play some role in maintaining or reducing the fear.

When cognitive restructuring does not reduce a client's level of distress, depression, or anxiety, the counselor and client may need to redefine the problem and goals. Perhaps the focus of treatment may need to be more on external psychosocial stressors rather than on internal events. The therapist should consider the possibility that his/her assessment has been inaccurate, and that there are, in fact, no internal sentences that are functionally tied to this particular client's problem. Remember that assessment of initial problems may not always turn out to be valid or accurate.

Assuming that the original problem assessment is accurate, perhaps a change in parts of the cognitive restructuring procedure is necessary. Here are some possible revisions:

1. The amount of time the client uses to shift from self-defeating to coping thoughts and to imagine coping thoughts can be increased.

2. The coping statements selected by client may not be very helpful; you may need to help the client change the type of coping statements.

3. Cognitive restructuring should be supplemented either with additional coping skills, such as deep breathing or relaxation, or with skill training.

Another reason for failure of cognitive restructuring may be that the client's problem behaviors result from errors in encoding rather than errors in reasoning. In the next chapter, we describe a strategy—reframing—that is designed to alter encoding or perceptual errors.

MODEL DIALOGUE: COGNITIVE RESTRUCTURING

We demonstrate with Joan, whom we met earlier and who is having problems in math class. The interview will be directed toward helping Joan replace self-defeating thoughts with coping thoughts. This is the "nuts and bolts" of cognitive restructuring.

1. *Counselor:* Good to see you again, Joan. How did your week go?
   Client: Pretty good. I did a lot of practice. I also tried to do this in math class. It helped some, but I still felt nervous. Here are my logs.

*In response 2, the counselor gives a **rationale** for cognitive restructuring, **explains the purpose of "coping" thoughts to Joan,** and gives an **overview** of the strategy.*

2. *Counselor:* Today we're going to work on having you learn to use some more constructive thoughts. I call these *coping thoughts.* You can replace the negative thoughts with coping thoughts that will help you when you're anticipating your class, in your class itself, and when things happen in your class that are especially hard for you—like taking a test or going to the board. What questions do you have about this?
   Client: I don't think any—although I don't know if I know exactly what you mean by a coping thought.

*The counselor, in response 3, will **explain and give some examples of coping thoughts** and particular times or phases when Joan might need to use them.*

3. *Counselor:* OK, let me explain about these and give you some examples. Then perhaps you can think of your own examples. The first thing is that there are probably different times when you could use coping thoughts—like before math class when you're anticipating it. Only, instead of worrying about it, you can use this time to prepare to handle it. For example, some coping thoughts you might use before math class are "No need to get nervous. Just think about doing OK" or "You can manage this situation" or "Don't worry so much—you've got the ability to do OK." Then, during math class, you can use coping thoughts to get through the class and to concentrate on what

you're doing, such as "Just psych yourself up to get through this" or "Look at this class as a challenge, not a threat" or "Keep your cool, you can control your nervousness." Then, if there are certain times during math class that are especially hard for you, like taking a test or going to the board, there are coping thoughts you can use to help you deal with really hard things, like "Think about staying very calm now" or "Relax, take a deep breath" or "Stay as relaxed as possible. This will be over shortly." After math class, or after you cope with a hard situation, then you can learn to encourage yourself for having coped by thinking things like "You did it" or "You were able to control your negative thoughts" or "You're making progress." Do you get the idea?
   Client: Yes, I think so.

*Next, in responses 4 through 7, the counselor will instruct Joan **to select and practice coping thoughts at each critical phase,** starting with **preparing for class.***

4. *Counselor:* Joan, let's take one thing at a time. Let's work just with what you might think before your math class. Can you come up with some coping thoughts you could use when you're anticipating your class?
   Client: Well. [Pauses] I could think about just working on my problems and not worrying about myself. I could think that when I work at it, I usually get it even if I'm slow.

5. *Counselor:* OK, good. Now just to get the feel for these, practice using them. Perhaps you could imagine you are anticipating your class—just say these thoughts aloud as you do.
   Client: Well, I'm thinking that I could look at my class as a challenge. I can think about just doing my work. When I concentrate on my work, I usually do get the answers.

6. *Counselor:* OK—good! How did that feel?
   Client: Well, OK. I can see how this might help. Of course, I don't usually think these kinds of things.

7. *Counselor:* I realize that, and later on today we will practice actually having you use these thoughts. You'll get to the point where you can use your nervousness as a signal to cope. You can stop the self-defeating thoughts and use these coping thoughts instead. Let's practice this some more. [Additional practice ensues.]

*In responses 8, 9, and 10, the counselor asks Joan **to select and practice verbalizing coping thoughts** she can use **during class.***

8. *Counselor:* OK, Joan, now you seem to have several kinds of coping thoughts that might help you when you're anticipating math class. What about some coping thoughts you could use during the class? Perhaps some of these could help you concentrate on your work instead of your tenseness.
   Client: Well, I could tell myself to think about what I need to do—like to get the problems. Or I could think—just take one situation at a time. Just psych myself up 'cause I know I really can do well in math if I believe that.

9. *Counselor:* OK, it sounds like you've already thought of several coping things to use during class. This time, why don't you

pretend you're sitting in your class? Try out some of these coping thoughts. Just say them aloud.

Client: OK. Well, I'm sitting at my desk, my work is in front of me. What steps do I need to take now? Well, I could just think about one problem at a time; not worry about all of them. If I take it slowly, I can do OK.

10. *Counselor:* OK, that seemed pretty easy for you. Let's do some more practice like this just so these thoughts don't seem unfamiliar to you. As you practice, try hard to think about the meaning of what you're saying to yourself. [More practice occurs.]

Next, Joan **selects and practices coping thoughts** *to help her deal with especially* **stressful or critical situations** *that come up in math class (responses 11, 12, and 13).*

11. *Counselor:* This time, let's think of some particular coping statements that might help you if you come up against some touchy situations in your math class—things that are really hard for you to deal with, like taking a test, going to the board, or being called on. What might you think at these times that would keep the lid on your nervousness?

Client: Well, I could think about just doing what is required of me—maybe, as you said earlier taking a deep breath and just

LEARNING ACTIVITY 44 *Cognitive Restructuring*

I. Listed below are eight statements. Read each statement carefully and decide whether it is a self-defeating or a self-enhancing statement. Remember, a self-defeating thought is a negative, unproductive way to view a situation; a self-enhancing thought is a realistic, productive interpretation of a situation or of oneself. Write down your answers. Feedback is given after the learning activities on page 388.

1. "Now that I've had this accident, I'll *never* be able to do anything I want to do again."
2. "How can I ever give a good speech when I don't know what I want to say?"
3. "Using a wheelchair is not as hard as it looks. I can get around wherever I want to go."
4. "I had to come to this country without my son and now that he is coming here too, I know he won't want to have anything to do with me."
5. "What I need to think about is what I want to say, not what I think I *should* say."
6. "If I just weren't a diabetic, a lot more opportunities would be available to me."
7. "Why bother? She probably wouldn't want to go out with me anyway."
8. "Of course I would prefer that my daughter marries a man but if she chooses to be single or be with another woman, I'm okay with that too. It's her life and I love her no matter what."

II. This learning activity is designed to help you personalize cognitive restructuring in some way by using it yourself.

1. Identify a problem situation for yourself—a situation in which you don't do what you want to, not because you don't have the skills, but because of your negative, self-defeating thoughts. Some examples:

   a. You need to approach your boss about a raise, promotion, or change in duties. You know what to

say, but you are keeping yourself from doing it because you aren't sure it would have any effect and you aren't sure how the person might respond.

   b. You have the skills to be an effective helper, yet you constantly think that you aren't.

   c. You continue to get positive feedback about the way you handle a certain situation, yet you are constantly thinking you don't do this very well.

2. For about a week, every time this situation comes up, monitor all the thoughts you have *before, during,* and *after* the situation. Write these thoughts in a log. At the end of the week

   a. Identify which of the thoughts are self-defeating.

   b. Identify which of the thoughts are self-enhancing.

   c. Determine whether the greatest number of self-defeating thoughts occur before, during, or after the situation.

3. In contrast to the self-defeating thoughts you have, identify some possible coping or self-enhancing thoughts you could use. On paper, list some you could use before, during, and after the situation, with particular attention to the time period when you tend to use almost all self-defeating thoughts. Make sure that you include in your list some positive or self-rewarding thoughts, too—for coping.

4. Imagine the situation—before, during, and after it. As you do this, stop any self-defeating thoughts and replace them with coping and self-rewarding thoughts. You can even practice this in role play. This step should be practiced until you can feel your coping and self-rewarding thoughts taking hold.

5. Construct a homework assignment for yourself that encourages you to apply this as needed when the self-defeating thoughts occur.

FEEDBACK
Cognitive Restructuring

I. 1. *Self-defeating:* the word *never* indicates the person is not giving himself or herself any chance for the future.

2. *Self-defeating:* the person is doubting both his or her ability to give a good speech and knowledge of the subject.

3. *Self-enhancing:* the person is *realistically* focusing on what she or he can do.

4. *Self-defeating:* the person is saying with certainty, as evidenced by the word *know,* that there is no chance to regain a relationship with her son. This is said without supporting evidence or data.

5. *Self enhancing:* the client is realistically focusing on his or her own opinion, not on the assessment of others.

6. *Self-defeating:* the person is viewing the situation only from a negative perspective.

7. *Self-defeating:* the person predicts a negative reaction without supporting evidence.

8. *Self-enhancing:* the person recognizes a preference yet focuses on her love for her daughter.

---

thinking about staying calm, not letting my anxiety get the best of me.

12. *Counselor:* OK, great. Let's see—can you practice some of these aloud as if you were taking a test or had just been asked a question or were at the board in front of the class?

Client: OK. Well, I'm at the board, I'm just going to think about doing this problem. If I start to get really nervous, I'm going to take a deep breath and just concentrate on being calm as I do this.

13. *Counselor:* OK, let's practice this several times. Maybe this time you might use another tense moment, like being called on by your teacher. [Further practice goes on.]

Next, the counselor **points out how Joan may discourage or punish herself after class** *(responses 14 and 15). Joan selects and* **practices encouraging or self-rewarding thoughts** *(responses 16, 17, and 18).*

14. *Counselor:* OK, Joan, there's one more thing I'd like you to practice. After math class, what do you usually think?

Client: I feel relieved. I think about how glad I am it's over. Sometimes I think about the fact that I didn't do well.

15. *Counselor:* Well, those thoughts are sort of discouraging, too. What I believe might help is if you could learn to encourage

yourself as you start to use these coping thoughts. In other words, instead of thinking about not doing well, focus on your progress in coping. You can do this during class or after class is over. Can you find some more positive things you could think about to encourage yourself—like giving yourself a pat on the back?

Client: You mean like I didn't do as bad as I thought?

16. *Counselor:* Yes, anything like that.

Client: Well, it's over, it didn't go too badly. Actually I handled things OK. I can do this if I believe it. I can see progress.

17. *Counselor:* OK, now, let's assume you've just been at the board. You're back at your seat. Practice saying what you might think in that situation that would be encouraging to you.

Client: Well, I've just sat down. I might think that it went fast and I did concentrate on the problem, so that was good.

18. *Counselor:* OK. Now let's assume class is over. What would you say would be positive, self-encouraging thoughts after class?

Client: Well, I've just gotten out. Class wasn't that bad. I got something out of it. If I put my mind to it, I can do it. [More practice of positive self-statements occurs.]

In response 19, the counselor instructs Joan **to practice the entire sequence** *of stopping a self-defeating thought and using a coping thought before, during, and after class. Usually the client practices this by* **imagining the situation.**

19. *Counselor:* So far we've been practicing these coping thoughts at the different times you might use them so you can get used to these. Now let's practice this in the sequence that it might actually occur—like before your class, during the class, coping with a tough situation, and encouraging yourself after class. We can also practice this with your thought stopping. If you imagine the situation and start to notice any self-defeating thoughts, you can practice stopping these. Then switch immediately to the types of coping thoughts that you believe will help you most at that time. Concentrate on the coping thoughts. How does this sound?

Client: OK, I think I know what you mean. [Looks a little confused.]

Sometimes long instructions are confusing. Modeling may be better. In responses 20 and 21, the counselor **demonstrates how Joan can apply coping thoughts in practice.**

20. *Counselor:* Well, I just said a lot, and it might make more sense if I showed this to you. First, I'm going to imagine I'm in English class. It's almost time for the bell, then it's math class. Wish I could get out of it. It's embarrassing. *Stop!* That's a signal to use my coping thoughts. I need to think about math class as a challenge. Something I can do OK if I work at it. [Pauses.] Joan, do you get the idea?

Client: Yes, now I do.

21. *Counselor:* OK, I'll go on and imagine now I'm actually in the class. He's given us a worksheet to do in 30 minutes. Whew!

How will dumb me ever do that! Wait a minute. I know I can do it, but I need to go slowly and concentrate on the work, not on me. Just take one problem at a time.

Well, now he wants us to read our answers. What if he calls on me? I can feel my heart pounding. Oh well, if I get called on, just take a deep breath and answer. If it turns out to be wrong, it's not the end of the world.

Well, the bell rang. I am walking out. I'm glad it's over. Now, wait a minute—it didn't go that badly. Actually I handled it pretty well. Now, why don't you try this? [Joan practices the sequence of coping thoughts several times, first with the counselor's assistance, gradually in a completely self-directed manner.]

Before terminating the session, the counselor **assigns daily homework practice.**

22. *Counselor:* This week I'd like you to practice this several times each day—just like you did now. Keep track of your practices on your log. And you can use this whenever you feel it would be helpful—such as before, during, or after math class. Jot these times down too, and we'll go over this next week.

SUMMARY

Various cognitive change procedures of cognitive modeling, problem solving, reframing, stress inoculation, and cognitive restructuring are being used more frequently in counseling and psychotherapy. An individual's construction of a particular situation is like a photograph. The individual's influence or bias can blur or distort the mental picture. Cognitive restructuring or replacement presented in this chapter is like providing the client with other pictures or a different construction or mental picture of a situation. Cognitive structural change can be more than modifying habitual cognitions, rules, formulas, assumptions, and imperatives that skew perceptions of situations; it can also provide emotional relief. As with any intervention, adaptations need to be made depending on the client's gender and culture. For example, some clients may benefit if cognitive restructuring is offered in a group setting. Counselors using cognitive restructuring also need to be sensitive to the diverse perspectives reflected by the client's culture. We present two related cognitive change procedures, reframing and stress inoculation, in the next chapter.

P O S T E V A L U A T I O N

Part One

Objective One asks you to identify and describe the six major components of cognitive restructuring in a client case. Using the case described here, explain briefly how you would use the steps and components of cognitive restructuring with *this* client. You can use the six questions following the client case to describe your use of this procedure. Feedback follows on page 394.

Description of client: Doreen is a junior in college, majoring in education and getting very good grades. She reports that she has an active social life and has some good close friendships with both males and females. Despite obvious "pluses," the client reports constant feelings of being worthless and inadequate. Her standards for herself seem to be unrealistically high: Although she has almost a straight-A average, she still chides herself that she does not have all A's. Although she is attractive and has an active social life, she thinks that she should be more attractive and more talented. At the end of the initial session, she adds that as an African American woman she always has felt as though she has to prove herself more than the average person.

1. How would you explain the rationale for cognitive restructuring to this client?
2. Give an example you might use with this client to point out the difference between a self-defeating and a self-enhancing thought. Try to base your example on the client's self-description.
3. How would you have the client identify her thoughts about herself—her grades, appearance, social life, and so on?
4. What are some coping thoughts this client might use?
5. Explain how, in the session, you would help the client practice shifting from self-defeating to coping thoughts.
6. What kind of homework assignment would you use to help the client increase her use of coping thoughts about herself?

Part Two

Objective Two asks you to teach the six components of cognitive restructuring to someone else or to demonstrate these components with a role-play client. Use the Interview Checklist for Cognitive Restructuring on pages 390–393 as a teaching and evaluation guide.

(continued)

POSTEVALUATION   (continued)

INTERVIEW CHECKLIST FOR COGNITIVE RESTRUCTURING

Instructions to observer: Determine whether the counselor demonstrated the lead listed in the checklist. Check which leads the counselor used.

Item	Examples of counselor leads
I. Rationale and Overview	
_____ 1. Counselor explains purpose and rationale of cognitive restructuring.	"You've reported that you find yourself getting anxious and depressed during and after these conversations with the people who have to evaluate your work. This procedure can help you identify some things you might be thinking in this situation that are just beliefs, not facts, and are unproductive. You can learn more realistic ways to think about this situation that will help you cope with it in a way that you want to."
_____ 2. Counselor provides brief overview of procedure.	"There are three things we'll do in using this procedure. *First,* this will help you identify the kinds of things you're thinking before, during, and after these situations that are self-defeating. *Second,* this will teach you how to stop a self-defeating thought and replace it with a coping thought. *Third,* this will help you learn how to give yourself credit for changing these self-defeating thoughts."
_____ 3. Counselor explains difference between self-enhancing thoughts (facts) and self-defeating thoughts (beliefs) and provides culturally relevant examples of each.	"A self-defeating thought is one way to interpret the situation, but it is usually negative and unproductive, like thinking that the other person doesn't value you or what you say. In contrast, a self-enhancing thought is a more constructive and realistic way to interpret the situation—like thinking that what you are talking about has value to you."
_____ 4. Counselor explains influence of irrational and self-defeating thoughts on emotions and performance.	"When you're constantly preoccupied with yourself and worried about how the situation will turn out, this can affect your feelings and your behavior. Worrying about the situation can make you feel anxious and upset. Concentrating on the situation and not worrying about its outcome can help you feel more relaxed, which helps you handle the situation more easily."
_____ 5. Counselor confirms client's willingness to use strategy.	"Are you ready to try this now?"
II. Identifying Client Thoughts in Problem Situations	
_____ 6. Counselor asks client to describe problem situations and identify examples of self-enhancing thoughts and of self-defeating thoughts client typically experiences in these situations.	"Think of the last time you were in this situation. Describe for me what you think before you have a conversation with your evaluator. . . . What are you usually thinking during the conversation? . . . What thoughts go through your mind after the conversation is over? Now, let's see which of those thoughts are actual facts about the situation or are constructive ways to interpret the situation. Which ones are your beliefs about the situation that are unproductive or self-defeating?"
_____ 7. If client is unable to complete step 6, counselor models examples of thoughts or "links" between event and client's emotional response.	"OK, think of the thoughts that you have while you're in this conversation as a link between this event and your feelings afterward of being upset and depressed. What is

(continued)

Item	Examples of counselor leads

II. Identifying Client Thoughts in Problem Situations (continued)

	the middle part? For instance, it might be something like 'I'll never have a good evaluation, and I'll lose this position' or 'I always blow this conversation and never make a good impression.' Can you recall thinking anything like this?''
_____ 8. Counselor instructs client to monitor and record content of thoughts *before, during,* and *after* stressful or upsetting situations before next session.	"One way to help you identify this link or your thoughts is to keep track of what you're thinking in these situations as they happen. This week I'd like you to use this log each day. Try to identify and write down at least three specific thoughts you have in these situations each day and bring this in with you next week.''
_____ 9. Using client's monitoring, counselor and client identify client's self-defeating thoughts.	"Let's look at your log and go over the kinds of negative thoughts that seem to be predominant in these situations. We can explore how these thoughts affect your feelings and performance in this situation—and whether you feel there is any evidence or rational basis for these.''

III. Introduction and Practice of Coping Thoughts

_____10. Counselor explains purpose and potential use of ''coping thoughts'' and gives some examples of coping thoughts to be used:	"Up to this point, we've talked about the negative or unproductive thoughts you have in these situations and how they contribute to your feeling uncomfortable, upset, and depressed. Now we're going to look at some alternative, more constructive ways to think about the situation—using coping thoughts. These thoughts can help you prepare for the situation, handle the situation, and deal with feeling upset or overwhelmed in the situation. As long as you're using some coping thoughts, you avoid giving up control and letting the old self-defeating thoughts take over. Here are some examples of coping thoughts.''
_____a. Before the situation—preparing for it	
_____b. During the situation	
_____1. Focusing on task	
_____2. Dealing with feeling overwhelmed	
_____11. Counselor instructs client to think of additional coping thoughts client could use or has used before.	"Try to think of your own coping thoughts—perhaps ones you can remember using successfully in other situations, ones that seem to fit for you.''
_____12. Counselor instructs client to practice verbalizing selected coping statements.	"At first you will feel a little awkward using coping statements. It's like learning to drive a stick shift after you've been used to driving an automatic. So one way to help you get used to this is for you to practice these statements aloud.''
_____a. Counselor instructs client first to practice coping statements individually. Coping statements to use before a situation are practiced, then coping statements to use during a situation.	"First just practice each coping statement separately. After you feel comfortable with saying these aloud, practice the ones you could use before this conversation. OK, now practice the ones you could use during this conversation with your evaluator.''
_____b. Counselor instructs client to practice sequence of coping statements as they would be used in actual situation.	"Now let's put it all together. Imagine it's an hour before your meeting. Practice the coping statements you could use then. We'll role-play the meeting. As you feel aroused or overwhelmed, stop and practice coping thoughts during the situation.''

(continued)

POSTEVALUATION   **(continued)**

Item	Examples of counselor leads
III. Introduction and Practice of Coping Thoughts (continued)	
_____c. Counselor instructs client to become actively involved and to internalize meaning of coping statements during practice.	"Try to really put yourself into this practice. As you say these new things to yourself, try to think of what these thoughts really mean."
IV. Shifting from Self-Defeating to Coping Thoughts	
_____13. Counselor models shift from recognizing a self-defeating thought and stopping it to replacing it with a coping thought.	"Let me show you what we will practice today. First, I'm in this conversation. Everything is going OK. All of a sudden I can feel myself starting to tense up. I realize I'm starting to get overwhelmed about this whole evaluation process. I'm thinking that I'm going to blow it. No, I stop that thought at once. Now, I'm just going to concentrate on calming down, taking a deep breath, and thinking only about what I have to say."
_____14. Counselor helps client practice shift from self-defeating to coping thoughts. Practice consists of four steps:	"Now let's practice this. You will imagine the situation. As soon as you start to recognize the onset of a self-defeating thought, stop it. Verbalize the thought aloud, and tell yourself to stop. Then verbalize a coping thought in place of it and imagine carrying on with the situation."
_____ a. Having client imagine situation or carry it out in a role play (behavior rehearsal).	
_____ b. Recognizing self-defeating thought (which could be signaled by a hand or finger).	
_____ c. Stopping self-defeating thought (which could be supplemented with a hand clap).	
_____ d. Replacing thought with coping thought (possibly supplemented with deep breathing).	
_____15. Counselor helps client practice using shift for each problem situation until anxiety or stress felt by client while practicing situation is decreased to a reasonable or negligible level and client can carry out practice and use coping thoughts in self-directed manner.	"Let's keep working with this situation until you feel pretty comfortable with it and can shift from self-defeating to coping thoughts without my help."
V. Introduction and Practice of Positive, or Reinforcing, Self-Statements	
_____16. Counselor explains purpose and use of positive self-statements and gives some examples of these to client.	"You have really made a lot of progress in learning to use coping statements before and during these situations. Now it's time to learn to reward or encourage yourself. After you've coped with a situation, you can pat yourself on the back for having done so by thinking a positive or rewarding thought like 'I did it' or 'I really managed that pretty well.' "

(continued)

Item	Examples of counselor leads

V. Introduction and Practice of Positive, or Reinforcing, Self-Statements (continued)

_____17. Counselor instructs client to think of additional positive self-statements and to select some to try out.

"Can you think of some things like this that you think of when you feel good about something or when you feel like you've accomplished something? Try to come up with some of these thoughts that seem to fit for you."

_____18. Counselor models application of positive self-statements as self-reinforcement for shift from self-defeating to coping thoughts.

"OK, here is the way you reward yourself for having coped. You recognize the self-defeating thought. Now you're in the situation using coping thoughts, and you're thinking things like 'Take a deep breath' or 'Just concentrate on this task.' Now the conversation is finished. You know you were able to use coping thoughts, and you reward yourself by thinking 'Yes, I did it' or 'I really was able to manage that.' "

_____19. Counselor instructs client to practice use of positive self-statements in interview following practice of shift from self-defeating to coping thoughts. This should be practiced in sequence (coping *before* and *during* situation and reinforcing oneself *after* situation).

"OK, let's try this out. As you imagine the conversation, you're using the coping thoughts you will verbalize. . . . Now, imagine the situation is over, and verbalize several reinforcing thoughts for having coped."

VI. Homework and Follow-Up

_____20. Counselor instructs client to use cognitive restructuring procedure (identifying self-defeating thought, stopping it, shifting to coping thought, reinforcing with positive self-statement) in situations outside the interview.

"OK, now you're ready to use the entire procedure whenever you have these conversations in which you're being evaluated—or any *other* situation in which you recognize your negative interpretation of the event is affecting you. In these cases, you recognize and stop any self-defeating thoughts, use the coping thoughts before the situation to prepare for it, and use the coping thoughts during the situation to help focus on the task and deal with being overwhelmed. After the situation is over, use the positive self-thoughts to reward your efforts."

_____21. Counselor instructs client to monitor and record on log sheet number of times client uses cognitive restructuring outside the interview.

"I'd like you to use this log to keep track of the number of times you use this procedure and to jot down the situation in which you're using it. Also rate your tension level on a 1-to-5 scale before and after each time you use this."

_____22. Counselor arranges for follow-up.

"Do this recording for the next two weeks. Then let's get together for a follow-up session."

Observer comments: _____

SUGGESTED READINGS

Arean, P. (1993). Cognitive behavioral therapy with older adults. *The Behavior Therapist, 16,* 236–239.

Beck, J. S. (1995). *Cognitive therapy.* New York: Guilford.

Bedrosian, R. C., & Bozicas, G. D. (1994). *Treating family of origin problems: A cognitive approach.* New York: Guilford.

Deffenbacher, J. L., Oetting, E. R., Huff, M. E., & Thwaites, G. A. (1995). Fifteen-month follow-up of social skills and cognitive-relaxation approaches to anger reduction. *Journal of Counseling Psychology, 42,* 400–405.

Foa, E. B., Rothbaum, B. O., Riggs, D. S., & Murdock, T. B. (1991). Treatment of posttraumatic stress disorder in rape victims: A comparison between cognitive-behavioral approaches and counseling. *Journal of Consulting and Clinical Psychology, 59,* 715–723.

Freeman, A., & Dattillio, F. (Eds.). (1992). *Comprehensive casebook of cognitive therapy.* NY: Plenum.

Grant, L. D., & Haverkamp, B. E. (1995). A cognitive-behavioral approach to chronic pain management. *Journal of Counseling and Development, 74,* 25–31.

Greenberger, D., & Padesky, C. (1995). Mind over mood: A cognitive therapy treatment manual for clients. New York: Guilford.

Hays, P. (1995). Multicultural applications of cognitive behavior therapy. *Professional Psychology, 26,* 309–315.

Horan, J. (1996). Effects of computer-based cognitive restructuring on rationally mediated self-esteem. *Journal of Counseling Psychology, 43,* 371–375.

Johnson, W. B., & Ridley, C. R. (1992). Brief Christian and non-Christian rational-emotive therapy with depressed Christian clients: An exploratory study. *Counseling and Values, 36,* 220–229.

Leahy, R. (1996). *Cognitive therapy: Basic principles and implications.* Northvale, NJ: Aronson.

Mahoney, M. J. (Ed.). (1995). *Cognitive and constructive psychotherapies.* NY: Springer.

Nedate, Y., & Tagami, F. (1994). Effects of instruction through writing on subjective well-being modification: When adopting a cognitive restructuring approach. *Japanese Journal of Counseling Science, 27,* 21–26.

Sookman, D., Pinard, G. B., & Beauchemin, N. (1994). Multidimensional schematic restructuring treatment for obsessions: Theory and practice. *Journal of Cognitive Psychotherapy, 8,* 175–194.

Thompson, L. W. (1996). Cognitive-behavioral therapy and treatment for late-life depression. *Journal of Clinical Psychiatry, 57,* 29–32.

FEEDBACK
POSTEVALUATION

Part One

1. The overall goal would be for Doreen to feel more empowered about herself and to feel less pressure to have to constantly prove herself as a Black woman. You can explain that CR would help her identify some of her thoughts about herself that are beliefs, not facts, and are unrealistic thoughts, leading to feelings of depression and worthlessness. In addition, CR would help her learn to think about herself in more realistic, self-enhancing ways. See the Interview Checklist for Cogni-

tive Restructuring on pages 390–393 for another example of the CR rationale.

2. A core issue for Doreen to challenge is her belief system about her race and gender—that as an African American and as a woman she must constantly prove herself in order to be a worthy person. Thinking that she is not good enough is self-defeating. Self-enhancing or positive thoughts about herself are more realistic interpretations of her experiences—good grades, close friends, active social life, and so on. Recognition that she is intelligent and attractive is a self-enhancing thought.

3. You could ask the client to describe different situations and the thoughts she has about herself in them. She could also observe this during the week. You could model some possible thoughts she might be having. See leads 6, 7, 8, and 9 in the Interview Checklist for Cognitive Restructuring.

4. There are many possible coping thoughts she could use. Here are some examples: "Hey, I'm doing pretty well as it is." "Don't be so hard on myself. I don't have to be perfect." "That worthless feeling is a sign to cope—recognize my assets." "What's more attractive anyway? I am attractive." "Don't let that one B get me down. It's not the end of the world." "I'm an African American woman and I'm proud of it. I feel okay about myself the way I am. I don't have to prove my worth to anyone."

5. See leads 13 through 16 on the Interview Checklist for Cognitive Restructuring.

6. Many possible homework assignments might help. Here are a few examples:

   a. Every time Doreen uses a coping thought, she could record it on her log.

   b. She could cue herself to use a coping thought by writing these down on note cards and reading a note before doing something else, like getting a drink or making a phone call, or by using a phone-answering device to report and verbalize coping thoughts.

   c. She could enlist the aid of a close friend or roommate. If the roommate notices that the client starts to "put herself down," she could interrupt her. Doreen could then use a coping statement.

Part Two

Use the Interview Checklist for Cognitive Restructuring on pages 390–393 to assess your teaching or counseling demonstration of this procedure.

(continued)

CHAPTER 16

REFRAMING AND STRESS INOCULATION

Reframing and stress inoculation assume that maladaptive emotions and thinking are influenced or mediated by one's core beliefs, schemas (see Chapter 15), perceptions, and cognitions. Both procedures have components of cognitive restructuring and coping skills. These two procedures help clients to access and to determine the relations among their perceptions, cognitions, emotions, and actions; to identify how clients interpret their situations and experiences; and to urge clients to substitute a new interpretive frame (Gendlin, 1996, p. 242).

OBJECTIVES

1. Demonstrate 8 out of the 11 steps of reframing in a role-play interview.
2. Using a simulated client case, describe how the five components of stress inoculation would be used with a client.
3. Demonstrate 17 out of 21 steps of stress inoculation in a role-play interview.

THE PROCESS OF REFRAMING

People who grow up in chaotic, abusive, rejecting, or disorganized attachment style environments will generally display distorted attribution processes. These people often perceive personal problems in ways that appear to have no workable solution for emotional relief. For example, a man who has experienced rejection from his parents while growing up may say, "I don't have the skills to work with people." From this man's frame of reference, he views "lack of interpersonal skills" as unchangeable, and he experiences a sense of hopelessness about working with people. It is his myth, personal script, belief, or frame of reference about his interpersonal skills. This habitual schema leads to self-limiting patterns of feeling, thinking, and behaving with people. If the man in the example continues the cycle of

self-indictment, he will experience a sense of despair, and become regressive and withdrawn from social interactions. The man in the above example feels stuck in his schema, which was created by the state-dependent memory, learning, and behaving he experienced in his family. The meaning and emotions he experiences in social interactions lock him in a trancelike schema which only limit his perceptions, beliefs, and options for alternative ways of behaving. Reframing might be one intervention that could help this man revise his way to perceive and modify his interpersonal skills.

Reframing means to explore how an incident or situation is typically perceived and to offer another view or frame for the situation. Reframing helps a client to change emotions, meaning, and perceived options. The reframe can change clients' everyday conscious sets (trance) and perceptions of their personal limitations. But as Gendlin (1996) points out, reframing "sometimes works, and sometimes not. To determine whether the reframing is effective, you must sense whether it has brought a *bodily* change or not. You must sense what actually comes in your body in response to a reframing. A real change is a shift in the concrete bodily way you have the problem, and not *only* a new way of thinking" (p. 9). A therapist can assess the validity of the reframe by asking the client what bodily responses or sensations he or she feels while experiencing the reframe. One technique to bring awareness to the body is to have the client engage in a body scan (see Chapter 17) while thinking about the reframe. Embodiment of an effective reframe "has a directly sensed effect. A weight is lifted; there is a physical freeing and new bodily energy. The client feels able to do what seemed impossible before" (Gendlin, 1996, p. 243). Sometimes the initial effect of the reframe is small. If this is the case, the counselor encourages the client to stay with the reframe longer so that it can become embodied more fully (Gendlin, 1996, p. 243).

Reframing (sometimes also called relabeling) is an approach that modifies or restructures a client's perceptions

TABLE 16-1. Research on Reframing

AIDS prevention

Citizens Commission on AIDS for New York City and Northern New Jersey. (1991). AIDS prevention and education: Reframing the message. *AIDS Education and Prevention, 3,* 147–163.

Anxiety

Ishiyama, F. I. (1991). A Japanese reframing technique for brief social anxiety treatment: An exploratory study of cognitive and therapeutic effects of Morita therapy. *Journal of Cognitive Psychotherapy, 5,* 55–70.

Kass, R. G., & Fish, J. M. (1991). Positive reframing and the test performance of test anxious children. *Psychology in the Schools, 28,* 43–52.

Children and adolescents

Komori, Y., Miyazato, M., & Orii, T. (1991). The Family Journal Technique: A simple, positive-reframing technique in Japanese pediatrics. *Family Systems Medicine, 9,* 19–24.

Ritchie, M. H. (1994). Counseling difficult children. Special Issue: Perspective on working with difficult clients. *Canadian Journal of Counseling, 28,* 58–68.

Rodriguez, C., & Moore, N. (1995). Perceptions of pregnant/parenting teens: Reframing issues for an integrated approach to pregnancy problems. *Adolescence, 30,* 685–706.

Cognitive change

Guterman, J. T. (1992). Disputation and reframing: Contrasting cognitive-change methods. *Journal of Mental Health Counseling, 14,* 440–456.

Decision making

Mozdzierz, G. J., & Greenblatt, R. L. (1992). Clinical paradox and experimental decision making. *Individual Psychology: Journal of Adlerian Theory, Research and Practice, 48,* 301–312.

Depression

Boer, C. (1992). Reframing depression: A systems perspective. *Family Systems Medicine, 10,* 405–411.

Brack, G., LaClave, L., & Wyatt, A. S. (1992). The relationship of problem solving and reframing to stress and depression in female college students. *Journal of College Student Development, 33,* 124–131.

Swoboda, J. S., Dowd, E. T., & Wise, S. L. (1990). Reframing and restraining directives in the treatment of clinical depression. *Journal of Counseling Psychology, 37,* 254–260.

Failure of reframe

Brack, G., Brack, C. J., & Hartson, D. (1991). When a reframe fails: Explorations into students' ecosystems. *Journal of College Student Development, 6,* 103–118.

Family therapy

Lawson, D. M. (1991). Reframing family change rate. *Journal of Family Psychotherapy, 2,* 75–87.

Prinz, R. J. (1992). Overview of behavioural family interventions with children: Achievements, limitations, and challenges. *Behaviour Change, 9,* 120–125.

(continued)

---

or views of a problem or a behavior. Reframing is used frequently in family therapy as a way to redefine presenting problems so as to shift the focus away from an "identified patient" or "scapegoat" and onto the family as a whole, as a system in which each member is an interdependent part. When used in this way, reframing changes the way a family encodes an issue or a conflict.

With individual clients, reframing has a number of uses as well. By changing or restructuring what clients encode and perceive, reframing can reduce defensiveness and mobilize the client's resources and forces for change. Second, it can shift the focus from overly simplistic trait attributions of behavior that clients are inclined to make ("I am lazy" or "I am not assertive") to analyses of important contextual and situational cues associated with the behavior. Finally, reframing can be a useful strategy for helping with a client's self-efficacy (see Chapter 20).

Uses of Reframing

Table 16-1 presents a sample of reframing research. A quick glance at the listing in the table reveals that reframing has been used for a number of purposes, such as these: to change the message about AIDS; to reduce social and test anxiety; in dealing with difficult children; to effect cognitive change; and to help clients cope with depression, illness, pain, and panic.

Meaning Reframes

Counselors reframe whenever they ask or encourage clients to see an issue from a different perspective. In this chapter, we propose a more systematic way for counselors to help clients reframe a problem behavior. The most common method of reframing—and the one that we illustrate in this

TABLE 16-1. Research on Reframing (continued)

Illness
Kleinman, A. (1992). Local worlds of suffering: An interpersonal focus for ethnographies of illness experience. *Qualitative Health Research, 2,* 127–134.

Lupus
Braden, C. J., McGlone, K., & Pennington, F. (1993). Specific psychosocial and behavioral outcomes from the Systemic Lupus Erythematosus Self-Help course. Special issue: Arthritis health education. *Health Education Quarterly, 20,* 29–41.

Mental health
Johnson, G. B., & Werstlein, P. O. (1990). Reframing: A strategy to improve care of manipulative patients. *Issues in Mental Health Nursing, 11,* 237–241.

Pesut, D. J. (1991). The art, science, and techniques of reframing in psychiatric mental health nursing. Special Issue: Psychiatric nursing for the 1990s: New concepts, new therapies. *Issues in Mental Health Nursing, 12,* 9–18.

Older adults
Dressel, P. L., & Barnhill, S. K. (1994). Reframing gerontological thought and practice: The case of grandmothers with daughters in prison. *Gerontologist, 34,* 685–691.

Motenko, A. K., & Greenberg, S. (1995). Reframing dependence in old age. *Social Work, 40,* 382–390.

Pain
Shutty, M. S., & Sheras, P. (1991). Brief strategic psychotherapy with chronic pain patients: Reframing and problem resolution. *Psychotherapy, 28,* 636–642.

Panic
Neeleman, J. (1992). The therapeutic potential of positive reframing in panic. *European Psychiatry, 7,* 135–139.

Parenting
Lam, J. A., Rifkin, J., & Townley, A. (1989). Reframing conflict: Implications for fairness in parent-adolescent mediation. *Mediation Quarterly, 7,* 15–31.

Resistance
Robinson, T., & Ward, J. V. (1991). "A belief in self far greater than anyone's disbelief": Cultivating resistance among African American female adolescents, Special Issue: Women, girls and psychotherapy: Reframing resistance. *Women and Therapy, 11* (3–4), 87–103.

Stern, L. (1991). Disavowing the self in female adolescence. Special Issue: Women, girls and psychotherapy: Reframing resistance. *Women and Therapy, 11* (3–4), 105–117.

Smoking relapse
Haaga, D. A. F., & Allison, M. L. (1994). Thought suppression and smoking relapse: A secondary analysis of Haaga (1989). *British Journal of Clinical Psychology, 33,* 327–331.

Supervision
Masters, M. A. (1992). The use of positive reframing in the context of supervision. *Journal of Counseling and Development, 70,* 387–390.

Terminally ill
Baack, C. M. (1993). Nursing's role in the nutritional care of the terminally ill: Weathering the storm. Special Issue: Nutrition and hydration in hospice care: Needs, strategies, ethics. *Hospice Journal, 9,* 1–13.

Trauma
Malon, D., & Hurley, W. (1994). Ericksonian utilization of depressive self-blame in a traumatized patient. *Journal of Systemic Therapies, 13,* 38–46.

chapter—is to reframe the *meaning* of a problem situation or behavior. When you reframe meaning, you are challenging the meaning that the client (or someone else) has assigned to a given problem behavior. Usually, the longer a particular meaning (or label) is attached to a client's behavior, the more necessary the behavior itself becomes in maintaining predictability and equilibrium in the client's functioning. Also, when meanings are attached to client behavior over a long period of time, clients are more likely to develop "functional fixity"—that is, seeing things in only one way or from one perspective or being fixated on the idea that this particular situation or attribute is *the* issue. Reframing helps clients by providing alternative ways to view a problem behavior without directly challenging the behavior itself and by loosening a client's perceptual frame, thus setting the stage for other kinds of therapeutic work. Once the *meaning* of a behavior or a situation changes, the person's response to the situation or the person's typical behavior usually also changes, providing the reframe is valid and acceptable to the client. The essence of a meaning reframe is to give a situation or a behavior a new label or a new name that has a different meaning. This new meaning always has a different connotation, and usually it is a positive one. For example, client "stubbornness" might be reframed as "independence," or "greediness" might be reframed as "ambition."

Reframing involves six steps:

1. Rationale: purpose and overview of the procedure
2. Identification of client perceptions and feelings in problem situations
3. Deliberate enactment of selected perceptual features

4. Identification of alternative perceptions
5. Modification of perceptions in problem situations
6. Homework and follow-up

A detailed description of the steps associated with these components can be found in the Interview Checklist for Reframing at the end of this chapter.

Treatment Rationale

The rationale used to present reframing attempts to strengthen the client's belief that perceptions or attributions about the problem situation can cause emotional distress. Here is a rationale that can be used to introduce reframing:

> When we think about or when we are in a problem situation, we automatically attend to selected features of the situation. As time goes on, we tend to get fixated on these same features of the situation and ignore other aspects of it. This can lead to some uncomfortable emotions, such as the ones you're feeling now. In this procedure, I will help you identify what you are usually aware of during these problem situations. Then we'll work on increasing your awareness of other aspects of the situation that you usually don't notice. As this happens, you will notice that your feelings about and responses to this problem will start to change. Do you have any questions?

Identification of Client Perceptions and Feelings

Assuming that the client accepts the rationale the counselor provides, the next step is to help clients become aware of what they automatically attend to in problem situations. Clients are often unaware of what features or details they attend to in a problem situation and what information about these situations they encode. For example, clients who have a fear of water may attend to how deep the water is because they cannot see the bottom and encode the perception that they might drown. Clients who experience test anxiety might attend to the large size of the room or how quickly the other people seem to be working. These features are encoded and lead to the clients' feeling overwhelmed, anxious, and lacking in confidence. In turn, these feelings can lead to impaired performance in or avoidance of the situation.

Within the interview, the therapist helps clients discover what they typically attend to in problem situations. The therapist can use imagery or role play to help clients reenact the situation(s) in order to become aware of what they notice and encode. While engaging in role play or in imagining the problem situation, the therapist can help the client become more aware of typical encoding patterns by asking questions like these:

"What are you attending to now?"
"What are you aware of now?"
"What are you noticing about the situation?"

In order to link feelings to individual perceptions, these questions can be followed with further inquiries, such as these:

"What are you feeling at this moment?"
"What do you feel in your body?"

The counselor may have to help clients engage in role play or imagery several times so that they can reconstruct the situation and become more aware of salient encoded features. The therapist may also suggest what the client might have felt and what the experience appears to mean to the client in order to bring these automatic perceptions into awareness. The therapist also helps clients notice "marginal impressions"—fleeting images, sounds, feelings, and sensations that were passively rather than deliberately processed by the client yet affect the client's reaction to the situation.

Deliberate Enactment of Selected Perceptual Features

After clients become aware of their rather automatic attending, they are asked to reenact the situation and intentionally attend to the selected features that they have been processing automatically. For example, the water-phobic client reenacts (through role play or imagery) approach situations with the water and deliberately attends to the salient features such as the depth of the water and inability to see the bottom of the pool. By deliberately attending to these encoded features, clients are able to bring these habitual attentional processes fully into awareness and under direct control. This sort of "dramatization" seems to sharpen the client's awareness of existing perceptions. When these perceptions are uncovered and made more visible through this deliberate reenactment, it is harder for the client to maintain old illusions. This step may need to be repeated several times during the session or assigned as a homework task.

Identification of Alternative Perceptions

The counselor can help the client change his or her attentional focus by selecting other features of the problem situation to attend to rather than ignore. For example, the water-phobic client who focuses on the depth of the water might be instructed to focus on how clear, clean, and wet the water appears. For the test-anxious client who attends to the size of the room, the counselor can redirect the client's attention to how roomy the testing place is or how comfortable the seats are. Both clients and counselors can suggest other features of the problem situation or person to utilize that have a positive or at least a neutral connotation. The counselor can ask the client what features provide a felt sense of relief.

For reframing to be effective, the alternative perceptions you identify must be acceptable to the client. The best reframes are the ones that are accurate and are as valid a way of looking at the world as the way the person sees things now. All reframes or alternative perceptions have to be tailored to the clients' values, style, and sociocultural context, and they have to fit the clients' experience and model of their world. The alternative perceptions or reframes you suggest obviously also need to match the external reality of the situation enough to be plausible. If, for example, a husband is feeling very angry with his wife because of her extramarital affair, reframing his anger as "loving concern" is probably not representative enough of the external situation to be plausible to the client. A more plausible reframe might be something like "frustration over not being able to control your wife's behavior" or "frustration from not being able to protect the (marital) relationship."

The delivery of a reframe is also important. When suggesting alternative perceptions to clients, it is essential that the counselor's nonverbal behavior be congruent with the tone and content of the reframe. It is also important to use your voice effectively in delivering the reframe by emphasizing key words or phrases.

Modification of Perceptions in Problem Situations

Modifying what clients attend to can be facilitated with role play or imagery. The therapist instructs the client to attend to other features of the problem situations during the role play or imagery enactment. This step may need to be repeated several times. Repetition is designed to embody new perceptual responses so that the client gradually experiences a felt sense of relief (Gendlin, 1996).

Homework and Follow-up

The therapist can suggest to the client that during in vivo situations, the client follow the same format used in therapy. The client is instructed to become more aware of salient encoded features of a stressful or arousing situation, to link these to uncomfortable feelings, to engage in deliberate enactments or practice activities, and to try to make "percep-

tual shifts" during these situations to other features of the situation previously ignored.

As the client becomes more adept at this process, the therapist will need to be alert to slight perceptual shifts and point these out to clients. Typically, clients are unskilled at detecting these shifts in encoding. Helping the client discriminate between old and new encodings of a problem situation can be very useful in promoting the overall goal of the reframing strategy—to alleviate and modify encoding or perceptual errors and bias.

Context Reframes

Although the steps we propose for reframing involve reframing of meaning, another way you can also reframe is to reframe the *context* of a problem behavior. Reframing the context helps a client to explore and decide *when, where,* and *with whom* a given problem behavior is *useful* or appropriate. Context reframing is based on the assumption that every behavior is useful in *some* but not all contexts or conditions. Thus, when a client states "I'm too lazy," a context reframe would be "In what situations (or with what people) is it useful or even helpful to be lazy?" The client may respond by noting that it is useful to be lazy whenever she wants to play with the children. At this point the counselor can help the client sort out and contextualize a given problem behavior so that clients can see where and when they do and do not want the behavior to occur. Context reframes are most useful when dealing with client generalizations—for example "I'm *never* assertive," "I'm *always* late," and so on.

MULTICULTURAL APPLICATIONS OF REFRAMING

As we mentioned earlier, for a reframe to be effective, it must be plausible and acceptable to the client. Client demographic factors such as age, gender, race, and ethnicity are very important components to consider in developing reframes with diverse groups of clients. An excellent example of a culturally relevant meaning reframe is given by Oppenheimer (1992) in working with a severely depressed 67-year-old Latina woman. The woman was able to improve only after she could reframe her depression within the context of her Latino spiritualist beliefs. Her therapist, also a Latina woman, not only refrained from labeling as pathological the client's belief in the supernatural, but she actually used these beliefs to create a valid meaning reframe by taking the client's references to "intranquil spirits or ghosts" to reframe her pain surrounding a loss. Similar examples of culturally sensitive reframes have been used with Asian American adolescents to negotiate conflicting cultural values (Huang, 1994), with HIV positive men to reframe stress around the threat of AIDS (Leserman, Perkins, & Evans, 1992), and with elderly people to reframe dependence (Motenko & Greenberg, 1995).

A good example of a multiculturally useful reframe is the feminist notion of the meaning of *resistance*. In this case, rather than referring to resistance as the client's conscious and unconscious attempts "to avoid the truth and evade change," it means "the refusal to merge with dominant cultural norms and to attend to one's own voice and integrity" (Brown, 1994, p. 15). The meaning of resistance shifts from something that is pathological to something that is healthy and desirable. In this sense, resistance means "learning the ways in which each of us is damaged by our witting or unwitting participation in dominant norms or by the ways in which such norms have been thrust upon us" (Brown, 1994, p. 25). An example might be what we do when our loyal, conscientious, and trustworthy office manager is in danger of losing her or his job because of organizational restructuring and "downsizing": do we look the other way, do we ignore it, or do we speak the truth on her or his behalf to persons holding the power in the dominant social structure of the organization? And as the office manager's therapist, do we attempt to soothe the manager and have her or him adjust, or do we attempt to empower the person to give voice to his or her anger and outrage?

In this sense, to resist means to tell the truth about what is actually happening and what is possible and available to each client as "avenues for change" (Brown, 1994, p. 26). Gay, lesbian, and bisexual clients are confronted with this sort of resistance daily in the "coming out" process. Smith (1992) notes that as an African American girl she was brought up to be a "resister," to be honest and self-reliant, and to stand up for herself. Notably absent from her experience were unrealistic fairy tale expectations such as being rescued by a "Prince Charming." Recent research by Taylor, Gilligan, and Sullivan (1995) also confirms that many African American adolescent girls have a strong sense of healthy resistance. Robinson and Ward (1992) have made an important contribution to the notion of healthy resistance by distinguishing between resistance strategies for *survival* versus resistance strategies for *liberation*. Resistance strategies for survival are crisis oriented and are short-term methods that include self-denigration and "racelessness," *excessive* autonomy at the expense of connectedness to one's collective culture, and "quick fixes" such as early and unplanned pregnancies, substance use, and school failure and/or dropping-out. Resistance strategies for liberation are strategies in which problems of oppression are acknowledged and demands for change are empowered. Strategies for liberation for African Americans are strengthened with an Afrocentric perspective. Robinson and Ward (1992) base their strategies for liberation on an African-centered Nguzo Saba value system (Karenga, 1980), as summarized in Table 16-2. Their model is relevant for all of us seeking to resist and empower rather than to submit and adjust.

MODEL CASE: REFRAMING

The following model case illustrates the process of the reframing intervention. In it, the therapist reframes the respective roles of a mother and her son, in relationship contexts. Note how the reframes are grounded in the family's cultural values. The therapist considers Asian American family ties in reframing Mrs. Kim's responsibilities as a mother, and her son's duties as a son and a parent.

THE KIM FAMILY*

Some months ago the local hospital asked the author to help with an elderly Korean lady who was about to be discharged. It appeared that the lady was very likeable and a good patient. She loved to be pampered by the staff, which concerned them because she appeared to be settling in for a long hospital stay. She was generally reluctant to use her weekend pass to go home, preferring to stay at the hospital and acting as if she were going to be there forever. She was becoming dependent on the staff and asking for more medication and care than she needed.

The hospitalization of 67-year-old Mrs. Kim had been precipitated by her attempted suicide (by cutting her throat with a butcher knife). She had been in the hospital for about a month, and her wound was healing nicely, a fact which the patient tended to minimize. The initially positive relationship between the patient and the staff appeared to be souring. Puzzled and frustrated by the lack of progress, the staff called the author, saying that there seemed to be a "family problem."

The staff had been encouraging Mrs. Kim to be independent and not to lean on her oldest son but to rely more on her husband. To their dismay, the husband seldom visited, but the oldest son came to see her every day even though he had a grueling schedule going to school, working full time, and raising two children as a single parent. They said they had talked to both mother and son to no avail and were getting

*SOURCE: From "Different and Same: Family Therapy with Asian-American Families," by I. K. Berg and A. Jaya. Reprinted from Volume 19, pp. 36–38 of the *Journal of Marital and Family Therapy*. Copyright © 1993 by the American Association for Marriage and Family Therapy. Reprinted with permission.

TABLE 16-2. Resistance strategies

Survival/Oppression	Liberation/Empowerment
Isolation and disconnectedness from the larger African American community	Unity with African people that transcends age, socioeconomic status, geographic origin, and sexual orientation (Umoja)
Self defined by others (the media, educational system) in a manner that oppresses and devalues blackness	Self-determination through confrontation and repudiation of oppressive attempts to demean self. New models used to make active decisions that empower and affirm self and racial identity (Kujichagulia)
Excessive individualism and autonomy; racelessness	Collective work and responsibility; the self seen in connection with the larger body of African people, sharing a common destiny (Ujima)
"I've got mine, you get yours" attitude	Cooperative economics advocating a sharing of resources through the convergence of the "I" and the "we" (Ujaama)
Meaninglessness in life, immediate gratification to escape life's harsh realities; the use of "quick fixes"	Purpose in life that benefits the self and the collective, endorses delaying gratification as a tool in resistance (Nia)
Maintaining status quo, replicating existing models, although they may be irrelevant	Creativity through building new paradigms for the community through dialogue with other resisters (Kuumba)
Emphasis on the here and now, not looking back and not looking forward; myopic vision	Faith through an intergenerational perspective where knowledge of the history of Africa and other resisters and care for future generations gives meaning to struggle and continued resistance (Imani)

SOURCE: From "A Belief in Self Far Greater than Anyone's Belief: Cultivating Resistance among African American Female Adolescents," by T. Robinson and J. Ward. In *Women, Girls, and Psychotherapy: Reframing Resistance,* by C. Gilligan, A. Rogers, and D. Tolman (Eds.), p. 99, Table 1. Copyright © 1992 by The Haworth Press, Inc., Binghamton, NY. Reprinted by permission.

somewhat frustrated with the dependent posture she was taking with her son.

The author decided that a joint session would be unproductive since mediated negotiation is more productive than confrontation in working with Asian-American families. The therapeutic task for this family was two-fold: to find a way to help Mrs. Kim get back to live with her husband without losing face and to help her son find a way to get back to being a responsible father and a loyal son without "killing himself" in the process.

The separate interviews with the mother and son revealed the following. The son, the oldest of three sons, had come to this country about 12 years ago to seek honor for the family by earning a college degree and getting an important job. Instead, he fell in love with an American woman, married, had two children, and had to drop out of school in order to support the family. Only recently had he resumed his pursuit of his education part-time. The marriage had ended in divorce, and he had custody of the children. In order to show responsibility to his family, he had managed to bring his two younger brothers and his aging parents to the U.S. However unhappy and difficult their lives were here, none of the Kim family was in a position to return to Korea since the shame of failure in this country would be intolerable to face.

When the parents first came to this country a couple of years ago, they had moved in with the son and his children. It appears that the grandchildren were not properly respectful to their grandparents; they were noisy, expressed their opinions freely, and did not appreciate the cultural heritage or customs, at times making rude remarks about the "old world" ways of the grandparents. They complained loudly to their father about their grandparents' "unreasonable" expectations, strange food, and strange way of doing things. In order to make things easier for everyone, the son made arrangements for his parents to move into a housing project for the elderly.

In order to pay deference to her status and age, the author met with Mrs. Kim alone first. Using the authority and status of a consultant brought in by the hospital, the author talked to her about how impressed she was with her dutiful son and what a fine job she had done raising such a bright, responsible son. The therapist and Mrs. Kim talked about her family and how much she must be suffering in a strange land so far away from home. The therapist commiserated with her about the young people nowadays, particularly those children who were born and raised in American ways, about how selfish they were and so on. This was designed to align the author with Mrs. Kim and to lay the groundwork for reframing her suicide attempt as a selfless sacrifice. In order to do so, the author joined with Mrs. Kim in her view of the situation she had tried to leave.

Mrs. Kim was reluctant to talk at first, but she gradually warmed up. By the second visit with her in the hospital, she was more willing to open up and talk about her sense of failure as a mother and humiliation at being pushed out of her son's home by the grandchildren. Seizing on this opportunity of her willingness to talk about the circumstances, the author reframed the son's behavior as an expression of his misguided but well-intentioned sense of responsibility toward his own children, which the author was sure that the mother had instilled in him. It was pointed out that no mother would want to see her son fail as a parent since that would surely mean that she had failed as a mother. Besides, had she succeeded in her suicide attempt, she would have left her son to fail permanently in his children's eyes, and she certainly cared about her son more than that. On the other hand, the author told her how impressed she was that she took her duty as a mother seriously enough to want to end her life.

The therapist reminded her that—as she well knew already—young people not only have a responsibility toward parents but we, as the older generation, have a duty to protect and encourage our children to succeed. The therapist added that this duty included being happy with our spouse so that we free our children from worrying about us. It is our responsibility to give our children all the chances they need to succeed. Mrs. Kim agreed and thanked me for coming.

Mrs. Kim's view was supported and reframed as her attempt to fulfill her cultural role as an elder of the family. Moreover, her return to her husband was depicted as important and vital—a necessity for her son as he carried out the family honor in raising his children.

When talking with the son, the author complimented him on being a dutiful son and doing so much for his parents and his younger brothers. The author suggested that he must have been taught well by his mother since he was not only dutiful to his parents and to his brothers but also responsible to his children. Since a failure to take care of his children would indeed be a dishonor to his parents, he was urged to continue to work hard in order to help his own children to succeed. His world view was supported by the therapist and viewed as necessary.

Mrs. Kim was discharged to her husband's care soon after this session. A 1-year follow-up indicated that she was doing well and that there had been no recurrence of the depression.

The two interventions are interwoven and clearly address the norm of close-knit family ties often found in Asian-American families. Techniques of reframing in relationship contexts were used to respect the world view presented by each client. The tasks implied and suggested were designed to respect the client's view and failed attempts to solve the problem.

STRESS INOCULATION: PROCESS AND USES

Stress inoculation is an approach to teaching both physical and cognitive coping skills. Meichenbaum (1993) states that stress inoculation training "helps clients acquire sufficient knowledge, self-understanding, and coping skills to facilitate better ways of handling expected stressful encounters. Stress inoculation training combines elements of Socratic and didactic teaching, client self-monitoring, cognitive restructuring, problem solving, self-instructional and relaxation training, behavioral and imagined rehearsal, and environmental change" (p. 381). Eliot and Eisdorfer (1982) and Meichenbaum (1993) classify stressful events in the following types:

1. One event that is time limited and not chronic, such as a medical biopsy, surgery, a dental procedure, an oral examination.
2. One event that triggers a series of stressful reactions, such as job loss, divorce, death of loved one, natural or man-made disaster, or sexual assault.
3. Chronic and intermittent events, such as musical performances, athletic competitions, military combat, recurrent headaches.
4. Chronic and continual events, such as chronic medical or mental illness, marital conflict, chronic physical-emotional or psychological abuse, some professions—nursing, teaching, or policework. (Meichenbaum, 1993, p. 373)

Table 16-3 presents a sample of stress inoculation research used with a variety of concerns. These include academic performance, anger management and reduction, childhood asthma, athletic performance, burnout among nurses, dental treatment, hypertension, and unemployment.

SEVEN COMPONENTS OF STRESS INOCULATION

Stress inoculation* has seven major components:

1. Rationale
2. Information giving
3. Acquisition and practice of direct-action coping skills
4. Acquisition and practice of cognitive coping skills
5. Application of all coping skills to problem-related situations

---

*We wish to acknowledge the work of Meichenbaum (1993, 1994) in our presentation of stress management.

LEARNING ACTIVITY **45** *Reframing*

This activity is designed to help you use the reframing procedure with yourself.

1. Identify a situation that typically produces uncomfortable or distressing feelings for you. Examples:
   a. You are about to initiate a new relationship with someone.
   b. You are presenting a speech in front of a large audience.

2. Try to become aware of what you rather automatically attend to or focus on during this situation. Role-play it with another person or pretend you're sitting in a movie theater and project the situation onto the screen in front of you. As you do so, ask yourself:
   "What am I aware of now?"
   "What am I focusing on now?"
   Be sure to notice fleeting sounds, feelings, images, and sensations.

3. Establish a link between these encoded features of the situation and your resulting feelings. As you reenact the situation, ask yourself: "What am I feeling at this moment?" "What am I experiencing now?"

4. After you have become aware of the most salient features of this situation, reenact it either in role play or in imagination. This time, deliberately attend to these selected features during the reenactment. Repeat this process until you feel that you have awareness and control of the perceptual process you engage in during this situation.

5. Next, select other features of the session (previously ignored) that you could focus on or attend to during this situation that would allow you to view and handle the situation differently. Consider images, sounds, and feelings as well as people and objects. Ask yourself questions such as "What other aspects of this same situation aren't readily apparent to me that would provide me with a different way to view the situation?" You may wish to query another person for ideas. After you have identified alternative features, again reenact the situation in role play or imagination—several times if necessary—in order to break old encoding patterns.

6. Construct a homework assignment for yourself that encourages you to apply this process as needed for use during actual situations.

---

6. Application of all coping skills to potential problem situations
7. Homework and follow-up

A detailed description of each step associated with these seven parts can be found in the Interview Checklist for Stress Inoculation at the end of this chapter.

Treatment Rationale

Here is an example of a rationale that a counselor might use for stress inoculation.

Purpose. The counselor might explain as follows the purpose of stress inoculation for a client having trouble with hostility:

> You find yourself confronted with situations in which your temper gets out of hand. You have trouble managing your anger, especially when you feel provoked. This procedure can help you learn to cope with provoking situations and can help you manage the intensity of your anger when you're in these situations so it doesn't control you.

Overview. Then the counselor can give the client a brief overview of the procedure:

> First, we will try to help you understand the nature of your feelings and how certain situations may provoke your feelings and lead from anger to hostility. Next you will learn some ways to manage this and to cope with situations in which you feel this way. After you learn these coping skills, we will set up situations where you can practice using these skills to help you control your anger. How does this sound to you?

Information Giving

In this procedure, before learning and applying various coping strategies, the client should be given some information about the nature of a stress reaction and the possible coping strategies that might be used. It is helpful for the client to understand the nature of a stress reaction and how various coping strategies can help manage the stress. The education phase of stress inoculation helps the client conceptualize reactions to stressful events and builds a foundation for the components.

TABLE 16-3. Stress inoculation research

Academic performance

Kiselica, M. S., Baker, S. B., Thomas, R. N., & Reedy, S. (1994). Effects of stress inoculation training on anxiety, stress, and academic performance among adolescents. *Journal of Counseling Psychology, 41,* 335–342.

Anger

Deffenbacher, J., McNamara K., Stark, R., & Sabadell, P. (1991). A combination of cognitive relaxation and behavioral coping skills in the reduction of anger. *Journal of College Student Development, 26,* 114–212.

Wilcox, D., & Dowrick, P. W. (1992). Anger management with adolescents. *Residential Treatment for Children and Youth, 9,* 29–39.

Anxiety

Burnley, M. C., Cross, P. A., & Spanos, N. P. (1993). The effects of stress inoculation training and skills training on the treatment of speech anxiety. *Imagination, Cognition and Personality, 12,* 355–366.

Register, A. C., Beckham, J. C., May, J. G., & Gustafson, D. J. (1991). Stress inoculation bibliotherapy in the treatment of test anxiety. *Journal of Counseling Psychology, 38,* 115–119.

Salovey, P., & Haar, M. D. (1990). The efficacy of cognitive-behavior therapy and writing process training for alleviating writing anxiety. *Cognitive Therapy and Research, 14,* 513–526.

Saunders, T., Driskell, J., Johnston, J. H., & Salas, E. (1996). The effect of stress inoculation training on anxiety and performance. *Journal of Occupational Health Psychology, 1,* 170–186.

Schneider, W. J., & Nevid, J. S. (1993). Overcoming math anxiety: A comparison of stress inoculation training and systematic desensitization. *Journal of College Student Development, 34,* 283–288.

Asthma

Benedito-Monleon, C., & Lopez-Andreu, J. A. (1994). Psychological factors in childhood asthma. *Behavioural and Cognitive Psychotherapy, 22,* 153–161.

Athletic performance

Kerr, G., & Leith, L. (1993). Stress management and athletic performance. *Sport Psychologist, 7,* 221–231.

Burnout among nurses

Freedy, J. R., & Hobfoll, S. E. (1994). Stress inoculation for reduction of burnout: A conservation of resources approach. *Anxiety, Stress and Coping: An International Journal, 6,* 311–325.

Cancer

Elsesser, K., Van Berkel, M., Sartory, G., & Biermann-Gocke, W. (1994). The effects of anxiety management training on psychological variables and immune parameters in cancer patients: A pilot study. *Behavioural and Cognitive Psychotherapy, 22,* 13–23.

Children and adolescents

Maag, J. W., & Kotlash, J. (1994). Review of stress inoculation training with children and adolescents: Issues and recommendations. *Behavior Modification, 18,* 443–469.

Dental treatment

Law, A., Logan, H., & Baron, R. S. (1994). Desire for control, felt control, and stress inoculation training during dental treatment. *Journal of Personality and Social Psychology, 67,* 926–936.

Disaster work

Dunning, C. (1990). Mental health sequelae in disaster workers: Prevention and intervention. *International Journal of Mental Health, 19,* 91–103.

(continued)

Three things should be explained to the client: a framework for the client's emotional reaction, information about the phases of reacting to stress, and examples of types of coping skills and strategies.

Framework for Client's Reaction. In setting a framework, the counselor should first explain the nature of the client's reaction to a stressful situation. Although understanding one's reaction may not be sufficient for changing it, the conceptual framework lays some groundwork for beginning the change process. Usually an explanation of some kind of stress (anxiety, hostility, pain) involves describing the stress as having two components: physiological arousal and covert self-statements or thoughts that provoke anxiety, hostility, or pain. This explanation may help the client realize that coping strategies must be directed toward the arousal behaviors *and* the cognitive processes. For example, to describe this framework to a client who has trouble controlling hostility, the counselor could say something like this:

> Perhaps you could think about what happens when you get very angry. You might notice that certain things happen to you physically—perhaps your body feels tight, your face may feel warm, you may experience more rapid breathing, or your heart may pound. This is the physical part of your anger. However, there is another thing that probably goes on while you're very angry—that is, what you're thinking. You might be thinking such things as "He had no right to attack me; I'll get back at him; Boy, I'll show him who's boss; I'll teach her to keep her mouth shut," and so on. These kinds of thoughts only intensify your anger. So the way you interpret and think about an

TABLE 16-3. Stress inoculation research (continued)

Exercise

Long, B. C. (1993). Aerobic conditioning (jogging) and stress inoculation interventions. An exploratory study of coping. Special Issue: Exercise and psychological well-being. *International Journal of Sport Psychology, 24,* 94–109.

Headache

Cruzado-Rodriguez, J. A., & Labrador-Encinas, F. J. (1990). Effects of stress inoculation training on tension headaches. *Revista de Psicologia General y Aplicada, 43,* 353–367.

Hypertension

Amigo, I., Buceta, J. M., Becona, E., & Bueno, A. M. (1991). Cognitive behavioural treatment for essential hypertension: A controlled study. *Stress Medicine, 7,* 103–108.

Older adults

Hayslip, B. (1989). Alternative mechanisms for improvements in fluid ability performance among older adults. *Psychology and Aging, 4,* 122–124.

Lopez, M. A., & Silber, A. (1991). Stress management for the elderly: A preventive approach. *Clinical Gerontologist, 10,* 73–76.

Pain

Ross, M., & Berger, R. (1996). Effects of stress inoculation training on athletes' postsurgical pain and rehabilitation after orthopedic injury. *Journal of Consulting and Clinical Psychology, 64,* 406–410.

Whitmarsh, B. G., & Alderman, R. B. (1993). Role of psychological skills training in increasing athletic pain tolerance. *Sport Psychologist, 7,* 388–399.

Parents

Jay, S. M., & Elliott, C. H. (1990). A stress inoculation program for parents whose children are undergoing painful medical procedures. *Journal of Consulting and Clinical Psychology, 58,* 799–804.

Prevention

Hains, A. A., & Ellmann, S. W. (1994). Stress inoculation training as a preventative intervention for high school youths. *Journal of Cognitive Psychotherapy, 8,* 219–232.

Rape victims

Foa, E. B., Rothbaum, B. O., Riggs, D. S., & Murdock, T. B. (1991). Treatment of posttraumatic stress disorder in rape victims: A comparison between cognitive-behavioral procedures and counseling. *Journal of Consulting and Clinical Psychology, 59,* 715–723.

Roth, S., & Newman, E. (1991). The process of coping with sexual trauma. *Journal of Traumatic Stress, 4,* 279–297.

Staff workers

Keyes, J. B., & Dean, S. F. (1988). Stress inoculation training for direct contact staff working with mentally retarded persons. *Behavioral Residential Treatment, 3,* 315–323.

Safety

Hytten, K., Jensen, A., & Skauli, G. (1990). Stress inoculation training for smoke divers and free fall lifeboat passengers. *Aviation, Space, and Environmental Medicine, 61,* 983–988.

Stepcouples

Fausel, D. (1995). Stress inoculation training for stepcouples. *Marriage and Family Review, 21,* 137–155.

Teacher stress

Cecil, M. A., & Forman, S. G. (1990). Effects of stress inoculation training and coworker support groups on teachers' stress. *Journal of School Psychology, 28,* 105–118.

Unemployment

Caplan, R. D., Vinokur, A. D., Price, R. H., & van Ryn, M. (1989). Job seeking, reemployment, and mental health: A randomized field experiment in coping with job loss. *Journal of Applied Psychology, 74,* 10–20.

---

anger-provoking situation also contributes to arousing hostile feelings.

(Note that in the case of this and related examples, we are differentiating between appropriate, legitimate feelings of anger and hostility that leads to abuse or damage.)

Phases of Stress Reactions. After explaining a framework for emotional arousal, the counselor should describe the kinds of times or phases when the client's arousal level may be heightened. For example, phobic clients may view their anxiety as one "massive panic reaction." Similarly, clients who are angry, depressed, or in pain may interpret their feelings as one large, continuous reaction that has a certain beginning and end. Clients who interpret their reactions this

way may perceive the reaction as too difficult to change because it is so massive and overwhelming.

One way to help the client see the potential for coping with feelings is to describe the feelings by individual stages or phases of reacting to a situation. Meichenbaum (1993, 1994) used four stages to help the client conceptualize the various critical points of a reaction: (1) preparing for a stressful, painful, or provoking situation; (2) confronting and handling the situation or the provocation; (3) coping with critical moments or with feelings of being overwhelmed or agitated during the situation; and (4) rewarding oneself after the stress for using coping skills in the first three phases. Explanation of these stages in the preliminary part of stress inoculation helps the client understand the sequence of coping strategies to be learned. To explain the client's reaction as a series of phases, the counselor might say something like this:

> When you think of being angry, you probably just think of being angry for a continuous period of time. However, you might find that your anger is probably not just one big reaction but comes and goes at different points during a provoking situation. The first critical point is when you anticipate the situation and start to get angry. At this point you can learn to prepare yourself for handling the situation in a manageable way. The next point may come when you're in the middle of the situation and you're very angry. Here you can learn how to confront a provoking situation in a constructive way. There might also be times when your anger really gets intense and you can feel it starting to control you—and perhaps feel yourself losing control. At this time, you can learn how to cope with intense feelings of agitation. Then, after the situation is over, instead of getting angry with yourself for the way you handled it, you can learn to encourage yourself for trying to cope with it. In this procedure, we'll practice using the coping skills at these especially stressful or arousing times.

Information about Coping Skills and Strategies.   Finally, the counselor should provide some information about the kinds of coping skills and strategies that can be used at these critical points. The counselor should emphasize that there is a *variety* of useful coping skills; clients' input in selecting and tailoring these for themselves is *most* important. Allow clients to choose coping strategies that reflect their own preferences. In using stress inoculation, both "direct action" and "cognitive" coping skills are taught (Meichenbaum, 1993). *Direct-action* coping strategies are designed to help the client use coping behaviors to handle the stress; *cognitive* coping skills are used to give the client coping thoughts (self-statements) to handle the stress. The client should understand that *both* kinds of coping skills are important and that the two serve different functions, although some clients may prefer to rely more on one type than another, depending on their gender and culture. To provide the client with information about the usefulness of these coping skills, the counselor might explain them this way:

> In the next phase of this procedure, you'll be learning a lot of different ways to prepare for and handle a provoking situation. Some of these coping skills will help you learn to cope with provoking situations by your actions and behaviors; others will help you handle these situations by the way you interpret and think about the situation. Not all the strategies you learn may be useful or necessary for you, so your input in selecting the ones you prefer to use is important.

Acquisition and Practice of Direct-Action Coping Skills

In this phase of stress inoculation, the client acquires and practices some direct-action coping skills. The counselor first discusses and models possible action strategies; the client selects some to use and practices them with the counselor's encouragement and assistance. As you may recall, direct-action coping skills are designed to help the client acquire and apply coping behaviors in stressful situations. The most commonly used direct-action coping strategies are these:

1. Collecting objective or factual information about the stressful situation
2. Identifying short-circuit or escape routes or ways to decrease the stress
3. Palliative coping strategies
4. Mental relaxation methods
5. Physical relaxation methods

Information Collection.   Collecting objective or factual information about a stressful situation may help the client evaluate the situation more realistically. The process of problem identification is very helpful in collecting information (see Chapter 9). Moreover, information about a situation may reduce the ambiguity for the client and indirectly reduce the level of threat. For example, for a client who may be confronted with physical pain, information about the source and expected timing of pain can reduce stress. This coping method is widely used in childbirth classes. The women and their "labor coaches" are taught and shown that the experienced pain is actually a uterine contraction. They are given information about the timing and stages of labor and the timing and intensity of contractions so that when labor occurs, their anxiety will not be increased by misunderstanding or lack of information about what is happening in their bodies.

Collecting information about an anxiety- or anger-engendering situation serves the same purpose. For example, in using stress inoculation to help clients control anger, collecting information about the people who typically provoke them may help. Clients can collect information that can help them view provocation as a *task* or a problem to be solved, rather than as a *threat* or a personal attack.

Identification of Escape Routes.   Identifying escape routes is a way to help the client cope with stress before it gets out of hand. The idea of an escape route is to short-circuit the explosive or stressful situation or to deescalate the stress before the client behaves in a way that may "blow it." This coping strategy may help abusive clients learn to identify cues that elicit their physical or verbal abuse and to take some preventive action before striking out. This is similar to the stimulus-control self-management strategy discussed in Chapter 20. These escape or prevention routes can be very simple things that the client can *do* to prevent losing control or losing face in the situation. An abusive client could perhaps avoid striking out by counting to 60, leaving the room, or talking about something humorous.

Palliative Coping Strategies.   Meichenbaum (1993, 1994) describes palliative coping strategies that may be particularly useful for aversive or stressful situations that cannot be substantially altered or avoided, such as chronic or life-threatening illnesses:

> Train emotionally focused palliative coping skills, especially when the client has to deal with unchangeable and uncontrollable stressors (e.g., perspective taking, selective attention diversion procedures, as in the case of chronic pain patients; adaptive modes of affective expression such as humor, relaxation, and reframing the situation). (Meichenbaum, 1993, p. 384)

Mental Relaxation.   Mental relaxation can also help clients cope with stress. This technique may involve attention-diversion tactics: angry clients can control their anger by concentrating on a problem to solve, counting floor tiles in the room, thinking about a funny or erotic joke, or thinking about something positive about themselves. Attention-diversion tactics are commonly used to help people control pain. Instead of focusing on the pain, the person may concentrate very hard on an object in the room or on the repetition of a word (a mantra) or a number. Again, in the Lamaze method of childbirth, women are taught to concentrate on a "focal point" such as an object in the room or, as the authors used, a picture of a sailboat. In this way, the woman's attention is directed to an object instead of to the tightening sensations in her lower abdomen.

Some people find that mental relaxation is more successful when they use imagery or fantasy. People who enjoy daydreaming or who report a vivid imagination may find imagery a particularly useful way to promote mental relaxation. Generally, imagery as a coping method helps the client go on a fantasy trip instead of focusing on the stress, the provocation, or the pain. For example, instead of thinking about how anxious or angry he feels, the client might learn to fantasize about lying on a warm beach, being on a sailboat, making love, or eating a favorite food (see "Guided Imagery" in Chapter 13). For pain control, the person can imagine different things about the pain. A woman in labor can picture the uterus contracting like a wave instead of thinking about pain. Or a person who experiences pain from a routine source, such as extraction of a wisdom tooth, can use imagery to change the circumstances producing the pain. Instead of thinking about how terrible and painful it is to have a tooth pulled, the person can imagine that the pain is only the aftermath of intense training for a marathon race or is from being the underdog who was hit in the jaw during a fight with the world champion.

Physical Relaxation.   Physical relaxation methods are particularly useful for clients who report physiological components of anxiety and anger, such as sweaty palms, rapid breathing or heartbeat, or nausea. Physical relaxation is also a very helpful coping strategy for pain control, because body tension will heighten the sensation of pain. Physical relaxation may consist of muscle relaxation or breathing techniques. Chapters 17 and 18 describe these procedures in more detail.

Each direct-action strategy should first be explained to the client with discussion of its purpose and procedure. Several sessions may be required to discuss and model all the possible direct-action coping methods. After the strategies have been described and modeled, the clients should select the particular methods to be used. The number of direct-action strategies used by a client will depend on the intensity of the reaction, the nature of the stress, and the client's preferences. With the counselor's assistance, the client should practice using each skill in order to be able to apply it in simulated and in vivo situations.

Acquisition and Practice of Cognitive Coping Skills

This part of stress inoculation—the acquisition and practice of cognitive coping skills—is very similar to the cognitive restructuring strategy described earlier in this chapter. The counselor models some examples of coping thoughts the client can use during stressful phases of problem situations; then the client selects and practices substituting coping thoughts for negative or self-defeating thoughts.

Description of Four Phases of Cognitive Coping.   As you remember from our discussion of information giving, the counselor helps the client understand the nature of an emotional reaction by conceptualizing the reaction by phases. In helping the client acquire cognitive coping skills, the counselor may first wish to review the importance of learning to cope at crucial times. The counselor can point out that the client can learn a set of cognitive coping skills for each important phase: preparing for the situation, confronting and handling the situation, coping with critical moments in the situation, and stroking himself or herself after the situation. Note that the first phase concerns coping skills *before* the situation; the second and third phases, coping *during* the situation; and the fourth phase, coping *after* the situation. The counselor can describe these four phases to the client with an explanation similar to this:

> Earlier we talked about how your anger is not just one giant reaction but something that peaks at certain stressful points when you feel provoked or attacked. Now you will learn a method of cognitive control that will help you control any negative thoughts that may lead to hostility and also help you use coping thoughts at stressful points. There are four times that are important in your learning to use coping thoughts, and we'll work on each of these four phases. First is how you interpret the situation initially, and how you think about responding or preparing to respond. Second is actually dealing with the situation. Third is coping with anything that happens during the situation that *really* provokes you. After the situation, you learn to encourage yourself for dealing with your feelings in a way that is not hurtful.

Modeling Coping Thoughts. After explaining the four phases of using cognitive coping skills to the client, the counselor would model examples of coping statements that are especially useful for each of the four phases.

Meichenbaum (1994) and Meichenbaum and Turk (1976) have provided an excellent summary of the coping statements used by Meichenbaum and Cameron (1973) for anxiety control, by Novaco (1975) for anger control, and by Turk (1975) for pain control. These statements, presented in Table 16-4, are summarized for each of the four coping phases: preparing for the situation, confronting the situation, coping with critical moments, and reinforcing oneself for coping. The counselor would present examples of coping statements for each of the four phases of a stress reaction.

Client Selection of Coping Thoughts. After the counselor models some possible coping thoughts for each phase, the client should add some or select those that fit. The counselor should encourage the client to "try on" and adapt the thoughts in whatever way feels most natural. The client might look for coping statements he or she has used in other stress-related situations. At this point in the procedure, the counselor should be helping to tailor a coping program *specifically* for this client. If the client's self-statements are too general, they may lead only to "rote repetition" and not function as effective self-instructions. Also, specific coping

statements are more likely to be culturally relevant. The counselor might explain the importance of the client's participation like this:

> You know, your input in finding coping thoughts that work for you is very important. I've given you some examples. Some of these you might feel comfortable with, and there may be others you can think of too. What we want to do now is to come up with some specific coping thoughts you can and will use during these four times that fit for *you,* not me or someone else.

Client Practice of Coping Thoughts. After the client selects coping thoughts to use for each phase, the counselor will instruct the client to practice these self-statements by saying them aloud. This verbal practice is designed to help the client become familiar with the coping thoughts and accustomed to the words. After this practice, the client should also practice the selected coping thoughts in the sequence of the four phases. This practice helps the client learn the timing of the coping thoughts in the application phase of stress inoculation. The counselor can say something like this:

> First I'd like you to practice using these coping thoughts just by saying them aloud to me. This will help you get used to the words and ideas of coping. Next, let's practice these coping thoughts in the sequence in which you would use them when applying them to a real situation. Here, I'll show you. OK, first I'm anticipating the situation, so I'm going to use coping statements that help me prepare for the situation, like "I know this type of situation usually upsets me, but I have a plan now to handle it" or "I'm going to be able to control my anger even if this situation is rough." Next, I'll pretend I'm actually into the situation. I'm going to cope so I can handle it. I might say something to myself like "Just stay calm. Remember who I'm dealing with. This is her style. Don't take it personally" or "Don't overreact. Just relax."
>
> OK, now the person's harassment is continuing. I am going to cope with feeling more angry. I might think "I can feel myself getting more upset. Just keep relaxed. Concentrate on this" or "This is a challenging situation. How can *I* handle myself in a way I don't have to apologize for?" OK, now afterward I realize I didn't get abusive or revengeful. So I'll think something to encourage myself, like "I did it" or "Gee, I really kept my cool."
>
> Now you try it. Just verbalize your coping thoughts in the sequence of preparing for the situation, handling it, coping with getting really agitated, and then encouraging yourself.

Application of All Coping Skills to Problem-Related Situations

The next part of stress inoculation involves having the client apply both the direct-action and the cognitive coping skills in

TABLE 16-4. Examples of coping thoughts used in stress inoculation

Anxiety	Anger	Pain
I. *Preparing for a stressor* (Meichenbaum & Cameron, 1973) What is it you have to do? You can develop a plan to deal with it. Just think about what you can do about it. That's better than getting anxious. No negative self-statements; just think rationally. Don't worry; worry won't help anything. Maybe what you think is anxiety is eagerness to confront it.	*Preparing for a provocation* (Meichenbaum, 1994) This is going to upset me, but I know how to deal with it. What is it that I have to do? I can work out a plan to handle this. If I find myself getting upset, I'll know what to do. There won't be any need for an argument. Try not to take this too seriously. This could be a testy situation, but I believe in myself. Time for a few deep breaths of relaxation. Feel comfortable, relaxed, and at ease. Easy does it. Remember to keep your sense of humor.	*Preparing for the painful stressor* (Turk, 1975) What is it you have to do? You can develop a plan to deal with it. Just think about what you have to do. Just think about what you can do about it. Don't worry; worrying won't help anything. You have lots of different strategies you can call upon.
II. *Confronting and handling a stressor* (Meichenbaum & Cameron, 1973) Just "psych" yourself up—you can meet this challenge. One step at a time; you can handle the situation. Don't think about fear; just think about what you have to do. Stay relevant. This anxiety is what the doctor said you would feel. It's a reminder to use your coping exercises. This tenseness can be an ally, a cue to cope. Relax; you're in control. Take a slow deep breath. Ah, good.	*Impact and confrontation* (Meichenbaum, 1994) Stay calm. Just continue to relax. As long as I keep my cool, I'm in control. Just roll with the punches; don't get bent out of shape. Think of what you want to get out of this. You don't need to prove yourself. There is no point in getting mad. Don't make more out of this than you have to. I'm not going to let him get to me. Look for the positives. Don't assume the worst or jump to conclusions. It's really a shame that he has to act like this. For someone to be that irritable, he must be awfully unhappy. If I start to get mad, I'll just be banging my head against the wall. So I might as well relax. There is no need to doubt myself. What he says doesn't matter. I'm on top of this situation and it's under control.	*Confronting and handling the pain* (Turk, 1975) You can meet the challenge. One step at a time; you can handle the situation. Just relax, breathe deeply and use one of the strategies. Don't think about the pain, just what you have to do. This tenseness can be an ally, a cue to cope. Relax. You're in control; take a slow deep breath. Ah. Good. This anxiety is what the trainer said you might feel. That's right; it's the reminder to use your coping skills.

(continued)

TABLE 16-4. Examples of coping thoughts used in stress inoculation (continued)

Anxiety	Anger	Pain
III. *Coping with the feeling of being over-whelmed* (Meichenbaum & Cameron, 1973) When fear comes, just pause. Keep the focus on the present; what is it you have to do? Label your fear from 0 to 10 and watch it change. You should expect your fear to rise. Don't try to eliminate fear totally; just keep it manageable. You can convince yourself to do it. You can reason your fear away. It will be over shortly. It's not the worst thing that can happen. Just think about something else. Do something that will prevent you from thinking about fear. Describe what is around you. That way you won't think about worrying.	*Coping with arousal* (Meichenbaum, 1994) My muscles are starting to feel tight. Time to relax and slow things down. Getting upset won't help. It's just not worth it to get so angry. I'll let him make a fool of himself. I have a right to be annoyed, but let's keep the lid on. Time to take a deep breath. Let's take the issue point by point. My anger is a signal of what I need to do. Time to instruct myself. I'm not going to get pushed around, but I'm not going haywire either. Try to reason it out. Treat each other with respect. Let's try a cooperative approach. Maybe we are both right. Negatives lead to more negatives. Work constructively. He'd probably like me to get really angry. Well I'm going to disappoint him. I can't expect people to act the way I want them to. Take it easy, don't get pushy.	*Coping with feelings at critical moments* (Turk, 1975) When pain comes just pause; keep focusing on what you have to do. What is it you have to do? Don't try to eliminate the pain totally; just keep it manageable. You were supposed to expect the pain to rise; just keep it under control. Just remember, there are different strategies; they'll help you stay in control. When the pain mounts you can switch to a different strategy; you're in control.

(continued)

the face of stressful, provoking, or painful situations. Before the client is instructed to apply the coping skills in vivo, she or he practices applying coping skills under simulated conditions with the counselor's assistance. The application phase of stress inoculation appears to be important for the overall efficacy of the procedure. Simply having a client rehearse coping skills *without* opportunities to apply them in stressful situations seems to result in an improved but limited ability to cope.

The application phase involves modeling and rehearsing to provide the client with exposure to simulations of problem-related situations. For example, the client who wanted to manage hostility would have opportunities to practice coping in a variety of hostility-provoking situations. During this application practice, the client needs to be faced with a stressful situation and also practice the skills in a coping manner. In other words, the application should be arranged and conducted as realistically as possible. The hostile client can be encouraged to practice feeling very agitated and to rehearse even starting to lose control—but

then applying the coping skills to gain control (Novaco, 1975). This type of application practice is viewed as the client's providing a self-model of how to behave in a stressful situation. By imagining faltering or losing control, experiencing anxiety, and then coping with this, the person practices the thoughts and feelings as they are likely to occur in a real-life situation (Meichenbaum, 1994). In the application phase of stress inoculation, the client's anxiety, hostility, or distressful emotions are used as a cue or reminder to cope.

Modeling of Application of Coping Skills. The counselor should first model how the client can apply the newly acquired skills in a coping manner when faced with a stressful situation. Here is an example of a counselor demonstration of this process with a client who is working toward hostility control (in this case, with his family):

> I'm going to imagine that the police have just called and told me that my 16-year-old son was just picked up again for breaking and entering. I can feel myself start to get really hot. Whoops, wait a minute. That's a signal [arousal cue for

TABLE 16-4. Examples of coping thoughts used in stress inoculation (continued)

Anxiety	Anger	Pain
IV. *Reinforcing self-statements* (Meichenbaum & Cameron, 1973) It worked; you did it. Wait until you tell your therapist about this. It wasn't as bad as you expected. You made more out of the fear than it was worth. Your damn ideas—that's the problem. When you control them, you control your fear. It's getting better each time you use the procedures. You can be pleased with the progress you're making. You did it!	*Reflecting on the provocation* (Meichenbaum, 1994) a. *When conflict is unresolved* Forget about the aggravation. Thinking about it only makes you upset. These are difficult situations, and they take time to straighten out. Try to shake it off. Don't let it interfere with your job. I'll get better at this as I get more practice. Remember relaxation. It's a lot better than anger. Don't take it personally. Take a deep breath. b. *When conflict is resolved or coping is successful* I handled that one pretty well. It worked! That wasn't as hard as I thought. It could have been a lot worse. I could have gotten more upset than it was worth. I actually got through that without getting angry. My pride can sure get me into trouble, but when I don't take things too seriously, I'm better off. I guess I've been getting upset for too long when it wasn't even necessary. I'm doing better at this all the time.	*Reinforcing self-statements* (Turk, 1975) Good, you did it. You handled it pretty well. You knew you could do it! Wait until you tell the trainer about which procedures worked best.

SOURCE: From *A Clinical Handbook: Practical Therapist Manual for Assessing and Treating Adults with Post Traumatic Stress Disorder,* by D. Meichenbaum, pp. 407–408. Copyright 1994 by Institute Press. Originally published in "The Cognitive-Behavioral Management of Anxiety, Anger, and Pain," by D. Meichenbaum and D. Turk, in *The Behavioral Management of Anxiety, Depression, and Pain,* by P. O. Davidson (Ed.). Copyright 1976 by Brunner/Mazel. Reprinted by permission of the author.

coping]. I'd better start thinking about using my relaxation methods to stay calm and using my coping thoughts to prepare myself for handling this situation constructively.

OK, first of all, sit down and relax. Let those muscles loosen up. Count to ten. Breathe deeply [direct-action coping methods]. OK, now I'll be seeing my son shortly. What is it I have to do? I know it won't help to lash out or to hit him. That won't solve anything. So I'll work out another plan. Let him do most of the talking. Give him the chance to make amends or find a solution [cognitive coping: preparing for the situation]. OK, now I can see him walking in the door. I feel sort of choked up. I can feel my fists getting tight. He's starting to explain. I want to interrupt and let him have it. But wait [arousal cue for

coping]. Concentrate on counting and on breathing slowly [direct-action coping]. Now just tell myself—keep cool. Let him talk. It won't help now to blow up [cognitive coping: confronting situation]. Now I can imagine myself thinking back to the last time he got arrested. Why in the hell doesn't he learn? No son of mine is going to be a troublemaker [arousal]. Whew! I'm pretty damn angry. I've got to stay in control, especially now [cue for coping]. Just relax, muscles! Stay loose [direct-action coping]. I can't expect him to live up to my expectations. I can tell him I'm disappointed, but I'm not going to blow up and shout and hit [cognitive coping: feelings of greater agitation]. OK, I'm doing a good job of keeping my lid on [cognitive coping: self-reinforcement].

Client Application of Coping Skills in Imaginary and Role-Play Practice. After the counselor modeling, the client should practice a similar sequence of both direct-action and cognitive coping skills. The practice can occur in two ways: imagination and role play. We find it is often useful to have the client first practice the coping skills while imagining problem-related situations. This practice can be repeated until the client feels very comfortable in applying the coping strategies to imagined situations. Then the client can practice the coping skills with the counselor's aid in a role play of a problem situation. The role-play practice should be similar to the in vivo situations the client encounters. For instance, our angry client could identify particular situations and people with whom he or she is most likely to blow up or lose control. The client can imagine each situation (starting with the most manageable one) and imagine using the coping skills. Then, with the counselor taking the part of someone else such as a provoker, the client can practice the coping skills in role play.

Application of All Coping Skills to Potential Problem Situations

Any therapeutic procedure should be designed not only to help clients deal with current problems but also to help them anticipate constructive handling of potential problems. In other words, an adequate therapeutic strategy should prevent future problems as well as resolve current ones. The prevention aspect of stress inoculation is achieved by having clients apply the newly learned coping strategies to situations that are not problematic now but could be in the future. If this phase of stress inoculation is ignored, the effects of the inoculation may be very short-lived. In other words, if clients do not have an opportunity to apply the coping skills to situations other than the current problem-related ones, their coping skills may not generalize beyond the present problem situations.

Application of coping skills to other potentially stressful situations is accomplished in the same way as application to the present problem areas. First, after explaining the usefulness of coping skills in other areas of the client's life, the counselor demonstrates the application of coping strategies to a potential, hypothetical stressor. The counselor might select a situation the client has not yet encountered, one that would require active coping of anyone who might encounter it, such as not receiving a desired job promotion or raise, facing a family crisis, moving to a new place, anticipating retirement, or being very ill. After the counselor has modeled

application of coping skills to these sorts of situations, the client would practice applying the skills in these situations or in similar ones that she or he identifies. The practice can occur in imagination or in role-play enactments. A novel way to practice is to switch roles—the client plays the counselor and the counselor plays the role of the client. The client helps or trains the counselor to use the coping skills. Placing the client in the role of a helper or a trainer may provide another kind of application opportunity that may also have benefits for the client's acquisition of coping strategies, and bolster the client's self-efficacy (see Chapter 20).

Homework and Follow-up

When the client has learned and used stress inoculation within the interviews, she or he is ready to use coping skills in vivo. The counselor and client should discuss the potential application of coping strategies to actual situations. The counselor might caution the client not to expect to cope beautifully with every problematic situation initially. The client should be encouraged to use a daily log to record the particular situations and the number of times the coping strategies are used. The log data can be used in a later follow-up as one way to determine the client's progress.

In our opinion, stress inoculation training is one of the most comprehensive therapeutic treatments presently in use. Teaching clients both direct-action and cognitive coping skills that can be used in current and potential problematic situations provides skills that are under the clients' own control and are applicable to future as well as current situations.

Cultural Variations in Stress

The model of stress and the particular intervention that we present in this chapter are decidedly European American in nature and format, and therefore not optimally relevant to every client. Cultural differences exist in both the concept and the management of stress. Among some cultural groups in India, stress is not viewed as a major problem and certainly not a problem to bring to an "expert" (Laungani, 1993). Instead, stress is viewed as a part of life. There is a widespread acceptance of a magical explanation of stress, and persons particularly qualified to remove spells and exorcise evil spirits are often utilized. People seeking relief from various problems also make token offerings to deities and pilgrimages to shrines. Further, there is a greater reliance on self-healing forms of managing stress such as meditation and yoga. We describe these in the following chapters.

LEARNING ACTIVITY **46** *Stress Inoculation*

I. Listed below are 12 examples of various direct-action coping skills. Using the coding system that precedes the examples, identify on paper the *type* of direct-action coping skill displayed in each example. Feedback follows on page 414.

> *Code*
> Information (I)
> Escape route (ER)
> Social support network (SSN)
> Ventilation (V)
> Perspective taking (PT)
> Attention diversion (AD)
> Imagery manipulations (IM)
> Muscle relaxation (MR)
> Breathing techniques (B)

Examples

1. "Learn to take slow, deep breaths when you feel especially tense."
2. "Instead of thinking just about the pain, try to concentrate very hard on one spot on the wall."
3. "Imagine that it's a very warm day and the warmth makes you feel relaxed."
4. "If it really gets to be too much, just do the first part only—leave the rest for a while."
5. "You can expect some pain, but it is really only the result of the stitches. It doesn't mean that something is wrong."
6. "Just tighten your left fist. Hold it and notice the tension. Now relax it—feel the difference."
7. "Try to imagine a strong, normal cell attacking the weak, confused cancer cells when you feel the discomfort of your treatment."

8. "When it gets very strong, distract yourself—listen hard to the music or study the picture on the wall."
9. "If you talk about it and express your feelings about the pain, you might feel better."
10. "Your initial or intuitive reaction might cause you to see only selected features of the situation. There are also some positive aspects we need to focus on."
11. "It would be helpful to have your family and neighbors involved to provide you feedback and another perspective."
12. "Social skills are important for you to learn in order to develop the support you need from other people. Others can lessen the effects of the aversive situation."

II. Listed below are eight examples of cognitive coping skills used at four phases: preparing for a situation, confronting or handling the situation, dealing with critical moments in the situation, and self-encouragement for coping. On paper, identify which phase is represented by each example. Feedback follows.

1. "By golly, I did it."
2. "What will I need to do?"
3. "Don't lose your cool even though it's tough now. Take a deep breath."
4. "Think about what you want to say—not how people are reacting to you now."
5. "Relax, it will be over shortly. Just concentrate on getting through this rough moment now."
6. "Can you feel this—the coping worked!"
7. "When you get in there, just think about the situation, not your anxiety."
8. "That's a signal to cope now. Just keep your mind on what you're doing."

There are also variations in the prevalence of stress and stressors among various cultural groups. For example, in the United States, a much higher number of women than men are diagnosed with anxiety and stress-related disorders (Barlow & Durand, 1995). Post-traumatic stress syndromes are also seen in non-western countries where highly patriarchal systems of social order may be maintained by violence against women. Also, people who feel marginalized from the mainstream cultural group, such as persons of color, gays and lesbians, the elderly, and the physically and mentally challenged, "are more likely to experience increased numbers of stressful events in the form of discrimination, poverty, humiliation, and harassment" (Castillo, 1997, p. 173). Stress

syndromes are also highly prevalent among refugee groups because of related traumas.

Specific presentations of anxiety and stress in other countries, as summarized by Castillo (1997), include

1. The *dhat* syndrome in India, characterized by somatic symptoms and anxiety over a loss of semen (which is thought to result in loss of spiritual power).
2. *Nervios* in Latin American countries, characterized by generalized anxiety along with a range of principal symptoms.
3. *Hwa-byung* syndrome in Korea, characterized by anxiety and somatic symptoms often associated with the suppression of anger.

F E E D B A C K 46
Stress Inoculation

I. 1. B
   2. AD
   3. IM
   4. ER
   5. I
   6. MR
   7. IM
   8. AD
   9. V
   10. PT
   11. SSN
   12. SSN

If this was difficult for you, you might review the information presented in the text on direct-action coping skills.

II. 1. Encouraging phase
   2. Preparing for the situation
   3. Dealing with a critical moment
   4. Confronting the situation
   5. Dealing with a critical moment
   6. Encouragement for coping
   7. Preparing for the situation
   8. Confronting the situation

If you had trouble identifying the four phases of cognitive coping skills, you may want to review Table 16-4.

(See Castillo's book for a complete discussion of these cultural variations.)

MODEL DIALOGUE: STRESS INOCULATION

Session 1

In this session, the counselor will teach Joan some direct-action coping skills for mental and physical relaxation to help her cope with her physical sensations of nervousness about her math class. Imagery manipulations and slow, deep breathing will be used.

1. *Counselor:* Hi, Joan. How was your week?
   Client: Pretty good. You know, this, well, whatever you call it, it's starting to help. I took a test this week and got an 85—I usually get a 70 or 75.

The counselor introduces the **idea of other coping skills to deal with Joan's nervousness.**

2. *Counselor:* That really is encouraging. And that's where the effects of this count—on how you do in class. Because what we did last week went well for you, I believe today we might work with some other coping skills that might help you decrease your nervous feelings.
   Client: What would this be?

In responses 3 and 4, the counselor explains and **models possible direct-action coping skills.**

3. *Counselor:* Well, one thing we might do is help you learn how to imagine something that gives you very calm feelings, and while you're doing this to take some slow, deep breaths—like this [counselor models closing eyes, breathing slowly and deeply]. When I was doing that, I thought about curling up in a chair with my favorite book—but there are many different things you could think of. For instance, to use this in your math class, you might imagine that you are doing work for which you will receive some prize or award. Or you might imagine that you are learning problems so you'll be in a position to be a helper for someone else. Do you get the idea?
   Client: I think so. I guess it's like trying to imagine or think about math in a pretend kind of way.

4. *Counselor:* Yes—and in a way that reduces rather than increases the stress of it for you.
   Client: I think I get the idea. It's sort of like when I imagine that I'm doing housework for some famous person instead of just at my house—it makes it more tolerable.

In response 5, the counselor asks Joan to **find some helpful imagery manipulations to promote calm feelings.**

5. *Counselor:* That's a good example. You imagine that situation to prevent yourself from getting too bored. Here, you find a scene or scenes to think about to prevent yourself from getting too nervous. Can you take a few minutes to think about one or two things you could imagine—perhaps about math—that would help you feel calm instead of nervous?
   Client: (Pauses) Well, maybe I could pretend that the math class is part of something I need in order to do something exciting, like being an Olympic downhill skier.

In responses 6 and 7, the counselor instructs Joan to **practice these direct-action coping skills.**

6. *Counselor:* OK, good. We can work with that, and if it doesn't help, we can come up with something else. Why don't you try first to practice imagining this while you also breathe slowly and deeply, as I did a few minutes ago? [Joan practices.]

7. *Counselor:* OK. How did that feel?
   Client: OK—it was sort of fun.

In response 8, the counselor gives **homework—** *asks Joan to engage in* **self-directed practice** *of these coping skills before the next session.*

8. *Counselor:* Good. Now this week I'd like you to practice this in a quiet place two or three times each day. Keep track of your practice sessions in your log and also rate your tension level before and after you practice. Next week we will go over this log and then work on a way you can apply what we did today—and the coping thoughts we learned in our two previous sessions. So I'll see you next week.

Session 2

In this session, the counselor helps Joan integrate the strategies of some previous sessions (coping thoughts, and imagery and breathing coping skills). Specifically, Joan learns to apply all these coping skills in imagery and role-play practices of some stressful situations related to math class. Application of coping skills to problem-related situations is a part of stress inoculation and helps the client to generalize the newly acquired coping skills to in vivo situations as they occur.

In responses 1 and 2, the counselor will **review Joan's use of the direct-action skills homework.**

1. *Counselor:* How are things going, Joan?
   Client: OK. I've had a hard week—one test and two pop quizzes in math. But I got 80s. I also did my imagination and breathing practice. You know, that takes a while to learn.

2. *Counselor:* That's natural. It does take a lot of practice before you really get the feel of it. So it would be a good idea if you continued the daily practice again this week. How did it feel when you practiced?
   Client: OK—I think I felt less nervous than before.

The counselor introduces the idea of **applying all the coping skills in practice situations** *and* **presents a rationale for this application phase.**

3. *Counselor:* That's good. As time goes on, you will notice more effects from it. Up to this point, we've worked on some things to help you in your math class—stopping self-defeating thoughts and using imagination and slow breathing to help you cope and control your nervousness. What I think might help now is to give you a chance to use all these skills in practices of some of the stressful situations related to your math class. This will help you use the skills when you need to during the class or related situations. Then we will soon be at a point where we can go through some of these same procedures for the other situations in which you want to express yourself differently and more frequently, such as with your folks. Does this sound OK?
   Client: Yes.

Next, the counselor **demonstrates (models) how Joan can practice her skills in an imaginary practice.**

4. *Counselor:* What I'd like you to do is to imagine some of the situations related to your math class and try to use your coping thoughts *and* the imagination scene and deep breathing to control your nervousness. Let me show you how you might do this. OK, I'm imagining that it's almost time for math class. I'm going to concentrate on thinking about how this class will help me train for the Olympic downhill program. If I catch myself thinking I wish I didn't have to go, I'm going to use some coping thoughts. Let's see—class will go pretty fast. I've been doing better. It can be a challenge. Now, as I'm in class, I'm going to stop thinking about not being able to do the work. I'm going to just take one

problem at a time. One step at a time. Oops! Mr. Lamborne just called on me. Boy, I can feel myself getting nervous. Just take a deep breath. . . . Do the best I can. It's only one moment anyway. Well, it went pretty well. I can feel myself starting to cope when I need to. OK, Joan, why don't you try this several times now? [Joan practices applying coping thoughts and direct action with different practice situations in imagination.]

In response 5, the counselor **checks Joan's reaction** *to applying the skills in practice through imagination.*

5. *Counselor:* Are you able to really get into the situation as you practice this way?
   Client: Yes, although I believe I have to work harder to use this when it really happens.

Sometimes **role play makes the practice more real.** *The counselor introduces this next. Note that the counselor will add a stress element by calling on Joan at unannounced times.*

6. *Counselor:* That's right. This kind of practice doesn't always have the same amount of stress as the "real thing." Maybe it would help if we did some role-play practice. I'll be your teacher this time. Just pretend to be in class. I'll be talking, but at an unannounced time, I'm going to call on you to answer a question. Just use your coping thoughts and your slow breathing as you need to when this happens. [Role-play practice of this and related scenarios occurs.]

The counselor **assesses Joan's reaction** *to role-play practice and* **asks Joan to rate her level of nervousness** *during the practice.*

7. *Counselor:* How comfortable do you feel with these practices? Could you rate the nervousness you feel as you do this on a 1-to-5 scale, with 1 being not nervous and 5 being very nervous?
   Client: Well, about a 2.

The counselor encourages Joan to **apply coping statements in the math-related problem situations** *as they occur, assigns* **homework,** *and schedules a* **follow-up.**

8. *Counselor:* Well, I think you are ready to use this as you need to during the week. Remember, any self-defeating thought or body tenseness is a cue to cope, using your coping thoughts and imagination and breathing skills. I'd like you to keep track of the number of times you use this on your log sheets. Also rate your level of nervousness before, during, and after math class on the log sheet. How about coming back in two weeks to see how things are going?
   Client: Fine.

SUMMARY

Reframing is a useful way to help the client develop alternative ways to view a problem. To be effective the reframe must be plausible, acceptable, and culturally meaningful to the client. An example of a meaning reframe is the newer notion of "healthy resistance," meaning clients' empowerment to resist oppression and to stand up for themselves and their beliefs.

The prevalence of stress and stressors varies among cultural groups. The stress inoculation training we have presented and the corresponding view of stress in this model is primarily Eurocentric in that stress is viewed as a major problem requiring the assistance of an expert and some "training" to be alleviated or managed. In Eastern cultures, stress is viewed as a part of life and self-healing practices such as meditation and yoga are utilized more often than outside help. We describe meditation and yoga in the following chapters.

P O S T E V A L U A T I O N

Part One

Objective One asks you to demonstrate at least 8 out of 11 steps of the reframing procedure with a role-play client. Assess this activity using the Interview Checklist for Reframing on pages 416–417.

Part Two

Objective Two asks you to describe how you would apply the five major components of stress inoculation with a client case. Using the client description below, respond to the five questions following the case description as if you were using stress inoculation with this client. Feedback follows.

Description of client: The client has been referred to you by Family Services. He is unemployed, is receiving welfare support, and has three children. He is married to his second wife; the oldest child is hers by another marriage. He has been referred because of school complaints that the oldest child, a seventh grader, has arrived at school several times with obvious facial bruises and cuts. The child has implicated the stepfather in this matter. After a long period of talking, the client reports that he has little patience with this boy and sometimes does strike him in the face as his way of disciplining the child. He realizes that maybe, on occasion, he has gone too far. Still, he gets fed up with the boy's "irresponsibility" and "lack of initiative" for his age. At these times, he reports, his impatience and anger get the best of him.

1. Explain the purpose of stress inoculation as you would to this client.

INTERVIEW CHECKLIST FOR REFRAMING

Instructions to observer: Determine whether the counselor demonstrated the lead listed in the checklist. Check which leads were used.

2. Briefly give an overview of the stress inoculation procedure.

3. Describe and explain one example of each of the following kinds of direct-action coping skills that might be useful to this client.

    a. Information about the situation
    b. An escape route
    c. An attention-diversion tactic
    d. An imagery manipulation
    e. Physical relaxation
    f. A palliative coping strategy (perspective taking, social support, or ventilation)

4. Explain, as you might to this client, the four phases of an emotional reaction and of times for coping. For each of the four phases, give two examples of cognitive coping skills (thoughts) that you would give to this client. The four phases are preparing for a disagreement or argument with the boy; confronting the situation; dealing with critical, very provoking times; and encouraging himself for coping.

5. Describe how you would set up practice opportunities in the interview with this client to help him practice applying the direct-action and cognitive coping skills in simulated practices of the provoking situations.

Part Three

Objective Three asks you to demonstrate 17 out of 21 steps of the stress inoculation procedure with a role-play client. Assess this activity using the Interview Checklist for Stress Inoculation on pages 418–421.

Item	Examples of counselor leads
I. Rationale for Reframing	
_____ 1. Counselor explains purpose of reframing.	"Often when we think about a problem situation, our initial or intuitive reaction can lead to emotional distress. For example, we focus only on the negative features of the situation and overlook other details. By focusing only
	(continued)

Item	Examples of counselor leads
	on the selected negative features of a situation, we can become nervous or anxious about the situation."
_____ 2. Counselor provides overview of reframing.	"We'll identify what features you attend to when you think of the problem situation. Once you become aware of these features, we will look for other neutral or positive aspects of the situation that you may ignore or overlook. Then we will work on incorporating these other things into your perceptions of the problem situation."
_____ 3. Counselor confirms client's willingness to use the strategy.	"How does this all sound? Are you ready to try this?"

II. Identification of Client Perceptions and Feelings in Problem Situations

_____ 4. Counselor has client identify features typically attended to during problem situation. (May have to use imagery with some clients.)	"When you think of the problem situation or one like it, what features do you notice or attend to? What is the first thing that pops into your head?"
_____ 5. Counselor has client identify typical feelings during problem situation.	"How do you usually feel?" "What do you experience [or are you experiencing] during this situation?"

III. Deliberate Enactment of Selected Perceptual Features

_____ 6. Counselor asks client to reenact situation (by role play or imagery) and to deliberately attend to selected features. (This step may need to be repeated several times.)	"Let's set up a role play [or imagery] in which we act out this situation. This time I want you to deliberately focus on these aspects of the situation we just identified. Notice how you attend to _____."

IV. Identification of Alternative Perceptions

_____ 7. Counselor instructs client to identify positive or neutral features of problem situation. The new reframes are plausible and acceptable to the client and fit the client's values and age, gender, race, and ethnicity.	"Now, I want us to identify other features of the problem situation that are neutral or positive. These are things you have forgotten about or ignored. Think of other features." "What other aspects of this situation that aren't readily apparent to you could provide a different way to view the situation?"

V. Modification of Perceptions in Problem Situations

_____ 8. Counselor instructs client to modify perceptions of problem situation by focusing on or attending to the neutral or positive features. (Use of role play or imagery can facilitate this process for some clients.) (This step may need to be repeated several times.)	"When we act out the problem situation, I want you to change what you attend to in the situation by thinking of the neutral or positive features we just identified. Just focus on these features."

VI. Homework and Follow-Up

_____ 9. Counselor encourages client to practice modifying perceptions during in vivo situations.	"Practice is very important for modifying your perceptions. Every time you think about or encounter the problem situation, focus on the neutral or positive features of the situation."
_____10. Counselor instructs client to monitor aspects of the strategy on homework log sheet.	"I'd like you to use this log to keep track of the number of times you practice or use this. Also record your initial and resulting feelings before and after these kinds of situations."
_____11. Counselor arranges for a follow-up. (During follow-up, counselor comments on client's log and points out small perceptual shifts.)	"Let's get together in two weeks. Bring your log sheet with you. Then we can see how this is working for you."

Observer comments: _____

(continued)

POSTEVALUATION   (continued)

INTERVIEW CHECKLIST FOR STRESS INOCULATION

Instructions to observer: Determine which of the following steps the counselor demonstrated in using stress inoculation with a client or in teaching stress inoculation to another person. Check any step the counselor demonstrated in the application of the procedure.

Item	Examples of counselor leads
I. Rationale	
_____ 1. Counselor explains purpose of stress inoculation.	"Stress inoculation is a way to help you cope with feeling anxious so that you can manage your reactions when you're confronted with these situations."
_____ 2. Counselor provides brief overview of stress inoculation procedure.	"First we'll try to understand how your anxious feelings affect you now. Then you'll learn some coping skills that will help you relax physically—and help you use coping thoughts instead of self-defeating thoughts. Then you'll have a chance to test out your coping skills in stressful situations we'll set up."
_____ 3. Counselor checks to see whether client is willing to use strategy.	"How do you feel now about working with this procedure?"
II. Information Giving	
_____ 4. Counselor explains nature of client's emotional reaction to a stressful situation.	"Probably you realize that when you feel anxious, you are physically tense. Also, you may be thinking in a worried way—worrying about the situation and how to handle it. Both the physical tenseness and the negative or worry thoughts create stress for you."
_____ 5. Counselor explains possible *phases* of reacting to a stressful situation.	"When you feel anxious, you probably tend to think of it as one giant reaction. Actually, you're probably anxious at certain times or phases. For example, you might feel very uptight just anticipating the situation. Then you might feel uptight during the situation, especially if it starts to overwhelm you. After the situation is over, you may feel relieved—but down on yourself, too."
_____ 6. Counselor explains specific kinds of coping skills to be learned in stress inoculation and importance of client's input in tailoring coping strategies.	"We'll be learning some action kinds of coping strategies—like physical or muscle relaxation, mental relaxation, and just commonsense ways to minimize the stress of the situation. Then also you'll learn some different ways to view and think about the situation. Not all these coping strategies may seem best for you, so your input in selecting the ones you feel are best for you is important."
III. Acquisition and Practice of Direct-Action Coping Skills	
_____ 7. Counselor discusses and models direct-action coping strategies (or uses a symbolic model):	"First, I'll explain and we can talk about each coping method. Then I'll demonstrate how you can apply it when you're provoked."
_____a. Collecting objective or factual information about stressful situation	"Sometimes it helps to get any information you can about things that provoke and anger you. Let's find out the types of situations and people that can do this to you. Then we can see whether there are other ways to view the provocation. For example, what if you looked at it as a situation to challenge your problem-solving ability rather than as a personal attack?"

(continued)

Item	Examples of counselor leads
_____b. Identifying short-circuit or escape routes—alternative ways to deescalate stress of situation	"Suppose you're caught in a situation. You feel it's going to get out of hand. What are some ways to get out of it or to deescalate it *before* you strike out? For example, little things like counting to 60, leaving the room, using humor, or something like that."
Mental relaxation: _____c. Attention diversion	"OK, one way to control your anger is to distract yourself—take your attention away from the person you feel angry with. If you have to stay in the same room, concentrate very hard on an object in the room. Think of all the questions about this object you can."
_____d. Imagery manipulations	"OK, another way you can prevent yourself from striking out is to use your imagination. Think of something very calming and very pleasurable, like listening to your favorite record or being on the beach with the hot sun."
Physical relaxation: _____e. Muscle relaxation	"Muscle relaxation can help you cope whenever you start to feel aroused and feel your face getting flushed or your body tightening up. It can help you learn to relax your body, which can, in turn, help you control your anger."
_____f. Breathing techniques	"Breathing is also important in learning to relax physically. Sometimes, in a tight spot, taking slow, deep breaths can give you time to get yourself together before saying or doing something you don't want to."
Palliative coping strategies: _____g. Perspective taking	"Let's try to look at this situation from a different perspective—what else about the situation might you be overlooking?"
_____h. Social support network	"Let's put together some people and resources you could use as a support system."
_____i. Ventilation of feelings	"Perhaps it would be helpful just to spend some time getting your feelings out in the open." [You could use Gestalt dialoguing as one ventilation tool in addition to discussion.]
_____ 8. Client selects most useful coping strategies and practices each under counselor's direction.	"We've gone over a lot of possible methods to help you control your anger so it doesn't result in abusive behavior. I'm sure that you have some preferences. Why don't you pick the methods that you think will work best for you? We'll practice with these so you can get a feel for them."

IV. Acquisition and Practice of Cognitive Coping Skills

_____ 9. Counselor describes four phases of using cognitive coping skills to deal with a stressful situation.	"As you may remember from our earlier discussion, we talked about learning to use coping procedures at important points during a stressful or provoking situation. Now we will work on helping you learn to use coping thoughts during these four important times—preparing for the situation, handling the situation, dealing with critical moments during the situation, and encouraging yourself after the situation."

(continued)

POSTEVALUATION   **(continued)**

Item	Examples of counselor leads

IV. Acquisition and Practice of Cognitive Coping Skills (continued)

_____10. For each phase, counselor models examples of coping statements.

"I'd like to give you some ideas of some possible coping thoughts you could use during each of these four important times. For instance, when I'm trying to psych myself up for a stressful situation, here are some things I think about."

_____11. For each phase, client selects most natural coping statements.

"The examples I gave may not feel natural for you. I'd like you to pick or add ones that you could use comfortably, that wouldn't seem foreign to you."

_____12. Counselor instructs client to practice using these coping statements for each phase.

"Sometimes, because you aren't used to concentrating on coping thoughts at these important times, it feels a little awkward at first. So I'd like you to get a feel for these just by practicing aloud the ones you selected. Let's work first on the ones for preparing for a provoking situation."

_____13. Counselor models and instructs client to practice sequence of all four phases and verbalize accompanying coping statements.

"OK, next I'd like you to practice verbalizing the coping thoughts aloud in the sequence that you'll be using when you're in provoking situations. For example, [counselor models]. Now you try it."

V. Application of All Coping Skills to Problem-Related Situations

_____14. Using coping strategies and skills selected by client, counselor models how to apply these in a coping manner while imagining a stressful (problem-related) situation.

"Now you can practice using all these coping strategies when confronted with a problem situation. For example, suppose I'm you and my boss comes up to me and gives me criticism based on misinformation. Here is how I might use my coping skills in that situation."

_____15. Client practices coping strategies while imagining problem-related stressful situations. (This step is repeated as necessary.)

"OK, this time why don't you try it? Just imagine this situation—and imagine that each time you start to lose control, that is a signal to use some of your coping skills."

_____16. Client practices coping strategies in role play of problem-related situation. (This step is repeated as necessary.)

"We could practice this in role play. I could take the part of your boss and initiate a meeting with you. Just be yourself and use your coping skills to prepare for the meeting. Then, during our meeting, practice your skills whenever you get tense or start to blow up."

VI. Application of All Coping Skills to Potential Problem Situations (Generalization)

_____17. Counselor models application of client-selected coping strategies to non-problem-related or other potentially stressful situations.

"Let's work on some situations now that aren't problems for you but could arise in the future. This will give you a chance to see how you can apply these coping skills to other situations you encounter in the future. For instance, suppose I just found out I didn't get a promotion that I believe I really deserved. Here is how I might cope with this."

_____18. Client practices, as often as needed, applying coping strategies to potentially stressful situations by

"OK, you try this now."

_____a. Imagining a potentially stressful situation

"Why don't you imagine you've just found out you're being transferred to a new place? You are surprised by this. Imagine how you would cope."

(continued)

Item	Examples of counselor leads

VI. Application of All Coping Skills to Potential Problem Situations (Generalization) (continued)

| _____b. Taking part in a role-play practice | "This time let's role-play a situation. I'll be your husband and tell you I've just found out I am very ill. You practice your coping skills as we talk." |
| _____c. Taking part of a teacher in a role play and teaching a novice how to use coping strategies for stressful situations | "This time I'm going to pretend that I have chronic arthritis and am in constant pain. It's really getting to me. I'd like you to be my trainer or helper and teach me how I could learn to use some coping skills to deal with this chronic discomfort." |

VII. Homework and Follow-Up

_____19. Counselor and client discuss application of coping strategies to in vivo situations.	"I believe now you could apply these coping skills to problem situations you encounter during a typical day or week. You may not find that these work as quickly as you'd like, but you should find yourself coping more and not losing control as much."
_____20. Counselor instructs client how to use log to record uses of stress inoculation for in vivo situations.	"Each time you use the coping skills, mark it down on the log and briefly describe the situation in which you used them."
_____21. Counselor arranges for a follow-up.	"We could get together next week and go over your logs and see how you're doing."

Observer comments: _____

_____

_____

_____

_____

_____

_____

_____

_____

_____

_____

SUGGESTED READINGS

Bedrosian, R. C., & Bozicas, G. D. (1994). *Treating family of origin problems: A cognitive approach.* New York: Guilford.

Castillo, R. (1997)., *Culture and mental illness.* Pacific Grove, CA.; Brooks/Cole.

Fausel, D. (1995). Stress inoculation training for stepcouples. *Marriage and Family Review, 21,* 137–155.

Foa, E. B., Rothbaum, B. O., Riggs, D. S., & Murdock, T. B. (1991). Treatment of posttraumatic stress disorder in rape victims: A comparison between cognitive-behavioral procedures and counseling. *Journal of Consulting and Clinical Psychology, 59,* 715–723.

Gendlin, E. T. (1996). *Focusing-oriented psychotherapy: A manual of the experiential method.* New York: Guilford.

Guterman, J. T. (1992). Disputation and reframing: Contrasting cognitive-change methods. *Journal of Mental Health Counseling, 14,* 440–456.

Huang, L. N. (1994). An integrative approach to clinical assessment and intervention with Asian-American adolescents. *Journal of Clinical Child Psychology, 23,* 21–31.

Laungani, P. (1993). Cultural differences in stress and its management. *Stress Medicine, 9,* 37–43.

Lehrer, P. M., & Woolfolk, R. L. (Eds.). (1993). *Principles and practice of stress management* (2nd ed.). New York: Guilford.

Leserman, J., Perkins, D., & Evans, D. (1992). Coping with the threat of AIDS: The role of social support. *American Journal of Psychiatry, 149,* 1514–1520.

Meichenbaum, D. (1993). Stress inoculation training: A 20-year update. In P. M. Lehrer & R. L. Woolfolk (Eds.), *Principles and practice of stress management* (2nd ed., pp. 373–406). New York: Guilford.

Meichenbaum, D. (1994). *A clinical handbook: Practical therapist manual for assessing and treating adults with post-traumatic stress disorder.* Waterloo, Ontario: Institute Press.

Midori-Hanna, S., & Brown, J. (1995). *The practice of family therapy.* Pacific Grove, CA: Brooks/Cole.

Motenko, A. K., & Greenberg, S. (1995). Reframing dependence in old age. *Social Work, 40,* 382–390.

O'Hanlon, W. H., & Weiner-Davis, M. (1989). *In search of solutions: A new direction in psychotherapy.* New York: Norton.

Oppenheimer, M. (1992). Alma's bedside ghost: Or the importance of cultural similarity. *Hispanic Journal of Behavioral Sciences, 14,* 496–501.

Robinson, T., & Ward, J. (1992). "A belief in self far greater than anyone's disbelief": Cultivating resistance among African American female adolescents. In C. Gilligan, A. Rogers, & D. Tolman (Eds.), *Women, girls and psychotherapy: Reframing resistance* (pp. 87–104). Binghamton, NY: Haworth Press.

Rodriguez, C., & Moore, N. (1995). Perceptions of pregnant/parenting teens: Reframing issues for an integrated approach to pregnancy problems. *Adolescence, 30,* 685–706.

Ross, M., & Berger, R. (1996). Effects of stress inoculation training on athletes' postsurgical pain and rehabilitation after orthopedic injury. *Journal of Consulting and Clinical Psychology, 64,* 406–410.

Weeks, G., & Treat, S. (1992). *Couples in treatment.* New York: Brunner/Mazel.

F E E D B A C K
POSTEVALUATION

Part One

Use the Interview Checklist for Reframing to assess your interview.

Part Two

1. Your rationale to this client might sound something like this:

   "You realize that there are times when your anger and impatience do get the best of you. This procedure can help you learn to control your feelings at especially hard times—when you're very upset with this child—so that you don't do something you will regret later."

2. Here is a brief overview of stress inoculation:

   "First, we'll look at the things the child can do that really upset you. When you realize you're in this type of situation, you can learn to control how upset you are—through keeping yourself calm. This procedure will help you learn different ways to keep calm and not let these situations get out of hand."

3. Information—See lead 7, part a, on the Interview Checklist for Stress Inoculation for some examples.
   Escape route—See lead 7, part b.
   Attention diversion—See lead 7, part c.
   Imagery manipulations—See lead 7, part d.
   Physical relaxation—See lead 7, parts e and f.
   Palliative coping—See lead 7, parts g, h, and i.

4. Here are some examples of a possible explanation of the four coping phases and of cognitive coping skills you might present to this client.

Phase	Explanation	Cognitive coping
Preparing for a provoking situation	Before you have a disagreement or discussion, you can plan how you want to handle it.	"What do I want to say to him that gets my point across?" "I can tell him how I feel without shouting."
Confronting a provoking situation	When you're talking to him, you can think about how to stay in control.	"Just keep talking in a normal voice, no yelling." "Let him talk, too. Don't yell a lot; it doesn't help."

(continued)

FEEDBACK: POSTEVALUATION   (continued)

Dealing with a very provoking moment	If you feel very angry, you really need to think of some things to keep you from blowing your cool.	"Wait a minute. Slow down. Don't let the lid off." "Keep those hands down. Stay calm now."
Encouraging self for coping	Recognize when you do keep your cool. It's important to do this, to give yourself a pat on the back for this.	"I kept my cool that time!" "I could feel myself getting angry, but I kept in control then."

5. Practice opportunities can be carried out by the client in imagination or by you and the client in role play. In a role-play practice, you could take the part of the child. See leads 14, 15, and 16 on the Interview Checklist for Stress Inoculation for some examples of this type of practice.

Part Three

Use the Interview Checklist for Stress Inoculation to assess your role-play interview.

MEDITATION AND MUSCLE RELAXATION

Feel anxious? Stressful? Wired?

Do you have tension headaches?

Do you abuse "soft" drugs—alcohol, or tobacco?

Do you feel chronic fatigue?

Are you irritable with low frustration tolerance?

Do you have high blood pressure in certain situations or at certain times?

Does your immune system seem to be not working well?

A great number of people would respond yes to one or more of these questions. Anxiety is one of the most common problems reported by clients, and stress is related to physiological discomfort such as headaches and indigestion. Stress is also correlated with heart disease, cancer, and other serious diseases. Perhaps as a consequence of the "stress syndrome," the last several years have produced an explosion in procedures for management of stress or anxiety, originally introduced in 1929 as "progressive relaxation" (Jacobson, 1929). Many books related to stress management have appeared in book stores in the health and self-help sections (Benson, 1976; Benson & Stuart, 1992; Borysenko, 1988; Kabat-Zinn, 1990, 1994; Ornish, 1990). In the same period, a flurry of research has explored the relative strengths and weaknesses of stress-management approaches (Lehrer & Woolfolk, 1993).

This chapter presents five stress-management or relaxation strategies. Three meditation procedures are described: Kabat-Zinn's (1990, 1994) mindfulness meditation, Benson (1987) and Benson and Stuart's (1992) relaxation response, and Borysenko and Borysenko's lovingkindness or *metta* meditation (1994). In addition, two approaches to muscle-relaxation training are discussed. As you will see, muscle relaxation (body scan) can be one of the components of meditation, and meditating (awareness) by focusing (like a laser beam of light) on different muscle groups is a component of muscle relaxation. These five strategies are typically used to treat both cognitive and physiological indexes of stress.

OBJECTIVES

1. Identify which step of the relaxation response is reflected by each of nine counselor responses, accurately identifying at least seven of the nine examples.

2. Identify which step of the mindfulness meditation procedure is reflected by at least eight counselor responses.

3. Select either mindfulness meditation or relaxation response and teach the procedure to another person. Audiotape your teaching and assess your steps with the Interview Checklist for Mindfulness Meditation or the Interview Checklist for Relaxation Response, or have an observer evaluate your teaching, using the checklist.

4. Describe how you would apply the seven major components of the muscle-relaxation procedure, given a simulated client case.

5. Demonstrate 13 out of 15 steps of muscle relaxation with a role-play client, using the Interview Checklist for Muscle Relaxation to assess your performance.

6. Using one of the two body scan scripts, demonstrate the body scan with a role-play client.

MEDITATION: PROCESSES AND USES

Several practitioners and researchers define the meditation strategy in different ways. Fontana (1991) describes what meditation is not and what it is: "Meditation *isn't:* falling asleep, going into a trance, shutting yourself off from reality and becoming unworldly, being selfish, doing something 'unnatural,' becoming lost in thought, forgetting where you are. Meditation *is:* keeping the mind alert and attentive, keeping the mind focused and concentrated, becoming aware of the world, becoming more human, knowing where you are" (p. 17). Levine (1991) describes meditation as "a means to an endlessness. Meditation enters the mind to heal the mind, but continues to those levels so deep and universal, only the word 'heart' will suffice as description. Meditation allows us to directly participate in our lives instead of living life as an afterthought" (pp. 8–9). Dean Ornish (1990) describes meditation as "concentration," "focusing your awareness," "paying attention," and "one-pointedness" (p. 238). The Zen master and French/Vietnamese Buddhist, Thich Nhat Hanh (1976) describes mindfulness in mindful meditation as "keeping one's consciousness alive to the present reality (p. 11); . . . Mindfulness is like that—it is the miracle which can call back in a flash our dispersed mind and restore it to wholeness so that we can live each minute of life" (p. 14). LeShan (1974) says that "we meditate to find, to recover, to come back to something of ourselves we once dimly and unknowingly had and have lost without knowing what it was or where or when we lost it" (p. 1). "Meditation has infiltrated our culture. Millions of Americans have tried meditation, and many have incorporated it into their busy lives" (Goleman, 1988, p. *xxii*), probably because the process helps break "through stereotyped perception" (p. 20). "Meditation is now a standard tool used in medicine, psychology, education, and self-development" (Goleman, 1988, p. *xxii*).

"Meditation is one of the techniques for eliciting the relaxation response. In the broadest sense of the word, it is natural and familiar to everyone—it is simply a process of focusing the mind on an object or activity, something you do most of the time. When you use meditation to elicit the relaxation response, you turn your attention inward, concentrating on a repetitive focus such as breathing or a word or a prayer. Your body and your mind begin to quiet down. A state of physiological and mental rest ensues. But, as we all know too well, the mind is usually very active and difficult to focus" (Benson & Stuart, 1992, p. 46). Joan Borysenko (1988) says that "mindfulness is meditation in action and involves a 'be here now' approach that allows life to unfold without the limitation of prejudgment. It means being open to an awareness of the moment as it is and to what the moment could hold. It is a relaxed state of attentiveness to both the inner world of thoughts and feelings and the outer world of actions and perceptions" (p. 91).

Patel (1993) describes meditation as a practice of "taking a comfortable position—either sitting, lying down, or standing, although sitting is the most usual posture. It then involves being in a quiet environment, regulating the breath, adopting a physically relaxed and mentally passive attitude, and dwelling single-mindedly upon an object. The object of meditation does not have to be physical. It can be an idea, image, or happening; it can be mental repetition of a word or phrase, as in mantra meditation; it can be observing one's own thoughts, perception, or reaction; or it can be concentrating on some bodily generated rhythm (e.g., breathing)" (p. 127). Barry Long (1995) says that "meditation is constructive because it gradually dissolves the false you—false self you think you are, but are not" (p. 1).

Finally, Borysenko (1987) says that "meditation is any activity that keeps the attention pleasantly anchored in the present moment. . . . To develop a state of inner awareness, to witness and to let go of the old dialogues, you need an observation point. If you went out in a boat to view offshore tides but neglected to put down an anchor, you would soon be carried off to sea. So it is with the mind. Without an anchor to keep the mind in place, it will be carried away by the torrent of thoughts. Your ability to watch what is happening will be lost. The practice of meditation, which calms the body through the relaxation response and fixes the mind through dropping the anchor of attention, is the most important tool of self-healing and self-regulation" (p. 36). We agree with Benson (1987) that the "entire process of mindfulness meditation can provide what's been called a 'primer for psychotherapy.' The mental doors are open to greater insights and creativity—and the way is paved to escape from past obsessions, compulsions, or bad habits. The connection of this process of meditation with the techniques and goals of psychotherapy is obvious" (p. 90). Carrington (1993) says that "it is the *experience* of meditation that brings about therapeutic change, rather than the *techniques* used to evoke this experience" (p. 139).*

Carrington (1993) summarizes the primary symptoms or difficulties for which a person might benefit from meditation: "tension and/or anxiety states, psychophysiological disorders, chronic fatigue states, insomnias and hypersomnias, alcohol, drug, or tobacco abuse, excessive self-blame, chronic low-grade depressions or subacute reaction depressions, irritability, low frustration tolerance, strong submissive trends, poor developed psychological differentiation, difficulties with self-assertion, pathological bereavement reactions, separation anxiety, blocks to productivity or creativity, inadequate contact with affective life, a need to shift emphasis from client's reliance on therapist to reliance on self—of particular use when terminating psychotherapy" (pp. 150–151). Murphy and Donovan (1997) provide a thorough documentation of the physical and psychological effects of

---

*SOURCE: This and all following quotations from "Modern Forms of Meditation," by P. Carrington, in *Principles and Practice of Stress Management,* 2nd Ed., by P. M. Lehrer and R. L. Woolfolk, copyright 1993 by The Guilford Press, are reprinted by permission.

meditation. Also, as you can see from the listing in Table 17-1, recent research on meditation reflects a wide range of applications: anxiety reduction, treatment of hypertension, enhancement of longevity, relapse prevention, respiratory rehabilitation, control of stressful situations and stress in the workplace, and treatment for different types of substance abusers.

APPLICATIONS OF MEDITATION AND RELAXATION WITH DIVERSE CLIENTS

We have found only a few instances in which meditation practices and relaxation have been used with diverse groups of clients. Ruben (1989) used relaxation training as a part of a classroom guidance program for Hispanic and Black fifth-grade students. Relaxation training enhanced self esteem and reduced the drop-out rate of these students. Ibanez-Ramirez, Delgado-Morales, and Pulido-Diez (1989) reported the use of relaxation training as part of a therapeutic program that was used successfully to treat impotence in young Hispanic gay men. Meditation has also been used with the elderly in various ways, including the extension of longevity (Alexander, Robinson, Orme-Johnson, & Schneider, 1994), the management of respiratory problems (Connolly, 1993), and the reduction of hypertension (Alexander, et al., 1994). Generally, meditation has been more successful than relaxation in achieving these outcomes with elderly persons.

BASIC MEDITATION

Patel (1993, p. 130) describes seven steps for meditation.*

1. Meditate in a place free of distracting noise, movement, light, telephones, and activity of other people.
2. Make sure you are comfortable and the room is warm; wear loose clothes, empty your bladder and bowel, and do not practice for at least two hours after a meal.
3. Make sure your back is straight, your body is relaxed, and your eyes are half or fully closed.
4. Breathe through the nostrils and down into the abdomen. Make sure that your breathing is regular, slow, and rhythmical.
5. Dwell on a single object, word, phrase, or your breath.
6. Develop a passive and relaxed attitude toward distractions.
7. Practice regularly.

Three variations of meditation are illustrated in Table 17-2 and described below: (1) mindfulness, (2) relaxation responses, and (3) lovingkindness. We have used all three variations with our clients who seek mental relief, with participants in fitness centers, and as complementary strategies for individuals with medical concerns.

STEPS IN MINDFULNESS MEDITATION

Kabat-Zinn's (1990) mindfulness meditation is "to embark upon a journey of self-development, self-discovery, learning, and healing" (p. 1).

Rationale for Treatment

Jon Kabat-Zinn (1990) offers the following rationale for engaging in mindfulness meditation:

> We live immersed in a world of constant doing. Rarely are we in touch with who is doing the doing or, put otherwise, with the world of being. To get back in touch with being is not that difficult. We only need to remind ourselves to be mindful. Moments of mindfulness are moments of peace and stillness, even in the midst of activity. When your whole life is driven by doing, formal meditation practice can provide a refuge of sanity and stability that can be used to restore some balance and perspective. It can be a way of stopping the headlong momentum of all the doing and giving yourself some time to dwell in a state of deep relaxation and well-being and to remember who you are. The formal practice can give you the strength and the self-knowledge to go back to the doing and do it from out of your being. Then at least a certain amount of patience and inner stillness, clarity and balance of mind, will infuse what you are doing, and the busyness and pressure will be less onerous. In fact they might just disappear entirely (p. 60). . . . This is why we make a special time each day for formal meditation practice. It is a way of stopping, a way of "re-minding" ourselves, of nourishing the domain of being for a change. (p. 61)*

Mindfulness meditation is about watching or witnessing whatever comes up from one moment to the next moment. The client is told about selecting a quiet place to meditate, with eyes closed, focusing on the breath, allowing thoughts to flow freely for a period of about ten to twenty minutes. If the mind "trips out," we can bring it back by focusing on the

*SOURCE: This and all following quotations from "Yoga-Based Therapy," by C. Patel, in *Principles and Practices of Stress Management,* 2nd Ed., by P. M. Lehrer and R. L. Woolfolk, pp. 89–137, copyright © 1993 by The Guilford Press, are reprinted by permission.

*SOURCE: This and all following quotations from *Full Catastrophe Living,* by Jon Kabat-Zinn, copyright © 1990 by J. Kabat-Zinn, are used by permission of Dell Books, a division of Bantam Doubleday Dell Publishing Group, Inc.

TABLE 17-1. Research on meditation

Anxiety reduction

Eppley, K. R., Abrams, A. T., & Shear, J. (1989). Differential effects of relaxation techniques on trait anxiety: A meta-analysis. *Journal of Clinical Psychology, 45,* 957–974.
Sakairi, Y. (1988). The effect of Transcendental Meditation in reducing anxiety level. *Japanese Journal of Hypnosis, 33,* 8–14.

Blood circulation and metabolic changes

Jevning, R., Wallace, R. K., & Beidebach, M. (1992). The physiology of meditation: A review: A wakeful hypometabolic integrated response. *Neuroscience and Biobehavioral Reviews, 16,* 415–424.

Depression

Teasdale, J., Segal, Z., Williams, J. & Mark, G. (1995). How does cognitive therapy prevent depressive relapse and why should attentional control (mindfulness) training help? *Behaviour Research and Therapy, 33,* 25–39.

Elderly

Alexander, C. N., Robinson, P., Orme-Johnson, D., & Schneider, R. H. (1994). The effects of transcendental meditation compared to other methods of relaxation and meditation in reducing risk factors, morbidity, and mortality. *Homeostasis in Health and Disease, 35,* 243–263.

Hypertension

Sothers, K., & Anchor, K. N. (1989). Prevention and treatment of essential hypertension with meditation-relaxation methods. *Medical Psychotherapy: An International Journal, 2,* 137–156.

Marksmanship

Hall, E. G., & Hardy, C. J. (1991). Ready, aim, fire . . . relaxation strategies for enhancing pistol marksmanship. *Perceptual and Motor Skills, 72,* 775–786.

Relapse prevention

O'Connell, D. F. (1991). The use of transcendental meditation in relapse prevention counseling. *Alcoholism Treatment Quarterly, 8,* 58–63.

Taub, E., Steiner, S. S., Weingarten, E., & Walton, K. G. (1994). Effectiveness of broad spectrum approaches to relapse prevention in severe alcoholism: A long-term, randomized, controlled trial of Transcendental Meditation, EMG biofeedback and electronic neurotherapy. Special Issue: Self-recovery: Treating addictions using Transcendental Meditation and Maharishi Ayur-Veda: I. *Alcoholism Treatment Quarterly, 11,* 187–220.

Respiratory rehabilitation

Connolly, M. J. (1993). Respiratory rehabilitation in the elderly patient. *Reviews in Clinical Gerontology, 3,* 281–294.

School use

Laselle, K. M., & Russell, T. T. (1993). To what extent are school counselors using meditation and relaxation techniques? *School Counselor, 40,* 178–183.

Stress

Alexander, C. N., Swanson, G. C., Rainforth, M. V., & Carlisle, T. W. (1993). Effects of the transcendental meditation program on stress reduction, health, and employee development: A prospective study in two occupational settings. *Anxiety, Stress and Coping: An International Journal, 6,* 245–262.
Saito, Y., & Sasaki, Y. (1993). The effect of transcendental meditation training on psychophysiological reactivity to stressful situations. *Japanese Journal of Hypnosis, 38,* 20–26.
Staggers, F., Alexander, C. N., & Walton, K. G. (1994). Importance of reducing stress and strengthening the host in drug detoxification: The potential offered by Transcendental Meditation. Special Double Issue: Self-recovery: Treating addictions using Transcendental Meditation and Maharishi Ayur-Veda: II. *Alcoholism Treatment Quarterly, 11,* 297–331.

Substance abuse

Gelderloos, P., Walton, K. G., Orme-Johnson, D. W., & Alexander, C. N. (1991). Effectiveness of the transcendental meditation program in preventing and treating substance misuse: A review. *International Journal of the Addictions, 26,* 293–325.

---

breath. To facilitate and enhance the practice of meditation, a participant needs attitudinal foundations.

Instructions on the Attitudinal Foundations for Mindfulness Practice

Kabat-Zinn describes seven attitudinal foundations for mindfulness meditative practice:

1. *Nonjudging* means that mindfulness is facilitated by being an impartial witness or observer to one's own experience. We have a habit of categorizing or judging our experiences, which locks us into unaware "knee jerk" or mechanical reactions that often do not have an objective basis. For example, you can be practicing and think about all the things you have to do and how boring the practice is. These are judgments that take you away from observing whatever comes up. If you pursue these thoughts, it takes you away from moment by moment awareness. To remedy this, just watch your breathing.

2. *Patience* means that we often have to allow things to unfold in their own time. Practicing patience means that we don't have to fill our lives with moments of doing and activity.

3. *Expectations of the beginner's mind* is often based on our past experiences or cognitive schema, but prevent us

TABLE 17-2. Steps for mindfulness meditation of Kabat-Zinn (1990), the relaxation response of Benson (1987) and Benson and Stuart (1992), and lovingkindness (*metta*) meditation of Borysenko and Borysenko (1994)

Mindfulness Mediation	Relaxation Response	Lovingkindness Meditation
1. Rationale—about nondoing, watching whatever comes up in the mind; provides energy and self-knowledge. Give overview of procedure. Confirm client's willingness to use.	1. Rationale—give purpose and overview; meditation is a way to elicit the relaxation response; sit quietly, get a focus word or phrase, and focus on breathing. Confirm willingness to try.	1. Rationale—meditating about bringing lovingkindness awareness and equanimity as a continuously available source of resolution.
2. Instruct the client about attitudinal foundations for mindfulness practice: nonjudging, patience, beginner's mind, trust, nonstriving, acceptance, and letting go.	2. Instruct about when, where, and how long to practice.	2. Instruct in breathing.
3. Instruct about commitment, self-discipline, and energy.	3. Instruct about focus word, phrase, or prayer. Give examples.	3. Instruct to imagine a Great Star of Light.
4. Instruct about preparations for meditation.	4. Instruct the client about position for meditation, and eyes.	4. Provide invocation for self: peace, heart open, true nature, be healed, and source of healing for all.
5. Do a quick body scan to relax the muscles.	5. Request the client to relax her or his muscles—or do a quick body scan.	5. Instruct to bring loved ones to mind and to imagine loving light shining down on them.
6. Give client breathing instructions.	6. Provide instructions for breathing.	6. Give invocation for loved ones.
7. Instruct about a wandering mind, focus on breathing to control the mind.	7. Instruct about a passive attitude when meditating.	7. Instruct to imagine persons by whom client feels wronged, and use Light to wash away negativity.
8. Instruct to sit quietly, close eyes, be present in the moment, 10 to 20 minutes, come out slowly.	8. Instruct the client to meditate for about ten to twenty minutes.	8. Invoke peace for the persons who are judged.
9. Inquire about the just completed meditation experience. How did it feel? How did you handle distracting thoughts?	9. Inquire about the just-completed meditation experience for the client. How did it feel? Did client handle distracting thoughts?	9. Instruct to imagine spaciousness of planet.
10. Homework—instruct client to meditate every day or six days a week, for eight weeks, and about fifteen to thirty minutes a day. Instruct about informal meditation.	10. Homework—instruct the client to practice meditation once or twice daily for the next week. Not within one hour after eating; use quiet place, several hours before bedtime. Instruct how to apply in vivo.	10. Instruct to invoke peace and openness.
		11. Instruct client to meditate.
		12. Discuss lovingkindness type of meditation experience.
		13. Homework—Do daily, and as often as client wishes throughout the day; find a quiet place; don't meditate within one hour after eating; good way to start the day.

from seeing things as they really are. It is important for beginning meditators to be open to moment-by-moment experiences without framing the moment with expectations of how we think the moment will be.

4. *Trust* is about developing trust in your feelings and intuition. Clients are instructed to trust their feelings and wisdom, and not discount them. For example, if you are sitting in a particularly uncomfortable posture while meditating, change to another posture that feels better. If your intuition says do this, follow what your intuition is telling you and experiment to find a way that matches your needs. The message is to obey and trust what your body or feelings are telling you.

5. *Non-striving* means that mindfulness meditation is about the process of practice and not about striving to achieve something or get somewhere. Instruct the client to experience the moments; they do not have to get anywhere—just attend to or be with whatever comes up.

6. *Acceptance* means don't worry about results. Instruct the client to just focus on seeing and accepting things as they are, moment by moment, and in the present.

7. *Letting go* means nonattachment or not holding on to thoughts. Often we perseverate or hold on. If for example, a client becomes judgmental, instruct the client to let go and just observe the judging mind. (pp. 33–40)

Instruction About Commitment, Self-discipline, and Energy

Kabat-Zinn (1990) asks his clients to make a commitment to the practice of mindfulness meditation similar to what would be required in athletic training. In addition to the seven attitudes, clients are instructed to make a strong commitment to working on themselves, to conjure up enough self-discipline to persevere, and to generate enough energy to develop a strong meditative practice and a high degree of mindfulness (p. 41). Jon tells his clients, "You don't have to like it; you just have to do it" (pp. 41–42). Then, after eight weeks of practice, the client can say whether the practice was useful.

Preparations for Meditation

Clients are instructed to set aside a particular *block of time* every day—at least six days a week and at least eight consecutive weeks—to practice. In using mindfulness meditation, we find that a three week period is necessary for the practice to take hold. Clients are given instruction about making a *special place* in their homes to meditate. The place should be comfortable and free of interruptions. The recommended pose for mindfulness meditation is a *sitting posture* on either a chair or the floor. Clients are instructed to sit erect with their head, neck, and back aligned vertically; this posture allows the breath to flow easily. If sitting in a chair, sit away from the back of the chair so the spine is self-supporting (pp. 61–62). If sitting on the floor, sit on a cushion to raise the buttocks off the floor. Some clients feel more comfortable and prefer to meditate lying on their backs. We have found that some clients who meditate lying down often fall asleep. They associate relaxation with sleep, and they lose consciousness. After meditating for a period of time (a couple of weeks), these clients start to maintain awareness and decrease their urge to sleep.

Body Scan to Relax the Muscles

Kabat-Zinn (1990) describes the body scan as a purification process for different parts of the body. The body scan process helps the client to discover his or her body and to bring mind/body awareness to the moment. (We include a body scan script in the next section of this chapter on muscle relaxation.)

Breathing Instructions

The client is instructed to observe the breath as it flows in and out (see Chapter 18). Ask the client to notice the difference in temperature of the out breath and in breath, and to feel the sensations of the air as it goes in and out of the nostrils.

Instructions about a Wandering Mind

Attention is often carried away by thoughts cascading through the mind. Instruct clients that when this happens they are to return their attention to the flow of breathing and to let go of the thoughts. Kabat-Zinn (1990) tells his patients that thinking is not bad nor is it even undesirable during meditation. What matters is whether you are aware of your thoughts and feelings during meditation and how you handle them (p. 69). "Meditation is not so concerned with how much thinking is going on as it is with how much room you are making for it to take place within the field of your awareness from one moment to the next" (p. 69). If the client gets stuck in a thought, feeling, sensation, sound, pain, or discomfort, instruct the client to bring his or her attention back to breathing and to let go by exhaling these distractions.

Instructions on Meditating

Instruct clients to close their eyes and to meditate for 10 to 20 minutes. After they have concluded the meditation, they may wish to sit quietly, then to move or stretch, and just to relax for a few moments before opening their eyes.

Discussion of the Meditation Experience

Discuss or probe the client's reaction to meditation. Clients may be unsure of themselves because they are judging the process. Discuss with them how they felt with their first meditative experience. The counselor should be nonjudgmental about what clients experience. For example, if clients say that most of their experience was chasing after thoughts, encourage them to continue meditating because every practice will be different and it is the process of the experience that is important.

Homework

The counselor should instruct the client to select a quiet place and time and to determine how often to meditate, using the instructions from the previous mindfulness meditation experience. Also, the counselor should instruct the client not to meditate within one hour after eating; if meditation is done in the evening, it should occur several hours before bedtime. Finally, encourage the client to bring mindful awareness throughout the day while, for example, eating, stuck in traffic, doing everyday tasks, and interacting with people. The mantra is "to be here now," "be in the moment," "be present with what you are experiencing—in feeling and thought."

Kabat-Zinn (1990) describes four other types of mindfulness meditation in addition to the breath meditation described above:

1. Sitting with the breath and the body as a whole
2. Sitting with sound from the environment, nature, or music—it is not about listening for sounds; rather hearing what is in the moment
3. Sitting with thoughts and feelings, which means perceiving them as events in your mind and noting their content and their change
4. Sitting with choiceless awareness—just being open and receptive to whatever comes into your field of awareness, allowing it to come and to go. (pp. 72–74)

STEPS FOR THE RELAXATION RESPONSE

The following steps illustrate the relaxation response (Benson, 1987; Benson & Stuart, 1992); they are summarized in Table 17-2.

Rationale for Treatment

Here is an example of a rationale for meditation used by Benson and Stuart (1992):

Meditation is one of the techniques for eliciting the relaxation response. In the broadest sense of the word, it is natural and familiar to everyone—it is simply a process of focusing the mind on an object or activity, something you do most of the time. When you use meditation to elicit the relaxation response, you turn your attention inward, concentrating on a repetitive focus such as breathing or a word or a prayer. Your body and your mind begin to quiet down. A state of physiological and mental rest ensues. But, as we all know too well, the mind is usually very active and difficult to focus. (p. 46)*

Instructions About When, Where, and How Long to Practice

Benson and Stuart (1992) recommend a period before breakfast as the best time to meditate because meditating before the daily schedule begins sets a positive tone for the day; this time is usually uncluttered with events and activities. Choosing a regular time is best; so that a routine is developed. The place where you practice is very important. Benson and

Stuart (1992) recommend selecting a place that is attractive and feels safe. The place that one selects should be quiet and free from distractions and interruptions. Ideally, a person should practice 10 to 20 minutes once or twice a day. In the beginning, clients may have difficulty setting aside a routine time to meditate and committing to meditate for a particular length of time. They should experiment with different approaches to learn what works best for them.

Instructions about a Focus Word, Phrase, or Prayer

One way to focus your mind is to link the focusing to breathing, either by concentrating on the breath or by attending closely to something. You can focus on a word, phrase, or prayer. You might draw the focal object from your belief system. "For example, a Christian might select the opening of Psalm 23, *The Lord is my shepherd;* a Jew, *shalom;* a Muslim, *Allah.* Or, a client may prefer a neutral word or phrase such as *one* or *peace* or *love.* Or, the client may be instructed to focus on [his or her] breathing—the air comes and the air goes out" (Benson & Stuart, 1992, p. 45).

Instructions about Position and Eyes

Any posture that is comfortable is appropriate for meditation—lying down on one's back or sitting up. We prefer to meditate while sitting up comfortably with the back straight and having good support. Clients are requested to close their eyes because they can attend or be in the present moment more easily when their eyes are closed.

Request Client to Relax Muscles

With the client focused on breathing, the counselor can do a quick body scan starting from head to feet, or from feet to head. Any of the muscle relaxation procedures described in the next section of the chapter can be used as a prelude to meditation. Basically, the client focuses on different muscle groups and breathes tension out as she or he exhales.

Breathing Instructions

When you instruct clients in breathing, ask them to notice the rising and falling of the abdomen. Have them focus on the air coming in and going out through the nostrils; notice the slightly cooler air entering the nose and the warmer air leaving. As their breathing becomes quieter and more regular, clients may notice a subtle pause when the inhalation ends

---

*SOURCE: This and all following quotations from *The Wellness Book,* by Herbert Benson, M. D., Eileen Stuart, R. N., M. S., and the staff of the Mind/Body Medical Institute, copyright © 1992 by the Mind/Body Medical Institute, are used by arrangement with Carol Publishing Group. A Birch Lane Press Book.

and before the exhalation begins, and vice versa (Benson & Stuart, 1992, p. 43). Chapter 18 describes different breathing exercises we have used for relaxation, body awareness, with meditation, and with Hatha Yoga.

Instructions About a Passive Attitude

Benson and Stuart (1992) say that meditation can be like going to a movie: you can choose to become emotionally involved in the movie or you can pull back and say, "It's only a movie" (p. 47). The practice of meditation allows people to observe or witness their thoughts, feelings, or bodily sensations. Mark Epstein (1995) in his *Thoughts Without a Thinker* describes attention during meditation as "diminishing reactivity. . . . Separating out the reactive self from the core experience, the practice of bare attention eventually returns the meditator to a state of unconditional openness" (p. 117). It is like watching a train—just standing or sitting and looking at one car at a time go by without changing your position. You are simply witnessing or observing without using the judging part of your mind. As quickly as one car of the train enters your right, you let it go and attend to the next car that comes into view. Also, you can instruct clients not to worry about whether the meditation is going "correctly" or how they are doing. Tell them just to maintain a passive attitude, to let go, and return to their word or breath if judgmental thoughts come to mind.

Instructions About Length of Meditation

Clients are instructed to meditate for 10 to 20 minutes. The counselor and client may want to select the length of time for the first meditation before the client begins. We have found that ten minutes may be long for the first in-session meditation, but we have also found that some clients prefer a longer period.

Discussion of Meditation Experience

The counselor gets the client's reaction to the just-completed meditative experience. For example, how did the client feel? How did the client handle the distractions?

Homework and Follow-Up

The counselor instructs the client to meditate twice or at least once daily for the next week. He or she can help the client identify a quiet place and time to practice during the next week, reminding the client that regular meditation is an important part of the therapeutic and healing process.

STEPS FOR LOVINGKINDNESS

Borysenko and Borysenko (1994) have developed a different type of meditation called "lovingkindness meditation" that includes the steps listed in Table 17-2. Their system is based on Dossey's (1993) *Healing Words* and his description of nonlocal or transpersonal medicine in which the "mind is a factor in healing both *within* and *between* persons" (p. 42).

Rationale for Treatment

Borysenko and Borysenko claim that meditating about bringing lovingkindness awareness and equanimity to the people in your life can be a continuously available source of resolution. They suggest meditating in this way about yourself, your loved ones, and the negativity you have for other people, opening your heart to yourself and others. You can use the imagery of light in this meditation to bring psychological, emotional, and spiritual peace.

Instruction for Breathing

In preparation for the meditation, the client is instructed to begin by "taking a few letting-go breaths and start to quiet down by focusing on either diaphragmatic or belly breathing or breath of bridging earth and heaven" [explained below] (Borysenko & Borysenko, 1994, p. 128). The client can use the following instructions for the breath of bridging the earth and heaven as awareness training. The instructions can be tape-recorded so the client can use them for homework during the week.

1. Sit with your feet flat on the ground and your back straight, yet relaxed. . . . Take a minute to stretch so that you feel as comfortable as possible in your body. . . . Now allow your eyes to close.
2. Request the client to take a big letting-go breath, and then switch into belly breathing, feeling or imagining your belly expanding as you breathe in, and relaxing as you breathe out. . . . Continue until you feel relaxed and present in your body. . . . This is one of many ways to breathe. Now we will learn another way called the breath of bridging earth and heaven. We will imagine breathing energy from the sky and the earth simultaneously into the heart.
3. Let's start with breathing in the sky energy. Feel or imagine the energy of the sun above you—it doesn't matter whether it happens to be day or night as you do this. . . . As you inhale, draw this energy in through the top of your head and into your heart. Breathe out a sense of spaciousness as if your breath could move out to the edges of the universe. . . . Try this for several breaths. . . .

4. Now we will breathe in the energy of the earth. Feel or imagine the earth energy beneath your feet.... As you inhale, draw this energy in through your feet and up into your heart. Breathe out a sense of spacious awareness into the universe. Try this for a minute or two until you begin to get comfortable with it....

5. Now open your eyes and look out at nature. See the earth and the sky as you continue to do the breathing exercise. This breath brings you into the present moment and blesses the universe with peaceful presence. (Borysenko & Borysenko, 1994, pp. 110–111)*

Instructions for Light Imagery

Borysenko and Borysenko offered the following instructions for using the imagery of light:

Imagine a Great Star of Light above you and slightly in front of you, pouring a waterfall of love and light over you. Let the light enter the top of your head and wash through you, revealing the purity of your own heart, which expands and extends beyond you, merges with the Divine Light. See yourself totally enclosed in the Egg of Light and then repeat these lovingkindness blessings for yourself, with all the respect and love that you would have for your only child. (p. 128)

Make Invocation for Self

The client can use the following mantra or invocation for a self-blessing:

May I be at peace,
May my heart remain open,
May I awaken to the light of my own true nature,
May I be healed,
May I be a source of healing for all beings. (p. 128)

Instructions to Imagine Loved Ones

At this point, the client is instructed to bring his or her loved ones into the imagery:

Next, bring one or more loved ones to mind. See them in as much detail as possible, imagining the loving Light shining down on them and washing through them, revealing the light within their own hearts. Imagine this light growing brighter, merging with the Divine Light and enclosing them in the Egg of Light. Then bless them. (p. 129)

---

*SOURCE: This and all following quotations from *The Power of the Mind to Heal,* by J. Borysenko and M. Borysenko, copyright © 1994 by Hay House, are reprinted by permission.

Give Invocation for Loved Ones

The counselor instructs the client to use the following invocation:

May you be at peace,
May your heart remain open,
May you awaken to the light of your own nature,
May you be healed,
May you be a source of healing for all beings. (p. 129)

Instructions for Imagining Persons by Whom Client Feels Wronged

Now the attention turns to individuals who have somehow wronged the client. The client is instructed to address their image in this way:

Next, think of a person or persons who may have been unjust to you and to whom you're ready to begin extending forgiveness. Place them in the Egg of Light, and see the Light washing away all their negativity and illusions, just as it did for you and your loved ones. (p. 129)

Instructions to Invoke Peace for Negative Individuals

To these individuals the client is instructed to direct the following blessing:

May you be at peace,
May your heart remain open,
May you awaken to the light of your own true nature,
May you be healed,
May you be a source of healing for all beings. (p. 129)

Instructions to Imagine the Spaciousness of the Planet

The counselor now instructs the client to capture the spaciousness of Earth:

See our beautiful planet as it appears from outer space, a delicate jewel spinning in the starry vastness of space. Imagine the earth surrounded by light—the green continents, the blue waters, the white polar caps, the two-leggeds, and four-leggeds, the fish that swim, and the birds that fly. Earth is a realm of opposites. Of day and night, good and evil, sickness and health, riches and poverty, up and down, male and female. Be spacious enough to hold it all as you offer these: (p. 130)

Instructions to Invoke Peace and Openness

The client offers the following invocation:

May there be peace on earth,
May the hearts of all people be open to themselves and to each other,
May all people awaken to the light of their own nature,
May all creation be blessed and be a blessing to All That Is.
(p. 130)

Instruction to Meditate

Instruct the client to use the above visualizations and invocations (with the counselor's help) to meditate. Continue the meditation for 15 to 20 minutes.

Discussion of Meditation

Discuss with clients how they felt with their first meditation. The counselor should be nonjudgmental about their experience. If this type of meditation is not appealing to the client, the counselor might suggest an alternative.

Homework

Instruct the client to do the meditation at least once a day, and to repeat it throughout the day as often as desired. Remind the client to meditate in a quiet place, before eating, or at least one hour after eating, and to continue the meditation for 15 to 20 minutes. This type of lovingkindness meditation may be a good way to start and end the day.

CONTRAINDICATIONS AND ADVERSE EFFECTS OF MEDITATION

Carrington (1993) and Goleman (1988) offer some cautions about using meditation, noting that it may not be appropriate for some clients. Carrington (1993) has found that some people may be "hypersensitive to meditation" and may need a shorter period of time to meditate than other people (p. 153). Some may release emotional material that is difficult for them to handle because they meditate for a very long period of time—three or four hours. For example, patients with an active psychiatric history may experience adverse effects by meditating for an extended period (Carrington, 1993, p. 154). Goleman (1988) points out that some clients with a schizoid disorder may become overly absorbed in inner realities and less connected with reality if they meditate (p. 171). People "in acute emotional states may be too agitated to begin meditation, and obsessive-compulsive clients might become overzealous with the new experience of meditation" (Goleman, 1988, pp. 171–172).

Finally, for some clients, the action of certain drugs may be enhanced by meditation. Carrington (1993) recommends monitoring patients practicing meditation if they are also using antianxiety, antidepressant, antihypertensive, or thyroid-regulating drugs (p. 154). The continued practice of meditation may allow some of these patients to lower their dosage of some drugs. The counselor should be aware of any medication the client is taking and know what potential interaction might occur with a particular drug or medication and the intervention strategy. It is important for the counselor to individualize the meditative process to address client needs and concerns.

MODEL EXAMPLE: MINDFULNESS MEDITATION

In this model example, we present a narrative account of how mindfulness meditation might be used with a 49-year-old Japanese American male client. Yoshi, an air traffic controller, has reported that he would like to decrease the stress he experiences in his job. He believes that decreasing his stress will help to heal his ulcer and allow him to cope better with the demands of his job. In addition to the physical signs of stress (hypertension), Yoshi also reports that he worries constantly about making mistakes at work.

1. Rationale

First, we explain to Yoshi that mindfulness meditation has been used to help people cope with job-related stress. We tell him that the procedure has also been used to help people with high blood pressure and anxiety as well as those who want to feel more alert. We give him an overview of mindfulness meditation, telling him that the procedure is a process of focusing on breathing, in a quiet place, with eyes closed, allowing thoughts to flow freely. We explain that he should focus on breathing if his thoughts become too distracting, and tell him that most people using this technique usually meditate for 10 to 20 minutes a day. We tell Yoshi that to help the practice of meditation, participants need a foundation of attitudes on which to build a meditative practice. Finally, we confirm his willingness to use meditation, and we answer any questions he may have about the process.

2. Attitudinal Foundations for Practice

We explain to Yoshi that there seven attitudes that will help mindfulness meditation practice. We tell him that mindfulness is helped by being nonjudgmental, and when we meditate, we want to be an impartial witness or observer of our experience. The mind has a tendency to categorize experiences, and we want to avoid that habit. We explain to Yoshi that we don't have to fill our lives with moments of doing and activity, and we ask him to be patient by allowing things to unfold in their own time while he is meditating. We tell him that beginners of meditation usually have expectations about how they think the moments will be while meditating, but that he should just be open to the moment-by-moment experiences without injecting expectations based on past experience. We tell him to trust his feeling and intuition; there is no "right" or "correct" way to meditate. We ask him to experiment with the process to learn what fits his needs, and to obey and trust what his feelings and intuition tell him. We explain that mindfulness meditation is about the process of practice and not about striving to achieve something or get somewhere. All he has to do is experience, to be in the moment, and attend to or be with whatever comes up. We talk to him about acceptance and tell him not to worry about the outcome—just to see and accept the way things are, moment by moment, and in the present. Finally, we talk to Yoshi about letting go and experiencing nonattachment or not holding on to thoughts and feelings; we ask him just to observe whatever comes up and observe the judging mind.

3. Instruction About Commitment, Self-Discipline, and Energy

We tell Yoshi to commit to practicing mindfulness in much the same way an athlete would commit to training. Yoshi understands that he must make a firm commitment, discipline himself to persevere, and generate enough energy so that he can develop a strong meditative practice.

4. Preparations for Meditation

We ask Yoshi to set aside a block of time every day and meditate for at least six days a week, to find a quiet and special place to meditate without interruption, and to meditate in a sitting position with his back straight. We tell him to meditate for eight weeks so that he can become adjusted to the process. We instruct him not to meditate within one hour after eating, and to wear comfortable clothing during the practice time.

5. Body Scan to Relax the Muscles

We conduct a body scan (see next section) with Yoshi, and tell him that we will scan different muscle groups of his body as a purification process. We tell him that this relaxes his body, and brings mind/body awareness to the moment.

6. Breathing Instructions

We ask him to breathe deeply and notice how his belly expands on the in breath and falls on the out breath. We ask him to feel the difference in temperature of the out breath and the in breath, and to feel the sensation of the air as it goes in and out of the nostrils.

7. Instructions About the Mind Wandering

Yoshi says that his attention is often carried away by cascading thoughts. We tell him that there is nothing wrong with that; when it happens, he is just to return his attention to the flow of breathing and let the thoughts go. We tell him to be aware of his thoughts and feelings during meditation. If he gets stuck in a thought, feeling, bodily sensation, sound, pain, or discomfort, he is to bring his attention back to breathing, and exhale these distractions.

8. Instructions to Meditate

We instruct Yoshi to sit quietly and get relaxed for about a minute. Then he is to close his eyes and focus his breathing; the air comes in and goes out. He is to just "ride" the tide of his breath. We tell Yoshi that mindfulness meditation is not an exercise and requires no effort; he can't force it. We mention to him that if distracting thoughts, feelings, sensations, or sounds occur, he should allow them to come and not try to influence them—just observe them and return to his breathing: "the air comes in and goes out." We tell Yoshi that he will meditate for 10 to 20 minutes. When the time is up, we ask him to come out of the meditation slowly by sitting with his eyes closed for about a minute; he may want to move and stretch. We instruct Yoshi to absorb what he is experiencing and then to open his eyes slowly.

9. Inquire About the Just Completed Meditation Experience

We ask Yoshi a series of questions about his experience of meditation: "How did you feel about the experience? How did you handle distractions? What are your feelings right now?"

10. Homework

We instruct Yoshi to meditate once a day, preferably in the morning just after he wakes up. We remind him of the things to do to prepare: find a quiet environment; select a special time; do a quick body scan; remember the seven attitudes and commitment, discipline, and energy. We tell him not to meditate within an hour after eating, and instruct him to try to be mindful and aware throughout the day, moment to moment. (See also Learning Activity #47.)

MUSCLE RELAXATION: PROCESS AND USES

In muscle relaxation, a person is taught to relax by becoming aware of the sensations of tensing and relaxing major muscle groups. Take a few moments to feel and to become aware of some of these sensations. Make a fist with your preferred (dominant) hand. Clench your fist of that hand. Clench it tightly and study the tension in your hand and forearm. Become aware and feel those sensations of tension. Now let the tension go in your fist, hand, and forearm. Relax your hand and rest it. Note the difference between the tension and the relaxation. Do the exercise once more, only this time close your eyes. Clench your fist tightly; become aware of the tension in your hand and forearm; then relax your hand and let the tension flow out. Note the different sensations of relaxing and tensing your fist. Try it.

If you did this exercise, you may have noticed that your hand and forearm *cannot* be tense and relaxed at the same time. In other words, relaxation is incompatible with tension. You may also have noted that you instructed your hand to tense up and then to relax. You sent messages from your head to your hand to impose tension and then to create relaxation. You can cue a muscle group (the hand and forearm, in this case) to perform or respond in a particular manner (tense up and relax). This exercise was perhaps too brief for you to notice changes in other bodily functions. For example, tension and relaxation can affect one's blood pressure, heart rate, and respiration rate and can also influence covert processes and the way one performs or responds overtly. "The long-range goal of muscle relaxation is for the body to monitor instantaneously all of its numerous control signals, and automatically to relieve tensions that are not desired" (McGuigan, 1993, p. 21).

Relaxation training is not new, but it has recently become a popular technique for dealing with a variety of client concerns. Jacobson (1929, 1964) developed an extensive procedure called "progressive relaxation." Later, Wolpe (1958) described muscle relaxation as an anxiety-inhibiting procedure with his systematic desensitization strategy (see Chapter 19). Bernstein and Borkovec (1973) wrote a thorough relaxation manual entitled *Progressive Relaxation Training*.

Muscle relaxation has been used to address many emotional and physical states (see Table 17-3). Among these are anger, asthma, depression, headaches, herpes, hypertension, insomnia, irritable bowel syndrome, panic disorder, and urinary incoordination. Also, body scan relaxation, a tech-

---

LEARNING ACTIVITY 47 *Meditation (Relaxation Response and Mindfulness Meditation)*

I. Teaching mindful meditation or the relaxation response to a client is a psychoeducational process. The counselor provides the instructions, and the client engages in meditation in a self-directed manner. To practice giving instructions to someone about meditation, select a partner or a role-play client and give instructions as described in the Interview Checklist for Mindfulness Meditation or the Interview Checklist for the Relaxation Response at the end of the chapter. Then assess how well your partner was able to implement your instructions. If you wish, reverse roles so that you can experience being instructed by another person.

II. This learning activity provides an opportunity to try formal meditation. Do this in a quiet, restful place when you will not be interrupted for 20 minutes. Do *not* do this within one hour *after* a meal or within two hours of going to sleep.

1. Get in a comfortable sitting position and close your eyes.

2. Relax your entire body. Think about all the tension draining out of your body.

3. Meditate for 15 to 20 minutes.
   a. Breathe easily and naturally through your nose.
   b. Focus on your breathing with the thought of a number (one) or a word. Say (think) your word silently each time you inhale and exhale.
   c. If other thoughts or images appear, don't dwell on them but don't force them away. Just relax and focus on your word or breathing.

4. Try to assess your reactions to your meditative experience:
   How do you feel about it?
   How do you feel afterward?
   What sorts of thoughts or images come into your mind?
   How much difficulty did you have with distractions?

5. Practice the relaxation response systematically—twice daily for a week, if possible.

TABLE 17-3. Muscle relaxation research

Anger

Schloss, P. J., Smith, M., Santora, C., & Bryant, R. (1989). A respondent conditioning approach to reducing anger responses of a dually diagnosed man with mild mental retardation. *Behavior Therapy, 20,* 459–464.

Asthma

Lehrer, P., Sargunaraj, D., & Hochron, S. M. (1992). Psychological approaches to the treatment of asthma. Special Issue: Behavioral medicine: An update for the 1990s. *Journal of Consulting and Clinical Psychology, 60,* 639–643.

Cancer

Baider, L., Uziely, B., & Kaplan De Nour, A. (1994). Progressive muscle relaxation and guided imagery in cancer patients. *General Hospital Psychiatry, 16,* 340–347.

Depression

Broota, A., & Dhir, R. (1990). Efficacy of two relaxation techniques in depression. *Journal of Personality and Clinical Studies, 6,* 83–90.

Headache

Applebaum, K. A., Blanchard, E. B., Nicholson, N. L., & Radnitz, C. (1990). Controlled evaluation of the addition of cognitive strategies to a home-based relaxation protocol for tension headache. *Behavior Therapy, 21,* 292–303.

Blanchard, E. B., Appelbaum, K. A., Radnitz, C. L., & Michultka, D. (1990). Placebo-controlled evaluation of abbreviated progressive muscle relaxation and of relaxation combined with cognitive therapy in the treatment of tension headache. *Journal of Consulting and Clinical Psychology, 58,* 210–215.

Blanchard, E. B., Kim, M., Hermann, C. U., & Steffeck, B. D. (1993). Preliminary results of the effects on headache relief of perception of success among tension headache patients receiving relaxation. *Headache Quarterly, 4,* 249–253.

Blanchard, E. B., Nicholson, N. L., Taylor, A. E., & Steffeck, B. D. (1991). The role of regular home practice in the relaxation treatment of tension headache. *Journal of Consulting and Clinical Psychology, 59,* 467–470.

Arena, J., Bruno, G., Hannah, S., & Meador, K. (1995). A comparison of frontal electromyographic biofeedback training, trapezius electromyographic biofeedback training, and progressive muscle relaxation therapy in the treatment of tension headache. *Headache, 35,* 411–419.

Herpes

Burnette, M. M., Koehn, K. A., Kenyon-Jump, R., & Hutton, K. (1991). Control of genital herpes recurrences using progressive muscle relaxation. *Behavior Therapy, 22,* 237–247.

Hypertension

Broota, A., Varma, R., & Singh, A. (1995). Role of relaxation in hypertension. *Journal of the Indian Academy of Applied Psychology, 21,* 29–36.

Haaga, D. A. F., Davison, G. C., Williams, M. E., & Dolezal, S. L. (1994). Mode-specific impact of relaxation training for hypertensive men with Type A behavior pattern. *Behavior Therapy, 25,* 209–223.

Larkin, K. T., Knowlton, G. E., & D'Alessandri, R. (1990). Predicting treatment outcome to progressive relaxation training in essential hypertensive patients. *Journal of Behavioral Medicine, 13,* 605–618.

Insomnia

Gustafson, R. (1992). Treating insomnia with a self-administered muscle relaxation training program: A follow-up. *Psychological Reports, 70,* 124–126.

Irritable bowel syndrome

Blanchard, E. B., Green, B., Scharff, L., & Schwarz-McMorris, S. P. (1993). Relaxation training as a treatment for irritable bowel syndrome. *Biofeedback and Self-Regulation, 18,* 125–132.

Panic disorder

Ost, L-G., Westling, B. E., & Hellstrom, K. (1993). Applied relaxation, exposure *in vivo* and cognitive methods in the treatment of panic disorder with agoraphobia. *Behaviour Research and Therapy, 31,* 383–394.

Review of abbreviated progressive muscle relaxation

Carlson, C. R., & Hoyle, R. H. (1993). Efficacy of abbreviated progressive muscle relaxation training: A quantitative review of behavioral medicine research. *Journal of Consulting and Clinical Psychology, 61,* 1059–1067.

Urinary incoordination

Philips, H. C., Fenster, H. N., & Samsom, D. (1992). An effective treatment for functional urinary incoordination. *Journal of Behavioral Medicine, 15,* 45–63.

---

nique for muscle relaxation, is an integral part of meditation and yoga training (Kabat-Zinn, 1990; LePage, 1994; Patel, 1993).

The effects of muscle relaxation, like those of any other strategy, are related to satisfactory problem assessment, client characteristics, and the therapist's ability to apply the procedure competently and confidently. Therapists should also heed other cautions: they should not apply relaxation training indiscriminately without first exploring the causes of the client's reported tension. The counselor would probably have made a reasonable determination of these root causes during problem assessment (see Chapters 8 and 9). For example, is muscle relaxation a logical strategy for alleviating the client's problem? If the client is experiencing tension in a job situation, the counselor and client may need to deal first with the client's external situation (the job). If the client is experiencing tension as a result of oppression and discrimination, this condition will need to be targeted for change. Bernstein and Borkovec (1973) point out the difference between dealing with the tension of daily problems and handling the tension of someone who is on the verge of financial disaster. In the latter case, combinations of therapeutic strategies may be necessary.

STEPS OF MUSCLE RELAXATION

Muscle relaxation consists of the following seven steps:

1. Rationale
2. Instructions about dress
3. Creation of a comfortable environment
4. Counselor modeling of the relaxation exercises
5. Instructions for muscle relaxation
6. Posttraining assessment
7. Homework and follow-up

These steps are described in detail in the Interview Checklist for Muscle Relaxation at the end of the chapter.

Treatment Rationale

Here is an example of one way a counselor might explain the *purpose* of relaxation: "This process, if you practice it regularly, can help you become relaxed. The relaxation benefits you derive can help you sleep better at night." An *overview* of the procedure might be this: "The procedure involves learning to tense and relax different muscle groups in your body. By doing this, you can contrast the difference between tenseness and relaxation. This will help you to recognize tension so you can instruct yourself to relax."

In addition, the counselor should explain that muscle relaxation is a *skill*. The process of learning will be gradual and will require regular practice. Finally, the counselor might explain that some discomfort may occur during the relaxation process. If so, the client can just move his or her body to a more comfortable position. Finally, the client may experience some floating, warming, or heavy sensations typical for some people learning muscle relaxation. The counselor should inform the client about these possible sensations. The explanation of the rationale for muscle relaxation should be concluded with a probe of the client's willingness to try the procedure.

Instructions About Dress

Before the actual training session, clients should be instructed about appropriate clothing. They should wear comfortable clothes such as slacks, a loose-fitting blouse or shirt, or any apparel that will not be distracting during the exercises. Clients who wear contact lenses should be told to wear their regular glasses for the training. They can take off the glasses while going through the exercises. It is uncomfortable to wear contact lenses when your eyes are closed.

Creation of a Comfortable Environment

A comfortable environment is necessary for effective muscle-relaxation training. The training environment should be quiet and free of distracting noises such as ringing telephones, street repair work outside, and airplane sounds. A padded recliner chair should be used if possible. If the counseling facility cannot afford one, an aluminum lawn chair or recliner covered with a foam pad may be satisfactory. If relaxation training is to be applied to groups, pads or blankets can be placed on the floor, with pillows to support each client's head. The clients can lie on the floor on their backs, with their legs stretched out and their arms along their sides with palms down.

Counselor Modeling of the Relaxation Exercises

Just before the relaxation training begins, the counselor should model briefly at least a few of the muscle exercises that will be used in training. The counselor can start with either the right or the left hand (make a fist, then relax the hand, opening the fingers; tense and relax the other hand; bend the wrists of both arms and relax them; shrug the shoulders and relax them) and continue demonstrating some of the rest of the exercises. The counselor should tell the client that the demonstration is going much faster than the speed at which the client will perform the exercises. The counselor should also punctuate the demonstration with comments like "When I clench my biceps like this, I feel the tension in my biceps muscles, and now, when I relax and drop my arms to my side, I notice the difference between the tension that was in my biceps and the relative relaxation I feel now." These comments are used to show clients how to discriminate between tension and relaxation.

Instructions for Muscle Relaxation

Muscle-relaxation training can start after the counselor has given the client the rationale for the procedure, answered any questions about relaxation training, instructed the client about what to wear, created a comfortable environment for the training, and modeled some of the various muscle-group exercises. In delivering (or reading) the instructions for the relaxation training exercises, the counselor's voice should be conversational, not dramatic. We recommend that the counselor practice along with the client during the beginning exercises. Practicing initial relaxation exercises with the client can give the counselor a sense of timing for delivering the verbalizations of relaxation and tension and may decrease any awkwardness the client feels about doing "body type" exercises.

In instructing the client to tense and relax muscles, remember that you do *not* want to instruct the client to tense up as hard as possible. You do not want the client to strain a muscle. Be careful of your vocabulary when giving instructions. Do not use phrases like "as hard as you can," "sagging or drooping muscles," or "tense the muscles until they feel like they could snap." Sometimes you can supplement instructions to tense and relax with comments about the client's breathing or the experiencing of warm or heavy sensations. These comments may help the client to relax.

The various muscle groups used for client training can be categorized into 17 groups, 7 groups, or 4 groups. These sets of muscle groups, adapted from Bernstein and Borkovec (1973), are listed in Table 17-4. Generally, in initial training sessions, the counselor instructs the client to go through all 17 muscle groups. When the client can alternately tense and relax any of the 17 muscle groups on command, you can abbreviate this somewhat long procedure and train the client in relaxation using 7 muscle groups. After this process, the client can practice relaxation using only four major muscle groups. Start with either 17 or 7 muscle groups. This may help the client to discriminate sensations of tension and relaxation in different parts of the body. Then the number of muscle groups involved in the relaxation can be reduced gradually. When the client gets to the point of using the relaxation in vivo, 4 muscle groups are much less unwieldy than 17!

The following section illustrates how the counselor can instruct the client in relaxation using all 17 muscle groups. First, the counselor instructs the client to settle back as comfortably as possible—either in the recliner chair or on the floor with the head on a pillow. The arms can be alongside the body, resting on the arms of the chair or on the floor with the palms of the hands down. The counselor then instructs the client to close her or his eyes. In some instances, a client may not wish to do this; at other times, the counselor and the client may decide that it might be more therapeutic to keep the eyes open during the training. In such cases, the client can focus on some object in the room or on the ceiling. Tell the client to *listen* and to *focus* on your instructions. When presenting instructions for each muscle group, direct the

client's attention to the tension, which is held for five to seven seconds, and then to the feelings of relaxation that follow when the client is instructed to relax. Allow about ten seconds for the client to enjoy the relaxation associated with each muscle group before delivering another instruction. Intermittently throughout the instructions, make muscle-group comparisons—for example, "Is your forehead as relaxed as your biceps?" While delivering the instructions, gradually lower your voice and slow the pace of delivery. Usually in initial training sessions, each muscle group is presented twice.

Here is a way the counselor might proceed with initial training in muscle relaxation, using the list of 17 muscle groups in Table 17-4:

1. *Fist of dominant hand.* "First think about your right arm, your right hand in particular. Clench your right fist. Clench it tightly and study the tension in the hand and in the forearm. Study those sensations of tension. [Pause.] Now let go. Just relax the right hand and let it rest on the arm of the chair [or floor]. [Pause.] And note the difference between the tension and the relaxation." [Ten-second pause.]

2. *Fist of nondominant hand.* "Now we'll do the same with your left hand. Clench your left fist. Notice the tension [five-second pause] and now relax. Enjoy the difference between the tension and the relaxation." [Ten-second pause.]

3. *Wrist of one or both arms.* The counselor can instruct the client to bend the wrists of both arms at the same time or to bend each separately. You might start with the dominant arm if you instruct the client to bend the wrists one at a time. "Now bend both hands back at the wrists so that you tense the muscles in the back of the hand and in the forearm. Point your fingers toward the ceiling. Study the tension, and now relax. [Pause.] Study the difference between tension and relaxation." [Ten-second pause.]

4. *Biceps.* The counselor can instruct the client to work with both biceps or just one at a time. If you train the client to do one at a time, start with the dominant biceps. The instructions for this exercise are "Now clench both your hands into fists and bring them toward your shoulders. As you do this, tighten your biceps muscles, the ones in the upper part of your arm. Feel the tension in these muscles. [Pause.] Now relax. Let your arms drop down to your sides. See the difference between the tension and the relaxation." [Ten-second pause.]

5. *Shoulders.* Usually the client is instructed to shrug both shoulders. However, the client could be instructed to

shrug one shoulder at a time. "Now we'll move to the shoulder area. Shrug your shoulders. Bring them up to your ears. Feel and hold the tension in your shoulders. [Pause.] Now, let both shoulders relax. Note the contrast between the tension and the relaxation that's now in your shoulders." [Ten-second pause.]

6. *Forehead.* This and the next three exercises are for the facial muscles. The instructions for the forehead are "Now we'll work on relaxing the various muscles of the face. First, wrinkle up your forehead and brow. Do this until you feel your brow furrow. [Pause.] Now relax. Smooth out the forehead. Let it loosen up." [Ten-second pause.]

7. *Eyes.* The purpose of this exercise is for the client to contrast the difference between tension and relaxation for the muscles that control the movements of the eyes. "Now close your eyes tightly. Can you feel tension all around your eyes? [Five-second pause.] Now relax those muscles, noting the difference between the tension and the relaxation." [Ten-second pause.]

8. *Tongue or jaws.* You can instruct some clients to clench their jaws: "Now clench your jaws by biting your teeth together. Pull the corners of your mouth back. Study the tension in the jaws. [Five-second pause.] Relax your jaws now. Can you tell the difference between tension and relaxation in your jaw area?" [Ten-second pause.] This exercise may be difficult for some clients who wear dentures. An alternative exercise is to instruct them: "Press your tongue into the roof of your mouth. Note the tension within your mouth. [Five-second pause.] Relax your mouth and tongue now. Just concentrate on the relaxation." [Ten-second pause.]

9. *Pressing the lips together.* The last facial exercise involves the mouth and chin muscles. "Now press your lips together tightly. As you do this, notice the tension all around the mouth. [Pause.] Now relax those muscles around the mouth. Enjoy this relaxation in your mouth area and your entire face. [Pause.] Is your face as relaxed as your biceps [intermuscle-group comparison]?"

10. *The head.* "Now we'll move to the neck muscles. Press your head back against your chair. Can you feel the tension in the back of your neck and in your upper back? Hold the tension. [Pause.] Now let your head rest comfortably. Notice the difference. Keep on relaxing." [Pause.]

11. *Chin in chest.* This exercise focuses on the muscles in the neck, particularly the front of the neck. "Now continue to concentrate on the neck area. Bring your head forward. See whether you can bury your chin into your chest. Note the tension in the front of your neck. Now relax and let go." [Ten-second pause.]

12. *The back.* Be careful here—you don't want the client to get a sore back. "Now direct your attention to your upper back area. Arch your back as if you were sticking out your chest and stomach. Can you feel tension in your back? Study that tension. [Pause.] Now relax. Note the difference between the tension and the relaxation." [Ten-second pause.]

13. *Chest muscles.* Inhaling (filling the lungs) and holding the breath focuses the client's attention on the muscles in the chest and down into the stomach area. "Now take a deep breath, filling your lungs, and hold it. Feel the tension all through your chest and into your stomach area. Hold that tension. [Pause.] Now relax and let go. Let your breath out naturally. Enjoy the pleasant sensations." [Ten-second pause.]

14. *Stomach muscles.* "Now think about your stomach. Tighten up the muscles in your abdomen. Hold this. Make the stomach like a knot. Now relax. Loosen those muscles now. [Ten-second pause.] Is your stomach as relaxed as your back and chest [muscle-group comparison]?" An alternative instruction is to tell the client to "pull in your stomach" or "suck in your stomach."

15. *The buttocks.* Moving down to other areas of the body, the counselor instructs or coaches the client to tighten the buttocks. This muscle group is optional; with some clients, the counselor may delete it and move on to the legs. The model instructions are "Now tighten [tense or contract] your buttocks by pulling them together and pushing them into the floor [or chair]. Note the tension. And now relax. Let go and relax." [Ten-second pause.]

16. *Legs.* "I'd like you now to focus on your legs. Stretch both legs. Feel tension in the thighs. [Five-second pause.] Now relax. Study the difference again between tension in the thighs and the relaxation you feel now." [Ten-second pause.]

17. *Toes.* "Now concentrate on your lower legs and feet. Tighten both calf muscles by pointing your toes toward your head. Pretend a string is pulling your toes up. Can you feel the pulling and the tension? Note that tension. [Pause.] Now relax. Let your legs relax deeply. Enjoy the difference between tension and relaxation." [Ten-second pause.]

After each muscle group has been tensed and relaxed twice, the counselor usually concludes relaxation training with a summary and review. The counselor goes through the review by listing each muscle group and asking the client to dispel any tension that is noted as the counselor names the muscle area. Here is an example:

> Now, I'm going to go over once more the muscle groups that we've covered. As I name each group, try to notice whether there is any tension in those muscles. If there is any, try to concentrate on those muscles and tell them to relax. Think of draining the tension completely out of your body as we do this. Now relax the muscles in your feet, ankles, and calves. [Pause.] Get rid of tension in your knees and thighs. [Five-second pause.] Loosen your hips. [Pause.] Let the muscles of your lower body go. [Pause.] Relax your abdomen, waist, lower back. [Pause.] Drain the tension from your upper back, chest, and shoulders. [Pause.] Relax your upper arms, forearms, and hands. Loosen the muscles of your throat and neck. [Pause.] Relax your face. [Pause.] Let all the tension drain out of your body. [Pause.] Now just sit quietly with your eyes closed.

The therapist can conclude the training session by evaluating the client's level of relaxation on a scale from 0 to 5 or by counting aloud to the client to instruct him or her to become successively more alert. For example:

> Now I'd like you to think of a scale from 0 to 5, where 0 is complete relaxation and 5 is extreme tension. Tell me where you would place yourself on that scale now.
> I'm going to count from 5 to 1. When I reach the count of 1, open your eyes. 5 . . . 4 . . . 3 . . . 2 . . . 1. Open your eyes now.

Posttraining Assessment

After the session of relaxation training has been completed, the counselor asks the client about the experience. The counselor can ask "What is your reaction to the procedure?" "How do you feel?" "What reaction did you have when you focused on the tension?" "What about relaxation?" or "How did the contrast between the tension and relaxation feel?" The counselor should be encouraging about the client's performance, praise the client, and build a positive expectancy set about the training and practice.

People experiencing relaxation training may have several problems (Bernstein & Borkovec, 1973). Some of these potential problem areas are cramps, excessive laughter or talking, spasms or tics, intrusive thoughts, falling asleep, inability to relax individual muscle groups, unfamiliar sensations and holding the breath. If the client experiences muscle cramps, possibly too much tension is being created in the particular muscle group. In this case, the counselor can instruct the client to decrease the amount of tension. If spasms and tics occur in certain muscle groups, the counselor can mention that these occur commonly, as in one's sleep,

and possibly the reason the client is aware of them now is that he or she is awake. Excessive laughter or talking would most likely occur in group-administered relaxation training. Possibly the best solution is to ignore it or to discuss how such behavior can be distracting.

The most common problem is for the client to fall asleep during relaxation training. The client should be informed that continually falling asleep can impede learning the skills associated with muscle relaxation. By watching the client throughout training, the counselor can confirm whether the client is awake. The counselor also can tell the client to "stay awake" during the muscle relaxation process.

If the client has difficulty or is unable to relax a particular muscle group, the counselor and client might work out an alternative exercise for that muscle group. If intrusive client thoughts become too distracting, the counselor might suggest changing the focus of the thought to something less distracting or to more positive or pleasant thoughts. It might be better for some clients to gaze at a picture of their choosing placed on the wall or ceiling throughout the training. Another strategy for dealing with interfering or distracting thoughts is to help the client use task-oriented coping statements or thoughts (see Chapter 16), that would facilitate focusing on the relaxation training.

Another potential problem is the occurrence of unfamiliar sensations, such as floating, warmth, and headiness. The counselor should point out that these sensations are common and that the client should not fear them. The last problem is that some clients have a tendency to hold their breath while tensing a muscle. The counselor needs to observe the client's breathing during muscle relaxation. If the client is holding the breath, the counselor can instruct the client to breathe freely and easily. The counselor can also use a variation of this procedure that we describe later.

Homework and Follow-Up

The last step in muscle relaxation is assigning homework. Four or five therapist training sessions with two daily home practice sessions between therapy sessions are probably sufficient. Some therapists have found that a minimal therapist contract with the client and home-based relaxation training using manuals and audiotapes with telephone consultation were just as effective in reducing tension headaches as six hours of therapist training. Regardless of the amount of time or number of therapist training sessions with the client, the therapist should inform the client that relaxation training, like learning any skill, requires a great deal of practice.

The more the client practices the procedure, the more proficient he or she will become in gaining control over tension, anxiety, or stress. The client should be instructed to select a quiet place for practice, free from distracting noise. The client should be encouraged to practice the muscle-relaxation exercises about 15 to 20 minutes twice a day. The exercises should be done when there is no time pressure. Some clients may not be willing to practice twice a day. The therapist can encourage these clients to practice several times or as often as they can during the week. The exercises can be done in a recliner chair or on the floor with a pillow supporting the head.

The client should be encouraged to complete a homework log after each practice. Figure 17-1 is an example of a homework log. Clients can rate their reactions on a scale from 1 (little or no tension) to 5 (extremely tense) before and after each practice. They can practice the relaxation exercises using a tape recording of the relaxation instructions or from memory. After client homework practices, a follow-up session should be scheduled.

A therapist can use several techniques to promote client compliance with relaxation homework assignments. One technique is to ask the client to demonstrate during the therapy session how the exercises for the muscles in the neck or the calf, for example, were done during last week's home practice. The counselor can select randomly from four or five muscle groups for the client to demonstrate. If the exercises are demonstrated accurately, the client probably practiced.

CONTRAINDICATIONS AND ADVERSE EFFECTS OF MUSCLE RELAXATION

Generally, muscle relaxation is benign and pleasant, but for some people it can have adverse side effects. For example, clients with generalized anxiety disorder, panic disorder, or a history of hyperventilation may experience adverse side effects (Bernstein & Carlson, 1993, p. 66). Some clients with certain muscles or connective tissues that have been damaged or are chronically weak will have difficulty in tensing and relaxing a particular muscle group. Also, some clients are incapable of exercising voluntary control over all muscles in the body because of a neuromuscular disability. Finally, with some medication such as insulin for diabetes or propanolol for hypertension, clients receiving relaxation training may

TABLE 17-4. Relaxation exercises for 17, 7, and 4 muscle groups

17 muscle groups	7 muscle groups	4 muscle groups
1. Clenching *fist* of dominant *hand* 2. Clenching *fist* of nondominant *hand* 3. Bending *wrist* of one or both arms 4. Clenching *biceps* (one at a time or together) 5. Shrugging *shoulders* (one at a time or together) 6. Wrinkling *forehead* 7. Closing *eyes* tightly 8. Pressing *tongue* or clenching *jaws* 9. Pressing *lips* together 10. Pressing *head* back (on chair or pillow) 11. Pushing *chin* into chest 12. Arching *back* 13. Inhaling and holding *chest muscles* 14. Tightening *stomach* muscles 15. Contracting *buttocks*[a] 16. Stretching *legs* 17. Pointing *toes* toward head	1. Hold *dominant arm* in front with elbow bent at about 45-degree angle while making a *fist* (hand, lower arm, and biceps muscles). 2. Same exercise with *nondominant arm.* 3. Facial muscle groups. Wrinkle *forehead* (or frown), squint *eyes*, wrinkle up *nose*, clench *jaws* or press *tongue* on roof of mouth, press *lips* or pull corners of mouth back. 4. Press or bury *chin* in chest (neck and throat). 5. *Chest, shoulders, upper back,* and *abdomen.* Take deep breath, hold it, pull shoulder blades back and together, while making stomach hard (pulling in). 6. *Dominant thigh, calf,* and *foot.* Lift foot off chair or floor slightly while pointing toes and turning foot inward. 7. Same as 6, with *nondominant thigh, calf,* and *foot.*	1. Right and left *arms, hands,* and *biceps* (same as 1 and 2 in 7-muscle group) 2. *Face* and *neck* muscles. Tense all *face* muscles (same as 3 and 4 in 7-muscle group) 3. *Chest, shoulders, back* and *stomach* muscles (same as 5 in 7-muscle group) 4. Both left and right upper *leg, calf,* and *foot* (combines 6 and 7 in 7-muscle group)

[a]This muscle group can be eliminated; its use is optional. *SOURCE:* From *Progressive Relaxation Training: A Manual for Helping Professions,* by D. A. Bernstein and T. D. Borkovec. Copyright 1973 by Research Press. Used by permission.

require a change in the amount of medication they need (Bernstein & Carlson, 1993, p. 67). A medical consultation may be necessary before beginning muscle relaxation training with clients who are taking certain types of medication. Be sure to seek such a consultation if there is a question about medications such as those described above. In cases of anxiety or panic disorders, the counselor can start with breathing exercises described in the first section of the next chapter. If a client has a problem with a particular muscle group, the counselor can avoid that group or do a body scan. Herman (1994) notes that relaxation is often contraindicated for clients who present with severe trauma histories because of their need to maintain some degree of vigilance in order to feel safe.

VARIATIONS OF MUSCLE RELAXATION

There are several variations of the muscle-relaxation training procedure as we've described it. These variations, which include recall, counting, differential relaxation, and body scan are arranged and designed in successive approximations from the counselor assisting the client to acquire the skills to the client applying the relaxation skills in real-life situations. The four-muscle-group exercises listed in Table 17-4 can be used in combination with the recall and counting procedures described by Bernstein and Borkovec (1973).

Recall

Recall proceeds according to the relaxation exercises for the four muscle groups (Table 17-4) without muscular tension. The counselor first instructs the client about the rationale for using this variation of relaxation training: "to increase your relaxation skills without the need to tense up the muscles." The client is asked to focus on each muscle group. Then the counselor instructs the client to focus on one of the four muscle groups (arms; face and neck; chest, shoulders, back, and stomach; legs and feet) and to relax and just recall what it was like when the client released the tension (in the previous session) for that particular muscle group. The counselor might suggest that if there is tension in a particular muscle group, the client should just relax or send a message for the muscle to relax and should allow what tension there is to "flow out." The counselor gives similar instructions for all four muscle groups. Again, the client is to recall what the relaxation felt like for each muscle group. Clients can generally use recall after first learning the tension/relaxation-contrast procedure for the four muscle groups. Gradually, the client can use recall to induce relaxation in self-directed

practices. Recall can also be used in combination with counting.

Counting

The rationale for counting is that it helps the client become very deeply relaxed. Again, the counselor explains the rationale for using counting. The counselor says that she or he will count from one to ten and that this will help the client to become more relaxed after each number. The counselor might say slowly:

> One—you are becoming more relaxed; two—notice that your arms and hands are becoming more and more relaxed; three—feel your face and neck becoming more relaxed; four, five—more and more relaxed are your chest and shoulders; six—further and further relaxed; seven—back and stomach feel deeply relaxed; eight—further and further relaxed; nine—your legs and feet are more and more relaxed; ten—just continue to relax as you are—relax more and more.

The counselor can use this counting procedure with recall. The client can also be instructed to use counting in real situations that provoke tension. For a more detailed presentation of counting, see Bernstein and Borkovec (1973). As you may remember from Chapter 16, counting is one type of direct-action coping skill used in stress inoculation. Counting can increase relaxation and decrease tension, and the client should be encouraged to practice it outside the session.

Differential Relaxation

This variation—differential relaxation—may contribute to generalization of the relaxation training from the treatment session to the client's world. The purpose of differential relaxation is to help the client recognize what muscles are needed in various situations, body positions, and activities in order to differentiate which muscle groups are used and which are not. Table 17-5 illustrates some possible levels for the differential-relaxation procedure.

TABLE 17-5. Levels of differential-relaxation procedure

Situation	Body position	Activity level
Quiet	Sitting	Low—inactive
Noisy	Standing	High—routine movements

HOMEWORK LOG SHEET						
DATE	TAPE NUMBER	NOTE WHICH MUSCLE GROUPS EXERCISED	PRACTICE SESSION NUMBER	LOCATION OF SESSION	LEVEL OF TENSION (1–5)	
					BEFORE SESSION	AFTER SESSION

NOTE: 1 = slightly or not tense; 2 = somewhat tense; 3 = moderately tense; 4 = very tense; 5 = extremely tense.

FIGURE 17-1. Example of homework log sheet for relaxation training

As an example of differential relaxation, the counselor might have the client sit in a regular chair (not a recliner) and ask the client to identify which muscles are used and which are not when sitting. If the client feels tension in muscles that are not used (face, legs, and stomach), he or she is instructed to induce and to maintain relaxation in the muscles not required for what the client is doing (sitting). The counselor can instruct the client to engage in different levels of the differential-relaxation procedure—for example, sitting down in a quiet place while inactive, or standing up. After several practice sessions, the client can be assigned homework to engage in various levels of these activities. Examples might be sitting in a quiet cafeteria, sitting in a noisy cafeteria while eating, standing in line for a ticket to some event, or walking in a busy shopping center. In practicing differential relaxation, the client tries to recognize whether any tension exists in the nonessential muscle groups. If there is tension in the nonengaged muscles, the client concentrates on dispelling it.

Body Scan

A body scan can be a powerful technique for reestablishing contact with the body because it is a way to develop concentration and flexibility of attention simultaneously (Kabat-Zinn, 1990, p. 77). The purpose of scanning your body is to feel each region or muscle group. You can focus the breath on each region of the body by breathing into the region and breathing out of the region. If the client feels tension in one region, the counselor instructs the client to breathe the tension out on the exhalation. We offer two scripts for body scanning. The first provides instructions for deep muscle relaxation with little focus on breathing (Patel, 1993, pp. 123–124). The second body scan (LePage, 1994, pp. 6.15–6.16) uses cues to focus on the breath throughout the instructions. Body scans are very useful variations of the traditional method of muscle relaxation for clients who tend to hold their breath while tensing muscles.

INSTRUCTIONS FOR DEEP MUSCLE RELAXATION BODY SCAN*

1. Close your eyes. Very slowly fill your lungs, starting at the diaphragm and working right up to the top of the chest, then very slowly breathe out. After three slow breaths, allow your breathing to become normal and regular. Breathe in and out gently and rhythmically, using your diaphragm. Don't force your breath. Don't try to make it slow deliberately. Just keep your own rhythm. Be completely aware of your breathing pattern. Feel the subtle difference in the temperature of the air you are inhaling and the air you are exhaling. The air you breathe in is cooler, and the air you breathe out is warmer.

*SOURCE: From *Integrative Yoga Therapy Manual,* by Joseph LePage, pp. 6.22–6.23. Copyright © 1994 by Integrative Yoga Therapy, Aptos, CA. Used by permission.

2. Now you are consciously going to relax each part of the body in turn. Relaxation means the complete absence of movement, since even the slightest movement means that some of your muscles are contracting. It also excludes holding any part rigid. Concentrate on the part you are relaxing.

3. Now take your mind to your right foot and relax your toes, instep, heel, and the ankle; stay there for a few seconds. Now move your attention slowly up, relaxing your leg, calf, knee, thigh, and the hip. Feel all the muscles, joints, and tissues of your right leg becoming completely relaxed. Relax as deeply as you can. Just keep your awareness on this feeling of deep relaxation in your right leg for a few moments.

4. Now take your mind to your left foot and repeat the process, working up the leg, knee, thigh, and the hip as before. Let all the tension ease away and enjoy the feeling of relaxation for a few seconds.

5. Next concentrate on your right hand. Relax the fingers, thumb, palm, and wrist. Move your attention up to your forearm, elbow, upper arm, and shoulder. Feel every muscle, joint, and tissue in your right arm becoming deeply relaxed. Fix your attention on the sensation of relaxation in your entire right arm for a few moments.

6. Now become aware of your left hand and relax the fingers, thumb, palm, wrist, forearm, elbow, upper arm, and the shoulder. Let all the tension ease away from the left arm.

7. Now concentrate on the base of your spine, vertebra by vertebra, relaxing each vertebra and the muscles on either side of the spine into the floor. Relax your back—first the lower back, then the middle back, and finally the upper back. Release all the tension from your back. Let the relaxation become deeper and deeper. Feel your back merging with the floor.

8. Let the muscles in your neck relax next. Let all the muscles in the front of your neck relax. Let your head rest gently and feel the back of the neck relaxing. Let the relaxation become as deep as possible.

9. Relax your chest. Every time you breathe out, relax a little more. Let your body sink into the floor a little more each time. Let all the nerves, muscles, and organs in your chest relax completely. Now relax the muscles of your stomach. Let all the nerves, muscles , and organs in your stomach relax completely. Just feel them relaxing.

10. Now concentrate on your jaw. Let it relax so that it drops slightly, your lips are just touching each other, and your teeth are apart. Relax your tongue; relax the muscles around your cheekbones. Relax your eyes and muscles around your eyes. Feel them becoming relaxed. Your eyes must become very still. Now relax your forehead; let all the muscles in your forehead become completely relaxed. There is no tension in your facial muscles at all. Now relax your scalp and all the muscles around your head.

11. Your body is now completely relaxed. Keep it relaxed for a few more minutes.

To come out of relaxation, take one deep breath, feeling the energy coming down into your arms and legs. Move your arms and legs slowly. Open your eyes without reacting to the light, and slowly sit up and stretch your body, feeling refreshed and re-energized.

BODY AWARENESS RELAXATION

Allow the body to begin to completely relax. . . . Inhale and feel the breath flow from the soles of the feet to the crown of the head, like a gentle slow motion wave. With each exhalation allow tension to flow out of the body. Bring your awareness to the fingers of the left hand. Inhale breath and awareness through the fingers and up the left arm. Exhale, release the arm into the support of the earth. . . . Allow relaxation to deepen with each exhalation. Now bring your awareness to the right fingertips and inhale the breath up the arm, exhale and completely relax. As you relax the arms, become more aware of all the feelings and sensations. . . . Focus all of your awareness into these sensations and then relax into them.

Now bring your awareness down to the toes of the left foot, drawing the wave of breath up to the top of the leg, and on exhalation relax the leg fully. Now bring your awareness into the right leg, and allow the wave of breath to flow up the right leg, and with the exhalation completely surrender the weight of the leg. Feel and see both legs now and with each breath become more aware of all the sensations in the legs, with each exhalation relax even more deeply. Listen to the sound of the waves as the breath flows through the body.

Now bring the breath and your awareness up into the hips, pelvis and buttocks. On the inhalation feel the pelvis area naturally expand and exhaling allow it rest down into the earth. . . . With each inhalation feel the pelvic floor being drawn gently up into the abdomen and with each exhalation allow it to completely release. Feel the wave of breath rising up from the pelvis filling the abdomen. Feel the abdomen rise and fall, and explore the abdomen with your awareness. With each exhalation the abdomen becomes softer and softer. Feel this softness touch the lower back and feel the breath there. Explore the sensations in the low back and then allow this area to soften into the earth.

Now allow breath and awareness to flow up the spine. With each inhalation, the spine fills with sensation. With each exhalation, the spine relaxes into the earth. Feeling the breath now through the entire back. Inhaling, sensing; exhaling, completely relaxing.

Now bring your awareness again to the rising and falling of the abdomen. As you inhale draw the breath up into the solar plexus filling that area fully with breath and awareness. And as you exhale, relax into the center of that awareness. Now focus the breath up into the heart and lungs, and with each exhalation relax deeper and deeper into the center of the heart.

Draw the breath into the neck and throat. Exhale, allowing any tension to be released. . . . Allow the breath to flow up through the head, with each inhalation become more aware of the sensations, with each exhalation, relax. Relax the jaw, the eyes, the forehead and the back of the head, soften the inner ears, and relax into the earth.

Feel the entire body now washed by a gentle wave of breath from the soles of the feet and the tips of the fingers, all the way to the crown of the head. Feel the peace and complete relaxation as the breath becomes softer and softer. Feel the sensations in the body becoming softer and more subtle and relax into them.

Now allow the wave of breath to be felt a little more strongly, rising up through the soles of the feet, and rising and falling in the abdomen. As the breath becomes stronger allow the sensations in the body to increase. Let the body gently begin to move with the breath. Move the toes and fingers. . . . allow the whole body to begin to gently stretch. Remaining with the eyes closed, begin to gently roll over onto the right side. Let every movement be an experience of awareness. Over the next minute come up into a seated position. And as you come to the seated position, feel the deep three part breath and experience how the body, breath and mind are in balance.

MODEL DIALOGUE: MUSCLE RELAXATION

In this dialogue, the counselor demonstrates relaxation training to help Joan deal with her physical sensations of nervousness.

First, the counselor gives Joan a **rationale** *for relaxation. The counselor explains the* **purpose** *of muscle relaxation and gives Joan a brief* **overview** *of the procedure.*

1. *Counselor:* Basically, we all learn to carry around some body tension. Some is OK. But in a tense spot, usually your body is even more tense, although you may not realize this. If you can learn to recognize muscle tension and relax your muscles, this state of relaxation can help to decrease your nervousness or anxiety. What we'll do is to help you recognize when your body is relaxed and when it is tense, by deliberately tensing and relaxing different muscle groups in your body. We should get to the point where, later on, you can recognize the sensations that mean tension and use these as a signal to yourself to relax. Does this make sense?
   Client: I think so. You'll sort of tell me how to do this?

Next, the counselor **will "set up" the relaxation by attending to details about the room** *and the client's comfort.*

2. *Counselor:* Yes. At first I'll show you so you can get the idea of it. One thing we need to do before we start is for you to get as comfortable as possible. So that you won't be distracted by light, I'm going to turn off the light. If you are wearing your contact lenses, take them out if they're uncomfortable, because you may feel more comfortable if you go through this with your eyes closed. Also, I use a special chair for this. You know the straight-backed chair you're sitting on can seem like a rock after a time. That might distract, too. So I have a padded chaise you can use for this. [Gets lounge chair out.]
   Client (sits in chaise): Umm. This really is comfortable.

Next the counselor begins **to model the muscle relaxation** *for Joan. This shows Joan how to do it and may alleviate any embarrassment on her part.*

3. *Counselor:* Good. That really helps. Now I'm going to show you how you can tense and then relax your muscles. I'll start first with my right arm. [Clenches right fist, pauses and notes tension, relaxes fist, pauses and notes relaxation; models several other muscle groups.] Does this give you an idea?
   Client: Yes. You don't do your whole body?

The counselor provides **further information about muscle relaxation, describes sensations** *Joan might feel, and checks to see whether Joan is completely clear on the procedure before going ahead.*

4. *Counselor:* Yes, you do. But we'll take each muscle group separately. By the time you tense and relax each muscle group, your whole body will feel relaxed. You will feel like you are "letting go," which is very important when you tense up—to let go rather than to tense even more. Now, you might not notice a lot of difference right away—but you might. You might even feel like you're floating. This really depends on the person. The most important thing is to remain as comfortable as possible while I'm instructing you. Do you have any questions before we begin, anything you don't quite understand?
   Client: I don't think so. I think that this is maybe a little like yoga.

The counselor proceeds with **instructions to alternately tense and relax** *each of 17 muscle groups.*

5. *Counselor:* Right. It's based on the same idea—learning to soothe away body tension. OK, get very comfortable in your chair and we'll begin. [Gives Joan several minutes to get comfortable, then uses the relaxation instructions. Most of the session is spent in instructing Joan in muscle relaxation as illustrated on pp. 438–439.]

After the relaxation, the counselor **queries Joan** *about her feelings during and after the relaxation. It is important to find out how the relaxation affected the client.*

6. *Counselor:* OK, Joan, how do you feel now?
   Client: Pretty relaxed.
7. *Counselor:* How did the contrast between the tensed and relaxed muscles feel?
   Client: It was pretty easy to tell. I guess sometimes my body is pretty tense and I don't think about it.

LEARNING ACTIVITY 48 — *Muscle Relaxation*

Because muscle relaxation involves the alternate tensing and relaxing of a variety of muscle groups, learning the procedure well enough to use it with a client is sometimes difficult. We have found that the easiest way to learn muscle relaxation is to do it yourself. Using it not only helps you learn what is involved but also may have some indirect benefits for you—increased relaxation!

In this learning activity, you will apply the muscle-relaxation procedure you've just read about to yourself. You can do this by yourself or with a partner. You may wish to try it out alone and then with someone else.

By Yourself

1. Get in a comfortable position, wear loose clothing, and remove your glasses or contact lenses.
2. Use the written instructions in this chapter to practice muscle relaxation. You can do this by putting the instructions on tape or by reading the instructions to yourself. Go through the procedure quickly to get a feel for the process; then do it again slowly without trying to rely too much on having to read the instructions. As you go through the relaxation, study the differences between tension and relaxation.
3. Try to assess your reactions after the relaxation. On a scale from 0 to 5 (0 being very relaxed and 5 being very tense), how relaxed do you feel? Were there any particular muscle groups that were hard for you to contract or relax?
4. One or two times through muscle relaxation is not enough to learn it or to notice any effects. Try to practice this procedure on yourself once or twice daily over the next several weeks.

With a Partner

One of you can take the role of a helper; the other can be the person learning relaxation. Switch roles so you can practice helping someone else through the procedure and trying it out on yourself.

1. The helper should begin by giving an explanation and a rationale for muscle relaxation and any instructions about it before you begin.
2. The helper can read the instructions on muscle relaxation to you. The helper should give you ample time to tense and relax each muscle group and should encourage you to note the different sensations associated with tension and relaxation.
3. After going through the process, the helper should query you about your relaxation level and your reactions to the process.

The counselor assigns **relaxation practice** *to Joan as* **daily homework.**

8. *Counselor:* As I mentioned before, this takes regular practice in order for you to use it when you need it—and to really notice the effects. I have put these instructions on this audiotape, and I'd like you to practice with this tape two times each day during the next week. Do the practice in a quiet place at a time when you don't feel pressured, and use a comfortable place when you do practice. Do you have any questions about the practice?

Client: No, I think I understand.

Counselor **explains the use of the log.**

9. *Counselor:* Also, I'd like you to use a log sheet with your practice. Mark down where you practice, how long you practice, what muscle groups you use, and your tension level before and after each practice on this 5-point scale. Remember, 0 is complete relaxation and 5 is complete or extreme tension. Let's go over an example of how you use the log. . . . Now, any questions?

Client: No. I can see this will take some practice.

Finally, the counselor arranges a **follow-up.**

10. *Counselor:* Right, it really is like learning any other skill—it doesn't just come automatically. Why don't you try this on your own for two weeks and then come back, OK?

SUMMARY

In this chapter we described three meditation strategies, a procedure for muscle relaxation, and two scripts on scanning the body. Any one of the meditation strategies can be used in vivo. Counselors must be aware that meditation and muscle relaxation can have contraindications and adverse effects for some clients. The muscle-relaxation strategy can be used with 17, 7, or 4 muscle groups. All these strategies are often used as a single treatment to prevent stress and to deal with stress-related situations. In addition, these strategies are used in Hatha Yoga (Chapter 18); muscle relaxation may also be used with systematic desensitization, eye movement desensitization, and reprocessing (see Chapter 19).

P O S T E V A L U A T I O N

Part One

For Objective One, you will be able to identify accurately the steps of the relaxation response procedure represented by at least seven out of nine examples of counselor instructive responses. On paper, for each of the following counselor responses, identify which part of the meditation procedure is being implemented. There may be more than one counselor response associated with a part. These examples are not in any particular order. The eight major parts of meditation are as follows:

1. Rationale
2. Instructions about when, where, and how long to practice
3. Instruction about focusing on a word, phrase, or prayer
4. Breathing instructions
5. Instruction about passive attitude
6. Meditating for 10 to 20 minutes
7. Probing about the meditative experience
8. Homework and practice

Feedback for the Postevaluation follows on pages 457–458.

1. "It is very important that you practice this at home regularly. Usually there are no effects without regular practice—about twice daily."
2. "Find a comfortable place in your home to practice, one free of interruptions and noise."
3. "This procedure has been used to help people with high blood pressure and people who have trouble sleeping and just as a general stress-reduction process."
4. "Breathe through your nose and focus on your breathing. If you can concentrate on one word as you do this, it may be easier."
5. "Be sure to practice at a quiet time when you don't think you'll be interrupted. And do not practice within two hours after a meal or within two hours before going to bed."
6. "Just continue now to meditate like this for 10 or 15 minutes. Sit quietly then for several minutes after you're finished."
7. "How easy or hard was this for you to do?"
8. "There may be times when other images or thoughts come into your mind. Try to just maintain a passive attitude. If you're bothered by other thoughts, don't dwell on them, but don't force them away. Just focus on your breathing and your word."
9. "Pick a word like *one* or *zum* that you can focus on—something neutral to you."

Part Two

Objective Two asks you to identify accurately the steps for the mindfulness meditation procedure represented by at least eight out of ten counselor instructive responses. Do this on paper. There may be more than one response for a given step. The counselor examples are not in order. The eight major steps of mindfulness meditation are as follows:

1. Rationale
2. Instructions about attitude
3. Preparation for meditation
4. Instructions about commitment, self-discipline, and energy
5. Breathing instructions
6. Body scan instructions
7. Discussion of client's reaction to first meditation
8. Homework and practice

Feedback follows.

1. "Meditation has benefited people by reducing tension, anxiety, stress, and headaches."
2. "Meditate for eight weeks and at the same times once or twice a day."
3. "Allow distracting thoughts to flow. Allow memories, images, and thoughts to occur. Don't try to influence them."
4. "Find a comfortable position in which to meditate."
5. "Notice as the air you breathe enters and exits your nostrils."
6. "Come out of meditation slowly. Sit with your eyes closed for two minutes. Slowly open your eyes. How do you feel?"
7. "Allow the wave of your breath to enter your body from the tip of your toes to the crown of your head."
8. "At first it will be like you are in training."

Part Three

Objective Three asks you to teach the process of meditation to another person. Select either relaxation response or mindfulness meditation to teach. You can have an observer evaluate you, or you can audiotape your teaching session and rate yourself. You can use the Interview Checklist for Mindfulness Meditation or the Interview Checklist for Relaxation Response that follows as a teaching guide and evaluation tool.

(continued)

POSTEVALUATION (continued)

Part Four

Objective Four asks you to describe how you would apply the seven major parts of the muscle-relaxation procedure. Using this client description and the seven questions following it, describe how you would use certain parts of the procedure with this person. You can check your responses with the feedback that follows.

Description of client: The client is a middle-age man who is concerned about his inability to sleep at night. He has tried sleeping pills but does not want to rely on medication.

1. Give an example of a rationale you could use about the procedure. Include the purpose and an overview of the strategy.
2. Give a brief example of instructions you might give this client about appropriate dress for relaxation training.
3. List any special environmental factors that may affect the client's use of muscle relaxation.
4. Describe how you might model some of the relaxation exercises for the client.

5. Describe some of the important muscle groups that you would instruct the client to tense and relax alternately.
6. Give two examples of probes you might use with the client after relaxation to assess his use of and reactions to the process.
7. What instructions about a homework assignment (practice of relaxation) would you give to this client?

Part Five

Objective Five asks you to demonstrate 13 out of 15 steps of muscle relaxation with a role-play client. An observer or the client can assess your performance, or you can assess yourself, using the Interview Checklist for Muscle Relaxation on pages 453–456.

Part Six

Objective Six asks you to demonstrate the body scan procedure with a role-play client. An observer can assess you or you can tape-record this activity using the script on pp. 443–445.

Interview Checklist for Mindfulness Meditation

Instructions: Determine which of the following counselor leads or questions were demonstrated in the interview. Check each of the leads used by the counselor. Some examples of counselor leads are provided in the right column.

Item	Examples of counselor leads
I. Rationale	
_____ 1. Counselor describes purpose of procedure.	"I would like to teach you mindfulness meditation. This type of meditation has been used to relieve fatigue caused by anxiety, to decrease stress that leads to high blood pressure, and to bring balance and focus in your life. Meditation helps you become more relaxed and deal more effectively with your tension and stress. It may bring you new awareness about yourself and a new way of seeing and doing in your life."
_____ 2. Counselor gives client an overview.	"First we will select a quiet place in which to meditate. You will then get into a relaxed and comfortable position. With your eyes closed, you will focus on your breathing, and allow your thoughts to flow freely. If your mind wanders off, you can bring it back by focusing on the breath. You will meditate for 10 to 20 minutes. Then, we will talk about the experience."
_____ 3. Counselor confirms client's willingness to use strategy.	"How do you feel now about practicing meditation?"
II. Instructions About Attitudinal Foundations for Meditation	
_____ 4. Counselor instructs client about attitudes to help the practice of meditation.	"There are seven attitudes that will help with your practice of meditation.

(continued)

Item	Examples of counselor leads

II. Instructions About Attitudinal Foundations for Meditation (continued)

_____ 5. Counselor instructs the client about being non-judging.

"First, it is best to be nonjudging. We have a tendency to categorize or judge people, things, or our experiences. These judgments take you away from observing whatever comes up while you are meditating. Judging steals energy from the moment-by-moment awareness. To remedy this, focus on your breathing."

_____ 6. Counselor instructs the client about patience.

"Second, have patience, which means just allow things to unfold in their own time. We don't have to fill our lives with moments of doing and activity."

_____ 7. Counselor instructs the client about beginner's mind and basing moment-by-moment awareness on past experiences.

"Third, as a beginner, what we experience in the moment is often based on our past experiences and ways of doing things. Just be open to moment-by-moment experience. Don't let past experiences judge and steal energy from moment-by-moment awareness."

_____ 8. Counselor instructs client about trusting feelings and intuition.

"Fourth, trust your feelings and intuition while meditating. For example, if your body tells you that your posture for meditating is not comfortable, change to another posture that feels better.

_____ 9. Counselor instructs the client to be nonstriving.

"Fifth, try to be nonstriving, which means mindfulness meditation is about the process of practice; every practice will be different. You don't want a mind-set that requires you to achieve something or get somewhere. Just be in the moment, and attend to whatever comes up."

_____ 10. Counselor instructs the client about acceptance.

"Sixth, just focus on seeing and accepting things as they are, moment by moment, and in the present."

_____ 11. Counselor instructs the client about letting go.

"Seventh, just let go, which means nonattachment or not holding on to thoughts."

III. Instructions about Commitment, Self-Discipline, and Energy

_____ 12. Counselor instructs client about commitment, self-discipline, and energy.

"You want to make the kind of commitment required in athletic training. This strong commitment is about working on your self. You have to summon enough self-discipline to generate enough energy that you can develop a strong meditative practice and a high degree of mindfulness. You don't have to like it; you just have to do it. Then, at the end of eight weeks of practice, we can see whether the practice was useful."

IV. Instructions About Preparations for Meditation

_____ 13. Counselor instructs the client about time, place, and posture.

"Select a particular time every day to meditate. Meditate for at least six days a week, and for eight weeks. Find a place to meditate that will be free of interruptions and that will be comfortable. When you meditate, sit erect in a chair or on the floor. Try to have your back so that it is self-supporting."

V. Body Scan Instructions

_____ 14. Counselor instructs client to do a body scan.

"Allow the body to begin to relax completely. Inhale and feel the breath flow from the soles of the feet to the crown of the head, like a gentle slow motion wave. With each exhalation allow tension to flow out of the body." (Continue this script on page 444.)

(continued)

Item	Examples of counselor leads

VI. Breathing Instruction

_____15. Counselor instructs the client about breathing.

"Observe your breathing as it flows in and flows out. Notice the difference in temperature of the out breath and in breath. Feel the sensation of the air as it goes in and out of the nostrils."

VII. Instructions about Wandering Mind

_____16. Counselor instructs client about what to do with cascading thoughts, feelings, sensations, sounds, pain, or discomfort.

"If you find yourself getting stuck in thoughts, feelings, sensations, sounds, pain, or discomfort, this is normal; just bring your attention to breathing, and let go by exhaling these distractions."

VIII. Instructions to Meditate

_____17. Counselor instructs client about sitting quietly and relaxed for a minute. You can do a quick body scan to facilitate relaxation.

"Sit quietly for a while; just relax; focus on your breathing."

_____18. Counselor instructs client to close eyes, focus on breathing, and get in a comfortable position.

"Close your eyes, get in a comfortable position, focus on your breathing; the air comes in and flows out."

_____19. Counselor instructs client to be in the moment, to have awareness and observe what comes up, and not give distractions any energy.

"Just be in the moment; be aware and observe whatever comes to mind. If distractions of thoughts, feelings, sounds, pain, or discomfort steal energy, just breathe them out and continue to observe and not move with the flow of these distractions."

_____20. Counselor tells the client that she or he will meditate for 10 to 20 minutes.

"Meditate for 10 to 20 minutes. I will keep time, and tell you when to stop.

_____21. Counselor instructs client to come out of meditation slowly.

"I want you to come out of the meditation slowly. Just sit there with your eyes closed for a while; take time to absorb what you experienced. You may wish to stretch and open your eyes slowly."

IX. Discussion of Client's Reaction to Mindfulness Meditation

_____22. Counselor asks client about experience with mindfulness meditation.

"What was the experience like for you?"
"How did you handle distractions?"
"How did you feel about mindfulness meditation?"

X. Homework

_____23. Counselor instructs client to meditate at home once a day. Reminds client about preparation for meditation.

"Practice mindfulness meditation once a day at least five days a week. Remember to select a quiet environment without distractions. Do not take any alcoholic beverages or nonprescription drugs at least 24 hours before meditating. Wait for an hour before meditating after eating solid foods or drinking beverages containing caffeine. Be in the moment when you meditate; just observe what comes up without being carried away."

_____24. Counselor instructs client about informal meditation.

"You can meditate informally when you are stressed out in stressful situations that may occur daily. Just relax and focus on your breathing; be aware and observe what is going on without giving energy to stress, and be peaceful in the situation."

Observer comments: _____

_____

_____

Interview Checklist for Relaxation Response

Instructions: Determine which of the following counselor leads or questions were demonstrated in the interview. Check each of the leads used by the counselor. Some examples of counselor leads are provided in the right column.

Item	Examples of counselor leads

I. Rationale

_____ 1. Counselor describes purpose of meditation.

"The relaxation response has been used to relieve anxiety, to decrease stress that can lead to high blood pressure, and to become more relaxed. It may give you a new focus and awareness about your self and your world."

_____ 2. Counselor gives client an overview.

"You will select a focus word, then you will get in a comfortable position, relax your body, and focus on your breathing. You will maintain a passive attitude while meditating, which will elicit the relaxation response. You will meditate for 10 to 20 minutes. Then, we will talk about the experience."

_____ 3. Counselor confirms client's willingness to use relaxation response.

"How do you feel about working with meditation that will elicit the relaxation response?"

II. Instructions about When, Where, and How Long to Practice

_____ 4. Counselor instructs client about when, where, and how long to practice.

"One of the best times to practice is before breakfast because it sets a positive tone for the day. This time is uncluttered with events and activities of the day. Relaxation works best if a regular time is chosen for it so that a routine is developed. Select a place to practice that is quiet and free from distractions and interruptions. Try to practice 10 to 20 minutes at least once and better twice a day."

III. Instructions about Focus Word, Phrase, or Prayer

_____ 5. Counselor provides rationale for mental word, phrase, or prayer.

"One major way to focus your mind is link it to breathing either by concentrating on the breath or by focusing on something. You can focus on a word, phrase, or prayer."

_____ 6. Counselor gives examples of focus word, phrase, or prayer.

"You might prefer a neutral word or phrase such as 'one' or 'peace' or 'love'; or you can use a phrase—'the air flows in and the air flows out.' Or you could focus on a prayer consistent with your belief system. For example, a Christian might select the opening of Psalm 23, *The Lord is my shepherd;* a Jew, *shalom;* a Muslim, *Allah.*"

IV. Instructions about Body Position and Eyes

_____ 7. Counselor instructs client about body posture and eyes.

"There are several ways to meditate. I'll show you one. I want you to get into a comfortable position while you are sitting there."

V. Body Scan

_____ 8. Counselor does quick body scan with the client.

"Relax all the muscles in your body; relax (said slowly) your head, face, neck, shoulders, chest, your torso, thighs, calves, and your feet. Relax your body." (You can use the body scan script on pages 443–444).

VI. Breathing Instructions

_____ 9. Counselor gives instructions about how to breathe.

"Breathe through your nose and focus on (or become aware of) your breathing. It is sometimes difficult to be natural when you are doing this. Just let the air come to you. Breathe easily and naturally. As you do this, say your focus word for each inhalation and exhalation. Say your focus word silently to yourself each time you breathe in and out."

(continued)

POSTEVALUATION **(continued)**

Item	Examples of counselor leads
VII. Instructions about Passive Attitude	
_____10. Counselor instructs client about passive attitude.	"Be calm and passive. If distracting thoughts or images occur, attempt to be passive about them by not dwelling on them. Return to repeating your focus word, phrase, or prayer. Try to achieve effortless breathing. After more practice, you will able to examine these thoughts or images with greater detachment. Do not attempt to keep the thoughts out of your mind; just let them pass through. Keep your mind open; don't try to solve problems or think things over. Allow thoughts to flow smoothly into your mind and then drift out. Say your focus word and relax. Don't get upset with distracting thoughts. Just return to your focus word, phrase, or prayer."
VIII. Instruct the Client to Meditate for about Ten to Twenty Minutes	
_____11. Counselor instructs client to meditate 10 to 20 minutes.	"Now, meditate for 10 to 20 minutes. You can open your eyes to check on the time. After you have finished, sit quietly for several minutes. You may wish to keep your eyes closed for a couple of minutes and later open them. You may not want to stand up for a few minutes."
IX. Obtain Client Reaction to Relaxation Response	
_____12. Counselor asks client about experience with relaxation response.	"How do you feel about the experience?" "What sorts of thoughts or images flowed through your mind?" "What did you do when the distracting thoughts drifted in?" "How did you feel about your focus word, phrase, or prayer?"
X. Homework	
_____13. Counselor instructs client to meditate daily for the next week.	"Practice the relaxation response two times a day. Get comfortable in your relaxation response position. Practice in a quiet place away from noise and interruptions. Do not meditate within two hours after eating or within a couple of hours before bedtime."
_____14. Counselor instructs client to apply relaxation response in vivo.	"Also, it would be helpful for you to apply an informal meditation in problem or stressful situations that may occur daily. You can do this by becoming detached and passive in the stressful situation. Observe yourself and focus on being calm and on your breathing. Be relaxed in situations that evoke stress."

Observer Comments: _____

_____

_____

(continued)

Interview Checklist for Muscle Relaxation

Instructions: Indicate with a check mark each counselor lead demonstrated in the interview. Some example leads are provided in the right column.

Item	Examples of counselor leads
I. Rationale	
_____ 1. Counselor explains purpose of muscle relaxation.	"The name of the strategy that I believe will be helpful is *muscle relaxation.* Muscle relaxation has been used very effectively to benefit people who have a variety of concerns like insomnia, high blood pressure, anxiety, or stress or for people who are bothered by everyday tension. Muscle relaxation will be helpful in decreasing your tension. It will benefit you because you will be able to control and to dispel tension that interferes with your daily activities."
_____ 2. Counselor gives overview of how muscle relaxation works.	"I will ask you to tense up and relax various muscle groups. All of us have some tensions in our bodies—otherwise we could not stand, sit, or move around. Sometimes we have too much tension. By tensing and relaxing, you will become aware of and contrast the feelings of tension and relaxation. Later we will train you to send a message to a particular muscle group to relax when nonessential tension creeps in. You will learn to control your tension and relax when you feel tension."
_____ 3. Counselor describes muscle relaxation as a skill.	"Muscle relaxation is a skill. And, as with any skill, learning it well will take a lot of practice. A lot of repetition and training are needed to acquire the muscle-relaxation skill."
_____ 4. Counselor instructs client about moving around if uncomfortable and informs client of sensations that may feel unusual.	"At times during the training and muscle exercises, you may want to move while you are on your back on the floor [or on the recliner]. Just feel free to do this so that you can get more comfortable. You may also feel heady sensations as we go through the exercise. These sensations are not unusual. Do you have any questions concerning what I just talked about? If not, do you want to try this now?"
II. Client Dress	
_____ 5. Counselor instructs client about what to wear for training session.	"For the next session, wear comfortable clothing." "Wear regular glasses instead of your contact lenses."
III. Comfortable Environment	
_____ 6. Counselor uses quiet environment, padded recliner chair, or floor with a pillow under client's head.	"During training, I'd like you to sit in this recliner chair. It will be more comfortable and less distracting than this wooden chair."
IV. Modeling the Exercises	
_____ 7. Counselor models some exercises for muscle groups.	"I would like to show you [some of] the exercises we will use in muscle relaxation. First, I make a fist to create tension in my right hand and forearm and then relax it."
V. Instructions for Muscle Relaxation	
_____ 8. Counselor reads or recites instructions from memory in conversational tone and practices along with client.	
_____ 9. Counselor instructs client to get comfortable, close eyes, and listen to instructions.	"Now, get as comfortable as you can, close your eyes, and listen to what I'm going to be telling you. I'm going to make

(continued)

POSTEVALUATION **(continued)**

Item	Examples of counselor leads
V. Instructions for Muscle Relaxation (continued)	you aware of certain sensations in your body and then show you how you can reduce these sensations to increase feelings of relaxation."
_____10. Counselor instructs client to tense and relax alternately each of the 17 muscle groups (*two* times for each muscle group in initial training). Also occasionally makes muscle-group comparisons.	
_____a. Fist of dominant hand	"First study your right arm, your right hand in particular. Clench your right fist. Clench it tightly and study the tension in the hand and in the forearm. Study those sensations of tension. [Pause.] And now let go. Just relax the right hand and let it rest on the arm of the chair. [Pause.] And note the difference between the tension and the relaxation." [Ten-second pause.].
_____b. Fist of nondominant hand	"Now we'll do the same with your left hand. Clench your left fist. Notice the tension [five-second pause] and now relax. Enjoy the difference between the tension and the relaxation." [Ten-second pause.]
_____c. One or both wrists	"Now bend both hands back at the wrists so that you tense the muscles in the back of the hand and in the forearm. Point your fingers toward the ceiling. Study the tension, and now relax. [Pause.] Study the difference between tension and relaxation." [Ten-second pause.]
_____d. Biceps of one or both arms	"Now, clench both your hands into fists and bring them toward your shoulders. As you do this, tighten your bicep muscles, the ones in the upper part of your arm. Feel the tension in these muscles. [Pause.] Now relax. Let your arms drop down again to your sides. See the difference between the tension and the relaxation." [Ten-second pause.]
_____e. Shoulders	"Now we'll move to the shoulder area. Shrug your shoulders. Bring them up to your ears. Feel and hold the tension in your shoulders. Now, let both shoulders relax. Note the contrast between the tension and the relaxation that's now in your shoulders. [Ten-second pause.] Are your shoulders as relaxed as your arms?"
_____f. Forehead	"Now we'll work on relaxing the various muscles of the face. First, wrinkle up your forehead and brow. Do this until you feel your brow furrow. [Pause.] Now relax. Smooth out the forehead. Let it loosen up." [Ten-second pause.]
_____g. Eyes	"Now close your eyes tightly. Can you feel tension all around your eyes? [Five-second pause.] Now relax those muscles, noting the difference between the tension and the relaxation." [Ten-second pause.]

(continued)

Item	Examples of counselor leads
V. Instructions for Muscle Relaxation (continued)	
_____h. Tongue or jaw	"Now clench your jaw by biting your teeth together. Pull the corners of your mouth back. Study the tension in the jaws. [Five-second pause.] Relax your jaws now. Can you tell the difference between tension and relaxation in your jaw area?" [Ten-second pause.]
_____i. Lips	"Now, press your lips together tightly. As you do this, notice the tension all around the mouth. [Pause.] Now relax those muscles around the mouth. Just enjoy the relaxation in your mouth area and your entire face." [Pause.]
_____j. Head backward	"Now we'll move to the neck muscles. Press your head back against your chair. Can you feel the tension in the back of your neck and in the upper back? Hold the tension. Now let your head rest comfortably. Notice the difference. Keep on relaxing." [Pause.]
_____k. Chin in chest	"Now continue to concentrate on the neck area. See whether you can bury your chin into your chest. Note the tension in the front of your neck. Now relax and let go." [Ten-second pause.]
_____l. Back	"Now direct your attention to your upper back area. Arch your back as if you were sticking out your chest and stomach. Can you feel tension in your back? Study that tension. [Pause.] Now relax. Note the difference between the tension and the relaxation."
_____m. Chest muscles	"Now take a deep breath, filling your lungs, and hold it. See the tension all through your chest and into your stomach area. Hold that tension. [Pause.] Now relax and let go. Let your breath out naturally. Enjoy the pleasant sensations. Is your chest as relaxed as your back and shoulders?" [Ten-second pause.]
_____n. Stomach muscles	"Now think about your stomach. Tighten the abdomen muscles. Hold this tension. Make your stomach like a knot. Now relax. Loosen these muscles now." [Ten-second pause.]
_____o. Buttocks	"Focus now on your buttocks. Tense your buttocks by pulling them in or contracting them. Note the tension that is there. Now relax—let go." [Ten-second pause.]
_____p. Legs	"I'd like you now to focus on your legs. Stretch both legs. Feel tension in the thighs. [Five-second pause.] Now relax. Study the difference again between the tension in the thighs and the relaxation you feel now." [Ten-second pause.]
_____q. Toes	"Now concentrate on your lower legs and feet. Tighten both calf muscles by pointing your toes toward your head. Pretend a string is pulling your toes up. Can you feel the pulling and the tension? Note that tension. [Pause.] Now relax. Let your legs relax deeply. Enjoy the difference between tension and relaxation." [Ten-second pause.]

(continued)

POSTEVALUATION (continued)

Item	Examples of counselor leads
V. Instructions for Muscle Relaxation (continued)	
_____11. Counselor instructs client to review and relax all muscle groups.	"Now, I'm going to go over again the different muscle groups that we've covered. As I name each group, try to notice whether there is any tension in those muscles. If there is any, try to concentrate on those muscles and tell them to relax. Think of draining any residual tension out of your body. Relax the muscles in your feet, ankles, and calves. [Pause.] Let go of your knee and thigh muscles. [Pause.] Loosen your hips. [Pause.] Loosen the muscles of your lower body. [Pause.] Relax all the muscles of your stomach, waist, lower back. [Pause.] Drain any tension from your upper back, chest, and shoulders. [Pause.] Relax your upper arms, forearms, and hands. [Pause.] Let go of the muscles in your throat and neck. [Pause.] Relax your face. [Pause.] Let all the muscles of your body become loose. Drain all the tension from your body. [Pause.] Now sit quietly with your eyes closed."
_____12. Counselor asks client to rate relaxation level following training session.	"Now I'd like you to think of a scale from 0 to 5, where 0 is complete relaxation and 5 extreme tension. Tell me where you would place yourself on that scale now."
VI. Posttraining Assessment	
_____13. Counselor asks client about first session of relaxation training, discusses problems with training if client has any.	"How do you feel?" "What is your overall reaction to the procedure?" "Think back about what we did—did you have problems with any muscle group?" "What reaction did you have when you focused on the tension? What about relaxation?" "How did the contrast between the tension and relaxation feel?"
VII. Homework and Follow-up	
_____14. Counselor assigns homework and requests that client complete homework log for practice sessions.	"Relaxation training, like any skill, takes a lot of practice. I would like you to practice what we've done today. Do the exercises twice a day for 15 to 20 minutes each time. Do them in a quiet place in a reclining chair, on the floor with a pillow, or on your bed with a head pillow. Also, try to do the relaxation at a time when there is no time pressure—like arising, after school or work, or before dinner. Try to avoid any interruptions, like telephone calls and people wanting to see you. Complete the homework log I have given you. Make sure you fill it in for each practice session. Do you have any questions?"
_____15. Counselor arranges for follow-up session.	"Why don't you practice with this over the next two weeks and come back then?"

Notations for problems encountered or variations used: _____

_____

_____

SUGGESTED READINGS

Alexander, C. N., Robinson, P., Orme-Johnson, D., & Schneider, R. H. (1994). The effects of transcendental meditation compared to other methods of relaxation and meditation in reducing risk factors, morbidity, and mortality. *Homeostasis in Health and Disease, 35,* 243–263.

Arena, J., Bruno, G., Hannah, S., & Meador, K. (1995). A comparison of frontal electromyographic biofeedback training, trapezius electromyographic biofeedback training, and progressive muscle relaxation therapy in the treatment of tension headache. *Headache, 35,* 411–419.

Bankart, C. P. (1997). *Talking cures: A history of Western and Eastern psychotherapies.* Pacific Grove, Ca.: Brooks/Cole.

Benson, H., & Stuart, E. M. (1992). *The wellness book: The comprehensive guide to maintaining health and treating stress-related illness.* New York: Birch Lane Press.

Bernstein, D. A., & Carlson, C. R. (1993). Progressive relaxation: Abbreviated methods. In P. M. Lehrer & R. L. Woolfolk (Eds.), *Principles and practice of stress management* (2nd ed., pp. 53–87). New York: Guilford.

Borysenko, J. (1997). *Meditation for inner guidance and self-healing.* (Audio tape). Niles, IL: Nightingale Conant.

Borysenko, J., & Borysenko, M. (1994). *The power of the mind to heal.* Carson, CA: Hay House.

Broota, A., Varma, R. & Singh, A. (1995). Role of relaxation in hypertension. *Journal of the Indian Academy of Applied Psychology, 21,* 29–36.

Carlson, C. R., & Hoyle, R. H. (1993). Efficacy of abbreviated progressive muscle relaxation training: A quantitative review of behavioral medicine research. *Journal of Counseling and Clinical Psychology, 61,* 1059–1067.

Carrington, P. (1993). Modern forms of meditation. In P. M. Lehrer & R. L. Woolfolk (Eds.), *Principles and practice of stress management* (2nd ed., pp. 139–168). New York: Guilford.

Epstein, M. (1995). *Thoughts without a thinker.* New York: Basic Books.

Fontana, D. (1991). *The elements of meditation.* New York: Element.

Hanh, T. N. (1991). *Peace is every step.* New York: Bantam.

Ibanez-Ramirez, M., Delgado-Morales, F., & Pulido-Diez, J. (1989). Secondary situational impotence in case of homosexuality. *Analisis y Modificacion de Conducta, 15,* 297–304.

Johanson, G., & Kurtz, R. (1991). *Grace unfolding: Psychotherapy in the spirit of Tao-te Ching.* New York: Bell Tower.

Kabat-Zinn, J. (1990). *Full catastrophe living.* New York: Dell.

Kabat-Zinn, J. (1994). *Wherever you go, there you are.* New York: Hyperion.

Kabat-Zinn, J. (1995). *Mindfulness meditation.* (Audio tape). Niles, IL: Nightingale Conant.

Lehrer, P. M., & Woolfolk, R. L. (Eds.). (1993). *Principles and practice of stress management* (2nd ed.). New York: Guilford.

Levine, S. (1991). *Guided meditations, explorations and healings.* New York: Anchor.

Long, B. (1995). *Meditation* (rev. ed.). London: Barry Long Books.

Murphy, M., & Donovan, S. (Eds.). (1997). *The physical and psychological effects of meditation* (2nd ed.). Sausalito, CA: Institute of Noetic Sciences.

Simpkins, C. A., & Simpkins, A. M. (1996). *Principles of meditation.* Boston: Charles E. Tuttle.

Teasdale, J., Segal, Z., Williams, J. & Mark, G. (1995). How does cognitive therapy prevent depressive relapse and why should attentional control (mindfulness) training help? *Behaviour Research and Therapy, 33,* 25–39.

Weil, A. (1995). *Spontaneous healing.* New York: Ballantine.

Weil, A. (1997). *8 weeks to optimum health.* New York: Knopf.

FEEDBACK POSTEVALUATION

Part One

1. *Homework* (practice)
2. *Instruction* about where to practice
3. *Rationale*—telling the client how the procedure is used
4. *Breathing Instructions*
5. *Homework*—giving the client instructions about how to carry out the practice
6. *Instructing the client to meditate* for 10 to 20 minutes
7. *Probing* about the meditative experience—assessing the client's reactions
8. *Instruction about a passive attitude*
9. *Instruction about focusing* on word or phrase

Part Two

1. *Rationale*—reason
2. *Homework*—when to practice
3. *Instructions about attitude*

(continued)

FEEDBACK: POSTEVALUATION   (continued)

4. *Preparation* about position for meditating
5. *Breathing* instructions
6. *Discussion* about reactions to meditation
7. *Body scan*
8. *Instructions about self-discipline and commitment*

Part Three
Use the Interview Checklist for Relaxation Response or the Interview Checklist for Mindfulness Meditation as a guide to assess your teaching.

Part Four

1. Rationale for client:
   a. Purpose: "This procedure, if you practice it regularly, can help you become relaxed. The relaxation benefits you derive can help you sleep better."
   b. Overview: "This procedure involves learning to tense and relax different muscle groups in your body. By doing this, you can contrast the difference between tenseness and relaxation. This will help you to recognize tension so you can instruct yourself to relax."
2. Instructions about dress: "You don't want anything to distract you, so wear comfortable, loose clothes for training. You may want to remove your glasses or contact lenses."
3. Environmental factors:
   a. Quiet room with reclining chair
   b. No obvious distractions or interruptions
4. Modeling of exercises: "Let me show you exactly what you'll be doing. Watch my right arm closely. I'm going to clench my fist and tighten my forearm, studying the tension as I do this. Now I'm going to relax it like this [hand goes limp], letting all the tension just drain out of the arm and hand and fingertips."
5. Muscle groups used in the procedure include
   a. fist of each arm
   b. wrist of each arm
   c. biceps of each arm
   d. shoulders
   e. facial muscles—forehead, eyes, nose, jaws, lips
   f. head, chin, and neck muscles
   g. back
   h. chest
   i. stomach
   j. legs and feet
6. Some possible probes are these:
   a. "On a scale from 0 to 100, 0 being very relaxed and 100 very tense, how do you feel now?"
   b. "What is your overall reaction to what you just did?"
(continued)

c. "How did the contrast between the tensed and relaxed muscles feel?"
   d. "How easy or hard was it for you to do this?"
7. Homework instructions should include
   a. practice twice daily
   b. practice in a quiet place; avoid interruptions
   c. use a reclining chair, the floor, or a bed with pillow support for your head

Part Five
Use the Interview Checklist for Muscle Relaxation to assess your performance.

Part Six
Use the scripts on pp. 443–445 to assess the way you used the body scan with a role-play client.

CHAPTER 18

BREATHING AND HATHA YOGA

A 16-year-old boy was having difficulty in school because he could not stay awake, and he was always tired. His parents didn't understand why he was so tired. They said that he had more than eight hours of sleep every school night.

A 29-year-old computer programmer had lower back pain. She said that she worked long hours at the computer and she felt that she was under pressure all the time. She was fidgety and had difficulty falling asleep at night.

A former drug addict said that ever since his rehabilitation he was clean, but he was afraid that he might revert to his old habits if he couldn't find something positive and self-fulfilling to do.

An elderly man said that he had dizzy spells, shakiness, and trembling. An exam by his physician did not reveal any causes or abnormal functions to explain his symptoms.

A woman who was raped about six months ago reported that she was experiencing a great deal of discomfort in her body.

A counselor in a mental health center reported that he always felt uncentered and unbalanced. He knew he was experiencing stress, but he said he didn't know what to do. He was a single parent and he didn't have time to work out like he thought he should.

These are examples of people who might benefit from the two interventions presented in this chapter. For years, health professionals have treated the mind and the body as two independent systems. We are now changing our old ways of thinking about health and healing. For example,

- We no longer believe that there is a single cause for each illness. We now recognize that disease, like health, is the result of a web of circumstances that involve our mind, body, spirit, and environment.
- We no longer believe that there is only one way to heal. We now recognize that no singular approach, practice, or treatment has all the answers.
- We no longer believe that the power to heal is exclusively contained in external agents or treatments. We now recognize

the wealth of healing capacities that are built into our minds, bodies, and spirits.

- We no longer believe that health professionals have all the answers to our questions about health and disease. We recognize that there are answers we must seek ourselves. (Dacher, 1996, p. 12)*

Research in the field of psychoneuroimmunology (see Benson & Stuart, 1992; Goleman & Gurin, 1993; Moyers, 1993) has provided an understanding of the biochemical pathways that link the mind and the body. Our psychological state influences the physiology of our bodies. Within the last 15 years, scientists have found increasing evidence of the link between the mind and body and how it influences our mental and physical health. The interconnections of our thoughts, feelings, images, and biochemistry are mediated through a mobile neuropeptide messenger system—natural chemicals that transfer information between the mind and body. Reflective of this type of research is the work of Candace Pert (1993), who found that everything in the body is run by information molecules, and these peptides and receptors are the biochemistry of emotions. From this research, alternative health healing practices have emerged and are beginning to be viewed as worthy of consideration by mental health professionals and as complementary to traditional medical procedures and mental health interventions. Complementary procedures once considered exotic and strange are now common among many health care practitioners and facilities (see Benson & Stuart, 1992; Moyers, 1993). Each of these complimentary practices is what Dacher (1996) refers to as "something we can do for ourselves in contrast to something that is done *to* us" (pp. 25–26). There is an entire curriculum of practices for self-regulation and creating the mind/body

---

*SOURCE: From *Whole Healing,* by Elliott S. Dacher. Copyright © 1996 by Elliott S. Dacher, M.D. Used by permission of Dutton Signet, a division of Penguin Books USA Inc.

connection and consciousness. These practices include meditation, yoga, breathing techniques, tai chi, aikido, imagery, writing, and journaling. Imagery was presented in Chapter 13, meditation in the previous chapter. These practices are derived from what Bankart (1997) refers to as "Eastern psychotherapies," which

1. integrate physical health, spiritual well-being, interpersonal awareness, and emotional stability into a unified whole;
2. expand consciousness; and
3. are accomplished in relative silence (Bankart, 1997, pp. 482–483).

In this chapter, we present two powerful mind/body practices: breathing and Hatha Yoga. You can train clients in one or more breathing exercises. By experiencing yoga postures (by actual experience or in your imagery), you can enhance your awareness of yoga so that you can refer clients who can benefit from working with a trained Hatha Yoga therapist. Also, if you experience the benefits of breathing exercises and/or Hatha Yoga practice, you will find that you will become more aware, balanced, and focused in your relationship with clients and in other parts of your life. Breathing exercises and Hatha Yoga can be used either as a single treatment or in conjunction with other treatments.

A great deal of psychotherapy has traditionally focused on working with a client's emotions, cognitions, and behaviors. So much of what we do in many Western cultural groups is "in our head" without much awareness about the connection with our body. Many clients can benefit from the therapeutic breath work and yoga postures presented in this chapter. Within each section of breathing exercises and Hatha Yoga, we present a list of the kinds of clients who might benefit from either strategy. With the precautions described in the breathing section, you can teach breathing exercises to clients. You can *not* teach Hatha Yoga to a client unless you are a certified yoga instructor or therapist; but you can become familiar with the process of Hatha Yoga so that you can make knowledgeable referrals to a certified yoga therapist. If you feel that a client might benefit from mind/body work, you can train the person in breathing exercises. If the client likes and responds well to the breath work, you can refer him or her to a certified yoga therapist because breath work and diaphragmatic breathing are integral parts of Hatha Yoga. We have no definitive way of knowing what kinds of clients benefit from Hatha Yoga. It has been our experience, however, that most clients can benefit from either breathing exercises or Hatha Yoga, or both, particularly those in need of stress management.

OBJECTIVES

1. Assess your breathing by using the instructions for "breathing awareness." You might wish to audiotape the instructions before you do the exercise.
2. Demonstrate 10 of the 12 steps for diaphragmatic breathing. You may wish to audiotape the instructions, which might enhance your focus on the exercise.
3. Demonstrate 9 of 11 steps for alternate nostril breathing (audiotape the instructions).
4. Describe the benefits of 10 of the 13 Sun Salutation warm-up postures, and 11 of 14 Hatha Yoga postures. You can describe the benefits in writing, or audiotape them.
5. Describe 4 of the 6 rules for breathing during movement in the postures. Again, you can write or list the rules, or audiotape them.
6. Describe the location and one potential emotional and cognitive issue of 6 of the 7 energy centers or chakras. Describe these in writing or audiotape them.

BREATHING: PROCESSES AND USES

Andrew Weil (1995) declares that "if the breath is the movement of the spirit in the body—a central mystery that connects us to all creation—then working with the breath is a form of spiritual practice" (p. 204). Breathing is *pranayama,* which means control (*yama*) of life force or breath (*prana*), and breathing (pranayama) is part of a system that aims to integrate mind, body and spirit" (Stewart, 1994, p. 8). Prana or life force is more than breath or the gaseous exchange. For example, Patel (1993) says that "when someone gets a creative insight into something we call it an *inspiration*" (p. 117). . . . The more prana or life force one has, the more vital energy and mental alertness and awareness one also has" (p. 118). The oxygen we breathe is important for our metabolism; we cannot burn energy without it, and we need it to nourish, repair, and regenerate our bodies. Conscious breathing allows us to control and to distribute vital energy (prana) with the body.

"Breath is the bridge which connects life to consciousness, which unites your body to your thoughts. . . . Our breath is the bridge from our body to our mind, the element which reconciles our body and mind and which makes possible one-ness of body and mind. Breath is aligned to both body and mind and it alone is the tool which can bring them both together, illuminating both and bringing both peace and calm" (Hanh, 1976, pp. 15, 23). A brief summary of the research about disordered breathing and its effects on mind/body functioning is reported by Fried (1993) and Patel

(1993). Fried (1990) has used breathing exercises as a part of treatment for breathing, stress-related, and psychosomatic disorders. Among the specific problems he has treated with breathing exercises are tension and anxiety, burn-out syndrome, panic disorder, depression, tension headache, hypertension, Raynaud's disease, bruxism, asthma, gastritis, hypoglycemia, heart arrhythmias, diabetes, insomnia, and aerophagia (pp. 230–231).*

PHYSIOLOGY OF BREATHING AND STRESS

An overview of the physiology of breathing is helpful in explaining what happens to the mind/body when breathing is disordered. The function of breathing is to supply the body with oxygen in which respiration (inspiration) oxygenates body cells, and ventilation (expiration) removes excess carbon dioxide. Inhalation or respiration brings air into the lungs. Figure 18-1 shows a diagram of the respiratory system: the trachea (windpipe), heart, an aorta, two lungs, and bronchioles. Included in Figure 18-1 are two inserts: one with the bronchioles with clusters of alveoli enlarged, and the other showing a cluster of alveoli with a network of capillaries. The heart and lungs work together. The heart takes the oxygen-rich blood from the lungs and pumps the blood through the aorta to all parts of the body. The oxygen-poor blood—carrying carbon dioxide—is pumped to the lungs for the exchange of gases. From the lungs, some of the oxygen moves from the air into the bloodstream. At the same time, carbon dioxide moves from the blood into air and is breathed out (ventilation). Oxygen and carbon dioxide are exchanged in tiny air sacs called the alveoli. The alveoli are in direct contact with the capillaries carrying blood (see Figure 18-1). The walls of the air sacs and the capillaries are moist and very thin. Oxygen molecules flow from the alveoli through the membranes into the blood of the capillaries. The hemoglobin in the red cells picks up the oxygen as the blood plasma releases carbon dioxide to the air in the alveoli.

Metabolic activity provides oxygenation to the body, a process achieved by blood circulation or by the oxygen transport system that adjusts the amount of oxygen delivery needed. When there is an increase in metabolic demand, a homeostatic (stable state of equilibrium or balance) adjustment can contribute "to physiological by-products favoring a compromise that may result in chronic graded hypoxia" (Fried, 1993, p. 302). Fried distinguishes between hypoxia (decreased oxygen availability) and anoxia (no oxygen) (p. 302). Hyperventilation can be viewed as evidence of hypoxia, and as one extreme on a continuum of stress reactions. According to Fried (1993), hyperventilation does not describe behavior, but its outcome: hypocapnia, or decreased alveolar carbon dioxide (p. 305). "Hypocapnia impairs all organ systems, including muscle, myocardial tissue, and nerves, but especially the blood and arteries. . . . Hypocapnia causes profound cerebral and peripheral arterial vasoconstriction, reducing blood flow to body extremities and to the brain" (Fried, 1993, p. 306). One of the most important symptoms of stress is hyperventilation, or disordered breathing and hypocapnia (Fried, 1993, p. 306). Table 18-1 lists the various symptoms reported by stress sufferers that were identical to symptoms of the hyperventilation syndrome. As Table 18-1 reveals, disordered breathing influences emotions and can contribute to psychophysiological disorders.

The practice of diaphragmatic or abdominal breathing, in contrast to chest or shallow breathing, balances the sympathetic and parasympathetic nervous systems. These two systems govern the internal organs and blood vessels, affecting heart rate, blood pressure, unconscious breathing,

TABLE 18-1. Symptoms reported by stress sufferers that are identical symptoms of the hyperventilation syndrome.

-Tension (a "feeling of tension," muscle ache)
-Irritability, low frustration tolerance
-Anxiety (apprehension, heightened vigilance)
-Dyspnea (inability to catch one's breath, choking sensation, feeling of suffocation, frequent sighing, chest heaving, lump in throat)
-Fatigue, tiredness, burn-out
-Insomnia
-Heart palpitations (pounding in chest, seemingly accelerated pulse rate, sensation of heaviness or weight on the chest, diffuse chest pain)
-Depression, restlessness, nervousness
-Dizzy spells, shakiness, trembling
-Coldness of the hands and feet, and occasional tingling sensations
-Inability to concentrate
-Bloating

SOURCE: Adapted from "The Role of Respiration in Stress and Stress Control: Toward a Theory of Stress as a Hypoxic Phenomenon," by R. Fried. In *Principles and Practice of Stress Management*, 2nd Ed., by P. M. Lehrer & R. L. Woolfolk, pp. 310–311. Copyright © 1993 by The Guilford Press. Used by permission.

---

*SOURCE: This and all following quotations from "The Role of Respiration in Stress and Stress Control: Toward a Theory of Stress as a Hypoxic Phenomenon," by R. Fried, in *Principles and Practice of Stress Management*, 2nd Ed., by P. M. Lehrer & R. L. Woolfolk, copyright © 1993 by The Guilford Press, are reprinted by permission.

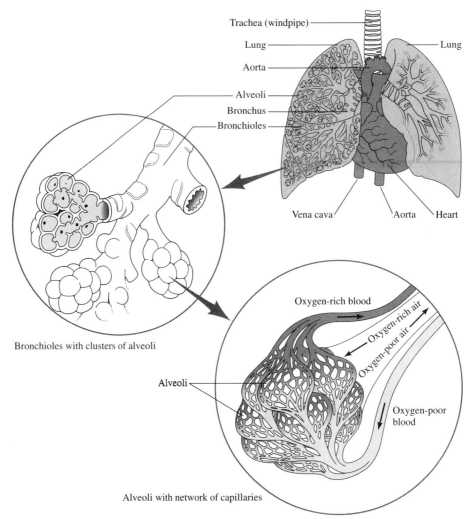

FIGURE 18-1. Diagram of the respiratory system—trachea, aorta, bronchia, lungs, and heart. Also shown are an insert of enlarged bronchioles with clusters of alveoli, and an insert of further enlarged alveoli with a network of capillaries.

and digestion. The sympathetic nervous system (SNS) activates the stress response and disordered breathing. The SNS is critically important for self-protection and the altered breathing is a normal part of the stress response. The SNS responds to threatening situations. The parasympathetic nervous system (PNS) produces the relaxation response that can evoke deep, abdominal breathing. The PNS performs a homeostatic and balancing function.

The pattern of breathing influences the body and mind, as shown in a person's physical, emotional, psychological, and spiritual well-being (Patel, 1993, p. 119). Chest or shallow breathing created by stress or learned breathing patterns causes psychological and/or physiological states of disequilibrium, or lack of homeostasis with the body and mind systems. Because there is a reciprocal relationship between breathing and the mind, diaphragmatic or abdominal breathing leads to mental relaxation.

STEPS FOR BREATHING

Take a little time to answer the following questions:

Do you inflate your chest when you take a deep breath?
Do you experience queasy sensations in your chest or stomach?

Is your breath mostly up in your chest?

Is your breathing shallow?

Do you often feel that you are not getting a full breath?

Do you get mild or more severe headaches, often in the afternoon?

Do you sometimes have painful sensations in your rib cage or shooting pains that make you want to hold your breath?

Are your muscles often tense or sore to the touch?

Do you sigh often?

Do you feel breathless fairly often?

Do you tire easily or wake up tired?

When you are calm and restful, do you breathe more than 15 times a minute?

(Hendricks 1995, p. IX)

With a yes answer to any of the above questions, a client may benefit from one of the five breathing exercises in Table 18-2. You can discuss with the client whether breathing exercises might help with his or her presenting concerns.

Table 18-2 offers an orientation to breathing awareness plus five breathing exercises: diaphragmatic breathing, interval breathing, single-nostril breathing, alternate-nostril breathing, and Kapalabhati (cleansing) breathing. The benefits of any of these five conscious breathing exercises can be one or a combination of the following: releases stress and tension, builds energy and endurance, contributes to emotional mastery, prevents and heals physical problems, manages pain, contributes to graceful aging, enhances mental concentration and physical performance, and promotes psychospiritual transformation (Hendricks, 1995, pp. 7–31).

When you are teaching the awareness orientation to breathing or any of the five breathing exercises, consider the client's age, gender, class, and ethnicity. Also, determine whether the client has any medical or physical condition that could make these breathing exercises inappropriate. If you or the client has any doubts, instruct the client to confer with his or her primary care physician and receive approval to engage in the exercise *before* you begin the instruction. Also, refer to the section on Contraindications and Adverse Effects of Diaphragmatic Breathing later in the chapter.

Awareness of Breathing

The awareness of breathing orientation (first column, Table 18-2) is a technique used to help a person become aware and conscious of her or his breathing. The orientation process has five steps:

1. Ask client to get in a comfortable position (in a recliner chair or whatever is comfortable) with legs apart, the feet relaxed and off to the side; one arm is bent at the elbow and placed on the belly button; the other arm and hand are relaxed and alongside the body.

2. Instruct the client to breathe slowly through the nose and to become aware, conscious; he or she is to feel the movement of the abdomen on inhaling and exhaling.

3. Instruct the client to keep breathing slowly and deeply, and to be relaxed.

4. After several minutes of breathing while lying down, instruct the client to sit up straight with the spine erect; he or she is to place one hand below the navel and the other hand on the upper chest.

5. Ask the client to notice the movement of the hands on inhalation and exhalation, and to assess which hand moves more. If the hand on the upper chest moves more than the other hand, it probably means the client is breathing from the chest rather than the abdomen, in which case the breathing is shallow and not as deep as diaphragmatic or abdominal breathing.

For some clients, this exercise brings awareness or consciousness of breathing before the rationale for diaphragmatic breathing and training is presented. We have found that providing this awareness exercise will facilitate and enhance diaphragmatic breathing training for some clients. For other clients, the counselor can begin with the rationale and overview for breathing, and start the diaphragmatic breathing training without the awareness exercise.

How to Work with the Breath

Table 18-3 presents five dimensions of the breath: (1) The counselor can observe the location of the client's breath—chest for problem breathing, belly for relaxed breathing. (2) The counselor can bring awareness to the client of the depth of breath—shallow for problem breathing, deep for relaxed breathing. (3) The counselor can determine the rate of breathing—fast for problem breathing, slow for relaxed breathing. (4) The counselor can observe whether the client holds or has a rapid exhalation of the breath—a problem. Finally, (5) The counselor can observe whether the client's inhalation is choppy and shallow—also a problem indicator. For any of these problems, the counselor can teach one of the appropriate exercises in Table 18-2.

Empowered Consent for Breathing Exercises

The counselor needs to act in good faith to protect clients' rights and welfare by informing them adequately about the breathing exercises. The therapist should obtain clients'

TABLE 18-2. Orientation to breathing awareness and five breathing exercises

Breathing Awareness	Diaphragmatic Breathing	Interval Breathing	Single-Nostril Breathing	Alternate-Nostril Breathing	Kapalabhati—Cleansing Breath
1. Get in a comfortable position. Place one hand over the belly button. Keep your legs apart, feet relaxed and off to the side. Place the other arm and hand alongside the body, relaxed.	1. Rationale—purposes and benefits.	1. Rationale—purposes and benefits.	1. Rationale—purposes and benefits.	1. Rationale—purposes and benefits.	1. Rationale—purposes and benefits.
2. Breathe slowly and gently. Breathe through the nose. Become aware and feel the movement of your abdomen. Notice the movement of your abdomen as you inhale and exhale.	2. Get in a comfortable position, as in a lounge chair. Breathe through your nose. Just notice your breath.	2. On the inhalation, breathe in slowly and relax. Do this while sitting erect.	2. Sit up in comfortable, spine-erect pose. If you sit on the floor with your legs crossed, place a cushion under your buttocks.	2. Sit up with your spine erect. Make a gentle fist with one of your hands.	2. Do this while sitting with back straight, or standing while bending at the waist, with hands on thighs above the knees. Relax, breathe gently.
3. Keep breathing through your nose. Breathe slowly, deeply; be relaxed.	3. Bend your arms and place your thumbs gently below the rib cage, with the rest of your hand and fingers perpendicular to the body. Visualize the movement of the diaphragm. As you inhale, the muscular fibers of the diaphragm contract and are drawn downward. As you exhale, the diaphragm is drawn upward as the air is pushed out of the lungs and forms a cone shape.	3. Breathe out a little and then pause.	3. Make a mudra (hand position for breathing).	3. Extend your thumb, ring finger and/or little finger, leaving your middle two fingers tucked into your palm. Bring the fist in front of your nose.	3. On the inhalation, relax the abdominal muscles, allowing the abdomen to rise, and visualize the diaphragm descending. Inhale deeply.
4. Sit up straight, place one hand on your abdomen (belly button), and place the other hand on your upper chest.	4. Simulate the movement with your hands. As you inhale, straighten the fingers out; as you exhale, curve the fingers so that they become cone shaped.	4. Continue to pause for about one second after each exhalation and inhalation.	4. Close your right nostril with the mudra hand.	4. Inhale into both nostrils. Now block the right nostril with the thumb and slowly exhale from the left nostril using deep breathing.	4. On the exhalation, pull in the abdominal muscles forcefully as though you might be hit or punched in the belly. Notice that your diaphragm will move up quickly and become cone shaped; the air will be pushed out of your lungs.
5. Notice the movement of your hands as you inhale and as you exhale. Which hand moves more?	5. Sit up with spine erect, eyes closed; visualize the abdomen rising and falling like the tide, the diaphragm moving.	5. On the exhalation, breathe out and then breathe in a little. Use the pause, breathe in, then pause sequence. Pause for one second and with equal intervals of inhalation.	5. Exhale completely through the left nostril. Inhale to a count of 4, then exhale to a count of 8.	5. Now, inhale into the left nostril, block it with the last two fingers (or, the ring or little finger) and exhale through the right nostril.	5. Do 15 rapid pumpings for a round. Do two or three rounds.
	6. Do homework, practice, and in vivo use.	6. Pause at the end of the breath and then breathe normally.	6. Repeat this series 5 times.	6. When you get accustomed to alternate-nostril breathing, establish a count whereby your exhalation is twice as long as your inhalation.	6. Do homework and practice.
		7. Do homework.	7. Close your left nostril and do the series.	7. Do this about 3 minutes to start and gradually increase to 6 or more minutes. Visualize breath.	
			8. Repeat the series 5 times.	8. Do homework and practice.	
			9. Do homework.		

TABLE 18-3. How to work with the breath

Dimensions of the breath	Problems in breathing	Relaxed breathing	Levels of interaction
Location of breath	Chest	Belly	Observe only
Depth of breath	Shallow	Deep	Bring to awareness
Rate of breath	Fast	Slow	Teach or use one of the breath exercises in Table 18-2 to modify problems in breathing
Exhalation of breath	Holding or rapid	Released (often with sound) slowly	Teach or use one of the breath exercises in Table 18-2 to modify problems in breathing
Inhalation of breath	Choppy and shallow	Deep and smooth	Teach or use one of the breath exercises in Table 18-2 to modify problems in breathing

Note: All the above apply to both the counselor and the client

consent and willingness to use the exercise(s) prior to offering any instruction. As part of the consent process the counselor should give clients the following information:

1. Description of breathing exercises, the activities involved
2. Rationale, purpose, and potential benefits of the breathing exercises
3. Description of the therapist's role
4. Description of the client's role
5. Description of possible risks or discomforts
6. Description of expected benefits
7. Estimated time needed for the exercise
8. Answers to client's questions about the breathing exercises
9. Explanation of client's right to discontinue the exercises at any time
10. Summary and clarification/exploration of client's reactions

You can use these items to construct a written consent form that clients will sign indicating that they have received all the above elements of information. Also, refer to the contraindications on page 470 for use of diaphragmatic breathing. If you or any clients have doubts about performing these exercises, clients should consult with their primary care physician.

Rationale and Overview for Breathing Training

The rationale and overview for breathing training could be presented like this:

A lot of people do not breathe deeply enough. Learning to breathe with greater use of your diaphragm or abdomen can increase your oxygen intake. This way of breathing utilizes the full capacity of the lungs and enables you to inhale about seven times more oxygen than with normal shallow breathing. Also, you can practice this any time during the day—while you are waiting in a line or in your car stuck in traffic. Increased oxygen capacity can have many mental and physical benefits. Breathing training stimulates the PNS or relaxation response (see Chapter 17), and calms the central nervous system. In the beginning you might experience a little discomfort or lightheadedness, but such discomfort is rare. The benefits you could derive are numerous and include the following: releases stress and tension, builds energy and endurance, contributes to emotional mastery, prevents and heals physical problems, helps with pain management, contributes to graceful aging, and enhances mental concentration and physical performance. There are no known risks in learning and practicing breathing exercises, although for some clients there may be contraindications; refer to page 470. If you feel any pain or discomfort, which is very unlikely, we will stop immediately. The process will include training you in abdominal breathing, or breathing with your diaphragm. We can explore other breathing exercises that you might find helpful. You will practice the breathing exercises during the week. If you practice once a day, you will spend five or six minutes. The time is doubled if you practice twice a day. Also, you will learn to use deep breathing when you are in stressful situations, or when you need a break, or for a refreshing relief to reenergize yourself. You can be trained to use the breathing exercises in one session. You can practice during the week, and if you feel the need, we can check in by phone to see how you are doing. Practice helps you incorporate deep breathing in your everyday life and in stressful situations. How does that sound?

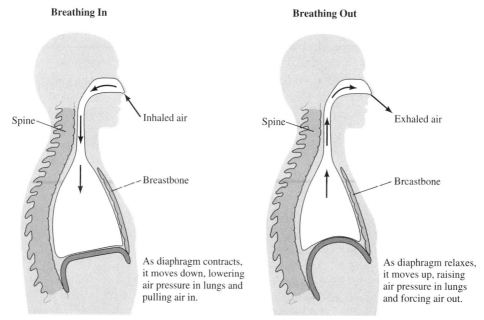

Breathing In

Spine

Inhaled air

Breastbone

As diaphragm contracts, it moves down, lowering air pressure in lungs and pulling air in.

Breathing Out

Spine

Exhaled air

Brcastbone

As diaphragm relaxes, it moves up, raising air pressure in lungs and forcing air out.

FIGURE 18-2. The diaphragm below the lungs. The drawing on the left shows that the diaphragm contracts on inhalation and moves down, lowering the air pressure in the lungs. The drawing on the left shows the diaphragm almost flat. The drawing on the rights shows the diaphragm on exhalation. As the diaphragm relaxes, it moves up and becomes cone shaped, raising the pressure in the lungs.

Diaphragmatic Breathing

Usually we start training in diaphragmatic breathing with a diagram like the one in Figure 18-2. The drawing helps to explain this exercise, also called belly or abdominal breathing because the movement of the abdomen is out during inhalation and in during exhalation. As Figure 18-2 shows, the diaphragm is below the lungs; it is a wide, fan-shaped muscle that separates the chest cavity or thorax and its contents from the abdominal cavity and its contents. If the client's breathing is too shallow, diaphragmatic breathing should help him or her acquire a deeper breathing pattern. You can use the following steps as a guide in teaching this exercise.

1. The counselor provides the client with a rationale. You can say to the client, "Notice that at the start of each breath your stomach rises and the lower edge of your rib cage expands. If your breathing is too shallow, diaphragmatic breathing should help you acquire a deeper pattern of breathing." You can use the rationale similar to the example presented on page 465. The counselor explains to the client that when one inhales, the muscle fibers of the diaphragm contract and are drawn downward toward

the abdomen, as illustrated in the left drawing in Figure 18-2. The picture on the right in Figure 18-2 illustrates the cone-shaped position of the diaphragm when one is exhaling. We use a picture like this because we want the client to visualize the movement of the diaphragm at the beginning of diaphragmatic training. As presented in Table 18-2, diaphragmatic training includes six steps (column 2). This exercise is also called belly or abdominal breathing because the movement of the abdomen is out during inhalation and in during exhalation.

2. The counselor instructs the client to get in a comfortable position and to breathe through the nose. Legs are comfortably apart with feet relaxed and off to the side.

3. The client is told to bend the arms at the elbow and place the thumbs gently below the rib cage; the rest of the hand and fingers are pointing toward each other and perpendicular to the body. Then ask the client to visualize the diaphragm on inhaling to see its muscular fibers contract and draw downward. Instruct the client on exhalation to visualize the diaphragm being drawn upward as the air is pushed out of the lungs and forms a cone shape. "As you inhale, extend your abdomen outward and as you exhale, allow your abdomen to come inward. You may wish to

exaggerate this movement. Always breathe through your nostrils in an even and smooth manner. When you start, it is helpful to focus on the rise and fall of your hand as you inhale and exhale. Focus on each inhalation: Your hand goes up and your belly comes out. On your exhalation, press in and back on your abdominal muscles (allowing the abdomen to draw inward toward the spine), and your hand goes down. Remember the diaphragm is a muscle, and like any muscle it can be trained and strengthened."

4. The client is instructed to simulate the movement of the diaphragm with the hands. The thumbs are just below the rib cage on the abdomen and the fingers are slightly interlaced. As the client inhales, the interlaced fingers are flattened to simulate the diaphragm being drawn downward (see the left picture in Figure 18-3). On the exhalation, the client is instructed to curve the fingers so they are cone shaped, simulating the movement of the diaphragm being drawn upward as the air is being pushed out of the lungs. On the inhalation the muscle fibers of the diaphragm contract and are drawn downward. The diaphragm becomes less cone shaped and almost flat. As the diaphragm descends, it presses on the stomach, liver, and other organs, and gently massages and stimulates them. "When you exhale, the muscles contract and the diaphragm becomes cone shaped and pushes upward toward the lungs. Visualize your diaphragm working and exercising the internal organs of digestion and elimination, massaging and kneading them for each inhalation and exhalation, forcing the blood into the organs and then squeezing the blood out. This way of breathing uses the full capacity of your lungs" (Birch, 1995, p. 45).

5. The client is instructed to sit up with the spine erect and eyes closed, to visualize the diaphragm's position while inhaling and exhaling, and to feel the abdomen rising and falling like the tide.

6. The counselor instructs the client to practice diaphragmatic breathing for homework at least twice daily. The counselor reminds the client to find a time and place to practice that will be free of distractions and interruptions. The client is also instructed to use abdominal or diaphragmatic breathing when in a stressful situation or to reenergize the mind or body. This way of breathing uses the full capacity of your lungs. It enables you to inhale about seven times more oxygen than with normal shallow breathing. Instruct the client to use diaphragmatic breathing anytime during the day.

Bill often uses this breathing exercise to coordinate diaphragmatic breathing with subtle body movements in preparation for Hatha Yoga. Instruct the client to lie down on the back with knees bent, feet on the floor, and arms alongside the body. (Note that for some clients who have experienced physical and sexual abuse, you should not use this exercise which involves laying down, use the former one which involves sitting.) On the exhalation, lift the tail bone (coccyx) slightly off the floor so that the small of the back is

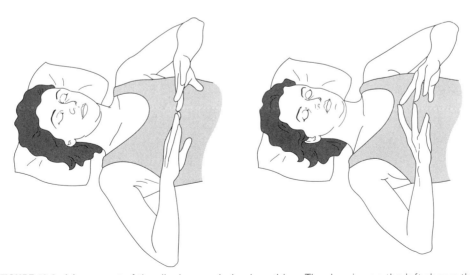

FIGURE 18-3. Movement of the diaphragm during breathing. The drawing on the left shows the fingers flat, some of fingers touching slightly, and simulating the shape of the diaphragm during inhalation. The drawing on the right shows the fingers cone shaped, most of the fingers slightly touching, and simulating the shape of the diaphragm during exhalation.

on the floor. On the inhalation, arch the small of the back slightly so that the client can slide a finger between it and the floor; keep the tail bone on the floor. Repeat slowly and in a relaxed fashion. Instruct the client to be mindful of the movements while visualizing the movement of the diaphragm.

Interval Breathing

1. The counselor presents the rationale for this breathing exercise, which is to teach clients how to slow their breathing and to experience what the feeling is like when they breathe in short intervals.
2. The client is instructed to get into a comfortable sitting position, with the spine erect, and first to breathe in slowly and relax.
3. The client is instructed on the exhalation to *breathe out* a little and then pause.
4. Continue to exhale a little and then pause for about one second until the lungs are empty.
5. On the inhalation, the client is told to breathe out, and then breathe in a little at a time. Again use the pause sequence when inhaling. The client is instructed to pause for one second and with equal intervals of inhalation.
6. Pause at the end of the breath and then breathe normally. The client can be instructed to repeat the sequence, alternating the pause sequence for the exhalation and inhalation as often as comfortable. We have found four repetitions for each sequence to be helpful. Again, remember that this is an optional breathing exercise for clients who wish to slow their breathing. If the counselor and client feel that little benefit will come from this exercise, single-nostril or alternate-nostril breathing may be an appropriate substitute.
7. If the client wishes to use this exercise at home, ask him or her to practice twice a day and to monitor the rate of exhalation and inhalation.

Single-Nostril Breathing

Single-nostril breathing (Table 18-2, column four) is a good preparation for pranayama (control of the breath) and for alternate-nostril breathing.

1. The counselor presents the rationale for single-nostril breathing, which is that the exercise is an excellent preparation for alternate-nostril breathing.
2. The client sits in a comfortable, spine-erect posture. If he or she sits on the floor with legs crossed, place a cushion under the buttocks for comfort, and keep the spine straight.

FIGURE 18-4. Diagram of mudra or hand gesture for breathing exercise. Thumb, ring finger, and little finger are extended; the other two fingers are folded down.

3. Instruct the client to make a mudra (hand gesture for breathing exercise) by extending the thumb, ring finger, and little finger, folding down the other two fingers (see Figure 18-4).
4. Instruct the client to close the right nostril with the mudra hand.
5. Then, ask the client to exhale completely through the left nostril, then inhale to a count of 4, and then exhale to a count of 8.
6. Request that the client repeat this series five times.
7. Instruct the client to close the left nostril, exhale completely through right nostril, then inhale to a count of 4, and then exhale to a count of 8.
8. Instruct to repeat this series five times.
9. Instruct the client to practice this exercise at home during the week.

Alternate-Nostril Breathing

1. In offering a rationale for alternate-nostril breathing, the counselor says that it is reputed to balance the mind and body and to bring lightness to the body, to instill mental alertness and focus, and to calm the nervous system. Additional benefits of alternate-nostril breathing are that it improves the appetite, benefits digestion, and helps bring sound sleep. The two hemispheres of the brain are connected by a thick bundle of fibers. The right hemisphere is said to control feeling, emotion, imagery, and intuition. The left hemisphere is said to deal with logic, mathematics, words, and linear thinking. The left side of the nose is connected to the right side of the brain, and the right nostril to the left side of the brain. Shifting between the brain hemispheres is what gives this breathing exercise power.
2. Instruct the client that this breathing exercise is done while sitting up in a comfortable, spine-erect pose.

3. Instruct the client to form a mudra with the preferred hand, and bring the hand to the nose, then to extend the thumb of the preferred hand, ring finger, and little finger, leaving the middle two fingers tucked into the palm of the hand (see Figure 18-4).

4. Ask the client to inhale into both nostrils, then, to block the right nostril with the thumb and slowly exhale from the left nostril, using deep breathing.

5. Instruct the client to inhale into the left nostril, block it with the last two fingers (or the ring, or little finger), and exhale through the right nostril.

6. When the client is accustomed to alternate-nostril breathing, instruct him or her to make the exhalation twice as long as the inhalation.

7. Instruct the client to do alternate-nostril breathing for three minutes to start, and then gradually increase to six or more minutes.

8. Finally, instruct the client to practice at home between sessions one time a day. Ask the client to visualize the action of the diaphragm as shown in Figures 18-2 and 18-3 throughout the exercise.

An advanced variation of this exercise is to inhale through the left nostril for the count of 4, retain or hold by closing both nostrils for the count of 16, exhale through the right nostril for a count of 8, inhale through the right for a count of 4, retain by closing both nostrils for a count of 16, and exhale through the left for a count of 8.

Kapalabhati (Cleansing) Breathing

1. The counselor provides the rationale for this advanced breathing exercise—that it offers a cleansing effect for the mind/body. The cleansing breath is known as the "breath of fire:" *kapal* means "skull," and *bhati* means "to cleanse, light, or luster" (Farhi, 1996, p. 158). The benefits are exercising of the diaphragm by having abdominal organs pushed repeatedly against it. The exercise strengthens the abdominal muscles, and tones and massages all the organs in the abdominal cavity. The most profound impact of the exercise is that it soothes and calms the entire nervous system. The exercise also oxygenates the brain by increasing the amount of blood to the brain. The increase of blood to the brain can improve concentration and memory, and can stimulate intellectual functioning. Counselors should, however, observe this cautionary note: This exercise *should not* be practiced by clients who have high blood pressure, a heart condition, epilepsy, an ear or eye problem, or a hernia. Also, the exercise should not be practiced during menstruation or pregnancy. If you or the client has any doubts, instruct the client to consult and get permission from her or his physician before beginning the exercise.

Here is an overview of the exercise. Sit in a comfortable position (you can practice this exercise while standing) with your back and spine straight. Relax your arms and hands, close your eyes, and relax your jaws. Then, inhale slowly, smoothly, and as deeply as you feel comfortable. Exhale briskly, while flattening and tightening your abdomen. Then, inhale naturally—almost involuntarily. Repeat: Exhale briskly (flatten and tighten your abdomen), and inhale naturally; do this cycle in rapid succession for several times. Another way to explain Kapalabhati is that it involves short, forceful exhalations (a pumping action) of air followed by passive inhalation. The exhalation is the active phase which is focused on, and the inhalation is passive by just allowing it to happen. Finally, resume normal breathing as you rest. The steps for directing the exercise follow.

2. Instruct the client to do this exercise while sitting with back straight (spine erect) *or* standing with slightly bent knees and bending forward from the waist, with hands on thighs above the knees. You can also do this exercise in a standing position. Instruct the client to relax and breathe gently.

3. On the inhalation, tell the client to relax the abdominal muscles, allowing the abdomen to rise, and to visualize the diaphragm movement. Client is to inhale deeply but not to force the inhalation.

4. Ask the client on the exhalation to pull in the abdominal muscles forcefully as though he or she has been hit or punched in the belly. Tell the client to notice that the diaphragm will move quickly and become cone shaped, and the air will be pushed out through the lungs. Instruct the client to establish a speed of inhalation and exhalation that is comfortable.

5. Instruct the client to repeat the inhalation/exhalation series for 15 to 20 rapid pumpings for a round, and to do two or three rounds. The client can be instructed to visualize the diaphragm while doing this exercise.

6. If the client finds this exercise helpful, provide homework instruction about when, where, and how often to practice.

CONTRAINDICATIONS AND ADVERSE EFFECTS OF DIAPHRAGMATIC BREATHING

Deep abdominal/diaphragmatic breathing may not be helpful for everyone. For example, some people doing these exercises may develop cramps. Fried (1993) offers clients the following instructions:

If an exercise causes pain or discomfort, stop it immediately. Also, do not do any exercise if you have any physical or medical condition or any injury that would contraindicate its safety. Among such conditions are the following:

1. Muscle or other tissue or organ malformation or injury—for example, sprained or torn muscles, torticolis, fractures, or recent surgery.
2. Any condition causing metabolic acidosis, where hyperventilation may be compensatory such as diabetes, kidney disease, heart disease, severe hypoglycemia, etc. If you are in doubt, please bring your condition to my attention.
3. Low blood pressure or any related condition, such as syncope (fainting). Deep abdominal breathing may cause a significant decrease in blood pressure.
4. Insulin-dependent diabetes. If you are an insulin-dependent diabetic, you should not do this or any other deep relaxation exercise without the expressed approval by your physician and his or her close monitoring of your insulin needs. (Fried, 1993, p. 323)

Pregnancy may also affect the client's capacity to do some of the breathing. Smokers also have more trouble with breathing exercises and Hendricks (1995) does not use breathing until after a person stops smoking.

MODEL EXAMPLE: DIAPHRAGMATIC BREATHING

In this model example, we present a narrative account of how two breathing exercises might be used with Yoshi, a 49-year-old male Japanese American, who is an air traffic controller.

1. *Rationale*

First, we explain to Yoshi that learning to breathe with his diaphragm or abdomen can increase his oxygen intake. We tell Yoshi that this way of breathing utilizes the full capacity of the lungs and enables him to inhale about seven times more oxygen than with normal shallow breathing. We tell him that the benefits of diaphragmatic breathing include releasing stress and tension and enhancing mental concentration. We provide an overview by telling him that the process will include training in diaphragmatic breathing. We tell Yoshi that he can practice during the week and use deep breathing in stressful situations. We confirm Yoshi's willingness to try diaphragmatic breathing and answer any questions he may have about the procedure. We provide all the elements of information illustrated in the example of a rationale on page 465.

2. *Show Diagram of Diaphragm and Instructions About Position*

We show Yoshi a diagram of the position of the diaphragm for exhalation and inhalation (Figure 18-2). Yoshi is instructed to get into a comfortable position and to breathe through his nose. He is to place one hand on his belly button and the other hand just below his rib cage. We request that he breathe through his nose, notice his breath (while breathing through his nose), and observe what happens to his hand as he exhales and inhales. We ask Yoshi what the movement is like for the hand on his belly button and the hand below his rib cage. For example, is the movement of the hands the same or does one hand move more than the other? If one hand moves more, is it the hand on the rib cage or the belly button? We explain to him that if he is breathing diaphragmatically, the hand on the belly button is moving more on the exhalation and inhalation than the hand below the rib cage. We ask Yoshi to relax and just notice his breathing.

3. *Visualize the Movement of the Diaphragm and Placement of the Hands*

We ask Yoshi to bend his arms and place his thumbs gently below his rib cage (we model the position of the hands for him as shown in Figure 18-3). Yoshi is instructed to place the rest of his fingers on each hand perpendicular to his body. We ask him to visualize the diagram (Figure 18-2) of the diaphragm, and as he inhales, remind him that the muscular fibers of the diaphragm contract and are drawn downward. We describe—as he exhales—the diaphragm being drawn upward as the air is pushed out of the lungs and forming a cone shape.

4. *Simulation of the Diaphragm with the Fingers*

Yoshi is instructed to simulate the movement of the diaphragm with his hands. We tell him, as he inhales, to straighten the fingers out flat. As he exhales, he is to curve the fingers so that they become cone shaped. We ask him to do this for several breaths, simulating the movement of the diaphragm while his breathing is gentle and relaxed.

5. *Diaphragmatic Breathing While Sitting*

Yoshi is instructed to sit up with his spine erect, his eyes closed, with his hands on his lap, his knees, or wherever is comfortable for him. We ask him to visualize his abdomen expanding or rising and falling like the tide, and to breathe in this relaxed fashion for a couple of minutes.

6. *Homework Practice and in Vivo Use*

Yoshi is instructed to practice diaphragmatic breathing twice a day for 5 minutes, once after getting up in the morning and once in the late afternoon or early evening. We remind Yoshi to find a quiet place to practice that is free of distractions and interruptions. Yoshi is instructed to visualize the movement of his diaphragm while he is practicing and when he uses abdominal breathing in stressful situations.

MODEL EXAMPLE: ALTERNATE-NOSTRIL BREATHING

We use the same client, Yoshi, to present a narrative account of how alternate-nostril breathing might be used.

1. *Rationale*

We explain the purpose and overview of alternate-nostril breathing. We say that alternate-nostril breathing is said to balance the mind and body, and claims to bring lightness of the body as well as mental alertness and focus, and to calm the nervous system. Additional benefits of alternate-nostril breathing are that it improves the appetite, benefits digestion, helps sound sleep, and helps to cleanse the nervous system. "You will use your hand to close one nostril, and breathe alternately through each nostril. How does that sound? Do you have any questions?"

2. *Position and Calm Breathing*

Yoshi is instructed to sit up straight in a comfortable position with the spine erect. We tell him that if he sits on the floor with his legs crossed, he should place a cushion under his buttocks for comfort.

3. *Mudra Hand Position for Alternate-Nostril Breathing*

Yoshi is instructed to make a gentle fist with the preferred hand. "Extend your thumb, ring finger, and your little finger, and leave your two middle fingers tucked into your palm" (see Figure 18-4). "Bring the fist in front of your nose."

4. *Right Nostril*

Yoshi is instructed to inhale into both nostrils, then to close his right nostril with his mudra hand, and to make sure the nostril is completely closed. We tell him to exhale slowly from the left nostril using deep breathing.

5. *Alternating Nostrils*

We say to Yoshi, "Now, inhale into the left nostril, block it with the mudra hand, and exhale through the right nostril." Yoshi is instructed to continue alternate-nostril breathing.

6. *Exhalation Twice as Long as Inhalation*

When Yoshi gets accustomed to alternate-nostril breathing, he is instructed to establish a count whereby his exhalation is twice as long as his inhalation.

7. *Instructions About Length of Time for Breathing Exercise*

Yoshi is instructed to do alternate-nostril breathing for about three minutes to start and gradually increase the time to six minutes or more. We ask him to visualize the breath as he does the exercise.

8. *Homework and Practice*

Yoshi is instructed to do alternate-nostril breathing twice a day at a time that is convenient for him and in a quiet place free of distractions and interruptions. If appropriate, Yoshi is instructed to use alternate breathing when he needs a refreshing distraction from a stress situation at work.

There are several benefits the counselor and Yoshi might expect following a week or two of Yoshi's using the diaphragmatic and alternate-nostril breathing exercises. With respect to Yoshi's presenting problem, he will probably experience less stress, more energy and endurance, greater mental concentration, and fewer feelings of being overwhelmed as he handles his responsibilities as an air traffic controller during peak periods. In other parts of his life, he will probably experience more calmness and mental alertness, more energy, and less fatigue. All these mental states are attributed to an increase in oxygen intake, which is seven times greater than he experienced before he started using the diaphragmatic and alternate-nostril breathing exercises. More oxygen is distributed throughout his body than before when

LEARNING ACTIVITY 49   *Breathing Exercises*

The emphasis in this chapter is on mind/body strategies. In this learning activity, you will use the breathing exercises so that your practice or experience with conscious breathing will facilitate your teaching others.

1. Try the breathing awareness exercise and the five different breathing exercises presented in this section of the chapter. Do the five exercises for several days so that you become familiar with them and experience their effects.
2. Practice diaphragmatic and alternate-nostril breathing for two weeks. Select a regular time and place where you can practice without interruptions and distractions.

Also, try to use the diaphragmatic breathing in real, stressful situations. Any time that you feel the need to become energized, in addition to your regular practice, engage in alternate-nostril breathing.

3. During the two-week period, while engaging in the diaphragmatic and alternate-nostril breathing, keep a journal or log of each practice session describing what you experienced from your practice. Record your overall energy and level of relaxation in what you perceive as a consequence of your practice with the breathing exercises. Also, in your journal, record on what occasions you used either diaphragmatic or alternate-nostril breathing in vivo.

---

he was using shallow breathing and not using the full capacity of his lungs.

HATHA YOGA: PROCESSES AND BENEFITS

Remember our goals for presenting Hatha Yoga: (1) to alert you to the potential benefits of Hatha Yoga so that you can refer to a certified yoga therapist those clients who might benefit from this kind of body work; (2) to provide you with a self-administered stress-reduction strategy. For mental health care providers, the practice of yoga is an excellent way to relax and reduce potential job stress and burn-out.

Yoga is not another fitness technique, a calisthenics workout, or an aerobic routine. It is not a religion. It is a mind/body technology aimed at the transformation of consciousness or awareness. Almost 5,000 years ago, yoga was created as a complete system of mind/body training to cultivate a state of awareness and relaxation. Today, yoga is recognized by Western science and health professionals as a potent tool for stress reduction, fitness, mental and physical health, and heightened consciousness. The word *yoga* means union. In our complex and chaotic world, the practice of yoga unifies all the different aspects of ourselves, bringing us in contact with the depth of our true inner nature. There are many approaches to yoga, and each uses different ways to achieve this common goal. The most popular form of yoga in the United States is the discipline known as Hatha (pronounce "Hatta") Yoga. Iyengar (1988) says the *Ha* in Hatha means sun which is the sun or soul in your body, and *tha* means moon or your consciousness. Hatha expresses the balance of opposing forces, sun-moon, male-female,

positive-negative, and yin-yang. Iyengar (1988) says that the energy of the sun never fades, but the moon fades every month and after fading comes to fullness. The sun (our soul) is in us all the time and never fades, and our mind or consciousness draws energy from our soul. The consciousness has fluctuations, modulations, moods, and ups and downs just like the phases of the moon. When consciousness and the body are brought into union with one another (by the practice of Hatha Yoga), the energy of consciousness becomes still (Iyengar, 1988, p. 4). "Yoga attempts to bring within one perspective all three sides of human life—the body, the mind, and the soul; the physical side, the social side (life in action), and the spiritual side" (Patel, 1993, p. 96). Yoga is a "unitive discipline"—the discipline that leads to inner and outer union, harmony, and joy (Feuerstein & Bodian, 1993, p. 3). Yoga is the discipline of conscious living because you take responsibility for your life.

In the last chapter, we presented muscle relaxation and meditation, which is typically incorporated in the practice of Hatha Yoga. Also, mindful breathing is an integral part of yoga. Yoga involves the asanas or postures and breathing (pranayama—control of breath) with the goal of integrating mind, body, and spirit. The yoga postures (asanas) stretch, extend, and flex the spine, and exercise the muscles and joints. Yoga asanas use gravity to stretch, soak, and squeeze the muscles and internal organs of the body. Because the postures are done in conjunction with breathing, the interactive effect is to stimulate circulation, digestion, and the nervous and endocrine systems. The practice of Hatha Yoga includes using elaborate physical postures to detoxify and strengthen the body (and mind) in preparation for meditation.

Practice with the physical postures is a form of meditation in motion, calming the mind and cultivating a state of relaxed but alert concentration. The chaos of living in the 21st century introduces stressful blocks that take us away from our true nature. Yoga is a technology that can remove these blocks.

Research provides evidence that the regular practice of yoga increases strength and flexibility, regulates blood pressure, improves cardiovascular efficiency and blood flow, strengthens the immune and endocrine systems, and improves functioning of the parasympathetic nervous system that activates the relaxation response. Most of the controlled studies have demonstrated the beneficial effects of yoga on hypertension and coronary heart disease, and potential positive effects on anxiety states, asthma, and diabetes. An extensive bibliography of 1,600 references to scientific studies of yoga and meditation appears in Monro, Ghosh, and Kalish's (1989) *Yoga Research Bibliography: Scientific Studies on Yoga and Meditation.* More recently, breathing exercises, muscle relaxation, yoga, and meditation have become complementary features of medical programs at the Mind/Body Medical Institute of New England Deaconess Hospital and Harvard Medical School (Benson & Stuart, 1992), the Stress Reduction Clinic at the University of Massachusetts Medical Center in Worcester (Kabat-Zinn, 1990, 1993), and the Preventive Medicine Research Institute at the School of Medicine, University of California, San Francisco and Sausalito, California (Ornish, 1990).

Ward (1994) suggests that the practice of yoga is important, but it is not for perfection. Yoga practice requires dedication, consistency, enthusiasm, and an attitude that one has the ultimate responsibility for his or her health. Self-efficacy is important in the practice of yoga (see Chapter 20). Asanas or the postures of yoga can have a powerful effect on the body, but becoming aware of the psychological implications of this process will help to make the change more permanent. Awareness and understanding are increased by observing and dealing with the mental-emotional processes while engaged in the postures. When the interconnectedness of the physical and mental states is increased, better understanding and insight are created and the mind can function as its own therapist by shifting focus and reprogramming the mind as the intended and conscious choice (Radha, 1987, p. 4).

Caution About Yoga for Your Clients and You

The purpose of the breathing and yoga asanas is to bring about a harmonious balance in and between your body and mind. Everyone can do yoga—young (Stewart & Philips, 1995) and older adults (Stewart, 1994, Ward, 1994). No one needs to be excluded from doing yoga. If you have any concerns about referral of a client to a certified yoga therapist, request the client to consult a physician before beginning yoga. You might want to give the client names of several physicians who are familiar with and informed about Hatha Yoga.

Yoga is an excellent way for mental health therapists to become balanced and relieve stress. You can imagine doing the yoga postures and think about the potential benefits derived from each posture. Or, you can start gently to practice the postures presented in this chapter at your own pace. Although you can do the postures on your own, we strongly recommend that you start your practice with a yoga instructor.

Place, Clothing, and Time for Practice

Generally, most people will practice yoga indoors. Practice in a place with a lot of space (at least 64 square feet), not cluttered by furniture or things, and comfortably heated and well ventilated. If the space or room does not have a carpet, use a small rug or a nonstick mat. The place to practice should be a room or space that will be free of distractions and interruptions. A pillow can be used as a cushion for sitting while doing breathing exercises and possibly for some asanas. Also, a small blanket might be useful for the relaxation or corpse posture.

Yoga practitioners use clothing that is comfortable and that will allow the maximum range of movement. A loose-fitting cotton T-shirt, sweatshirt, or a turtleneck, comfortable sweatpants, leotards, or shorts are excellent choices. Do not wear shoes for yoga; they restrict movement and prevent you from being "grounded" with the floor or rug. Bare feet are the preference for doing yoga asanas. If your feet get cold, wear socks.

Set aside a particular time during the day in which there will not be any interruption in the practice of yoga. Use the same time each day. Some people prefer early morning as a good way to start the day. Some do their yoga practice during their lunch break whereas others prefer to practice yoga before dinner. Never practice within two hours after eating. Practice in the evening relieves the tensions of the day.

Allow an hour to an hour and a half for practice. This may seem like a long period, but it is well worth the time and energy. When you are busy, do a shorter session with fewer asanas. We subscribe to the Jon Kabat-Zinn (1990, 1994) belief that "you don't have to like it, just do it," and after six to eight weeks you can determine the benefits of your practice.

YOGA AND YOU

We feel that if you are to help potential clients with yoga, breathing, and meditation, you must have your own practice and do it almost daily. If you can experience the benefits from Hatha Yoga, clients will experience your energy and enthusiasm when you refer them to a certified yoga therapist. If you decide to engage in your own Hatha Yoga practice, it is perhaps best to be supervised by a qualified yoga teacher who can gently guide and demonstrate for you the asanas and breathing. Even better is to get training as a yoga instructor or therapist. Bill received his training from Joseph LePage, director of the Integrative Yoga Therapy Program (2975 Pacific Heights Drive, Aptos, California 95003).

As Ward (1994) suggests, a commitment to engage in a yoga practice requires dedication, consistency, enthusiasm, and an attitude that you have the ultimate responsibility for your health. Before you begin yoga asanas, look over the asanas and ascertain your body's capabilities. Progress may be slow; go at your own rate, and be gentle and relaxed when you hold an asana. Be aware of your body, and breathe consciously while doing the asanas. The practice of yoga with breathing in the movement facilitates the full flow of energy throughout the seven energy centers in the body. Remember, you are not qualified to teach Hatha Yoga to clients, but you can heighten your awareness about the benefits of yoga, or engage in your own practice of yoga.

THE CHAKRAS: THE SEVEN ENERGY CENTERS OF THE BODY

The purpose in describing the energy centers or chakras of the body is to increase your awareness of this approach to body work and to the types of clients who might benefit from working with a qualified yoga therapist who incorporates the energy centers into the therapeutic practice of Hatha Yoga. In the yogic tradition, there are seven energy centers in the body. They are called *chakras,* which means nerve center. The positions of these nerve or energy centers correspond to points at different levels along the spinal cord. According to yogic thought, the energy centers or chakras create energy fields (Birch, 1995; Myss, 1996a; Rama, Ballentine, & Ajaya, 1976; Joy, 1979). Table 18-4 lists the number of the energy center or chakra, the label for the energy center, and the psychological characteristics in the first column, and the body location in the second column. Column three

presents the parts of the body associated with each energy center. Columns four and five present for each chakra the psychological and emotional functioning for balanced flow of energy, and the psychological, emotional, and physical conditions for imbalanced flow of energy throughout the chakras. Finally, the last column names the yoga asanas or postures that, according to yogis, are effective for removing the blocks that impede the flow of energy for each chakra to the other chakras or energy centers.

A brief overview of the dynamics of the energy centers is helpful for understanding how the chakra system works. The first energy center, according to yogic tradition, is the seat of security and stability. The chakra is also associated with behaviors defined by the "tribe," family, region, nation, and/or culture to which the person belongs. Basic fears and insecurities are supposed to stem from this chakra or energy center. When we are confident and grounded with ourselves and without dependency on societal or external forces or attachment to external objects for our security, we feel calm, secure, balanced, and grounded to the earth. According to the yogic tradition, the second energy center or chakra is the source of our physical, sexual, and creative energy; it extends from around the pubic bone to the sacrum or lumbar region (lower or middle part of the abdomen). Blocked or imbalanced flow of energy in this chakra can contribute to frustration, blame, guilt, jealousy, sexual codependency, and overattachment to objects or people. In such cases, creative energy is consumed by these unhealthy emotions and processes.

The third chakra or energy center is located in the navel region between the base of the sternum and navel. It corresponds to the solar plexus. This energy center is associated with personal power, self-motivation or initiative, decision-making skills, good self-system, and ego boundaries. An imbalanced flow or block in the third chakra can yield powerlessness, greed, aggression, inhibition, difficulty in decision making, intimidation, and a fractured self-system. The fourth chakra is located around the area of the heart and includes the lungs and diaphragm. This center is associated with love, compassion, acceptance, fulfillment, forgiveness, hope, and trust. An imbalanced flow of energy in this center can result in one's being emotionally closed, insensitive, passive, lonely, self-centered, distrusting, pessimistic, lacking in compassion, being depressed, and bitter.

The fifth chakra is located at the base of the throat and is associated with verbal expression, will or capacity to make decisions, and ability to use personal knowledge to create. Blockage of energy in this area can create attachment and addiction problems, little consciousness of emotions and thoughts, and poor decision making. The sixth chakra is

TABLE 18-4. Chakras or energy centers, their locations, and functions

Energy center (chakra)	Body location	Parts and organs of body	Psychological and emotional functioning for balanced flow of energy	Psychological, emotional, and physical conditions for imbalanced flow of energy	Example asanas/postures
1st Tribe Security	Base of spine *Stability*	Base of the spine at the lower extreme of the spinal column (coccyx). Includes large intestine, rectum, male reproductive organs, prostate, kidneys, adrenal gland, legs, feet, bones, and immune system.	Safety and security, emotional and physical support. Trust and feelings of belongingness. Awareness and consciousness.	Self-centeredness, self-indulgence, insecurity, instability, grief. Phobia, anxiety, and fear related to survival and/or abandonment. Feeling of lack of physical and emotional support. Scarcity of consciousness. Problems with feet, legs, knees, base of spine, colon, kidneys, sympathetic nervous system. Hemorrhoids, constipation, sciatica.	Tree Forward Bend Head to Knee
2nd Power and control	Pubic bone to sacrum (last bone of the spine) *Sexuality, creativity, energy storage*	Pubic bone to the sacrum (last bone of the spine). Includes female reproductive glands, kidneys, bladder, hip area, and circulatory system.	Feeling and emotional energy center. Patience, endurance, self-confidence, well-being. Ethics and honor in relationships. Autonomy.	Frustration—blame and guilt; shame, doubt; attachment versus letting go; anxiety, fear. Sexual potency, frigidity; uterine, bladder, or kidney problems, lower back pain.	Triangle Cobra Forward Bend
3rd Self-socialization and social interaction	Navel region—between base of sternum and navel. Solar plexus *Personal power; social interaction*	Pancreas gland, upper intestines, digestive system, muscles in this region, stomach, liver, spleen, gall bladder, middle spine, nervous system.	Personal power, self-motivation (initiative), decision-making skills, good self-system (confidence, esteem, and respect), willfulness, good self-image, good ego boundaries.	Powerlessness, greed, aggression; anger, inhibition, guilt, inability to make good decisions; fear and intimidation; no good self-system; sensitivity to criticism. Stomach problems, digestive problems, diabetes, gastric, or duodenal ulcers, liver problems, adrenal dysfunction, pancreatitis, arthritis.	Bow Sun Salutation
4th Love, compassion	Area of the heart *Opens the heart, empathy*	Heart and circulatory system, thymus gland, diaphragm, ribs/breasts, lungs.	Compassion, acceptance, love, fulfillment, forgiveness, hope, and trust.	Insensitivity, emotionally closed, passivity, sadness, depression, resentment, loneliness, anger, self-centeredness, lack of compassion, distrust, pessimism, sullenness. Asthma, allergies, lung disease; heart disease—congestive heart failure, myocardial infarction, mitral value prolapse, high blood pressure; upper back discomfort, bronchial problems, lung disease.	Cobra Triangle Child

(continued)

TABLE 18-4. Chakras or energy centers, their locations, and functions (continued)

Energy center (chakra)	Body location	Parts and organs of body	Psychological and emotional functioning for balanced flow of energy	Psychological, emotional, and physical conditions for imbalanced flow of energy	Example asanas/postures
5th Will Communication and interaction inspired by a higher vision	Base of throat *Vocal expression*	Throat, trachea, neck, esophagus, vertebrae, thyroid, parathyroid mouth, shoulders, arms, hands.	Communication, expression, creativity—using personal power to create, interactions; inspiration; good decision-making ability.	Poor decision-making skills, little consciousness of emotions and cognitions, problem with expression, stagnation, obsession, inability to use personal knowledge or judgment to create, attachment and addiction problems. Sore throat; jaw joint problems; stiff neck; colds; speech disorders; hearing problems; asthma; mouth ulcers; gum problems; laryngitis; swollen glands; thyroid problems; hand, arm, and shoulder problems; upper back discomfort.	Neck Stretches Half and Full Shoulder Stand
6th Mind Experiencing and understanding	Between eyebrows *Third eye Visualization Intuitive*	Pineal gland, brain, nervous system, eyes, ears, nose.	Intellectual and emotional intelligence; visualization; imagination; perception; openness to experience and learning; nonattachment—letting go; awareness of spiritual dimension of existence; realistic—truthfulness; self-monitoring and evaluation; feeling adequate.	Difficulty focusing, aloof, detached emotionally, intellectually stagnant. Lack of awareness of spiritual dimension of existence. Problems with attachment and holding on. Stroke, brain tumor, seizures, headaches, eyestrain, blurred vision, hearing problems.	Half or Full Shoulder Stand Half Spinal Twist
7th Oneness Perception of reality beyond duality	Top of Head *Liberation Highest state of consciousness*	Top of the skull or the crown of the head, central nervous system, upper brain (cerebral cortex), the pituitary gland.	Selflessness, ability to trust the process of life. Capacity to see the big picture. Belongingness, moral values, and courage. Good energy; enlightenment; awareness; and consciousness.	Depression, confinement, closed mind, possible psychosis, worry, alienation, confusion, boredom, apathy, inability to learn and comprehend. Paralysis, genetic disorders, multiple sclerosis, amyotrophic lateral sclerosis (ALS), Alzheimer's/dementia, bone or brain cancer.	Corpse Inverted Postures

associated with the mind and the "third eye" of visualization, and with intuitive experiencing and understanding. This energy center is located between the eyebrows. Examples of an imbalance in the flow of energy in this center are difficulty in focusing, inability to learn from experience, aloofness and detachment, intellectual stagnation, and little awareness of a spiritual dimension. Finally, the seventh chakra is located at the top of the head or skull. This energy system is associated with oneness or the capacity to be a part of the cosmos, a perception of reality beyond duality, and a high level of consciousness. Blockage or imbalance around the energy of this center can result in depression, confinement, a closed mind, worry, alienation, confusion, boredom, apathy, and an inability to learn and comprehend.

Imbalanced or blocked energy flow of the chakras can relate to the developmental or attachment issues presented in Chapter 2. A yoga therapist might approach therapy by asking clients what types of daily activities they engage in. For example, clients might report that they work at their desks all day, processing information and making decisions, and spend very little time in physical activities. In addition, the yoga therapist could ask clients what types of concerns they have around each energy center, such as issues of (1) security and stability; (2) power and control, sexuality, creativity; (3) self, socialization, and social interactions, (4) compassion and love, (5) will, interactions, expression, creativity, inspiration, and decision making, (6) mind–emotional acceptance, perceptual openness, and intellectual fluidity, and (7) relatedness to others, and to wholeness and unblocked energy. The yoga therapist would assess where clients feel discomfort (blockage of energy) by having them mark the location on a diagram of the body. Then the yoga therapist could have clients focus on the postures or asanas associated with the blocked energy centers in addition to a routine of other Hatha Yoga postures.

So much of counseling and psychotherapy involves working with the client's cognitions, emotions, and behaviors. Clients do not often have the opportunity to work with or in their bodies. Yoga therapy offers a complementary mind/body approach for many different types of clients such as these:

- People who suffer from ills such as arthritis, cancer, chronic lung disease, chronic pain, coronary disease, fatigue, gastrointestinal disease, headaches, hypertension, insomnia
- Adolescents who have low self-esteem
- Elderly people in nursing homes or assisted-living facilities, or who belong to health clubs

- Substance abusers or people with addictions
- People with anxiety or who suffer from panic attacks
- People who suffer from some form of trauma
- People who experience stress because they do mostly mental work, with little physical activity or body movement
- People who seem to have assumed the role of victim and are holding on to psychological/emotional wounds
- Mental health professionals who feel unbalanced and uncentered because of stress and potential burn-out.

One of the important facets of Hatha Yoga is the mind/body connection. Meditation about the seven energy centers or chakras facilitates this connection. In addition to the different meditations presented in Chapter 17, we provide the following meditation that might be used by a yoga therapist who incorporates the chakras or energy centers in the practice of Hatha Yoga:

Close your eyes and get in a comfortable position. Take a couple of deep breaths and relax; breathe diaphragmatically. Focus on each of your seven chakras or energy centers. Draw your attention on the first chakra and feel yourself connected to your life in the region of your body at the base of your spine. Breathe energy into this area . . . feel the energy of your breath go into this area and experience the energy of security this area provides. Think about power and control, sexuality and creativity while focusing on the second chakra. Bring your attention to the area of your pubic bone, to the sacrum—the last bone of your spine. Experience the energy and sense the quality of the energy. Breathe restfully . . . focus on the third chakra . . . the area between the base of the sternum and the navel. Think about your self and your social interactions. Experience the energy this center provides. Examine the heart area of the fourth chakra and experience the love, compassion, and empathy energy this chakra brings to your body and you. If at any time you feel the amount of energy or quality is not sufficient, breathe into the energy center and allow the energy of the breath to refresh the center. Draw the attention of your will to the base of throat and experience the quality and amount of energy this center provides. Think about your interactions as being inspired by a higher vision. Think of your interactions as sending positive energy and thoughts to others. Bring your attention to the area between your eyebrows that brings from your mind understanding, wisdom . . . if you feel confused, breathe energy into this center, requesting it to give you insight and wisdom. On the seventh chakra at the top of your head, focus on oneness; allow the energy to let go of unfinished business and breathe in and send energy to this area for spiritual awareness. Now pause . . . and review each chakra. If you feel that a particular chakra needs more energy, like a beam of light draw your attention to that particular one . . . and consciously breathe and send quality energy to that chakra . . . experience the energy activity increasing in that part of your body. Feel the balance and flow of all the energy centers or chakras throughout your body.

This meditation can be done at the same time as the asanas or postures of Hatha Yoga, or the meditation can be used in the resting or *sponge* (corpse) posture.

STEPS FOR HATHA YOGA

In introducing Hatha Yoga, the first step is to provide the client with a rationale and overview of the practice. The counselor will instruct the client in abdominal or diaphragmatic breathing. Next, he or she will provide guidelines for breathing while engaging in the particular asanas or postures. At the conclusion of the sequence of asanas, the instructor will instruct the client to engage in a body scan (see the last section of Chapter 17); the Hatha Yoga session ends with a meditation (see the first part of Chapter 17).

Rationale for Hatha Yoga

Benson and Stuart (1992) offer the following rationale for Hatha Yoga: "Hatha-yoga exercises offer one means of developing body awareness and eliciting the relaxation response. Hatha-yoga can easily be modified to an individual's physical restrictions. In addition to some of the more obvious benefits of exercise, yoga can help you realign your bodily posture, release muscle tension, and develop a more subtle control of your body. Yoga is also helpful because it reinforces your experience of maintaining a basic 'resting state' in the musculature and the mind while carrying on daily activities" (p. 79).

Ornish (1990) says that "Hatha yoga is more than just a collection of stretching and breathing techniques. Ultimately, the purpose of Hatha yoga is to begin rebalancing these opposing forces and to experience the equilibrium, peace, and common unity that underlie these dualities. For most of us, the duality of our muscles—contracting and relaxing—is out of balance, for our muscles are chronically tensed and contracted. The first step toward experiencing inner peace and healing is to quiet down and relax the body" (pp. 146–147).

Kabat-Zinn (1990) offers the following rationale: "Yoga is a wonderful form of exercise for a number of reasons. To begin with, it is gentle. It can be beneficial at any level of physical conditioning and, if practiced regularly, counteracts the process of disuse atrophy. . . . Yoga is also good exercise because it is a type of full-body conditioning. It improves strength and flexibility in the entire body. . . . Perhaps the

most remarkable thing about yoga is how much energy you feel *after* you do it. . . . When you practice yoga, you should be on the lookout for the many ways, some quite subtle, in which your perspective on your body, your thoughts, and your whole sense of self can change when you adopt different postures on purpose and stay in them for a time, paying full attention from moment to moment. Practicing in this way enriches the inner work enormously and takes it far beyond the physical benefits that come naturally with the stretching and strengthening" (pp. 100, 102, 103).

Another brief rationale for and overview of Hatha Yoga is this: "Yoga is a process of self-integration. Yoga is a process of achieving equanimity of the mind. Yoga is a science of the soul and a method through which you can liberate yourself from the cycle of chaos. Yoga enriches and nourishes the meditative process. You will do movements or postures. There is no *precise* way of doing them correctly. You will measure the success of your postures not by their exactness, but by your awareness of them. The correctness of the postures is the degree to which they make you feel present, alive, and whole. No special equipment is required for yoga other than a mat or carpeted floor and comfortable clothing. You will learn to breathe diaphragmatically or from your belly. You will learn warm-up postures, and then, 10 to 14 asanas or postures. Also, you will use body scan and meditation technique. How does that sound? Do you have any questions?"

Breathing During Movement

A yoga instructor would teach the client about abdominal breathing and any other breathing technique presented in the first section of this chapter. Miller (1993) offers some general guidelines for breathing during movement:

1. Forward bending occurs most easily during exhalation.
2. Backward bending is generally executed during inhalation.
3. Twisting movements occur most readily during exhalation.
4. Movements that expand the chest and thorax are initiated during inhalation.
5. Movements that compress the abdominal area and thorax region are initiated during exhalation.
6. Generally, the exhalations are 1 to 1.5 times longer than the inhalation during asana practice. (pp. 28, 30)*

---

*SOURCE: From "Working with the Breath," by R. C. Miller, in *Living Yoga*, by G. Feuerstein and S. Bodian (Eds.). Copyright © 1993 by Yoga Journal. Reprinted by permission of the Putnam Publishing Group/Jeremy P. Tarcher, Inc.

There is an influential interaction between bodily movement and the breath. Often beginners of Hatha Yoga have a tendency to hold their breath while engaging in the asanas. A general guideline for holding the breath is to do so after an inhalation when the body is in extension. Holding the breath after an exhalation may occur during either extension or flexion of the body (Miller, 1993). As Miller (1993) suggests, clients can also be instructed to visualize the breath moving upward along the spine from the coccyx (tailbone) to the crown of the head on the exhalation and reverse the visualization on the inhalation, imagining the breath descending from the crown of the head to the base of the spine. Finally, the client is instructed to synchronize these movements with the breath. Exhale with movements that compress the abdomen, bend the body forward, or twist the body. The client visualizes the particular asana and the movement of the breath harmonizing with the flow of the movements in the posture or asana. The client is instructed that on the inhalation for asanas that extends, bend backwards, or expands to the body, visualize the particular asana and the downward movement of the breath harmonizing with the flow of the movements in the posture (Miller, 1993).

Sun Salutation—Warm-up

Figure 18-5 presents an overview of the asanas or postures used for the Sun Salutation or warm-up for the yoga postures. The Sun Salutation limbers up the mind and body in preparation for the sequence of postures that follows. Each of the 13 postures in Figure 18-5 provides a different movement

FIGURE 18-5. Sun Salutation—Warm-up

for the spinal column, intended to bring feelings of balance and connectedness to mind and body. Clients are instructed to start with four to six sequences to thoroughly warm up in preparation for the sequence of yoga asanas, and to do the Sun Salutation gently and smoothly.

The sequence of the 13 postures goes clockwise starting with the Prayer Posture (Namaste—the spirit within me salutes the spirit within you), and continues with the standing backward arch, forward bend, the lunge forward with right leg back, inclined plane, inverted V, modified push-up, cobra asana, inverted V again, lunge repeated with the other leg back and knee bent, forward bend, backward arch, and finally the mountain posture.

Another diagram of each Sun Salutation asana, instructions for each asana, and the potential benefits of each are presented in Table 18-5. For each asana, the client is instructed to visualize the movement to get into the posture, and to visulize the breath flowing with the movement or while holding the posture. The client is reminded of the movement guidelines for breathing presented in the previous section. Clients can use the Sun Salutation chart that appears in Chapter 20 to self-monitor their progress for each asana. The asanas in the Sun Salutation are used for a warm-up. The purpose is to heat the body. Each asana in Figure 18-5 is performed immediately after the preceding one so that the entire sequence of the 13 postures is done gracefully in a continuous flow. In the following section, each asana can be held for a longer period of time so that we can experience the squeezing, soaking, and stretching each posture provides.

Hatha Yoga Asanas

Diagrams for 14 asanas, with instructions for each asana and the potential benefits derived from each, appear in Table 18-6. Included are the following asanas: the triangle—with and without the hand on the floor, half-moon for both the left and right sides, the tree posture with the arch of the foot on the calf or inside the thigh of the standing leg, cobra asana, half and full locust, bow posture, half spinal twist—both sides, sitting knee to head for each leg, sitting forward bending, half shoulder stand, the child pose, cobbler, and finally the sponge. Again, the client is instructed to visualize the movement of the particular asana, and the appropriate flow of the breath. (See also Learning Activity #50.)

Informed Consent for Hatha Yoga

If you refer a client to a trained yoga therapist, you might offer the client an informed consent document that he or she can share with the therapist. The contents of the document can be the same as those on page 465 for the breathing exercises. The yoga therapist may use your informed consent document or one he or she has prepared. In either case, the client's welfare will be protected.

MODEL EXAMPLE: HATHA YOGA

In this model example, we show how Hatha Yoga might be used with Yoshi by a yoga therapist to reduce the stress he experiences. The yoga therapist presents the rationale for Hatha Yoga and determines that Yoshi has been cleared by his physician to participate in yoga. The Sun Salutation warm-up is introduced to Yoshi. He is informed about breathing while doing the postures or asanas; he is instructed to engage in the 14 asanas presented in Table 18-6 and to do a body scan in the corpse asana, as described in the previous chapter. Yoshi is instructed to meditate using any one of three techniques described in the previous chapter; and finally, the yoga therapist gives instructions for homework practice.

1. *Rationale*

The yoga therapist explains the rationale to Yoshi: "Hatha Yoga is a wonderful form of exercise for a number of reasons. To begin with, it is gentle. It can be beneficial at any level of physical conditioning and, if practiced regularly, counteracts the process of disuse atrophy. Yoga is also good exercise because it is a type of full-body conditioning. In addition to reducing stress and helping with your ulcer, it improves strength and flexibility in the entire body. Perhaps the most remarkable thing about yoga is how much energy you feel *after* you do it. When you practice yoga, you should be on the lookout for the many ways, some quite subtle, in which your perspective on your body, your thoughts, and your whole sense of self can change when you adopt different postures on purpose and stay in them for a time, paying full attention from moment to moment. Practicing the postures will enrich the inner work enormously and take it far beyond the physical benefits that come naturally with stretching, soaking, and strengthening the muscles and internal organs of the body. Soaking means the bathing of the cells with oxygen-rich blood. Yoga brings the mind and body together."

The yoga therapist confirms Yoshi's willingness to engage in the practice of yoga and answers any questions he has. In addition, the yoga therapist confirms with Yoshi that he has received approval from his physician to engage in the practice of yoga.

TABLE 18-5. Asanas and potential benefits of Sun Salutation

Diagram of asana or posture	Instructions for asana	Potential benefits
1	**Prayer Pose**—Stand straight, feet width of shoulders, arms bent at elbows, and hands in front of chest, head straight. Focus on your breathing.	Develops strength in the legs and flexibility of the pelvis and lower back. Helps teach basic alignment.
2	**Standing Backward Bend**—Inhale and lift your arms from your side over your head. While your head is between your arms, look at your hands. When arching your back, find your level of comfort.	Opens the chest and can stimulate the nervous system; can increase vitality.
3	**Forward Bend**—Keep your head between your arms and bend forward from the waist; keep your back straight. Bring your hands beside your feet, or touching your ankles, or dangle your arms. Keep your head down.	Soothes the nervous system. Can quiet the mind. If you have tight hamstrings or lower back problems, you can do Wide Spread Forward Bend in which the feet are spread wide apart.
4	**Lunge Posture**—Take a large step back with your right foot. Lower your knee to the floor. Palms or fingers are flat on the mat or floor. Leave the left foot between the hands and the left knee to the chest toward the left shoulder. Look up.	Brings strength and flexibility to the lower back and legs.
5	**Inclined Plane**—Move your left leg back, toes tucked on the floor, arms straight. Keep your head in line with your body.	Strengthens shoulders, arms, and legs.
6	**Downward Facing Dog**—Bring your feet closer to your hands and form a triangle by lifting your buttocks. Create a straight line from your buttocks along your back, stretching your spine. Keep your legs straight by pressing heels toward the floor, your arms alongside your ears while looking back at your feet.	Extends your spine and shoulders. Strengthens weak knees by stretching and contracting the hamstrings, quadriceps, shin, and calf muscles. Considered one of the most invigorating postures.
7	**Modified Push-up Posture**—Come down so that your knees, chest and chin or forehead are touching the floor. Your pelvis is slightly raised off the floor. Your palms are beneath your shoulders.	Works the lower back, abdominal, and pelvis areas.
8	**Cobra Asana**—Lower your pelvis to the floor. Stretch your head, neck, and chest up. Keep your elbows slightly bent and in toward your body so that the weight rests on your back. Top of toes are on the floor.	Exercises the back muscles, spine, and cranial nerves; expands the chest; can relieve backache, constipation, and gas.

(continued)

TABLE 18-5. Asanas and potential benefits of Sun Salutation (continued)

Diagram of asana or posture	Instructions for asana	Potential benefits
9	**Downward Facing Dog**—Tuck toes so that they are on the floor. Lift your buttocks to form a triangle. Push up and back into an upside-down V position. Stretch the backs of your legs through your heels. Form a straight line from your buttocks along your spine, shoulders, and arms, head looking back at your feet. Again, make sure your legs are straight by pressing your heels toward the floor. Make sure feet are parallel and about eight to twelve inches apart, lined up with your hip bones, your hands about the same distance apart and lined up with your shoulder blades, palms flat and fingers spread. Look toward your navel.	Same as 6 above.
10	**Lunge Asana**—Come out of Downward Facing Dog by placing your right foot forward between your hands, your knee is toward your right shoulder. Your left knee is on the floor.	Same as 4 above.
11	**Forward Bend**—Bring left foot between hands, straighten knees and bring head to knees.	Same as 3 above.
12	**Standing Backward Bend**—Lift from the waist—at your hips with your back flat—to a standing position. Inhale and lift arms from your sides over your head, with your head between your arms, arching your back with your head looking at your hands. Find your comfort zone while arching your back.	Same as 2 above.
13	**Mountain Posture**—Lower your arms to your sides and stand straight, feet width of shoulders. Close your eyes and focus on your breathing.	Develops strength in the legs and flexibility of the pelvis and lower back. Helps teach basic alignment.

2. Introduce the 13 Sun Salutation Warm-up Positions and Comfortable Clothing

The 13 Sun Salutation warm-up positions are demonstrated for Yoshi (see Figure 18-5). Yoshi starts with Namaste—the Prayer posture—and continues clockwise with all the postures until ending with the Mountain pose. Yoshi is instructed about each posture, and its potential benefits as he does each one. The instructor tells him to wear comfortable clothing but no shoes while doing yoga.

3. Instructions About Breathing while Doing the Sun Salutation Warm-up and Hatha Yoga

Yoshi is given the guidelines for breathing while doing the Sun Salutation warm-up postures and the Hatha Yoga asanas. The yoga instructor says, "Exhalation is easily done with forward bending; backward bending is generally done during inhalation; and twisting movements occur mostly during exhalation." Yoshi is reminded that he does not have to remember all this now, and that the instructor will tell him

TABLE 18-6. Hatha Yoga asanas

Diagram of asana or posture	Instruction for asana and potential benefits

Triangle

Instructions—Triangle: In a standing position, visualize doing the posture as you turn your right foot out to 90 degrees, and turn your left foot inward about 45 degrees. Align the arch of your left foot with the heel of your right foot. Spread your arms out parallel to the ground. The ankles of both feet should be comfortably under the wrist of each arm, so that, for example, the left ankle is under the wrist of your left arm. On an exhalation, slide your right hand down your right leg while lifting your left arm toward the ceiling. Bend at your waist. Turn your head so that you are looking up at your left hand. Hold the posture for about 30 seconds. You can gradually extend the time the posture is held to about a minute. Do each side 2 times.

Potential Benefits: In the Triangle, the abdominal cavity is stretched along the sides while the liver, colon, stomach, spleen, and kidneys are massaged. The intercostal muscles as well as the diaphragm are massaged.

Half Moon

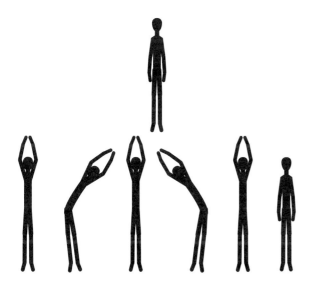

Instructions—Half Moon: From standing in a Mountain Pose, place your feet together and your arms at your sides. Take a couple of minutes to calm your mind and body by breathing diaphragmatically. Contract your quadriceps, gently firm your buttocks, and draw your abdominal area up and in. Find a spot at eye level and focus on it. Again, breathe deeply, and bring your awareness to your entire body. Visualize getting into this asana, inhale and bring your arms from your sides to over your head. You can lace your fingers or just keep your fingers and hands separated above your head and in the same plane as your body. Exhale and stretch to your right side. Try to keep the underside of your rib cage as long as possible. Feel the stretch along the curved outside part of your body. Come back to center on the inhalation. Take a breath in, then exhale and bend to the left side. When ready, return to center. You can do a couple of rounds more of the Half Moon asana making sure you do both sides for each round. You can close your eyes and visualize the movement and breathing as you do the asana.

Potential Benefits: The Half Moon posture provides lateral flexion of the spine. Raising the arms above the head helps with flexibility of the shoulder joints, tones the muscles of the arms, strengthens the musculature of the torso and abdomen. The Half Moon helps the kidneys, liver, spleen, and digestive system.

Tree

Instructions—Tree: In the Mountain Pose, look at a spot on a wall at eye level. Shift your weight to one leg, and gently lift the other foot slightly off the floor. Hold the foot off the floor for three breaths. Then, stand on one leg and bend the knee of the other leg. Place the arch of the lifted foot on the calf of the leg on which you are standing—or place it on the inside thigh of the leg you are standing on. Inhale, and lift both arms above your head. You can lace your fingers above your head if you like. Hold the posture for about 30 seconds, and don't forget to breathe. Now, let's do the other side. You can do 2 to 4 rounds for each leg.   (continued)

TABLE 18-6. Hatha Yoga asanas (continued)

Diagram of asana or posture	Instruction for asana and potential benefits

Potential Benefits: The Tree strengthens both legs. It improves concentration, balance, and respiration.

Cobra

Instructions—Cobra: Lie face down, place your chin and chest on the floor, and your hands under your shoulders with fingers facing toward your front. Exhale; make sure your hips are firm on the floor, and the tops of your feet are on the floor. The fingers or palms of your hands are under your shoulders. Now, on the exhalation, lift your chest off the floor with your head following. Lengthen your entire spine as you breathe out, and lift up to your level of comfort. Avoid the tendency to lift with your arms; the backbending movement comes from your spine and upper back. Gradually, raise the chest, bending the back vertebrae one by one. Tilt your head back slightly. Hold the posture while breathing to fit your level of comfort. Now, relax and slowly drop your chest down first, followed by your chin to the floor; turn your head to one side and rest it on the floor. Take your arms from beneath your shoulders, and place them down along each side of your body. Experience the relaxation, and study the sensation you experience in your body. Repeat the posture.

Potential Benefits: The Cobra exercises the back muscles, spine, and cranial nerves. It expands the chest and can relieve backache, constipation, and gas.

Head to Knee

Instructions—Head to Knee: Lie on your back, with your arms stretched overhead. Lock your thumbs and sit up with as much control as possible. Try to bend from the waist. Your arms are above your head when you sit up. Now, lower your arms and bend the left leg; place the sole of the foot against the upper, inside right thigh, with your heel pressing inward. Again, stretch the arms up, lock your thumbs, and fold forward over the right leg, bending at the waist. Grasp the leg as far down as comfortable. If you can, grasp the big toe with both hands and bring your face down toward and to your knee. Try to keep your leg straight. If this is not comfortable, bend the knee slightly. Then, to come up, lock the thumbs, stretch out and raise up as straight as you can. Repeat with the opposite leg. Do 2 to 3 times for each leg, holding the Head to Knee for about 10 seconds.

Potential Benefits: The Head to Knee increases your level of energy.

Forward Bending

Instructions—Forward Bending: Lie on your back, with your arms stretched overhead. As with the previous postures, visualize in your mind's eye the movement and breath of the asana. Now, lock your thumbs together and sit up on the exhalation. Try to lift your back straight, your arms remaining overhead. Then, fold forward over your legs and grasp as far down as possible. Eventually try to hold your big toes—with left first finger on your left toes and right finger on your right toes. If you can, bring your face down to your knees. Also, try not to

TABLE 18-6. Hatha Yoga asanas (continued)

Diagram of asana or posture	Instruction for asana and potential benefits
	allow your knees to bend. If it is too much of a strain not to bend them, bend them slightly with the idea that with future practice you will ultimately not bend them. Try to relax in the asana without bouncing or straining. Do Forward Bending 3 to 6 times, holding the posture for about 10 seconds each time. **Potential Benefits:** This posture stretches and exercises nearly all the posterior muscles and tones the abdominal viscera.
Half Shoulder Stand 	**Instructions—Half Shoulder Stand:** This is an inverted and invigorating posture. First, lie on your back, with your hands alongside your body and the palms of your hands down on the floor. On the exhalation, visualize while lifting your legs up to a 90-degree angle; try to keep your knees straight and bend at your waist. Press on the palms, swing the legs slightly over your head, and bring your hands to support your back. Place your hands with the bottom of your palms at the end of each pelvis bone on your back, your fingers alongside the pelvis. Your upper arm is parallel to and on the floor and your forearm is almost perpendicular to the floor, again with hand support at the end of the pelvis bone. Your elbows are fairly close together, just about under your back. Bring your chest as close to your chin as comfortable. (If this feels uncomfortable, you can use a small blanket or folded mat to elevate your shoulders, with your head resting on the floor.) Notice that your legs are bent, allowing your feet to be over or beyond your head. Hold the posture as long as comfortable. **Potential Benefits:** The Half Shoulder Stand tones the thyroid glands, regulates metabolism, and improves circulation.
Half Locust 	**Instructions—Half Locust:** Lie down with chin on the floor. Tuck your arms underneath your body, elbows close together so that your palms are facing up on the upper part of your thighs. Your legs are straight and the tops of your feet are resting on the floor. Remember to visualize doing the posture in your mind's eye as I instruct you with the movement of the posture and while you are doing the movement. Now, slowly raise the right leg, keeping your pelvis on your wrists. Take a couple of breaths while holding the leg up. Then slowly lower the right leg. Again, take a couple of breaths, and raise your left leg. Do this 2 times for each leg; hold for about 10 seconds each time. **Potential Benefits:** The Half Locust and the Full Locust exercise back, pelvis, and abdomen. They tone the sympathetic nervous system and can help the liver.
Full Locust 	**Instructions—Full Locust:** Follow the instructions for the Half Locust, but instead of lifting one leg at a time, lift both legs. The Full Locust is a little more rigorous than lifting one leg at a time. **Potential Benefits:** The possible benefits for the Full Locust are the same as for the Half Locust. (continued)

TABLE 18-6. Hatha Yoga asanas (continued)

Diagram of asana or posture	Instruction for asana and potential benefits
Bow 	**Instruction—** Bow: Lie face down with your chin or forehead on the floor. Bend your knees and grasp your ankles; visualize your movements as we do them. On inhalation, raise your head, chest, and thighs. Arch your back to allow your entire weight to fall on the abdomen. Keep your arms straight. A less rigorous variation of the Bow is to raise just your thighs while holding your chest on the floor, or just raise your chest while allowing your thighs to stay on the floor. Do the posture 3 to 6 times, holding the posture for about 10 seconds, gradually during your practice working up to holding it for one minute. **Potential Benefits:** The Bow gives the same benefits as the Cobra and Locust plus it reduces abdominal fat and tones the pancreas. It relieves chronic constipation, and can improve functioning of the liver, kidneys, spleen, stomach, and intestines. The Bow strengthens the back and thighs, aligns the vertebrae, and increases vigor and vitality.
Half Spinal Twist	**Instructions—**Half Spinal Twist: Sit with your knees drawn to your chest and your feet flat on the floor. Stretch out your left leg. Place the right foot on the outside of your left leg—knee or calf. Place the right hand on the floor behind or near your buttocks, fingers pointing away from your body. Bring your left arm around to the outside or the right side of your right knee. Look over the right shoulder, on the exhalation, twist the spine to the right, and to your level of comfort. Pause there for a couple of breaths. Then, slowly unwind on an inhalation. Reverse the asana by doing it on the other side. Do the posture 1 to 3 times, holding the asana for about 30 seconds and gradually building up to 3 minutes. **Potential Benefits:** The Half Spinal Twist helps the liver, spleen, kidneys, and adrenals. The asana tones the sympathetic nervous system, and strengthens the deep and superficial back muscles.
Child	**Instructions—**Child or Resting Posture: Get on the floor on your hands and knees. Visualize the movement as you take your arms off the floor, push back, and sit on your feet. On exhalation, place your forehead on the floor. Place your arms back alongside your body or straight out in front of you. Breathe gently, and remember that the Child Pose is a resting and relaxing asana. **Potential Benefits:** The Child Posture creates flexibility in the back, hips, knees, and ankles. It is highly restorative. It is supposed to balance the autonomic nervous system and improve digestion; it may be good for headaches and stomachaches. The posture is cooling, relaxing.
Cobbler	**Instructions—**Cobbler: Sit on the floor with your back straight, your knees bent, and your feet together. Breathe out and let your thighs go down and out as you stretch upward, forward, and lean toward the floor. The Cobbler's or Tailor's Posture is one of the oldest known postures.

TABLE 18-6. Hatha Yoga asanas (continued)

Diagram of asana or posture	Instruction for asana and potential benefits
Sponge (Corpse) 	**Potential Benefits:** The outward turn of the thighs at the hip joints is worked and stretched, which can relieve stiffness. It is one of the basic yoga poses for many other yoga asanas. **Instructions**—Sponge (Corpse): Lie down on your back. Straighten your legs out and allow your feet to flop to the sides. Place your arms and hands alongside your body, but not too close. Place your palms up. Check your back to make sure your spine is as flat as possible. The back of your waist will be raised a little, forming a small arch. Breathe and experience your abdomen rise and fall like the tide. A variation of this pose is to place your feet and calves on a chair with the thighs vertical, and your sit bones or tailbone and spine flat on the floor with outstretched arms and hands. **Potential Benefits:** The Sponge Posture creates awareness and relaxation. With the focus on your breathing, it brings you to the present moment. We will do the body scan and meditation in this pose.

about breathing as he does each posture. Breathing instructions are given: Movements that expand the chest and thorax usually are done during inhalation and postures or movements that compress the abdominal area and thorax region are started during exhalation. Generally, exhalations—with practice—are about 1 to 1.5 times longer than the inhalation. Yoshi is instructed about inhaling or exhaling as he begins each posture. After instructions for the initial part of the posture, he is told to breathe gently, inhaling and exhaling at his normal rate.

4. Instructions for Each Asana of the Sun Salutation, Potential Benefits, Breathing, and Visualization

Yoshi is instructed to do each asana of the Sun Salutation. The yoga therapist informs Yoshi that the purpose of the Sun Salutation is to warm and heat the body in preparation for the sequence of Hatha Yoga asanas that follows. He is told that he will start slowly but have little pause between asanas; the Sun Salutation will be done in a continuous gentle flow of postures to warm and heat the body. Yoshi is instructed to visualize in his mind's eye each posture and his breath as he does it. He starts with the Prayer Posture. The yoga therapist tells him the instructions for each posture, all of which appear in Table 18-5:

1. *Prayer Posture*—"Stand straight, with your feet apart about shoulder width, arms bent at your elbows; hold your hands in front of your chest, with your fingers pointed up in a prayer position. Your head is straight with your eyes looking forward or closed. Focus on your breathing. This posture develops strength in your legs and flexibility of the pelvis and lower back. The posture is helpful for learning basic alignment."

2. *Standing Backward Bend*—"Inhale and lift your arms from your side over your head. While your head is between your arms, look at your hands. When arching your back, find your level of comfort; don't strain. Backbends open the chest, stimulate the nervous system, and increase energy and vitality. Remember: Visualize doing the posture and your breath."

3. *Forward Bend*—"On the exhalation, visualize bending forward from the waist, keeping your head between your arms. Bring your hands beside your feet, or touching your ankles, or dangle your arms—whichever is most comfortable for you. Keep your head down. If you want to stay in the forward bend for a few seconds, just keep breathing gently. The forward bend is soothing for the nervous system. Forward bends can quiet the mind. If your hamstrings are tight or your lower back is uncomfortable, you can do a widespread forward bend in which your feet are wide apart."

4. *Lunge Posture*—"Visualize what we are doing as you take a large step back with your right foot. Lower your knee to the floor. Place the palms of your hands or fingers

flat on the floor. Leave the left foot between your hands, and your left knee to the chest toward the left shoulder. Inhale and look up. This lunge posture brings strength and flexibility to the lower back and legs."

5. *Inclined Plane*—"Visualize, on the exhalation, moving your left leg back, toes tucked on the floor, arms straight. Keep your head in line with your body. The inclined plane strengthens your shoulders, arms, and legs."

6. *Inverted V or Downward Facing Dog*—"Visualize yourself making an inverted V: Place your feet closer to your hands by lifting your buttocks on the exhalation. Create a straight line from your buttocks along your back, stretching your spine. Keep your legs straight by pressing your heels toward the floor. Place your arms alongside your ears while looking back at your feet. Breathe gently. This asana is considered one of the most invigorating postures because it extends your spine and shoulders. It strengthens weak knees because they stretch and contract the hamstrings, quadriceps, and shin and calf muscles.

7. *Modified Push-up Pose*—"See yourself, on an exhalation, coming down so that your knees, chest and chin or forehead are touching the floor. Your pelvis is slightly raised off the floor. The palms of your hands are either beneath or next to your shoulders. The modified push-up works the lower back, abdominal, and pelvis areas."

8. *Cobra Posture*—"From the modified push-up, inhale and visualize getting into the posture as you lower your pelvis to the floor. Stretch up your head, neck, and chest. Keep your elbows slightly bent and in toward your body so the weight rests on your back and not on your arms. The tops of your toes are on the floor. The cobra exercises the back muscles, spine, and cranial nerves; it expands the chest and can relieve backache."

9. *Inverted V or Downward Facing Dog*—"Visualize the posture on the inhalation; tuck your toes so that they are on the floor. Lift your sit bones or buttocks to form a V. Push up and stretch the backs of your legs through the heels. Form a straight line from your buttocks along your spine, shoulders, and arms. Place your head so that it is looking toward your feet. Again, make sure that your legs are straight by pressing your heels toward the floor. Make sure your feet are parallel, and about eight to twelve inches apart, lined up with your hip bones. Keep your hands about the same distance apart and lined up with the shoulder blades. Your palms are flat and fingers are spread. Look toward your feet or navel. This posture is very invigorating."

10. *Lunge Posture*—"Visualize yourself coming out of the inverted V by inhaling and placing your right foot forward between your hands, your knee toward your right shoulder. Your left knee is on the floor. The posture brings strength and flexibility to the lower back and legs."

11. *Forward Bend*—"Again, on the exhale, visualize bringing the left foot between the hands; straighten the knees and bring the head downward to the knees. Experience this posture soothing the nervous system, quieting the mind."

12. *Standing Backward Bend*—"On the inhalation, visualize lifting from the waist—at your hips, with your back flat—to a standing position. Lift your arms over your head, with your head between your arms, arching your back with your head back and looking at your hands. Find your comfort zone while arching your back. This posture opens your chest, can stimulate the nervous system, and can increase vitality and energy."

13. *Mountain Posture*—"On the exhalation, lower your arms to your sides, stand straight; your toes are pointing straight and your feet are shoulder width apart. Close your eyes and focus on your diaphragmatic breathing."

Yoshi is instructed to do three rounds of the Sun Salutation for homework. He is to increase gradually to as many as 10 repetitions, but not to overexert himself. "The benefits of this sequence are that it heats the body and includes combinations of the basic format of most asanas. If you do the sequence rapidly, you will become alert and energized. Going through the sequence at a slow pace can quiet your mind and body."

5. *Instructions for Hatha Yoga Asanas, Potential Benefits, Breathing, and Visualization*

Yoshi is told about the potential benefits of the sequence of Hatha Yoga postures that appear in Table 18-6. He is reminded of the rules for breathing during the movements and asked to visualize the posture mentally as he engages in each one. The yoga therapist says, "It is important to keep in mind while doing these postures that some may be challenging, but just let things happen rather than striving to achieve something." For each of the following postures, the yoga therapist gives Yoshi instructions:

1. *Triangle*—"Get in a standing position; visualize doing the posture as you turn your right foot out 90 degrees, and turn your left foot inward about 45 degrees. Line the arch

of your left foot with the heel of your right foot. Spread your arms parallel to the ground. The ankles of both feet should be comfortably under the wrist of each arm, so that, for example, the left ankle is under the wrist of your left arm. On an exhalation, slide your right hand down your right leg while lifting your left arm toward the ceiling. Bend at your waist. Turn your head so that you are looking up at your left hand. Hold the posture for about 30 seconds. You can gradually extend the time the posture is held to about a minute. Do each side 2 times. The benefits of the posture are that the abdominal cavity is stretched along the sides, massaging the liver, colon, stomach, spleen, and kidneys. The intercostal muscles as well as the diaphragm are massaged."

2. *Half Moon*—"Do the 'standing in a mountain' pose, with your feet together and your arms at your sides. Take a couple of minutes to calm your mind and body by breathing diaphragmatically. Contract your quadriceps, gently firm your buttocks, and draw your abdominal area up and in. Find a spot at eye level and focus on it. Again, breathe deeply, and bring your awareness to your entire body. Visualize getting into this asana; inhale and bring your arms from your sides to over your head. You can lace your fingers or just keep your fingers and hands separated above your head and in the same plane as your body. Exhale and stretch to your right side. Stretch to keep the underside of your rib cage as long as possible. Feel the stretch along the curved outside part of your body. Come back to center on the inhalation. Take a breath in, then exhale and bend to the left side. When ready, return to center. You can do a couple of rounds more of the Half Moon asana making sure you do both sides for each round. You can close your eyes and visualize the movement and breathing as you do the asana. The Half Moon posture provides lateral flexion of the spine. Raising the arms above the head helps with flexibility of the shoulder joints, tones the muscles of the arms, and strengthens the musculature of the torso and abdomen. The Half Moon helps the kidneys, liver, spleen, and digestive system."

3. *Tree*—"In the Mountain Pose, look at a spot on a wall at eye level. Shift your weight to one leg and gently lift the other foot slightly off the floor. Hold the foot off the floor for three breaths. Then, stand on one leg, and bend the knee of the other leg. Place the arch of the lifted foot on the calf of the leg on which you are standing, or on the inside thigh of that leg. Inhale, and lift both arms above your head. You can lace your fingers above your head if you like. Hold the posture for about 30 seconds, and don't

forget to breathe. Now, let's do the other side. You can do two to four rounds for each leg. The Tree strengthens both legs. It improves concentration, balance, and respiration."

4. *Cobra*—"Yoshi, lie face down, place your chin and chest on the floor, and put your hands under your shoulders with fingers facing toward the front. Exhale; make sure your hips are firm on the floor and the tops of your feet are on the floor. The fingers or palms of your hands are under your shoulders. Now, on the exhalation, lift your chest off the floor, with your head following. Lengthen your entire spine as you breathe out, and lift up to your level of comfort. Avoid the tendency to lift with your arms; the backbending movement comes from your spine and upper back. Gradually, raise the chest, bending the back vertebrae one by one. Tilt your head back slightly. Hold the posture while breathing to fit your level of comfort. Now, relax and slowly drop your chest down to the floor, followed by your chin; turn your head to one side and rest it on the floor. Take your arms from beneath your shoulders, and place them down by your side. Experience the relaxation, and study the sensation you experience in your body. Repeat the posture. The Cobra exercises the back muscles, spine, and cranial nerves. It expands the chest and can relieve backache, constipation, and gas."

5. *Head to Knee*—"Lie on your back, with your arms stretched overhead. Lock your thumbs and sit up with as much control as possible. Try to bend from the waist. Your arms are above your head when you sit up. Now, lower your arms and bend the left leg; place the sole of the foot against the upper, inside right thigh, with your heel pressing inward. Again, stretch the arms up, lock your thumbs, and fold forward over the right leg, bending at the waist. Grasp the leg as far down as comfortable. If you can, grasp the big toe with both hands and bring your face down toward and to your knee. Try to keep your leg straight. If this is not comfortable, bend the knee slightly. Then, to come up, lock the thumbs, stretch out, and rise up as straight as you can. Repeat with the opposite leg. Do two to three times for each leg, holding the head to the knee for about 10 seconds. The Head to Knee posture helps to increase your level of energy."

6. *Forward Bending*—"You need to lie on your back, with your arms stretched overhead. As with the previous postures, visualize the movement and breath of the asana. Now, lock your thumbs together and sit up on the exhalation. Try to lift your back straight with your arms remaining overhead. Then, fold forward over your legs and grasp as far down as possible. Eventually try to hold

your big toes—with left first finger on your left toes and right finger on your right toes. If you can, bring your face down to your knees. Try not to allow your knees to bend. If not bending them is too much of a strain, bend them slightly; work toward not bending them ultimately. Try to relax in the asana without bouncing or straining. Do Forward Bending three to six times, holding the posture for about 10 seconds each time. This posture stretches and exercises nearly all the posterior muscles and tones the abdominal viscera."

7. *Half Shoulder Stand*—"Yoshi, this is an inverted and invigorating posture. First, lie on your back, with your hands alongside your body and the palms of your hands down on the floor. On the exhalation, visualize while lifting your legs up to a 90 degree angle; try to keep your knees straight, and bend at your waist. Press on the palms, swing the legs slightly over your head, and bring your hands to support your back. Place your hands with the bottom of your palms at the end of each pelvis bone on your back, your fingers alongside the pelvis. Your biceps are on the floor and parallel to it; your forearm is almost perpendicular to the floor, again with hand support at the end of the pelvis bone. Your elbows are fairly close together, just about under your back. Bring your chest as close to your chin as comfortable. (If this feels uncomfortable, you can use a small blanket or folded mat to elevate your shoulders, with your head resting on the floor.) Notice that your legs are bent, allowing your feet to be over or beyond your head. Hold the posture as long as comfortable. The Half Shoulder Stand tones the thyroid glands, regulates metabolism, and improves circulation."

8. *Half Locust*—"Lie down with your chin on the floor. Tuck your arms underneath your body, elbows close together so that your palms are facing up, resting on the upper part of your thighs. Your legs are straight and the tops of your feet are resting on the floor. Visualize doing the posture as I instruct you and while you are doing the posture movement. Now, slowly raise the right leg, keeping your pelvis on your wrists. Take a couple of breaths while holding the leg up. Then, slowly lower your right leg. Again, take a couple of breaths and raise your left leg. Do this two times for each leg, holding for about 10 seconds each time. The Half Locust and the Full Locust exercise the back, pelvis, and abdomen. They tone the sympathetic nervous system and can help the liver."

9. *Full Locust*—"We can do the Full Locust by following the instructions for the Half Locust, but instead of lifting one leg at a time, you lift both legs. The Full Locust is a little more rigorous than lifting one leg at a time. The

possible benefits for the Full Locust are the same as for the Half Locust. Hold for 10 seconds."

10. *Bow*—"Lie face down with your chin or forehead on the floor. Bend your knees and grasp your ankles; visualize your movements as we do them. On inhalation, raise your head, chest, and thighs. Arch your back to allow your entire weight to fall on the abdomen. Keep your arms straight. A less rigorous variation of the Bow is to raise just your thighs while holding your chest on the floor, or raise just your chest while allowing your thighs to stay on the floor. Do the posture three to six times, holding it for about 10 seconds; gradually during your practice work up to holding it for one minute. The Bow gives the same benefits as the Cobra and Locust, plus reducing abdominal fat and toning the pancreas. It relieves chronic constipation and can improve functioning of the liver, kidneys, spleen, stomach, and intestines. The Bow strengthens the back and thighs, aligns the vertebrae, and increases vigor and vitality."

11. *Half Spinal Twist*—"Sit with your knees drawn to your chest and your feet flat on the floor. Stretch out your left leg. Place the right foot on the outside of your left leg—knee or calf. Place the right hand on the floor behind or near your buttocks, fingers pointing away from your body. Bring your left arm around to the outside or the right side of your right knee. Look over the right shoulder; on the exhalation, twist the spine to the right, within your level of comfort. Pause there for a couple of breaths. Then, slowly unwind on an inhalation. Reverse the asana by doing it on the other side. Do the posture one to three times, holding the asana for about 30 seconds and gradually building up to three minutes. The Half Spinal Twist helps the liver, spleen, kidneys, and adrenals. The asana tones the sympathetic nervous system and strengthens the deep and superficial back muscles."

12. *Child or Resting Posture*—"Get on the floor on your hands and knees. Visualize the movement as you take your arms off the floor and push back, to sit on your feet. On exhalation, place your forehead on the floor. Place your arms back along your body or straight out in front of you. Breathe gently, and remember that the Child Pose is a resting and relaxing asana. The Child Posture creates flexibility in the back, hips, knees, and ankles. It is highly restorative. It is supposed to balance the autonomic nervous system and improve digestion; it may be good for headaches and stomachaches. The posture is cooling and relaxing."

13. *Cobbler*—"Sit on the floor with your back straight, your knees bent, and your feet together. Breathe out and let your thighs go down and out as you stretch upward and forward, and lean toward the floor. The Cobbler's or Tailor's Posture is one of the oldest known postures. The outward turn of the thighs at the hip joints is worked and stretched; this movement can relieve stiffness. It is one of the basic yoga poses for many other yoga asanas."

14. *Sponge (Corpse)*—"Lie down on your back. Straighten your legs out and allow your feet to flop to the sides. Place your arms and hands by your sides, but not too close. Place your palms up. Check your back to make sure your spine is as flat as possible. The back of your waist will rise a little, forming a small arch. Breathe and experience your abdomen rising and falling like the tide. A variation of this pose is to place your feet and calves on a chair with the thighs vertical and your sit bones or tailbone and spine flat on the floor; your arms and hands are outstretched. The Sponge Posture creates awareness and relaxation. With focus on your breathing, it brings you to the present moment. We will do the body scan and meditation in this pose."

6. *Instructions for Body Scan*

While Yoshi is in the Sponge Posture, the yoga instructor has him do a body scan. The yoga therapist can use the one by Patel (1993) "Instructions for Deep Muscle Relaxation" or "Body Awareness Relaxation" by LePage (1994). Instructions for both body scans are presented in Chapter 17.

7. *Instructions for Meditation*

Following the body scan, Yoshi is instructed to engage in meditation. Usually, a variation of one of the three meditation procedures described in Chapter 17 is used. Alternatively, one is used that fits the client's and/or the yoga therapist's sensitivity.

8. *Homework*

Yoshi is instructed to select a place for his practice that has enough space for him to do the asanas and is not cluttered by furniture or things that might be distracting. The yoga therapist tells him to make sure the place he selects is well ventilated with a comfortable temperature. Yoshi is told to wear comfortable clothing that will give him full range of movement and to wear no shoes for yoga. He is to set aside a particular time during the day when he will not be interrupted in his practice of yoga. Yoshi is instructed to do the Sun Salutation warm-ups and the Hatha Yoga asanas daily. He is reminded about his breathing while doing each asana and the visualization of the particular asana. If Yoshi wants to keep a log of his yoga practice, he can use forms for self-monitoring the Sun Salutation warm-up and Hatha Yoga asanas like the ones that appear in Chapter 20.

The counselor might expect Yoshi to benefit in several ways from working with a certified yoga instructor or therapist. With respect to his presenting problem, yoga practice would probably help him feel more mentally relaxed but with alert concentration. Also, the discomfort from his

---

LEARNING ACTIVITY 50   *Sun Salutation and Hatha Yoga*

If you have any problem or medical concern, check with your physician before starting. (Note: If you are unable to do the postures, simply visualize doing each warm-up posture and the Hatha Yoga postures. Also, review the benefits as you visualize each posture.) Select a time and place to practice without interruptions or distractions, and with enough room to do all the movements. Wear comfortable clothing, without shoes or socks. Again, if possible, practice the asanas while receiving supervision from a certified yoga instructor or therapist in order to use the proper alignment for each asana.

You might find it helpful to tape-record the instructions for the postures. Do the Sun Salutation without much pause between postures. You want to heat your body.

Then, do the Hatha Yoga asanas, holding each posture at your own level of comfort. The process of being in awareness for each asana is more important than speeding through them. You may wish to start with five or seven asanas, and then extend the number as you gain more practice. Don't forget the rules for breathing when you are doing the asanas. Also, *visualize* the posture and think of the benefits as you do each one. Try to establish a routine for your yoga practice, and practice for four to six weeks to maximize the benefits. Keep a journal or log with brief entries for each practice. The self-monitoring charts for the Sun Salutation and Hatha Yoga asanas in Chapter 20 might be helpful to use.

ulcer should be relieved, and there is a high probability that the ulcer would be almost healed. In addition, the postures, breathing, and meditation associated with Hatha Yoga should provide Yoshi, for his life in general, better insight, understanding, and awareness. He will have better energy among the chakras or energy centers in his body. There will be improvement in his cardiovascular efficiency, his endocrine and immune systems will be strengthened, he will have more strength and flexibility both mentally and physically, and he will feel more energized.

SUMMARY

In this chapter, we described a way to become aware of the depth of your breathing and presented five breathing exercises: diaphragmatic breathing, interval breathing, single-nostril breathing, alternate-nostril breathing, and Kapalabhati (cleansing) breathing. We also described and illustrated 13 yoga postures for warming the body, and 14 Hatha Yoga asanas or postures. Our goals for presenting Hatha Yoga were these: (1) to heighten your awareness about a powerful body therapy and the kinds of clients who might benefit from your referral to a certified yoga therapist; (2) to provide a self-strategy you can use to manage the stress you might experience as a therapist. We included instructions for breathing while moving into a particular posture. We presented a brief overview of the seven energy centers or chakras and described how a yoga therapist might use these energy centers.

We illustrated a chakra meditation that can be used separately or while doing the asanas or postures. We described the use of breathing, meditation, and body scan as essential parts of Hatha Yoga. We alerted you to the possibility of contraindications and adverse effects of the breathing exercises and yoga postures for a small number of people.

Secure the client's informed consent if you instruct him or her in one or more of the breathing exercises. Ask clients to check with their primary care physician if you or they have questions, doubts, or concerns about the appropriateness of yoga for them. You will not train clients in Hatha Yoga, but before making a referral to a certified yoga instructor, you might wish to provide an informed consent document to the client to be completed by the yoga instructor. Also, you should have the client check with an informed physician before beginning the practice of yoga. Again, always consider the contraindications for breathing exercises and yoga for each client before you refer him or her to a yoga instructor.

The interventions presented in this chapter, once considered exotic, are now common complementary procedures among many mental health and physical health practitioners. The breathing exercises and yoga asanas presented here strengthen self-regulation, enhance consciousness, and reinforce the mind/body connection. The two strategies presented in this chapter may be used to complement other strategies presented in previous chapters, and used with systematic desensitization and eye movement desensitization and reprocessing described in the next chapter.

P O S T E V A L U A T I O N

Part One
Objective One asks you to assess your breathing by using the breathing awareness exercise in column one of Table 18-2. See whether you can do this exercise from memory. Use Table 18-2 to check whether you did all the steps.

Part Two
Objective Two asks you to demonstrate the steps for diaphragmatic breathing. Use the checklist at the end of the chapter to assess your performance (page 493).

Part Three
Objective Three asks you to demonstrate the steps for alternate-nostril breathing. Assess your performance by using the checklist at the end of the chapter on page 493.

Part Four
Objective Four asks you to describe the benefits for the Sun Salutation warm-up postures and the Hatha Yoga postures. See whether you can visualize the postures by drawing a stick figure illustrating each one; describe the benefits for each posture. Use the checklist at the end of the chapter to assess your descriptions of the benefits (pages 494–495). Use Tables 18-5 and 18-6 to check your accuracy in visualizing the postures for the Sun Salutation and Hatha Yoga.

Part Five
Objective Five asks you to describe the rules for breathing during movement. Check your answer on page 478 of this chapter.

(continued)

Part Six

Objective Six asks you to describe the location and one potential emotional and cognitive issue for each chakra or energy center. Check your answers using Table 18-4 on pages 475–476.

Checklist for Diaphragmatic Breathing
I. Rationale

_____ 1. If you have doubts, check with your primary care physician before starting.
_____ 2. Think about the purposes and benefits of abdominal breathing.

II. Comfortable position (lounge chair)

_____ 3. Get in a comfortable position so as to experience and feel the movement of the abdomen.
_____ 4. Breathe through your nose. Just notice your breath.
_____ 5. As you inhale, notice how cool the air feels, and how much warmer the air feels when you exhale.

III. Placement of hands; visualize the movement of the diaphragm, and simulate the movement of the diaphragm with the hands

_____ 6. Bend your arms and place the thumbs below your rib cage. Place the rest of the fingers of each hand perpendicular to your body.
_____ 7. Visualize the movement of the diaphragm; it contracts and is drawn downward as you inhale; when you exhale, the air is pushed out of the lungs, and the diaphragm is drawn upward.
_____ 8. Simulate the movement of the diaphragm with your fingers; the fingers straighten out as you inhale, and the fingers are curved and cone shaped.

IV. Diaphragmatic breathing

_____ 9. Continue to breathe diaphragmatically without your hand gestures or movement; place your hands in your lap or beside your body.
_____10. As you breathe diaphragmatically, visualize the movement of the diaphragm as the abdomen rises and falls like the tide.

V. Daily practice and use in stressful situations

_____11. Select a time and place to practice daily for a week.
_____12. When you are in a stress situation, start breathing diaphragmatically.

Checklist for Alternate-Nostril Breathing
I. Rationale

_____ 1. Review the purposes and benefits of alternate-nostril breathing. Again, if you have doubts about doing it, check with your primary care physician.
_____ 2. Think about the overview of the procedure.

II. Position and breathing

_____ 3. Sit up straight in a comfortable position with your spine erect.
_____ 4. Breathe calmly and diaphragmatically.

III. Hand mudra

_____ 5. Make a gentle fist with your preferred hand; extend your thumb, ring finger, and little finger; leave your middle two fingers tucked into your palm.

IV. Close right nostril with hand mudra

_____ 6. Bring your hand mudra in front of your nose. Inhale into both nostrils, then close your right nostril with hand mudra. Make sure the nostril is completely closed. Slowly exhale from the left nostril.

V. Alternating nostrils

_____ 7. Inhale into the left nostril, block it with your hand, and exhale through the right nostril. Continue to do alternate nostril breathing.

VI. Exhale/inhale ratio

_____ 8. After you get accustomed to alternate-nostril breathing, try to have your exhalation two times as long as your inhalation.

VII. Time and visualization

_____ 9. Breathe with alternating nostrils for three to four minutes.
_____10. Visualize the breath as you do this exercise.

VIII. Daily practice

_____11. Establish a time and place where you can practice without interruptions or distractions. Use alternate-nostril breathing as a refreshing distraction from a stress situation. If you wish, keep a journal of your practice sessions.

(continued)

POSTEVALUATION (continued)

Checklist for Hatha Yoga
I. Rationale

_____ 1. Give some purposes for your practice of yoga. You have received permission from your primary care physician to practice yoga.
_____ 2. Overview of four parts of yoga: warm-up postures, breathing, doing and visualizing the postures and meditation.

II. Clothing, Space, and Time

_____ 3. Wear comfortable clothing
_____ 4. Select a space free of distractions/interruptions
_____ 5. Select a convenient time for your yoga practice each day.

III. Breathing During Movement in the Postures

_____ 6. Exhale when bending forward.
_____ 7. Inhale bending backward.
_____ 8. Exhale during twisting movements.
_____ 9. Inhale with movements that expand the chest.
_____10. Exhale with movements that compress the abdominal area and thorax region.
_____11. Generally, exhale 1 to 1.5 times longer than you exhale during the practice of the postures.

IV. 13 Sun Salutation Warm-Up Postures and Benefits (Use the guidelines for breathing and visualize the posture as you do it.)

_____12. Prayer Posture; aligns the body.
_____13. Standing Backward Bend; opens the chest, stimulates the nervous system.
_____14. Standing Forward Bend; soothes the nervous system, quiets the mind.
_____15. Lunge; provides strength and flexibility to lower back and legs.
_____16. Inclined Plane; strengthens shoulders, arms, legs.
_____17. Inverted V (Downward Dog); invigorates; extends the spine and shoulders.
_____18. Modified Push-up; works the lower back, abdominal, and pelvis areas.
_____19. Cobra; exercises the back muscles, spine, and cranial nerves; expands the chest.
_____20. Inverted V or Downward Dog; see 17.
_____21. Lunge; see 15.
_____22. Standing Forward Bend; see 14.

_____23. Standing Backward Bend; see 13.
_____24. Mountain Pose; aligns the body.

V. Hatha Yoga Postures and Benefits (Use the guidelines for breathing and visualize the posture as you do it.)

_____25. Triangle—stretches the abdominal cavity along the sides, massages the liver, stomach, spleen, and kidneys; the intercostal muscles and the diaphragm are massaged.
_____26. Half Moon—flexes the spine laterally; helps flexibility of the shoulder joints; tones the arm muscles; strengthens the muscles of the torso and abdomen; helps the kidneys, liver, spleen, and digestive system.
_____27. Tree—strengthens both legs; improves concentration, balance, and respiration.
_____28. Cobra—exercises back muscles, spine, and cranial nerves; expands the chest.
_____29. Head to Knee—increases level of energy.
_____30. Forward Bending—stretches and exercises most posterior muscles; tones the abdominal viscera.
_____31. Half Shoulder Stand—tones the thyroid glands; helps to regulate metabolism; improves circulation.
_____32. Half Locust—exercises the back, pelvis, and abdomen; tones the sympathetic nervous system; may help the liver.
_____33. Full Locust—provides same benefits as Half Locust.
_____34. Bow—reduces abdominal fat; tones the pancreas; may improve functioning of the liver, kidneys, spleen, stomach, and intestines; strengthens the back and thighs; aligns the vertebrae; increases vigor and vitality.
_____35. Half Spinal Twist—helps the liver, spleen, kidneys, and adrenals; tones the sympathetic nervous system; strengthens the deep and superficial back muscles.
_____36. Child Pose—creates flexibility in the back, hips, knees, and ankles; may balance the autonomic nervous system; improves digestion; is a cooling, relaxing, and restorative posture.
_____37. Cobbler—stretches and relieves stiffness in the thighs and hip joints.
_____38. Sponge—creates awareness and relaxation; focus on your breathing brings you to the present moment.

(continued)

VI. Body Scan

_____39. Scan your body using one of the two instructions presented in Chapter 17.

VII. Meditation

_____40. Use any one of the three meditations presented in Chapter 17, or the Chakra Meditation presented in this chapter.

VIII. Daily Practice

_____41. Make a plan for daily practice of Hatha Yoga with the breathing guidelines, Sun Salutation warm-up, visualization of each posture, body scan, and meditation. You can use the self-monitoring charts for yoga in Chapter 20.

SUGGESTED READINGS

Baxter, K. (1994). *Pain release.* (Videotape). Lenox, Ma.: Kripalu Center.

Bell, L., & Seyfer, E. (1982). *Gentle yoga.* Berkeley, CA: Celestial Arts.

Birch, B. B. (1995). *Power yoga.* New York: Fireside.

Brennan, B. A. (1987). *Hands of light: A guide to healing through the human energy field.* New York: Bantam.

Christensen, A. (1987). *The American Yoga Association beginner's manual.* New York: Fireside.

Cohen, K. (1997). *Healthy breathing.* (Audio tape). Boulder, CO: Sounds True.

Dacher, E. S. (1996). *Whole healing.* New York: Dutton.

Farhi, D. (1996). *The breathing book.* New York: Holt.

Fried, R. (1990). *The breathing connection.* New York: Plenum.

Fried, R. (1993). The role of respiration in stress and stress control: Toward a theory of stress as a hypoxic phenomenon. In P. M. Lehrer & R. L. Woolfolk (Eds.), *Principles and practice of stress management* (2nd ed., pp. 301–331). New York: Guilford.

Feuerstein, G., & Bodian, S. (Eds.). (1993). *Living yoga.* New York: Jeremy P. Tarcher/Perigee Books.

Goleman, D., & Gurin, J. (1993). *Mind/body medicine.* New York: Consumer Report Books.

Hendricks, G. (1995). *Conscious breathing.* New York: Bantam.

Iyengar, B. K. S. (1988). *The tree of yoga.* Boston: Shambhala.

Joy, W. B. (1979). *Joy's way.* New York: Tarcher/Putnam.

Karagulia, S., & Kunz, D. (1989). *The chakras and the human energy field.* London: Quest Books.

Kent, H. (1993). *Yoga made easy.* Allentown, PA: The People's Medical Society.

Kroner, H., Hebing, G., Van-Rijn-Kalkmann, U., & Frenzel, A. (1995). The management of chronic tinnitus: Comparison of a cognitive-behavioral group training with yoga. *Journal of Psychosomatic Research, 39,* 153–165.

Lidell, L., Rabinovitch, N., & Rabinovitch, G. (1983). *The Sivananda companion to yoga.* New York: Fireside.

Monro, R., Ghosh, A. K., & Kalish, D. (1989). *Yoga research bibliography: Scientific studies on yoga and meditation.* Cambridge, England: Yoga Biomedical Trust.

Myss, C. (1996a). *Energy anatomy: The science of personal power, spirituality, and health.* Audio tapes. Boulder, CO: Sounds True Audio.

Myss, C. (1996b). *Anatomy of the spirit: The seven stages of power and healing.* New York: Harmony Books.

Patel, C. (1993). Yoga-based therapy. In P. M. Lehrer & R. L. Woolfolk (Eds.), *Principles and practice of stress management* (2nd ed., pp. 89–137). New York: Guilford.

Pierce, M. D., & Pierce, M. G. (1996). *Yoga for your life.* Portland, OR: Rudra Press.

Radha, S. S. (1987). *Hatha yoga.* Spokane, WA: Timeless Books.

Rama, S., Ballentine, R., & Ajaya, S. (1974). *Yoga and psychotherapy.* Honesdale, PA: The Himalayan International Institute of Yoga Science and Philosophy.

Steward, M. (1994). *Yoga over 50.* New York: Fireside.

Steward, M., & Philips, K. (1992). *Yoga for children.* New York: Simon & Schuster.

Telles, S., Nagarathna, R., & Nagendra, H. R. (1995). Improvement in visual perception following yoga training. *Journal of Indian Psychology, 13,* 30–32.

Ward, S. W. (1994). *Yoga for the young at heart.* Santa Barbara, CA: CAPRA Press.

Weller, S. (1995). *Yoga therapy.* London: Thorsons.

Zi, N. (1994). *The art of breathing.* Glendale, CA: Vivi.

FEEDBACK
POSTEVALUATION

Part One

For Objective One, you were to assess another person's breathing by providing instructions in the orientation awareness for the breathing exercise. Use the steps listed in Table 18-2 to assess your performance.

Part Two

Objective Two asks you to teach another person diaphragmatic and alternate-nostril breathing. Evaluate your instructions by using the Interview Checklists for Diaphragmatic and Alternate-Nostril Breathing on page 493.

Part Three

For Objective Three, evaluate your performance in teaching the Sun Salutation Warm-Up by using the Interview Checklist for the Sun Salutation Warm-Up on page 494.

Part Four

Use the Interview Checklist for Hatha Yoga, Body Scan, Meditation, and Homework on pages 494–495 to assess your instructions in teaching another person these techniques.

CHAPTER 19

SYSTEMATIC DESENSITIZATION*
by Cynthia R. Kalodner

Consider the following case:

> A high school student gets good grades on homework and self-study assignments, but whenever he takes a test, he "freezes." Some days, if he can, he avoids or leaves the class because he feels so anxious about the test, even the day before. When he takes a test, he feels overcome with anxiety; he cannot remember very much, and his resulting test grades are quite low.

This case description reflects an instance in which a person has learned an anxiety response to a situation. According to Bandura (1969), anxiety is a persistent, learned maladaptive response resulting from stimuli that have acquired the capacity to elicit very intense emotional reactions. In addition, the student described in this case felt fear in a situation where there was no obvious external danger (sometimes called a *phobia;* Morris, 1991, p. 161). Further, to some degree, he managed to avoid the nondangerous feared situation (sometimes called a *phobic reaction;* p. 161). Individuals with phobias often require counseling or therapy.

In contrast, in the next two cases, a person is prevented from learning an anxiety response to a certain situation.

> A child is afraid to learn to swim because of a prior bad experience with water. The child's parent or teacher gradually introduces the child to swimming, first by visiting the pool, dabbling hands and feet in the water, getting in up to the knees, and so on. Each approach to swimming is accompanied by a pleasure—being with a parent, having a toy or an inner tube, or playing water games.

> A person has been in a very bad car accident. The person recovers and learns to get back in a car by sitting in it, going for short distances first, often accompanied by a friend or hearing pleasant music on the radio.

In the two descriptions you just read, the situation never got out of hand; that is, it never acquired the capacity to elicit a persistent anxiety response, nor did the persons learn to avoid the situation continually. Why? See whether you can identify common elements in these situations that prevented these two persons from becoming therapy candidates. Go over the last two cases again. Do you notice that, in each one, some type of stimulus or emotion was present that counteracted the fear or anxiety? The parent used pleasurable activities to create enjoyment for the child while swimming; the person in the car took a friend or listened to music. In addition, these persons learned gradually to become more comfortable with a potentially fearful situation. Each step of the way represented a larger or more intense dose of the feared situation.

In a simplified manner, these elements reflect some of the processes that seem to occur in the procedure of *systematic desensitization,* a widely used anxiety-reduction strategy. According to Wolpe (1990, p. 150),

> Systematic desensitization is one of a variety of methods used to break down neurotic anxiety-response habits in piecemeal fashion. . . . After a physiological state inhibiting anxiety has been induced in the patient by means of muscle relaxation, the patient is exposed to a weak anxiety-arousing stimulus for a few seconds. If the exposure is repeated, the stimulus progressively loses its ability to evoke anxiety. Successively stronger stimuli are then similarly treated.

OBJECTIVES

1. Using written examples of four sample hierarchies, identify at least three hierarchies by type (conventional or idiosyncratic).
2. Given a written client case description, identify and describe at least 9 of the following 11 procedural steps of desensitization:
   a. A rationale
   b. An overview

*Cynthia R. Kalodner is director of doctoral clinical training, Department of Counseling Psychology, West Virginia University. Material on eye movement desensitization and reprocessing (EMDR) was contributed by Bill Cormier.

    c. A method for identifying client emotion-provoking situations

    d. A type of hierarchy appropriate for this client

    e. The method of ranking hierarchy items the client could use

    f. An appropriate counterconditioning response

    g. The method of imagery assessment

    h. The method of scene presentation

    i. The method of client signaling during scene presentation

    j. A written notation method to record scene-presentation progress

    k. An example of a desensitization homework task

3. Demonstrate at least 22 out of 28 steps of systematic desensitization in several role-play interviews.

REPORTED USES OF DESENSITIZATION

Systematic desensitization was used widely as early as 1958 by Wolpe. In 1961, Wolpe reported its effectiveness in numerous case accounts, which were substantiated by successful case reports cited by Lazarus (1967). Since 1963, when Lang and Lazovik conducted the first controlled study of systematic desensitization, its use as a therapy procedure has been the subject of numerous empirical investigations and case reports.

    Desensitization has been used to treat various kinds of phobias (Hoffman, Herzog-Bronsky, & Zim, 1994; King, 1993; Kelly & Cooper, 1993) as well as anxiety related to public speaking (Motley & Molloy, 1994), mathematics (Foss & Hadfield, 1993; Schneider & Nevid, 1993), dietary restraint (Pitre & Nicki, 1994), and test taking (Strumpf & Foder, 1993). It also has been used to treat obsessive-compulsive disorder (Cox, Swinson, Morrison, & Lee, 1993). Increasingly, desensitization is being used in behavioral medicine—in areas such as preparing patients for invasive medical procedures (Horne, Vatmanidis, & Careri, 1994; Korth, 1993) (see Table 19-1). The most recent application of desensitization involves posttraumatic stress disorder (PTSD) in which a new application of desensitization called eye movement desensitization and reprocessing (EMDR, Shapiro, 1995) is used. (We describe EMDR in greater detail in the section on variations of desensitization.) Several recent studies have compared desensitization to other treatment interventions. Desensitization was compared to social skills training for 41 institutionalized aggressive children (Schneider, 1991). Both groups improved in cooperative play and reduced aggressive behavior although neither condition

produced changes in teacher ratings. Desensitization has also been compared to stress inoculation training (SIT) in the treatment of math anxiety of college students (Schneider & Nevid, 1993). Both treatments were effective compared to a control condition in significantly reducing math anxiety ratings. Vicarious and in vivo desensitization were compared in reducing fear of water for children three to eight years old (Menzies & Clarke, 1993). The in vivo form was superior to the vicarious form of desensitization.

    Of course, one should not apply desensitization automatically whenever a client reports "anxiety." In some cases, the anxiety may be the logical result of another problem. For example, a person who continually procrastinates on work deadlines may feel anxious. Using this procedure would only help the person become desensitized or numb to the possible consequences of continued procrastination. A more logical approach might be to help the client reduce the procrastination behavior that is clearly the antecedent for the experienced anxiety. This illustration reiterates the importance of thorough problem assessment (Chapters 8 and 9) as a prerequisite for selecting and implementing counseling strategies.

    Generally, desensitization is the superior treatment for phobias and is most appropriate when a client has the capability or the skills to handle a situation or perform an activity but avoids the situation or performs less than adequately because of anxiety (Kleinknecht, 1991). For example, in the case described at the beginning of this chapter, the high school student had the ability to do well and possessed adequate study skills, yet his performance on tests was not up to par because of his response. In contrast, if a person avoids a situation because of skill deficits, then desensitization will be inappropriate and ineffective (Rimm & Masters, 1979). As you may recall from Chapter 12, modeling procedures work very well with many kinds of skill-deficit problems. People with many fears or with general, pervasive anxiety may benefit more from cognitive change strategies (Chapters 15 and 16) or from combinations of strategies in which desensitization may play some role. In addition, some anxiety may be maintained by the client's maladaptive self-verbalizations. In such instances, cognitive restructuring, reframing, or stress inoculation (Chapter 16) may be a first treatment choice or may be used in conjunction with desensitization (see Berman, Miller, & Massman, 1985). Van Hout, Emmelkamp, and Scholing (1994) evaluated the role of negative self-statements during in vivo desensitization of eight adult clients with phobia and agoraphobia. Four of the clients were most improved and four were least im-

TABLE 19-1. Uses of systematic desensitization

Anxiety
Dietary restraint anxiety
Pitre, A., & Nicki, R (1994). Desensitization of dietary restraint anxiety and its relationship to weight loss. *Journal of Behavioral Therapy and Experimental Psychiatry, 25,* 153–154.

Math anxiety
Foss, D., & Hadfield, O. (1993). A successful clinic for the reduction of mathematics anxiety among college students. *College Student Journal, 27,* 157–165.
Schneider, W., & Nevid, J. (1993). Overcoming math anxiety: A comparison of stress inoculation training and systematic desensitization. *Journal of College Student Development, 34,* 283–288.

Speech anxiety
Motley, M., & Molloy, J. (1994). An efficacy test of a new therapy for public speaking anxiety. *Journal of Applied Communication Research, 22,* 48–58.

Test anxiety
Strumpf, J., & Fodor, I. (1993). The treatment of test anxiety in elementary school-age children: Review and recommendations. *Child and Family Behavior Therapy, 15,* 19–42.

Behavioral medicine
Preparation for invasive medical procedures
Horne, D., Vatmanidis, P., & Careri, A. (1994). Preparing patients for invasive medical and surgical procedures: II. Using psychological interventions with adults and children. *Behavioral Medicine, 20,* 15–21.

Korth, E. (1993). Psychological operation-preparation with children: Preparation of a five-year old boy for the amputation of his left leg with play, family, and behavior therapy. *Zeitschrift fur Klinische Psychologie, 22,* 62–76.

Obsessive-compulsive disorder
Cox, B., Swinson, R., Morrison, B., & Lee, P. (1993). Clomipramine, fluoxetine, and behavior therapy in the treatment of obsessive-compulsive disorder: A meta-analysis. *Journal of Behavior Therapy and Experimental Psychiatry, 24,* 149–153.

Phobias
Childhood phobias
King, N. (1993). Simple and social phobias. *Advances in Clinical Child Psychology, 15,* 305–341.

Claustrophobia
Hoffman, S., Herzog-Bronsky, R., & Zim, S. (1994). Dialectical psychotherapy of phobias: A case study. *International Journal of Short Term Psychotherapy, 9,* 229–233.

Panic and agoraphobia
Kelly, C., & Cooper, S. (1993). Panic and agoraphobia associated with a cerebral arteriovenous malformation. *Irish Journal of Psychological Medicine, 10,* 94–96.

Acrophobia
Menzies, R., & Clarke, J. (1995). Individual response patterns, treatment matching, and the effects of behavioural and cognitive interventions for acrophobia. *Anxiety, Stress, and Coping, 8,* 141–160.

---

proved. Results showed that the total frequency of negative self-statements at the beginning, middle, and end of desensitization differentiated best between the most and least improved clients. These authors suggest that, therapeutically, it seems wise to continue desensitization not only until both subjective and physiological measures of anxiety are reduced but also until the frequency of negative self-statements is reduced to zero (Van Hout, Emmelkamp, & Scholing, 1994). Desensitization should not be used when the client's anxiety is nonspecific, or free floating (Foa, Steketee, & Ascher, 1980). Biofeedback, meditation, and muscle relaxation may also be useful supplemental strategies (Meyer & Deitsch, 1996).

At the same time, desensitization should not be overlooked as a possible treatment strategy for client problems that do not involve anxiety. Marquis, Morgan, and Piaget (1973) suggest that desensitization can be used with any conditioned emotion such as anger and loss.

MULTICULTURAL APPLICATIONS OF SYSTEMATIC DESENSITIZATION

To date, the use of desensitization with diverse groups of clients has been limited. Traditional desensitization has been used successfully to treat Japanese undergraduates with test phobia and speech anxiety (Kamimura & SaSaki, 1991). In vivo desensitization was used to treat multiple phobias of a 42-year-old Latina woman (Acierno, Tremont, Last, & Montgomery, 1994). Eye movement desensitization and reprocessing has been used to treat PTSD for a 68-year-old World War II Native American veteran and also several African American Vietnam combat veterans (Thomas & Gafner, 1993; Lipke & Botkin, 1992).

All these cases report the successful use of desensitization—either imagined, in real situations, or through EMDR; Note, however, that although anxiety by and large appears to be a universal phenomenon, "the specific social events antecedent to many emotional states and the responses to those events are, in part, culture-specific" (Kleinknecht, Dinnel, Tanouye-Wilson, & Lonner, 1994, p. 175). Kleinknecht et al. (1994), for example, distinguish between social phobia as described in Western countries and *Taijin Kyofusho* (TKS), a Japanese condition referred to as social phobia—although this condition has been reported in China and Korea as well as Japan. The Western version of social phobia, described in the *Diagnostic and Statistical Manual of Mental Disorders* (DSM-IV; American Psychiatric Association, 1994), begins in childhood or early adolescence and appears to be equal across genders; the anxiety begins with the concern that the person will embarrass *himself* or *herself*. In TKS, which affects more males than females, the concern is that the person will do something that in some way will embarrass or offend *someone else* (Kleinknecht et al., 1994). This difference reflects the variation in cultural values between some Western cultures that focus on independence and individuality ("The squeaky wheel gets the grease") and some of the Asian cultures that focus on interdependence or collectivity ("The nail that sticks out gets pounded down") (Triandis, 1990). As Kleinknecht et al. (1994) observe, different cultural values affect how anxiety is presented by the client, what the client's major concern about the anxiety is, and what cognitive and behavioral patterns the therapist must focus on (p. 178). Therapists who are not aware of these variations may miss the crucial component of an anxiety-related client problem and focus the intervention on a secondary rather than the primary area. In Chapter 16 we discussed some other cultural variations of anxiety.

EXPLANATIONS OF DESENSITIZATION

Although desensitization has enjoyed substantial empirical support, a great deal of controversy surrounds its current status. There is general agreement that desensitization is effective in reducing fears and neurotic behavior. The controversy centers on how and why the procedure works, or what processes surrounding desensitization are responsible for its results. Connor-Greene (1993) found that nonspecific factors such as the therapeutic alliance and the client's expectation for change appeared to account for some of the therapeutic effects of desensitization. Ford (cited in Wolpe, 1990), asserts that there is evidence to the contrary to indicate that expectations of change were not a reliable predictor of long-term improvement.

We briefly summarize here two of the possible theoretical explanations of the desensitization procedure. This discussion should help you understand the theoretical basis of both the counterconditioning and the extinction rationales for implementing desensitization.

Desensitization by Reciprocal Inhibition

In 1958, Wolpe explained the way in which desensitization works by the principle of *reciprocal inhibition*. When reciprocal inhibition occurs, a response such as fear is inhibited by another response or activity that is stronger than and incompatible with the fear response (or any other response to be inhibited). In other words, if an incompatible response occurs in the presence of fear of a stimulus situation, and if the incompatible response is stronger than the fear, desensitization occurs, and the stimulus situation loses its capacity to evoke fear. The reciprocal inhibition theory is based on principles of classical conditioning. For desensitization to occur, according to the reciprocal inhibition principle, three processes are required:

1. A strong anxiety-competing or counterconditioning response must be present. Usually this competing or inhibiting response is deep muscle relaxation. Although other responses (such as eating, assertion, and sexual responses) can be used, Wolpe (1990) believes relaxation is most helpful.
2. A graded series of anxiety-provoking stimuli is presented to the client. These stimulus situations are typically arranged in a hierarchy with low-intensity situations at the bottom and high-intensity situations at the top.
3. Contiguous pairing must occur involving one of these aversive stimulus situations and the competing or counterconditioning response (relaxation). This pairing is usually accomplished by having the client achieve a state of deep relaxation and then imagine an aversive stimulus (presented as a hierarchy item) while relaxing. The client stops imagining the situation whenever anxiety (or any other emotion to be inhibited) occurs. After additional relaxation, the situation is presented again several times.

In recent years, some parts of the reciprocal inhibition principle have been challenged, both by personal opinion and by empirical explorations. There is some doubt that relaxation behaves in the manner suggested by Wolpe—as a response that is inherently antagonistic to anxiety. As Kazdin and Wilcoxon (1976) observe, some research indicates that desensitization is not dependent on muscle relaxation or a hierarchical arrangement of anxiety-provoking stimuli or the pairing of these stimuli with relaxation as an incompatible response (p. 731). These research results have led some people to abandon a reciprocal inhibition explanation for desensitization.

Desensitization by Extinction

Lomont (1965) proposed that extinction processes account for the results of desensitization. In other words, anxiety responses diminish when conditioned stimuli are presented repeatedly without reinforcement. This theory is based on principles of operant conditioning. Wolpe (1990) agrees that desensitization falls within this operational definition of extinction and that extinction may play a role in desensitization. Similarly, Wilson and Davison (1971) have argued that desensitization reduces a client's anxiety level sufficiently that the client gradually approaches the feared stimuli and the fear is then extinguished. Figure 19-1 shows the seven major components of systematic desensitization. A summary of the procedural steps associated with each component is found in the Interview Checklist for Systematic Desensitization at the end of the chapter.

COMPONENTS OF DESENSITIZATION

Treatment Rationale

The purpose and overview given to the client about desensitization are important because they introduce the client to the principles of desensitization. Further, the outcomes of desen-

sitization may be enhanced when the client is given very clear instructions and a positive expectancy set.

Rationale and Overview of Counterconditioning Model. With the counterconditioning model, you would present a rationale that explains how the client's fear or other conditioned emotion can be counterconditioned using desensitization. Your overview of the procedure would emphasize the use of an anxiety-free response to replace the conditioned emotion, the construction of a hierarchy consisting of a graduated series of items representing the emotion-provoking situations, and the pairing of these hierarchy items with the anxiety-free response (such as relaxation).

MODEL DIALOGUE: RATIONALE FOR DESENSITIZATION

Here is an example of a rationale the counselor could use to explain to Joan how desensitization can help her with her fear and avoidance of math class:

Joan, we've talked about how you get very nervous before and during your math class. Sometimes you try to skip it. But you realize you haven't always felt this way about math. You've *learned* to feel this way. There is a procedure called desensitization that can help you replace your tension with relaxation. Eventually, the things about math class you now fear will not be tense situations for you. This procedure has been used very

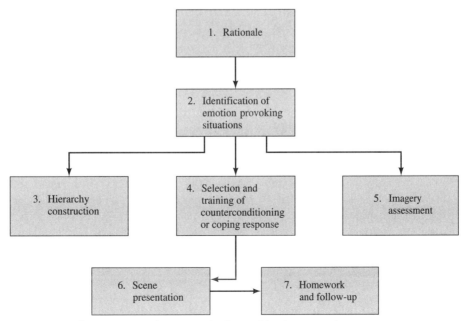

FIGURE 19-1. Components of systematic desensitization procedure

successfully to help other people reduce their fear of a certain situation.

In desensitization, you will learn how to relax. After you're relaxed, I'll ask you to imagine some things about your math class—starting with not too stressful things and gradually working with more stressful things. As we go along, the relaxation will start to replace the anxiety. These stressful situations then will no longer seem so stressful to you.

What questions do you have about this?

Identification of Emotion-Provoking Situations

If the counselor and client have defined the problem thoroughly, there will already be some indications about the dimensions or situations that provoke anxiety (or any other emotional arousal) in the client. However, the counselor and client must be sure to isolate the most crucial situations in which the client should become less anxious or upset (Goldfried & Davison, 1976, p. 114). This is not always an easy task, as first appearances can be deceiving. Wolpe (1990) describes a case in which the initial theme seemed to be fear of social places but was determined actually to be fear of rejection or criticism. In another case, an unexpected source of anxiety concerning impotence was found to be a fear of inflicting physical trauma. Goldfried and Davison recommend that counselors ask themselves and their clients what the consequences may be of desensitization to one thing or another (p. 115).

The emotion-provoking situations must be defined idiosyncratically for each client. Marquis, Ferguson, and Taylor (1980) observe that, even among clients who have the same type of fear or phobia, the specific anxiety-provoking situations associated with the fear can vary greatly.

Counselors can use at least three ways to try to identify past and present situations that are anxiety-provoking to the client. These three methods are the interview assessment, client self-monitoring, and client completion of related self-report questionnaires.

Interview Assessment.   The interview assessment will be similar to the one proposed in Chapters 8 and 9 on problem assessment. The counselor should use leads that will establish the particular circumstances and situations that elicit the conditioned emotion. For instance, does the client feel anxious in all social situations or only those with strangers present? Does the client's anxiety vary with the number of people present? With whether the client is accompanied or alone? Does the client experience more anxiety with people of the same or the other sex? These are examples of the kind of information the interview assess-

ment could provide about a client's anxiety in certain social situations.

Client Self-Monitoring.   In addition to the information obtained during the session, the counselor may obtain even more data by having the client observe and record on a log the emotion-provoking situations as they occur during the week. The client would observe and note what was going on, where, with whom, and when the emotion, such as anxiety, was detected. The client also might rate the level of anxiety felt during the situation on a scale of 1 (low) to 10 (high) or on a scale of 0 (no anxiety) to 100 (panic).

Self-Report Questionnaires.   Some counselors find that additional data about particular emotion-provoking situations can be gained by having the client complete one or more self-report questionnaires. A commonly used questionnaire is the Wolpe and Lang (1964) Fear Survey Schedule (FSS), which contains 87 items that may cause fear or unpleasant feelings. The FSS is available in Appendix C of the Wolpe (1990) text. Also available in the Wolpe book is the Willoughby questionnaire, which lists anxiety responses in interpersonal contexts.

The counselor should persist in this process until specific emotion-provoking situations are identified. Marquis et al. (1973, p. 2) indicate that information gathering is not complete until the counselor knows the factors related to the onset and maintenance of the client's problem and until the client believes that all pertinent information has been shared with the counselor. At this point, the counselor and client are ready to construct a hierarchy.

MODEL DIALOGUE: IDENTIFYING EMOTION-PROVOKING SITUATIONS

Counselor: Joan, we've already discussed some of the situations about your math class that make you feel anxious. What are some of these?

Client: Well, before class, just thinking about having to go bothers me. Sometimes at night I get anxious—not so much doing math homework but studying for tests.

Counselor: OK. Can you list some of the things that happen during your math class that you feel anxious about?

Client: Well, always when I take a test. Sometimes when I am doing problems and don't know the answers—having to ask Mr. Lamborne for help. And, of course, when he calls on me or asks me to go to the board.

Counselor: OK, good. And I believe you said before that you feel nervous about volunteering answers, too.

Client: Right—that, too.

Counselor: And yet these sorts of situations don't seem to upset you in your other classes?

Client: No. And really math class has never been as bad as it is this year. I guess part of it is the pressure of getting closer to graduating and, well, my teacher makes me feel dumb. I felt scared of him the first day of class. And I've always felt somewhat nervous about working with numbers.

Counselor: So some of your fear centers on your teacher, too—and then perhaps there's some worry about doing well enough to graduate.

Client: Right. Although I know I won't do *that* badly.

Counselor: OK, good. You realize that, even with not liking math and worrying about it, you won't get a bad grade.

Client: Not worse than a C.

Counselor: There's one thing I'd like you to do this week. Could you make a list of anything about the subject of math—and your math class— that has happened and has made you nervous? Also, write down anything about math or your class that *could* happen that you would feel anxious about.

Client: OK.

Counselor: Earlier, too, you had mentioned that sometimes these same feelings occur in situations with your parents, so after we work with math class, we'll come back and work on the situations with you and your parents.

Hierarchy Construction

A hierarchy is a list of stimulus situations to which the client reacts with graded amounts of anxiety or some other emotional response (Wolpe, 1990). Hierarchy construction can consume a good deal of interview time because of the various factors involved in constructing an adequate hierarchy. These factors include selection of a type of hierarchy, the number of hierarchies (single or multiple), identification of hierarchy items, identification of control items, and ranking and spacing of items. (See Learning Activity #51.)

Types of Hierarchies. On the basis of the stimulus situations that evoke anxiety (or any emotion to be counterconditioned), the counselor should select an appropriate type of hierarchy in which to cast individual items or descriptions of the aversive situations. Wolpe (1990) distinguishes between conventional and idiosyncratic hierarchies. Conventional hierarchies are unidimensional and incremental in terms of time or distance from a feared object. Most simple phobias can be treated with the development of a conventional hierarchy. On the other hand, idiosyncratic hierarchies are highly individualized as they are based on past conditioning of an individual. The type of hierarchy used will depend on the nature of the client's problem. A conventional hierarchy is developed by using items that represent physical or time dimensions, such as distance from one's house or time before taking an exam. In either case, anxiety seems to vary with proximity to the feared object or situation. Someone who is afraid to leave the house will become more anxious as the distance from the house increases. A client who is anxious about an exam will become more anxious as the exam draws closer.

Here is an example of a conventional hierarchy used with a client who was anxious about taking a test. You will see that the items are arranged according to time:

1. Your instructor announces on the first day of class that the first exam will be held in one month. You know that the month will go quickly.
2. A week before the exam, you are sitting in class and the instructor reminds the class of the exam date. You realize you have a lot of studying to do during the week.
3. You are sitting in the class and the instructor mentions the exam, scheduled for the next class session, two days away. You realize you still have a lot of pages to read.
4. Now it is one day before the exam. You are studying in the library. You wonder whether you have studied as much as everyone else in the class.
5. It is the night before the test. You are in your room studying. You think about the fact that this exam grade is one-third of your final grade.
6. It is the night before the exam—late evening. You have just finished studying and have gone to bed. You're lying awake going over your reading in your mind.
7. You wake up the next morning and your mind flashes that this is exam day. You wonder how much you will remember of what you read the night and day before.
8. It is later in the day, one hour before the exam. You do some last-minute scanning of your lecture notes. You start to feel a little hassled—even a little sick. You wish you had more time to prepare.
9. It is 15 minutes before the class—time to walk over to the classroom. As you're walking over, you realize how important this grade will be. You hope you don't "blank out."

10. You go into the building, stop to get a drink of water, and then enter the classroom. You look around and see people laughing. You think that they are more confident and better prepared than you.

11. The instructor is a little late. You are sitting in class waiting for the teacher to come and pass out the tests. You wonder what will be on the test.

12. The instructor has just passed out tests. You receive your copy. Your first thought is that the test is so long—will you finish in time?

13. You start to work on the first portion of the test. There arc some questions you aren't sure of. You spend time thinking and then see that people around you are writing. You skip these questions and go on.

14. You look at your watch. The class is half over—only 25 minutes left. You feel you have dawdled on the first part of the test. You wonder how much your grade will be pulled down if you don't finish.

15. You continue to work as fast as you can; occasionally you worry about the time. You glance at your watch—five minutes left. You still have a lot of unanswered questions.

16. Time is just about up. There are some questions you had to leave blank. You worry again because this test accounts for one-third of your grade.

An idiosyncratic hierarchy focuses on a particular issue that is associated with anxiety for an individual client. For example, Wolpe (1990) describes several cases of clients who are concerned about illnesses in themselves and others. Items on the "Illness in Others" list include the sight of a physical deformity, the sight of bleeding, automobile accidents, and nurses in uniform. A list of "Illnesses in Self" includes a tight sensation in the head, clammy feet, perspiring hands, dizziness, and rapid breathing. These items are developed with each client individually and pertain to individual fears and concerns. There is no sequential ordering; rather, the counselor and the client must work together to develop a hierarchy that contains the feared items in increasing increments of fear-producing potential. An example of this kind of hierarchy (Wolpe, 1990, p. 167) is provided below:

EXTERNAL SERIES (ILLNESS IN OTHERS)

1. Child with two wasted legs
2. Man walking slowly—short of breath owing to a weak heart
3. Blind man working elevator
4. Child with one wasted leg
5. A hunchback

6. A person groaning with pain
7. A man with a club foot
8. A one-armed man
9. A one-legged man
10. A person with a high temperature owing to a relatively nondangerous disease such as influenza

Number of Hierarchies. Whether you use one or several hierarchies also depends on the client's problem and preferences and on your preferences. Some therapists believe separate hierarchies should be constructed for different themes or different parameters of one theme. Using multiple hierarchies may be less confusing but can require more time for hierarchy construction and presentation. Generally, up to four different hierarchies can be used in an individual desensitization session (Wolpe, 1990).

Identification of Hierarchy Items. The counselor must initiate a method of generating the items for the hierarchy. The client's role in this process is extremely important. Generally, the counselor can ask the client to aid in identifying hierarchy items by interview questions or by a homework assignment. The counselor can question the client about particular emotion-provoking scenes during the interview. However, questioning the client about the scenes should not occur simultaneously with the client's relaxation training. If the client is queried about hierarchy items after engaging in a period of deep relaxation, her or his responses may be altered.

If the client has difficulty responding concretely to interview questions, the counselor can assign homework for item identification. The counselor can give the client a stack of blank 3 × 5 index cards. The client is asked to generate items during the week and to write down each item on a separate note card. Often this homework assignment is useful even with clients who respond very thoroughly to the interview questions. During the week, the client has time to add items that were not apparent during the session.

The counselor should continue to explore and generate hierarchy items until a number have been identified that represent a range of aversive situations and varying degrees of emotional arousal. A hierarchy typically contains 10 to 20 items but occasionally may have more or fewer. Goldfried and Davison (1976) and Marquis et al. (1973) suggest some criteria to use in constructing adequate hierarchy items:

1. Some of the items should represent situations that, if carried out by the client in vivo, are under the client's control (do not require instigation from others).

2. An item must be concrete and specific. Embellishing the item description with sufficient details may help the client

obtain a clear and vivid visualization of the item during scene presentation. As an example, an item that reads "Your best friend disapproves of you" is too general. A more concrete item would be "Your best friend disapproves of your boyfriend and tells you that you are stupid for going out with him."

3. Items should be similar to or represent actual situations the client has faced or may have to face in the future. If dialogue is written into an item, the language used should be adapted to the client.

4. Items selected should reflect a broad range of situations in which the client's fear (or other emotion) does or could occur.

5. Items should be included that reflect all different levels of the emotion, ranging from low to high intensity.

After the hierarchy items are identified, the client and counselor can identify several control items.

Identification of Control Items. A control item consists of a relaxing or neutral scene to which the client is not expected to have any strong emotional reaction. Control scenes are placed at the bottom of the hierarchy and represent a zero or "no anxiety" ranking. Some examples of control items are to "imagine a colored object," "imagine you're sitting in the sun on a day with a completely blue sky," or "imagine you're looking at a field of vivid yellow daffodils." A control item is often used to test the client's ability to visualize anxiety-free material and to give the client a relaxing or pleasant scene to imagine during scene presentation in order to enhance the level of relaxation. After all the hierarchy and control items have been identified, the client can arrange the items in order of increasing emotional arousal through a ranking method.

Ranking and Spacing of Hierarchy Items. The counselor and client work together to identify an appropriate order for the items in the hierarchy. Generally, the client plays the major role in ranking, but the counselor must ensure that the spacing between items is satisfactory. The hierarchy items are ranked in order of increasing difficulty, stress, or emotional arousal. The control items are placed at the bottom of the hierarchy, and each item that represents more difficulty or greater anxiety is placed in a successively higher position in the hierarchy. Items at the top of the hierarchy represent the situations that are most stressful or anxiety-producing for the client.

The counselor should explain how the hierarchy items are arranged before asking the client to rank them. The counselor should also explain the purpose of this type of hierarchy arrangement so the client fully understands the necessity of spending time to rank the items. The counselor can point out that desensitization occurs gradually and that the function of a hierarchy is to identify low-stress items to which the client will be desensitized before higher-stress items. The client's learning to face or cope with a feared situation will begin with more manageable situations first and gradually extend to more difficult situations. The counselor may emphasize that at no point will the client be asked to imagine or cope with a scene or situation that is very stressful before learning to deal successfully with less stressful scenes. This point is often reassuring to an occasional client whose anxiety is so great that the desensitization procedure itself is viewed with great trepidation.

The *sud* scale is used to quantify the items on the hierarchies. A sud is a *subjective unit of disturbance* (Wolpe, 1990) and can be explained to clients in the following way: "Think of the worst anxiety you can imagine and assign it the number 100. Then think of being absolutely calm—that is, no anxiety at all—and call this zero. Now you have a scale of anxiety. At every waking moment of your waking life you must be somewhere between zero and 100. How do you rate yourself at this moment?" (Wolpe, 1990, p. 91).

When a client uses the sud scale to arrange items; each item is assigned a number representing the amount of stress it generates for the client. If the item doesn't generate much stress, the client may assign it 10, 15, or 20 suds. Average amounts of stress might be assigned 35, 40, 45, or 50; whereas 85, 90, 95, and 100 suds represent situations that produce much anxiety or stress.

After the items are arranged according to the assigned suds, the counselor should make sure that no item is separated from the previous item by more than 10 suds; at the high end of the scale, spacing of no more than 5 suds between items is often necessary (Wolpe, 1990). If there are large gaps (greater than 10 or 5 suds), the counselor and client should write additional, intermediate items to fill in. If there are too many items around the same level, particularly at the lower end of the hierarchy, some may be deleted. The sud system makes it easy to determine whether there is too much or too little space between items. Second, the use of the sud scale at this point in desensitization introduces the client to a way to discriminate and label varying degrees of relaxation and tension. Often this kind of labeling system is useful during relaxation training and scene presentation.

Although we have described how to construct a hierarchy for an individual client, hierarchy construction can also be adapted for groups of clients. For some clients, standardized hierarchies may work as well as individualized ones.

It is even possible to have machines conduct the desensitization process. Lang, Melamed, and Hart (1970) described the use of two tape recorders, one with the hierarchy items and one with relaxation items, to treat snake phobia. Others have also described the use of tape-recorded material instead of the counselor's participation (Denholtz & Mann, 1974; Kahn & Baker, 1968). The use of tapes may be especially helpful in cases when the client is too anxious to relax in the presence of the therapist (Wolpe, 1990). Recent developments also present hierarchies on computers using "virtual realities" (Nelissen, Muris, & Merckelbach, 1995).

MODEL DIALOGUE: HIERARCHY CONSTRUCTION USING SUD SCALE

Counselor: Hi, Joan. I see you brought your list with you. That's great, because today we're going to work on a list that's called a hierarchy, which is like a staircase or a ladder. In your case, it will center on the theme or the idea of math anxiety. It's a list of all the situations about math that are anxiety-producing for you. We'll list these situations and then I'll ask you to assign each one a sud value. A sud is a number that describes how anxious you feel. The higher the sud, the more anxious you feel. It works like this. Think of the worst anxiety you can imagine and assign it the number 100. Then think of being absolutely calm—that is, having no anxiety at all—and call this zero. Now you have a scale of anxiety. At every waking moment of your waking life you must be somewhere between zero and 100. Now we are going to go through your list and assign suds to each item. Does this seem clear?

Client: Yes. Actually I did something like that this week in making my list, didn't I?

Counselor: Right, you did. Now what we want to do is take your list, add any other stressful situations that aren't on here, and make sure each item on this list is specific. We may need to add some details to some of the items. The idea is to get a close description of the way the situation actually is or actually does or could happen. Let's take a look at your list now.

1. Sitting in English thinking about math class
2. On way to math class
3. At home, doing math homework
4. At home, studying for a math test
5. In math class, teacher giving out test
6. In math class, taking test
7. In math class, teacher asking me question
8. In math class, at board, having trouble
9. In math class, working problems at desk, not knowing some answers
10. Asking teacher for help
11. Volunteering an answer
12. Getting test or assignment back with low grade
13. Hearing teacher tell me I'll flunk or barely pass
14. Doing anything with numbers, even outside math class, like adding up a list of numbers
15. Talking about math with someone

Counselor: Well, it looks like you've really worked hard at putting down some math situations that are stressful for you and indicating just how stressful they are. OK, let's go over this list and fill in some details. For each item here, can you write in one or two more details about the situation? What exactly happens that you feel nervous about? For instance, when you're sitting in English, what is it you're thinking that makes you nervous?

Client: OK, I see what you mean.

Counselor: Let's go over each item here, and as you tell me the details, I'll jot these down. [This step proceeds until a concrete description is obtained for each item. Counselor checks them to see whether items meet necessary criteria, which, with added details, these do. The criteria are these: some items are under client's control; items are concrete; items represent past, present, or future anxiety-provoking scenes; items sample a broad range of situations; items represent varying levels of anxiety.]

Counselor: What else can you think of about math that is or could be stressful?

Client: Nothing right now. Not everything on my list has happened, but like, if my teacher did tell me I was going to flunk, that would be very tense.

Counselor: You've got the idea. Now can you think of something not related to math that would be pleasant or relaxing for you to think about—like skiing down a slope or lying on the beach?

Client: Well, what about sitting in front of a campfire roasting marshmallows?

Counselor: Good. Now later on, as we proceed, I might ask you to relax by imagining a pleasant scene. Then you could imagine something like that.

Client: OK.

Counselor: I'd like you to take a look at the items we've listed on these cards and assign a sud value to each item. We can start with the pleasant item having to do with the campfire at the bottom of the list and assign a sud score of 0. Now we have to assign a sud value to each of the other situations we described.

Client: OK, so the next thing should be something that only bothers me a little bit, right? Then, I'll pick the one about me sitting at home and doing problems that are pretty easy. A sud score could be 10.

Counselor: Great! You have the right idea. Which one would be next on the hierarchy?

Client: I think that when I think about going to math class, I get anxious, so I'll pick the one about being in English class and knowing that math class is next, and I'll give it a sud of 20.

Counselor: OK, now what would be next?

Client: On the way to math class. Going in and having the teacher look at homework. This makes me more anxious. I'll give it a sud of 30.

Counselor: You are doing great. We have more items to rate. Which would be next?

Client: Talking to my friend about a test that is coming up. Worrying that I might not pass it. This is a sud of about 35. And, the next one would be the one about seeing a long list of numbers and having to see if the addition is right. That would be about 40 on the sud scale.

Counselor: OK. What would be next?

Client: I get anxious thinking about being in math class, at my desk and having trouble doing my work. That one would be a sud of 50. The next one would also be in math class and would involve not being able to do some of the problems. I'd have to ask my teacher for help. That would be 60.

Counselor: Looks like we are about half done. Let's finish the list and continue to assign sud values to each item. [The process continues until each of the 16 cards has been assigned a sud value.] OK, now it seems as if you have ordered each of these situations in terms of the sud value. We can see that there are no large gaps between items, so we can begin with this hierarchy. Later we can move items around or reassign sud values, if we need to do so. I'm going to lay each card out to see what you've got here, starting at the bottom.

Card 1: Sitting in front of a campfire on a cool night
(sud 0)   with friends, singing songs and roasting marshmallows [control item].

Card 2: Sitting in my room at my desk doing routine
(sud 10)   math homework over fairly easy material.

Card 3: Sitting in English about ten minutes before
(sud 20)   the bell. Thinking about going to math class next and wondering if I can hide or if I'll get called on.

Card 4: Walking down the hall with a couple of
(sud 30)   classmates to math class. Walking in the door and seeing the teacher looking over our homework. Wondering how I did.

Card 5: A girlfriend calls up and talks about our up-
(sud 35)   coming test in math—wonder if I'll pass it.

Card 6: Seeing a big list of numbers, like on a store
(sud 40)   receipt, and having to check the total to make sure it's OK.

Card 7: In math class, sitting at my desk; having to
(sud 50)   work on problems and coming across some that are confusing. Don't have much time to finish.

Card 8: Working on problems at my desk. I'm
(sud 60)   stumped on a couple. Nothing I try works. Having to go up and ask Mr. Lamborne for help. He tries to do it for me; I feel dumb.

Card 9: Sitting in my room at home the night before
(sud 65)   a big math test; studying for the test and wondering if I'll blank out on it.

Card 10: In math class taking a test and coming across
(sud 75)   some problems I don't know how to do.

Card 11: Waiting to get a test or an assignment back
(sud 80)   and worrying about a low grade.

Card 12: Sitting in math class and the teacher asks for
(sud 85)   the answer; raising my hand to volunteer it; wonder if it's right.

Card 13: Sitting in math class waiting for a big test to
(sud 90)   be passed out. Wondering what's on it and if I'll be able to remember things.

Card 14: Sitting in math class and suddenly the teacher
(sud 95)   calls on me and asks me for an answer. I feel unprepared.

Card 15: Sitting in math class and the teacher sends
(sud 98)   me to the board. I'm at the board trying to work a problem in front of the class. I'm getting stuck.

Card 16: The teacher calls me in for a conference after
(sud   school. Mr. Lamborne is telling me I'm in
100)   big trouble and barely passing. There's a good chance I could flunk math.

Counselor: OK, now it seems like each of these items represents a somewhat more stressful situation. Do you feel that there are any large jumps between

LEARNING ACTIVITY 51   *Hierarchy Construction*

This learning activity is designed to give you some practice in constructing hierarchies. You can do this activity by yourself or with another person.

Part One: Conventional Hierarchy

Think of for yourself, or have your partner identify, a situation you fear and avoid. This should be a situation in which the fear increases as the distance or time proximity toward the feared object or situation gets closer. For example, you might fear and avoid approaching certain kinds of animals or high places (distance). Or you might get increasingly anxious as the time before an exam, a speech, or an interview diminishes (time). For this situation, identify the specific situations that are anxiety provoking. Try to identify all the relevant parameters of the situation.

For example, does your anxiety vary if you're alone or with another person, if you're taking a midterm or a quiz, if you're speaking before a large or a small group? List each anxiety-provoking situation that could be a hierarchy item

on a separate index card. Also list one control (pleasant) item on a card. After you or your partner believes all the relevant items are listed, take the cards and assign a sud value to each item. The control item will be at the bottom, followed by items that evoke successively greater anxiety. Check to make sure there are no differences between sud values that exceed 10 units.

Part Two: Idiosyncratic Hierarchy

See whether you or your partner can identify a situation about which you have painful or unpleasant memories. Such situations might include loss of a prized object, loss of a job or friend, or termination of a close relationship. Generate emotion-provoking situations associated with pleasant memories and unpleasant memories. List each situation on a separate card. When all the items are identified and listed, assign sud values to each. Check the sud values to make sure there are no differences between sud values that exceed 10 units.

---

items—like going from a very low-stress situation to a higher-stress one suddenly?

Client (looks over list): No, I don't think so.

Counselor: OK, we'll stick with this list and this order for now. Of course, this is tentative. Later on, if we feel something needs to be moved around or added, we will do so.

Selection and Training of Counterconditioning Response

According to the principles of reciprocal inhibition and counterconditioning, for desensitization to take place, the client must respond in a way that inhibits (or counterconditions) the anxiety or other conditioned emotion. The counselor selects, and trains the client to use, a response that can be considered either an alternative to anxiety or incompatible with anxiety.

Selection of a Response.   The counselor's first task is to select an appropriate counterconditioning response for the client to use. Typically, the anxiety-inhibiting or counterconditioning response used in desensitization is deep muscle relaxation (Wolpe, 1990). Muscle relaxation has some advantages. Levin and Gross (1985) found that relaxation height-

ens the vividness of imagery. Also, as you may remember from Chapter 17, its use in anxiety reduction and management is well documented. Wolpe (1990) prefers muscle relaxation because it doesn't require any sort of motor activity to be directed from the client toward the sources of anxiety (p. 154). Muscle relaxation is easily learned by most clients and easily taught in the interview. It is also adaptable for daily client practice. However, an occasional client may have difficulty engaging in relaxation. Further, relaxation is not always applicable to in vivo desensitization, in which the client carries out rather than imagines the hierarchy items.

When deep muscle relaxation cannot be used as the counterconditioning response, the counselor may decide to proceed without this sort of response or to substitute an alternative response. In some cases, clients have been desensitized without relaxation. However, with a client who is very anxious, it may be risky to proceed without any response to counteract the anxiety.

If muscle relaxation is not suitable for a client, guided imagery (Chapter 13), meditation (Chapter 17), and coping thoughts (Chapter 16) may be reasonable substitutes that are practical to use in the interview and easy to teach. For example, if the counselor selects guided imagery, the client can focus on pleasant scenes during desensitization. If meditation is selected, the client can focus on breathing and counting. In the case of coping thoughts, the client can whisper or subvocalize coping statements.

Explanation of Response to the Client. The client will be required to spend a great deal of time in the session and at home learning the response. Usually a large amount of client time will result in more payoffs if the client understands how and why this sort of response should be learned.

In emphasizing that the response is for counterconditioning, the counselor can explain that one of the ways desensitization can help the client is by providing a substitute for anxiety (or other emotions). The counselor should emphasize that this substitute response is incompatible with anxiety and will minimize the felt anxiety so that the client does not continue to avoid the anxiety-provoking situations.

Training in the Response. The counselor will need to provide training for the client in the particular response to be used. The training in muscle relaxation or any other response may require at least portions of several sessions to complete. The training in a counterconditioning response can occur simultaneously with hierarchy construction. Half the interview can be used for training; the rest can be used for hierarchy construction. Remember, though, that identifying hierarchy items should not occur simultaneously with relaxation. The counselor can follow portions of the interview protocol for cognitive restructuring (Chapter 15) for training in coping statements; the interview checklists for guided imagery (Chapter 13), muscle relaxation, and meditation (Chapter 17) can be used to provide training in these responses.

Before and after each training session, the counselor should ask the client to rate the felt level of stress or anxiety. This is another time the sud scale is very useful. The client can use the 0-to-100 scale and assign a numerical rating to the level of anxiety. Generally, training in the counterconditioning response should be continued until the client can discriminate different levels of anxiety and can achieve a state of relaxation after a training session equivalent to 10 or less on the 100-point sud scale. If, after successive training sessions, the client has difficulty using the response in a nonanxious manner, another response may need to be selected.

After the client has practiced the response with the counselor's direction, daily homework practice should be assigned. An adequate client demonstration of the counterconditioning response is one prerequisite for actual scene presentation. A second prerequisite involves a determination of the client's capacity to use imagery.

MODEL DIALOGUE: SELECTION OF AND TRAINING IN COUNTERCONDITIONING RESPONSE

Counselor: Joan, perhaps you remember that when I explained desensitization to you, I talked about replacing anxiety with something else, like relaxation. What I'd like to do today is show you a relaxation method you can learn. How does that sound?

Client: OK, is it like yoga?

Counselor: Well, it's carried out differently from yoga, but it is a skill you can learn with practice and it has effects similar to those of yoga. This is a process of body relaxation. It involves learning to tense and relax different muscle groups in your body. Eventually you will learn to recognize when a part of you starts to get tense, and you can signal to yourself to relax.

Client: Then how do we use it in desensitization?

Counselor: After you learn this, I will ask you to imagine the items on your hierarchy—but only when you're relaxed, like after we have a relaxation session. You'll be imagining something stressful, only you'll be relaxed. After you keep working with this, the stressful situations become less and less anxiety provoking for you.

Client: That makes sense to me, I think. The relaxation can help the situation to be less tense.

Counselor: Yes, it plays a big role—which is why I consider the time we'll spend on learning the relaxation skill so important. Now, one more thing, Joan. Before and after each relaxation session, I'll ask you to tell me how tense or how relaxed you feel at that moment. You can do this by using a number from 0 to 100—0 would be total relaxation and 100 would be total anxiety or tenseness. How do you feel right now, on that scale?

Client: Well, not totally relaxed, but not real tense. Maybe around a 30.

Counselor: OK. Would you like to begin with a relaxation-training session now?

Client: Sure. [Training in muscle relaxation following the interview checklist presented in Chapter 17 is given to Joan. An abbreviated version of this is also presented in the model dialogue on scene presentation later in this chapter.]

Imagery Assessment

The typical administration of desensitization relies heavily on client imagery. The relearning (counterconditioning) achieved in desensitization occurs during the client's visualization of the hierarchy items. This, of course, assumes that imagination of a situation is equivalent to a real situation and that the learning that occurs in the imagined situation generalizes to the real situation. M. J. Mahoney (1974) notes evidence that there may be considerable variability in the degree to which these assumptions about imagery really operate. Still, if desensitization is implemented, the client's capacity to generate images is vital to the way this procedure typically is used.

Explanation to the Client.   The counselor can explain that the client will be asked to imagine the hierarchy items as if the client were a participant in the situation. The counselor might say that imagining a feared situation can be very similar to actually being in the situation. If the client becomes desensitized while imagining the aversive situation, then the client will also experience less anxiety when actually in the situation. The counselor can suggest that because people respond differently to using their imaginations, it is a good idea to practice visualizing several situations.

Assessment of Client Imagery.   The client's capacity to generate clear and vivid images can be assessed by use of practice (control) scenes or by a questionnaire, as described in Chapter 13. Generally, it is a good idea to assess the client's imagery for desensitization at two times—when the client is deliberately relaxed and when the client is not deliberately relaxed. According to Wolpe (1990), imagery assessment of a scene under relaxation conditions serves two purposes. First, it gives the therapist information about the client's ability to generate anxiety-free images. Second, it suggests whether any factors are present that may inhibit the client's capacity to imagine anxiety-free material. For example, a client who is concerned about losing self-control may have trouble generating images of a control item (Wolpe, 1990). After each visualization, the counselor can ask the client to describe the details of the imagined scene aloud. Clients who cannot visualize scenes may have to be treated with an alternative strategy for fear reduction that does not use imagery, such as participant modeling or in vivo desensitization (see also page 517).

Criteria for Effective Imagery.   In the typical administration of desensitization, the client's use of imagery plays a major role. A client who is unable to use imagery may not benefit from a hierarchy that is presented in imagination. From the results of the client's imagery assessment, the counselor should determine whether the client's images meet the criteria for effective therapeutic imagery. These four

criteria have been proposed by Marquis et al. (1973, p. 10):

1. The client must be able to imagine a scene concretely, with sufficient detail and evidence of touch, sound, smell, and sight sensations.
2. The scene should be imagined in such a way that the client is a participant, not an observer.
3. The client should be able to switch a scene image on and off upon instruction.
4. The client should be able to hold a particular scene as instructed without drifting off or changing the scene.

If these or other difficulties are encountered during imagery assessment, the counselor may decide to continue to use imagery and provide imagery training or to add a dialogue or a script; to present the hierarchy in another manner (slides, role plays, or actual experience); or to terminate desensitization and use an alternative therapeutic strategy. Whenever the client is able to report clear, vivid images that meet most of the necessary criteria, the counselor can initiate the "nuts and bolts" of desensitization—presentation of the hierarchy items.

MODEL DIALOGUE: IMAGERY ASSESSMENT

The following assessment should be completed two times: once after a relaxation session and once when Joan is not deliberately relaxed.

Counselor: Joan, I will be asking you in the procedure to imagine the items we've listed in your hierarchy. Sometimes people use their imaginations differently, so it's a good idea to see how you react to imagining something. Could you just sit back and close your eyes and relax? Now get a picture of a winter snow scene in your mind. Put yourself in the picture, doing something. [Pauses.] Now, can you describe exactly what you imagined?

Client: Well, it was a cold day, but the sun was shining. There was about a foot of snow on the ground. I was on a toboggan with two friends going down a big hill very fast. At the bottom of the hill we rolled off and fell in the snow. That was cold!

Counselor: So you were able to imagine sensations of coldness. What colors do you remember?

Client: Well, the hill, of course, was real white and the sky was blue. The sun kind of blinded you. I had on a bright red snow parka.

Counselor: OK, good. Let's try another one. I'll describe a scene and ask you to imagine it for a certain amount of time. Try to get a clear image as soon as I've de-scribed the scene. Then, when I say "Stop the image," try to erase it from your mind. OK, here's the scene. It's a warm, sunny day with a good breeze. You're out on a boat on a crystal-clear lake. OK—now imagine this—put in your own details. [Pauses.] OK, Joan, stop the image. Can you tell me what you pictured? [Joan describes the images.] How soon did you get a clear image of the scene after I described it?

Client: Oh, not long. Maybe a couple of seconds.

Counselor: Were you able to erase it when you heard me say *stop?*

Client: Pretty much. It took me a couple of seconds to get completely out of it.

Counselor: Did you always imagine being on a boat, or did you find your imagination wandering or revising the scene?

Client: No, I was on the boat the entire time.

Counselor: How do you feel about imagining a scene now?

Client: These are fun. I don't think imagination is hard for me anyway.

Counselor: Well, you do seem pretty comfortable with it, so we can go ahead.

Joan's images meet the criteria for effective imagery: The scenes are imagined concretely; she sees herself in a scene as a participant; she is able to turn the image on and off fairly well on instruction; she holds a scene constant; there is no evidence of any other difficulties.

Hierarchy Scene Presentation and Signaling Method

Scenes in the hierarchy are presented after the client has been given training in a counterconditioning response and after the client's imagery capacity has been assessed. Each scene presentation is paired with the counterconditioning response so that the client's anxiety (or other emotion) is countercon-ditioned, or decreased. There are different ways to present scenes to the client. Wolpe (1990) described the following method for presenting scenes to clients. The person is instructed to imagine the scene as described by the counselor and is told to raise an index finger to indicate to the counselor that the scene is clear. The counselor allows the client to hold the image of the scene for five to seven seconds and terminates it by saying "Stop the scene." Next, the counselor asks the client how much anxiety was generated in terms of suds; the counselor might ask, "How much did imagining this scene increase your sud level?"

In the signaling method suggested by Wolpe (1990), the client raises a finger to inform the counselor that the scene is clear. Other signaling methods have been described else-where (Marquis et al., 1973).

Format of a Scene-Presentation Session. Scene presen-tation follows a fairly standardized format. Each scene-presentation session should be preceded by a training session involving the designated counterconditioning response. As you will recall, the idea is to present the hierarchy items concurrently with some counterconditioning or coping re-sponse. For example, the counselor can inform the client that the first part of the session will be a period of relaxation, after which the counselor will ask the client to imagine a scene from the hierarchy. Each desensitization session should begin with a brief period of muscle relaxation, meditation, or guided imagery. The client's relaxation rating following this period should be 10 or less on the 100-point sud scale before the counselor presents a hierarchy item.

At this point, the counselor begins by describing a hierarchy item to the client and instructing the client to evoke the scene in imagination. The initial session begins with the first (least anxiety-provoking) item in the hierarchy. Succes-sive scene presentations always begin with the last item successfully completed at the preceding session. This helps to make a smooth transition from one session to the next and checks on the client's learning retention. Starting with the last successfully completed item may also prevent spontane-ous recovery of the anxiety response (Marquis et al., 1973, p. 11). Sometimes relapse between two scene-presentation sessions does occur, and this procedure is a way to check for it.

In presenting the item, the counselor should describe it and ask the client to imagine it. Usually the counselor presents an item for a specified amount of time before asking the client to stop the image. The duration of a scene is usually five to seven seconds. There are reasons to vary the amount of time that a scene is imagined by a client. First, if the client indicates that he or she is experiencing strong anxiety, the scene should be stopped. If the counselor believes the client might have a large increase in anxiety with a particular scene, the counselor might limit the time imag-ining that scene to less than two seconds. Wolpe (1990) indicates that early presentations of scenes tend to be shorter and that later ones tend to be longer.

If the client holds the scene for the specified duration and does not report any tension, the counselor can instruct the client to stop the scene and to take a little time to relax. This relaxation time serves as a breather between item presentations. During this time, the counselor can cue the onset of relaxation with descriptive words such as "let all your muscles relax" or with the presentation of a control item. There is no set time for a pause between items. Generally a pause of 10 to 30 seconds is sufficient, although some clients may need as much as two or three minutes (Wolpe, 1990). It is suggested that the counselor check with the client regarding the sud level between scene presentations.

If the client experienced anxiety during the visualization, the counselor will instruct the client to imagine the scene again and check on the level of anxiety generated. It is often necessary to present a scene three or four times until the presentation of the scene does not lead to an increase in anxiety. Wolpe (1990) indicates that at times as many as ten presentations are necessary to reduce the anxiety to zero. Anxiety must be reduced to zero suds with each item before the next item in the hierarchy is presented.

An item that continues to elicit anxiety after three presentations may indicate some trouble and a need for adjustment. Continued anxiety for one item may indicate a problem in the hierarchy or in the client's visualization. There are at least three things a counselor can try to alleviate continual anxiety resulting from the same item: a new, less anxiety-provoking item can be added to the hierarchy; the same item can be presented to the client more briefly; or the client's visualization can be assessed to determine whether the client is drifting from or revising the scene.

The counselor should be careful to use standardized instructions at all times during scene-presentation sessions. Standardized instructions are important regardless of whether the client signals anxiety or reports a high or a low anxiety rating on the sud scale. Rimm and Masters (1979) observe that a counselor can inadvertently reinforce a client for not signaling anxiety by saying "Good." The client, often eager to please the counselor, may learn to avoid giving reports of anxiety because these are not similarly reinforced.

Each scene-presentation session should end with an item that evokes no anxiety or receives a low sud rating, as the last item of a series is well remembered. At times, the counselor may need to return to a lower item on the hierarchy so that a non-anxiety-provoking scene can end the session. Systematic desensitization sessions should be terminated after 15 to 30 minutes (Wolpe, 1990), although some clients may be able to work longer, especially after the first few sessions have been completed. Limiting the presentation may allow time for up to 10 scene presentations, and at advanced stages of the process, some clients will be able to go through more than 30 scenes in a single session. A session may be terminated sooner if the client seems restless. Desensitization requires a great deal of client concentration, and the counselor should not try to extend a session beyond the client's concentration limits.

Identify Notation Method.   Desensitization also requires some concentration and attention on the counselor's part. Just think about the idea of conducting perhaps four or five scene-presentation sessions with one client and working with one or more hierarchies with 10 to 20 items per hierarchy! The counselor has a great deal of information to note and recall. Most counselors use some written recording method during the scene-presentation sessions. There are several ways to make notations of the client's progress in each scene-presentation session. We describe two. These methods are only suggestions; you may discover a notation system that parallels more closely your own procedural style of desensitization.

Marquis et al. (1973) use a "Desensitization Record Sheet" to record the hierarchy item numbers and the anxiety or sud rating associated with each item presentation. Their record sheet is shown in Figure 19-2, with a sample notation provided at the top of the sheet.

Goldfried and Davison (1994) use a notation system written on the 3 × 5 index card that contains the description of the hierarchy item and the item number. Under the item description is space for the counselor to note the duration of the item presentations and whether item presentation elicited anxiety. An example is presented in Figure 19-3. In this example, the numbers refer to the time in seconds that the client visualized each presentation of the item. The plus sign indicates a no-anxiety or low-sud visualization, and the minus sign indicates an anxiety or high-sud visualization. Note that there were two successive no-anxiety visualizations (+5 and +7) before the item was terminated.

MODEL DIALOGUE: SCENE PRESENTATION

Counselor: Joan, after our relaxation session today, we're going to start working with your hierarchy. I'd like to explain how this goes. After you've relaxed, I'll ask you to imagine the first item on the low end of your hierarchy—that is, the pleasant one. It will help you relax even more. Then I'll describe the next item. I will show you a way to let me know if you feel any anxiety while you're imagining it. If you do, I'll ask you to stop or erase the image and to relax. You'll have some time to relax before I give you an item again. Does this seem clear?

Client: I believe so.

Counselor: One more thing. If at any point during the time you're imagining a scene you feel nervous or anxious

Subject's Name: _Jane Doe_

Theme of Hierarchy: _Criticism_

Time needed to relax at the beginning of the session: _15 minutes_

Time needed to visualize the scene presented: _10 sec./8 sec./9 sec./5 sec._

Date and Total Time Spent in Session	Item Hierarchy Number	Anxiety + or − or Sud Rating		Time Between Items	Comments, Observations, Changes in Procedure, or Other Special Treatment
7-14-97 45 minutes	4	+ 8 + 20	- 15 + 30	60 sec./ 60 sec./ 30 sec./ 60 sec./	

FIGURE 19-2. Desensitization record sheet. *SOURCE:* From *A Guidebook for Systematic Desensitization,* 3rd Ed., by J. Marquis, W. Morgan, and G. Piaget, 1973. Veterans' Workshop, Palo Alto, CA. Reprinted with permission of the authors.

Item No. 6
Date 7-14-97

ITEM DESCRIPTION
You are walking to class thinking about the upcoming exam. Your head feels crammed full of details. You are wondering whether you've studied the right material.

+5 −7 +5 +7

FIGURE 19-3. Notation card. *SOURCE:* Adapted from *Clinical Behavior Therapy* (expanded edition), by M. R. Goldfried and G. C. Davison. Copyright © 1994 by Wiley Interscience. Reprinted by permission of the authors.

about it, just raise your finger. This will signal that to me. I'll also ask you to report the sud level that you feel after each scene is imagined.

Client: OK.

Counselor: Just to make sure we're on the same track, could you tell me what you believe will go on during this part of desensitization?

Client: Well, after relaxation you'll ask me to imagine an item at the bottom of the hierarchy. If I feel any anxiety, I'll raise my finger and you'll ask me to erase the

scene and relax. And I'll also tell you how anxious I feel then.

Counselor: Good. And even if you don't signal anxiety after a little time of imagining an item, I'll tell you to stop and relax. This gives you a sort of breather. Ready to begin?

Client: Yep.

Counselor: OK, first we'll begin with some relaxation. Just get in a comfortable position and close your eyes and relax. . . . Let the tension drain out of your body. . . .

Now, to the word *relax,* just let your arms go limp. . . . Now relax your face. . . . Loosen up your face and neck muscles. . . . As I name each muscle group, just use the word *relax* as the signal to let go of all the tension. . . . Now, Joan, you'll feel even more relaxed by thinking about a pleasant situation. . . . Just imagine you're sitting around a campfire on a cool winter night. . . . You're with some good friends, singing songs and roasting marshmallows. [Presentation of item 1, or control item]. [Gives Joan about 30 seconds for this image.] Now I'd like you to imagine you're sitting in your room at your desk doing math homework that's pretty routine and is fairly easy. [Presentation of item 2 in hierarchy]. [Counselor notes duration of presentation on stopwatch. Counselor allows the scene to be imagined for five to seven seconds and watches to see whether Joan signals any anxiety by raising her finger. Joan does not respond, indicating that she is not feeling anxious while imagining this item.] OK, Joan, stop that image and erase it from your mind. Just concentrate on feeling very relaxed. [Pauses 10 to 30 seconds.] Now I'd like you to again imagine you're in your room sitting at your desk doing math homework that is routine and fairly simple. [Second presentation of item 2] [Counselor allows the scene to be imagined again for five to seven seconds and watches to see whether Joan signals any anxiety by raising her finger. Joan does not respond, indicating that she is not feeling anxious while imagining this item.] OK, Joan, now just erase the image from your mind and relax. Let go of all your muscles. [Pause 10 to 30 seconds. As two successive presentations of this item did not elicit any anxiety, the counselor will move on to item 3.]

Now I'd like you to imagine you're sitting in English class. It's about ten minutes before the bell. Your mind drifts to math class. You wonder if anything will happen, like getting called on. [Presentation of item 3 in hierarchy]. [Counselor notes duration of presentation with stopwatch. Counselor allows the scene to be imagined again for five to seven seconds and watches to see whether Joan signals any anxiety by raising her finger. This time, Joan raises her finger at six seconds, indicating that she is feeling anxious while imagining this item.] Joan, what is the sud level that you are feeling right now?

Client: I feel about 20 suds thinking about this.

Counselor: OK, Joan, just erase that image from your mind. . . . Now relax. Let relaxation flood your body. . . . Think again about being in front of a campfire. [Pauses for 10 to 30 seconds for relaxation.]

Now I'd like you to again imagine you're sitting in English class. It's almost time for the bell. You think about math class and wonder if you'll be called on. [Second presentation of item 3 in the hierarchy]. [Counselor notes duration with stopwatch. Counselor allows the scene to be imagined again for five to seven seconds and watches to see whether Joan signals any anxiety by raising her finger. Joan does not respond, indicating that she is not feeling anxious while imagining this item.] OK, Joan, now just erase that image and concentrate on relaxing. [Pauses 10 to 30 seconds.] OK, again imagine yourself sitting in English class. It's just a few minutes before the bell. You think about math class and wonder if you'll be called on. [Third presentation of item 3]. [Counselor allows the scene to be imagined again for five to seven seconds and watches to see whether Joan signals any anxiety by raising her finger. Joan does not respond, indicating that she is not feeling anxious while imagining this item.] As the last two presentations of this item did not evoke anxiety, the counselor can move on to item 4. OK, Joan, stop imagining that scene. Think about a campfire. . . . Just relax. [Another control item can be used for variation. After 10 to 30 seconds, item 4 is presented or session is terminated.]

If this session had been terminated after the successful completion of item 3, the next scene-presentation session would begin with item 3. Other hierarchy items would be presented in the same manner as in this session. If Joan reported anxiety for three successive presentations of one item, the session would be interrupted, and an adjustment in the hierarchy would be made. (See Learning Activity #52.)

Homework and Follow-Up

Homework is essential to the successful completion of desensitization! Homework may include daily practice of the selected relaxation procedure, visualization of the items completed in the previous session, and exposure to in vivo situations.

Assignment of Homework Tasks. Most counselors instruct clients to practice once or twice daily the relaxation method being used. Practice is especially critical in the early sessions, in which training in the counterconditioning response occurs. In addition, a counselor can assign the client to practice visualizing the items covered in the last session

LEARNING ACTIVITY **52** *Scene Presentation*

This learning activity is designed to familiarize you with some of the procedural aspects of scene presentation. You can complete this activity by yourself or with a partner who can serve as your client.

1. Select one of the hierarchies you or your partner developed in Learning Activity #51 on hierarchy construction.
2. Administer relaxation or imagery to yourself or to your partner.
3. If you have a partner to act as a role-play client, tell the client to signal anxiety by raising a finger and that the sud numbers will be used to indicate anxiety when the client raises a finger.
4. By yourself or with your partner, start by presenting the lowest item in the hierarchy. If no anxiety is signaled

after a specified duration, instruct your partner to remove the image and relax; then re-present the same scene. Remember, two successive no-anxiety presentations are required before the next item is presented. If anxiety is signaled, instruct yourself or your partner to remove the image and relax. After 10 to 30 seconds, re-present the same item.

5. Select a notation system to use. Record at least the number of times each item was presented, the duration of each presentation, and whether each presentation did or did not evoke anxiety. If anxiety was indicated, be sure to use the sud scale to note the amount of anxiety the client felt.

---

after the relaxation session. Cassette tapes or computerized "virtual realities" can also be used for this purpose. Gradually, in vivo homework tasks can be added. As desensitization progresses, the client should be encouraged to participate in real-life situations that correspond to the situations covered in hierarchy-item visualization during the sessions. The shift to real life is very important in facilitating generalization from imagined to real anxiety-producing situations. However, there may be some risk in the client's engaging in a real situation corresponding to a hierarchy item that has not yet been covered in the scene-presentation sessions.

Homework Log Sheets and Follow-Up.   The client should record completion of all homework assignments on daily log sheets. After all desensitization sessions are completed, a follow-up session or contact should be arranged.

MODEL DIALOGUE: HOMEWORK AND FOLLOW-UP

Counselor: Joan, you've been progressing through the items on your list very well in our session. I'd like you to try some practice on your own similar to what we've been doing.

Client: OK, what is it?

Counselor: Well, I'm thinking of an item that's near the middle of your list. It's something you've been able to imagine last week and this week without reporting any

nervousness. It's the one on your volunteering an answer in class.

Client: You think I should do that?

Counselor: Yes, although there is something I'd like to do first. I will put this item and the two before it on tape. Each day after you practice your relaxation, I'd like you to use the tape and go over these three items just as we do here. If this goes well for you, then next week we can talk about your actually starting to volunteer a bit more in class.

Client: OK with me.

Counselor: One more thing. Each time you use the tape this week, make a notation on a blank log sheet. Also note your tension level before and after the practice with the tape on the 0-to-100 scale. Then I'll see you next week.

Figure 19-4 summarizes all the components of systematic desensitization. You may find this to be a useful review of procedural aspects of this strategy.

PROBLEMS ENCOUNTERED DURING DESENSITIZATION

Although desensitization can be a very effective therapeutic procedure, occasionally problems are encountered that make it difficult or impossible to administer. Sometimes these problems can be minimized or alleviated. At other times, a problem may require the counselor to adopt an alternative strategy.

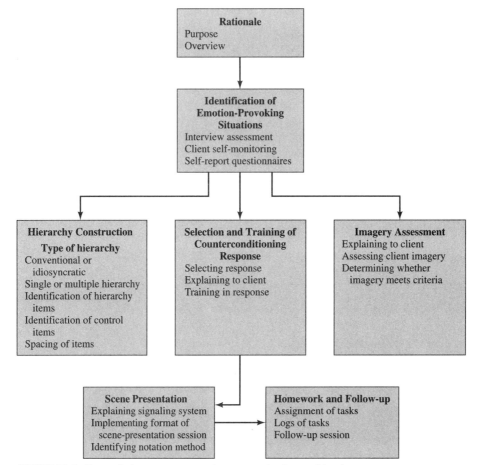

FIGURE 19-4. Expanded components of systematic desensitization

Wolpe (1990) identifies three major barriers to effective implementation of desensitization. These include difficulties of relaxation, misleading or irrelevant hierarchies, and inadequacies of imagery (p. 181). When a client is unable to lower anxiety to less than 10 suds, relaxation is not working well. According to Wolpe (1990), it may be effective to use drugs such as diazepam or codeine before the systematic desensitization session. An additional alternative is to try inducing relaxation through hypnosis. Another reason that desensitization may not proceed is related to the hierarchies being used. Sometimes, despite the time spent developing the hierarchies, the client and counselor have missed the core problem, and the hierarchy does not address the problem. In this case, the solution is to consider redeveloping the hierarchy. Finally, a client may be unable to visualize adequately the images that are required for the procedure to work. In this case, the counselor and client can work together to develop more complete and vivid descriptions of the items on the hierarchy. Another possibility to consider is that the item is generating so much anxiety that the client detaches from the imagery. Additional hierarchy items may be necessary before moving on to the item that preceded an inability to visualize.

Occasionally, clients will benefit from a different form of desensitization. Two of the possible variations of desensitization are discussed in the next section.

VARIATIONS OF SYSTEMATIC DESENSITIZATION: IN VIVO DESENSITIZATION AND EYE MOVEMENT DESENSITIZATION AND REPROCESSING

The desensitization procedure described in this chapter reflects the traditional procedure applied over a series of sessions to an individual client by a counselor, using an individualized hierarchy imagined by the client. This section briefly describes two possible variations of this method of desensitization: in vivo and eye movement desensitization and reprocessing (EMDR). For more detailed information, we encourage you to consult the references mentioned in this section and those listed in the suggested readings at the end of the chapter.

In Vivo Desensitization

In vivo desensitization involves actual client exposure to the situations in the hierarchy. The client engages in the graded series of situations instead of imagining each item. This variation is used when a client has difficulty using imagery or does not experience anxiety during imagery or when a client's actual exposure to the situations will have more therapeutic effects. If the client can actually be exposed to the feared stimuli, then in vivo desensitization is preferable to imagined exposure because it will produce more rapid results and will foster greater generalization. At times the counselor may accompany the client to the feared situation. In vivo desensitization resembles participant modeling (Chapter 12), in which the client performs a graduated series of difficult tasks with the help of induction aids. In vivo desensitization has been used to treat noise phobia and spider phobias in young children (McGrath, Tsui, Humphries, & Yule, 1990; Nelissen, Muris, & Merckelbach, 1995). It has also been successfully used to treat chronic paruesis (inability to urinate in the presence of others) of an adult man (McCracken & Larkin, 1991); panic attacks in adults (Harris, 1991; Van Hout, Emmelkamp, & Scholing, 1994); varying kinds of phobias in adults, including claustrophobia (Edinger & Radtke, 1993); and balloon phobia (Kraft, 1994). More recently in vivo desensitization was used to treat primary vaginismus in adult women (Kennedy, Doherty, & Barnes, 1995).

The main procedural problem associated with in vivo desensitization involves adequate use and control of a counterconditioning response. Sometimes it is difficult for a client to achieve a state of deep relaxation while simultaneously performing an activity. McGrath, Tsui, Humphries, and Yule (1990) reported using conversation and a play therapy environment as counterconditioning responses when using in vivo desensitization for a nine-year-old girl. It is not always necessary to use a counterconditioning response to decrease the client's anxiety in threatening situations. Often exposure alone will result in sufficient anxiety reduction, particularly if the exposure occurs in graduated amounts and with induction aids. Nelissen et al. (1995) suggest that during in vivo exposure, information that is discordant with the irrational elements of anxiety is more available and more easily incorporated by the client, resulting in a reduction of the feared stimuli.

Eye Movement Desensitization and Reprocessing

Within the last ten years, a new therapy known as eye movement desensitization and reprocessing (EMDR) has become very popular throughout the United States, Europe, Canada, and Australia. The approach was developed by Francine Shapiro (1995) and is designed to help clients deal with traumatic memories. EMDR has been used with clients who have experienced single traumatic events or multiple traumatic episodes. The therapy has also been used for accident and burn victims; for clients dealing with addictions, anxiety, depression, grief, or phobia; and for victims of crime and sexual assault (Shapiro, 1995).

After reviewing the research, Greenwald (1996) concluded that the mixed results and controversy concerning the effectiveness of EMDR have several causes, rooted in the way the therapy is presented: (1) There is an information gap between practitioners who have received formal, supervised training provided by Shapiro's EMDR Institute and those who have not received such training. Therapy conducted by these two groups will often produce quite different results. (2) For the therapy to be conducted appropriately, supervised training is mandatory. (3) Even so, being trained in EMDR is not always a guarantee that practitioners will adhere closely to the EMDR protocol. In spite of these conclusions, Greenwald (1996) reports that EMDR does have an encouraging research base for treatment of posttraumatic stress disorder (Silver, Brooks, & Obenchain, 1995; Vaughan et al., 1994; Wilson, Becker, & Tinker, 1995).

Accelerated Information Processing Model. Eye movement desensitization and reprocessing (EMDR) is a specific method operating within what Shapiro (1995) calls an accelerated information processing (AIP) model, which can also include a variety of other treatment strategies. According to the AIP model, there is a neurological balance that allows information to be processed. When a person has neurological balance, he or she processes information with appropriate associations and integrates the experience into a positive emotional and cognitive schema (Shapiro, 1995, p. 29). For example, an adolescent may be embarrassed by a parent in front of her peers. She may be troubled by the experience at first, but later she overcomes her embarrassment and is able to use the information to guide her actions in the future. In this example, the information the adolescent obtains from the episode becomes integrated, over time, into a positive schema, creating a neurological balance.

But what happens to a person who experiences severe trauma? According to Shapiro's (1995) AIP model, the traumatic event appears to create an *imbalance* or a *block* in the nervous system with respect to processing the information. The block may be caused by changes in the psychoneuroendocrinological systems (i.e., neurotransmitters, adrenaline, etc.). Information acquired during the traumatic event is held neurologically in a disturbed state that includes images, sounds, emotions, cognitions, and physical sensations associated with the event. Shapiro (1995) hypothesizes that the eye movements used in EMDR trigger a physiological mechanism that activates the imbalanced or blocked information processing system. She suggests that when the client brings up a memory of the trauma, a link may be established "between consciousness and the site where the information is stored in the brain" (Shapiro, 1995, p. 30). She conjectures that the eye movements rebalance or remove the blocks from the information processing system. If the information processing system continues to be blocked and imbalanced, the person will exhibit maladaptive patterns of responding. The AIP model suggests that the psychoneuroendocrinological disturbance of the information processing system, caused by the trauma, creates "frozen information (stored in the form in which it was experienced), isolated in its own neuro network, and stored in its originally disturbing state-specific form" (Shapiro, 1995, p. 40). For the client to learn or to integrate new information is very difficult when he or she is stranded and blocked in a static/imbalanced information processing system. The resolution of the imbalance and the processing of the disturbing and dysfunctional material is achieved, according to Shapiro (1995), through the "stimulation of the client's inherent self-healing processes" (p. 44).

Eye Movements. In EMDR therapy, the clinician holds up the first two fingers with the thumb over the last two fingers and the palm facing the client at a distance of 12 to 14 inches from the client's face. The clinician moves her/his fingers horizontally from the extreme left to the extreme right of the client's visual field, asking the client to follow the motion of her fingers. The client and clinician evaluate the comfort of the client's eye movement in terms of the distance, speed, and height of the clinician's fingers. In addition to the left-right eye movement, the clinician may also introduce vertical, circular, or diagonal eye movements (Shapiro, 1995, pp. 63–67). Clinicians report that a client must experience a

24-movement set to be able to process cognitive material to a new level of adaptation (Shapiro, 1995, p. 65). For some clients, the process may require more movements. After each reprocessing set, the clinician asks the client, "What comes up?" or "What do you get now?" If the client indicates increased adaptiveness, eye movements continue for the next set in the same direction of finger movement, speed, and distance. If nothing comes up for a client after an eye movement set, adaptiveness can be achieved by changing the direction of the finger movement. However, many other interventions in addition to change of direction are necessary to achieve full resolution. EMDR consists of much more than eye movement.

The clinician starts the eye movement reprocessing by instructing the client to elicit different aspects of the traumatic event and hold them in consciousness while eye movements are induced. This process will elicit other associations in the memory network. As each new event is elicited from the client's memory, the clinician leads the client through a set of eye movements. The sets are used one at a time with each memory, thought, image, emotion, person, and sensation that comes into the client's mind. These associative channels are the client's memory network and are stored and linked to one another as a consequence of the traumatic event. The dysfunctionally stored memories are statically associated. The material is disturbing because the memories cannot form new and more functional connections, and for new learning to occur around this dysfunctional material is difficult. The key to the AIP model in general and the EMDR strategy specifically is to access dysfunctional information and to process it, facilitating appropriate connections by linking the targeted blocked memory network to healthier associations.

The specific targets that clients identify in their associative channels help them gain access to the dysfunctional stored and blocked memory network. Clinicians trained to use EMDR have reported that, as a result of this therapy, their clients' negative images, emotions, cognitions, and sensations become diffuse and are replaced with positive images, emotions, and cognitions (Shapiro, 1995, p. 39). The accelerated information processing, which occurs as a result of the clients' eye movements, transforms the dysfunctional information. After the blocks to the imbalanced information processing system are removed, the information can be adapted and integrated into a more functional and positive emotional and cognitive schema.

Shapiro (1995) has found that in addition to eye movements, other stimuli can activate the dysfunctional information processing system. Hand taps and the repetition of an auditory cue can help clients activate and process stored disturbing material. For hand taps, the client sits with the palms of the hands turned up on the knees; then the counselor taps the client's palms, alternating right and left at about the same speed as an eye movement set. Auditory cues require the counselor to snap his or her fingers next to each of the client's ears, alternating between ears at a rate comparable to the one used with sets of eye movements (1995, p. 67).

Eight Phases of the EMDR Approach. The EMDR approach, delineated below, integrates many of the features of the therapeutic processes and interventions presented earlier in this book.

1. A *client history* is obtained first to *determine whether EMDR is appropriate* for this client. If the client can handle the disturbance created by accessing and processing the dysfunctional information evoked by the EMDR, the clinician *obtains target material from the narrative* history that needs to be reprocessed. Features in this book included in the first phase are listening and action responses; conceptualizing and defining the client problem; and identifying, defining, and evaluating outcome goals.
2. *Preparation* involves building a therapeutic alliance with the client, explaining the theory of EMDR to the client, testing his or her comfort level for eye movements, creating a place where the client can feel safe with emotive imagery, and setting expectations. This phase requires a self-aware counselor, ingredients of an effective helping relationship, relationship enhancement variables and interpersonal influences, and guided imagery intervention.
3. *Assessment* includes evaluating the subjective unit of disturbance (SUD) scale of images, negative and positive cognitions, validity of cognitions, and physical sensations associated with the traumatic event. The therapist will help the client create a positive cognition about the traumatic event and rate the positive cognition on the validity of cognitions (VOC) scale. Processes and interventions included in this book that are part of phase three are conceptualizing and defining client problems, doing cognitive restructuring, and using a body scan to assess body sensations.
4. *Desensitization* with eye movements and other appropriate interventions allows the emergence of insight and pattern recognition of negative and positive client emotions as reflected in the SUD scale. Features of this phase are presented in this chapter.
5. The *installation* phase means increasing the strength of the positive cognitions as a replacement for the negative

cognition of the trauma. The chapter in this book notable for this phase is Chapter 20—building self-efficacy (confidence) to use positive cognitions.
6. The *body scan* phase is completed when the client scans his or her body and feels little tension while visualizing the target event and experiencing the positive cognition (see Chapter 15 and desensitization in this chapter).
7. The *closure* phase consists of five parts: building feelings of self-efficacy; using the safe place image; learning posttreatment material; logging any disturbing material; and using relaxation, meditation, or imagery. Guided imagery, self-monitoring, meditation, muscle relaxation, breathing exercises, and Hatha Yoga, all of which are presented earlier in this book, can be appropriate interventions for this phase of the EMDR approach.
8. The last phase is *reevaluation.* Here the SUD and VOC scales are used to evaluate any material from the past as well as present persons or events that elicit disturbing material from the past; any anticipated event, person, or situation that might access past dysfunctional material; and follow-up after termination of treatment. The last phase of EMDR can include the following processes and interventions presented in this book: conceptualizing and defining client problems, establishing therapeutic goals, desensitizing, doing cognitive restructuring, self-monitoring, and outcome evaluation.

With some cases a clinician may be able to get through the first seven phases during the first session, or about an hour and-a-half.

Eye movement desensitization and reprocessing (EMDR) integrates a wide range of therapeutic interventions, all of which contribute to the complexity and therapeutic power of this approach. The therapy, however, should be used only by persons who have received *clinical training* and *supervision* in its application. There are ethical issues in using EMDR without proper training and supervision with any client, and especially with clients who suffer from some form of trauma. Again, we caution you *not* to use the procedure *unless* you receive proper training and supervision from a person trained and clinically experienced in using EMDR. For information regarding training, contact the EMDR Institute, P.O. Box 51010, Pacific Grove, CA 93950.

SUMMARY

Systematic desensitization is one of the oldest of the behavioral strategies developed to alleviate anxiety. There does not seem to be a great deal of ongoing research testing this procedure, perhaps because research conducted until the 1970s demonstrated the effectiveness of desensitization through repeated studies (Wolpe, 1990). Although desensitization is time-consuming to implement, it is clearly an effective way to help clients alleviate anxiety and specifically to treat phobias. In practice, many counselors find that the combination of desensitization and other strategies may be used to help clients lessen anxiety and learn new ways of handling situations that are anxiety provoking. Clients seem to benefit most when they have multiple ways of controlling anxiety and fear through desensitization, and additional strategies they can continue to use after the counseling process has ended.

Desensitization has been used in limited ways with diverse groups of clients who present with phobias and anxiety problems. Remember that although anxiety by and large seems to be a universal phenomenon, the specific presentation of it as well as the cues surrounding it are often very culturally specific.

The newest variation of desensitization has been developed by Shapiro, as summarized in her recent book (1995). It is called eye movement desensitization and reprocessing or EMDR and involves the use of rhythmic eye movements to treat traumatic stress and memories of clients. The most commonly reported use of EMDR is with posttraumatic stress disorders.

POSTEVALUATION

Part One

Objective One states that you should be able to identify accurately at least three out of four hierarchies by type. Read each hierarchy carefully and then identify on a piece of paper whether the hierarchy is conventional or idiosyncratic. Feedback is provided at the end of the postevaluation.

Hierarchy 1 (fear of heights)

1. You are walking along the sidewalk. It is on a completely level street.
2. You are walking along the sidewalk, ascending. At the top of the street, you look down and realize you've climbed a hill.
3. You are climbing a ladder up to a second-story window.
4. You are riding in a car, and the road curves higher and higher.
5. You are riding in a car and you look outside. You notice you are driving on the edge of a good-sized hill.
6. You are starting to climb to the top of a fire tower. You are halfway up. You look down and see how far you've climbed.
7. You are climbing a ladder to the roof of a three-story house.
8. You have climbed to the top of a fire tower and look down.
9. You are riding in a car and are at the edge of a cliff on a mountain.
10. You are at the very top of a mountain, looking down into the surrounding valley.

Hierarchy 2 (fear of being rejected)

1. You speak to a stranger on the street. He doesn't hear you.

2. You go into a department store and request some information from one of the clerks. The clerk snaps at you in response.
3. You ask a stranger to give you change. She gives you a sarcastic reply.
4. You ask a casual acquaintance to lend you a book. He refuses.
5. You ask a friend over to dinner. The friend is too busy to come.
6. You apply for a membership in a social club, and your application is denied.
7. You are competing for a job. You and another person are interviewed. The other person is hired; you are not chosen.
8. You have an argument with your best friend. She leaves suddenly. You don't hear from her for a while.
9. You have an argument with your partner. Your partner says he would rather do things alone than with you.
10. Your partner says he doesn't love you any more.

Hierarchy 3 (loss of a close relationship)

1. You remember a warm, starry night. You ask this woman you love to marry you. She accepts. You are very happy.
2. The two of you are traveling together soon after your marriage, camping out and traveling around in a van.
3. The two of you are running in the water together at the beach and having a good time being together.
4. You and this person are eating dinner together at home.
5. The two of you are disagreeing over how to spend money. She wants to save it; you are arguing to use some of it for camping supplies.

(continued)

6. The two of you are arguing over your child. She wants the child to go with you on all trips; you want a babysitter occasionally.
7. The two of you are starting to eat some meals apart. You are working late to avoid coming home for dinner.
8. She is wrapped up in her social activities; you, in your work. On the weekends you go your separate ways.
9. You have a discussion about your relationship and separate activities. You start sleeping on the couch.
10. The two of you go to see a lawyer to initiate discussion about a separation.

Hierarchy 4 (fear of giving speeches)

1. Your instructor casually mentions a required speech to be given by the end of the course.
2. Your instructor passes around a sign-up sheet for the speeches. You sign up.
3. You talk about the speech with some of your classmates. You aren't sure what to say.
4. You go to the library to look up some resource material for your speech. You don't find too much.
5. Some of your classmates start to give speeches. You think about how good their speeches are and wonder how yours will be.
6. It is a week before the speech. You're spending a lot of time working on it.
7. It is the day before the speech. You're going over your notes religiously.
8. It is the night before the speech. You lie awake thinking about it.
9. It is the next morning. You wake up and remember it is speech day. You don't feel hungry at breakfast.
10. Later that morning you're walking to speech class. A classmate comes up and says "Well, I guess you're on today."
11. You're sitting in speech class. The instructor will call on you any moment. You keep going over your major points.

Part Two

Objective Two asks you to identify and describe at least 9 out of 11 procedural steps of desensitization, using a written client case description. Read this case description carefully; then respond by identifying and describing the 11 items listed after the description.

Your client is a fifth-grade boy at a local elementary school. This year, the client's younger sister has entered first grade at the same school. After a few weeks at school, your client, Ricky, began to complain about school to his teacher and parents. He would come to school and get sick. His parents would come and take him home. After a medical check-up, the doctor can find nothing physically wrong with Ricky. Yet Ricky continues either to get sick at school or to wake up sick in the morning. He appears to be better on weekends. He says he hates school and it makes him sick to his stomach to have to go. On occasion, he has vomited in the morning. The parents report that it is getting harder and harder to get Ricky to attend school. Suppose you were to use desensitization as one strategy in this case to help Ricky overcome his tension and avoidance of school. Identify and describe how you would implement the following 11 steps of desensitization with Ricky. Adapt your language to words that a 10-year-old could understand.

1. Your rationale of desensitization
2. Your description of an overview of desensitization
3. A method for helping Ricky identify the anxiety-provoking situations about school
4. The type of hierarchy that would be used with Ricky
5. The sud method Ricky would use to arrange the hierarchy items
6. An appropriate counterconditioning response you could train Ricky to use
7. The method of assessing Ricky's imagery capacity
8. The method of scene presentation you would use with Ricky
9. The method Ricky could use for signaling during scene presentation
10. A notation method you might use to keep track of hierarchy presentation
11. An example of one homework task associated with desensitization that you might assign to Ricky to complete

Feedback follows the Postevaluation.

Part Three

Objective Three asks you to demonstrate at least 22 out of 28 steps of systematic desensitization with a role-play client. Several role-play interviews may be required for you to include all the major procedural components of desensitization. Use the Interview Checklist for Systematic Desensitization on pages 524–529 as an assessment tool.

SUGGESTED READINGS

Acierno, K., Tremont, G., Last, C., & Montgomery, D. (1994). Tripartite assessment of the efficacy of eye-movement desensitization in a multi-phobic patient. *Journal of Anxiety Disorders, 8,* 259–276.

Edinger, J., & Radtke, R. (1993). Use of *in vivo* desensitization to treat a patient's claustrophobic response to nasal CPAP. *Sleep, 16,* 678–680.

Foss, D., & Hadfield, O. (1993). A successful clinic for the reduction of mathematics anxiety among college students. *College Student Journal, 27,* 157–165.

Greenwald, R. (1996). The information gap in the EMDR controversy. *Professional Psychology: Research and Practice, 27,* 67–72.

Guglielmi, S., Cox, D., & Spyker, D. (1994). Behavioral treatment of phobic avoidance in multiple chemical sensitivity. *Journal of Behavior Therapy and Experimental Psychiatry, 25,* 197–209.

Kamimura, E., & SaSaki, Y. (1991). Fear and anxiety reduction in systematic desensitization and imaging strategies: A comparison of response and stimulus oriented imaging. *Japanese Journal of Behavior Therapy, 17,* 29–38.

King, N. (1993). Simple and social phobias. *Advances in Clinical Child Psychology, 15,* 305–341.

Hoffman, S., Herzog-Bronsky, R., & Zim, S. (1994). Dialectical psychotherapy of phobias: A case study. *International Journal of Short Term Psychotherapy, 9,* 229–233.

Horne, D., Vatmanidis, P., & Careri, A. (1994). Preparing patients for invasive medical and surgical procedures: II. Using psychological interventions with adults and children. *Behavioral Medicine, 20,* 15–21.

Menzies, R., & Clarke, J. (1993). A comparison of *in vivo* and vicarious exposure in the treatment of childhood water phobia. *Behavior Research and Therapy, 31,* 9–15.

Morris, R. J. (1991). Fear reduction methods. In F. H. Kanfer & A. P. Goldstein (Eds.), *Helping people change* (4th ed., pp. 161–201). New York: Pergamon Press.

Pitre, A., & Nicki, R. (1994). Desensitization of dietary restraint anxiety and its relationship to weight loss. *Journal of Behavior Therapy and Experimental Psychiatry, 25,* 153–154.

Schneider, W., & Nevid, J. (1993). Overcoming math anxiety: A comparison of stress inoculation training and systematic desensitization. *Journal of College Student Development, 34,* 283–288.

Shapiro, F. (1995). *Eye movement desensitization and reprocessing: Basic principles, protocols and procedures.* New York: Guilford.

Silver, S. M., Brooks, A., & Obenchain, J. (1995). Treatment of Vietnam war veterans with PTSD: A comparison of eye movement desensitization and reprocessing, biofeedback, and relaxation training. *Journal of Traumatic Stress, 8,* 337–342.

Strumpf, J., & Fodor, I. (1993). The treatment of test anxiety in elementary school-age children: Review and recommendations. *Child and Family Behavior Therapy, 15,* 19–42.

Thomas, R., & Gafner, G. (1993). PTSD in an elderly male: Treatment with eye movement desensitization and reprocessing (EMDR). *Clinical Gerontologist, 14,* 57–59.

VanHout, W., Emmelkamp, P., & Scholing, A. (1994). The role of negative self-statements during exposure *in vivo:* A process study of eight panic disorder patients with agoraphobia. *Behavior Modification, 18,* 389–410.

Vaughan, K., Armstrong, M. S., Gold, R., O'Connor, N., Jenneke, W., & Tarrier, N. (1994). A trial of eye movement desensitization compared to image habituation training and applied muscle relaxation in post-traumatic stress disorder. *Journal of Behavior Therapy and Experimental Psychiatry, 25,* 283–291.

Wilson, S. A., Becker, L. A., & Tinker, R. H. (1995). Eye movement desensitization and reprocessing (EMDR) treatment for psychologically traumatized individuals. *Journal of Consulting and Clinical Psychology, 63,* 928–937.

Wolpe, J. (1990). *The practice of behavior therapy* (4th ed.). New York: Pergamon Press.

FEEDBACK 19
POSTEVALUATION

Part One

1. Conventional. Items are arranged by increasing height off the ground.
2. Idiosyncratic. Items are arranged around the theme of rejection.
3. Idiosyncratic. Items are arranged from pleasant to unpleasant memories of an ex-spouse.
4. Conventional. Items are arranged by time; as the time approaching the situation diminishes, the fear intensifies.

Part Two

Here are some possible descriptions of the 11 procedural steps of desensitization you were asked to identify and

(continued)

describe. See whether your responses are in some way similar to these.

1. Rationale: "Ricky, it seems that it's very hard for you to go to school now or even think about school without feeling sick. There are some things about school that upset you this much. We can work together to find out what bothers you, and I can help you learn to be able to go to school without feeling so upset or sick to your stomach, so you can go to school again and feel OK about it. How does that sound?"

2. Overview: "There are several things you and I will do together. First we'll talk about the things about school that upset you. I'll ask you to think about these situations, only instead of feeling upset when you do, I'll show you a way to stay calm and keep the butterflies out of your stomach. It will take a lot of practice in this room, but after a while you will be able to go back to your class and feel calm and OK about it!"

3. Method for identifying the anxiety-provoking situations:
   a. Use of interview leads such as "Ricky, what do you do in school that makes you feel sick? What about school makes you want to stay at home? What happens at school that bothers you? When do you feel most upset about school?"
   b. Use of client self-monitoring: "Ricky, could you keep a chart for me this week? Each time you feel upset about school, mark down what has happened or what you're thinking about that makes you feel upset or sick."

4. Type of hierarchy: An idiosyncratic hierarchy would be used. One hierarchy might consist of school-related anxiety-provoking situations. Depending on the anxiety-provoking situations identified, another idiosyncratic hierarchy may emerge, dealing with jealousy. It is possible that the avoidance of school is a signal that Ricky really fears being upstaged by his younger sister.

5. Ranking method: The sud method can be used with Ricky. The counselor will need to take additional time to explain the method to Ricky and provide examples of things that might be as high as 90 suds and items that would be as low as 10 suds.

6. Counterconditioning response: Muscle relaxation can be used easily with a child Ricky's age as long as you show him (by modeling) the different muscle groups and the way to tighten and relax a muscle.

(continued)

7. Method of imagery assessment: Ask Ricky to tell you some daydreams he has or some things he loves to do. Before and after a relaxation-training session, ask him to imagine or pretend he is doing one of these things. Then have him describe the details of his imagined scene. Children often have a capacity for more vivid and descriptive imagery than adults.

8. Scene presentation: Ricky is told to imagine the scene as described by the counselor and to raise an index finger to indicate to the counselor that the scene is clear in his mind. The counselor should allow Ricky to hold the image of the scene for five to seven seconds and terminate it by saying "Stop the scene." Next, the counselor asks Ricky how much anxiety was generated in terms of suds; the counselor might ask "How much did imagining this scene increase your sud level?"

9. Signaling method: The counselor should tell Ricky to raise a finger to inform the counselor that he sees the scene clearly.

10. Notation method: The easiest notation method might be to use each hierarchy card and note the number of times each item is presented, the duration of each presentation, and an indication of whether Ricky did or did not report being "tense" during or after the item. This notation system looks like this:

Item No. _____Date_____
Item description
+5 −7 +5 +7

The item was presented four times; the numbers 5, 7, 5, and 7 refer to the duration of each presentation; the + indicates no anxiety report; the − indicates a "tense" signal.

11. Examples of possible homework tasks:
   a. A list of anxiety-related situations
   b. Practice of muscle relaxation
   c. Practice of items covered in the interview
   d. Exposure to certain school-related in vivo situations

Part Three
You or an observer can rate your desensitization interviews using the Interview Checklist for Systematic Desensitization that follows.

(continued)

FEEDBACK: POSTEVALUATION   (continued)

Interview Checklist for Systematic Desensitization
Instructions to observer: Listed below are some procedural steps of systematic desensitization. Check which of these steps were used by the counselor in implementing this procedure. Some possible examples of these leads are described in the right column of the checklist.

Item	Examples of counselor leads
I. Rationale	
_____ 1. Counselor gives client rationale for desensitization, clearly explaining how it works.	"This procedure is based on the idea that you can learn to replace your fear (or other conditioned emotion) in certain situations with a better or more desirable response, such as relaxation or general feelings of comfort." "You have described some situations in which you have learned to react with fear (or some other emotion). This procedure will give you skills to help you cope with these situations so they don't continue to be so stressful."
_____ 2. Counselor describes brief overview of desensitization procedure.	"There are three basic things that will happen—first, training you to relax; next, constructing a list of situations in which you feel anxious; and finally, having you imagine scenes from this list, starting with low-anxiety scenes, while you are deeply relaxed." "First you will learn how to relax and how to notice tension so you can use it as a signal to relax. Then we'll identify situations that, to varying degrees, upset you or make you anxious. Starting with the least discomforting situations, you will practice the skill of relaxation as a way to cope with the stress."
_____ 3. Counselor checks to see whether client is willing to use strategy.	"Are you ready to try this now?"
II. Identification of Emotion-Provoking Situations	
_____ 4. Counselor initiates at least one of the following means of identifying anxiety-provoking stimulus situations:	
_____ a. Interview assessment through problem leads	"When do you notice that you feel most _____?" "Where are you when this happens?" "What are you usually doing when you feel _____?" "What types of situations seem to bring on this feeling?"
_____ b. Client self-monitoring	"This week I'd like you to keep track of any situation that seems to bring on these feelings. On your log, write down where you are, what you're doing, whom you're with, and the intensity of these feelings."
_____ c. Self-report questionnaires	"One way that we might learn more about some of the specific situations that you find stressful is for you to complete this short questionnaire. There are no right or wrong answers—just describe how you usually feel or react in the situations presented."
_____ 5. Counselor continues to assess anxiety-provoking situations until client identifies some specific situations.	"Let's continue with this exploration until we get a handle on some things. Right now you've said that you get nervous and upset around certain kinds of people. Can you tell me some types or characteristics of people that bother you or make you anxious almost always?"

(continued)

Item	Examples of counselor leads
	"OK, good, so you notice you're always very anxious around people who can evaluate or criticize you, like a boss or teacher."

III. Hierarchy Construction

_____ 6. Counselor identifies a type of hierarchy to be constructed with client:

"Now we're going to make a list of these anxiety-provoking situations and fill in some details and arrange these in an order, starting with the least anxiety-provoking situation all the way to the most anxiety-provoking one."

_____a. Conventional

"Because you get more and more anxious as the time for the speech gets closer and closer, we'll construct these items by closer and closer times to the speech."

_____b. Idiosyncratic

"We'll arrange these items according to the different kinds of situations in which people criticize you—depending on who does it, what it's about, and so on."

_____ 7. Counselor identifies the number of hierarchies to be developed:

_____a. Single hierarchy

"We will take all these items that reflect different situations that are anxiety-producing for you and arrange them in one list."

_____b. Multiple hierarchies

"Because you find several types of situations stressful, we'll construct one list for situations involving criticism and another list for situations involving social events."

_____ 8. Counselor initiates identification of hierarchy items through one or more methods:

"I'd like us to write down some items that describe each of these anxiety-provoking scenes with quite a bit of detail."

_____a. Interview questions (*not* when client is engaged in relaxation)

"Describe for me what your mother could say that would bother you most. How would she say it? Now who, other than your mother, could criticize you and make you feel worse? What things are you most sensitive to being criticized about?"

_____b. Client completion of note cards (homework)

"This week I'd like you to add to this list of items. I'm going to give you some blank index cards. Each time you think of another item that makes you get anxious or upset about criticism, write it down on one card."

_____ 9. Counselor continues to explore hierarchy items until items are identified that meet the following criteria:

_____a. Some items, if carried out in vivo, are under client's control (do not require instigation from others).

"Can you think of some items that, if you actually were to carry them out, would be things you could initiate without having to depend on someone else to make the situation happen?"

_____b. Items are concrete and specific.

"OK, now just to say that you get nervous at social functions is a little vague. Give me some details about a social function in which you might feel pretty comfortable and one that could make you feel extremely nervous."

_____c. Items are similar to or represent past, present, or future situations that *have* provoked or *could* provoke the emotional response from client.

"Think of items that represent things that have made you anxious before or currently—and things that could make you anxious if you encountered them in the future."

(continued)

FEEDBACK: POSTEVALUATION   (continued)

Item	Examples of counselor leads

III. Hierarchy Construction (continued)

_____d. Items have sampled broad range of situations in which emotional response occurs.

"Can you identify items representing different types of situations that seem to bring on these feelings?"

_____e. Items represent different levels of emotion aroused by representative stimulus situations.

"Let's see if we have items here that reflect different amounts of the anxiety you feel. Here are some items that don't make you too anxious. What about ones that are a little more anxiety-provoking, up to ones where you feel panicky?"

_____10. Counselor asks client to identify several control items (neutral, non-emotion-arousing).

"Sometimes it's helpful to imagine some scenes that aren't related to things that make you feel anxious. Could you describe something you could imagine that would be pleasant and relaxing?"

_____11. Counselor explains purpose of ranking and spacing items according to increasing levels of arousal.

"It may take a little time, but you will rank these hierarchy items from least anxiety-producing to most anxiety-producing. This gives us an order to the hierarchy that is gradual, so we can work just with more manageable situations before moving on to more stressful ones."

_____12. Counselor asks client to arrange hierarchy items in order of increasing arousal, using sud method; explains method to client:

"Now I would like you to take the items and arrange them in order of increasing anxiety, using the following method."

_____a. Sud scale

"I'd like you to arrange these items using a 0-to-100 scale. 0 represents total relaxation and 100 is comparable to complete panic. If an item doesn't give you any anxiety, give it a 0. If it is just a little stressful, may be a 15 or 20. Very stressful items would get a higher number, depending on how stressful they are."

_____13. Counselor adds or deletes items if necessary to achieve reasonable spacing of items in hierarchy.

"Let's see, at the lower end of the hierarchy you have many items. We might drop out a few of these. But you have only three items at the upper end, so we have some big gaps here. Can you think of a situation provoking a little bit more anxiety than this item but not quite as much as this next one? We can add that in here."

IV. Selection and Training of Counterconditioning Response

_____14. Counselor selects appropriate counterconditioning response to use to countercondition anxiety (or other conditioned emotion):

_____a. Deep muscle relaxation   (contrasting tensed and relaxed muscles)

_____b. Guided imagery   (evoking pleasurable scenes in imagination)

_____c. Meditation   (focusing on breathing and counting)

_____d. Coping thoughts or statements   (concentrating on coping or productive thoughts incompatible with self-defeating ones)

_____15. Counselor explains purpose of particular response selected and describes its role in desensitization.

"This response is like a substitute for anxiety. Learning it will take time, but it will help to decrease your anxiety so that you can face rather than avoid these feared situations."

(continued)

Item	Examples of counselor leads
	"This training will help you recognize the onset of tension. You can use these cues you learn as a signal to relax away the tension."
_____16. Counselor trains client in use of counterconditioning response and suggests daily practice of this response.	"We will spend several sessions learning this so you can use it as a way to relax. This relaxation on your part is a very important part of this procedure. After you practice this here, I'd like you to do this at home two times each day over the next few weeks. Each practice will make it easier for you to relax."
_____17. Counselor asks client before and after each training session to rate felt level of anxiety or arousal.	"Using a scale from 0 to 100, with 0 being complete relaxation and 100 being intense anxiety, where would you rate yourself now?"
_____18. Counselor continues with training until client can discriminate different levels of anxiety and can use nonanxiety response to achieve 10 or lower rating on 0-to-100 scale.	"Let's continue with this until you feel this training really has an effect on your relaxation state after you use it."

V. Imagery Assessment

_____19. Counselor explains use of imagery in desensitization.	"In this procedure, I'll ask you to imagine each hierarchy item as if you were actually there. We have found that imagining a situation can be very similar to actually being in the situation. Becoming desensitized to anxiety you feel while imagining an unpleasant situation will transfer to real situations, too."
_____20. Counselor assesses client's capacity to generate vivid images by	"It might be helpful to see how you react to using your imagination."
_____a. Presenting control items when client is using a relaxation response	"Now that you're relaxed, get a picture in your mind of sitting in the sun on a warm day. The sky is very blue, not a cloud in it. The grass and trees are green. You can feel the warmth of the sun on your body."
_____b. Presenting hierarchy items when client is not using a relaxation response	"OK, just imagine that you're at this party. You don't know anyone. Get a picture of yourself and the other people there. It's a very large room."
_____c. Asking client to describe imagery evoked in *a* and *b*	"Can you describe what you imagined? What were the colors you saw? What did you hear or smell?"
_____21. Counselor, with client's assistance, determines whether client's imagery meets the following criteria and, if so, decides to continue with desensitization:	
_____a. Client is able to imagine scene concretely with details.	"Were you able to imagine the scene clearly? How many details can you remember?"
_____b. Client is able to imagine scene as participant, not onlooker.	"When you imagined the scene, did you feel as if you were actually there and involved—or did it seem as if you were just an observer, perhaps watching it happen to someone else?"
_____c. Client is able to switch scene on and off when instructed to.	"How soon were you able to get an image after I gave it to you? When did you stop the image after I said *Stop?*"
_____d. Client is able to hold scene without drifting off or revising it.	"Did you ever feel as if you couldn't concentrate on the scene and started to drift off?" "Did you ever change anything about the scene during the time you imagined it?"

(continued)

FEEDBACK: POSTEVALUATION   (continued)

Item	Examples of counselor leads
_____e. Client shows no evidence of other difficulties.	"What else did you notice that interfered with getting a good picture of this in your mind?"

VI. Hierarchy Scene Presentation

Item	Examples of counselor leads
_____22. Counselor explains method of scene presentation.	"I am going to present an item in the hierarchy and I'd like you to imagine the scene as clearly as you can. I will wait while you imagine the scene and watch to see whether you signal any anxiety."
_____23. Counselor checks client's anxiety level; allows the client to hold the image of the scene for five to seven seconds and terminates it by saying, "Stop the scene."	"How much did imagining this scene increase your sud level?"
_____24. For each session of scene presentation:	
_____ a. Counselor precedes scene presentation with muscle relaxation or other procedures to help client achieve relaxation before scenes are presented.	"Let your whole body become heavier and heavier as all your muscles relax. . . . Feel the tension draining out of your body. . . . Relax the muscles of your hands and arms. . . ."
_____ b. Counselor begins initial session with lowest (least anxiety-provoking) item in hierarchy and for successive sessions begins with last item successfully completed at previous session.	"I'm going to start this first session with the item that is at the bottom of the hierarchy." "Today we'll begin with the item we ended on last week for a review."
_____ c. Counselor describes item and asks client to imagine it for five to seven seconds.	"Just imagine you are sitting in the classroom waiting for the test to be passed to you, wondering how much you can remember." [Counts five to seven seconds.]
_____(1) If client held image and did not signal anxiety, counselor instructs client to stop image and relax for 10 to 30 seconds.	"Now, stop visualizing this scene and just take a little time to relax. Think of sitting in the sun on a warm day, with blue sky all around you."
_____(2) If client indicated anxiety during or after visualizing scene, counselor asks for a sud rating and tells the client to erase the scene and relax for 10 to 30 seconds. Counselor then represents that same scene for five to seven seconds and watches to see whether the client signals any anxiety by raising her finger.	"What is your sud rating?"
_____d. After pause of 10 to 30 seconds between items, counselor presents each item to client a second time.	"Now I want you to imagine the same thing. Concentrate on being very relaxed, then imagine that you are sitting in the classroom waiting for the test to be passed to you, wondering how much you can remember."
_____e. Each item is successfully completed (with no anxiety) at least two successive times (more for items at top of hierarchy) before new item is presented.	"I'm going to present this scene to you once more now. Just relax, then imagine that. . ."

(continued)

Item	Examples of counselor leads
_____f. If an item elicits anxiety after three presentations, counselor makes some adjustments in hierarchy or in client's visualization process.	"Let's see what might be bogging us down here. Do you notice that you are drifting away from the scene while you're imagining it—or revising it in any way? Can you think of a situation we might add here that is just a little bit less stressful for you than this one?"
_____g. Standardized instructions are used for each phase of scene presentation; reinforcement of *just* the no-anxiety items is avoided.	"OK, I see that was not stressful for you. Just concentrate on relaxing a minute." "What was your feeling of anxiety on the 0-to-100 scale? 20. OK, I want you to just relax for a minute, then I'll give you the same scene."
_____h. Each scene-presentation session ends with a successfully completed item (no anxiety for at least two successive presentations).	"OK, let's end today with this item we've just been working on, since you reported 5 suds during the last two presentations."
_____i. Each session is terminated _____(1) After 15 to 20 minutes of scene presentation _____(2) After indications of client restlessness or distractibility	"We've done quite a bit of work today. Just spend a few minutes relaxing, and then we will stop."
_____25. Counselor uses written recording method during scene presentation to note client's progress through hierarchy.	"As we go through this session, I'm going to make some notes about the number of times we work with each item and your anxiety rating of each presentation."

VII. Homework and Follow-Up

_____26. Counselor assigns homework tasks that correspond to treatment progress of desensitization procedure:	"There is something I'd like you to do this week on a daily basis at home."
_____a. Daily practice of selected relaxation procedure	"Practice this relaxation procedure two times each day in a quiet place."
_____b. Visualization of items successfully completed at previous session	"On this tape there are three items we covered this week. Next week at home, after your relaxation sessions, practice imagining each of these three items."
_____c. Exposure to in vivo situations corresponding to successfully completed hierarchy items.	"You are ready now to actually go to a party by yourself. We have gotten to a point where you can imagine doing this without any stress."
_____27. Counselor instructs client to record completion of homework on daily log sheets.	"Each time you complete a homework practice, record it on your log sheets."
_____28. Counselor arranges for follow-up session or check-in.	"Check in with me in two weeks to give me a progress report."

Observer comments: _____

_____
_____

CHAPTER 20

SELF-MANAGEMENT STRATEGIES: SELF-MONITORING, STIMULUS CONTROL, SELF-REWARD, SELF-AS-A-MODEL, AND SELF-EFFICACY

The counselor helps the client to accept increasing responsibility for her or his behavior by preparing the client *during* the therapy session. The client does most of the work *between* sessions. As most mental health professionals agree, the major goal of therapy is to assist clients to help themselves accept more responsibility and function more independently (Kanfer & Gaelick-Buys, 1991).

Definitions of self-management have still not attained wide agreement, partly because of the confusing terms they contain. For example, self-change methods have been referred to as self-control (Cautela, 1969; Thoresen & Mahoney, 1974), self-regulation (Kanfer & Gaelick-Buys, 1991), and self-management (Mahoney, 1971, 1972). We prefer the label *self-management* because it suggests conducting and handling one's life with some skill. Also, the term *self-management* avoids the concepts of inhibition and restraint often associated with the words *control* and *regulation* (Thoresen & Mahoney, 1974). Whatever label is used, there is agreement about the processes involved in self-management. According to Bandura (1977, 1986, 1989, 1991), the social cognitive theory holds that human behavior is extensively motivated and regulated by the ongoing exercise of self-influence. The major self-management processes operate through four principal subfunctions: (1) self-monitoring of one's behavior, its components, and its effects; (2) judgment of one's behavior; (3) affective self-reactions; and (4) self-efficacy. Self-efficacy plays a central role in the exercise of personal agency by its strong impact on thought, affect, motivation, and action. In therapy, self-efficacy is very important in helping clients achieve treatment goals and enhancing their confidence and ability to execute the self-management strategies. The therapeutic alliance between the client and counselor is necessary to motivate the client "as a cooperative observer, reporter and change agent" (Kanfer & Gaelick-Buys, 1991, p. 306). Behavioral change is very often challenging and frequently not pleasant. If clients are involved in negotiating treatment planning and setting goals,

they are much more likely to implement the strategies and to achieve the therapeutic goals successfully. Finally, as Kanfer and Gaelick-Buys (1991) maintain, situation problems or symptoms often cannot be changed; the client must then learn coping strategies to handle these intractable situations.

Self-management strategies have several client outcomes that may include the following: (1) to use more effective task, interpersonal, cognitive, and emotional behaviors; (2) to alter perceptions of and judgmental attitudes toward problem situations or persons; and (3) either to change or learn to cope with a stress-inducing situation (Kanfer & Gaelick-Buys, 1991, p. 307). This chapter describes four self-management strategies:

Self-monitoring—Observing and recording your own particular behaviors (thoughts, feelings, and actions) about yourself and your interactions with environmental events

Stimulus control—Prearranging antecedents or cues to increase or decrease your performance of a target behavior

Self-reward—Giving yourself a positive stimulus following a desired response

Self-as-a-model—Using yourself as the model; seeing yourself performing the goal behavior in the desired manner

These four strategies may be viewed as self-management because in each procedure the client, in a self-directed fashion, monitors, alters, rewards, models, and possesses confidence (self-efficacy) to perform a specific task to produce the desired behavioral changes. Self-efficacy is a process in which the counselor attempts to empower confidence in the client. We describe the self-efficacy process later in this chapter. As we have indicated so often in previous chapters, none of these strategies is entirely independent of the client's personal history, gender, age, culture, ethnicity, and environmental variables.

In addition to these four self-management procedures, a client can use virtually any helping strategy in a self-directed manner. For example, a client could apply relaxation training to manage anxiety by using a relaxation-training audiotape without the assistance of a counselor. In fact, some degree of client self-management may be a necessary component of every successful therapy case. For example, in all the other helping strategies described in this book, some elements of self-management are suggested in the procedural guidelines for strategy implementation. These self-managed aspects of any therapy procedure typically include the following:

1. Client self-directed practice in the interview
2. Client self-directed practice in the in vivo setting (often through homework tasks)
3. Client self-observation and recording of target behaviors or of homework
4. Client self-reward (verbal or material) for successful completion of action steps and homework assignments

OBJECTIVES

1. Given a written client case description, describe the use of self-monitoring and stimulus control for this client.
2. Teach another person how to engage in self-monitoring as a self-change strategy.
3. Given a written client case description, be able to describe the use of a culturally relevant self-management program for this client.
4. Teach another person how to use self-reward, stimulus control, self-monitoring, and self-as-a-model.

CLINICAL USES OF SELF-MANAGEMENT STRATEGIES

Self-management strategies have been used for a wide range of clinical concerns (see Table 20-1). They have been applied to several health problems including AIDS, arthritis, asthma, cardiac disease, cystic fibrosis, diabetes, epilepsy, headaches, and self-health care. Among the psychological problems for which self-management strategies have been applied are autism, mood disorders, behavior disorders, bulimia, depression, emotional disturbance in adolescents, and insomnia. Self-management, in combination with social support, has been used to manage pain, to decrease heroin and alcohol abuse, to help compensate for developmental disabilities, and to improve disruptive students' classroom behavior. A self-management scale has been developed and self-efficacy scores are used as a predictor of self-management behavior.

MULTICULTURAL APPLICATIONS OF SELF-MANAGEMENT

Self-management has been used with diverse groups of clients in the areas of behavioral medicine, and HIV education and prevention. Jacob, Penn, Kulik, and Spieth (1992) researched the effects of self-management and positive reinforcement on the self-reported compliance rate of African American women who performed breast self-examinations over a nine-month period. Both self-management and positive reinforcement were associated with high compliance rates, especially for women who were designated initially as "monitors" (e.g., more likely to "track" things about themselves).

Roberson (1992) also explored the role of compliance and self-management in adult rural African Americans who had been diagnosed with chronic health conditions. She found that the patients and their health professionals had different notions of compliance and also different treatment goals. The patients defined compliance in terms of apparent "good health" and wanted a treatment that was manageable, viable, and effective. They developed systems of self-management to cope with their illness that were suitable to their lifestyles, belief patterns, and personal priorities. They believed they were managing their illness and treatment regimen effectively. Roberson (1992) suggested that as professionals we need to focus less on noncompliance rates and more on enhancing clients' efforts to manage their own illnesses and to live effectively with them.

In a related study, Haire-Joshu, Fisher, Munro, and Wedner (1993) explored differences in attitudes toward self-management between African American adults with asthma who received care in a public acute care setting and those who received care in a private setting stressing self-management. Those persons in the acute care setting were more likely to engage in self-treatment or to avoid or delay care compared to the patients who had learned preventive asthma self-management techniques. Rao and Kramer (1993) found self-control to be an important aspect of stress reduction and coping among African American mothers who had infants with sickle cell conditions.

Self-management has also been found an effective component of HIV infection risk reduction training with gay men (Kelly & St. Lawrence, 1990; Martin, 1993). In a recent study of several hundred African American youth, self-management was part of an eight-week HIV risk reduction program that compared information and skills training versus both information and skills training (St. Lawrence, Brasfield,

TABLE 20-1. Research on self-management

AIDS

Hagopian, L. P., Weist, M. D., & Ollendick, T. H. (1990). Cognitive-behavior therapy with an 11-year-old girl fearful of AIDS infection, other diseases, and poisoning: A case study. *Journal of Anxiety Disorders, 4,* 257–265.

Alcohol abuse

Sobell, L. C., Sobell, M. B., Toneatto, T., & Leo, G. I. (1993). What triggers the resolution of alcohol problems without treatment? *Alcoholism Clinical and Experimental Research, 17,* 217–224.

Anger

Smith, L. L., Smith, J. N., & Beckner, B. M. (1994). An anger-management workshop for women inmates. *Families in Society, 75,* 172–175.

Arthritis

Allegrante, J. P., Kovar, P. A., MacKenzie, C. R., & Peterson, M. G. (1993). A walking education program for patients with osteoarthritis of the knee: Theory and intervention strategies, Special Issue: Arthritis health education. *Health Education Quarterly, 20,* 63–81.

Lorig, K., & Gonzales, V. (1992). The integration of theory with practice: A 12-year case study. Special Issue: Roles and uses of theory in health education practice. *Health Education Quarterly, 19,* 355–368.

Lorig, K., & Holman, H. (1993). Arthritis self-management studies: A twelve-year review. Special Issue: Arthritis health education. *Health Education Quarterly, 20,* 17–28.

Taal, E., Riemsma, R. P., Brus, H. L., & Seydel, E. R. (1993). Group education for patients with rheumatoid arthritis. Special Issue: Psychosocial aspects of rheumatic diseases. *Patient Education and Counseling, 20,* 177–187.

Asthma

Clark, N. M., Evans, D., Zimmerman, B. J., & Levison, M. J. (1994). Patient and family management of asthma: Theory-based techniques for the clinician. *Journal of Asthma, 31,* 427–435.

Vazquez, M. I., & Buceta, J. M. (1993). Effectiveness of self-management programmes and relaxation training in the treatment of bronchial asthma: Relationships with trait anxiety and emotional attack triggers. *Journal of Psychosomatic Research, 37,* 71–81.

Vazquez, M. I., Fontan, B. J., & Buceta, J. M. (1992). Self-perception of asthmatic children and modification through self-management programmes. *Psychological Reports, 71,* 903–913.

Autism

Pierce, K. L., & Schreibman, L. (1994). Teaching daily living skills to children with autism in unsupervised settings through pictorial self-management. *Journal of Applied Behavior Analysis, 27,* 471–481.

Stahmer, A. C., & Schreibman, L. (1992). Teaching children with autism appropriate play in unsupervised environments using a self-management treatment package. *Journal of Applied Behavior Analysis, 25,* 447–459.

Behavior disorders

Lazarus, B. D. (1993). Self-management and achievement of students with behavior disorders. *Psychology in the Schools, 30,* 67–74.

Bulimia nervosa

Viens, M. J., & Hranchuk, K. (1992). The treatment of bulimia nervosa following surgery using a stimulus control procedure: A case study. *Journal of Applied Behavior Analysis, 23,* 313–317.

Cardiology

Dodge, J. A., Janz, N. K., & Clark, N. M. (1994). Self-management of the health care regimen: A comparison of nurses' and cardiac patients' perceptions. *Patient Education and Counseling, 23,* 73–82.

Starker, S. (1994). Self-care materials in the practice of cardiology: An explorative study among American cardiologists. *Patient Education and Counseling, 24,* 91–94.

Classroom behavior

Brigham, T. A. (1992). A brief commentary on the future of self-management interventions in education. *School Psychology Review, 21,* 264–268.

Cole, C. L. (1992). Self-management interventions in the schools. *School Psychology Review, 21,* 188–192.

Fantuzzo, J. W., & Rohrbeck, C. A. (1992). Self-managed groups: Fitting self-management approaches into classroom systems. *School Psychology Review, 21,* 255–263.

Seabaugh, G. O., & Schumaker, J. B. (1994). The effects of self-regulation training on the academic productivity of secondary students with learning problems. *Journal of Behavioral Education, 4,* 109–133.

Skinner, C. H., & Smith, E. S. (1992). Issues surrounding the use of self-management interventions for increasing academic performance. *School Psychology Review, 21,* 202–210.

(continued)

Jefferson, Alleyne, O'Bannon, & Shirley, 1995). Youth who received both information and skills training lowered their risk to a greater degree, maintained risk reduction changes better, and deferred the onset of sexual activity to a greater extent than those who received only one component of training.

In the majority of these studies on self-management as well as a study on self-monitoring (Carr & Punzo, 1993), the participants were African American. There may be something about the use of self-management strategies that is consistent with the values and beliefs and behavioral patterns of some

TABLE 20-1. Research on self-management (continued)

Cystic fibrosis

Parcel, G. S., Swank, P. R., Mariotto, M. J., & Bartholomew, L. K. (1994). Self-management of cystic fibrosis: A structural model for educational and behavioral variables. *Social Science and Medicine, 38,* 1307–1315.

Depression

Rokke, P. D., & Kozak, C. D. (1989). Self-control deficits in depression: A process investigation. *Cognitive Therapy and Research, 13,* 609–621.

Developmental disabilities

Harchik, A. E., Sherman, J. A., & Sheldon, J. B. (1992). The use of self-management procedures by people with developmental disabilities: A brief review. *Research in Developmental Disabilities, 13,* 211–227.

Diabetes

Ruggiero, L., Spirito, A., Coustan, D., & McGarvey, S. T. (1993). Self-reported compliance with diabetes self-management during pregnancy. *International Journal of Psychiatry in Medicine, 23,* 195–207.

Emotionally disturbed adolescents

Ninness, H. C., Fuerst, J., Rutherford, R. D., & Glenn, S. S. (1991). Effects of self-management training and reinforcement on the transfer of improved conduct in the absence of supervision. *Journal of Applied Behavior Analysis, 24,* 499–508.

Epilepsy

Dilorio, C., Faherty, B., & Manteuffel, B. (1994). Epilepsy self-management: Partial replication and extension. *Research in Nursing and Health, 17,* 167–174.

Self-graphing

DiGangi, S. A., Maag, J. W., & Rutherford, R. B. (1991). Self-graphing of on-task behavior: Enhancing the reactive effects of self-monitoring on on-task behavior and academic performance. *Learning Disability Quarterly, 14,* 221–230.

Headaches

Beams, L., Sanders, M. R., & Bor, W. (1992). The role of parent training in the cognitive behavioral treatment of children's headaches. *Behavioural Psychotherapy, 20,* 167–180.

Health care

Marks, I. M. (1994). Behavior therapy as an aid to self-care. *Current Directions in Psychological Science, 3,* 19–22.

Heroin addiction

Van-Bilsen, H. P., Henck, J. G., & Whitehead, B. (1994). Learning controlled drugs use: A case study. *Behavioural and Cognitive Psychotherapy, 22,* 87–95.

Insomnia

Gustafson, R. (1992). Treating insomnia with a self-administered muscle relaxation training program: A follow-up. *Psychological Reports, 70,* 124–126.

Mood

Thayer, R. E., Newman, J. R., & McClain, T. M. (1994). Self-regulation of mood: Strategies for changing a bad mood, raising energy, and reducing tension. *Journal of Personality and Social Psychology, 67,* 910–925.

Pain

Payne, T. J., Johnson, C. A., Penzien, D. B., & Porzelius, J. (1994). Chest pain self-management training for patients with coronary artery disease. *Journal of Psychosomatic Research, 38,* 409–418.

Polydipsia

Turnbull, J. (1993). Treatment of polydipsia using a contingency management procedure incorporating a self-management approach: A single case study. *Behavioural and Cognitive Psychotherapy, 21,* 275–279.

Self-management scale

Williams, R. L., Moore, C. A., Pettibone, T. J., & Thomas, S. P. (1992). Construction and validation of a brief self-report scale of self-management practices. *Journal of Research in Personality, 26,* 216–234.

Social support

Conn, V. S., Taylor, S. G., & Hayes, V. (1992). Social support, self-esteem, and self-care after myocardial infarction. *Health Values: The Journal of Health Behavior, Education and Promotion, 16,* 25–31.

Sobell, L. C., Sobell, M. B., Toneatto, T., & Leo, G. I. (1993). What triggers the resolution of alcohol problems without treatment? *Alcoholism Clinical and Experimental Research, 17,* 217–224.

Taal, E., Rasker, J. J., Seydel, E. R., & Wiegman, O. (1993). Health status, adherence with health recommendations, self-efficacy and social support in patients with rheumatoid arthritis. Special Issue: Psychosocial aspects of rheumatic diseases. *Patient Education and Counseling, 20,* 63–76.

Self-efficacy as a predictor of self-management

Parcel, G. S., Swank, P. R., Mariotto, M. J., & Bartholomew, L. K. (1994). Self-management of cystic fibrosis: A structural model for educational and behavioral variables. *Social Science and Medicine, 38,* 1307–1315.

African American clients. For example, in the chapter on reframing, we noted that some African American women report having learned to be self-reliant as a general strategy for living (although caution should be used in generalizing this to all African American women). Such value and belief systems and behavioral patterns as well as the belief that therapy should promote action and not just "talk" and self-exploration (Sue & Sue, 1990) would be very consistent with the use of a self-management intervention. It would seem that self-management could be incorporated as a culturally effective intervention with many clients from diverse groups, especially as self-management is time limited, deals with the present, and focuses on "concrete resolution of problems" (Sue & Sue, 1990, p. 200). Self-management also may appeal to some clients of color who do not like or feel comfortable with traditional mental health services. Keep in mind that in self-management efforts the work is client-managed and most of it occurs outside counseling sessions. Caution must be used, however, in selecting self-management as an appropriate intervention for all clients from diverse cultural groups. McCafferty (1992) has suggested that the process of self-regulation varies among cultures and societies. Many of the notions involved in self-management strategies are decidedly Eurocentric. Casas (1988) asserts that the basic notion underlying self-management may "not be congruent with the life experiences of many racial/ethnic minority persons. More specifically, as a result of life experiences associated with racism, discrimination, and poverty, people may have developed a cognitive set (e.g., an external locus of control, an external locus of responsibility, and learned helplessness) that . . . is antithetical to any self-control approaches" (pp. 109–110). Recall from the discussion of worldviews in Chapter 11 on treatment planning that there are four quadrants of Sue and Sue's (1990) cultural identity model based on the dimensions of locus of control and locus of responsibility (IC-IR, IC-ER, EC-IR, EC-ER) and these quadrants range from internal control to external control and from internal responsibility to external responsibility. Thus, locus of control and locus of responsibility seem to be mediating variables that affect the appropriateness of using self-management for some women clients and for some clients of color. In an innovative study conducted by St. Lawrence (1993), African American female and male youth completed measures of knowledge related to AIDS, attitudes toward the use of condoms, vulnerability to HIV infection, peer sexual norms, personal sexual behavior, contraceptive preferences, and locus of control. Condom use as prevention was associated with greater internal locus of control, which was higher for the African American girls than for the boys. In addition to the mediating variables of locus of control and locus of responsibility, the client's identification with his or her cultural (collective) identity, acculturation status, and assimilation may also be mediating variables that affect the use of self-management for non-Euro-American clients.

Guidelines for Using Self-Management with Diverse Groups of Clients

We recommend the following guidelines in using self-management approaches with diverse groups of clients.

1. Consider the client's lifestyle, beliefs, behavioral patterns, and personal priorities in assessing the usefulness of self-management. For example, if the client is interested in following the progress of events, a strategy such as self-monitoring may be relevant to his or her personal and cognitive style. For clients who have no interest in such tracking, self-monitoring may appear to be a waste of time, an activity that is personally and culturally irrelevant.

2. Adapt the intervention to the client's culture. Some clients have been socialized to be very private and would feel most uncomfortable in publicly displaying their self-monitoring data. Other clients would be unlikely to discipline themselves to go to one place to obtain stimulus control, such as using a smoking chair to help control smoking. And, depending on the client's history, the idea of using self-rewards may be irrelevant. At the very least, the rewards must be tailored to the client's gender and culture.

3. Discover the client's worldview (see Chapter 11) and consider the relevance of self-management based on this perception of the world. For clients whose cultural identity is shaped by external locus of control and external locus of responsibility, self-management is not as useful.

4. Consider the relevance of self-management against the client's goals for counseling and also the context of the client's life. If the client is also struggling with problems in living; aversive external structures; the presence of racism, discrimination, and oppression, self-management is not likely to be very useful. Consider how it would feel if you were a low-income woman with no social support and few resources and were regularly beaten by your live-in male partner—and your therapist told you to engage in some form of self-management!

CHARACTERISTICS OF AN EFFECTIVE SELF-MANAGEMENT PROGRAM

Well-constructed and well-executed self-management programs have some advantages that are not so apparent in counselor-administered procedures. For instance, use of a self-management procedure may increase a person's perceived control over the environment and decrease dependence on the counselor or others. Perceived control over the environment often motivates a person to take some action. Second, self-management approaches are practical—inexpensive and portable (Thoresen & Mahoney, 1974, p. 7). Third, such strategies are usable. By this we mean that occasionally a person will refuse to go "into therapy" to stop drinking or to lose weight (for example) but will agree to use the self-administered instructions that a self-management program provides. This may be particularly advantageous with some clients who are mistrustful of therapy. In fact, one study found that people who had never received counseling were more agreeable than clients to the idea of using self-management. Finally, self-management strategies may enhance generalization of learning—both from the interview to the environment and from problematic to nonproblematic situations (Thoresen & Mahoney, 1974, p. 7). These are some of the possible advantages of self-management that have spurred both researchers and practitioners to apply and explore some of the components and effects of successful self-management programs. Although many questions remain unanswered, we can say tentatively that the following factors may be important in an effective self-management program:

1. A combination of strategies, some focusing on antecedents of behavior and others on consequences
2. Consistent use of strategies over a period of time
3. Evidence of client self-evaluation, goal setting, and self-efficacy
4. Use of covert, verbal, or material self-reinforcement
5. Some degree of external, or environmental, support

Combination of Strategies

A combination of self-management strategies is usually more useful than a single strategy. In a weight-control study, Mahoney, Moura, and Wade (1973) found that the addition of self-reward significantly enhanced the procedures of self-monitoring and stimulus control. Further, people who combined self-reward and self-punishment lost more weight than those who used just one of the procedures. Greiner and

Karoly (1976) found that students who used self-monitoring, self-reward, and planning strategies improved their study behavior and academic performance more than students who used only one strategy. Mitchell and White (1977) found that the frequency of clients' reported migraine headaches was reduced in direct proportion to the number of self-management skills they used. Similarly, Perri and Richards (1977) and Heffernan and Richards (1981) discovered that successful self-controllers reported using a greater number of techniques for a longer time than unsuccessful self-controllers. (In these studies, successful self-controllers were defined as persons who had increased or decreased the target behavior at least 50% and had maintained this level for several months.) Problem areas for which comprehensive self-management programs have been developed include weight control (Mahoney & Mahoney, 1976), interpersonal skills training (McFall & Dodge, 1982), developmental disabilities (Litrownik, 1982), anxiety (Deffenbacher & Suinn, 1982), addictive disorders (Marlatt & Parks, 1982), depression (Rehm, 1982), insomnia (Bootzin, 1977), and academic performance (Neilans & Israel, 1981). Finally, self-reinforcement, stimulus control, and self-monitoring were used as maintenance programs on the long-term management of obesity.

Consistent Use of Strategies

Consistent, regular use of the strategies is a very important component of effective self-management. Seeming ineffectiveness may be due not to the impotence of the strategy but to its inconsistent or sporadic application (Thoresen & Mahoney, 1974, p. 107). It is probable that successful self-controllers use strategies more frequently and more consistently than unsuccessful self-controllers. Similarly, "failures" in a self-management program may be lax in using the procedures. Also, if self-management efforts are not used over a certain period of time, their effectiveness may be too limited to produce any change.

Self-Evaluation, Standard Setting, and Self-Efficacy

Self-evaluation in the form of standard setting (or goal setting) and intention statements seems to be an important component of a self-management program. Some evidence also suggests that self-selected stringent standards affect performance more positively than lenient standards (Bandura, 1971). Self-efficacy or the level of confidence a person has in his or her ability to develop intentions and set behavioral goals for himself or herself is also an important factor in an effective self-management program. Successful self-controllers usually set higher goals and criteria for change than unsuccessful self-controllers. However, the standards set should be realistic and within reach, or it is unlikely that self-reinforcement will ever occur.

Use of Self-Reinforcement

Self-reinforcement, either covert, verbal, or material, appears to be an important ingredient of an effective self-management program. Being able to praise oneself covertly or to note positive improvement seems to be correlated with self-change. In contrast, self-criticism (covert and verbal) seems to militate against change (Mahoney & Mahoney, 1976). Material self-reward (such as money) may be more effective than either self-monitoring or self-punishment. Self-reinforcement must also be relevant to the client's gender and culture.

Environmental Support

Some degree of external support is necessary to effect and maintain the changes resulting from a self-management program. For example, public display of self-monitoring data and the help of another person provide opportunities for social reinforcement that often augment behavior change. Successful self-controllers may report receiving more positive feedback from others about their change efforts than unsuccessful self-controllers. To maintain any self-managed change, there must be some support from the social and physical environment, although clients from some cultural backgrounds may not want to have their changes publicly acknowledged.

STEPS IN DEVELOPING A CLIENT SELF-MANAGEMENT PROGRAM

We have incorporated these five characteristics of effective self-management into a description of the steps associated with a self-management program. The steps are applicable to any program in which the client uses stimulus control, self-monitoring, or self-reward. Figure 20-1 summarizes the steps associated with developing a self-management program; the characteristics of effective self-management reflected in the steps are noted in the left column of the figure.

In developing a self-management program, steps 1 and 2 both involve aspects of standard setting and self-evaluation. In step 1, the client identifies and records the target behavior

and its antecedents and consequences. This step involves self-monitoring in which the client collects baseline data about the behavior to be changed. If baseline data have not been collected as part of problem assessment (Chapter 9), it is imperative that such data be collected now, before using any self-management strategies. In step 2, the client explicitly identifies the desired behavior, conditions, and level of change. As you may remember from Chapter 10, the behavior, conditions, and level of change are the three parts of a counseling outcome goal. Defining the goal is an important part of self-management because of the possible motivating effects of standard setting. Establishing goals may interact with some of the self-management procedures and contribute to the desired effects.

Steps 3 and 4 are directed toward helping the client select a combination of self-management strategies to use. The counselor will need to explain all the possible self-management strategies to the client (step 3). The counselor should emphasize that the client should select some strategies that involve prearrangement of the antecedents and some that involve manipulation and self-administration of consequences. Ultimately, the client is responsible for selecting which self-management strategies should be used (step 4). Client selection of the strategies is an important part of the overall *self-directed* nature of self-management.

Steps 5 through 9 all involve procedural considerations that may strengthen client commitment and may encourage consistent use of the strategies over time. First, the client commits himself or herself verbally by specifying what and how much change is desired and the action steps (strategies) the client will take to produce the change (step 5). Next, the counselor will instruct the client in how to carry out the selected strategies (step 6). (The counselor can follow the guidelines listed later in the chapter for self-monitoring, those for stimulus control, and the ones for self-reward.) Explicit instructions and modeling by the counselor may encourage the client to use a procedure more accurately and effectively. The instructional set given by a counselor may contribute to some degree to the overall treatment outcomes. The client also may use the strategies more effectively if there is an opportunity to rehearse the procedures in the interview under the counselor's direction (step 7). Finally, the client applies the strategies in vivo (step 8) and records (monitors) the frequency of use of each strategy and the level of the target behavior (step 9). Some of the treatment effects of self-management may also be a function of the client's self-recording.

Steps 10 and 11 involve aspects of self-evaluation, self-reinforcement, and environmental support. The client

**Characteristics of an Effective
Self-Management Program**

**Steps in Developing a
Self-Management Program**

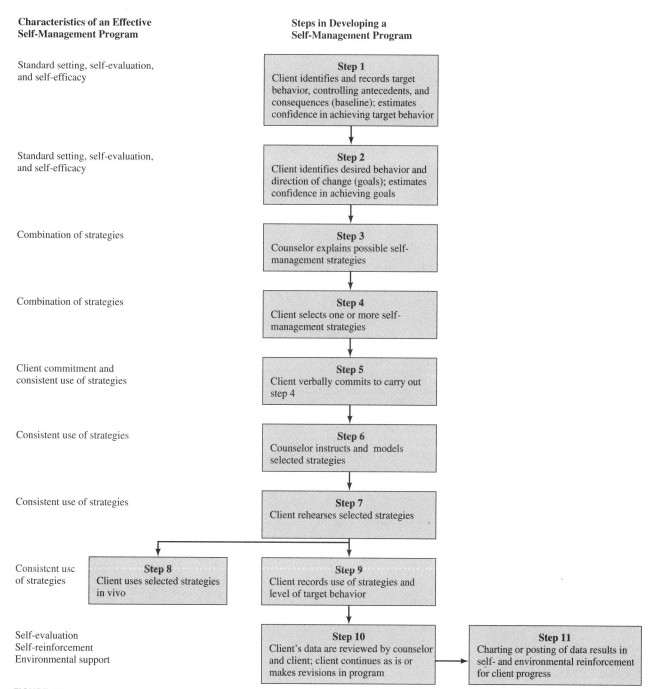

Standard setting, self-evaluation,
and self-efficacy

> **Step 1**
> Client identifies and records target
> behavior, controlling antecedents, and
> consequences (baseline); estimates
> confidence in achieving target behavior

Standard setting, self-evaluation,
and self-efficacy

> **Step 2**
> Client identifies desired behavior and
> direction of change (goals); estimates
> confidence in achieving goals

Combination of strategies

> **Step 3**
> Counselor explains possible self-
> management strategies

Combination of strategies

> **Step 4**
> Client selects one or more self-
> management strategies

Client commitment and
consistent use of strategies

> **Step 5**
> Client verbally commits to carry out
> step 4

Consistent use of strategies

> **Step 6**
> Counselor instructs and models
> selected strategies

Consistent use of strategies

> **Step 7**
> Client rehearses selected strategies

Consistent use
of strategies

> **Step 8**
> Client uses selected strategies
> in vivo

> **Step 9**
> Client records use of strategies and
> level of target behavior

Self-evaluation
Self-reinforcement
Environmental support

> **Step 10**
> Client's data are reviewed by counselor
> and client; client continues as is or
> makes revisions in program

> **Step 11**
> Charting or posting of data results in
> self- and environmental reinforcement
> for client progress

FIGURE 20-1. Developing an effective self-management program

has an opportunity to evaluate progress toward the goal by reviewing the self-recorded data collected during strategy implementation (step 10). Review of the data may indicate that the program is progressing smoothly or that some adjustments are needed. When the data suggest that some progress toward the goal is being made, the client's self-evaluation may set the occasion for self-reinforcement. Charting or posting the data (step 11) can enhance self-reinforcement and can elicit important environmental support for long-term maintenance of client change.

The following section describes how self-monitoring can be used to record the target behavior. Such recording can occur initially for problem assessment and goal setting, or it can be introduced later as a self-change strategy. We will discuss specifically how self-monitoring can be used to promote behavior change.

SELF-MONITORING: PURPOSES, USES, AND PROCESSES

Purposes of Self-Monitoring

In Chapter 9 we defined self-monitoring as a process in which clients observe and record things about themselves and their interactions with environmental situations. Self-monitoring is a useful adjunct to problem assessment because the observational data can verify or change the client's verbal report about the problem behavior. We recommend that clients record their daily self-observations over a designated time period on a behavior log. Usually the client observes and records the problem behavior, controlling antecedents, and resulting consequences. Thoresen and Mahoney (1974) assert that self-monitoring is a major *first* step in any self-change program (as in any change program!). The client must be able to discover what is happening *before* implementing a self-change strategy, just as the counselor must know what is going on before using any other therapeutic procedure. In other words, any self-management strategy, like any other strategy, should be preceded by a baseline period of self-observation and recording. During this period, the client collects and records data about the behavior to be changed (B), the antecedents (A) of the behavior, and the consequences (C) of the behavior. In addition, the client may wish to note how much or how often the behavior occurs. For example, a client might record the daily amount of study time or the number of times he or she left the study time and place to do something else. The behavior log presented in Chapter 9 for problem-assessment data can also be used by a client to collect baseline data before implementing a self-management

program. If the counselor introduces self-management strategies *after* problem assessment, these self-observation data should be already available.

As we discussed in Chapter 10, self-monitoring is also very useful for evaluation of goals or outcomes. When a client self-monitors the target behavior either before or during a treatment program, "the primary utility of self-monitoring lies in its assessment or data collection function" (Ciminero et al., 1977, p. 196). In recent years, however, practitioners and researchers have realized that the mere act of self-observation can produce change. As one collects data about oneself, the data collection may influence the behavior being observed. We now know that self-monitoring is useful not only to collect data but also to promote client change. If properly structured and executed, self-monitoring can be used as one type of self-management strategy.

Clinical Uses of Self-Monitoring

A number of research reports and clinical studies have explored self-monitoring as the major treatment strategy. Table 20-2 indicates the variety of subjects for which self-monitoring has been used. These include academic productivity, career mobility, classroom behaviors, cross-cultural relationship adjustment, and risk behaviors for health concerns. Other topics are the immune system, mood states, panic attacks, problem-solving performance, sleep onset, and weight loss.

Factors Influencing the Reactivity of Self-Monitoring

As you may recall from Chapter 9, two issues involved in self-monitoring are the reliability of the self-recording and its reactivity. Reliability, the accuracy of the self-recorded data, is important when self-monitoring is used to evaluate the goal behaviors. However, when self-monitoring is used as a change strategy, the accuracy of the data is less crucial. From a counseling perspective, the reactivity of self-monitoring makes it suitable for a change strategy. As an example of reactivity, Kanfer and Gaelick-Buys (1991) noted that a married couple using self-monitoring to observe their frequent arguments reported that whenever the monitoring device (a tape-recorder) was turned on, the argument was avoided.

Although the reactivity of self-monitoring can be a problem in data collection, it can be an asset when self-monitoring is used intentionally as a helping strategy. In using self-monitoring as a change strategy, try to maximize

TABLE 20-2. Research on self-monitoring

Academic productivity

Carr, S., & Punzo, R. (1993). The effects of self-monitoring of academic accuracy and productivity on the performance of students with behavioral disorders. *Journal of Black Psychology, 19,* 155–168.

Maag, J. W., Reid, R., & DiGangi, S. A. (1993). Differential effects of self-monitoring attention, accuracy, and productivity. *Journal of Applied Behavior Analysis, 26,* 329–344.

Performance of children with learning disabilities

Harris, K. R., Graham, S., Reid, R., & McElroy, K. (1994). Self-monitoring of attention versus self-monitoring of performance: Replication and cross-task comparison studies. *Learning Disability Quarterly, 17,* 121–139.

Marshall, K. J., Lloyd, J. W., & Hallahan, D. P. (1993). Effects of training to increase self-monitoring accuracy. *Journal of Behavioral Education, 3,* 445–459.

McDougall, D., & Brady, M. (1995). Using audio-cued self-monitoring for students with severe behavior disorders. *Journal of Educational Research, 88,* 309–317.

Reid, R., & Harris, K. R. (1993). Self-monitoring of attention versus self-monitoring of performance: Effects on attention and academic performance. *Exceptional Children, 60,* 29–40.

Attitudes toward learning statistics

Lan, W. Y., Bradley, L., & Parr, G. (1993). The effects of a self-monitoring process on college students' learning in an introductory statistics course. *Journal of Experimental Education, 62,* 26–40.

Autism

Strain, P. S., Kohler, F. W., Storey, K., & Danko, C. D. (1994). Teaching preschoolers with autism to self-monitor their social interactions: An analysis of results in home and school settings. *Journal of Emotional and Behavioral Disorders, 2,* 78–88.

Blood glucose

Bernbaum, M., Albert, S. G., McGinnis, J., & Brusca, S. (1994). The reliability of self blood glucose monitoring in elderly diabetic patients. *Journal of the American Geriatrics Society, 42,* 779–781.

Career mobility

Kilduff, M., & Day, D. V. (1994). Do chameleons get ahead? The effects of self-monitoring on managerial careers. *Academy of Management Journal, 37,* 1047–1060.

Classroom

Brown, D., & Frank, A. R. (1990). "Let me do it": Self-monitoring in solving arithmetic problems. *Education and Treatment of Children, 13,* 239–248.

Dunlap, G., Clarke, S., Jackson, M., & Wright, S. (1995). Self-monitoring of classroom behaviors with students exhibiting emotional and behavioral challenges. *School Psychology Quarterly, 10,* 165–177.

Hughes, C. A., & Boyle, J. R. (1991). Effects of self-monitoring for on-task behavior and task productivity on elementary students with moderate mental retardation. *Education and Treatment of Children, 14,* 96–111.

Maag, J. W., Rutherford, R. B., & DiGangi, S. A. (1992). Effects of self-monitoring and contingent reinforcement on on-task behavior and academic productivity of learning-disabled students: A social validation study. *Psychology in the Schools, 29,* 157–172.

Malone, L. D., & Matropieri, M. A. (1992). Reading comprehension instruction: Summarization and self-monitoring training for students with learning disabilities. *Exceptional Children, 58,* 270–279.

McCurdy, B. L., & Shapiro, E. S. (1992). A comparison of teacher-, peer-, and self-monitoring with curriculum-based measurement in reading among students with learning disabilities. *Journal of Special Education, 26,* 162–180.

Prater, M. A., Joy, R., Chilman, B., & Temple, J. (1991). Self-monitoring of on-task behavior by adolescents with learning disabilities. *Learning Disability Quarterly, 14,* 164–177.

Cross-cultural studies

Gudykunst, W. B., Gao, G., Nishida, T., & Bond, M. H. (1989). A cross-cultural comparison of self-monitoring. *Communication Research Reports, 6,* 7–12.

Hamid, P. N. (1993). Self-monitoring and ethnic group membership. *Psychological Reports, 72,* 1347–1350.

Goodwin, R., & Soon, A. P. Y. (1994). Self-monitoring and relationship adjustment: A cross-cultural analysis. *Journal of Social Psychology, 134,* 35–39.

Cardiovascular disease–risk behaviors

Adachi, Y., Nakasone, H., & Meno, T. (1991). Behavioral group therapy for hypercholesterolemia in a public health center. *Japanese Journal of Behavior Therapy, 17,* 1–11.

Clark, N. M., Janz, N. K., Dodge, J. A., & Sharpe, P. A. (1992). Self-regulation of health behavior: The "take PRIDE" program. Special Issue: Roles and uses of theory in health education practice. *Health Education Quarterly, 19,* 341–354.

Madsen, J., Sallis, J. F., Rupp, J. W., & Senn, K. L. (1993). Relationship between self-monitoring of diet and exercise change and subsequent risk factor changes in children and adults. *Patient Education and Counseling, 21,* 61–69.

Estimations of blood alcohol concentration

Martin, C. S., Rose, R. J., & Obremski, K. M. (1991). Estimation of blood alcohol concentrations in young male drinkers. *Alcoholism Clinical and Experimental Research, 15,* 494–499.

Tucker, J. A., Vuchinich, R. E., Harris, C. V., & Gavornik, M. G. (1991). Agreement between subject and collateral verbal reports of alcohol consumption in older adults. *Journal of Studies on Alcohol, 52,* 148–155.

(continued)

TABLE 20-2. Research on self-monitoring (continued)

Emotional expressive cues

Friedman, H. S., & Miller-Herringer, T. (1991). Nonverbal display of emotion in public and in private: Self-monitoring, personality, and expressive cues. *Journal of Personality and Social Psychology, 61,* 766–775.

Hair and fiber ingestion

Kaushik, S. S., & Singh, A. K. (1993). Modification of trichobazoar through behaviour modification techniques: A case report. *Indian Journal of Clinical Psychology, 20,* 57–59.

Immune system

Halley, F. M. (1991). Self-regulation of the immune system through biobehavioral strategies. *Biofeedback and Self-Regulation, 16,* 55–74.

Inflammatory bowel disease and stress

Greene, B. R., Blanchard, E. B., & Wan, C. K. (1994). Long-term monitoring of psychosocial stress and symptomatology in inflammatory bowel disease. *Behaviour Research and Therapy, 32,* 217–226.

Marital antecedents and consequences of stressful interactions

Halford, W. K., Gravestock, F. M., Lowe, R., & Scheldt, S. (1992). Toward a behavioral ecology of stressful marital interactions. *Behavioral Assessment, 14,* 199–217.

Mood states

Reynolds, M., & Salkovskis, P. M. (1992). Comparison of positive and negative intrusive thoughts and experimental investigation of the differential effects of mood. *Behaviour Research and Therapy, 30,* 273–281.

Panic attacks

de Beurs, E., Lange, A., & Van Dyck, R. (1992). Self-monitoring of panic attacks and retrospective estimates of panic: Discordant findings. *Behaviour Research and Therapy, 30,* 411–413.

de Beurs, E., Garssen, B., Buikhuisen, M., & Lange, A. (1994). Continuous monitoring of panic. *Acta Psychiatrica Scandinavica, 90,* 38–45.

Pain perception

Dar, R., & Leventhal, H. (1993). Schematic processes in pain perception. *Cognitive Therapy and Research, 17,* 341–357.

Perceived physical exertion

Dunbar, C. C., Robertson, R. J., Baun, R., & Blandin, M. F. (1992). The validity of regulating exercise intensity by ratings of perceived exertion. *Medicine and Science in Sports and Exercise, 24,* 94–99.

Preschoolers

de Haas-Warner, S. J. (1992). The utility of self-monitoring for preschool on-task behavior. *Topics in Early Childhood Special Education, 12,* 478–495.

Problem solving

Delclos, V. R., & Harrington, C. (1991). Effects of strategy monitoring and proactive instruction on children's problem-solving performance. *Journal of Educational Psychology, 83,* 35–42.

Positive and negative intrusive thoughts

Reynolds, M., & Salkovskis, P. M. (1992). Comparison of positive and negative intrusive thoughts and experimental investigation of the differential effects of mood. *Behaviour Research and Therapy, 30,* 273–281.

Reactivity

Kirby, K. C., Fowler, S. A., & Baer, D. M. (1991). Reactivity in self-recording: Obtrusiveness of recording procedure and peer comments. *Journal of Applied Behavior Analysis, 24,* 487–498.

Self-graphing

DiGangi, S. A., Maag, J. W., & Rutherford, R. B. (1991). Self-graphing of on-task behavior: Enhancing the reactive effects of self-monitoring on on-task behavior and academic performance. *Learning Disability Quarterly, 14,* 221–230.

Trammel, D. L., Schloss, P. J., & Alper, S. (1994). Using self-recording, evaluation, and graphing to increase completion of homework assignments. *Journal of Learning Disabilities, 27,* 75–81.

Sleep onset

Downey, R., & Bonnet, M. H. (1992). Training subjective insomniacs to accurately perceive sleep onset. *Sleep, 15,* 58–63.

Suicide ideation

Clum, G. A., & Curtin, L. (1993). Validity and reactivity of a system of self-monitoring suicide ideation. *Journal of Psychopathology and Behavioral Assessment, 15,* 375–385.

Swimming

Critchfield, T. S., & Vargas, E. A. (1991). Self-recording, instructions, and public self-graphing: Effects on swimming in the absence of coach verbal interaction. *Behavior Modification, 15,* 95–112.

Tourette syndrome

Peterson, A. L., & Azrin, N. H. (1992). An evaluation of behavioral treatments for Tourette Syndrome. *Behaviour Research and Therapy, 30,* 167–174.

Weight loss

Baker, R. C., & Kirschenbaum, D. S. (1993). Self-monitoring may be necessary for successful weight control. *Behavior Therapy, 24,* 377–397.

Foreyt, J. P., & Goodrick, G. K. (1994). Attributes of successful approaches to weight loss and control. *Applied and Preventive Psychology, 3,* 209–215.

Workers' health

Fox, M. L., & Dwyer, D. (1995). Stressful job demands and worker health: An investigation of the effects of self-monitoring. *Journal of Applied Social Psychology, 25,* 1973–1995.

the reactive effects of self-monitoring—at least to the point of producing desired behavioral changes. Self-monitoring for *long* periods of time maintains reactivity.

A number of factors seem to influence the reactivity of self-monitoring. A summary of these factors suggests that self-monitoring is most likely to produce positive behavioral changes when change-motivated subjects continuously monitor a limited number of discrete, positively valued target behaviors; when performance feedback and goals or standards are made available and are unambiguous; and when the monitoring act is both salient and closely related in time to the target behaviors.

Nelson (1977) has identified eight variables that seem to be related to the occurrence, intensity, and direction of the reactive effects of self-monitoring:

1. *Motivation.* Clients who are interested in changing the self-monitored behavior are more likely to show reactive effects when they self-monitor.
2. *Valence of target behaviors.* Behaviors a person values positively are likely to increase with self-monitoring; negative behaviors are likely to decrease; neutral behaviors may not change.
3. *Type of target behaviors.* The nature of the behavior that is being monitored may affect the degree to which self-monitoring procedures effect change.
4. *Standard setting (goals), reinforcement, and feedback.* Reactivity is enhanced for people who self-monitor in conjunction with goals and the availability of performance reinforcement or feedback.
5. *Timing of self-monitoring.* The time when the person self-records can influence the reactivity of self-monitoring. Results may differ depending on whether self-monitoring occurs before or after the target response.
6. *Devices used for self-monitoring.* More obtrusive or visible recording devices seem to be more reactive than unobtrusive devices.
7. *Number of target responses monitored.* Self-monitoring of only one response increases reactivity. As more responses are concurrently monitored, reactivity decreases.
8. *Schedule for self-monitoring.* The frequency with which a person self-monitors can affect reactivity. Continuous self-monitoring may result in more behavior change than intermittent self-recording.

Three factors may contribute to the reactive effects of self-monitoring:

1. *Client characteristics.* Client intellectual and physical abilities may be associated with greater reactivity when self-monitoring.

2. *Expectations.* Clients seeking help may have some expectations for desirable behavior changes. However, it is probably impossible to separate client expectations from implicit or explicit therapeutic "demands" to change the target behavior.
3. *Behavior change skills.* Reactivity may be influenced by the client's knowledge and skills associated with behavior change. For example, the reactivity of addictive behaviors may be affected by the client's knowledge of simple, short-term strategies such as fasting or abstinence. These are general guidelines and their effects may vary with gender, class, race, and ethnicity of each specific client.

STEPS OF SELF-MONITORING

Self-monitoring involves at least six important steps: *rationale* for the strategy, *discrimination* of a response, *recording* of a response, *charting* of a response, *display* of data, and *analysis* of data (Thoresen & Mahoney, 1974, pp. 43–44). Each of these six steps and guidelines for their use are discussed here and summarized in Table 20-3. Remember that the steps are all interactive, and the presence of all of them may be required for a person to use self-monitoring effectively. Also, remember that any or all of these steps may need to be adapted, depending on the client's gender and culture.

Treatment Rationale

First, the therapist explains the rationale for self-monitoring. Before using the strategy, the client should be aware of what the self-monitoring procedure will involve and how the procedure will help with the client's problem. An example, adapted from Benson and Stuart (1992), follows:

> The purpose of self-monitoring is to increase your awareness of your sleep patterns. Research has demonstrated that people who have insomnia benefit from keeping a self-monitoring diary. Each morning for a week you will record the time you went to bed the previous night; approximately how many minutes it took you to fall asleep; if you awakened during the night, how many minutes you were awake; the total number of hours you slept; and the time you got out of bed in the morning. Also, on a scale you will rate how rested you feel in the morning, how difficult it was to fall asleep the previous night, the quality of sleep, your level of physical tension when you went to bed the previous night, your level of mental activity when you went to bed, and how well you think you were functioning the previous day. The diary will help us evaluate your sleep and remedy any problems. This kind of

TABLE 20-3. Steps of self-monitoring

1. *Rationale* for self-monitoring	A. Purpose
	B. Overview of procedure
2. *Discrimination* of a response	A. Selection of target response to monitor
	1. Type of response
	2. Valence of response
	3. Number of responses
3. *Recording* of a response	A. Timing of recording
	1. Prebehavior recording to decrease a response; postbehavior recording to increase a response
	2. Immediate recording
	3. Recording when no competing responses distract recorder
	B. Method of recording
	1. Frequency counts
	2. Duration measures
	a. Continuous recording
	b. Time sampling
	C. Devices for recording
	1. Portable
	2. Accessible
	3. Economical
	4. Somewhat obtrusive
4. *Charting* of a response	A. Charting and graphing of daily totals of recorded behavior
5. *Displaying* of data	A. Chart for environmental support
6. *Analysis* of data	A. Accuracy of data interpretation
	B. Knowledge of results for self-evaluation and self-reinforcement

awareness facilitates correcting any problems that might contribute to your insomnia. How does that sound?

Discrimination of a Response

When a client engages in self-monitoring, first an observation, or discrimination, of a response is required. For example, a client who is monitoring fingernail biting must be able to discriminate instances of nail biting from instances of other behavior. Discrimination of a response occurs when the client is able to identify the presence or absence of the behavior, and whether overt, like nail biting, or covert, like a positive self-thought. Thoresen and Mahoney (1974, p. 43) point out that making behavioral discriminations can be thought of as the "awareness" facet of self-monitoring.

Discrimination of a response involves helping the client identify *what* to monitor. Often this decision will require counselor assistance. The type of the monitored response may affect the results of self-monitoring. For example, self-monitoring may produce greater weight loss for people who recorded their daily weight and daily caloric intake than for those who recorded only daily weight. As McFall (1977) has observed, it is not very clear why some target responses seem to be better ones to self-monitor than others; at this point, the selection of target responses remains a pragmatic choice. Mahoney (1977, pp. 244–245) points out that there may be times when self-monitoring of certain responses could detract from therapeutic effectiveness, as in asking a suicidal client to monitor depressive thoughts.

The effects of self-monitoring also vary with the valence of the target response. There are always "two sides" of a behavior that could be monitored—the positive and the negative (Mahoney & Thoresen, 1974, p. 37). There seem to be times when one side is more important for self-monitoring than the other (Mahoney & Thoresen, 1974, p. 37).

Unfortunately, there are very few data to guide a decision about the exact type and valence of responses to monitor. Because the reactivity of self-monitoring is affected by the value assigned to a behavior (Watson & Tharp, 1997), one guideline might be to have the client monitor the behavior that she or he cares *most* about changing. Generally, it is a good idea to encourage the client to limit monitoring to one response, at least initially. If the client engages in self-monitoring of one behavior with no problems, then more items can be added.

Recording of a Response

After the client has learned to make discriminations about a response, the counselor can provide instructions and examples about the method for recording the observed response. Most clients have probably never recorded their behavior *systematically*. Systematic recording is crucial to the success of self-monitoring, so it is imperative that the client understand the importance and methods of recording. The client needs instructions in when and how to record and devices for recording. The timing, method, and recording devices all can influence the effectiveness of self-monitoring.

Timing of Self-Monitoring: When to Record.   One of the least understood processes of self-monitoring involves timing, or the point when the client actually records the target behavior. Instances have been reported of both prebehavior and postbehavior monitoring. In prebehavior monitoring, the client records the intention or urge to engage in the behavior *before* doing so. In postbehavior monitoring, the client records each completed instance of the target behavior—*after* the behavior has occurred. Kazdin (1974, p. 239) points out that the precise effects of self-monitoring may depend on the point at which monitoring occurs in the chain of responses relative to the response being recorded. Kanfer and Gaelick-Buys (1991) conclude that existing data are insufficient to judge whether pre- or postbehavior monitoring will have maximal effects. Nelson (1977) indicates that the effects of the timing of self-monitoring may depend partly on whether other responses are competing for the person's attention at the time the response is recorded. Another factor influencing the timing of self-monitoring is the amount of time between the response and the actual recording. Most people agree that delayed recording of the behavior weakens the efficacy of the monitoring process (Kanfer & Gaelick-Buys, 1991; Kazdin, 1974).

We suggest four guidelines that may help the counselor and client decide when to record. First, if the client is using monitoring as a way to *decrease* an undesired behavior, prebehavior monitoring may be more effective, as this seems to interrupt the response chain early in the process. An example of the rule of thumb for self-monitoring an undesired response would be to record whenever you have the urge

to smoke or to eat. Prebehavior monitoring may reduce smoking behavior. Prebehavior monitoring may result in more change than postbehavior monitoring. If the client is using self-monitoring to *increase* a desired response, then postbehavior monitoring may be more helpful. Postbehavior monitoring can make a person more aware of a low frequency, desirable behavior. Third, recording instances of a desired behavior as it occurs or immediately after it occurs may be most helpful. The rule of thumb is to "record *immediately* after you have the urge to smoke—or *immediately* after you have covertly praised yourself; do not wait even for 15 or 20 minutes, as the impact of recording may be lost." Fourth, the client should be encouraged to self-record the response when not distracted by the situation or by other competing responses. The client should be instructed to record the behavior in vivo as it occurs, if possible, rather than at the end of the day, when he or she is dependent on retrospective recall. In vivo recording may not always be feasible, and in some cases the client's self-recording may have to be performed later.

Method of Self-Monitoring: How to Record.   The counselor also needs to instruct the client in a *method* for recording the target responses. McFall (1977) points out that the method of recording can vary in a number of ways:

> It can range from a very informal and unstructured operation, as when subjects are asked to make mental notes of any event that seems related to mood changes, to something fairly formal and structured, as when subjects are asked to fill out a mood-rating sheet according to a time-sampling schedule. It can be fairly simple, as when subjects are asked to keep track of how many cigarettes they smoke in a given time period; or it can be complex and time-consuming, as when they are asked to record not only how many cigarettes they smoke, but also the time, place, circumstances, and affective response associated with lighting each cigarette. It can be a relatively objective matter, as when counting the calories consumed each day; or it can be a very subjective matter, as when recording the number of instances each day when they successfully resist the temptation to eat sweets. (p. 197)

Ciminero et al. (1977, p. 198) suggest that the recording method should be "easy to implement, must produce a representative sample of the target behavior, and must be sensitive to changes in the occurrence of the target behavior." Keep the method informal and unstructured for clients who are not "monitors" or who do not value "tracking" in such a systematic way.

As you may remember from our description of outcome evaluation in Chapter 10, frequency, duration, and intensity can be recorded with either a continuous recording or a time-sampling method. Selection of one of these methods will depend mainly on the type of target response and the frequency of its occurrence. To record the *number* of target responses, the client can use a frequency count. Frequency counts are most useful for monitoring responses that are discrete, do not occur all the time, and are of short duration (Ciminero et al., 1977, p. 190). For instance, clients might record the number of times they have an urge to smoke or the number of times they praise or compliment themselves covertly.

Other kinds of target responses are recorded more easily and accurately by duration. Any time a client wants to record the amount or length of a response, a duration count can be used. Ciminero et al. (1977, p. 198) recommend the use of a duration measure whenever the target response is not discrete and varies in length. For example, a client might use a duration count to note the amount of time spent reading textbooks or practicing a competitive sport. Or a client might want to keep track of the length of time spent in a "happy mood."

Sometimes a client may want to record two different responses and use both the frequency and duration methods. For example, a client might use a frequency count to record each urge to smoke and a duration count to monitor the time spent smoking a cigarette. Watson and Tharp (1997) suggest that the counselor can recommend frequency counts whenever it is easy to record clearly separate occurrences of the behavior and duration counts whenever the behavior continues for long periods.

Clients can also self-record intensity of responses whenever data are desired about the relative severity of a response. For example, a client might record the intensity of happy, anxious, or depressed feelings or moods.

Format of Self-Monitoring Instruments. There are many formats of self-monitoring instruments a client can use to record frequency, duration, or severity of the target response as well as information about contributing variables. The particular format of the instrument can affect reactivity and can increase client compliance with self-monitoring. The format of the instrument should be tailored to the client problem and to the client. Figure 20-2 shows examples of four formats for monitoring instruments. Example 1 in the figure illustrates a thought record for client mood problems (Greenberger & Padesky, 1995). The client records episodes of situations, moods, automatic thoughts and images, evidence that supports the hot thoughts, evidence that does not support the hot thoughts, alternative or balanced thoughts, and a new rating of moods. Example 2 illustrates a self-

monitoring format used for Sun Salutation warm-up and Hatha Yoga postures and for each day of the week. Clients can check off each posture they do on a daily basis, or they can rate their daily experience with each posture on a scale from 1—little benefit from posture—to 5—feelings of extraordinary benefit. Example 3 shows a format we use with couples in marital therapy for self-monitoring of content and quality of marital interactions. In this format, each person records the content of the interaction with the spouse (for example, having dinner together, talking about finances, discussing work, going to movies) and self-rates the quality of that interaction. Example 4 shows a format we use with clients to self-record aspects of anxiety responses. This format can be adapted to other covert (internal) responses. Each of these formats can use a variety of self-recording devices.

Devices for Self-Monitoring. Often clients report that one of the most intriguing aspects of self-monitoring involves the device or mechanism used for recording. For recording to occur systematically and accurately, the client must have access to some recording device. A variety of devices have been used to help clients keep accurate records. Note cards, daily log sheets, and diaries can be used to make written notations. A popular self-recording device is a wrist counter, such as a golf counter. The golf counter can be used for self-recording in different settings. If several behaviors are being counted simultaneously, the client can wear several wrist counters or use knitting tallies. A wrist counter with rows of beads permits the recording of several behaviors. Audio- and videotapes, toothpicks, or small plastic tokens can also be used as recording devices. Watson and Tharp (1997) report the use of pennies to count: a client can carry pennies in one pocket and transfer one penny to another pocket each time a behavior occurs. Children can record frequencies by pasting stars on a chart or by using a countoon, which has pictures and numbers for three recording columns: "What do I do," "My count," and "What happens." Clocks, watches, and kitchen timers can be used for duration counts.

The counselor and client select a recording device. Here is an opportunity to be inventive! There are several practical criteria to consider in helping a client select a recording device. The device should be portable and accessible so that it is present whenever the behavior occurs (Watson & Tharp, 1997). It should be easy and convenient to use and economical. The obtrusiveness of the device should also be considered. The recording device can function as a cue (discriminative stimulus) for the client to self-monitor, so it should be noticeable enough to remind the client to engage in

1. Thought record

1. Situation Who? What? When? Where?	2. Moods a. What did you feel? b. Rate each mood (0–100%).	3. Automatic Thoughts (Images) a. What was going through your mind just before you started to feel this way? Any other thoughts? Images? b. Circle the hot thought.	4. Evidence That Supports The Hot Thought	5. Evidence That Does *Not* Support Hot Thought	6. Alternative/ Balanced Thoughts a. Write an alternative or balanced thought. b. Rate how much you believe in each alternative or balanced thought (0–100%).	7. Rate Moods Now Rerate moods listed in column 2 as well as any new moods (0–100%).

SOURCE: From *Mind over Mood: A Cognitive Therapy Treatment Manual for Clients,* by D. Greenberger and C. A. Padesky, p. 37. Copyright 1995 by The Guilford Press. Reprinted by permission.

2. Yoga postures
Sun Salutation Warm-Up

Asana/Posture Days	1 Mountain	2 Back Bend	3 Forward Bend	4 Lunge	5 Inclined Plane	6 Downward Dog	7 Modified Push-up	8 Cobra	9 Downward Dog	10 Lunge	11 Forward Bend	12 Back Bend	13 Mountain
Monday													
Tuesday													
Wednesday													
Thursday													
Friday													
Saturday													
Sunday													

Self-Monitoring Chart—Comments:

(continued)

FIGURE 20-2. Four examples of formats for self-monitoring instruments

Hatha Yoga Asanas

Asanas	1 Triangle 	2 Half Moon 	3 Tree 	4 Cobra 	5 Half Locust 	6 Full Locust 	7 Bow	8 Half Spinal Twist 	9 Head to Knee 	10 Forward Bend	11 Half Shoulder Stand 	12 Child	13 Cobbler 	14 Sponge (Corpse) 
Days														
Monday														
Tuesday														
Wednesday														
Thursday														
Friday														
Saturday														
Sunday														

Comments:

3. Content and quality of marital interactions

Record the type of interaction under "Contents." For each interaction circle one category that best represents the quality of that interaction.

Time	Content of interaction	Quality of interaction				
		Very pleasant	Pleasant	Neutral	Unpleasant	Very unpleasant
6:30 A.M.		++	+	0	−	−−
7:00		++	+	0	−	−−
7:30		++	+	0	−	−−
8:00		++	+	0	−	−−

(continued)

FIGURE 20-2. Four examples of formats for self-monitoring instruments (continued)

self-monitoring. However, a device that is too obtrusive may draw attention from others who could reward or punish the client for self-monitoring (Ciminero et al., 1977, p. 202). Finally, the device should be capable of giving cumulative frequency data so that the client can chart daily totals of the behavior (Thoresen & Mahoney, 1974).

After the client has been instructed in the timing and method of recording, and after a recording device has been selected, the client should practice using the recording system. Breakdowns in self-monitoring often occur because a client did not understand the recording process clearly. Rehearsal of the recording procedures may ensure that the client

4. Self-monitoring log for recording anxiety responses

Instructions for recording:

Date and time	Frequency of anxiety response	External events	Internal dialogue (self-statements)	Behavioral factors	Degree of arousal	Skill in handling situation
Record day and time of incident	Describe each situation in which anxiety occurred	Note what triggered the anxiety	Note your thoughts or things you said to yourself when this occurred	Note how you responded—what you did	Rate the intensity of the anxiety: (1) a little intense (2) somewhat intense (3) very intense (4) extremely intense	Rate the degree to which you handled the situation effectively: (1) a little (2) somewhat (3) very (4) extremely

FIGURE 20-2. Four examples of formats for self-monitoring instruments (continued)

will record accurately. Generally, a client should engage in self-recording for three to four weeks. Usually the effects of self-monitoring are not apparent in only one or two weeks' time.

Charting of a Response

The data recorded by the client should be translated onto a more permanent storage record such as a chart or graph that enables the client to inspect the self-monitored data visually. This type of visual guide may provide the occasion for client self-reinforcement (Kanfer & Gaelick-Buys, 1991), which, in turn, can influence the reactivity of self-monitoring. The data can be charted by days, using a simple line graph. For example, a client counting the number of urges to smoke a cigarette could chart these by days, as in Figure 20-3. A client recording the amount of time spent studying daily could use the same sort of line graph to chart duration of study time. The vertical axis would be divided into time intervals such as 15 minutes, 30 minutes, 45 minutes, or 1 hour.

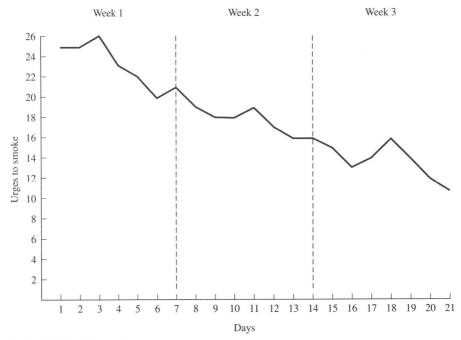

FIGURE 20-3. Self-monitoring chart

The client should receive either oral or written instructions on a way to chart and graph the daily totals of the recorded response. The counselor can assist the client in interpreting the chart in the sessions on data review and analysis. If a client is using self-monitoring to increase a behavior, the line on the graph should go up gradually if the self-monitoring is having the desired effect; if self-monitoring is influencing an undesired response to decrease, the line on the graph should go down gradually.

Display of Data

After the graph has been made, the client has the option of displaying the completed chart. If the chart is displayed in a "public" area, this display may prompt environmental reinforcement, a necessary part of an effective self-management program. The effects of self-monitoring are usually augmented when the data chart is displayed as a public record. However, some clients will not want to make their data public for reasons of confidentiality or shame avoidance.

Analysis of Data

If the client's recording data are not reviewed and analyzed, the client may soon feel as if he or she was told to make a graph just for practice in drawing straight lines! A very important facet of self-monitoring is the information it can provide to the client. There is some evidence that people who receive feedback about their self-recording change more than those who do not. The recording and charting of data should be used *explicitly* to provide the client with knowledge of results about behavior or performance. Specifically, the client should bring the data to weekly counseling sessions for review and analysis. In these sessions, the counselor can encourage the client to compare the data with the desired goals and standards. The client can use the recorded data for self-evaluation and determine whether the data indicate that the behavior is within or outside the desired limits. The counselor can also aid in data analysis by helping the client interpret the data correctly. As Thoresen and Mahoney observe, "Errors about what the charted data represent can seriously hinder success in self-control" (1974, p. 44).

MODEL EXAMPLE: SELF-MONITORING

As you may recall from Joan's goal chart in Chapter 10, one of Joan's goals was to increase her positive thoughts (and simultaneously decrease her negative thoughts) about her ability to do well with math. This goal lends itself well to application of self-management strategies for several reasons. First, the goal represents a covert behavior (positive thoughts), which is observable only by Joan. Second, the "flip side" of the goal (the negative thoughts) represents a very well-learned habit. Probably most of these negative thoughts occur *outside* the counseling sessions. To change this thought pattern, Joan will need to use strategies she can apply frequently (as needed) in vivo, and she will need to use strategies she can administer to herself.

Here is a description of the way self-monitoring could be used to help Joan achieve this goal.

1. *Treatment rationale.* The counselor would provide an explanation of what Joan will self-monitor and why, emphasizing that this is a strategy she can apply herself, can use with a "private" behavior, and can use as frequently as possible in the actual setting.
2. *Discrimination of a response.* The counselor would need to help Joan define the target response explicitly. One definition could be "Any time I think about myself doing math or working with numbers successfully." The counselor should provide some possible examples of this response, such as "Gee, I did well on my math homework today" or "I was able to balance the checkbook today." Joan should also be encouraged to identify some examples of the target response. Because Joan wants to increase this behavior, the target response would be stated in the "positive."
3. *Recording of a response.* The counselor should instruct Joan in timing, a method, and a device for recording. In this case, because Joan is using self-monitoring to increase a desired behavior, she would use postbehavior monitoring. Joan should be instructed to record *immediately* after a target thought has occurred. She is interested in recording the *number* of such thoughts, so she could use a frequency count. A tally on a note card or a wrist counter could be selected as the device for recording. After these instructions, Joan should practice recording before actually doing it. She should be instructed to engage in self-monitoring for about four consecutive weeks.
4. *Charting of a response.* After each week of self-monitoring, Joan can add her daily frequency totals and chart them by days on a simple line graph, as shown in Figure 20-4.

   Joan is using self-monitoring to increase a behavior; as a result, if the monitoring has the desired effect, the line on her graph should gradually rise. It is just starting to do so here; additional data for the next few weeks should

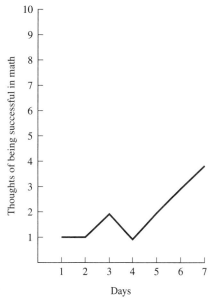

FIGURE 20-4. Simple line graph for self-monitoring

show a greater increase if the self-monitoring is influencing the target behavior in the desired direction.
5. *Display of data.* After Joan has made a data chart, she may wish to post it in a place such as her room, although this is a very idiosyncratic decision.
6. *Analysis of data.* During the period of self-monitoring, Joan should bring in her data for weekly review sessions with the counselor. The counselor can provide reinforcement and help Joan interpret the data accurately. Joan can use the data for self-evaluation by comparing the "story" of the data with her stated desired behavior and level of change.

STIMULUS CONTROL

Kanfer and Gaelick-Buys (1991, p. 335) define stimulus control as the predetermined arrangement of environmental conditions that makes it impossible or unfavorable for an undesired behavior to occur. Stimulus-control methods emphasize rearranging or modifying environmental conditions that serve as cues or antecedents of a particular response. As you may recall from the discussion of the ABC model of behavior in Chapter 8, a behavior often is guided by certain things that precede it (antecedents) and is maintained by positive or negative events that follow it (consequences). You may also remember that both antecedents and consequences

LEARNING ACTIVITY 53   *Self-Monitoring*

This learning activity is designed to help you use self-monitoring yourself. The instructions describe a self-monitoring plan for you to try out.

1. Discrimination of a target response:
   a. Specify one target behavior you would like to change. Pick either the positive or the negative side of the behavior to monitor—depending on which you value more and whether you want to increase or decrease this response.
   b. Write down a definition of this behavior. How clear is your definition?
   c. Can you write some examples of this behavior? If you had trouble with these, try to tighten up your definition—or contrast positive and negative instances of the behavior.
2. Recording of the response:
   a. Specify the *timing* of your self-recording. Remember the rules of thumb:
      1. Use prebehavior monitoring to decrease an undesired response.
      2. Use postbehavior monitoring to increase a desired response.
      3. Record immediately—don't wait.
      4. Record when there are no competing responses. Write down the timing you choose.
   b. Select a *method* of recording (frequency, duration, and so on). Remember:
      1. Frequency counts for clearly separate occurrences of the response
      2. Duration or latency measures for responses that occur for a period of time
      3. Intensity measures to determine the severity of a response
   c. Select a *device* to assist you in recording. Remember that the device should be
      1. Portable
      2. Accessible
      3. Economical
      4. Obtrusive enough to serve as a reminder to self-record
   d. After you have made these determinations, engage in self-monitoring for at least a week (preferably two). Then complete steps 3, 4, and 5.
3. Charting of response: Take your daily self-recording data and chart them on a simple line graph for each day that you self-monitored.
4. Displaying of data: Arrange a place [that you feel comfortable with] to display your chart.
5. Analysis of data: Compare your chart with your stated desired behavior change. What has happened to the behavior?

---

can be external (overt) or internal (covert). For example, an antecedent could be a situation, an emotion, a cognition, or an overt or covert verbal instruction.

Clinical Uses of Stimulus Control

Stimulus-control procedures have been used for a wide range of concerns; see Table 20-4 for a sample of research on stimulus control. This procedure has been used to treat bulimia nervosa, childhood and adolescent obesity, and anxiety about pelvic examinations. Other applications have been in toilet training, to teach behavior, and to change noncompliant behavior. If number of studies is an indicator, an important use of stimulus control is in the treatment of insomnia.

How Antecedents Acquire Stimulus Control

When antecedents are consistently associated with a behavior that is reinforced in the *presence* (not the absence) of these antecedent stimuli, they gain control over the behavior. You might think of this as "stimulus control," as an antecedent is a stimulus for a certain response. When an antecedent gains stimulus control over the response, there is a high probability that the response will be emitted in the presence of these particular antecedent events. For example, most of us automatically slow down, put our foot on the brake, and stop the car when we see a red traffic light. The red light is a stimulus that has gained control over our stopping-the-car behavior. Generally, the fact that antecedents exert stimulus control is helpful, as it is in driving when we go ahead with a green light and stop at the sight of a red light.

Inappropriate Stimulus Control in Problem Behavior

Client problem behaviors may occur because of *inappropriate* stimulus control. Inappropriate stimulus control may be related to obesity. Eating responses of overweight people

TABLE 20-4. Research on stimulus control

Bulimia

Viens, M. J., & Hranchuk, K. (1992). The treatment of bulimia nervosa following surgery using a stimulus control procedure: A case study. *Journal of Behavior Therapy and Experimental Psychiatry, 23,* 313–317.

Effective teaching

Martens, B. K., & Kelly, S. Q. (1993). A behavioral analysis of effective teaching. *School Psychology Quarterly, 8,* 10–26.

Fitness

Estabrooks, P., Courneya, K., & Nigg, C. (1996). Effect of a stimulus control intervention on attendance at a university fitness center. *Behavior Modification, 20,* 202–215.

Insomnia

Engle-Friedman, M., Bootzin, R. R., Hazlewood, L., & Tsao, C. (1992). An evaluation of behavioral treatments for insomnia in the older adult. *Journal of Clinical Psychology, 48,* 77–90.

France, K. G., & Hudson, S. M. (1990). Behavior management of infant sleep disturbance. *Journal of Applied Behavior Analysis, 23,* 91–98.

Horvath, A. O., & Goheen, M. D. (1990). Factors mediating the success of defiance- and compliance-based interventions. *Journal of Counseling Psychology, 37,* 363–371.

Jacobs, G. D., Benson, H., & Friedman, R. (1993). Home-based central nervous system assessment of a multifactor behavioral intervention for chronic sleep-onset insomnia. *Behavior Therapy, 24,* 159–174.

Jacobs, G. D., Rosenberg, P. A., Friedman, R., & Matheson, J. (1993). Multifactor behavioral treatment of chronic sleep-onset insomnia using stimulus control and the relaxation response: A preliminary study. *Behavior Modification, 17,* 498–509.

Nakano, K. (1988). The combined effects of stimulus control and relaxation for sleep onset insomnia. *Japanese Journal of Behavior Therapy, 14,* 21–24.

Morawetz, D. (1989). Behavioral self-help treatment for insomnia: A controlled evaluation. *Behavior Therapy, 20,* 365–379.

Sloan, E. P., Hauri, P., Bootzin, R., & Morin, C. (1993). The nuts and bolts of behavioral therapy for insomnia. *Journal of Psychosomatic Research, 37,* 19–37.

In natural settings

Halle, J. W., & Holt, B. (1991). Assessing stimulus control in natural settings: An analysis of stimuli that acquire control during training. *Journal of Applied Behavior Analysis, 24,* 579–589.

Non-compliance behavior

Cameron, M. J., Luiselli, J. K., McGrath, M., & Carlton, R. (1992). Stimulus control analysis and treatment of noncompliant behavior. *Journal of Developmental and Physical Disabilities, 4,* 141–150.

Obesity

Haddock, C. K., Shadish, W. R., Klesges, R. C., & Stein, R. J. (1994). Treatments for childhood and adolescent obesity. *Annals of Behavioral Medicine, 16,* 235–244.

Pelvic examination

Williams, J. G., Park, L. I., & Kline, J. (1992). Reducing distress associated with pelvic examinations: A stimulus control intervention. *Women and Health, 18,* 41–53.

Toilet training

Taylor, S., Cipani, E., & Clardy, A. (1994). A stimulus control technique for improving the efficacy of an established toilet training program. *Journal of Behavior Therapy and Experimental Psychiatry, 25,* 155–160.

Writing tasks

Stromer, R., MacKay, H., Howell, S., & McVay, A. (1996). Teaching computer-based spelling to individuals with developmental and hearing disabilities. Transfer of stimulus control to writing tasks. *Journal of Applied Behavior Analysis, 29,* 25–42.

---

tend to be associated with many environmental cues. If a person eats something not only at the dining table but also when working in the kitchen, watching television, walking by the refrigerator, and stopping at a Dairy Queen, the sheer number of eating responses could soon result in obesity. Too many environmental cues often are related to other client problems, particularly "excesses" such as substance use. In these cases, the primary aim of a self-management stimulus-control method is to reduce the number of cues associated with the undesired response, such as eating or smoking.

Other problem behaviors have been observed that seem to involve excessively narrow stimulus control. At the opposite pole from obesity are people who eat so little that their physical and psychological health suffers (a condition called

"anorexia nervosa"). For these people, there are too few eating cues. Lack of exercise can be a function of too narrow stimulus control. For some people, the paucity of environmental cues associated with exercise results in very little physical activity. In these cases, the primary aim of a stimulus-control strategy is to establish or increase the number of cues that will elicit the desired behavior.

To summarize, stimulus-control self-management involves reducing the number of antecedent stimuli associated with an undesirable behavior and simultaneously increasing the antecedent cues associated with a desirable response (Mahoney & Thoresen, 1974; Thoresen & Mahoney, 1974). Table 20-5 shows the principal methods of stimulus control and some examples.

TABLE 20-5. Principles and examples of stimulus-control strategies

Principle of change	Example
To decrease a behavior: Reduce or narrow the frequency of cues associated with the behavior.	
1. Prearrange or alter cues associated with the place of the behavior:	
a. Prearrange cues that make it hard to execute the behavior.	Place fattening foods in high, hard-to-reach places.
b. Prearrange cues so that they are controlled by others.	Ask friends or family to serve you only one helping of food and to avoid serving fattening foods to you.
2. Alter the time or sequence (chain) between the antecedents cues and the resulting behaviors:	
a. Break up the sequence.	Buy and prepare food only on a full stomach.
b. Change the sequence.	Substitute and engage in nonfood activities when you start to move toward snacking (toward refrigerator, cupboard, or candy machine).
c. Build pauses into the sequence.	Delay second helpings of food or snacks for a predetermined amount of time.
To increase a behavior: Increase or prearrange the cues associated with the response.	
1. Seek out these cues deliberately to perform the desired behavior.	Initially arrange only one room with a desk to study. When you need to study, go to this place.
2. Concentrate on the behavior when in the situation.	Concentrate only on studying in the room. If you get distracted, get up and leave. Don't mix study with other activities, such as listening to records or talking.
3. Gradually extend the behavior to other situations.	When you have control over studying in one room, extend the behavior to another conducive room or place.
4. Promote the occurrence of helpful cues by other people or by self-generated reminders.	Ask your roommate to remind you to leave the desk when you are talking or distracted; remind yourself of good study procedures by posting a list over your study desk or by using verbal or covert self-instructions.

Using Stimulus Control to Decrease Behavior

To decrease the rate of a behavior, the antecedent cues associated with the behavior should be reduced in frequency or prearranged or altered in terms of time and place of occurrence. When cues are separated from the habitual behavior by alteration or elimination, the old, undesired habit can be terminated (Mahoney & Thoresen, 1974, p. 42). Many behavioral "excesses," such as eating, smoking, drinking, or self-criticism, are tied to a great number of antecedent situations. Reducing these cues can restrict the occurrence of the undesired behavior. Existing cues can be prearranged to make the target behavior so hard to execute that the person is unlikely to do it. An example would be altering the place of smoking by moving one's smoking chair to an inconvenient place like the basement. The smoker would have to go downstairs each time she or he wanted a cigarette. Cues can also be prearranged by

placing their control in the hands of another person. Giving your pack of cigarettes to a friend is an example of this method. The friend should agree to help you reduce smoking and should agree not to reinforce or punish any instances of your smoking behavior (the undesired response).

A behavior can also be reduced through stimulus control by interrupting the learned pattern or sequence that begins with one or more antecedent cues and results in the undesired response. This sequence may be called a *chain*. A problem behavior is often the result of a long chain of events. For example, a variety of behaviors make up the sequence of smoking. Before puffing on a cigarette, a person has to go to a cigarette machine, put money in the machine, take out a pack of cigarettes, take out one cigarette from the pack, and light the cigarette.

This chain might be interrupted in one of three ways: breaking up the chain of events, changing the chain, or building pauses into the chain (Watson & Tharp, 1997, pp. 145–149). All these methods involve prearranging or altering the time of occurrence of the behavior. A chain of events can be broken up by discovering and interrupting an event early in the sequence or by scrambling the typical order of events. For example, the smoker could break up the chain by not carrying the change required for a cigarette machine. Or, if the smoker typically smokes at certain times, the usual order of events leading up to smoking could be mixed up. The smoker could also change the typical chain of events. People who start to light up a cigarette whenever they are bored, tense, or lacking something to do with their hands could perform a different activity at this point, such as calling a friend when bored, relaxing when tense, or knitting or playing cards to provide hand activity. Finally, smokers could interrupt the chain by deliberately building pauses into it. As you may recall, when antecedents exert control over a behavior, the behavior occurs almost automatically. As Watson and Tharp point out, one way to deal with this automatic quality is to pause before responding to a cue (p. 147). For instance, whenever the smoker has an urge to light up in response to a stress cue, a deliberate pause of ten minutes can be built in before the person actually does light up. Gradually this time interval can be increased. Watson and Tharp (1997) note that the use of pauses is good for "indulgent behaviors," and a two-minute pause is often long enough for the "urge" to pass (p. 146). Sometimes you can even strengthen the pause procedure by covertly instructing yourself on what you want to do or by thinking about the benefits of not smoking. The pause itself can then become a new antecedent (Watson & Tharp, 1997, p. 146).

Using Stimulus Control to Increase Behavior

Stimulus-control methods can also be used to increase a desired response. As noted in Table 20-5, to increase the rate of a response, a person increases or prearranges the antecedent cues associated with the desired behavior. The person deliberately seeks out these cues to perform the behavior and concentrates only on this behavior when in the situation. Competing or distracting responses must be avoided. Gradually, as stimulus control over the behavior in one situation is achieved, the person can extend the behavior by performing it in another, similar situation. This process of stimulus generalization means that a behavior learned in one situation can be performed in different but similar situations (Watson & Tharp, 1997, p. 156). The person can promote the occurrence of new antecedent cues by using reminders from others, self-reminders, or overt or covert self-instructions

(Watson & Tharp, 1997). The rate of a desired response is increased by increasing the times and places in which the person performs the response.

As an example, suppose you are working with a client who wants to increase his or her amount of daily exercise. First, more cues would be established to which the person would respond with isometric or physical activity. For example, the person might perform isometric activities whenever sitting in a chair or waiting for a traffic light. Or the person might perform physical exercises each morning and evening on a special exercise mat. The client would seek out these prearranged cues and concentrate on performing the activity while in the situation. Other behaviors should not be performed while in these situations, as a competing response could interfere with the exercise activity (Watson & Tharp, 1997, p. 153). Gradually, the client could extend the exercise activities to new but similar situations—for example, doing isometrics while sitting on the floor or waiting for a meeting to start. The person could also promote exercise behavior in these situations by reminders—posting an exercise chart on the wall or carrying it around in a pocket or wallet, displaying an exercise list, and so forth.

Stimulus-control instructions also have been used to increase sleep. For example, clients were instructed: (1) Go to bed or lie down to sleep only when sleepy. (2) Do not read, watch TV, or eat in bed—use the bed only for sleeping and/or sexual activities. (3) If unable to fall asleep after 10 to 20 minutes, get out of bed and engage in some activity. Return to bed only when sleepy—and continue this procedure throughout the night as necessary. (4) Set the alarm clock and get up at the same time every morning regardless of the amount of sleep obtained during the night. (5) Do not take naps during the day.

According to Kanfer and Gaelick-Buys (1991, p. 335), one advantage of stimulus control is that only minimal self-initiated steps are required to trigger environmental changes that effect desired or undesired responses. However, stimulus-control methods often are insufficient to modify behavior without the support of other strategies. As Mahoney and Thoresen (1974) observe, stimulus-control methods are not usually sufficient for long-term self-change unless accompanied by other self-management methods that exert control over the *consequences* of the target behavior. One self-management method that involves self-presented consequences is discussed in the following section.

MODEL EXAMPLE: STIMULUS CONTROL

This model example will illustrate how stimulus control can be used as *one way* to help Joan achieve her goal of

LEARNING ACTIVITY 54 *Stimulus Control*

As the emphasis in this chapter is on self-management, the learning activities in this chapter are designed to help you use these strategies yourself! The purpose of this learning activity is to help you reduce an unwanted behavior, using stimulus-control methods.

1. Specify a behavior that you find undesirable and you wish to decrease. It can be an overt one, such as smoking, eating, biting your nails, or making sarcastic comments, or it can be a covert behavior, such as thinking about yourself in negative ways or thinking how great food or smoking tastes.
2. Select one or more stimulus-control methods to use for behavior reduction from the list and examples given in Table 20-5. Remember, you will be reducing the number of cues or antecedent events associated with this

behavior by altering the times and places the undesired response occurs.
3. Implement these stimulus-control methods daily for two weeks.
4. During the two weeks, engage in self-monitoring of your target response. Record the type and use of your method and amount of your target behavior, using frequency or duration methods of recording.
5. At the end of two weeks, review your recording data. Did you use your selected method consistently? If not, what contributed to your infrequent use? If you used it consistently, did you notice any gradual reduction in the target behavior by the end of two weeks? What problems did you encounter in applying a stimulus-control method with yourself? What did you learn about stimulus control that might help you when using it with clients?

---

increasing positive thoughts about her math ability. Recall that the principle of change in using stimulus control to increase a behavior is to increase the cues associated with the behavior. Here's how we would implement this principle with Joan.

1. Establish at least one cue that Joan could use as an antecedent for positive thoughts. We might suggest something like putting a piece of tape over her watch.
2. Develop a list of several positive thoughts about math. Each thought could be written on a blank card that Joan could carry with her.
3. Instruct Joan to read or think about a thought on one card *each* time she looks at her watch. Instruct her to seek out the opportunity deliberately by looking at her watch frequently and then concentrating on one of these positive thoughts.
4. When Joan gets to the point that she automatically thinks of a positive thought after looking at her watch, other cues can be established that she can use in the same way. For instance, she can put a ☺ on her math book. Each time she gets out her math book and sees the "smiley face," she can use this cue to concentrate on another positive thought.
5. She can promote more stimulus control over these thoughts by using reminders. For instance, Joan could put a list of positive thoughts on the mirror or the closet door in her room. Each time she sees the list, it serves as a reminder. Or she can ask a friend or classmate to remind

her to "think positively" whenever the subject of math or math class is being discussed.

SELF-REWARD: PROCESSES AND USES

Self-monitoring and stimulus-control procedures may be enough to maintain the desired goal behavior for many people. However, for some people with low self-esteem, depression, strong emotional reactions, environmental consequences, or low self-efficacy, self-monitoring may not always be effective in regulating behavior (Kanfer & Gaelick-Buys, 1991). In such cases, self-reward procedures are used to help clients regulate and strengthen their behavior with the aid of self-produced consequences. Many actions of an individual are controlled by self-produced consequences as much as by external consequences.

According to Bandura (1971), there are several necessary conditions of self-reinforcement, or self-reward:

1. The individual (rather than someone else) determines the criteria for adequacy of her or his performance and for resulting reinforcement.
2. The individual (rather than someone else) controls access to the reward.
3. The individual (rather than someone else) is his or her own reinforcing agent and administers the rewards.

Note that self-reward involves *both* the self-determination and the self-administration of a reward. This distinction has, at times, been overlooked in self-reinforcement research and application. Nelson, Hayes, Spong, Jarrett, and McKnight (1983, p. 565) propose that "self-reinforcement is effective primarily because of its stimulus properties in cuing natural environmental consequences."

As a self-management procedure, self-reward is used to strengthen or increase a desired response. The operations involved in self-reward are assumed to parallel those that occur in external reinforcement. In other words, a self-presented reward, like an externally administered reward, is defined by the function it exerts on the target behavior. A reinforcer (self- or external) is something that, when administered following a target response, tends to maintain or increase the probability of that response in the future. A major advantage of self-reward over external reward is that a person can use and apply this strategy independently.

Self-rewards can be classified into two categories: positive and negative. In positive self-reward, one presents oneself with a positive stimulus (to which one has free access) *after* engaging in a specified behavior. Examples of positive reward include praising yourself after you have completed a long and difficult task, buying yourself a new compact disc after you have engaged in a specified amount of piano practice, or imagining that you are resting on your favorite spot after you have completed your daily exercises. Negative self-reward involves the removal of a negative stimulus after execution of a target response. For example, taking down an uncomplimentary picture or chart from your wall after performing the target response is an example of negative self-reward.

Our discussion of self-reward as a therapeutic strategy is limited to the use of positive self-reward for several reasons. First, there has been very little research to validate the negative self-reward procedure. Second, by definition, negative self-reward involves an aversive activity. It is usually unpleasant for a person to keep suet in the refrigerator or to put an ugly picture on the wall. Many people will not use a strategy that is aversive. Second, we do not recommend that counselors suggest strategies that seem aversive, because the client may feel that terminating the counseling is preferable to engaging in an unpleasant change process.

Uses of Self-Reward

Self-reward has been used as one component with developmental disabilities (Harchik, Sherman, & Sheldon, 1992); with the academic productivity of secondary students with learning problems (Seabaugh & Schumaker, 1994); and to treat an 11-year-old girl fearful of AIDS infection, other

diseases, and poisoning (Hagopian, Weist, & Ollendick, 1990). Research has reported the effective use of self-reward to facilitate weight loss, improve study skills, enhance dating skills, and increase the activity levels of depressed persons. Other research has explored moderating variables with the use of self-reward. For example, Enzle, Roggeveen, and Look (1991) found that ambiguous standards of performance coupled with self-administration of rewards reduced intrinsic motivation whereas clear standards with self-administration of rewards maintained high levels of intrinsic motivation. In a similar study, extrinsically motivated young children (ages 3 to 6 years) set higher performance standards and arranged a leaner schedule of self-reinforcement under stringent and lenient instructional demands than intrinsically motivated children (Switzky & Haywood, 1992). People who have a self-defeating attitude and are depressed were observed to use a lower frequency of self-reinforcement (Schill & Kramer, 1991). Rokke and Kozak (1989) found that depressed people had lower levels of self-evaluation and self-reward than nondepressed people, but there were also differences in the level of self-reward between the two groups when self-evaluation ratings were controlled. Finally, Field and Steinhardt (1992) reported that a lower external locus of control score was associated with higher positive self-reinforcement for university students' orientations toward exercise and wellness behavior. The last group of studies indicates the importance of client state variables and environmental demand characteristics in the application of a self-reward strategy.

Some of the clinical effects typically attributed to the self-reinforcement procedure may also be due to certain external factors, including a client's previous reinforcement history, client goal setting, the role of client self-monitoring, surveillance by another person, external contingencies in the client's environment, and the instructional set given to the client about the self-reward procedure. The exact role these external variables may play in self-reward is still relatively unknown. However, a counselor should acknowledge and perhaps try to capitalize on some of these factors to heighten the clinical effects of a self-reward strategy.

COMPONENTS OF SELF-REWARD

Self-reward involves planning by the client of appropriate rewards and of the conditions in which they will be used. Self-reward can be described by four major components: (1) selection of appropriate self-rewards, (2) delivery of self-rewards, (3) timing of self-rewards, and (4) planning for self-change maintenance. These components are described in this portion of the chapter and are summarized in the following list. Although these components are discussed separately, remember that all of them are integral parts of an effective self-reward procedure.

1. Selection of appropriate rewards

   a. Individualize the reward.
   b. Use accessible rewards.
   c. Use several rewards.
   d. Use different types of rewards (verbal/symbolic, material, imaginal, current, potential).
   e. Use potent rewards.
   f. Use rewards that are not punishing to others.
   g. Match rewards to target response.
   h. Use rewards that are relevant to the client's culture, gender, age, class, and so on.

2. Delivery of self-rewards

   a. Self-monitor for data of target response.
   b. Specify what and how much is to be done for a reward.
   c. Specify frequent reinforcement in small amounts for different levels of target response.

3. Timing of self-reward

   a. Reward should come after, not before, behavior.
   b. Rewards should be immediate.
   c. Rewards should follow performance, not promises.

4. Planning for self-change maintenance.

   a. Enlist help of others in sharing or dispensing rewards (if desired).
   b. Review data with counselor.

Selection of Appropriate Rewards

In helping a client to use self-reward effectively, some time and planning must be devoted to selecting rewards that are appropriate for the client and the desired target behavior. Selecting rewards can be time-consuming. However, effective use of self-reward is somewhat dependent on the availability of events that are truly reinforcing to the client. The counselor can assist the client in selecting appropriate self-rewards; however, the client should have the major role in determining the specific contingencies.

Rewards can take many different forms. A self-reward may be verbal/symbolic, material, or imaginal. One verbal/symbolic reward is self-praise, such as thinking or telling oneself "I did a good job." This sort of reward may be especially useful with a very self-critical client (Kanfer &

Gaelick-Buys, 1991). A material reward is something tangible—an event (such as a movie), a purchase (such as a banana split), or a token or point that can be exchanged for a reinforcing event or purchase. An imaginal reinforcer is the covert visualization of a scene or situation that is pleasurable and produces good feelings. Imaginal reinforcers might include picturing yourself as a thin person after losing weight or imagining that you are water-skiing on a lake you have all to yourself.

Self-rewards can also be classified as current or potential. A current reward is something pleasurable that happens routinely or occurs daily, such as eating, talking to a friend, or reading a newspaper. A potential reward is something that would be new and different if it happened, something a person does infrequently or anticipates doing in the future. Examples of potential rewards include going on a vacation or buying a "luxury" item (something you love but rarely buy for yourself, not necessarily something expensive). Engaging in a "luxury" activity—something you rarely do—can be a potential reinforcer. For a person who is very busy and constantly working, "doing nothing" might be a luxury activity that is a potential reinforcer.

In selecting appropriate self-rewards, a client should consider the availability of these various kinds of rewards. We believe that a well-balanced self-reward program involves a *variety* of types of self-rewards. A counselor might encourage a client to select *both* verbal/symbolic and material rewards. Relying only on material rewards may ignore the important role of positive self-evaluations in a self-change program. Further, material rewards have been criticized for overuse and misuse. Imaginal reinforcers may not be so powerful as verbal/symbolic and material ones. However, they are completely portable and can be used to supplement verbal/symbolic and material rewards when it is impossible for an individual to use these other types (Watson & Tharp, 1997).

In selecting self-rewards, a client should also consider the use of both current and potential rewards. One of the easiest ways for a client to use current rewards is to observe what daily thoughts or activities are reinforcing and then to rearrange these so that they are used in contingent rather than noncontingent ways (Watson & Tharp, 1997). However, whenever a client uses a current reward, some deprivation or self-denial is involved. For example, agreeing to read the newspaper only after cleaning the kitchen involves initially denying oneself some pleasant, everyday event in order to use it to reward a desired behavior. As Thoresen and Mahoney (1974) point out, this initial self-denial introduces an aversive element into the self-reward strategy. Some people do not respond well to any aversiveness associated with self-change or self-directed behavior. One of the authors, in fact, consistently "abuses" the self-reward principle by doing the reward before the response (reading the paper before cleaning the kitchen)—precisely as a reaction against the aversiveness of this "programmed" self-denial. One way to prevent self-reward from becoming too much like programmed abstinence is to have the client select novel or potential reinforcers to use in addition to current ones.

There are several ways a counselor can help a client identify and select various kinds of self-rewards. One way is simply with verbal report. The counselor and client can discuss current self-reward practices and desired luxury items and activities (Kanfer & Gaelick-Buys, 1991). The client can also identify rewards by using in vivo observation. The client should be instructed to observe and list current consequences that seem to maintain some behaviors. Finally, the client can identify and select rewards by completing preference and reinforcement surveys. A preference survey is designed to help the client identify preferred and valued activities. Here is an example of one that Watson and Tharp (1997, p. 211) recommend:

1. What will be the rewards of achieving your goal?
2. What kind of praise do you like to receive, from yourself or from others?
3. What kinds of things do you like to have?
4. What are your major interests?
5. What are your hobbies?
6. What people do you like to be with?
7. What do you like to do with those people?
8. What do you do for fun?
9. What do you do to relax?
10. What do you do to get away from it all?
11. What makes you feel good?
12. What would be a nice present to receive?
13. What kinds of things are important to you?
14. What would you buy if you had an extra $20? $50? $100?
15. On what do you spend your money each week?
16. What behaviors do you perform every day? (Don't overlook the obvious or the commonplace.)
17. Are there any behaviors that you usually perform instead of the target behavior?
18. What would you hate to lose?
19. Of the things you do every day, which would you hate to give up?
20. What arc your favorite daydreams and fantasies?
21. What are the most relaxing scenes you can imagine?

The client can complete this sort of preference survey in writing or in a discussion. Clients who find it difficult to identify rewarding events might also benefit from completing a more formalized reinforcement survey, such as the Reinforcement Survey Schedule or the Children's Reinforcement Survey Schedule, written by Cautela (1977). The client can be given homework assignments to identify possible verbal/symbolic and imaginal reinforcers. For instance, the client might be asked to make a daily list for a week of positive self-thoughts or of the positive consequences of desired change. Or the client could make a list of all the things about which she or he likes to daydream or of some imagined scenes that would be pleasurable (Watson & Tharp, 1997).

Sometimes a client may seem thwarted in initial attempts to use self-reward because of difficulties in identifying rewards. Watson and Tharp (1997) note that people whose behavior consumes the reinforcer (such as smoking or eating), whose behavior is reinforced intermittently, or whose avoidance behavior is maintained by negative reinforcement may not be able to identify reinforcing consequences readily. Individuals who are "locked into" demanding schedules cannot find daily examples of reinforcers. Sometimes depressed people have trouble identifying reinforcing events. In these cases, the counselor and client have several options that can be used to overcome difficulties in selecting effective self-rewards.

A client who does not have the time or money for material rewards might use imaginal rewards. Imagining pleasant scenes following a target response has been described by Cautela (1970) as *covert positive reinforcement* (CPR). In the CPR procedure, the client usually imagines performing a desired behavior, followed by imagination of a reinforcing scene. A counselor might consider use of imaginal reinforcers only when other kinds are not available.

A second available option for problem cases is to use a client's everyday activity as a self-reward. Some clinical cases have used a mundane activity such as answering the phone or opening the daily mail as the self-reward. (Actually, such an activity may work more as a cuing device than a reinforcer; see Thoresen and Mahoney, 1974). If a frequently occurring behavior is used as a self-reward, it should be a desirable or at least a neutral activity. As Watson and Tharp (1997) note, clients should not use as a self-reward any high-frequency behavior that they would stop immediately if they could. Using a negative high-frequency activity as a reward may seem more like punishment than reinforcement.

With depressed clients, selecting self-rewards is often difficult, because many events lose their reinforcing value for someone who is depressed. Before using self-reward with a depressed client, it might be necessary to increase the reinforcing value of certain events. Anton, Dunbar, and Friedman (1976) describe the procedure of "anticipation training" designed to increase depressed clients' positive anticipations of events. In anticipation training, a client identifies and schedules several pleasant events to perform and then constructs three positive anticipation statements for each activity. The client imagines engaging in an activity and imagines the anticipation statements associated with the activity. An example adapted from Anton et al. of some anticipation statements for one activity appears below:

Activity planned: *Spending an afternoon at the lake*
Date to be carried out: *Tuesday; Wednesday if it rains Tuesday*
I will enjoy: *sitting on the beach reading a book*
I will enjoy: *getting in the water on a hot day*
I will enjoy: *getting fresh air*

No thought, event, or imagined scene is reinforcing for everyone. Often what one person finds rewarding is very different from the rewards selected by someone else. In using self-reward, it is important to help clients choose rewards that will work well for *them*—not for the counselor, a friend, or a spouse. Kanfer and Gaelick-Buys (1991) note the importance in considering the client's history, of also taking into account the client's gender, culture, age, and class.

The counselor should use the following guidelines to help the client determine some self-rewards that might be used effectively.

1. *Individualize* the reward to the client.
2. The reward should be *accessible* and *convenient* to use after the behavior is performed.
3. *Several* rewards should be used interchangeably to prevent satiation (a reward can lose its reinforcing value because of repeated presentations).
4. Different *types* of rewards should be selected (verbal/symbolic, material, imaginal, current, potential).
5. The rewards should be *potent* but not so valuable that an individual will not use them contingently.
6. The rewards should not be *punishing* to others. Watson and Tharp (1997) suggest that if a reward involves someone else, the other person's agreement should be obtained.
7. The reward should be *compatible* with the desired response (Kanfer & Gaelick-Buys, 1991). For instance, a person losing weight might use new clothing as a reward or thoughts of a new body image after weight loss. Using eating as a reward is not a good match for a weight-loss target response.

8. The rewards should be relevant to the client's culture, gender, age, class, race, and ethnicity.

Delivery of Self-Reward

The second part of working out a self-reward strategy with a client involves specifying the conditions and method of delivering the self-rewards. First, a client cannot deliver or administer a self-reward without some database. Self-reward delivery is dependent on systematic data gathering; self-monitoring is an essential first step.

Second, the client should determine the precise conditions under which a reward will be delivered. The client should, in other words, state the rules of the game. The client should know *what* and *how much* has to be done before administering a self-reward. Usually self-reward is more effective when clients reward themselves for small steps of progress. In other words, performance of a subgoal should be rewarded. Waiting to reward oneself for demonstration of the overall goal usually introduces too much of a delay between responses and rewards.

Finally, the client should indicate how much and what kind of reward will be given for performing various responses or different levels of the goals. The client should specify that doing so much of the response results in one type of reward and how much of it. Usually reinforcement is more effective when broken down into smaller units that are self-administered more frequently. The use of tokens or points provides for frequent, small units of reinforcement; these can be exchanged for a "larger" reinforcer after a certain amount of points or tokens is accumulated.

Timing of Self-Reward

The counselor also needs to instruct the client about the timing of self-reward—when a self-reward should be administered. There are three ground rules for the timing of a self-reward.

1. A self-reward should be administered *after* performing the specified response, not before.
2. A self-reward should be administered *immediately* after the response. Long delays may render the procedure ineffective.
3. A self-reward should follow *actual performance,* not promises to perform.

Planning for Self-Change Maintenance

Self-reward, like any self-change strategy, needs environmental support for long-term maintenance of change (Kanfer

& Gaelick-Buys, 1991; Mahoney, 1974). The last part of using self-reward involves helping the client find ways to plan for self-change maintenance. First, the counselor can give the client the option of enlisting the help of others in a self-reward program. Other people can share in or dispense some of the reinforcement if the client is comfortable with this (Watson & Tharp, 1997). Some evidence indicates that certain people may benefit more from self-reward if initially in the program they have received their rewards from others (Mahoney & Thoresen, 1974). Second, the client should plan to review the data collected during self-reward with the counselor. The review sessions give the counselor a chance to reinforce the client and to help the client make any necessary revisions in the use of the strategy. Therapist expectations and approval for client progress may add to the overall effects of the self-reward strategy if the therapist serves as a reinforcer to the client.

Some Cautions in Using Rewards

The use of rewards as a motivational and informational device is a controversial issue (Eisenberger & Cameron, 1996). Using rewards, especially material ones, as incentives has been criticized on the grounds that tangible rewards are overused, are misused, and often discourage rather than encourage the client.

As a therapy technique, self-reward should not be used indiscriminately. Before suggesting self-reward, the counselor should carefully consider the individual client, the client's previous reinforcement history, and the client's desired change. Self-reward may not be appropriate for clients from cultural backgrounds in which the use of rewards is considered "undesirable or immodest" (Kanfer & Gaelick-Buys, 1991, p. 338). When a counselor and client do decide to use self-reward, two cautionary guidelines should be followed. First, material rewards should not be used solely or promiscuously. The therapist should seek ways to increase a person's intrinsic satisfaction in performance before automatically resorting to extrinsic rewards as a motivational technique. Second, the counselor's role in self-reward should be limited to providing instructions about the procedure and encouragement for progress. The client should be the one who selects the rewards and determines the criteria for delivery and timing of reinforcement. When the target behaviors and the contingencies are specified by someone other than the person using self-reward, the procedure can hardly be described accurately as a self-change operation.

MODEL EXAMPLE: SELF-REWARD

This example will illustrate how self-reward could be used to help Joan increase her positive thoughts about her ability to do well in math.

1. *Selection of rewards:* First the counselor would help Joan select some appropriate rewards to use for reaching her predetermined goal. The counselor would encourage Joan to identify some self-praise she could use to reward herself symbolically or verbally ("I did it"; "I can gradually see my attitude about math changing"). Joan could give herself points for daily positive thoughts. She could accumulate and exchange the points for material rewards, including current rewards (such as engaging in a favorite daily event) and potential rewards (such as a purchase of a desired item). These are suggestions; Joan should be responsible for the actual selection. The counselor could suggest that Joan identify possible rewards through observation or completion of a preference survey. The counselor should make sure that the rewards Joan selects are accessible and easy to use. Several rewards should be selected to prevent satiation. The counselor should also make sure that the rewards selected are potent, compatible with Joan's goal, not punishing to anyone else, and relevant to Joan.

2. *Delivery of rewards:* The counselor would help Joan determine guidelines for delivery of the rewards selected. Joan might decide to give herself a point for each positive thought. This allows for reinforcement of small steps toward the overall goal. A predetermined number of daily points, such as 5, might result in delivery of a current reward, such as watching TV or going over to her friend's house. A predetermined number of weekly points could mean delivery of a potential self-reward, such as going to a movie or purchasing a new item. Joan's demonstration of her goal beyond the specified level could result in delivery of a bonus self-reward.

3. *Timing of rewards:* The counselor would instruct Joan to administer the reward *after* the positive thoughts or after the specified number of points is accumulated. The counselor can emphasize that the rewards follow performance, not promises. The counselor should encourage Joan to engage in the rewards as soon as possible after the daily and weekly target goals have been met.

4. *Planning for self-change maintenance:* The counselor can help Joan find ways to plan for self-change maintenance. One way is to schedule periodic "check-ins" with the counselor. In addition, Joan might select a friend who

LEARNING ACTIVITY **55**   *Self-Reward*

This learning activity is designed to have you engage in self-reward.

1. Select a target behavior you want to increase. Write down your goal (the behavior to increase, desired level of increase, and conditions in which behavior will be demonstrated).
2. Select several types of self-rewards to use and write them down. The types to use are verbal/symbolic, material (both current and potential), and imaginal. See whether your selected self-rewards meet the following criteria:
   a. Individually tailored to you?
   b. Accessible and convenient to use?
   c. Several self-rewards?
   d. Different types of self-rewards?

   e. Are rewards potent?
   f. Are rewards not punishing to others?
   g. Are rewards compatible with your desired goal?
   h. Are rewards relevant to your gender and culture?

3. Set up a plan for delivery of your self-reward: What type of reinforcement and how much will be administered? How much and what demonstration of the target behavior are required?
4. When do you plan to administer a self-reward?
5. How could you enlist the aid of another person?
6. Apply self-reward for a specified time period. Did your target response increase? To what extent?
7. What did you learn about self-reward that might help you in suggesting its use to diverse groups of clients?

---

could help her share in the reward by watching TV or going shopping with her or by praising Joan for her goal achievement.

SELF-AS-A-MODEL

The self-as-a-model procedure uses the client as the model. The procedure as we present it in this chapter has been developed primarily by Hosford (1974). Hosford and de Visser (1974) have described self-modeling as a procedure in which the client sees himself or herself as the model—performing the goal behavior in the desired manner. The client also practices with an audiotape. Successful practices are rewarded and errors are corrected. Note that this procedure involves not only modeling but also practice and feedback.

Why have the client serve as the model? The literature indicates that such model characteristics as prestige, status, age, sex, and ethnic identification have differential influence on clients (Bandura, 1969, 1971). For some people, observing another person—even one with similar characteristics—may produce negative reactions. Some people may attend and listen better when they see or hear themselves on a tape or in a movie (Hosford, Moss, & Morrell, 1976). For example, when we perform in front of a video camera or a tape recorder, we have to admit there is a little exhibitionism and "ham" in each of us.

Several studies have investigated the effects of self-as-a-model with different populations. Winfrey and Weeks (1993) used the procedure with 11 female gymnasts, ages 8 to 13 years, to improve their balance beam performance. Two case studies found success in using self-as-a-model with electively mute six-year-old children (Holmbeck-Grayson & Lavigne, 1992; Kehle, Owen, & Cressy, 1990). Woltersdorf (1992) videotaped self-modeling of four attention deficit hyperactivity disorder (ADHD) nine- and ten-year-old boys to enhance their math performance and reduce fidgeting, distractibility, and vocalization. Walker and Clement (1992) used self-as-a-model and stress inoculation training with six first- and second-grade ADHD boys. Four mothers were trained with self-as-a-model to issue more direct, clear commands and praise to their four- to seven-year-old sons, contingent on the boys' compliance (Meharg & Lipsker, 1991). Cognitive-behavior therapy, relaxation training, and self-modeling were equally effective in decreasing depression and increasing self-esteem for 68 sixth, seventh, and eighth graders who were moderately to severely depressed (Kahn, Kehle, Jenson, & Clark, 1990). A Dutch study discusses the use of self-as-a-model film as part of stress-reduction training with a 12-year-old boy with leukemia (Begeer-Kooy, 1988). Johnson (1989) used self-as-a-model with counselor trainees (age 22 to 50 years), and the results showed that this group experienced less physiological and experiential anxiety and less preoccupation with how they performed than the comparison group. Videotaped self-modeling has been used effectively to teach skills to children with physical disabilities (Dowrick & Raeburn, 1995).

The behavior to be modeled should have coherence with the client's age, gender, and culture. We have adopted five steps associated with the self-as-a-model procedure from Hosford and deVisser (1974). These five components, which are illustrated in the Checklist for Self-as-a-Model on p. 562, are the following:

1. Rationale about the strategy
2. Recording the desired behavior on tape
3. Editing the tape
4. Demonstrating with the edited tape
5. Homework: client self-observation and practice

Treatment Rationale

After the client and counselor have reviewed the problem behaviors and the goal behaviors for counseling, the counselor presents a treatment rationale for the self-as-a-model procedure to the client. The counselor might say something like this:

> The procedure we are going to use is based on the idea that people learn new habits or skills by observing other people in various situations. [reason] The way this is done is that people watch other people doing things or they observe a film or tape of people doing things. What we are going to do is vary this procedure a little by having you observe yourself rather than someone else. The way we can do this is to videotape [or audiotape] your desired behavior, and then you can see [hear] yourself on the tape performing the behavior. After that, you will practice the behavior that you saw [heard] on the tape, and I will give you feedback about your practice performance. I think that seeing yourself perform and practice these behaviors will help you acquire these skills. [overview] How does this sound to you? [client's willingness]

Of course, this is only one version of the rationale for the self-as-a-model procedure a counselor might use. A counselor could add "Seeing yourself perform these behaviors will give you confidence in acquiring these skills." This statement emphasizes the cognitive component of the self-as-a-model strategy: by using oneself as the model, one sees oneself coping with a formerly anxiety-arousing or difficult situation.

Recording the Desired Behaviors

The desired goal behaviors are recorded on audio- or videotape first. For example, one particular client may need to acquire several assertion skills, such as expression of personal opinions using a firm and strong voice tone, delivery of opinions without errors in speech, and delivery of the assertive message without response latency (within five seconds after the other person's message). For this example, the counselor and client might start with voice tone and record the client expressing opinions to another person in a firm, strong voice. The counselor might have to coach the client so that at least some portion of the taped message reflects this desired response. The tape should be long enough that the client will later be able to hear himself or herself expressing opinions in a firm voice throughout several verbal exchanges with the other person. The counselor might have to spend a great deal of time staging the recording sessions in order to obtain tapes of the client's goal behavior. A dry run might be helpful before the actual tape is made.

Sometimes the counselor can instruct clients to obtain recordings of their behavior in vivo. For example, clients who stutter could be asked to audiotape their interactions with others during designated times of the week. We have also suggested such recordings to people who felt incompetent in talking with those of the other sex. The advantage of in vivo recordings is that examples of the client's actual behavior in real-life situations are obtained. However, it is not always possible or desirable to do this, particularly if the client's baseline level of performing the desired skill is very low. Whether tapes are made in vivo or in the session, the recording is usually repeated until a sample of the desired behavior is obtained.

Editing the Tape

Next, the counselor will edit the audio- or videotape recordings so that the client will see or hear *only* the appropriate (goal) behavior. Hosford et al. (1976) recommend that the "inappropriate" behaviors be deleted from the tape, leaving a tape of only the desired responses. The purpose in editing out the inappropriate behaviors is to provide the client with a positive, or self-enhancing, model. It is analogous to weeding out the dandelions in a garden and leaving the daffodils. In our example, we would edit out portions of the tape when the client did not express opinions in a strong voice and leave in all the times when the client did use a firm voice tone. For the stutterer, the stuttering portions of the tape would be deleted so that the edited tape included only portions of conversations in which stuttering did not occur.

Demonstrating with the Edited Tape

After the tape has been edited, the counselor plays it for the client. First, the client is told what to observe on the tape. For our examples of stuttering and assertion training, the counselor might say "Listen to the tape and notice that, in these conversations you have had, you are able to talk without stuttering," or "Note that you are maintaining eye contact when you are delivering a message to the other person."

After these instructions, the counselor and client play back the tape. If the tape is long, it can be stopped at various points to obtain the client's reaction. At these points, or after the tape playback, it is important for the counselor to give encouragement or positive feedback to the client for demonstrating the desired behavior.

After the tape playback, the client should practice behaviors that were demonstrated on the tape. The counselor can facilitate successful practice by coaching, rewarding successes, and correcting errors. This component of self-as-a-model relies heavily on practice and feedback.

Homework: Client Self-Observation and Practice

The client may benefit more from the self-as-a-model strategy when the edited tape is used in conjunction with practice outside the interview. The counselor can instruct the client to use a self-model audiotape as a homework aid by listening to it daily. (For homework purposes, the use of a videotape may not be practical.) After each daily use of the taped playback, the client should practice the target behavior covertly or overtly. The client could also be instructed to practice the behavior without the tape. Gradually, the client should be instructed to use the desired responses in actual instances outside the interview setting. In addition, the client should record the number of practice sessions and the measurement of the goal behaviors on a log sheet. And, as with any homework assignment, the counselor should arrange for a follow-up after the client completes some portion of the homework. The following checklist for self-as-a-model illustrates these steps.

CHECKLIST FOR SELF-AS-A-MODEL

Rationale About Strategy

_____ 1. Counselor provides rationale about strategy.
_____ 2. Counselor provides overview of strategy.
_____ 3. Counselor determines client's willingness to try strategy.

Recording Desired Behavior

_____ 4. Counselor breaks desired behaviors into subskills.
_____ 5. For each subskill, counselor coaches client about ways to perform successfully.

_____ 6. Client performs skill, using counselor or tape for feedback.
_____ 7. Client records self demonstrating skill on video- or audiotape. Recording occurs in the interview or in vivo situation outside the session. Recording is repeated until sample of desired behavior is obtained.

Editing the Tape

_____ 8. Counselor edits tape so that a clear picture of client's desired behavior is evident and instances of undesired behavior are deleted.

Demonstrating with the Edited Tape

_____ 9. Client is instructed about what to look or listen for during tape playback.
_____10. Counselor initiates playback of edited tape for client observation.
_____11. Counselor provides positive feedback to client for demonstration of desired behavior on tape.
_____12. Counselor initiates client practice of taped behaviors; successes are rewarded and errors corrected.

Homework: Client Self-Observation and Practice

_____13. Client is requested to observe or listen to the model tape and to practice the goal responses daily—both overtly and covertly—for a certain period of time.
_____14. Counselor gives some kind of self-directed prompts, such as cue cards.
_____15. Counselor asks client to record number of practice sessions and to rate performance of goal behaviors on a homework log sheet.
_____16. Counselor initiates a face-to-face or telephone follow-up to assess client's use of homework and to provide encouragement.

MODEL DIALOGUE: SELF-AS-A-MODEL

To assist you in identifying the steps of a self-as-a-model strategy, the following dialogue is presented with our client Joan. In this dialogue, the strategy is used to help Joan achieve one of her counseling goals described in Chapter 10, increasing her initiating skills in her math class.

Session 1

In response 1, the counselor provides Joan with a **rationale** *for the self-as-a-model strategy. One initiating skill, that of volunteering*

LEARNING ACTIVITY **56** *Self-as-a-Model*

You may recall from the case of Ms. Weare and Freddie that Ms. Weare wanted to eliminate the assistance she gave Freddie in getting ready for school in the morning. One of Ms. Weare's concerns is to find a way to tell Freddie about the new ground rules—mainly that she will not help him get dressed and will not remind him when the bus is five minutes away. Ms. Weare is afraid that after she does this, Freddie will either pout or talk back to her. She is concerned that she will not be able to follow through with her plan or else will not be firm in the way she delivers the ground rules to him.

Describe how you could use the five components of the self-as-a-model strategy to help Ms. Weare accomplish these four things:

1. Deliver clear instructions to Freddie.
2. Talk in a firm voice.
3. Maintain eye contact while talking.
4. Avoid talking down, giving in, or changing her original instructions.

Feedback follows on p. 564.

---

answers to questions, will be targeted using this strategy. Note that the counselor presents a **rationale** *and also confirms the* **client's willingness** *to try the strategy.*

1. *Counselor:* One of the things we discussed that is a problem for you now in your math class is that you rarely volunteer answers or make comments during class. As we talked about before, you feel awkward doing this and unsure about how to do it in a way that makes you feel confident. One thing we might try that will help you build up your skills for doing this is called "self-as-a-model." It's sort of a fun thing because it involves not only you but also this tape recorder. It's a way for you to actually hear how you come across when volunteering answers. It can help you do this the way you want to and also can build up your confidence about this. What do you think about trying this?

   Client: Well, I've never heard myself on tape too much before. Other than that, it sounds OK.

In response 2, the counselor **responds to Joan's concern** *about the tape recorder and initiates a period of using it so it doesn't interfere with the strategy.*

2. *Counselor:* Sometimes the tape recorder does take a little time to get used to, so we'll work with it first so you are accustomed to hearing your voice on it. We might spend some time doing that now. [Joan and the counselor spend about 15 minutes recording and playing back their conversation.]

In response 3, the counselor gives Joan an **overview** *of what is involved in the self-as-a-model strategy.*

3. *Counselor:* You seem to feel more comfortable with the recorder now. Let me tell you what this involves so you'll have an idea of what to expect. After we work out the way you want to volunteer answers, you'll practice this and we'll tape several practice sessions. Then I'll give you feedback, and we'll use the tape as feedback. We'll take the one practice that really sounds good to you, and you can take that and the recorder home so you can listen to it each day. Does that seem pretty clear?

   Client: I think so. I guess the tape is a way for me to find out how I really sound.

In response 4, the counselor emphasizes the cognitive or **coping part** *of self-as-a-model.*

4. *Counselor:* That's right. The tape often indicates you can do something better than you think, which is the reason it does help.

   Client: I can see that. Just hearing myself a little while ago helped. My voice doesn't sound as squeaky as I thought.

In this case, the client's verbal participation has already been defined by three behaviors. One behavior, volunteering answers, will be worked with at this point. The other two can be added later. In response 5, the counselor will **coach** *Joan on ways to perform this skill.*

5. *Counselor:* OK, Joan, let's talk about what you might do to volunteer answers in a way that you would feel good about. What comes to your mind about this?

   Client: Well, I just hardly ever volunteer in the class now. I just wait until Mr. Lamborne calls on me. It's not that I don't know the answer, because lots of times I do. I guess I just need to raise my hand and give out the answer. See, usually he'll say, "OK, who has the answer to this problem?" So all I need to do is raise my hand and give the answer, like 25 or 40 or whatever. I don't know why I don't do it. I guess I'm afraid I will sound silly or maybe my voice will sound funny.

In the next response, the counselor uses a **clarification** *to determine Joan's particular concern about this skill.*

6. *Counselor:* So are you more concerned with the way you sound than with what you have to say?

   Client: I think so.

In response 7, the counselor continues to **coach** *Joan about ways to perform the desired skill, volunteering answers, and also initiates a* **trial practice.**

7. *Counselor:* Well, let's try this. Why don't I pretend to be Mr. Lamborne and then you raise your hand and give me an answer? Just try to speak in a firm voice that I can easily hear. Maybe even take a deep breath at first. OK? [Counselor turns on tape

FEEDBACK
Self-as-a-Model

1. *Rationale for strategy*
   First, you would explain to Ms. Weare how the self-as-a-model procedure could help her (rationale) and what is involved in the procedure (overview). Then, ask Ms. Weare how she feels about trying this procedure (client's willingness).

2. *Recording the desired behavior*
   According to the case description, there are four things Ms. Weare wants to do in talking to Freddie.
   The counselor will probably need to coach Ms. Weare on a successful way to perform the skills before recording her demonstration of it, and a dry run may be necessary. When the counselor believes Ms. Weare can demonstrate the skills at least sometimes, a video or audio recording will be made. (For eye contact, a videotape would be necessary.) Because Ms. Weare presently is not engaging in these behaviors with Freddie, an in-session tape would be more useful than an in vivo tape at this point. The counselor can role-play the part of Freddie during the taping. The taping should continue until an adequate sample of each of the four skills is obtained.

3. *Editing the tape*
   After the tape has been recorded, the counselor will edit it. Only inappropriate examples of the skill would be deleted. For example, instances when Ms. Weare looks away would be erased from the tape. The edited tape would consist only of times when she maintains eye contact. A final tape in which she uses all four skills would consist only of times when she was using the desired skills.

4. *Demonstrating with the edited tape*
   After the edited tape is ready, it would be used for demonstration and practice with Ms. Weare. The counselor would instruct Ms. Weare about what to look for and then play back the tape. The counselor would give positive feedback to Ms. Weare for instances of demonstrating eye contact and the other three skills. After the playback, Ms. Weare would practice the skill and receive feedback from the counselor about the practice performance.

5. *Homework: Client self-observation and practice*
   After Ms. Weare was able to practice the skills with the counselor, she could use the self-modeling tape as homework. Specifically, the counselor would instruct her to listen to or view the tape on her own if possible. She could also practice the skills—first covertly and later overtly—with Freddie. This practice could occur with or without the tape. A follow-up should be arranged to check on her progress.

recorder.] (as Mr. Lamborne) Who has the answer to this problem?
[Joan raises her hand.]
Counselor (as Mr. Lamborne, looks around room, pauses): Joan?
Client (in a pretty audible voice): 25.

After the dry run, the counselor, in responses 8, 9, and 10, gives **feedback** *(using tape playback) about Joan's performance of the target behavior.*

8. *Counselor:* OK, let's stop. What did you think about that?
   Client: Well, it wasn't really that hard. I took a deep breath.

9. *Counselor:* Your voice came across pretty clear. Maybe it could be just a little stronger. OK. Let's hear this on tape. [Playback of tape.]

10. *Counselor:* How do you feel about what you just heard?
    Client: Well, I sound fine. I mean my voice didn't squeak.

In response 11, the counselor initiates **tape recordings** *of Joan's demonstration of the skill (volunteering answers). This tape will be edited and used as a modeling tape.*

11. *Counselor:* No, it was pretty strong. Let's do this several times now. Just take a deep breath and speak firmly. [Practice ensues and is recorded.]

In response 12, the counselor explains the **tape-editing process;** *the tape is edited before their next session.*

12. *Counselor:* OK, I'm going to need to go over this tape before we use it for feedback. So maybe that's enough for today. We can get together next week, and I'll have this tape ready by then. Basically, I'm just going to edit it so you can hear the practice examples in which your voice sounded clear and firm. [Before the next session, the counselor erases any portions of the tape in which Joan's answers were inaudible or high-pitched, leaving only audible, firm, level-pitched answers.]

Session 2

After a brief warm-up period in this session, the counselor **instructs** *Joan about what to listen for in the* **demonstration with the edited tape playback.**

1. *Counselor:* Well, Joan, I've got your tape ready. I'd like to play back the tape. When I do, I'd like you to note how clearly and firmly you are able to give the answers. [Tape is played.]

2. *Counselor:* What did you think?
   Client: You're right. I guess I don't sound silly, at least not on that tape.

In response 3, the counselor gives **positive feedback** *to Joan about demonstrating the skill.*

3. *Counselor:* You really sounded like you felt very confident about the answers. It was very easy to hear you.
   Client: But will I be able to sound like that in class?

In response 4, the counselor instructs Joan on how to use the tape as **daily homework** *in conjunction with practice. Note that the homework assignment specifies* **what** *and* **how much** *Joan will do.*

4. *Counselor:* Yes, and we'll be working on that as our next step. In the meantime, I'd like you to work with this tape during the week. Could you set aside a certain time each day when you could listen to the tape, just like we did today? Then after you listen to the tape, practice again. Imagine Mr. Lamborne asking for the answer. Just raise your hand, take a deep breath, and speak firmly. Do you understand how to use this now?

Client: Yes, just listen to it once a day and then do another round of practice.

In response 5, the counselor asks Joan to **record her use of homework on log sheets.**

5. *Counselor:* As you do this, I'd like you to use these log sheets and mark down each time you do this homework. Also, rate on this five-point scale how comfortable you feel in doing this before and each time you practice.

Client: That doesn't sound too difficult, I guess.

In response 6, the counselor encourages Joan to **reinforce herself** *for progress and* **arranges for follow-up** *on homework at their next session.*

6. *Counselor:* Well, this recording on your log sheet will help you see your progress. You've made a lot of progress, so give yourself a pat on the back after you hear the tape this week. Next week we can talk about how this worked out and then we'll see whether we can do the same type of thing in your classes.

The next step would be to obtain some tape-recorded samples of Joan's volunteering in a class situation. A non-threatening class in which Joan presently does volunteer might be used first, followed by her trying this out in math class. The biggest problem in this step is to arrange for tape-recorded samples in a way that is not embarrassing to Joan in the presence of her classmates.

SELF-EFFICACY

Self-efficacy is described by Bandura, Adams, and Beyer (1977) as a cognitive process that mediates behavioral change. Self-efficacy is defined as a concept about our personal beliefs of how capable we are of exercising control over events in our lives. In other words, self-efficacy is an estimate of a person's skill in dealing with some task. Self-esteem is enhanced, our self-efficacy is high, when we feel capable of doing what is required to achieve success, or when we realize that we have reached our goal. Underlying self-efficacy is optimism and hope that yields high efficacy, or helplessness and despair that contributes to low efficacy; both these extremes can be learned. Self-efficacy belongs to a constellation of similar concepts such as self-agency, expectation, self-confidence, self-empowerment, optimism, hopefulness, and enthusiasm.

Over the last several years research about self-efficacy has flourished. Table 20-6 lists a sample of these studies. The research can be grouped by achievement and performance, career decision making, health concerns, substance abuse, divorce, performance in sports, exercise, legal compliance, pain, panic, problem solving, rehabilitation, relapse prevention, safety, and shyness.

Several sources contribute to self-efficacy and to similar concepts in the constellation of one's personality or personal constructs. According to Bandura (1977), four primary sources influence efficacy expectations: (1) actual performance accomplishments, (2) mind/body states, (3) environmental experiences, and (4) verbal installation—as a prelude to treatment. Verbal installation is the process the therapist uses to enhance a client's confidence or self-efficacy expectations about performing specific tasks. A therapist might install positive self-efficacy by talking with the client about his or her past success in performing similar or related tasks. Or, the therapist can attempt to build confidence by discussing any of the client's successful experiences if the client has not engaged in tasks for which self-confidence or efficacy was needed.

Performance Accomplishments

There is huge variability in how people perform. People who have a high degree of self-efficacy recoup very quickly from failure. These types of people are motivated, energized, and risk taking despite the possibility of failure. At the other extreme are those people who fall into a state of learned helplessness (Seligman, 1990). These people are plagued with depressed feelings that contribute to pessimism, a low level of energy, negative internal dialogue, vulnerability, and hopelessness. When these people experience failure, their automatic explanation is, "It's my fault," "I always screw things up," "This is just another example that I can never do anything right," or "This is going to ruin my life." This learned behavior for dealing with failures creates a pessimistic and catastrophizing mode for feeling and thinking.

Most of us fall somewhere between these two extremes. Perceived self-efficacy is a major determinant in whether people engage in a task, the amount of effort they exert if they do engage, and how long they will persevere with the task if they encounter adverse circumstances. For example, people who frequently surf the Internet on their personal computers in search of a particular web page feel competent in performing this task. They may feel quite confident in their abilities and persevere for some time in their search, even when they are unsuccessful in locating a specific web page. The same people, however, may feel less competent in programming software and may avoid programming tasks.

TABLE 20-6. Research on self-efficacy

Abortion
Cozzarelli, C. (1993). Personality and self-efficacy as predictors of coping with abortion. *Journal of Personality and Social Psychology, 65,* 1224–1236.

AIDS
Denson, D. R., Voight, R., & Eisenman, R. (1994). Self-efficacy and AIDS prevention for university students. *International Journal of Adolescence and Youth, 5*(1–2), 105–113.
Weeks, K., Levy, S., Zhu, C., & Perhats, C. (1995). Impact of a school-based AIDS prevention program in young adolescents' self-efficacy skills. *Health Education Research, 10,* 329–344.

Alcohol use
Aas, H., Klepp, K., Laberg, J., & Edvard, L. (1995). Predicting adolescents' intentions to drink alcohol: Outcome expectancies and self-efficacy. *Journal of Studies on Alcohol, 156,* 293–299.
Moss, H. B., Kirisci, L., & Mezzich, A. C. (1994). Psychiatric comorbidity and self-efficacy to resist heavy drinking in alcoholic and nonalcoholic adolescents. *American Journal of Addictions, 3,* 204–212.

Battered women
Mancoske, R. J., Standifer, D., & Cauley, C. (1994). The effectiveness of brief counseling services for battered women. *Research on Social Work Practice, 4,* 53–63.

Birth control
Haydem, J., (1993). The condom race. *Journal of American College Health, 42,* 133–136.
Levinson, R. A. (1995). Reproductive and contraceptive knowledge, contraceptive self-efficacy, and contraceptive behavior among teenage women. *Adolescence, 30,* 65–85.

Breast self-examination
Friedman, L. C., Nelson, D. V., Webb, J. A., & Hoffman, L. P. (1994). Dispositional optimism, self-efficacy, and health beliefs as predictors of breast self-examination. *American Journal of Preventive Medicine, 10,* 130–135.

Cancer
Hertog, J. K., Finnegan, J. R., Rooney, B., & Viswanath, K. (1993). Self-efficacy as a target population segmentation strategy in a diet and cancer risk reduction campaign. *Health Communication, 5,* 21–40.

Cardiovascular
Wright, R. A., Wadley, V. G., Pharr, R. P., & Butler, M. (1994). Interactive influence of self-reported ability and avoidant task demand on anticipatory cardiovascular responsivity. *Journal of Research in Personality, 28,* 68–86.

Career
Luzzo, D. A. (1995). The relative contributions of self-efficacy and locus of control to the prediction of career maturity. *Journal of College Student Development, 36,* 61–66.

Ryan, N., Solberg, V., & Brown, S. (1996). Family dysfunction, parental attachment, and career search self-efficacy among community college students. *Journal of Counseling Psychology, 43,* 84–89.

Childbirth
Lowe, N. K. (1993). Maternal confidence for labor: Development of the Childbirth Self-Efficacy Inventory. *Research in Nursing and Health, 16,* 141–149.

Compliance with the law
Jenkins, A. L. (1994). The role of managerial self-efficacy in corporate compliance with the law. *Law and Human Behavior, 18,* 71–88.

Decision making
Stone, D. N. (1994). Overconfidence in initial self-efficacy judgments: Effects on decision processes and performance. *Organizational Behavior and Human Decision Processes, 59,* 452–474.

Dental health
Tedesco, L. A., Keffer, M. A., Davis, E. L., & Christersson, L. A. (1993). Self-efficacy and reasoned action: Predicting oral health status and behaviour at one, three, and six month intervals. Special Issue: Dental health psychology. *Psychology and Health, 8,* 105–121.

Diabetes
Kavanagh, D. J., Gooley, S., & Wilson, P. H. (1993). Prediction of adherence and control of diabetes. *Journal of Behavioral Medicine, 16,* 509–522.

Dietary self-efficacy and marital interaction
Schafer, R. B., Keith, P. M., & Schafer, E. (1994). The effects of marital interaction, depression and self-esteem on dietary self-efficacy among married couples. *Journal of Applied Social Psychology, 24,* 2209–2222.

Divorce and perceived self-efficacy of children
Kurtz, L., & Derevensky, J. L. (1993). The effects of divorce on perceived self-efficacy and behavioral control in elementary school children. *Journal of Divorce and Remarriage, 20,* 75–90.

Epilepsy
Dilorio, C., Faherty, B., & Manteuffel, B. (1994). Epilepsy self-management: Partial replication and extension. *Research in Nursing and Health, 17,* 167–174.

Exercise
DuCharme, K., & Brawley, L. (1995). Predicting the intentions and behavior of exercise initiates using two forms of self-efficacy. *Journal of Behavioral Medicine, 18,* 479–497.
Rodgers, W., & Brawley, L. (1996). The influence of outcome expectancy and self-efficacy on the behavioral intentions of novice exercisers. *Journal of Applied Social Psychology, 26,* 618–634.

(continued)

TABLE 20-6. Research on self-efficacy (continued)

Gymnastics

Winfrey, M. L., & Weeks, D. L. (1993). Effects of self-modeling on self-efficacy and balance beam performance. *Perceptual and Motor Skills, 77,* 907–913.

Headache

Martin, N. J., Holroyd, K. A., & Rokicki, L. A. (1993). The Headache Self-Efficacy Scale: Adaptation to recurrent headaches. *Headache, 33,* 244–248.

Health

Eachus, P. (1993). Development of the Health Student Self-Efficacy Scale. *Perceptual and Motor Skills, 77,* 670.

Grembowski, D., Patrick, D., Diehr, P., & Durham, M. (1993). Self-efficacy and health behavior among older adults. *Journal of Health and Social Behavior 34,* 89–104.

Job procurement

Wenzel, S. L. (1993). The relationship of psychological resources and social support to job procurement self-efficacy in the disadvantaged. *Journal of Applied Social Psychology, 23,* 1471–1497.

Learning

Moriarty, B., Douglas, G., Punch, K., & Hattie, J. (1995). The importance of self-efficacy as a mediating variable between learning environments and achievement. *British Journal of Educational Psychology, 65,* 73–84.

Maternal self-efficacy

Gross, D., Conrad, B., Fogg, L., & Wothke, W. (1994). A longitudinal model of maternal self-efficacy, depression, and difficult temperament during toddlerhood. *Research in Nursing and Health, 17,* 207–215.

Math self-efficacy

Randhawa, B. S. (1994). Self-efficacy in mathematics, attitudes, and achievement of boys and girls from restricted samples of two countries. *Perceptual and Motor Skills, 79,* 1011–1018.

Myocardial infarction

Coyne, J. C., & Smith, D. A. J. (1994). Couples coping with a myocardial infarction: Contextual perspective on patient self-efficacy. *Journal of Family Psychology, 8,* 43–54.

Opiate addicts relapse risk

Reilly, P., Sees, K., Shopshire, M., & Hall, S. (1995). Self-efficacy and illicit opoid use in a 180-day methadone detoxification treatment. *Journal of Consulting and Clinical Psychology, 63,* 158–162.

Osteoarthritis

Allegrante, J. P., Kovar, P. A., MacKenzie, C. R., & Peterson, M. G. (1993). A walking education program for patients with osteoarthritis of the knee: Theory and intervention strategies. Special Issue: Arthritis health education. *Health Education Quarterly, 20,* 63–81.

Pain

Bickelew, S., Murray, S., Hewett, J., & Johnson, J. (1995). Self-efficacy, pain, and physical activity among fibromyalgia subjects. *Arthritis Care and Research, 8,* 43–50.

Parenting

Rodrigue, J. R., Geffken, G. R., Clark, J. E., & Hunt, F. (1994). Parenting satisfaction and efficacy among caregivers of children with diabetes. *Children's Health Care, 23,* 181–191.

Problem solving

Huet, N., Marine, C., & Escribe, C. (1994). Auto-evaluation des competences en resolution de problems chez des adultes peu qualifies. *International Journal of Psychology, 29,* 273–289.

Rehabilitation

Guthrie, S., & Harvey, A. (1994). Motivation and its influence on outcome in rehabilitation. *Reviews in Clinical Gerontology, 4,* 235–243.

Relapse for substance abuse

Rawson, R. A., Obert, J. L., McCann, M. J., Marinelli-Casey, P. (1993). Relapse prevention models for substance abuse treatment. Special Issue: Psychotherapy for the addictions. *Psychotherapy, 30,* 284–298.

Safety

Celuch, K. G., Lust, J. A., & Showers, L. S. (1995). An investigation of the relationship between self-efficacy and the communication effectiveness of product manual formats. *Journal of Business and Psychology, 9,* 241–252.

Shyness

Kuzuu, S., (1994). The effects of self-observation on self-efficacy of shy students. *Japanese Journal of Counseling Science, 27,* 97–104.

Smoking

De Vries, H., & Backbier, E. (1994). Self-efficacy as an important determinant of quitting among pregnant women who smoke: The o-pattern. *Preventive Medicine: An International Journal Devoted to Practice and Theory, 23,* 167–174.

Mudde, A., Kok, G., & Strecher, V. (1995). Self-efficacy as a predictor for the cessation of smoking. *Psychology and Health, 10,* 353–367.

Students—medical students

Tresolini, C. P., & Stritter, F. T. (1994). An analysis of learning experiences contributing to medical students' self-efficacy in conducting patient education for health promotion. *Teaching and Learning in Medicine, 6,* 247–254.

Traumatized adolescents

Saigh, P., Mroueh, M., Zimmerman, B., & Fairbank, J. (1995). Self-efficacy expectations among traumatized adolescents. *Behaviour Research and Therapy, 33,* 701–704.

Sometimes people feel competent at performing a task, but do not perform it because there is no incentive for doing so. Also, some people may have unrealistic expectations about performing a task simply because they are unfamiliar with the task. For example, some people may feel *over*confident about doing something and others may experience *less* confidence. Generally, when we develop competence of any kind, we enhance and strengthen our self-efficacy, confidence, self-esteem, willingness to take risk, and ability to perform the task.

Mind/Body Link to Self-Efficacy

Self-efficacy influences our emotions, thinking, perceptions, motivation, and biochemistry. There is a chain reaction and interaction within the mind/body information processing system. Everything in our bodies is run by messenger molecules, many of which are called peptides. There is sufficient scientific evidence to suggest that information molecules, peptides, and receptors are the biochemicals of emotion (Pert, 1993). These messenger molecules or neuropeptides influence self-efficacy. For example, a person who has a high degree of self-confidence and high level of self-esteem, compared with someone of lower self-confidence, is inclined to attempt more difficult tasks, use more energy, persevere longer at problem solutions when faced with adversity, and refuse to blame himself or herself when encountering failure. This type of person has feelings of control. One hypothesis is that the production of endogenous morphine (endorphins) in the brain is high and the production of catecholamines is low. A relatively high production of endorphins is correlated with confidence. It has an analgesic effect, which spreads throughout the body and reduces sensitivity to pain (Bandura, Cioffi, Taylor, & Brouillard, 1988; Pert, 1993).

In contrast, those people without feelings of control have low levels of perceived self-efficacy, avoid difficult tasks, have lower expectancies, and have a weak commitment to achieving a goal. They dwell on their personal inadequacies, allow negative self-talk, put forth less effort, have little energy for a task, and take longer to recover from failure on some task. They are very susceptible to stress and depression. For these people who feel that they are not in control, their feelings of low self-efficacy can possibly increase the production of catecholamines or stress hormones (Bandura, Taylor, Williams, Mefford, & Barchas, 1985). Our emotional state affects our perceived level of self-efficacy; this state, in turn, causes information molecules or neuropeptides to produce either stress hormones or brain opioids, depending on our emotional state. The degree of perceived self-efficacy, self-confidence, and biochemical reactivity have their origins in the family and the environmental and cultural context in which the family resides.

Environmental Influences

A person's self-efficacy is influenced by reciprocal interactions of cognitive, affective, behavioral, relational, and environmental and/or cultural variables (Bandura, 1989). Family of origin, culture, and environmental setting mold a person's perceived self-efficacy, which contributes to cognitive development and functioning. Bandura (1993) proposed that perceived self-efficacy exerts a powerful influence on four major developmental processes: cognitive, motivational, affective, and perceptual selection processes. For example, Bandura (1993) asserts that there are four different levels at which perceived self-efficacy operates as an important contributor to academic development: (1) students' beliefs in their efficacy to regulate their own learning and to master academic activities determine their level of motivation and academic accomplishments; (2) teachers' beliefs in their efficacy to motivate and promote learning affect the types of learning environments they create and the degree of their students' academic progress; (3) faculties' beliefs in their collective instructional efficacy contribute significantly to their schools' level of achievement; and (4) student body characteristics influence school-level achievement more strongly by altering faculties' beliefs in their collective efficacy. Note that these four levels of perceived self-efficacy are also affected by the client's worldview.

Self-Efficacy as a Prelude for Treatment

A client's expectations, based on his or her perceived self-efficacy, determine the underlying cognitive process that accounts for changes in therapeutic outcomes and achievement of treatment goals. Clients must acquire self-efficacy (confidence) to perform the specific skills associated with a particular therapeutic strategy so they can achieve their therapeutic goals. Bandura (1982) has demonstrated that there is a strong relationship between an individual's level of perceived self-efficacy and his or her later level of performance. Self-efficacy is a mental precondition that has a striking influence on the successful application of treatment. We believe that if the counselor can maximize the client's perceived self-efficacy and expectancies about treatment, the client will be confident about using the specific steps associated with the treatment protocol and enhance the potential benefits and effectiveness of treatment. Rosenthal and Steffek (1991) say that "all psychotherapies—whatever techniques they comprise—must raise patients' self-efficacy if patients are to attempt, or persist at, activities they formerly avoided or inhibited" (p. 80).

The strength of self-efficacy assessment involves the specification of the goals for therapy discussed in Chapter 10. Clients can be provided with a series of specific goal behaviors or outcome goals for treatment. They can be asked to rate how much they believe they can perform these behaviors and how confident they are of their judgment. These ratings are usually measured on a scale ranging from 0 (uncertain) to 100 (certain). For example, Ozer and Bandura (1990) offered 45 women volunteers a program to deal effectively with the risk of sexual assault. All the women had limited their cultural, educational, and social activities because they lacked a sense of ability (self-efficacy) to handle potential dangers that might occur to them from going out alone. Self-efficacy training was assessed in several areas: (1) competence in self-defense was measured on scales of perceived capacity to retaliate with disabling blows, for example, if assaulted; (2) ability to increase activities was measured on scales of perceived comfort in resuming nighttime outdoor recreation, attending concerts, and going to restaurants and nightclubs—all activities that take place after dark; (3) efficacy for interpersonal encounters was measured on scales of the women's perceived capacity for assertive response if they were threatened or harassed in social contexts such as at work, on dates, on public transportation, or at parties; (4) efficacy to control troublesome cognitions was measured by a scale of perceived capacity to stop ruminating about possible attacks; and (5) a variety of ratings of thoughts about, vulnerability to, risk of facing, and fear concerning sexual assault were measured (Ozer & Bandura, 1990). The success of achieving therapeutic goals may be largely a function of the client's self-efficacy, with each strategy designed to reach a specific goal. Also, the client's age, gender, social class, and cultural and ethnic background will influence the degree of efficacy or confidence he or she feels in performing a specific task.

MULTICULTURAL APPLICATIONS OF SELF-EFFICACY

Increasingly, self-efficacy is being identified as a factor in prevention and risk reduction programs and in academic jobs and career performance of diverse groups of clients (see Table 20-7). In the area of prevention and risk reduction, self-efficacy has been examined as a factor in prevention of HIV infection, drug use, and cardiovascular disease.

A number of studies have examined how self-efficacy impacts HIV risk-reduction behaviors among African American and Hispanic men and also among gay and bisexual men. A higher level of self-efficacy among heterosexual African American men using cocaine was associated with increased condom use (Malow, Corrigan, Cunningham, & West, 1993) and also with HIV prevention behaviors among African American incarcerated young men (Belgrave, Randolph, Carter, & Braithwaite, 1993). Rotheram-Borus, Rosario, Reid, and Koopman (1995) examined the role of self-efficacy and other predictor variables of HIV sexual risk acts (SRAs) in a longitudinal study of 136 gay and bisexual African American and Hispanic male adolescents. Patterns of change in these HIV SRAs were assessed at intake and at 3-, 6-, and 12-month follow-ups. There were five patterns of SRAs: protected, improved, relapsed, variable, and unprotected. The protected and improved patterns of SRAs were associated with low levels of anxiety, depression, and substance use, and a high level of self-efficacy.

Self-efficacy has been suggested as an important component of HIV prevention programs for gay and bisexual men (Antoni, 1991; Aspinwall, Kemeny, Taylor, & Schneider, 1991). As a result of all this research, self-efficacy theory, as well as the health belief model, are being used increasingly to predict HIV-related risk behaviors. Cochran and Mays (1993) discuss some potential issues in the application of these models to predict risk behaviors of African Americans. They note that these models emphasize the importance of individualistic, direct control of behavioral choices and deemphasize external factors such as racism and poverty that are particularly relevant to those within the African American community at highest risk for HIV infection. They contend that applications of these models without consideration of the unique issues associated with behavioral choices within the African American community may not capture the most relevant determinants of risk behaviors. Others have also discussed the need for AIDS education and prevention that mobilizes the will of members of the African American community (Gasch, Poulson, Fullilove, & Fullilove, 1991). These authors note that such a prevention program is based on three assumptions:

1. The environment plays a critical role in conditioning behavior.
2. The experiences of community deterioration differ between African American men and women.
3. African Americans in urban areas are struggling to make sense of the threatening ecological and social environment with which they come into contact on a daily basis.

TABLE 20-7. Multicultural applications of self-efficacy

HIV risk prevention

Antoni, M. (1991). Psychosocial stressors and behavioral interventions in gay men with HIV infection. *International Review of Psychiatry, 3,* 383–399.

Aspinwall, L., Kemeny, M., Taylor, S., & Schneider, S. (1991). Psychosocial predictors of gay men's AIDS risk-reduction behavior. *Health Psychology, 10,* 432–444.

Belgrave, F., Randolph, S., Carter, C., & Braithwaite, N. (1993). The impact of knowledge, norms, and self-efficacy on intentions to engage in AIDS-preventive behaviors among young incarcerated African American males. *Journal of Black Psychology, 19,* 155–168.

Cochran, S., & Mays, V. (1993). Applying social psychological models to predicting HIV-related sexual risk behaviors among African Americans. *Journal of Black Psychology, 19,* 142–154.

Gasch, H., Poulson, D., Fullilove, R., & Fullilove, M. (1991). Shaping AIDS education and prevention programs for African Americans amidst community decline. *Journal of Negro Education, 60,* 85–96.

Malow, R., Corrigan, S., Cunningham, S., & West, J. (1993). Psychosocial factors associated with condom use among African American drug abusers in treatment. *AIDS Education and Prevention, 5,* 244–253.

Rotherman-Borus, M. J., Rosario, M., Reid, H., & Koopman, C. (1995). Predicting patterns of sexual acts among homosexual and bisexual youths. *American Journal of Psychiatry, 152,* 588–595.

Drug use prevention

Epstein, J., Botvin, G., Diaz, T., & Toth, V. (1995). Social and personal factors in marijuana use and intentions to use drugs among inner city minority youth. *Journal of Developmental and Behavioral Pediatrics, 16,* 14–20.

Krepcho, M., Fernandez-Esquer, M., Freeman, A., & Magee, E. (1993). Predictors of bleach use among current African-American injecting drug users: A community study. *Journal of Psychoactive Drugs, 25,* 135–141.

Van-Hasselt, V., Hersen, M., Null, J., & Ammerman, R. (1993). Drug abuse prevention for high risk African American children and their families: A review and model program. *Addictive Behaviors, 18,* 213–234.

Cardiovascular disease prevention

Winkleby, M., Flora, J., & Kraemer, H. (1994). A community-based heart disease intervention: Predictors of change. *American Journal of Public Health, 84,* 767–772.

Academic performance and career selection

Bryan, T., & Bryan, J. (1991). Positive mood and math performance. *Journal of Learning Disabilities, 24,* 490–494.

Church, A., Teresa, J., Rosebrook, R., & Szendre, D. (1992). Self-efficacy for careers and occupational consideration in minority high school equivalency students. *Journal of Counseling Psychology, 39,* 498–508.

Lauver, P., & Jones, R. (1991). Factors associated with perceived career options in American Indian, White and Hispanic rural high school students. *Journal of Counseling Psychology, 38,* 159–166.

Morrow, S., Gore, P., & Campbell, B. (1996). The application of a socio-cognitive framework to the career development of lesbian women and gay men. *Journal of Vocational Behavior, 48,* 136–148.

Solberg, V., O'Brien, K., Villareal, P., & Kennel, R. (1993). Self-efficacy and Hispanic college students: Validation of the College Self-Efficacy Instrument. *Hispanic Journal of Behavioral Sciences, 15,* 80–95.

Job performance

Orpen, C. (1995). Self-efficacy beliefs and job performance among Black managers in South Africa. *Psychology Reports, 76,* 649–650.

Prostate cancer

Boehm, S., Coleman-Burns, P., Schlenk, E., & Funnell, M. (1995). Prostate cancer in African American men: Increasing knowledge and self-efficacy. *Journal of Community Health Nursing, 12,* 161–169.

Exercise

Clark, D. O., Patrick, D., Grembowski, D., & Durham, M. (1995). Socio-economic status and exercise self-efficacy in late life. *Journal of Behavioral Medicine, 18,* 355–376.

McAuley, E., Shaffer, S., & Rudolph, D. (1995). Affective responses to acute exercise in elderly impaired males: The moderating effects of self-efficacy and age. *International Journal of Aging and Human Development, 41,* 13–27.

Disability and low back disorders

Lackner, J., Carosella, A., & Feuerstein, M., (1996). Pain expectancies, pain, and functional self-efficacy expectancies as determinants of disability in patients with chronic low back disorders. *Journal of Consulting and Clinical Psychology, 64,* 212–220.

Gasch and associates propose a model of AIDS education that focuses on social responsibility for African American men and *contextual* self-efficacy for African American women.

Self-efficacy was examined as a predictor in marijuana use and in intentions to use drugs among 757 African American and Hispanic seventh-grade, low-income youth (Epstein, Botvin, Diaz, & Toth, 1995). Social influences including tolerance of marijuana use by adults, friends, and role models predicted marijuana use. Lack of self-efficacy and low academic performance were related to the youths' intentions to use cocaine, crack, and other drugs. Self-efficacy figured in another drug use study—one of 117 African American adults who used drugs through needle injections (Krepcho, Fernandez-Esquer, Freeman, & Magee, 1993). Specifically, self-efficacy was explored as one of a number of predictor variables for cleaning needles with bleach during drug use. Out of all ten predictor variables, the best predictors of bleach use were self-efficacy, expectations about bleach use, and age of the adult. Focusing on preventing drug use, Van-Hasselt, Hersen, Null, and Ammerman (1993) describe a drug use prevention program for African American children that focuses on the development of family-based alternative activities to promote self-efficacy, achievement, and self-esteem.

Winkleby, Flora, and Kraemer (1994) examined self-efficacy as one of a number of factors related to changes in cardiovascular disease risk-factor scores (RFSS) of 411 adult men and women of varied racial and ethnic groups who had received a six-year cardiovascular disease risk-reduction educational program. The subgroup with the highest proportion of positive changes was composed of people over 55 years of age with the highest perceived risk, highest health media use, and highest blood pressure and cholesterol levels. The subgroup with the lowest proportion of positive changes consisted of those who were less educated, reported less health knowledge, and had lower self-efficacy scores.

Self-efficacy is also an important variable in academic success. Bryan and Bryan (1991) found self-efficacy and positive mood induction to be related to performance of both junior high and high school students with a learning disability. At the postsecondary level, college self-efficacy—or the degree of confidence that one can successfully complete college—was an important determinant of student adjustment for Mexican American and Latino American college students (Solberg, O'Brien, Villareal, & Kennel, 1993).

Bandura's (1986) self-efficacy model has also been applied to career selection. Morrow, Gore, and Campbell (1996) investigated the effects of self-efficacy on the career development patterns of lesbian women and gay men in another study. With an ethnically mixed group of high school equivalency students (predominantly Hispanic and Native American girls and boys) from seasonal farm work backgrounds, students' interests, perceived incentives, and self-efficacy expectations (beliefs about their ability to learn to engage in an occupation successfully) predicted their willingness to consider the occupations (Church, Teresa, Rosebrook, & Szendre, 1992).

Among an ethnically mixed high school population of over 800 rural girls and boys (Hispanic, Native American, and White youth), Lauver and Jones (1991) found differences in self-efficacy estimates for career choice among varied ethnic groups, with efficacy lowest for 7 of the 18 occupations studied among the Native American youth.

Gender also is a variable that affects self-efficacy in career selection. Church et al. (1992) found that both Hispanic and Native American male and female high school equivalency students reported greater willingness to consider occupations dominated by their own gender, with the women students showing a greater tendency to reject occupations dominated by the opposite gender. This finding suggests less self-efficacy related to expectations of success in nontraditional occupations. In the Lauver and Jones (1991) study of Hispanic, Native American, and White rural youth, gender differences were found in both interest and self-efficacy estimates for same-gender and cross-gender occupations, with the boys considering fewer cross-gender options than the girls. Self-efficacy has also been explored as a moderating variable in the exercise habits of elderly persons (McAuley, Shaffer, & Rudolph, 1995).

Guidelines for Using Self-Efficacy with Diverse Groups of Clients

From a multicultural standpoint, the goal of strengthening self-efficacy, when used alone, suffers from some of the criticisms we have made of the relevance of self-management approaches for clients of color. As Cochran and Mays (1993) have so aptly noted, the self-efficacy model stresses the importance of individualistic, direct control of behavioral choices. This reflects a worldview that is high in both internal locus of control and internal locus of responsibility. People from cultures that stress collectivity and unity may not feel comfortable with this model, nor may individuals who have experienced unjust societal conditions such as oppression, racism, and poverty. Still, improved self-efficacy is potentially useful as an empowerment tool for women and clients of color if it is sought within the context of additional change processes (Gutierrez, 1990). Gutierrez (1990) proposes that an empowerment process for women and clients of color include not only ways to increase self-efficacy but also the development of group and cultural consciousness and the reduction of personal and self-blame. Her perspective is highly useful because it focuses on external social-contextual factors as well as the internal sense of confidence in oneself.

MODEL EXAMPLE: SELF-EFFICACY

In Chapter 10, we presented the goals for Joan. The first goal—asking questions and making reasonable requests—had four subgoals: (1) to decrease anxiety ratings associated with anticipation of failure in math class and rejection by parents, (2) to increase positive self-talk and thoughts that "girls are capable" in math class and other competitive situations from zero or two times a week to four or five times a week over the next two weeks, (3) to increase attendance in math class from two or three times a week to four or five times a week during treatment, and (4) to increase verbal participation and initiation in math class and with her parents from none or once a week to three or four times a week over the next two weeks during treatment. Verbal participation is defined as asking and answering questions posed by teacher or parents, volunteering answers or offering opinions, or going to the chalkboard.

The counselor can determine the extent of Joan's self-efficacy (confidence) for each of the goal behaviors. Self-efficacy can be measured by asking Joan to give a verbal rating of her confidence for each goal on a scale from 0—no confidence—to 100—a great deal of confidence. Alternatively, the counselor can design a rating scale and ask Joan to circle her rating of confidence for each goal. Possible scales for all the goals are shown below.

Goal One for Math Class:
Confidence in *decreasing anxiety* (from 70 to 50) about possible failure in *math class*
0　10　20　30　40　50　60　70　80　90　100
Uncertain　　　　　　　　　　　　　　　Total
　　　　　　　　　　　　　　　　　　　　Certainty

Goal One for Parents:
Confidence in *decreasing anxiety* (from 70 to 50) about possible *rejection by parents*
0　10　20　30　40　50　60　70　80　90　100
Uncertain　　　　　　　　　　　　　　　Total
　　　　　　　　　　　　　　　　　　　　Certainty

Goal Two for Math Class:
Confidence to *increase positive self-talk and thoughts*—"girls are capable"—to 4 or 5 times a week in *math class*
0　10　20　30　40　50　60　70　80　90　100
Uncertain　　　　　　　　　　　　　　　Total
　　　　　　　　　　　　　　　　　　　　Certainty

Goal Two for Other Situations:
Confidence to *increase positive self-talk and thoughts*—"girls are capable"—to 4 or 5 times a week in *other competitive situations*

0　10　20　30　40　50　60　70　80　90　100
Uncertain　　　　　　　　　　　　　　　Total
　　　　　　　　　　　　　　　　　　　　Certainty

Goal Three:
Confidence to *increase attendance* in *math class* to 4 or 5 times a week
0　10　20　30　40　50　60　70　80　90　100
Uncertain　　　　　　　　　　　　　　　Total
　　　　　　　　　　　　　　　　　　　　Certainty

Goal Four for Math Class:
Confidence in *asking* questions in *math class*
0　10　20　30　40　50　60　70　80　90　100
Uncertain　　　　　　　　　　　　　　　Total
　　　　　　　　　　　　　　　　　　　　Certainty

Confidence in *answering* questions in *math class*
0　10　20　30　40　50　60　70　80　90　100
Uncertain　　　　　　　　　　　　　　　Total
　　　　　　　　　　　　　　　　　　　　Certainty

Confidence in *volunteering answers* in *math class*
0　10　20　30　40　50　60　70　80　90　100
Uncertain　　　　　　　　　　　　　　　Total
　　　　　　　　　　　　　　　　　　　　Certainty

Confidence in *going to the chalkboard* in *math class*
0　10　20　30　40　50　60　70　80　90　100
Uncertain　　　　　　　　　　　　　　　Total
　　　　　　　　　　　　　　　　　　　　Certainty

Goal Four for Parents:
Confidence in *asking parents questions*
0　10　20　30　40　50　60　70　80　90　100
Uncertain　　　　　　　　　　　　　　　Total
　　　　　　　　　　　　　　　　　　　　Certainty

Confidence in *answering questions asked by parents*
0　10　20　30　40　50　60　70　80　90　100
Uncertain　　　　　　　　　　　　　　　Total
　　　　　　　　　　　　　　　　　　　　Certainty

Confidence in *offering opinions to parents*
0　10　20　30　40　50　60　70　80　90　100
Uncertain　　　　　　　　　　　　　　　Total
　　　　　　　　　　　　　　　　　　　　Certainty

There are seven different behaviors and settings listed in objective four. We are very specific about assessing the degree of Joan's self-efficacy for each *behavior* and *setting* described in objective four. As Joan becomes more successful in achieving her goal behaviors, measures of her self-efficacy or confidence will increase.

SUMMARY

Self-management is a process in which clients direct their own behavior change with any one therapeutic strategy or a

LEARNING ACTIVITY **57** *Self-Efficacy*

In this learning activity, you are to determine and assess your self-efficacy.

1. Review Chapter 10 and select some goals you would like to achieve. Your general goal may have to be divided into subgoals.
2. Write down your goals and/or subgoals. Make sure your written goals comply with the guidelines presented in Chapter 10.
3. For the goal you would like to achieve, make a scale (0 to 100) to measure your self-efficacy (confidence) in

performing your goal behaviors, thoughts, or feelings for each subgoal, and for each person with whom and setting in which the goal is to be performed.
4. Assess your self-efficacy by circling the number on each scale that reflects your degree of uncertainty or certainty (confidence) in performing each goal.
5. You might wish to use the self-efficacy scales to self-monitor your confidence over a period of time as you gain more experience in performing the goal behaviors.

---

combination of strategies. The self-monitoring strategy provides a method by which a client can become more aware of his or her overt behavior or internal responses such as thoughts or feelings. Self-monitoring may also provide information about the social and environmental context that influences these behaviors. Stimulus-control procedures require predetermined arrangement of environmental conditions that are antecedents of a target behavior, or cues to increase or decrease that behavior. As a self-management strategy, self-reward involves presenting oneself with a positive stimulus *after* engaging in a specified behavior. Self-modeling is a method in which the client can acquire desired behaviors using himself or herself as a model. Self-efficacy, a measure of confidence, mediates all behavioral change. These four strategies typically are classified as self-management because in each procedure the client prompts, alters, or controls antecedents and consequences to produce desired changes in behavior in a self-directed fashion. Promoting client commitment to use self-management strategies can be achieved by introducing these strategies later in therapy, assessing the client's motivation for change, creating a social support system to aid the client in the use of the strategy, and maintaining contact with the client while self-management strategies are being used. All of these self-management strategies, as well as all of the interventions we describe in this book, are affected by self-efficacy or self-empowerment, a cognitive process that mediates behavioral change.

Self-management interventions have been used with diverse groups of clients primarily in two areas: behavioral medicine, and education about and prevention of HIV infection and drug use. Self-management may apply to some clients of color who value self-reliance and who desire interventions that are time limited, focused on the present,

and designed to resolve identified problems. However, the process of self-regulation appears to vary across cultures and societies and many of the notions underlying self-management are decidedly Eurocentric, specifically the notions of internal locus of control and internal locus of responsibility. These two variables, as well as the client's cultural (collective) identity and acculturation and assimilation status, are mediating variables that affect the appropriateness of self-management. We concur with Gutierrez (1990) who proposes that an empowerment process for clients from diverse groups includes not only ways to increase the client's sense of self-efficacy but also ways to develop group and cultural consciousness and the reduction of personal self-blame and denigration.

A FINAL NOTE

As we terminate this journey, we would like to leave you with three thoughts. First, set careful and realistic personal and professional limits for yourself as a therapist. Learn what you can and cannot do for and with clients. A critical part of effective help giving is to know when to back off and turn over some responsibility to the person in front of you. The more you do for your clients, the less they will do for themselves. Recently there has been a focus in the help giving professions on "compassion fatigue"—a state of physical and/or mental exhaustion that occurs in professional helpers when they don't have good boundaries for themselves and care too much for others at the expense of themselves. "Take good care of yourself" is an important axiom for helpers.

Second, examine your expectations for yourself and your clients. Therapists often find that their expectations for change differ markedly from those of clients, particularly reluctant or pessimistic clients. In such instances, therapists often are meeting more of their own needs for change and success than pursuing the needs and issues of clients. Therapy is endangered when the counselor wants more for clients than they want for themselves. In such cases, the therapist fights with the client (on the client's behalf), but the client loses an important ally. Recognize that change efforts by all of us are difficult, especially with long-standing behavior patterns. Also, remember that clients' behaviors to resist change are often positive because they are protective.

Finally, above all, be flexible. The skills and strategies offered in this book are simply methodology that is more or less effective depending on the creativity and intuition of the user. Remember that there is great variance among individual clients—not only in terms of cultural and gender factors but also in terms of covert and overt behavioral patterns, life and developmental histories, and sociopolitical milieus. We have spoken a great deal in this book about subgroups of clients— men, women, young, old, gay, lesbian, learning disabled, physically challenged, and people of all different races and ethnic groups. Remember, however, that each subgroup is made up of individuals who cannot and should not be categorized or stereotyped by their status in a group. Therapists who are flexible regard each client as unique, and each helping strategy as a tool to be used or set aside depending on its effectiveness in producing client-generated outcomes.

POSTEVALUATION

Part One
For Objective One, describe the use of self-monitoring and stimulus control for the following client case.

Case Description
The client, Maria, is a thirty-something Puerto Rican woman who was physically separated from her husband of 15 years when they came to the United States in separate trips. Although they were reunited about a year ago, she reports that during the last year she has had "*ataques de nervios*"—which she describes as trembling and faintness. She worries that her husband will die young and she will be left alone. Her history reveals no evidence of *early* loss or abandonment; however, she has experienced losses with her immigration. Also, she seems to be self-sacrificing and dependent on Juan, her husband. She reports being a very religious person and praying a lot about this.

She asks assistance in gaining some control over these "*ataques de nervios.*" How would you use self-monitoring and stimulus control to help her decrease the "*ataques de nervios*"?

What else would you focus on in addition to the use of these two strategies, given her cultural background and the case description?

Feedback follows.

Part Two
Objective Two asks you to teach someone else how to engage in self-monitoring. Your teaching should follow the six guidelines listed in Table 20-3: rationale, response discrimination, self-recording, data charting, data display, and data analysis on p. 542. Feedback follows.

Part Three
Objective Three asks you to describe the application of a culturally relevant self-management program for a given client case (self-efficacy, self-modeling, self-monitoring, self-reward, and stimulus control).

Case Description
The client, Thad, is a young African American man who has recently identified himself as gay. Thad has been working with you in coming to terms with his sexual orientation and now is at a point where he has visited some gay bars and has participated in some gay activities. However, he has not asked anyone out and wants to find a way to do so. You have discussed the use of self-monitoring and self-reward as possible interventions for this goal. Thad is interested in this and would like to go out at least once a week with a male partner.

How would you use and adapt the interventions of self-monitoring, self-reward, self-modeling, stimulus control, and self-efficacy with this particular client?

Feedback follows.

Part Four
Objective Four asks you to teach someone else how to use self-reward, self-monitoring, self-control, and self-as-a model. You can use the steps for self-monitoring on page 542, the stimulus-control principles on page 552, the components for self-reward on page 556, and the checklist for self-as-a-model on page 562.

SUGGESTED READINGS

Self-Management

Cole, C. L. (1992). Self-management interventions in the schools. *School Psychology Review, 21,* 188–192.

Lazarus, B. D. (1993). Self-management and achievement of students with behavior disorders. *Psychology in the Schools, 30,* 67–74.

Marks, I. M. (1994). Behavior therapy as an aid to self-care. *Current Directions in Psychological Science, 3,* 19–22.

Parcel, G. S., Swank, P. R., Mariotto, M. J., & Bartholomew, L. K. (1994). Self-management of cystic fibrosis: A structural model for educational and behavioral variables. *Social Science and Medicine, 38,* 1307–1315.

Rokke, P. D., & Kozak, C. D. (1989). Self-control deficits in depression. A process investigation. *Cognitive Therapy and Research, 13,* 609–621.

Smith, L. L., Smith, J. N., & Beckner, B. M. (1994). An anger-management workshop for women inmates. *Families in Society, 75,* 172–175.

Taal, E., Riemsma, R. P., Brus, H. L., & Seydel, E. R. (1993). Group education for patients with rheumatoid arthritis. Special Issue: Psychosocial aspects of rheumatic diseases. *Patient Education and Counseling, 20,* 177–187.

Watson, D. L., & Tharp, R. G. (1997). *Self-directed behavior* (7th ed.). Pacific Grove, CA: Brooks/Cole.

Williams, R. L., Moore, C. A., Pettibone, T. J., & Thomas, S. P. (1992). Construction and validation of a brief self-report scale of self-management practices. *Journal of Research in Personality, 26,* 216–234.

Self-Monitoring

Fox, M., & Deyer, D. (1995). Stressful job demands and worker health: An investigation of the effects of self-monitoring. *Journal of Applied Social Psychology, 25,* 1973–1995.

Goodwin, R., & Soon, A. P. Y. (1994). Self-monitoring and relationship adjustment: A cross-cultural analysis. *Journal of Social Psychology, 134,* 35–39.

Halford, W. K., Gravestock, F. M., Lowe, R., & Scheldt, S. (1992). Toward a behavioral ecology of stressful marital interactions. *Behavioral Assessment, 14,* 199–217.

Harris, K. R., Graham, S., Reid, R., & McElroy, K. (1994). Self-monitoring of attention versus self-monitoring of performance: Replication and cross-task comparison studies. *Learning Disability Quarterly, 17,* 121–139.

Madsen, J., Sallis, J. F., Rupp, J. W., & Senn, K. L. (1993). Relationship between self-monitoring of diet and exercise change and subsequent risk factor changes in children and adults. *Patient Education and Counseling, 21,* 61–69.

McDougall, D., & Brady, M. (1995). Using audio-cued self-monitoring for students with severe behavior disorders. *Journal of Educational Research, 88,* 309–317.

Reynolds, M., & Salkovskis, P. M. (1992). Comparison of positive and negative intrusive thoughts and experimental investigation of the differential effects of mood. *Behaviour Research and Therapy, 30,* 273–281.

Stimulus Control

Estabrooks, P., Courneya, K., & Nigg, C. (1996). Effects of a stimulus control intervention on attendance at a university fitness center. *Behavior Modification, 20,* 202–215.

Haddock, C. K., Shadish, W. R., Klesges, R. C., & Stein, R. J. (1994). Treatments for childhood and adolescent obesity. *Annals of Behavioral Medicine, 16,* 235–244.

Jacobs, G. D., Benson, H., & Friedman, R. (1993). Home-based central nervous system assessment of a multifactor behavioral intervention for chronic sleep-onset insomnia. *Behavior Therapy, 24,* 159–174.

Jacobs, G. D., Rosenberg, P. A., Friedman, R., & Matheson, J. (1993). Multifactor behavioral treatment of chronic sleep-onset insomnia using stimulus control and the relaxation response: A preliminary study. *Behavior Modification, 17,* 498–509.

Stromer, R., MacKay, H., Howell, S., & McVay, A. (1996). Teaching computer-based spelling to individuals with developmental and hearing disabilities. *Journal of Applied Behavior Analysis, 29,* 25–42.

Taylor, S., Cipani, E., & Clardy, A. (1994). A stimulus control technique for improving the efficacy of an established toilet training program. *Journal of Behavior Therapy and Experimental Psychiatry, 25,* 155–160.

Williams, J. G., Park, L. I., & Kline, J. (1992). Reducing distress associated with pelvic examinations: A stimulus control intervention. *Women and Health, 18,* 41–53.

Self-Reward

Eisenberger, R., & Cameron, J. (1996). Detrimental effects of reward: Reality or myth? *American Psychologist, 51,* 1153–1166.

Enzle, M. E., Roggeveen, J. P., & Look, S. C. (1991). Self- versus other- reward administration and intrinsic motivation. *Journal of Experimental Social Psychology, 27,* 468–479.

Field, L. K. & Steinhardt, M. A. (1992). The relationship of internally directed behavior to self-reinforcement, self-esteem, and expectancy values for exercise. *American Journal of Health Promotion, 7,* 21–27.

Kanfer, F. H., & Gaelick-Buys, L. (1991). Self-management methods. In F. H. Kanfer & A. P. Goldstein (Ed.). *Helping people change* (4th ed., pp. 305–360). New York: Pergamon Press.

Schill, T., & Kramer, J. (1991). Self-defeating personality, self-reinforcement, and depression. *Psychological Reports, 69,* 137–138.

Self-as-a-Model

Dowrick, P., & Raeburn, J. (1995). Self-modeling: Rapid skill training for children with physical disabilities. *Journal of Developmental and Physical Disabilities, 7,* 25–37.

Houlihan, P., Miltenberger, R., Trench, B., & Larson, M. (1995). A videotaped peer self-modeling program to increase community involvement. *Child and Family Behavior Therapy, 17,* 1–11.

Meharg, S. S., & Lipsker, L. E. (1991). Parent training using videotape self-modeling. *Child and Family Behavior Therapy, 13,* 1–27.

Winfrey, M. L., & Weeks, D. L. (1993). Effects of self-modeling on self-efficacy and balance beam performance. *Perceptual and Motor Skills, 77,* 907–913.

Woltersdorf, M. A. (1992). Videotape self-modeling in the treatment of attention-deficit hyperactivity disorder. *Child and Family Behavior Therapy, 14,* 53–73.

Self-Efficacy

Bandura, A. (1977). Self-efficacy: Toward a unifying theory of behavioral change. *Psychological Review, 84,* 191–215.

Bandura, A. (1982). Self-efficacy mechanism in human agency. *American Psychologist, 37,* 122–147.

Bandura, A. (1989). Human agency in social cognitive theory. *American Psychologist, 44,* 1175–1185.

Bandura, A., Taylor, C. B., Williams, S. L., Mefford, I. N., & Barchas, J. D. (1985). Catecholamine secretion as function of perceived coping self-efficacy. *Journal of Consulting and Clinical Psychology, 53,* 406–415.

Friedman, L. C., Nelson, D. V., Webb, J. A., & Hoffman, L. P. (1994). Dispositional optimism, self-efficacy, and health beliefs as predictors of breast self-examination. *American Journal of Preventive Medicine, 10,* 130–135.

Luzzo, D. A. (1995). The relative contributions of self-efficacy and locus of control to the prediction of career maturity. *Journal of College Student Development, 36,* 61–66.

Mancoske, R. J., Standifer, D., & Cauley, C. (1994). The effectiveness of brief counseling services for battered women. *Research on Social Work Practice, 4,* 53–63.

Ozer, R. L., & Bandura, A. (1990). Mechanisms governing empowerment effects: A self-efficacy analysis. *Journal of Personality and Social Psychology, 58,* 472–486.

Rodgers, W., & Brawley, L. (1996). The influence of outcome expectancy and self-efficacy on the behavioral intentions of novice exercisers. *Journal of Applied Social Psychology, 26,* 618–634.

Ryan, N., Solberg, V., & Brown, S. (1996). Family dysfunction, parental attachment, and career search self-efficacy among community college students. *Journal of Counseling Psychology, 43,* 84–89.

Self-Management and Diverse Groups of Clients

Belgrave, F., Randolph, S., Carter, C., & Braithwaite, N. (1993). The impact of knowledge, norms, and self-efficacy on intentions to engage in AIDS-preventive behaviors among young incarcerated African American males. *Journal of Black Psychology, 19,* 155–168.

Carr, S., & Punzo, R. (1993). The effects of self-monitoring of academic accuracy and productivity on the performance of students with behavioral disorders. *Behavioral Disorders, 18,* 241–250.

Church, A., Teresa, J., Rosebrook, R., & Szendre, D. (1992). Self-efficacy for careers and occupational consideration in minority high school equivalency students. *Journal of Counseling Psychology, 39,* 498–508.

Cochran, S., & Mays, V. (1993). Applying social psychological models to predicting HIV-related sexual risk behaviors among African Americans. *Journal of Black Psychology, 19,* 142–154.

Epstein, J., Botvin, G., Diaz, T., & Toth, V. (1995). Social and personal factors in marijuana use and intentions to use drugs among inner city minority youth. *Journal of Developmental and Behavioral Pediatrics, 16,* 14–20.

Gutierrez, L. (1990). Working with women of color: An empowerment perspective. *Social Work, 35,* 149–153.

Jacob, T., Penn, N., Kulik, J., & Spieth, L. (1992). Effects of cognitive style and maintenance strategies of breast self-examination (BSE) practice by African American women. *Journal of Behavioral Medicine, 15,* 589–609.

Lackner, J., Carosella, A., & Feuerstein, M. (1996). Pain expectancies, pain, and functional self-efficacy expectancies as determinants of disability in patients with chronic low back disorders. *Journal of Consulting and Clinical Psychology, 64,* 212–220.

McAuley, E., Shaffer, S., & Rudolph, D. (1995). Affective responses to acute exercise in elderly impaired males: The moderating effects of self-efficacy and age. *International Journal of Aging and Human Development, 41,* 13–27.

McCafferty, S. (1992). The use of private speech by adult second language learners: A cross-cultural study. *Modern Language Journal, 76,* 179–189.

Malow, R., Corrigan, S., Cunningham, S., & West, J. (1993). Psychosocial factors associated with condom use among African American drug abusers in treatment. *AIDS Education and Prevention, 5,* 244–253.

Martin, D. (1993). Coping with AIDS-risk reduction efforts among gay men. *AIDS Education and Prevention, 5,* 104–120.

Roberson, M. (1992). The meaning of compliance: Patient perspectives. *Qualitative Health Research, 2,* 7–26.

Rotheram-Borus, M. J., Rosario, M., Reid, H., & Koopman, C. (1995). Predicting patterns of sexual acts among homosexual and bisexual youths. *American Journal of Psychiatry, 152,* 588–595.

St. Lawrence, J. (1993). African American adolescents' knowledge, health-related attitudes, sexual behavior, and contraceptive decisions: Implications for the prevention of adolescent HIV infection. *Journal of Consulting and Clinical Psychology, 61,* 104–112.

Solberg, V., O'Brien, K., Villareal, P., & Kennel, R. (1993). Self-efficacy and Hispanic college students: Validation of the College Self-Efficacy Instrument. *Hispanic Journal of Behavioral Sciences, 15,* 80–95.

Van-Hasselt, V., Hersen, M., Null, J., & Ammerman, R. (1993). Drug abuse prevention for high risk African American children and their families: A review and model program. *Addictive Behaviors, 18,* 213–234.

Winkleby, M., Flora, J., & Kraemer, H. (1994). A community-based heart disease intervention: Predictors of change. *American Journal of Public Health, 84,* 767–772.

FEEDBACK
POSTEVALUATION

Part One

Self-Monitoring

1. In the rationale, you would emphasize how this strategy could help obtain information about the client's *ataques de nervios* as well as modify them in the desired direction. You would also explain that she would be recording defined *ataques de nervios* in vivo on a daily basis for several weeks. You need to be careful to frame the rationale in a way that respects her cultural values.

2. Response-discrimination training would involve selecting, defining, and giving examples of the response to be monitored. The counselor should model some examples of the defined behavior and elicit some others from the client. Specifically, you would help the client define the nature and content of the behaviors she would be recording, such as feeling faint.

3. *Timing of the recording:* Because this client is using self-monitoring to decrease an undesired behavior, she would engage in prebehavior monitoring: each time she felt faint or worried, she would record.

   Method of recording: The client would be instructed to use a frequency count and record the number of times she felt faint or worried. If she was unable to discern when these started and ended, she could record with time sampling. For example, she could divide a day into equal time intervals and use the "all or none" method. If such thoughts occurred during an interval, she would record *yes;* if they did not, she would record *no.* Or, during each interval, she could rate the approximate frequency of these behaviors on a numerical scale, such as 0 for "never occurring," 1 for "occasionally," 2 for "often," and 3 for "very frequently."

   Device for recording: There is no one right device to assist this client in recording. She could count the frequency using a tally on a note card or a golf wrist counter. Or she could use a daily log sheet to keep track of interval occurrences.

4. A simple chart might have days along the horizontal axis and frequency of behaviors along the vertical axis.

5. This client may not wish to display the data in a public place at home. She could carry the data in her purse or knapsack.

(continued)

FEEDBACK: POSTEVALUATION   (continued)

6. The client could engage in data analysis by reviewing the data with the counselor or by comparing the data with the baseline or with her goal (desired level of behavior change). The latter involves self-evaluation and may set the stage for self-reinforcement.

Stimulus Control

You can explain the use of stimulus control as another way to help her gain some feeling of personal control surrounding these "*ataques de nervios*" by confining them to particular places and times so that they don't occur so randomly and unpredictably. You could suggest the use of a worry spot or worry chair that she goes to at a designated time to do her worrying and that she is to stop worrying when she leaves this place or chair.

Also, in addition to these two self-management interventions, it would be useful to explore her feelings of loss and safety surrounding her immigration experience, the adaptations she is having to make to a different culture, and the conflicts she may be experiencing between the two cultures.

Part Two

Use Table 20-3 as a guide to assist your teaching. You might also determine whether the person you taught implemented self-monitoring accurately.

Part Three

First, you would need to determine how well the use of self-management "fits" with Thad's beliefs, values, worldview, and lifestyle. Assuming that Thad is receptive to the use of self-management and that he is oriented more toward an internal than an external locus of control and responsibility, you can proceed. (However, it is also important to explore whether there may be any external social factors that are contributing to his sense of discomfort.) You may first wish to assess and work with self-efficacy, or Thad's confidence in himself and the contacts he will make with other men. Note that in this case his sense of self-efficacy is related to both his identity development as a gay male and as an African American. We anticipate that as Thad uses various self-management tools, his sense of self-efficacy will increase.

Self-modeling could be used to videotape or audiotape role plays in which Thad initiates social contacts with other men.

Self-reward can be used in conjunction with times he actually makes social contacts with men and goes out with a man.

1. The *verbal symbolic rewards* used by Thad could consist of self-praise or covert verbalizations about the

positive consequences of his behavior. Here are some examples:

"I did it! I asked him out."
"I did just what I wanted to do."
"Wow! What a good time I'll have with _____."

Material rewards would be things or events Thad indicates he prefers or enjoys. These might include watching TV, listening to music, or playing sports. Both current and potential rewards should be used. Of course, these activities are only possibilities; he has to decide whether they are reinforcing.

Examples of an *imaginal* reward may include either pleasant scenes or scenes related to going out:

Imagining oneself on a raft on a lake
Imagining oneself on a football field
Imagining oneself with one's partner at a movie
Imagining oneself with one's partner lying on a warm beach

Self-monitoring can be used to help Thad track the number of social contacts he has with other men.

Stimulus control can be used to help Thad increase the number of cues associated with increasing his social contacts with other men. For example, he might start in one place or with one activity where he feels most comfortable; gradually he can increase his visits to other places and activities where he will find other gay men.

Part Four

Use the following:

Self-monitoring steps on page 542
Principles of stimulus control on page 552
Components of self-reward on page 556
Checklist for self-as-a-model on page 562

(continued)

OPERATIONALIZATION OF THE
MULTICULTURAL COUNSELING COMPETENCIES

Patricia Arredondo, Rebecca Toporek,
Sherlon Pack Brown, Janet Jones,
Don C. Locke, Joe Sanchez,
Holly Stadler

For the past 20 years, the Association for Multicultural Counseling and Development (AMCD) has provided leadership for the American counseling profession in major sociocultural and sociopolitical domains. Through our vision of the centrality of culture and multiculturalism to the counseling profession, we have created new directions and paradigms for change. One of our major contributions has been the development of the Multicultural Counseling Competencies (Sue, Arredondo, & McDavis, 1992).

For the first time in the history of the profession, competencies to guide interpersonal counseling interactions with attention to culture, ethnicity, and race have been articulated. Through the leadership of Thomas Parham, president of the Association for Multicultural Counseling and Development (AMCD) 1991–1992, the Professional Standards and Certification Committee was charged to develop multicultural counseling competencies. On the direction of President Marlene Rutherford-Rhodes (1994–1995), the Committee was asked to provide additional clarification to the revised com-

Patricia Arredondo is president-elect of the Association of Multicultural Counseling and Development (AMCD), past chair of AMCD's Professional Standards and Certification Committee, and president of Empowerment Workshops, Boston. **Rebecca Toporek, Sherlon Pack Brown, Janet Jones, Don C. Locke, Joe Sanchez,** and **Holly Stadler** are also members of AMCD's Professional Standards and Certification Committee. Correspondence regarding this article should be sent to Patricia Arredondo, 251 Newbury Street, Boston, MA 02116.

SOURCE: From "Operationalization of the Multicultural Counseling Competencies," by P. Arredondo, R. Toporek, S. P. Brown, J. Jones, D. C. Locke, J. Sanchez, and H. Stadler, in *Journal of Multicultural Counseling and Development, 24,* pp. 42–78. Copyright © 1996 by ACA. No further reproduction authorized without written permission of the American Counseling Association.

petencies and to specify enabling criteria as well. This objective has been addressed through this document.

Multicultural counseling refers to preparation and practices that integrate multicultural and culture-specific awareness, knowledge, and skills into counseling interactions. The term *multicultural,* in the context of counseling preparation and application, refers to five major cultural groups in the United States and its territories: African/Black, Asian, Caucasian/European, Hispanic/Latino and Native American or indigenous groups who have historically resided in the continental United States and its territories. It can be stated that the U.S. is a pluralistic or multicultural society and that all individuals are ethnic, racial, and cultural beings.

All persons can point to one or more of these macrolevel, cultural groups as sources of their cultural heritage. For the aforementioned five major cultural groups, race and ethnicity are further identifiers, although oftentimes the terms are interchanged with culture, introducing confusion. What is noteworthy about cultural groupings is that they point to historical and geographic origins as well as to racial heritage.

African Americans and Haitians might similarly claim African heritage with etiology tracing back to the African continent. Some individuals might prefer to self-identify in racial terms—Black—or based on their country of origin. Thus the terms *ethnicity* or *nationality* come into play, and self-descriptors could include Haitian, Nigerian, Afro-American, and so forth.

Individuals of Asian cultural background—Chinese, Japanese, Korean, Vietnamese—can point to roots on the continent of Asia, but all speak different languages and dialects. East Asians are another group coming primarily from India, Pakistan, and Iran and other countries not part of the Orient geographically. The term *Oriental* is considered pejorative and no longer used in multicultural counseling literature.

Persons racially listed as White or Caucasian, are usually of European heritage. In the United States, men of European background have held, and continue to hold, economic, political, and educational power. This is an important factor in the development of multicultural counseling competencies and the domain of multicultural counseling. As the normative cultural group in the United States, Euro-Americans have been the yardstick by which individuals of other cultural groups and women have been measured.

Hispanics/Latinos are similar to and a bit different from the other cultural groups. Generally speaking, they can point to both the North and the South American continents for their roots. Central America, although not a continent, is the homeland of many who are classified as Hispanics. Racially, Hispanics are biracial by birth, representing as though this were the only way to identify the person. How might one feel to be told, "Oh, you're a man. What do you understand about sexual harassment?" or "We are so glad you are now working here. We need the viewpoint of a Latina," or "Why should you care about affirmative action, you're not Black." Few of us escape the tendency to buy into the labels of identity, although they may be limiting.

The Dimensions of Personal Identity model can be used as a paradigm to see people more completely, as well as an educational tool. It provides a reference point for recognizing the complexity of all persons. The model highlights our different identity-based affiliations, memberships, and subcultures and, therefore, complements the discussion of multiculturalism.

A Dimension. The A Dimension is a listing of characteristics that serve as a profile of all people. The majority of the dimensions we are born with or born into, making most "fixed" and less changeable. For example, our age, gender, culture, ethnicity, race, and language are predetermined. We have no control over these when we are born, and there is very little we can do to change most of these dimensions. Some research suggests that sexual orientation is biologically based, whereas other data promote a sociocultural explanation. In the model, sexual orientation appears as an A Dimension characteristic. For some individuals, it has been possible to transcend economic roots. Social class status, however, may persist for generations based on one's culture or society. For example, in India this may occur through the caste system, whereby individuals are born into a caste, complete with its privileges and limitations. In the U.S., social class may play out differently based on historical and familial lineage. One artifact of social class status is the social register, which accords a listing for some at the time of birth. For better or worse, attributions and judgments are made about all of us based on our social status. At times, this is less visible or known. Nevertheless, appearances are often used to make assessments of individuals' "value." How

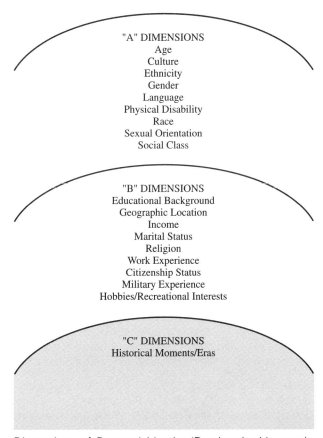

Dimensions of Personal Identity (Reprinted with permission of Empowerment Workshops, Inc.)

someone dresses, their "attractiveness" in terms of height, weight, and other physical criteria also interact with A Dimensions. Would a counselor respond similarly to an overweight, White woman as he or she would to a professional Black woman? How might his or her previous experiences, or lack thereof, with these types of women affect his or her assumptions, comfort, and behavior in a counseling encounter?

Note that a number of the A Dimension characteristics also hold "protected class" status based on government classification/Equal Employment Opportunity (EEO) and Title VII of the Civil Rights Act of 1964. The other noteworthy feature of the A Dimension list is that these are characteristics that most readily engender stereotypes, assumptions, and judgments, both positively and negatively. For example, an African-American woman may be assumed (simply because of her race and gender) to be strong and direct in sharing her thoughts and feelings. Although some people in the general society may view strength and direct behavior as desirable personal characteristics, others may view these characteristics as intimidating and overpowering. An African-American woman shared an experience that led to misassumptions based on a lack of cultural awareness:

> I came into work one day feeling pretty tired from staying up all night with my sick 3-year-old. I guess you can say I was not very communicative. No one asked how I was or if something was wrong. I didn't think anything of it at the time but later I learned why I was left alone. My coworkers thought my behavior was a statement of my racial identity. They assumed that this was Black behavior. I was floored. Of course, when I asked them what made my behavior racial, they could not explain.

Misattributions can readily occur, even among people who work together, because of a lack of cultural awareness and knowledge and interracial discomfort. In counseling environments, professional counselors, different by ethnicity and race and often the only person of color, are more often scrutinized because of one or more A Dimensions. Individuals report that they feel the pressure to perform at higher levels than their White counterparts because of their visible difference.

Another example from the A Dimension is that a person speaking English with an accent might be assumed to be less intelligent, more difficult to deal with, or viewed in other negative ways. Oftentimes, immigrant adults experience the impatience and even the ridicule of monolingual English speakers when they seek services at a human service agency. Individuals from Jamaica speak English, but it might be heard as new and not discernible to a front desk attendant. Chinese language speakers have a different tone when speaking, which often causes the impatience of medical personnel. One can only wonder how the verbal and nonverbal behavior of the professional staff, from whom clients expect to receive help, will influence an individual's discontinuation of treatment. Literature indicates the extraordinary incidence of dropout among clients of color after an initial visit.

Continuing with the theme of accents, there also are accents typically perceived more positively by Americans. These are British and Australian accents and in turn these individuals may be perceived as more highly desirable and valuable. Lack of awareness, knowledge, experience of, and respect for cultures that are not of European heritage easily introduces interpersonal barriers to counseling transactions.

Advocates of the Americans With Disabilities Act of 1990 remind the public that everyone is "one accident away from being disabled." Because the effects of an injury are usually irreversible, this is also considered an A Dimension characteristic. Consider the perception of a physically chal-lenged woman of Puerto Rican heritage who uses a wheelchair. Is she written off as a welfare case or assumed to have assets such as bilingual and bicultural experiences?

If placed on a continuum, all of the A Dimension characteristics can bring positive and negative reactions. Because they tend to be more visible, they invite feedback, both wanted and unwanted, from others, thus contributing to self-concept and self-esteem. It is the A Dimension characteristics that invite and challenge counselors to operate from a framework of multiculturalism and cultural competence. Because each person holds a cultural identity from one or more of the five groups cited, and because individuals embody all of the A Dimension characteristics, to be culturally effective counselors need to see individuals holistically, not in terms of color, ethnicity, culture, or accent alone. People are complete packages, as is described in the B and the C Dimensions.

C Dimension. The C Dimension is discussed next because it also encompasses universal phenomena. This dimension indicates first of all that all individuals must be seen in a context; we do not exist in a vacuum. The C Dimension grounds us in historical, political, sociocultural, and economic contexts indicating that events of a sociopolitical, global, and environmental form have a way of affecting one's personal culture and life experiences. The time one is born is an historical moment that will never happen again. In presentations, participants are encouraged to think about the following: How was your family life at the time of your birth? What was taking place in the local community or in your home country? What was going on in the world? Reflecting on the questions and the data that emerge provides individuals with a landscape for their personal history.

Individuals who were born during the pre- or post-World War II period, have different recollections of their families. Many talk about their fathers being away during their childhood and the relief felt when they returned. African Americans also share recollections of their parents or grandparents serving in all-Black units in the military. Many Americans are unaware that Native Americans, because of their linguistic abilities, were involved in decoding communication among enemy camps. Because U.S. history books are written from a Euro-American, monocultural perspective, the experiences of individuals and families of color are often unreported or minimized. One example is the mode of entry historically experienced in territories recognized today as the United States. Blacks were brought as slaves to the North and South American continents. Native Americans and Mexicans populated the continental territory, but were subsumed under American jurisdiction as a result of treaties. In other words, they became conquered people. The island of Puerto Rico was colonized by the Spaniards and today it is an American commonwealth. The Japanese and Chinese were brought as laborers to the Northwest to build the new frontier. For millions of people, the U.S. did not represent the land of freedom.

This lack of knowledge can place a counselor at a disadvantage because all he or she can reference is his or her personal experience. "Revisionist" versions of American history, as they are called, are providing missing information that again highlight some of the experiences of Americans marginalized based on cultural, ethnic, and racial differences.

The time at which one was born also indicates the significant political and environmental incidents that may also affect personal identity. Women and persons of color often point to the importance of the Civil Rights Act of 1964 and its consequences for employment, education, and housing. Before this legislation, men and women of color had limited access to the same institutions, job opportunities, and housing compared with that of their White counterparts. For counselors, knowledge about historical and political realities faced by persons of color should be known when providing career counseling or some other interventions. The work experiences of immigrants must also be inquired about more carefully. Often, accents of limited-English ability may lead to a reaction of surprise when a counselor learns that Mr. Wu had actually been an accountant in Taiwan or that Senora Garcia had been a school teacher.

Some women, particularly Caucasian, also remark about the women's era of the 1970s. The historical, cultural, and sociopolitical contexts into which women are born, however, influence how women of diverse ethnocultural heritages behave, think, and feel. For example, African-American women have historically worked outside the home and their history, culture, and sociopolitical realities have influenced their perceptions about working. Often this perception is one of "I will/must work." Because the role of gender is not highlighted in the multicultural texts and literature, it is underscored here. Women, particularly women of color, have been portrayed in roles of servitude and secondary status to men. This historical, political, and sociocultural reality has marginalized women and this continues to be evident in many contexts, including the counseling profession. For example, despite increases in the number of women who earn doctorates, concerns have been raised about the feminization of the profession and the consequences on earning power. Another example can be pointed to in the field of multicultural counseling. There are very few women who have authored texts or books with a multicultural focus.

In more recent presentations, the Gulf War has been cited as an historical and political event that affected and continues to affect individuals and their families. At the time, it led to many anti-Arab public sentiments, lumping all Middle Eastern peoples into a category labeled *terrorists.* This is also a reminder of the internment of American citizens of Japanese heritage during World War II. Although many of these individuals were contributors to the American society, they were deemed suspect when the war commenced.

Because of technology and the emphasis on acquisitions, many persons who were born or grew up in the late 1970s and 1980s probably had greater material options than those who were products of the Depression era. There is a different type of story for immigrants and refugees as well. The relationship between their country and the United States, their socioeconomic circumstances, and their racial heritage will all have a bearing on their status, adjustment, and acceptance in this country. For example, a Chicano psychologist shared how he was stopped and questioned as he crossed the Mexican border into San Diego, California, his hometown. Another incident was shared by a Latina professor who was returning from an educational trip from Colombia and entering the country through Puerto Rico. Officials of the Immigration and Naturalization Service insisted on questioning her because she looked "Spanish." Their concern was that she was entering illegally.

The C Dimension suggests that there are many factors that surround us over which we have no control as individuals, but which will, nevertheless, affect us both positively and negatively. These contextual factors, although they may not seem to have a direct impact, do affect the way people are treated and perceived. From historical, political, and sociocultural perspectives, persons of color have experienced more incidents of oppression, disenfranchisement and legislated discrimination for racial reasons alone. This may also help to explain the phenomenon of learned helplessness and why individuals of color may suffer from experiences that make them the victim of an unfair decision or practice. Counseling from a multicultural perspective indicates developing knowledge of how American history has been experienced differently by persons of color, by those who were of lower socioeconomic status, by people with less education and access to power, and by women. Oppression is a dynamic factor that emerges in these discussions, because historically, power has been used based on an A Dimension to oppress others with different A Dimensions. Where there are victims of homophobia, racism, and sexism, there are also beneficiaries. The C Dimension also invites exploration of institutional oppression and how it continues to occur in contemporary society and counseling sites. One example may be calling the Department of Social Services and announcing yourself as a counselor to the receptionist in a situation in which the client has previously had difficulty reaching his or her caseworker, has been put on hold indefinitely, or has been told that the caseworker will return the call. By announcing the identity as a counselor to the receptionist, the call may be expedited so the client is able to take care of his or her business.

Another hypothetical example may be that of a student of color in a large educational institution who approaches a counselor with a complaint about a faculty member, citing racial discrimination and sexual harassment. The student complains that he or she went through the grievance process but did not feel that it was satisfactorily resolved and was told that the incidents were minor and not adequate to justify taking action. The culturally competent counselor, having been approached in the past by other students with similar complaints against the same faculty member, may choose to intervene within the institution's policies and procedures for such situations.

B Dimension. The B Dimension is discussed last because theoretically it may represent the "consequences" of the A and C dimensions. What occurs to individuals relative to their B Dimension is influenced by some of the immutable characteristics of the A Dimension and the major historical, political, sociocultural, and economic legacies of the C Dimension.

Educational experience is one example. Many more women and people of color have pursued higher education in the past 25 years as opportunities and access have become more possible because of Title VII of the Civil Rights Act. As a result of this legislation, colleges and universities can no longer discriminate based on gender, race, religion, and so forth. With increased levels of education, the work experience and parental status for women looks more varied than it did 25 years ago, although in terms of earning power, it does not equal that of White men. Education and socioeconomic conditions can enhance or limit a person seen only through the lens of an A Dimension. What happens to individuals is not totally within their control despite the American myths about self-control and self-reliance. Enabling conditions also play a role in what one can access or even think about. Laurence Graham, a Black author and lawyer, spoke to the Boston Human Resources Association in November 1994, about the reactions of his White peers when they learned he had been accepted to Yale. Rather than congratulating him, they stated that he had achieved entry as a result of affirma-

tive action, thus diminishing his achievement. Graham also reported that many of his peers, not his par academically, were still admitted to prestigious universities. Why? As Graham noted, they had their own form of affirmative action, entry based on family legacy, connections, and so forth. The difference was that no one questioned this practice; things just worked that way.

One wonders how a high school counselor might see these two different sets of experiences. How aware and knowledgeable are counselors about the issues of access as they relate to the B Dimensions? For most people of color in this country, access has been restricted legislatively and has been based on interpersonal discomfort and racism. In the CEO ranks, women and persons of color are sparse. Even going to the right schools may not always help. Why? As the literature indicates, organizations tend to hire in their own image, and historically this has been a certain type of White man. How might this information assist a therapist who sees a woman or a person of color who has again been passed over for a promotion? Chances are that encouraging the person to "try harder" will not provide the appropriate empathy. To be culturally competent, counselors need to understand the political power dynamics of the workplace and how they perpetuate the dominance of certain groups over others. With this knowledge, counselors can respond more effectively to the reality of the client's experience.

A contemporary example also can be seen when considering individuals who are gay or lesbian. In most work settings, there remains a lack of openness and comfort to be yourself ostensibly denying one the freedom to be totally real. Focusing on only one dimension of a person's totality limits understanding. For example, an Irish-American gay man shares with his family that he is gay. The family expresses their discomfort with their son's lifestyle. The man goes on to reveal to his family that his significant other is an African American. The family asserts that they are uncomfortable with the gay lifestyle but threatens to disown him if he does not give up his African-American partner. When counseling from a multicultural perspective, a counselor would know that the racial and gay identity of the men cannot be isolated as two independent dimensions. Both have to be recognized and respected. Counselors would also have to be aware of their own feelings and judgments about interracial and homosexual relationships. To say that one can offer unconditional positive regard does not necessarily mean that one is comfortable or culturally competent to assist the client.

The B Dimension also represents possible shared experiences that might not be observable if one were to focus exclusively on the A Dimension. You cannot tell whether a person is from Ohio, whether a single mother, or an avid reader of poetry by looking at that individual out of context. If you see an Asian woman with a child, you might assume she is the mother, although you may not be able to discern her relationship status. Is she straight or gay, an unmarried mother, divorced or married? There are many possibilities for self-definition that go beyond our A Dimensions and in counseling these possibilities need to be recognized.

The B Dimension can be a point of connection. In presentations, participants are usually asked about how the B Dimensions relate to them. People from the same organization are invariably surprised when they learn that others attended the same university, were also in the military, or have children under the age of 5. There are more ways that categories (B Dimensions) can actually foster rapport-building between client and counselor than seems apparent. This may depend, however, on the counselor's position of self-disclosure. Multicultural counseling leaders have reported that rapport building is critical in counseling with persons of color, and at times, a counselor's sharing that he or she has been to the client's home country or knows someone from there may facilitate the connection. Counseling with college-age Latina students experiencing homesickness and guilt for being away from family may be more soothing if the counselor can emphasize with the students about culturally based expectations for Latinas. This demonstration of "recognizing" the client's dilemma can contribute to relationship development with the client.

A, B, and C Dimension summary. The purpose of this model is to demonstrate the complexity and holism of individuals. It suggests that despite the categories we may all fit into or that are assigned to us, the combination of these affiliations is what makes everyone unique. Personal culture is composed of these different dimensions of identity. By definition and in reality everyone is a "multicultural person." The sum is not greater than the parts.

In reference to the Multicultural Counseling Competencies, this model introduces the many ways in which individuals, clients and counselors alike, may self-define. It further suggests that when counseling with persons perceived as different from self in terms of culture, ethnicity, or race, working from stereotyped assumptions and focusing only on one A or B Dimension without consideration of the C Dimension, will be likely to lead to miscommunication and the undermining of a potential counseling relationship. Concomitantly, even if one is counseling with a person seemingly more like oneself, these perceived dimensions of sameness, that is, gender or race, may not be the most relevant point of connection. There are many dimensions that influence us and assumptions must be questioned regarding which ones are most salient for an individual. It is ultimately up to the client to self-define this. When counseling from a multicultural perspective, culturally competent counselors would know that culture is not to be "blamed" for a person's problems, nor does the presenting problem have to be based on culture or race for a person of color. For example, if counseling an African-American couple, a culturally competent counselor would primarily assess the relationship issues and not necessarily ascribe their concerns to race. Over time, the topic of race might be introduced, but it should neither be imposed nor ignored.

An assumption is often made, although not verbalized, that multicultural counseling is for poor persons of color who use public services. Not only is this erroneous but it does group all persons of color into one economic class, which is not the reality at all. Admittedly, there is an overrepresentation of Blacks, Latinos, and Native Americans, particularly women and children, living in poverty, but not everyone fits the profile. When counseling at college counseling centers, mental health agencies, or in private practice, counselors need to check their assumptions about persons of color. They come from varied backgrounds just like White people.

By stepping back and using the Personal Dimensions of Identity (PDI) model as objective criteria, counselors can more readily "see" the range of human potentiality every person possesses. I draw from my experiences, values, and individuality as a woman from Mexican-American heritage, who grew up in Ohio, who has lived in Boston for nearly 25 years, who has a doctoral degree in psychology, who is a former university professor, married without children, and the owner of a management consulting business. To categorize and see me through one or two A dimensions may only limit the contributions I can make as a counseling professional in an organization.

From an institutional perspective, the PDI model can assist leaders to become more aware of how the culture of their organization may alienate, marginalize, or lose people of color, women, and other minority groups, if cultural competency is not valued and practiced. The multicultural competencies are designed to promote culturally effective relationships, particularly in interpersonal counseling. But by applying this paradigm in institutional settings, it may become possible to understand the relationship between an organization and its clients or customers. Is the environment friendly to culturally and economically different people?

Increased multicultural competence can enable counselors to provide culturally appropriate counseling as well as to use the PDI model more effectively. Through increasing awareness, counselors are better able to understand how their own personal dimensions affect their ability to perceive and understand the personal dimension of their clients. Similarly, greater knowledge enhances the counselor's ability to more accurately understand the various cultures or elements that make up their clients' personal dimensions. Developing greater multicultural counseling skills allows for appropriate interventions, advocacy, and an effective use of culturally appropriate models, such as the PDI model. The shift in counseling paradigms will require counselors to continue to develop themselves, their profession, and institutions along with a much broader spectrum of society.

The revised Multicultural Counseling Competencies and accompanying Explanatory Statements further clarify and define the three domains of awareness, knowledge, and skills. In this format, they go beyond the original Multicultural Counseling Competencies document (Sue, Arredondo, & McDavis, 1992) and take the profession further along in the process of institutionalizing counselor training and practices to be multicultural at the core. With the Explanatory Statements are examples and anecdotes that give life to the competencies. They are operationalized through language that describes the means of achieving and demonstrating a said competency. This new format further underscores the opportunity to use multicultural perspectives generically, with all interpersonal counseling.

MULTICULTURAL ORGANIZATIONS

The emphasis throughout the competencies is on individual change. Yet, it is obvious that if only individuals change and not the systems in which they work, the textbooks that are used for teaching, the practicum experiences that are provided, the ethical standards and competencies that guide professional practice, and the institutions that set up policies and influence legislation, the status quo will remain. We will continue to be segregated professionally and societally.

It is recommended that institutions of higher education, public and private schools, mental health facilities, and other settings where counseling is practiced engage in a self-examination process. Assess the cultural appropriateness and relevance of your organizational systems, policies, and practices. We encourage organizational leaders to search out the literature that will guide their process to becoming multicultural. It can be done.

MULTICULTURAL COUNSELING COMPETENCIES

I. Counselor Awareness of Own Cultural Values and Biases
  A. Attitudes and Beliefs
    1. Culturally skilled counselors believe that cultural self-awareness and sensitivity to one's own cultural heritage is essential.

       Explanatory Statements
       - Can identify the culture(s) to which they belong and the significance of that membership including the relationship of individuals in that group with individuals from other groups institutionally, historically, educationally, and so forth (include A, B, and C Dimensions as do the other suggestions in this section).
       - Can identify the specific cultural group(s) from which counselor derives fundamental cultural heritage and the significant beliefs and attitudes held by those cultures that are assimilated into their own attitudes and beliefs.
       - Can recognize the impact of those beliefs on their ability to respect others different from themselves.
       - Can identify specific attitudes, beliefs, and values from their own heritage and cultural learning that support behaviors that demonstrate respect and valuing of differences and those that impede or hinder respect and valuing of differences.
       - Actively engage in an ongoing process of challenging their own attitudes and beliefs that do not support respecting and valuing of differences.
       - Appreciate and articulate positive aspects of their own heritage that provide them with strengths in understanding differences.
       - In addition to their cultural groups, can recognize the influence of other personal dimensions of identity (PDI) and their role in cultural self-awareness.

    2. Culturally skilled counselors are aware of how their own cultural background and experiences have influenced attitudes, values, and biases about psychological processes.

       Explanatory Statements
       - Can identify the history of their culture in relation to educational opportunities and its impact on their current worldview (includes A and some B Dimensions).

- Can identify at least five personal, relevant cultural traits and can explain how each has influenced the cultural values of the counselor.
- Can identify social and cultural influences on their cognitive development and current information processing styles and can contrast that with those of others (includes A, B, and C Dimensions).
- Can identify specific social and cultural factors and events in their history that influence their view and use of social belonging, interpretations of behavior, motivation, problem-solving and decision methods, thoughts and behaviors (including subconscious) in relation to authority and other institutions and can contrast these with the perspectives of others (A and B Dimensions).
- Can articulate the beliefs of their own cultural and religious groups as these relate to sexual orientation, able-bodiedness, and so forth, and the impact of these beliefs in a counseling relationship.

3. Culturally skilled counselors are able to recognize the limits of their multicultural competency and expertise.

   Explanatory Statements
   - Can recognize in a counseling or teaching relationship, when and how their attitudes, beliefs, and values are interfering with providing the best service to clients (primarily A and B Dimensions).
   - Can identify preservice and in-service experiences which contribute to expertise and can identify current specific needs for professional development.
   - Can recognize and use referral sources that demonstrate values, attitudes, and beliefs that will respect and support the client's developmental needs.
   - Can give real examples of cultural situations in which they recognized their limitations and referred the client to more appropriate resources.

4. Culturally skilled counselors recognize their sources of discomfort with differences that exist between themselves and clients in terms of race, ethnicity, and culture.

   Explanatory Statements
   - Able to recognize their sources of comfort/discomfort with respect to differences in terms of race, ethnicity, and culture.
   - Able to identify differences (along A and B Dimensions) and are nonjudgmental about those differences.
   - Communicate acceptance of and respect for differences both verbally and nonverbally.
   - Can identify at least five specific cultural differences, the needs of culturally different clients, and how these differences are handled in the counseling relationship.

B. Knowledge
   1. Culturally skilled counselors have specific knowledge about their own racial and cultural heritage and how it personally and professionally affects their definitions of and biases about normality/abnormality and the process of counseling.

      Explanatory Statements
      - Have knowledge regarding their heritage. For example, A Dimensions in terms of ethnicity, language, and so forth, and C Dimensions in terms of knowledge regarding the context of the time period in which their ancestors entered the established United States or North American continent.
      - Can recognize and discuss their family's and culture's perspectives of acceptable (normal) codes of conduct and what are unacceptable (abnormal) and how this may or may not vary from those of other cultures and families.
      - Can identify at least five specific features of culture of origin and explain how those features affect the relationship with culturally different clients.

   2. Culturally skilled counselors possess knowledge and understanding about how oppression, racism, discrimination, and stereotyping affect them personally and in their work. This allows individuals to acknowledge their own racist attitudes, beliefs, and feelings. Although this standard applies to all groups, for White counselors it may mean that they understand how they may have directly or indirectly benefited from individual, institutional, and cultural racism as outlined in White identity development models.

Explanatory Statements

- Can specifically identify, name, and discuss privileges that they personally receive in society due to their race, socioeconomic background, gender, physical abilities, sexual orientation, and so on.
- Specifically referring to White counselors, can discuss White identity development models and how they relate to one's personal experiences.
- Can provide a reasonably specific definition of racism, prejudice, discrimination, and stereotype. Can describe a situation in which they have been judged on something other than merit. Can describe a situation in which they have judged someone on something other than merit.
- Can discuss recent research addressing issues of racism, White identity development, antiracism, and so forth, and its relation to their personal development and professional development as counselors.

3. Culturally skilled counselors possess knowledge about their social impact on others. They are knowledgeable about communication style differences, how their style may clash with or foster the counseling process with persons of color or others different from themselves based on the A, B, and C Dimensions, and how to anticipate the impact it may have on others.

Explanatory Statements

- Can describe the A and B Dimensions of Identity with which they most strongly identify.
- Can behaviorally define their communication style and describe both their verbal and nonverbal behaviors, interpretations of others' behaviors, and expectations.
- Recognize the cultural bases (A Dimension) of their communication style, and the differences between their style and the styles of those different from themselves.
- Can describe the behavioral impact and reaction of their communication style on clients different from themselves. For example, the reaction of an older (1960s) Vietnamese male recent immigrant to continuous eye contact from the young, female counselor.

- Can give examples of an incident in which communication broke down with a client of color and can hypothesize about the causes.
- Can give three to five concrete examples of situations in which they modified their communication style to compliment that of a culturally different client, how they decided on the modification, and the result of that modification.

C. Skills

1. Culturally skilled counselors seek out educational, consultative, and training experiences to improve their understanding and effectiveness in working with culturally different populations. Being able to recognize the limits of their competencies, they (a) seek consultation, (b) seek further training or education, (c) refer to more qualified individuals or resources, or (d) engage in a combination of these.

Explanatory Statements

- Can recognize and identify characteristics or situations in which the counselor's limitations in cultural, personal, or religious beliefs, or issues of identity development require referral.
- Can describe objectives of at least two multicultural-related professional development activities attended over the past 5 years and can identity at least two adaptations to their counseling practices as a result of these professional development activities.
- Have developed professional relationships with counselors from backgrounds different from their own and have maintained a dialogue regarding multicultural differences and preferences.
- Maintain an active referral list and continuously seek new referrals relevant to different needs of clients along A and B Dimensions.
- Understand and communicate to the client that the referral is being made because of the counselor's limitations rather than communicating that it is caused by the client.
- On recognizing these limitations, the counselor actively pursues and engages in professional and personal growth activities to address these limitations.
- Actively consult regularly with other professionals regarding issues of culture to receive

feedback about issues and situations and whether or where referral may be necessary.

2. Culturally skilled counselors are constantly seeking to understand themselves as racial and cultural beings and are actively seeking a nonracist identity.

   Explanatory Statements

   - Actively seek out and participate in reading and in activities designed to develop cultural self-awareness, and work toward eliminating racism and prejudice.

   - Maintain relationships (personal and professional) with individuals different from themselves and actively engage in discussions allowing for feedback regarding the counselor's behavior (personal and professional) concerning racial issues. (For example, a White counselor maintains a personal/professional relationship with a Latina counselor that is intimate enough to request and receive honest feedback regarding behaviors and attitudes and their impact on others, "I seem to have difficulty retaining Latina students in my class, given how I run my class, can you help me find ways that I may make it a more appropriate environment for Latina students?" or "When I said _____, how do you think others perceived that comment?") This requires the commitment to develop and contribute to a relationship that allows for adequate trust and honesty in very difficult situations.

   - When receiving feedback, the counselor demonstrates a receptivity and willingness to learn.

(See Appendix A, p. 594, for strategies to achieve the competencies and objectives for Area I.)

II. Counselor Awareness of Client's Worldview
   A. Attitudes and Beliefs
      1. Culturally skilled counselors are aware of their negative and positive emotional reactions toward other racial and ethnic groups that may prove detrimental to the counseling relationship. They are willing to contrast their own beliefs and attitudes with those of their culturally different clients in a nonjudgmental fashion.

         Explanatory Statements

         - Identify their common emotional reactions about individuals and groups different from themselves and observe their own reactions in encounters. For example, do they feel fear when approaching a group of three young African-American men? Do they assume that the Asian-American clients for whom they provide career counseling will be interested in a technical career?

         - Can articulate how their personal reactions and assumptions are different from those who identify with that group. (e.g., If the reaction on approaching three young African-American men is fear, what is the reaction of a young African-American man or woman in the same situation? What might be the reaction of an African-American woman approaching a group of White young men?)

         - Identify how general emotional reactions observed in oneself could influence effectiveness in a counseling relationship. (Reactions may be regarding cultural differences as well as along A and B Dimensions.)

         - Can describe at least two distinct examples of cultural conflict between self and culturally different clients, including how these conflicts were used as "content" for counseling. For example, if a Chicana agrees to live at home rather than board at a 4-year college to support her mother, can a counselor be nonjudgmental?

      2. Culturally skilled counselors are aware of their stereotypes and preconceived notions that they may hold toward other racial and ethnic minority groups.

         Explanatory Statements

         - Recognize their stereotyped reactions to people different from themselves. (e.g., silently articulating their awareness of a negative stereotypical reaction, "I noticed that I locked my car doors when that African-American teenager walked by.")

         - Consciously attend to examples that contradict stereotypes.

         - Can give specific examples of how their stereotypes (including "positive" ones) referring to the A and B Dimensions can affect the counselor-client relationship.

         - Recognize assumptions of those in a similar cultural group but who may differ based on A or B Dimension.

B. Knowledge
  1. Culturally skilled counselors possess specific knowledge and information about the particular group with which they are working. They are aware of the life experiences, cultural heritage, and historical background of their culturally different clients. This particular competency is strongly linked to the minority identity development models available in the literature.

     Explanatory Statements
     - Can articulate (objectively) differences in non-verbal and verbal behavior of the five major cultural groups most frequently seen in their experience of counseling.
     - Can describe at least two different models of minority identity development and their implications for counseling with persons of color or others who experience oppression or marginalization.
     - Understand and can explain the historical point of contact with dominant society for various ethnic groups and the impact of the type of contact (enslaved, refugee, seeking economic opportunities, conquest, and so forth) on current issues in society.
     - Can identify within-group differences and assess various aspects of individual clients to determine individual differences as well as cultural differences. For example, the counselor is aware of differences within Asian Americans: Japanese Americans, Vietnamese Americans, and so forth; differences between first generation refugees versus second or third generation; differences between Vietnamese refugees coming in the "first wave" in 1975 versus Vietnamese refugees coming to the United States in 1990.
     - Can discuss viewpoints of other cultural groups regarding issues such as sexual orientation, physical ability or disability, gender, and aging.

  2. Culturally skilled counselors understand how race, culture, ethnicity, and so forth may affect personality formation, vocational choices, manifestation of psychological disorders, help-seeking behavior, and the appropriateness or inappropriateness of counseling approaches.

     Explanatory Statements
     - Can distinguish cultural differences and expectations regarding role and responsibility in family, participation of family in career decision making, appropriate family members to be involved when seeking help, culturally acceptable means of expressing emotion and anxiety and so forth (primarily along A Dimension and portions of B Dimension).
     - Based on literature about A Dimensions, can describe and give examples of how a counseling approach may or may not be appropriate for a specific group of people based primarily on an A Dimension.
     - Understand and can explain the historical point of contact with dominant society for various ethnic groups and the impact of the type of contact (e.g., enslaved, refugee, seeking economic opportunities, conquest) on potential relationships and trust when seeking help from dominant culture institutions.
     - Can describe one system of personality development, the population(s) on which the theory was developed, and how this system relates or does not relate to at least two culturally different populations.
     - Can identify the role of gender, socioeconomic status, and physical disability as they interact with personality formation across cultural groups.

  3. Culturally skilled counselors understand and have knowledge about sociopolitical influences that impinge on the life of racial and ethnic minorities. Immigration issues, poverty, racism, stereotyping, and powerlessness may affect self-esteem and self-concept in the counseling process.

     Explanatory Statements
     - Can identify implications of concepts such as internalized oppression, institutional racism, privilege, and the historical and current political climate regarding immigration, poverty, and welfare (public assistance).
     - Can explain the relationship between culture and power. Can explain dynamics of at least two cultures and how factors such as poverty and powerlessness have influenced the current conditions of individuals of those cultures.

- Understand the economic benefits and contributions gained by the work of various groups, including migrant farm workers, to the daily life of the counselor and the country at large.
- Can communicate an understanding of the unique position, constraints, and needs of those clients who experience oppression based on an A or B Dimension alone (and families of clients) who share this history.
- Can identify current issues that affect groups of people (A and B Dimensions) in legislation, social climate, and so forth, and how that affects individuals and families to whom the counselor may be providing services.
- Are aware of legal legislation issues that affect various communities and populations (e.g., in California it is essential for a counselor to understand the ramifications of the recent passage of Proposition 187 and how that will affect not only undocumented individuals, but also families and anyone that has Chicano features, a Mexican-American accent, and speaks Spanish. In addition, the counselor must be aware of how this will affect health issues, help-seeking behaviors, participation in education, and so forth.)
- Counselors are aware of how documents such as the book *The Bell Curve* and affirmative action legislation affect society's perception of different cultural groups.

C. Skills
   1. Culturally skilled counselors should familiarize themselves with relevant research and the latest findings regarding mental health and mental disorders that affect various ethnic and racial groups. They should actively seek out educational experiences that enrich their knowledge, understanding, and cross-cultural skills for more effective counseling behavior.

      Explanatory Statements
      - Can discuss recent research regarding such topics as mental health, career decision making, education and learning, that focuses on issues related to different cultural populations and as represented in A and B Dimensions.
      - Complete (at least 15 hours per year) workshops, conferences, classes, in-service training regarding multicultural counseling skills and knowledge. These should span a variety of topics and cultures and should include discussions of wellness rather than focusing only on negative issues (medical model) related to these cultures.
      - Can identify at least five multicultural experiences in which counselor has participated within the past 3 years.
      - Can identify professional growth activities and information that are presented by professionals respected and seen as credible by members of the communities being studied (e.g., the book *The Bell Curve* may not represent accurate and helpful information regarding individuals from non-White cultures).
      - Can describe in concrete terms how they have applied varied information gained through current research in mental health, education, career choices, and so forth, based on differences noted in A Dimension.

   2. Culturally skilled counselors become actively involved with minority individuals outside the counseling setting (e.g., community events, social and political functions, celebrations, friendships, neighborhood groups) so that their perspective of minorities is more than an academic or helping exercise.

      Explanatory Statements
      - Can identify at least five multicultural experiences in which the counselor has participated within the past 3 years. These include various celebrations, political events, or community activities involving individuals and groups from racial and cultural backgrounds different from their own, such as political fund-raisers, Tet celebrations, and neighborhood marches against violence.
      - Actively plan experiences and activities that will contradict negative stereotypes and preconceived notions they may hold.

(See Appendix B, p. 594, for strategies to achieve the competencies and objectives for Area II.)

III. Culturally Appropriate Intervention Strategies
   A. Beliefs and Attitudes
      1. Culturally skilled counselors respect clients' religious and spiritual beliefs and values, including attributions and taboos, because these affect worldview, psychosocial functioning, and expressions of distress.

Explanatory Statements
- Can identify the positive aspects of spirituality (in general) in terms of wellness and healing aspects.
- Can identify in a variety of religious and spiritual communities the recognized form of leadership and guidance and their client's relationship (if existent) with that organization and entity.

2. Culturally skilled counselors respect indigenous helping practices and respect help-giving networks among communities of color.
- Can describe concrete examples of how they may integrate and cooperate with indigenous helpers when appropriate.
- Can describe concrete examples of how they may use intrinsic help-giving networks from a variety of client communities.

3. Culturally skilled counselors value bilingualism and do not view another language as an impediment to counseling ("monolingualism" may be the culprit).
   Explanatory Statements
   - Communicate to clients and colleagues values and assets of bilingualism (if client is bilingual).

B. Knowledge
1. Culturally skilled counselors have a clear and explicit knowledge and understanding of the generic characteristics of counseling and therapy (culture bound, class bound, and monolingual) and how they may clash with the cultural values of various cultural groups.
   Explanatory Statements
   - Can articulate the historical, cultural, and racial context in which traditional theories and interventions have been developed.
   - Can identify, within various theories, the cultural values, beliefs, and assumptions made about individuals and contrast these with values, beliefs, and assumptions of different racial and cultural groups.
   - Recognize the predominant theories being used within counselor's organization and educate colleagues regarding the aspects of those theories and interventions that may clash with the cultural values of various cultural and racial minority groups.
   - Can identify and describe primary indigenous helping practices in terms of positive

and effective role in at least five A or B Dimensions, relevant to counselor's client population.

2. Culturally skilled counselors are aware of institutional barriers that prevent minorities from using mental health services.
   Explanatory Statements
   - Can describe concrete examples of institutional barriers within their organizations that prevent minorities from using mental health services and share those examples with colleagues and decision-making bodies within the institution.
   - Recognize and draw attention to patterns of usage (or non-usage) of mental health services in relation to specific populations.
   - Can identify and communicate possible alternatives that would reduce or eliminate existing barriers within their institutions and within local, state, and national decision-making bodies.

3. Culturally skilled counselors have knowledge of the potential bias in assessment instruments and use procedures and interpret findings in a way that recognizes the cultural and linguistic characteristics of the clients.
   Explanatory Statements
   - Demonstrate ability to interpret assessment results including implications of dominant cultural values affecting assessment/interpretation, interaction of cultures for those who are bicultural, and the impact of historical institutional oppression.
   - Can discuss information regarding cultural, racial, gender profile of normative group used for validity and reliability on any assessment used by counselor.
   - Understand the limitations of translating assessment instruments as well as the importance of using language that includes culturally relevant connotations and idioms.
   - Use assessment instruments appropriately with clients having limited English skills.
   - Can give examples, for each assessment instrument used, of the limitations of the instrument regarding various groups represented in A and B Dimensions.
   - Recognize possible historical and current sociopolitical biases in *DSM (Diagnostic &*

Statistical Manual of Mental Disorders) system of diagnosis based on racial, cultural, sexual orientation, and gender issues.

4. Culturally skilled counselors have knowledge of family structures, hierarchies, values, and beliefs from various cultural perspectives. They are knowledgeable about the community where a particular cultural group may reside and the resources in the community.

Explanatory Statements

- Are familiar with and use organizations that provide support and services in different cultural communities.
- Can discuss the traditional ways of helping in different cultures and continue to learn the resources in communities relevant to those cultures.
- Adequately understand the client's religious and spiritual beliefs to know when and what topics are or are not appropriate to discuss regarding those beliefs.
- Understand and respect cultural and family influences and participation in decision making.

5. Culturally skilled counselors should be aware of relevant discriminatory practices at the social and the community level that may be affecting the psychological welfare of the population being served.

Explanatory Statements

- Are aware of legal issues that affect various communities and populations (e.g., in Proposition 187 California described earlier).

C. Skills

1. Culturally skilled counselors are able to engage in a variety of verbal and nonverbal helping responses. They are able to send and receive both verbal and nonverbal messages accurately and appropriately. They are not tied down to only one method or approach to helping, but recognize that helping styles and approaches may be culture bound. When they sense that their helping style is limited and potentially inappropriate, they can anticipate and modify it.

Explanatory Statements

- Can articulate what, when, why, and how they apply different verbal and nonverbal helping responses based on A and B Dimensions.

- Can give examples of how they may modify a technique or intervention or what alternative intervention they may use to more effectively meet the needs of a client.
- Can identify and describe techniques in which they have expertise for providing service that may require minimal English language skills (e.g., expressive therapy).
- Can communicate verbally and nonverbally to the client the validity of the client's religious and spiritual beliefs.
- Can discuss with the client (when appropriate) aspects of their religious or spiritual beliefs that have been helpful to the client in the past.

2. Culturally skilled counselors are able to exercise institutional intervention skills on behalf of their clients. They can help clients determine whether a "problem" stems from racism or bias in others (the concept of healthy paranoia) so that clients do not inappropriately personalize problems.

Explanatory Statements

- Can recognize and discuss examples in which racism or bias may actually be imbedded in an institutional system or in society.
- Can discuss a variety of coping and survival behaviors used by a variety of individuals from their A and B Dimensions to cope effectively with bias or racism.
- Communicate to clients an understanding of the necessary coping skills and behaviors viewed by dominant society as dysfunctional that they may need to keep intact.
- Can describe concrete examples of situations in which it is appropriate and possibly necessary for a counselor to exercise institutional intervention skills on behalf of a client.

3. Culturally skilled counselors are not averse to seeking consultation with traditional healers or religious and spiritual leaders and practitioners in the treatment of culturally different clients when appropriate.

Explanatory Statements

- Participate or gather adequate information regarding indigenous or community helping resources to make appropriate referrals (e.g., be familiar with the American Indian community enough to recognize when, how, and to whom it may be appropriate to refer a client for indigenous healers).

4. Culturally skilled counselors take responsibility for interacting in the language requested by the client and, if not feasible, make appropriate referrals. A serious problem arises when the linguistic skills of the counselor do not match the language of the client. This being the case, counselors should (a) seek a translator with cultural knowledge and appropriate professional background or (b) refer to a knowledgeable and competent bilingual counselor.

   Explanatory Statement

   - Are familiar with resources that provide services in languages appropriate to clients.
   - Will seek out, whenever necessary, services or translators to ensure that language needs are met.
   - If working within an organization, actively advocate for the hiring of bilingual counselors relevant to client population.

5. Culturally skilled counselors have training and expertise in the use of traditional assessment and testing instruments. They not only understand the technical aspects of the instruments but are also aware of the cultural limitations. This allows them to use test instruments for the welfare of culturally different clients.

   Explanatory Statements

   - Demonstrate ability to interpret assessment results including implications of dominant cultural values affecting assessment and interpretation, interaction of cultures for those who are bicultural, and the impact of historical institutional oppression.
   - Can discuss information regarding cultural, racial, and gender profile of norm group used for validity and reliability on any assessment used by counselor.
   - Understand that although an assessment instrument may be translated into another language, the translation may be literal without an accurate contextual translation including culturally relevant connotations and idioms.

6. Culturally skilled counselors should attend to, as well as work to eliminate, biases, prejudices, and discriminatory contexts in conducting evaluations and providing interventions, and should develop sensitivity to issues of oppression, sexism, heterosexism, elitism, and racism.

Explanatory Statements

- Recognize incidents in which clients, students, and others are being treated unfairly based on such characteristics as race, ethnicity, and physical ableness and take action by directly addressing incident or perpetrator, filing informal complaint, filing formal complaint, and so forth.
- Work at an organizational level to address, change, and eliminate policies that discriminate, create barriers, and so forth.
- If an organization's policy created barriers for advocacy, the counselor works toward changing institutional policies to promote advocacy against racism, sexism, and so forth.

7. Culturally skilled counselors take responsibility for educating their clients to the processes of psychological intervention, such as goals, expectations, legal rights, and the counselor's orientation.

   Explanatory Statements

   - Assess the client's understanding of and familiarity with counseling and mental health services and provide accurate information regarding the process, limitations, and function of the services into which the client is entering.
   - Ensure that the client understands client rights, issues, and definitions of confidentiality, and the expectations placed on that client. In this educational process, counselors adapt information to ensure that all concepts are clearly understood by the client. This may include defining and discussing these concepts.

(See Appendix C for strategies to achieve competencies and objectives in Area III. See Appendix D for strategies to achieve competencies and objectives in all three areas.)

Adapted from "Multicultural Counseling Competencies and Standards: A Call to the Profession" (Sue, Arredondo, & McDavis (1992)).

REFERENCES

Arredondo, P., & Glauner, T. (1992). *Personal Dimensions of Identity Model.* Boston, MA: Empowerment Workshops.

Helms, J. (1990). *White identity development.* New York: Greenwood Press.

Packer, A. H., & Johnston, W. B. (1987). *Workforce 2000: Work and Workers for the 21st Century.* Indiana: Hudson Institute.

Sue, D. W., Arredondo, P., & McDavis, R. J. (1992). Multicultural counseling competencies and standards: A call to the profession. *Journal of Counseling & Development, 70,* 477–483.

APPENDIX A
Strategies to Achieve the Competencies and Objectives (I)

Read materials regarding identity development. For example, a European-American counselor may read materials on White or Majority Identity Development or an African American may read materials on Black Identity Development to gain an understanding of their own development. Additionally, reading about others' identity development processes is essential. The following are some resources specifically for European-American or White counselors:

> Carter, R. T. (1990). The relationship between racism and racial identity among White Americans: An exploratory investigation. *Journal of Counseling & Development, 69,* 46–50.
>
> Corvin, S., & Wiggins, F. (1989). An anti-racism training model for White professionals. *Journal of Multicultural Counseling and Development, 17,* 105–114.
>
> Helms, J. (1990). *White identity development.* New York: Greenwood Press.
>
> Pedersen, P. B. (1988). *A handbook for development of multicultural awareness.* Alexandria, VA: American Association for Counseling and Development.
>
> Pope-Davis, D. B., & Ottavi, T. M. (1992). The influence of White racial identity attitudes on racism among faculty members: A preliminary examination. *Journal of College Student Development, 33,* 389–394.
>
> Sabnani, H. B., Ponterotto, J. G., & Borodovsky, L. G. (1991). White racial identity development and cross-cultural training. *The Counseling Psychologist, 19,* 76–102.
>
> Wrenn, C. G. (1962). The culturally encapsulated counselor. *Harvard Educational Review, 32,* 444–449.

Other Professional Activities

- Attend annual conferences and workshops such as:

> Annual Conference on Race and Ethnicity in Higher Education sponsored by the Center for Southwest Studies Oklahoma (1995, Santa Fe)
> Third World Counselor's Association Annual Conference (Palm Springs, 1995)
> AMCD Annual Western Summit

- Engage a mentor from your own culture who you identify as someone who has been working toward becoming cross-culturally competent and who has made significant strides in ways you have not.
- Engage a mentor or two from cultures different from your own who are willing to provide honest feedback regarding your behavior, attitudes, and beliefs. Be willing to listen and work toward change!
- Film: *The Color of Fear* by Lee Mun Wah
- Film: *A Class Divided* produced by PBS for "Frontline"
- Film: "True Colors"—"20/20" Special
- Video: *The Trial Model* by Paul Pederson

APPENDIX B
Strategies to Achieve the Competencies and Objectives (II)

- The following reading list may be helpful for counselors to broaden their understanding of different worldviews (some of these materials would also be helpful in developing culturally appropriate intervention strategies):

> Atkinson, D., Morten, G., & Sue, D. W. (1989). *Counseling American minorities: A cross-cultural perspective.* Dubuque, IA: Brown.
>
> Collins, P. (1990). *Black feminist thought: Knowledge, consciousness and the politics of empowerment.* Boston, MA: Unwin Hyman.
>
> Sue, D. W., & Sue, D. (1990). *Counseling the culturally different: Theory and practice* (2nd ed.). New York: Wiley.

- Attend annual conferences and workshops such as:

> Annual Conference on Race and Ethnicity in Higher Education sponsored by the Center for Southwest Studies Oklahoma (1995, Santa Fe)
> Third World Counselor's Association Annual Conference (Palm Springs, 1995)
> AMCD Annual Western Summit

- Enroll in ethnic studies courses at local community colleges or universities that focus on cultures different from your own (if none are offered, communicate to that school your expectation that they will offer them in the future).
- Spend time in communities different from your own (e.g., shopping in grocery stores, attending churches, walking in marches).
- Read newspapers and other periodicals targeting specific populations different from your own (i.e., Spanish language newspapers, *Buffalo Soldier, Lakota Times*).
- Engage in activities and celebrations within communities different from your own (e.g., Juneteenth, Tet, Cinco de Mayo).

- Engage a mentor or two from cultures different from your own who are also working toward cross-cultural competency (be sure to discuss with them your contribution to the relationship).
- Accept that it is your responsibility to learn about other cultures and implications in counseling and do not expect or rely on individuals from those cultures to teach you.
- Learn a second or third language relevant to clients to begin to understand the significance of that language in the transmission of culture.
- Seek out and engage in consultation from professionals from cultures relevant to your client population.
- Spend time in civil service offices observing service orientation toward individuals of color (Chicano/Latino, African American, Asian American, Native American) and contrast that with service orientation toward White individuals. Also observe any differences in service orientation that may be based on class issues (e.g., someone alone and well dressed versus a woman with children wearing older clothing, somewhat disheveled).
- Film: *The Color of Fear* by Lee Mun Wah
- Film: *El Norte*
- Film: *Stand and Deliver*
- Film: *Roots*
- Film: *Lakota Woman*
- Film: *Daughters of the Dust*

APPENDIX C
Strategies to Achieve the Competencies and Objectives (III)

- The following reading list may be helpful for building a foundation to develop and apply culturally appropriate interventions:

> Atkinson, D., Morten, G., & Sue, D. W. (1989). *Counseling American minorities: A cross-cultural perspective*. Dubuque, IA: Brown.
>
> Ibrahim, F. A., & Arredondo, P. M. (1990). Ethical issues in multicultural counseling. In B. Herlihy & L. Golden (Eds.), *Ethical standards casebook* (pp. 137–145). Alexandria, VA: American Association for Counseling and Development.
>
> Katz, J. (1978). *White awareness: Handbook for anti-racism training*. Norman, OK: Oklahoma.
>
> LaFromboise, T. D., & Foster, S. L. (1990). Cross-cultural training: Scientist-practitioner model and methods. *The Counseling Psychologist, 20,* 472–489.
>
> LaFromboise, T. D., & Foster, S. L. (1989). Ethics in multicultural counseling. In P. B. Pedersen, W. J. Lonner, & J. E.

> Trimble (Eds.), *Counseling across cultures* (3rd ed., pp. 115–136). Honolulu, HI: University of Hawaii Press.

- Meet with leaders and heads of organizations that specifically focus on providing service to individuals of certain cultural groups (for example in San Jose, CA, AACI-Asian Americans for Community Involvement) to discuss how you may work cooperatively together and what support you may provide the organization.
- Conduct informal research of your clientele, your organizations' clientele, to determine if there are patterns of use or non use along cultural and/or racial lines.

APPENDIX D
Overall Strategies for Achieving Competencies and Objectives in all Three Areas

- Assess self in terms of cross-cultural counseling competencies either by reviewing the competencies and giving examples in each area and/or using any of the following resources regarding assessment instruments:

> Ho, M. K. (1992). *Minority children and adolescents in therapy*. Newbury Park: Sage (see Appendix)
>
> LaFromboise, T. D., Coleman, H. L. K., & Hernandez, A. (1991). Development and factor structure of the Cross Cultural Counseling Inventory-Revised. *Professional Psychology: Research and Practice, 22,* 380–388.
>
> Ponterotto, J. G., Rieger, B. P., Barrett, A., & Sparks, R. (1994). Assessing multicultural counseling competence: A review of instrumentation. *Journal of Counseling & Development, 72,* 316–322.

- Learn a second or third language relevant to clients.
- Communicate to conference organizers and workshop providers that you will attend only if the activity addresses cross-cultural aspects of the topic.
- Actively communicate in your organization the need for training in cross-cultural training relevant to that organization.
- Speak up in your organization when you observe that clients, students or others are being treated unfairly based on such characteristics as race, ethnicity, or physical ableness.
- Become a member of AMCD, Division 45/APA, or state and local organizations that provide cross-cultural exchanges.

GUIDELINES FOR AUTHORS

The *Journal of Multicultural Counseling and Development* invites articles concerned with research, theory, and program applications pertinent to multicultural and ethnic minority interests or experiences in all areas of counseling and related areas of human development. If you wish to contribute to the journal, follow these guidelines.

- Manuscripts should be well organized and concise so that the development of ideas is logical. Avoid dull, stereotyped writing and aim at the clear and interesting communication of ideas. Limit manuscript length to 15 pages. Include an abstract of not more than 50 words.
- The article title should appear on a separate page accompanying the manuscript. Include on this page the names of the authors and a paragraph that repeats the names of the authors and gives the professional title and institutional affiliation of each.
- Article titles and headings within the articles should be as short as possible.
- Authors must provide the final version of accepted manuscripts on a microcomputer floppy diskette (5¼″ or 3½″) in **ASCII or WordPerfect** format for IBM compatible computers or in **Text file** format for Macintosh computers. The diskette should be labeled with the first author's name, the journal title, and the language in which the article was written.
- Send an original and two clean copies of all material on regular 8½″ by 11″ white bond paper.
- Double-space all material, including references and extensive quotations.
- Do not use footnotes.
- Check all references for completeness, including volume number, issue number, and pages for journal citations. Make sure all references mentioned in the text are listed in the reference section and vice versa.
- Tables should be kept to a minimum. Include only the essential data and combine tables wherever possible. Refer to a recent copy of the journal for style of tabular presentations. Each table should be on a separate sheet of paper following the reference section of the article. Final placement of tables is at the discretion of the production editor; in all cases, tables will be placed near the first reference of the table in the text.
- Figures (graphs, illustrations, line drawings) should be supplied as camera-ready art. Titles for the figures may be attached to the art and will be set in the appropriate type.
- Lengthy quotations (generally 300–500 cumulative words or more from one source) require *written* permission from the copyright holder for reproduction. Adaptations of tables and figures also require reproduction approval from the copyrighted source. It is the **author's responsibility** to secure such permission and a copy of the publisher's written

permission must be provided to the journal editor immediately on acceptance of the article for publication by ACA.

- Manuscript style is that of the *Publication Manual of the American Psychological Association,* fourth edition, which is available from the Order Department, APA, 750 First St. N.E., Washington, DC 20002-4242. Only manuscripts conforming to this style will be reviewed by the Editorial Board.
- Never submit material that is under consideration by another periodical.
- Submit manuscripts to Thomas S. Gunnings, Room A229, East Fee Hall, Michigan State University, East Lansing, MI 48824-1316.

ACA journal articles are edited within a uniform style for correctness and consistency of grammar, spelling, and punctuation. In some cases, portions of manuscripts have been reworded for conciseness or clarity of expression. Computer printouts of edited manuscripts are sent to the senior authors of each article in the journal. Changes at this stage must be limited to correcting inaccuracies and typographical errors. Authors must bear responsibility for the accuracy of references, tables, and figures. On publication of an article, a complimentary copy is sent to all authors.

ASSOCIATION FOR MULTICULTURAL COUNSELING AND DEVELOPMENT

1995–1996 OFFICERS

President
Sherlon Pack Brown
Bowling Green State
  University

President-Elect
Patricia Arredondo
Empowerment Workshops,
  Boston

Past President
Marlene Rutherford-Rhodes
St. Louis Community
  College at Forest Park

*Vice President for
African-American Concerns*
Ruth Lewis
St. Louis Community
  College at Forest Park

*Vice President for
Hispanic/Latino Concerns*
Bernal Baca
Yakima Community College

Secretary
Susan Cameron
The University of New
  Mexico

Treasurer
J. Otis Smith
Stand By Systems II, Inc.,
  Philadelphia

President Emeritus
Samuel Johnson, NSSFNS

Parliamentarian
Seymour Bryson
Southern Illinois
  University-Carbondale

*Representative to ACA
Governing Council*
Queen D. Fowler
St. Louis Public Schools

EXECUTIVE COUNCIL

*North Atlantic Regional
Representative*
M. Nartel Green
St. Croix, U.S. Virgin
  Islands

*Southern Regional
Representative*
Victor E. Bilbins, Sr.
Elizabeth City State
  University

*Midwest Regional
Representative*
Florida Bosh
Washington University

*Western Regional
Representative*
Marcelett C. Henry
California Department of
  Education

NEWSLETTER EDITOR

Wanda Harewood-Jones
Ohio Department of Education

CODE OF ETHICS, AMERICAN COUNSELING ASSOCIATION

(Approved by the ACA Governing Council, April, 1995)

Preamble

The American Counseling Association is an educational, scientific, and professional organization whose members are dedicated to the enhancement of human development throughout the lifespan. Association members recognize diversity in our society and embrace a cross-cultural approach in support of the worth, dignity, potential, and uniqueness of each individual.

The specification of a code of ethics enables the association to clarify to current and future members, and to those served by members, the nature of the ethical responsibilities held in common by its members. As the code of ethics of the association, this document establishes principles that define the ethical behavior of association members. All members of the American Counseling Association are required to adhere to the *Code of Ethics* and the *Standards of Practice.* The Code of Ethics will serve as the basis for processing ethical complaints initiated against members of the association.

Section A: The Counseling Relationship

A.1. Client Welfare
   a. *Primary Responsibility.* The primary responsibility of counselors is to respect the dignity and to promote the welfare of clients.
   b. *Positive Growth and Development.* Counselors encourage client growth and development in ways that foster the clients' interest and welfare; counselors avoid fostering dependent counseling relationships.

c. *Counseling Plans.* Counselors and their clients work jointly in devising integrated, individual counseling plans that offer reasonable promise of success and are consistent with abilities and circumstances of clients. Counselors and clients regularly review counseling plans to ensure their continued viability and effectiveness, respecting clients' freedom of choice. (See A.3.b.)
d. *Family Involvement.* Counselors recognize that families are usually important in clients' lives and strive to enlist family understanding and involvement as a positive resource, when appropriate.
e. *Career and Employment Needs.* Counselors work with their clients in considering employment in jobs and circumstances that are consistent with the client's overall abilities, vocational limitations, physical restrictions, general temperament, interest and aptitude patterns, social skills, education, general qualifications, and other relevant characteristics and needs. Counselors neither place nor participate in placing clients in positions that will result in damaging the interest and the welfare of clients, employers, or the public.

A.2. Respecting Diversity
   a. *Nondiscrimination.* Counselors do not condone or engage in discrimination based on age, color, culture, disability, ethnic group, gender, race, religion, sexual orientation, marital status, or socioeconomic status. (See C.5.a., C.5.b., and D.1.i.)
   b. *Respecting Differences.* Counselors will actively attempt to understand the diverse cultural backgrounds of the clients with whom they work. This includes, but is not limited to, learning how the counselor's own cultural/ethnic/racial identity impacts her/his values and beliefs about the counseling process. (See E.8. and F.2.i.)

A.3. Client Rights

a. *Disclosure to Clients.* When counseling is initiated, and throughout the counseling process as necessary, counselors inform clients of the purposes, goals, techniques, procedures, limitations, potential risks and benefits of services to be performed, and other pertinent information. Counselors take steps to ensure that clients understand the implications of diagnosis, the intended use of tests and reports, fees, and billing arrangements. Clients have the right to expect confidentiality and to be provided with an explanation of its limitations, including supervision and/or treatment team professionals; to obtain clear information about their case records; to participate in the ongoing counseling plans; and to refuse any recommended services and be advised of the consequences of such refusal. (See E.5.a. and G.2.)

b. *Freedom of Choice.* Counselors offer clients the freedom to choose whether to enter into a counseling relationship and to determine which professional(s) will provide counseling. Restrictions that limit choices of clients are fully explained. (See A.1.c.)

c. *Inability to Give Consent.* When counseling minors or persons unable to give voluntary informed consent, counselors act in these clients' best interests. (See B.3.)

A.4. Clients Served by Others

If a client is receiving services from another mental health professional, counselors, with client consent, inform the professional persons already involved and develop clear agreements to avoid confusion and conflict for the client. (See C.6.c.)

A.5. Personal Needs and Values

a. *Personal Needs.* In the counseling relationship, counselors are aware of the intimacy and responsibilities inherent in the counseling relationship, maintain respect for clients, and avoid actions that seek to meet their personal needs at the expense of clients.

b. *Personal Values.* Counselors are aware of their own values, attitudes, beliefs, and behaviors and how these apply in a diverse society, and avoid imposing their values on clients. (See C.5.a.)

A.6. Dual Relationships

a. *Avoid When Possible.* Counselors are aware of their influential positions with respect to clients, and they avoid exploiting the trust and dependency of clients. Counselors make every effort to avoid dual relationships with clients that could impair professional judgment or increase the risk of harm to clients. (Examples of such relationships include, but are not limited to, familial, social, financial, business, or close personal relationships with clients.) When a dual relationship cannot be avoided, counselors take appropriate professional precautions such as informed consent, consultation, supervision, and documentation to ensure that judgment is not impaired and no exploitation occurs. (See F.1.b.)

b. *Superior/Subordinate Relationships.* Counselors do not accept as clients superiors or subordinates with whom they have administrative, supervisory, or evaluative relationships.

A.7. Sexual Intimacies with Clients

a. *Current Clients.* Counselors do not have any type of sexual intimacies with clients and do not counsel persons with whom they have had a sexual relationship.

b. *Former Clients.* Counselors do not engage in sexual intimacies with former clients within a minimum period of two years after terminating the counseling relationship. Counselors who engage in such relationship after two years following termination have the responsibility to thoroughly examine and document that such relations did not have an exploitative nature, based on factors such as duration of counseling, amount of time since counseling, termination circumstances, client's personal history and mental status, adverse impact on the client, and actions by the counselor suggesting a plan to initiate a sexual relationship with the client after termination.

A.8. Multiple Clients

When counselors agree to provide counseling services to two or more persons who have a relationship (such as husband and wife, or parents and children), counselors clarify at the outset which person or persons are clients and the nature of the relationship they will have with each involved person. If it becomes apparent that counselors may be called upon to perform potentially conflicting roles, they clarify, adjust, or withdraw from roles appropriately. (See B.2. and B.4.d.)

A.9. Group Work

a. *Screening.* Counselors screen prospective group counseling/therapy participants. To the extent possible, counselors select members whose needs and goals are compatible with goals of the group, who will not impede the group process, and whose well-being will not be jeopardized by the group experience.

b. *Protecting Clients.* In a group setting, counselors take reasonable precautions to protect clients from physical or psychological trauma.

A.10. Fees and Bartering (See D.3.a. and D.3.b.)

a. *Advance Understanding.* Counselors clearly explain to clients, prior to entering the counseling relationship, all financial arrangements related to professional services including the use of collection agencies or legal measures for nonpayment. (A.11.c.)

b. *Establishing Fees.* In establishing fees for professional counseling services, counselors consider the financial status of clients and locality. In the event that the established fee structure is inappropriate for a client, assistance is provided in attempting to find comparable services of acceptable cost. (See A.10.d., D.3.a., and D.3.b.)

c. *Bartering Discouraged.* Counselors ordinarily refrain from accepting goods or services from clients in return for counseling services because such arrangements create inherent potential for conflicts, exploitation, and distortion of the professional relationship. Counselors may participate in bartering only if the relationship is not exploitive, if the client requests it, if a clear written contract is established, and if such arrangements are an accepted practice among professionals in the community. (See A.6.a.)

d. *Pro Bono Service.* Counselors contribute to society by devoting a portion of their professional activity to services for which there is little or no financial return (pro bono).

A.11. Termination and Referral

a. *Abandonment Prohibited.* Counselors do not abandon or neglect clients in counseling. Counselors assist in making appropriate arrangements for the continuation of treatment, when necessary, during interruptions such as vacations, and following termination.

b. *Inability to Assist Clients.* If counselors determine an inability to be of professional assistance to clients, they avoid entering or immediately terminate a counseling relationship. Counselors are knowledgeable about referral resources and suggest appropriate alternatives. If clients decline the suggested referral, counselors should discontinue the relationship.

c. *Appropriate Termination.* Counselors terminate a counseling relationship, securing client agreement when possible, when it is reasonably clear that the client is no longer benefiting, when services are no longer required, when counselor no longer serves the client's needs or interests, when clients do not pay fees charged, or when agency or institution limits do not allow provision of further counseling services. (See A.10.b. and C.2.g.)

A.12. Computer Technology

a. *Use of Computers.* When computer applications are used in counseling services, counselors ensure that: (1) the client is intellectually, emotionally, and physically capable of using the computer application; (2) the computer application is appropriate for the needs of the client; (3) the client understands the purpose and operation of the computer applications; and (4) a follow-up of client use of a computer application is provided to correct possible misconceptions, discover inappropriate use, and assess subsequent needs.

b. *Explanation of Limitations.* Counselors ensure that clients are provided information as a part of the counseling relationship that adequately explains the limitations of computer technology.

c. *Access to Computer Applications.* Counselors provide for equal access to computer applications in counseling services. (See A.2.a.)

Section B: Confidentiality

B.1. Right to Privacy

a. *Respect for Privacy.* Counselors respect their clients' right to privacy and avoid illegal and unwarranted disclosures of confidential information. (See A.3.a. and B.6.a.)

b. *Client Waiver.* The right to privacy may be waived by the client or their legally recognized representative.

c. *Exceptions.* The general requirement that counselors keep information confidential does not apply when disclosure is required to prevent clear and imminent danger to the client or others or when legal requirements demand that confidential information be revealed. Counselors consult with other professionals when in doubt as to the validity of an exception.

d. *Contagious, Fatal Diseases.* A counselor who receives information confirming that a client has a disease commonly known to be both communicable and fatal is justified in disclosing information to an identifiable third party, who by his or her relationship with the client is at a high risk of contracting the disease. Prior to making a disclosure the counselor should ascertain that the client has not already informed the third party about his or her disease and that the client is not intending to inform the third party in the immediate future. (See B.1.c. and B.1.f.)

e. *Court Ordered Disclosure.* When court ordered to release confidential information without a client's permission, counselors request to the court that the disclosure not be required due to potential harm to the client or counseling relationship. (See B.1.c.)

f. *Minimal Disclosure.* When circumstances require the disclosure of confidential information, only essential information is revealed. To the extent possible, clients are informed before confidential information is disclosed.

g. *Explanation of Limitations.* When counseling is initiated and throughout the counseling process as necessary, counselors inform clients of the limitations of confidentiality and identify foreseeable situations in which confidentiality must be breached. (See G.2.a.)

h. *Subordinates.* Counselors make every effort to ensure that privacy and confidentiality of clients are maintained by subordinates including employees, supervisees, clerical assistants, and volunteers. (See B.1.a.)

i. *Treatment Teams.* If client treatment will involve a continued review by a treatment team, the client will be informed of the team's existence and composition.

B.2. Groups and Families

a. *Group Work.* In group work, counselors define confidentiality and the parameters for the specific group being entered, explain its importance, and discuss the difficulties related to confidentiality involved in group work. The fact that confidentiality cannot be guaranteed is clearly communicated to group members.

b. *Family Counseling.* In family counseling, information about one family member cannot be disclosed to another member without permission. Counselors

protect the privacy rights of each family member. (See A.8., B.3., and B.4.d.)

B.3. Minor or Incompetent Clients

When counseling clients who are minors or individuals who are unable to give voluntary, informed consent, parents or guardians may be included in the counseling process as appropriate. Counselors act in the best interests of clients and take measures to safeguard confidentiality. (See A.3.c.)

B.4. Records

a. *Requirements of Records.* Counselors maintain records necessary for rendering professional services to their clients and as required by laws, regulations, or agency or institution procedures.

b. *Confidentiality of Records.* Counselors are responsible for securing the safety and confidentiality of any counseling records they create, maintain, transfer, or destroy whether the records are written, taped, computerized, or stored in any other medium. (See B.1.a.)

c. *Permission to Record or Observe.* Counselors obtain permission from clients prior to electronically recording or observing sessions. (See A.3.a.)

d. *Client Access.* Counselors recognize that counseling records are kept for the benefit of clients, and therefore provide access to records and copies of records when requested by competent clients, unless the records contain information that may be misleading and detrimental to the client. In situations involving multiple clients, access to records is limited to those parts of records that do not include confidential information related to another client. (See A.8., B.1.a., and B.2.b.)

e. *Disclosure or Transfer.* Counselors obtain written permission from clients to disclose or transfer records to legitimate third parties unless exceptions to confidentiality exist as listed in Section B.1. Steps are taken to ensure that receivers of counseling records are sensitive to their confidential nature.

B.5. Research and Training

a. *Data Disguise Required.* Use of data derived from counseling relationships for purposes of training, research, or publication is confined to content that is disguised to ensure the anonymity of the individuals involved. (See B.1.g. and G.3.d.)

b. *Agreement for Identification.* Identification of a client in a presentation or publication is permissible only when the client has reviewed the material and has agreed to its presentation or publication. (See G.3.d.)

B.6. Consultation

a. *Respect for Privacy.* Information obtained in a consulting relationship is discussed for professional purposes only with persons clearly concerned with the case. Written and oral reports present data germane to the purposes of the consultation, and every effort is made to protect client identity and avoid undue invasion of privacy.

b. *Cooperating Agencies.* Before sharing information, counselors make efforts to ensure that there are defined policies in other agencies serving the counselor's clients that effectively protect the confidentiality of information.

Section C: Professional Responsibility

C.1. Standards Knowledge

Counselors have a responsibility to read, understand, and follow the *Code of Ethics* and the *Standards of Practice.*

C.2. Professional Competence

a. *Boundaries of Competence.* Counselors practice only within the boundaries of their competence, based on their education, training, supervised experience, state and national professional credentials, and appropriate professional experience. Counselors will demonstrate a commitment to gain knowledge, personal awareness, sensitivity, and skills pertinent to working with a diverse client population.

b. *New Specialty Areas of Practice.* Counselors practice in specialty areas new to them only after appropriate education, training, and supervised experience. While developing skills in new specialty areas, counselors take steps to ensure the competence of their work and to protect others from possible harm.

c. *Qualified for Employment.* Counselors accept employment only for positions for which they are qualified by education, training, supervised experience, state and national professional credentials, and appropriate professional experience. Counselors hire for professional counseling positions only individuals who are qualified and competent.

d. *Monitor Effectiveness.* Counselors continually monitor their effectiveness as professionals and take steps to improve when necessary. Counselors in private practice take reasonable steps to seek out peer supervision to evaluate their efficacy as counselors.

e. *Ethical Issues Consultation.* Counselors take reasonable steps to consult with other counselors or related professionals when they have questions regarding their ethical obligations or professional practice. (See H.1.)

f. *Continuing Education.* Counselors recognize the need for continuing education to maintain a reasonable level of awareness of current scientific and professional information in their fields of activity. They take steps to maintain competence in the skills they use, are open to new procedures, and keep current with the diverse and/or special populations with whom they work.

g. *Impairment.* Counselors refrain from offering or accepting professional services when their physical, mental, or emotional problems are likely to harm a client or others. They are alert to the signs of impairment, seek assistance for problems, and, if necessary, limit, suspend, or terminate their professional responsibilities. (See A.11.c.)

C.3. Advertising and Soliciting Clients

a. *Accurate Advertising.* There are no restrictions on advertising by counselors except those that can be specifically justified to protect the public from deceptive practices. Counselors advertise or represent their services to the public by identifying their credentials in an accurate manner that is not false, misleading, deceptive, or fraudulent. Counselors may only advertise the highest degree earned which is in counseling or a closely related field from a college or university that was accredited when the degree was awarded by one of the regional accrediting bodies recognized by the Council on Postsecondary Accreditation.

b. *Testimonials.* Counselors who use testimonials do not solicit them from clients or other persons who, because of their particular circumstances, may be vulnerable to undue influence.

c. *Statements by Others.* Counselors make reasonable efforts to ensure that statements made by others about them or the profession of counseling are accurate.

d. *Recruiting Through Employment.* Counselors do not use their places of employment or institutional affiliation to recruit or gain clients, supervisees, or consultees for their private practices. (See C.5.3.)

e. *Products and Training Advertisements.* Counselors who develop products related to their profession or conduct workshops or training events ensure that the advertisements concerning these products or events are accurate and disclose adequate information for consumers to make informed choices.

f. *Promoting to Those Served.* Counselors do not use counseling, teaching, training, or supervisory relationships to promote their products or training events in a manner that is deceptive or would exert undue influence on individuals who may be vulnerable. Counselors may adopt textbooks they have authored for instruction purposes.

g. *Professional Association Involvement.* Counselors actively participate in local, state, and national associations that foster the development and improvement of counseling.

C.4. Credentials

a. *Credentials Claimed.* Counselors claim or imply only professional credentials possessed and are responsible for correcting any known misrepresentations of their credentials by others. Professional credentials include graduate degrees in counseling or closely related mental health fields, accreditation of graduate programs, national voluntary certification, government-issued certifications or licenses, ACA professional membership, or any other credential that might indicate to the public specialized knowledge or expertise in counseling.

b. *ACA Professional Membership.* ACA professional members may announce to the public their membership status. Regular members may not announce their ACA membership in a manner that might imply they are credentialed counselors.

c. *Credential Guidelines.* Counselors follow the guidelines for use of credentials that have been established by the entities that issue the credentials.

d. *Misrepresentation of Credentials.* Counselors do not attribute more to their credentials than the credentials represent, and do not imply that other counselors are not qualified because they do not possess certain credentials.

e. *Doctoral Degrees From Other Fields.* Counselors who hold a master's degree in counseling or a closely related mental health field, but hold a doctoral degree from other than counseling or a closely related field do not use the title, "Dr." in their practices and do not announce to the public in relation to their practice or status as a counselor that they hold a doctorate.

C.5. Public Responsibility

a. *Nondiscrimination.* Counselors do not discriminate against clients, students, or supervisees in a manner that has a negative impact based on their age, color, culture, disability, ethnic group, gender, race, religion, sexual orientation, or socioeconomic status, or for any other reason. (See A.2.a.)

b. *Sexual Harassment.* Counselors do not engage in sexual harassment. Sexual harassment is defined as sexual solicitation, physical advances, or verbal or nonverbal conduct that is sexual in nature, that occurs in connection with professional activities or roles, and that either: (1) is unwelcome, is offensive, or creates a hostile workplace environment, and counselors know or are told this; or (2) is sufficiently severe or intense to be perceived as harassment to a reasonable person in the context. Sexual harassment can consist of a single intense or severe act or multiple persistent or pervasive acts.

c. *Reports to Third Parties.* Counselors are accurate, honest, and unbiased in reporting their professional activities and judgments to appropriate third parties including courts, health insurance companies, those who are the recipients of evaluation reports, and others. (See B.1.g.)

d. *Unjustified Gains.* Counselors do not use their professional positions to seek or receive unjustified personal gains, sexual favors, unfair advantage, or unearned goods or services. (See C.3.d.)

C.6. Responsibility to Other Professionals

a. *Different Approaches.* Counselors are respectful of approaches to professional counseling that differ from their own. Counselors know and take into account the traditions and practices of other professional groups with which they work.

b. *Personal Public Statements.* When making personal statements in a public context, counselors clarify that they are speaking from their personal perspectives and that they are not speaking on behalf of all counselors or the profession. (See C.5.d.)

c. *Clients Served by Others.* When counselors learn that their clients are in a professional relationship with another mental health professional, they request release from clients to inform the other professionals and strive to establish positive and collaborative professional relationships. (See A.4.)

Section D: Relationships With Other Professionals

D.1. Relationships with Employers and Employees

a. *Role Definition.* Counselors define and describe for their employers and employees the parameters and levels of their professional roles.

b. *Agreements.* Counselors establish working agreements with supervisors, colleagues, and subordinates regarding counseling or clinical relationships, confidentiality, adherence to professional standards, distinction between public and private material, maintenance and dissemination of recorded information, workload, and accountability. Working agreements in each instance are specified and made known to those concerned.

c. *Negative Conditions.* Counselors alert their employers to conditions that may be potentially disruptive or damaging to the counselor's professional responsibilities or that may limit their effectiveness.

d. *Evaluation.* Counselors submit regularly to professional review and evaluation by their supervisor or the appropriate representative of the employer.

e. *In-Service.* Counselors are responsible for in-service development of self and staff.

f. *Goals.* Counselors inform their staff of goals and programs.

g. *Practices.* Counselors provide personnel and agency practices that respect and enhance the rights and welfare of each employee and recipient of agency services. Counselors strive to maintain the highest levels of professional services.

h. *Personnel Selection and Assignment.* Counselors select competent staff and assign responsibilities compatible with their skills and experiences.

i. *Discrimination.* Counselors, as either employers or employees, do not engage in or condone practices that are inhumane, illegal, or unjustifiable (such as considerations based on age, color, culture, disability, ethnic group, gender, race, religion, sexual orientation, or socioeconomic status) in hiring, promotion, or training. (See A.2.a. and C.5.b.)

j. *Professional Conduct.* Counselors have a responsibility both to clients and to the agency or institution within which services are performed to maintain high standards of professional conduct.

k. *Exploitive Relationships.* Counselors do not engage in exploitive relationships with individuals over whom they have supervisory, evaluative, or instructional control or authority.

l. *Employer Policies.* The acceptance of employment in an agency or institution implies that counselors are in agreement with its general policies and principles. Counselors strive to reach agreement with employers as to acceptable standards of conduct that allow for changes in institutional policy conducive to the growth and development of clients.

D.2. Consultation (See B.6.)

a. *Consultation as an Option.* Counselors may choose to consult with any other professionally competent person about their clients. In choosing consultants, counselors avoid placing the consultant in a conflict of interest situation that would preclude the consultant being a proper party to the counselor's efforts to help the client. Should counselors be engaged in a work setting that compromises this consultation standard, they consult with other professionals whenever possible to consider justifiable alternatives.

b. *Consultant Competency.* Counselors are reasonably certain that they have or the organization represented has the necessary competencies and resources for giving the kind of consulting services needed and that appropriate referral resources are available.

c. *Understanding with Clients.* When providing consultation, counselors attempt to develop with their clients a clear understanding of problem definition, goals for change, and predicted consequences of interventions selected.

d. *Consultant Goals.* The consulting relationship is one in which client adaptability and growth toward self-direction are consistently encouraged and cultivated. (See A.1.b.)

D.3. Fees for Referral

a. *Accepting Fees from Agency Clients.* Counselors refuse a private fee or other remuneration for rendering services to persons who are entitled to such services through the counselor's employing agency or institution. The policies of a particular agency may make explicit provisions for agency clients to receive counseling services from members of its staff in private practice. In such instances, the clients must be informed of other options open to them should they seek private counseling services. (See A.10.a., A.11.b., and C.3.d.)

b. *Referral Fees.* Counselors do not accept a referral fee from other professionals.

D.4. Subcontractor Arrangements

When counselors work as subcontractors for counseling services for a third party, they have a duty to inform clients of the limitations of confidentiality that the organization may place on counseling services to clients. The limits of such confidentiality ordinarily are discussed as part of the intake session. (See B.1.e. and B.1.f.)

Section E: Evaluation, Assessment, and Interpretation

E.1. General

a. *Appraisal Techniques.* The primary purpose of educational and psychological assessment is to provide measures that are objective and interpretable in either comparative or absolute terms. Counselors recognize the need to interpret the statements in this section as applying to the whole range of appraisal techniques, including test and nontest data.

b. *Client Welfare.* Counselors promote the welfare and best interests of the client in the development, publication, and utilization of educational and psychological assessment techniques. They do not misuse assessment results and interpretations and take reasonable steps to prevent others from misusing the information these techniques provide. They respect the client's right to know the results, the interpretations made, and the bases for their conclusions and recommendations.

E.2. Competence to Use and Interpret Tests

a. *Limits of Competence.* Counselors recognize the limits of their competence and perform only those testing and assessment services for which they have been trained. They are familiar with reliability, validity, related standardization, error of measurement, and proper application of any technique utilized. Counselors using computer-based test interpretation are trained in the construct being measured and the specific instrument being used prior to using this type of computer application. Counselors take reasonable measures to ensure the proper use of psychological assessment techniques by persons under their supervision.

b. *Appropriate Use.* Counselors are responsible for the appropriate application, scoring, interpretation, and use of assessment instruments, whether they score and interpret such tests themselves or use computerized or other services.

c. *Decisions Based on Results.* Counselors responsible for decisions involving individuals or policies that are based on assessment results have a thorough understanding of educational and psychological measurement, including validation criteria, test research, and guidelines for test development and use.

d. *Accurate Information.* Counselors provide accurate information and avoid false claims or misconceptions when making statements about assessment instruments or techniques. Special efforts are made to avoid unwarranted connotations of such terms as IQ and grade equivalent scores. (See C.5.c.)

E.3. Informed Consent

a. *Explanation to Clients.* Prior to assessment, counselors explain the nature and purposes of assessment and the specific use of results in language the client (or other legally authorized person on behalf of the client) can understand, unless an explicit exception to this right has been agreed upon in advance. Regardless of whether scoring and interpretation are completed by counselors, by assistants, or by computer or any other outside services, counselors take reasonable steps to ensure that appropriate explanations are given to the client.

b. *Recipients of Results.* The examinee's welfare, explicit understanding, and prior agreement determine the recipients of test results. Counselors include accurate and appropriate interpretations with any release of individual or group test results. (See B.1.a. and C.5.c.)

E.4. Release of Information to Competent Professionals

a. *Misuse of Results.* Counselors do not misuse assessment results, including test results, and interpretations, and take reasonable steps to prevent the misuse of such by others. (See C.5.c.)

b. *Release of Raw Data.* Counselors ordinarily release data (e.g. protocols, counseling or interview notes, or questionnaires) in which the client is identified only with the consent of the client or the client's legal representative. Such data are usually released only to persons recognized by counselors as competent to interpret the data. (See B.1.a.)

E.5. Proper Diagnosis of Mental Disorders

    a. *Proper Diagnosis.* Counselors take special care to provide proper diagnosis of mental disorders. Assessment techniques (including personal interview) used to determine client care (e.g., locus of treatment, type of treatment, or recommended follow-up) are carefully selected and appropriately used. (See A.3.a. and C.5.c.)

    b. *Cultural Sensitivity.* Counselors recognize that culture affects the manner in which clients' problems are defined. Clients' socioeconomic and cultural experience is considered when diagnosing mental disorders.

E.6. Test Selection

    a. *Appropriateness of Instruments.* Counselors carefully consider the validity, reliability, psychometric limitations, and appropriateness of instruments when selecting tests for use in a given situation or with a particular client.

    b. *Culturally Diverse Populations.* Counselors are cautious when selecting tests for culturally diverse populations to avoid inappropriateness of testing that may be outside of socialized behavioral or cognitive patterns.

E.7. Conditions of Test Administration

    a. *Administration Conditions.* Counselors administer tests under the same conditions that were established in their standardization. When tests are not administered under standard conditions or when unusual behavior or irregularities occur during the testing session, those conditions are noted in interpretation, and the results may be designated as invalid or of questionable validity.

    b. *Computer Administration.* Counselors are responsible for ensuring that administration programs function properly to provide clients with accurate results when a computer or other electronic methods are used for test administration. (See A.12.b.)

    c. *Unsupervised Test-Taking.* Counselors do not permit unsupervised or inadequately supervised use of tests or assessments unless the tests or assessments are designed, intended, and validated for self-administration and/or scoring.

    d. *Disclosure of Favorable Conditions.* Prior to test administration, conditions that produce most favorable test results are made known to the examinee.

E.8. Diversity in Testing

Counselors are cautious in using assessment techniques, making evaluations, and interpreting the performance of populations not represented in the norm group on which an instrument was standardized. They recognize the effects of age, color, culture, disability, ethnic group, gender, race, religion, sexual orientation, and socioeconomic status on test administration and interpretation and place test results in proper perspective with other relevant factors. (See A.2.a.)

E.9. Test Scoring and Interpretation

    a. *Reporting Reservations.* In reporting assessment results, counselors indicate any reservations that exist regarding validity or reliability because of the circumstances of the assessment or the inappropriateness of the norms for the person tested.

    b. *Research Instruments.* Counselors exercise caution when interpreting the results of research instruments possessing insufficient technical data to support respondent results. The specific purposes for the use of such instruments are stated explicitly to the examinee.

    c. *Testing Services.* Counselors who provide test scoring and test interpretation services to support the assessment process confirm the validity of such interpretations. They accurately describe the purpose, norms, validity, reliability, and applications of the procedures and any special qualifications applicable to their use. The public offering of an automated test interpretations service is considered a professional-to-professional consultation. The formal responsibility of the consultant is to the consultee, but the ultimate and overriding responsibility is to the client.

E.10. Test Security

Counselors maintain the integrity and security of tests and other assessment techniques consistent with legal and contractual obligations. Counselors do not appropriate, reproduce, or modify published tests or parts thereof without acknowledgment and permission from the publisher.

E.11. Obsolete Tests and Outdated Test Results

Counselors do not use data or test results that are obsolete or outdated for the current purpose. Counselors make every effort to prevent the misuse of obsolete measures and test data by others.

E.12. Test Construction

Counselors use established scientific procedures, relevant standards, and current professional knowledge for test design in the development, publication, and utilization of educational and psychological assessment techniques.

Section F: Teaching, Training, and Supervision

F.1. Counselor Educators and Trainers

a. *Educators as Teachers and Practitioners.* Counselors who are responsible for developing, implementing, and supervising educational programs are skilled as teachers and practitioners. They are knowledgeable regarding the ethical, legal, and regulatory aspects of the profession, are skilled in applying that knowledge, and make students and supervisees aware of their responsibilities. Counselors conduct counselor education and training programs in an ethical manner and serve as role models for professional behavior. Counselor educators should make an effort to infuse material related to human diversity into all courses and/or workshops that are designed to promote the development of professional counselors.

b. *Relationship Boundaries with Students and Supervisees.* Counselors clearly define and maintain ethical, professional, and social relationship boundaries with their students and supervisees. They are aware of the differential in power that exists and the student's or supervisee's possible incomprehension of that power differential. Counselors explain to students and supervisees the potential for the relationship to become exploitive.

c. *Sexual Relationships.* Counselors do not engage in sexual relationships with students or supervisees and do not subject them to sexual harassment. (See A.6. and C.5.b.)

d. *Contributions to Research.* Counselors give credit to students or supervisees for their contributions to research and scholarly projects. Credit is given through coauthorship, acknowledgment, footnote statement, or other appropriate means, in accordance with such contributions. (See G.4.b. and G.4.c.)

e. *Close Relatives.* Counselors do not accept close relatives as students or supervisees.

f. *Supervision Preparation.* Counselors who offer clinical supervision services are adequately prepared in supervision methods and techniques. Counselors who are doctoral students serving as practicum or internship supervisors to master's level students are adequately prepared and supervised by the training program.

g. *Responsibility for Services to Clients.* Counselors who supervise the counseling services of others take reasonable measures to ensure that counseling services provided to clients are professional.

h. *Endorsement.* Counselors do not endorse students or supervisees for certification, licensure, employment, or completion of an academic or training program if they believe students or supervisees are not qualified for the endorsement. Counselors take reasonable steps to assist students or supervisees who are not qualified for endorsement to become qualified.

F.2. Counselor Education and Training Programs

a. *Orientation.* Prior to admission, counselors orient prospective students to the counselor education or training program's expectations, including but not limited to the following: (1) the type and level of skill acquisition required for successful completion of the training, (2) subject matter to be covered, (3) basis for evaluation, (4) training components that encourage self-growth or self-disclosure as part of the training process, (5) the type of supervision settings and requirements of the sites for required clinical field experiences, (6) student and supervisee evaluation and dismissal policies and procedures, and (7) up-to-date employment prospects for graduates.

b. *Integration of Study and Practice.* Counselors establish counselor education and training programs that integrate academic study and supervised practice.

c. *Evaluation.* Counselors clearly state to students and supervisees, in advance of training, the levels of competency expected, appraisal methods, and timing of evaluations for both didactic and experiential components. Counselors provide students and supervisees with periodic performance appraisal and evaluation feedback throughout the training program.

d. *Teaching Ethics.* Counselors make students and supervisees aware of the ethical responsibilities and standards of the profession and the students' and supervisees' ethical responsibilities to the profession. (See C.1. and F.3.e.)

e. *Peer Relationships.* When students or supervisees are assigned to lead counseling groups or provide clinical supervision for their peers, counselors take steps to ensure that students and supervisees placed in these roles do not have personal or adverse relationships with peers and that they understand they have the same ethical obligations as counselor educators, trainers, and supervisors. Counselors make every effort to ensure that the rights of peers are not compromised when students or supervisees are assigned to lead counseling groups or provide clinical supervision.

f. *Varied Theoretical Positions.* Counselors present varied theoretical positions so that students and supervisees may make comparisons and have opportunities to develop their own positions. Counselors provide information concerning the scientific bases of professional practice. (See C.6.a.)

g. *Field Placements.* Counselors develop clear policies within their training program regarding field placement and other clinical experiences. Counselors provide clearly stated roles and responsibilities for the student or supervisee, the site supervisor, and the program supervisor. They confirm that site supervisors are qualified to provide supervision and are informed of their professional and ethical responsibilities in this role.

h. *Dual Relationships as Supervisors.* Counselors avoid dual relationships such as performing the role of site supervisor and training program supervisor in the student's or supervisee's training program. Counselors do not accept any form of professional services, fees, commissions, reimbursement, or remuneration from a site for student or supervisee placement.

i. *Diversity in Programs.* Counselors are responsive to their institution's and program's recruitment and retention needs for training program administrators, faculty, and students with diverse backgrounds and special needs. (See A.2.a.)

F.3. Students and Supervisees

a. *Limitations.* Counselors, through ongoing evaluation and appraisal, are aware of the academic and personal limitations of students and supervisees that might impede performance. Counselors assist students and supervisees in securing remedial assistance when needed, and dismiss from the training program supervisees who are unable to provide competent service due to academic or personal limitations. Counselors seek professional consultation and document their decision to dismiss or refer students or supervisees for assistance. Counselors assure that students and supervisees have recourse to address decisions made, to require them to seek assistance, or

to dismiss them.

b. *Self-Growth Experiences.* Counselors use professional judgment when designing training experiences conducted by the counselors themselves that require student and supervisee self-growth or self-disclosure. Safeguards are provided so that students and supervisees are aware of the ramifications their self-disclosure may have, on counselors whose primary role as teacher, trainer, or supervisor requires acting on ethical obligations to the profession. Evaluative components of experiential training experience explicitly delineate predetermined academic standards that are separate and not dependent on the student's level of self-disclosure. (See A.6.)

c. *Counseling for Students and Supervisees.* If students or supervisees request counseling, supervisors or counselor educators provide them with acceptable referrals. Supervisors or counselor educators do not serve as counselor to students or supervisees over whom they hold administrative, teaching, or evaluative roles unless this is a brief role associated with a training experience. (See A.6.b.)

d. *Clients of Students and Supervisees.* Counselors make every effort to ensure that the clients at field placements are aware of the services rendered and the qualifications of the students and supervisees rendering those services. Clients receive professional disclosure information and are informed of the limits of confidentiality. Client permission is obtained in order for the students and supervisees to use any information concerning the counseling relationship in the training process. (See B.1.e.)

e. *Standards for Students and Supervisees.* Students and supervisees preparing to become counselors adhere to the *Code of Ethics* and the *Standards of Practice.* Students and supervisees have the same obligations to clients as those required of counselors. (See H.1.)

Section G: Research and Publication

G.1. Research Responsibilities

a. *Use of Human Subjects.* Counselors plan, design, conduct, and report research in a manner consistent with pertinent ethical principles, federal and state laws, host institutional regulations, and scientific standards governing research with human subjects. Counselors design and conduct research that reflects cultural sensitivity appropriateness.

b. *Deviation from Standard Practices.* Counselors seek consultation and observe stringent safeguards to protect the rights of research participants when a research problem suggests a deviation from standard acceptable practices. (See B.6.)

c. *Precautions to Avoid Injury.* Counselors who conduct research with human subjects are responsible for the subjects' welfare throughout the experiment and take reasonable precautions to avoid causing injurious psychological, physical, or social effects to their subjects.

d. *Principal Research Responsibility.* The ultimate responsibility for ethical research practice lies with the principal researcher. All others involved in the research activities share ethical obligations and full responsibility for their own actions.

e. *Minimal Interference.* Counselors take reasonable precautions to avoid causing disruptions in subjects' lives due to participation in research.

f. *Diversity.* Counselors are sensitive to diversity and research issues with special populations. They seek consultation with appropriate professionals. (See A.2.a. and B.6.)

G.2. Informed Consent

a. *Topics Disclosed.* In obtaining informed consent for research, counselors use language that is understandable to research participants and that: (1) accurately explains the purpose and procedures to be followed; (2) identifies any procedures that are experimental or relatively untried; (3) describes the attendant discomforts and risks; (4) describes the benefits or changes in individuals or organizations that might be reasonably expected; (5) discloses appropriate alternative procedures that would be advantageous for subjects; (6) offers to answer any inquiries concerning the procedures; (7) describes any limitations on confidentiality; and (8) instructs that subjects are free to withdraw their consent and to discontinue participation in the project at any time. (See B.1.f.)

b. *Deception.* Counselors do not conduct research involving deception unless alternative procedures are not feasible and the prospective value of the research justifies the deception. When the methodological requirements of a study necessitate concealment or deception, the investigator is required to explain clearly the reasons for this action as soon as possible.

c. *Voluntary Participation.* Participation in research is typically voluntary and without any penalty for refusal to participate. Involuntary participation is appropriate only when it can be demonstrated that participation will have no harmful effects on subjects and is essential to the investigation.

d. *Confidentiality of Information.* Information obtained about research participants during the course of an investigation is confidential. When the possibility exists that others may obtain access to such information, ethical research practice requires that the possibility, together with the plans for protecting confidentiality, be explained to participants as a part of the procedure. (See B.1.e.)

e. *Persons Incapable of Giving Informed Consent.* When a person is incapable of giving informed consent, counselors provide an appropriate explanation, obtain agreement for participation and obtain appropriate consent from a legally authorized person.

f. *Commitments to Participate.* Counselors take reasonable measures to honor all commitments to research participants.

g. *Explanations After Data Collection.* After data are collected, counselors provide participants with full clarification of the nature of the study to remove any misconceptions. Where scientific or human values justify delaying or withholding information, counselors take reasonable measures to avoid causing harm.

h. *Agreement to Cooperate.* Counselors who agree to cooperate with another individual in research or publication incur an obligation to cooperate as promised in terms of punctuality of performance and with regard to the completeness and accuracy of the information required.

i. *Informed Consent for Sponsors.* In the pursuit of research, counselors give sponsors, institutions, and publication channels the same respect and opportunity for giving informed consent that they accord to individual research participants. Counselors are aware of their obligation to future research workers and ensure that host institutions are given feedback information and proper acknowledgment.

G.3. Reporting Results

a. *Information Affecting Outcome.* When reporting research results, counselors explicitly mention all variables and conditions known to the investigator that may have affected the outcome of a study or the interpretation of data.

b. *Accurate Results.* Counselors plan, conduct, and report research accurately and in a manner that minimizes the possibility that results will be misleading. They provide thorough discussions of the limitations of their data and alternative hypotheses. Counselors do not engage in fraudulent research, distort data, misrepresent data, or deliberately bias their results.

c. *Obligation to Report Unfavorable Results.* Counselors communicate to other counselors the results of any research judged to be of professional value. Results that reflect unfavorably on institutions, programs, services, prevailing opinions, or vested interests are not withheld.

d. *Identity of Subjects.* Counselors who supply data, aid in the research of another person, report research results, or make original data available take due care to disguise the identity of respective subjects in the absence of specific authorization from the subjects to do otherwise. (See B.1.g. and B.5.a.)

e. *Replication Studies.* Counselors are obligated to make available sufficient original research data to qualified professionals who may wish to replicate the study.

G.4. Publication

a. *Recognition of Others.* When conducting and reporting research, counselors are familiar with and give recognition to previous work on the topic, observe copyright laws, and give full credit to those to whom credit is due. (See F.1.d. and G.4.c.)

b. *Contributors.* Counselors give credit through joint authorship, acknowledgment, footnote statements, or other appropriate means to those who have contributed significantly to research or concept development in accordance with such contributions. The principal contributor is listed first and minor technical or professional contributions are acknowledged in notes or introductory statements.

c. *Student Research.* For an article that is substantially based on a student's dissertation or thesis, the student is listed as the principal author. (See F.1.d. and G.4.a.)

d. *Duplicate Submission.* Counselors submit manuscripts for consideration to only one journal at a time. Manuscripts that are published in whole or in substantial part in another journal or published work are not submitted for publication without acknowledgment and permission from the previous publication.

e. *Professional Review.* Counselors who review material submitted for publication, research, or other scholarly purposes respect the confidentiality and proprietary rights of those who submitted it.

Section H: Resolving Ethical Issues

H.1. Knowledge of Standards

Counselors are familiar with the *Code of Ethics* and the *Standards of Practice* and other applicable ethics codes from other professional organizations of which they are members, or from certification and licensure bodies. Lack of knowledge or misunderstanding of an ethical responsibility is not a defense against a charge of unethical conduct. (See F.3.e.)

H.2. Suspected Violations

a. *Ethical Behavior Expected.* Counselors expect professional associates to adhere to Code of Ethics. When counselors possess reasonable cause that raises doubts as to whether a counselor is acting in an ethical manner, they take appropriate action. (See H.2.d. and H.2.e.)

b. *Organization Conflicts.* If the demands of an organization with which counselors are affiliated pose a conflict with Code of Ethics, counselors specify the nature of such conflicts and express to their supervisors or other responsible officials their commitment to Code of Ethics. When possible, counselors work toward change within the organization to allow full adherence to Code of Ethics.

c. *Informal Resolution.* When counselors have reasonable cause to believe that another counselor is violating an ethical standard, they attempt to first resolve the issue informally with the other counselor if feasible, providing that such action does not violate confidentiality rights that may be involved.

d. *Reporting Suspected Violations.* When an informal resolution is not appropriate or feasible, counselors, upon reasonable cause, take action such as reporting the suspected ethical violation to state or national ethics committees, unless this action conflicts with confidentiality rights that cannot be resolved.

e. *Unwarranted Complaints.* Counselors do not initiate, participate in, or encourage the filing of ethics complaints that are unwarranted or intend to harm a counselor rather than to protect clients or the public.

H.3. Cooperation with Ethics Committees

Counselors assist in the process of enforcing Code of Ethics. Counselors cooperate with investigations, proceedings, and requirements of the ACA Ethics Committee or ethics committees of other duly constituted associations or boards having jurisdiction over those charged with a violation. Counselors are familiar with the ACA Policies and Procedures and use it as a reference in assisting the enforcement of the Code of Ethics.

ETHICAL PRINCIPLES OF PSYCHOLOGISTS, AMERICAN PSYCHOLOGICAL ASSOCIATION

INTRODUCTION

The American Psychological Association's (APA's) Ethical Principles of Psychologists and Code of Conduct (hereinafter referred to as the Ethics Code) consists of an Introduction, a Preamble, six General Principles (A-F), and specific Ethical Standards. The Introduction discusses the intent, organization, procedural considerations, and scope of application of the Ethics Code. The Preamble and General Principles are *aspirational* goals to guide psychologists toward the highest ideals of psychology. Although the Preamble and General Principles are not themselves enforceable rules, they should be considered by psychologists in arriving at an ethical course of action and may be considered by ethics bodies in interpreting the Ethical Standards. The Ethical Standards set forth *enforceable* rules for conduct as psychologists. Most of the Ethical Standards are written broadly, in order to apply to psychologists in varied roles, although the application of an Ethical Standard may vary depending on the context. The Ethical Standards are not exhaustive. The fact that a given conduct is not specifically addressed by the Ethics Code does not mean that it is necessarily either ethical or unethical.

Membership in the APA commits members to adhere to the APA Ethics Code and to the rules and procedures used to implement it. Psychologists and students, whether or not they are APA members, should be aware that the Ethics Code may be applied to them by state psychology boards, courts, or other public bodies.

This Ethics Code applies only to psychologists' work-related activities, that is, activities that are part of the psychologists' scientific and professional functions or that are psychological in nature. It includes the clinical or counseling practice of psychology, research, teaching, supervision of trainees, development of assessment instruments, conducting assessments, educational counseling, organizational consulting, social intervention, administration, and other activities as well. These work-related activities can be distinguished from the purely private conduct of a psychologist, which ordinarily is not within the purview of the Ethics Code.

The Ethics Code is intended to provide standards of professional conduct that can be applied by the APA and by other bodies that choose to adopt them. Whether or not a psychologist has violated the Ethics Code does not by itself determine whether he or she is legally liable in a court action, whether a contract is enforceable, or whether other legal consequences occur. These results are based on legal rather than ethical rules. However, compliance with or violation of the Ethics Code may be admissible as evidence in some legal proceedings, depending on the circumstances.

In the process of making decisions regarding their professional behavior, psychologists must consider this Ethics Code, in addition to applicable laws and psychology board regulations. If the Ethics Code establishes a higher standard of conduct than is required by law, psychologists must meet the higher ethical standard. If the Ethics Code standard appears to conflict with the requirements of law, then psychologists make known their commitment to the Ethics Code and take steps to resolve the conflict in a responsible manner. If neither law nor the Ethics Code resolves an issue, psychologists should consider other professional materials and the dictates of their own conscience, as well as seek consultation with others within the field when this is practical.

The procedures for filing, investigating, and resolving complaints of unethical conduct are described in the current Rules and Procedures of the APA Ethics Committee. The actions that APA may take for violations of the Ethics Code include actions such as reprimand, censure, termination of APA membership, and referral of the matter to other bodies. Complainants who seek remedies such as monetary damages in alleging ethical violations by a psychologist must resort to private negotiation, administrative bodies, or the courts. Actions that violate the Ethics Code may lead to the imposition of sanctions on a psychologist by bodies other than APA, including state psychological associations, other professional groups, psychology boards, other state or federal agencies, and payors for health services. In addition to actions for violation of the Ethics Code, the APA Bylaws provide that APA may take action against a member after his or her conviction of a felony, expulsion or suspension from an affiliated state psychological association, or suspension or loss of licensure.

PREAMBLE

Psychologists work to develop a valid and reliable body of scientific knowledge based on research. They may apply that knowledge to human behavior in a variety of contexts. In doing so, they perform many roles, such as researcher, educator, diagnostician, therapist, supervisor, consultant, administrator, social interventionist, and expert witness. Their goal is to broaden knowledge of behavior and, where appropriate, to apply it pragmatically to improve the condition of both the individual and society. Psychologists respect the central importance of freedom of inquiry and expression in research, teaching, and publication. They also strive to help the public in developing informed judgments and choices concerning human behavior. This Ethics Code provides a common set of values upon which psychologists build their professional and scientific work.

This Code is intended to provide both the general principles and the decision rules to cover most situations encountered by psychologists. It has as its primary goal the welfare and protection of the individuals and groups with whom psychologists work. It is the individual responsibility of each psychologist to aspire to the highest possible standards of conduct. Psychologists respect and protect human and civil rights, and do not knowingly participate in or condone unfair discriminatory practices.

The development of a dynamic set of ethical standards for a psychologist's work-related conduct requires a personal commitment to a lifelong effort to act ethically; to encourage ethical behavior by students, supervisees, employees, and colleagues, as appropriate; and to consult with others, as needed, concerning ethical problems. Each psychologist supplements, but does not violate, the Ethics Code's values and rules on the basis of guidance drawn from personal values, culture, and experience.

GENERAL PRINCIPLES

Principle A: Competence

Psychologists strive to maintain high standards of competence in their work. They recognize the boundaries of their particular competencies and the limitations of their expertise. They provide only those services and use only those techniques for which they are qualified by education, training, or experience. Psychologists are cognizant of the fact that the competencies required in serving, teaching, and/or studying groups of people vary with the distinctive characteristics of those groups. In those areas in which recognized professional standards do not yet exist, psychologists exercise careful judgment and take appropriate precautions to protect the welfare of those with whom they work. They maintain knowledge of relevant scientific and professional information related to the services they render, and they recognize the need for ongoing education. Psychologists make appropriate use of scientific, professional, technical, and administrative resources.

Principle B: Integrity

Psychologists seek to promote integrity in the science, teaching, and practice of psychology. In these activities psychologists are honest, fair, and respectful of others. In describing or reporting their qualifications, services, products, fees, research, or teaching, they do not make statements that are false, misleading, or deceptive. Psychologists strive to be aware of their own belief systems, values, needs, and limitations and the effect of these on their work. To the extent feasible, they attempt to clarify for relevant parties the roles they are performing and to function appropriately in accordance with those roles. Psychologists avoid improper and potentially harmful dual relationships.

Principle C: Professional and Scientific Responsibility

Psychologists uphold professional standards of conduct, clarify their professional roles and obligations, accept appropriate responsibility for their behavior, and adapt their methods to the needs of different populations. Psychologists consult with, refer to, or cooperate with other professionals and institutions to the extent needed to serve the best interests of their patients, clients, or other recipients of their services. Psychologists' moral standards and conduct are personal matters to the same degree as is true for any other person, except as psychologists' conduct may compromise their professional responsibilities or reduce the public's trust in psychology and psychologists. Psychologists are concerned about the ethical compliance of their colleagues' scientific and professional conduct. When appropriate, they consult with colleagues in order to prevent or avoid unethical conduct.

Principle D: Respect for People's Rights and Dignity

Psychologists accord appropriate respect to the fundamental rights, dignity, and worth of all people. They respect the rights of individuals to privacy, confidentiality, self-determination, and autonomy, mindful that legal and other obligations may lead to inconsistency and conflict with the exercise of these rights. Psychologists are aware of cultural, individual, and role differences, including those due to age, gender, race, ethnicity, national origin, religion, sexual orientation, disability, language, and socioeconomic status. Psychologists try to eliminate the effect on their work of biases based on those factors, and they do not knowingly participate in or condone unfair discriminatory practices.

Principle E: Concern for Others' Welfare

Psychologists seek to contribute to the welfare of those with whom they interact professionally. In their professional actions, psychologists weigh the welfare and rights of their patients or clients, students, supervisees, human research participants, and other affected persons, and the welfare of animal subjects of research. When conflicts occur among psychologists' obligations or concerns, they attempt to resolve these conflicts and to perform their roles in a responsible fashion that avoids or minimizes harm. Psychologists are sensitive to real and ascribed differences in power between themselves and others, and they do not exploit or mislead other people during or after professional relationships.

Principle F: Social Responsibility

Psychologists are aware of their professional and scientific responsibilities to the community and the society in which they work and live. They apply and make public their knowledge of psychology in order to contribute to human welfare. Psychologists are concerned about and work to mitigate the causes of human suffering. When undertaking research, they strive to advance human welfare and the science of psychology. Psychologists try to avoid misuse of their work. Psychologists comply with the law and encourage the development of law and social policy that serve the interests of their patients and clients and the public. They are encouraged to contribute a portion of their professional time for little or no personal advantage.

ETHICAL STANDARDS

1. General Standards

These General Standards are potentially applicable to the professional and scientific activities of all psychologists.

1.01 Applicability of the Ethics Code
The activity of a psychologist subject to the Ethics Code may be reviewed under these Ethical Standards only if the activity is part of his or her work-related functions or the activity is psychological in nature. Personal activities having no connection to or effect on psychological roles are not subject to the Ethics Code.

1.02 Relationship of Ethics and Law
If psychologists' ethical responsibilities conflict with law, psychologists make known their commitment to the Ethics Code and take steps to resolve the conflict in a responsible manner.

1.03 Professional and Scientific Relationship
Psychologists provide diagnostic, therapeutic, teaching, research, supervisory, consultative, or other psychological services only in the context of a defined professional or scientific relationship or role. (See also Standards 2.01, Evaluation, Diagnosis, and Interventions in Professional Context, and 7.02, Forensic Assessments.)

1.04 Boundaries of Competence
   (a) Psychologists provide services, teach, and conduct research only within the boundaries of their competence, based on their education, training, supervised experience, or appropriate professional experience.
   (b) Psychologists provide services, teach, or conduct research in new areas or involving new techniques only after first undertaking appropriate study, training, supervision, and/or consultation from persons who are competent in those areas or techniques.
   (c) In those emerging areas in which generally recognized standards for preparatory training do not yet exist, psychologists nevertheless take reasonable steps to ensure

the competence of their work and to protect patients, clients, students, research participants, and others from harm.

1.05 Maintaining Expertise

Psychologists who engage in assessment, therapy, teaching, research, organizational consulting, or other professional activities maintain a reasonable level of awareness of current scientific and professional information in their fields of activity, and undertake ongoing efforts to maintain competence in the skills they use.

1.06 Basis for Scientific and Professional Judgments

Psychologists rely on scientifically and professionally derived knowledge when making scientific or professional judgments or when engaging in scholarly or professional endeavors.

1.07 Describing the Nature and Results of Psychological Services

(a) When psychologists provide assessment, evaluation, treatment, counseling, supervision, teaching, consultation, research, or other psychological services to an individual, a group, or an organization, they provide, using language that is reasonably understandable to the recipient of those services, appropriate information beforehand about the nature of such services and appropriate information later about results and conclusions. (See also Standard 2.09, Explaining Assessment Results.)

(b) If psychologists will be precluded by law or by organizational roles from providing such information to particular individuals or groups, they so inform those individuals or groups at the outset of the service.

1.08 Human Differences

Where differences of age, gender, race, ethnicity, national origin, religion, sexual orientation, disability, language, or socioeconomic status significantly affect psychologists' work concerning particular individuals or groups, psychologists obtain the training, experience, consultation, or supervision necessary to ensure the competence of their services, or they make appropriate referrals.

1.09 Respecting Others

In their work-related activities, psychologists respect the rights of others to hold values, attitudes, and opinions that differ from their own.

1.10 Nondiscrimination

In their work-related activities, psychologists do not engage in unfair discrimination based on age, gender, race, ethnicity, national origin, religion, sexual orientation, disability, socioeconomic status, or any basis proscribed by law.

1.11 Sexual Harassment

(a) Psychologists do not engage in sexual harassment. Sexual harassment is sexual solicitation, physical advances, or verbal or nonverbal conduct that is sexual in nature, that occurs in connection with the psychologist's activities or roles as a psychologist, and that either: (1) is unwelcome, is offensive, or creates a hostile workplace environment, and the psychologist knows or is told this; or (2) is sufficiently severe or intense to be abusive to a reasonable person in the context. Sexual harassment can consist of a single intense or severe act or of multiple persistent or pervasive acts.

(b) Psychologists accord sexual-harassment complainants and respondents dignity and respect. Psychologists do not participate in denying a person academic admittance or advancement, employment, tenure, or promotion, based solely upon their having made, or their being the subject of, sexual-harassment charges. This does not preclude taking action based upon the outcome of such proceedings or consideration of other appropriate information.

1.12 Other Harassment

Psychologists do not knowingly engage in behavior that is harassing or demeaning to persons with whom they interact in their work based on factors such as those persons' age, gender, race, ethnicity, national origin, religion, sexual orientation, disability, language, or socioeconomic status.

1.13 Personal Problems and Conflicts

(a) Psychologists recognize that their personal problems and conflicts may interfere with their effectiveness. Accordingly, they refrain from undertaking an activity when they know or should know that their personal problems are likely to lead to harm to a patient, client, colleague, student, research participant, or other person to whom they may owe a professional or scientific obligation.

(b) In addition, psychologists have an obligation to be alert to signs of, and to obtain assistance for, their personal problems at an early stage, in order to prevent significantly impaired performance.

(c) When psychologists become aware of personal problems that may interfere with their performing work-related duties adequately, they take appropriate measures, such as obtaining professional consultation or assistance, and determine whether they should limit, suspend, or terminate their work-related duties.

1.14 Avoiding Harm

Psychologists take reasonable steps to avoid harming their patients or clients, research participants, students, and others with whom they work, and to minimize harm where it is foreseeable and unavoidable.

1.15 Misuse of Psychologists' Influence

Because psychologists' scientific and professional judgments and actions may affect the lives of others, they are alert to and guard against personal, financial, social, organizational, or political factors that might lead to misuse of their influence.

1.16 Misuse of Psychologists' Work

(a) Psychologists do not participate in activities in which it appears likely that their skills or data will be misused by others, unless corrective mechanisms are available. (See also Standard 7.04, Truthfulness and Candor.)

(b) If psychologists learn of misuse or misrepresentation of their work, they take reasonable steps to correct or minimize the misuse or misrepresentation.

1.17 Multiple Relationships

(a) In many communities and situations, it may not be feasible or reasonable for psychologists to avoid social or other nonprofessional contacts with persons such as patients, clients, students, supervisees, or research participants. Psychologists must always be sensitive to the potential harmful effects of other contacts on their work and on those persons with whom they deal. A psychologist refrains from entering into or promising another personal, scientific, professional, financial, or other relationship with such persons if it appears likely that such a relationship reasonably might impair the psychologist's objectivity or otherwise interfere with the psychologist's effectively performing his or her functions as a psychologist, or might harm or exploit the other party.

(b) Likewise, whenever feasible, a psychologist refrains from taking on professional or scientific obligations when preexisting relationships would create a risk of such harm.

(c) If a psychologist finds that, due to unforeseen factors, a potentially harmful multiple relationship has arisen, the psychologist attempts to resolve it with due regard for the best interests of the affected person and maximal compliance with the Ethics Code.

1.18 Barter (With Patients or Clients)

Psychologists ordinarily refrain from accepting goods, services, or other nonmonetary remuneration from patients or clients in return for psychological services because such arrangements create inherent potential for conflicts, exploita-

tion, and distortion of the professional relationship. A psychologist may participate in bartering only if (1) it is not clinically contraindicated, and (2) the relationship is not exploitative. (See also Standards 1.17, Multiple Relationships, and 1.25, Fees and Financial Arrangements.)

1.19 Exploitative Relationships

(a) Psychologists do not exploit persons over whom they have supervisory, evaluative, or other authority such as students, supervisees, employees, research participants, and clients or patients. (See also Standards 4.05–4.07 regarding sexual involvement with clients or patients.)

(b) Psychologists do not engage in sexual relationships with students or supervisees in training over whom the psychologist has evaluative or direct authority, because such relationships are so likely to impair judgment or be exploitative.

1.20 Consultations and Referrals

(a) Psychologists arrange for appropriate consultations and referrals based principally on the best interests of their patients or clients, with appropriate consent, and subject to other relevant considerations, including applicable law and contractual obligations. (See also Standards 5.01, Discussing the Limits of Confidentiality, and 5.06, Consultations.)

(b) When indicated and professionally appropriate, psychologists cooperate with other professionals in order to serve their patients or clients effectively and appropriately.

(c) Psychologists' referral practices are consistent with law.

1.21 Third-Party Requests for Services

(a) When a psychologist agrees to provide services to a person or entity at the request of a third party, the psychologist clarifies to the extent feasible, at the outset of the service, the nature of the relationship with each party. This clarification includes the role of the psychologist (such as therapist, organizational consultant, diagnostician, or expert witness), the probable uses of the services provided or the information obtained, and the fact that there may be limits to confidentiality.

(b) If there is a foreseeable risk of the psychologist's being called upon to perform conflicting roles because of the involvement of a third party, the psychologist clarifies the nature and direction of his or her responsibilities, keeps all parties appropriately informed as matters develop, and resolves the situation in accordance with this Ethics Code.

1.22 Delegation to and Supervision of Subordinates

(a) Psychologists delegate to their employees, supervisees, and research assistants only those responsibilities that such persons can reasonably be expected to perform competently, on the basis of their education, training, or experience, either independently or with the level of supervision being provided.

(b) Psychologists provide proper training and supervision to their employees or supervisees and take reasonable steps to see that such persons perform services responsibly, competently, and ethically.

(c) If institutional policies, procedures, or practices prevent fulfillment of this obligation, psychologists attempt to modify their role or to correct the situation to the extent feasible.

1.23 Documentation of Professional and Scientific Work

(a) Psychologists appropriately document their professional and scientific work in order to facilitate provision of services later by them or by other professionals, to ensure accountability, and to meet other requirements of institutions or the law.

(b) When psychologists have reason to believe that records of their professional services will be used in legal proceedings involving recipients of or participants in their work, they have a responsibility to create and maintain documentation in the kind of detail and quality that would be consistent with reasonable scrutiny in an adjudicative forum. (See also Standard 7.01, Professionalism, under Forensic Activities.)

1.24 Records and Data

Psychologists create, maintain, disseminate, store, retain, and dispose of records and data relating to their research, practice, and other work in accordance with law and in a manner that permits compliance with the requirements of this Ethics Code. (See also Standard 5.04, Maintenance of Records.)

1.25 Fees and Financial Arrangements

(a) As early as is feasible in a professional or scientific relationship, the psychologist and the patient, client, or other appropriate recipient of psychological services reach an agreement specifying the compensation and the billing arrangements.

(b) Psychologists do not exploit recipients of services or payors with respect to fees.

(c) Psychologists' fee practices are consistent with law.

(d) Psychologists do not misrepresent their fees.

(e) If limitations to services can be anticipated because of limitations in financing, this is discussed with the patient, client, or other appropriate recipient of services as early as is feasible. (See also Standard 4.08, Interruption of Services.)

(f) If the patient, client, or other recipient of services does not pay for services as agreed, and if the psychologist wishes to use collection agencies or legal measures to collect the fees, the psychologist first informs the person that such measures will be taken and provides that person an opportunity to make prompt payment. (See also Standard 5.11, Withholding Records for Nonpayment.)

1.26 Accuracy in Reports to Payors and Funding Sources

In their reports to payors for services or sources of research funding, psychologists accurately state the nature of the research or service provided, the fees or charges, and where applicable, the identity of the provider, the findings, and the diagnosis. (See also Standard 5.05, Disclosures.)

1.27 Referrals and Fees

When a psychologist pays, receives payment from, or divides fees with another professional other than in an employer–employee relationship, the payment to each is based on the services (clinical, consultative, administrative, or other) provided and is not based on the referral itself.

2.   Evaluation, Assessment, or Intervention

2.01 Evaluation, Diagnosis, and Interventions in Professional Context

(a) Psychologists perform evaluations, diagnostic services, or interventions only within the context of a defined professional relationship. (See also Standard 1.03, Professional and Scientific Relationship.)

(b) Psychologists' assessments, recommendations, reports, and psychological diagnostic or evaluative statements are based on information and techniques (including personal interviews of the individual when appropriate) sufficient to provide appropriate substantiation for their findings. (See also Standard 7.02, Forensic Assessments.)

2.02 Competence and Appropriate Use of Assessments and Interventions

(a) Psychologists who develop, administer, score, interpret, or use psychological assessment techniques, interviews, tests, or instruments do so in a manner and for purposes that are appropriate in light of the research on or evidence of the usefulness and proper application of the techniques.

(b) Psychologists refrain from misuse of assessment techniques, interventions, results, and interpretations and take reasonable steps to prevent others from misusing the information these techniques provide. This includes refraining from releasing raw test results or raw data to persons, other than to patients or clients as appropriate, who are not qualified to use such information. (See also Standards 1.02, Relationship of Ethics and Law, and 1.04, Boundaries of Competence.)

2.03 Test Construction

Psychologists who develop and conduct research with tests and other assessment techniques use scientific procedures and current professional knowledge for test design, standardization, validation, reduction or elimination of bias, and recommendations for use.

2.04 Use of Assessment in General and With Special Populations

(a) Psychologists who perform interventions or administer, score, interpret, or use assessment techniques are familiar with the reliability, validation, and related standardization or outcome studies of, and proper applications and uses of, the techniques they use.

(b) Psychologists recognize limits to the certainty with which diagnoses, judgments, or predictions can be made about individuals.

(c) Psychologists attempt to identify situations in which particular interventions or assessment techniques or norms may not be applicable or may require adjustment in administration or interpretation because of factors such as individuals' gender, age, race, ethnicity, national origin, religion, sexual orientation, disability, language, or socioeconomic status.

2.05 Interpreting Assessment Results

When interpreting assessment results, including automated interpretations, psychologists take into account the various test factors and characteristics of the person being assessed that might affect psychologists' judgments or reduce the accuracy of their interpretations. They indicate any significant reservations they have about the accuracy or limitations of their interpretations.

2.06 Unqualified Persons

Psychologists do not promote the use of psychological assessment techniques by unqualified persons. (See also Standard 1.22, Delegation to and Supervision of Subordinates.)

2.07 Obsolete Tests and Outdated Test Results

(a) Psychologists do not base their assessment or intervention decisions or recommendations on data or test results that are outdated for the current purpose.

(b) Similarly, psychologists do not base such decisions or recommendations on tests and measures that are obsolete and not useful for the current purpose.

2.08 Test Scoring and Interpretation Services

(a) Psychologists who offer assessment or scoring procedures to other professionals accurately describe the purpose, norms, validity, reliability, and applications of the procedures and any special qualifications applicable to their use.

(b) Psychologists select scoring and interpretation services (including automated services) on the basis of evidence of the validity of the program and procedures as well as on other appropriate considerations.

(c) Psychologists retain appropriate responsibility for the appropriate application, interpretation, and use of assessment instruments, whether they score and interpret such tests themselves or use automated or other services.

2.09 Explaining Assessment Results

Unless the nature of the relationship is clearly explained to the person being assessed in advance and precludes provision of an explanation of results (such as in some organizational consulting, preemployment or security screenings, and forensic evaluations), psychologists ensure that an explanation of the results is provided using language that is reasonably understandable to the person assessed or to another legally authorized person on behalf of the client. Regardless of whether the scoring and interpretation are done by the psychologist, by assistants, or by automated or other outside services, psychologists take reasonable steps to ensure that appropriate explanations of results are given.

2.10 Maintaining Test Security

Psychologists make reasonable efforts to maintain the integrity and security of tests and other assessment techniques consistent with law, contractual obligations, and in a manner that permits compliance with the requirements of this Ethics Code. (See also Standard 1.02, Relationship of Ethics and Law.)

3. Advertising and Other Public Statements

3.01 Definition of Public Statements

Psychologists comply with this Ethics Code in public statements relating to their professional services, products, or publications or to the field of psychology. Public statements include but are not limited to paid or unpaid advertising, brochures, printed matter, directory listings, personal resumes or curricula vitae, interviews or comments for use in media, statements in legal proceedings, lectures and public oral presentations, and published materials.

3.02 Statements by Others

(a) Psychologists who engage others to create or place public statements that promote their professional practice, products, or activities retain professional responsibility for such statements.

(b) In addition, psychologists make reasonable efforts to prevent others whom they do not control (such as employers, publishers, sponsors, organizational clients, and representatives of the print or broadcast media) from making deceptive statements concerning psychologists' practice or professional or scientific activities.

(c) If psychologists learn of deceptive statements about their work made by others, psychologists make reasonable efforts to correct such statements.

(d) Psychologists do not compensate employees of press, radio, television, or other communication media in return for publicity in a news item.

(e) A paid advertisement relating to the psychologist's activities must be identified as such, unless it is already apparent from the context.

3.03   Avoidance of False or Deceptive Statements

(a) Psychologists do not make public statements that are false, deceptive, misleading, or fraudulent, either because of what they state, convey, or suggest or because of what they omit, concerning their research, practice, or other work activities or those of persons or organizations with which they are affiliated. As examples (and not in limitation) of this standard, psychologists do not make false or deceptive statements concerning (1) their training, experience, or competence; (2) their academic degrees; (3) their credentials; (4) their institutional or association affiliations; (5) their services; (6) the scientific or clinical basis for, or results or degree of success of, their services; (7) their fees; or (8) their publications or research findings. (See also Standards 6.15, Deception in Research, and 6.18, Providing Participants With Information About the Study.)

(b) Psychologists claim as credentials for their psychological work, only degrees that (1) were earned from a regionally accredited educational institution or (2) were the basis for psychology licensure by the state in which they practice.

3.04   Media Presentations

When psychologists provide advice or comment by means of public lectures, demonstrations, radio or television programs, prerecorded tapes, printed articles, mailed material, or other media, they take reasonable precautions to ensure that (1) the statements are based on appropriate psychological literature and practice, (2) the statements are otherwise consistent with this Ethics Code, and (3) the recipients of the information are not encouraged to infer

that a relationship has been established with them personally.

3.05   Testimonials

Psychologists do not solicit testimonials from current psychotherapy clients or patients or other persons who because of their particular circumstances are vulnerable to undue influence.

3.06   In-Person Solicitation

Psychologists do not engage, directly or through agents, in uninvited in-person solicitation of business from actual or potential psychotherapy patients or clients or other persons who because of their particular circumstances are vulnerable to undue influence. However, this does not preclude attempting to implement appropriate collateral contacts with significant others for the purpose of benefiting an already engaged therapy patient.

4.   Therapy

4.01   Structuring the Relationship

(a) Psychologists discuss with clients or patients as early as is feasible in the therapeutic relationship appropriate issues, such as the nature and anticipated course of therapy, fees, and confidentiality. (See also Standards 1.25, Fees and Financial Arrangements, and 5.01, Discussing the Limits of Confidentiality.)

(b) When the psychologist's work with clients or patients will be supervised, the above discussion includes that fact, and the name of the supervisor, when the supervisor has legal responsibility for the case.

(c) When the therapist is a student intern, the client or patient is informed of that fact.

(d) Psychologists make reasonable efforts to answer patients' questions and to avoid apparent misunderstandings about therapy. Whenever possible, psychologists provide oral and/or written information, using language that is reasonably understandable to the patient or client.

4.02   Informed Consent to Therapy

(a) Psychologists obtain appropriate informed consent to therapy or related procedures, using language that is reasonably understandable to participants. The content of informed consent will vary depending on many circumstances; however, informed consent generally implies that the person (1) has the capacity to consent, (2) has been informed of significant information concerning the procedure, (3) has freely and without undue influence expressed consent, and (4) consent has been appropriately documented.

(b) When persons are legally incapable of giving informed consent, psychologists obtain informed permission from a legally authorized person, if such substitute consent is permitted by law.

(c) In addition, psychologists (1) inform those persons who are legally incapable of giving informed consent about the proposed interventions in a manner commensurate with the persons' psychological capacities, (2) seek their assent to those interventions, and (3) consider such persons' preferences and best interests.

4.03 Couple and Family Relationships

(a) When a psychologist agrees to provide services to several persons who have a relationship (such as husband and wife or parents and children), the psychologist attempts to clarify at the outset (1) which of the individuals are patients or clients and (2) the relationship the psychologist will have with each person. This clarification includes the role of the psychologist and the probable uses of the services provided or the information obtained. (See also Standard 5.01, Discussing the Limits of Confidentiality.)

(b) As soon as it becomes apparent that the psychologist may be called on to perform potentially conflicting roles (such as marital counselor to husband and wife, and then witness for one party in a divorce proceeding), the psychologist attempts to clarify and adjust, or withdraw from, roles appropriately. (See also Standard 7.03, Clarification of Role, under Forensic Activities.)

4.04 Providing Mental Health Services to Those Served by Others

In deciding whether to offer or provide services to those already receiving mental health services elsewhere, psychologists carefully consider the treatment issues and the potential patient's or client's welfare. The psychologist discusses these issues with the patient or client, or another legally authorized person on behalf of the client, in order to minimize the risk of confusion and conflict, consults with the other service providers when appropriate, and proceeds with caution and sensitivity to the therapeutic issues.

4.05 Sexual Intimacies With Current Patients or Clients

Psychologists do not engage in sexual intimacies with current patients or clients.

4.06 Therapy With Former Sexual Partners

Psychologists do not accept as therapy patients or clients persons with whom they have engaged in sexual intimacies.

4.07 Sexual Intimacies With Former Therapy Patients

(a) Psychologists do not engage in sexual intimacies with a former therapy patient or client for at least two years after cessation or termination of professional services.

(b) Because sexual intimacies with a former therapy patient or client are so frequently harmful to the patient or client, and because such intimacies undermine public confidence in the psychology profession and thereby deter the public's use of needed services, psychologists do not engage in sexual intimacies with former therapy patients and clients even after a two-year interval except in the most unusual circumstances. The psychologist who engages in such activity after the two years following cessation or termination of treatment bears the burden of demonstrating that there has been no exploitation, in light of all relevant factors, including (1) the amount of time that has passed since therapy terminated, (2) the nature and duration of the therapy, (3) the circumstances of termination, (4) the patient's or client's personal history, (5) the patient's or client's current mental status, (6) the likelihood of adverse impact on the patient or client and others, and (7) any statements or actions made by the therapist during the course of therapy suggesting or inviting the possibility of a posttermination sexual or romantic relationship with the patient or client. (See also Standard 1.17, Multiple Relationships.)

4.08 Interruption of Services

(a) Psychologists make reasonable efforts to plan for facilitating care in the event that psychological services are interrupted by factors such as the psychologist's illness, death, unavailability, or relocation or by the client's relocation or financial limitations. (See also Standard 5.09, Preserving Records and Data.)

(b) When entering into employment or contractual relationships, psychologists provide for orderly and appropriate resolution of responsibility for patient or client care in the event that the employment or contractual relationship ends, with paramount consideration given to the welfare of the patient or client.

4.09 Terminating the Professional Relationship

(a) Psychologists do not abandon patients or clients. (See also Standard 1.25e, under Fees and Financial Arrangements.)

(b) Psychologists terminate a professional relationship when it becomes reasonably clear that the patient or client no longer needs the service, is not benefiting, or is being harmed by continued service.

(c) Prior to termination for whatever reason, except where precluded by the patient's or client's conduct, the psychologist discusses the patient's or client's views and needs, provides appropriate pretermination counseling, suggests alternative service providers as appropriate, and takes other reasonable steps to facilitate transfer of responsibility to another provider if the patient or client needs one immediately.

5. Privacy and Confidentiality

These Standards are potentially applicable to the professional and scientific activities of all psychologists.

5.01   Discussing the Limits of Confidentiality

(a) Psychologists discuss with persons and organizations with whom they establish a scientific or professional relationship (including, to the extent feasible, minors and their legal representatives) (1) the relevant limitations on confidentiality, including limitations where applicable in group, marital, and family therapy or in organizational consulting, and (2) the foreseeable uses of the information generated through their services.

(b) Unless it is not feasible or is contraindicated, the discussion of confidentiality occurs at the outset of the relationship and thereafter as new circumstances may warrant.

(c) Permission for electronic recording of interviews is secured from clients and patients.

5.02   Maintaining Confidentiality

Psychologists have a primary obligation and take reasonable precautions to respect the confidentiality rights of those with whom they work or consult, recognizing that confidentiality may be established by law, institutional rules, or professional or scientific relationships. (See also Standard 6.26, Professional Reviewers.)

5.03   Minimizing Intrusions on Privacy

(a) In order to minimize intrusions on privacy, psychologists include in written and oral reports, consultations, and the like, only information germane to the purpose for which the communication is made.

(b) Psychologists discuss confidential information obtained in clinical or consulting relationships, or evaluative data concerning patients, individual or organizational clients, students, research participants, supervisees, and employees, only for appropriate scientific or professional purposes and only with persons clearly concerned with such matters.

5.04   Maintenance of Records

Psychologists maintain appropriate confidentiality in creating, storing, accessing, transferring, and disposing of records under their control, whether these are written, automated, or in any other medium. Psychologists maintain and dispose of records in accordance with law and in a manner that permits compliance with the requirements of this Ethics Code.

5.05   Disclosures

(a) Psychologists disclose confidential information without the consent of the individual only as mandated by law, or where permitted by law for a valid purpose, such as (1) to provide needed professional services to the patient or the individual or organizational client, (2) to obtain appropriate professional consultations, (3) to protect the patient or client or others from harm, or (4) to obtain payment for services, in which instance disclosure is limited to the minimum that is necessary to achieve the purpose.

(b) Psychologists also may disclose confidential information with the appropriate consent of the patient or the individual or organizational client (or of another legally authorized person on behalf of the patient or client), unless prohibited by law.

5.06   Consultations

When consulting with colleagues, (1) psychologists do not share confidential information that reasonably could lead to the identification of a patient, client, research participant, or other person or organization with whom they have a confidential relationship unless they have obtained the prior consent of the person or organization or the disclosure cannot be avoided, and (2) they share information only to the extent necessary to achieve the purposes of the consultation. (See also Standard 5.02, Maintaining Confidentiality.)

5.07   Confidential Information in Databases

(a) If confidential information concerning recipients of psychological services is to be entered into databases or systems of records available to persons whose access has not been consented to by the recipient, then psychologists use coding or other techniques to avoid the inclusion of personal identifiers.

(b) If a research protocol approved by an institutional review board or similar body requires the inclusion of personal identifiers, such identifiers are deleted before the information is made accessible to persons other than those of whom the subject was advised.

(c) If such deletion is not feasible, then before psychologists transfer such data to others or review such data collected by others, they take reasonable steps to determine that appropriate consent of personally identifiable individuals has been obtained.

5.08   Use of Confidential Information for Didactic or Other Purposes

(a) Psychologists do not disclose in their writings, lectures, or other public media, confidential, personally identifiable information concerning their patients, individual or organizational clients, students, research participants, or other recipients of their services that they obtained during the course of their work, unless the person or organization has consented in writing or unless there is other ethical or legal authorization for doing so.

(b) Ordinarily, in such scientific and professional presentations, psychologists disguise confidential information concerning such persons or organizations so that they are not individually identifiable to others and so that discussions do not cause harm to subjects who might identify themselves.

5.09   Preserving Records and Data

A psychologist makes plans in advance so that confidentiality of records and data is protected in the event of the psychologist's death, incapacity, or withdrawal from the position or practice.

5.10   Ownership of Records and Data

Recognizing that ownership of records and data is governed by legal principles, psychologists take reasonable and lawful steps so that records and data remain available to the extent needed to serve the best interests of patients, individual or organizational clients, research participants, or appropriate others.

5.11   Withholding Records for Nonpayment

Psychologists may not withhold records under their control that are requested and imminently needed for a patient's or client's treatment solely because payment has not been received, except as otherwise provided by law.

6.   Teaching, Training Supervision, Research, and Publishing

6.01   Design of Education and Training Programs

Psychologists who are responsible for education and training programs seek to ensure that the programs are competently designed, provide the proper experiences, and meet the requirements for licensure, certification, or other goals for which claims are made by the program.

6.02   Descriptions of Education and Training Programs

(a) Psychologists responsible for education and training programs seek to ensure that there is a current and accurate description of the program content, training goals and objectives, and requirements that must be met for satisfactory completion of the program. This information must be made readily available to all interested parties.

(b) Psychologists seek to ensure that statements concerning their course outlines are accurate and not misleading, particularly regarding the subject matter to be covered, bases for evaluating progress, and the nature of course experiences. (See also Standard 3.03, Avoidance of False or Deceptive Statements.)

(c) To the degree to which they exercise control, psychologists responsible for announcements, catalogs, brochures, or advertisements describing workshops, seminars, or other non-degree-granting educational programs ensure that they accurately describe the audience for which the program is intended, the educational objectives, the presenters, and the fees involved.

6.03   Accuracy and Objectivity in Teaching

(a) When engaged in teaching or training, psychologists present psychological information accurately and with a reasonable degree of objectivity.

(b) When engaged in teaching or training, psychologists recognize the power they hold over students or supervisees and therefore make reasonable efforts to avoid engaging in conduct that is personally demeaning to students or supervisees. (See also Standards 1.09, Respecting Others, and 1.12, Other Harassment.)

6.04   Limitation on Teaching

Psychologists do not teach the use of techniques or procedures that require specialized training, licensure, or expertise, including but not limited to hypnosis, biofeedback, and projective techniques, to individuals who lack the prerequisite training, legal scope of practice, or expertise.

6.05   Assessing Student and Supervisee Performance

(a) In academic and supervisory relationships, psychologists establish an appropriate process for providing feedback to students and supervisees.

(b) Psychologists evaluate students and supervisees on the basis of their actual performance on relevant and established program requirements.

6.06   Planning Research

(a) Psychologists design, conduct, and report research in accordance with recognized standards of scientific competence and ethical research.

(b) Psychologists plan their research so as to minimize the possibility that results will be misleading.

(c) In planning research, psychologists consider its ethical acceptability under the Ethics Code. If an ethical issue is unclear, psychologists seek to resolve the issue through consultation with institutional review boards, animal care and use committees, peer consultations, or other proper mechanisms.

(d) Psychologists take reasonable steps to implement appropriate protections for the rights and welfare of human participants, other persons affected by the research, and the welfare of animal subjects.

6.07   Responsibility

(a) Psychologists conduct research competently and with due concern for the dignity and welfare of the participants.

(b) Psychologists are responsible for the ethical conduct of research conducted by them or by others under their supervision or control.

(c) Researchers and assistants are permitted to perform only those tasks for which they are appropriately trained and prepared.

(d) As part of the process of development and implementation of research projects, psychologists consult those with expertise concerning any special population under investigation or most likely to be affected.

6.08   Compliance With Law and Standards

Psychologists plan and conduct research in a manner consistent with federal and state law and regulations, as well as professional standards governing the conduct of research, and particularly those standards governing research with human participants and animal subjects.

6.09   Institutional Approval

Psychologists obtain from host institutions or organizations appropriate approval prior to conducting research, and they provide accurate information about their research proposals. They conduct the research in accordance with the approved research protocol.

6.10   Research Responsibilities

Prior to conducting research (except research involving only anonymous surveys, naturalistic observations, or similar research), psychologists enter into an agreement with participants that clarifies the nature of the research and the responsibilities of each party.

6.11   Informed Consent to Research

(a) Psychologists use language that is reasonably understandable to research participants in obtaining their appropriate informed consent (except as provided in Standard 6.12, Dispensing With Informed Consent). Such informed consent is appropriately documented.

(b) Using language that is reasonably understandable to participants, psychologists inform participants of the nature of the research; they inform participants that they are free to participate or to decline to participate or to withdraw from the research; they explain the foreseeable consequences of declining or withdrawing; they inform participants of significant factors that may be expected to influence their willingness to participate (such as risks, discomfort, adverse effects, or limitations on confidentiality, except as provided in Standard 6.15, Deception in Research); and they explain other aspects about which the prospective participants inquire.

(c) When psychologists conduct research with individuals such as students or subordinates, psychologists take special care to protect the prospective participants from adverse consequences of declining or withdrawing from participation.

(d) When research participation is a course requirement or opportunity for extra credit, the prospective participant is given the choice of equitable alternative activities.

(e) For persons who are legally incapable of giving informed consent, psychologists nevertheless (1) provide an appropriate explanation, (2) obtain the participant's assent, and (3) obtain appropriate permission from a legally authorized person, if such substitute consent is permitted by law.

6.12   Dispensing With Informed Consent

Before determining that planned research (such as research involving only anonymous questionnaires, naturalistic observations, or certain kinds of archival research) does not require the informed consent of research participants, psychologists consider applicable regulations and institutional review board requirements, and they consult with colleagues as appropriate.

6.13   Informed Consent in Research Filming or Recording

Psychologists obtain informed consent from research participants prior to filming or recording them in any form, unless the research involves simply naturalistic observations in public places and it is not anticipated that the recording will be used in a manner that could cause personal identification or harm.

6.14   Offering Inducements for Research Participants

(a) In offering professional services as an inducement to obtain research participants, psychologists make clear the nature of the services, as well as the risks, obligations, and limitations. (See also Standard 1.18, Barter [With Patients or Clients].)

(b) Psychologists do not offer excessive or inappropriate financial or other inducements to obtain research participants, particularly when it might tend to coerce participation.

6.15   Deception in Research

(a) Psychologists do not conduct a study involving deception unless they have determined that the use of deceptive techniques is justified by the study's prospective scientific, educational, or applied value and that equally effective alternative procedures that do not use deception are not feasible.

(b) Psychologists never deceive research participants about significant aspects that would affect their willingness to participate, such as physical risks, discomfort, or unpleasant emotional experiences.

(c) Any other deception that is an integral feature of the design and conduct of an experiment must be explained to participants as early as is feasible, preferably at the conclusion of their participation, but no later than at the conclusion of the research. (See also Standard 6.18, Providing Participants With Information About the Study.)

6.16   Sharing and Utilizing Data

Psychologists inform research participants of their anticipated sharing or further use of personally identifiable research data and of the possibility of unanticipated future uses.

6.17   Minimizing Invasiveness

In conducting research, psychologists interfere with the participants or milieu from which data are collected only in a manner that is warranted by an appropriate research design and that is consistent with psychologists' roles as scientific investigators.

6.18   Providing Participants With Information About the Study

(a) Psychologists provide a prompt opportunity for participants to obtain appropriate information about the nature, results, and conclusions of the research, and psychologists attempt to correct any misconceptions that participants may have.

(b) If scientific or humane values justify delaying or withholding this information, psychologists take reasonable measures to reduce the risk of harm.

6.19   Honoring Commitments

Psychologists take reasonable measures to honor all commitments they have made to research participants.

6.20   Care and Use of Animals in Research

(a) Psychologists who conduct research involving animals treat them humanely.

(b) Psychologists acquire, care for, use, and dispose of animals in compliance with current federal, state, and local laws and regulations, and with professional standards.

(c) Psychologists trained in research methods and experienced in the care of laboratory animals supervise all procedures involving animals and are responsible for ensuring appropriate consideration of their comfort, health, and humane treatment.

(d) Psychologists ensure that all individuals using animals under their supervision have received instruction in research methods and in the care, maintenance, and handling of the species being used, to the extent appropriate to their role.

(e) Responsibilities and activities of individuals assisting in a research project are consistent with their respective competencies.

(f) Psychologists make reasonable efforts to minimize the discomfort, infection, illness, and pain of animal subjects.

(g) A procedure subjecting animals to pain, stress, or privation is used only when an alternative procedure is unavailable and the goal is justified by its prospective scientific, educational, or applied value.

(h) Surgical procedures are performed under appropriate anesthesia; techniques to avoid infection and minimize pain are followed during and after surgery.

(i) When it is appropriate that the animal's life be terminated, it is done rapidly, with an effort to minimize pain, and in accordance with accepted procedures.

6.21   Reporting of Results

(a) Psychologists do not fabricate data or falsify results in their publications.

(b) If psychologists discover significant errors in their published data, they take reasonable steps to correct such errors in a correction, retraction, erratum, or other appropriate publication means.

6.22   Plagiarism

Psychologists do not present substantial portions or elements of another's work or data as their own, even if the other work or data source is cited occasionally.

6.23   Publication Credit

(a) Psychologists take responsibility and credit, including authorship credit, only for work they have actually performed or to which they have contributed.

(b) Principal authorship and other publication credits accurately reflect the relative scientific or professional contributions of the individuals involved, regardless of their relative status. Mere possession of an institutional position, such as Department Chair, does not justify authorship credit. Minor contributions to the research or to the writing for publications are appropriately acknowledged, such as in footnotes or in an introductory statement.

(c) A student is usually listed as principal author on any multiple-authored article that is substantially based on the student's dissertation or thesis.

6.24 Duplicate Publication of Data

Psychologists do not publish, as original data, data that have been previously published. This does not preclude republishing data when they are accompanied by proper acknowledgment.

6.25 Sharing Data

After research results are published, psychologists do not withhold the data on which their conclusions are based from other competent professionals who seek to verify the substantive claims through reanalysis and who intend to use such data only for that purpose, provided that the confidentiality of the participants can be protected and unless legal rights concerning proprietary data preclude their release.

6.26 Professional Reviewers

Psychologists who review material submitted for publication, grant, or other research proposal review respect the confidentiality of and the proprietary rights in such information of those who submitted it.

7. Forensic Activities

7.01 Professionalism

Psychologists who perform forensic functions, such as assessments, interviews, consultations, reports, or expert testimony, must comply with all other provisions of this Ethics Code to the extent that they apply to such activities. In addition, psychologists base their forensic work on appropriate knowledge of and competence in the areas underlying such work, including specialized knowledge concerning special populations. (See also Standards 1.06, Basis for Scientific and Professional Judgments; 1.08, Human Differences; 1.15, Misuse of Psychologists' Influence; and 1.23, Documentation of Professional and Scientific Work.)

7.02 Forensic Assessments

(a) Psychologists' forensic assessments, recommendations, and reports are based on information and techniques (including personal interviews of the individual, when appropriate) sufficient to provide appropriate substantiation for their findings. (See also Standards 1.03, Professional and Scientific Relationship; 1.23, Documentation of Professional and Scientific Work; 2.01, Evaluation, Diagnosis, and Interventions in Professional Context; and 2.05, Interpreting Assessment Results.)

(b) Except as noted in (c), below, psychologists provide written or oral forensic reports or testimony of the psychological characteristics of an individual only after they have conducted an examination of the individual adequate to support their statements or conclusions.

(c) When, despite reasonable efforts, such an examination is not feasible, psychologists clarify the impact of their limited information on the reliability and validity of their reports and testimony, and they appropriately limit the nature and extent of their conclusions or recommendations.

7.03 Clarification of Role

In most circumstances, psychologists avoid performing multiple and potentially conflicting roles in forensic matters. When psychologists may be called on to serve in more than one role in a legal proceeding—for example, as consultant or expert for one party or for the court and as a fact witness—they clarify role expectations and the extent of confidentiality in advance to the extent feasible, and thereafter as changes occur, in order to avoid compromising their professional judgment and objectivity and in order to avoid misleading others regarding their role.

7.04 Truthfulness and Candor

(a) In forensic testimony and reports, psychologists testify truthfully, honestly, and candidly and, consistent with applicable legal procedures, describe fairly the bases for their testimony and conclusions.

(b) Whenever necessary to avoid misleading, psychologists acknowledge the limits of their data or conclusions.

7.05 Prior Relationships

A prior professional relationship with a party does not preclude psychologists from testifying as fact witnesses or from testifying to their services to the extent permitted by applicable law. Psychologists appropriately take into account ways in which the prior relationship might affect their professional objectivity or opinions and disclose the potential conflict to the relevant parties.

7.06 Compliance With Law and Rules

In performing forensic roles, psychologists are reasonably familiar with the rules governing their roles. Psychologists are aware of the occasionally competing demands placed upon them by these principles and the requirements of the court system, and attempt to resolve these conflicts by making known their commitment to this Ethics Code and taking steps to resolve the conflict in a responsible manner. (See also Standard 1.02, Relationship of Ethics and Law.)

8. Resolving Ethical Issues

8.01 Familiarity With Ethics Code

Psychologists have an obligation to be familiar with this Ethics Code, other applicable ethics codes, and their application to psychologists' work. Lack of awareness or misunderstanding of an ethical standard is not itself a defense to a charge of unethical conduct.

8.02 Confronting Ethical Issues

When a psychologist is uncertain whether a particular situation or course of action would violate this Ethics Code, the psychologist ordinarily consults with other psychologists knowledgeable about ethical issues, with state or national psychology ethics committees, or with other appropriate authorities in order to choose a proper response.

8.03 Conflicts Between Ethics and Organizational Demands

If the demands of an organization with which psychologists are affiliated conflict with this Ethics Code, psychologists clarify the nature of the conflict, make known their commitment to the Ethics Code, and to the extent feasible, seek to resolve the conflict in a way that permits the fullest adherence to the Ethics Code.

8.04 Informal Resolution of Ethical Violations

When psychologists believe that there may have been an ethical violation by another psychologist, they attempt to resolve the issue by bringing it to the attention of that individual if an informal resolution appears appropriate and the intervention does not violate any confidentiality rights that may be involved.

8.05 Reporting Ethical Violations

If an apparent ethical violation is not appropriate for informal resolution under Standard 8.04 or is not resolved properly in that fashion, psychologists take further action appropriate to the situation, unless such action conflicts with confidentiality rights in ways that cannot be resolved. Such action might include referral to state or national committees on professional ethics or to state licensing boards.

8.06 Cooperating With Ethics Committees

Psychologists cooperate in ethics investigations, proceedings, and resulting requirements of the APA or any affiliated state psychological association to which they belong. In doing so, they make reasonable efforts to resolve any issues as to confidentiality. Failure to cooperate is itself an ethics violation.

8.07 Improper Complaints

Psychologists do not file or encourage the filing of ethics complaints that are frivolous and are intended to harm the respondent rather than to protect the public.

CODE OF ETHICS
OF THE
NATIONAL ASSOCIATION OF SOCIAL WORKERS

As adopted by the Delegate Assembly of August 1996
West Virginia Chapter
National Association of Social Workers
1608 Virginia Street East
Charleston, West Virginia 25311

OVERVIEW

The National Association of Social Workers Code of Ethics is intended to serve as a guide to the everyday professional conduct of social workers. This code includes four sections. Section one, "Preamble," summarizes the social work profession's mission and core values. Section two, "Purpose of the Code of Ethics," provides an overview of the Code's main functions and a brief guide for dealing with ethical issues or dilemmas in social work practice. Section three, "Ethical Principles," presents broad ethical principles, based on social work's core values, that inform social work practice. The final section, "Ethical Standards," includes specific ethical standards to guide social workers' conduct and to provide a basis for adjudication.

PREAMBLE

The primary mission of the social work profession is to enhance human well-being and help meet basic human needs of all people, with particular attention to the needs and empowerment of people who are vulnerable, oppressed and living in poverty. An historic and defining feature of social work is the profession's focus on individual well-being in a social context and the well-being of society. Fundamental to social work is attention to the environmental forces that create, contribute to, and address problems in living.

Social workers promote social justice and social change with and on behalf of clients. 'Clients' is used inclusively to refer to individuals, families, groups, organizations, and communities. Social workers are sensitive to cultural and ethnic diversity and strive to end discrimination, oppression, poverty, and other forms of social injustice. These activities may be in the form of direct practice, community organizing, supervision, consultation, administration, advocacy, social and political action, policy development and implementation, education, and research and evaluation. Social workers seek to enhance the capacity of people to address their own needs. Social workers also seek to promote the responsiveness of organizations, communities, and other social institutions to individuals' needs and social problems.

The mission of the social work profession is rooted in a set of core values. These core values, embraced by social workers throughout the profession's history, are the foundation of social work's unique purpose and perspective:

• Service
• Social justice
• Dignity and worth of the person
• Importance of human relationships
• Integrity
• Competence

The constellation of these core values reflects what is unique to the social work profession. Core values, and the principles which flow from them, must be balanced within the context and complexity of the human experience.

PURPOSE OF THE CODE OF ETHICS

Professional ethics are at the core of social work. The profession has an obligation to articulate its basic values, ethical principles, and ethical standards. The NASW Code of Ethics sets forth these values, principles, and standards to guide social workers' conduct. The code of ethics is relevant to all social workers and social work students, regardless of their professional functions, the settings in which they work, or the populations they serve.

This NASW Code of Ethics serves six purposes:

- The code identifies core values on which social work's mission is based.
- The code summarizes broad ethical principles that reflect the profession's core values and establishes a set of specific ethical standards that should be used to guide social work practice.
- The code of ethics is designed to help social workers identify relevant considerations when professional obligations conflict or ethical uncertainties arise.
- The code provides ethical standards to which the general public can hold the social work profession accountable.
- The code socializes practitioners new to the field to social work's mission, values, ethical principles, and ethical standards.
- The code articulates standards that the social work profession itself can use to assess whether social workers have engaged in unethical conduct. NASW has formal procedures to adjudicate ethics complaints filed against its members.* In subscribing to this code social workers are required to cooperate in its implementation, participate in NASW adjudication proceedings, and abide by any NASW disciplinary rulings or sanctions based on it.

This code offers a set of values, principles, and standards to guide decision making and conduct when ethical issues arise. It does not provide a set of rules that prescribe how social workers should act in all situations. Specific applications of the code must take into account the context in which it is being considered and the possibility of conflicts among the code's values, principles, and standards. Ethical responsibilities flow from all human relationships, from the personal and familial to the social and professional.

Further, the code of ethics does not specify which values, principles, and standards are most important and ought to outweigh others in instances when they conflict. Reasonable differences of opinion can and do exist among social workers with respect to the ways in which values, ethical principles, and ethical standards should be rank-ordered when they conflict. Ethical decision making in a given situation must apply the informed judgment of the individual social worker and should also consider how the issues would be judged in a peer review process where the ethical standards of the profession would be applied.

Ethical decision making is a process. There are many instances in social work where simple answers are not available to resolve complex ethical issues. Social workers should take into consideration all the values, principles, and standards in this code that are relevant to any situation in which ethical judgment is warranted. Social workers' decisions and actions should be consistent with the spirit as well as the letter of this code.

In addition to this code, there are many other sources of information about ethical thinking that may be useful. Social workers should consider ethical theory and principles generally, social work theory and research, laws, regulations, agency policies, and other relevant codes of ethics, recognizing that among codes of ethics social workers should consider the NASW Code of Ethics as their primary source. Social workers also should be aware of the impact on ethical decision-making of their clients' and their own personal values, cultural and religious beliefs, and practices. They should be aware of any conflicts between personal and professional values and deal with them responsibly. For additional guidance social workers should consult relevant literature on professional ethics and ethical decision making, and seek appropriate consultation when faced with ethical dilemmas. This may involve consultation with an agency-based or social work organization's ethics committee, regulatory body, knowledgeable colleagues, supervisors, or legal counsel.

Instances may arise where social workers' ethical obligations conflict with agency policies, relevant laws or regulations. When such conflicts occur, social workers must make a responsible effort to resolve the conflict in a manner that is consistent with the values, principles, and standards expressed in this code. If a reasonable resolution of the conflict does not appear possible, social workers should seek proper consultation before making a decision.

This code of ethics is to be used by NASW and by other individuals, agencies, organizations, and bodies (such as licensing and regulatory boards, professional liability insurance providers, courts of law, agency boards of directors, government agencies, and other professional groups) that choose to adopt it or use it as a frame of reference. Violation of standards in this code does not automatically imply legal liability or violation of the law. Such determination can only be made in the context of legal and judicial proceedings. Alleged violations of the code would be subject to a peer review process. Such processes are generally separate from legal or administrative procedures and insulated from legal review or proceedings in order to allow the profession to counsel and/or discipline its own members.

---

*For information on NASW adjudication procedures, see *NASW Procedures for the Adjudication of Grievances.*

A code of ethics cannot guarantee ethical behavior. Moreover, a code of ethics cannot resolve all ethical issues or disputes, or capture the richness and complexity involved in striving to make responsible choices within a moral community. Rather a code of ethics sets forth values, ethical principles and ethical standards to which professionals aspire and by which their actions can be judged. Social workers' ethical behavior should result from their personal commitment to engage in ethical practice. This code reflects the commitment of all social workers to uphold the profession's values and to act ethically. Principles and standards must be applied by individuals of good character who discern moral questions and, in good faith, seek to make reliable ethical judgments.

ETHICAL PRINCIPLES

The following broad ethical principles are based on social work's core values of: service, social justice, dignity and worth of the person, importance of human relationships, integrity, and competence. These principles set forth ideals to which all social workers should aspire.

VALUE: *Service*
Ethical Principle: *Social workers' primary goal is to help people in need and to address social problems.*

Social workers elevate service to others above self-interest. Social workers draw on their knowledge, values, and skills to help people in need and to address social problems. Social workers are encouraged to volunteer some portion of their professional skills with no expectation of significant financial return (pro bono service).

VALUE: *Social Justice*
Ethical Principle: *Social workers challenge social injustice.*

Social workers pursue social change, particularly with and on behalf of vulnerable and oppressed individuals and groups of people. Social workers' social change efforts are focused primarily on issues of poverty, unemployment, discrimination, and other forms of social injustice. These activities seek to promote sensitivity to and knowledge about oppression, and cultural and ethnic diversity. Social workers strive to ensure equality of opportunity, access to needed information, services, resources, and meaningful participation in decision making for all people.

VALUE: *Dignity and Worth of the Person*
Ethical Principle: *Social workers respect the inherent dignity and worth of the person.*

Social workers treat each person in a caring and respectful fashion, mindful of individual differences and cultural and ethnic diversity. Social workers promote clients' socially responsible self-determination. Social workers seek to enhance clients' capacity and opportunity to change and to address their own needs. Social workers are cognizant of their dual responsibility to clients and to the broader society. They seek to resolve conflicts between clients' and the broader society's interests in a socially responsible manner consistent with the values, ethical principles, and ethical standards of the profession.

VALUE: *Importance of Human Relationships*
Ethical Principle: *Social workers recognize the central importance of human relationships.*

Social workers understand that relationships between and among people are an important vehicle for change. Social workers engage people as partners in the helping process. Social workers seek to strengthen relationships among people in a purposeful effort to promote, restore, maintain, and enhance the well-being of individuals, families, social groups, organizations, and communities.

VALUE: *Integrity*
Ethical Principle: *Social workers behave in a trustworthy manner.*

Social workers are continually aware of the profession's mission, values, ethical principles, and ethical standards, and practice in a manner consistent with them. Social workers act honestly and responsibly and promote ethical practices on the part of the organizations with which they are affiliated.

VALUE: *Competence*
Ethical Principle: *Social workers practice within their areas of competence and develop and enhance their professional expertise.*

Social workers continually strive to increase their professional knowledge and skills and to apply them in practice. Social workers should aspire to contribute to the knowledge base of the profession.

ETHICAL STANDARDS

The following ethical standards are relevant to the professional activities of all social workers. These standards concern: (1) social workers' ethical responsibilities to clients, (2) social workers' ethical responsibilities to colleagues, (3) social workers' ethical responsibilities in practice settings, (4) social workers' ethical responsibilities as professionals, (5) social workers' ethical responsibilities to the profession, and (6) social workers' ethical responsibilities to the broader society.

Some of the standards that follow are enforceable guidelines for professional conduct and some are more aspirational in nature. The extent to which each standard is enforceable is a matter of professional judgment to be exercised by those responsible for reviewing alleged violations of ethical standards.

1. Social Workers' Ethical Responsibilities to Clients

1.01 Commitment to Clients
Social workers' primary responsibility is to promote the well-being of clients. In general, clients' interests are primary. However, social workers' responsibility to the larger society or specific legal obligations may on limited occasions supersede the loyalty owed clients and clients should be so advised. (Examples include when a social worker is required by law to report that a client has abused a child or has threatened to harm self or others.)

1.02 Self-Determination
Social workers respect and promote the right of clients to self-determination and assist clients in their efforts to identify and clarify their goals. Social workers may limit clients' right to self-determination when, in their professional judgment, clients' actions or potential actions pose a serious, foreseeable, and imminent risk to themselves or others.

1.03 Informed Consent
a. Social workers should provide services to clients only in the context of a professional relationship based, when appropriate, on valid informed consent. Social workers should use clear and understandable language to inform clients of the purpose of the service, risks related to the service, limits to service because of the requirements of a third-party payor, relevant costs, reasonable alternatives, clients' right to refuse or withdraw consent, and the time frame covered by the consent. Social workers should provide clients with an opportunity to ask questions.

b. In instances where clients are not literate or have difficulty understanding the primary language used in the practice setting, social workers should take steps to ensure clients' comprehension. This may include providing clients with a detailed verbal explanation or arranging for a qualified interpreter and/or translator whenever possible.

c. In instances where clients lack the capacity to provide informed consent, social workers should protect clients' interests by seeking permission from an appropriate third party, informing clients consistent with their level of understanding. In such instances social workers should seek to ensure that the third party acts in a manner consistent with

clients' wishes and interests. Social workers should take reasonable steps to enhance such clients' ability to give informed consent.

d. In instances where clients are receiving services involuntarily, social workers should provide information about the nature and extent of services, and of the extent of clients' right to refuse service.

e. Social workers who provide services via electronic mediums (such as computers, telephone, radio, and television) should inform recipients of the limitations and risks associated with such services.

f. Social workers should obtain clients' informed consent before audiotaping or videotaping clients, or permitting third party observation of clients who are receiving services.

1.04 Competence
a. Social workers should provide services and represent themselves as competent only within the boundaries of their education, training, license, certification, consultation received, supervised experience, or other relevant professional experience.

b. Social workers should provide services in substantive areas or use intervention techniques or approaches that are new to them only after engaging in appropriate study, training, consultation, and/or supervision from persons who are competent in those interventions or techniques.

c. When generally recognized standards do not exist with respect to an emerging area of practice, social workers should exercise careful judgment and take responsible steps—including appropriate education, research, training, consultation, and supervision—to ensure the competence of their work and to protect clients from harm.

1.05 Cultural Competence and Social Diversity
a. Social workers should understand culture and its function in human behavior and society, recognizing the strengths that exist in all cultures.

b. Social workers should have a knowledge base of their clients' cultures and be able to demonstrate competence in the provision of services that are sensitive to clients' culture and to differences among people and cultural groups.

c. Social workers should obtain education about and seek to understand the nature of social diversity and oppression with respect to race, ethnicity, national origin, color, sex, sexual orientation, age, marital status, political belief, religion, and mental or physical disability.

1.06 Conflicts of Interest
a. Social workers should be alert to and avoid conflicts of interest that interfere with the exercise of professional

discretion and impartial judgment. Social workers should inform clients when a real or potential conflict of interest arises and take reasonable steps to resolve the issue in a manner that makes the clients' interests primary and protects clients' interests to the greatest extent possible. In some cases, protecting clients' interests may require termination of the professional relationship with proper referral of the client.

b.   Social workers should not take unfair advantage of any professional relationship or exploit others to further their personal, religious, political, or business interests.

c.   Social workers should not engage in dual or multiple relationships with clients or former clients in which there is a risk of exploitation or potential harm to the client. In instances when dual or multiple relationships are unavoidable, social workers should take steps to protect clients and are responsible for setting clear, appropriate, and culturally sensitive boundaries. (Dual or multiple relationships occur when social workers relate to clients in more than one relationship, whether professional, social, or business. Dual or multiple relationships can occur simultaneously or consecutively.)

d.   When social workers provide services to two or more persons who have a relationship with each other (for example, couples, family members), social workers should clarify with all parties which individuals will be considered clients and the nature of social workers' professional obligations to the various individuals who are receiving services. Social workers who anticipate a conflict of interest among the individuals receiving services, or who anticipate having to perform in potentially conflicting roles (for example, when a social worker is asked to testify in a child custody dispute or divorce proceedings involving clients), should clarify their role with the parties involved and take appropriate action to minimize any conflict of interest.

1.07 Privacy and Confidentiality

a.   Social workers should respect clients' right to privacy. Social workers should not solicit private information from clients unless it is essential to providing service or conducting social work evaluation or research. Once private information is shared, standards of confidentiality apply.

b.   Social workers may disclose confidential information when appropriate with a valid consent from a client, or a person legally authorized to consent on behalf of a client.

c.   Social workers should protect the confidentiality of all information obtained in the course of professional service, except for compelling professional reasons. The general expectation that social workers will keep information confidential does not apply when disclosure is necessary to prevent serious, foreseeable, and imminent harm to a client

or other identifiable person or when laws or regulations require disclosure without a client's consent. In all instances, social workers should disclose the least amount of confidential information necessary to achieve the desired purpose; only information that is directly relevant to the purpose for which the disclosure is made should be revealed.

d.   Social workers should inform clients, to the extent possible, about the disclosure of confidential information and the potential consequences and, when feasible, before the disclosure is made. This applies whether social workers disclose confidential information as a result of a legal requirement or based on client consent.

e.   Social workers should discuss with clients and other interested parties the nature of confidentiality and limitations of clients' right to confidentiality. Social workers should review with clients circumstances where confidential information may be requested and where disclosure of confidential information may be legally required. This discussion should occur as soon as possible in the social worker-client relationship and as needed throughout the course of the relationship.

f.   When social workers provide counseling services to families, couples, or groups, social workers should seek agreement among the parties involved concerning each individual's right to confidentiality and obligation to preserve the confidentiality of information shared by others. Social workers should inform participants in family, couples, or group counseling that social workers cannot guarantee that all participants will honor such agreements.

g.   Social workers should inform clients involved in family, couples, marital, or group counseling of the social worker's, employer's, and/or agency's policy concerning the social worker's disclosure of confidential information among the parties involved in the counseling.

h.   Social workers should not disclose confidential information to third party payors, unless clients have authorized such disclosure.

i.   Social workers should not discuss confidential information in any setting unless privacy can be assured. Social workers should not discuss confidential information in public or semi-public areas (such as hallways, waiting rooms, elevators, and restaurants).

j.   Social workers should protect the confidentiality of clients during legal proceedings to the extent permitted by law. When a court of law or other legally authorized body orders social workers to disclose confidential or privileged information without a client's consent and such disclosure could cause harm to the client, social workers should request that the court withdraw or limit the order as narrowly as possible and/or maintain the records under seal, unavailable for public inspection.

k. Social workers should protect the confidentiality of clients when responding to requests from members of the media.

l. Social workers should protect the confidentiality of clients' written and electronic records and other sensitive information. Social workers should take reasonable steps to ensure that clients' records are stored in a secure location and that clients' records are not available to others who are not authorized to have access.

m. Social workers should take precautions to ensure and maintain the confidentiality of information transmitted to other parties through the use of computers, electronic mail, facsimile machines, telephones and telephone answering machines, and other electronic or computer technology. Disclosure of identifying information should be avoided whenever possible.

n. Social workers should transfer or dispose of clients' records in a manner that protects clients' confidentiality and is consistent with state statutes governing records and social work licensure.

o. Social workers should take reasonable precautions to protect client confidentiality in the event of the social worker's termination of practice, incapacitation, or death.

p. Social workers should not disclose identifying information when discussing clients for teaching or training purposes, unless the client has consented to disclosure of confidential information.

q. Social workers should not disclose identifying information when discussing clients with consultants, unless the client has consented to disclosure of confidential information or there is a compelling need for such disclosure.

r. Social workers should protect the confidentiality of deceased clients consistent with the preceding standards.

1.08 Access to Records

a. Social workers should provide clients with reasonable access to records concerning them. Social workers who are concerned that clients' access to their records could cause serious misunderstanding or harm to the client should provide assistance in interpreting the records and consultation with the client regarding the records. Social workers should limit client access to social work records, or portions of clients' records, only in exceptional circumstances when there is compelling evidence that such access would cause serious harm to the client. Both the client's request and the rationale for withholding some or all of the record should be documented in the client's file.

b. When providing clients with access to their records, social workers should take steps to protect the confidentiality of other individuals identified or discussed in such records.

1.09 Sexual Relationships

a. Social workers should under no circumstances engage in sexual activities or sexual contact with current clients, whether such contact is consensual or forced.

b. Social workers should not engage in sexual activities or sexual contact with clients' relatives or other individuals with whom clients maintain a close, personal relationship where there is a risk of exploitation or potential harm to the client. Sexual activity or sexual contact with clients' relatives or other individuals with whom clients maintain a personal relationship has the potential to be harmful to the client and may make it difficult for the social worker and client to maintain appropriate professional boundaries. Social workers—not their clients, their clients' relatives or other individuals with whom the client maintains a personal relationship—assume the full burden for setting clear, appropriate and culturally sensitive boundaries.

c. Social workers should not engage in sexual activities or sexual contact with former clients because of the potential for harm to the client. If social workers engage in conduct contrary to this prohibition or claim that an exception to this prohibition is warranted due to extraordinary circumstances, it is social workers—not their clients—who assume the full burden of demonstrating that the former client has not been exploited, coerced, or manipulated, intentionally or unintentionally.

d. Social workers should not provide clinical services to individuals with whom they have had a prior sexual relationship. Providing clinical services to a former sexual partner has the potential to be harmful to the individual and is likely to make it difficult for the social worker and individual to maintain appropriate professional boundaries.

1.10 Physical Contact

Social workers should not engage in physical contact with clients where there is a possibility of psychological harm to the client as a result of the contact (such as cradling or caressing clients). Social workers who engage in appropriate physical contact with clients are responsible for setting clear, appropriate, and culturally sensitive boundaries that govern such physical contact.

1.11 Sexual Harassment

Social workers should not sexually harass clients. Sexual harassment includes sexual advances, sexual solicitation, requests for sexual favors, and other verbal or physical conduct of a sexual nature.

1.12 Derogatory Language

Social workers should not use derogatory language in their written or verbal communications to or about clients. Social workers should use accurate and respectful language in all communications to and about clients.

1.13 Payment for Services

a.   When setting fees, social workers should ensure that the fees are fair, reasonable, and commensurate with the service performed. Consideration should be given to the client's ability to pay.

b.   Social workers should avoid accepting goods or services from clients as payment for professional services. Bartering arrangements, particularly involving services, create the potential for conflicts of interest, exploitation, and inappropriate boundaries in social workers' relationships with clients. Social workers should explore and may participate in bartering only in very limited circumstances where it can be demonstrated that such arrangements are an accepted practice among professionals in the local community, considered to be essential for the provision of service, negotiated without coercion and entered into at the client's initiative and with the client's informed consent. Social workers who accept goods or services from clients as payment for professional services assume the full burden of demonstrating that this arrangement will not be detrimental to the client or the professional relationship.

c.   Social workers should not solicit a private fee or other remuneration for providing services to clients who are entitled to such available services through the social workers' employer or agency.

1.14 Clients Who Lack Decision-Making Capacity

When social workers act on behalf of clients who lack the capacity to make informed decisions, social workers should take reasonable steps to safeguard the interests and rights of those clients.

1.15 Interruption of Services

Social workers should make reasonable efforts to ensure continuity of services in the event that they are interrupted by factors such as unavailability, relocation, illness, disability, or death.

1.16 Termination of Services

a.   Social workers should terminate services to clients, and professional relationships with them, when such services

and relationships are no longer required or no longer serve the clients' needs or interests.

b.   Social workers should take reasonable steps to avoid abandoning clients who are still in need of services. Social workers should withdraw services precipitously only under unusual circumstances, giving careful consideration to all factors in the situation and taking care to minimize possible adverse effects. Social workers should assist in making appropriate arrangements for continuation of services when necessary.

c.   Social workers in fee-for-service settings may terminate services to clients who are not paying an overdue balance if the financial contractual arrangements have been made clear to the client, if the client does not pose an imminent danger to self or others, and if the clinical and other consequences of the current non-payment have been addressed and discussed with the client.

d.   Social workers should not terminate services to pursue a social, financial, or sexual relationship with a client.

e.   Social workers who anticipate the termination or interruption of services to clients should notify clients promptly and seek the transfer, referral, or continuation of services in relation to the clients' needs and preferences.

f.   Social workers who are leaving an employment setting should inform clients of appropriate options for the continuation of service and their benefits and risks.

2. Social Workers' Ethical Responsibilities to Colleagues

2.01 Respect

a.   Social workers should treat colleagues with respect, and represent accurately and fairly the qualifications, views, and obligations of colleagues.

b.   Social workers should avoid unwarranted negative criticism of colleagues with clients or with other professionals. Unwarranted negative criticism may include demeaning comments that refer to colleagues' level of competence or to individuals' attributes such as race, ethnicity, national origin, color, age, religion, sex, sexual orientation, marital status, political belief, mental or physical disability, or any other preference, personal characteristic, or status.

c.   Social workers should cooperate with social work colleagues and with colleagues of other professions when it serves the well-being of clients.

2.02 Confidentiality with Colleagues

Social workers should respect confidential information shared by colleagues in the course of their professional relationships and transactions. Social workers should ensure that such colleagues understand social workers' obligation to respect confidentiality and any exceptions related to it.

2.03 Interdisciplinary Collaboration

a. Social workers who are members of an interdisciplinary team should participate in and contribute to decisions that affect the well-being of clients by drawing on the perspectives, values, and experiences of the social work profession. Professional and ethical obligations of the interdisciplinary team as a whole and of its individual members should be clearly established.

b. Social workers for whom a team decision raises ethical concerns should attempt to resolve the disagreement through appropriate channels. If the disagreement cannot be resolved social workers should pursue other avenues to address their concerns, consistent with client well-being.

2.04 Disputes Involving Colleagues

a. Social workers should not take advantage of a dispute between a colleague and employer to obtain a position or otherwise advance the social worker's own interests.

b. Social workers should not exploit clients in a dispute with a colleague or engage clients in any inappropriate discussion of a social worker's conflict with a colleague.

2.05 Consultation

a. Social workers should seek advice and counsel of colleagues whenever such consultation is in the best interests of clients.

b. Social workers should keep informed of colleagues' areas of expertise and competencies. Social workers should seek consultation only from colleagues who have demonstrated knowledge, expertise and competence related to the subject of the consultation.

c. When consulting with colleagues about clients, social workers should disclose the least amount of information necessary to achieve the purposes of the consultation.

2.06 Referral for Services

a. Social workers should refer clients to other professionals when other professionals' specialized knowledge or expertise is needed to serve clients fully, or when social workers believe they are not being effective or making reasonable progress with clients and additional service is required.

b. Social workers who refer clients to other professionals should take appropriate steps to facilitate an orderly transfer of responsibility. Social workers who refer clients to

other professionals should disclose, with clients' consent, all pertinent information to the new service providers.

c. Social workers are prohibited from giving or receiving payment for a referral when no professional service is provided by the referring social worker.

2.07 Sexual Relationships

a. Social workers who function as supervisors or educators should not engage in sexual activities or contact with supervisees, students, trainees, or other colleagues over whom they exercise professional authority.

b. Social workers should avoid engaging in sexual relationships with colleagues where there is potential for a conflict of interest. Social workers who become involved in, or anticipate becoming involved in, a sexual relationship with a colleague have a duty to transfer professional responsibilities, when necessary, in order to avoid a conflict of interest.

2.08 Sexual Harassment

Social workers should not engage in any sexual harassment of supervisees, students, trainees, or colleagues. Sexual harassment includes sexual advances, sexual solicitation, requests for sexual favors, and other verbal or physical conduct of a sexual nature.

2.09 Impairment of Colleagues

a. Social workers who have direct knowledge of a social work colleague's impairment which is due to personal problems, psychosocial distress, substance abuse, or mental health difficulties, and which interferes with practice effectiveness, should consult with that colleague when feasible and assist the colleague in taking remedial action.

b. Social workers who believe that a social work colleague's impairment interferes with practice effectiveness and that the colleague has not taken adequate steps to address the impairment should take action through appropriate channels established by employers, agencies, NASW, licensing and regulatory bodies, and other professional organizations.

2.10 Incompetence of Colleagues

a. Social workers who have direct knowledge of a social work colleague's incompetence should consult with that colleague when feasible and assist the colleague in taking remedial action.

b. Social workers who believe that a social work colleague is incompetent and has not taken adequate steps to address the incompetence should take action through appropriate channels established by employers, agencies, NASW, licensing and regulatory bodies, and other professional organizations.

2.11 Unethical Conduct of Colleagues

a. Social workers should take adequate measures to discourage, prevent, expose, and correct the unethical conduct of colleagues.

b. Social workers should be knowledgeable about established policies and procedures for handling concerns about colleagues' unethical behavior. Social workers should be familiar with national, state, and local procedures for handling ethics complaints. These include policies and procedures created by NASW, licensing and regulatory bodies, employers, agencies, and other professional organizations.

c. Social workers who believe that a colleague has acted unethically should seek resolution by discussing their concerns with the colleague when feasible and when such discussion is likely to be productive.

d. When necessary, social workers who believe that a colleague has acted unethically should take action through appropriate formal channels (such as contacting a state licensing board or regulatory body, NASW committee on inquiry, or other professional ethics committees).

e. Social workers should defend and assist colleagues who are unjustly charged with unethical conduct.

3. Social Workers' Ethical Responsibilities in Practice Settings

3.01 Supervision and Consultation

a. Social workers who provide supervision or consultation should have the necessary knowledge and skill to supervise or consult appropriately and should do so only within their areas of knowledge and competence.

b. Social workers who provide supervision or consultation are responsible for setting clear, appropriate, and culturally sensitive boundaries.

c. Social workers should not engage in any dual or multiple relationships with supervisees in which there is a risk of exploitation of or potential harm to the supervisee.

d. Social workers who provide supervision should evaluate supervisees' performance in a manner that is fair and respectful.

3.02 Education and Training

a. Social workers who function as educators, field instructors for students, or trainers should provide instruction only within their areas of knowledge and competence, and should provide instruction based on the most current information and knowledge available in the profession.

b. Social workers who function as educators or field instructors for students should evaluate students' performance in a manner that is fair and respectful.

c. Social workers who function as educators or field instructors for students should take reasonable steps to ensure that clients are routinely informed when services are being provided by students.

d. Social workers who function as educators or field instructors for students should not engage in any dual or multiple relationships with students in which there is a risk of exploitation or potential harm to the student. Social work educators and field instructors are responsible for setting clear, appropriate, and culturally sensitive boundaries.

3.03 Performance Evaluation

Social workers who have the responsibility for evaluating the performance of others should fulfill such responsibility in a fair and considerate manner, and on the basis of clearly stated criteria.

3.04 Client Records

a. Social workers should take reasonable steps to ensure that documentation in records is accurate and reflective of the services provided.

b. Social workers should include sufficient and timely documentation in records to facilitate the delivery of services and to ensure continuity of services provided to clients in the future.

c. Social workers' documentation should protect clients' privacy to the extent that is possible and appropriate, and should include only that information that is directly relevant to the delivery of services.

d. Social workers should store records following the termination of service to ensure reasonable future access. Records should be maintained for the number of years required by state statutes or relevant contracts.

3.05 Billing

Social workers should establish and maintain billing practices that accurately reflect the nature and extent of services provided, and by whom the service was provided in the practice setting.

3.06 Client Transfer

a. When an individual who is receiving services from another agency or colleague contacts a social worker for services, the social worker should carefully consider the client's needs before agreeing to provide services. In order to minimize possible confusion and conflict, social workers should discuss with potential clients the nature of their current relationship with other service providers and the implications, including possible benefits or risks, of entering into a relationship with a new service provider.

b. If a new client has been served by another agency or colleague, social workers should discuss with the client whether consultation with the previous service provider is in the client's best interest.

3.07 Administration

a. Social work administrators should advocate within and outside of their agencies for adequate resources to meet clients' needs.

b. Social workers should advocate for resource allocation procedures that are open and fair. When not all clients' needs can be met, an allocation procedure should be developed that is non-discriminatory and based on appropriate and consistently applied principles.

c. Social workers who are administrators should take reasonable steps to ensure that adequate agency or organizational resources are available to provide appropriate staff supervision.

d. Social work administrators should take reasonable steps to ensure that the working environment for which they are responsible is consistent with and encourages compliance with the NASW Code of Ethics. Social work administrators should take reasonable steps to eliminate any conditions in their organizations that violate, interfere with, or discourage compliance with the Code of Ethics.

3.08 Continuing Education and Staff Development

Social work administrators and supervisors should take reasonable steps to provide or arrange for continuing education and staff development for all staff for whom they are responsible. Continuing education and staff development should address current knowledge and emerging developments related to social work practice and ethics.

3.09 Commitments to Employers

a. Social workers generally should adhere to commitments made to employers and employing organizations.

b. Social workers should work to improve employing agencies' policies and procedures, and the efficiency and effectiveness of their services.

c. Social workers should take reasonable steps to ensure that employers are aware of social workers' ethical obligations as set forth in the NASW Code of Ethics and their implications for social work practice.

d. Social workers should not allow an employing organization's policies, procedures, regulations, or administrative orders to interfere with their ethical practice of social work. Social workers should take reasonable steps to ensure that their employing organizations' practices are consistent with the NASW Code of Ethics.

e. Social workers should act to prevent and eliminate discrimination in the employing organization's work assignments and in its employment policies and practices.

f. Social workers should accept employment or arrange student field placements only in organizations where fair personnel practices are exercised.

g. Social workers should be diligent stewards of the resources of their employing organizations, wisely conserving funds where appropriate, and never misappropriating funds or using them for unintended purposes.

3.10 Labor-Management Disputes

a. Social workers may engage in organized action, including the formation of and participation in labor unions, to improve services to clients and working conditions.

b. The actions of social workers who are involved in labor-management disputes, job actions, or labor strikes should be guided by the profession's values, ethical principles, and ethical standards. Reasonable differences of opinion exist among social workers concerning their primary obligation as professionals during an actual or threatened labor strike or job action. Social workers should carefully examine relevant issues and their possible impact on clients before deciding on a course of action.

4. Social Workers' Ethical Responsibilities as Professionals

4.01 Competence

a. Social workers should accept responsibility or employment only on the basis of existing competence or the intention to acquire the necessary competence.

b. Social workers should strive to become and remain proficient in professional practice and the performance of professional functions. Social workers should critically examine, and keep current with, emerging knowledge relevant to social work. Social workers should routinely review professional literature and participate in continuing education relevant to social work practice and social work ethics.

c. Social workers should base practice on recognized knowledge, including empirically-based knowledge, relevant to social work and social work ethics.

4.02 Discrimination

Social workers should not practice, condone, facilitate, or collaborate with any form of discrimination on the basis of race, ethnicity, national origin, color, age, religion, sex, sexual orientation, marital status, political belief, or mental or physical disability.

4.03 Private Conduct

Social workers' should not permit their private conduct to interfere with their ability to fulfill their professional responsibilities.

4.04 Dishonesty, Fraud, and Deception

Social workers should not participate in, condone, or be associated with dishonesty, fraud, or deception.

4.05 Impairment

a. Social workers should not allow their own personal problems, psychosocial distress, legal problems, substance abuse, or mental health difficulties to interfere with their professional judgment and performance or jeopardize the best interests of those for whom they have a professional responsibility.

b. Social workers whose personal problems, psychosocial distress, legal problems, substance abuse, or mental health difficulties interfere with their professional judgment and performance should immediately seek consultation and take appropriate remedial action by seeking professional help, making adjustments in workload, terminating practice, or taking any other steps necessary to protect clients and others.

4.06 Misrepresentation

a. Social workers should make clear distinctions between statements made and actions engaged in as a private individual and as a representative of the social work profession, a professional social work organization, or of the social worker's employing agency.

b. Social workers who speak on behalf of professional social work organizations should accurately represent the official and authorized positions of the organizations.

c. Social workers should ensure that their representations to clients, agencies, and the public of professional qualifications, credentials, education, competence, affiliations, services provided, or results to be achieved are accurate. Social workers should claim only those relevant professional credentials they actually possess and take steps to correct any inaccuracies or misrepresentations of their credentials by others.

4.07 Solicitations

a. Social workers should not engage in uninvited solicitation of potential clients who, because of their circumstances, are vulnerable to undue influence, manipulation, or coercion.

b. Social workers should not engage in solicitation of testimonial endorsements (including solicitation of consent to use a client's prior statement as a testimonial endorsement) from current clients or from other persons who, because of their particular circumstances, are vulnerable to undue influence.

4.08 Acknowledging Credit

a. Social workers should take responsibility and credit, including authorship credit, only for work they have actually performed and to which they have contributed.

b. Social workers should honestly acknowledge the work of and the contributions made by others.

5. Social Workers' Ethical Responsibilities to the Social Work Profession

5.01 Integrity of the Profession

a. Social workers should work toward the maintenance and promotion of high standards of practice.

b. Social workers should uphold and advance the values, ethics, knowledge, and mission of the profession. Social workers should protect, enhance, and improve the integrity of the profession through appropriate study and research, active discussion, and responsible criticism of the profession.

c. Social workers should contribute time and professional expertise to activities that promote respect for the value, integrity, and competence of the social work profession. These activities may include teaching, research, consultation, service, legislative testimony, presentations in the community and participation in their professional organizations.

d. Social workers should contribute to the knowledge base of social work and share with colleagues their knowledge related to practice, research, and ethics. Social workers should seek to contribute to the profession's literature and to share their knowledge at professional meetings and conferences.

e. Social workers should act to prevent the unauthorized and unqualified practice of social work.

5.02 Evaluation and Research

a. Social workers should monitor and evaluate policies, the implementation of programs, and practice interventions.

b. Social workers should promote and facilitate evaluation and research in order to contribute to the development of knowledge.

c. Social workers should critically examine and keep current with emerging knowledge relevant to social work and fully utilize evaluation and research evidence in their professional practice.

d. Social workers engaged in evaluation or research should consider carefully possible consequences and should follow guidelines developed for the protection of evaluation and research participants. Appropriate institutional review boards should be consulted.

e. Social workers engaged in evaluation or research should obtain voluntary and written informed consent from participants, when appropriate, without any implied or actual deprivation or penalty for refusal to participate, without undue inducement to participate, and with due regard for participants' well-being, privacy and dignity. Informed consent should include information about the nature, extent, and duration of the participation requested and disclosure of the risks and benefits of participation in the research.

f. When evaluation or research participants are incapable of giving informed consent, social workers should provide an appropriate explanation to them, obtain the participant's assent, and obtain consent from an appropriate proxy.

g. Social workers should never design or conduct evaluation or research that does not use consent procedures, such as certain forms of naturalistic observation and/or archival research, unless rigorous and responsible review of the research has found it to be justified because of its prospective scientific yield, educational, or applied value and unless equally effective alternative procedures that do not involve waiver of consent are not feasible.

h. Social workers should inform participants of their rights to withdraw from evaluation and research at any time without penalty.

i. Social workers should take appropriate steps to ensure that participants in evaluation and research have access to appropriate supportive services.

j. Social workers engaged in evaluation or research should protect participants from unwarranted physical or mental distress, harm, danger, or deprivation.

k. Social workers engaged in the evaluation of services should discuss collected information only for professional purposes and only with persons professionally concerned with this information.

l. Social workers engaged in evaluation or research should ensure the anonymity or confidentiality of participants and the data obtained from them. Social workers should inform participants of any limits of confidentiality, the measures that will be taken to ensure confidentiality, and when any records containing research data will be destroyed.

m. Social workers who report evaluation and research results should protect participants' confidentiality by omitting identifying information unless proper consent has been obtained authorizing disclosures.

n. Social workers should report evaluation and research findings accurately. They should not fabricate or falsify results and should take steps to correct any errors later found in published data using standard publication methods.

o. Social workers engaged in evaluation or research should be alert to and avoid conflicts of interest and dual relationships with participants, should inform participants when a real or potential conflict of interest arises, and should take steps to resolve the issue in a manner that makes participants' interests primary.

p. Social workers should educate themselves, their students, and colleagues about responsible research practices.

6. Social Workers' Ethical Responsibilities to the Broader Society

6.01 Social Welfare
Social workers should promote the general welfare of society, from local to global levels, and the development of people, their communities, and their environment. Social workers should advocate for living conditions conducive to the fulfillment of basic human needs and promote social, economic, political, and cultural values and institutions that are compatible with the realization of social justice.

6.02 Public Participation
Social workers should facilitate informed participation by the public in shaping social policies and institutions.

6.03 Public Emergencies
Social workers should provide appropriate professional services in public emergencies, to the greatest extent possible.

6.04 Social and Political Action
a. Social workers should engage in social and political action that seeks to ensure that all persons have equal access to the resources, employment, services, and opportunities that they require in order to meet their basic human needs and to develop fully. Social workers should be aware of the impact of the political arena on practice, and should advocate for changes in policy and legislation to improve social conditions in order to meet basic human needs and promote social justice.

b. Social workers should act to expand choice and opportunity for all persons, with special regard for vulnerable, disadvantaged, oppressed, and exploited persons and groups.

c. Social workers should promote conditions that encourage respect for the diversity of cultures and social diversity within the United States and globally. Social workers should promote policies and practices that demonstrate respect for difference, support the expansion of cultural knowledge and resources, advocate for programs and institutions that demonstrate cultural competence, and promote policies that safeguard the rights of and confirm equity and social justice for all people.

d. Social workers should act to prevent and eliminate domination, exploitation, and discrimination against any person, group, or class on the basis of race, ethnicity, national origin, color, age, religion, sex, sexual orientation, marital status, political belief, mental or physical disability, or any other preference, personal characteristic, or status.

MANUALS FOR EMPIRICALLY VALIDATED TREATMENTS

A Project of the Task Force on Psychological Interventions Division of Clinical Psychology, American Psychological Association
William C. Sanderson, PhD & Sheila Woody, PhD (Editors)
June 1995

Development of this resource

The Division 12 (Clinical Psychology) Task Force on Promotion and Dissemination of Psychological Interventions issued a report in October of 1993 establishing criteria for judging whether a treatment may be considered to be empirically validated as well as forming an initial list of well-established and probably efficacious treatments. Please see that report (Task Force on Promotion and Dissemination of Psychological Procedures, 1995) for complete details. In order to facilitate the dissemination of these treatments, the Task Force on Psychological Interventions has now collected a list of manuals detailing the well-established treatments. (One exception will be noted by careful readers: Cognitive Behavior Therapy for Bulimia was inadvertently omitted from the initial Task Force list but is included in this resource.)

Of course, this list will be updated as new treatments are empirically evaluated. In addition, several treatments that are in fact empirically validated but were overlooked in the initial report will be added to the updated version. We want to emphasize that this list is inclusive, not exclusive.

In order to compile this resource listing of manuals, we wrote to leading investigators in the respective areas of treatment research, particularly those whose work formed the basis for judging a particular treatment to be efficacious. These investigators provided citations for those published manuals.

Many of them offered to provide copies of unpublished manuals to other clinicians, although requests must be accompanied by a check to cover the costs of photocopying and postage.

What qualifies as a manual? In building this resource, we attempted to locate materials that provide sufficient detail to allow a trained clinician to replicate the treatment. Of course, no treatment manual is adequate in the absence of solid theoretical grounding and supervised training in the particular approach. Recognizing this, we have also included, when available, information about training in these approaches. We specifically excluded conference workshops as a training resource, because these workshops typically do not offer the opportunity for supervised experience.

Comments and feedback

This listing is intended as a growing resource for researchers and clinicians. We welcome your feedback on the materials listed here, as well as your suggestions for materials we have overlooked.

Please contact either of us:

William C. Sanderson, Ph.D.
Department of Psychiatry
Albert Einstein College of Medicine
Montefiore Medical Center
Bronx, NY 10467-2490
e-mail: SANDERSO@aecom.yu.edu

Sheila Woody, Ph.D.
Department of Psychology
Yale University
P.O. Box 208205
New Haven, CT 06520
e-mail: SWOODY@minerva.cis.yale.edu

Reference:

Task Force on Promotion and Dissemination of Psychological Procedures. (1995). Training in and dissemination of empirically-validated psychological treatments: Report and recommendations: the Clinical Psychologist, 48, 3–23.

BULIMIA

Cognitive Behavioral Therapy, Treatment References/Manuals:

Fairburn, C. G. (1985). Cognitive-behavioral treatment for bulimia. In D. M. Garner & P. E. Garfinkel (Eds.), *Handbook of psychotherapy for anorexia nervosa and bulimia*. New York: Plenum Press.

Fairburn, C. G., Marcus, M. D., & Wilson, G. T. (1993). Cognitive-behavioral therapy for binge eating and bulimia nervosa. In C. G. Fairburn & G. T. Wilson (Eds.), *Binge eating: nature, assessment, and treatment*. New York: Guilford Press.

Interpersonal Therapy, Treatment References/Manuals:

Fairburn, C. G. (1993). Interpersonal psychotherapy for bulimia nervosa. In G. L. Klerman & M. M.. Weissman (Eds.), *New applications of interpersonal therapy*. Washington, DC: American Psychiatric Press.

CHRONIC HEADACHE

Behavioral Treatment, Treatment References/Manuals:

Blanchard, E. B., & Andrasik, F. (1985). *Management of chronic headache: A psychological approach*. Elmsford, NY: Pergamon Press.

CHRONIC PAIN

Cognitive Behavioral Treatment, Treatment References/Manuals:

Cognitive Behavioral Treatment for Arthritis Pain (Contact: Francis Keefe, PhD, Pain Management Program, Duke Medical Center, Box 3159, Durham, NC 27710. Cost=$30.00).

Turk, D. C., Meichenbaum, D., & Genest, M. (1983). *Pain and behavioral medicine: A cognitive-behavioral perspective*. New York: Guilford Press.

Training Available:

Francis J. Keefe, PhD
Director, Pain Management Program
Duke Medical Center - Box 3159
Durham, NC 27710
Telephone: 919-684-6212

Pain Evaluation and Treatment Institute
University of Pittsburgh School of Medicine
4601 Baum Boulevard
Pittsburgh, PA 15213

CHRONICALLY MENTALLY ILL

Token Economy Programs, Treatment References/Manuals:

Ayllon, T., & Azrin, N. (1968). *The token economy: A motivational system for therapy and rehabilitation*. New York: Appleton-Century-Crofts.

DEPRESSION

Cognitive Therapy, Treatment References/Manuals:

Beck, A. T., Rush, A. J., Shaw, B. F., & Emery, G. (1979). *Cognitive therapy of depression*. New York: Guilford.

Training Available:

Cory Newman, PhD
Center for Cognitive Therapy
University of Pennsylvania Medical School
3600 Market Street, Room 754
Philadelphia, PA 19104
Telephone: 215-898-4100

Judy S. Beck, PhD
Beck Institute for Cognitive Therapy
GSB Building - Suite 700
City Line & Belmont Avenues
Bala Cynwyd, PA 19004
Telephone: 610-664-3020

Cognitive Therapy Training Program
Cognitive Therapy Center of New York
3 East 80th Street
New York, NY 10021
Telephone: 212-717-1052

Interpersonal Therapy, Treatment References/Manuals:

Klerman, G. L., Weissman, M. M.., Rounsaville, B. J., & Chevron, E. S. (1984). *Interpersonal psychotherapy of depression*. New York: Basic Books.

Training Available:

Cleon Cornes, MD & Ellen Frank, PhD
Western Psychiatric Institute and Clinic
3811 O'Hara Street
Pittsburgh, PA 15213
Telephone: 412-624-2211

Myrna Weissman
Psychosocial Therapeutic Systems
Graywind Publications
Stuyvesant Plaza

Executive Park Drive
Albany, NY 12203
Telephone: 518-438-3231

Videotape: Interpersonal Therapy of Depression
IPT Educational Foundation
5307 Cherokee
Houston, TX 77005
Telephone: 800-782-0015

DISCORDANT COUPLES

Behavior Therapy, Treatment References/Manuals:

Jacobson, N. S., & Margolin, G. (1979). *Marital therapy: Strategies based on social learning and behavior exchange principles.* New York: Brunner/Mazel.

Baucom, D. H., & Epstein, N. (1990). Cognitive-Behavioral Marital Therapy. New York: Brunner/Mazel.

ENURESIS

Behavioral Treatment, Treatment References/Manuals:

Azrin, N. H. & Besalel, V. B. (1979). *A parent's guide to bedwetting control.* New York: Pocket Books.

Full Spectrum Home Training for Nocturnal Enuresis. (Contact: Arthur C. Houts, Department of Psychology, University of Memphis, Memphis, TN 38152. Cost=$5.00.)

GENERALIZED ANXIETY DISORDER

Anxiety Management, Treatment References/Manuals:

Anxiety Management for Generalized Anxiety. (Contact: Secretary, Department of Psychology, Warnerford Hospital, Headington, Oxford, OX3 7JX. Cost = 2 pounds, prepayment required).

Cognitive Behavior Therapy, Treatment References/ Manuals:

Controlling Anxiety. (Contact: Secretary, Department of Psychology, Warnerford Hospital, Headington, Oxford, OX3 7JX. Cost = 2 pounds, prepayment required).

Brown, T., O'Leary, T., & Barlow, D. H. (1994). Generalized anxiety disorder. In D. H. Barlow (Ed.), *Clinical handbook of psychological disorders.* New York: Guilford.

Training Available:

Cory Newman, PhD
Center for Cognitive Therapy
University of Pennsylvania Medical School
3600 Market Street, Room 754

Philadelphia, PA 19104
Telephone: 215-898-4100

Judy S. Beck, PhD
Beck Institute for Cognitive Therapy
GSB Building - Suite 700
City Line & Belmont Avenues
Bala Cynwyd, PA 19004
Telephone: 610-664-3020

OBSESSIVE COMPULSIVE DISORDER

Behavioral Treatment, Treatment References/Manuals:

Steketee, G. (1993). *Treatment of obsessive compulsive disorder.* New York: Guilford Press.

Riggs, D. S. & Foa, E. B. Obsessive compulsive disorder. In D. H. Barlow (Ed.), *Clinical handbook of psychological disorders.* New York: Guilford Press.

PANIC DISORDER

Cognitive Therapy, Treatment References/Manuals:

Barlow, D. H., & Cerny, J. A. (1988). *Psychological treatment of panic.* New York: Guilford Press.

Barlow, D., & Craske, M. (1994). *Mastery of your anxiety and panic - II.* Albany, NY: Graywind Publications. (Both therapist and client versions are available - To order call 518-438-3231)

Clark, D. M. (1989). Anxiety states: Panic and generalized anxiety. In K. Hawton, P. Salkovskis, J. Kirk, & D. M. Clark (Eds.), *Cognitive behavior therapy for psychiatric problems.* Oxford: Oxford University Press.

Salkovskis, P. M., & Clark, D. M. (1991). Cognitive treatment of panic disorder. *Journal of cognitive psychotherapy, 3,* 215-226.

Training Available:

Graywind Publications
Executive Park Drive
Albany, NY 12203
Telephone: 518-438-3231

POST TRAUMATIC STRESS DISORDER

Treatment References/Manuals:

Clinical Handbook/Therapist Manual on PTSD. (Contact: Donald Meichenbaum, University of Waterloo, Department of Psychology, Waterloo, Ontario, Canada N2L 3G1, Phone: 519-885-1211, ext. 2551, cost: $40 plus $5 shipping.)

SOCIAL PHOBIA

Cognitive Behavioral Group Therapy, Treatment References/Manuals:

Cognitive Behavioral Group Therapy for Social Phobia by R. Heimberg (Contact: Karen Law, Center for Stress and Anxiety Disorders, Pine West Plaza, Building 4, Washington Avenue Extension, Albany, NY 12205. Cost = $20.00).

Social Effectiveness Therapy: A Program for Overcoming Social Anxiety and Phobia (Contact: Samuel M. Turner, Ph.D. or Deborah C. Beidel, Ph.D., Turndel Inc., Suite 200, 615 Wesley Drive Charleston, SC 29464. Cost = $39.00).

Training Available:

Richard Heimberg, Ph.D. & Harlan Justen, Ph.D.
Center for Stress and Anxiety Disorders

Pine West Plaza, Building 4
Washington Avenue Extension
Albany, NY 12205
Telephone: (518) 464-0241 or 869-2033
E-mail: rh188@albnyvms.bitnet

SPECIFIC PHOBIA

Systematic Desensitization, Treatment References/Manuals:

Wolpe, J. (1990). *Practice of behavior therapy* (4th Edition). New York: Pergamon Press.

Exposure Therapy, Treatment References/Manuals:

Marks, I. (1978). *Living with fear*. New York: McGraw Hill.

REFERENCES

Achterberg, J. (1985). *Imagery in healing: Shamanism and modern medicine.* Boston: Shambhala.

Acierno, K., Tremont, G., Last, C., & Montgomery, D. (1994). Tripartite assessment of the efficacy of eye-movement desensitization in a multi-phobic patient. *Journal of Anxiety Disorders, 8,* 259–276.

Adler, A. (1964). *Social interest: A challenge to mankind.* New York: Capricorn Books.

Agency for Health Care Policy and Research (AHCPR). (1990). *Clinical guideline development.* (AHCPR Program Note, Publication No. 0m90-0086). Silver Spring, MD: AHCPR Publications Clearinghouse.

Agency for Health Care Policy and Research (AHCPR). (1993). *Clinical practice guideline quick reference guide for clinicians: Depression in primary care, detection, diagnosis and treatment.* AHCPR Publication No. 93-0552. Rockville, MD: AHCPR Publications.

Ahijevych, K., & Wewers, M. (1993). Factors associated with nicotine dependence among African American women cigarette smokers. *Research in Nursing and Health, 16,* 283–292.

Ainsworth, M. D. S. (1989). Attachment beyond infancy. *American Psychologist, 44,* 709–716.

Ainsworth, M. D. S., & Bowlby, J. (1991). An ethological approach to personality development. *American Psychologist, 46,* 333–341.

Alagna, F. J., Whitcher, S. J., Fisher, J. D., & Wicas, E. A. (1979). Evaluative reaction to interpersonal touch in a counseling interview. *Journal of Counseling Psychology, 26,* 265–272.

Alexander, C. N., Robinson, P., Orme-Johnson, D. W., & Schneider, R. (1994). The effects of transcendental meditation compared to other methods of relaxation and meditation in reducing risk factors, morbidity, and mortality. *Homeostasis in Health and Disease, 35,* 243–263.

Allison, K., Crawford, I., Echemendia, R., Robinson, L., & Knepp, D. (1994). Human diversity and professional competence. *American Psychologist, 49,* 792–796.

Altmaier, E. M., Ross, D. L., Leary, M. R., & Thornbrough, M. (1982). Matching stress inoculation's treatment components to clients' anxiety mode. *Journal of Counseling Psychology, 29,* 331–334.

Alvarez, J., & Jason, L. (1993). The effectiveness of legislation, education, and loaners for child safety in automobiles. *Journal of Community Psychology, 21,* 280–284.

American Counseling Association. (1994). Proposed revision to American Counseling Association Code of Ethics and Standards of Practice. *Counseling Today, 37,* 3, 20–28.

American Counseling Association. (1995). *Code of ethics.* Alexandria, VA: Author.

American Psychiatric Association. (1993). Practice guidelines for major depressive disorder in adults. *American Journal of Psychiatry, 150* (Suppl.), 1–26.

American Psychiatric Association. (1994). *Diagnostic and statistical manual of mental disorders* (4th ed.). Washington, DC: Author.

American Psychological Association. (1990). *Guideline for providers of psychological services to ethnic, linguistic, and culturally diverse populations.* Washington, DC: Author.

American Psychological Association. (1992). *Ethical principles of psychologists.* Washington, DC: Author.

American Psychological Association, Division 12. (1993). *Task force on promotion and dissemination of psychological procedures.* Washington, DC: Author.

Anderson, E. A. (1987). Preoperative preparation for cardiac surgery facilitates recovery, reduces psychological distress, and reduces the incidence of acute postoperative hypertension. *Journal of Consulting and Clinical Psychology, 55,* 513–550.

Anderson, R., & McMillion, P. (1995). Effects of similar and diversified modeling on African American women's efficacy expectations and intentions to perform breast self-examination. *Health Communication, 7,* 327–343.

Andrada, P., & Korte, A. (1993). *En aquellas tiempos:* A reminiscing group with Hispanic elderly. *Journal of Gerontological Social Work, 20,* 25–42.

Ankuta, G., & Abeles, N. (1993). Client satisfaction, clinical significance, and meaningful change in psychotherapy. *Professional Psychology, 24,* 70–74.

Anton, J. L., Dunbar, J., & Friedman, L. (1976). Anticipation training in the treatment of depression. In J. D. Krumboltz & C. E. Thoresen (Eds.), *Counseling methods* (pp. 67–74). New York: Holt, Rinehart and Winston.

Antoni, M. (1991). Psychosocial stressors and behavioral interventions in gay men with HIV infection. *International Review of Psychiatry, 3,* 383–399.

Antonuccio, D., Danton, W., & DeNelsky, G. (1995). Psychotherapy versus medication for depression: Challenging the conventional wisdom with data. *Professional Psychology, 26,* 574–585.

Arean, P. A. (1993). Cognitive behavioral therapy with older adults. *The Behavior Therapist, 16,* 236–239.

Armour-Thomas, E., Bruno, K., & Allen, B. (1992). Toward an understanding of higher-order thinking among minority students. *Psychology in the Schools, 29,* 273–280.

Arnow, B. A., Taylor, C. B., Agras, W. S., & Telch, M. J. (1985). Enhancing agoraphobia treatment outcome by changing couple communication pattern. *Behavior Therapy, 16,* 452–467.

Arredondo, P., Toporek, R., Brown, S. P., Jones, J., Locke, D. C., Sanchez, J., & Stadler, H. (1996). Operationalization of the multicultural counseling competencies. *Journal of Multicultural Counseling and Development, 24,* 42–78.

Aspinwall, L., Kemeny, M., Taylor, S., & Schneider, S. (1991). Psychosocial predictors of gay men's AIDS risk-reduction behavior. *Health Psychology, 10,* 432–444.

Atkinson, D. R., Brown, M. T., Parham, T. A., Matthews, L. G., Landrum-Brown, J., & Kim, A. U. (1996). African American client skin tone and clinical judgments of African American and European American psychologists. *Professional Psychology, 27,* 500–505.

Atkinson, D. R., & Hackett, G. (1995). *Counseling diverse populations.* Madison, WI: Brown & Benchmark.

Atkinson, D. R., Morten, G., & Sue, D. W. (1993). *Counseling American minorities.* Madison, WI: Brown & Benchmark.

Auerbach, R., & Kilmann, P. R. (1977). The effects of group systematic desensitization on secondary erectile failure. *Behavior Therapy, 8,* 330–339.

Axelson, J. A, (1993). *Counseling and development in a multicultural society* (2nd ed.). Pacific Grove, CA: Brooks/Cole.

Azrin, N. H., Bersalel, A., Bechtel, R., Michalicek, A., Mancera, M., Carroll, D., Shuflord, D., & Cox, J. (1980). Comparison of reciprocity and discussion-type counseling for marital problems. *American Journal of Family Therapy, 8,* 21–28.

Azrin, N. H., Nunn, R. G., & Frantz, S. E. (1980). Habit reversal vs. negative practice treatment of nailbiting. *Behaviour Research and Therapy, 18,* 281–285.

Azrin, N. H., Nunn, R. G., & Frantz-Renshaw, S. (1980). Habit reversal treatment of thumbsucking. *Behaviour Research and Therapy, 18,* 395–399.

Bachelor, A. (1995). Clients' perceptions of the therapeutic alliance: A qualitative analysis. *Journal of Counseling Psychology, 42,* 323–337.

Baider, L., Uziely, B., & Kaplan-DeNour, A. (1994). Progressive muscle relaxation and guided imagery in cancer patients. *General Hospital Psychiatry, 16,* 340–347.

Baker, J., & Krugh, M. (1996, April). *Do I say Hispanic or Latino/Latina?* Paper presented at the Ohio University Multicultural Counselor Education Conference, Athens, OH.

Baker, S. B. (1981). *Cleaning up our thinking: A unit in self-improvement.* Unpublished manuscript, Division of Counseling and Educational Psychology, Pennsylvania State University, University Park.

Bandler, R., & Grinder, J. (1975). *The structure of magic I: A book about language and therapy.* Palo Alto, CA: Science and Behavior Books.

Bandura, A. (1969). *Principles of behavior modification.* New York: Holt, Rinehart & Winston.

Bandura, A. (1971). Vicarious and self-reinforcement processes. In R. Glaser (Ed.), *The nature of reinforcement.* New York: Academic Press.

Bandura, A. (1976). Effecting change through participant modeling. In J. D. Krumboltz & C. E. Thoresen (Ed.), *Counseling methods* (pp. 248–265). New York: Holt, Rinehart & Winston.

Bandura, A. (1977). Self-efficacy: Toward a unifying theory of behavior change. *Psychological Review, 84,* 191–215.

Bandura, A. (1982). Self-efficacy mechanism in human agency. *American Psychologist, 37,* 122–147.

Bandura, A. (1986). *Social foundations of thought and action: A social cognitive theory.* Englewood Cliffs, NJ: Prentice-Hall.

Bandura, A. (1989). Human agency in social cognitive theory. *American Psychologist, 44,* 1175–1185.

Bandura, A. (1991). Social cognitive theory of self-regulation. Special Issue: Theories of cognitive self-regulation. *Organizational Behavior and Human Decision Processes, 50* (2), 248–287.

Bandura, A. (1993). Perceived self-efficacy in cognitive development and functioning. *Educational Psychologist, 28,* 117–148.

Bandura, A., Adams, N. E., & Beyer, J. (1977). Cognitive processes mediating behavioral change. *Journal of Personality and Social Psychology, 35,* 125–139.

Bandura, A., Cioffi, D., Taylor, C., & Brouillard, M. E. (1988). Perceived self-efficacy in coping with cognitive stressors and opioid activation. *Journal of Personality and Social Psychology, 55,* 477–488.

Bandura, A., & Simon, K. (1977). The role of proximal intentions in self-regulation of refractory behavior. *Cognitive Therapy and Research, 1,* 177–193.

Bandura, A., Taylor, C., Williams, S. L., Mefford, I. N., & Barchas, J. D. (1985). Catecholamine secretion as function of perceived coping self-efficacy. *Journal of Consulting and Clinical Psychology, 53,* 406–415.

Bankart, C. P. (1995). *Talking cures: A history of Western and Eastern psychotherapies.* Pacific Grove, CA: Brooks/Cole.

Banken, D., & Wilson, G. (1992). Treatment acceptability of alternative therapies for depression: A comparative analysis. *Psychotherapy, 29,* 610–618.

Barak, A., Patkin, J., & Dell, D. M. (1982). Effects of certain counselor behaviors in perceived expertness and attractiveness. *Journal of Counseling Psychology, 29,* 261–267.

Barkham, M. (1988). Empathy in counseling and psychotherapy. Present status and future directions. *Counseling Psychology Quarterly, 1,* 407–428.

Barlow, D. H. (Ed.). (1994). *Clinical handbook of psychosocial disorders* (2nd ed.). New York: Guilford.

Barlow, D. H., Craske, M. G., Cerny, J. A., & Klosko, J. S. (1989). Behavioral treatment of panic disorder. *Behavior Therapy, 20,* 261–282.

Barlow, D. H., & Durand, V. M. (1995). *Abnormal psychology: An integrative approach.* Pacific Grove, CA: Brooks/Cole.

Barlow, D. H., Hayes, S. C., & Nelson, R. O. (1984). *The scientist practitioner.* New York: Pergamon Press.

Barrett-Lennard, G. T. (1981). The empathy cycle: Refinement of a nuclear concept. *Journal of Counseling Psychology, 28,* 91–100.

Bartholomew, K. (1990). Avoidance of intimacy: An attachment perspective. *Journal of Social and Personal Relationships, 7,* 147–178.

Bartholomew, K., & Horowitz, L. M. (1991). Attachment styles among young adults: A test of a four-category model. *Journal of Personality and Social Psychology, 61,* 226–244.

Basic Behavioral Science Research for Mental Health. (1995). Emotion and motivation. *American Psychologist, 50,* 838–845.

Basic Behavioral Science Task Force of the National Advisory Mental Health Council. (1996). Thought and communication. *American Psychologist, 51,* 181–189.

Baucom, D..H., & Epstein, N. (1990). *Cognitive-behavioral marital therapy.* New York: Brunner/Mazel.

Bauermeister, J., Berrios, B., Jiminez, A., Acevedo, L., & Gordon, M. (1990). Some issues and instruments for the assessment of attention-deficit hyperactivity disorder in Puerto Rican children. *Journal of Clinical Child Psychology, 19,* 9–16.

Beck, A. T. (1970). Cognitive therapy: Nature and relation to behavior therapy. *Behavior Therapy, 1,* 184–200.

Beck, A. T. (1967). *Depression.* New York: Hoeber.

Beck, A. T. (1972). *Depression: Causes and treatment.* Philadelphia: University of Pennsylvania Press.

Beck, A. T. (1976). *Cognitive therapy and the emotional disorders.* New York: International Universities Press.

Beck, A. T. (1993). Cognitive therapy: Past, present, and future. *Journal of Consulting and Clinical Psychology, 62,* 194–198.

Beck, A. T., & Emery, G. (1979). *Cognitive therapy of anxiety.* Philadelphia: Center for Cognitive Therapy.

Beck, A. T., Sokol, L., Clark, D. A., Berchick, B., & Wright, F. (1992). A crossover study of focused cognitive therapy for panic disorder. *American Journal of Psychiatry, 149,* 778–783.

Beck, A. T., Steer, R. A., & Brown, G. K. (1996). *The Beck Depression Inventory II.* San Antonio: The Psychological Corporation, Harcourt-Brace.

Beck, J. S. (1995). *Cognitive therapy.* New York: Guilford.

Becker, D., & Lamb, S. (1994). Sex bias in the diagnosis of borderline personality disorders and post-traumatic stress disorder. *Professional Psychology, 25,* 55–61.

Bedrosian, R. C., & Bozicas, G. D. (1994). *Treating family of origin problems: A cognitive approach.* New York: Guilford.

Begeer-Kooy, J. Y. (1988). Stress-reduction training with child oncology patients: Introduction of a self-modeling film. *Tijdschrift voor Psychotherapie, 14,* 247–257.

Belenky, M. F., Clinchy, B. M., Goldbeyer, N. R., & Tarule, J. M. (1986). *Women's ways of knowing.* New York: Basic Books.

Belgrave, F., Randolph, S., Carter, C., & Braithwaite, N. (1993). The impact of knowledge, norms, and self-efficacy on intentions to engage in AIDS-preventive behaviors among young incarcerated African American males. *Journal of Black Psychology, 19,* 155–168.

Bell, L., & Seyfer, E. (1982). *Gentle yoga.* Berkeley, CA: Celestial Arts.

Bell, Y., Brown, R., & Bryant, A. (1993). Traditional and culturally relevant presentations of a logical reasoning task and performance among African-American students. *Western Journal of Black Studies, 17,* 173–178.

Bemis, K. (1980). Personal communication. Cited in P. C. Kendall & S. D. Hollon (Eds.), *Assessment strategies for cognitive behavioral interventions.* New York: Academic Press.

Benson, H. (1976). *The relaxation response.* New York: Avon.

Benson, H. (1987). *Your maximum mind.* New York: Avon.

Benson, H., & Stuart, E. M. (Eds.). (1992). *The wellness book: The comprehensive guide to maintaining health and treating stress-related illness.* New York: Birch Lane Press.

Berenson, B. C., & Mitchell, K. M. (1974). *Confrontation: For better or worse.* Amherst, MA: Human Resource Development Press.

Berg, I., & Jaya, A. (1993). Different and same: Family therapy with Asian-American families. *Journal of Marital and Family Therapy, 19,* 31–38.

Berman, J. S., Miller, R. C., & Massman, P. J. (1985). Cognitive therapy vs. systematic desensitization: Is one treatment superior? *Psychological Bulletin, 97,* 451–461.

Berne, E. (1964). *Games people play.* New York: Grove Press.

Bernstein, D. A., & Borkovec, T. D. (1973). *Progressive relaxation training: A manual for helping professions.* Champaign, IL: Research Press.

Bernstein, D. A., & Carlson, C. R. (1993). Progressive relaxation: Abbreviated methods. In P. M. Lehrer & R. L. Woolfolk (Eds.), *Principles and practice of stress management* (2nd ed., pp. 53–87). New York: Guilford.

Betancourt, H., & Lopéz, S. (1993). The study of culture, ethnicity, and race in American psychology. *American Psychologist, 48,* 629–637.

Beutler, L., & Clarkin, J. (1990). *Systematic treatment selection.* New York: Brunner/Mazel.

Bhargava, S. C. (1988). Participant modeling in stuttering. *Indian Journal of Psychiatry, 30,* 397–405.

Bijou, S. W., & Baer, D. M. (1976). *Child development I: A systematic and empirical theory.* Englewood Cliffs, NJ: Prentice-Hall.

Birch, B. B. (1995). *Power yoga.* New York: Fireside.

Birdwhistell, R. L. (1970). *Kinesics and context.* Philadelphia: University of Pennsylvania Press.

Blanchard, E. B., Andrasik, F., Ahles, T. A., Teders, S. J., & O'Keefe, D. (1980). Migraine and tension headache: A meta-analytic review. *Behavior Therapy, 11,* 613–631.

Blanchard, E. B., Schwartz, S. P., & Radnitz, C. (1987). Psychological assessment and treatment of irritable bowel syndrome. *Behavior Modification, 11,* 348–372.

Bloom, S. L., & Bills, L. (April, 1995). *The new trauma paradigm: Implications for institutions.* Paper presented to the Traumatology Conference, West Virginia University, Morgantown.

Bly, R. (1996). *The sibling society.* Reading, MA: Addison-Wesley.

Bootzin, R. R. (1977). Effects of self-control procedures for insomnia. In R. B. Stuart (Ed.), *Behavioral self-management: Strategies, techniques, and outcomes* (pp. 176–195). New York: Brunner/Mazel.

Bordin, E. S. (1979). The generalizability of the psychoanalytic concept of the working alliance. *Psychotherapy: Theory, Research and Practice, 16,* 256–260.

Borkovec, T. D., Mathews, A. M., Chambers, A., Ebrahimi, S., Lytle, R., & Nelson, R. (1987). The effects of relaxation training with cognitive or nondirective therapy and the role of relaxation-induced anxiety in the treatment of generalized anxiety. *Journal of Consulting and Clinical Psychology, 55,* 883–888.

Borysenko, J. (1987). *Minding the body, mending the mind.* New York: Bantam.

Borysenko, J. (1988). *Minding the body, mending the mind.* New York: Bantam.

Borysenko, J. (1997). *Meditation for inner guidance and self-healing.* (Audio tape). Niles, IL: Nightingale Conant.

Borysenko, J., & Borysenko, M. (1994). *The power of the mind to heal.* Carson, CA: Hay House.

Boszormenyi-Nagy, I., & Sparks, G. (1973). *Invisible loyalties.* New York: Harper & Row.

Botvin, G., Baker, E., Botvin, E., & Dusenbury, L. (1993). Factors promoting cigarette smoking among Black youth: A causal modeling approach. *Addictive Behaviors, 18,* 397–405.

Bowen, M. (1978). *Family therapy in clinical practice.* New York: Jason Aronson.

Bowlby, J. (1988). *A secure base: Parent-child attachments and healthy human development.* New York: Basic Books.

Boyd-Franklin, N. (1989). *Black families in therapy: A multi systems approach.* New York: Guilford.

Bracho-de-Carpio, A., Carpio-Cedraro, F., & Anderson, L. (1990). Hispanic families and teaching about AIDS: A participatory approach at the community level. *Hispanic Journal of Behavioral Sciences, 12,* 165–176.

Bradford, E., & Lyddon, W. J. (1994). Assessing adolescent and adult attachment: An update. *Journal of Counseling and Development, 73,* 215–219.

Brammer, L. M., Shostrom, E. L., & Abrego, P. J. (1989). *Therapeutic psychology: Fundamentals of counseling and psychotherapy* (5th ed.). Englewood Cliffs, NJ: Prentice-Hall.

Brigham, D. (1994). *Imagery for getting well.* New York: Norton.

Bronfenbrenner, U. (1979). *The ecology of human development.* Cambridge, MA: Harvard University Press.

Bronfenbrenner, U. (1993). The ecology of cognitive development: Research models and fugitive findings. In R. H. Wozniak & K. W. Fischer (Eds.), *Development in context: Acting and thinking in specific environments* (The Jean Piaget Symposium Series, pp. 3–44). Hillsdale, NJ: Erlbaum.

Brooks, J., & Scarano, T. (1985). Transcendental meditation in the treatment of post-Vietnam adjustment. *Journal of Counseling and Development, 64,* 212–215.

Brothers, D. (1995). *Falling backwards: An exploration of trust and self-experience.* New York: Norton.

Brown, L. M., & Gilligan, C. (1992). *Meeting at the crossroads.* New York: Ballantine Books.

Brown, L. S. (1992). Introduction in L. S. Brown & M. Ballou (Ed.), *Personality and psychopathology: Feminist reappraisals* (pp. 111–115). New York: Guilford.

Brown, L. S. (1994). *Subversive dialogues: Theory in feminist therapy.* New York: Basic Books.

Brown, L. S., & Ballou, M. (1992). *Personality and psychopathology: Feminist reappraisals.* New York: Guilford.

Bryan, T., & Bryan, J. (1991). Positive mood and math performance. *Journal of Learning Disability, 24,* 490–494.

Burnett, J. W., Anderson, W. P., & Heppner, P. P. (1995). Gender roles and self-esteem: A consideration of environmental factors. *Journal of Counseling and Development, 73,* 323–326.

Butler, G., Fennell, M., Robson, P., & Gelder, M. (1991). Comparison of behavior therapy and cognitive behavior therapy in the treatment of generalized anxiety disorder. *Journal of Consulting and Clinical Psychology, 59,* 167–175.

Carkhuff, R. R. (1969a). *Helping and human relations.* Vol. 1: *Practice and research.* New York: Holt, Rinehart and Winston.

Carkhuff, R. R. (1969b). *Helping and human relations.* Vol. 2: *Practice and research.* New York: Holt, Rinehart and Winston.

Carkhuff, R. R. (1987). *The art of helping.* (6th ed.). Amherst, MA: Human Resource Development Press.

Carkhuff, R. R. (1993). *The art of helping.* (8th ed.). Amherst, MA: Human Resource Development Press.

Carkhuff, R. R., & Pierce, R. M. (1975). *Trainer's guide: The art of helping.* Amherst, MA: Human Resource Development Press.

Carkhuff, R. R., Pierce, R. M., & Cannon, J. R. (1977). *The art of helping III.* Amherst, MA: Human Resource Development Press.

Carlson, C. R., & Hoyle, R. H. (1993). Efficacy of abbreviated progressive muscle relaxation training: A quantitative review of behavioral medicine research. *Journal of Consulting and Clinical Psychology, 61,* 1059–1067.

Carr, S., & Punzo, R. (1993). The effects of self-monitoring of academic accuracy and productivity on the performance of students with behavioral disorders. *Behavioral Disorders, 18,* 241–250.

Carrington, P. (1993). Modern forms of meditation. In P. M. Lehrer & R. L. Woolfolk (Eds.), *Principles and practice of stress management* (2nd ed., pp. 139–168). New York: Guilford.

Carter, R. E. (1990). The relationship between racism and racial identity among White Americans: An exploratory investigation. *Journal of Counseling and Development, 69,* 47.

Casas, J. M. (1988). Cognitive behavioral approaches: A minority perspective. *The Counseling Psychologist, 16,* 106–110.

Caspar, F. (1995). *Plan analysis: Toward optimizing psychotherapy.* Seattle, WA: Hogrefe & Huber.

Cass, V. C. (1979). Homosexual identity formation: A theoretical model. *Journal of Homosexuality, 4,* 219–235.

Castillo, R. J. (1997). *Culture and mental illness.* Pacific Grove, CA: Brooks/Cole.

Cautela, J. R. (1966). The treatment of compulsive behavior by covert sensitization. *Psychological Record, 16,* 33–41.

Cautela, J. R. (1969). Behavior therapy and self-control: Techniques and implications. In C. Franks (Ed.), *Behavior therapy: Appraisal and status* (pp. 323–340). New York: McGraw-Hill.

Cautela, J. R. (1970). Covert reinforcement. *Behavior Therapy, 1,* 33–50.

Cautela, J. R. (1977). *Behavior analysis forms for clinical intervention* (Vol. 2). Champaign, IL: Research Press.

Cavaliere, F. (October, 1995). Payers demand increased provider documentation. *American Psychological Association Monitor,* p. 41.

Chambless, D. L., & Gillis, M. M. (1993). Cognitive therapy of anxiety disorders. *Journal of Consulting and Clinical Psychology, 61,* 248–260.

Cheatham, H., Ivey, A., Ivey, M. B., & Simek-Morgan, L. (1993). Multicultural counseling and therapy. In A. Ivey, M. B. Ivey, & L. Simek-Morgan (Eds.), *Counseling and psychotherapy: A multicultural perspective* (pp. 114–115). Needham Heights, MA: Allyn & Bacon.

Cheatham, H., & Stewart, J. (1990). *Black families: Interdisciplinary perspective.* New Brunswick, NJ: Transactional Publishers.

Cheston, S. E. (1991). *Making effective referrals: The therapeutic process.* New York: Gardner Press.

Chojnacki, J. T., & Gelberg, S. (1995). The facilitation of gay/lesbian/bisexual support-therapy group by heterosexual counselors. *Journal of Counseling and Development, 73,* 352–354.

Chopra, D. (1991). *Perfect health.* New York: Harmony Books.

Christensen, A. (1987). *The American Yoga Association beginner's manual.* New York: Fireside.

Church, A., Teresa, J., Rosebrook, R., & Szendre, D. (1992). Self-efficacy for careers and occupational consideration in minority high school equivalency students. *Journal of Counseling Psychology, 39,* 498–508.

Ciminero, A. R., Nelson, R. O., & Lipinski, D. P. (1977). Self-monitoring procedures. In A. R. Ciminero, K. S. Calhoun, & H. E. Adams (Eds.), *Handbook of behavioral assessment* (pp. 195–232). New York: Wiley.

Cimmarusti, R. (1996). Exploring aspects of Filipino-American families. *Journal of Marital & Family Therapy, 22,* 205–217.

Claiborn, C. D. (1979). Counselor verbal interventions, nonverbal behavior, and social power. *Journal of Counseling Psychology, 26,* 378–383.

Claiborn, C. D. (1982). Interpretation and change in counseling. *Journal of Counseling Psychology, 29,* 439–453.

Claiborn, C. D., & Dowd, E. T. (1985). Attributional interpretations in counseling: Content versus discrepancy. *Journal of Counseling Psychology, 32,* 188–196.

Claiborn, C. D., Ward, S. R., & Strong, S. R. (1981). Effects of congruence between counselor interpretations and client beliefs. *Journal of Counseling Psychology, 28,* 101–109.

Clark, D. M., & Salkovskis, P. M. (1989). *Cognitive therapy for panic and hypochondriasis.* New York: Pergamon Press.

Clark, D. M., Salkovskis, P. M., Hackman, A., Middleton, H., Anastasiades, P., & Gelder, M. (1994). A comparison of cognitive therapy, applied relaxation, and imipramine in the treatment of panic disorder. *British Journal of Psychiatry, 32,* 188–196.

Clinton, J., McCormick, K., & Besteman, J. (1994). Enhancing clinical practice. The role of practice guidelines. *American Psychologist, 49,* 30–33.

Cochran, S., & Mays, V. (1993). Applying social psychological models to predicting HIV-related sexual risk behaviors among African Americans. *Journal of Black Psychology, 19,* 142–154.

Cohen, K. (1997). Healthy breathing. (Audio tape). Boulder, CO: Sounds True.

Cohen, R. E., Creticso, P. S., & Norman, P. S. (1993–1994). The effects of guided imagery (GI) on allergic subjects' responses to ragweed-pollen nasal challenge: An exploratory investigation. *Imagination, Cognition and Personality, 13,* 259–269.

Cole, C. L. (1992). Self-management interventions in the schools. *School Psychology Review, 21,* 188–192.

Coleman, E. (Ed.). (1988). *Integrated identity for gay men and lesbians.* New York: Harrington Park.

Coleman, E., & Remafedi, G. (1989). Gay, lesbian and bisexual adolescents: A critical challenge to counselors. *Journal of Counseling and Development, 68,* 36–40.

Coleman, H. (1995). Strategies for coping with cultural diversity. *The Counseling Psychologist, 23,* 722–740.

Connolly, M. J. (1993). Respiratory rehabilitation in the elderly patient. *Reviews in Clinical Gerontology, 3,* 281–294.

Connor-Greene, P. (1993). The therapeutic context: Preconditions for change in psychotherapy. *Psychotherapy, 30,* 375–382.

Consumer Reports. (1995, November). Does therapy help? (pp. 734–739).

Cooper, J. F. (1995). *A primer of brief therapy.* New York: Norton.

Corey, G., Corey, M., & Callanan, P. (1993). *Issues and ethics in the helping professions* (4th ed.). Pacific Grove, CA: Brooks/Cole.

Corrigan, J. D., Dell, D. M., Lewis, K. N., & Schmidt, L. D. (1980). Counseling as a social influence process: A review. *Journal of Counseling Psychology, 27,* 395–441.

Cox, B., Swinson, R., Morrison, R., & Lee, P. (1993). Clomipramine, fluoxetine, and behavior therapy in the treatment of obsessive-compulsive disorder: A meta-analysis. *Journal of Behavior Therapy and Experimental Psychiatry, 24,* 149–153.

Cozzarelli, C. (1993). Personality and self-efficacy as predictors of coping with abortion. *Journal of Personality and Social Psychology, 65,* 1224–1236.

Craig, R. J. (Ed.). (1989). *Clinical and diagnostic interviewing.* Northvale, NJ: Aronson.

Craigie, F. C., Jr., & Ross, S. M. (1980). The case of a video-tape pretherapy training program to encourage treatment-seeking among alcohol detoxification patients. *Behavior Therapy, 11,* 141–147.

Cronbach, L. J. (1990). *Essentials of psychological testing.* (5th ed.). New York: Harper & Row.

Cross, W. E. (1971). The Negro-to-Black conversion experience: Toward a psychology of Black liberation. *Black World, 20,* 13–27.

Csikszentmihalyi, M. (1990). *Flow: The psychology of optimal experience.* New York: Harper Perennial.

Cullen, C. (1983). Implications of functional analysis. *British Journal of Clinical Psychology, 22,* 137–138.

Cusumano, J., & Robinson, S. (1993). The short-term psychophysiological effects of hatha yoga and progressive relaxation on female Japanese students. *Applied Psychology: An International Review, 42,* 77–90.

Dacher, E. S. (1996). *Whole healing.* New York: Dutton.

Das, A. K. (1995). Rethinking multicultural counseling: Implications for counselor education. *Journal of Counseling and Development, 74,* 45–74.

Day, J. (1995). Obligation and motivation: Obstacles and resources for counselor well-being and effectiveness. *Journal of Counseling & Development, 73,* 108–110.

Day, R. W., & Sparacio, R. T. (1980). Structuring the counseling process. *Personnel and Guidance Journal, 59,* 246–250.

DeAngelis, T. (1994, August). Women less likely to pursue technology-related careers. *American Psychological Association Monitor,* p. 30.

Deffenbacher, J. L., & Suinn, R. M. (1982). The self-control of anxiety. In P. Karoly & F. H. Kanfer (Eds.), *Self-management and behavior change* (pp. 393–442). New York: Pergamon Press.

DeJong, W. (1994). Relapse prevention. *International Journal of the Addictions, 29,* 681–705.

Denholtz, M., & Mann, E. (1974). An audiovisual program for group desensitization. *Journal of Behavior Therapy and Experimental Psychiatry, 5,* 27–29.

Derogatis, L. R. (1983). *SCL-90 R administration, scoring and procedures manual-II.* Towson, MD: Clinical Psychometric Research.

DiMascio, A., Weissmand, M. M., Prusoff, B. A., Neu, C., Zwilling, M., & Klerman, G. L. (1979). Differential symptom reduction by drugs and psychotherapy in acute depression. *Archives of General Psychiatry, 36,* 1450–1456.

Dixon, D. N., & Glover, J. A. (1984). *Counseling: A problem-solving approach.* New York: Wiley.

Dobson, K. S. (1989). A meta-analysis of the efficacy of cognitive therapy for depression. *Journal of Consulting and Clinical Psychology, 57,* 414–419.

Dorn, F. J. (1984). The social influence model: A social psychological approach to counseling. *Personnel and Guidance Journal, 62,* 342–345.

Dorn, F. J., & Day, B. J. (1985). Assessing change in self-concept: A social psychological approach. *American Mental Health Counselors Association Journal, 7,* 180–186.

Dorris, M. (1995, Spring). Heroic possibilities. *Teaching tolerance* (pp. 11–15).

Dossey, L. (1993). *Healing words.* New York: Harper/Collins.

Dostalek, C. (1994). Physiological bases of yoga techniques in the prevention of diseases. CIANS-ISBM Satellite Conference Symposium: Lifestyle changes in the prevention and treatment of diseases. *Homeostasis in Health and Disease, 35,* 205–208.

Dowrick, P., & Raeburn, J. (1995). Self-modeling: Rapid skill training for children with physical disabilities. *Journal of Developmental and Physical Disabilities, 7,* 25–37.

Duan, C., & Hill, C. (1996). The current state of empathy research. *Journal of Counseling Psychology, 43,* 261–274.

Duhl, F. J., Kantor, D., Duhl, B. S. (1973). Learning, pace and action in family therapy: A primer of sculpture. In D. A. Bloch (Ed.), *Techniques of family psychotherapy.* New York: Grune & Stratton.

Duley, S. M., Cancelli, A. A., Kratochwill, T. R., Bergan, J. R., & Meredith, K. E. (1983). Training and generalization of motivational analysis interview assessment skills. *Behavioral Assessment, 5,* 281–293.

Duncan, S. P., Jr. (1972). Some signals and rules for taking speaking turns in conversations. *Journal of Personality and Social Psychology, 23,* 283–292.

Duncan, S. P., Jr. (1974). On the structure of speaker-auditor interaction during speaking turns. *Language and Society, 2,* 161–180.

D'Zurilla, T. J. (1986). *Problem-solving therapy: A social competence approach to clinical intervention.* New York: Springer.

D'Zurilla, T. J. (1988). Problem-solving therapies. In K. D. Dobson (Ed.), *Handbook of cognitive-behavioral therapies* (pp. 85–135). New York: Guilford.

D'Zurilla, T. J. & Goldfried, M. R. (1971). Problem-solving and behavior modification. *Journal of Abnormal Psychology, 78,* 107–126.

D'Zurilla, T. J., & Nezu, C. M. (Eds.). (1990). Development and preliminary evaluation of the Social Problem-Solving Inventory (SPSI). *Psychological Assessment: A Journal of Consulting and Clinical Psychology, 2,* 156–163.

Eckert, P. (1994). Cost control through quality improvement: The new challenge for psychology. *Professional Psychology, 25,* 3–8.

Edinger, J., & Radtke, R. (1993). Use of in vivo desensitization to treat a patient's claustrophobic response to nasal CPAP. *Sleep, 16,* 678–680.

Edwards, C., & Murdock, N. (1994). Characteristics of therapist self-disclosure in the counseling process. *Journal of Counseling & Development, 72,* 384–389.

Egan, G. (1975). *The skilled helper.* Pacific Grove, CA: Brooks/Cole.

Egan, G. (1990). *The skilled helper* (4th ed.). Pacific Grove, CA: Brooks/Cole.

Egan, G. (1994). *The skilled helper* (5th ed.). Pacific Grove, CA: Brooks/Cole.

Eisenberger, R., & Cameron, J. (1996). Detrimental effects of reward: Reality or myth? *American Psychologist, 51,* 1153–1166.

Ekman, P. (1964). Body position, facial expression and verbal behavior during interviews. *Journal of Abnormal and Social Psychology, 68,* 295–301.

Ekman, P. (1982). Methods for measuring facial action. In K. R. Scherer, & P. Ekman (Eds.), *Handbook of methods in nonverbal behavior research* (pp. 45–135). New York: Cambridge University Press.

Ekman, P. (Ed.). (1983). Emotion in the human face (2nd ed.). Cambridge, MA: Cambridge University Press.

Ekman, P., & Friesen, W. V. (1967). Head and body cues in the judgment of emotion: A reformulation. *Perceptual and Motor Skills, 24,* 711–724.

Ekman, P., & Friesen, W. V. (1969). Nonverbal leakage and clues to deception. *Psychiatry, 32,* 88–106.

Eliot, G. R., & Eisdorfer, C. (1982). *Stress and human health.* New York: Springer-Verlag.

Elkin, I., Shea, M. T., Watkins, J. T., Imber, S. D., Sotsky, S. M., Collins, J. F., Glass, D. R., Pilkonis, P. A., Leber, W. R., Docherty, J. P., Fiester, S. J., & Parloff, M. B. (1989). National Institute of Mental Health Treatment of Depression Collaborative Research Program. *Archives of General Psychiatry, 46,* 971–982.

Ellis, A. (1962). *Reason and emotion in psychotherapy.* New York: Lyle Stuart.

Ellis, A. (1984). *Rational-emotive therapy and cognitive behavior therapy.* New York: Springer.

Enns, C. (1993). Twenty years of feminist counseling and therapy. *The Counseling Psychologist, 21,* 33–87.

Enzle, M. E., Roggeveen, J. P., & Look, S. C. (1991). Self- versus other-reward administration and intrinsic motivation. *Journal of Experimental Social Psychology, 27,* 468–479.

Epstein, J., Botvin, G., Diaz, T., & Toth, V. (1995). Social and personal factors in marijuana use and intentions to use drugs among inner city minority youth. *Journal of Developmental and Behavioral Pediatrics, 16,* 14–20.

Epstein, M. (1995). *Thoughts without the thinker.* New York: Basic Books.

Erasmus, U. (1992). Behavioral treatment of needle phobia. *Tijdschrift voor Psychotherapie, 18,* 335–347.

Erickson, F. (1975). One function of proxemic shifts in face-to-face interaction. In A. Kendon, R. M. Harris, & M. R. Keys (Eds.), *Organization of behavior in face to face interactions* (pp. 175–187). Chicago, IL: Aldine.

Erickson, M. H., Rossi, E., & Rossi, S. (1976). *Hypnotic realities.* New York: Irvington.

Erikson, E. (1968). *Identity: Youth and crisis.* New York: Norton.

Espin, O. M. (1993). Psychological impact of migration on Latinas. In D. R. Atkinson, G. Morten, & D. W. Sue (Eds.), *Counseling American minorities* (pp. 279–296). Madison, WI: Brown & Benchmark.

Essandoh, P. K. (1995). Counseling issues with African college students in U.S. colleges and universities. *The Counseling Psychologist, 23,* 348–360.

Estes, C. (1992). *Women who run with the wolves.* New York: Ballantine.

Exline, R. V., & Winters, L. C. (1965). Affective relations and mutual glances in dyads. In S. S. Tompkins & C. E. Izard (Eds.), *Affect, cognition, and personality.* New York: Springer.

Fairburn, C. G., Jones, R., Peveler, R. C., Hope, R. A., & O'Conner, M. (1993). Psychotherapy and bulimia nervosa: Longer-term effects of interpersonal psychotherapy, behavior therapy, and cognitive behavior therapy. *Archives of General Psychiatry, 50,* 419–428.

Falco, K. L. (1991). *Psychotherapy with lesbian clients: Theory into practice.* New York: Brunner/Mazel.

Falloon, R. H., Boyd, J. L., McGill, C. W., Williamson, M., Razani, A., Moss, H. B., Giulderman, A. M., & Simpson, G. M. (1985). Family management in the prevention of morbidity of schizophrenia: Clinical outcome of a two-year longitudinal study. *Archives of General Psychiatry, 42,* 887–896.

Faludi, S. (1991). *Backlash: The undeclared war against American women.* New York: Crown Books.

Farhi, D. (1996). *The breathing book.* New York: Holt.

Faust, J., Olson, R., & Rodriguez, H. (1991). Same-day surgery preparation: Reduction of pediatric patient arousal and distress through participating modeling. *Journal of Consulting and Clinical Psychology, 59,* 475–478.

Feeney, J. A., Noller, P., & Hanrahan, M. (1994). Assessing adult attachment. In M. B. Sperling & W. H. Berman (Eds.), *Attachment in adults* (pp. 128–152). New York: Guilford.

Feinbach, B. E., Winstead, B. A., & Derlega, V. J. (1989). Sex differences in diagnosis and treatment recommendations for anti-social personality and somatization disorders. *Journal of Social and Clinical Psychology, 8,* 238–255.

Feindler, E. L., Ecton, R. B., Kingsley, D., & Dubey, D. R. (1986). Group anger-control training for institutionalized psychiatric male adolescents. *Behavior Therapy, 17,* 109–123.

Fensterheim, H. (1983). Basic paradigms, behavioral formulation and basic procedures. In H. Fensterheim & H. Glazer (Eds.), *Behavioral psychotherapy: Basic principles and case studies in an integrative clinical model* (pp. 40–87). New York: Brunner/Mazel.

Feuerstein, G., & Bodian, S. (Eds.). (1993). *Living yoga.* New York: Jeremy P. Tarcher/Perigee Books.

Fezler, W. (1989). *Creative imagery: How to visualize in all five senses.* New York: Simon & Schuster.

Field, L. K., & Steinhardt, M. A. (1992). The relationship of internally directed behavior to self-reinforcement, self-esteem, and expectancy values for exercise. *American Journal of Health Promotion, 7,* 21–27.

Fisch, R., Weakland, J., & Segal, L. (1982). *The tactics of change: Doing therapy briefly.* San Francisco, CA: Jossey-Bass.

Fischer, J., & Corcoran, K. (1994). *Measures for clinical practice* (2nd ed.). New York: Free Press.

Fishelman, L. (1991, March). Algorithm construction and use. In *Developing algorithms and treatment guidelines in mental health.* Institute sponsored by Harvard Community Health Plan and Harvard Medical School, Boston, MA.

Fishman, S. T., & Lubetkin, B. S. (1983). Office practice of behavior therapy. In M. Hersen (Ed.), *Outpatient behavior therapy: A clinical guide.* New York: Grune & Stratton.

Foa, E. B., Rothbaum, B. O., Riggs, D. S., & Murdock, T. B. (1991). Treatment of posttraumatic stress disorder in rape victims: A comparison between cognitive-behavioral procedures and counseling. *Journal of Consulting and Clinical Psychology, 59,* 715–723.

Foa, E. B., Steketee, G. S., & Ascher, L. M. (1980). Systematic desensitization. In A. Goldstein & E. B. Foa (Eds.), *Handbook of behavioral interventions: A clinical guide* (pp. 38–91). New York: Wiley.

Fodor, I. G. (1992). The agoraphobic syndrome: From anxiety neurosis to panic disorder. In L. S. Brown & M. Ballou (Eds.), *Personality and psychopathology: Feminist reappraisals* (pp. 177–205). New York: Guilford.

Fong, M. L., & Cox, B. G. (1983). Trust as an underlying dynamic in the counseling process: How clients test trust. *Personnel and Guidance Journal, 62,* 163–166.

Fontana, D. (1991). *The elements of meditation.* New York: Element.

Ford, M., & Widiger, T. A. (1989). Sex bias in the diagnosis of histrionic and antisocial personality disorders. *Journal of Consulting and Clinical Psychology, 57,* 301–305.

Foss, D., & Hadfield, O. (1993). A successful clinic for the reduction of mathematics anxiety among college students. *College Student Journal, 27,* 157–165.

Freiberg, P. (1995, May). Older people thrive with the right therapy. *American Psychological Association Monitor,* p. 38.

Freire, P. (1972). *Pedagogy of the oppressed.* New York: Herder & Herder.

Fretz, B. R., Corn, R., Tuemmler, J. M., & Bellet, W. (1979). Counselor nonverbal behaviors and client evaluations. *Journal of Counseling Psychology, 26,* 304–311.

Fried, R. (1990). *The breathing connection.* New York: Plenum.

Fried, R. (1993). The role of respiration in stress and stress control: Toward a theory of stress as a hypoxic phenomenon. In P. M. Lehrer & R. L. Woolfolk (Eds.), *Principles and practice of stress management* (2nd ed., pp. 301–331). New York: Guilford.

Friedman, H. S., & Miller-Herringer, T. (1991). Nonverbal display of emotion in public and in private: Self-monitoring, personality, and expressive cues. *Journal of Personality and Social Psychology, 61,* 766–775.

Friedman, L. C., Nelson, D. V., Webb, J. A., & Hoffman, L. P. (1994). Dispositional optimism, self-efficacy, and health beliefs as predictors of breast-self-examination. *American Journal of Preventive Medicine, 10,* 130–135.

Gaines, R., & Price-Williams, P. (1990). Dreams and imaginative processes in American and Balinese artists. *Psychiatric Journal of the University of Ottawa, 15,* 107–110.

Galassi, J. P., & Gersh, T. (1993). Myths, misconceptions, and missed opportunity: Single-case designs and counseling psychology. *Journal of Counseling Psychology, 40,* 525–531.

Gambrill, E. D. (1977). *Behavior modification: Handbook of assessment, intervention, and evaluation.* San Francisco, CA: Jossey-Bass.

Gambrill, E. D. (1990). *Critical thinking in clinical practice.* San Francisco, CA: Jossey-Bass.

Gasch, H., Poulson, D., Fullilove, R., & Fullilove, M. (1991). Shaping AIDS education and prevention programs for African Americans amidst community decline. *Journal of Negro Education, 60,* 85–96.

Gazda, G. M., Asbury, F. S., Balzer, F., Childers, W. C., Phelps, R. E., & Walters, R. P. (1995). *Human relations development: A manual for educators* (5th ed.). Needham, MA: Allyn & Bacon.

Gelso, C. J., & Carter, J. A. (1985). The relationship in counseling and psychotherapy: Components, consequences, and theoretical antecedents. *The Counseling Psychologist, 13,* 155–243.

Gelso, C. J., & Carter, J. A. (1994). Components of the psychotherapy relationship: Their interaction and unfolding during treatment. *Journal of Counseling Psychology, 41,* 296–306.

Gelso, C. J., & Fretz, B. R. (1992). *Counseling psychology.* New York: Harcourt Brace Jovanovich.

Gendlin, E. T. (1996). *Focusing-oriented psychotherapy: A manual of the experiential method.* New York: Guilford.

Gilbert, B. O., Johnson, S. B., Spillar, R., McCallum, M., Silverstein, J. H., & Rosenbloom, A. (1982). The effects of a peer-modeling film on children learning to self-inject insulin. *Behavior Therapy, 13,* 186–193.

Gilliland, B. E., James, R. K., & Bowman, J. T. (1994). *Theories and strategies in counseling and psychotherapy* (3rd ed.). Englewood Cliffs, NJ: Prentice-Hall.

Gladstein, G. (1983). Understanding empathy: Integrating counseling, development, and social psychology perspectives. *Journal of Counseling Psychology, 30,* 467–482.

Goldfried, M., & Wolfe, B. (1996). Psychotherapy practice and research: Repairing a strained alliance. *American Psychologist, 51,* 1007–1016.

Goldfried, M. R. (1983). The behavior therapist in clinical practice. *Behavior Therapy, 6,* 45–46.

Goldfried, M. R. (1995). *From cognitive behavior therapy to psychotherapy integration.* New York: Springer.

Goldfried, M. R., & Davison, G. C. (1976). *Clinical behavior therapy.* New York: Holt, Rinehart and Winston.

Goldfried, M. R., & Davison, G. C. (1994). *Clinical behavior therapy.* (Expanded ed.). New York: Wiley Interscience.

Goldfried, M. R., Greenberg, L. S., & Marmer, C. (1990). Individual psychotherapy: Process and outcome. *Annual Review of Psychology, 41,* 659–688.

Goldiamond, I., & Dyrud, J. E. (1967). Some applications and implications of behavioral analysis in psychotherapy. In J. Schlein (Ed.), *Research in psychotherapy* (Vol. 3). Washington, DC: American Psychological Association.

Goldstein, A. P. (1971). *Psychotherapeutic attraction.* New York: Pergamon Press.

Goldstein, A. P., & Higginbotham, H. N. (1991). Relationship-enhancement methods. In F. H. Kanfer & A. P. Goldstein (Eds.), *Helping people change* (4th ed., pp. 20–69). New York: Pergamon Press.

Goleman, D. (1988). *The meditative mind.* New York: Jeremy P. Tarcher/Perigee.

Goleman, D., & Gurin, J. (1993). *Mind/body medicine.* New York: Consumer Report Books.

Goodwin, R., & Soon, A. P. Y. (1994). Self-monitoring and relationship adjustment: A cross-cultural analysis. *Journal of Social Psychology, 134,* 35–39.

Gorrell, J. (1993). Cognitive modeling and implicit rules: Effects on problem-solving performance. *American Journal of Psychology, 106,* 51–65.

Gottlieb, L. (1991, March). Algorithm-based clinical quality improvement. *Developing algorithm and treatment guidelines in mental health.* Institute sponsored by Harvard Community Health Plan and Harvard Medical School, Boston, MA.

Gottman, J. M., & Leiblum, S. R. (1974). *How to do psychotherapy and evaluate it.* New York: Holt, Rinehart and Winston.

Goulding, M., & Goulding, R. (1978). *The power is in the patient: A TA/Gestalt approach to psychotherapy.* San Francisco: TA Press.

Graves, J. R., & Robinson, J. D. (1976). Proxemic behavior as a function of inconsistent verbal and nonverbal messages. *Journal of Counseling Psychology, 23,* 333–338.

Greenberg, L. S., & Goldman, R. L. (1988). Training in experiential therapy. *Journal of Consulting and Clinical Psychology, 56,* 696–702.

Greenberg, L. S., Rice, L. N., & Elliott, R. (1993). *Facilitating emotional change.* New York: Guilford.

Greenberger, D., & Padesky, C. A. (1995). *Mind over mood: A cognitive therapy treatment manual for clients.* New York: Guilford.

Greenson, R. R. (1967). *The technique and practice of psychoanalysis* (Vol. 1). Madison, CT: International University Press.

Greenwald, R. (1996). The information gap in the EMDR controversy. *Professional Psychology: Research and Practice, 27,* 67–72.

Greiner, J., & Karoly, P. (1976). Effects of self-control training on study activity and academic performance: An analysis of self-monitoring, self-reward and systematic-planning components. *Journal of Counseling Psychology, 23,* 494–502.

Gresham, F. M., & Nagle, R. J. (1980). Social skills training with children: Responsiveness of modeling and coaching as a function of peer orientation. *Journal of Consulting and Clinical Psychology, 48,* 718–729.

Guglielmi, S., Cox, D., & Spyker, D. (1994). Behavioral treatment of phobic avoidance in multiple chemical sensitivity. *Journal of Behavior Therapy and Experimental Psychiatry, 25,* 197–209.

Guidano, V. F. (1995). Self-observation in constructivist psychotherapy. In R. A. Neimeyer & M. J. Mahoney (Eds.), *Constructivism in psychotherapy* (pp. 155–168). Washington, DC: American Psychological Association.

Gutierrez, L. (1990). Working with women of color: An empowerment perspective. *Social Work, 35,* 149–153.

Gutkin, T. B. (1993). Cognitive modeling: A means for achieving prevention in school-based consultation. *Journal of Educational and Psychological Consultation, 4,* 179–183.

Guterman, J. T. (1992). Disputation and reframing: Contrasting cognitive-change methods. *Journal of Mental Health Counseling, 14,* 440–456.

Hackney, H., & Cormier, L. S. (1994). *Counseling strategies and interventions* (4th ed.). Needham Heights, MA: Allyn & Bacon.

Hackney, H., & Cormier, L. S. (1996). *The professional counselor* (3rd ed.). Needham Heights, MA: Allyn & Bacon.

Haddock, C. K., Shadish, W. R., Klesges, R. C., & Stein, R. J. (1994). Treatment of childhood and adolescent obesity. *Annals of Behavioral Medicine, 16,* 235–244.

Hagopian, L. P., Weist, M. D., & Ollendick, T. H. (1990). Cognitive-behavior therapy with an 11-year old girl fearful of AIDS infection, other diseases, and poisoning: A case study. *Journal of Anxiety Disorders, 4,* 257–265.

Hains, A. (1989). An anger-control intervention with aggressive delinquent youths. *Behavioral Residential Treatment, 4,* 213–230.

Haire-Joshu, D., Fisher, E., Munro, J., & Wedner, J. (1993). A comparison of patient attitudes toward asthma self-management among acute and preventive care settings. *Journal of Asthma, 30,* 359–371.

Haley, J. (1987). *Problem-solving therapy* (2nd ed.). San Francisco: Jossey-Bass.

Halford, W. K., Gravestock, F. M., Lowe, R., & Scheldt, S. (1992). Toward a behavioral ecology of stressful marital interactions. *Behavioral Assessment, 14,* 199–217.

Hall, E. T. (1966). *The hidden dimension.* Garden City, NY: Doubleday.

Halley, F. M. (1991). Self-regulation of the immune system through biobehavioral strategies. *Biofeedback and Self-Regulation, 16,* 55–74.

Hammond, R., & Yung, B. (1991). Preventing violence in at-risk African American youth. *Journal of Health Care for the Poor and Underserved, 2,* 359–373.

Hanh, T. N. (1976). *The miracle of mindfulness.* Boston: Beacon Press.

Hanh, T. N. (1991). *Peace is every step.* New York: Bantam.

Hanna, F. J., & Ritchie, M. H. (1995). Seeking the active ingredients of psychotherapeutic change: Within and outside the context of therapy. *Professional Psychology, 26,* 176–183.

Hansen, J. C., Rossberg, R., & Cramer, S. (1994). *Counseling theory and process* (5th ed.). Needham Heights, MA: Allyn & Bacon.

Harchik, A. E., Sherman, J. A., & Sheldon, J. B. (1992). The use of self-management procedures by people with developmental disabilities: A brief review. *Research in Developmental Disabilities, 13(3),* 211–227.

Harris, G. (1991). Hypnotherapy for agoraphobia: A case study. *International Journal of Psychosomatics, 38,* 92–94.

Harris, K. R., Graham, S., Reid, R., & McElroy, K. (1994). Self-monitoring of attention versus self-monitoring of performance: Replication and cross-task comparison studies. *Learning Disability Quarterly, 17,* 121–139.

Hatch, M., & Paradis, C. (1993). Panic disorder with agoraphobia. A focus on group treatment with African Americans. *The Behavior Therapist, 16,* 240–242.

Hays, P. A. (1995). Multicultural applications of cognitive-behavior therapy. *Professional Psychology, 26,* 309–315.

Heffernan, T., & Richards, C. S. (1981). Self-control of study behavior: Identification and evaluation of natural methods. *Journal of Counseling Psychology, 28,* 361–364.

Heimberg, R. G., Dodge, C. S., Hope, D. A., Kennedy, C. R., & Zollo, L. J. (1990). Cognitive behavioral group treatment for social phobia: Comparison with credible placebo control. *Cognitive Therapy and Research, 14,* 1–23.

Hein, E. C. (1980). *Communication in nursing practice* (2nd ed.). Boston, MA: Little, Brown.

Helms, J. E. (Ed.). (1990a). *Black and white racial identity: Theory, research and practice.* New York: Greenwood Press.

Helms, J. E. (1990b). Counseling attitudinal and behavioral predispositions: The black/white interaction model. In J. E. Helms (Ed.), *Black and white racial identity: Theory, research, and practice* (pp. 145–163). New York: Greenwood Press.

Hendricks, G. (1995). *Conscious breathing.* New York: Bantam.

Heppner, P. P. (1988). *The problem solving inventory.* Palo Alto, CA: Consulting Psychologist Press.

Heppner, P. P., & Claiborn, C. D. (1989). Social influence research in counseling: A review and critique. *Journal of Counseling Psychology, 36,* 365–387.

Heppner, P. P., & Heesacker, M. (1982). Interpersonal influence process in real-life counseling: Investigating client perceptions, counselor experience level, and counselor power over time. *Journal of Counseling Psychology, 29,* 215–223.

Heppner, P. P., & Heesacker, M. (1983). Perceived counselor characteristics, client expectations, and client satisfaction with counseling. *Journal of Counseling Psychology, 30,* 31–39.

Herman, J. L. (1994, April). *Women's pathways to healing.* Paper presented at the Learning from Women Conference, Boston, MA.

Hermansson, G. L., Webster, A. C., & McFarland, K. (1988). Counselor deliberate postural lean and communication of facilitative conditions. *Journal of Counseling Psychology, 35,* 149–153.

Herring, R., & Meggert, S. (1994). The use of humor as a counselor strategy with Native American Indian children. *Elementary School Guidance and Counseling, 29,* 67–76.

Herron, W. G., Javier, R. A., Primavera, L. H., & Schultz, C. L. (1994). The cost of psychotherapy. *Professional Psychology, 25,* 106–110.

Hess, E. H. (1975). *The tell-tale eye.* New York: Van Nostrand Reinhold.

Hill, C. E., & Corbett, M. (1993). A perspective on the history of process and outcome research in counseling psychology. *Journal of Counseling Psychology, 40,* 3–24.

Hill, C. E., Siegelman, L., Gronsky, B. R., Sturniolo, F., & Fretz, B. R. (1981). Nonverbal communication and counseling outcome. *Journal of Counseling Psychology, 28,* 203–212.

Ho, D. Y. F. (1995). Internalized culture, culturocentrism, and transcendence. *The Counseling Psychologist, 23,* 4–24.

Hoffman, S., Herzog-Bronsky, R., & Zim, S. (1994). Dialectical psychotherapy of phobias: A case study. *International Journal of Short Term Psychotherapy, 9,* 229–233.

Hoffman-Graff, M. A. (1977). Interviewer use of positive and negative self-disclosure and interviewer-subject sex pairing. *Journal of Counseling Psychology, 24,* 184–190.

Hogarty, G. E., Anderson, C. M., Reiss, D. J., Kornblith, S. J., Greenwald, D. P., Janva, C. D., & Madonia, M. J. (1986). Family psychoeducation, social skills training, and maintenance chemotherapy in the aftercare treatment of schizophrenia: 1. One-year effects of a controlled study on relapse and expressed emotion. *Archives of General Psychiatry, 43,* 633–642.

Hollon, S. (1996). The efficacy and effectiveness of psychotherapy relative to medications. *American Psychologist, 51,* 1025–1030.

Hollon, S. D., & Kendall, P. C. (1981). In vivo assessment techniques for cognitive-behavioral processes. In P. C. Kendall & S. D. Hollon (Eds.), *Assessment strategies for cognitive-behavioral interventions* (pp. 319–362). New York: Academic Press.

Holmbeck-Grayson, N., & Lavigne, J. V. (1992). Combining self-modcling and stimulus fading in the treatment of an electively mute child. *Psychotherapy, 29,* 661–667.

Homme, L. E. (1965). Perspectives in psychology: XXIV. Control of coverants, the operants of the mind. *Psychological Record, 15,* 505–511.

Horne, D., Vatmanidis, P., & Careri, A. (1994). Preparing patients for invasive medical and surgical procedures: II. Using psychological interventions with adults and children. *Behavioral Medicine, 20,* 15–21.

Horvath, A. O., & Symonds, B. D. (1991). Relation between working alliance and outcome in psychotherapy: A meta-analysis. *Journal of Counseling Psychology, 38,* 139–149.

Hosford, R. E. (1974). *Counseling techniques: Self-as-a-model film.* Washington, DC: American Personnel and Guidance Press.

Hosford, R. E., & de Visser, L. (1974). *Behavioral approaches to counseling: An introduction.* Washington, DC: American Personnel and Guidance Press.

Hosford, R. E., Moss, C., & Morrell, G. (1976). The self-as-a-model technique: Helping prison inmates change. In J. D. Krumboltz & C. E. Thoresen (Eds.), *Counseling methods* (pp. 487–495). New York: Holt, Rinehart and Winston.

Howard, K. I., Kopta, S. M., Krause, M. S., & Orlinsky, D. E. (1986). The dose-effect relationship in psychotherapy. *American Psychologist, 41,* 159–164.

Howath, A. O., & Symonds, B. D. (1991). Relation between working alliance and outcome in psychotherapy: A meta-analysis. *Journal of Counseling Psychology, 38,* 139–149.

Hoyt, W. T. (1996). Antecedents and effects of perceived therapist credibility: A meta-analysis. *Journal of Counseling Psychology, 43,* 430–447.

Huang, L. N. (1994). An integrative approach to clinical assessment and intervention with Asian-American adolescents. *Journal of Clinical Child Psychology, 23,* 21–31.

Hubble, M. A., Noble, F. C., & Robinson, S. E. (1981). The effect of counselor touch in an initial counseling session. *Journal of Counseling Psychology, 28,* 533–535.

Hughes, D. (1990). Participant modeling as a classroom activity. *Teaching of Psychology, 17,* 238–240.

Hull, C. L. (1980). *A behavior system.* New Haven, CT: Yale University Press.

Hurd, E., Moore, C., & Rogers, R. (1995). Quiet success: Parenting strengths among African Americans. *Families in Society, 76,* 434–443.

Hutchins, D., & Cole, C. (1992). *Helping relationships and strategies* (2nd ed.). Pacific Grove, CA: Brooks/Cole.

Hutchins, J. (1995). AAMFT readers meet, set agenda. *Family Therapy News,* June 5.

Ibanez-Ramirez, M., Delgado-Morales, F., & Pulido-Diez, J. (1989). Secondary situational impotence in case of homosexuality. *Analisis y Modificacion de Conducta, 15,* 297–304.

Ilacqua, G. E. (1994a). Migraine headaches: Coping efficacy of guided imagery training. *Headache, 34,* 99–102.

Ilacqua, G. E. (1994b). *In the mind's eye: The power of imagery for personal enrichment.* New York: Guilford.

Itai, G., & McRae, G. (1994). Counseling older Japanese American clients: An overview and observation. *Journal of Counseling and Development, 72,* 373–377.

Ivey, A. E. (1994). *Intentional interviewing and counseling* (3rd ed.). Pacific Grove, CA: Brooks/Cole.

Ivey, A. E., & Gluckstern, N. (1984). *Basic influencing skills: Participant manual.* Amherst, MA: Microtraining Associates.

Ivey, A. E., Gluckstern, N. B., & Ivey, M. B. (1993). *Basic attending skills* (3rd ed.). North Amherst, MA: Microtraining Associates.

Ivey, A. E., Ivey, M. B., & Simek-Downing, L. (1987). *Counseling and psychotherapy: Skills, theories, and practice* (2nd ed.). Englewood Cliffs, NJ: Prentice-Hall.

Ivey, A. E., Ivey, M. B., & Simek-Morgan, L. (1993). *Counseling and psychotherapy: A multicultural perspective* (3rd ed.). Needham Heights, MA: Allyn & Bacon.

Iwamasa, G. Y. (1993). Asian Americans and cognitive behavioral therapy. *The Behavior Therapist, 16,* 233–235.

Iyengar, B. K. S. (1988). *The tree of yoga.* Boston, MA: Shambhala.

Jackson, G. G. (1987). Cross-cultural counseling with Afro-Americans. In P. Pedersen (Ed.), *Handbook of cross-cultural counseling and therapy* (pp. 231–237). New York: Praeger.

Jacob, T., Penn, N., Kulik, J., & Spieth, L. (1992). Effects of cognitive style and maintenance strategies of breast self-examination (BSE) practice by African American women. *Journal of Behavioral Medicine, 15,* 589–609.

Jacobs, G. D., Benson, H., & Friedman, R. (1993). Home-based central nervous system assessment of a multifactor behavioral intervention for chronic sleep-onset insomnia. *Behavior Therapy, 24,* 159–174.

Jacobs, G. D., Rosenberg, P. A., Friedman, R., & Matheson, J. (1993). Multifactor behavioral treatment of chronic sleep-onset insomnia using stimulus control and the relaxation response: A preliminary study. *Behavior Modification, 17,* 498–509.

Jacobson, E. (1929). *Progressive relaxation.* Chicago: University of Chicago Press.

Jacobson, E. (1964). *Anxiety and tension control.* Philadelphia: Lippincott.

Jacobson, N. & Hollon, S. (1996). Cognitive behavior therapy vs. pharmacotherapy: Now that the jury's returned it's verdict, it is time to present the rest of the evidence. *Journal of Consulting and Clinical Psychology, 64,* 74–80.

Jacobson, N. S., & Follette, W. C. (1985). Clinical significance of improvement resulting from two behavioral marital therapy components. *Behavior Therapy, 16,* 133–145.

Jacobson, N. S., Follette, W. C., & Revenstorf, D. (1984). Psychotherapy outcome research: Methods of reporting variability and evaluating clinical significance. *Behavior Therapy, 15,* 336–352.

Jacobson, N. S., & Truax, P. (1991). Clinical significance: A statistical approach to defining meaningful change in psychotherapy research. *Journal of Consulting and Clinical Psychology, 59,* 12–19.

Jaffee v. Redmond., 116 s. ct. 1923, (1996).

Jay, S. M., Elliott, C. H., Ozolins, M., Olson, R. A., & Pruitt, S. D. (1985). Behavioral management of children's distress during painful medical procedures. *Behaviour Research & Therapy, 23,* 513–520.

Johanson, J., & Kurtz, R. (1991). *Grace unfolding: Psychotherapy in the spirit of the Tao-te Ching.* New York: Bell Tower, Division of Crown.

Johnson, D. W. (1993). *Reaching out: Interpersonal effectiveness and self-actualization* (5th ed.). Needham Heights, MA: Allyn & Bacon.

Johnson, M. E. (1989). Effects of self-observation and self-as-a-model on counselor trainees' anxiety and self-evaluation. *Clinical Supervision, 7,* 59–70.

Johnson, S. M., & Greenberg, L. S. (1985). Differential effects of experiential and problem-solving interventions in resolving marital conflict. *Journal of Consulting and Clinical Psychology, 53,* 138–147.

Johnson, W. B., & Ridley, C. R. (1992). Brief Christian and non-Christian rational-emotive therapy with depressed Christian clients: An exploratory study. *Counseling and Values, 36,* 220–229.

Jones, A. S., & Gelso, C. (1988). Differential effects of style of interpretation: Another look. *Journal of Counseling Psychology, 35,* 363–369.

Jones, E. E. (1993). Introduction to special section: Single-case research in psychotherapy. *Journal of Consulting and Clinical Psychology, 61,* 371–372.

Jones, E. E., Kanouse, D., Kelley, H. H., Wisbett, R. E., Valins, S., & Weiner, B. (Eds.). (1972). *Attribution: Perceiving the causes of behavior.* Morristown, NJ: General Learning Press.

Josselson, R. (1992). *The space between us.* San Francisco: Jossey-Bass.

Joy, W. B. (1979). *Joy's way.* New York: Tarcher/Putnam.

Kabat-Zinn, J. (1990). *Full catastrophe living.* New York: Dell.

Kabat-Zinn, J. (1993). Meditation. In B. Moyers (Ed.), *Healing and the mind* (pp. 115–144). New York: Doubleday.

Kabat-Zinn, J. (1994). *Wherever you go there you are.* New York: Hyperion.

Kabat-Zinn, J. (1995). *Mindfulness meditation.* (Audio tape). Niles, IL: Nightingale Conant.

Kahn, J. S., Kehle, T. J., Jenson, W. R., & Clark, E. (1990). Comparison of cognitive-behavioral, relaxation, and self-modeling interventions for depression among middle-school students. *School Psychology Review, 19,* 196–211.

Kahn, M. (1991). *Between therapist and client.* New York: W. H. Freeman.

Kahn, M., & Baker, B. L. (1968). Desensitization with minimal therapist contact. *Journal of Abnormal Psychology, 73,* 198.

Kamimura, E., & SaSaki, Y. (1991). Fear and anxiety reduction in systematic desensitization and imaging strategies: A comparison of response and stimulus oriented imaging. *Japanese Journal of Behavior Therapy, 17,* 29–38.

Kanfer, F. H. (1990). The scientist-practitioner connection: A bridge in need of constant attention. *Professional Psychology, 21,* 264–270.

Kanfer, F. H., & Gaelick-Buys, L. (1991). Self-management methods. In F. H. Kanfer & A. P. Goldstein (Eds.), *Helping people change* (4th ed., pp. 305–360). New York: Pergamon Press.

Kanfer, F. H., & Goldstein, A. P. (Eds.). (1991). *Helping people change.* New York: Pergamon Press.

Kanfer, F. H., & Saslow, G. (1969). Behavioral diagnosis. In C. M. Franks (Ed.), *Behavior therapy: Appraisal and status.* (pp. 417–444). New York: McGraw-Hill.

Kantor, J. R. (1970). An analysis of the experimental analysis of behavior (TEAB). *Journal of the Experimental Analysis of Behavior, 13,* 101–108.

Kantrowitz, R., & Ballou, M. (1992). A feminist critique of cognitive-behavioral therapy. In L. S. Brown & M. Ballou (Eds.), *Personality and psychopathology: Feminist reappraisals* (pp. 70–87). New York: Guilford.

Kaplan, H. I., & Sadock, B. J. (1995). *Modern synopsis of comprehensive textbook of psychiatry III* (6th ed.). Baltimore, MD: Williams & Wilkins.

Karasu, T., Docherty, J., Gelenberg, A., Kupfer, D., Merriam, A., & Shadoan, R. (1993). Practice guidelines for major depressive disorder in adults. *American Journal of Psychiatry, 150* (Supplement), 1–26.

Karenga, M. (1980). *Kawaida theory.* Los Angeles: Kawaida Publications.

Katz, P. A., & Walsh, P. V. (1991). Modification of children's gender-stereotyped behavior. *Child Development, 62,* 338–351.

Kazdin, A. E. (1974). Self-monitoring and behavior change. In M. J. Mahoney & C. E. Thoresen (Eds.), *Self-control: Power to the person* (pp. 218–246). Pacific Grove, CA: Brooks/Cole.

Kazdin, A. E. (1976). Developing assertive behavior through covert modeling. In J. D. Krumboltz & C. E. Thoresen (Eds.), *Counseling methods* (pp. 475–486). New York: Holt, Rinehart and Winston.

Kazdin, A. E. (1977). Assessing the clinical or applied importance of behavior change through social validation. *Behavior Modification, 1,* 427–452.

Kazdin, A. E. (1993). Evaluation in clinical practice: Clinically sensitive and systematic methods of treatment delivery. *Behavior Therapy, 24,* 11–15.

Kazdin, A. E., & Bass, D. (1989). Power to detect differences between alternative treatments in comparative psychotherapy outcome research. *Journal of Consulting and Clinical Psychology, 57,* 138–147.

Kazdin, A. E., & Wilcoxon, L. A. (1976). Systematic desensitization and nonspecific treatment effects: A methodological evaluation. *Psychological Bulletin, 83,* 729–758.

Keane, T. M., Black, J. L., Collins, F. L., Jr., & Venson, M. C. (1982). A skills training program for teaching the behavioral interview. *Behavioral Assessment, 4,* 53–62.

Keefe, F. J., Dunsmore, J., & Burnett, R. (1992). Behavioral and cognitive-behavioral approaches to chronic pain: Recent advances and future directions. *Journal of Consulting and Clinical Psychology, 60,* 528–536.

Kehle, T. J., Owen, S. V., & Cressy, E. T. (1990). The use of self-modeling as an intervention in school psychology: A case study of an elective mute. *School Psychology Review, 19,* 115–121.

Kelley, C. R. (1979). Freeing blocked anger. *The Radix Journal, 1*(3), 19–33.

Kelly, C., & Cooper, S. (1993). Panic and agoraphobia associated with a cerebral arteriovenous malformation. *Irish Journal of Psychological Medicine, 10,* 94–96.

Kelly, G. A. (1955). *The psychology of personal construct* (2 vols.). New York: Norton.

Kelly, J., & St. Lawrence, J. (1990). The impact of community-based groups to help persons reduce HIV infection risk behaviors. *AIDS Care, 2,* 25–36.

Kennedy, P., Doherty, N., & Barnes, J. (1995). Primary vaginismus: A psychometric study of both partners. *Sexual and Marital Therapy, 10,* 9–22.

Kent, H. (1993). *Yoga made easy.* Allentown, PA: The People's Medical Society.

Kern, J. M. (1982). The comparative external and concurrent validity of three role-plays for assessing heterosocial performance. *Behavior Therapy, 13,* 666–680.

Kiesler, D. J. (1966). Some myths of psychotherapy research and the search for a paradigm. *Psychological Bulletin, 65,* 110–136.

Kim, Y. O. (1995). Cultural pluralism and Asian-Americans: Culturally sensitive social work practice. *International Social Work, 38,* 69–78.

King, N. (1993). Simple and social phobias. *Advances in Clinical Child Psychology, 15,* 305–341.

Kiresuk, T. J., & Sherman, R. E. (1968). Goal attainment scaling: A general method for evaluating comprehensive mental health programs. *Community Mental Health Journal, 4,* 443–453.

Kleinknecht, R. (1991). *Mastering anxiety: The nature and treatment of anxious conditions.* New York: Plenum.

Kleinknecht, R., Dinnel, D., Tanouye-Wilson, S., & Lonner, W. (1994). Cultural variation in social anxiety and phobia: A study of *Taijin Kyofusho. The Behavior Therapist, 17,* 175–178.

Kliewer, W., & Lewis, H. (1995). Family influences on coping processes in children and adolescents with sickle cell disease. *Journal of Pediatric Psychology, 20,* 511–525.

Klorman, R., Hilpert, P. L., Michael, R., LaGana, C., & Sveen, O. B. (1980). Effects of coping and mastery modeling on experienced and inexperienced pedodontic patients' disruptiveness. *Behavior Therapy, 11,* 156–168.

Knapp, M. L., & Hall, J. (1992). *Nonverbal communication in human interaction* (3rd ed.). Orlando, FL: Holt, Rinehart & Winston.

Kohut, H. (1971a). *The analysis of the self.* New York: International Universities Press.

Kohut, H. (1971b). *The restoration of the self.* Madison, CT: International Universities Press.

Kohut, H. (1984). *How does analysis cure?* Chicago: University of Chicago Press.

Korth, E. (1993). Psychological operation-preparation with children: Preparation of a five-year old boy for the amputation of his left leg with play, family, and behavior therapy. *Zeitschrift fur Klinische Psychologie, 22,* 62–76.

Kraft, T. (1994). The combined use of hypnosis and *in vivo* desensitization in the successful treatment of a case of balloon phobia. *Contemporary Hypnosis, 11,* 71–76.

Krepcho, M., Fernandez-Esquer, M., Freeman, A., & Magee, E. (1993). Predictors of bleach use among current African-American injecting drug users: A community study. *Journal of Psychoactive Drugs, 25,* 135–141.

Krivonos, P. D., & Knapp, M. L. (1975). Initiating communication: What do you say when you say hello? *Central States Speech Journal, 26,* 115–125.

Krumboltz, J. D., & Thoresen, C. E. (Eds.). (1976). *Counseling methods.* New York: Holt, Rinehart and Winston.

Kuehlwein, K. T. (1992). Working with gay men. In A. Freeman & F. M. Dattillio (Eds.), *Comprehensive casebook of cognitive therapy* (pp. 249–255). New York: Plenum.

Kupfersmid, J. (1989). Treatment of nocturnal enuresis: A status report. *The Psychiatric Forum, 14,* 37–46.

Kutz, L., & Derevensky, J. L. (1993). The effects of divorce on perceived self-efficacy and behavioral control in elementary school children. *Journal of Divorce and Remarriage, 20,* 75–90.

LaCrosse, M. B. (1980). Perceived counselor social influence and counseling outcomes: Validity of the counselor rating form. *Journal of Counseling Psychology, 27,* 320–327.

LaFromboise, T., Coleman, H. L. K., & Gerton, J. (1993). Psychological impact of biculturalism: Evidence and theory. *Psychological Bulletin, 114,* 395–412.

LaFromboise, T., & Dixon, D. N. (1981). American Indian perception of trustworthiness in a counseling interview. *Journal of Counseling Psychology, 28,* 135–139.

LaFromboise, T., & Low, K. (1989). American Indian adolescents. In J. Gibbs & L. Hwang (Eds.), *Children of color* (pp. 114–147). San Francisco, CA: Jossey-Bass.

LaFromboise, T., Trimble, T. D., & Mohatt, G. (1993). Counseling intervention and American Indian tradition: An integrative approach. In D. Atkinson, G. Morten, & D. W. Sue (Eds.), *Counseling American minorities* (pp. 145–170). Madison, WI: Brown & Benchmark.

Lambert, M., & Bergin, A. (1992). Achievements and limitations of psychotherapy research. In D. K. Freedheim (Ed.), *History of psychotherapy: A century of change* (pp. 360–390). Washington, DC: American Psychological Association.

Lambert, M. J., Ogles, B. M., & Masters, K. S. (1992). Choosing outcome assessment devices: An organizational and conceptual scheme. *Journal of Counseling and Development, 70,* 527–532.

Lang, P. J., & Lazovik, A. (1963). Experimental desensitization of a phobia. *Journal of Abnormal and Social Psychology, 66,* 519–525.

Lang, P. J., Melamed, B. G., & Hart, J. (1970). A psychophysiological analysis of fear modification using an automated desensitization procedure. *Journal of Abnormal Psychology, 76,* 221.

Lankton, S. R. (1980). *Practical magic: A translation of basic neurolinguistic programming into clinical psychotherapy.* Cupertine, CA: Meta Publications.

Larsen, D. L., Attkisson, C. C., Hargreaves, W. A., & Nguyen, T. D. (1979). Assessment of client/patient satisfaction: Development of a general scale. *Evaluation and Program Planning, 2,* 197–207.

Larson, J. D. (1992). Anger and aggression management techniques through the Think First Curriculum. *Journal of Offender Rehabilitation, 18,* 101–117.

Laungani, P. (1993). Cultural differences in stress and its management. *Stress-Medicine, 9,* 37–43.

Lauver, P., & Jones, R. (1991). Factors associated with perceived career options in American Indian, White and Hispanic rural high school students. *Journal of Counseling Psychology, 38,* 159–166.

Lazarus, A. A. (1967). In support of technical eclecticism. *Psychological Reports, 21,* 415–416.

Lazarus, A. A. (1976). *Multimodal behavior therapy.* New York: Springer.

Lazarus, A. A. (1984). *In the mind's eye: The power of imagery for personal enrichment.* New York: Guilford.

Lazarus, A. A. (1989). *The practice of multimodal therapy.* Baltimore, MD: Johns Hopkins University Press.

Lazarus, B. D. (1993). Self-management and achievement of students with behavior disorders. *Psychology in the Schools, 30,* 67–74.

Leahy, R. (1996). *Cognitive therapy: Basic principles and implications.* Northvale, NJ: Aronson.

Lecomte, C., Bernstein, B. L., & Dumont, F. (1981). Counseling interactions as a function of spatial-environmental conditions. *Journal of Counseling Psychology, 28,* 536–539.

Lehrer, P. M., & Woolfolk, R. (1982). Self-report assessment of anxiety: Somatic, cognitive, and behavioral modalities. *Behavioral Assessment, 4,* 167–177.

Lehrer, P. M., & Woolfolk, R. L. (Eds.). (1993). *Principles and practice of stress management* (2nd ed.). New York: Guilford.

LePage, J. (1994). *Integrative yoga therapy manual.* Aptos, CA: Integrative Yoga Therapy.

Lerman, H. (1992). The limits of phenomenology: A feminist critique of the humanistic personality theories. In L. S. Brown & M. Ballou (Eds.), *Personality and psychotherapy: Feminist reappraisals.* New York: Guilford.

Leserman, J., Perkins, D., & Evans, D. (1992). Coping with the threat of AIDS: The role of social support. *American Journal of Psychiatry, 149,* 1514–1520.

LeShan, L. (1974). *How to meditate.* New York: Bantam.

LeShan, L. (1995). Mobilizing the life force, treating the individual. *Alternative Therapies, I,* 63–69.

Levin, R., & Gross, A. (1985). The role of relaxation in systematic desensitization. *Behaviour Research and Therapy, 23,* 187–196.

Levine, S. (1991). *Guided meditations, explorations and healings.* New York: Anchor.

Lewinsohn, P. M., Hoberman, H. M., & Clarke, G. N. (1989). The coping with depression course: Review and future directions. *Canadian Journal of Behavioural Science, 21,* 470–493.

Liberman, R. P. (1972). Behavioral modification of schizophrenia: A review. *Schizophrenia Bulletin, 16,* 37–48.

Lin, Meei-Ju, Kelly, K. R., & Nelson, R. C. (1996). A comparative analysis of the interpersonal process in school-based counseling and consultation. *Journal of Counseling Psychology, 43,* 389–393.

Lindh, R. (1987). Suggestive covert modeling as a method with disturbed pupils. *Jyaskyla Studies in Education, Psychology and Social Research, 60,* 1–194.

Linehan, M. M., Armstrong, H. E., Suarez, A., Allmon, D., & Heard, H. L. (1991). Cognitive-behavioral treatment of chronically parasuicidal borderline patients. *Archives of General Psychiatry, 48,* 1060–1064.

Lipke, H., & Botkin, A. (1992). Case studies of eye movement desensitization and reprocessing (EMDR) with chronic post-traumatic stress disorder. *Psychotherapy, 29,* 591–595.

Litrownik, A. J. (1982). Special considerations in the self-management training of the developmentally disabled. In P. Karoly & F. H. Kanfer (Eds.), *Self-management and behavior change* (pp. 315–352). New York: Pergamon Press.

Lomont, J. F. (1965). Reciprocal inhibition or extinction? *Behaviour Research and Therapy, 3,* 209–219.

Long, B. (1995). *Meditation.* London: Barry Long Books.

Long, L., & Prophit, P. (1981). *Understanding/responding: A communication manual for nurses.* Monterey, CA: Wadsworth Health Sciences.

Lopez, F. G. (1995). Contemporary attachment theory: An introduction with implications for counseling psychology. *The Counseling Psychologist, 23,* 395–415.

Lopez, M., & Mermelstein, R. (1995). A cognitive-behavioral program to improve geriatric rehabilitation outcome. *Gerontologist, 35,* 696–700.

LoPiccolo, J., & Stock, W. E. (1986). Treatment of sexual dysfunction. *Journal of Consulting and Clinical Psychology, 54,* 158–167.

Luborsky, L. (1984). *Principles of psychoanalytic psychotherapy: A manual for supportive expressive treatment.* New York: Basic Books.

Luborsky, L., Crits-Christoph, P., & Mellon, J. (1986). Advent of objective measures of the transference concept. *Journal of Consulting and Clinical Psychology, 54,* 39–47.

Luborsky, L., & DeRubeis, R. (1984). The use of psychotherapy treatment manuals: A small revolution in psychotherapy research style. *Clinical Psychology Review, 4,* 5–14.

Lum, D. (1996). *Social work practice and people of color* (3rd ed.). Pacific Grove, CA: Brooks/Cole.

Lum, L. C. (1976). The syndrome of habitual chronic hyperventilation. In O. Hill (Ed.), *Modern trends in psychosomatic medicine.* (Vol. 3). Boston: Butterworths.

Luzzo, D. A. (1995). The relative contributions of self-efficacy and locus of control to the prediction of career maturity. *Journal of College Student Development, 36,* 61–66.

Lyddon, W. J. (1995). First- and second-order change: Implications for rationalist and constructivist cognitive therapies. *Journal of Counseling & Development, 69,* 122–127.

MacKay, P., & Marlatt, G. (1990). Maintaining sobriety. *International Journal of the Addictions, 25,* 1257–1276.

Maddux, J. E. (Ed.). (1995). *Self-efficacy, adaptation, and adjustment.* New York: Plenum.

Madsen, J., Sallis, J. F., Rupp, J. W., & Senn, K. L. (1993). Relationship between self-monitoring of diet and exercise change and subsequent risk factor changes in children and adults. *Patient Education and Counseling, 21,* 61–69.

Mahoney, K., & Mahoney, M. J. (1976). Cognitive factors in weight reduction. In J. D. Krumboltz & C. E. Thoresen (Eds.), *Counseling methods* (pp. 99–105). New York: Holt, Rinehart and Winston.

Mahoney, M. J. (1971). The self-management of covert behavior: A case study. *Behavior Therapy, 2,* 575–578.

Mahoney, M. J. (1972). Research issues in self-management. *Behavior Therapy, 3,* 45–63.

Mahoney, M. J. (1974). *Cognition and behavior modification.* Cambridge, MA: Ballinger.

Mahoney, M. J. (1977). Some applied issues in self-monitoring. In J. Cone & R. Hawkins (Eds.), *Behavioral assessment: New directions in clinical psychology* (pp. 241–254). New York: Brunner/Mazel.

Mahoney, M. J. (1989). Scientific psychology and radical behaviorism: Important distinctions based in scientism and objectivism. *American Psychologist, 44,* 1372–1377.

Mahoney, M. J. (1991). *Human change processes.* New York: Basic Books.

Mahoney, M. J. (1995). Theoretical developments in the cognitive psychotherapies. In M. J. Mahoney (Ed.). *Cognitive and constructive psychotherapies* (pp. 3–19). New York: Springer.

Mahoney, M. J., Moura, N. G., & Wade, T. C. (1973). Relative efficacy of self-reward, self-punishment, and self-monitoring techniques for weight loss. *Journal of Consulting and Clinical Psychology, 40,* 404–407.

Mahoney, M. J., & Thoresen, C. E. (Eds.). (1974). *Self-control: Power to the person.* Pacific Grove, CA: Brooks/Cole.

Maibach, E., & Flora, J. A. (1993). Symbolic modeling and cognitive rehearsal: Using video to promote AIDS prevention self-efficacy. Special Issue: The role of communication in health promotion. *Communication Research, 20,* 517–545.

Mail, P. (1995). Early modeling of drinking behavior by Native American elementary school children playing drunk. *International Journal of the Addictions, 30,* 1187–1197.

Malgady, R., Rogler, T., & Costantino, G. (1990). Culturally sensitive psychotherapy for Puerto Rican children and adolescents: A program of treatment outcome research. *Journal of Consulting and Clinical Psychology, 58,* 704–712.

Mallinckrodt, B. (1991). Clients' representations of childhood emotional bonds with parents, social support, and formation of working alliance. *Journal of Counseling Psychology, 38,* 401–409.

Mallinckrodt, B. (1993). Session impact, working alliance, and treatment outcome in brief counseling. *Journal of Counseling Psychology, 40,* 25–32.

Mallinckrodt, B. (1996). Change in working alliance, social support, and psychological symptoms in brief therapy. *Journal of Counseling Psychology, 43,* 448–455.

Mallinckrodt, B., Coble, H., & Gantt, D. (1995). Working alliance, attachment memories, and social competencies of women in brief therapy. *Journal of Counseling Psychology, 42,* 79–84.

Mallinckrodt, B., Gantt, D., & Coble, H. (1995). Attachment patterns in the psychotherapy relationship: Development of the client attachment to therapist scale. *Journal of Counseling Psychology, 42,* 307–317.

Malow, R., Corrigan, S., Cunningham, S., & West, J. (1993). Psychosocial factors associated with condom use among African American drug abusers in treatment. *AIDS Education and Prevention, 5,* 244–253.

Mancoske, R. J., Standifer, D., & Cauley, C. (1994). The effectiveness of brief counseling services for battered women. *Research on Social Work Practice, 4,* 53–63.

Marks, I. (1994). Behavior therapy as an aid to self-care. *Current Directions in Psychological Science, 3,* 19–22.

Marks, I., & O'Sullivan, G. (1988). Drugs and psychological treatments for agoraphobia/panic and obsessive-compulsive disorders: A review. *British Journal of Psychiatry, 153,* 650–658.

Marlatt, G., & Gordon, J. R. (1985). *Relapse prevention.* New York: Guilford.

Marlatt, G., & Parks, G. A. (1982). Self-management of addictive behaviors. In P. Karoly & F. H. Kanfer (Eds.), *Self-management and behavior change* (pp. 443–448). New York: Pergamon Press.

Marquis, J. N., Morgan, W., & Piaget, G. (1973). *A guidebook for systematic desensitization* (3rd ed.). Palo Alto, CA: Veterans' Workshop.

Marsella, A., Friedman, M., Gerrity, E., & Scarsfield, R. C. (1996). *Ethnocultural aspects of posttraumatic stress disorder.* Washington, DC: American Psychological Association.

Marshall, W. L., Jones, R., Ward, T., Johnston, P., & Barbaree, H. E. (1991). Treatment outcome with sex offenders. *Clinical Psychology Review, 11,* 465–485.

Marten, P., & Heimberg, R. (1995). Toward an integration of independent practice and clinical research. *Professional Psychology, 26,* 48–53.

Martin, D. (1993). Coping with AIDS-risk reduction efforts among gay men. *AIDS Education and Prevention, 5,* 104–120.

Martin, K., & Hall, C. (1995). Using mental imagery to enhance intrinsic motivation. *Journal of Sport and Exercise Psychology, 17,* 54–65.

Mattick, R. P., Andrews, G., Hadzi-Pavlovic, D., & Christensen, H. (1990). Treatment of panic and agoraphobia: An integrative review. *Journal of Nervous and Mental Disease, 178,* 567–576.

Mattick, R. P., & Peters, L. (1988). Treatment of severe social phobia: Effects of guided exposure with and without cognitive restructuring. *Journal of Consulting and Clinical Psychology, 56,* 251–260.

Mattick, R. P., Peters, L., & Clarke, J. C. (1989). Exposure and cognitive restructuring for social phobia. *Behavior Therapy, 29,* 3–23.

Maurer, R. E., & Tindall, J. H. (1983). Effect of postural congruence on client's perception of empathy. *Journal of Counseling Psychology, 30,* 158–163.

McAuley, E., Shaffer, S., & Rudolph, D. (1995). Affective responses to acute exercise in elderly impaired males: The moderating effects of self-efficacy and age. *International Journal of Aging and Human Development, 41,* 13–27.

McCafferty, S. (1992). The use of private speech by adult second language learners: A cross-cultural study. *Modern Language Journal, 76,* 179–189.

McCarthy, P. (1982). Differential effects of counselor self-referent responses and counselor status. *Journal of Counseling Psychology, 29,* 125–131.

McCarthy, P., & Betz, N. (1978). Differential effects of self-disclosing versus self-involving counselor statements. *Journal of Counseling Psychology, 25,* 251–256.

McConnaughy, E., Prochaska, J., & Velicer, W. (1983). Stages of change in psychotherapy: Measurement and sample profiles. *Psychotherapy, 20,* 368–375.

McCracken, L., & Larkin, K. (1991). Treatment of paruresis with *in vivo* desensitization: A case report. *Journal of Behavior Therapy and Experimental Psychiatry, 22,* 57–62.

McFall, R. M. (1977). Parameters of self-monitoring. In R. B. Stuart (Ed.), *Behavioral self-management: Strategies, techniques and outcomes* (pp. 196–214). New York: Brunner/Mazel.

McFall, R. M., & Dodge, K. A. (1982). Self-management and interpersonal skills learning. In P. Karoly & F. H. Kanfer (Eds.), *Self-management and behavior change* (pp. 353–392). New York: Pergamon Press.

McGill, D. W. (1992). The cultural story in multicultural family therapy. *Families in Society, 73,* 339–349.

McGrath, T., Tsui, E., Humphries, S., & Yule, W. (1990). Successful treatment of a noise phobia in a nine-year old girl with systematic desensitization *in vivo. Educational Psychology, 10,* 79–83.

McGuigan, F. J. (1984). Progressive relaxation: Origins, principles, and clinical applications. In R. L. Woolfolk & P. M. Lehrer (Eds.), *Principles and practice of stress management* (pp. 12–42). New York: Guilford.

McGuigan, F. J. (1993). Progressive relaxation: Origins, principles, and clinical applications. In P. M. Lehrer & R. L. Woolfolk (Eds.), *Principles and practice of stress management* (pp. 17–52). New York: Guilford.

McNamee, S., & Gergen, K. J. (Eds.) (1992). *Therapy as social construction.* London: Sage.

McNeill, B. W., May, R. J., & Lee, V. J. (1987). Perceptions of counselor source characteristics by premature and successful terminators. *Journal of Counseling Psychology, 34,* 86–89.

Meador, B., & Rogers, C. (1984). Person-centered therapy. In R. J. Corsini (Ed.), *Current psychotherapies* (pp. 142–195). Itasca, IL: Peacock.

Mehan, H., Hubbard, L., & Villanueva, I. (1994). Forming academic identities: Accommodations without assimilation among involuntary minorities. *Anthropology and Education Quarterly, 25,* 91–117.

Meharg, S. S., & Lipsker, L. E. (1991). Parent training using videotape self-modeling. *Child and Family Behavior Therapy, 13,* 1–27.

Mehrabian, A. (1976). *Public places and private spaces.* New York: Basic Books.

Meichenbaum, D. H. (1985). *Stress-inoculation training.* New York: Pergamon Press.

Meichenbaum, D. H. (1993). Stress inoculation training: A 20-year update. In P. M. Lehrer & R. L. Woolfolk (Eds.), *Principles and practice of stress management* (2nd ed., pp. 373–406). New York: Guilford.

Meichenbaum, D. H. (1994). *A clinical handbook: Practical therapist manual for assessing and treating adults with post-traumatic stress disorder.* Waterloo, Ontario, Canada: Institute Press.

Meichenbaum, D. H., & Cameron, R. (1973). *Stress inoculation: A skills training approach to anxiety management.* Unpublished manuscript, University of Waterloo, Waterloo, Ontario, Canada.

Meichenbaum, D. H., & Cameron, R. (1983). Stress inoculation training: Toward a general paradigm on training coping skills. In D. H. Meichenbaum & M. E. Jaremko (Eds.), *Stress reduction and prevention* (pp. 115–157). New York: Plenum.

Meichenbaum, D. H., & Goodman, J. (1971). Training impulsive children to talk to themselves: A means of developing self-control. *Journal of Abnormal Psychology, 77,* 115–126.

Meichenbaum, D. H., & Turk, D. (1976). The cognitive-behavioral management of anxiety, anger, and pain. In P. O. Davidson (Ed.), *The behavioral management of anxiety, depression and pain.* New York: Brunner/Mazel.

Melamed, B. G., & Siegel, L. J. (1975). Reduction of anxiety in children facing hospitalization and surgery by use of filmed modeling. *Journal of Consulting and Clinical Psychology, 43,* 511–521.

Menzies, R., & Clarke, J. (1993). A comparison of *in vivo* and vicarious exposure in the treatment of childhood water phobia. *Behaviour Research and Therapy, 31,* 9–15.

Mesquita, B., & Frijda, N. (1992). Cultural variations in emotions: A review. *Psychological Bulletin, 112,* 179–204.

Meyer, R. G., & Deitsch, S. (1996). *The clinician's handbook* (4th ed.). Needham Heights, MA: Allyn & Bacon.

Middleton, M., & Cartledge, G. (1995). The effects of social skills instruction and parental involvement on the aggressive behavior of African American males. *Behavior Modification, 19,* 192–210.

Midori-Hanna, S., & Brown, J. (1995). *The practice of family therapy.* Pacific Grove, CA: Brooks/Cole.

Miller, J. B. (1991). The development of women's sense of self. In J. Jordan, A. Kaplan, J. B. Miller, I. Stiver, & J. Surrey (Eds.), *Women's growth in connection* (pp. 11–26). New York: Guilford.

Miller, N., & Magruder, K. (Eds.) (In press). *The cost-effectiveness of psychotherapy: A guide for practitioners, researchers, and policymakers.* New York: Wiley.

Miller, R. C. (1993). Working with the breath. In G. Feuerstein & S. Bodian (Eds.), *Living yoga* (pp. 27–39). New York: Tarcher/Perigee - Putnam.

Milne, C. R., & Dowd, E. T. (1983). Effect of interpretation style on counselor social influence. *Journal of Counseling Psychology, 51,* 603–606.

Mindess, H. (1988). *Makers of psychology: The personal factor.* New York: Insight Books.

Minuchin, S. (1974). *Families and family therapy.* Cambridge, MA: Harvard University Press.

Miranda, J., & Dwyer, E. V. (1993). Cognitive behavioral therapy for disadvantaged medical patients. *The Behavior Therapist, 16,* 226–228.

Mischel, W. (1973). *Personality and assessment* (2nd ed.). New York: Wiley.

Mitchell, K. R., & White, R. G. (1977). Behavioral self-management: An application to the problem of migraine headaches. *Behavior Therapy, 8,* 213–221.

Mitchell, R. W. (1991). *Documentation in counseling records.* Washington, DC: American Counseling Association.

Moncher, M., & Schinke, S. (1994). Group intervention to prevent tobacco use among Native American youth. *Research on Social Work Practice, 4,* 160–171.

Monro, R., Ghosh, A. K., & Kalish, D. (1989). *Yoga research bibliography: Scientific studies on yoga and meditation.* Cambridge, England: Yoga Biomedical Trust.

Moras, K., Telfer, L. A., & Barlow, D. H. (1993). Efficacy and specific effects data on new treatments: A case study strategy with mixed anxiety-depression. *Journal of Consulting and Clinical Psychology, 61,* 412–420.

Morganstern, K. P. (1988). Behavioral interviewing. In A. S. Bellack & M. Hersen (Eds.), *Behavioral assessment: A practical handbook* (3rd ed., pp. 86–118). New York: Pergamon Press.

Morran, D. K., Kurpius, D. J., Brack, G., & Rozecki, T. G. (1994). Relationship between counselors' clinical hypotheses and client ratings of counselor effectiveness. *Journal of Counseling and Development, 72,* 655–660.

Morran, K., Kurpius, D., Brack, C., & Brack, G. (1995). A cognitive-skills model for counselor training and supervision. *Journal of Counseling & Development, 73,* 384–389.

Morris, R. J. (1991). Fear reduction methods. In F. H. Kanfer & A. P. Goldstein (Eds.), *Helping people change* (4th ed., pp. 161–201). New York: Pergamon Press.

Morrison, J. (1995). *The first interview: A guide for clinicians.* New York: Guilford.

Morrow, S., Gore, P., & Campbell, B. (1996). The application of a socio-cognitive framework to the career development of lesbian women and gay men. *Journal of Vocational Behavior, 48,* 136–148.

Motenko, A. K., & Greenberg, S. (1995). Reframing dependence in old age. *Social Work, 40,* 382–390.

Motley, M., & Molloy, J. (1994). An efficacy test of a new therapy for public-speaking anxiety. *Journal of Applied Communication Research, 22,* 48–58.

Munoz, R. F., Hollon, S. P., McGrath, E., Rehm, L. P., & VandenBos, G. (1994). On the AHCPR depression in primary care guidelines: Further considerations for practitioners. *American Psychologist, 49,* 41–61.

Murphy, M., & Donovan, S. (Eds.). (1997). *The physical and psychological effects of meditation* (2nd ed.). Sausalito, CA: Institute of Noetic Sciences.

Mwaba, K., & Pedersen, P. (1990). Relative importance of intercultural, interpersonal, and psychopathological attributions in judging critical incidents by multicultural counselors. *Journal of Multicultural Counseling and Development, 18,* 107–117.

Myrick, R. D., & Myrick, L. S. (1993). Guided imagery: From mystical to practical. Special Issue: Counseling and children's play. *Elementary School Guidance and Counseling, 28,* 62–70.

Myss, C. (1996a). *Anatomy of the spirit: The seven stages of power and healing.* New York: Harmony Books.

Myss, C. (1996b). *Energy anatomy: The science of personal power, spirituality, and health.* Audio tapes. Boulder, CO: Sounds True Audio.

Napier, A. Y. (1988). *The fragile bond.* New York: Harper & Row.

National Association of Social Workers. (1996). *Code of ethics.* Washington, DC: Author.

Neilans, T. H., & Israel, A. C. (1981). Towards maintenance and generalization of behavior change: Teaching children self-regulation and self-instructional skills. *Cognitive Therapy and Research, 5,* 189–195.

Neill, K. (1993). Ethnic pain styles in acute myocardial infarction. *Western Journal of Nursing Research, 15,* 531–543.

Neimeyer, R. A., & Mahoney, M. J. (Eds.). (1995). *Constructivism in psychotherapy.* Washington, DC: American Psychological Association.

Nelissen, I., Muris, P., & Merckelbach, H. (1995). Computerized exposure and in vivo exposure treatments of spider fear in children: Two case reports. *Journal of Behavior Therapy and Experimental Psychiatry, 26,* 153–156.

Nelson, R. O. (1977). Methodological issues in assessment via self-monitoring. In J. D. Cone & R. P. Hawkins (Eds.), *Behavioral assessment: New directions in clinical psychology* (pp. 217–254). New York: Brunner/Mazel.

Nelson, R. O. (1983). Behavioral assessment: Past, present, and future. *Behavioral Assessment, 5,* 195–206.

Nelson, R. O., & Barlow, D. H. (1981). Behavioral assessment: Basic strategies and initial procedures. In D. H. Barlow (Ed.), *Behavioral assessment of adult disorders* (pp. 13–43). New York: Guilford.

Nelson, R. O., Hayes, S. C., Spong, R. T., Jarrett, R. B., & McKnight, D. L. (1983). Self-reinforcement: Appealing misnomer or effective mechanism. *Behaviour Research and Therapy, 21,* 557–566.

Newman, F., & Tejeda, M. (1996). The need for research that is designed to support decisions in the delivery of mental health services. *American Psychologist, 51,* 1040–1049.

Nezu, A. M., & Nezu, C. M. (Eds.). (1989). *Clinical decision making in behavior therapy: A problem-solving perspective.* Champaign, IL: Research Press.

Nezu, A. M., Nezu, C. M., & Perri, M. B. (1989). *Problem-solving therapy for depression: Theory, research, and clinical guidelines.* New York: Wiley.

Nichols, M. P. (1995). *The lost art of listening.* New York: Guilford.

Nickerson, K., Helms, J., & Terrell, F. (1994). Cultural mistrust, opinions about mental illness, and black students' attitudes toward seeking psychological help from white counselors. *Journal of Counseling Psychology, 41,* 378–385.

Nietzel, M. T., Bernstein, D. A., & Russell, R. L. (1988). Assessment of anxiety and fear. In A. S. Bellack & M. Hersen (Eds.), *Behavioral assessment: A practical handbook* (3rd ed., pp. 280–312). New York: Pergamon Press.

Nilsson, D., Strassberg, D., & Bannon, J. (1979). Perceptions of counselor self-disclosure: An analogue study. *Journal of Counseling Psychology, 26,* 399–404.

Nolen-Hoeksema, S., Girgus, J. S., & Seligman, M. (1992). Predictors and consequences of childhood depressive symptoms: A 5-year longitudinal study. *Journal of Abnormal Psychology, 101,* 405–422.

Novaco, R. W. (1975). *Anger control: The development and evaluation of an experimental treatment.* Lexington, MA: Heath.

Ogles, B., Lambert, M., & Masters, K. (1996). *Assessing outcome in clinical practice.* Boston: Allyn & Bacon.

O'Hanlon, W. H., & Weiner-Davis, M. (1989). *In search of solutions: A new direction in psychotherapy.* New York: Norton.

Okun, B. F. (1992a). *Effective helping: Interviewing and counseling techniques* (4th ed.). Pacific Grove, CA: Brooks/Cole.

Okun, B. F. (1992b). Objective relations and self psychology: Overview and feminist perspective. In L. S. Brown & M. Ballou (Eds.), *Personality and psychotherapy: Feminist reappraisals* (pp. 20–45). New York: Guilford.

Okun, B. F. (1997). *Effective helping* (5th ed.). Pacific Grove, CA: Brooks/Cole.

O'Leary, K. D., & Wilson, G. T. (1987). *Behavior therapy* (2nd ed.). Englewood Cliffs, NJ: Prentice-Hall.

Oppenheimer, M. (1992). Alma's bedside ghost: Or the importance of cultural similarity. *Hispanic Journal of Behavioral Sciences, 14,* 496–501.

Organista, K., Dwyer, E. V., & Azocar, F. (1993). Cognitive behavioral therapy with Latino outpatients. *The Behavior Therapist, 16,* 229–232.

Ornish, D. (1990). *Reversing heart disease.* New York: Ballantine.

Ost, L. (1988). Applied relaxation vs. progressive relaxation in the treatment of panic disorder. *Behaviour Research and Therapy, 26,* 13–22.

Ost, L., & Westling, B. E. (1991). *Treatment of panic disorder by applied relaxation versus cognitive therapy.* Paper presented at the meeting of the European Association of Behaviour Therapy, Oslo, Norway.

Ostaseski, F. (1994). Stories of lives lived and now ending. *The Sun,* Dec. #228, 10–13.

Ottens, A., Shank, G., & Long, R. (1995). The role of abductive logic in understanding and using advanced empathy. *Counselor Education and Supervision, 34,* 199–211.

Overholser, J. C. (1990). Passive relaxation training with guided imagery: A transcript for clinical use. *Phobia Practice and Research Journal, 3 (2),* 107–122.

Overholser, J. C. (1991). The use of guided imagery in psychotherapy: Modules for use with passive relaxation training. *Journal of Contemporary Psychotherapy, 21,* 159–172.

Oz, S. (1995). A modified balance-sheet procedure for decision making in therapy: Cost-cost comparison. *Professional Psychology, 25,* 78–81.

Ozer, E. M., & Bandura, A. (1990). Mechanisms governing empowerment effects: A self-efficacy analysis. *Journal of Personality and Social Psychology, 58,* 472–486.

Papp, P. (1976). Family choreography. In P. J. Guerin, Jr. (Ed.), *Family therapy: Theory and practice* (pp. 465–479). New York: Gardner Press.

Parcel, G. S., Swank, P. R., Mariotto, M. J., & Bartholomew, L. K. (1994). Self-management of cystic fibrosis: A structural model for educational and behavioral variables. *Social Science and Medicine, 38,* 1307–1315.

Passons, W. R. (1975). *Gestalt approaches in counseling.* New York: Holt, Rinehart and Winston.

Patel, C. (1993). Yoga-based therapy. In P. M. Lehrer & R. L. Woolfolk (Eds.), *Principles and practice of stress management* (2nd ed., pp. 89–137). New York: Guilford.

Patterson, L. E., & Welfel, E. R. (1994). *The counseling process* (4th ed.). Pacific Grove, CA: Brooks/Cole.

Paul, G. L. (1967). Strategy of outcome research in psychotherapy. *Journal of Consulting Psychology, 31,* 109–118.

Paye, B. D., & Manning, B. H. (1990). The effect of cognitive self-instructions on preservice teacher's anxiety about teaching. *Contemporary Educational Psychology, 15,* 261–267.

Payton, C. R. (1994). Implications of the 1992 ethics code for diverse groups. *Professional Psychology, 25,* 317–320.

Peck, S. (1978). *The road less traveled.* New York: Simon & Schuster.

Pecsok, E. H., & Fremouw, W. J. (1988). Controlling laboratory binging among restrained eaters through self-monitoring and cognitive restructuring procedures. *Addictive Behaviors, 13,* 37–44.

Pedersen, P. (1987). Ten frequent assumptions of cultural bias in counseling. *Journal of Multicultural Counseling and Development, 15,* 16–22.

Perls, F. S. (1973). *The Gestalt approach and eyewitness to therapy.* Palo Alto, CA: Science and Behavior Books.

Perri, M. G., & Richards, C. S. (1977). An investigation of naturally occurring episodes of self-controlled behaviors. *Journal of Counseling Psychology, 24,* 178–183.

Pert, C. (1993). The chemical communicators. Interview by B. Moyers with C. Pert, in *Healing and the mind* (pp. 177–193). New York: Doubleday.

Peterson, L., & Shigetomi, C. (1981). The use of coping techniques to minimize anxiety in hospitalized children. *Behavior Therapy, 12,* 1–14.

Picucci, M. (1992). Planning an experiential weekend workshop for lesbians and gay males in recovery. *Journal of Chemical Dependency Treatment, 5,* 119–139.

Pierce, M. D., & Pierce, M. G. (1996). *Yoga for your life.* Portland: Rudra Press.

Pietrofesa, J. J., Hoffman, A., Splete, H. H., & Pinto, D. V. (1978). *Counseling: Theory, research, and practice.* Chicago, IL: Rand McNally.

Pilkington, N. W., & Cantor, J. M. (1996). Perceptions of heterosexual bias in professional psychology programs: A survey of graduate students. *Professional Psychology, 27,* 604–612.

Pinsof, W., & Wynne, L. (Eds.). (1995). Family therapy effectiveness: Current research and theory. (Special Issue) *Journal of Marital and Family Therapy, 21*(4).

Piper, W. E., Azim, H. F., McCallum, M., & Joyce, A. S. (1990). Patient suitability and outcome in short-term individual psychotherapy. *Journal of Consulting and Clinical Psychology, 58,* 475–481.

Pipher, M. (1994). *Reviving Ophelia: Saving the selves of adolescent girls.* New York: Putnam.

Pistole, M. C., & Watkins, C. E., Jr. (1995). Attachment theory, counseling process, and supervision. *The Counseling Psychologist, 23,* 457–478.

Pitre, A., & Nicki, R. (1994). Desensitization of dietary restraint anxiety and its relationship to weight loss. *Journal of Behavior Therapy and Experimental Psychiatry, 25,* 153–154.

Pollard, D. (1993). Gender, achievement, and African-American students' perceptions of their school experience. *Educational Psychologist, 28,* 341–356.

Pope-Davis, D. B., Menefee, L. A., & Ottavi, T. M. (1993). The comparison of White racial identity attitudes among faculty and students: Implications for professional psychologists. *Professional Psychology, 24,* 443–449.

Prochaska, J., & DiClemente, C. (1982). Transtheoretical therapy: Towards a more integrative model of change. *Psychotherapy, 19,* 276–278.

Prochaska, J., & DiClemente, C. (1985). Common processes of self-change of smoking: Toward an integrative model of change. *Journal of Consulting and Clinical Psychology, 51,* 390–395.

Prochaska, J., & DiClemente, C., & Norcross, J. (1992). In search of how people change: Applications to addictive behaviors. *American Psychologist, 47,* 1102–1114.

Pulos, L. (1996). *The power of visualization.* (Audiotapes). Niles, IL: Nightingale/Conant.

Radha, S. S. (1987). *Hatha yoga.* Spokane, WA: Timeless Books.

Rama, S., Ballentine, R., & Ajaya, S. (1974). *Yoga and psychotherapy.* Honesdale, PA: Himalayan International Institute of Yoga Science and Philosophy.

Rappaport, A. F., & Cammer, L. (1977). Breath meditation in the treatment of essential hypertension. *Behavior Therapy, 8,* 269–270.

Rao, R., & Kramer, L. (1993). Stress and coping among mothers of infants with a sickle cell condition. *Children's Health Care, 22,* 169–188.

Raskin, N., & Rogers, C. (1995). Person-centered therapy. In J. Corsini & D. Wedding (Eds.), *Current psychotherapies* (5th ed., pp. 128–161). Itasca, IL: Peacock.

Reed, J. R., Patton, M. J., & Gold, P. B. (1993). Effects of turn-taking sequences in vocational test interpretation interviews. *Journal of Counseling Psychology, 40,* 144–155.

Reeder, C., & Kunce, J. (1976). Modeling techniques, drug-abstinence behavior, and heroin addicts: A pilot study. *Journal of Counseling Psychology, 23,* 560–562.

Rehm, L. P. (1982). Self-management in depression. In P. Karoly & F. H. Kanfer (Eds.), *Self-management and behavior change* (pp. 522–567). New York: Pergamon Press.

Reichelova, E., & Baranova, E. (1994). Training program for the development of prosocial behavior in children. *Psychologia a Patopsychologia Dietata, 29,* 41–50.

Reid, W. H. (1995). *Treatment of the DSM-IV psychiatric disorders.* New York: Brunner/Mazel.

Renfrey, G. S. (1992). Cognitive-behavior therapy and the Native American client. *Behavior Therapy, 23,* 321–340.

Resnick, R. J. (1995). How come outcome? *The American Psychological Association Monitor, 26* (6), 2.

Reyes, M., Routh, D., Jean-Gilles, M., & Sanfilippo, M. (1991). Ethnic differences in parenting children in fearful situations. *Journal of Pediatric Psychology, 16,* 717–726.

Reynolds, M., & Salkovskis, P. M. (1992). Comparison of positive and negative intrusive thoughts and experimental investigation of the differential effects of mood. *Behaviour Research and Therapy, 30,* 273–281.

Rhodes, F., & Humfleet, G. (1993). Using goal-oriented counseling and peer support to reduce HIV/AIDS risk among drug users not in treatment. *Drugs and Society, 7,* 185–204.

Richardson, B., & Stone, G. L. (1981). Effects of cognitive adjunct procedure within a microtraining situation. *Journal of Counseling Psychology, 28,* 168–175.

Richardson, T., & Helms, J. (1994). The relationship of the racial identity attitudes of Black men to perceptions of "parallel" counseling dyads. *Journal of Counseling & Development, 73,* 172–177.

Riley, W., Barenie, J., Mabe, P., & Myers, D. (1990). Smokeless tobacco use in adolescent females: Prevalence and psychosocial factors among racial/ethnic groups. *Journal of Behavioral Medicine, 13,* 207–220.

Riley, W., Barenie, J., Mabe, P., & Myers, D. (1991). The role of race and ethnic status on the psychosocial correlates of smokeless tobacco use in adolescent males. *Journal of Adolescent Health, 12,* 15–21.

Rimm, D. C., & Masters, J. C. (1979). *Behavior therapy: Techniques and empirical findings* (2nd ed.). New York: Academic Press.

Ritchie, E. C. (1992). Treatment of gas mask phobia. *Military Medicine, 157,* 104–106.

Roberson, M. (1992). The meaning of compliance: Patient perspectives. *Qualitative Health Research, 2,* 7–26.

Robinson, T., & Ward, J. (1992). A belief in self far greater than anyone's disbelief: Cultivating resistance among African American female adolescents. In C. Gilligan, A. Rogers, & D. Tolman (Eds.), *Women, girls and psychotherapy: Reframing resistance* (pp. 87–104). Binghamton, NY: Haworth Press.

Rogers, C. (1942). *Counseling and psychotherapy.* Boston, MA: Houghton Mifflin.

Rogers, C. (1951). *Client-centered therapy.* Boston, MA: Houghton Mifflin.

Rogers, C. (1957). The necessary and sufficient conditions of therapeutic personality change. *Journal of Consulting Psychology, 21,* 95–103.

Rogers, C., Gendlin, E., Kiesler, D., & Truax, C. (1967). *The therapeutic relationship and its impact: A study of psychotherapy with schizophrenics.* Madison, WI: University of Wisconsin Press.

Rogoff, B., & Chavajay, P. (1995). What's become of research on the cultural basis of cognitive development? *American Psychologist, 50,* 859–877.

Rokke, P. D., & Kozak, C. D. (1989). Self-control deficits in depression: A process investigation. *Cognitive Therapy and Research, 13,* 609–621.

Root, M. P. (1992). Reconstructing the impact of trauma on personality. In L. S. Brown & M. Ballou (Eds.), *Personality and psychopathology: Feminist reappraisals* (pp. 229–265). New York: Guilford.

Root, M. P. (1993). Guidelines for facilitating therapy with Asian American clients. In D. Atkinson, G. Morten, & D. W. Sue (Eds.), *Counseling American minorities* (pp. 211–224). Madison, WI: Brown & Benchmark.

Rosado, J. W., Jr., & Elias, M. J. (1993). Ecological and psychocultural mediators in the delivery of services for urban, culturally diverse Hispanic clients. *Professional Psychology, 24,* 450–459.

Rose, S. D. & LeCroy, C. W. (1991). Group methods. In F. H. Kanfer & A. P. Goldstein (Eds.), *Helping people change* (pp. 422–453). New York: Pergamon Press.

Rosenthal, T. L. (1993). To soothe the savage beast. *Behaviour Research and Therapy, 31,* 439–462.

Rosenthal, T. L., & Steffek, B. D. (1991). Modeling methods. In F. H. Kanfer & A. P. Goldstein (Eds.), *Helping people change* (4th ed., pp. 70–121). New York: Pergamon Press.

Ross, C. A. (1990). *Multiple personality disorder: Diagnosis, clinical features and treatment.* New York: Wiley.

Rossi, E. (1987). Mind/body connections and the new language of human facilitation. In J. Zeig (Ed.), *The evolution of psychotherapy.* New York: Brunner/Mazel.

Rothbard, J. C., & Shaver, P. R. (1994). Continuity of attachment across the life span. In M. B. Sperling & W. H. Berman (Eds.), *Attachment in adults* (pp. 31–71). New York: Guilford.

Rotherman-Borus, M. J., Rosario, M., Reid, H., & Koopman, C. (1995). Predicting patterns of sexual acts among homosexual and bisexual youths. *American Journal of Psychiatry, 152,* 588–595.

Rotter, J. B. (1966). Generalized expectancies for internal versus external control of reinforcement. *Psychological Monographs: General and Applied, 80*(1, Whole No. 609).

Rounds, K. A., Weil, M., & Bishop, K. (1994). Practice and culturally diverse families of young children with disabilities. *Families in Society, 75,* 3–15.

Ruben, A. (1989). Preventing school dropouts through classroom guidance. *Elementary School Guidance and Counseling, 24,* 21–29.

Ruiz, A. S. (1990). Ethnic identity: Crisis and resolution. *Journal of Multicultural Counseling and Development, 18,* 29–40.

Safran, J. C., Crocker, P., McMain, S., & Murray, P. (1990). Therapeutic alliance rupture as a therapy event for empirical investigation. *Psychotherapy, 27,* 154–165.

Saha, G. B., Palchoudhury, S., & Mardal, M. K. (1982). A study on facial expression of emotion. *Psychologia, 25,* 255–259.

St. Lawrence, J. (1993). African American adolescents' knowledge, health-related attitudes, sexual behavior, and contraceptive decisions: Implications for the prevention of adolescent HIV infection. *Journal of Consulting and Clinical Psychology, 61,* 104–112.

St. Lawrence, J., Brasfield, T., Jefferson, K., Alleyne, E., O'Bannon, R., & Shirley, A. (1995). Cognitive-behavioral intervention to reduce African American adolescents' risk for HIV infection. *Journal of Consulting and Clinical Psychology, 63,* 221–237.

Samson, D., & Rachman, S. (1989). The effect of induced mood on fear reduction. *British Journal of Clinical Psychology, 28,* 227–238.

Sanders, M. R., & Jones, L. (1990). Behavioural treatment of injection, dental and medical phobias in adolescents: A case study. *Behavioural Psychotherapy, 18,* 311–316.

Santiago-Rivera, A. (1995). Developing a culturally sensitive treatment modality for bilingual Spanish-speaking clients. *Journal of Counseling and Development, 74,* 12–17.

Sarason, I. G., & Sarason, B. R. (1981). Teaching cognitive and social skills to high school students. *Journal of Consulting and Clinical Psychology, 49,* 908–918.

Satir, V. (1972). *Peoplemaking.* Palo Alto, CA: Science and Behavior Books.

Schill, T., & Kramer, J. (1991). Self-defeating personality, self-reinforcement, and depression. *Psychological Reports, 69,* 137–138.

Schinke, S., Botvin, G., Orlandi, M., & Schilling, R. (1990). African American and Hispanic American adolescents, HIV infection, and prevention intervention. *AIDS Education and Prevention, 2,* 305–312.

Schinke, S., Orlandi, M., Gordon, A., & Weston, E. (1989). AIDS prevention via computer-based intervention. *Computers in Human Services, 5,* 147–156.

Schlossberger, E., & Hecker, L. (1996). HIV and family therapist's duty to warn: A legal and ethical analysis. *Journal of Marital & Family Therapy, 22,* 27–40.

Schneider, B. (1991). A comparison of skill-building and desensitization strategies for intervention with aggressive children. *Aggressive Behavior, 17,* 301–311.

Schneider, W., & Nevid, J. (1993). Overcoming math anxiety: A comparison of stress inoculation training and systematic desensitization. *Journal of College Student Development, 34,* 283–288.

Schoefield, W. (1964). *Psychotherapy: The purchase of friendship.* Englewood Cliffs, NJ: Prentice-Hall.

Schulz, R., & Barefoot, J. (1974). Nonverbal responses and affiliative conflict theory. *British Journal of Social and Clinical Psychology, 13,* 237–243.

Scotti, J. R., Evans, I. M., Meyer, L. H., & Walker, P. (1991). A meta-analysis of intervention research with problem behavior: Treatment validity and standards of practice. *American Journal of Mental Retardation, 96,* 233–256.

Seabaugh, G. O., & Schumaker, J. B. (1994). The effects of self-regulation training on the academic productivity of secondary students with learning problems. *Journal of Behavioral Education, 4,* 109–133.

Sedney, M., Baker, J., & Gross, E. (1994). "The story" of a death: Therapeutic considerations with bereaved families. *Journal of Marital and Family Therapy, 20,* 287–296.

Selby, J. W., & Calhoun, L. G. (1980). Psychodidactics: An undervalued and underdeveloped treatment tool of psychological intervention. *Professional Psychology, 11,* 236–241.

Seligman, M. (1982). *Helplessness: On depression, development and death.* San Francisco: Freeman.

Seligman, M. (1990). *Learned optimism.* New York: Pocket Books.

Seligman, M. (1994). *What you can change and what you can't.* New York: Knopf.

Seligman, M. (1995). The effectiveness of psychotherapy: The *Consumer Reports* survey. *American Psychologist, 50,* 965–974.

Seligman, M. (1996). Science as an ally of practice. *American Psychologist, 51,* 1072–1088.

Senour, M. (1982). How counselors influence clients. *Personnel and Guidance Journal, 60,* 345–350.

Sexton, T. L., & Whiston, S. C. (1994). The status of the counseling relationship: An empirical review, theoretical implications, and research directions. *The Counseling Psychologist, 22,* 6–78.

Shainberg, D. (1993). *Healing in psychotherapy.* Langhorne, PA: Gordon & Breach Science Publishers USA.

Shapiro, F. (1995). *Eye movement desensitization and reprocessing: Basic principles, protocols and procedures.* New York: Guilford.

Shapiro, F., & Forrest, M. S. (1997). *EMDR.* New York: Basic Books.

Shefler, G., & Dasberg, H. (1989, June). *A randomized controlled outcome and follow-up study of James Mann's time-limited psychotherapy in a Jerusalem community mental health center.* Paper presented at the meeting of the Society for Psychotherapy Research, Toronto.

Sheikh, A. A. (Ed.). (1986). *Anthology of imagery techniques.* Milwaukee, WI: American Imagery Institute.

Silver, S. M., Brooks, A., & Obenchain, J. (1995). Treatment of Vietnam war veterans with PTSD: A comparison of eye movement desensitization and reprocessing, biofeedback, and relaxation training. *Journal of Traumatic Stress, 8,* 337–342.

Simon, O. C., & Hensen, R. (1992). *The healing journey.* New York: Bantam.

Simon, S. (1995). *Values clarification* (2nd ed.). New York: Warner Books.

Simpkins, C. A., & Simpkins, A. M. (1996). *Principles of meditation.* Boston, MA: Charles E. Tuttle.

Sinacore-Guinn, A. L. (1995). The diagnostic window: Culture- and gender-sensitive diagnosis and training. *Counselor Education and Supervision, 35,* 18–31.

Singer, E. (1970). *New concepts in psychotherapy.* New York: Basic Books.

Singer, J. L. (1975). Navigating the stream of consciousness: Research in daydreaming and related inner experience. *American Psychologist, 30,* 727–738.

Sitton, S. C., & Griffin, S. T. (1981). Detection of deception from clients' eye contact patterns. *Journal of Counseling Psychology, 28,* 269–271.

Skovholt, T., & Ronnestad, M. (1995). *The evolving professional self.* New York: Wiley.

Smart, J. F., & Smart, D. (1995a). Acculturative stress of Hispanics: Loss and challenge. *Journal of Counseling and Development, 73,* 390–396.

Smart, J. F., & Smart, D. (1995b). Acculturative stress: The experience of the Hispanic immigrant. *The Counseling Psychologist, 23,* 25–42.

Smith, B. (1992). Raising a resister. In C. Gilligan, A. Rogers, & D. Tolman (Eds.), *Women, girls and psychotherapy: Reframing resistance* (pp. 137–148). Binghamton, NY: Haworth.

Smith, D., & Fitzpatrick, M. (1995). Patient-therapist boundary issues. *Professional Psychology, 26,* 499–506.

Smith, E. W. (1985). *The body in psychotherapy.* Jefferson, NC: McFarland & Company.

Smith, K. J., Subich, L., & Kalodner, C. (1995). The transtheoretical model's stages and processes of change and their relation to premature termination. *Journal of Counseling Psychology, 42,* 34–39.

Smith, L. L., Smith, J. N., & Becker, B. M. (1994). An anger-management workshop for women inmates. *Families in Society, 75,* 172–175.

Snowden, L. R., & Cheung, F. K. (1990). Use of inpatient mental health services by members of ethnic minority groups. *American Psychologist, 45,* 347–355.

Snyder, C. R. (1995). Conceptualizing, measuring and nurturing hope. *Journal of Counseling and Development, 73,* 355–360.

Solberg, V., O'Brien, K., Villareal, P., & Kennel, R. (1993). Self-efficacy and Hispanic college students: Validation of the College Self-Efficacy Instrument. *Hispanic Journal of Behavioral Sciences, 15,* 80–95.

Sommers-Flanagan, J., & Sommers-Flanagan, S. (1995). Intake interviewing with suicidal patients: A systematic approach. *Professional Psychology, 26,* 41–47.

Spiegel, S. B., & Hill, C. E. (1989). Guidelines for research on therapist interpretation: Toward greater methodological rigor and relevance to practice. *Journal of Counseling Psychology, 36,* 121–129.

Spitzer, R. L., Gibbon, M., Skodol, A. E., Williams, J. B., & First, M. B. (1994). *DSM IV Casebook.* Washington, DC: American Psychiatric Association.

Spivack, G., Platt, J. J., & Shure, M. B. (1976). *The problem-solving approach to adjustment.* San Francisco: Jossey-Bass.

Steenbarger, B. N. (1994). Duration and outcome of psychotherapy: An integrative review. *Professional Psychology, 25,* 111–119.

Steketee, G., Foa, E. B., & Grayson, J. B. (1982). Recent advances in the behavioral treatment of obsessive compulsives. *Archives of General Psychiatry, 39,* 1365–1371.

Stern, D. N. (1985). *The interpersonal world of the infant: A view from psychoanalysis and development psychology.* New York: Basic Books.

Stevenson, H. C., & Renard, G. (1993). Trusting ole wise ones: Therapeutic use of cultural strengths in African-American families. *Professional Psychology, 24,* 433–442.

Stewart, M. (1994). *Yoga over 50.* New York: Fireside.

Stewart, M., & Philips, K. (1995). *Yoga for children.* New York: Fireside.

Stiles, W. B., & Snow, J. S. (1984). Counseling session impact as seen by novice counselors and their clients. *Journal of Counseling Psychology, 31,* 3–12.

Stricker, G. (1992). The relationship of research to clinical practice. *American Psychologist, 47,* 543–549.

Stroebe, M., Gergen, M., Gergen, K., & Stroebe, W. (1992). Broken hearts or broken bonds. *American Psychologist, 47,* 1205–1212.

Strong, S. R. (1968). Counseling: An interpersonal influence process. *Journal of Counseling Psychology, 15,* 215–224.

Strong, S. R., & Claiborn, C. (1982). *Change through interaction: Social psychological processes of counseling and psychotherapy.* New York: Wiley-Interscience.

Strumpf, J., & Fodor, I. (1993). The treatment of test anxiety in elementary school-age children: Review and recommendations. *Child and Family Behavior Therapy, 15,* 19–42.

Strupp, H. H. (1980). Success and failure in time-limited psychotherapy: A systematic comparison of two cases: Comparison I. *Archives of General Psychiatry, 37,* 595–603.

Strupp, H. H. (1988). What is therapeutic change? *Journal of Cognitive Psychotherapy, 2,* 75–82.

Strupp, H. H. (1989). Psychotherapy: Can the practitioner learn from the research? *American Psychologist, 44,* 717–724.

Strupp, H. H., & Binder, J. L. (1984). *Psychotherapy in a new key.* New York: Basic Books.

Strupp, H. H., Fox, R., & Lessler, K. (1969). *Patients view their psychotherapy.* Baltimore, MD: Johns Hopkins University Press.

Strupp, H. H., & Hadley, S. W. (1977). A tripartite model of mental health and therapeutic outcomes with special reference to negative effects in psychotherapy. *American Psychologist, 32,* 187–195.

Sue, D., & Sue, D. W. (1993). Ethnic identity: Cultural factors in the psychological development of Asians in America. In D. Atkinson, G. Morten, & D. W. Sue (Eds.), *Counseling American minorities* (pp. 199–210). Madison, WI: Brown & Benchmark.

Sue, D. W. (1994). Asian-American mental health and help-seeking behavior. *Journal of Counseling Psychology, 41,* 292–295.

Sue, D. W., Arredondo, P., & McDavis, R. (1992). Multicultural counseling competencies and standards: A call to the profession. *Journal of Counseling & Development, 70,* 477–487.

Sue, D. W., Ivey, A. E., & Pedersen, P. B. (1996). *A theory of multicultural counseling and therapy.* Pacific Grove, CA: Brooks/Cole.

Sue, D. W., & Sue, D. (1990). *Counseling the culturally different* (2nd ed.). New York: Wiley.

Sue, S. (1991). Ethnicity and culture in psychological research and practice. In J. D. Goodchilds (Ed.), *Psychological perspectives on human diversity in America* (pp. 51–85). Washington, DC: American Psychological Association.

Sue, S., & Sue, D. W. (1971). Chinese-American personality and mental health. *Amerasia Journal, 1,* 36–49.

Suinn, R. M. (1986). *Seven steps to peak performance: The mental training manual for athletes.* Lewiston, NY: Huber.

Suinn, R. M., & Richardson, F. (1971). Anxiety management training: A nonspecific behavior therapy program for anxiety control. *Behavior Therapy, 3,* 308–310.

Sweeney, M. A., Cottle, W. C., & Kobayashi, M. J. (1980). Nonverbal communication: A cross-cultural comparison of American and Japanese counseling students. *Journal of Counseling Psychology, 27,* 150–156.

Swenson, L. C. (1997). *Psychology and law.* (2nd ed.). Pacific Grove, CA: Brooks/Cole.

Switzky, H. N., & Haywood, H. C. (1992). Self-reinforcement schedules in young children: Effects of motivational orientation and instructional demands. *Learning and Individual Differences, 4*(1), 59–71.

Szapocznik, J., & Kurtines, W. (1993). Family psychology and cultural diversity. *American Psychologist, 48,* 400–407.

Szapocznik, J., & Kurtines, W. (1980). Acculturation, biculturalism and adjustment among Cuban Americans. In A. Padilla (Ed.), *Recent advances in acculturation research: Theory, models, and some new findings* (pp. 139–157). Boulder, CO: Westview.

Taal, E., Riemsma, R. P., Brus, H. L., & Seydel, E. R. (1993). Group education for patients with rheumatoid arthritis. Special Issue: Psychosocial aspects of rheumatic diseases. *Patient Education and Counseling, 20,* 177–187.

Taylor, C. B. (1983). DSM-III and behavioral assessment. *Behavioral Assessment, 5,* 5–14.

Taylor, J., Gilligan, C., & Sullivan, A. (1995). *Between voice and silence: Women and girls, race and relationships.* Cambridge, MA: Harvard University Press.

Taylor, J., & Schneider, B. (1992). The sport-clinical intake protocol: A comprehensive interviewing instrument for applied sport psychology. *Professional Psychology, 23,* 318–325.

Taylor, S., Cipani, E., & Clardy, A. (1994). A stimulus control technique for improving the efficacy of an established toilet training program. *Journal of Behavior Therapy and Experimental Psychiatry, 25,* 155–160.

Teders, S. J., Blanchard, E. B., Andrasik, F., Jurish, S. E., Neff, D. F., & Arena, J. G. (1984). Relaxation training for tension headache: Comparative efficacy and cost-effectiveness of minimal therapist contact versus a therapist-delivered procedure. *Behavior Therapy, 15,* 59–70.

Terr, L. (1990). *Too scared to cry: Psychic trauma in childhood.* New York: Harper & Row.

Teyber, E. (1997). *Interpersonal processes in psychotherapy* (3rd ed.). Pacific Grove, CA: Brooks/Cole.

Thomas, R., & Gafner, G. (1993). PTSD in elderly male: Treatment with eye movement desensitization and reprocessing (EMDR). *Clinical Gerontologist, 14,* 57–59.

Thomas, T. C. (1993). Counseling Native Americans: An introduction for non-Native American counselors. In D. Atkinson, G. Morten, & D. W. Sue (Eds.), *Counseling American minorities* (pp. 171–191). Madison, WI: Brown & Benchmark.

Thompson, L. W. (1996). Cognitive-behavioral therapy and treatment for late-life depression. *Journal of Clinical Psychiatry, 57,* 29–37.

Thompson, L. W., Gallagher, D., & Breckenridge, J. S. (1987). Comparative effectiveness of psychotherapies for depressed elders. *Journal of Consulting and Clinical Psychology, 55,* 385–390.

Thoresen, C. E., & Mahoney, M. J. (1974). *Behavioral self-control.* New York: Holt, Rinehart and Winston.

Tisdelle, D. A., & St. Laurence, J. S. (1988). Adolescent interpersonal problem-solving skill training: Social validation and generalization. *Behavior Therapy, 19,* 171–182.

Todd, D. M., Jacobus, S. I., & Boland, J. (1992). Uses of computer database to support research-practice integration in a training clinic. *Professional Psychology, 23,* 52–58.

Trager, G. L. (1958). Paralanguage: A first approximation. *Studies in Linguistics, 13,* 1–12.

Treviño, J. (1996). Worldview and change in cross-cultural counseling. *The Counseling Psychologist, 24,* 198–215.

Triandis, H. (1990). Cross-cultural studies of individualism and collectivism. In J. Berry, J. Draguns, & M. Cole (Eds.), *Cross-cultural perspectives.* Lincoln, NE: University of Nebraska Press.

Triandis, H. (1996). The psychological measurement of cultural syndromes. *American Psychologist, 51,* 407–415.

Triandis, H., Lambert, W., Berry, J., Lonner, W., Heron, A., Brislin, R., & Draguns, J. (Eds.). (1980). *Handbook of cross-cultural psychology,* Vols. 1–6. Boston, MA: Allyn & Bacon.

Truax, C. B., & Mitchell, K. M. (1971). Research on certain therapist interpersonal skills in relation to process and outcome. In A. Bergin & S. Garfield (Eds.), *Handbook of psychotherapy and behavior change: An empirical analysis* (pp. 299–344). New York: Wiley.

Trull, T. J., Nietzel, M. T., & Main, A. (1988). The use of meta-analysis to assess the clinical significance of behavior therapy for agoraphobia. *Behavior Therapy, 19,* 527–538.

Turk, D. (1975). *Cognitive control of pain: A skills training approach for the treatment of pain.* Unpublished master's thesis, University of Waterloo. Waterloo, Ontario, Canada.

Turner, W. L. (1993). Identifying African-American family strengths. *Family Therapy News, 24* (2), 9, 14.

Turock, A. (1980). Immediacy in counseling: Recognizing clients' unspoken messages. *Personnel and Guidance Journal, 59,* 168–172.

Uhlemann, M., Lee, Y. D., & Martin, J. (1994). Client cognitive responses as a function of quality of counselor verbal responses. *Journal of Counseling and Development, 73,* 198–203.

Uhlemann, M. R., & Koehn, C. V. (1989). Effects of covert and overt modeling on the communication of empathy. *Canadian Journal of Counseling, 23,* 372–381.

Ussher, J. (1990). Cognitive behavioral couples therapy with gay men referred for counseling in an AIDS setting: A pilot study. *AIDS-Care, 2,* 43–51.

Vaccaro, F. J. (1990). Application of social skills training in a group of institutionalized aggressively elderly subjects. *Psychology and Aging, 5,* 369–378.

Valentich, M. (1986). Feminism and social work practice. In F. Turner (Ed.), *Social work treatment: Interlocking theoretical approaches* (pp. 564–589). New York: Free Press.

VandenBos, G. R. (Ed.). (1996). Outcome assessment of psychotherapy (Special Issue). *American Psychologist, 51*(10).

Van Hasselt, V., Hersen, M., Null, J., & Ammerman, R. (1993). Drug abuse prevention for high risk African American children and their families: A review and model program. *Addictive Behaviors, 18,* 213–234.

Van Hoose, W. H., & Kottler, J. A. (1985). *Ethical and legal issues in counseling and psychotherapy* (2nd ed.). San Francisco, CA: Jossey-Bass.

Van Hout, W., Emmelkamp, P., & Scholing, A. (1994). The role of negative self-statements during exposure *in vivo:* A process study of eight panic disorder patients with agoraphobia. *Behavior Modification, 18,* 213–234.

Vargas, A. M., & Borkowski, J. G. (1982). Physical attractiveness and counseling skills. *Journal of Counseling Psychology, 29,* 246–255.

Vasquez, M. (1993). The 1992 ethics code: Implications for the practice of psychotherapy. *Texas Psychologist, 45,* 11.

Vasquez, M. (1994). Implications of the 1992 ethics code for the practice of individual psychotherapy. *Professional Psychology, 25,* 321–328.

Vaughan, K., Armstrong, M. S., Gold, R., O'Connor, N., Jenneke, W., & Tarrier, N. (1994). A trial of eye movement desensitization compared to image habituation training and applied muscle relaxation in post-traumatic stress disorder. *Journal of Behavior Therapy and Experimental Psychiatry, 25,* 283–291.

Waehler, C. A., & Lenox, R. A. (1994). A concurrent (versus stage) model for conceptualizing and representing the counseling process. *Journal of Counseling & Development, 73,* 17–22.

Wahler, R. G., & Fox, J. J. (1981). Setting events in applied behavior analysis: Toward a conceptual and methodological expansion. *Journal of Applied Behavior Analysis, 14,* 327–338.

Walker, C. J., & Clement, P. W. (1992). Treating inattentive, impulsive, hyperactive children with self-modeling and stress inoculation training. *Child and Family Behavior Therapy, 14,* 75–85.

Waller, R. J. (1995). *Border music.* New York: Warner Books.

Walter, H. I., & Gilmore, S. K. (1973). Placebo versus social learning effects in parent training procedures designed to alter the behavior of aggressive boys. *Behavior Therapy, 4,* 361–377.

Walters, R. P. (1980). Amity: Friendship in action. Boulder, CO: Christian Helpers, Inc.

Ward, S. W. (1994). *Yoga for the young at heart.* Santa Barbara, CA: Capra Press.

Watkins, C. E. (1990). The effects of counselor self disclosure: A research review. *The Counseling Psychologist, 18,* 477–500.

Watson, B., Bell, P., & Chavez, E. (1994). Conflict handling skills used by Mexican American and White non-Hispanic students in the educational system. *High School Journal, 78,* 35–39.

Watson, D. L., & Tharp, R. G. (1997). *Self-directed behavior* (7th ed.). Pacific Grove, CA: Brooks/Cole.

Watson, O. M. (1970). *Proxemic behavior: A cross-cultural study.* The Hague: Mouton.

Watts-Jones, D. (1990). Toward a stress scale for African-American women. *Psychology of Women Quarterly, 14,* 271–275.

Watzlawick, P. (1993). *The language of change: Elements of therapeutic communication.* New York: Norton.

Waugh, D. (1991, September 2). Facts about group names: Ethnic groups change names with the times. *San Francisco Examiner,* p. H-6.

Webster-Stratton, C. (1981a). Modification of mothers' behaviors and attitudes through a videotape modeling group discussion program. *Behavior Therapy, 12,* 634–642.

Webster-Stratton, C. (1981b). Videotape modeling: A method of parent education. *Journal of Clinical Child Psychology, 10,* 93–98.

Webster-Stratton, C., Kolpacoff, M., & Hollinsworth, T. (1988). Self-administered videotape therapy for families with conduct-problem children: Comparison with two cost-effective treatments and a control group. *Journal of Consulting and Clinical Psychology, 56,* 558–566.

Weeks, G., & Treat, S. (1992). *Couples in treatment.* New York: Brunner/Mazel.

Wehrly, B. (1995). *Pathways to multicultural counseling competence.* Pacific Grove, CA: Brooks/Cole.

Weil, A. (1995). *Spontaneous healing.* New York: Ballantine.

Weil, A. (1997). *8 weeks to optimum health.* New York: Knopf.

Weiss, J. (1995). Cognitive therapy and life-review therapy. *Journal of Mental Health Counselors, 17,* 157–172.

Welch, B. (1992, August). The best care: Integrated, not managed. *The American Psychological Association Monitor,* p. 30.

Weller, S. (1995). *Yoga therapy.* London: Thorsons.

Wells, K. C., & Egan, J. (1988). Social learning and systems family therapy for childhood oppositional disorder: Comparative treatment outcome. *Comprehensive Psychiatry, 29,* 138–146.

Wells, M., & Glickauf-Hughes, C. (1986). Techniques to develop object constancy with borderline clients. *Psychotherapy, 23,* 460–468.

Whiston, S., & Sexton, T. (1993). An overview of psychotherapy outcome research: Implications for practice. *Professional Psychology, 24,* 43–51.

Whitaker, C., & Keith, D. (1981). Symbolic-experimental therapy. In A. Gurman & D. Kniskern (Eds.), *Handbook of family therapy* (pp. 187–225). New York: Brunner/Mazel.

White, M., & Epston, P. (1990). *Narrative means to therapeutic ends.* New York: Norton.

Whitman, J. (1995). Providing training about sexual orientation in counselor education. *Counselor Education and Supervision, 35,* 168–176.

Whittrock, D. A., Blanchard, E. B., & McCory, G. C. (1988). Three studies on the relation of process to outcome in the treatment of essential hypertension with relaxation and thermal biofeedback. *Behaviour Research and Therapy, 26,* 53–66.

Wilfley, D. E., Agras, W. S., Telch, C. F., Rossiter, E. M., Schneider, J. A., Cole, A. G., Sifford, L., & Raeburn, S. D. (1993). Group cognitive-behavioral therapy and group interpersonal psychotherapy for the nonpurging bulimic individual: A controlled comparison. *Journal of Consulting and Clinical Psychology, 61,* 296–305.

Williams, J. G., Park, L. I., & Kline, J. (1992). Reducing distress associated with pelvic examinations: A stimulus control intervention. *Women and Health, 18,* 41–53.

Williams, O. (1994). Group work with African American men who batter: Toward more ethnically sensitive practice. *Journal of Comparative Family Studies, 25,* 91–103.

Williams, R. L., Moore, C. A., Pettibone, T. J., & Thomas, S. P. (1992). Construction and validation of a brief self-report scale of self-management practices. *Journal of Research in Personality, 26,* 216–234.

Wilson, G. T. (1995). Behavior therapy. In R. J. Corsini & D. Wedding (Eds.), *Current psychotherapies* (pp. 197–228). Itasca, IL: Peacock.

Wilson, G. T. (1996). Manual-based treatments: The clinical application of research findings. *Behaviour Research and Therapy, 34,* 197–212.

Wilson, G. T., & Agras, W. S. (1992). The future of behavior therapy. *Psychotherapy, 29,* 39–43.

Wilson, G. T., & Davison, G. C. (1971). Processes of fear reduction in systematic desensitization: Animal studies. *Psychological Bulletin, 76,* 1–14.

Wilson, G. T., Rossiter, E., Kleifield, E. I., & Lindholm, L. (1986). Cognitive behavioral treatment of bulimia nervosa: A controlled evaluation. *Behaviour Research and Therapy, 24,* 277–288.

Wilson, L. L., & Stith, S. M. (1993). Culturally sensitive therapy with Black clients. In D. R. Atkinson, G. Morten, & D. W. Sue (Eds.), *Counseling American minorities* (pp. 101–122). Madison, WI: Brown & Benchmark.

Wilson, S. A., Becker, L. A., & Tinker, R. H. (1995). Eye movement desensitization and reprocessing (EMDR) treatment for psychologically traumatized individuals. *Journal of Consulting and Clinical Psychology, 63,* 928–937.

Winfrey, M. L., & Weeks, D. L. (1993). Effects of self-modeling on self-efficacy and balance beam performance. *Perceptual and Motor Skills, 77,* 907–913.

Wing, R. R., Epstein, L. H., Norwalk, M. P., & Scott, N. (1988). Self-regulation in the treatment of Type II diabetes. *Behavior Therapy, 19,* 11–23.

Winkleby, M., Flora, J., & Kraemer, H. (1994). A community-based heart disease intervention: Predictors of change. *American Journal of Public Health, 84,* 767–772.

Winnicott, D. W. (1958). *The maturational processes and the facilitating environment.* New York: International Universities Press.

Winston, A., Pollack, J., McCullough, L., Flegenheimer, W., Kestenbaum, R., & Trujillo, M. (1991). Brief dynamic psychotherapy of personality disorders. *Journal of Nervous and Mental Disease, 179,* 188–193.

Wolfe, J. L. (1992). Working with gay women. In A. Freeman & F. M. Dattilio (Eds.), *Comprehensive casebook of cognitive therapy* (pp. 249–255). New York: Plenum.

Wolpe, J. (1958). *Psychotherapy by reciprocal inhibition.* Stanford, CA: Stanford University Press.

Wolpe, J. (1961). The systematic desensitization treatment of neuroses. *Journal of Nervous and Mental Disease, 132,* 189–203.

Wolpe, J. (1990). *The practice of behavior therapy* (4th ed.). New York: Pergamon Press.

Wolpe, J., & Lang, P. J. (1964). A fear survey schedule for use in behavior therapy. *Behaviour Research and Therapy, 2,* 27–30.

Woltersdorf, M. A. (1992). Videotape self-modeling in the treatment of attention-deficit hyperactivity disorder. *Child and Family Behavior Therapy, 14,* 53–73.

Woodman, M. (1992). *Leaving my father's house.* Boston, MA: Shambhala.

Woodman, M. (1993). The eternal feminine: Mirror and container. *New Dimensions, 20,* 8–13.

Woody, G. E., Luborsky, L., McLellan, A. T., & O'Brien, C. P. (1990). Corrections and revised analyses for psychotherapy in methadone maintenance patients. *Archives of General Psychiatry, 47,* 788–789.

Woody, G. E., Thompson, L., Gallagher, D., & Zitrin, C. (1991). Meta-analysis of therapist effects in psychotherapy outcome studies. *Psychotherapy Research, 1,* 81–91.

Woolfolk, R. L., & Lehrer, P. M. (Eds.). (1984). *Principles and practice of stress management.* New York: Guilford.

Wright, J., & Davis, D. (1994). The therapeutic relationship in cognitive-behavioral therapy: Patient perceptions and therapist responses. *Behavior Therapy, 24,* 25–45.

Wright, L., & Walker, C. E. (1978). A simple behavioral treatment program for psychogenic encopresis. *Behaviour Research and Therapy, 16,* 209–212.

Wurtele, S. K., Marrs, S. R., & Miller-Perrin, C. L. (1987). Practice makes perfect? The role of participant modeling in sexual abuse prevention programs. *Journal of Consulting and Clinical Psychology, 55,* 599–602.

Wynd, C. A. (1992). Personal power imagery and relaxation techniques use in smoking cessation programs. *American Journal of Health Promotion, 6,* 184–189, 196.

Yang, B., & Clum, G. (1994). Life stress, social support, and problem-solving skills predictive of depression symptoms, hopelessness, and suicide ideation in an Asian student population: A test model. *Suicide and Life Threatening Behavior, 24,* 127–139.

Yoder, J. D., & Kahn, A. S. (1993). Working toward an inclusive psychology of women. *American Psychologist, 48,* 846–850.

Yoshimi, Y., Yonezawa, S., Sugiyama, K., & Matsui, H. (1989). The effects of the viewpoints and observers: Traits on the observational learning of altruistic behavior. *Japanese Journal of Psychology, 60,* 98–104.

Youell, K. J., & McCullough, J. P. (1975). Behavioral treatment of mucous colitis. *Journal of Consulting and Clinical Psychology, 43,* 740–745.

Young-Eisendrath, P. (1984). *Hags and heroes: A feminist approach to Jungian psychotherapy with couples.* Toronto, Canada: Inner City Books.

Zamostny, K. P., Corrigan, J. D., & Eggert, M. A. (1981). Replication and extension of social influence processes in counseling: A field study. *Journal of Counseling Psychology, 28,* 481–489.

Zaragoza, R. E., & Navarro-Humanes, J. F. (1988). Behavioral treatment of blood phobia: Participant modeling vs. gradual "in vivo" exposure. *Analisis y Modicacion de Conducta, 14,* 119–134.

Zi, N. (1994). *The art of breathing.* Glendale, CA: Vivi.

Zlotlow, S. F., & Allen, G. J. (1981). Comparison of analogue strategies for investigating the influence of counselors' physical attractiveness. *Journal of Counseling Psychology, 28,* 194–202.

AUTHOR INDEX

SUBJECT INDEX

TO THE OWNER OF THIS BOOK:

We hope that you have found *Interviewing Strategies for Helpers: Fundamental Skills and Cognitive Behavioral Interventions*, Fourth Edition, useful. So that this book can be improved in a future edition, would you take the time to complete this sheet and return it? Thank you.

School and address: _____

Department: _____

Instructor's name: _____

1. What I like most about this book is: _____

_____

_____

2. What I like least about this book is: _____

_____

_____

3. My general reaction to this book is: _____

_____

4. The name of the course in which I used this book is: _____

_____

5. Were all of the chapters of the book assigned for you to read? _____

   If not, which ones weren't?   _____

 6. In the space below, or on a separate sheet of paper, please write specific suggestions for improving this book and anything else you'd care to share about your experience in using the book.

_____

_____

_____

_____

Optional:

Your name: _____ Date: _____

May Brooks/Cole quote you, either in promotion for *Interviewing Strategies for Helpers: Fundamental Skills and Cognitive Behavioral Interventions,* Fourth Edition, or in future publishing ventures?

Yes: _____ No: _____

Sincerely,

Sherry Cormier
Bill Cormier

FOLD HERE

- -

BUSINESS REPLY MAIL

FIRST CLASS   PERMIT NO. 358   PACIFIC GROVE, CA

POSTAGE WILL BE PAID BY ADDRESSEE

ATT: *Sherry Cormier & Bill Cormier* _____

Brooks/Cole Publishing Company
511 Forest Lodge Road
Pacific Grove, California 93950-9968

- -

FOLD HERE

Brooks/Cole is dedicated to publishing quality publications for education in the human services fields. If you are interested in learning more about our publications, please fill in your name and address and request our latest catalogue, using this prepaid mailer.

Name: _____

Street Address: _____

City, State, and Zip: _____

FOLD HERE

BUSINESS REPLY MAIL
FIRST CLASS   PERMIT NO. 358   PACIFIC GROVE, CA

POSTAGE WILL BE PAID BY ADDRESSEE

ATT: _Human Services Catalogue_____

Brooks/Cole Publishing Company
511 Forest Lodge Road
Pacific Grove, California 93950-9968

FOLD HERE

IN-BOOK SURVEY

At Brooks/Cole, we are excited about creating new types of learning materials that are interactive, three-dimensional, and fun to use. To guide us in our publishing/development process, we hope that you'll take just a few moments to fill out the survey below. Your answers can help us make decisions that will allow us to produce a wide variety of videos, CD-ROMs, and Internet-based learning systems to complement standard textbooks. If you're interested in working with us as a student Beta-tester, be sure to fill in your name, telephone number, and address. We look forward to hearing from you!

In addition to books, which of the following learning tools do you currently use in your counseling/human services/social work courses?

_____ **Video** _____ in class _____ school library _____ own VCR

_____ **CD-ROM** _____ in class _____ in lab _____ own computer

_____ **Macintosh disks** _____ in class _____ in lab _____ own computer

_____ **Windows disks** _____ in class _____ in lab _____ own computer

_____ **Internet** _____ in class _____ in lab _____ own computer

How often do you access the Internet? _____

My own home computer is:

_____ Macintosh _____ DOS _____ Windows _____ Windows 95

The computer I use in class for counseling/human services/social work courses is:

_____ Macintosh _____ DOS _____ Windows _____ Windows 95

If you are NOT currently using multimedia materials in your counseling/human services/social work courses, but can see ways that video, CD-ROM, Internet, or other technologies could enhance your learning, please comment below:

_____

_____

_____

Other comments (optional): _____

_____

Name _____ Telephone _____

Address _____

School _____

Professor/Course_____

You can fax this form to us at (408) 375-6414; e:mail to: info@brookscole.com; or detach, fold, secure, and mail.